Tables of Physical Data

TABLE 21-3 MEASURED SPEED OF SOUND IN AIR, 444

TABLE 21-4 TYPICAL VALUES OF INTENSITY LEVEL ABOVE THRESHOLD, 450

PART IV Electromagnetism

TABLE 24-1 TYPICAL VALUES OF DIELECTRIC CONSTANT AND DIELECTRIC STRENGTH, 519

TABLE 25-1 RESISTIVITY AND TEMPERATURE COEFFICIENT OF COMMON METALS AND ALLOYS, 535

TABLE 30-1 MAGNETIC SUSCEPTIBILITIES OF NONFERROMAGNETIC METALS AT 300° K, 639

PART V Light

TABLE 34-1 INDICES OF REFRACTION FOR YELLOW LIGHT OF WAVELENGTH 589 nm, 717

TABLE 36-1 REFRACTIVE INDICES AT 20° C FOR VARIOUS WAVELENGTHS, 755

PART VI Quantum and Relativistic Properties of Radiation and Matter

TABLE 42-1 THE PERIODIC TABLE, 874

TABLE 45-1 'EXTERNAL' PROPERTIES OF CERTAIN NUCLEI, 913

TABLE 46-1 MESONS AND BARYONS: MEAN LIFETIMES, 942

TABLE 46-2 A PARTIAL LIST OF PARTICLE–ANTIPARTICLE PAIRS, 945

Greek Alphabet

Because physicists, mathematicians, and engineers invariably use certain letters of the Greek alphabet as symbols to denote particular physical and geometrical quantities, we follow this convenient practice. For reference, the letters of the Greek alphabet, in lower and upper case, together with their names, are given below:

α	A	alpha	ι	I	iota	ρ	P	rho
β	B	beta	κ	K	kappa	σ	Σ	sigma
γ	Γ	gamma	λ	Λ	lambda	τ	T	tau
δ	Δ	delta	μ	M	mu	υ	Υ	upsilon
ε	E	epsilon	ν	N	nu	ϕ	Φ	phi
ζ	Z	zeta	ξ	Ξ	xi	χ	X	chi
η	H	eta	o	O	omicron	ψ	Ψ	psi
θ	Θ	theta	π	Π	pi	ω	Ω	omega

George Shortley B.E.E., PH.D. CONSULTANT, BOOZ, ALLEN APPLIED RESEARCH INC.;
FORMERLY PROFESSOR OF PHYSICS, THE OHIO STATE UNIVERSITY

Dudley Williams A.B., PH.D. REGENTS' DISTINGUISHED PROFESSOR OF PHYSICS,
KANSAS STATE UNIVERSITY; FORMERLY PROFESSOR OF PHYSICS, THE OHIO STATE UNIVERSITY

ELEMENTS OF PHYSICS

FIFTH EDITION

For Students of Science and Engineering

Prentice-Hall, Inc. Englewood Cliffs, New Jersey

PRENTICE-HALL INTERNATIONAL, INC.	*London*
PRENTICE-HALL OF AUSTRALIA, PTY. LTD.	*Sydney*
PRENTICE-HALL OF CANADA, LTD.	*Toronto*
PRENTICE-HALL OF INDIA PRIVATE LIMITED	*New Delhi*
PRENTICE-HALL OF JAPAN, INC.	*Tokyo*

13-268383-0

Current printing (last digit):

10   9   8   7   6   5   4   3   2   1

© 1971, 1965, 1961, 1955, 1953 by
Prentice-Hall, Inc.
Englewood Cliffs, New Jersey

ALL RIGHTS RESERVED

No part of this book may be reproduced in any form, by mimeograph or any other means, without permission in writing from the publisher.

Library of Congress Catalog Card Number: 78-150397

PRINTED IN THE UNITED STATES OF AMERICA

PREFACE TO THE FIFTH EDITION

We again begin the preface to this new, fifth edition of *Elements of Physics* by expressing our pleasure at the wide, favorable reception accorded the earlier editions and by thanking the users for their many helpful and constructive suggestions. Without altering the general mathematical level of the text or adding appreciably to the length of the text, we have been able to incorporate most of these suggestions in the present edition.

This book is primarily designed as a text for the introductory physics course normally taken by students of physical science and engineering—students who are concurrently enrolled in a course in the calculus. We attempt to give a *rigorous* introductory treatment that takes full advantage of the student's background. In particular, we have attempted to take advantage of recent improvements in the teaching of science and mathematics in the high schools and the recent advances being made in college mathematics courses. We have made optimum use of the comments elicited from student users of the text by the four reviewers who provided our publishers with detailed criticisms of the fourth edition.

Our primary objectives are to provide the student with an understanding of physics as a quantitative science based on *observation* and *experiment* and with an appreciation of the *experimental laws* and *fundamental principles* that describe the behavior of the physical world. We attempt to show how *experimental laws* develop as limited generalizations based directly on observation and experiment, and how *fundamental principles* develop, in turn, as generalizations of higher order. In the text, the most widely applicable generalizations are carefully designated as *principles,* while experimental relations of restricted applicability are as carefully called *laws.*

An operational approach to the formulation of the definitions of physical quantities has been employed throughout. The concept of physical dimensions is formally introduced in the first chapter; thereafter, the basic ideas of dimensional analysis are taught through meticulously careful attention to units. As in the third and fourth editions, we employ the coherent system of metric units now known as the *Système International* and the shortened abbreviations for these SI units now used in all languages; correspondingly shortened abbreviations are used for the coherent system of British gravitational units employed in mechanics.

In the first half of the present edition, we continue to follow the 'great tradition' in the order of topics: *Mechanics, Heat and Molecular Physics,* followed by *Wave Motion and Sound.* However, in the second half of the new edition, we have departed from the order previously followed by taking up *Electromagnetism* before *Light.* This rearrangement has the advantage of providing an immediate interpretation of many optical phenomena in terms of electromagnetic waves similar to those produced by oscillating circuits; the altered order of text materials has been tested in the course at Kansas State University and elsewhere with satisfactory results. Although we have continued to treat much of what is conventionally regarded as 'modern physics' within the framework of classical physics in the body of the text and in the extensive sets of 'graded' problems at the ends of the chapters, we have, at the suggestion of various users of earlier editions, introduced a sixth division of the book entitled *Quantum and Relativistic Properties of Radiation and Matter.* In this sixth division we show how quantum and relativity theories have developed and how they have been applied in atomic, nuclear, and high-energy physics.

In the present edition, the part of the book dealing with *Light* has been reorganized. The emphasis on *wave properties* has been retained, but more geometrical optics is presented than in the preceding edition. In particular, we have added a short chapter on optical instruments in response to the suggestions of some users of the fourth edition; these colleagues point out that the introductory physics course constitutes the sole exposure to optics for many engineering students, who will need to make use of various optical devices in their later professional work. The inclusion of this chapter may also facilitate the understanding of the subsequent treatment of diffraction limitations of optical instruments.

The general mathematical level of the text has not been altered, since the level used in earlier editions has proved generally satisfactory. Although we use some derivatives and a few integrals in early chapters, the instructor whose students are not yet fully prepared can readily postpone this mathematical treatment until later in his course. Without presupposing prior knowledge of vector analysis, we introduce vector addition and subtraction in connection with *displacement,* the scalar product of vectors in connection with *work,* the vector product of vectors in connection with *torque,* and vector calculus in connection with specific problems in dynamics and electromagnetism.

As indicated above, we have benefited greatly from the helpful comments of many colleagues who have used earlier editions of this book. We wish to express our special thanks to PROFESSOR DONALD LANG for advice regarding our revised treatment of nuclear and high-energy physics. We also gratefully acknowledge the assistance we have received from PROFESSORS WILLY HAEBERLI, F. R. BIARD, and J. B. GREENE, who gave detailed criticisms of the original manuscript of the present edition. We also wish to thank the editorial staff of Prentice-Hall for its continued helpfulness.

GEORGE SHORTLEY
DUDLEY WILLIAMS

CONTENTS

1   Physics, 1

1. The scientific method, 2
2. Measurement and definition, 4
3. Physical quantities: units and dimensions, 6
4. Systems of units, 8
5. Order of topics, 9

PART I   Mechanics

2   Displacement, 13

1. Measurement of lengths and angles, 14
2. Methods of specifying positions, 17
3. Displacement, 19
4. Addition of vectors, 22
5. Subtraction of vectors, 24
6. Resolution of a vector, 25

3   Kinematics of a Particle, 30

1. Time intervals, 31
2. Speed, 32
3. Velocity, 33
4. Relative velocity, 37
5. Acceleration, 39
6. Rectilinear motion with constant acceleration, 41
7. Motion in a circle at constant speed, 42
8. Motion when acceleration is given, 44

4 Dynamics of a Particle, 51

1. Newton's first principle, 52
2. Mass, 53
3. Newton's second principle, 56
4. Newton's third principle, 58
5. Planetary motion; universal gravitation, 60
6. Weight; gravitational acceleration, 62
7. British gravitational system of units, 66
8. Standard weight, 68
9. Density, 69
10. Freely falling bodies: vertical motion, 71
11. Freely falling bodies: projectiles, 73
12. The Newtonian principle of relativity, 75

5 Systems of Forces; Friction, 83

1. Treatment of forces as vectors, 84
2. Equilibrium of forces, 85
3. Friction between solid surfaces, 87
4. Examples of accelerated motion, 93

6 Work, Energy, and Power, 102

1. Work, 102
2. The scalar product of two vectors, 104
3. Energy, 107
4. Kinetic energy, 108
5. Gravitational potential energy, 110
6. Transformations of mechanical energy, 112
7. Conservation of energy, 114
8. Work and energy in moving coordinate systems, 117
9. Power, 120
10. Mechanical power transmission, 121

7 Momentum, 129

1. Momentum and impulse, 129
2. Principle of conservation of momentum, 132
3. Elastic collisions, 134
4. Perfectly inelastic collisions; impulse and reaction forces, 137
5. Acceleration of rockets, 138

8 Rotational Motion, 145

1. Kinematics of pure rotation, 146
2. Motion of a point in a rigid body in pure rotation, 148
3. Work, power, torque, 151
4. Kinetic energy, rotational inertia, 153
5. Dynamics of pure rotation, 155
6. Angular momentum, 157
7. Power transmission, 158

9 Statics, 164

1. Translational equilibrium, 165
2. Rotational equilibrium, 165
3. Center of mass, 167
4. Equilibrium of a rigid body, 171
5. The torque vector: vector product of two vectors, 173

10 Dynamical Systems, 181

1. The momentum of a dynamical system, 182
2. Effect of external forces on a dynamical system, 183
3. Equilibrium of a dynamical system, 184
4. Two-dimensional motion of a rigid body, 186
5. Rolling bodies, 191
6. Angular momentum, 195
7. Gyroscopic motion, 198
8. Applications of gyroscopes, 201
9. Motion of a particle relative to a spherical earth, 202
10. Earth satellites, 206
11. Ballistic missiles, 209

11 Elastic Properties of Solids and Liquids, 219

1. Hooke's law, elastic potential energy, 219
2. Longitudinal stress and strain: generalization of Hooke's law, 223
3. Volume elasticity: bulk modulus, 226
4. Elasticity of shape: shear modulus, 227
5. Relations among elastic constants, 230
6. Elastic limit and ultimate strength of materials, 231
7. The atomic structure of solids and liquids, 232
8. Internal forces in solids, 235

12 Periodic Motion, 242

1. Simple harmonic motion, 243
2. The reference circle, 248
3. Energy relations in simple harmonic motion, 249
4. Angular simple harmonic motion; torsional oscillation, 251
5. The motion of a pendulum, 252
6. Forced oscillations, 254

13 Mechanics of Fluids, 260

1. Fluid pressure, 260
2. Fluid statics, 264
3. The barometer; pressure gauges, 267
4. Fluid dynamics; Bernoulli's law, 269
5. Applications of Bernoulli's law, 273
6. The lift of an airfoil, 276
7. Fluid viscosity, 278

PART II Heat and Molecular Physics

14 Temperature; Thermal Expansion, 289

1. The common temperature scales, 290
2. Thermal equilibrium, 292
3. The absolute temperature scale, 292
4. Thermal expansion of solids, 294
5. Thermal expansion of liquids, 297

15 Heat and Thermal Energy, 303

1. Quantity of heat, 304
2. Heat capacity; specific heat capacity, 306
3. Latent heats, 308
4. Thermal energy, 310
5. Solids, liquids, and gases, 312
6. Heat of combustion, 313

16 Heat Transfer, 318

1. Methods of heat transfer, 318
2. Laws of heat conduction, 319
3. Convection and radiation; Newton's law of cooling, 322
4. Thermal radiation, 323

17 Ideal Gases, 331

1. Atoms and molecules, 332
2. Avogadro's law; the kilomole; Avogadro's number, 335
3. The gas laws, 337
4. External work; thermal energy, 340
5. Heat capacity per kilomole at constant volume, 342
6. Work done by an expanding gas; heat capacity per kilomole at constant pressure, 343
7. The kinetic theory of gases, 346
8. Pressure in the interior of a gas, 351
9. The Maxwellian distribution; Brownian motion; Avogadro's constant, 352

18 Solids, Liquids, and Gases, 359

1. Fusion; freezing, 359
2. Vaporization; condensation, 361
3. Sublimation; the triple point, 366
4. The critical point, 367
5. Real gases, 368
6. Boiling, 371
7. Mixtures of gases and vapors; hygrometry, 372
8. Surface phenomena of liquids, 375

19 Thermodynamics, 381

1. The first principle of thermodynamics, 382
2. The second principle of thermodynamics, 383
3. Isothermal expansion and compression, 383
4. Reversible adiabatic expansion and compression, 385
5. The Carnot cycle, 390
6. Carnot's theorems, 393
7. The absolute thermodynamic temperature scale, 395
8. Entropy, 397
9. The steam engine, 399
10. Refrigerators; heat pumps, 401
11. Liquefaction of gases, 403

PART III Wave Motion and Sound

20 Wave Motion, 411

1. Mechanical waves, 412
2. Sinusoidal wave motion, 415
3. Derivation of the speed of a transverse wave on a string, 419
4. Energy of sinusoidal wave motion, 421
5. Interference phenomena; the superposition principle, 424
6. Standing waves, 427
7. Production of standing waves, 430
8. Reflection of waves, 431

21 Sound, 438

1. Production of sound by vibrating solids, 439
2. Speed of sound in solids, liquids, and gases, 442
3. Production of sound by vibrating air columns, 445
4. Intensity of sound waves, 449
5. Pitch and quality, 451
6. Response of the ear to sound waves, 452
7. Interference of sound waves, 454

PART IV Electromagnetism

22 Electrostatics, 465

1. Coulomb's principle, 466
2. Electric fields, 471
3. Difference of potential, 477
4. Potential resulting from a charge distribution, 480
5. Line-integral relation for $\mathcal{E}$, 483
6. The charge of the electron, 484
7. The electron-volt as a unit of energy, 484
8. The nuclear model of the atom, 485
9. Electric dipoles, molecular fields, 488

23 Electrostatic Fields, 493

1. Vector lines, vector tubes, vector flux, 494
2. Gauss's relation, 496
3. Charges on conductors; shielding, 499
4. Charging by induction; the electroscope, 504
5. The Faraday ice-pail experiment, 506

24 Capacitance, 513

1. Capacitance, 513
2. The force between charged plates, 515
3. Dielectric constant, 517
4. Energy of a charged capacitor, 521
5. Capacitors in parallel and in series, 521
6. Dielectric strength, 522

25 Electric Currents, 529

1. Current arising from a capacitor discharge, 529
2. Constant currents, 531
3. Resistance; Ohm's law, 532
4. Resistors in parallel and in series, 533
5. Resistivity, 534
6. Resistivity of nonmetals, 538
7. Semiconductors, rectifiers, 538
8. The electron; thermionic emission, 541

26 Direct Electric Currents, 546

1. Terminal voltages, 546
2. Simple circuits, 549
3. Electrical networks; Kirchhoff's rules, 550
4. Measuring instruments, 551
5. Charge and discharge of a capacitor, 554
6. Nonlinear circuit elements; the thermionic vacuum tube and the transistor, 557

27 Electrochemistry; Thermoelectricity, 566

1. Electrolysis, 566
2. Charge transport in electrolysis, 568
3. Voltage necessary for electrolysis, 571
4. Cells in present use as sources of EMF, 574
5. Fuel cells, 577
6. Thermoelectric effects; the thermocouple, 577

28 Magnetic Forces, 582

1. Magnetic forces, 583
2. Magnetic intensity, 584
3. The earth's magnetic field, 587
4. Magnetic force on a conductor carrying current, 590
5. The moving-coil galvanometer, 593
6. Magnetic force on a moving charged particle, 594
7. The mass of the electron, 596
8. The mass spectrograph; nuclidic masses, 596
9. The cyclotron, 598

29 Magnetic Fields, 603

1. The magnetic field of an electric current, 604
2. Ampère's principle, 606
3. Ampère's line-integral relation; magnetic poles, 608
4. Properties of magnetic fields, 613
5. The field of a solenoidal coil, 615
6. Forces on solenoids, 619
7. The current balance; determination of ε_0; electrical standards, 621
8. Magnetic trapping of charged particles, 622

30 Magnetic Properties of Matter, 628

1. Permanent magnets, 629
2. Magnetization of a ferromagnetic toroid, 632
3. The 'magnetic circuit' 636
4. Real magnetic materials, 636
5. Magnetic susceptibility, 638
6. Magnetic properties of atoms; paramagnetism, 639
7. Diamagnetism, 641

31 Electromagnetic Induction, 646

1. Motion of a wire in a magnetic field, 647
2. Relation between EMF and rate of change of flux, 649
3. Eddy currents, 652
4. Induction by changing current, 655
5. The betatron, 656
6. Mutual inductance, 658
7. Self inductance, 659

32 Alternating Currents, 667

1. The series circuit, 668
2. Resistance, 668
3. Capacitance, 670
4. Inductance, 673
5. General series circuits, 675
6. Series resonance, 676
7. The transformer, 679
8. Measuring instruments, 681

33 Oscillating Circuits; Electromagnetic Waves, 686

1. Oscillations in a circuit containing inductance and capacitance, 687
2. The damped oscillating circuit; continuous oscillations, 688
3. Electromagnetic radiation, 690
4. Microwaves, 693

PART V Light

34 The Wave Nature of Light, 699

1. The propagation of light, 701
2. Interference resulting from thin films, 704
3. The Michelson interferometer, 709
4. Speed of light, 713
5. Index of refraction, 717
6. Transmission of energy by light waves; radiometry, 720
7. Photometry, 722

35 Shadows: Fresnel Diffraction, 729

1. Huygens's hypotheses, 730
2. Young's experiment, 732
3. Diffraction by a slit and by a straight-edge, 735
4. Diffraction by a circular aperture, 737

36 Reflection and Refraction, 743

1. Reflection at plane surfaces, 743
2. Image formation by plane mirrors, 746
3. Refraction of light, 748
4. Fermat's principle, 752
5. Dispersion by refraction, 753
6. Transmission and absorption of light waves, 755

37 Mirrors and Lenses, 760

1. Ellipsoidal and paraboloidal mirrors, 761
2. Concave spherical mirrors, 763
3. Reflection by convex spherical mirrors, 770
4. Treatment of convex mirrors by wave optics, 772
5. The ideal lens, 775
6. Thin lenses, 777
7. Treatment of lenses by wave optics, 784
8. Aberrations of mirrors and lenses, 788

38 Optical Instruments, 792

1. Combinations of lenses, 792
2. Magnifying power of an optical instrument, 795
3. The simple microscope, 798
4. The compound microscope, 799
5. The astronomical telescope, 801
6. Terrestrial telescopes, 802

39 Image Formation: Fraunhofer Diffraction, 806

1. Diffraction resulting from an aperture, 807
2. Resolving power of optical instruments, 810
3. The diffraction grating, 815

40 Polarization, 819

1. Polarization by selective absorption, 820
2. Polarization by reflection, 823
3. Polarization by scattering, 825
4. Double refraction, 827
5. Magneto-optical and electro-optical effects, 833

PART VI Quantum and Relativistic Properties of Radiation and Matter

41 Quantum Properties of Radiation, 839

1. Emission and absorption spectra, 840
2. Black-body radiation; Planck's quantum principle, 843
3. The photoelectric effect, 847
4. Atomic spectra, 848
5. The Bohr theory of the structure of hydrogen, 852
6. Emission and absorption processes; lasers, 855
7. The inverse photoelectric effect; X rays, 859
8. Photons, 861

42 Quantum Mechanics: Atomic Structure, 866

1. Waves associated with material particles, 867
2. Quantum mechanics, 868
3. Structure of atoms, 872
4. Magnetic properties of atoms, 877
5. Applications of quantum mechanics, 880

43 Electromagnetic Radiation, 883

1. Maxwell's equations in integral form, 883
2. Maxwell's equations for free space in differential form, 887
3. Character and speed of a plane electromagnetic wave, 890

44 Einstein's Theory of Relativity, 895

1. The speed of light, 896
2. Relativistic kinematics, 897
3. Relativistic dynamics, 899
4. Accelerators for high-energy particles, 904
5. Energy and momentum transfer; quantum and relativistic considerations, 906

45 Nuclear Physics, 911

1. The 'external' properties of nuclei, 912
2. Natural radioactivity, 914
3. Nuclear reactions, 916
4. The positron; annihilation and creation of matter, 919
5. Transformation of matter to energy in nuclear reactions, 921
6. Nuclear energy, 925
7. Nuclear models, 929

46 High-Energy Physics: Elementary Particles, 935

1. Electromagnetic interactions, 936
2. Strong interactions: pions, 936
3. Weak interactions, 938
4. Mesons and baryons, 939
5. Detection devices for high-energy particles, 941
6. Antimatter, 944
7. Elementary particles, 946

Appendix, i

1. SYSTEMS OF MECHANICAL AND ELECTRICAL UNITS, i
2. FUNDAMENTAL PHYSICAL CONSTANTS, iii
3. ASTRONOMICAL DATA, iv
4. THE ELEMENTS, iv
5. TABLES OF CONVERSION FACTORS, vii
6. NATURAL TRIGONOMETRIC FUNCTIONS, xii
7. TABLE OF LOGARITHMS TO THE BASE 10, xvi
8. TABLES OF EXPONENTIALS, xviii
9. TABLE OF SQUARE ROOTS, xx
10. A SHORT TABLE OF CUBE ROOTS, xxiv

Index, i

PHYSICS 1

Physics is a *science*—a *quantitative* science—a science of measurement, experiment, and of systematization of the results of experiment. We shall presently discuss the *scientific method,* which furnishes the fundamental basis of physics as well as of other sciences. And we shall describe how *measurement* underlies scientific *definition*.

Among the quantitative sciences are astronomy, chemistry, metallurgy, physics, and certain geological and biological sciences. It is impossible to give accurate, mutually exclusive definitions of any of these sciences, in particular of *physics*. The hard core of 'classical' physics as it developed prior to 1900 was fairly sharply defined and consisted of the subjects: *mechanics, heat, sound, light,* and *electromagnetism*. But during the twentieth century research physicists, as well as other types of research scientists, have been studying the structure of matter, of the atom, and of the atomic nucleus—and these subjects are not the exclusive province of physics. They are equally the province of chemistry and, within these areas, many subdivisions have come to be recognized as 'primarily physics' or 'primarily chemistry' principally because they happened to be first studied in laboratories occupied by professional 'physicists' or 'chemists,' respectively. That there are no accurate boundaries defining the different sciences is exemplified by the fact that there are scientists and periodicals dedicated to subjects with such compound names as 'physical chemistry,' 'chemical physics,' 'astrophysics,' 'geophysics,' 'biophysics,' 'biochemistry,' 'medical physics,' and 'molecular biology.'

Before turning to a discussion of the scientific method, we might say a word about *engineering*. Although much of the work that goes on in engineering research laboratories is truly *science,* the *practice* of engineering is *applied science* (application of the bodies of knowledge of the various natural sciences), supplemented as necessary by *art* ('know-how' built up and handed down from past experience). In the same way, we can say that the *practice* of medicine is *applied biological science* supplemented, perhaps to a greater degree than is engineering, by *art*.

1. The Scientific Method

The term *science* refers to much more than a body of knowledge considered as a mere collection of facts. It refers to a body of models and generalizations that systematizes and correlates observed facts and from which predictions can be made that may be compared with later observation or experiment. An accepted scientific conceptual scheme is usually called a *theory*. A theory is never *proved*. A theory is considered to be a valid model of reality if it correlates in fertile fashion a considerable body of facts and if no one happens to *disprove* it by finding a fact in contradiction with its predictions. Because of this characteristic of a scientific theory, there is always the chance that the best-established theory will someday be confronted with facts inconsistent with its predictions and will have to be revised, corrected, refined, or even abandoned.

> The **scientific method** is the systematic attempt to construct theories that correlate wide groups of observed facts and are capable of predicting the results of future observations. Such theories are tested by controlled experimentation and are accepted only so long as they are consistent with all observed facts.

Science starts, then, with the systematic recording of facts—accurate, well-defined, and, in physics, quantitative facts. The scientist then essentially speculates; he mulls over the facts, trying to make order out of them, trying to relate them to each other and to other known facts. He develops alternative *working hypotheses* capable of prediction. These hypotheses are provisional conjectures designed to guide further investigation. A hypothesis is incorporated into scientific theory only when it has been empirically verified in many ways. It falls if its predictions are contradicted by observation.

Thus, in the 16th century, the Danish astronomer TYCHO BRAHE observed and plotted the positions of the planets in great detail, collecting facts. From these data, the German astronomer JOHANN KEPLER showed that the planets moved in ellipses with the sun at one focus and with speeds predictable by definite *laws*. This was scientific progress since future positions of the planets could be predicted, and the predictions were verified. Then, late in the 17th century, the British scientist SIR ISAAC NEWTON made much more substantial progress with his beautifully simple *principles* of mechanics and gravitation which enabled prediction of the motion not only of the planets but of the moon and of the apples that fell from his tree, and which would have been refuted did not the planets move in ellipses at Kepler's speeds. Newton's principles stood the test of time until this century, when discrepancies from their predictions led to the development of relativity theory and of quantum theory, which are in a sense refinements of Newton's theories since they give Newton's results under ordinary circumstances, but depart from Newton's predictions in the cases of the extremely high speeds and the extremely small particles encountered in atomic and nuclear physics. These later theories are in no sense yet completely satisfactory; minor discrepancies still exist in the theory of the atom. But Newton's principles will always be useful in predicting the motions of machines and of vehicles, since predictions from these principles agree with experiment on bodies of ordinary size and speed within any accuracy of observation obtainable at present.

The general procedure of science is excellently illustrated by the above example. The first step is the collection of *facts* by careful observation or experiment, as exemplified by the work of Tycho Brahe. The second step is a discovery of relations among the facts and the formulation of *laws,* as exemplified by Kepler's laws of planetary motion. The third step is the development of general *principles,* as exemplified by Newton's principles of mechanics. From Newton's principles one can derive not only Kepler's laws of planetary motion but also the 'laws of falling bodies' developed by Galileo and other laws governing the mechanical behavior of rockets, earth satellites, automobiles, aircraft, gyroscopes, and a host of other mechanical devices unknown in the time of Newton.

We must add one important note to this discussion: Science never considers the question *'why?'*—it considers only the question *'how?'* It is not in the slightest concerned with *why* the universe behaves as it does, it is only concerned with *how* the universe behaves. It attempts, however, to understand how it behaves in terms of the fewest, simplest, and most general principles that it can find.

2. Measurement and Definition

Physics is a quantitative science concerned with relations between careful measurements of well-defined quantities. In physics, *measurement* comes first:

> The **definition of a physical quantity** is the description of the operational procedure for measuring the quantity.

A definition of this type is called an *operational definition*.

Let us consider the definition of a familiar physical quantity: *average speed*—for example the average speed of an automobile in a 500-mile race. This was reported as 156 miles per hour for the winner of the 1970 Memorial-Day race at Indianapolis. How do we *define* this average speed? We define it as the distance traversed (in this case 500 miles) divided by the total elapsed time (3.21 hours). But then how do we define the distance? We measure the length of the track with a calibrated tape, calibrated in terms of a certain standard length, that of a platinum bar kept in Washington. *We have defined length by telling how to measure it.* What about the elapsed time? We measure this with a stopwatch calibrated so that it records 24 hours per day, the day being the average time between *noons* (when the sun is at the meridian, due north or south or directly overhead). This definition is not complete; we must define north, south, etc.; but again we do so by telling how to determine these directions. *We have defined elapsed time by telling how to measure it.*

Note that in defining distance we have told how to measure it in terms of a number and a unit—*500 miles*. We have defined time by telling how to measure it in terms of a number and a unit—*3.21 hours*. We have defined average speed as the ratio

$$\text{average speed} = \frac{\text{length}}{\text{time}} = \frac{500 \text{ mile}}{3.21 \text{ hour}} = 156 \frac{\text{mile}}{\text{hour}}.$$

In this calculation we divide the two numbers to obtain the number 156. We divide the two units to obtain the new *derived* unit *mile/hour*. The *magnitude* of the average speed is *156 miles per hour*.

> The **magnitude of a physical quantity** is specified by a *number* and a *unit*.

The same magnitude, 156 mile/hour, can be specified in other units, such as foot/second. By definition,

$$1 \text{ mile} = 5280 \text{ foot}, \qquad 1 \text{ hour} = 3600 \text{ second}.$$

We write the right sides of these equations in the singular because *in mathematical equations* such expressions are always to be understood to represent 5280×(1 foot) and 3600×(1 second), a number *multiplied by a unit*. These equations again represent operational definitions, as the reader will realize if he considers their physical implication in terms of the *calibration* of a one-mile tape in feet, or the *calibration* of the face of a watch.

We can find the relation between the derived units, mile/hour and foot/second, by direct substitution, as follows:

$$1\frac{\text{mile}}{\text{hour}} = \frac{1 \text{ mile}}{1 \text{ hour}} = \frac{5280 \text{ foot}}{3600 \text{ second}} = 1.47 \frac{\text{foot}}{\text{second}}.$$

By use of this relation, the average speed discussed earlier becomes

$$\text{average speed} = 156 \times 1\frac{\text{mile}}{\text{hour}} = 156 \times 1.47 \frac{\text{foot}}{\text{second}} = 229 \frac{\text{foot}}{\text{second}}.$$

In the above computations, we have 'rounded-off' all numbers to three significant figures. This practice, which simplifies arithmetic computation and permits the use of the slide rule, will be followed in almost all of the examples and problems in this text. We must, however, realize that more than three significant figures should often be carried in real scientific computations. But one must *never* express an answer to *more* significant figures than (a) are justified by the accuracy of the observed data entering the computation, and (b) are useful for the purposes that the answer is to serve. Violation of (a) would give an answer that is physically incorrect; violation of (b) is merely a waste of time and effort. In most of the examples in this book, answers serve only the purpose of indicating that we know how to solve correctly a problem involving hypothetical data; giving more than three significant figures would not further this purpose.

A physical quantity is frequently represented by a single letter symbol. For example, the average speed might be denoted by v, so that in the Indianapolis race,

$$v = 156 \frac{\text{mile}}{\text{hour}} = 229 \frac{\text{foot}}{\text{second}}.$$

We must remember that *a letter such as 'v' that represents a physical quantity does not stand for merely a number, but a number and a unit*, and that in substitution in equations such as

$$v = D/t,$$
$$(\text{speed}) = (\text{distance})/(\text{time})$$

we must substitute both a number *and* a unit for each symbol.

Sometimes derived units are given special names. The mile/hour has no such special name; we prefer the complexity of saying 60 *miles per hour* rather than, say, 60 *spartas*. But the *nautical mile per hour*

(1 nautical mile=6076.10 foot), which is used in describing the speed of ships and airplanes, does have such a special name, the *knot:*

$$1 \text{ knot} = 1 \text{ nautical mile/hour}.$$

Example. *Express the average speed in the Indianapolis race mentioned above in knots.*

To three significant figures,

$$1 \text{ nautical mile} = 6080 \text{ foot} = 6080 \text{ (1 foot)},$$
$$1 \text{ foot} = \tfrac{1}{6080} \text{ nautical mile}.$$

Then
$$v = 229 \frac{\text{foot}}{\text{second}} = 229 \frac{\tfrac{1}{6080} \text{ nautical mile}}{\tfrac{1}{3600} \text{ hour}}$$
$$= \frac{229 \cdot 3600}{6080} \frac{\text{nautical mile}}{\text{hour}} = \frac{229 \cdot 3600}{6080} \text{ knot}$$
$$= 136 \text{ knot}.$$

For convenience, the common physical units are given short abbreviations, and these abbreviations are generally used in writing equations. Thus, *mile* is abbreviated as mi, *foot* as f, *hour* as h, *second* as s; the mi/h and the f/s are units of speed, with 1 mi/h = 1.47 f/s, to three significant figures. The abbreviations of units are always printed in roman type, to distinguish them from symbols representing physical quantities, which are printed in italic type. As a matter of convention, units named after men, such as the *volt* and the *watt,* have abbreviations beginning with capital letters (V and W in these instances), while units not named after men, such as those mentioned earlier, have abbreviations beginning with small letters.

3. Physical Quantities: Units and Dimensions

Any physical quantity used in the formulation of laws and principles must be operationally defined by a specification of the procedure to be used for measuring the quantity. It is desirable to keep the number of arbitrary units employed in measurement as small as possible. Therefore, it is convenient to regard certain physical quantities as *fundamental quantities;* these are measured in terms of arbitrary but internationally accepted *fundamental units.* All other physical quantities are defined in terms of the fundamental quantities and are therefore called *derived quantities.*

Three fundamental quantities are needed for the treatment of the part of physics known as *mechanics;* the fundamental quantities employed in the international metric system are *length, mass,* and *time.* These quantities are measured in terms of the arbitrarily chosen fundamental units: the *meter,* the *kilogram,* and the *second.* All other mechanical quantities are defined in terms of the fundamental quantities, just as average speed was defined in terms of length and time in Sec. 2, and are measured in terms of *derived units* defined in terms of the fundamental units; e.g., speed is measured in the unit *meter/second* in the international metric system.

Only three *additional* fundamental quantities and units are needed in all other branches of physics: in *heat* we define the quantity *temperature difference* and its unit, the *kelvin;* in *electromagnetism* we define *electric current* and its unit, the *ampere;* in *light* we define *luminous intensity* and its unit, the *candela*. From just six fundamental quantities, all other quantities employed in physics can be derived.

At this point, we should briefly mention the concept of the *dimensions* of physical quantities. The physical quantity *speed* can be expressed in mi/h, or in f/s, or in *yards per minute,* or even in *inches per day*. The common thing about all of these derived units is that they represent *a unit of length divided by a unit of time;* a speed can be expressed in any unit so derived, and cannot be expressed in any unit not so derived. We compute a speed by dividing a length by a time; *speed* is said to have the *dimensions* of *(length)/(time)*.

Similarly, *acceleration,* which will be defined on p. 39, is suitably measured in f/s^2, or mi/h·s, or even in *(inches per day) per minute*. These units all represent a length divided by the product of two times, and *acceleration* is said to have the *dimensions* of $(length)/(time)^2$. Power, defined on p. 120, is expressed in the metric system in the complex unit kilogram-meter^2 per second^3, called the *watt,* and has the *dimensions* of $(mass) \times (length)^2/(time)^3$.

> The **dimensions** of a physical quantity are specified by giving the powers of the fundamental physical quantities that occur in its derivation—alternatively, the dimensions are specified by giving the powers of the fundamental units that enter the derived unit suitable for its measurement.

Recognition of the concept of dimensions is of importance because

> *One can add or subtract two physical quantities if, and only if, they have the same physical dimensions.*

Thus one *can* add 6 f/s and 1 mi/h because these physical quantities have the same dimensions; the sum is seen to be 7.47 f/s. One *can* add 3 f^3 and 10 gallons because these quantities both represent volumes, with the dimensions of (length)^3; since 1 gallon=0.134 f^3, the sum is 4.34 f^3. But one *cannot* add 1 mi/h and 10 gallons. The last statement is so obvious that it needs no explanation—the important thing to realize is that the fundamental reason that these quantities cannot be added is that they have *different* physical dimensions, namely (length)/(time) and (length)^3, respectively.

As a generalization of the above discussion, we conclude that

In any equation, the physical dimensions of each term must be identical.

This conclusion follows from the fact that if all terms of the equation are placed on one side and equated to zero, we are adding (or subtracting) the terms, and this process represents nonsense unless all terms have the same dimensions.

We shall frequently be dealing with equations involving derived quantities having complex dimensions. The reader will find it very helpful to consider the dimensions of all physical quantities with which he works, and to check equations for consistency of the dimensions of the different terms. Section 1 of the Appendix lists the derived metric units that we shall employ, and expresses them in terms of the fundamental metric units; from these expressions, the dimensions of the various physical quantities can be conveniently determined.

4. Systems of Units

Until the nineteenth century, each country of the world had its own particular system of units of length, mass, volume, and other physical quantities. And these systems, like our own common system of units, were not decimal systems. Thus, we divide the inch into eighths, 16ths, and 64ths; there are 12 inches to the foot, three feet to the yard, and 1760 yards to the mile. Also there are 16 ounces to the pound, 112 pounds to the hundredweight (cwt), and 20 cwt to the long ton. Computation with such units is just as inconvenient as the practice of accounting with pounds, shillings, and pence, which the British are now changing to a decimal monetary system like our own.

An international decimal system for the units of length and mass was conceived as early as the sixteenth century, but implementation waited until 1790, when the French National Assembly requested the French Academy of Sciences to establish such a system and to create the basic standards. The resulting so-called metric system, based on the meter as the unit of length and the gram as the unit of mass, was soon adopted by scientists throughout the world. It was also adopted by most of the countries of the world. It was 'legalized' in the United States in 1866, and in 1893 our common units were redefined in terms of the international metric units.

The permanent International Bureau of Weights and Measures was established in Paris in 1875 to maintain the international standards and calibrate the national standards, such as those maintained by our National Bureau of Standards, against the international standards.

The metric system has had a considerable evolution in the past century. At first it was a CG (centimeter, gram) system. Then the second was added as a unit of time and we had a CGS system. Some of the derived units were *coherent,* in that they were products of powers of the fundamental units, but some of them, notably in electromagnetism, were incoherent. A completely coherent system of derived units was adopted in 1954, and in 1960 was given the formal name of *International System of Units* and the abbreviation SI (for *Système International*), to be used in all languages.

The SI system is based on six fundamental units, the *meter* as the unit of length, the *kilogram* as the unit of mass, the *second* as the unit of time, the *ampere* as the unit of electric current, the *kelvin* as the unit of temperature difference, and the *candela* as the unit of luminous intensity.

The great advantage of the SI system, as we employ it in this book, is its *coherence.* All derived units are simply multiples of the more fundamental units. Thus a force of *1 newton* gives a mass of *1 kilogram* an acceleration of *1 meter/second^2*; a force of *1 newton* applied for a distance of *1 meter* does *1 joule* of work; work at the rate of *1 joule/second* represents a power of *1 watt;* and so forth throughout the entire body of derived units.

Because, in this book, we place more emphasis on teaching the *principles of physics* than on teaching a system of units, and also because SI units have not been generally adopted in this country for either common or engineering units (except for electromagnetic quantities), we also employ a system of English units in mechanics so that the reader sees some of the results expressed in familiar rather than unfamiliar terms.

It seems clear that to compete in world markets and to be in tune with the rest of the world, both our engineering and our common usage must eventually go metric. South Africa, New Zealand, Australia, and Great Britain have deciced to go metric, leaving just the United States and Canada as major countries not employing metric units. Great Britain expects to complete her 'metrication,' as she calls it, by 1975.

5. Order of Topics

A word should be said regarding the order of topics in this text. We shall follow what is frequently called the 'traditional' order: *mechanics, heat, sound, electromagnetism,* and *light.* There are valid reasons for this particular order of topics; physics has a definite 'structure' and is best understood when it is examined in a logical manner.

In mechanics, we introduce the fundamental quantities: *mass, length,* and *time.* Derived quantities such as force, work, energy, momentum, pressure, etc., are defined in terms of the three fundamental quantities of mechanics and are thereafter used not only in mechanics but in the other branches of physics as well. The fourth fundamental quantity, *temperature,* introduced in the second part of the book dealing with heat, is defined in terms of certain thermodynamic cycles, which can be understood only after the principles of mechanics are clearly understood. An understanding of the transmission of *sound* waves, discussed in the third part of the book, involves not only an understanding of fluid mechanics but also an understanding of thermal phenomena. The fifth fundamental quantity, *electric current,* is defined in the fourth part of the book in terms of forces and lengths, while other electrical quantities such as voltage are defined in terms of work. The fifth part of the book dealing with *light* depends on the wave concepts introduced earlier in the treatment of sound waves; in this part of the book the *candela* is introduced as the fundamental photometric unit. The waves involved in light transmission are electromagnetic in character, as elucidated in a later chapter dealing with electromagnetic radiation.

This 'traditional' order of topics largely parallels the historical development of classical physics. However, we shall not approach the subject from a historical point of view but shall introduce the new concepts associated with twentieth-century physics at appropriate points. The atomic nature of matter is accepted as the basis for certain discussions in mechanics; modern atomic theory is widely employed thereafter. The basic concepts of quantum mechanics are introduced in connection with emission and absorption of light. The last part of the book is devoted to phenomena that can be interpreted only on the basis of *quantum* and *relativistic* considerations; however, all of the theory introduced in earlier parts is involved in these considerations. The final chapter is devoted to high-energy physics and includes a survey of our present knowledge of elementary particles.

One final word of caution should be given: Because of the logical structure of physics, one cannot 'skip about' at will. In the field of geography, one can understand the geography of Australia without prior knowledge of the geography of Europe. In physics, anyone wishing to understand electromagnetism must begin by understanding the basic principles of mechanics and heat. Similarly, anyone wishing to understand quantum and relativistic mechanics should have a clear understanding of classical physics.

PART I

MECHANICS

DISPLACEMENT 2

Mechanics involves the study of *motions* of material particles (including, of course, the special case in which the particles are at rest). Motion involves change of position in space. A change in position is called a *displacement*. Displacement is a *vector* quantity; its specification requires a *direction* as well as a *magnitude*. In this chapter, we learn first to specify position and then change in position, or displacement. Using displacement as a convenient example, we show how to describe *vector quantities* graphically and analytically, and how to *add* vector quantities.

What we shall learn about vectors in discussing displacement will be directly applicable to later study of many other physical quantities having vector properties. The basic quantities used in the treatment of electromagnetism are vector quantities, as are the angular momenta of electrons, protons, neutrons, and the other 'elementary particles' of modern physics. Thus, it is important to develop an early understanding of vectors and 'vector mathematics,' which includes certain types of multiplication in addition to the rules for addition given in the present chapter. The use of vector mathematics affords a method of writing many physical laws and principles in a simple, easily understandable form that is independent of the choice of any particular coordinate system.

1. Measurement of Lengths and Angles

The description of the position of an object in space is one of the first steps in the analysis of its motion. Such a description is made in terms of lengths of lines and of angles between lines. The standard unit of length used for the calibration of most scientific instruments is the *meter*.

Until October 1960, the international standard meter was the distance between two fine lines ruled on a platinum-iridium bar kept at the International Bureau of Weights and Measures in Sèvres, France, the laboratory set up by international treaty in 1875 to prepare, measure, and preserve the standards of the international metric system. However, there has long been concern that this international prototype meter bar might be lost or destroyed, or that its metal might, in time, change in length in an unpredictable manner. Of more recent concern was the fact that secondary standards could be calibrated against the international standard only to an accuracy of about one part in 10 million, whereas scientific measurements were being made with greater accuracy and even industrial tolerances in precision instrumentation were beginning to approach this limit. As a result, the 1960 International General Conference on Weights and Measures redefined the meter as exactly 1 650 763.73 times the wavelength of the orange light emitted by a pure nuclide, or isotope, that of mass number 86, of krypton gas. Since the operations involved in employing this new definition require knowledge of optics, spectroscopy, and atomic structure, we shall postpone further discussion of these operations until Chapter 41. We note, however, that a length can be measured in terms of this wavelength more accurately than 1 part in 100 000 000, and that, based as it is on a fundamental atomic property, this wavelength is immutable and inviolable. Hence it should serve indefinitely as a standard of length immune to the failings of a concrete standard such as a metal bar.

The first of the three fundamental units of the international metric system is thus defined as follows:

> The **meter** is a length that is exactly 1 650 763.73 times the wavelength of the orange light emitted when a gas consisting of the pure krypton nuclide of mass number 86 is excited in an electrical discharge, the wavelength being measured in a vacuum.

In the metric system, the following prefixes are used to indicate decimal multiples of the basic units:

$$\begin{array}{llll} \text{kilo- (k)} & \text{for} \quad 1000 \quad = 10^3, & \text{giga- (G)} & \text{for} \quad 10^9, \\ \text{mega- (M)} & \text{for} \quad 1\,000\,000 = 10^6, & \text{tera- (T)} & \text{for} \quad 10^{12}. \end{array}$$

One sometimes also sees the prefixes deka- (dk) for 10 and hecto- (h) for 100. Of the units that are multiples of the meter, the kilometer (km) is the most commonly used.

For the prefixes used to indicate submultiples, the following list, approved in 1962 by the International Committee on Weights and Measures, is now in common scientific usage:

deci- (d) for 10^{-1}, nano- (n) for 10^{-9},
centi- (c) for 10^{-2}, pico- (p) for 10^{-12},
milli- (m) for 10^{-3}, femto- (f) for 10^{-15},
micro- (μ) for 10^{-6}, atto- (a) for 10^{-18}.

The centimeter (cm) and the millimeter (mm) are in common popular usage in all non-English-speaking countries; the micrometer (μm) is widely used in biology; while the still smaller units, nm, pm, and fm, have important applications in measurements related to light waves and to atomic and nuclear structure. Until recent years, the micrometer (μm) was called simply the *micron* and designated by μ, while the nanometer (nm) was called the *millimicron* and designated by mμ; these special terms, which do not fit into the rational system described above, are falling into disuse.

The unit of length used in most commercial transactions and in much engineering practice in English-speaking countries is the foot (f), which is 12 inches (i) or one-third of a yard (yd). The foot is defined by recent (1959) international agreement as exactly 0.3048 m. This definition leads to the following relations* among British and metric units of length:

1 f = 30.48 cm = 0.3048 m, 1 m = 39.37 i = 3.281 f,
1 i = 2.54 cm = 0.0254 m, 1 km = 3281 f = 0.6214 mi.

The statute mile (mi) is defined as exactly 5280 f.

Measurement of angles. The angle between two lines is readily defined if any circle is drawn with its center at the intersection of the two lines as shown in Fig. 1. The magnitude of the angle is proportional to the fraction of the circumference of the circle lying between the two lines. This fraction is independent of the radius of the circle. We shall have occasion to use the following two types of angular units (*see* Fig. 2):

The **degree** (°) is the angle subtended at the center of a circle by $\frac{1}{360}$ of the circumference. The **minute** (′) is $\frac{1}{60}$°. The **second** (″) is $\frac{1}{60}$′.

The **radian** (rad) is the angle subtended at the center of a circle by a portion of the circumference equal in length to the radius of the circle, that is, by the fraction $1/2\pi$ of the circumference.

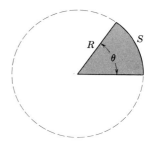

Fig. 1. Definition of an angle θ.

---

*More complete sets of relations among units will be found in the "Tables of Conversion Factors" in Sec. 5 of the Appendix. The reader will find these tables extremely useful in working problems and for general reference purposes.

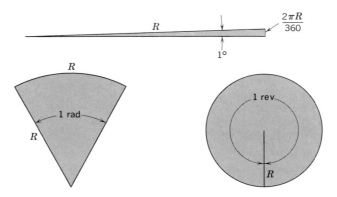

Fig. 2. The angular units: 1 degree, 1 radian, 1 revolution.

The angle θ in Fig. 1 is given in *radians* by the relation

$$\theta = S/R,$$

where S is the arc length on the circumference and R is the radius of the reference circle. It follows that $\theta = 1$ rad when $S = R$, and $\theta = 2\pi$ rad when S is one circumference. The angle subtended by the full circumference is commonly called 1 *revolution* (rev). Thus,

$$1 \text{ rev} = 360° = 2\pi \text{ rad}, \qquad 1° = \tfrac{1}{360} \text{ rev},$$
$$1 \text{ rad} = 57°.3 = (1/2\pi) \text{ rev}, \qquad 1° = (2\pi/360) \text{ rad} = 0.01745 \text{ rad}.$$

The physical quantity *angle,* defined by $\theta = S/R$, is seen to be *dimensionless;* it is the *ratio* of two lengths and hence is independent of the unit used to measure length. The *radian* is considered as the *fundamental* unit of angular measure; the *degree* is more useful in calibrating instruments for angular measurement because it is integrally related to the circle.

Example. *If the arc length S in Fig. 1 is* 20 cm *and the radius R is* 1 m, *what is the angle θ in radians, degrees, and revolutions?*

From the definition of angle in radians we have

$$\theta = \frac{S}{R} = \frac{20 \text{ cm}}{1 \text{ m}} = \frac{20 \text{ cm}}{100 \text{ cm}} = 0.2.$$

The result is a pure number. However, to make clear what angular unit we are using, it is convenient to insert the dimensionless term rad, and write

$$\theta = 0.2 \text{ rad}.$$

We convert this angle to degrees and revolutions by using the equations immediately preceding this example. Thus,

$$\theta = 0.2 \text{ rad} = 0.2 \,(57°.3) = 11°.5; \qquad \theta = 0.2 \text{ rad} = 0.2 \,(1/2\pi) \text{ rev} = 0.0318 \text{ rev}.$$

The answers here have been rounded in accordance with the practice we shall follow throughout this book: angles in degrees to the nearest $\frac{1}{10}°$, all other results to three significant figures.

Let us now compute S from R and θ, using the equation $S = R\theta$. In this equation, θ *must* be in radians. We substitute:

$$S = R\theta = (1 \text{ m}) \times (0.2 \text{ rad}) = 0.2 \text{ m} \cdot \text{rad}$$

Here, we should *drop* the dimensionless unit rad, just as we decided to *add* it before, writing, more sensibly, $S = 0.2 \text{ m} = 20 \text{ cm}$.

In equations employing letter symbols, angles are always understood to be in radians unless otherwise specified. It is only the angular unit *rad* that can be dropped or inserted at will as in the example above.

2. Methods of Specifying Positions

Coordinate systems are required to specify the position of a point in space. A coordinate system has a fixed reference point called the origin O and fixed reference lines relative to which the positions of points can be specified.

A *rectangular coordinate system* like that in Fig. 3(a) can be used, for example, to specify the position of a ball on a football field if one takes the coordinate origin O at the southwest corner of the playing field, the X-axis along the south goal line, and the Y-axis along the west side of the field. When the ball is on the ground at the 40-yard line and is 30 yards from the west side of the field, its position P has the coordinates

$$X = 30 \text{ yd}, \quad Y = 40 \text{ yd}.$$

A third coordinate is needed for specifying positions when motion occurs in three dimensions. If the football is in the air, its position can be given in terms of an XYZ-coordinate system whose Z-axis points vertically upward from the southwest corner of the field. When the ball is 5 yd above a point on the ground that is on the 40-yd line and 30 yd from the west side of the field, its position has the coordinates

$$X = 30 \text{ yd}, \quad Y = 40 \text{ yd}, \quad Z = 5 \text{ yd}.$$

The position of a point in a plane can also be specified in terms of a *polar coordinate system* as in Fig. 3(b). Here one gives the distance R of the point P from the coordinate origin and the angle θ that the line OP makes with a fixed reference line, usually taken as the X-axis. For the same point P whose rectangular coordinates are X and Y, as indicated in Fig. 3(a), the polar coordinates indicated in Fig. 3(b) are given by

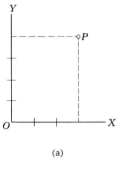

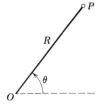

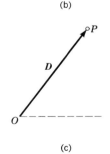

Fig. 3. (a) Rectangular coordinates of a point P. (b) Plane polar coordinates of the point P. (c) Displacement vector from O to P.

$$X = R\cos\theta, \qquad Y = R\sin\theta.$$
Conversely,
$$R = \sqrt{X^2 + Y^2}, \qquad \theta = \arctan(Y/X).$$

In the rectangular coordinate system of Fig. 3(a), X is taken to be positive when the point P is to the right of the origin, negative when P is to the left. The direction to the right is called the *positive X*-direction; that to the left is called the *negative X*-direction. Similarly Y is positive for a point above the origin, negative for a point below. These relations are illustrated in Fig. 4. The plane is said to be divided into four quadrants, called the first, second, third, and fourth, as specified in Fig. 4. When both X and Y are positive, the point is in the first quadrant; when X is negative but Y positive, it is in the second quadrant; and so forth.

In polar coordinates, *R is always positive*—it is the distance from

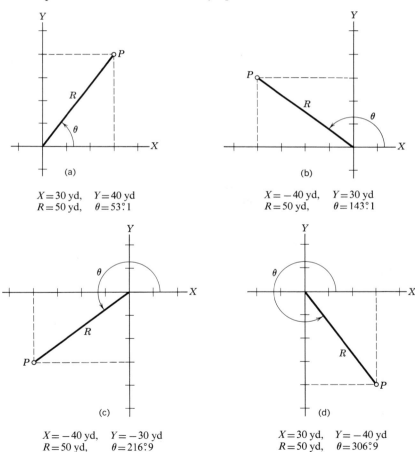

(a)
$X = 30$ yd, $Y = 40$ yd
$R = 50$ yd, $\theta = 53°.1$

(b)
$X = -40$ yd, $Y = 30$ yd
$R = 50$ yd, $\theta = 143°.1$

(c)
$X = -40$ yd, $Y = -30$ yd
$R = 50$ yd, $\theta = 216°.9$

(d)
$X = 30$ yd, $Y = -40$ yd
$R = 50$ yd, $\theta = 306°.9$

Fig. 4. Examples of coordinates of points in (a) the first quadrant, (b) the second quadrant, (c) the third quadrant, (d) the fourth quadrant.

the origin. The quadrant is determined by the size of the angle θ, which is measured counterclockwise from the positive X-axis and goes from $0°$ to $360°$ to cover the entire plane.

3. Displacement

If an object, such as a football, moves from O to P in Fig. 3, its change in position is measured by a physical quantity called the *displacement*. The displacement is represented by the arrow-tipped line segment D drawn from O to P in Fig. 3(c). The directed line segment D has two essential properties:

1. *Magnitude,* as indicated by the length of the line segment.
2. *Direction,* as indicated by the angular orientation of the line and the placement of the arrowhead on a particular end of the line.

Displacement is an example of the type of physical quantity called a *vector quantity,* or simply a *vector*. A vector requires for its specification both a magnitude and a direction. Throughout physics we find that a great many quantities are characterized by both magnitude and direction and have a common method of 'addition,' identical with the method of addition of displacements that we shall discuss in Section 4; hence, the usefulness of the vector concept.

> A **vector quantity** is one that is specified by a magnitude and a direction and that has a rule for addition identical with the rule for addition of displacements.

If a man has his home in Flushing, on Long Island, and his office in Wall Street, on Manhattan Island, his task each evening is to effect a displacement of himself from office to home. In accomplishing this displacement it is immaterial, so far as net physical result is concerned, whether he takes the subway, rides the buses, or uses his automobile. The route he follows will be quite different in these three cases but the change in position will be the same— a displacement from office to home that can be represented by the vector D in Fig. 5.

Displacement is a vector quantity that is a measure of *change in position*. When a particle moves from point P to point Q, its displacement is given by the length and direction of the vector whose tail is at P and whose head is at Q. The displacement

Figure 5

is independent of the actual path that may have been followed in the motion from P to Q.

A quantity that has no associated direction, and that can be completely specified by giving merely a magnitude, is called a *scalar quantity* or simply a *scalar*. A scalar is completely specified by the statement of a numeral and a unit—for example, a change in volume of *minus ten gallons*. On the other hand, *a vector requires for its specification a numeral, a unit, and a direction*—for example, a displacement of *two feet eastward*.

A **scalar quantity** is one that is completely specified by a magnitude.

Two vectors are *equal* if they have the same magnitude and direction. Thus, an airplane that moves 10 mi northeastward from the Chicago airport and one that moves 10 mi northeastward from the Washington airport have the *same* displacement; the *change* in position is the same in each case, represented by a vector 10 mi long pointing northeastward. Thus, the three vectors in Fig. 6 are considered equal; in fact, the three arrows represent the *same* vector.

A vector is usually denoted in print by boldface type. A vector can be denoted conveniently in handwriting by underscoring the letter or by putting an arrow over it. The *magnitude* of a vector is indicated in ordinary type; thus, D denotes the magnitude of the vector **D**. The magnitude of a vector is always taken as a *positive* quantity. Thus, if the displacement

$$\mathbf{D} = 10 \text{ mi northeastward,}$$

its magnitude $\qquad D = 10 \text{ mi.}$

A negative sign before the symbol indicating a vector merely changes the sense of the direction; that is, it interchanges the head (arrow tip) and tail without changing the length or orientation of the line segment. This convention is illustrated in Fig. 7. A numerical factor before the symbol indicating a vector changes the magnitude of the vector by that

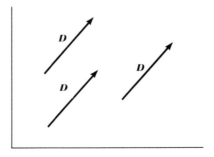

Fig. 6. Equal vectors.

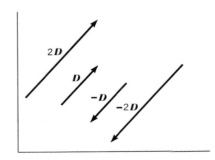

Fig. 7. The vectors $-\mathbf{D}, 2\mathbf{D}, -2\mathbf{D}$.

factor without a change in direction; thus, $2\mathbf{D}$ indicates a vector whose magnitude is twice that of $\mathbf{D}$ but whose direction is the same. It follows that $-2\mathbf{D}$ indicates a vector parallel to $\mathbf{D}$ with the opposite sense of direction and twice the magnitude.

It is at once evident from comparison of Figs. 3(b) and 3(c) that the vector representing a displacement from O to P can be specified in terms of the polar coordinates of P relative to O. The polar coordinate R is the magnitude D of the vector $\mathbf{D}$, and the angle θ defines the direction of $\mathbf{D}$ in the plane of the diagram.

The displacement vector $\mathbf{D}$ could also be specified by giving rectangular coordinates of P relative to O. In this method of specifying a vector $\mathbf{D}$, the projections of the vector on the X- and Y-axes are called the *scalar components* of the vector $\mathbf{D}$. These components are denoted by the symbols D_X and D_Y, as in Fig. 8.

The **scalar components** D_X and D_Y of a vector $\mathbf{D}$ specify the X- and Y-coordinates of the head of the arrow representing $\mathbf{D}$ when the tail is placed at the origin.

It is seen that with θ measured counterclockwise from the $+X$-axis as in Fig. 8, the components are correctly given by the relations

$$D_X = D\cos\theta, \qquad D_Y = D\sin\theta,$$

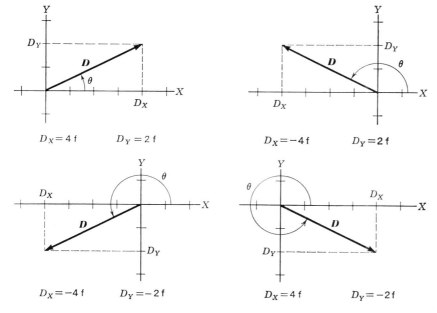

Fig. 8. Illustrating the signs attached to the scalar components D_X and D_Y of a displacement vector.

where D, the magnitude of the vector $\boldsymbol{D}$, is always a positive number.

The discussion above has been confined to vectors lying in a given plane. If we desire to represent a vector pointing in an arbitrary direction in space, we choose a rectangular XYZ-coordinate system and represent the vector by three components, D_X, D_Y, and D_Z, which are its projections, taken with proper signs, on the three coordinate axes.

Example. *Find the magnitude D and the angle θ for each vector in Fig. 8.*

In each case the magnitude

$$D = \sqrt{(4\text{ f})^2 + (2\text{ f})^2} = \sqrt{20}\text{ f} = 4.47\text{ f}.$$

The *acute* angle between $\boldsymbol{D}$ and the $+$ or $-X$-axis has its tangent equal to $2/4 = 0.500$. The angle whose tangent is 0.500 is $26°.6$. Placing this angle in the proper quadrant gives, in the four successive cases, $\theta = 26°.6$; $\theta = 180° - 26°.6 = 153°.4$; $\theta = 180° + 26°.6 = 206°.6$; $\theta = 360° - 26°.6 = 333°.4$.

4. Addition of Vectors

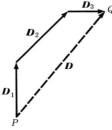

Figure 9

Let us now consider how several successive displacements are combined to find the resultant displacement. Suppose that a morning commuter first drives 3 mi north, effecting the displacement $\boldsymbol{D}_1$ in Fig. 9; then he drives 4 mi on a street running northeast, effecting the displacement $\boldsymbol{D}_2$; and finally drives 2 mi east, effecting the displacement $\boldsymbol{D}_3$. Having started at point P, he ends up at point Q. His *resultant* displacement from home is represented by the vector $\boldsymbol{D}$ connecting P to Q. The vector $\boldsymbol{D}$ is called the *sum* of the vectors $\boldsymbol{D}_1$, $\boldsymbol{D}_2$, and $\boldsymbol{D}_3$. It expresses the net displacement resulting from the several successive displacements. Mathematically, $\boldsymbol{D}$ is written as

$$\boldsymbol{D} = \boldsymbol{D}_1 + \boldsymbol{D}_2 + \boldsymbol{D}_3.$$

The **+** signs here are made heavy to indicate that this is not ordinary addition but a special kind of addition called *vector addition* that is illustrated in Fig. 9.

> The **sum of several vectors** is the single vector leading from the tail of the first vector to the head of the last when the vectors are placed tail to head in any order.

No matter whether the vectors represent displacements or any other type of vector quantity, the sum is defined in this way.

To add the vectors $\boldsymbol{D}_1$, $\boldsymbol{D}_2$, $\boldsymbol{D}_3$, $\boldsymbol{D}_4$ shown at the left of Fig. 10, start at the origin and place the vectors tail to head in any order. The sum $\boldsymbol{D}$ is then the single vector leading from the origin to the head of the last vector in the series, as indicated at the right of Fig. 10. That the

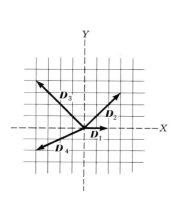
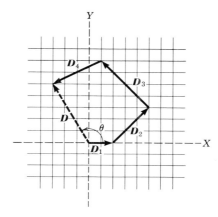

Fig. 10. Addition of displacement vectors (coordinate units in feet):

$D_{1X}=$ 2 f	$D_{1Y}=$ 0 f	$\tan\theta = D_Y/D_X = (5\text{ f})/(-3\text{ f}) = -1.667$
$D_{2X}=$ 3	$D_{2Y}=$ 3	$\theta = 121.^\circ 0$
$D_{3X}=-4$	$D_{3Y}=$ 4	$D = \sqrt{D_X^2 + D_Y^2} = \sqrt{34\text{ f}^2}$
$D_{4X}=-4$	$D_{4Y}=-2$	$D = 5.83$ f.
$D_X = -3$ f	$D_Y =$ 5 f	

vector $\boldsymbol{D}$ obtained by this procedure is independent of the order in which the vectors are placed tail to head will be demonstrated presently.

It is easy to show that the components of the vector sum $\boldsymbol{D}$ are just the sums of the components of the vectors that are added. Thus, if

$$\boldsymbol{D} = \boldsymbol{D}_1 + \boldsymbol{D}_2 + \boldsymbol{D}_3 + \boldsymbol{D}_4, \tag{1}$$

we can show that
and
$$\left. \begin{array}{l} D_X = D_{1X} + D_{2X} + D_{3X} + D_{4X}, \\ D_Y = D_{1Y} + D_{2Y} + D_{3Y} + D_{4Y}. \end{array} \right\} \tag{2}$$

To prove this statement, consider the diagram at the right of Fig. 10. Since $D_{1X}=2$ f, the head of $\boldsymbol{D}_1$ is 2 f to the right of the Y-axis. Since $D_{2X}=3$ f, the head of $\boldsymbol{D}_2$ is 3 f further to the right, or 2 f + 3 f = 5 f to the right of the Y-axis. Since $D_{3X}=-4$ f, the head of $\boldsymbol{D}_3$ is back 4 f, or 2 f + 3 f − 4 f = 1 f to the right of the Y-axis; and since $D_{4X}=-4$ f, the head of $\boldsymbol{D}_4$ is back 4 f more, or 2 f + 3 f − 4 f − 4 f = −3 f to the right of the Y-axis (this means 3 f to the left). But the position of the head of $\boldsymbol{D}_4$ is the same as the position of the head of $\boldsymbol{D}$, and the position of the head of $\boldsymbol{D}$ relative to the Y-axis is the X-component D_X. Hence $D_X = 2\text{ f} + 3\text{ f} - 4\text{ f} - 4\text{ f} = -3\text{ f} = D_{1X} + D_{2X} + D_{3X} + D_{4X}$. A similar argument may be applied to the Y-component of $\boldsymbol{D}$. Thus we see that the geometrical definition that we have given for the vector sum (1) implies the analytical relations (2) between the components. Since the relations (2) give values of

D_X and D_Y that are independent of the order in which the components of $\boldsymbol{D}_1$, $\boldsymbol{D}_2$, $\boldsymbol{D}_3$, and $\boldsymbol{D}_4$ are added, the geometrical construction must give a vector $\boldsymbol{D}$ that is independent of the order in which $\boldsymbol{D}_1$, $\boldsymbol{D}_2$, $\boldsymbol{D}_3$, and $\boldsymbol{D}_4$ are placed tail to head. To illustrate this statement, the reader should make diagrams like that at the right of Fig. 10, but in which the same vectors are placed tail to head in different orders.

In most cases we shall wish to use analytical rather than graphical methods for the addition of vectors, employing the rules in equations (2):

The X-component of the sum of a number of vectors is the sum of the X-components of the vectors.

The Y-component of the sum of a number of vectors is the sum of the Y-components of the vectors.

After we have found the components D_X and D_Y, we can find the magnitude D and direction θ from Pythagoras' theorem and simple trigonometry, as in the legend of Fig. 10.

For the addition of just two vectors, the following rule is frequently given:

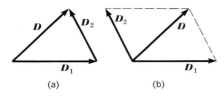

Fig. 11. Addition of vectors $\boldsymbol{D}_1$ and $\boldsymbol{D}_2$ by (a) laying them off end to end, (b) completing the parallelogram.

Lay off the vectors from a common origin and complete the parallelogram. The sum is the directed diagonal from the origin to the opposite corner of the parallelogram.

This procedure is illustrated in Fig. 11(b). Comparison with Fig. 11(a) shows that this rule is equivalent to the one we have previously given.

If we wish to add vectors that are not coplanar, we must work in three-dimensional space. One can imagine the vectors as represented by actual material arrows placed end to end just as we placed our directed lines in two dimensions. The sum is, by definition, the vector leading directly from the tail of the first arrow to the head of the last. Three relations, similar to those in (2), give the three components of the sum.

5. Subtraction of Vectors

In Fig. 11, we see that

$$\boldsymbol{D} = \boldsymbol{D}_1 + \boldsymbol{D}_2. \tag{3}$$

From this equation, it would be convenient to write

$$\boldsymbol{D}_2 = \boldsymbol{D} - \boldsymbol{D}_1, \tag{4}$$

with $\boldsymbol{D}_1$ *subtracted* from the sum $\boldsymbol{D}$ to obtain $\boldsymbol{D}_2$. Equation (4) is valid if we *define subtraction*, just as in ordinary algebra, as *addition of the negative:*

$$D_2 = D + (-D_1), \qquad (5)$$

where we have already defined, as in Fig. 7, the negative of D_1 as a vector of the same magnitude but opposite in direction. That equation (5) is valid for the vectors of Fig. 11 is demonstrated graphically in Fig. 12.

We see that if the vector D has components D_X and D_Y, the vector $-D$ has components $-D_X$ and $-D_Y$. Thus, for any vector quantities A and B, the vector difference,

$$C = A - B,$$

is represented graphically by

$$C = A + (-B)$$

and analytically by

$$C_X = A_X - B_X \qquad C_Y = A_Y - B_Y.$$

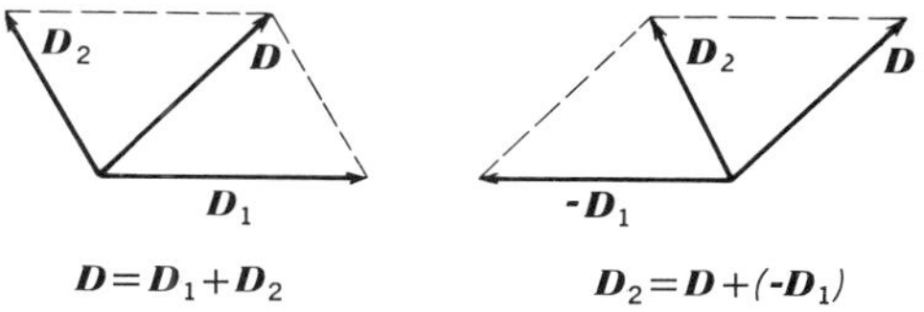

$D = D_1 + D_2$ $\qquad D_2 = D + (-D_1)$

Figure 12

We shall have frequent occasion to employ vector subtraction when we define velocity and acceleration in the following chapter.

6. Resolution of a Vector

The process of resolution of a vector is the inverse of the process of vector addition in that it *replaces* a vector by two or more vectors which have the given vector as their sum. The most useful type of resolution is that in which the vector is replaced by its vector projections on a set of mutually perpendicular axes. If attention is confined to a plane, we use two such axes, but if we are working in three dimensions we must use three. These vector projections are known as *vector components*. Figure 13 shows four examples of the resolution of a vector D into vector components D_1 and D_2, such that

$$D = D_1 + D_2.$$

These *vector components* are related to, but must be distinguished from, the *scalar components* previously introduced.

> The **vector components** D_1 and D_2 (*see* Fig. 13) are two *vectors* lying along the X- and Y-axes whose sum is D and which can together replace D.

It must be emphasized that the vector components can be used on a diagram only to *replace* the original vector, *not to supplement it*. For

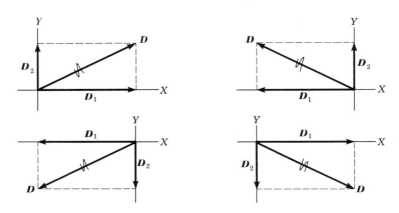

Fig. 13. Resolution of a vector into vector components.

this reason the original vector is shown crossed out in Fig. 13 after it has been replaced by two vector components.

PROBLEMS

NOTE: *The numerical data given in the problems in this book should be regarded as exact numbers, valid to as many significant figures as may be required. Unless otherwise instructed, give all answers to three significant figures. Angles should be given in degrees and tenths.*

1. A boy is 5 f tall. What is his height in meters? Ans: 1.52 m.

2. Find the length of a football field in meters.

3. If the distance between two cities is 750 km, what is their separation in miles? Ans: 466 mi.

4. A living room is 15 f wide and 24 f long. What is the area of the room in square meters?

5. If the arc length S in the figure is 3.0 m and the radius R is 12 m, what is the angle θ in radians? in degrees? in revolutions? Ans: 0.250 rad; 14°.3; 0.0398 rev.

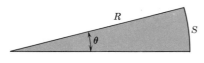

Problem 5

6. The diameter of the moon is 2160 mi, and its mean distance from the earth is 239 000 mi. The diameter of the sun is 864 000 mi, and its mean distance from the earth is 92 900 000 mi. Determine the apparent angular diameters of the moon and the sun, as seen from the earth when they are at their mean distances, in radians. Express these angular diameters in minutes, and show that they are both approximately equal to 0°.5.

7. The rectangular coordinates of a point are $X = 4$ f, $Y = 3$ f. Give the polar coordinates of this point. Ans: $R = 5$ f; $\theta = 36°.9$.

8. The rectangular coordinates of a point are $X = -4$ f, $Y = 8$ f. Give the polar coordinates of this point.

9. The polar coordinates of a point are $R = 20$ f, $\theta = 225°$. What are the rectangular coordinates of this point? Ans: $X = -14.1$ f; $Y = -14.1$ f.

10. The polar coordinates of a point are $R = 9$ m, $\theta = 300°$. Find its rectangular coordinates.

11. A man travels 6 mi eastward, and then 4.5 mi northward. Find the magnitude and direction of the resultant displacement from his starting point.
Ans: 7.5 mi; $\theta = 36°.9$ (N of E).

12. A man travels first 10 mi northeastward, then 12 mi southward, and finally 8 mi in a direction 30° N of W. What are the magnitude and direction of his resultant displacement?

13. An airplane flies 50 mi in a direction 30° N of E. What is the eastward component of the plane's displacement? the northward component? Ans: 43.3 mi; 25.0 mi.

14. An automobile travels 3 mi uphill along a roadway. If the roadway makes an angle of 6° with the horizontal, find the magnitude of the automobile's horizontal and vertical components of displacement.

15. In the figure, $D_{1X} = 4$ f, $D_{1Y} = 0$; $D_{2X} = -1$ f, $D_{2Y} = 5$ f. Find the vector difference, $D_3 = D_2 - D_1$. Verify your result by showing, graphically and algebraically, that $D_1 + D_3 = D_2$. Ans: $D_{3X} = -5$ f, $D_{3Y} = 5$ f.

Problem 15

16. Find the sum of the three displacements, $A + B + C$, in the figure.

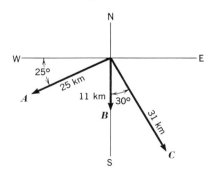

Problems 16, 18

17. Find the magnitude R and the angle θ for the sum of the three displacement vectors in the figure. Ans: $R = 65.6$ km; $\theta = 69°.2$.

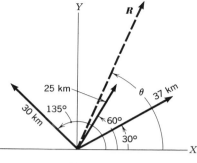

Problem 17

18. In the figure, find the magnitude and direction of the vector $A - B + C$.

19. Find the sum of five displacements having the following magnitudes and making the following angles, measured counterclockwise from the horizontal X-axis: A, 37 m, 350°; B, 15 m, 80°; C, 14.5 m, 170°; D, 31.5 m, 90°; E, 7.5 m, 180°. Ans: 45.8 m, 67°.8.

20. A mark is put on the uppermost portion of the rim of a wheel of 3-f radius, mounted on a fixed horizontal axle. What is the displacement of this mark when the wheel turns clockwise through ¼ rev? When it subsequently turns through ½ rev? Then when it makes 1 more whole revolution? What is the sum of these three displacements? Draw a diagram showing the three displacement vectors and their sum.

21. What is the displacement of the point of a wheel initially in contact with the ground when the wheel *rolls* forward ½ rev? Take the radius of the wheel as R and the X-axis in the forward direction.
Ans: $3.72 R$ at $32°.5$ with the X-axis.

22. By means of the rules for vector addition, show that $C^2 = A^2 + B^2 + 2AB \cos\phi$, a relation

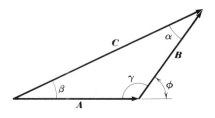

Problem 22

called the 'law of cosines,' and $A/\sin\alpha = B/\sin\beta = C/\sin\gamma$, a relation called the 'law of sines.' Note that, in the figure, $\mathbf{C} = \mathbf{A} + \mathbf{B}$.

23. The length of the minute hand on the north face of a large tower clock is 5 f. What is the magnitude of the displacement of a point at the very end of the hand between 4 P.M. and 4:15 P.M.? What is the length of the path actually traversed by the point in question? What are the northward, the eastward, and the upward components of the displacement?
Ans: 7.05 f; 7.85 f; 0, -5 f, -5 f.

24. Find the magnitude of the displacement of the point on the clock hand described in the preceding problem during the time interval 4:30 P.M. to 4:50 P.M. What is the length of the path actually traversed? Find the northward, eastward, and upward components of the displacement.

25. In working with vectors in three-dimensional space, it is necessary to give *three* numbers in order to specify the magnitude and direction of the vector. For example, the displacement $\mathbf{D}$ in the figure can be defined by giving its scalar components D_X, D_Y, and D_Z. (a) From two successive applications of the Pythagorean theorem, show that $D^2 = D_X^2 + D_Y^2 + D_Z^2$. (b) Show that $D_X = D \cos\phi_X$, $D_Y = D \cos\phi_Y$, and $D_Z = D \cos\phi_Z$; these relations make it possible to describe a vector completely by giving its magnitude D and its *direction angles* ϕ_X, ϕ_Y, ϕ_Z, or their cosines, which are known as *direction cosines*. (c) However, since only three numbers are needed, the angles are not completely independent. Show that they are related by the equation $\cos^2\phi_X + \cos^2\phi_Y + \cos^2\phi_Z = 1$.

26. A vector can also be specified in terms of its magnitude D, the polar angle θ, and the azimuthal angle ϕ, which gives the angle between the X-axis and the projection of $\mathbf{D}$ on the XY-plane (*see figure*). Show that $D_X = D \sin\theta \cos\phi$, $D_Y = D \sin\theta \sin\phi$, and $D_Z = D \cos\theta$.

27. Approximating the earth by a sphere of 4000-mi radius, determine the magnitude of the displacement of an airplane that flies from a point at 40° N latitude on the prime meridian to a point at 40° N latitude on the 180° meridian. What is the length of the path actually traversed by the plane if the pilot follows the 40° N latitude parallel? If he flies directly over the north pole? Employ a coordinate system attached to the earth in determining the displacement. Ans: 6130 mi; 9620 mi; 6990 mi.

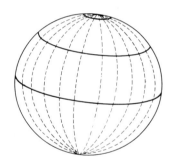

Problems 27–28

28. An airplane starts at the point where the equator intersects the prime meridian, flies 4000 mi due north, 4000 mi due east, 4000 mi due south, and finally 4000 mi due west. Find (a) the latitude and longitude of the plane's final position and (b) the magnitude of the plane's final displacement from its starting point. Approximate the earth by a sphere of 4000-mi radius, and use a coordinate system attached to the earth in determining the displacement. What other coordinate systems might be chosen?

29. Resolve a displacement of 100 mi in a direction 60° s of E into two vector components, one of which is directed eastward.
Ans: 86.6 mi, s; 50 mi, E.

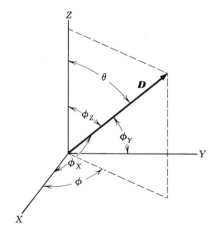

Problems 25–26

30. A ship sails from its home port and after 10 h is 60 mi north of its starting point. Resolve the displacement of the ship into two vector components—one northeastward and the other northwestward. How much further would the ship travel in reaching its final position if it actually sailed northeastward and then northwestward, rather than due northward?

31. An aircraft takes off in the northeast direction at an angle of 10° with the horizontal and flies in a straight line for five miles. Determine the eastward, the northward, and the upward components of its displacement during this five-mile flight. Determine the direction angles of the displacement vector, as specified in Prob. 25, if the X-axis is eastward, the Y-axis is northward, and the Z-axis is vertical. Verify that the sum of the squares of the direction cosines equals unity.

Ans: 3.48, 3.48, 0.868 mi; 45°.9, 45°.9, 80°.

3 KINEMATICS OF A PARTICLE

Kinematics is the branch of science concerned with the quantitative description of motion.

Motion of a body of any type can be described as a continuous change in the position of each particle of the body. The scope of kinematics is limited to a mere *description of motion* and does not include a treatment of the causes of the motion; treatment of these causes is the subject matter of *dynamics*.

In this chapter we shall confine our attention to the description of the motion of a *particle*. This description is fundamental to later consideration of the motion of extended bodies, since in the last analysis such bodies are collections of particles. In Chapters 8 through 10, we shall consider the motion of *rigid bodies,* while in Chapter 13 we shall consider the motion of fluids.

At any instant, the position of an isolated particle or the position of a particular particle of a fluid or of a rigid body is represented by the position of a geometrical point. To describe the motion of this point, and hence of the physical particle, we shall define and learn to compute the vector quantities *velocity* and *acceleration*.

1. Time Intervals

A kinematical description of motion gives position as a function of *time;* hence, it is necessary to define the units used to measure time intervals.

Astronomical observations were used for many centuries to define the time unit. In its orbital motion, the earth *revolves* around the sun in one *year*. The length of the year is determined by observing the apparent position of the sun with respect to the 'fixed' stars; i.e., the stars that are so distant that they have no apparent motion with respect to one another. During the course of one year the sun appears to make a complete circuit around the sky with respect to the fixed stars. Thus, if the position of the sun relative to the stars is noted at a given time, the time required for the sun to circle the sky and return to the same position relative to the stars is one year. Until recently, the *year* furnished the basic unit of time interval.

However, because the earth's orbital motion may be subject to some slight variation over a period of many years, a time interval based on astronomical 'events' has proved unsatisfactory as a standard on which modern science and technology can be based. To establish a more reproducible and 'permanent' standard of time interval, the International Bureau of Weights and Measures adopted, in October 1946, an 'atomic' standard, in line with the similar standard of length. The cesium atom, which has only one isotope (all atoms are exactly alike), has a convenient frequency of radiation in the microwave radar region. In terms of this frequency,

1 **second** = time of 9 192 631 770 Cs 'vibrations.'

In terms of the second, other common units of time are defined as follows:
$$
\begin{aligned}
1 \text{ minute} &= 60 \text{ seconds} \\
1 \text{ hour} = 60 \text{ min} &= 3\ 600 \text{ s} \\
1 \text{ day} = 24 \text{ h} = 1440 \text{ min} &= 86\ 400 \text{ s}.
\end{aligned}
$$

The abbreviations for hour, minute, and second are h, min, and s, respectively. The definition of the *second,* as given above, is so chosen that the *day* corresponds to a *mean solar day* and very closely approximates the time interval between successive local noons averaged over a long period of years. Local noon is the instant the sun passes over the local observer's geographical *meridian*. Because the earth's orbit is not exactly circular, the length of a solar day varies appreciably during a given year and may show slight variations from year to year—hence, the necessity of speaking of a *mean* solar day. In terms of the day, astronomical observations, actually extending over several millenia, have given the observed value:

$$1 \text{ year} = 365.242\ 198\ 79 \text{ day}.$$

The units, year and day, are abbreviated y and d respectively.*

One accurate type of clock is furnished by the apparent revolution of the fixed stars around the earth. The time between successive meridian passages of a star is called a *sidereal day*. Although the rotational speed of the earth is very nearly constant, sidereal days do not furnish a convenient practical time unit because there is one more sidereal day per year than there are solar days. The word *day*, used without qualification, denotes the *solar day*.

We note that, in astronomical terminology, the earth and the other planets *rotate* about their axes and *revolve* about the sun. The period of rotation of the earth is one sidereal day $= {}^{365}\!/_{366}$ day; the period of revolution is one year.

2. Speed

Let us consider the motion of an automobile along a level roadway that is curved as in Fig. 1. The average speed of the car in moving from point O to point A is defined as the ratio of the total path length or distance traversed *along the road* to the total time elapsed while the motion is taking place:

The **average speed** of traversing a given path is a *scalar quantity* defined as the total length of the path divided by the time elapsed.

The path length and the elapsed time are both scalar quantities, and hence the ratio of these quantities is also a scalar. If the total path length from O to A is L and the total elapsed time is t, the average speed between O and A will be L/t. To determine the average speed during the portion of the trip between points B and E, we would compute $\Delta l/\Delta t$ where Δt is the time elapsed during the motion of the car from B to E.

By considering successively smaller path lengths Δl, always including the point C, we obtain

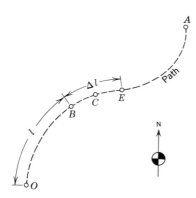

Figure 1

*The official international abbreviation for *year* is a, for *annus*.

values for average speed that approach a limit called the *instantaneous speed of the car at point C:*

> The **instantaneous speed** at a point is defined as the limit of the average speed over a path length that approaches zero but always includes the point.

In applying this definition, we note that as the path length Δl becomes smaller and smaller, the time interval Δt by which Δl is to be divided also becomes smaller and smaller, and the ratio approaches a finite limit. The instantaneous speed v is defined in calculus notation by the expression

$$v = \lim_{\Delta t \to 0} \frac{\Delta l}{\Delta t} = \frac{dl}{dt}. \tag{1}$$

This definition of instantaneous speed implies that the traversed path length l is a function of *time,* which can be regarded as the independent variable. The length l of path measured from O is indeed a function of the time t elapsed after the car has left point O.

Speed is a scalar quantity. The units in which it is commonly expressed include f/s, m/s, mi/h, and km/h. The conventional speedometer gives, to a good approximation, a direct reading of 'instantaneous' speed. It can be shown that the average speed, as we have defined it, is the *time* average of the instantaneous speed.*

3. Velocity

The general problem of kinematics is one of finding the *position* of a body as a function of time. Hence it is necessary to define a physical quantity that gives not only the *speed* of a body but also the *direction* in which the body is moving. The quantity that has these properties is the vector quantity called *velocity*. The *average velocity* of a particle is defined as the ratio of its *displacement* to the length of the time interval in which the displacement occurs:

---

*The proof of this statement involves a simple application of calculus. The time average of the instantaneous speed, between times t_0 and t_1, is, by the definition of time average:

$$\frac{1}{t_1 - t_0} \int_{t_0}^{t_1} v\, dt.$$

Since $v = dl/dt$, this becomes

$$\frac{1}{t_1 - t_0} \int_{t_0}^{t_1} \frac{dl}{dt} dt = \frac{1}{t_1 - t_0} \int_{l_0}^{l_1} dl = \frac{l_1 - l_0}{t_1 - t_0},$$

if $l_1 - l_0$ is the path length traversed in time $t_1 - t_0$. But the last expression is just our definition of average speed.

The **average velocity** of a particle during a specified time interval is a *vector quantity* defined as the displacement of the particle during that time interval divided by the time interval.

Since displacement is a vector and time is a scalar, the ratio is a vector quantity that has the same direction as the displacement.

Figure 2(a) shows again the path of Fig. 1. The actual path of the car is given by the dotted curve; the displacement $\boldsymbol{D}$ of the car in making the complete trip from O to A is given by the arrow. Denoting the total elapsed time for the trip by t, we may write the average velocity $\bar{\boldsymbol{v}}$ as

$$\bar{\boldsymbol{v}} = \boldsymbol{D}/t. \qquad (O \rightarrow A)$$

The average velocity $\bar{\boldsymbol{v}}$ can be represented by the arrow shown in (b); the length of the arrow gives a measure of the magnitude of the average velocity and the direction of the arrow is the same as that of the arrow representing $\boldsymbol{D}$.

Let us now write the expression for the average velocity of the car in moving from B to E. As measured from the starting point, the displacement of the car at B is represented in Fig. 2(c) by the arrow labeled $\boldsymbol{D}_B$ and the displacement of the car at E is given by the arrow labeled $\boldsymbol{D}_E$. The change in displacement of the car in going from B to E is given by the arrow labeled $\Delta \boldsymbol{D}$. This vector $\Delta \boldsymbol{D}$ must be added to the displacement $\boldsymbol{D}_B$ to obtain the displacement $\boldsymbol{D}_E$. Since $\boldsymbol{D}_E = \boldsymbol{D}_B + \Delta \boldsymbol{D}$, $\Delta \boldsymbol{D}$ is the vector difference (*see* pp. 24–25):

$$\Delta \boldsymbol{D} = \boldsymbol{D}_E - \boldsymbol{D}_B.$$

The average velocity of the car in moving from B to E is given by

$$\bar{\boldsymbol{v}} = \Delta \boldsymbol{D}/\Delta t, \qquad (B \rightarrow E)$$

where Δt is the time required for the car to move from B to E. This average velocity can be represented by the arrow shown in Fig. 2(d).

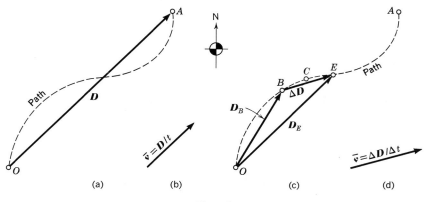

Figure 2

The direction of this arrow is parallel to the arrow representing $\Delta \boldsymbol{D}$; the length of the arrow gives a measure of the magnitude of the average velocity.

> The **instantaneous velocity** at a point is defined as the limit of the average velocity over a path length that approaches zero but always includes the point.

In calculus notation, considering the displacement $\boldsymbol{D}$ from O as a function of time, we can write the instantaneous velocity as

$$\boldsymbol{v} = \lim_{\Delta t \to 0} \frac{\Delta \boldsymbol{D}}{\Delta t} = \frac{d\boldsymbol{D}}{dt}. \tag{2}$$

Just as in the case of speed, it can be shown that the average velocity, as we have defined it, is the *time* average of this instantaneous velocity.*

The units in which the magnitude of the velocity is measured are the same as those in which speed is measured, but the statement of the velocity must include a direction. For example, the velocity of a particle might be stated as 30 mi/h eastward or 12 m/s upward. Hereafter, unless otherwise indicated, the term *velocity* means *instantaneous velocity*.

Now let us see how we can compute the velocity of a particle if we are given its X-, Y-, and Z-coordinates as functions of time. Suppose that the particle traverses the space path indicated in Fig. 3 and we are given functions $X(t)$, $Y(t)$, and $Z(t)$ that give its coordinates at any time t. We desire to determine the velocity of the particle at a certain point P on its path. Starting from P, the particle will undergo a displacement $\Delta \boldsymbol{D}$ in a time Δt, as indicated on Fig. 3. The vector $\Delta \boldsymbol{D}$ has rectangular components ΔX, ΔY, and ΔZ as indicated. The velocity at P, defined by (2), is a vector which has components

$$v_X = \lim_{\Delta t \to 0} \frac{\Delta X}{\Delta t}; \qquad v_Y = \lim_{\Delta t \to 0} \frac{\Delta Y}{\Delta t}; \qquad v_Z = \lim_{\Delta t \to 0} \frac{\Delta Z}{\Delta t}.$$

But, by definition, these three limits are just the derivatives of the three variables X, Y, and Z, and so we obtain the components of the velocity by differentiating these three functions of time:

$$v_X = dX/dt; \qquad v_Y = dY/dt; \qquad v_Z = dZ/dt. \tag{3}$$

By considering what happens in Fig. 3 as the magnitude of $\Delta \boldsymbol{D}$ becomes smaller and smaller, we see that

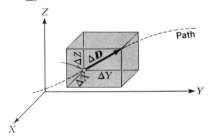

Figure 3

---

*The proof is like that for the case of speed, except that vector relations are involved. The time average of the instantaneous velocity is

$$\frac{1}{t_1 - t_0} \int_{t_0}^{t_1} \boldsymbol{v} \, dt = \frac{1}{t_1 - t_0} \int_{t_0}^{t_1} \frac{d\boldsymbol{D}}{dt} dt = \frac{1}{t_1 - t_0} \int_{D_0}^{D_1} d\boldsymbol{D} = \frac{\boldsymbol{D}_1 - \boldsymbol{D}_0}{t_1 - t_0},$$

which is the average velocity as we have defined it.

The magnitude of the instantaneous velocity of a particle at point P is equal to the instantaneous speed of the particle at P, and the direction of the instantaneous-velocity vector at P is the direction of the tangent to the path at that point.

Since the magnitude of the velocity is the speed, we are justified in our choice of the symbols $\mathbf{v}$ for velocity and v for speed.

The simplest case in which to apply these ideas is the motion of a particle along a straight line. Let us take the straight-line path as the X-axis of our coordinate system. Then there is only one nonvanishing velocity component,

$$v_X = dX/dt.$$

This can be evaluated at any point or at any time, provided the relation between X and t is known. For example, suppose that the position of the particle as a function of time is given by the relation

$$X = a + bt + ct^2, \tag{4}$$

where a, b, and c are constants that must have the dimensions of length, length/time, and length/(time)^2, respectively (*why?*). Then the X-component of the velocity of the particle is given as a function of time by

$$v_X = dX/dt = b + 2ct. \tag{5}$$

In the case of the motion of a particle along the X-axis, the velocity is a vector directed along this axis, in the positive sense if the component v_X is positive, in the negative sense if v_X is negative.

Example. *The position of a particle moving along the X-axis is given by*

$$X = 6 \text{ m} + [8 \text{ m/s}] t - [2 \text{ m/s}^2] t^2.$$

Find the position and velocity of this particle at $t = 0$, 1 s, 2 s, and 3 s. Find the velocity when $X = 13.5$ m.

Differentiation of the above equation gives

$$v_X = dX/dt = 8 \text{ m/s} - [4 \text{ m/s}^2] t.$$

Straightforward substitution gives, at $t = 0$, $X = 6$ m, $v_X = 8$ m/s; at $t = 1$ s, $X = 12$ m, $v_X = 4$ m/s; at $t = 2$ s, $X = 14$ m, $v_X = 0$; at $t = 3$ s, $X = 12$ m, $v_X = -4$ m/s. Thus the velocity vector points in the positive sense along the X-axis from $t = 0$ to $t = 2$ s; at $t = 2$ s the particle is at rest; at later times the X-component of velocity is negative, the velocity vector points in the negative sense, and the particle moves to the left if the positive sense is toward the right. It is apparent from the above computation that the particle is at $X = 13.5$ m twice, once between $t = 1$ s and $t = 2$ s when it is moving toward the right, and once between $t = 2$ s and $t = 3$ s when it is moving toward the left. Substitution of $X = 13.5$ m in the quadratic equation that gives X as a function of t, and solution, gives $t = 1.5$ s and $t = 2.5$ s as these two values of t. Substitution of these values of t in the velocity equation gives $v_X = 2$ m/s and $v_X = -2$ m/s respectively. In this example, which represents a case of constant acceleration that we shall study later in this chapter, when the particle moves to the left through a given point on the X-axis it has a velocity equal and opposite to the velocity it had when it moved through the same point to the right. The reader should verify that when the particle moves back through its initial position ($X=6$ m) it has the velocity component $v_X = -8$ m/s.

4. Relative Velocity

All velocities are relative.

We shall show in the next chapter that it is impossible in principle to determine what would be called an *absolute* velocity of any body. The velocity of a body is always specified *relative* to some other body. When the other body is not mentioned, it is understood to be the earth.

Consider a man walking at 3 mi/h down the aisle of a train. This is his velocity relative to the train, but the train may be moving at 60 mi/h relative to the surface of the earth. In turn, the earth is moving relative to the sun at about 66 000 mi/h because of the motion of the earth in its yearly orbit, on which is superposed a surface velocity, resulting from the daily rotation of the earth, that varies in magnitude from 1000 mi/h at the equator to zero at the poles. The sun itself is moving at about 540 000 mi/h around the center of our galaxy—while the center of our galaxy is moving relative to the other galaxies in the universe! It would be quite a chore to determine the absolute velocity of the man walking down the aisle of the train—if this had a meaning. However, one of the principles of physics (the *principle of relativity*) asserts that absolute velocity has no meaning whatsoever.

Relative velocities add vectorially according to the following rule:

The velocity of body A relative to body C, denoted by v_{AC}, is the vector sum of the velocity of body A relative to body B, v_{AB}, and the velocity of body B relative to body C, v_{BC}, that is,

$$v_{AC} = v_{AB} + v_{BC}. \tag{6}$$

This rule for addition of velocities follows from the fact that displacements add vectorially. Before giving a formal proof, let us consider the particular example illustrated in Fig. 4. Here a river flows north at velocity v_{WE}, the velocity of the water relative to the earth. A boat keeps a heading directly east and is propelled by a motor at velocity v_{BW}, the velocity of the boat relative to the water. We desire to find v_{BE}, the velocity of the boat relative to the earth. Let us take the particular case where the river is one mile wide and is flowing at 8 mi/h, while the boat is propelled relative to the water at 6 mi/h. The boat then takes ten minutes to make the one mile across the river. So long as it keeps its heading due east, it takes ten minutes to cross the river whether the water is stationary or flowing. During these ten minutes, the motor displaces the boat an

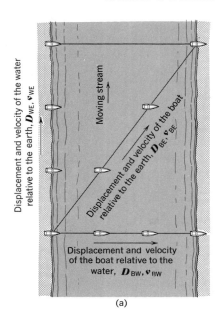 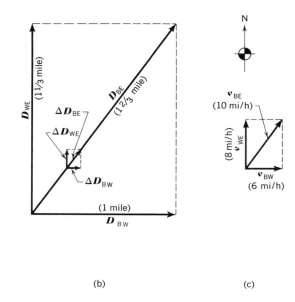

Figure 4

amount D_{BW} relative to the water. But during these same ten minutes the water is displaced $1\frac{1}{3}$ mi to the north. The boat is carried with the water this distance to the north during these ten minutes independent of the progress of the boat across the stream. The net result is that the boat travels the path indicated in Fig. 4(a); is displaced $1\frac{2}{3}$ mi as indicated by D_{BE} in (b); and has velocity relative to the earth of magnitude 10 mi/h in the direction indicated by v_{BE} in (c). (The magnitudes D_{BE} and v_{BE} that we have just given are directly computed from the Pythagoras theorem.)

The proof of (6) follows from arguments similar to the above. In any short period of time Δt,

$$\Delta D_{BE} = \Delta D_{BW} + \Delta D_{WE},$$

since displacements add vectorially. Although Fig. 4 shows the case in which ΔD_{BW} is perpendicular to ΔD_{WE}, this vectorial relation is true for any heading of the boat. Dividing the above equation through by Δt,

$$\Delta D_{BE}/\Delta t = \Delta D_{BW}/\Delta t + \Delta D_{WE}/\Delta t,$$

and going to the limit $\Delta t \rightarrow 0$, we find $v_{BE} = v_{BW} + v_{WE}$, as in (6).

Example. *An airplane flies at an airspeed of* 200 mi/h. *At the altitude at which it is flying, there is a wind of* 100 mi/h *from due west. The plane desires* to proceed due north relative to the ground. In what direction should it head, and what will be its ground speed?

See Fig. 5, in which P stands for plane, A for air, and G for ground. We are given that $v_{AG} = 100$ mi/h east; $v_{PA} = 200$ mi/h at an unknown angle θ; v_{PG}, which by (6) is the vector sum of these, is of unknown magnitude but points due north. It is apparent that the plane must head into the wind as in Fig. 5. From the parallelogram in this figure we see that $200 \sin\theta = 100$, $\sin\theta = \frac{1}{2}$, $\theta = 30°$. Hence the ground speed will be

$$v_{PG} = (200 \text{ mi/h}) \cos 30° = (200 \text{ mi/h})(0.866) = 173 \text{ mi/h}.$$

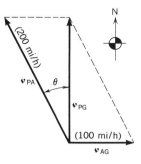

Figure 5

5. Acceleration

In most cases, the velocity of a moving body changes as the motion proceeds, and the body is said to move with *accelerated motion* or to have an *acceleration*. We first define the *average acceleration* of a particle:

The **average acceleration** of a particle during a specified time interval is a *vector quantity* defined as the change in velocity of the particle during that time interval divided by the time interval.

Since the change in velocity is a vector and time is a scalar, the ratio is a vector. If v_0 is the velocity of a body at time t_0 and v_1 is the velocity of the body at time t_1, the average acceleration $\bar{a}$ during the time interval $t_1 - t_0$ is given by

$$\bar{a} = \frac{v_1 - v_0}{t_1 - t_0} = \frac{\Delta v}{\Delta t}.$$

For example, in Fig. 6 the vector Δv is the change in velocity $v_1 - v_0$ that has occurred in the time interval $\Delta t = t_1 - t_0$. The average acceleration vector $\bar{a} = \Delta v / \Delta t$ has the same *direction* as Δv. If Δv is measured in f/s or m/s, the magnitude of the average acceleration is measured in (f/s) *per second* or (m/s) *per second*, that is, in f/s² or m/s².

The *instantaneous acceleration* of a body is defined in a manner analogous to that in which instantaneous velocity was defined:

The **instantaneous acceleration** of a particle at a given point in its path is defined as the limit of the average acceleration over a path length that approaches zero but always includes the given point.

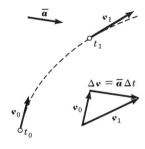

Figure 6

In calculus notation, $$a = \lim_{\Delta t \to 0} \frac{\Delta v}{\Delta t} = \frac{dv}{dt}. \qquad (7)$$

As in the case of speed and velocity, the average acceleration as defined above is the *time* average of the instantaneous acceleration.

Since the vector Δv has components Δv_X, Δv_Y, and Δv_Z, we conclude that, in the motion of a particle along a space curve as in Fig. 3, the components of the acceleration vector are given by

$$a_X = \lim_{\Delta t \to 0} \frac{\Delta v_X}{\Delta t} = \frac{dv_X}{dt} = \frac{d^2 X}{dt^2};$$

$$a_Y = \lim_{\Delta t \to 0} \frac{\Delta v_Y}{\Delta t} = \frac{dv_Y}{dt} = \frac{d^2 Y}{dt^2}; \qquad (8)$$

$$a_Z = \lim_{\Delta t \to 0} \frac{\Delta v_Z}{\Delta t} = \frac{dv_Z}{dt} = \frac{d^2 Z}{dt^2}.$$

In the case of motion along a straight line, the acceleration is directed along the path and has magnitude equal to the rate of change of speed, but in the case of motion in a curved path *neither of these statements is true,* since the *velocity* changes, even if the *speed* does not change, and there is a component of the acceleration normal to the path and directed toward the concave side, as will seem reasonable by study of Fig. 6. We shall consider acceleration in motion along curved paths in Section 7 and in Chapter 8.

Figure 7

Let us first consider the simplest case of motion along a straight line, which we shall take as the X-axis of our coordinate system as in Fig. 7. In this diagram, the particle is at point A at time t_0 and has velocity component v_{X0} in the direction of the positive X-axis. At a later time t_1, the particle is at B and has velocity component v_{X1}. The X-component of the average acceleration during this interval is given by the relation

$$\bar{a}_X = \frac{v_{X1} - v_{X0}}{t_1 - t_0} = \frac{\Delta v_X}{\Delta t}. \qquad (9)$$

This is positive or negative according to whether v_{X1} is greater than or less than v_{X0}. If the velocity is toward the right (v_X positive) but the speed is decreasing ($v_{X1} < v_{X0}$), $\bar{a}_X$ is negative and is said to represent a *deceleration.*

If the velocity is given by equation (5), the acceleration given by (8) has the value

$$a_X = dv_X/dt = 2c;$$

in other words, for the motion of (4), the acceleration is constant. In the example given on p. 36, the reader should verify that the acceleration has the constant magnitude 4 m/s² in the negative X-direction. The problem of rectilinear motion with *constant* acceleration is an important one which we shall next consider in detail.

6. Rectilinear Motion with Constant Acceleration

The simplest type of accelerated motion is that of a particle moving along a straight line with *constant* acceleration. We shall take the X-axis along this line, so that v_X will be positive when the velocity vector points in the $+X$-direction, otherwise negative; a_X will be positive when the acceleration vector points in the $+X$-direction, otherwise negative.

From (9) we see that when a_X is constant, the X-component of velocity of the particle increases or decreases by the same amount in each unit of time. The change Δv_X in any time interval Δt is given by

$$\Delta v_X = a_X \Delta t. \tag{10}$$

Let the body be at the origin $X=0$ at time $t=0$, and let the initial velocity component be v_{X0}. At any later time t the velocity component will have changed by $\Delta v_X = a_X t$, as we see from (10), so that v_X will be

$$v_X = v_{X0} + \Delta v_X = v_{X0} + a_X t. \tag{11}$$

Similarly, the position X of the particle at any time t is given by

$$X = \bar{v}_X t, \tag{12}$$

where $\bar{v}_X$ is the average velocity component during the entire time of motion t. Now, since v_X is changing at a linear rate in (11), the average value during the time interval t equals one-half the sum of the value v_{X0} at the beginning of the interval and the value $v_{X0} + a_X t$ at the end. Thus,

$$\bar{v}_X = \tfrac{1}{2}[v_{X0} + (v_{X0} + a_X t)], \quad \text{or} \quad \bar{v}_X = v_{X0} + \tfrac{1}{2} a_X t. \tag{13}$$

Substitution of this value in (12) gives, for the position X of the particle at any time t,

$$X = v_{X0} t + \tfrac{1}{2} a_X t^2. \tag{14}$$

It is sometimes useful to employ a relation that gives directly the velocity component v_X at any position X. This relation can be obtained by solving equation (11) for t and substituting this expression for t in equation (14). The resulting expression is

$$v_X^2 = v_{X0}^2 + 2 a_X X, \tag{15}$$

where v_X is the velocity component when the particle is at point X. This expression will be derived in a more physically meaningful way in Chapter 6.

Let us now verify the correctness of (14) by differentiation. From (3),

$$v_X = dX/dt = v_{X0} + a_X t,$$

in agreement with (11). A second differentiation, in accordance with (8), gives
$$a_X = d^2X/dt^2 = dv_X/dt = a_X,$$
the assumed constant acceleration.

Example. *A car starting from rest has a constant acceleration of* 4 f/s², *southward, along a straight horizontal roadway. What is its velocity at the end of* 10 s? *What is the car's position at the end of* 10 s? *What is the average velocity of the car during the first* 10 s *of motion?*

On the basis of (11) and the fact that the initial velocity of the car was zero, we obtain, at the end of 10 s,
$$v_X = 0 + (4 \text{ f/s}^2)(10 \text{ s}) = 40 \text{ f/s}$$
as the velocity of the car along a $+X$-axis directed southward.

The displacement of the car in this time interval can be obtained directly from (14):
$$X = \tfrac{1}{2}(4 \text{ f/s}^2)(100 \text{ s}^2) = 200 \text{ f};$$
hence the displacement is 200 f, southward.

The average velocity during the first 10 s is, by definition,
$$v_X = X/t = 200 \text{ f}/10 \text{ s} = 20 \text{ f/s}$$
in the southward direction. This result is seen to be one-half the sum of the initial and final velocities, in agreement with (13).

Example. *A car is traveling at* 60 mi/h *along a straight road. The driver sees a* 30-mi/h *speed-limit sign* 150 f *ahead. What constant deceleration is required to reduce the speed to* 30 mi/h *in this distance?*

We take the $+X$-direction in the direction of travel and use (15) with $v_X = 30$ mi/h = 44 f/s, $v_{X0} = 60$ mi/h = 88 f/s, and $X = 150$ f:
$$(44 \text{ f/s})^2 = (88 \text{ f/s})^2 + 2a_X(150 \text{ f}),$$
$$1936 \text{ f}^2/\text{s}^2 = 7744 \text{ f}^2/\text{s}^2 + (300 \text{ f})\, a_X,$$
from which
$$a_X = -19.4 \text{ f/s}^2.$$
The negative result corresponds to a *deceleration*. The speed *decreases* at the rate of 19.4 f/s (or 13.2 mi/h) *per second*.

The most common example of rectilinear motion with very nearly constant acceleration is the vertical free fall of a body near the earth's surface. The kinematics and dynamics of this problem will be treated in Chapter 4.

7. Motion in a Circle at Constant Speed

In rectilinear motion, the acceleration is tangent to the path and results in change of magnitude of the velocity without change in direction (except for possible *reversal* of the velocity vector).

We have remarked that, for motion on a curved path, the acceleration vector is *not* tangent to the path but must have a component perpendicular to the path pointing toward the concave side, to accomplish the change in direction of the velocity.

We consider first the simplest case, that of *motion in a circle at constant speed*. In this case we shall find that the acceleration has no component tangent to the circular path; it is a vector perpendicular to the path, directed toward the center of the circle, whose whole action is to change the direction of the velocity vector *without* changing its magnitude.

Consider a particle moving at *constant* speed v in a circular path of radius R as shown in Fig. 8(a). *During the time Δt required for the particle to move from A to B*, the body experiences a displacement $\Delta \boldsymbol{D}$ of magnitude equal to the chord of the circle joining points A and B. Now let us consider the velocity-vector diagram shown in Fig. 8(b). Here $\boldsymbol{v}_A$ and $\boldsymbol{v}_B$ represent instantaneous velocities of the particle at points A and B, respectively, and $\Delta \boldsymbol{v}$ represents the change in the velocity of the particle in the time interval Δt. The magnitudes of $\boldsymbol{v}_A$ and $\boldsymbol{v}_B$ are both equal to v, but vectorially $\boldsymbol{v}_B = \boldsymbol{v}_A + \Delta \boldsymbol{v}$.

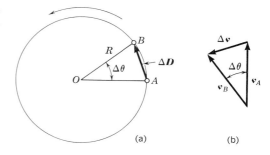

Fig. 8. (a) Particle moving at constant speed in a circular path of radius R, showing the displacement $\Delta \boldsymbol{D}$ in the motion from A to B. (b) Velocity diagram: $\boldsymbol{v}_A$ and $\boldsymbol{v}_B$ give the velocities of the particle at points A and B, respectively.

In circular motion the instantaneous velocity is always tangent to the circle and hence is perpendicular to the radius connecting the particle to the center of the circle. Hence, in the diagrams in (a) and (b), $\boldsymbol{v}_A$ is perpendicular to OA and $\boldsymbol{v}_B$ is perpendicular to OB; hence the angle between $\boldsymbol{v}_A$ and $\boldsymbol{v}_B$ is equal to the angle $\Delta\theta$ between OA and OB. Further, since $v_A = v_B = v$ and $OA = OB = R$, the triangles in (a) and (b) are similar isosceles triangles. Thus, Δv is perpendicular to the chord $\Delta \boldsymbol{D}$ and therefore in the limit as $\Delta\theta \to 0$ is directed toward the center of the circle. Since $\bar{\boldsymbol{a}} = \Delta\boldsymbol{v}/\Delta t$, where $\bar{\boldsymbol{a}}$ is the average acceleration during time Δt, the average acceleration during any time interval Δt has the same direction as Δv and is also directed toward the center of the circle in the limit $\Delta\theta \to 0$.

The instantaneous-acceleration vector is the limit of the average-acceleration vector as $\Delta t \to 0$ and hence $\Delta\theta \to 0$; therefore *the acceleration vector is directed toward the center of the circle at every point in the motion.*

So much for the direction of $\boldsymbol{a}$. We can now easily compute its magnitude. For small $\Delta\theta$, we see from Fig. 8(b) that the magnitude of $\Delta \boldsymbol{v}$ is $v\,\Delta\theta$, with $\Delta\theta$ in radians, since, for small angles, the chord can be set equal to the arc. Hence, from the definition (7), we can write the magnitude of the acceleration as

$$a = \lim_{\Delta t \to 0} \frac{v\,\Delta\theta}{\Delta t} = v\frac{d\theta}{dt}, \qquad (16)$$

since v is a constant. Similarly, we can write the magnitude of the velocity itself as

$$v = \lim_{\Delta t \to 0} \frac{AB}{\Delta t} = \lim_{\Delta t \to 0} \frac{R\, \Delta\theta}{\Delta t} = R\frac{d\theta}{dt}. \tag{17}$$

Dividing (16) by (17) so that $d\theta/dt$ cancels, we find

$$\frac{a}{v} = \frac{v}{R}, \quad \text{or} \quad a = \frac{v^2}{R}. \tag{18}$$

A particle moving in a circle of radius R at constant speed v has acceleration $\mathbf{a}$ which at each instant is directed toward the center of the circle and has magnitude v^2/R.

This is called the *centripetal* ('center-seeking') acceleration. It is this acceleration that accomplishes the continuous turning of the velocity vector in direction, without change of speed.

8. Motion When Acceleration Is Given

Since, as we shall see in the next chapter, the forces acting on a particle determine its acceleration, it is important to learn to compute velocity and position as a function of time *when the acceleration is given.*

We have already solved the particularly simple case of rectilinear motion with *constant* acceleration. In the more general case in which the acceleration is not constant, the methods of the integral calculus are directly applicable in determining velocities and displacements. Let us suppose that the acceleration $\mathbf{a}$ is expressed as a function of time; then, from equation (7),

$$\mathbf{a} = d\mathbf{v}/dt,$$

we find

$$d\mathbf{v} = \mathbf{a}\, dt.$$

The velocity $\mathbf{v}$ is determined by integration:

$$\int_0^t d\mathbf{v} = \int_0^t \mathbf{a}\, dt,$$

or

$$\mathbf{v} - \mathbf{v}_0 = \int_0^t \mathbf{a}(t)\, dt,$$

where $\mathbf{v}_0$ is the velocity at $t=0$. This relation gives

$$\mathbf{v} = \mathbf{v}_0 + \int_0^t \mathbf{a}\, dt. \tag{19}$$

The integral on the right is a *vector* that represents the sum of infinitesimal vectors. Just as a particular component of the sum of several finite vectors equals the sum of that component of the individual vectors (p. 24), so in the limit represented by an integral, a component of a vector integral equals the integral of that component. Hence the scalar components of the velocity are given by

$$\left.\begin{array}{l} v_X = v_{X0} + \displaystyle\int_0^t a_X\, dt, \\[6pt] v_Y = v_{Y0} + \displaystyle\int_0^t a_Y\, dt, \\[6pt] v_Z = v_{Z0} + \displaystyle\int_0^t a_Z\, dt. \end{array}\right\} \quad (20)$$

When the velocity has been determined as a function of time, the displacement can be determined by a second integration based on (2). Since

$$d\mathbf{D} = \mathbf{v}\, dt,$$

we may write

$$\int_0^t d\mathbf{D} = \int_0^t \mathbf{v}\, dt,$$

whence

$$\mathbf{D} - \mathbf{D}_0 = \int_0^t \mathbf{v}\, dt,$$

where we write $\mathbf{D}_0$ for the initial displacement from the origin of coordinates. Thus we have

$$\mathbf{D} = \mathbf{D}_0 + \int_0^t \mathbf{v}\, dt. \quad (21)$$

The scalar components of the displacement are

$$\left.\begin{array}{l} X = X_0 + \displaystyle\int_0^t v_X\, dt, \\[6pt] Y = Y_0 + \displaystyle\int_0^t v_Y\, dt, \\[6pt] Z = Z_0 + \displaystyle\int_0^t v_Z\, dt. \end{array}\right\} \quad (22)$$

The equations presented in this section are of general validity and can be used to derive the equations for the simple case of rectilinear motion with constant acceleration considered in Section 6, as in the following example.

Example. *Employ* (20) *and* (22) *to derive* (11) *and* (14), *the equations giving the velocity and displacement of a particle moving along the X-axis with initial displacement* $X_0 = 0$ *and constant acceleration* a_X.

Since a_x is constant, (20) assumes the simple form

$$v_x = v_{x0} + a_x \int_0^t dt = v_{x0} + a_x t,$$

which is the required equation (11). Substitution of this value of v_x in (22) gives

$$X = X_0 + \int_0^t v_{x0}\, dt + \int_0^t a_x t\, dt,$$

or, since v_{x0} and a_x are constants and since $X_0 = 0$,

$$X = v_{x0} \int_0^t dt + a_x \int_0^t t\, dt = v_{x0} t + \tfrac{1}{2} a_x t^2,$$

the required equation (14).

PROBLEMS

1. Explain in detail why there is one more sidereal day in a year than there are solar days; hence, that the length of the sidereal day and the period of rotation of the earth are each $365/366$ day.

2. A pilot desires to fly his aircraft so that he remains continuously at solar noon. How fast and in what direction should he fly, with respect to the earth's surface, if he is (a) at the equator, (b) at 30° N latitude, and (c) at 60° N latitude? After he flies once around the earth, what time and day is it? Answer the same questions for a pilot who starts at 10 P.M. and keeps the same star continuously due south of him.

3. A motorist makes a round trip from city A to city B, which are 120 mi apart by road. If the time required for the trip from A to B is 2 h and the time required for the return trip is 3 h, find (a) the average speed of the car in going from A to B, (b) the average speed of the car on the return trip, and (c) the average speed of the car for the entire trip.
Ans: (a) 60 mi/h; (b) 40 mi/h; (c) 48 mi/h.

4. An automobile makes a round trip between two points 200 mi apart by highway. If the outward trip requires 5 h and the return trip 10 h, find the average speeds for the outward trip, the return trip, and the complete round trip.

5. Over a 2-s period, the distance l traveled by a certain body is accurately described by the equation

$$l = (10 \text{ m/s}) t + (8 \text{ m/s}^2) t^2 - (2 \text{ m/s}^3) t^3.$$

Find the instantaneous speed of the body when $t = 2$ s.   Ans: 18 m/s.

6. To illustrate the definition of instantaneous speed as the limit of average speed, compute accurately the average speed of the body in Prob. 5, over the time intervals 0.5 to 1.5 s, 0.9 to 1.1 s, and 0.99 to 1.01 s, and compare with the instantaneous speed at $t = 1$ s.

7. The motion of a steel ball rolling down an inclined plane is noted at 1-second intervals, and the data obtained are recorded in the accompanying table. From the data given, calculate the average speed during (a) the first second, (b) the fifth second, and (c) the first 5 seconds.
Ans: (a) 2 cm/s; (b) 18 cm/s; (c) 10 cm/s.

Time t (s)	Position l (cm)	Time t (s)	Position l (cm)
0	0	6	72
1	2	7	98
2	8	8	128
3	18	9	162
4	32	10	200
5	50		

Problems 7–10

8. From the data given in the accompanying table, calculate the average speed of the steel ball during (*a*) the sixth second, (*b*) the tenth second, and (*c*) the last 5 seconds of motion.

9. To the data shown in the accompanying table, fit a simple equation giving the position *l* as a function of time. Using this equation, find the instantaneous speed of the ball at the end of the fifth second. Ans: 20 cm/s.

10. As in Prob. 9, find the instantaneous speed of the steel ball at $t=3$ s, $t=6$ s, and $t=9$ s.

11. A particle moving along the X-axis has the position

$$X = 4 \text{ cm} + (4 \text{ cm/s}) \, t - (1 \text{ cm/s}^2) \, t^2.$$

Find (*a*) the X-component of the average velocity of the particle during the first 5 seconds and (*b*) the X-component of the instantaneous velocity of the particle at the end of the fifth second. Ans: (*a*) -1 cm/s; (*b*) -6 cm/s.

12. For the particle mentioned in Prob. 11, find the X-component of the average velocity during the first 3-s interval and during the second 3-s interval, and find the X-component of the velocity when $t=3$ s.

13. A car is traversing the circular racetrack shown in the figure at a constant speed of 75 mi/h. If the circumference of the racetrack is 1.25 mi and if the car started at point *O*, find the instantaneous velocity at the end of 15 s, 30 s, 45 s, and 60 s. What is the average velocity during the first minute? Ans: 75 mi/h, E; 75 mi/h, S; 75 mi/h, W; 75 mi/h, N; zero.

14. For the car described in Prob. 13, find the average velocity during the first 15 s, the first 30 s, and the first 45 s. What is the average velocity during the first 2½ min? During the time interval $t=30$ s to $t=60$ s?

15. An airplane has a cruising speed of 240 mi/h. On a day when there is no wind this plane travels due eastward for 3 h and then travels northward for 4 h. What is the average velocity of this plane during the first 3 h? during the second interval of 4 h? during the entire 7-h interval? What is the total displacement of the plane at the end of 7 h? Ans: 240 mi/h, E; 240 mi/h, N; 171 mi/h, 53°1 N of E; 1200 mi, 53°1 N of E.

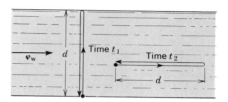

Problem 16

16. A boat travels at speed v_{BW} relative to the water in a river of width *d*. The water is moving at speed v_W. Show that the time required to cross the river to a point exactly opposite the starting point and then to return is

$$t_1 = \frac{2d}{\sqrt{v_{BW}^2 - v_W^2}} = \frac{2d/v_{BW}}{\sqrt{1 - v_W^2/v_{BW}^2}}.$$

Show that the time for the boat to travel a distance *d* exactly downstream and then to return is

$$t_2 = \frac{2d/v_{BW}}{1 - v_W^2/v_{BW}^2}.$$

Compare these times with the time $t_0 = 2d/v_{BW}$ required to traverse the distance $2d$ in still water, for the special cases $v_W/v_{BW} = 9/10, 5/10, 1/10$. (The fact that $t_1 \neq t_2$ in this computation when the speed v_{BW} is relative to a moving medium, whereas Michelson and Morley, in 1887, could find no difference in times for *light* propagated parallel and perpendicular to the earth's orbital velocity, disproved the concept that light is propagated *relative* to a medium. Prior to this period, it was generally assumed that light, like all mechanical waves, was propagated through a medium, which was

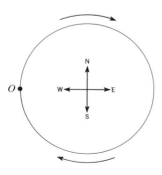

Problems 13–14

given the name *luminiferous ether*. Exploration of the logical consequences of the Michelson-Morley experiment led to Einstein's theory of relativity—*see* Chapter 44.)

17. In order to cross a stream in which the water moves at 3 mi/h, in a boat that travels at 5 mi/h, at what angle upstream should the boat be headed in order to reach the point directly opposite? What is the speed of the boat relative to the ground?
Ans: 36°.9; 4.00 mi/h.

18. An airplane with an airspeed of 280 knots desires to proceed due northwest when a wind of 80 knots is blowing from due west. In what direction should the plane head and what will be its ground speed?

19. The water in a river 1 mi wide flows due north at 8 mi/h. A motorboat travels at 6 mi/h relative to the water. A man starts from the west bank in the boat and desires to reach the point on the east bank directly opposite. The boat is incapable of landing at this point because it cannot travel as fast as the water. The man must land downstream and walk back along the bank. He walks at 3 mi/h. (a) At what angle should he *head* the boat to reach his objective in *minimum time*. What are the times of crossing, of walking along the bank, and the total time? (b) Compare with the case in which he *heads* due east so as to minimize the time of crossing. (c) Compare with the case in which he heads due southeast.
Ans: (a) 33°.0 s of E; 11.9, 18.7, 30.6 min; (b) 10.0, 26.7, 36.7 min; (c) 14.2, 17.8, 32.0 min.

20. Repeat Prob. 19(a) for the case in which the boat will go 9 mi/h, but the man still desires to minimize his time by walking along the bank. Compare the total time with the time required if the boat proceeds directly to the opposite point on the bank.

21. A car is equipped with a special speedometer that registers speed in f/s. While the car is moving eastward along a straight street, an occupant of the front seat observes the speed at various times and records the data shown in the table. From these data calculate the average acceleration (a) during the first 6 seconds of motion, (b) during the interval $t=4$ s to $t=6$ s, and (c) during the interval $t=6$ s to $t=8$ s.
Ans: (a) 3 f/s² E; (b) 4 f/s² E; (c) −1 f/s² E or 1 f/s² W.

Time (s)	Speed (f/s)	Time (s)	Speed (f/s)
0	0	8	16
2	4	10	8
4	10	12	4
6	18	14	0

Problems 21–22

22. From the data given in the table, find the average acceleration (a) during the first 4 seconds, (b) during the interval $t=8$ s to $t=12$ s, and (c) during the entire period of motion.

23. An automobile starting from rest and traveling northward attains a speed of 15 mi/h in 10 s. What is the change in the velocity of the car during this 10-s interval? What is the average acceleration of the car? Ans: 15 mi/h or 22 f/s, N; 1.5 mi/h·s or 2.2 f/s², N.

24. If the driver of the automobile in Prob. 23 immediately applies his brakes when the speedometer reaches 15 mi/h and brings the car to rest in 12 s, what is the change in velocity during the 12-s interval? What is the average acceleration during this interval?

25. A car starting from rest experiences a constant acceleration of 6 f/s², northward. What is the velocity of the car at the end of 4 s? What is the average velocity of the car during this 4-s interval? How far does the car move during this interval?
Ans: 24 f/s, N; 12 f/s, N; 48 f.

26. If a car starting from rest has a constant acceleration of 2.2 f/s², northward, how long will it take the car to acquire a velocity of 60 mi/h, northward? How far will the car move during this time?

27. Two sleds are initially at the top of a steep slope. Two seconds after starting, the first sled has moved a distance of 16 f down the slope, and the second sled begins its descent. If both sleds start from rest and have the same constant acceleration, what is the distance between the sleds 3 s after the second sled starts down the slope? Ans: 64 f.

28. An automobile moving along a straight road at a constant speed of 45 mi/h passes a motor patrolman. Just as the automobile passes, the patrolman starts his motorcycle. If the motorcycle has a constant acceleration of 1.5 f/s², how long will it take the patrolman to overtake the speeding motorist? What will be the speed and total displacement of the motorcycle at the time the motorist is overtaken?

29. A motorcycle moving with an initial velocity of 15 m/s experiences a constant deceleration of 3 m/s². How far does the motorcycle move during the first 4 s after its velocity begins to decrease?   Ans: 36 m.

30. A car has an initial velocity of 15 m/s and experiences a constant deceleration of 2 m/s². How far has the car moved by the time its velocity is reduced to 5 m/s?

31. A particle starting from rest moves with a constant acceleration of 4 f/s², N, for 3 s, then at constant velocity for 4 s, and finally with an acceleration of −2 f/s², N, until it comes to rest. How long does the body move? What is its total displacement before it comes to rest? What is its average velocity during the entire motion? its average acceleration?
Ans: 13 s; 102 f, N; 7.85 f/s, N; 0.

32. A particle starting from rest moves with a constant acceleration of 3 m/s², E, for 4 s, then with constant velocity for 5 s, and finally with a constant acceleration of −3 m/s², E, until it comes to rest. Answer all questions in Prob. 31 for this case.

33. According to Einstein's theory of relativity, no particle can have a speed greater than the speed of light (3×10^8 m/s). However with *particle accelerators* it is possible to give particles such as protons and electrons speeds approaching that of light. For example, in an electrostatic accelerator having an acceleration tube 1.5 m long, it is possible to give protons a speed equal to 0.8 the speed of light. If the protons enter with practically zero speed and leave with a speed of 2.4×10^8 m/s, calculate the acceleration (assumed constant) of the protons as they pass through the tube.
Ans: 1.92×10^{16} m/s².

34. A baseball is observed to describe a parabola having the parametric equations

$X = (20 \text{ f/s}) t,$
$Y = 4 \text{ f} + (30 \text{ f/s}) t - (16 \text{ f/s}^2) t^2,$

where t is the time after the ball is struck, the X-axis is horizontal, and the Y-axis is vertically upward, as in the figure. Plot the parabola accurately until the ball strikes the ground ($Y=0$). Determine the magnitude and direction of the acceleration *a* of the ball.

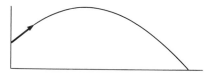

Problem 34

35. An automobile moves at a constant speed of 60 f/s around a curve having a radius of 800 f. What is the magnitude of the car's acceleration? What is its direction?
Ans: 4.5 f/s², toward center of curve.

36. What is the speed at a point on the circumference of a long-playing phonograph record of 9-inch diameter when the record is turning at the prescribed rate of 33⅓ rev/min? What is the magnitude of the centripetal acceleration of the point in question?

37. A wheel having a diameter of 2 i is mounted on the shaft of a motor turning at 1800 rev/min. What is the total path length traversed during 1 min by a point on the circumference of the wheel? What is its speed? What is the magnitude of its acceleration?
Ans: 942 f; 15.7 f/s; 2960 f/s².

38. If the 2-i wheel described in Prob. 37 were replaced by a 1-f wheel, what would be the speed and the acceleration of a point at the circumference when the motor shaft is turning at 1800 rev/min?

39. A particle starting from rest moves with a variable acceleration given by the expression $a = kt^2$, where k is a constant. The motion is rectilinear. Derive (*a*) an expression for the velocity as a function of time and (*b*) an expression for the displacement as a function of time.

40. If the magnitude of the acceleration given by the expression $a = kt^2$ in Prob. 39 is given in f/s², what are the units in which the constant k must be expressed? Verify the dimensional validity of the expressions obtained in Prob. 39.

41. A particle moving in the XY-plane starts from the origin at $t = 0$, with initial velocity components $v_{X0} = 12$ f/s, $v_{Y0} = 0$. The particle has constant acceleration components $a_X = 0$, $a_Y = 4$ f/s². Find the displacement, velocity, and acceleration vectors at $t = 4$ s.
Ans: 57.7 f, 33°.7; 20 f/s, 53°.1; 4 f/s², 90°.

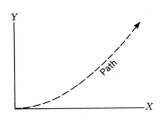

Problems 41–42

42. Show that the path of the particle in Prob. 41 is a parabola; determine its equation.

43. The acceleration of a particle starting from rest at the origin at $t=0$ is given by the expression $a = (2 \text{ f/s}^4) t^2$, southward. Find (a) the acceleration of the particle at $t=3$ s, (b) the velocity of the particle at $t=3$ s, and (c) the position at $t=3$ s.
Ans: 18 f/s²; 18 f/s; 13.5 f.

44. A particle starts from rest at the origin of coordinates and moves in the XY-plane under the influence of forces that give it the following acceleration components:

$$a_X = (2 \text{ f/s}^3) t. \qquad a_Y = 6 \text{ f/s}^2.$$

Determine the velocity components v_X, v_Y, and the coordinates X, Y, of the particle as a function of time. Compute these four quantities at $t=1$ s and at $t=2$ s.

DYNAMICS OF A PARTICLE 4

Dynamics is the study of motion in terms of the forces that produce the motion.

All of the vast body of knowledge called *classical mechanics,* including *statics, dynamics, hydrodynamics, aerodynamics, theory of elasticity, wave motion,* and *sound,* has as its logical foundation three *principles of motion* formulated by SIR ISAAC NEWTON and published in 1687.

Since Newton's principles play such an important role in the subject of mechanics, we shall discuss in detail the reasoning involved in their formulation. Briefly, we can say that *the first step was experimental,* involving observation of the motion of interacting bodies; the directly observable kinematic factors were *positions* and *times,* from which velocities and accelerations could be calculated. From accelerations occurring during interactions between pairs of bodies, Newton showed that it is possible to attribute to every body a property called its *mass.* Newton found it desirable to describe the observed interaction in terms of the three fundamental quantities *mass, length,* and *time. The second step involved the formulation of a set of principles* giving a consistent account of the experimental observations. These principles involve certain assumptions and the introduction of a fourth quantity called *force,* which is *defined* in terms of the three fundamental quantities by Newton's second principle.

Newton formulated his principles in order to give a single coherent account of observations and the results of experiments actually performed. In Newton's own words, "These laws must be considered as resting on convictions drawn from observation and experiment, not on intuitive perception." The real justification or 'proof' of Newton's principles is the success with which they and the relations based upon them can be applied to the practical problems encountered in physics and engineering. It was not until the beginning of the twentieth century that it was discovered that there were types of problems that could not be handled by the classical dynamics based upon Newton's principles. One of these types involves motion at high speeds, comparable with the speed of light (3×10^8 m/s or 186 000 mi/s), and must be treated by Einstein's *relativistic mechanics;* the other type involves atomic and sub-atomic particles, to which *quantum mechanics* must be applied. It should be pointed out that relativistic and quantum mechanics both reduce to classical Newtonian mechanics when applied to bodies of the size and speed of automobiles, bullets, baseballs, aircraft, and even rockets. Hence Newton's principles will continue to be of extreme usefulness in spite of the fact that under certain recognized conditions they require refinement.

1. Newton's First Principle

By a series of experiments (employing, for example, small spherical bodies rolling on a horizontal table top), GALILEO GALILEI (1564–1642) and Newton were led to make the following postulate in regard to the motion of a particle:

NEWTON'S FIRST PRINCIPLE: *A particle left to itself has constant velocity.*

This postulate includes the case of a stationary particle, whose velocity remains zero as long as the particle is 'left to itself'; here the postulate seems obvious. In the case of moving particles, the postulate is an idealization from experimental observations. For example, a marble rolling across a rug quickly comes to rest as a result of interaction with the rug; the same marble rolling across a bare floor will move for a much longer time before coming to rest because the horizontal interaction between the marble and floor is smaller than the interaction between the marble and the rug; in neither case is the marble really 'left to itself.' In no way is it possible to perform an experiment in which a body is left completely to itself, and hence the above postulate is an idealization which cannot be verified by a particular experiment.

The property of a particle that results in its maintaining its velocity unchanged unless it interacts with other particles was termed *inertia* by Newton, and the principle we have just stated was called by Newton the *law of inertia*. According to this principle, if we were to throw a ball from a rocket ship in free space where there is no air and no appreciable gravitational effect, the ball would go on indefinitely along a straight line with the same speed that it had when it left our hand.

Before the days of Galileo and Newton it was frequently assumed that the 'natural' state of a body was one of *rest*, and that interaction with other bodies was required to keep a body in motion. Newton's first principle asserts that such interaction is only required to *change* the velocity, that is, to produce *acceleration*.

2. Mass

According to the first principle, whenever a given particle experiences an *acceleration,* the acceleration must be attributed to the influence of other particles. Newton first studied the acceleration produced by interactions between *pairs* of particles. He noted that *whenever there was an acceleration of one particle, there was also an acceleration of the other particle*. The accelerations were found to be oppositely directed, as shown in Fig. 1.

As for the magnitudes of the two accelerations, a_1 and a_2, Newton concluded *as a result of his observations* that their ratio was always a constant for a given pair of particles, provided there were no interactions with other bodies. This conclusion is found to be valid regardless of the type of interaction; whether the interaction is of elastic, electric, magnetic, or gravitational character, the ratio of the magnitudes of the accelerations experienced by the two interacting bodies is always the same, provided interactions with other bodies can be ignored.

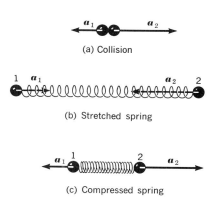

Fig. 1. Whenever a particular pair of particles interact, their accelerations are oppositely directed and have a constant magnitude ratio.

All evidence points to the fact that Newton's conclusion is correct. This conclusion may be stated in the form

$$a_2/a_1 = \text{constant} = R_{12},$$

where the ratio R_{12} has a definite constant value for a particular pair of particles, as indicated in Fig. 1. Since the accelerations are always oppositely directed, a vector equation may be written in the form

$$\boldsymbol{a}_2 = -R_{12}\,\boldsymbol{a}_1,$$

where the constant R_{12} is a positive number. In interpreting R_{12}, Newton proposed to assign a number m_1 to one particle and to define a corresponding number m_2 for the other particle by the equation

$$R_{12} = m_1/m_2.$$

Therefore, $\quad \dfrac{a_2}{a_1} = \dfrac{m_1}{m_2},\quad$ or $\quad m_1 a_1 = m_2 a_2,$

and
$$m_2 = (a_1/a_2)\,m_1, \tag{1a}$$

where the acceleration ratio (a_1/a_2) can be determined from experimental observation. In this manner, if a number m_1, called the *mass*, is arbitrarily assigned to particle 1, the value m_2 of mass to be assigned to particle 2 is determined. Similarly, by an experiment in which a particle 1 is made to interact in any manner with a third particle 3, a mass is determined for 3:

$$m_3 = (a_1/a_3)\,m_1. \tag{1b}$$

In this manner, a value of mass may be assigned to any particle in terms of the mass assigned to the 'standard' particle 1.

We now need to introduce another generalization from experiment that was made by Newton. If particles 2 and 3 are assigned masses m_2 and m_3 by experiments in which first one and then the other interacts with particle 1, and then particles 2 and 3 are made to interact with each other in any manner, the same values of masses may be used to determine the ratios of the accelerations in this interaction. The relation is

$$\frac{a_3}{a_2} = \frac{m_2}{m_3}, \quad \text{or} \quad m_2 a_2 = m_3 a_3. \tag{2}$$

This experimental fact shows that the mass values determined in terms of a standard particle to which an arbitrary mass is assigned are universally applicable in determining the ratios of the accelerations in the interaction between *any* two particles.

The mass is said to be a *measure of the inertia* of a particle since, in any interaction, the particle of greater mass has the lesser acceleration. The simple demonstration indicated in Fig. 2(a) illustrates this point. We might place on a billiard table two balls of the same diameter but composed of different materials, say wood and lead. Attempts to set the balls in motion with the cue make it evident immediately that the mass (inertia) of the lead ball is greater than that of the wooden ball, since the lead ball has a greater tendency to remain at rest and is more difficult to accelerate than the wooden ball. In a sense, we use the cue as a 'standard body' to interact with the two balls. We might compare the

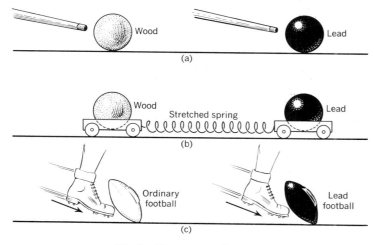

Fig. 2. Comparison of masses.

masses of the two balls by a more direct interaction between them with the arrangement shown in Fig. 2(b). We place the two balls in small cars connected by a light spring. If the cars are separated by stretching the spring and then released, the magnitude of the acceleration of the wooden ball will be much greater than that of the lead ball; again we conclude that the mass of the lead ball is greater. A third demonstration we might perform is indicated in Fig. 2(c). Here we have an ordinary football and another football made of lead; a kicker would have no difficulty in telling which ball has the greater mass and the lesser acceleration.

> The **mass** of a body is defined by the equation $m = (a_S/a) m_S$ when an experiment is performed in which the body interacts with a 'standard' body assigned mass m_S, and the body experiences acceleration a, while the standard body experiences acceleration a_S.

We shall later describe convenient and accurate methods of comparing masses. However, in principle, equation (1) enables us to determine the mass m of any body in terms of the mass m_1 of some standard body. It is, of course, important that some one standard mass be selected as a unit in terms of which all other masses can be measured. The *standard unit mass is the kilogram* (kg), equal to 1000 *grams* (g):

> The **kilogram** is the mass of a cylinder of platinum-iridium kept at the International Bureau of Weights and Measures.

Such a concrete standard* of course suffers from the deficiencies discussed

> *A 'concrete' standard is one that can be 'put away on the shelf.'

on p. 14 in connection with the old concrete standard of length, and may eventually be replaced by an atomic standard of mass. At present, however, the mass of the prototype kilogram can be compared with other masses with much higher accuracy than can the mass of an atomic particle.

3. Newton's Second Principle

When two particles interact in such a way that they are accelerated, they are said to exert *forces* on each other. If a body of mass m experiences acceleration $\boldsymbol{a}$, the vector $m\boldsymbol{a}$ is called the force acting on this body. *This is the fundamental definition of force.* It is convenient because the above discussion shows that, when two particles interact, $m_1\,\boldsymbol{a}_1 = -m_2\,\boldsymbol{a}_2$, so that the two particles exert equal and opposite forces on each other.

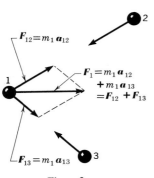

Figure 3

Fundamentally, all forces are exerted by individual particles on each other. However, a given particle may experience forces from a large number of other particles simultaneously, and we need to know how these forces add. Here we have recourse to a further generalization that Newton made from experimental observation. Let particle 1 in Fig. 3 be attracted by both particle 2 and particle 3. Suppose that if particle 3 were not present, particle 1 would experience acceleration $\boldsymbol{a}_{12}$ owing to particle 2 alone, and that if particle 2 were not present, particle 1 would have acceleration $\boldsymbol{a}_{13}$ owing to particle 3. Then it is concluded from experiment that with both particles present, particle 1 will have acceleration

$$\boldsymbol{a}_1 = \boldsymbol{a}_{12} + \boldsymbol{a}_{13},$$

the vector sum. Multiplying this equation through by m_1, we obtain

$$m_1\,\boldsymbol{a}_1 = m_1\,\boldsymbol{a}_{12} + m_1\,\boldsymbol{a}_{13}.$$

Here $m_1\,\boldsymbol{a}_{12}$ is $\boldsymbol{F}_{12}$, the force exerted by 2 on 1; $m_1\,\boldsymbol{a}_{13}$ is $\boldsymbol{F}_{13}$, the force exerted by 3 on 1; and $m_1\,\boldsymbol{a}_1$ is *defined* as the *resultant* force $\boldsymbol{F}_1$ acting on particle 1. Hence with this definition, forces add vectorially and the preceding equation gives

$$\boldsymbol{F}_1 = \boldsymbol{F}_{12} + \boldsymbol{F}_{13}. \tag{3}$$

The resultant force acting on a particle is the vector sum of the various forces exerted on that particle by other particles.

If the resultant force $\boldsymbol{F}$ acts on a particle of mass m, the particle experiences an acceleration $\boldsymbol{a}$ given by

$$\boldsymbol{F} = m\boldsymbol{a}, \quad \text{or} \quad \boldsymbol{a} = \frac{\boldsymbol{F}}{m}, \tag{4}$$

which, stated in words, gives

NEWTON'S SECOND PRINCIPLE: *The acceleration of a particle is directly proportional to the resultant force acting on the particle, is inversely proportional to the mass of the particle, and has the same direction as the resultant force.*

It is seen that Newton's first principle is really a special case of the second principle in which the force, and hence the acceleration, vanish.

In the metric system of units, the unit of length is the meter, the unit of mass is the kilogram, the unit of time is the second, and the unit of force defined by equation (4) is the *newton* (N):

> The **newton** is the resultant force that imparts to a one-kilogram mass an acceleration of one meter per second per second.

The newton is a derived unit. From (4) we see that the metric force unit is the kg·m/s^2, and *newton* is merely a convenient *name* for this complex unit:

$$1 \text{ N} = 1 \text{ kg} \cdot \text{m/s}^2. \tag{5}$$

In solving dynamical problems in metric units, forces come out initially in kg·m/s^2, and this fact can be used as a check on the correctness of the work. *Force is a derived quantity of physical dimensions mass × length/time^2.*

The newton is called an *absolute unit* of force because its value is defined in terms of its effect on an object quite independently of the position of the object in the universe. In Section 7, we shall define the *gravitational* unit of force used with British units. This unit, the *pound*, is the *earth's* force of gravity on a standard mass at a particular location.

Example. *An automobile of mass* 900 kg *accelerates at a constant rate from rest to* 15 m/s *in* 6 s. *What is the resultant force on the car and what body exerts this force?*

From (10), p. 41, we see that the acceleration has the magnitude

$$a = \frac{15 \text{ m/s}}{6 \text{ s}} = 2.5 \text{ m/s}^2.$$

Hence, from (4) and (5), the resultant force has the magnitude

$$F = ma = (900 \text{ kg})(2.5 \text{ m/s}^2) = 2250 \text{ kg} \cdot \text{m/s}^2 = 2250 \text{ N}.$$

The direction of the resultant force is *forward,* in the direction of the acceleration, by Newton's second principle. There are only three sources of force on the car—there is the downward gravitational pull of the earth which we shall discuss in Section 6; there is the road, which is touching the car wheels; and there is the air, also touching the car but exerting negligible force at low speeds. The road must push on the car with a force that has an upward component equal in magnitude to the downward force of gravity, since *the resultant force has no vertical component* (if it did the car would accelerate vertically). The force of the road must also have a *forward* component of 2250 N. *The force that accelerates the car is exerted*

4. Newton's Third Principle

Now that we have defined force by Newton's second principle, let us return to the relation

$$m_1 \mathbf{a}_1 = -m_2 \mathbf{a}_2 \tag{6}$$

that Newton used in comparing masses. By the second principle,

$$\mathbf{F}_1 = m_1 \mathbf{a}_1 \tag{7}$$

is the force exerted on particle 1 by particle 2. Similarly,

$$\mathbf{F}_2 = m_2 \mathbf{a}_2 \tag{8}$$

is the force exerted on particle 2 by particle 1. Substitution of (7) and (8) in equation (6) leads to the result

$$\mathbf{F}_1 = -\mathbf{F}_2, \tag{9}$$

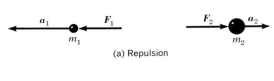

(a) Repulsion

(b) Attraction

Fig. 4. When two particles interact, the magnitude of the force exerted on the first particle by the second particle is equal to the magnitude of the force exerted on the second particle by the first. The forces are oppositely directed.

which states that the force exerted on particle 1 by particle 2 is equal in magnitude and opposite in direction to the force exerted on particle 2 by particle 1. Figure 4(a) gives a diagram showing the results of a repulsive interaction between 'particle' 1 and 'particle' 2; Fig. 4(b) shows the results of an attractive interaction. Equation (9) expresses

NEWTON'S THIRD PRINCIPLE: *If particle 1 exerts a force on particle 2, then particle 2 exerts an equal and opposite force on particle 1.*

It can be readily shown that if this principle is true for each pair of 'particles,' it is true for interactions between any two bodies, as illustrated in Fig. 5.

Newton's third principle holds rigorously no matter what the state of motion of the bodies may be. This principle also holds in cases of action at a distance such as occur with gravitational, electric, and magnetic forces. Thus, the earth exerts a downward gravitational force on any object near its surface, for example on a flying airplane. The airplane exerts an equal and opposite upward force on the earth. Although the force of the earth on the plane is of great practical importance, that of the plane on the earth is of no detectable importance in connection with the motion of the earth. The gravitational pull of the earth on the moon

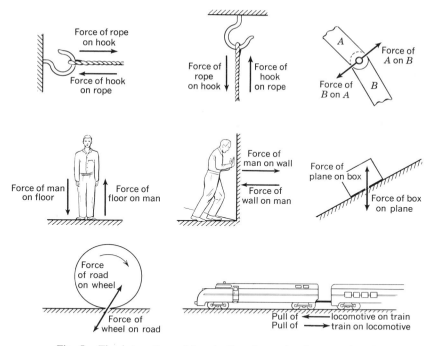

Fig. 5. The interaction of two bodies always involves equal and opposite forces.

keeps the moon revolving around the earth just as the pull of the sun keeps the earth revolving around the sun. In the case of the earth-moon system, the reaction force exerted by the moon does introduce a detectable wobble into the earth's regular motion around the sun, and the gravitational pull of the moon on the waters of the earth is the principal tide-producing force.

Let us now give a résumé of the way in which Newton's three principles were formulated. There were four conclusions based on idealization of experiment:

(a) The velocity of a particle is constant as long as there are no interactions with other particles.

(b) When two particles interact, they both experience accelerations. The accelerations are oppositely directed, and the ratio of the magnitudes of the accelerations is a constant for a given pair of particles, independent of the method by which they interact.

(c) Characteristic masses can be assigned to each particle, by comparison with a standard particle, such that when *any* two particles interact the ratio of the magnitudes of their accelerations is the inverse of the ratio of their preassigned masses.

(d) When a particle interacts with several other particles simultaneously, the acceleration it experiences is the vector sum of the accelerations it would experience by interacting with each of the other particles individually.

Conclusion (a) is the *first principle.* Conclusions (b) and (c) permit the definition of *mass,* and hence of *force* as *ma.* Conclusion (d) shows that forces obey the law of vector addition and hence permits formulation of the *second principle* in terms of resultant force. Conclusion (b) and the definition of force lead directly to the *third principle.*

5. Planetary Motion; Universal Gravitation

As pointed out in Chapter 3, a particle describing motion in a circle at constant speed experiences a centripetal acceleration of magnitude $a = v^2/R$, where v is the speed of the particle and R is the radius of the circular path. According to Newton's second principle, if the particle has mass m, it must be acted upon by a force $\boldsymbol{F} = m\boldsymbol{a}$ if it is to have this acceleration. This force is called the *centripetal force* and its magnitude is

$$F = mv^2/R. \tag{10}$$

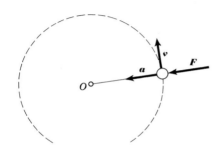

Fig. 6. A constant resultant force of magnitude $F = mv^2/R$ must act at all times on a particle moving with constant speed v in a circular path of radius R. This force is directed toward the center of the circle and is called the *centripetal force.*

Since the centripetal acceleration is directed toward the center of the circle, the centripetal force is also directed toward the center of the circle, as in Fig. 6. One of Newton's earliest applications of the relation (10) was to the motions of the moon and of the planets, since these bodies move, to a close approximation, in circles at constant speed.

To the approximation in which we consider the orbits of the planets (which are actually slightly elliptical) as circles with the sun at the center, one of the empirical laws of planetary motion formulated by Kepler states that the speeds v_P of the different planets are inversely proportional to the square roots of the radii R_P of their orbits. Thus if P1 and P2 represent any two of the planets,

$$\frac{v_{P1}}{v_{P2}} = \sqrt{\frac{R_{P2}}{R_{P1}}}, \quad \text{or} \quad \frac{v_{P1}^2}{v_{P2}^2} = \frac{R_{P2}}{R_{P1}}. \tag{11}$$

A planet moving at speed v_P in a circle of radius R_P around the sun has acceleration v_P^2/R_P toward the sun. The sun must therefore attract planets P1 and P2 with the centripetal forces given by (10) as

$$F_{P1} = m_{P1} v_{P1}^2 / R_{P1}, \quad F_{P2} = m_{P2} v_{P2}^2 / R_{P2},$$

where m_{P1} and m_{P2} are the masses of the planets. If we take the ratio F_{P1}/F_{P2} and substitute the observed speed ratio (11), we find

$$\frac{F_{P1}}{F_{P2}} = \frac{m_{P1}/R_{P1}^2}{m_{P2}/R_{P2}^2}. \tag{12}$$

The force exerted by the sun on a planet is thus found to be one of attraction, proportional to the mass of the planet, inversely proportional to the square of the distance to the planet, and directed along the line connecting the planet and the sun.

Since the force between the sun and a planet is proportional to the mass of the planet, it is reasonable, in extending this force law to other situations, such as the earth-moon system, to suspect that the force is proportional to both of the masses in the system. By these considerations, Newton was led to formulate the principle of universal gravitation, which has been amply confirmed by detailed astronomical observation:

PRINCIPLE OF UNIVERSAL GRAVITATION: *Every particle of matter in the universe attracts every other particle with a gravitational force. The magnitude of the gravitational force between two particles is proportional to the product of the masses of the particles and inversely proportional to the square of the distance between them; the gravitational forces of attraction between two particles act along the line joining the two particles.*

The equation expressing this relation for a given pair of particles of masses m_1 and m_2 is

$$F = G\frac{m_1 m_2}{R^2}, \tag{13}$$

where R is the distance between the two particles and G is a fixed proportionality constant called the *gravitation constant*. The value of the gravitation constant G can be determined experimentally in terms of the arbitrary fundamental units, meter, kilogram, and second. If we solve (13) for G we see that the unit in which it is expressed is the $N \cdot m^2/kg^2$ or, equivalently, the $m^3/kg \cdot s^2$.

Because the standard of mass exists only in the laboratory, G cannot be determined by astronomical observation; it was first measured accurately by Cavendish in 1798 with the apparatus in Fig. 7. Two small balls, each of mass m, are attached to the ends of a light rod. The resulting 'dumbbell' is suspended in a horizontal position by a fine quartz fiber. Heavy lead balls can be moved close to the ends of the dumbbell by sliding them along sup-

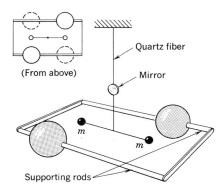

Fig. 7. The experiment of Cavendish for determining the gravitation constant G.

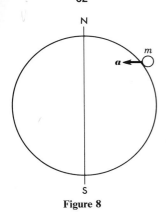

Figure 8

Fig. 9. Spring scale.

porting rods. When the large balls are in the position shown, the gravitational forces they exert on the small balls tend to twist the fiber in one direction. When they are moved into the position indicated by broken lines, the forces tend to twist the fiber the other way. The angle through which the fiber is twisted when the balls are moved from one position to the other is measured by observing the deflection of a beam of light reflected from the small mirror attached to the fiber. With a knowledge of the geometry of the apparatus and the elastic properties of the supporting fiber, the value of G can be determined from the observed angle of twist. The most accurate value of G is

$$G = 6.670 \times 10^{-11} \text{ N} \cdot \text{m}^2/\text{kg}^2,$$

as determined by P. R. HEYL and P. CHRZANOWSKI at the U.S. National Bureau of Standards in 1942 by a dynamical method not different in principle from that used by Cavendish.

The extremely small magnitude of G shows that gravitational forces *between* bodies on the earth's surface are extremely small and quite negligible for all ordinary purposes. The Cavendish experiment is a very delicate experiment indeed. The large and important gravitational force of the earth on all bodies near its surface is accounted for by the extremely large mass of the earth, which is computed in an example in the following section.

6. Weight; Gravitational Acceleration

The concept of *weight* is very useful in the consideration of the static equilibrium of bodies at rest near the surface of the earth, and in the consideration of the dynamics of bodies in motion *over short distances* near the earth's surface. The concept of weight is not useful in the consideration of the dynamics of long-range ballistic missiles or of earth satellites because the earth is both round and rotating, and because the gravitational attraction varies with position relative to the earth. We shall consider the latter types of problems in Chapter 10. Here we shall consider problems in which the earth can be considered flat and the gravitational attraction can be considered constant.

A particle at rest relative to the surface of the earth actually has a centripetal acceleration a resulting from the earth's rotation; this centripetal acceleration is perpendicular to the earth's axis, as in Fig. 8. Consider the same particle supported by a spring scale as in Fig. 9. The forces now acting on the particle are shown in Fig. 10, where $\boldsymbol{F}_\text{G}$ is the true gravitational attraction of the earth and $\boldsymbol{F}_\text{S}$ is the force exerted by the spring. As indicated in Fig. 10, Newton's second principle requires that $\boldsymbol{F}_\text{G} + \boldsymbol{F}_\text{S} = m\boldsymbol{a}$.

By definition, $-\boldsymbol{F}_\text{S}$ in Fig. 10 is called the *weight* $\boldsymbol{w}$ of the particle. It is the opposite of the force required to support the particle near the earth's surface. The weight w is called the *apparent force of gravity* on the particle.

> The **weight** of a particle is the opposite of the force required to support the particle at rest near the earth's surface. The **vertical**, at a point near the earth's surface, is the direction of the weight of a particle; the **horizontal** is the plane normal to the vertical.

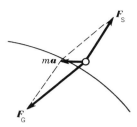

Figure 10

The length of the vector $m\mathbf{a}$ is greatly exaggerated in Fig. 10; actually the weight of a particle differs very little from the force $\mathbf{F}_G$ of gravity since the vector $m\mathbf{a}$, even at the equator where it is the greatest, has only 0.003 times the magnitude of $\mathbf{F}_G$.

It is apparent from the above discussion that in problems of the equilibrium of forces on *stationary* particles we can ignore the fact that the earth's surface is accelerated and treat such problems as if our coordinate system were unaccelerated and the effect of gravity contributed a force w, the weight of the particle, in the downward vertical direction, as illustrated in Fig. 11. We can show that we can do the same in dynamic problems in a restricted region near the earth's surface.

Fig. 11. The effective downward force on a particle near the surface of the earth is its weight w.

Let us first consider the acceleration of a *freely falling particle* near the earth's surface, relative to a coordinate system attached to the earth.

> A **freely falling particle** is one that experiences no force except that of the earth's gravity.

The freely falling particle in Fig. 12 is acted on only by the force $\mathbf{F}_G$, so its true acceleration is $\mathbf{F}_G/m$. We see that we can write

$$\mathbf{F}_G/m = \mathbf{a} + \mathbf{w}/m, \tag{14}$$

the acceleration $\mathbf{a}$ *of the rotating coordinate system* attached to the earth, plus the acceleration $\mathbf{w}/m$ *relative to this coordinate system*. It is the relative acceleration $\mathbf{w}/m$ that we measure when we stand in our rotating coordinate system and observe the acceleration of a freely falling body. This apparent acceleration is called the *gravitational acceleration*, and is denoted by $\mathbf{g}$:

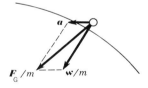

Fig. 12. Acceleration of a freely falling particle.

$$\mathbf{w}/m = \mathbf{g}; \qquad \mathbf{w} = m\mathbf{g}. \tag{15}$$

By similar arguments it can be shown that if a body is at rest near the earth's surface the combined effect of the earth's gravitational attraction and the centripetal acceleration can still be represented by a force $\mathbf{w}$ in a coordinate system attached to the earth, and that, by doing so, the earth's rotation can be ignored.

For moving particles, the above considerations are still valid but represent only a *very good approximation* because the coordinate system attached to the earth's surface is actually *rotating*, as well as undergoing a linear acceleration. Consider, for example, the coordinate system at the pole, which is unaccelerated but has a pure rotation every 24 hours.

The correction for rotation leads to an additional system of apparent forces called *Coriolis forces* which are instrumental in causing the cyclonic motions of large-scale meteorology.

Since the force of gravity F_G is proportional to the mass m of a particle, by the principle of universal gravitation, the ratio F_G/m in (14) is a constant vector. Since a is also a constant vector at a particular location, the ratio $g = w/m$ is also a constant vector independent of the mass m of the body. Hence,

At a particular location near the earth, the gravitational acceleration g is constant in magnitude and direction.

This result is confirmed by experiment. Up to the Middle Ages it was thought that heavy bodies fall more rapidly than light bodies, and, indeed, it is a matter of ordinary casual observation that a lead shot, a feather, a snowflake, and a water droplet fall through the air at different velocities. It remained for Galileo, who might be termed the first modern experimental physicist, to give a correct analysis of the problem. Galileo introduced the clearly defined concept of acceleration and, as a result of indirect experimental measurements, concluded that *in vacuum all freely falling bodies would experience the same constant acceleration.* This conclusion, which can easily be verified in a laboratory by modern experimental methods, was based on a series of experiments that were hampered by the fact that the only devices available for the measurement of time were crude water clocks, which had not been improved appreciably since their invention by the Egyptians and Babylonians thousands of years earlier.

Table 4-1 VALUES OF THE GRAVITATIONAL ACCELERATION

At sea level			At 40° latitude			
Latitude	f/s²	m/s²	Altitude (f)	f/s²	Altitude (m)	m/s²
0°	32.0878	9.780 39	0	32.1578	0	9.801 71
10°	32.0929	9.781 95	500	32.1563	500	9.800 17
20°	32.1076	9.796 41	1 000	32.1547	1 000	9.798 64
30°	32.1302	9.793 29	2 000	32.1516	2 000	9.795 54
40°	32.1578	9.801 71	4 000	32.1454	4 000	9.789 37
50°	32.1873	9.810 71	8 000	32.1331	8 000	9.777 02
60°	32.2151	9.819 18	16 000	32.1084	16 000	9.752 33
70°	32.2377	9.826 08	32 000	32.0608	32 000	9.702 96
80°	32.2525	9.830 59				
90°	32.2577	9.832 17				

Near the earth's surface the magnitude of g is approximately 9.8 m/s². In solving problems we shall ordinarily use the value

$$g = 9.8 \text{ m/s}^2 = 32 \text{ f/s}^2,$$

but it should be noted that accurate measurements show a slight dependence of g on latitude and altitude. In Table 4-1 are given the accurate values of g for sea-level stations at various latitudes and for stations at various altitudes at 40°. latitude. Even these values are only mean values, since there are local variations of smaller magnitude that depend on the particular character of the underlying earth. Measurement of such local variations is a useful tool in geophysical prospecting for oil.

Example. *What is the weight of a man of* 100-kg *mass* (a) *on a mountain near the equator where* $g = 9.76 \text{ m/s}^2$? (b) *at sea level and* 60° *latitude where* $g = 9.82 \text{ m/s}^2$?

The weights in newtons are obtained directly from equations (15) and (5):

(a) $w = mg = (100 \text{ kg})(9.76 \text{ m/s}^2) = 976 \text{ N}$;
(b) $w = mg = (100 \text{ kg})(9.82 \text{ m/s}^2) = 982 \text{ N}$.

These are the magnitudes of the weights; the directions are of course along the downward verticals.

We are now in position to compute the mass of the earth from equation (13), using the observed value of the gravitation constant G and the observed value of the gravitational acceleration g. In this computation we shall need to use the following theorem, which will occur as a problem in Chapter 23:

The resultant gravitational force between two spherically symmetric bodies is the same as if the whole mass of each body were concentrated in a small particle at its center.

The approximate mass of the earth is computed in the following example:

Example. *The approximate radius of the earth is* 6.4×10^6 m. *Compute the approximate mass of the earth.*

To a very good approximation the earth can be assumed to be a sphere, since the polar and equatorial radii differ by only ⅓ per cent. We shall make this assumption and use the approximate value 9.8 m/s^2 for the acceleration of a body near the earth's surface. To this approximation, we can consider that the weight w of a body represents the true force F_G of gravity. The preceding theorem then shows that the force between the earth, of mass m_E, and a particle of mass m near its surface is

$$F_G = w = G \frac{m m_E}{R_E^2},$$

where R_E is the distance from the center of the earth to the particle. This force is the weight of the particle, and results in the acceleration

$$g = w/m = G m_E / R_E^2.$$

Substitution of the known quantities in this equation gives

$$m_E = \frac{gR_E^2}{G} = \frac{(9.8 \text{ m/s}^2)(6.4 \times 10^6 \text{ m})^2}{6.67 \times 10^{-11} \text{ N} \cdot \text{m}^2/\text{kg}^2} = 6.0 \times 10^{24} \text{ kg},$$

where we have substituted, from (5), $1 \text{ N} = 1 \text{ kg} \cdot \text{m/s}^2$ in converting the units. This figure is the approximate mass of the earth.

From a consideration of Newton's law of gravitation, it is easy to see one reason for the variations of g with latitude and altitude as given in Table 4-1. If the earth were a perfect sphere of uniform density, the magnitude of the gravitational force exerted on a particle of mass m above the earth's surface would decrease with the distance of the particle from the center of the earth. This variation with *altitude* is observed. The observed variation of g with *latitude* is readily understood when we recall that the earth is not a perfect sphere but is an oblate spheroid, and the distance from sea level to the center of the earth becomes less as one proceeds from the equator toward the poles. Therefore, g would be expected to show a slight increase with increasing latitude, as it does. Actually, the earth's rotation, which decreases the gravitational acceleration at the equator, accounts for approximately one-half of the latitude change.

7. British Gravitational System of Units

The British system of force and mass units, still used in most engineering computations in some English-speaking countries, is, for historical reasons, defined in a more complex fashion than is the newer metric system. The basic unit is a unit of *force,* the *pound* (p), rather than one of mass. Other common force units are the *ounce* ($\frac{1}{16}$p), and the *ton* (2000 p).

The pound was originally the weight of a standard cylinder kept in London. To avoid the use of a double standard, the mass of this standard cylinder was defined in 1959, by international agreement, as exactly 0.453 592 37 kg, and the pound is now defined as follows:

The **pound** is the weight of a body of mass 0.453 592 37 kg at a location where the gravitational acceleration has the standard value

$$g_S = 32.173 \ 98 \text{ f/s}^2 = 9.806 \ 65 \text{ m/s}^2. \tag{16}$$

We can readily use this definition to relate the pound to the newton. The weight of the given mass at the specified location is

$$w = mg_S = (0.4536 \text{ kg})(9.807 \text{ m/s}^2) = 4.448 \text{ N}.$$

But this weight is, by definition, 1 pound, so

$$1 \text{ p} = 4.448 \text{ N}; \quad 1 \text{ N} = 0.2248 \text{ p}. \tag{17}$$

The *mass* unit in the British gravitational system is called the *slug* (sl) and is defined in such a way that the equation $\boldsymbol{F} = m\boldsymbol{a}$ is valid when accelerations are expressed in f/s²:

The **slug** is the mass of a body that experiences an acceleration of 1 f/s² when a resultant force of 1 p acts upon it.

A procedure that involves the definition of a mass unit, the *slug*, in terms of a force unit, the *pound*, may at first seem rather involved, particularly in view of the fact that the *pound* was itself defined in terms of the gravitational force exerted upon another mass. However, this procedure is actually worthwhile because it permits us to use a force unit of *convenient* size in the *convenient* equation $\boldsymbol{F} = m\boldsymbol{a}$. Whenever British units are used in this equation, the force should be expressed in p, the mass in sl, and the acceleration in f/s².

From the equation $m = F/a$, it follows that the slug is a derived unit equal to the p·s²/f:

$$1 \text{ sl} = 1 \text{ p} \cdot \text{s}^2/\text{f}; \tag{18}$$

hence
$$1 \text{ p} = 1 \text{ sl} \cdot \text{f}/\text{s}^2. \tag{19}$$

To summarize, both of the systems of units that we shall use are based on the expression

$$\boldsymbol{F} = m\boldsymbol{a} \quad \begin{Bmatrix} F \text{ in N} \\ m \text{ in kg} \\ a \text{ in m/s}^2 \end{Bmatrix} \text{ or } \begin{Bmatrix} F \text{ in p} \\ m \text{ in sl} \\ a \text{ in f/s}^2 \end{Bmatrix} \tag{20}$$

for Newton's second principle. In either system, the weight of a body is

$$w = mg. \tag{21}$$

In this equation, the units must be the same as in (20).

From the definition on the preceding page, we see that the pound will give a mass of 0.4536 kg an acceleration $g_s = 32.17$ f/s². From the definition above, we see that the pound will give a mass of 1 sl an acceleration of only 1 f/s². Since, for the same force, acceleration is inversely proportional to mass, we see that $(1/32.17)$ sl = 0.4536 kg, or

$$\left.\begin{array}{l} 1 \text{ sl} = (32.17)(0.4536 \text{ kg}) = 14.59 \text{ kg}, \\ 1 \text{ kg} = 0.068\ 52 \text{ sl}. \end{array}\right\} \tag{22}$$

By (21), the weight of a 1-sl mass is approximately 32 p near the earth's surface, since g is approximately 32 f/s²; hence the mass of a body of 1-p weight is approximately $\frac{1}{32}$ sl. For many purposes these statements, based on an approximate value of g, are sufficiently accurate. *In working problems, if weights are given in pounds and the value of g is unspecified, we shall assume $g = 32$ f/s², so that the mass of a body in slugs is numerically $\frac{1}{32}$ times its weight in pounds.*

Example. *If an automobile weighs* 4000 p *at a high elevation near the equator where* $g=32.0$ f/s², *what is its mass? What resultant force is required to accelerate the automobile at* 16 f/s²?

We determine the mass of the automobile from (21):

$$m = w/g = (4000 \text{ p})/(32.0 \text{ f/s}^2) = 125 \text{ p} \cdot \text{s}^2/\text{f} = 125 \text{ sl},$$

where we use (18) to obtain the last equality. The magnitude of the force required to accelerate the automobile at 16 f/s² is thus

$$F = ma = (125 \text{ sl})(16 \text{ f/s}^2) = 2000 \text{ sl} \cdot \text{f/s}^2 = 2000 \text{ p}.$$

The direction of the force is the same as the direction of the acceleration.
The fact that this force equals just half the weight is not at all accidental. A force equal to the weight of a body, by definition of weight, would give the body an acceleration g; a force equal to half the weight would give it an acceleration $\tfrac{1}{2} g$. This statement is true no matter what the value of g may be. If $g = 32.0$ f/s², a force equal to half the weight would give an acceleration of 16.0 f/s².

Example. *What are the weight and mass of the automobile described in the preceding example at a location at* 55° *latitude where* $g=32.3$ f/s²? *What resultant force would be required to accelerate it at* 16 f/s² *at the latter location?*

The mass was computed in the preceding example as 125 sl; the *mass is independent* of the location of the automobile. The weight at 55° latitude is

$$w = mg = (125 \text{ sl})(32.2 \text{ f/s}^2) = 4025 \text{ p}.$$

The automobile *weighs* 25 p more at the 55°-latitude location, but it has the *same mass*, so the same resultant force as computed in the preceding example, 2000 p, would be required to accelerate it at 16 f/s². Half its weight, 2012.5 p, would accelerate it at $\tfrac{1}{2} g$, or 16.1 f/s².

The value of G to be used in the principle of universal gravitation (13), to obtain force in pounds when the masses m are expressed in slugs and the distance R in feet, is determined by transformation of units in the SI value given on p. 62:

$$G = 6.670 \times 10^{-11} \frac{\text{N} \cdot \text{m}^2}{\text{kg}^2} = 6.670 \times 10^{-11} \frac{(0.2248 \text{ p})(3.281 \text{ f})^2}{(6.852 \times 10^{-2} \text{ sl})^2}$$

$$= 3.438 \times 10^{-8} \frac{\text{p} \cdot \text{f}^2}{\text{sl}^2}.$$

8. Standard Weight

In everyday speech, the word *mass* is not used, and no clear distinction is made between the fundamentally distinct quantities *weight* and *mass*. The word *weight*, as commonly used, actually is usually intended to mean *mass,* as we have defined it. The 'weight' of a man, or of a load of coal, or of a bunch of bananas, is ordinarily determined on a scale correctly labeled *No Springs.* This scale is actually a beam balance (*see* Chapter 9) that compares masses and gives a reading quite independent of the value of g—it does not measure weight but mass. However, it purports to give weight in pounds. What it actually gives is a *standard weight;* it is calibrated to read the weight the mass *would have* at a location where g has its standard value g_S given in (16):

> The **standard weight** of a body is the weight it would have at a location where the gravitational acceleration has the standard value $g_S = 32.173\,98$ f/s².

Standard weight is a *measure* of the *mass* of a body, not of its weight, although it is expressed in weight units. Given the standard weight, we can determine the mass of the body but we cannot determine its weight. Standard weight, w_S, is given by

$$w_S = mg_S; \quad m = w_S/g_S;$$

hence we can determine mass in slugs by dividing standard weight in pounds by 32.173 98 f/s².

9. Density

> The **average density** of an object or of a quantity of matter is defined as the total mass divided by the total volume occupied by the object or by the quantity of matter.

From the above definition of average density, which is usually a directly measurable quantity, we go to the concept of density of matter at a point by going to the limit of collapsing the volume to an infinitesimal region around the point. Let a volume ΔV containing the point in question have the mass Δm. Then the density, denoted by ρ, at the point is defined by

$$\rho = \lim_{\Delta V \to 0} \frac{\Delta m}{\Delta V}, \tag{23}$$

the limit of *mass per unit volume.*

However, when we remember that matter is composed of atoms, and that an atom consists of a very small nucleus that has most of the mass, a number of electrons, and is mostly empty space, we see that there is a physical difficulty in applying equation (23) in the strict mathematical sense of going to an infinitesimal volume. The selected point may be inside a nucleus but is most likely in empty space. In either case we do not get the value of density that we are really interested in. This same type of difficulty will occur throughout the discussion of the mechanical, thermal, and electromagnetic properties of materials. The physicist is not really interested in the mathematician's *point;* he is interested in a *volume* that is macroscopically small but that still contains an enormous number of atoms—for example a volume of a cubic micrometer. One μm^3 of ordinary air still contains 3×10^7 molecules, with solids and liquids having about 1000 times this number of molecules.

A physicist then understands the $\Delta V \rightarrow 0$ in (23) to indicate that ΔV gets very small, *but not too small.* The physicist is interested in a *macroscopic* property of the material, not a *microscopic* (atomic scale) property. He thinks of ΔV as shrinking to a volume too small to see—such as, for example, $1 \, \mu\text{m}^3$—that still contains an enormous number of atoms. It is in this sense of going to a *physical infinitesimal,* rather than to a *mathematical infinitesimal,* that a physicist defines point functions, such as density, in material media, and it is in this way that the physicist understands the calculus limit $\Delta V \rightarrow 0$.

A *homogeneous* material will have a constant density throughout. Data on the densities of various homogeneous materials are given in Table 4–2. Water has the density of $1.000 \, \text{Mg/m}^3 = 1.000 \, \text{g/cm}^3$. The meter and kilogram were originally intended to make this density *exactly* $1 \, \text{g/cm}^3$ at the temperature $39°$ F or $4°$ C at which water has maximum density at normal atmospheric pressure; this objective was actually missed

Table 4–2 Typical Densities ρ of Liquids and Solids at 68° F and Normal Atmospheric Pressure

Liquids	Mg/m³ or g/cm³	sl/f³	Solid metals	Mg/m³ or g/cm³	sl/f³
Water (32–50°)	1.000	1.940	Aluminum	2.700	5.25
Sea water	1.030	2.00	Cast iron	7.200	14.0
Benzene	0.879	1.71	Copper	8.890	17.3
Carbon tetrachloride	1.594	3.09	Gold	19.300	37.5
Ethyl alcohol	0.789	1.53	Lead	11.340	22.0
Gasoline	0.680	1.32	Magnesium	1.740	3.39
Kerosene	0.800	1.55	Nickel	8.850	17.2
Lubricating oil	0.900	1.75	Silver	10.500	20.4
Methyl alcohol	0.792	1.54	Steel	7.800	15.1
Sulphuric acid, 100%	1.831	3.55	Tungsten	19.000	37.0
Turpentine	0.873	1.69	Zinc	7.140	13.9
Mercury (32° F)	13.595	26.38	Brass or bronze	8.700	16.9
Nonmetallic solids			Woods		
Ice (32° F)	0.922	1.79	Balsa	0.130	0.25
Concrete	2.300	4.48	Pine	0.480	0.93
Earth, packed	1.500	2.92	Maple	0.640	1.24
Glass	2.600	4.97	Oak	0.720	1.40
Granite	2.700	5.25	Ebony	1.200	2.33

by 3 parts in 100 000. The ratio of the density of a substance to that of water at this temperature and pressure is called the *specific gravity* of the substance.

The density will vary from point to point in an object composed of inhomogeneous materials, or even in an object composed entirely of the same material if the pressure or the temperature varies from point to point in the object. A good example of a highly heterogeneous object is the earth, whose *average* density is computed in the following example:

Example. *In the example on p. 65, we computed the mass of the earth as 6.0×10^{24} kg. Assuming the earth to be a sphere of radius 6.4×10^6 m, compute its average density and specific gravity.*

The volume of the sphere is

$$\tfrac{4}{3}\pi R_E^3 = \tfrac{4}{3}\pi\,(6.4 \times 10^6 \text{ m})^3 = 11 \times 10^{20} \text{ m}^3.$$

The average density is then

$$\bar{\rho} = 6.0 \times 10^{24} \text{ kg}/11 \times 10^{20} \text{ m}^3 = 5500 \text{ kg/m}^3 = 5.5 \text{ Mg/m}^3,$$

or 5.5 times the density of water, which figure gives the average specific gravity. A more accurate value of the average density, given in Sec. 3 of the Appendix, is 5.52 Mg/m^3.

10. Freely Falling Bodies: Vertical Motion

Near the earth's surface, the magnitude of g varies only slowly with altitude, and g is directed vertically downward. Hence the motion of a body dropped from rest or projected vertically upward or downward is, to a good approximation, rectilinear motion with constant acceleration, provided air resistance is negligible. In this case the equations of motion developed on p. 41 are directly applicable.

We shall first consider the rectilinear motion of a freely falling body occurring along the Y-axis of a rectangular coordinate system with the positive Y-axis pointing *upward,* so that the Y-component of acceleration is $-g$. We choose the origin of our coordinate system at the initial position of the body, so that $Y=0$ at $t=0$. The initial velocity component will be denoted by v_{Y0}; this is positive if the initial velocity is upward, negative if it is downward. With this notation, equations (10)–(15) on p. 41 apply if we substitute Y for X and $-g$ for a_X, and give

$$v_Y = v_{Y0} - gt, \qquad Y = v_{Y0}\,t - \tfrac{1}{2}gt^2, \qquad v_Y^2 = v_{Y0}^2 - 2gY. \tag{24}$$

Example. *Suppose that a man on the top of a building 300 f high leans over the edge of the building and throws a ball vertically upward. If the magnitude of the initial upward velocity is 64 f/s, find the position and velocity of the ball at one-second intervals after it leaves the man's hand, and find the time required for the ball to reach the pavement on the ground at the side of the building. Take g as exactly 32 f/s^2.*

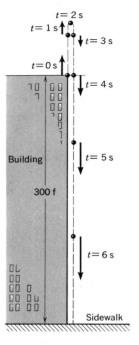

Fig. 13. The position and velocity of a ball at the end of successive one-second intervals after the ball is thrown upward from the top of a building at 64 f/s.

Substitution in (24) gives

$$Y = (64 \text{ f/s}) t - \tfrac{1}{2}(32 \text{ f/s}^2) t^2$$
$$= (64 \text{ f/s}) t - (16 \text{ f/s}^2) t^2, \tag{i}$$
$$v_Y = 64 \text{ f/s} - (32 \text{ f/s}^2) t. \tag{ii}$$

From these equations, the following table of positions and velocities can be obtained:

Time	Position	Velocity
0 s	0 f	+ 64 f/s
1	+ 48	+ 32
2	+ 64	0
3	+ 48	− 32
4	0	− 64
5	− 80	− 96
6	− 192	− 128
7	− 336	− 160

These values of position and velocity are shown graphically in Fig. 13.

From these results, we note that the time required for the ball to reach the ground at the side of the building ($Y = -300$ f) is between 6 s and 7 s. For this value of Y, (i) gives the quadratic equation

$$-300 \text{ f} = (64 \text{ f/s}) t - (16 \text{ f/s}^2) t^2$$

or

$$16 t^2 - (64 \text{ s}) t - 300 \text{ s}^2 = 0.$$

This equation has two solutions: $t = 6.77$ s and $t = -2.77$ s. Only the positive value can give the solution to the problem we are considering, so the time to reach the ground is 6.77 s. (The solution $t = -2.77$ s does have an interesting physical significance: It is the time at which the ball would have to have been projected from the ground with exactly the correct upward initial velocity to have reached the top of the wall at $t = 0$ with upward velocity 64 f/s so that the ball would continue the motion tabulated above.)

Two points might be noted in connection with the results shown in the table in the previous example. One of these is that the ball traveled upward for two seconds and then required two seconds to return to its original position. This result is quite general; the time of rise and time of descent to the original level are equal whenever an object is thrown upward if air resistance is negligible. The second point to be noted is that the downward velocity of the ball at the time it returns to its initial position has the same magnitude as the initial upward velocity of the ball; this result is also quite general.

It should be re-emphasized that the discussion we have given holds strictly only for the idealization of a *freely falling* body that experiences no air resistance. Air resistance is approximately proportional to the projected area of a body and increases rapidly with its speed. If a tennis ball, or a raindrop, were to fall from a considerable height, the free-fall approximation might be good during the first second or two of fall, but would become poorer as the speed increased. Eventually a constant limiting downward speed, known as the *terminal speed,* would be reached. The terminal speed is that at which the upward resisting force of the air equals the weight, so that the resultant force is zero, and no further acceleration is experienced.

11. Freely Falling Bodies: Projectiles

When a body is projected with an initial velocity having a horizontal component, it describes a curved path (*trajectory*) as in Figs. 14 and 16. If the range R and height h in these figures are very small in comparison with the radius of the earth, we can assume g to be constant along the trajectory. If, in addition, air resistance is negligible, the projectile is a freely falling body, acted on only by the downward force mg. Since there is no horizontal component of force, there is no horizontal component of acceleration, so *the horizontal component of velocity is constant.* The vertical component of velocity varies in accordance with (24).

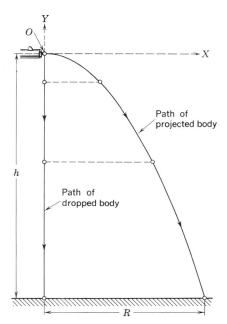

Fig. 14. A body projected horizontally has the same *vertical* motion as a body dropped from rest at the instant of projection, and reaches the ground at the same instant.

From these considerations we see that if, as shown in Fig. 14, a body is projected horizontally at a height h above the ground and a second body is released from rest at height h at the instant the first body is projected, and falls vertically, the two bodies are at any instant at the same vertical distance above the ground and reach the ground at the same time. This result can be easily verified in the laboratory.

To determine the range R of a body projected horizontally with initial speed v_0, as in Fig. 14, let us place the origin of our coordinate system at the point of projection. The projectile will reach the ground when its Y-coordinate is equal to $-h$. Substituting this value and $v_{Y0}=0$ in the center equation of (24), we may determine the time T required for the projectile to reach the ground (the *time of flight*). This time is $T=\sqrt{2h/g}$. In the case we are considering, v_X, the constant horizontal component of the velocity, is equal to the initial speed v_0. Therefore the horizontal component X of the total displacement is given by the equation $X=v_0 t$. The *range* R is given by substituting the value $T=\sqrt{2h/g}$ of the time of flight in this equation; that is, $R=v_0\sqrt{2h/g}$.

The velocity v of the shell at any time t during the flight can be determined from its horizontal and vertical components, which are given

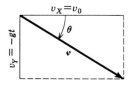

Figure 15

by the relations $v_X = v_0$; $v_Y = -gt$. The magnitude of the resultant velocity is $v = \sqrt{v_X^2 + v_Y^2} = \sqrt{v_0^2 + g^2 t^2}$, and the direction, given by the angle θ measured from the horizontal, is such that $\tan\theta = v_Y/v_X = -gt/v_0$, as shown in Fig. 15. The negative value of the tangent indicates that θ is negative and below the horizontal.

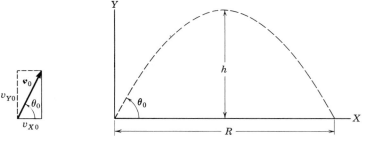

Fig. 16. Path of a body projected with initial velocity of magnitude v_0 at elevation angle θ_0.

Now let us consider a body projected from ground level as in Fig. 16, with initial velocity of magnitude v_0 at angle θ_0 above the horizontal. The initial velocity has components

$$v_{X0} = v_0 \cos\theta_0; \qquad v_{Y0} = v_0 \sin\theta_0. \tag{25}$$

Throughout the trajectory, the horizontal component v_X of the velocity remains constant at the initial value v_{X0}. If we assume the body to be projected at time $t = 0$, its horizontal position at any later time is given by the relation

$$X = v_{X0} \, t. \tag{26}$$

The vertical velocity component v_Y and the Y-position are given, at any time t, by (24) as

$$v_Y = v_{Y0} - gt, \qquad Y = v_{Y0} \, t - \tfrac{1}{2} g t^2. \tag{27}$$

The equation

$$v_Y^2 = v_{Y0}^2 - 2gY \tag{28}$$

gives the Y-component of velocity as a function of height Y.

The equations giving X and Y as functions of t are the parametric equations for a *parabola,* which is the shape of the trajectory. These equations, together with the associated velocity equations, describe the motion completely. By means of these equations such quantities as the total time of flight T, the range R, the maximum height h, and the velocity of the projectile at any time can be calculated.

Projectile motion in a parabola is an example of motion in a curved path, in which case the acceleration vector (vertically downward in this case) is not tangent to the path but points inward toward the concave side of the path.

Example. *A gun pointed* 30° *above the horizontal fires a shell with muzzle velocity of* 640 f/s. *Neglecting air resistance, find* (a) *the maximum height reached,* (b) *the time of flight,* (c) *the range. Take* $g=32$ f/s^2.

We have $v_{X0}=(640 \text{ f/s}) \cos 30° = 554$ f/s; $v_{Y0}=(640 \text{ f/s}) \sin 30° = 320$ f/s.

(a) At the maximum altitude, $v_Y=0$, since the velocity (always tangent to the path) is horizontal. Equation (28) then gives, if we set $v_Y=0$ and $Y=h$, $2\,gh=v_{Y0}^2$; thus $h=v_{Y0}^2/2g=(320 \text{ f/s})^2/(64 \text{ f/s}^2)=1600$ f.

(b) When the shell returns to the ground, $Y=0$. Putting $Y=0$ in the second equation of (27) and solving for t gives $T=20$ s, the time of flight.

(c) The range is the horizontal distance traveled in the 20-s time of flight, which from (26) is $R=v_{X0} \cdot 20 \text{ s} = 11\,100$ f.

As indicated in Fig. 17, both height and range of actual projectiles are reduced by air resistance, which acts in the direction opposite to the velocity and tends to decrease the magnitudes of both components of velocity. Determination of the air resistance is made by wind-tunnel and other types of experimental measurements and used in the computation of 'firing tables' for military guns. These computations are extremely complicated. Study of actual trajectories is included in the science of *ballistics*.

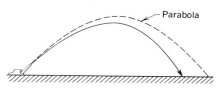

Figure 17

12. The Newtonian Principle of Relativity

We stated on p. 37 that all velocities are relative and that no mechanical experiment can determine an absolute velocity. Inside a closed room in a moving train or airplane, there is no way of determining the velocity of the room relative to the earth. So long as this velocity is constant, all mechanical experiments—the acceleration of a body down an inclined plane, the path of a projected body, etc.—give the same results as if the room were at rest on the ground. Inside the room we could easily measure the *acceleration* of the train or plane, as we know from common experience, but it is impossible to measure its velocity. These ideas were formulated by Newton as the

NEWTONIAN PRINCIPLE OF RELATIVITY: *The principles of mechanics that are valid in one coordinate system are equally valid in a coordinate system in motion at constant velocity relative to it; hence on the basis of mechanical phenomena there is no way of determining absolute velocity; only relative velocity can be measured.*

It is not difficult to see that if Newton's principles are applicable in one coordinate system, they are applicable in a second coordinate system moving at constant velocity relative to the first; for example, the coordinate system in the moving train or plane mentioned above. Of course, different velocities are assigned to a particle in the two coordinate systems, but if a particle moves at *constant* velocity in one it moves at constant velocity in the other and would be considered as acted on by no forces in either case. Hence, Newton's first principle is equally valid in the two coordinate systems. Although different velocities are assigned to particles in the two coordinate systems, these velocities always differ by a constant which is the relative velocity of the two systems. Hence, a particle has the *same acceleration* in the two coordinate systems, and Newton's second principle assigns the *same forces*. Since the forces are the same, Newton's third principle is valid in one system if it is valid in the other. These arguments prove the Newtonian principle of relativity.

A system of coordinates in which Newton's principles are valid is called an *inertial frame*. A system of coordinates that is accelerated relative to an inertial frame is not an inertial frame—for example, in the accelerated frame a particle acted on by no forces appears nevertheless to be accelerated.

The extension of the principle of relativity to the laws of electrodynamics led Einstein to the important conclusions embodied in the Einstein theory of relativity which we shall discuss in Chapter 44.

Example. *An elevator cage is moving upward at a constant speed of* 12 f/s. *A light bulb detaches itself from the ceiling and smashes on the floor* 9 f *below. Compute the time of flight from the point of view of the man in the cage. Verify, in this example, the Newtonian principle of relativity by computing the time of flight from the point of view of a man standing in the building looking into the open cage of the elevator.*

By the principle of relativity, the man in the elevator can completely ignore his motion relative to the earth and consider that the light bulb falls from rest and travels 9 f. The time required is given, as we have seen, by

$$T = \sqrt{2h/g} = \sqrt{2\,(9\text{ f})/(32\text{ f/s}^2)} = \tfrac{3}{4}\text{ s}.$$

During this $\tfrac{3}{4}$ s, the elevator has moved upward exactly 9 f, so to the man in the building it must look as though the bulb left the ceiling and hit the floor at the same Y-coordinate. The man in the building does not see the bulb fall from rest; at the moment it is detached from the ceiling, he sees it projected upward with initial velocity $v_{Y0} = 12$ f/s. A body projected upward at this speed rises for a time that is given by setting $v_Y = 0$ in the first equation of (29) as

$$v_{Y0}/g = (12\text{ f/s})/(32\text{ f/s}^2) = \tfrac{3}{8}\text{ s}.$$

The projectile returns to its starting point in twice this time, or ¾ s. In ¾ s, the elevator has moved up exactly 9 f; the floor of the elevator is where the ceiling was initially, so the man in the building predicts that the impact takes place at just this instant.

PROBLEMS

NOTE: *In solving problems involving gravitational acceleration, use $g = 32$ f/s^2 = 9.8 m/s^2 and neglect air resistance, unless otherwise indicated.*

1. A 'particle' with a mass of 1 kg is attached by means of a light spring to a second 'particle' A of unknown mass m_A. By compressing or stretching the spring and then releasing it in such a way that the particles interact only with each other, it is found that the acceleration of the second particle always has a mgnitude that is half the magnitude of the acceleration of the 1-kg particle. What can be said concerning the *directions* of the accelerations of the two particles? What is m_A? Ans: 2 kg.

Problems 1–3

2. The particle A with mass m_A mentioned in Prob. 1 is connected to another particle B, with unknown mass m_B, by means of a spring. In studying interactions between the particles, it is found that the following relation exists between the magnitudes of the accelerations of A and B: $a_A = \frac{2}{5} a_B$. What is m_B? If particle B interacted in any way with a 1-kg particle, what prediction could you make regarding the resulting accelerations?

3. Body A with mass 2 kg and body B with mass 3 kg are connected by means of a light stretched spring and are then released. By observing the resulting motion, an observer finds that body A has an initial acceleration of 3 m/s^2, eastward. What resultant force acts on body A? What is the initial acceleration of body B? What resultant force acts on body B?
Ans: 6 N, E; 2 m/s^2, w; 6 N, w.

4. A 2-kg body is subjected to one force of 8 N, eastward, and a second force of 6 N, northward. What resultant force acts on the body? What is the acceleration of the body? Find the northward and eastward components of the body's acceleration.

5. A 1000-kg automobile moves eastward along a horizontal road at a constant speed of 16 m/s. What resultant force acts on the automobile? If the brakes are applied and the car is brought to rest in 8 s with uniform deceleration, what constant resultant force acts on the car?
Ans: 0; 2000 N, w.

6. A 2000-kg car is moving southward along a straight horizontal roadway at a speed of 20 m/s when the brakes are applied in such a way as to produce constant deceleration. If the car travels 100 m along the road before coming to rest, find the magnitude and direction of the resultant external force acting on the car. What body exerts this force on the car?

7. A 1000-kg car moves around a circular race track at a constant speed of 20 m/s. If the radius of the track is 200 m, what is the centripetal acceleration of the car? What is the magnitude of the centripetal force exerted on the car? What exerts this force on the car?
Ans: 2 m/s^2; 2000 N; the track.

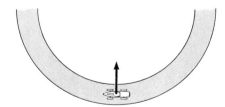

Problem 7

8. From the mass and radius of the sun (*see* Appendix), determine the gravitational acceleration at its surface.

9. From the mass and radius of the moon (*see* Appendix), determine the gravitational acceleration at its surface.   Ans: 1.62 m/s².

10. If the radius of the earth's orbit were doubled, what would be the effect on the gravitational force exerted on the earth by the sun? Assuming that the earth continued to move in a circular orbit with twice the radius of its present orbit, find its new orbital speed in terms of its present orbital speed and find the new length of the year. If the earth's orbit were so changed, what would be the effect on the weight of a body near the earth's surface?

11. Calculate the gravitational force that would be exerted on a 10-kg sphere by a second 10-kg sphere when the distance between the centers of the spheres is 60 cm. If these were the only forces acting on the spheres, what would be their accelerations?
    Ans: 1.85×10^{-8} N; 1.85×10^{-9} m/s².

12. At what point along the line from the earth to the moon is the gravitational pull of the moon equal to that of the earth? At this point a moon rocket would leave the 'influence' of earth gravity and enter the 'influence' of moon gravity. Express your result as a percentage of the distance from the center of the earth to the center of the moon. The mass of the moon is 0.0123 times that of the earth.

Problem 12

13. Compute the value of the mass of the sun in kg from the approximation that the earth moves in a circular orbit of radius 150 000 000 km and has a period of revolution of 365 days. Use the observed value of G.
    Ans: 1.99×10^{30} kg.

14. Compute the value of the earth's mass from the approximation that the moon completes one revolution about the earth every 27.3 days in a circular path of radius 384 000 km. Use the observed value of G. Compare your result with that obtained in the example in the text on p. 65.

15. Show that if two solid gold spheres of equal radii are in contact, their gravitational force of attraction is directly proportional to the fourth power of their radii.

16. Compute the average density of the moon in Mg/m³, approximating the moon by a sphere of radius 1700 km and mass 7.3×10^{22} kg. Compare with the average density of the earth.

17. What is the gravitational force exerted on a 10-kg solid gold ball by a second gold ball of the same mass when the two balls are in contact?   Ans: 6.67×10^{-7} N.

18. In analogy to Kepler's law which states that planetary speeds are inversely proportional to the square roots of the radii of the orbits, show that the speeds of different earth satellites in circular orbits are inversely proportional to the square roots of the radii of their orbits. Show also that the times of revolution are directly proportional to the $3/2$ power of these radii.

19. Given that the period of revolution of the moon is 27.3 days and that the orbit of the moon is approximately a circle of radius 239 000 mi, compute, from the result in Prob. 18, the period of an artificial earth satellite in a circular orbit of radius 4500 mi (about 500 mi above the earth's surface).   Ans: 101 min.

20. A 200-p man stands on a spring scale in an elevator. What does the scale read if the elevator ascends at a constant velocity of 8 f/s? if the elevator has an upward acceleration of $1/10\,g$? a downward acceleration of $1/5\,g$? if the cable breaks and the elevator falls freely?

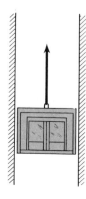

Problems 20–22

21. An elevator cage weighing 2000 p experiences an upward acceleration of 8 f/s². If friction is negligible, what upward force is exerted on the elevator cage by the supporting cable? Ans: 2500 p.

22. If the elevator cage in Prob. 21 experiences a downward acceleration of 8 f/s², what upward force is exerted by the supporting cable?

23. A body weighs 300 p at sea level. Using the radius of the earth as 4000 mi, find the height above sea level at which the body in question would weigh 150 p. What would be the initial downward acceleration of the body at that height if it were released and allowed to fall? Ans: 1640 mi; 16 f/s².

24. Find the magnitude of the acceleration of a body acted on by a resultant force whose magnitude is twice the magnitude of the body's weight; three times that of the body's weight; etc. Derive the expression $a/g = F/w$, which gives the magnitude of the acceleration a of a body of weight w when the body is subjected to a force of magnitude F. Note that this equation holds for any choice of units, provided only that the same units are used for F and w and the same units are used for a and g.

25. What is the weight of a body having a mass of 1 sl at a place at which the acceleration of gravity has its standard value $g_s = 32.173\ 98$ f/s²? What is the weight of a 1-kg body at this place? Ans: 32.173 98 p; 9.806 65 N.

26. Consider an earth satellite so positioned that it appears stationary to an observer on the earth and serves the purpose of a fixed relay station for intercontinental transmission of television and other communications. What is the location and diameter of the orbit of such a SYNCOM satellite? What is the direction of motion?

27. Compute the time of revolution of an earth satellite in a circular orbit 1000 mi above the surface of the earth; in a circular orbit 2000 mi above the surface. Approximate the earth by a sphere of radius 4000 mi with gravitational acceleration of 32 f/s² at the surface. Ans: 119 min; 157 min.

28. A dive bomber has an acceleration of 4 g_s upward (where g_s is the standard acceleration of gravity) at the lowest point of its dive when its velocity is 500 f/s horizontally. What is the

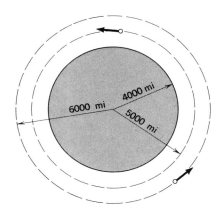

Problem 27

radius of curvature of its path at this point? Determine the force between the pilot and his seat expressed as a multiple of the pilot's weight.

Problem 28

29. Using the values for density given in Table 4–2, find the magnitude of the resultant force that will give 1 f³ of aluminum an acceleration of 3 f/s². What is the weight of 2 f³ of nickel? Ans: 15.8 p; 1100 p.

30. What is the mass of a 1-f³ block of ice? What resultant force would be required to give this block of ice an acceleration of magnitude 1 f/s²; 2 f/s²; 32 f/s²?

31. When a 7-kg body initially at rest is subjected to a resultant force equal to its own weight, what is its acceleration? What will be the velocity of the body at the end of 4 s? What will be the displacement of the body during the first 4 s? Ans: 9.8 m/s², downward; 39.2 m/s, downward; 78.4 m, downward.

32. A 2-kg body falls from rest under the influence of gravity. What is the magnitude of the gravitational force acting on the body? What is its acceleration? How far does the body move during the first second of motion? What is its average velocity during the first 6 seconds?

33. Using $g=32$ f/s², compute the maximum upward displacement of a ball that is thrown vertically upward with an initial velocity of 96 f/s, upward. What is its velocity at the highest point in its trajectory? What is its acceleration at the highest point in its trajectory? How long will the ball continue to rise? Ans: 144 f, upward; 0; 32 f/s², downward; 3 s.

34. A man at the top of a tall building leans over the edge and throws a ball downward. If the initial downward speed of the ball is 48 f/s, what is the downward displacement of the ball at the end of 2 s? If the point of release is 512 f above the ground, how long does it take the ball to reach the ground?

35. A bullet is fired horizontally from the top of a cliff 64 f above the surface of a lake. If the muzzle velocity of the projectile is 1200 f/s, what is the horizontal range of the projectile? Ans: 2400 f.

36. A bullet weighing 4 ounces is fired from ground level. The horizontal component of the muzzle velocity is 2000 f/s and the vertical component is 960 f/s. How long is required for the bullet to reach the highest point in its trajectory? What is the velocity of the bullet at the highest point in its trajectory? the acceleration? the resultant force on it? What is the time of descent from the highest point to ground level? the total time of flight? the total horizontal range of the bullet?

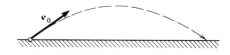

Problems 36–37

37. Answer all questions in Prob. 36 for a 400-g bullet for which the vertical component of the muzzle velocity is 98 m/s and the horizontal component is 200 m/s. Ans: 10 s; 200 m/s, horizontal; 9.8 m/s², downward; 3.92 N, downward; 10 s; 20 s; 4000 m.

38. Show that, as indicated in the figure, a freely falling body projected with *any* initial speed has the same range when projected at initial angles with the horizontal of 15° and 75°; at 30° and 60°; in general at θ_0 and $90° - \theta_0$. Derive an equation giving the range R in terms of v_0 and θ_0. By differentiation of R with respect to θ_0, show that R is a maximum for $\theta_0 = 45°$.

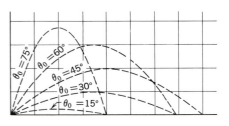

Problems 38–39

39. The shell fired from a trench mortar has a muzzle velocity of 320 f/s. Calculate two angles of elevation at which the mortar can be fired in order to hit a target 2770 f away and at the same level as the mortar. Calculate the maximum heights and the times of flight in 'direct' and 'indirect' fire at the target. Ans: $\theta_0 = 30°$ and 60°; $h = 400$ f and 1200 f; $t = 10$ s and 17.3 s.

40. As in the figure, a gun is 'bore-sighted' on a target. At the instant the projectile leaves the muzzle of the gun, the target starts to fall from rest, vertically. Show that for any angle of elevation of the gun, any muzzle velocity, and any distance to the target, the projectile will always hit the target if the effect of air resistance is negligible.

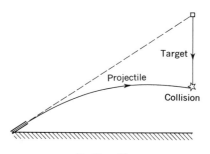

Problem 40

41. During the First World War, the Germans bombarded Paris with a specially constructed long-range gun (Big Bertha). The length of the gun barrel was approximately 120 f and the muzzle velocity was approximately 4800 f/s. For an elevation angle $\theta_0 = 45°$, calculate the range in the absence of air resistance. (The

maximum range actually attained was about 132 000 yd with $\theta_0 = 55°$; in the case of this very high muzzle velocity, air resistance is seen to have a very important effect.)
Ans: 242 000 yd.

42. In the example on pp. 76–77, the man in the building sees the light bulb rise for $3/8$ s, then fall for $3/8$ s, so that, at $t = 3/8$ s, it appears to him to be at rest. Compute the velocity of the light bulb at $t = 3/8$ s in the coordinates of the man in the elevator. By combining this relative velocity with that of the elevator in accordance with (6), p. 37, show that the velocity relative to the ground is indeed zero at this time.

43. The engineer of a train traveling at 80 mi/h sights a train one mile ahead of him traveling in the same direction on the same track, and applies his brakes. If the forward train is traveling at the constant speed of 40 mi/h, what is the minimum deceleration, in mi/h·s, of the rear train that will avoid a collision?
Ans: 0.222 mi/h·s.

44. Show that by flying along a parabolic trajectory at exactly the speed of a freely falling projectile, an aircraft pilot can, for a limited period, establish a 'zero-gravity' condition in which he and objects in the plane will be apparently weightless and a ball will float freely relative to the plane.

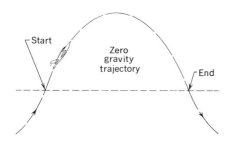

Problems 44-45

45. In a typical Air Force 'zero-gravity' experiment (*see* Prob. 44), a jet pilot starts a ballistic trajectory at an altitude of 15 000 f, while climbing with an initial velocity of 500 mi/h at an angle of 70° above the horizontal. When he returns to the 15 000-f altitude, he leaves the ballistic trajectory and begins to pull out of his dive. (*a*) For how long does he maintain the zero-gravity condition? (*b*) What is his speed and altitude at the highest point of the trajectory?
Ans: (*a*) 43.1 s; (*b*) 171 mi/h, 22 400 f.

46. When a certain body is suspended from a spring balance, the balance reads 40 p when the body is near sea level on the earth's surface. If the body were suspended from this balance near the surface of the moon, what would the balance read?

47. A lead brick has a mass of 3 sl. At what distance above the surface of the earth would this brick weigh 24 p? What would be its initial acceleration of free fall if it were released at this point?
Ans: 4000 mi; 8 f/s².

48. What constant resultant external force exerted on a body weighing 48 p near the earth's surface will give the body a velocity of 40 f/s at the end of 6 s if the body is initially at rest?

49. An automobile weighing 6400 p is moving southward at a speed of 60 f/s. What constant resultant force must be applied in order to bring the automobile to rest if the automobile is to move only 180 f after the force begins to act?
Ans: 2000 p, N.

50. It is found that a very small particle falling vertically through the air is subject to an upward resisting force directly proportional to its downward velocity. (*a*) Setting the resisting force equal to kv, where k is a proportionality constant, show that the body will attain a terminal velocity v_T and express this terminal velocity in terms of k, the body's mass m, and g. (*b*) The downward acceleration a can be determined from Newton's second principle in the form $ma = mg - kv$. Recalling that $a = dv/dt$, show by integration that the velocity v of a body falling from rest at $t = 0$ is given by the expression

$$v = \frac{mg}{k}(1 - e^{-(k/m)t}).$$

(*c*) Show that this expression for v reduces to the free-fall expression for sufficiently small values of k and t.

51. (*a*) In the course of his early experiments, Galileo tried to interpret his observations by a relation in which the vertical downward velocity was proportional to the distance of fall from rest: $v_Y = kY$. Show by integration that this leads to the relation $Y = Ce^{kt}$. This relation is clearly incorrect; why? (*b*) Huygens was also interested in obtaining a relation between v_Y and Y for a freely falling body and proposed the relation $v_Y = k\sqrt{Y}$. Show by integration that this is a valid expression, and find k in terms of g.

52. A pendulum, consisting of a heavy bob attached to the end of a light string, is suspended from the ceiling of a railway car moving along a straight, level track. At a certain time a passenger happens to notice that the pendulum is apparently stationary but does not hang vertically. It is inclined at an angle of 6°.5 to the vertical, the bob being displaced toward the rear of the train. What conclusions can the passenger make regarding the train's motion at the time of his observation? If the bob weighs 4 p, what force does the cord exert on the bob at the time of observation?

53. The dependence of the earth's gravitational acceleration g on the distance r from the center of the earth is, on the assumption that the earth is spherically symmetric, given by $g = k/r^2$, where k is a constant. From this formula, compute the value of dg/dr near the surface of the earth, and compare with the observed rate of change at 40° latitude obtained from the table on p. 65.

Ans: $dg/dr = -0.0030$ f/s² per 1000 f.

54. Compute the centripetal acceleration a in equation (14) at the equator, where it has its maximum value. Show that, as stated in the text, it is 0.003 times the gravitational acceleration.

55. Show that the centripetal acceleration a in equation (14) varies as the cosine of the latitude.

56. Calculate the gravitational force of the sun on a 150-p man on the earth. Is the man heavier by this amount when he is on the side of the earth away from the sun and correspondingly lighter when the sun is overhead? Explain.

57. A man in a car moving along a straight road with constant speed V throws a ball vertically upward with initial speed v_{Y0} and catches it on its return. Show how this straight line trajectory 'transforms' into the parabolic trajectory seen by an observer on the ground. Determine the initial velocity (magnitude and direction) as observed from the ground. Neglect air resistance.

SYSTEMS OF FORCES; FRICTION

5

In this chapter we shall consider particles and rigid bodies acted on by more than one force, so that the vector sum of the forces governs the acceleration of the body. We shall consider both the *static* case of *equilibrium,* in which the vector sum of the forces is zero and the body remains at rest or in motion at constant velocity, and the *dynamic* case where the vector sum is not zero and the body is accelerated. We consider only *translational motion* of rigid bodies, leaving discussion of *rotation* to Chapter 8:

> **Translational motion** of a rigid body is motion in which all particles of the body have the same velocity.

We state the empirical laws governing *friction* between solid bodies, since forces of friction play an important role, not only in engineering design, but in everyday life, as is apparent if one thinks of the troubles that arise when friction is too small, as on slippery floors and icy roads.

1. Treatment of Forces as Vectors

The resolution of a force vector into rectangular vector components is illustrated in Figs. 1, 2, and 3. Since the component forces take the place of the original force, the original force vector is shown crossed out

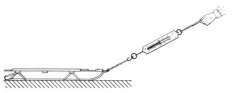

Fig. 1. A sled is pulled by a rope. The force is measured by the spring balance.

Fig. 2. Resolution of the force on the sled in Fig. 1 into horizontal and vertical components.

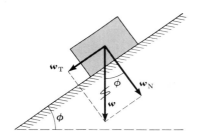

Fig. 3. Resolution of a vertical force into components parallel and normal to an inclined plane.

in Figs. 2 and 3 in favor of the two mutually perpendicular forces that replace it. The meaning of these component forces is easy to understand in a specific example such as that in which a sled is pulled along the ground. The horizontal force F_H in Fig. 2 is the effective force pulling the sled along the ground, and the vertical force F_V is the effective force tending to lift the sled off the ground.

The weight of an object is a force w directed vertically downward. It is frequently useful, in analyzing the effect of a body's weight on its motion along an inclined plane, to resolve w into vector components in directions parallel and normal to the inclined plane, as in Fig. 3. The tangential force w_T is the effective force that tends to pull the body downward along the incline, while w_N is the effective force that pulls the body normally against the inclined plane.

By *experiment* we can show that if several forces act at the same point of a body, as in Fig. 4(a), their net effect is in all respects identical with the effect of a single force that is their vector sum. Thus if the three forces of Fig. 4 have the magnitudes represented in (b), they are equivalent to the single force F of (c).

Example. *Find the single force F that is equivalent to the three forces F_1, F_2, F_3 in Fig. 4, when they act simultaneously at the same point of a body. F_1 has magnitude 3 N at an angle of 30° with the X-axis; F_2, 8 N at an angle of 60°; F_3, 9 N at an angle of 195°.*

The components are

$F_{1X} = (3\text{ N})(\cos 30°) = 2.60$ N, $F_{1Y} = (3\text{ N})(\sin 30°) = 1.50$ N,
$F_{2X} = (8\text{ N})(\cos 60°) = 4.00$ N, $F_{2Y} = (8\text{ N})(\sin 60°) = 6.93$ N,
$F_{3X} = (9\text{ N})(\cos 195°) = -8.69$ N, $F_{3Y} = (9\text{ N})(\sin 195°) = -2.33$ N.

The sum of these forces has components

$F_X = F_{1X} + F_{2X} + F_{3X} = (2.60 + 4.00 - 8.69)\text{ N} = -2.09$ N,
$F_Y = F_{1Y} + F_{2Y} + F_{3Y} = (1.50 + 6.93 - 2.33)\text{ N} = 6.10$ N.

Hence $\tan\phi = (2.09\text{ N})/(6.10\text{ N}) = 0.343$, $\phi = 18°.9$,

$\theta = 90° + 18°.9 = 108°.9$, $F = \sqrt{(2.09\text{ N})^2 + (6.10\text{ N})^2} = 6.45$ N.

The resultant force has a magnitude of 6.45 N at an angle of 108°.9 with the positive X-axis.

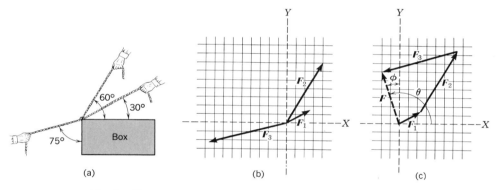

Fig. 4. Three men pull on three ropes attached to the corner of a box. In (b) and (c) the grid spacing represents one newton.

2. Equilibrium of Forces

We shall now consider the 'static' situation, in which the system of forces acting on a particle is such that the particle remains at rest or, equivalently, in motion at constant velocity. One type of 'particle' for which we shall consider the equilibrium of forces is a knot such as that in Fig. 5, which is *in equilibrium* under the action of the three forces $\mathbf{F}_1$, $\mathbf{F}_2$, and $\mathbf{F}_3$.

From Newton's second law, we have the

CONDITION FOR EQUILIBRIUM OF A PARTICLE: *The vector sum of all the forces acting on a particle in equilibrium must be zero.*

In this chapter we shall, for simplicity, confine our attention to cases in which all the force vectors can be drawn in the same plane (the forces have X- and Y-, but no Z-components). The condition for equilibrium is then most simply applied if we replace each force by its X- and Y-component forces. If the X-axis is horizontal and the Y-axis vertical, as would be appropriate in Fig. 5, the condition that the vector sum of the forces have zero horizontal component is expressed by the equation:

$$\left\{\begin{matrix}\text{the sum of the magnitudes of} \\ \text{the component forces acting} \\ \text{to the right}\end{matrix}\right\} = \left\{\begin{matrix}\text{the sum of the magnitudes of} \\ \text{the component forces acting} \\ \text{to the left}\end{matrix}\right\} ; \qquad \textbf{(1)}$$

and the condition that this vector sum have zero vertical component by:

$$\left\{\begin{matrix}\text{the sum of the magnitudes of} \\ \text{the component forces acting} \\ \text{vertically upward}\end{matrix}\right\} = \left\{\begin{matrix}\text{the sum of the magnitudes of} \\ \text{the component forces acting} \\ \text{vertically downward}\end{matrix}\right\} . \qquad \textbf{(2)}$$

These conditions will be illustrated by application to problems of various types. It is noted that sometimes it is convenient to take rectangular axes that are not horizontal and vertical, but, for example, parallel and perpendicular to an inclined plane; in such cases equations (1) and (2) undergo obvious modifications.

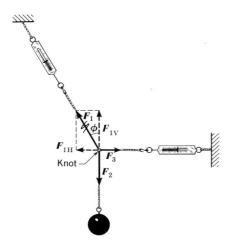

Fig. 5. A knot in equilibrium, under the action of three concurrent forces.

Example. *In the system of Fig. 5, determine the forces F_1 and F_3 in terms of the weight w of the suspended sphere and the angle ϕ.*

At the knot there are three strings. The vertical string pulls down on the knot with a force $\boldsymbol{F}_2$ equal to the weight $\boldsymbol{w}$ of the suspended sphere. The other two strings pull on the knot with forces that can be read on the spring scales. If we resolve $\boldsymbol{F}_1$ into vertical and horizontal component forces $\boldsymbol{F}_{1V}$ and $\boldsymbol{F}_{1H}$, equation (1) states that

$$F_3 = F_{1H}, \qquad \text{(i)}$$

while equation (2) gives

$$F_{1V} = F_2. \qquad \text{(ii)}$$

Since $F_{1V} = F_1 \cos\phi$, we can write equation (ii) in the form $F_1 \cos\phi = F_2$, or

$$F_1 = F_2/\cos\phi = w/\cos\phi \qquad \text{(iii)}$$

Since $F_{1H} = F_1 \sin\phi$, equation (i) becomes $F_3 = F_1 \sin\phi$. Substitution of (iii) in this equation gives

$$F_3 = \frac{F_2}{\cos\phi}\sin\phi = F_2 \tan\phi = w \tan\phi. \qquad \text{(iv)}$$

Equations (iii) and (iv) determine F_1 and F_3 in terms of w and ϕ.

Example. *If, in Fig. 5, $F_2 = 2.5$ N and $F_3 = 1.5$ N, determine F_1 and ϕ.*

From (i) and (ii) in the preceding example,

$$F_{1H} = F_3 = 1.5 \text{ N}, \qquad F_{1V} = F_2 = 2.5 \text{ N}.$$
Then
$$F_1 = \sqrt{F_{1H}^2 + F_{1V}^2} = \sqrt{(1.5 \text{ N})^2 + (2.5 \text{ N})^2} = 2.91 \text{ N},$$
and
$$\tan\phi = F_{1H}/F_{1V} = 1.5/2.5 = 0.6, \qquad \phi = 31°\!.0.$$

In the case of an extended rigid body acted on by forces that may not all be applied at the same point, the general conditions for equilibrium are more complex. For the body to remain at rest or, equivalently, in motion with all its particles having the same constant velocity, the so-called condition for *translational* equilibrium requires that the vector sum of the forces be zero. However there are, in addition, conditions for *rotational* equilibrium, which we shall study in Chapter 8. In many cases, such as the force system in Fig. 7 in the next section, where there is clearly no tendency to rotate, we need only apply the

CONDITION FOR TRANSLATIONAL EQUILIBRIUM: *A rigid body will remain at rest or in motion at constant velocity only if the vector sum of all the forces acting on it is zero.*

3. Friction between Solid Surfaces

Everyone is familiar with the friction that is encountered whenever the surface of one solid object moves or merely tends to move along the surface of another solid object with which it is in contact. It is friction that holds a nail in a board and keeps the individual fibers in a thread or rope from pulling apart. Friction makes it possible for us to walk; friction between the tires and the road is essential for the propulsion and the braking of an automobile.

The direction of the force of friction *on a sliding object* is always parallel to the surface along which motion occurs and opposite to the

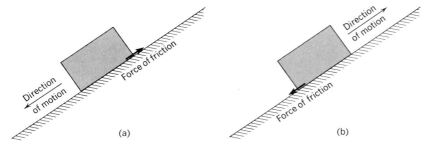

Fig. 6. The force of kinetic friction exerted *by a plane on a block*. This force is in the direction opposite to the motion of the block, as indicated. The force of friction that the *block* exerts *on the plane* would be opposite to the force shown here, by Newton's third law.

direction of motion, as illustrated in Fig. 6. The force of friction exerted *by the surface on the block* is tangent to the surface and opposite to the direction of motion of the block along the surface.

Two surfaces in contact exert tangential forces of friction on each other, not only when one slides past the other but also when the surfaces are stationary and one *tends* to slide past the other. If a man leans against a heavy table and the table does not move, a horizontal force of friction exerted on the table by the floor must oppose the horizontal force the man is exerting, if the table is to remain in equilibrium. Such a force is known as a force of *static friction* and is in such a direction as to oppose the tendency to move and to keep the body in equilibrium. If the man pushes hard enough on the table to make it actually move, the force of friction exerted by the floor while the table is moving is opposite to the direction of motion of the table and is known as a force of *kinetic friction.*

A **force of static friction** is a tangential force exerted by one solid surface on another when the two surfaces *are not* sliding past each other.

A **force of kinetic friction** is a tangential force exerted by one solid surface on another when the two surfaces *are* sliding past each other.

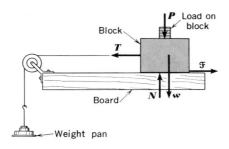

Fig. 7. Experiment to determine the force of friction.

We notice that the horizontal forces exerted by the tires of an automobile on a level road (and the equal and opposite forces exerted by the road on the tires) are usually forces of *static* friction, since the tires do not ordinarily *slide* on the road. However if the driver 'spins' the wheels in accelerating, or if the wheels are skidding, these forces become those of kinetic friction.

The laws governing frictional forces are determined by experiment. A simple experimental setup is shown in Fig. 7, in which the forces acting on the block are the following:

w = weight of the block,
P = downward force exerted by the load on the block,
N = normal force exerted by the board against the surface of the block,
T = tension in the string,
$\mathcal{F}$ = force of friction on the block.

The tension T in the string, which, if the block, and hence the weight pan, are unaccelerated, equals in magnitude the weight of the weight pan and its load,* tends to pull the block to the left along the surface

---

*This statement, which of course seems reasonable from experience, follows from Newton's principles, but we cannot discuss this matter rigorously until we study *rotational*

of the board, and the force $\mathcal{F}$ of friction, tending to oppose this motion, is directed to the right parallel to the surface of the board. When the block is in equilibrium, it either remains stationary or moves with constant velocity. The conditions (1) and (2) require that, in equilibrium,

$$N = w + P \quad \text{and} \quad T = \mathcal{F}.$$

Let us start the experiment with the weight pan detached. In this case $\mathcal{F} = 0$ because $T = 0$. Let us now attach the weight pan and start adding weights, gradually increasing T. Until T reaches a certain magnitude, the block does not move. During this period, the force of static friction is gradually increasing, since it must always equal T. Static friction always exerts just enough force to keep the block in equilibrium—up to a certain maximum. If T is increased sufficiently, the block will start to move, indicating that T has exceeded the maximum force of static friction. Thus *the force of static friction is a variable quantity,* depending on how much force is needed to keep the block from moving. But there is a definite maximum force of static friction; if the magnitude T of the applied force exceeds the maximum force of static friction, the block will move.

The maximum force of static friction $\mathcal{F}_\text{S}^\text{Max}$ can be found from the apparatus of Fig. 7 by gradually adding weights to the weight pan until the block starts to move. If we vary the load P on the block, we shall find that the values of $\mathcal{F}_\text{S}^\text{Max}$ so determined are *proportional* to the total weight $w + P$ of load and block, which equals the normal force N between the two surfaces. Hence, we can write

$$\mathcal{F}_\text{S}^\text{Max} = \mu_\text{S} N, \tag{3}$$

---

motion in Chapter 8. It will, however, be convenient to use certain results in advance of their formal proof.

    By the *tension* in a string, we mean the pull exerted by the string. This pull is exerted not only at the ends of the string, but by each section of the string on the adjoining sections—if we were to imagine cutting a string at any point and inserting a light spring scale, the scale would read the tension. Tension is discussed more fundamentally later, in Chapter 11.

    By a *frictionless pulley,* we mean one with no friction in its *bearings;* it is desirable that there be sufficient friction between the string and the pulley so that the string does not slip and the pulley rotates with movement of the string.

    By a *light pulley,* we mean one whose *inertia* can be neglected in the solution of a dynamical problem.

    With these definitions, we have the following results for the *static* case, and for the more general *dynamic* case, in which the string may be accelerated:

STATIC CASE: *An unaccelerated string passing over a frictionless pulley has the same tension on both sides of the pulley.*

DYNAMIC CASE: *A string passing over a light frictionless pulley has the same tension on both sides of the pulley.*

The pulley in Fig. 7 should be assumed to be both light and frictionless.

where the proportionality constant, μ_S, is called the *coefficient of static friction*. This coefficient is dimensionless (a pure number), since it is the ratio of two forces.

> **LAWS OF STATIC FRICTION:** *The maximum force of static friction between two surfaces is proportional to the normal force between the surfaces. The coefficient of proportionality μ_S depends on the materials and roughnesses of the surfaces but, over a wide range, is independent of the area of contact between the surfaces.*

The independence of area can be demonstrated by the apparatus of Fig. 7 by using various blocks of the same material but of different areas of contact with the plane.

The laws of static friction, as well as the laws of kinetic friction given below, are merely approximate expressions of the results of experimental observation. They are in no sense fundamental principles, but they are of great practical usefulness over the range of values for which they are approximately correct. They are valid for only a certain range of values of normal force. If the normal force is so great that one surface begins to dent or crush the other, the proportionalities no longer apply.

It is harder to get the block of Fig. 7 started than it is to keep it moving at constant velocity after it has once started moving. Once the weight pan has been loaded so that the block starts moving, it will be found that the block is *accelerated* to the left. Thus, T is greater than $\mathcal{F}_K$, the force of kinetic friction exerted by the board on the moving block. Once the block has started moving, T can be reduced somewhat and the block will still continue to move, indicating that $\mathcal{F}_K$ *is less than* $\mathcal{F}_S^{Max}$. One can determine $\mathcal{F}_K$ by loading the weight pan to a value somewhat less than $\mathcal{F}_S^{Max}$ and then giving the block a push with the finger to start it moving; if it slows down and comes to rest, one concludes that $T < \mathcal{F}_K$; if it speeds up, one concludes that $T > \mathcal{F}_K$. In this way one can adjust the weight until the block moves at constant velocity, once it has been started, in which case the weight is equal to $T = \mathcal{F}_K$.

By varying the load P on the block, it will be found that for a given pair of surfaces, $\mathcal{F}_K$ is proportional to the normal force N between the two surfaces. Hence, we can write

$$\mathcal{F}_K = \mu_K N, \tag{4}$$

where μ_K is a dimensionless constant called the *coefficient of kinetic friction*.

> **LAWS OF KINETIC FRICTION:** *The force of kinetic friction between two surfaces is proportional to the normal force between the surfaces. The proportionality constant μ_K depends on the materials and roughnesses of the surfaces but, over a wide range, is independent of the area of contact of the surfaces and of the relative velocity of the surfaces.*

Table 5-1 gives typical values of μ_S and μ_K. Note that the coefficient of static friction μ_S determines the *maximum* force of static friction and that $\mathcal{F}_S$ can have any magnitude from zero up to this maximum. The force of static friction is $\mathcal{F}_S^{Max}$ only when the surfaces are just on the point of starting to move relative to each other. On the other hand, μ_K determines *the* force of kinetic friction which exists *whenever* the surfaces are *moving* relative to each other. It should also be noted that, since frictional forces have their origin in both macroscopic 'surface roughness' and in interatomic forces (cohesion and adhesion), the values of the coefficients in Table 5-1 are not to be regarded seriously as physical constants; they are valid for use in (3) and (4) for contact between reasonably 'smooth' surfaces.

Table 5-1 TYPICAL VALUES OF THE COEFFICIENTS OF STATIC FRICTION, μ_S, AND OF KINETIC FRICTION, μ_K

Materials	μ_S	μ_K
Steel on steel	0.15	0.09
Steel on ice	0.03	0.01
Hemp rope on wood	0.5	0.4
Leather on wood	0.5	0.4
Oak on oak	0.5	0.3
Wrought iron on cast iron or bronze	0.19	0.18
Rubber tire on dry concrete road	1.0	0.7
Rubber tire on wet concrete road	0.7	0.5

Relations (3) and (4) are valid regardless of the inclination of the surfaces. For example, in Fig. 8 *the plane pushes on the block* with a force which has two components: the component normal to the plane is called the *normal force* or the *normal reaction* of the plane and is denoted by N; the component tangent to the plane is called the *force of friction* and is denoted by $\mathcal{F}$. Relation (3) gives the maximum force of static friction, and (4) gives the force of kinetic friction, with the same values of μ_S and μ_K as if the same block and plane were lying horizontally.

Example. *If a body is initially at rest on a horizontal plane, the body will remain at rest as the plane is gradually tilted until the angle of tilt reaches a characteristic value called the* limiting angle of repose *or the* angle of slip. *Once this angle has been attained, the body moves down the incline with constant acceleration. Show that the tangent of the angle of slip is equal to* μ_S.

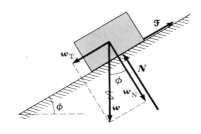

Figure 8

In Fig. 8, the force w of gravity can be resolved into components tangent and normal to the plane as shown. So long as the block does not move,

$$N = w_N = w \cos\phi, \qquad \mathcal{F} = w_T = w \sin\phi.$$

Thus, $\mathcal{F}$ increases as the inclination ϕ of the plane increases. The inclination ϕ cannot exceed a certain value ϕ^{Max} which makes $\mathcal{F}=\mathcal{F}_S^{\text{Max}}$, or else the block will slide. When the block is on the point of slipping, we have, from (3),

$$\mathcal{F}_S^{\text{Max}} = w \sin\phi^{\text{Max}} = \mu_S N;$$

also
$$w \cos\phi^{\text{Max}} = N.$$

Dividing the first equation by the second gives

$$\tan\phi^{\text{Max}} = \mu_S.$$

Thus, the plane can be raised to an angle of inclination whose tangent is μ_S before the block starts to slip.

Example. *The coefficient of static friction between the tires of a four-wheel-drive 'jeep' and a certain icy road is* 0.2, *while the coefficient of kinetic friction is* 0.1. *What is the maximum inclination of hill that the jeep can climb? What is the maximum inclination for the driver who spins his wheels?*

The tangential force the road exerts forward on the car is a force of friction. It will equal the backward force exerted by the tires on the road as the engine turns the wheels. It is a force of static friction if there is no slipping between the tires and the road. The system of forces *on the vehicle* is exactly that of Fig. 8. If the car is to move up the hill at constant speed, we must have $\mathcal{F}=w_T=w\sin\phi$. If $\sin\phi$ is too great, the required force of friction will exceed the maximum $\mathcal{F}_S^{\text{Max}}$, the wheels will spin, and the hill cannot be climbed because the force of kinetic friction between the spinning wheels and the road is always less than the maximum force of static friction. We determine the limiting angle ϕ^{Max} as in the preceding example:

$$\tan\phi^{\text{Max}} = \mu_S = 0.2, \qquad \phi^{\text{Max}} = 11°.3.$$

If engine power is increased so that the wheels spin, the forward force of the road on the tires will be that of kinetic friction, whatever the angle of the road. If the angle is sufficiently low, the car will accelerate forward; if the angle is too high, it will accelerate backward. The limiting angle, where the car moves at constant speed, is that in which (*see* Fig. 8)

$$w_T = \mathcal{F}_K = \mu_K N, \qquad \text{or} \qquad w\sin\phi = \mu_K w \cos\phi.$$

Hence
$$\tan\phi = \mu_K = 0.1, \qquad \phi = 5°.7.$$

At any greater inclination, w_T would exceed $\mathcal{F}_K$ and the moving car would slow down, stop, and slide back.

For an ordinary, two-wheel-drive automobile, these angles would be approximately half as great, since only approximately one-half the weight is 'on' the driving wheels. The reader should explain this conclusion.

The laws of friction that we have discussed above do not apply to lubricated surfaces, where there is a continuous film of liquid between solid surfaces. Friction between lubricated surfaces is essentially a problem in hydrodynamics that we shall not treat in this text. We also omit from consideration a slight variation in μ_K with speed; this variation probably involves a thin air film between the moving surfaces.

4. Examples of Accelerated Motion

When the vector sum of the forces acting on a body is not zero, the body accelerates in the direction of the resultant force in accordance with Newton's second principle.

Problems in accelerated motion are treated most simply by resolving all forces into rectangular component forces parallel and perpendicular to the *direction of the acceleration.* Then the component forces perpendicular to the acceleration balance out just as in the equilibrium case, but the component forces parallel to the acceleration do not balance but have a resultant equal to *ma.* Application of these ideas to the solution of problems is illustrated below by means of four examples. The reader will be assisted in the application of Newton's principles if he keeps in mind the following three requirements:

1. Select a definite well-defined body to which the principles are to be applied.
2. Be sure to include in the system of forces *all* the forces acting on the selected body. These will be the forces exerted by all other bodies that touch the selected body, the force of gravity, and possibly electric or magnetic forces.
3. Be sure *not* to include in the force system any forces that are acting on bodies other than the selected body. In particular do not include forces exerted by the selected body on other bodies—include only forces exerted by other bodies on the selected body.

Example. *Determine the acceleration of the body of mass m down the frictionless inclined plane in Fig. 9.*

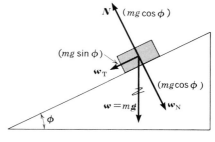

Figure 9

The gravitational force on the body is mg, which we can resolve into two components, one parallel to the inclined plane, and the other normal to the plane. The component normal to the plane, of magnitude $mg \cos\phi$, must be just balanced by the plane's normal reaction force N, since clearly there is no acceleration component normal to the plane. Hence, the resultant force on the body acts parallel to the plane and has the magnitude $mg \sin\phi$, as indicated in Fig. 9. Therefore, the body experiences an acceleration down the plane, of magnitude a given by

$$mg \sin\phi = ma, \quad \text{or} \quad a = g \sin\phi. \tag{i}$$

This equation gives the value of a in terms of g for any elevation angle ϕ of the inclined plane. We may check the results given by (i) for two limiting cases: For $\phi=90°$, equation (i) gives $a=g$. When $\phi=90°$, the frictionless plane is upright; the body becomes a freely falling body and by definition has acceleration g, as given by (i). For $\phi=0°$, the equation gives $a=0$. Since, when $\phi=0$, the plane is horizontal and there is no horizontal force on the body, we see that this result is also correct.

Example. *Determine the acceleration of the body in Fig. 10 if the coefficient of kinetic friction is μ_K and the body is sliding down the plane.*

As in the case of the frictionless plane, the component of the gravitational force perpendicular to the plane is just balanced by the reaction force exerted by the plane, because there is no acceleration normal to the plane. Hence we have drawn arrows only for force components acting parallel to the plane. As in Fig. 9, the gravitational force component down the plane is $mg \sin\phi$. Opposing this force is the frictional force $\mu_K N$, where N is the magnitude $mg \cos\phi$ of the normal force. The frictional force on the body is *up* the plane since it is specified that the body is sliding *down*. The resultant force F down the plane is

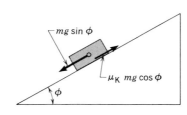

Figure 10

$$F = mg \sin\phi - \mu_K mg \cos\phi.$$

Then, by Newton's second principle, the acceleration a down the plane is given by

$$mg \sin\phi - \mu_K mg \cos\phi = ma, \quad \text{or} \quad a = g(\sin\phi - \mu_K \cos\phi).$$

The velocity of the body down the plane will increase if a is positive, that is, if

$$\sin\phi > \mu_K \cos\phi, \quad \text{or} \quad \tan\phi > \mu_K.$$

If the acceleration is zero, the velocity will remain unchanged; this condition occurs when

$$\sin\phi = \mu_K \cos\phi, \quad \text{or} \quad \tan\phi = \mu_K,$$

a result analogous to the angle of repose in the case of static friction. If the acceleration is negative, the velocity will decrease and the body will come to rest and remain at rest; this is the case when

$$\sin\phi < \mu_K \cos\phi, \quad \text{or} \quad \tan\phi < \mu_K.$$

The reader should explain why the body will *remain at rest* in this case.

Example. *If the string of the conical pendulum in Fig. 11 is 1.5 f long, what is the time of one revolution of the pendulum if the angle ϕ is to be 60°? Take $g = 32$ f/s². Show that the answer is independent of the mass at the end of the string, but does depend on the value of g.*

The reader can easily perform an experiment to verify the following computation by tying any small dense object to the end of a string about 1.5 f long, holding the other end in his fingers, and causing the object to rotate in a horizontal circle at any angle ϕ of about 60°. He will find that he can just about 'count seconds' in synchronism with the period of revolution.

As indicated in Fig. 11, the only forces acting on the mass are the tension T of the string and the downward force mg of gravity. Since the mass moves in a horizontal circle at constant speed, its acceleration is horizontal and directed toward the center of the circle—to the left in Fig. 11. The vertical component T_V of the tension T balances the weight, since there is no vertical acceleration. The horizontal component T_H of T is the only horizontal force, and gives the body its centripetal acceleration v^2/R, as in Fig. 6, p. 60. Expressed analytically, these statements become

$$T\cos\phi = mg; \qquad T\sin\phi = mv^2/R.$$

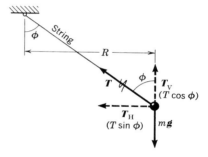

Fig. 11. The force system in a vertical plane through a mass attached to a string and moving in a horizontal circle at constant speed. (This is called a *conical pendulum* since the string generates a cone as the body moves.)

Dividing the second equation by the first, we obtain

$$\tan\phi = v^2/Rg, \quad \text{or} \quad v^2 = Rg\tan\phi.$$

We thus see that the required speed of the body is independent of the body's mass, but does depend on the value of g.

In our case,

$$R = (1.5 \text{ f})\sin\phi = (1.5 \text{ f})\sin 60° = (1.5 \text{ f})(0.866) = 1.30 \text{ f}$$

Hence $\quad v^2 = Rg\tan\phi = (1.30 \text{ f})(32 \text{ f/s}^2)(\tan 60°) = 72.05 \text{ f}^2/\text{s}^2;$

so $\qquad v = 8.49$ f/s.

The mass moves in a circle of radius $R = 1.30$ f. The circumference of this circle is 8.16 f. The time of one revolution is obtained by dividing this circumference by the speed v:

$$(8.16 \text{ f})/(8.49 \text{ f/s}) = 0.961 \text{ s},$$

in agreement with the observation that one can just about 'count seconds' with this conical pendulum.

Example. *What is the angle of 'bank' of a highway curve of 400-f radius in order that automobiles traveling at 60 mi/h shall have no tendency to skid sidewise on the road?*

Since the size of an automobile is small compared with 400 f, we can consider the automobile as a 'particle' in working this problem.

'No tendency to skid' means no sidewise force of friction for the car shown in Fig. 12; the only forces acting on the car are then its weight w and the normal force N of the road. We can resolve the force N into vertical and horizontal component forces, of magnitudes $N\cos\phi$ and $N\sin\phi$, respectively. Since there is no *vertical* acceleration,

$$w = N\cos\phi, \qquad N = w/\cos\phi = mg/\cos\phi.$$

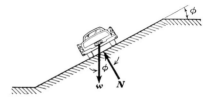

Figure 12

The horizontal acceleration to the left is

$$v^2/R = (88 \text{ f/s})^2/(400 \text{ f}) = 19.4 \text{ f/s}^2,$$

where we have made the transformation of units: 60 m/h=60 (5280 f)/(3600 s)=88 f/s. The only force to the left has the magnitude $N \sin\phi$, so, equating this force to the mass times the acceleration to the left, we obtain

$$m\,(19.4\text{ f/s}^2) = N \sin\phi = mg \sin\phi/\cos\phi = mg \tan\phi,$$

from which

$$\tan\phi = (19.4\text{ f/s}^2)/g = (19.4\text{ f/s}^2)/(32\text{ f/s}^2) = 0.606; \qquad \phi = 31°.2.$$

This is the required angle of bank.

PROBLEMS

1. A 2-kg body lies on a plane inclined at an angle of 30° with the horizontal. What is the weight of the body? Resolve the weight into two component forces: one parallel to the plane and the other normal to the plane.
Ans: 19.6 N; 9.8 N, 16.9 N.

2. A box with a mass of 4 sl lies on an inclined plane making an angle of 40° with the horizontal. What is the gravitational force exerted on the body by the earth? Resolve the gravitational force into components parallel to and perpendicular to the plane.

3. Three men pull on ropes attached to a single point on a trunk as shown in the figure. The ropes are in the same vertical XY-plane. The forces are as follows: $F_1=10$ p at an angle of 30° with the X-axis; $F_2=25$ p, upward; and $F_3=30$ p at an angle of 210° with the X-axis. Find the X-component of the resultant force; the Y-component of the resultant force; the magnitude and direction of the resultant force.
Ans: -17.3 p; 15.0 p; 22.9 p, 139°.1.

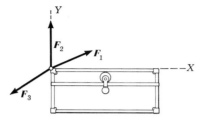

Problem 3

4. Three men push a piano across a level floor. The first man exerts a force of 50 p in the southward direction. The second man exerts a force of 40 p in a direction 15° east of south; the third man exerts a force of 75 p in a direction 30° west of south. Find the magnitude and direction of the resultant force exerted by the men.

5. A particle lies on a horizontal plane and is acted on by the three horizontal forces shown in the figure. Determine the angle θ of the 10-N force so that the resultant force will be exactly in the X-direction. What then is the magnitude of the resultant force?
Ans: 44°.4; 14.1 N.

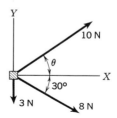

Problem 5

6. Three strings are attached, as in the figure, to a small ring on a horizontal circular 'force table' of the type used in physics demonstrations. The strings pass over pulleys to the edge of the table and support hangers to which laboratory 'weights' can be added. The load at the end of the first string is 200 g and that at the end of the second string is 400 g; the pulleys supporting these loads are 120° apart at the edge of the table. The load and direction for the third string are adjusted until the small ring is in equilibrium at the center of the table. Find the magnitude of the load for the third string, and the angle between its pulley and that of the first string.

Problem 6

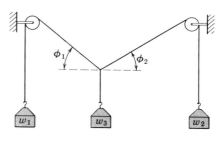

Problem 11

7. A 50-kg block of lead is attached by means of a hook to a light flexible nylon cable at a point midway between the points of support, which are 100 m apart and at the same height. After attachment of the lead block, the mid-point is 2 m below the points of support. What is the angle between the cable and the horizontal at the point of attachment? What total upward force does the cable exert on the block? What is the tension in the cable?
Ans: $2°3$; 490 N; 6100 N.

8. Solve Prob. 7 for a situation in which a 2-sl block of lead is attached to the center of an initially horizontal 80-f cable if the mid-point is 3 f below the points of support after the block is attached.

9. A 6-sl steel sphere is supported in equilibrium by two ropes that make angles of $30°$ and $45°$ with the horizontal. Find the magnitude of the force exerted by each rope at the knot where they are joined.   Ans: 141 p; 172 p.

Problems 9–10

10. If the sphere in the figure has a mass of 10 kg and the two ropes make angles of $25°$ and $60°$ with the horizontal, find the magnitude of the force exerted by each rope at the knot.

11. In the figure, show that if $w_1 = w_2$, then $\phi_1 = \phi_2$, and that if $w_1 > w_2$, then $\phi_1 > \phi_2$.

12. In Fig. 5, p. 86, the earth pulls down on the sphere with a force w. The force F_2 exerted by the vertical string on the knot was assumed to equal w. From the equilibrium conditions and Newton's third principle, show *rigorously* that indeed $F_2 = w$.

13. A trunk weighing 100 p lies at rest on a horizontal platform. A man pushes horizontally on the trunk with a gradually increasing force. No motion occurs until he exerts a force of 40 p, when the trunk suddenly begins to move. Once the trunk is moving, the man can keep it in motion at constant velocity by applying a force of 30 p. Find the coefficients of static and kinetic friction.   Ans: 0.4; 0.3.

14. A 50-kg box lies at rest on a platform. A horizontal force of 200 N must be applied to set the box in motion, but a horizontal force of 150 N will keep the trunk in motion with constant velocity once it has begun to move. Find the coefficients of static and kinetic friction.

15. A crate weighing 100 p lies at rest on a ramp making an angle of $20°$ with the horizontal. The coefficient of static friction between the crate and the ramp is 0.4; the coefficient of kinetic friction is 0.3. What is the normal force exerted by the ramp on the crate? What is the maximum value of the force of static friction that could be exerted on the crate by the ramp? What is the frictional force acting on the crate when it is pushed up the ramp at constant speed by a force parallel to the incline? down the ramp?
Ans: 94 p; 37.6 p; 28.2 p; 28.2 p.

16. A 3-kg block lies at rest on a board. One end of the board is raised gradually and the block begins to slide when the board is inclined at an angle of $40°$ with the horizontal. Just before the block slides, what is the normal force exerted against the board by the block? What is the component of the block's weight acting down the incline parallel to the board? What is the force of static friction? What is the coefficient of static friction?

17. A box with a mass of 0.5 sl lies on an inclined plane making an angle of 30° with the horizontal. If the coefficient of kinetic friction between the box and the plane is 0.4, what is the magnitude of the force that must be applied parallel to the incline to keep the box moving down the incline at constant speed? up the incline at constant speed?
Ans: 2.46 p, up incline; 13.5 p, up incline.

18. A 2-kg block is held at rest on an inclined plane that makes an angle of 60° with the horizontal. The coefficient of static friction between the block and the plane is 0.4; the coefficient of kinetic friction is 0.3. What upward force, parallel to the plane, must be applied to hold the block at rest? If the block is released what will be its acceleration in moving down the incline?

19. An automobile with half its weight on the driving wheels is on a level road. If the coefficient of static friction between the tires and the road is 0.3, what is the maximum acceleration attainable by the automobile?
Ans: 4.8 f/s².

20. An automobile equipped with a four-wheel drive is on a slippery road that goes uphill at an angle of 10° with the horizontal. If the coefficient of static friction between the tires and roadbed is 0.3, what is the maximum acceleration attainable in starting up the hill? in starting down the hill?

21. A body of mass 4 kg is projected at an initial speed of 5 m/s up a plane inclined at 20° to the horizontal, on which the coefficients of friction are $\mu_K = 0.2$ and $\mu_S = 0.3$.
(a) What is the deceleration of the block as it moves up?
(b) How far does it move along the plane before coming to rest?
(c) Show that the block will start back down after coming to rest.
(d) What is its acceleration down the plane?
(e) What is its speed when it again reaches the starting point? Ans: (a) 5.20 m/s²; (b) 2.40 m; (d) 1.54 m/s²; (e) 2.72 m/s.

22. As in the figure, a weight is suspended at the mid-point of a light rope 3 m long. The ends of the rope are attached to two rings, which are free to move on a horizontal rod. If the coefficient of static friction between the rings and the rod is 0.4, what is the maximum separation the rings can have without slipping?

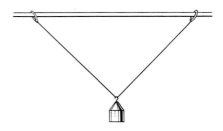

Problem 22

23. In the figure, if $\phi = 20°$, $\mu_S = 0.6$, and the block weighs 3 p, what horizontal push P is required to start the block moving up the plane? Ans: 3.69 p.

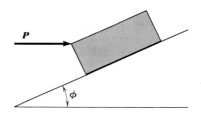

Problems 23–24

24. In the figure, show that the horizontal push P required to keep the block from starting to slide down the plane is given by

$$P = w \frac{\sin\phi - \mu_S \cos\phi}{\cos\phi + \mu_S \sin\phi}$$

Notice that this is positive only when ϕ is greater than the angle of repose, because only then is a force required to keep the body from sliding down the plane.

25. In the figure, if $\mu_S = 0.6$, $w = 25$ p, and $\phi = 30°$, find the magnitude of the pull P that is required to overcome static friction and start the block moving. Ans: 12.9 p.

26. In the figure, show that the magnitude of the pull P, acting at an angle ϕ above the

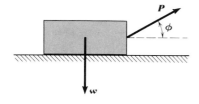

Problems 25-26

horizontal, that is required to overcome static friction and start the block moving is given by $P = \mu_s w / (\cos\phi + \mu_s \sin\phi)$.

27. In the figure, what push P acting at an angle of $\phi = 30°$ is required to prevent a 15-p block from sliding down a vertical wall if $\mu_s = 0.44$? What push is required to start the block moving up the wall?

Ans: 13.8 p; 23.2 p.

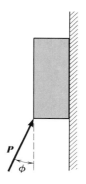

Problem 27

28. What *horizontal* push P will hold a 20-p block against a vertical wall and prevent it from sliding down if the coefficient of static friction between block and wall is 0.5?

29. In the figure, the weight of the body lying on the table top is 75 p and that of the sus-

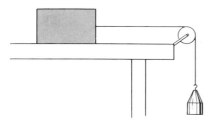

Problems 29-30

pended body is 25 p. Find the downward acceleration of the suspended body and the tension in the cord. (Ignore frictional effects.)

Ans: 8.0 f/s²; 18.8 p.

30. In the figure, the mass of the body on the table top is 4 kg and that of the suspended body is 2 kg. If the coefficient of kinetic friction between the table top and the body lying upon it is 0.3, calculate the tension in the cord and the magnitude of the resulting acceleration once the block is set in motion.

31. A man sits in a bosun's chair supported by a rope passing over a pulley, as in the figure. He pulls on the bitter end of the rope to lift himself up. If man and chair together weigh 200 p, with what force must he pull to raise himself at constant speed? at an acceleration of 0.1 g?

Ans: 100 p; 110 p.

Problems 31-32

32. In Prob. 31, if a single-sheave pulley is fastened to the top of the chair in addition to the one in the rigging, design a rope arrangement that will enable the man to lift himself with a pull of only 67 p. If each of the two pulleys has two sheaves, how low can the force be?

33. A light inextensible string hangs taut over a light frictionless pulley attached to the ceiling of a room. A 2-kg mass resting on the floor is attached to one end of the string. A 4-kg mass, attached to the other end of the string, falls from rest to the floor through a total distance of 2 m, thereby lifting the 2-kg mass, as in the figure. Describe in some detail the motion of the 2-kg mass. *See* figure.

Problem 33

34. In the 'airplane ride' at the carnival, cars are suspended from cables 14 f long attached to arms at points 20 f from the vertical axis of rotation. At what angular speed of rotation will the cars swing out so that the cables make an angle of 45° with the horizontal? an angle of 30°? Treat the cars as 'particles.'

Problem 34

35. A 1-kg lead ball is attached to a string hanging from the ceiling of a lecture hall, as in the figure. By pulling the ball to one side and then pushing the ball in the proper direction, the lecturer sets the ball in motion at constant speed in a horizontal circular path—the so-called 'conical pendulum.' If the radius of the circle is 100 cm and the speed of the ball is 90 cm/s, find the magnitude of the centripetal acceleration. What is the magnitude of the centripetal force?   Ans: 0.81 m/s²; 0.802 N

36. In Prob. 35, find the length of the cord.

37. If the string of a conical pendulum is 4 f long, what is the time of one revolution of the pendulum if angle $\phi = 45°$?   Ans: 1.86 s.

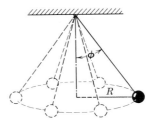

Problems 35–38

38. Find the time for one revolution of a conical pendulum if the length of the string is 10 f and angle ϕ is 30°.

39. There is a curve with a radius of 400 f in a level, unbanked roadway. What is the maximum speed at which a car can traverse the curve without slipping if the coefficient of static friction between the tires and the wet road surface is 0.5?   Ans: 80 f/s.

40. What is the maximum speed at which a car can traverse the curve mentioned in Prob. 39 on a dry day when the coefficient of static friction is 1.0? on an icy day when the coefficient is 0.10?

41. What is the angle of 'bank' of a highway curve of 1000-f radius if automobiles traveling at 80 mi/h have no tendency to skid?
   Ans: 23°.4.

42. Solve Prob. 41 for curves of 500-f and 2000-f radius.

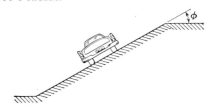

Problems 41–43

43. An automobile travels around a curve of 1000-f radius at 60 mi/h. What is the minimum coefficient of friction required to keep the car from skidding sidewise if the road is flat? if the road is banked 5° the 'right' way? 5° the 'wrong' way?   Ans: 0.242; 0.156; 0.330.

44. The frictional force of air resistance on a very small sphere of radius R falling at speed v relative to the air is given by the relation

$$\mathcal{F} = (10.9 \times 10^{-5} \text{ N} \cdot \text{s/m}^2)\, \pi v R$$

(at standard atmospheric conditions). Show that the terminal speed of a sphere of radius R and density ρ will be given by

$$v_T = (1.22 \times 10^4 \text{ m}^2/\text{N} \cdot \text{s}) R^2 \rho g,$$

so that speed of fall is proportional to the square of the radius.

45. From the last equation in the preceding problem determine the terminal speeds of fall of particles 1 and 10 μm in radius, having the density of water. This size range is that of smoke particles, fine pollens, and small cloud droplets, and the very slow terminal speeds account for the very slow settling of such fine particles. Ans: 0.0120 cm/s, 1.20 cm/s.

46. An oak block lies at rest near one end of a long oak board. The board is slowly tilted until the block begins to slide. If the board is held at rest at the angle at which sliding begins, what will be the velocity of the block 2 s after motion begins?

47. A wooden block with an initial speed of 8 f/s slides across a highly polished floor and comes to rest after moving a horizontal distance of 20 f. What is the coefficient of kinetic friction between the block and the floor?
Ans: 0.05.

48. Block A having a mass of 1 kg is attached to the wall by a string, as shown in the figure, and lies on top of block B, which has a mass of 3 kg. If the coefficient of static friction between block A and block B is 0.4 and that between block B and the floor is 0.2, what horizontal force F must be applied to block B

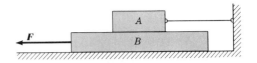

Problem 48

to set it in motion? At the instant motion is on the point of starting, what is the tension in the string attached to block A?

49. In the figure, block A has a mass of 0.25 sl, block B a mass of 0.75 sl. Between blocks A and B, the coefficient of kinetic friction is 0.2 and the coefficient of static friction is 0.3. Between block B and the floor, the corresponding coefficients of friction are 0.3 and 0.4, respectively. What horizontal force F applied to block B is required (a) to set the blocks in motion, (b) to keep the blocks in motion at constant velocity, and (c) to keep the blocks in motion with a common acceleration of 2 f/s²? Ans: 12.8 p; 9.6 p; 11.6 p.

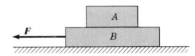

Problems 49–50

50. In the figure, what is the minimum value of applied force F that will cause block A to start to slide with respect to block B? If this force is continued after block A has fallen off block B, what will be the acceleration of block B?

6 WORK, ENERGY, AND POWER

In this chapter we shall define the quantities *work, energy,* and *power,* which are extremely important physical concepts. We shall develop the principle of *conservation of mechanical energy* for use in mechanical problems in which neither frictional forces nor collisions are involved. Then we shall introduce a more general *conservation-of-energy* principle applicable to problems in which heat is generated. The introduction of this principle at this point is somewhat premature, since the general principle was stated by Helmholtz and Mayer on the basis of experimental work that will not be discussed until Chapter 15. However, since conservation of energy is one of the most important principles in physics, it is desirable to introduce this principle immediately, even though we must postpone complete discussion of its experimental basis.

1. **Work**

In nonscientific language, we use the word *work* to denote any type of activity that requires the exertion of muscular or mental effort. In physics, however, the term *work* has an explicit, quantitative, operational definition, and is used *only* in the restricted sense of this definition. In order for physical work to be done, it is necessary for a force to act on a body and for the body to experience a displacement that has a component parallel to the direction in which the force is acting.

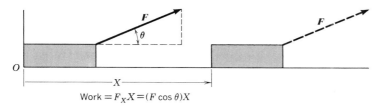

Work $= F_X X = (F\cos\theta) X$

Fig. 1. A body acted on by a constant force F moves to the right a distance X.

Figure 1 represents a body moving along a horizontal surface in a direction which we shall take as the X-axis. A constant force F acts on the body at an angle θ with the direction of motion. The work done by this force is *defined* in the following way:

> The **work** done by a constant force acting on a body while the body is displaced along a straight line is a scalar quantity defined as the product of the magnitude of the displacement by the component of the force in the direction of the displacement.

Thus, the work W done by the constant force in Fig. 1 when the body moves from the origin to the point X is the product of the component of the force in the direction of the motion, $F_X = F\cos\theta$, by the distance X:

$$W = (F\cos\theta)\, X. \tag{1}$$

In the special case in which the force has the same direction as the displacement, the work done by the force is equal to the product of the magnitudes of the force and the displacement; this is the case when the angle θ in Fig. 1 is zero and $\cos\theta = 1$. When $\theta = 90°$, the force has no component in the direction of the displacement ($\cos\theta = 0$), and no work is done by the force. If the angle θ between the force vector F and the direction of the displacement is greater than $90°$, $\cos\theta$ is negative and the work is negative. This is the case when a force is applied to retard the motion of a body already moving with a positive velocity in the X-direction.

We note that, regardless of how large a force may act, displacement must occur before work is done. Thus, although opposing teams may exert enormous forces on a rope during a tug of war, neither team performs any work unless the rope moves.

Since work is defined as the *product* of a force component by the magnitude of a displacement, the derived unit used in measuring work involves the product of a force unit and a length unit. In SI units, the force unit is the *newton* and the length unit is the *meter*. The work unit, the *meter-newton,* is called the *joule* (J):

> One **joule** is the work done by a constant force of one newton when the body on which the force is exerted moves a distance of one meter in the direction of the force.

Similarly, the work unit in the British gravitational system is the *foot-pound* (fp); there is no single abbreviation for the foot-pound:

> One **foot-pound** is the work done by a constant force of one pound when the body on which the force is exerted moves a distance of one foot in the direction of the force.

The SI unit of work is named for JAMES PRESCOTT JOULE (1818–1889), an English physicist who first determined the mechanical equivalent of heat.

The following relations between work units will be found useful:

$$1 \text{ J} = (1 \text{ m}) \times (1 \text{ N}) = (3.281 \text{ f}) \times (0.2248 \text{ p}) = 0.7376 \text{ fp},$$
$$1 \text{ fp} = (1 \text{ f}) \times (1 \text{ p}) = (0.3048 \text{ m}) \times (4.448 \text{ N}) = 1.356 \text{ J}.$$

2. The Scalar Product of Two Vectors

The definition of work given in the preceding section involves taking a type of product of two vector quantities, *force* and *displacement,* to obtain a scalar quantity, *work*. This type of product of two vector quantities is called a *scalar product:*

> The **scalar product of two vectors** is defined as a scalar quantity equal to the product of the magnitudes of the two vectors and the cosine of the angle between them.

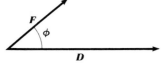

Figure 2

This particular product of two vectors is denoted by placing a dot between the two vectors. Thus, in Fig. 2, if a force $\boldsymbol{F}$ acts on a particle that has a displacement $\boldsymbol{D}$, the work done by the force is written as

$$W = \boldsymbol{F} \cdot \boldsymbol{D} = FD \cos\phi, \tag{2}$$

where the expression $\boldsymbol{F} \cdot \boldsymbol{D}$ is simply a shorthand notation for $FD \cos\phi$. Because of the use of a dot to denote this particular product of two vectors, the scalar product is sometimes called the *dot product*.

The scalar product of two vectors occurs so frequently in physics that it is desirable to develop further some of its algebraic properties.

From its definition, it is clearly *commutative,* i.e.,

$$A \cdot B = B \cdot A. \tag{3}$$

Furthermore (*see* Fig. 3), it is also *associative:*

$$(A+B) \cdot C = A \cdot C + B \cdot C. \tag{4}$$

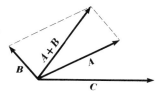

Figure 3

To prove this relation, note that if we take an X-axis in the direction of C, then the right side is $A_X C + B_X C$, while the left side is the X-component of $A+B$ multiplied by C. But we have already seen on p. 23 that the X-component of $A+B$ is the sum $A_X + B_X$. Hence (4) is correct.

Logical combination of the commutative property (3) and the associative property (4) shows that for any number of vectors:

$$(A+B+\cdots)\cdot(D+E+\cdots) = A\cdot D + A\cdot E + \cdots + B\cdot D + B\cdot E + \cdots. \tag{5}$$

Now let us consider a vector A (Fig. 4) resolved into rectangular vector components A_1, A_2, A_3, so that we can write

$$A = A_1 + A_2 + A_3. \tag{6}$$

Similarly, a vector B can be written as

$$B = B_1 + B_2 + B_3. \tag{7}$$

The scalar product of A and B becomes, by (5),

$$\begin{aligned}A \cdot B = & A_1 \cdot B_1 + A_2 \cdot B_2 + A_3 \cdot B_3 \\ & + A_1 \cdot B_2 + A_2 \cdot B_3 + A_3 \cdot B_1 \\ & + A_1 \cdot B_3 + A_2 \cdot B_1 + A_3 \cdot B_2.\end{aligned} \tag{8}$$

All of the scalar products in the second and third lines on the right vanish because in each case the two vector components are perpendicular and $\cos\phi = 0$ in the defining equation (2). Now let us consider the dot product

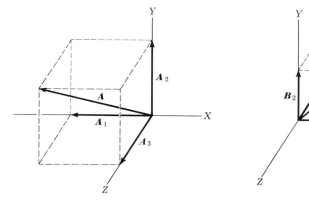

Figure 4

of the two collinear vectors A_1 and B_1. This product has the value $A_1 B_1$ if A_1 and B_1 point in the same direction, and has the value $-A_1 B_1$ if they point oppositely, as in Fig. 4, because in the first case $\cos\phi = 1$ and in the second case $\cos\phi = -1$. However, we note that because of the manner in which signs are attached to the scalar components A_X and B_X, it is always true that

$$A_1 \cdot B_1 = A_X B_X.$$

If A_1 and B_1 point in the same direction, A_X and B_X will have the same sign and the product will be positive; if A_1 and B_1 point oppositely, A_X and B_X will have different signs and the product will be negative. Similarly

$$A_2 \cdot B_2 = A_Y B_Y, \qquad A_3 \cdot B_3 = A_Z B_Z,$$

and from (8) we see that the scalar product of two vectors can be expressed in terms of the scalar components of the vectors by

$$A \cdot B = A_X B_X + A_Y B_Y + A_Z B_Z. \tag{9}$$

The expression (1) for *work* was stated for the simple case in which the motion was rectilinear and the force was constant. In the more general case of motion along an arbitrary path such as that in Fig. 5, with a force that may vary from point to point along the path, we define the element of work done during the element of displacement $d\mathbf{D}$ as

$$dW = \mathbf{F} \cdot d\mathbf{D} = F_X \, dX + F_Y \, dY + F_Z \, dZ, \tag{10}$$

where dX, dY, dZ are the components of $d\mathbf{D}$, as in Fig. 3, p. 35. The total work done by the force $\mathbf{F}$ in the motion from P_1 to P_2 is then the integral

$$W = \int_{P_1}^{P_2} \mathbf{F} \cdot d\mathbf{D}. \tag{11}$$

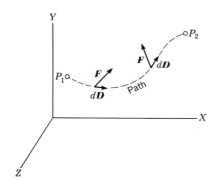

Figure 5

An integral such as this is called a *line integral* because it is evaluated along a particular path (line) from P_1 to P_2. In accordance with (10) this integral can be written as the sum of three integrals

$$W = \int_{X_1}^{X_2} F_X \, dX + \int_{Y_1}^{Y_2} F_Y \, dY + \int_{Z_1}^{Z_2} F_Z \, dZ, \tag{12}$$

where subscript 1 refers to the coordinates of P_1 and 2 refers to P_2.

Example. *The heavy spring in a railroad 'bumper,' Fig. 6, exerts a force on a car that is proportional to the distance X that the spring is compressed. The car exerts an equal and opposite force, say $F_X = KX$, to the right in Fig. 6, on the bumper. If a car compresses this bumper by an amount X_1 before* *it is brought to rest, compute the work done by the car on the bumper.*

We compute the work from expression (12). In this expression, only the first integral has a value since the force is in the X-direction. The initial value of X is 0, the final value is X_1; hence

$$W = \int_0^{X_1} F_X \, dX = \int_0^{X_1} KX \, dX = \tfrac{1}{2} KX^2 \Big|_0^{X_1} = \tfrac{1}{2} KX_1^2$$

is the work done by the car on the bumper.

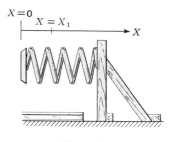

Figure 6

We should now make one further important observation, relating to the total work done if several forces act on a particle in its motion along the path in Fig. 5. If this path is that of a sled pulled along a surface, for example, the various forces would be the force of gravity, the normal reaction of the surface, the force of friction, and the force of the boy pulling the sled.

If several forces act on a particle in motion along a path, the total work done by all the forces can be computed by computing the work done by the vector sum of the forces.

The proof follows immediately from (11) and (5). If the individual forces are $\boldsymbol{F}_1, \boldsymbol{F}_2, \cdots$, with sum $\boldsymbol{F}$:

$$\boldsymbol{F} = \boldsymbol{F}_1 + \boldsymbol{F}_2 + \cdots,$$

we can write

$$\int_{P_1}^{P_2} \boldsymbol{F} \cdot d\boldsymbol{D} = \int_{P_1}^{P_2} (\boldsymbol{F}_1 + \boldsymbol{F}_2 + \cdots) \cdot d\boldsymbol{D} = \int_{P_1}^{P_2} \boldsymbol{F}_1 \cdot d\boldsymbol{D} + \int_{P_1}^{P_2} \boldsymbol{F}_2 \cdot d\boldsymbol{D} + \cdots,$$

which proves the assertion in italics.

3. Energy

In physics, the term *energy* has a precise technical meaning. The following will serve as a general definition:

> The **energy** of a body is a measure of the capacity or ability of the body to perform work. It is a scalar quantity and is measured in the same units as work.

Using this definition, we can immediately think of many examples of objects having energy because of their *motion*. For example, a moving freight car, a moving bullet, and a rotating flywheel all have the ability to do work during the process of being brought to rest. *The energy possessed by a body as a result of its motion is called kinetic energy.*

A body may also have energy as a result of its position or configuration. For example, a cubic foot of water at the top of a waterfall is capable, as a result of its *position,* of performing work in turning a water wheel at the bottom; while a clockspring, when wound, can do work in operating the clock mechanism as a result of its stressed *configuration.* *Energy possessed by a body as a result of its position or configuration is called potential energy.* The cubic foot of water has potential energy as a result of its position in the gravitational field of the earth; hence we refer to its energy as *gravitational potential energy.* In the case of the clock spring, the spring is an elastic body and, when distorted, tends to regain its original shape or configuration; we can therefore think of energy as being imparted to the spring when the clock is wound. This type of energy is referred to as *elastic potential energy.*

We can always think of energy as resulting from the performance of work. In the case of the cubic foot of water, the potential energy is imparted to the water when *work is done* in raising it to the top of the waterfall; in the case of the spring, the potential energy is produced when *work is done* in distorting the spring. Since work must be done in setting a body in motion, we see that *work is done* in giving kinetic energy to a body. These observations suggest operational definitions of kinetic and potential energy that are more satisfactory than the general definition given earlier, but which we shall show, by means of examples, are equivalent to the general definition:

> The **kinetic energy** of a body in motion is equal to the work that must be done by the forces acting on the body in order to change the body from a state of rest to its state of motion.
>
> The **change in gravitational or elastic potential energy,** when a body changes from one position or configuration to another, is the *negative* of the work done by the gravitational or elastic forces during the change in position or configuration.

4. Kinetic Energy

We shall first derive an expression giving the *kinetic* energy of a particle in terms of its mass m and its speed v. In order to obtain this expression, we must compute the work that is done in accelerating the particle, initially at rest, to velocity v. If a constant resultant force F acts on the particle, the acceleration a is constant and the displacement has the same direction as the force. Therefore, we may choose this direction as the X-axis of a coordinate system in which the particle is initially located at the origin, as in Fig. 7. The work W done in moving the particle to position X is

$$W = FX. \tag{13}$$

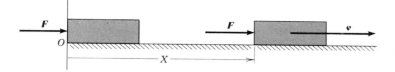

Fig. 7. The external work FX done in accelerating a body, initially at rest on a horizontal frictionless surface, is transformed into the kinetic energy $\frac{1}{2} mv^2$ of the body.

If the particle starts from rest at $t=0$, it reaches position X at time $t=T$ given by $X=\frac{1}{2} aT^2$. Substitution of this expression for X, ma for F, and finally $v=aT$, in (13) gives

$$W = FX = (ma)(\tfrac{1}{2} aT^2) = \tfrac{1}{2} m (aT)^2 = \tfrac{1}{2} mv^2. \tag{14}$$

Since the kinetic energy of a particle traveling at speed v is defined as the work W required to accelerate the body from rest to speed v, expression (14) is just the kinetic energy of the particle:

$$E_K = \tfrac{1}{2} mv^2. \quad \begin{Bmatrix} E_K \text{ in J} \\ m \text{ in kg} \\ v \text{ in m/s} \end{Bmatrix} \text{ or } \begin{Bmatrix} E_K \text{ in fp} \\ m \text{ in sl} \\ v \text{ in f/s} \end{Bmatrix} \tag{15}$$

As indicated, the units in the above equations must correspond to one of the two systems in which the equation $F=ma$ is valid.

The above simple derivation of (15) is for the special case in which the particle is accelerated at a constant rate along a straight line. However, it can be readily shown that the same result is obtained for a particle accelerated in any manner along any type of curve. Let us show that the work done by the varying resultant force $\mathbf{F}$ in the motion of a particle of mass m from P_1 to P_2 in Fig. 5 equals $\frac{1}{2} mv_2^2 - \frac{1}{2} mv_1^2$ and hence equals the change in kinetic energy given by the expression (15). The work done by $\mathbf{F}$ is

$$W = \int_{P_1}^{P_2} \mathbf{F} \cdot d\mathbf{D} = \int_{X_1}^{X_2} F_X \, dX + \int_{Y_1}^{Y_2} F_Y \, dY + \int_{Z_1}^{Z_2} F_Z \, dZ, \tag{16}$$

by (12). Since $F_X = m \, dv_X/dt$, the first integral on the right can be modified as follows:

$$\int_{X_1}^{X_2} F_X \, dX = \int_{X_1}^{X_2} m \frac{dv_X}{dt} dX = \int_{v_{X1}}^{v_{X2}} m \frac{dX}{dt} dv_X = m \int_{v_{X1}}^{v_{X2}} v_X \, dv_X$$

$$= \tfrac{1}{2} mv_X^2 \Big|_{v_{X1}}^{v_{X2}} = \tfrac{1}{2} mv_{X2}^2 - \tfrac{1}{2} mv_{X1}^2.$$

Similar manipulation of the last two integrals in (16) gives finally, as we set out to prove,

$$W = \tfrac{1}{2} m (v_{X2}^2 + v_{Y2}^2 + v_{Z2}^2) - \tfrac{1}{2} m (v_{X1}^2 + v_{Y1}^2 + v_{Z1}^2) = \tfrac{1}{2} mv_2^2 - \tfrac{1}{2} mv_1^2.$$

Expression (15) also gives the kinetic energy of an extended body in *translational motion,* in which all particles have the same velocity v. The kinetic energy of an extended body in more general motion involving rotation will be considered in Chapter 8.

5. Gravitational Potential Energy

Let us now obtain an expression for the change in gravitational potential energy when a particle changes its position relative to the earth's surface. Look first at the definition of this change as the *negative* of the work done by the gravitational force. If a stone moves from the top of a cliff to the bottom, it loses potential energy; that is, it loses the ability to do work by virtue of its position. In this motion the downward force of gravity does *positive* work, but the potential energy decreases; the *change* in potential energy is *negative*. But if we haul the stone at constant speed from the bottom to the top of the cliff, the force of gravity does *negative* work while the potential energy *increases*. We must pull up on the stone with a force that opposes the force of gravity. The force we exert does positive work; this is said to be work done *against* the force of gravity. When positive work is done *against* the force of gravity, the potential energy *increases;* when positive work is done *by* the force of gravity, the potential energy *decreases*. If the body is *projected* from the bottom to the top of the cliff, the force of gravity does *negative* work during the motion. The potential energy therefore *increases;* this time at the expense of the kinetic energy the body had when it was projected, as we shall see later.

A body has potential energy by virtue of its *position,* but, since the pull of gravity is directed *vertically* downward, the gravitational potential energy of the body changes only when the *height* of the body changes. In fact we need only find the work done in changing the height of the body by raising it vertically in order to find its change in potential energy. For example, let us consider the situation in Fig. 8 which shows a stone on a cliff at a height H above a lake; we wish to find the gravitational potential energy of the stone relative to the water surface as reference level. We may do this by finding the work done against the earth's gravitational pull when the stone is raised vertically a distance H at constant speed. The force F that must be exerted is equal to the weight mg of the stone and the work done is

$$W = F H = mg H = w H.$$

The stone's potential energy relative to its position at water level is equal to the work done in raising the stone:

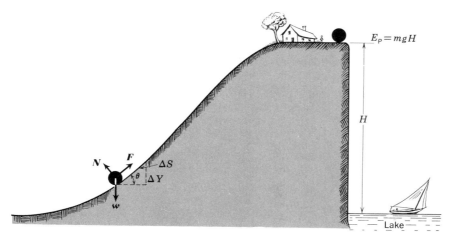

Fig. 8. At the top of the cliff, the stone has potential energy $mgH = wH$ with respect to the surface of the lake.

$$E_P = mg\,H = wH. \quad \begin{Bmatrix} E_P \text{ in J} \\ m \text{ in kg} \\ g \text{ in m/s}^2 \\ H \text{ in m} \\ w \text{ in N} \end{Bmatrix} \text{ or } \begin{Bmatrix} E_P \text{ in fp} \\ m \text{ in sl} \\ g \text{ in f/s}^2 \\ H \text{ in f} \\ w \text{ in p} \end{Bmatrix} \quad (17)$$

The units in (17) must again correspond to one of the two systems in which the equation $w = mg$ is valid.

It will be noted from this example that the potential energy is directly proportional to the height H *above some reference level,* which in this case we have chosen as the surface of the lake. The potential energy given by (17) represents only the ability of the stone to do work as it returns the whole distance to the reference level. For example, if the stone falls over the cliff, it can do work $W = mg\,H$ on a sailboat on the lake surface, but the stone has *no* ability to do work on the objects on the top of the cliff. The reference level of potential energy is arbitrary; *only differences in potential energy at two different levels have physical significance.*

We shall now verify that the work required to pull a body up any type of inclined path is (friction being absent) just the same as the work $mg\,H$ required to lift the body vertically the same vertical height. Suppose, as at the left of Fig. 8, we pull the body up a frictionless incline by applying a force just large enough to move the body without acceleration. Then, as we found in Chapter 5, we require $F = w\sin\theta$, where θ is the inclination of the hill at any stage. The work done when the body moves a short distance ΔS is $F\,\Delta S = w\,\Delta S\sin\theta$. But $\Delta S\sin\theta = \Delta Y$, the increment in vertical height. Hence the work required to move the body a vertical height ΔY is $w\,\Delta Y$ independent of the inclination. Thus, the total work required to move the body up the hill of vertical height H is $wH = mgH$, independent of the shape of the incline. If there is friction, more work will be required but, as we shall see in Section 7, only the work wH is done against gravity and goes into increase of gravitational potential energy; the additional work done against friction goes into heat. Note that the normal force N in Fig. 8 *does no work*.

> **Example.** *What is the kinetic energy of a* 2-ton *automobile traveling at* 60 mi/h? *At what height above the bottom of a cliff would the automobile have potential energy of this amount? Verify that if the automobile were to fall over the cliff from this height it would strike the bottom at* 60 mi/h. *Take* $g = 32$ f/s^2.
>
> The mass of the automobile is $m = w/g = (4000\text{ p})/(32\text{ f/s}^2) = 125$ sl. Since 60 mi/h = 88 f/s, substitution in (15) gives
>
> $$E_K = \tfrac{1}{2}\,mv^2 = \tfrac{1}{2}\,(125\text{ sl})(88\text{ f/s})^2 = 484\,000\text{ f}\cdot(\text{sl}\cdot\text{f/s}^2) = 484\,000\text{ fp}.$$
>
> To obtain the height of the cliff on which the automobile would have equal potential energy, we substitute this value for E_P in (17), and 4000 p for w, to obtain
>
> $$H = E_P/w = (484\,000\text{ fp})/(4000\text{ p}) = 121\text{ f}.$$
>
> Let us now obtain the speed of a body falling from rest a distance of 121 f. By (24), p. 71, this speed is given by
>
> $$v^2 = 2\,gH = 2\,(32\text{ f/s}^2)(121\text{ f}) = 64 \times 121\text{ f}^2/\text{s}^2,$$
>
> $$v = 8 \times 11\text{ f/s} = 88\text{ f/s} = 60\text{ mi/h}.$$
>
> This verification illustrates the transformation of potential energy into kinetic energy that we shall discuss in the next section.

Consideration of the computation of *elastic potential energy* will be postponed until Chapter 11, which is devoted to the study of elasticity.

6. Transformations of Mechanical Energy

The sum of the kinetic energy and the gravitational or elastic potential energy of a body is called the body's *mechanical energy*. It can be proved from Newton's principles that, *provided there are no frictional or other dissipative effects, the total mechanical energy of a system of bodies remains constant.* We can easily demonstrate this statement for a ball of mass m thrown vertically upward. From equation (24) p. 71, the velocity of the ball at height Y is given by

$$v^2 = v_0^2 - 2\,gY.$$

By multiplying both sides of this equation by $\tfrac{1}{2}\,m$, we obtain

$$\tfrac{1}{2}mv^2 = \tfrac{1}{2}mv_0^2 - mgY,$$

and, by transposition,

$$\tfrac{1}{2}mv^2 + mgY = \tfrac{1}{2}mv_0^2. \tag{18}$$

The quantities in equation (18) can be interpreted as follows:

$\tfrac{1}{2}mv^2 =$ kinetic energy of the ball at height Y,
$mgY =$ potential energy of the ball at height Y, relative to the origin,
$\tfrac{1}{2}mv_0^2 =$ initial kinetic energy of the ball.

The terms on the left side of equation (18) represent the *total mechanical energy* (kinetic + potential) of the ball at any point in its path, and the term on the right side represents the *total initial mechanical energy*, taking the potential energy of the ball at the origin as zero. Thus, equation (18) states that at all times *the total mechanical energy of the ball is constant*. Provided there is no air resistance, which is of the nature of a frictional force, mechanical energy is *conserved*. As indicated in Fig. 9, the initial kinetic energy is all changed to potential energy at the top of the path.

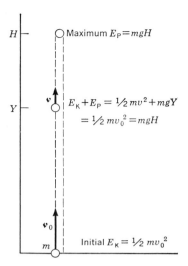

Fig. 9. The sum of the kinetic energy and the potential energy of a ball thrown vertically upward from a horizontal plane is at all times constant and equal to the initial kinetic energy of the ball.

To demonstrate conservation of mechanical energy in a slightly more complex case, consider the projectile of mass m shown in Fig. 10. Let the horizontal and vertical components of the initial velocity be v_{0X} and v_{0Y}, respectively. In the absence of air resistance, v_X will at all times be equal to v_{0X}:

$$v_X = v_{0X}, \tag{19}$$

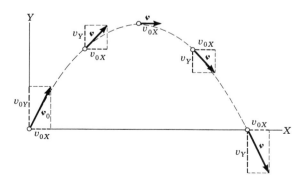

Fig. 10. Path of a projectile, with velocities indicated at several points along the trajectory.

while, from (28), p. 74, we may write

$$v_Y^2 = v_{0Y}^2 - 2gY, \quad \text{or} \quad v_Y^2 + 2gY = v_{0Y}^2. \tag{20}$$

Squaring both sides of equation (19) and adding the resulting equation to equation (20), we obtain

$$v_X^2 + v_Y^2 + 2gY = v_{0X}^2 + v_{0Y}^2.$$

Multiplication of both sides of this equation by $\tfrac{1}{2}m$ gives

$$\tfrac{1}{2} m (v_X^2 + v_Y^2) + mgY = \tfrac{1}{2} m (v_{0X}^2 + v_{0Y}^2),$$

or

$$\tfrac{1}{2} mv^2 + mgY = \tfrac{1}{2} mv_0^2. \tag{21}$$

The terms in equation (21) have the following interpretation:

$\tfrac{1}{2} mv^2 =$ kinetic energy of the projectile at height Y,
$mgY =$ potential energy of the projectile at height Y, relative to $Y=0$,
$\tfrac{1}{2} mv_0^2 =$ initial kinetic energy = total initial mechanical energy.

Hence, we see that equation (21), like equation (18), is an expression of the principle of conservation of mechanical energy.

This same principle can readily be demonstrated in the case of a body sliding on a frictionless incline, whether or not the inclined surface is plane. The work done by the force of gravity is, by definition of potential-energy change, the decrease in potential energy. Since the incline is frictionless, the force exerted on the body by the incline is normal to the direction of motion and does no work. Hence the work done by the force of gravity is by definition the increase in the kinetic energy. Therefore the increase in kinetic energy equals the decrease in potential energy, and the total mechanical energy is conserved. The same type of argument can be applied to a system of bodies connected by strings and pulleys, provided there is no friction.

7. Conservation of Energy

The examples above can be generalized to the following principle, which can be demonstrated rigorously from Newton's principles:

PRINCIPLE OF CONSERVATION OF MECHANICAL ENERGY: *For any system of bodies, connected together in any manner, the total mechanical energy of the system is conserved, provided there is no relative motion that involves friction, no collisions between the bodies, and all the forces are mechanical or gravitational in nature.*

In actual mechanical situations, frictional effects are usually present and collisions may occur; if these effects are small the above principle is a useful approximation. However, *whenever relative motion that involves friction or collisions occurs, mechanical energy is not conserved;* rather, the total mechanical energy always decreases. We say that mechanical energy is *dissipated*. The relative motion referred to in the principle may be between solid surfaces, between a solid object and the air or a liquid, or may be internal to a solid as in the case of the distortion that accompanies a *collision* between solid bodies. Mechanical energy is dissipated in collisions because of the presence of forces of the nature of internal friction: this dissipation will be discussed in some detail in Chapter 7.

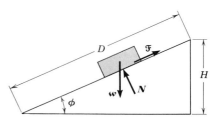

Figure 11

As an example of the way in which mechanical energy is dissipated when there is friction, consider the situation shown in Fig. 11. A box of mass m, initially at rest, slides down an inclined plane from a height H above a level floor. The potential energy of the box in the initial position (relative to the bottom of the plane) is given by mgH. If there were no friction between the box and the inclined plane, the kinetic energy $\tfrac{1}{2} mv^2$ of the box on reaching the lower end of the inclined plane would be exactly equal to the initial potential energy mgH. However, if friction is involved, some of the initial potential energy is dissipated in doing work against friction, and the kinetic energy of the box at the bottom of the inclined plane is less than the initial potential energy. Since the normal force $N = mg \cos\phi$, the frictional force $\mathcal{F}$ has the magnitude

$$\mathcal{F} = \mu_K N = \mu_K mg \cos\phi,$$

and is directed up the plane, opposite to the velocity. The work W done by the force of friction as the block moves the distance D down the plane is, since the direction of the force is *opposite* to the direction of the displacement,

$$W = -\mathcal{F} D = -\mu_K mg D \cos\phi. \tag{22}$$

We shall use this particular example to illustrate a general principle:

The change in mechanical energy resulting from kinetic friction equals the work done by the forces of friction. This work is always negative, so the mechanical energy always decreases.

In the example of Fig. 11, the resultant force down the plane is the difference between the component of the weight and the force of friction:

$$F = mg \sin\phi - \mathcal{F}. \tag{23}$$

This force results in an acceleration down the plane of magnitude $a = F/m$. The speed at the bottom of the plane is, by (15), p. 41,

$$v^2 = 2 aD = 2 FD/m.$$

The kinetic energy at the bottom is

$$\tfrac{1}{2} mv^2 = FD = mgD \sin\phi - \mathcal{F} D,$$

or $$\tfrac{1}{2}\,mv^2 = mgH - \mathfrak{F}D. \tag{24}$$

Thus, the kinetic energy at the bottom of the plane is not as great as the potential energy mgH at the top of the plane; it differs from this potential energy by the work $-\mathfrak{F}D$ done by friction. Mechanical energy $\mathfrak{F}D$ disappears during the motion and is said to be *dissipated in friction*.

Now let us consider the question: what becomes of the mechanical energy that is dissipated? It is recognized by everyone who has ever rubbed his hands together to warm them that work done against friction produces *heat*. However, not until more than one hundred years after Newton's death was it clearly understood that the amount of heat produced is directly proportional to the amount of mechanical energy dissipated and that *heat itself is a form of energy*. When work is done against friction, mechanical energy is converted into *thermal energy;* in heat engines, thermal energy is converted into mechanical energy. In both cases, one form of energy is transformed into another form of energy.

Recognition of this fact led to the formulation of one of the most important principles in physical science. This principle of *conservation of energy* can be stated as follows:

> PRINCIPLE OF CONSERVATION OF ENERGY: *Energy cannot be created or destroyed, although it can be changed from one form to another.*

If we rewrite equation (24) in the form

$$mgH = \tfrac{1}{2}\,mv^2 + \mathfrak{F}D,$$

we see that $\mathfrak{F}D$ must be the energy transformed into heat. Of the initial gravitational potential energy mgH at the top of the incline, $\tfrac{1}{2}\,mv^2$ appears as kinetic energy at the bottom of the incline, and the amount $\mathfrak{F}D$ of thermal energy has been generated. The heat produced by friction is sometimes readily observable. For example, a nail becomes noticeably warmer when driven into or suddenly withdrawn from a board. Similarly, when a bullet enters a piece of wood, the walls of the resulting hole become charred as a result of the high temperatures attained, and a lead bullet sometimes melts. It is also observed that fast-moving meteors and rockets are heated to incandescence on entering the earth's atmosphere.

In addition to mechanical kinetic and potential energy and thermal energy, there are still other forms of energy, which include chemical energy, electric energy, magnetic energy, and nuclear energy. Conversion of these other forms of energy into mechanical energy can *increase* the total *mechanical energy* of an isolated system; the total energy remains constant if all forms of energy are included. The principle of conservation of energy finds application in all branches of physics. There is no formal proof of the general principle, but we have no evidence that it is ever violated. The *total* energy in the *universe* seems to be constant! In the

Example. *In Fig. 11, let the angle ϕ be $30°$, the height H, 10 f, the slant height D, 20 f, the mass m, 3 sl, and the coefficient of friction μ_K, 0.25. If the mass starts from rest at the top of the plane, determine from energy considerations its speed at the bottom. Take $g=32$ f/s².*

At the top of the plane, the potential energy (relative to the bottom) is

$$mgH = (3 \text{ sl})(32 \text{ f/s}^2)(10 \text{ f}) = (96 \text{ p})(10 \text{ f}) = 960 \text{ fp}.$$

To get the kinetic energy at the bottom we must subtract from this initial potential energy the energy transformed into heat in friction—that is, the work done against friction. Since there is no acceleration perpendicular to the plane, $N = w \cos\phi = mg \cos\phi = (96 \text{ p})(0.866) = 83.1$ p. The force of friction is then $\mathfrak{F} = \mu_K N = (0.25)(83.1 \text{ p}) = 20.8$ p. The work done against friction is

$$\mathfrak{F}D = (20.8 \text{ p})(20 \text{ f}) = 416 \text{ fp}.$$

The kinetic energy at the bottom of the incline is the difference

$$\text{kinetic energy} = (960 - 416) \text{ fp} = 544 \text{ fp}.$$

We can equate this kinetic energy to $\frac{1}{2} mv^2$, where v is the speed at the bottom of the incline:

$$544 \text{ fp} = \tfrac{1}{2} mv^2 = \tfrac{1}{2} (3 \text{ sl}) v^2,$$

from which $v^2 = 363$ fp/sl $= 363$ f²/s², $v = 18.8$ f/s.

This is, of course, the same answer as would be obtained by direct application of the relation $F=ma$, as the reader can verify.

8. Work and Energy in Moving Coordinate Systems

By the Newtonian principle of relativity, we know that the principles of mechanics are equally valid for two different observers, one of whom is moving at constant velocity relative to the other. Hence the concepts of work and energy, and the principle of conservation of energy must be equally valid and useful for the two observers. However, as we shall see in the following example, both work and energy must be measured relative to a particular frame of reference since two observers in relative motion assign not only different kinetic energies but different amounts of work done. After working through the example, we shall examine the physical significances of these differences.

Example. *A particle of mass m rests on a smooth horizontal floor on a train moving forward with constant speed V. A constant force F is applied in the forward direction on the particle and gives it acceleration a for time t. Show* that the increase in kinetic energy of the particle equals the work done by the force, both as measured by an observer whose frame of reference is attached to the train and one whose frame is attached to the ground.

The force required, which is independent of the frame of reference, is given by Newton's second principle as $F=ma$. In time t the particle increases its speed by at.

The observer on the train determines the following quantities for the particle:

$$\text{initial speed } v_0 = 0$$
$$\text{final speed } v_1 = at$$
$$\text{average speed } \bar{v} = \tfrac{1}{2} at$$
$$\text{distance traveled } D = \bar{v}t = \tfrac{1}{2} at^2$$
$$\text{work done } W = FD = (ma)(\tfrac{1}{2} at^2) = \tfrac{1}{2} ma^2 t^2$$
$$\text{initial kinetic energy} = \tfrac{1}{2} mv_0^2 = 0$$
$$\text{final kinetic energy} = \tfrac{1}{2} mv_1^2 = \tfrac{1}{2} ma^2 t^2$$
$$\text{increase in kinetic energy} = \tfrac{1}{2} ma^2 t^2 = \text{work done.}$$

The corresponding quantities as determined by the observer on the ground are

$$\text{initial speed } v_0' = V$$
$$\text{final speed } v_1' = V + at$$
$$\text{average speed } \bar{v}' = V + \tfrac{1}{2} at$$
$$\text{distance traveled } D' = \bar{v}'t = Vt + \tfrac{1}{2} at^2$$
$$\text{work done } W' = FD' = maD' = maVt + \tfrac{1}{2} ma^2 t^2$$
$$\text{initial kinetic energy} = \tfrac{1}{2} mv_0'^2 = \tfrac{1}{2} mV^2$$
$$\text{final kinetic energy} = \tfrac{1}{2} mv_1'^2 = \tfrac{1}{2} mV^2 + maVt + \tfrac{1}{2} ma^2 t^2$$
$$\text{increase in kinetic energy} = maVt + \tfrac{1}{2} ma^2 t^2 = \text{work done.}$$

In connection with the above example, the question arises: If work has fundamental physical significance as energy expended, why does one get different results according to the frame of the observer? What is the energy actually expended in accelerating the particle?

The answer to these questions is that the energy expended is different according to who expends the energy, and in employing the principle of conservation of energy one should use the reference frame of the person doing the work. If the man on the train pulls on the particle to accelerate it, he exerts force F through distance D and does work FD. If, on the other hand, the man on the ground is imagined to have a string tied to the particle and to run along beside the train exerting force F to accelerate the particle, he must exert the force F through the much larger distance D' and do the greater work FD'.

Correspondingly, the differences in assigned kinetic energies have definite physical significances in the two frames of reference. We can think of the kinetic energy as measured by the train observer as the work the particle could do if it struck an object *stationary on the train* and was brought to rest *relative to the train*. The much larger kinetic energy assigned by the observer on the ground represents the much larger work the particle could do if it struck an object *stationary on the ground* and was brought to rest *relative to the ground*.

The case of work done against friction needs special consideration. Surely the quantity of heat generated in this case is independent of the velocity of the observer. It turns out that *work done against friction is to be computed by multiplying the force of friction $\mathcal{F}$ by the distance of relative motion of the two sliding surfaces*. This distance has a definite physical significance independent of the observer. Thus two observers would assign the same amount of mechanical energy transformed into heat in friction. The discussion following the next example should help clarify these statements.

Example. *What changes occur in the results of the preceding example if we assume that there is a coefficient of kinetic friction μ_K between the particle and the floor of the train? A force is to be applied to the particle such as to give it the same acceleration a for the same time t as in the previous case.*

In addition to the force $F=ma$, a force $\mathcal{F}_K=\mu_K mg$ is required to balance the opposing force of friction. The man on the train does additional work $\mathcal{F}_K D$; the man on the ground does additional work $\mathcal{F}_K D'$. There is no change in the initial, the final, or the increase in, kinetic energy. Increase in kinetic energy no longer equals work done.

In this example the additional work done by the man on the train is $\mathcal{F}_K D$, which is the amount of energy that goes into heat, since the distance of relative motion of the particle and the train floor on which it slides is D.

Of the additional work $\mathcal{F}_K D'$ done by the man on the ground, only the part $\mathcal{F}_K D$ goes into heat. Where does the rest of the energy, $\mathcal{F}_K (D'-D)$, go? Notice that $D'-D$ is the distance the train moves in the time t. One suspects that work is being done *on the train*. The man on the ground is indeed helping to pull the train, because the box is exerting on the train a *forward force* $\mathcal{F}_K$, and the work done by this force, $\mathcal{F}_K$ times the distance the train moves, relieves the engine of some of the burden of keeping the train moving at constant speed and represents work done on the train. The man on the train can do *no* work on the train; in his coordinate system the train *does not move!* We can break up the work done by the man on the ground into the several parts:

$$(F+\mathcal{F}_K) D' = FD' + \mathcal{F}_K D' = FD' + \mathcal{F}_K D + \mathcal{F}_K (D'-D),$$

of which the first term increases the kinetic energy of the box, the second term appears as heat between the surface of the box and the floor of the train, and the third term represents work done on the train itself.

9. Power

Time is not involved in any way in the definition of *work*. The same amount of work is done in raising a given weight through a given vertical distance regardless of the time required. However, the *time rate* at which work is done is an important quantity called power:

Power is the time rate at which work is done.

The unit used for the measurement of power involves the ratio of a work unit to a time unit. Thus, in the British system, we can use fp/s as a power unit. In the metric system, the corresponding unit is the *joule/second* which is given a special name, the *watt* (W):

$$1 \text{ watt} = 1 \text{ joule/second}. \tag{25}$$

The W and the fp/s are both inconveniently small units for many practical power measurements; hence it has been found desirable to define larger units. In the metric system, two multiples of the watt are commonly used: the *kilowatt* (1 kW = 1000 W) and the megawatt (1 MW = 1 000 000 W). In the British system, the commonly used power unit is the *horsepower* (hp), defined in the typically unwieldy fashion of the British system by

$$1 \text{ hp} = 550 \text{ fp/s} = 33\ 000 \text{ fp/min}. \tag{26}$$

From the relations between the newton, pound, meter, and foot, it can easily be shown that

$$1 \text{ hp} = 745.7 \text{ W} = 0.7457 \text{ kW}.$$

It is convenient to remember that 1 horsepower is about ¾ kilowatt.

A conveniently large *energy* unit in common usage is the kilowatt-hour:

One **kilowatt-hour** is the work done in one hour by a device working at a constant rate of one kilowatt.

Since 1 kW = 1000 W = 1000 J/s, and 1 h = 3600 s,

$$1 \text{ kWh} = (1 \text{ kW}) \times (1 \text{ h}) = (1000 \text{ J/s})(3600 \text{ s}) = 3.6 \times 10^6 \text{ J}. \tag{27}$$

It should be noted that the kilowatt-hour is a unit of *energy*, not of *power*. Thus, when an electric 'power company' presents a bill for a certain number of kWh, charges are actually made for *energy*, not power.

Because, for historical reasons, power requirements of electrical appliances are always stated in *watts*, while engines are always rated in *horsepower*, there is a popular misconception that there is something 'electrical' about a watt. This is not the case. This particular usage occurs because the more recently developed discipline of electrical engineering uses the *metric* system, while the older discipline of mechanical engineering employs British-system units such as horsepower. It would be quite permissible to state the power rating of incandescent lamps in horsepower. Similarly, it would be entirely proper to rate automobile engines in kilowatts. The common usage results purely from the historical fact that mechanical engineering was well developed before metric units were first introduced in 1795, while electrical engineering began *after* this event.

Let us now compute the power expended by a force acting on a moving body, that is, the *rate* at which the force does work. Suppose a force of magnitude F acting in the X-direction is exerted on a body as shown in Fig. 7 while the body undergoes a displacement dX in time dt. Then the work done is $dW = F\,dX$, and the instantaneous power is

$$P = dW/dt = F\,dX/dt = Fv. \tag{28a}$$

The rate at which a force $\mathbf{F}$ does work on a body moving at velocity $\mathbf{v}$ is Fv if the vectors $\mathbf{F}$ and $\mathbf{v}$ have the same direction. If the force $\mathbf{F}$ does not have the same direction as the velocity $\mathbf{v}$, it is easy to see that the rate at which the force does work is

$$P = \mathbf{F} \cdot \mathbf{v} = Fv\cos\theta, \quad \begin{Bmatrix} P \text{ in W} \\ F \text{ in N} \\ v \text{ in m/s} \end{Bmatrix} \text{ or } \begin{Bmatrix} P \text{ in fp/s} \\ F \text{ in p} \\ v \text{ in f/s} \end{Bmatrix} \tag{28b}$$

where θ is the angle between the vectors $\mathbf{F}$ and $\mathbf{v}$.

Example. *An automobile of mass* 110 sl *is accelerated from rest with the constant acceleration of* 5 f/s² *for* 10 s. *Determine the useful horsepower delivered by the engine as a function of time during this period.*

From $F = ma$, the required force is $F = 550$ p. From $v = at$, the speed as a function of time is $v = (5 \text{ f/s}^2)\,t$. From (28a), the power required is

$$P = Fv = (2750 \text{ fp/s}^2)\,t.$$

To change to horsepower we substitute 1 fp/s = $\tfrac{1}{550}$ hp, to get

$$P = (5 \text{ hp/s})\,t.$$

The power useful in accelerating the automobile increases linearly in time from zero at $t = 0$ to 50 hp at $t = 10$ s.

10. Mechanical Power Transmission

Power-transmission devices, sometimes called 'machines,' are devices used to transmit power from a *source* such as an engine or motor to the place where the power is used to do useful work. Examples of power-transmission devices are gear boxes, belt drives, pulley systems, jackscrews, and the like. Devices that are primarily rotational in character will be considered in Chapter 8.

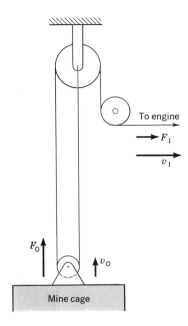

Fig. 12. Pulley system used to raise a mine cage.

Consider the simple pulley system of Fig. 12, in which the engine exerts a constant force F_I and moves the cable at speed v_I. The power $P_I = F_I v_I$ is known as the *input power*. Correspondingly, the *output power* is $P_O = F_O v_O$. If there were no power lost in friction, the output power would equal the input power, but in practice there are always power losses in a power-transmission device, so the output power is always less than the input power.

The **efficiency** of a power-transmission device is the ratio of output power to input power.

Thus in Fig. 12,

$$\text{efficiency} = \frac{\text{output power}}{\text{input power}} = \frac{F_O v_O}{F_I v_I}. \quad (29)$$

We notice that in this device, because of the geometrical arrangement of the cables, the cage moves at only half the speed at which the engine pulls the cable, hence

$$v_O = \tfrac{1}{2} v_I.$$

Therefore, if we computed F_O on the assumption that the machine had unit efficiency, we should have

$$F_O v_O = F_I v_I, \quad \text{or} \quad \frac{F_O}{F_I} = \frac{v_I}{v_O} = 2.$$

This value is known as the *ideal force ratio* or the *ideal mechanical advantage*:

The **ideal force ratio** of a power-transmission device is the ratio of output force to input force under the assumption of unit efficiency.

The actual force ratio (or mechanical advantage) is, from (29),

$$\frac{F_O}{F_I} = \text{efficiency} \times \frac{v_I}{v_O},$$

where efficiency is always less than unity. Hence in the case of Fig. 12, we must have a force ratio F_O/F_I that is less than 2.

Power-transmission devices having large force ratios frequently serve very useful purposes in hoisting and lifting, as in the case of the jackscrew and the differential chain hoist, which are considered in Probs. 32 and 36. In practical experience, it is usually found that the larger the force ratio of a device, the lower is its efficiency.

PROBLEMS

1. By exerting a constant force of 10 p on a rope attached to a sled, a man pulls the sled across the ice on the surface of a pond. If the rope makes an angle of 45° with the direction of motion, how much work does the man do in moving the sled a distance of 90 f?
Ans: 636 fp.

2. A sled weighing 30 p is dragged at constant velocity across a horizontal surface. If the coefficient of friction between the sled runners and the snow is 0.12, how much work is done in moving the sled a distance of 100 f if the applied force is in the direction of motion?

3. An elevator cage has a mass of 500 kg. If frictional effects are negligible, how much work is done in raising the cage 50 meters at constant speed?   Ans: 2.45×10^5 J.

4. A 30-kg trunk lies on a horizontal floor. If the coefficient of static friction between the trunk and the floor is 0.4 and that of kinetic friction is 0.3, what horizontal force must be exerted to set the trunk in motion? What horizontal force is required to keep the trunk in motion at constant velocity? How much work is done on the trunk by a man in moving the trunk at constant speed a distance of 12 m across the floor? While the trunk is being moved, how much work is done on the trunk by the force of friction exerted by the floor?

5. An automobile weighs 6400 p. What is its mass? How much kinetic energy does this automobile have when it is moving at a speed of 30 mi/h?   Ans: 194 000 fp.

6. A resultant external force of 4.8 N is applied to a 0.5-kg body initially at rest. What is the kinetic energy of the body after it has been displaced 8 m in the direction of the applied force? How much work is done on the body by the force?

7. What force must be applied parallel to the incline in order to pull a 2-kg body at constant velocity up an inclined plane making an angle of 30° with the horizontal if frictional effects are negligible? If the length of the inclined plane is 30 m, how much work is done against gravity in pulling the body to the top? What is the gravitational potential energy of the body at the top relative to the foot of the incline? What vertical force would be required to raise the body vertically to the same height?
Ans: 9.8 N; 294 J; 294 J; 19.6 N.

8. A resultant force of 20 N is applied to a 4-kg object initially at rest. What is the kinetic energy of the object at the end of 4 s? What would have been the kinetic energy of the object after the force had been applied for 6 s if its initial velocity had been (a) 10 m/s, in the direction of the force; (b) 10 m/s, in a direction opposite to that of the applied force?

9. A man weighing 200 p sits at rest in the club car of a passenger train moving at a speed of 60 mi/h. What is his kinetic energy relative to the other seated passengers? relative to the ties on the railroad track?   Ans: 0; 24 200 fp.

10. A 20-kg stone lies on top of a cliff at a height of 1200 m above the surface of a lake. What is its potential energy relative to the top of the cliff? relative to the lake surface?

11. What is the potential energy of 1 f³ of water at the top of a 45-f waterfall, relative to the surface of the river below the fall? What is the kinetic energy of the 1 f³ of water as it reaches the river provided there is no mechanical energy loss due to friction or turbulence? How much work could 1 f³ of water do to a turbine located at the top of the fall?
Ans: 2790 fp; 2790 fp; none.

12. What is the speed of water reaching the foot of the waterfall in Prob. 11 if friction can be neglected? How much work could 1 f³ of water do to a turbine located at the foot of the fall?

13. A 2-kg shell is fired with a muzzle velocity of 110 m/s from a howitzer aimed 60° above the horizontal. What is the initial kinetic energy of the shell? What is the kinetic energy of the shell at the highest point in its trajectory? the potential energy at the highest point? Neglect air resistance.
Ans: 12 100 J; 3020 J; 9080 J.

14. Show that if a particle of weight w, at the end of a string, is set in motion in a vertical circle, the tension in the string when the particle is at the lowest point is greater than the tension when the particle is at the highest point by 6 w.

15. A sled weighing 8 p slides from rest at the top of a hill 64 f high. The slope down which

the sled moves is inclined at an angle of 30° with the horizontal. What is the initial potential energy of the sled? If the coefficient of kinetic friction between the sled runners and the snow is 0.1, how much of the sled's initial mechanical energy is dissipated in doing work against friction? What is the final kinetic energy of the sled at the foot of the hill? What is the final speed of the sled?
   Ans: 512 fp; 88.7 fp; 423 fp; 58.2 f/s.

16. Repeat Prob. 15 for cases in which the sled moves down slopes having the following angles with the horizontal: 20°, 45°, 60°—all other data remaining unchanged.

17. An automobile weighing 6400 p moves at a speed of 60 f/s along a level roadway. When the car reaches an upward slope, the driver allows the car to coast uphill until it comes to rest. If the final position of the car on the hill is 45 f above the level of the approaching roadway, how much mechanical energy was dissipated during the car's ascent of the hill?
   Ans: 72 000 fp.

18. If the car in Prob. 17 were allowed to roll backward down the hill and the same amount of mechanical energy were lost during descent as during ascent, what would be the speed of the car when it reached the foot of the hill?

19. In the figure, body A has 1-sl mass, body B has ½-sl mass. Body C is a board of ³⁄₂-sl mass mounted on frictionless wheels. The coefficients of friction between B and C are $\mu_S=0.7$, $\mu_K=0.6$. The system starts from rest. Show that B will immediately start to slide along C. Determine (a) the accelerations of A, B, C; (b) the heat generated by friction in 1 s; (c) the loss of potential energy of body A in 1 s; (d) the kinetic energies of bodies A, B, C after 1 s; (e) verify that the loss of potential energy of A equals the gain in kinetic energy of A, B, and C plus the heat generated by friction.   Ans: (a) 14.9, 14.9, 6.40 f/s²;

(b) 40.7 fp; (c) 239 fp; (d) 111, 55.5, 31.2 fp.

20. In the figure, body A has 4-kg mass, body B has 5-kg mass; body C is a board of 12-kg mass mounted on frictionless wheels. The coefficients of friction between B and C are $\mu_S=0.4$, $\mu_K=0.3$. Make the same computations as in Prob. 19.

21. An automobile, starting from rest on a horizontal road, attains a speed of 96 f/s in 30 s. If the automobile weighs 6400 p, what is its final kinetic energy? At what average rate is work being done on the car? Assuming that there are no frictional losses, calculate the average power output of the engine and transmission of the car.
   Ans: 922 000 fp; 30 800 fp/s; 55.8 hp.

22. In Prob. 21, the average power involved in accelerating the automobile has been calculated for the entire 30-s period. Suppose that the engine is operated in such a way that the resultant force on the car is constant. What is the instantaneous power as a function of time? What are its minimum and maximum values?

23. A 250-kg elevator cage is raised at constant speed from the basement to the top floor of a building—a vertical distance of 100 m—in 1 min. What is the increase in the gravitational potential energy of the cage? What is the useful output power of the motor-and-hoist system? What is the resultant force on the cage? What vertical force is exerted on the cage by the cable if frictional effects are negligible?   Ans: 245 000 J; 4085 W; 0; 2450 N.

24. Assuming that, when the elevator cage in Prob. 23 is raised, 55 000 J of work are done against frictional forces of various kinds, find the average electric power that must be supplied to the motor. What is the total energy that must be supplied to the motor? If electric energy costs 4 cents per kWh, how much does it cost to raise the elevator? A workman can

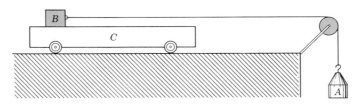

Problems 19–20

do work for a short period at a rate of 500 W (approximately ⅔ hp). If this workman were employed to raise the elevator by turning the shaft normally turned by the motor, how long would it take him to raise the elevator? If he were paid at the rate of $2.00 per hour, how much would he receive for raising the elevator?

25. A man weighing 198 p climbs a 10 000-f mountain peak in 8 h. How much work does he do in fp? in kWh? We now offer to pay the man for his day's work at the rate of 4 cents per kWh—the rate we would have to pay the 'power company' for supplying electrical energy. What is the man's total compensation? What is his hourly rate? Ans: 1 980 000 fp, 0.746 kWh; 3 cents; ⅜ cent/h.

26. A waterfall is 144 f high and 50 f³ of water pass over the fall each minute. If 80 per cent of the initial potential energy of the water can be utilized in operating a hydroelectric generator, what is the generator output in hp? in kW?

27. The force that resists the motion of an airplane at constant speed and altitude, called the *drag*, is to a good approximation proportional to the square of the speed. Find the ratio of the power required at 600 mi/h to that at 200 mi/h. Find the ratio of the energy expended per mile. Ans: 27; 9.

28. A diesel locomotive is pulling a train up a 4 per cent grade (the sine of the angle with the horizontal is 0.04). The total weight of engine and train is 1200 tons. A constant tractive force of 16 tons is required to overcome friction. The engine power is 5000 hp. What is the maximum constant speed at which the engine can pull the train up the grade?

29. (*a*) A particle is pushed up a *frictionless* incline of *constant slope* at *constant speed* by a force F which may be *variable* in magnitude and direction. Show that the work done by F in pushing the particle from P_1 to P_2 in the figure equals wH, the increase in gravitational potential energy. To demonstrate this, note that we can write $F+w+N=0$. Show that the line integral of N is zero; that of w is $-wH$, and hence that of F is $+wH$, independent of how F may vary.

(*b*) By arguments similar to those in (*a*), show that *if the speed is not constant*, the work done by a *variable* force F in pushing a particle up a *frictionless* incline of *constant slope* equals

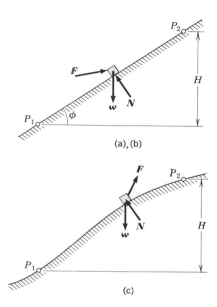

Problem 29

the increase in potential energy plus the increase in kinetic energy.

(*c*) Remove the restriction to constant slope and prove the statement of (*b*) for a *variable slope*, as in the figure. In this case one must use the information about normal and tangential accelerations given on pp. 150–151.

30. If a particle is pulled up a plane at constant speed by a force that is parallel to the plane, as in the figure, and the coefficient of friction is μ_K, show that the work done can be written in the form $wH + \mu_K wL$.

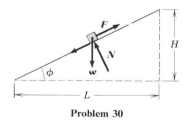

Problem 30

31. Since the angle ϕ does not occur explicitly in the expression derived in Prob. 30, one might expect the same relation to be valid even if the slope is variable, as in Prob. 29(*c*), so long as the speed is constant and the force is always applied parallel to the incline; show that the relation of Prob. 30 is *not valid* in this case of variable slope.

32. Show that the ideal force ratio of the jackscrew shown in the figure is $2\pi R/p$, where p is the pitch of the screw, that is, the distance between adjacent threads. If $p=0.5$ i, $R=4$ f, $F_O=5000$ p, and a force $F_I=50$ p is required, what is the actual force ratio and the efficiency of the jackscrew?

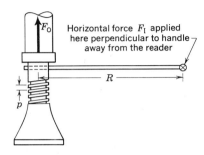

Problem 32

33. Determine the ideal force ratio of the pulley system in the figure, both by analyzing the tensions in the strings and by analyzing the geometrical motions. Assume all strings to be vertical. Ans: 3.

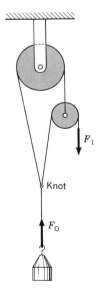

Problem 33

34. What is the ideal force ratio of the crank-and-axle type of hoist such as was used to raise the bucket in the old-fashioned well?

35. What is the ideal force ratio of the pulley system shown in the figure? If a force of 100 p is required to raise a load of 300 p at constant speed, what are the actual force ratio and the efficiency of the pulley system?

Ans: 4, 3.00, 75.0%.

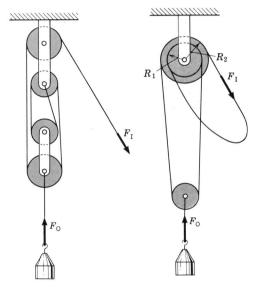

Problem 35 **Problem 36**

36. Show that the ideal force ratio of the differential chain hoist sketched in the figure is $F_O/F_I = 2R_2/(R_2 - R_1)$. The string shown in this drawing is actually a chain with teeth engaged in the pulleys so that it cannot slip. The two pulleys of radii R_1 and R_2 are fastened tightly together. If a chain hoist with an ideal force ratio of 80 is used to raise an automobile engine $2\frac{1}{2}$ f, and a man can pull the chain 1 yd at a time, how many times does the man have to pull down on the chain?

37. In the figure, a force **F** is used to pull the box at constant speed along a horizontal surface with coefficient μ_K of kinetic friction. Find the angle ϕ at which the magnitude F of the required force is a minimum. For the case $\mu_K = 1$, compute F at $\phi = 0°$, $30°$, $45°$, and $60°$, and verify the correctness of your solution in this particular case.

Ans: $\tan\phi = \mu_K$, $\phi = \arctan\mu_K$.

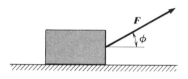

Problems 37-38

38. In Prob. 37, at what angle ϕ is the minimum *power* required to pull the box at a given constant speed? Explain what was probably an unexpected answer.

39. Show that the power required to give a block of mass m an acceleration a when it is moving at velocity v up a frictionless inclined plane making angle ϕ with the horizontal is

$$P = mva + mgv\sin\phi.$$

Derive this expression in two ways: (*a*) by computing the force required and using the relation $P = \mathbf{F} \cdot \mathbf{v}$; (*b*) by determining the time rate of change of mechanical energy by differentiation of $\tfrac{1}{2}mv^2 + mgH$. The direction of the force $\mathbf{F}$ is to be arbitrary, as in Prob. 29(*a*).

40. The expression mgH for the potential energy of a body of mass m at a height H above the earth's surface is an approximation that applies near the earth's surface. Noting that the actual force on a particle of mass m above the earth's surface at a distance R from the center of the earth is given by $Gm_E m/R^2$, where m_E is the mass of the earth, show by integration that the work done in lifting a particle vertically from the earth's surface ($R = R_E$) to a height H above the earth's surface ($R = R_E + H$) is given by $mgHR_E/(R_E + H)$, where mg is the gravitational force at the surface. Show that this expression reduces to mgH for $H \ll R_E$.

41. Taking the radius of the earth as 6400 km, use the results of Prob. 40 to compute the potential energy of a body of mass 12 kg at a height of 1600 km above the earth's surface. What initial vertical velocity should be given to a projectile if it is to reach a height of 1600 km? Neglect air resistance and rotational motion of the earth. Compare your answers with those obtained by using the formula mgH for potential energy.

Ans: 1.51×10^8 J; 5010 m/s.

42. The *average* gravitational force F_{Av} acting on a body of mass m moved from radius R_1 to radius R_2 from the center of the earth (both radii being greater than or equal to the earth's radius R_E) is defined by

$$F_{Av}(R_2 - R_1) = \int_{R_1}^{R_2} F\, dR.$$

Note that this is an average *with respect to distance*, not an average with respect to time such as we have been denoting by a bar. For the variation of force with radius given in Prob. 40, show that $F_{Av} = Gm_E m/R_1 R_2$. Show that change in potential energy is obtained by multiplying F_{Av} by $(R_2 - R_1)$.

43. A small car of mass m rolls down the incline of the 'loop-the-loop' shown in the figure from an initial height of $H = 4R$, where R is the radius of the loop. Find the magnitude of the car's velocity as it passes point A at the top of the loop. What downward force does the track exert on the car as it passes point A? Neglect friction. Ans: $2\sqrt{Rg}$; $3\,mg$.

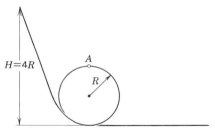

Problems 43-44

44. What is the minimum height H from which a car can roll and still maintain contact with the track at point A in the figure? Neglect friction.

45. A particle slides from rest down a plane of slant length l inclined at an angle ϕ with the horizontal. Derive an expression for its speed at the bottom of the plane in terms of l, ϕ, and the coefficient μ_K of kinetic friction.

46. A small oak block at the top of an incline slides from rest down an oak borad 12 f in length and inclined at an angle of $45°$ with the horizontal. What is its speed as it reaches the end of the incline? Take $\mu_K = 0.3$.

47. In the arrangement shown in the figure, $m_1 = 1$ kg, $m_2 = 2$ kg, and $h = 4$ m. Find the speed at which the larger body strikes the floor after falling from rest. (Ignore the effects of friction and the inertia of the pulley.) Solve this problem (*a*) by direct application of Newton's principles and (*b*) by the principle of energy conservation. Ans: 5.11 m/s.

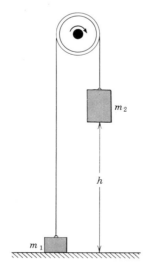

Problems 47–48

48. Keeping $m_1 = 1$ kg and $h = 4$ m in the figure, find the value of m_2 needed to cause m_2 to strike the floor 2 s after motion begins. Solve this problem by the two methods suggested in Prob. 47.

49. In the figure, a box of mass m and weight w rests on a long belt moving at constant speed V. The coefficient of friction between box and belt is μ_K. Starting at $t = 0$, a horizontal force F gives the block a constant acceleration a.

(*a*) Determine F and the power dissipated in friction as a function of time.

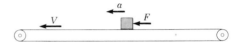

Problem 49

(*b*) If the force is exerted by a man walking on the belt, determine the power he expends as a function of time. Show that this power equals the power dissipated in friction plus the rate of increase of kinetic energy as he measures it.

(*c*) If the force is exerted by a man walking on the floor beside the belt, determine the power he expends as a function of time. Show that this power equals the power dissipated in friction plus the rate of increase of kinetic energy of the box as he computes it *plus the work he does on the moving belt*.

Ans: (*a*) $F = ma + \mu_K w$, $\mu_K wat$; (*b*) $\mu_K wat + ma^2 t$; (*c*) $\mu_K wat + ma(V + at) + \mu_K wV$.

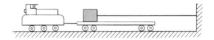

Problem 50

50. In the figure, a 1500-p box rests on a flatcar and is attached by a cable to a wall fixed to the ground. The engine moves slowly to the left, and thereby the box is dragged a distance of 28 f along the flatcar. If the coefficient of friction is 0.6, how much energy is dissipated in friction? From the standpoint of a man on the ground, what force does the work that supplies this energy? From the standpoint of a man riding on the flatcar, what force does the work that supplies this energy?

MOMENTUM 7

In this chapter we shall first treat problems involving collisions between bodies. In problems of this kind it is difficult, if not impossible, to determine the instantaneous values of the forces involved or even the exact time intervals during which the forces act. Therefore, it is difficult to specify numerical values in the equation $F=ma$. However, even though we cannot give definite values for force or acceleration, it is still possible to apply Newton's principles to these problems; we can do this conveniently by introducing the quantities *momentum* and *impulse*. We shall show how these quantities are used in the solution of impact and collision problems involving atomic particles as well as macroscopic bodies. The concept of a *dynamical system* will be introduced, and the *principle of conservation of momentum* demonstrated for such a system. This principle will be applied in a discussion of the propulsion of *jet aircraft* and *rockets*.

1. Momentum and Impulse

Before plunging into formal definitions of momentum and impulse, it might be well to discuss the general ideas involved. Consider a golf ball being driven from a tee. A large force acts on the ball during a short interval of time Δt, accelerating the ball from rest to a final velocity v. The magnitude of the force and the time interval Δt are difficult to determine, but the quantity $\boldsymbol{F}\Delta t$, where $\overline{\boldsymbol{F}}$ is the average force acting during the time Δt, has a value that is readily computed. From Newton's second principle, $\boldsymbol{F}=m\boldsymbol{a}$, we infer that $\overline{\boldsymbol{F}}=m\overline{\boldsymbol{a}}$, and hence that $\overline{\boldsymbol{F}}=m\boldsymbol{v}/\Delta t$, from the definition of average acceleration. Therefore,

$$\overline{\boldsymbol{F}}\Delta t = m\boldsymbol{v}.$$

This equation is so useful in problems involving impacts and collisions that the quantities occurring are given special names. The product $m\boldsymbol{v}$ is called the *momentum* of the golf ball. The product $\overline{\boldsymbol{F}}\Delta t$ is called the *impulse* required to change the momentum of the ball from zero to its final value $m\boldsymbol{v}$.

> The **momentum** $\boldsymbol{p}$ of a particle is a vector defined as the product of its mass m and its velocity $\boldsymbol{v}$:

$$\boldsymbol{p}=m\boldsymbol{v}. \tag{1}$$

Since mass is a scalar quantity and velocity is a vector quantity, *momentum is a vector quantity*. It is measured in units that are the product of a mass unit by a velocity unit, for example in kg·m/s or sl·f/s. From the definition given above, it is evident that the momentum of a particle changes whenever either the mass or the velocity changes. As we have already indicated, the mass of a given particle may be regarded as constant except for particles having enormous velocities, comparable with the speed of light. Hence, at ordinary velocities, any change in the momentum of a particle results from a change in its velocity. Newton's second principle, $\boldsymbol{F}=m\boldsymbol{a}$, states that force equals *mass times rate of change of velocity*, which is just the *rate of change of momentum* since, from (1), $d\boldsymbol{p}/dt = m\,d\boldsymbol{v}/dt$. Hence we can restate Newton's second principle in terms of momentum as follows:

> *The time rate of change of the momentum of a particle is equal to the resultant force acting on the particle and is in the direction of the resultant force:*

$$\frac{d\boldsymbol{p}}{dt}=\boldsymbol{F}. \tag{2}$$

This restatement turns out, as we shall see later, to be particularly useful when we consider 'reaction' devices such as turbines, jet-aircraft engines, and rockets.

Let us now return to a process of the type occurring when a golf ball is struck with a golf club. (We shall treat the golf ball as a 'particle' and neglect rotational effects.) The club is actually in contact with the ball for only a very short interval of time. During this short time interval a very large force is exerted on the golf ball; this force varies with time in a complex manner that is very difficult to determine. Forces of the kind exerted on the golf ball are called *impulsive forces*.

Let us assume that the graph in Fig. 1(a) shows the magnitude of the actual force exerted on the golf ball as a function of time for a case in which this force has a constant direction. At all times the relation $F=dp/dt$ applies and hence we may write

$$\int_{t_0}^{t_1} F\, dt = \int_{t_0}^{t_1} \frac{dp}{dt}\, dt = \int_{p_0}^{p_1} dp = p_1 - p_0 = \Delta p.$$

However, this integral can also be written as $\overline{F}\Delta t$, where $\overline{F}$ is the time-average force acting during the time interval $\Delta t = t_1 - t_0$. Hence, we may write

$$\Delta p = \overline{F}\Delta t. \qquad (3)$$

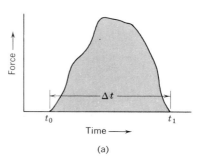
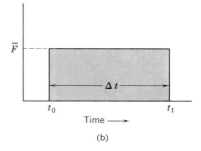

Fig. 1. An impulsive force.

By definition, the magnitude $\overline{F}$ of the time-average force is such that $\overline{F}\Delta t$ is the area under the force curve. The magnitude of $\overline{F}$ is shown in Fig. 1(b), where the rectangle is drawn so as to have the same area as the shaded area under the curve of Fig. 1(a).

Equation (3) states that the impulsive force causes a change in momentum of magnitude equal to the area under the force curve of Fig. 1(a). The vector $\overline{F}\Delta t$, having this magnitude, is called the *impulse*:

> The **impulse** is a vector defined as the time average of the force multiplied by the time interval during which the force acts.

Hence, from (3),

> *When an impulsive force acts on a particle, the change in momentum of the particle is equal to the impulse.*

By looking back over the preceding paragraph, we see that everything we have said about impulse applies equally well whether or not the *direction* of the impulsive force remains constant during the interval in which it acts.

Impulse is a vector quantity having the direction of the average force $\overline{F}$. It is measured in units of force × time, either N·s or p·s. These units are seen to be the same as the units in which momentum is measured, in accordance with the identities,

$$1\ \text{N·s} \equiv 1\ \text{kg·m/s}, \quad \text{and} \quad 1\ \text{p·s} \equiv 1\ \text{sl·f/s},$$

which follow from equations (5), p. 57, and (19), p. 67.

2. Principle of Conservation of Momentum

Now let us consider a collision between two particles, such as those of masses m_1 and m_2 shown in Fig. 2. Let us assume that during a collision these two particles exert forces on each other but that no other forces from outside this two-body system act on either particle. At any instant, F_1 is the force exerted on particle 1 by particle 2 and F_2 is the equal and opposite force exerted on particle 2 by particle 1. We may write the change in momentum that particle 1 suffers during the collision as

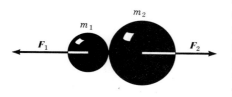

Fig. 2. Interaction between particles in collision.

$$\Delta p_1 = F_1 \, \Delta t,$$

where Δt is the time interval during which the impulsive forces act. Similarly, the change in momentum of particle 2 is

$$\Delta p_2 = F_2 \, \Delta t.$$

At each instant, $F_2 = -F_1$, hence $\overline{F}_2 = -\overline{F}_1$. Therefore

$$\Delta p_2 = -\Delta p_1. \tag{4}$$

Now let us consider the two particles as constituting a *system of particles* and define the *total momentum of the system* as

$$P = p_1 + p_2.$$

The change ΔP in the momentum of the *system* during the collision is

$$\Delta P = \Delta p_1 + \Delta p_2 = 0, \tag{5}$$

by (4). This equation expresses the important result that the total momentum of the system is not changed; this is a special case of the *principle of conservation of momentum*.

Example. *A man weighing* 200 p *and a boy weighing* 80 p *are on skates at rest facing each other. The man gives the boy a push that sends the boy backward at* 7.5 mi/h. *What happens to the man?* (*Neglect friction.*)

The forces between the man and the boy are like the forces we have considered in the collision in Fig. 2. Hence, the total momentum of the 'system' is conserved. It is initially zero; hence, after the impulsive push, the boy and man move off with momenta of equal magnitude but opposite direction:

$$m_{\text{Boy}} \, v_{\text{Boy}} + m_{\text{Man}} \, v_{\text{Man}} = 0.$$

We can use any units we please in this equation so long as we use the same units in both terms. Mass is proportional to weight, so we can substitute weight in p and speed in mi/h. If we take the velocity components in the direction of the boy's velocity, we find

$$(80 \text{ p})(7.5 \text{ mi/h}) + (200 \text{ p}) \, v_{\text{Man}} = 0.$$

Hence the man moves backward at the speed $v_{\text{Man}} = -3$ mi/h, where the minus sign indicates that the man's velocity is in the direction opposite to that of the boy.

The *principle of conservation of momentum* is a very general fundamental principle that can be derived for a *system* containing any number of particles by an extension of the method we have already used for two particles. Before discussing this principle further we shall define certain useful concepts:

> A **dynamical system** is any well-defined collection of matter.

The system may be composed of particles, rigid bodies, liquids, gases, or any combination, but by *well-defined* we mean that we have specified exactly which material is considered as part of the dynamical system; all other matter in the universe is *external* to the dynamical system.

> The **total momentum of a dynamical system** is the vector sum of the momenta of all the individual particles in the system.
>
> **Internal forces** are forces that the particles of a dynamical system exert *on each other*.
>
> **External forces** are forces exerted *on* particles in a dynamical system *by* particles external to the system.

If $p_1, p_2, p_3, \cdots$ represent the momenta of the individual particles or atoms in a dynamical system, we write the total momentum of the system as

$$P = p_1 + p_2 + p_2 + \cdots. \tag{6}$$

Then, *provided only internal forces are acting*, we can derive an equation exactly like (5) showing that, in any time interval Δt, the total momentum does not change:

$$\Delta P = \Delta p_1 + \Delta p_2 + \Delta p_3 + \cdots = 0. \tag{7}$$

Let $F_{12}, F_{13}, F_{14}, \cdots$ be the forces exerted on particle 1 by particles 2, 3, 4, $\cdots$. Similarly, let $F_{21}, F_{23}, F_{24}, \cdots$ be the forces on particle 2; $F_{31}, F_{32}, F_{34}, \cdots$ those on particle 3; etc. By (2) we see that, if these *internal* forces are the only forces acting,

$$\frac{dp_1}{dt} = F_{12} + F_{13} + F_{14} + \cdots,$$

$$\frac{dp_2}{dt} = F_{21} + F_{23} + F_{24} + \cdots,$$

$$\frac{dp_3}{dt} = F_{31} + F_{32} + F_{34} + \cdots, \quad \cdots.$$

When we add these equations for all the particles of the dynamical system, the left side becomes, from (6), just $d\mathbf{P}/dt$, while the sum on the right side vanishes, because for each force $\mathbf{F}_{ij}$ there also occurs the *equal and opposite force* (Newton's third principle) $\mathbf{F}_{ji}$. Hence the time rate of change of total momentum vanishes, and we obtain the

PRINCIPLE OF CONSERVATION OF MOMENTUM: *The total momentum of a dynamical system remains constant unless the system is acted on by external forces.*

This principle is true without exception since it is derived directly from Newton's principles. It even remains true in the modern quantum and relativistic modifications of dynamics.

3. Elastic Collisions

The problem of the collision of bodies is a difficult one because the conservation-of-momentum principle is the only mechanical principle applicable to the problem. This principle alone is insufficient to determine the motion after a collision from the motion before the collision, since an undetermined amount of energy may be converted into heat by deformation of the bodies during the collision, without violating conservation of momentum. *Momentum is always conserved,* but some *mechanical energy may be lost* during collision. Mechanical energy is *always* lost during collisions of ordinary bodies composed of large numbers of atoms; it may or may not be conserved during collisions of individual atomic or nuclear particles themselves.

There are two extreme cases in which the problem is soluble. One is the case in which we assume that no mechanical energy is dissipated—kinetic energy as well as momentum is conserved. Such collisions are called *perfectly elastic*. Collisions between ivory or glass balls fall approximately into this category. The other extreme category is that in which the bodies *stick together* after the collision. Such collisions are called *perfectly inelastic*. Collisions between two balls of putty are likely to be in this category, and a collision between a bullet and a block of wood, in which the bullet remains embedded in the wood, certainly is.

We shall first consider *perfectly elastic* collisions between two 'particles' in which the velocities both before and after collision are along the same straight line. To visualize such a collision, consider two smooth nonrotating spheres colliding exactly 'head-on' when they are moving horizontally. The instant after they separate they will be moving along the same straight line, still without rotation.

Such a collision is represented in Fig. 3, in which the masses of the bodies are m_1 and m_2, the velocity components before collision are u_1

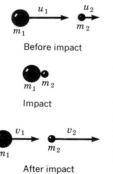

Fig. 3. Direct impact.

and u_2 and those after collision are v_1 and v_2. These velocity components and the corresponding momentum components are taken as positive when directed to the right. Necessarily, in the configuration of Fig. 3, $u_1 > u_2$ if the bodies are to collide, and $v_2 > v_1$ if the bodies are to separate.

The total momentum of the system is unaltered by the impact, hence

momentum before impact = momentum after impact

$$m_1 u_1 + m_2 u_2 = m_1 v_1 + m_2 v_2. \tag{8}$$

Since we have assumed that the collision is perfectly elastic, we have a second equation expressing the conservation of kinetic energy:

E_K before impact = E_K after impact

$$\tfrac{1}{2} m_1 u_1^2 + \tfrac{1}{2} m_2 u_2^2 = \tfrac{1}{2} m_1 v_1^2 + \tfrac{1}{2} m_2 v_2^2. \tag{9}$$

Rearranging this equation, we may write

$$m_1 (u_1^2 - v_1^2) = m_2 (v_2^2 - u_2^2),$$

or

$$m_1 (u_1 + v_1)(u_1 - v_1) = m_2 (v_2 + u_2)(v_2 - u_2). \tag{9'}$$

By rearranging equation (8), we may also write

$$m_1 (u_1 - v_1) = m_2 (v_2 - u_2). \tag{8'}$$

By dividing (9') by (8'), we get $u_1 + v_1 = v_2 + u_2$, or

$$u_1 - u_2 = v_2 - v_1. \tag{10}$$

This result indicates that, in an elastic collision in which the motion is confined to one dimension, *the relative velocity of approach before collision is equal to the relative velocity of separation after collision.*

Equations (8) and (10) are the simplest pair to use in determining the velocity components v_1 and v_2 after collision from the velocity components u_1 and u_2 before collision.

Several special cases are of interest. For example, if $m_1 = m_2$, equations (8) and (10) take the forms

$$u_1 + u_2 = v_1 + v_2$$

and

$$u_1 - u_2 = v_2 - v_1.$$

By adding and subtracting these two equations we can obtain the relations

$$u_1 = v_2 \quad \text{and} \quad u_2 = v_1, \tag{11}$$

which show that in direct, or 'head-on,' collision of two particles having equal masses, the particles exchange velocities during impact.

We shall now consider cases in which the body of mass m_2 is originally at rest. Since $u_2 = 0$, equations (8) and (10) become

$$m_1 u_1 = m_1 v_1 + m_2 v_2 \quad \text{and} \quad u_1 = v_2 - v_1.$$

These two equations are readily solved for v_1 and v_2 in terms of u_1, with the result

$$v_1 = \frac{m_1 - m_2}{m_1 + m_2} u_1, \qquad v_2 = \frac{2 m_1}{m_1 + m_2} u_1. \tag{12}$$

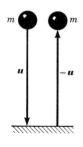

Figure 4

We note from these equations that if $m_2 \gg m_1$, $v_1 \approx -u_1$ and $v_2 \ll u_1$. In an elastic collision in which a particle strikes a much more massive particle at rest, the velocity of the light particle is approximately reversed and that of the massive particle is nearly unchanged. The extreme example is that of an elastic collision in which a particle strikes a smooth 'fixed' plane perpendicularly. By a fixed plane we mean one rigidly attached to the earth, so the mass factor associated with the plane is the whole mass of the earth. In this case the particle rebounds with exactly reversed velocity, as in Fig. 4.

Example. *The collisions between atomic nuclei and elementary particles such as neutrons are frequently perfectly elastic. In a nuclear reactor, such collisions are used to 'slow down' the high-speed neutrons emitted in atomic fission by passing the neutrons through a solid or a liquid called a 'moderator.' Both heavy-hydrogen nuclei (in heavy water) and carbon nuclei (in graphite) are used as moderators. Show that in a direct collision with a nucleus of heavy hydrogen, whose mass is twice that of the neutron, the neutron loses 8/9 of its kinetic energy. Determine the fraction lost in a direct collision with the heavier carbon nucleus, which has 12 times the mass of the neutron.*

The situation here is exactly the one that led to equations (12). The nucleus, of mass m_2, is initially at rest and moves off with velocity v_2. The neutron, of mass m_1, has initial velocity u_1 and initial kinetic energy $\tfrac{1}{2} m_1 u_1^2$. It moves off with velocity v_1, which in the case of collision with a nucleus of heavy hydrogen, where $m_2 = 2 m_1$, is, by (12),

$$v_1 = \frac{m_1 - 2 m_1}{m_1 + 2 m_1} u_1 = -\tfrac{1}{3} u_1.$$

The final kinetic energy of the neutron is

$$\tfrac{1}{2} m_1 v_1^2 = \tfrac{1}{9} (\tfrac{1}{2} m_1 u_1^2),$$

so 8/9, or 89 per cent, of the initial kinetic energy is transferred to the hydrogen nucleus. [We could check this by computing $\tfrac{1}{2} m_2 v_2^2$, which should come out to be 8/9 $(\tfrac{1}{2} m_1 u_1^2)$.]

In the case of collision with the heavier carbon nucleus, where $m_2 = 12 m_1$, we find

$$v_1 = \frac{m_1 - 12 m_1}{m_1 + 12 m_1} u_1 = -{}^{11}\!/_{13} u_1,$$

so

$$\tfrac{1}{2} m_1 v_1^2 = {}^{121}\!/_{169} (\tfrac{1}{2} m_1 u_1^2),$$

and only 48/169, or 28 per cent, of the kinetic energy of the neutron is lost. This computation shows why heavy water is a 'better' moderator than graphite in a nuclear reactor. (Although the above type of computation would indicate that a direct collision with ordinary light hydrogen would cause a neutron to lose *all* of its energy, light water has disadvantages as a moderator because light hydrogen strongly *absorbs* neutrons, to form heavy hydrogen, as we shall discuss in Chapter 45.)

4. Perfectly Inelastic Collisions; Impulse and Reaction Forces

Let us now consider perfectly inelastic collisions, in which the two bodies have the same velocity after the collision. Collisions of this type are ordinarily accompanied by pronounced deformation of one or both of the colliding bodies, and energy is lost in accomplishing this deformation. An inelastic collision may be visualized by imagining direct impact between two balls of putty. The two balls remain in contact after collision and therefore have a common final velocity v. This final velocity can be determined directly from the relation for conservation of momentum:

$$m_1 u_1 + m_2 u_2 = (m_1 + m_2) v. \tag{13}$$

When a stream of water impinges on a fixed flat plate, the impact closely approximates a perfectly inelastic collision. The stream spreads out during impact and the water flows along the surface of the plate as indicated at the left in Fig. 5. We can readily find the force $\boldsymbol{F}$ exerted by a steady jet of water on the plate. In time Δt, let mass Δm of water strike the plate. Before collision this water has X-component of momentum $\Delta m\, u$. After collision its X-component of momentum is zero. Therefore in the collision its X-component of momentum is *decreased* by $\Delta p = \Delta m\, u$. According to (3), this means that the plate must exert a force to the left:

$$F = \Delta p / \Delta t = u\, \Delta m / \Delta t.$$

But $\Delta m / \Delta t$ is the *mass of water striking the plate per unit time;* call this μ. Then

$$F = u\mu. \tag{14}$$

This is the magnitude of the force exerted to the left by the plate on

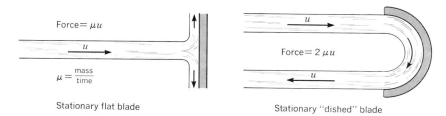

Fig. 5. Stream of water striking turbine blades. In the discussion, the blades are considered as fixed in position.

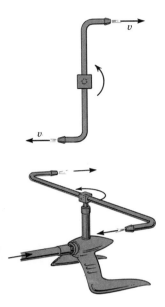

Fig. 6. Simple reaction turbine.

the water; by Newton's third principle, F is also the magnitude of *the force exerted to the right by the water jet on the plate.*

Engineers sometimes call this force an *impulse force*. The force applied in this way is employed in the *impulse turbine*, a water wheel operated by means of a high-speed water jet. The force exerted by a jet of water can be increased considerably, as in the Pelton wheel, by 'dishing' the blades as shown at the right in Fig. 5. The direction of the water stream is reversed when it strikes the curved surface of the dish, and if we neglect friction so that energy is conserved, the stream will leave the dish with a velocity equal in magnitude and opposite in direction to the velocity of the incident jet. The total force exerted on the dish is therefore given by $2u\mu$, since the change in velocity is $2u$. If the dished blades are moving to the right, as they would be in the case of a water wheel, we see from the principle of relativity that this expression still gives the force on the blade, provided that u is the speed *relative to the blade*. Engineers sometimes think of this total force as being made up of two parts: (*a*) the impulse force produced while the initial velocity to the right is reduced to zero and (*b*) the reaction force produced while the water is given a velocity toward the left. Such reaction forces are utilized in reaction turbines, of which the rotating sprinkler for lawns is a familiar example. A turbine of this type is shown in Fig. 6. Water flowing out from the center inside the turbine is ejected in the manner indicated, and the reaction force sets the turbine arms into rotation.

5. Acceleration of Rockets

For extremely long-range projectiles, rockets have proved more practicable than missiles propelled from guns. A rocket is propelled by reaction forces supplied from a high-speed jet of gas ejected toward the rear in much the same manner that reaction forces are used to set a reaction turbine in motion. In order to understand how the reaction is used, consider the sketch shown in Fig. 7(a), which shows a machine gun mounted on a car and firing a stream of bullets of mass m with muzzle velocity v. Each bullet receives momentum mv to the left as it leaves the gun. Hence, if n bullets are fired per second from the gun, they receive momentum $\Delta p = n\,mv$. From (3), with $\Delta t = 1$ s, we see that the average force exerted by the gun on these bullets is $F = n\,mv$ to the left. The magnitude of the reaction force acting on the gun is equal and opposite:

reaction = change of momentum of projectiles per second = $n\,mv$.

This reaction force, or 'recoil' force, accelerates the gun, and the car on which the gun is mounted, to the right.

An accelerating force is applied to the rocket in a similar manner, as indicated in the sketch in Fig. 7(b). In this case, we shall denote the

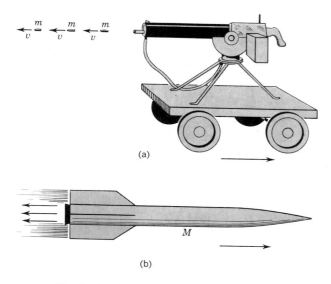

Fig. 7. Acceleration of bodies by reaction forces.

jet velocity *relative to the rocket* by v and shall denote the constant rate of discharge (mass per second) of hot gases by μ. Hence, the change of momentum per second of the material passing from the rocket into the jet is given by μv, and the magnitude of the reaction force which accelerates the rocket is given by the equation (derived exactly as in the case of the machine gun)

$$F = \mu v. \quad \begin{Bmatrix} F \text{ in N} \\ \mu \text{ in kg/s} \\ v \text{ in m/s} \end{Bmatrix} \text{ or } \begin{Bmatrix} F \text{ in p} \\ \mu \text{ in sl/s} \\ v \text{ in f/s} \end{Bmatrix} \quad (15)$$

This reaction force is called the *thrust*.

The true rocket is a reaction-propelled missile that carries all the materials from which the propulsion jet is formed, whereas the jet engine ordinarily used on manned aircraft carries only fuel and depends upon atmospheric oxygen to support combustion. The basic principles of propulsion are essentially the same for jet planes and for rockets. A true rocket operates best in the absence of an atmosphere.

Because of the occurrence in (15) of the factor v, the jet velocity, it is important to eject the jet at as high a speed as possible. In large rockets, the jet speeds are greater than a mile per second.

Example. *A rocket weighing 40 tons rises vertically from rest. It ejects gas at 6000 f/s at the rate of 30 sl/s for 40 s before the fuel is exhausted. Determine its initial upward acceleration and its acceleration at the end of 20 and 40 s. Neglect the variation of gravitational acceleration with altitude.*

From (15) we see the upward force is constant, with the value

$$\mu v = (30 \text{ sl/s})(6000 \text{ f/s}) = 180\,000 \text{ p}.$$

This force is opposed by the weight of the rocket, which is 80 000 p initially, 60 800 p after 20 s, and 41 600 p after 40 s. Hence the *net* upward force is

$$F_{t=0} = 100\ 000\ \text{p}; \qquad F_{t=20\ \text{s}} = 119\ 200\ \text{p}; \qquad F_{t=40\ \text{s}} = 138\ 400\ \text{p}.$$

The ratios F/w of the net upward force to the weight w of the rocket, at $t=0$, 20 s, and 40 s, respectively, are:

$$\frac{100\ 000}{80\ 000} = 1.25; \qquad \frac{119\ 200}{60\ 800} = 1.96, \qquad \frac{138\ 400}{41\ 600} = 3.33.$$

Since $F = ma$ and $w = mg$, $a/g = F/w$; hence the accelerations of the rocket are

$$1.25\ g, \qquad 1.96\ g, \qquad 3.33\ g$$

at these three times, where g is the gravitational acceleration.

PROBLEMS

1. A car of 2000-kg mass has a velocity of 15 m/s, eastward. What is the momentum of the car? What would be the velocity of a 150-g bullet if its momentum were the same as that of the car? Which would have the greater kinetic energy—the car or the bullet? Ans: 30 000 kg·m/s, E; 200 000 m/s, E; the bullet.

2. Compute the momentum of a 100-kg rocket traveling vertically upward (*a*) at the speed of sound (330 m/s) and (*b*) at twice the speed of sound. Find the ratio of the kinetic energies in the two cases.

3. A car weighs 4000 p and travels northward at a speed of 30 mi/h. Find the momentum and kinetic energy of the car.
Ans: 5500 sl·f/s, N; 121 000 fp.

4. A rocket of 2-sl mass moves vertically upward at the speed of sound (1080 f/s). what is the momentum of the rocket? its kinetic energy?

5. A 300-g rubber ball is dropped from an office window to a sidewalk. The speed of the ball just before impact is 30 m/s and just after impact is 20 m/s. Find the initial momentum just before impact, the momentum just after impact, and the total change in momentum during collision. Assuming that the ball was in contact with the sidewalk for 0.15 s, compute the average force exerted on the ball during impact.

Ans: 9 kg·m/s, downward; 6 kg·m/s, upward; 15 kg·m/s, upward; 100 N, upward.

6. Referring to Prob. 5, find the height of the office window from which the ball was initially dropped, the energy lost by the ball during collision, the height to which the ball rises on its first bounce, and the average force exerted on the sidewalk by the ball during its collision. Neglect air resistance.

7. An automobile weighing 4800 p has an initial velocity of 30 f/s, southward. What is the average value of the resultant external force acting on this car in the following cases: (*a*) the car is brought to rest in 5 s by application of the brakes; (*b*) the car is brought to rest in 1 s by collision with a large haystack; (*c*) the car collides with a concrete wall and rebounds with a northward speed of 10 f/s after an impact time of 0.1 s.
Ans: 900 p, N; 4500 p, N; 60 000 p, N.

8. A 96-p shell is fired from a gun with a 24-f barrel. If the muzzle velocity of the shell is 1200 f/s, calculate the time required for the shell to pass through the barrel on the assumption that its acceleration is constant. What is the magnitude of the resultant force acting on the shell during acceleration?

9. An 8-kg shell is fired horizontally westward with a muzzle velocity of 900 m/s from a 2000-kg gun mounted on a cliff. What is the recoil velocity of the gun? The gun is not rigidly attached to the cliff but is mounted on a track equipped with a recoil mechanism consisting of a set of springs arranged to bring the recoiling gun to rest. If the recoil mechanism is to stop the initial recoil in 1 s, what is the average force it must exert on the gun?

Ans: 3.6 m/s, E; 7200 N, w.

10. Referring to Prob. 9, compare the initial kinetic energy of the shell with the initial kinetic energy of the recoiling gun. Compute recoil velocities and kinetic energies for cases in which 1000-kg and 500-kg guns are used to launch the projectile of Prob. 9 in the manner specified. Why are guns provided with recoil mechanisms?

11. It is proposed to launch a shell having the same mass and velocity as that of the rocket of Prob. 4 from a gun with a 108-f barrel. If the shell has constant acceleration, how long does it stay in the gun barrel? What resultant force is exerted on the shell in the barrel? What should be the mass of the gun if its recoil velocity is not to exceed 10 f/s? How much would such a gun weigh? Discuss the advantages of rockets.

Ans: 0.2 s; 10 800 p; 216 sl; 6910 p.

12. By application of the principle of conservation of momentum, show that the kinetic energy of a recoiling gun is related to the kinetic energy of a shell in the following way: $(E_K)_{Gun} = (M_{Shell}/M_{Gun})(E_K)_{Shell}$. By a similar argument show that when a man throws a ball vertically upward, substantially no kinetic energy is imparted to the earth.

13. A 50-p boy dives from the stern of a 100-p boat. If the boy's horizontal component of velocity is 9 f/s, westward, what is the initial horizontal velocity component of the boat?

Ans: 4.5 f/s, E.

14. (a) Two men stand on a raft and toss a 10-p medicine ball back and forth. Discuss the effect of their play on the motion of the raft. (b) One of the authors heard the following problem propounded one evening by the driver of a leisurely bus: 'A 198-p man has a boat that will support exactly 200 p without sinking. He wishes to use the boat to transport three 1-p coconuts across a river and is able to accomplish this job in one crossing without getting the coconuts wet. How is this possible?'

After much discussion of the problem by the passengers, the bus driver presented his solution: 'The man was a juggler and was able to keep one coconut in the air at all times.' Discuss the validity of this solution.

15. A man tosses a 1-p body vertically into the air and then catches the body when it comes back downward. Suppose it takes the man 0.5 s to give the body a velocity of 32 f/s upward and 0.5 s to stop the body when it comes back down. What average *resultant* force acts on the body during the launching period? during the period required to stop its downward motion? Recalling that the man must also support the weight of the body during launching and stopping, find the average upward force he must exert in each case. What is the total time of 'free flight' of the body? What is the magnitude of the upward force exerted by the man averaged over the total time of launching, free flight, and stopping?

Ans: 2 p; 2 p; 3 p; 3 p; 2 s; 1 p.

16. A man gives a ball of mass m an initial upward velocity v_0 and then catches the ball when it comes back down. Show that the average upward force exerted by the man is always mg when the force is averaged over the total time of launching, free flight, and catching.

17. A 50-g steel ball moving with an initial velocity of 4 m/s eastward makes a perfectly elastic head-on collision with a second steel ball of mass m_2 initially at rest. considering the two balls as a dynamical system, find the initial and final momentum and kinetic energy of the system. What are the final velocities v_1 and v_2 of the two balls after collision for the following values of m_2: 10 kg, 50 g, 1 g?

Ans: 0.2 kg·m/s, 0.4 J; $v_1 = 3.96$ m/s, w, 0, 3.84 m/s, E; $v_2 = 0.0396$ m/s, E, 4 m/s, E, 7.84 m/s, E.

18. Calculate the fraction of the kinetic energy of the 50-g ball that is transferred to the second ball in each case in Prob. 17. Derive a general expression for the fraction of the initial energy transferred from the first to the second ball in terms of their masses m_1 and m_2, respectively.

19. Equation (10) indicates that in a perfectly elastic collision the relative velocity of approach $(u_1 - u_2)$ before collision is equal to the relative velocity of separation $(v_2 - v_1)$ after collision. It is found experimentally that, in collisions in which energy is lost, (10) must be replaced by the relation

$$v_2 - v_1 = e(u_1 - u_2),$$

where e is a constant called the *coefficient of restitution,* whose value ranges from unity for a perfectly elastic collision to zero for a perfectly inelastic collision. A steel ball is dropped from a height of 96 f to a concrete sidewalk. What is the height of rebound if the coefficient of restitution is 0.5? Ans: 24 f.

20. A steel ball falls from an initial height h_0 and strikes a large steel block. If it rebounds to a height h_1, show that the coefficient of restitution, as defined in Prob. 19, is given by the expression $e = \sqrt{h_1/h_0}$. Find the height h_2 of the second rebound in terms of e and h_0; find a general expression for the height h_n of the nth rebound in terms of e and h_0.

21. A 30-g bullet moving horizontally at an initial speed of 200 m/s strikes a 4-kg block of wood lying at rest on the surface of a frozen lake. The bullet becomes embedded in the wood and the block and bullet move together across the ice. Neglecting friction, calculate the speed of the block and bullet after the collision.
Ans: 1.49 m/s.

22. A 30-g bullet moving horizontally with a velocity of 300 m/s, eastward, strikes a 2-kg bird sitting in a tree. The bullet becomes embedded in the bird. If the branch on which the bird is sitting is 19.6 m above the ground, how far eastward will the bird move before striking the ground?

23. In a lecture-demonstration experiment, a 2-pound steel ball coated with beeswax makes a perfectly inelastic head-on collision with a 2-pound steel ball initially at rest. If the velocity of the first ball just before impact is 8 f/s, eastward, what is the common velocity of the balls after impact? How much mechanical energy is dissipated in the collision?
Ans: 4 f/s, E; 1 fp.

24. A particle of mass m_1 with initial velocity v_1 makes a perfectly inelastic collision with a second particle of mass m_2 at rest. Show that the ratio of the kinetic energy of the system after collision to its kinetic energy before collision is $m_1/(m_1+m_2)$. Show that when a bullet becomes embedded in the earth, substantially *all* of its kinetic energy is converted into heat.

25. A hunter carries an automatic rifle that shoots 1-ounce bullets with a muzzle velocity of 2000 f/s. A tiger weighing 160 p springs at the hunter. If the horizontal component of the tiger's velocity is 10 f/s, eastward, how many bullets must the man fire into the tiger in order to stop the tiger's horizontal motion? Assume that all bullets become embedded in the tiger. Would you say that the hunter is properly equipped for a tiger hunt? Ans: 13.

26. A car weighing 3200 p, moving southward at a speed of 40 f/s, makes a head-on collision with a north-bound car weighing 4800 p moving at a speed of 80 f/s. During collision the bumpers become locked. What is the common velocity of the cars immediately after collision? How much mechanical energy was dissipated during collision? What became of this energy?

27. A person leans over an elevator shaft and drops a light ball, which strikes the roof of an elevator moving upward at 10 f/s at the instant that the elevator is 40 f below. Assuming that the ball rebounds elastically, to what height does it rise? Use a coordinate system moving with the elevator and the Newtonian principle of relativity. Ans: 37.9 f above initial position.

28. A bullet of mass 100 g moving horizontally at 300 m/s imbeds itself in the center of a block of mass 3000 g initially at rest on a horizontal plane. If the coefficient of kinetic friction between the block and the plane is 0.3, how far does the block slide?

29. A fire hose discharges a stream of water against the north wall of a house. The stream is directed southward and the stream velocity has a magnitude of 60 f/s. If the hose discharges water at the rate of 1 f³/s, what force is exerted on the wall? Ans: 116 p, s.

30. It is claimed that the first jet-propelled vehicles were Chicago fire boats on Lake Michigan. These boats moved about by taking in lake water and then pumping it at high speed out of nozzles. For such a boat, determine the thrust attainable in terms of the jet speed and the volume of water pumped per second. If the jet speed is 200 f/s, how many cubic feet of water must be pumped each minute in order to obtain a thrust of 500 p?

31. A fueled rocket weighing 19 200 p ejects hot gases at 4000 f/s at the rate of 5 sl/s. What is the thrust exerted on the rocket? If burnout

takes place in 20 s, what is the impulse associated with the thrust?
Ans: 20 000 p; 400 000 p·s.

32. If the rocket of Prob. 31 were fired vertically, find the initial resultant force acting on the rocket and the final resultant force just before burnout. Find also the initial and final accelerations. Neglect air resistance.

33. A rocket with its load of fuel and oxidant is in free space where gravitational effects are negligible. The mass of the unfueled rocket is 8000 kg and that of the fuel and oxidant is 4000 kg. If the jet velocity has a magnitude of 2000 m/s and burnout occurs in 40 s, find the thrust exerted on the rocket. What is the impulse associated with the thrust? Calculate the change in speed when the rocket is fired, (*a*) on the assumption that this impulse changes the momentum of the fueled rocket and (*b*) on the assumption that only the empty rocket is accelerated (the actual speed is, of course, in between). Ans: 200 000 N; 8 000 000 N·s; 666 m/s, 1000 m/s.

34. The magnitude ΔV of the actual change in speed of the rocket in Prob. 33 is given by the expression

$$\Delta V = \frac{mv}{M} + \tfrac{1}{2} v \left(\frac{m}{M}\right)^2 + \tfrac{1}{3} v \left(\frac{m}{M}\right)^3 + \cdots,$$

where m is the mass of the fuel and oxidant, M the initial mass of the fueled rocket, and v the magnitude of the jet velocity. From this expression, estimate the magnitude of the rocket's change in velocity and compare your answers with those obtained in Prob. 33. Which of the assumptions in Prob. 33 gives the more nearly correct result?

35. What percentage of its kinetic energy is transferred to the stationary particle when a moving particle collides elastically with a stationary particle of 100 times its mass? of equal mass? of $\frac{1}{100}$ its mass?
Ans: 3.92 %; 100 %; 3.92 %.

36. Show that when a moving particle collides elastically with a stationary particle that is either k times as massive, or $1/k$ times as massive, it transfers to the stationary particle the fraction $4k/(1+k)^2$ of its kinetic energy.

37. The unmanned Surveyor missions involved the soft-landing of instrumentation on the moon's surface to determine the chemical composition of the surface material. One of the methods involved projecting helium nuclei of known kinetic energy (alpha particles from a radioactive source) at the moon's surface and measuring the kinetic energy of those alpha particles that were scattered exactly backward by elastic collision with nuclei in the surface material. (*a*) Show that an alpha particle that collides head-on with a nucleus of iron returns with 75 percent of its original energy. The relative masses of helium and iron nuclei are in the ratio 4.0 : 56. (*b*) Determine the fraction of its original energy that an alpha particle has after being scattered exactly backward from an oxygen nucleus. The relative masses of helium and oxygen nuclei are in the ratio 4.0 : 16.

38. As in the preceding problem, determine to two significant figures the fraction of its original kinetic energy that it retains when a helium nucleus is scattered exactly backward from a uranium nucleus, the mass ratio being 4.0 : 238.

39. A stream of water strikes a wall as shown at the left of Fig. 5. Find the force exerted on the wall when 5 f³ of water strike the wall each minute and the stream velocity u is 40 f/s. Find the force if the stream were directed against a stationary dished blade in the manner shown at the right of Fig. 5.
Ans: 6.45 p; 12.9 p.

40. If the stream of water mentioned in Prob. 39 were directed against the dished blades of a turbine, approximately what power is developed by the turbine if the speed of the water (with respect to the ground) is 10 f/s in the opposite direction from that at impact when it leaves the turbine, and the turbine blades are moving at 12 f/s at the point of water impact?

41. A railroad gondola car weighing 20 tons moves freely along a horizontal railroad track at a velocity of 40 f/s when 15 tons of snow fall vertically into the car from a snow bank overhanging the track. What is the final velocity of the car? Ans: 22.9 f/s.

42. Two stones, sliding on a smooth horizontal sheet of ice, collide. Before the collision, the first stone, of mass 3 kg, has translational velocity 4 m/s northward; the second stone, of mass 5 kg, has translational velocity 6 m/s eastward. After the collision, the first stone moves off at 7 m/sec eastward. What is the velocity of the second stone after the collision?

43. A jet engine is in a plane traveling at 600 f/s. Each second, the engine takes in 3000 f³ of air having a mass of 6 sl. The incoming air has of course the speed 600 f/s relative to the plane. This air is used to burn 0.3 sl of fuel each second, the energy being used to compress the products of combustion and eject them at the rear of the plane at 1200 f/s relative to the plane. What is the thrust of this jet engine and the horsepower it is delivering?   Ans: 3960 p, 4320 hp.

44. Show that when a very light particle makes a (head-on) elastic collision with a very heavy object moving toward it, it rebounds with its own speed *plus* twice the speed of the moving object, and hence gains kinetic energy. When it makes such a collision with an object moving away from it, how does it rebound and does it gain or lose energy? Consider these questions in two ways: first by employing (8) and (10); then by using a coordinate system moving with the heavy object and employing the Newtonian principle of relativity. The most important application of these results is in the case of gas molecules striking the face of a moving piston in a cylinder. They do work and lose energy if the piston is moving away (gas expanding); the piston does work on the gas if it is moving toward the colliding molecules and the gas is being compressed.

45. Water traveling at 100 f/s relative to the ground strikes the dished blades of a Pelton wheel. Assuming that the speed of the water is exactly reversed *relative to the blades,* find (*a*) the backward speed of the blades at which the force of the water is a maximum, (*b*) the speed at which the power developed is maximum, and the efficiency at this speed.
Ans: 0; 50 f/s, 100 %.

46. Using the notation employed in Prob. 34, with v for the jet speed, V for the rocket speed, and μ for the rate at which hot gases are ejected, we may by Newton's second principle set the thrust μv equal to the mass being accelerated times its acceleration $a = dV/dt$. However, the mass being accelerated is continuously decreasing as hot gases are ejected; the mass being accelerated is $M - \mu t$. Thus, we may write

$$(M - \mu t)\, a = \mu v,$$

or $\qquad (M - \mu t)\, dV/dt = \mu v.$

By integration show that

$$\Delta V = v \log_e [M/(M-m)],$$

after a mass m of propellant has been ejected. A series expansion of this expression gives the equation in Prob. 34. Recompute Prob. 34 from this expression.

ROTATIONAL MOTION 8

A rigid body is essentially a collection of particles constrained by forces that keep the particles in the body in place relative to each other. The dynamical behavior of the rigid body can be deduced from the dynamical behavior of the particles of which it is composed. In the present chapter we shall study one of the simpler, but nevertheless very important, types of rigid-body motion, namely the rotational motion of a rigid body mounted on a fixed axle.

We shall define *angular displacement, angular velocity, angular acceleration,* and *torque*. We shall see that *angular acceleration* is proportional to the *torque* exerted by the forces acting on a rigid body. Thus, for rotational motion, *torque* plays a role analogous to that of *force* in translational motion. The quantity having a corresponding analogy to *mass* is *rotational inertia;* we show how this quantity is determined. We close the chapter with a brief discussion of *power transmission* in rotational motion.

1. Kinematics of Pure Rotation

Pure rotation is the rotation of a rigid body about a fixed axis.

Let us consider Fig. 1, which shows the cross section of a rigid body mounted so that its only possible motion is one of rotation about a fixed axis through O perpendicular to the plane of the diagram. We use the term *axis* to denote the geometrical *line* through O that would be the *center line* of a real axle. All points in the body move in circles with centers located on the axis O. The position of every point in a body having this type of motion is determined by the angular position of a reference radius OA 'marked' *on the body* relative to some direction OX fixed in space. The angle between OA and OX can be used in determining the positions of all points in a rotating rigid body, since the *relative* positions of the points in the body do not change.

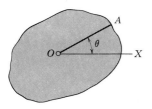

Fig. 1. Angular displacement. O is a fixed axis perpendicular to the paper; OX is a fixed direction in space; OA is a radial line 'painted' on the rigid body.

The angle between OX and OA describes the *orientation* of the rigid body as a whole. If OA initially coincides with OX and later has turned through angle θ, we can think of θ as the *angular displacement* of the body.

> **Angular displacement** of a rigid body in pure rotation is defined as the angle through which any radius of the body turns; it is usually taken as positive for a counterclockwise rotation and negative for a clockwise rotation.

Angular velocity, denoted by ω, is defined by means of the equation

$$\omega = \lim_{\Delta t \to 0} \frac{\Delta \theta}{\Delta t} = \frac{d\theta}{dt}, \qquad (1)$$

where $\Delta \theta$ is the angle through which the body turns in the short time interval Δt. If θ is measured in radians and t in seconds, ω is expressed in radians per second (rad/s).

> The **angular velocity** of a rigid body in pure rotation is the time rate of change of its angular displacement; the same sign convention is used as in the case of angular displacement.

When the angular velocity is constant, the angular displacement of the body in time t is given by

$$\theta = \omega\, t. \qquad (2)$$

This equation is analogous to the equation $X = v_X\, t$ which gives the linear displacement in time t of a particle moving with constant velocity v_X in translational motion along the X-axis.

If the angular velocity of a rotating body is changing, there is said to be an angular acceleration. If the change is $\Delta\omega$ in time Δt, the angular acceleration, denoted by α, is defined as

$$\alpha = \lim_{\Delta t \to 0} \frac{\Delta\omega}{\Delta t} = \frac{d\omega}{dt} = \frac{d^2\theta}{dt^2}. \tag{3}$$

If ω is measured in radians per second and time in seconds, angular acceleration α will be expressed in (rad/s) *per second*, that is, in rad/s².

> **Angular acceleration** of a rigid body in pure rotation is defined as the time rate of change of angular velocity.

If the body is rotating counterclockwise (ω positive) and has a positive angular acceleration, $\Delta\omega$ is positive and the speed of rotation is increasing; if it has a negative angular acceleration, $\Delta\omega$ is negative and the speed of rotation is decreasing.

We may now write equations describing the motion of a body experiencing *constant* angular acceleration α. If the angular velocity of the body is ω_0 at time $t=0$, its angular velocity ω at time t can be written as $\omega = \omega_0 + \Delta\omega$, where $\Delta\omega$ is the change in angular velocity during the time interval $\Delta t = t$. Since, for constant angular acceleration, $\Delta\omega = \alpha \Delta t = \alpha t$, the expression for ω becomes

$$\omega = \omega_0 + \alpha t. \tag{4}$$

In order to find the angular displacement θ experienced by the body during the time t, we write

$$\theta = \bar{\omega} t, \tag{5}$$

where $\bar{\omega}$ is the average angular velocity during this time. Just as in the case of uniformly accelerated rectilinear motion, we may express $\bar{\omega}$ in the form

$$\bar{\omega} = \tfrac{1}{2}(\omega_0 + \omega), \tag{6}$$

where ω is the final angular velocity given by equation (4). Substitution of (4) in (6) leads to $\bar{\omega} = \omega_0 + \tfrac{1}{2}\alpha t$, and substitution of this value in (5) gives

$$\theta = \omega_0 t + \tfrac{1}{2}\alpha t^2 \tag{7}$$

for the magnitude of the angular displacement during time t.

By elimination of t from equations (4) and (7), we obtain the relation

$$\omega^2 = \omega_0^2 + 2\alpha\theta. \tag{8}$$

Note that equations (4)–(8) are valid only if the angular acceleration α is *constant;* thus these equations are analogous to equations (11)–(15) on p. 41 for the special case of rectilinear motion with constant acceleration a_x.

Example. *A motor-driven grinding wheel starts from rest and receives a constant angular acceleration of 3 rad/s² for 12 s. Determine its angular velocity at the end of this interval and the angle through which it has turned.*

From (4), the angular velocity of the wheel at the end of 12 seconds is 36 rad/s. The average angular velocity of the wheel during these 12 seconds is 18 rad/s, as given by (6). The total angular displacement of the wheel can be obtained from (5):

$$\theta = (18 \text{ rad/s})(12 \text{ s}) = 216 \text{ rad},$$

or from (7): $\theta = \tfrac{1}{2}(3 \text{ rad/s}^2)(12 \text{ s})^2 = 216 \text{ rad}.$

Since there are 2π radians in one complete revolution, the wheel turns through $216/2\pi = 34.4$ revolutions during the first 12 seconds of motion.

There is an exact correspondence between rotational motion of a rigid body about a fixed axis and motion of a particle along a straight line (the X-axis) if we associate θ with X, ω with v_x, and α with a_x. The relations we have found among θ, ω, and α are exactly the same as the relations we found in Chapter 3 between X, v_x, and a_x. As we go into the dynamics of rotational motion about a fixed axis and enlarge this list of associated quantities, we shall find essentially that we shall not need to learn any new analytical relations if we remember the analytical relations governing the translational motion of a particle along a straight line and the fairly obvious system of correspondence between rotational and translational quantities.

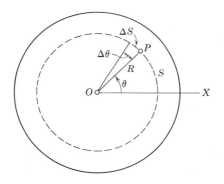

Figure 2

2. Motion of a Point in a Rigid Body in Pure Rotation

The angular displacement, angular velocity, and angular acceleration of a rigid body rotating about a fixed axis characterize the motion of the rigid body *as a whole*. Let us consider the displacement, velocity, and acceleration of a single point in the rotating body. Every point in a body in pure rotation moves in a circle whose center is on the axis. Let the broken circle in Fig. 2 represent the path of a point P at distance R from the axis O. When the radius OP makes angle θ with the fixed reference line OX, the arc length measured counterclockwise from the reference line to P will be

$$S = R\theta, \tag{9}$$

where θ must be expressed in radians.

In order to study the velocity and acceleration of the point P, we imagine that the polar angle θ in Fig. 3 is given as a function of time t. Then we can write the rectangular coordinates of P as

$$X = R\cos\theta, \qquad Y = R\sin\theta, \tag{10}$$

and derive rectangular components of velocity and acceleration by differentiation. We are not so much interested in X- and Y-components of the acceleration, as in *radial* and *tangential* components, so let us compute the X- and Y-components *at the angle $\theta = 0$, where the X-component is the outward radial component a_R and the Y-component is the tangential component a_T, taken as positive in the counterclockwise sense.* There is no loss of generality in choosing this particular point on the path because the X- and Y-axes in Fig. 3 can be oriented arbitrarily so that $\theta = 0$ can represent any desired point on the circle.

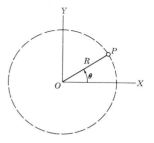

Figure 3

Differentiation of (10) with respect to t gives

$$v_X = \frac{dX}{dt} = -R\frac{d\theta}{dt}\sin\theta, \qquad v_Y = \frac{dY}{dt} = R\frac{d\theta}{dt}\cos\theta, \tag{11}$$

if θ is expressed in radians. At $\theta = 0$, these relations give

$$v_X = 0, \qquad v_Y = R(d\theta/dt). \qquad (\theta = 0) \tag{12}$$

The velocity is, of course, purely tangential. Since we intend to let $\theta = 0$ represent any arbitrary point on the path, we can generalize (12) by writing v_X as v_R, the outward radial component, and v_Y as v_T, the tangential component positive if counterclockwise. Then (12) becomes

$$v_R = 0, \qquad v_T = R(d\theta/dt) = R\omega. \tag{13}$$

We can similarly treat the acceleration components. Differentiation of (11) gives

$$\left. \begin{aligned} a_X &= \frac{dv_X}{dt} = \frac{d^2X}{dt^2} = -R\left(\frac{d\theta}{dt}\right)^2\cos\theta - R\frac{d^2\theta}{dt^2}\sin\theta, \\ a_Y &= \frac{dv_Y}{dt} = \frac{d^2Y}{dt^2} = -R\left(\frac{d\theta}{dt}\right)^2\sin\theta + R\frac{d^2\theta}{dt^2}\cos\theta. \end{aligned} \right\} \tag{14}$$

At $\theta = 0$, these expressions become

$$a_X = -R\left(\frac{d\theta}{dt}\right)^2, \qquad a_Y = R\frac{d^2\theta}{dt^2}. \qquad (\theta = 0) \tag{15}$$

More generally,

$$a_R = -R\left(\frac{d\theta}{dt}\right)^2 = -R\omega^2, \qquad a_T = R\frac{d^2\theta}{dt^2} = R\alpha. \tag{16}$$

Finally we note from (13) that

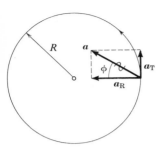

(a) Speed increasing

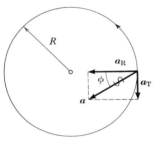

(b) Speed decreasing

Fig. 4. Centripetal and tangential component accelerations for particles *moving counterclockwise* in circles at variable speed.

$$\omega = \frac{1}{R} v_T, \qquad \text{hence} \qquad \alpha = \frac{d\omega}{dt} = \frac{1}{R} \frac{dv_T}{dt}. \tag{17}$$

Substitution of these relations in (16) gives

$$a_R = -R\omega^2 = -\frac{v_T^2}{R}, \qquad a_T = R\alpha = \frac{dv_T}{dt}. \tag{18}$$

Hence we have found that

A point P at radius R from the axis of rotation of a rigid body has speed $R\omega$, centripetal acceleration $R\omega^2$, tangential acceleration $R\alpha$. The velocity vector is tangent to the circular path of P and points counterclockwise if ω is positive, clockwise if ω is negative; similarly the tangential component of acceleration points counterclockwise or clockwise according to whether α is positive or negative (see Fig. 4). These relations hold only when the angular unit in ω and α is the radian.

Example. *A 'fun house' contains a circular horizontal rotating table that starts from rest and accelerates at the uniform rate of* 0.4 rad/s². *A person sits on the table at a distance of* 3 f *from the axis. If the coefficient of static friction is* 0.15, *at what time after the table starts to rotate does the person start to slide?*

The person will slide when the force required to accelerate him in his circular motion is greater than the maximum force of static friction, 0.15 mg, where mg is his weight. Let a^{Max} be the maximum acceleration this maximum force of static friction can cause. Then, by Newton's second principle,

$$0.15\ mg = ma^{\text{Max}} \qquad \text{and} \qquad a^{\text{Max}} = 0.15\ g = 0.15\ (32\ \text{f/s}^2) = 4.80\ \text{f/s}^2.$$

The tangential component of the acceleration of the person is constant at

$$a_T = R\ \alpha = (3\ \text{f})(0.4\ \text{rad/s}^2) = 1.20\ \text{f/s}^2.$$

Since the angular velocity at time t is

$$\omega = \alpha\ t = [0.4\ \text{rad/s}^2]\ t,$$

the radial component of acceleration is

$$a_R = R\ \omega^2 = [3\ \text{f}][0.16/\text{s}^4]\ t^2 = [0.480\ \text{f/s}^4]\ t^2.$$

The resultant acceleration is $a = \sqrt{a_T^2 + a_R^2}$. Equate the square of the resultant acceleration to the square of a^{Max} to determine when slipping will occur:

$$a_T^2 + a_R^2 = (a^{\text{Max}})^2.$$

Substitution gives

$$[1.20\ \text{f/s}^2]^2 + [0.480\ \text{f/s}^4]^2\ t^4 = [4.80\ \text{f/s}^2]^2,$$

from which $\qquad t^4 = 93.8\ \text{s}^4, \qquad t = 3.11\ \text{s}.$

The person will start to slip 3.11 s after the platform starts turning. At this instant, the angular velocity will be 1.24 rad/s or about 12 rev/min.

While the above derivation uses explicitly the fact that the motion is in a circle and R is constant, it turns out that, for motion on a plane curve that is not a circle, relations similar to (18) are applicable. Associated with any given point on a plane curve is a 'circle of curvature,' defined by passing a circle through the given point and two nearby points on either side, and then going to the limiting circle as the nearby points move closer and closer to the given point. In the limit, there is a definite 'center of curvature' and 'radius of curvature' R. Since, from its definition, the circle of curvature is tangent to the path, the center of curvature lies on the normal to the path. Relations (18) then correctly give the two component accelerations, one directed toward the center of curvature and one tangent to the curve. It is easy to visualize the meaning of the circle of curvature (Fig. 5) if one thinks of the path as a narrow road along which an automobile is driven. The circle of curvature at any point P of the path is the circle the automobile would traverse if the steering wheel were locked in position at that point.

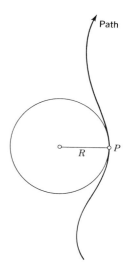

Fig. 5. Circle of curvature.

Example. *An automobile on a circular track of radius* 1500 f *is traveling at* 90 f/s *and increasing its speed at the rate of* 5 f/s² *. What is its acceleration?*

The situation is like that in Fig. 4(a). The tangential component of acceleration has magnitude equal to the rate of change of speed, that is, $a_T = 5$ f/s². The inward radial acceleration has magnitude

$$a_R = \frac{v^2}{R} = \frac{(90 \text{ f/s})^2}{1500 \text{ f}} = 5.4 \frac{\text{f}}{\text{s}^2}.$$

The acceleration therefore has the magnitude

$$a = \sqrt{5^2 + 5.4^2} \text{ f/s}^2 = \sqrt{54.16} \text{ f/s}^2 = 7.36 \text{ f/s}^2.$$

The angle ϕ between the acceleration vector and the radius [Fig. 4(a)] is such that

$$\tan\phi = \frac{a_T}{a_R} = \frac{5 \text{ f/s}^2}{5.4 \text{ f/s}^2} = 0.926; \qquad \phi = 42°.8.$$

3. Work, Power, Torque

We now consider the *dynamics* of rotational motion, that is, the relation between the forces acting on a rotating body and its angular acceleration. This relation can be derived directly from application of Newton's second principle to each of the particles in the body. It has a form analogous to Newton's relation $F = ma$ for rectilinear motion except that in place of F there occurs a quantity L called *torque*, and in place of m, a quantity I called *rotational inertia*. The resulting relation is $L = I\alpha$. We shall derive this relation by an alternative method in which we first obtain expressions for the power developed by a force acting on a rotating body, and for the kinetic energy of the body. Then we can derive the relation $L = I\alpha$ from the principle of conservation of energy; such a derivation is much simpler than the direct application of Newton's second principle to each particle of the body.

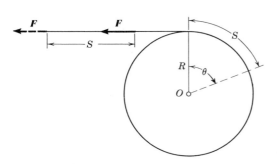

Fig. 6. Rotation of a wheel about a fixed axis.

Consider the wheel in Fig. 6, which has a string wound around the rim, pulled with a constant force F. When the string moves a distance S, work $W = FS$ is done by the force F. During this motion the wheel turns through an angle $\theta = S/R$, so $S = R\theta$. We can then write the work done by the force when the wheel turns through angle θ as

$$W = FS = FR\theta = L\theta, \tag{19}$$

where

$$L = FR \tag{20}$$

is called the *torque* exerted by the force F.

Torque is the product of force by the perpendicular distance from the line of action of the force to the axis of rotation. It has the dimensions of force times distance and is measured in such units as N·m or p·f.

Torques are usually designated as positive if they tend to cause counterclockwise rotation, negative if they tend to cause clockwise rotation. We shall discuss the above definition of torque in more detail in Chapter 9 when we consider the condition for rotational equilibrium.

In (19), W is in J when F is in N and S in m, while W is in fp when F is in p and S in f; hence the work done by torque L is given in the following units:

$$W = L\theta. \quad \begin{Bmatrix} W \text{ in J} \\ L \text{ in N·m} \\ \theta \text{ in rad} \end{Bmatrix} \text{ or } \begin{Bmatrix} W \text{ in fp} \\ L \text{ in p·f} \\ \theta \text{ in rad} \end{Bmatrix} \tag{21}$$

Now let us consider the rate of doing work, the *power* $P = dW/dt$. Since $d\theta/dt = \omega$, differentiation of (21) gives

$$P = L\omega. \quad \begin{Bmatrix} P \text{ in W} \\ L \text{ in N·m} \\ \omega \text{ in rad/s} \end{Bmatrix} \text{ or } \begin{Bmatrix} P \text{ in fp/s} \\ L \text{ in p·f} \\ \omega \text{ in rad/s} \end{Bmatrix} \tag{22}$$

Power is given by the product of torque and angular velocity.

Equation (22) is valid whether or not the force F and the torque L are constant, since we can write (19) as $dW = F\,dS = L\,d\theta$, from which (22) immediately follows.

4. Kinetic Energy, Rotational Inertia

Now consider the kinetic energy of a body in pure rotation. That a rotating body has energy by virtue of its motion is apparent if we try to stop a large flywheel by applying a brake shoe; the brake shoe becomes very hot as the rotational kinetic energy is transformed into heat. It is also a matter of experience that work must be done to set a wheel into rotation even when frictional effects are negligible.

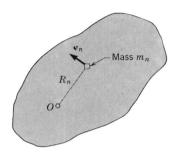

Figure 7

The kinetic energy of a rotating body is the sum of the kinetic energies of the particles of which it is composed. To obtain an expression for the kinetic energy, consider the body in Fig. 7, which is rotating at angular velocity ω about a fixed axis through O. A typical particle, of mass m_n at radius R_n, has kinetic energy $\frac{1}{2} m_n v_n^2$, where v_n is the speed of the particle in its circular path. Since $v_n = R_n \omega$, we may write the kinetic energy of the particle in the form $\frac{1}{2} m_n R_n^2 \omega^2$. Therefore, the total kinetic energy of the body is the sum

$$E_K = \Sigma \tfrac{1}{2} m_n R_n^2 \omega^2,$$

where the symbol Σ indicates summation over all the particles of the body. Since the angular velocity ω of all the particles is the same, we can factor out $\frac{1}{2} \omega^2$ and write

$$E_K = \tfrac{1}{2} [\Sigma m_n R_n^2] \omega^2.$$

This equation is analogous to the equation $E_K = \frac{1}{2} mv^2$ for the kinetic energy of translation of a particle. Just as m is said to be a measure of the *inertia* of the particle, the quantity in square brackets is called the *rotational inertia* (or *moment of inertia*) of the rigid body about the axis through O. Rotational inertia is ordinarily denoted by the letter I; thus, if we write

$$I = \Sigma m_n R_n^2, \tag{23}$$

we obtain $\qquad E_K = \tfrac{1}{2} I \omega^2 \qquad \begin{Bmatrix} E_K \text{ in J} \\ I \text{ in kg·m}^2 \\ \omega \text{ in rad/s} \end{Bmatrix} \text{ or } \begin{Bmatrix} E_K \text{ in fp} \\ I \text{ in sl·f}^2 \\ \omega \text{ in rad/s} \end{Bmatrix} \qquad (24)$

as the value of the kinetic energy of a body having rotational inertia I and rotating with angular velocity ω.

> The **rotational inertia** of a rigid body about an axis fixed relative to the body is defined as $\Sigma m_n R_n^2$, where m_n is the mass of a typical particle of the body, R_n its perpendicular distance from the axis, and the summation is extended over all particles of the body.

The summation involved in computing the rotational inertia can be performed readily for most regular solids by the methods of the calculus by expressing the sum as an integral over the volume elements of the body. We give in Fig. 8 expressions for the rotational inertia of several regular solids about axes through their centers. The case of the ring or thin-walled cylinder is simple: *all* of the mass is at radius R so the rotational inertia is just mR^2. For the solid cylinder or sphere, much of the mass is at radius less than R so the rotational inertia is less than mR^2.

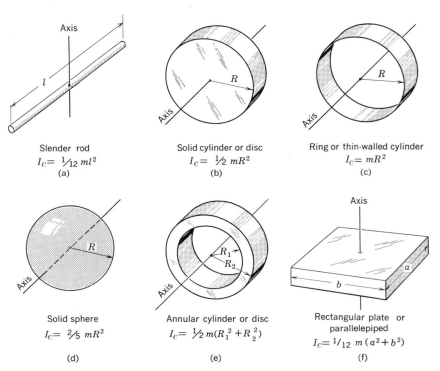

Fig. 8. The rotational inertias I_C of various rigid bodies of mass m about certain axes through their geometrical centers.

From the values given in Fig. 8 and from (23), it can be seen that the rotational inertia of a body depends upon two factors: (a) *the total mass of the body* and (b) *the distribution of the mass about the axis*. The second factor is determined by the *size* and *shape* of the body.

The principle of conservation of energy is a powerful tool for solving certain types of dynamical problems involving rotational motion. The following examples illustrate such problems:

Example. *Consider the frictionless arrangement shown in Fig. 9, in which a weight, suspended from a cord wrapped around a hub, is released and allowed to descend, thereby setting a wheel of rotational inertia I into rotation about* *a fixed axis O. Determine the linear velocity v of the suspended body and the angular velocity ω of the wheel after the suspended body, released from rest, has descended a distance h.*

We note that the final kinetic energy of the system will be equal to the decrease in the potential energy of the weight:

$$\tfrac{1}{2} I\omega^2 + \tfrac{1}{2} m_1 v^2 = wh,$$

where I is the rotational inertia of the wheel. In evaluating the final kinetic energy, it is convenient to use the relation $v = r\omega$ between the downward linear velocity v of the weight, the angular velocity ω of the wheel, and the radius r of the hub. Substituting this expression for v, we obtain

$$\tfrac{1}{2} I\omega^2 + \tfrac{1}{2} m_1 r^2 \omega^2 = wh,$$

or,

$$\omega^2 = \frac{2wh}{I + m_1 r^2},$$

from which ω and $v = r\omega$ can be computed.

Example. *If $w = 16$ p, $r = 4$ i, and $I = 3.12$ sl·f² in Fig. 9, what angular and linear velocities are acquired when the weight falls from rest through $h = 6$ f?*

By substitution of the above values in the last equation of the preceding example, we obtain

$$\omega^2 = \frac{2 \times 16 \text{ p} \cdot 6 \text{ f}}{3.12 \text{ sl} \cdot \text{f}^2 + \tfrac{1}{2} \text{ sl} \cdot \tfrac{1}{9} \text{f}^2} = 60.4 \frac{\text{p}}{\text{sl} \cdot \text{f}}$$

To interpret this last expression, we remember that $1 \text{ p} = 1 \text{ sl} \cdot \text{f}/\text{s}^2$, in accordance with Newton's second principle. If we make this substitution we find $\omega^2 = 60.4/\text{s}^2$. The square root of this value gives

$$\omega = 7.77 \text{ rad/s}.$$

The final downward velocity of the weight is therefore

$$v = r\omega = \tfrac{1}{3} \text{ f} \cdot 7.77 \text{ rad/s} = 2.59 \text{ f/s}.$$

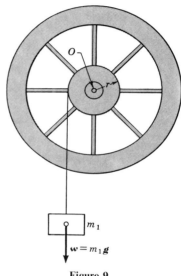

Figure 9

5. Dynamics of Pure Rotation

In this section, we derive the relation $L = I\alpha$, between torque and angular acceleration. We can start by demonstrating this relation for the simple system in Fig. 10, which shows a particle of mass m that is constrained by a support of negligible mass to move in a circular path of radius R about a vertical axis through O. Let us apply a tangential force of magnitude F_T to the particle of mass m. The particle will then experience a tangential acceleration of magnitude a_T given by Newton's second principle,

$$F_T = ma_T, \tag{25}$$

since F_T is the only force component tangent to the circle.

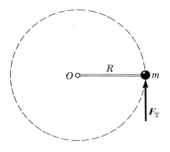

Figure 10

By use of equation (18) we may write

$$F_T = mR\alpha, \tag{26}$$

where α is the angular acceleration of the particle about the axis through O. This equation states that α is proportional to the magnitude of the tangential force. The torque L of this force about the axis through O is given by (20) as $F_T R$. Hence, by multiplying both sides of (26) by R, we obtain

$$F_T R = L = [mR^2]\alpha = I\alpha, \tag{27}$$

which gives the desired relation between L and α for this simple case.

Now consider a rigid body of any shape mounted so that it can rotate about an axis through O, as in Fig. 11. The body can be considered as divided up into small pieces, of which the nth typical piece, of mass m_n, is shown in the figure. If the rigid body is to have angular acceleration α, this typical piece must have acting on it a resultant tangential force given by (21) as $F_n = m_n r_n \alpha$, corresponding to a torque about O given by (22) as $L_n = [m_n r_n^2]\alpha$. The torque required to cause the angular acceleration must be supplied by some external force such as $\mathbf{F}$ at radius R, giving torque $L = FR$. We shall show from the principle of conservation of energy that *the total external torque needed to give the rigid body the angular acceleration α is just the sum over the pieces of the body of the torques* $L_n = [m_n r_n^2]\alpha$. Thus,

$$L = \sum L_n = \sum [m_n r_n^2]\alpha = I\alpha. \tag{28}$$

A rigorous demonstration of this relation can be obtained by equating the power furnished by the torque L to the rate of change of kinetic energy of the rotating body:

$$P = dE_K/dt.$$

Substitution of (22) for power and (24) for kinetic energy gives

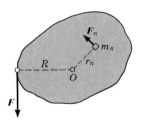

Figure 11

$$L\omega = \frac{d}{dt}(\tfrac{1}{2} I\omega^2) = I\omega\, d\omega/dt = I\omega\alpha,$$

or

$$L = I\alpha. \quad \begin{Bmatrix} L \text{ in N·m} \\ I \text{ in kg·m}^2 \\ \alpha \text{ in rad/s}^2 \end{Bmatrix} \text{ or } \begin{Bmatrix} L \text{ in p·f} \\ I \text{ in sl·f}^2 \\ \alpha \text{ in rad/s}^2 \end{Bmatrix} \tag{29}$$

If more than one force acts on the rotating body, it is easily seen that L in (29) should be the *algebraic sum* of the torques.

We note immediately the similarity between equation (29), $L = I\alpha$, and the equation $F = ma$ for Newton's second principle. The role of *torque L* in rotational motion is analogous to the role of *force F* in translational motion; the role of angular acceleration α is analogous to that of linear acceleration a; and the role of rotational inertia I is analogous to that of mass m.

Example. *Go back to Fig. 9 and the example on p. 155 and derive the expression for the angular velocity of the wheel, after the suspended body has fallen a distance h, from the dynamical equations $L=I\alpha$ and $F=ma$.*

We note that the resultant downward force on the suspended body is $(w-T)$, where T is the tension in the cord, and that the resultant counterclockwise torque acting on the wheel is $L=Tr$. If a is the downward acceleration of the suspended body, the equation $F=ma$ for this body becomes

$$w-T=m_1 a \quad \text{or} \quad w-T=m_1 r\alpha, \tag{i}$$

since the acceleration a equals the tangential acceleration of a point on the wheel at radius r, which, by (18), is $r\alpha$. For the wheel,

$$L=I\alpha \quad \text{or} \quad Tr=I\alpha. \tag{ii}$$

Elimination of T from the equations on the right in (i) and (ii) gives

$$\alpha = \frac{wr}{m_1 r^2 + I}.$$

While the weight falls a distance h, the wheel turns through the angle $\theta = h/r$. Equation (8) shows that a wheel starting from rest, having angular acceleration α, and turning through angle θ acquires angular velocity given by

$$\omega^2 = 2\alpha\theta = 2\frac{wr}{m_1 r^2 + I}\frac{h}{r} = \frac{2wh}{m_1 r^2 + I}.$$

This is the same result that we derived earlier, and somewhat more simply, from the principle of conservation of energy.

6. Angular Momentum

The rotational quantity that is analogous to linear momentum $p=mv$ is called angular *momentum* and is denoted by

$$J=I\omega. \tag{30}$$

Just as Newton's second principle can be restated as (force)=(time rate of change of momentum), the rotational analog $L=I\alpha$ can be restated as (torque)=(time rate of change of angular momentum). From (30), we see that

$$dJ/dt = I\, d\omega/dt = I\alpha = L, \tag{31}$$

which equates torque to time rate of change of angular momentum.

Again, we have a principle of conservation of angular momentum:

If no external torque acts on a rigid body mounted on a fixed axis, the angular momentum of the body is constant.

If an impulsive torque acts, such as might occur if the spoke of a wheel is struck with a hammer, again we may not be able to determine the actual torque L or the time Δt during which it acts, but we can, from (31) write

$$dJ = L\, dt,$$

at any instant, and hence

$$\Delta J = \overline{L}\, \Delta t. \tag{32}$$

This equation states that the change in angular momentum equals the average value of the impulsive torque multiplied by the time during which it acts. This equation is analogous to (3) on p. 131; just as $\overline{F}\,\Delta t$ in that equation is called the impulse the expression $\overline{L}\,\Delta t$ is called the *angular impulse*.

We have now defined all of the quantities and relations involved in discussing rotational motion about a fixed axis, and we note the complete analogy between these quantities and relations and those involved in the discussion of the motion of a particle along a straight line. The following table gives a convenient summary of the analogous quantities.

Quantity of relation	Rectilinear motion	Rotational motion
Displacement	X	θ (angular)
Velocity	$v = dX/dt$	$\omega = d\theta/dt$ (angular)
Acceleration	$a = dv/dt$	$\alpha = d\omega/dt$ (angular)
Inertia	m (mass)	I (rotational inertia)
Force, torque	F (force)	L (torque)
Newton's principle	$F = ma$	$L = I\alpha$
Element of work	$F\, dX$	$L\, d\theta$
Kinetic energy	$\tfrac{1}{2} mv^2$	$\tfrac{1}{2} I\omega^2$
Power	Fv	$L\omega$
Momentum	mv	$I\omega$ (angular)
Impulse	$\overline{F}\,\Delta t$	$\overline{L}\,\Delta t$ (angular)

7. Power Transmission

In the transmission of mechanical power from an engine or motor, one frequently introduces a device (a *transmission*) to change the rotational speed and correspondingly change the torque. We desire here to extend the treatment of pp. 121–122 to the case in which the power inputs and outputs involve rotational motion.

One type of transmission is a pair of gears such as those in Fig. 12. The input shaft turns at angular velocity ω_I, the output shaft at ω_O. The ratio of these speeds is fixed by the ratio of the numbers of teeth on the gears. The engine exerts torque L_I on the input shaft; the output or drive shaft exerts torque L_O on the machine being driven. If there is no loss of energy in heat in the gears or in the bearings supporting the shafts (not shown in Fig. 12), the power delivered to the machine, $L_O \omega_O$, will exactly equal the power delivered by the engine, $L_I \omega_I$; that is,

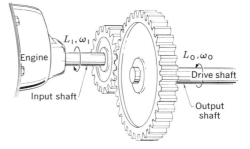

Figure 12

$$L_O \omega_O = L_I \omega_I, \quad \text{or} \quad \frac{L_O}{L_I} = \frac{\omega_I}{\omega_O}.$$

The *torque ratio* L_O/L_I would equal the *speed ratio* ω_I/ω_O. Since there is always energy loss, in an actual transmission,

$$L_O \omega_O < L_I \omega_I, \quad \text{or} \quad \frac{L_O}{L_I} < \frac{\omega_I}{\omega_O}. \tag{33}$$

Since the *efficiency* of a transmission is defined as the ratio of the output power to the input power,

$$\text{efficiency} = \frac{L_O \omega_O}{L_I \omega_I}, \quad \text{and} \quad \frac{L_O}{L_I} = \text{efficiency} \times \frac{\omega_I}{\omega_O}. \tag{34}$$

The efficiency is always less than 1. If the efficiency is 98 per cent, the torque ratio is 98 per cent of the speed ratio, and 2 per cent of the input energy goes into heat in the transmission.

For power transmission devices using gears, or nonslipping belts between pulleys of different sizes, the speed ratio is geometrically fixed and the torque ratio is determined by the efficiency as described above. In the case of *hydraulic torque converters,* in which the only contact between driving and driven members is by means of a liquid, neither the torque ratio nor the speed ratio is geometrically fixed, and they depend on each other in a complex fashion; however, the fundamental principle of conservation of energy implied in relations (33) and (34) must still be satisfied.

Example. *In accelerating an automobile equipped with a hydraulic torque converter, the engine shaft is turning at* 2000 rev/min, *the drive shaft at* 400 rev/min, *the torque ratio is* 3.0. *What is the efficiency of the converter under these conditions?*

We have $L_O/L_I = 3.0$, $\omega_I/\omega_O = 5.0$; hence from (34)

$$\text{efficiency} = \frac{3.0}{5.0} = 0.6 = 60 \text{ percent}.$$

PROBLEMS

1. The shaft of a motor is rotating at a constant angular velocity of 1800 rev/min. What is the angular velocity of the shaft in rad/s? What is the angular displacement of the shaft during an 8-s interval? Ans: 60π rad/s; 480π rad.

2. What is the angular velocity of the hour hand of a clock in rad/min? in rad/s? What is the angular velocity of the minute hand in the same units?

3. If it takes 10 s for the shaft of a motor to start from rest and attain its full speed of 1800 rev/min, what is the magnitude of the average angular acceleration of the shaft?
Ans: 6π rad/s^2.

4. If the shaft of the motor in Prob. 3 has constant angular acceleration, what is its angular displacement during the 10-s time interval required for it to attain full rotational speed?

5. A grinding wheel has an initial angular velocity of 200 rad/s when the driving motor is shut off. After 20 s, its angular velocity is 120 rad/s. If the angular acceleration is constant, what is the angular displacement of the wheel during this 20-s interval? What is the angular acceleration of the wheel? Assuming that the angular acceleration remains constant, find the total number of revolutions the wheel makes in coming to rest. How long will it take the angular velocity to drop from 120 rad/s to zero?
Ans: 3200 rad; -4 rad/s^2; 796 rev; 30 s.

6. A grinding wheel starts from rest and moves with a constant angular acceleration of 20 rad/s^2 for 30 s. What is its final angular velocity? What is its total angular displacement?

7. A grinding wheel 2 f in diameter has a constant angular velocity of 40 rad/s. What is the tangential velocity of a point on the rim? What is the radial acceleration of the point? What is the total path length traversed by the point in 20 s? What is the displacement of the point after the wheel has made 500 complete revolutions? Ans: 40 f/s; 1600 f/s^2; 800 f; 0.

8. At what angular velocity would a point on the rim of the grinding wheel of Prob. 7 have a radial acceleration of 9 g_s? What would be the tangential speed of the point under this condition?

9. A flywheel with 2-f radius has an angular velocity of 30 rad/s and an angular deceleration of 3 rad/s^2. Find the magnitudes of the radial and tangential components of the acceleration of a point on the rim of this wheel.
Ans: 1800 f/s^2; 6 f/s^2.

10. A flywheel has a radius of 60 cm. The radial and tangential components of acceleration of a point on the rim have equal magnitudes. If the magnitude of each component is equal to g_s, what are the angular velocity and angular acceleration of the wheel?

11. Consider the wheel shown in the figure. If the radius of the wheel is 1.5 f and the force $F=20$ p, what is the displacement of the end of the string while the wheel is making 1 rev counterclockwise? How much work is done by the force? What is the torque L? What is the angular displacement θ of the wheel during 1 rev? Verify arithmetically that the work done is given by the product $L\theta$.
Ans: 3π f; 60π fp; 30 fp; 2π rad.

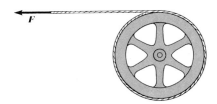

Problems 11–12

12. At what rate is work being done on the wheel in the figure when a force $F=5$ N is causing the end of the string to move toward the left at a velocity of 1.5 m/s? If the radius of the wheel is 0.3 m, what torque is associated with the force? What is the angular velocity of the wheel? Verify numerically that the rate at which work is being done is given by the product $L\omega$.

13. Show by integration that the rotational inertia of a solid disc about an axis through its center is, as shown in Fig. 8, $I=\frac{1}{2}mR^2$. Consider the disc as composed of thin rings of radius r, thickness dr, and mass dm. The contribution dI to I by each ring is $dI=r^2\,dm$; express dm in terms of the material density ρ and the volume of each ring.

14. Show by integration that the rotational inertia of a uniform slender rod about a transverse axis through its center is $I = \frac{1}{12}ml^2$ as shown in Fig. 8. Consider the contribution $dI = X^2\, dm$ for a typical mass element at a distance X from the axis, and express dm in terms of the mass per unit length. Show similarly that the rotational inertia of a slender rod about a transverse axis *through the end* is $\frac{1}{3}ml^2$. Verify the parallel-axis relation, given later, on p. 189.

15. What is the rotational inertia of a uniform slender rod 4 f long about an axis through its center and perpendicular to the rod if the mass of the rod is 0.3 sl? Ans: 0.399 sl·f².

16. A 50-kg flywheel with a rotational inertia of 115 kg·m² has an angular velocity of 50 rad/s. What is its rotational kinetic energy? What would be the velocity of a 50-g bullet having translational kinetic energy equal to the rotational kinetic energy of the flywheel?

17. The flywheel in the figure has a mass of 50 kg and a rotational inertia of 25 kg·m². A 3-kg block of steel is attached to the cord wound around a hub of radius 20 cm. The wheel is initially at rest when the steel block is released and allowed to move down a distance of 4 m. What is the resulting decrease in the potential energy of the system? In the absence of friction, what kinetic energy would the system acquire and what would be the angular velocity of the flywheel?

Ans: 118 J; 118 J; 3.07 rad/s.

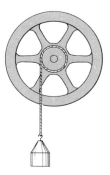

Problems 17–18

18. A flywheel with a rotational inertia of 150 kg·m² has an angular velocity of 10 rad/s. What is the kinetic energy of the flywheel? By means of a cord attached to the hub of this flywheel and a clutch arrangement, it is possible to use the kinetic energy of the flywheel to raise a 20-kg load. Through what vertical distance could the load be raised before the wheel comes to rest? Assume that frictional effects are negligible.

19. A grinding wheel with rotational inertia of 0.75 sl·f² has an angular velocity of 12 rad/s. After the driving motor is turned off, the wheel comes to rest in 1 min. Assuming a constant angular deceleration, determine the angular displacement of the wheel while its speed is decreasing, and the resultant frictional torque.

Ans: 360 rad; −0.159 p·f.

20. A 4-kg mass is joined to a 2-kg mass by a string that passes over a pulley of mass 7 kg, radius 14 cm, and a rotational inertia of 0.085 kg·m². If there is no friction at the bearings and no slipping of the string, find the acceleration of the masses and the tensions in the two parts of the string.

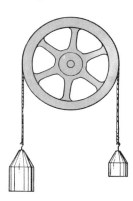

Problems 20–23

21. A light inextensible string hangs taut over a frictionless pulley of mass 12 kg, rotational inertia 0.2 kg·m², and radius 50 cm. A 0.5-kg mass resting on the floor is attached to one end of the string. A 1.5-kg mass is attached to the other end of the string at a point 2 m from the floor and allowed to fall from rest to the floor, thereby lifting the 0.5-kg mass. From the principle of conservation of energy, compute the speed with which the 1.5-kg mass strikes the floor. Ans: 3.74 m/s.

22. Repeat Prob. 21 for the case in which the string has a total mass of 0.5 kg.

23. Masses of 2 kg and 3 kg are connected by a light cord that passes over a pulley of 12-cm radius. The downward acceleration of the 3-kg mass is observed to be 1.2 m/s². Assuming that the cord does not slip and that the bearings of the pulley are frictionless, compute the rotational inertia of the pulley.
Ans: 0.0456 kg·m².

24. In the figure, a weight w is attached to a string wrapped around a solid cylinder of mass M mounted on a frictionless axle at O. If the weight starts from rest and falls a distance h, show that its speed is independent of the radius R of the cylinder.

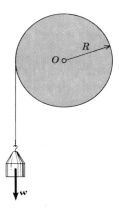

Problem 24

25. The driving belt exerts a constant resultant torque of 6.3 p·f on a flywheel with a mass of 4 sl and a rotational inertia of 4 sl·f². Bearing friction exerts a torque of magnitude 0.3 p·f. What will be the angular acceleration of the wheel? How long will be required for the wheel to start from rest and attain an angular velocity of 45 rad/s? Ans: 1.5 rad/s²; 30 s.

26. After the flywheel in Prob. 25 has attained an angular velocity of 45 rad/s, the driving belt slips off. Assuming that bearing friction remains unchanged, find the time required for the wheel to come to rest and the total angular displacement of the wheel during deceleration.

27. Two gears are meshed for transmission of mechanical power, as in Fig. 12. The gear attached to the motor shaft has 16 teeth; the gear attached to the load shaft has 64 teeth. The torque ratio is 3.5. What is the efficiency of this transmission? Ans: 87.5 percent.

28. In accelerating an automobile equipped with a hydraulic torque converter, the engine shaft is turning at 1800 rev/min and the drive shaft is turning at 400 rev/min. If the efficiency is 90 per cent, what is the torque ratio?

29. If a ½-hp motor operating at full load, with its shaft turning at 600 rev/min, supplies power to the transmission described in Prob. 27, what torque is applied to the input shaft? Find the rotational speed and the torque transmitted to the load. What power is supplied to the drive shaft?
Ans: 4.38 p·f; 150 rev/min; 15.3 p·f; ⁷⁄₁₆ hp.

30. If the automobile engine driving the torque converter described in Prob. 28 is delivering 70 hp, what torque is imparted to the engine shaft? to the drive shaft? How much power is transmitted to the drive shaft?

31. In the belt drive shown in the figure, the small pulley attached to the driving motor has a diameter of 4 i, the large pulley attached to the machine being driven has a diameter of 16 i. If the motor turns at 1600 rev/min, what is the angular speed of the large pulley? What is the ideal torque ratio? If the efficiency of the belt drive is 95 per cent and the motor delivers 20 hp, what torque is applied to the shaft of the machine being driven?
Ans: 400 rev/min; 4; 249 p·f.

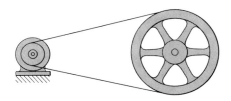

Problem 31

32. A motor turning at 3600 rev/min acts through a gear box and pulley system to lift an elevator weighing 5000 p at a constant speed of 18 f/s. If the efficiency of the power transmission system is 72 per cent, what are the required torque and horsepower of the motor?

33. In Sec. 7, the efficiency of a transmission is defined as the ratio of power output to power input. Show that the efficiency can also be expressed as the ratio of useful work done to the energy supplied. In the case of a transmission, no energy is stored; in the case of a spring clock drive or some other device in which energy *is* stored, which definition of efficiency should be used?

34. A rope brake is used to load a motor in a test of its power output. The motor is equipped with a flanged water-cooled pulley, and 1½ turns of rope are taken around the pulley. The two ends of the rope are brought vertically down to the floor where they are attached to spring balances. A suitable arrangement is provided to tighten the rope in order to load the motor.

(*a*) Draw a sketch of this rope brake.

(*b*) If the centers of the two vertical ropes are 18 i apart and the tensions are 125 and 325 p when the motor is turning at 480 rev/min, what horsepower is being delivered by the motor?

(*c*) If the electrical input to the motor is measured by electrical instruments as 12.2 kW, what is the efficiency of the motor?

35. An automobile on a circular race track of radius 2000 f is traveling at a speed of 100 f/s and is increasing its speed at the rate of 4 f/s². What is the magnitude of its acceleration? What is the angle represented by ϕ in Fig. 4(a)?
Ans: 6.40 f/s², $\phi = 38°6$.

36. A car on the racetrack described in Prob. 35 is moving at a speed of 80 f/s and is increasing its speed at such a rate that the radial and tangential components of acceleration are equal. At what rate is its speed increasing? What is the magnitude of the automobile's acceleration?

37. A car starts from rest on a circular race track of radius 1000 f and increases its speed at the rate of 3.6 f/s². How long will be required to attain the speed at which the tangential and radial components of the car's acceleration are equal? How far does the car move along the track before the radial and tangential components of its acceleration are equal?
Ans: 16.7 s; 500 f.

38. A point P on the rim of a wheel starting from rest has the position (*see* figure) $\theta = (\pi \text{ rad/s}^2) t^2$. If the radius of the wheel is ½ m, find the magnitude and direction of the velocity and acceleration of P at $t = 0$, ½, and 1 s.

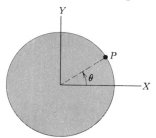

Problems 38–39

39. For the motion of point P described in Prob. 38, compute the magnitudes and directions of the velocity and acceleration at $t = ½$ s in the following two ways, and compare the results: (*a*) from the X- and Y-components computed from (11) and (14); (*b*) from the radial and tangential components computed from (13) and (16).

40. Back in the days when heavy naval guns had to be hand-cranked to change elevation, it was relatively easy to keep a gun that was facing broadside continuously aimed at a target in spite of the roll of the ship. Explain why this was so, in view of the very large rotational inertia of the gun about its elevation axis.

9 STATICS

Statics is the branch of mechanics that deals with the balance of forces on an object that remains at rest or in a state of uniform motion.

The principles of statics form the basis of much of the work of mechanical, civil, and architectural engineers engaged in machine, structure, and building design. These principles apply to cases in which forces act on a body in translational and rotational equilibrium—that is, on a body that has neither translational nor rotational acceleration.

In this chapter, we shall state the conditions for equilibrium and learn to apply them. In doing this we shall consider more carefully the computation of torques associated with forces. In computing the torques associated with gravitational forces, we introduce the concept of *center of mass* for an extended body and show how the position of the center of mass is determined. The chapter closes with a discussion of the vector properties of torque. The *torque vector* can be determined by computing the *vector product* of two vector quantities. This type of vector multiplication finds wide application in physics and engineering.

1. Translational Equilibrium

We are now in a position to give a rigorous proof of the

CONDITION FOR TRANSLATIONAL EQUILIBRIUM: *The vector sum of the external forces acting on a dynamical system at rest or in translational motion at constant velocity must be zero.*

This statement is slightly more general than the condition given on p. 87, since we have substituted *dynamical system* (see p. 133) for *rigid body*. For example, a quantity of liquid in a pail is a dynamical system but hardly a rigid body.

The Newtonian principle of relativity shows that translational motion at constant velocity is equivalent to being at rest, since there is always some observer relative to whom a dynamical system in translational motion *is* at rest, and he will observe the same set of forces.

Now consider an observer relative to whom the system *is* moving at constant velocity v. Let the *external forces* acting on the system be $F_1, F_2, F_3, \cdots$. Then, from (28b), p. 121, the power delivered by these external forces is

$$F_1 \cdot v + F_2 \cdot v + \cdots = (F_1 + F_2 + \cdots) \cdot v. \tag{1}$$

Since the internal forces (*see* p. 133) occur in equal and opposite pairs, the corresponding sum for the internal forces vanishes and they deliver no power. But *the external forces deliver no power either*, because the kinetic energy is not increasing; hence (1) must vanish. The expression (1) will vanish only if the sum of the external-force components in the direction of v vanishes. But since v can be chosen, by proper choice of observer, in any direction we please, the sum of the force components in any direction must vanish, and we conclude that, for translational equilibrium,

$$F_1 + F_2 + F_3 + \cdots = 0, \tag{2}$$

which expresses the condition given above.

2. Rotational Equilibrium

Even though the forces acting on a rigid body satisfy the condition for translational equilibrium, they may still be such as to cause the body to rotate. We shall now consider the *condition for rotational equilibrium*, which must be satisfied if a rigid body is to have no tendency to rotate if at rest or to change its angular velocity if rotating.

Let us first consider more carefully the definition of torque. Consider, in Fig. 1, a body mounted on a fixed axle and acted on by a force F, which is shown resolved into two vector components. The component

F_R, directed radially away from the axis, clearly has no tendency to cause rotation about O; only the component F_T tends to cause rotation. The torque of F_T is $L = F_T R$. Since $F_T = F\cos\phi$, and $R = l/\cos\phi$, we can write the torque of F as

$$L = F_T R = (F\cos\phi)(l/\cos\phi) = Fl.$$

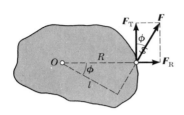

Fig. 1. Definition of torque: $L = F_T R = Fl$.

Note that l is the *perpendicular* distance from the line of action of F to the axis of rotation; hence we see why the word perpendicular occurs in the definition of torque given on p. 152, although the discussion that led up to that definition comprehended only forces that were normal to the radius at their point of application. A force F may have a third component, perpendicular to the paper in Fig. 1 and parallel to the axis. This component clearly has no effect on rotation about the axis. A complete definition of torque is thus:

> The **torque** of a force, about a given axis, is the product of the magnitude of the component force in a plane perpendicular to that axis by the *lever arm* of the force relative to the given axis.

Here lever arm is defined as follows:

> The **lever arm** of a force relative to a given axis is the perpendicular distance from the axis to the line of action of the force.

The distances l_1, l_2, l_3 shown in Fig. 2 are, respectively, the lever arms of forces F_1, F_2, F_3 applied to the hinged bar, relative to a horizontal axis through the hinge, perpendicular to the paper. A simple experiment would show that the forces required to keep the bar horizontal in the three cases are such that the *torques* $F_1 l_1$, $F_2 l_2$, and $F_3 l_3$ are equal.

If a rigid body is to have zero angular acceleration, the resultant torque must be zero. This gives the

> CONDITION FOR ROTATIONAL EQUILIBRIUM: *The sum of the counterclockwise torques about any axis must equal the sum of the clockwise torques about the same axis.*

When we come to apply this condition to determine the force required in Fig. 2 to hold up the horizontal bar, we see that the clockwise torque arises from gravitational forces *distributed* over the length of the bar. The resultant of these distributed gravitational torques can be computed if we locate the *center of mass* of the bar, to be defined in the next section.

When a rigid body is mounted on a real axle, the only forces tending to cause rotation are component forces perpendicular to the axle. There are many interesting problems in which a rigid body is not mounted on an axle but is in equilibrium under a system of forces all in the same plane (*coplanar forces*) as in Probs. 4, 7, 8, pp. 175–176. In these cases we can apply the condition for rotational equilibrium to any imaginary axis perpendicular to the plane of the forces. Since the body is in equilibrium, it does not tend to rotate about any axis whatsoever, so the torques must balance about any axis we choose. Examples will follow after we have discussed the torque arising from gravitational forces.

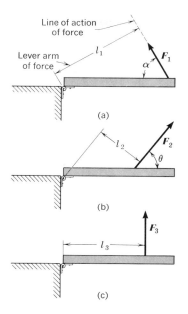

Fig. 2. The lines of action and the lever arms for several forces acting in a vertical plane on a horizontal bar.

3. Center of Mass

If an object is very small and can be regarded as a particle, its weight is simply a single force exerted vertically downward. If the object is extended, each of the constituent particles of the object experiences a downward gravitational force, and the entire object is subjected to a set of parallel downward forces.

When a rigid body is suspended by a single string, it is observed that the string hangs vertically. This familiar fact follows from the condition for translational equilibrium if we remember that the forces of gravity on all the particles of the body act vertically downward and have no horizontal components. Hence, in equilibrium the force exerted by the string must be vertically upward and can have no horizontal component.

An object suspended by a single string assumes an orientation in which it is in equilibrium. When the point of attachment of the string to the body is changed, the body assumes another equilibrium orientation. If the extended vertical line of the supporting string is marked in the body when it is in equilibrium for each of several points of attachment, all the lines so obtained are found to intersect in a common point called the *center of mass* (or the *center of gravity*) of the body. This point is particularly easy to find for a thin flat piece of material such as a piece of sheet metal.

The center of mass of an object is a point fixed relative to the object, but not necessarily inside the material of the object. The center of mass of a symmetric object such as a sphere, an automobile tire, or a meter stick lies at its geometrical center, whereas the center of mass of a tapered pole is a point on the symmetry axis between the ends but closer to the thicker than to the thinner end.

The position X_C of the center of mass of the tapered rod in Fig. 3 is at the point at which the rod will be in equilibrium under the action of the single upward force F and the downward forces of gravity on the particles of the rod. We consider the rod divided into a large number

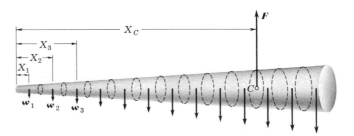

Fig. 3. A rod suspended by a single string attached at the center of mass so that the rod is 'balanced.'

of portions of weights w_1, w_2, w_3, $\cdots$ centered at distances X_1, X_2, X_3, $\cdots$ from the end of the rod, as in Fig. 3. The larger the number and the smaller the sizes of the portions into which the rod is divided, the more accurately does such a force system represent the action of gravity.

Now apply the equilibrium conditions to the rod of Fig. 3. Balancing the magnitudes of the vertical forces, we find that

$$F = w_1 + w_2 + w_3 + \cdots = W,$$

where we denote the whole weight of the rod by W. The condition of rotational equilibrium about an imaginary axis through the left end of the rod and perpendicular to the paper gives the equation

$$FX_C = WX_C = w_1 X_1 + w_2 X_2 + w_3 X_3 + \cdots. \tag{3}$$

This equation shows that

The resultant torque of the forces of gravity on the various parts of a body is the same as the torque of a single force equal to the whole weight concentrated at the center of mass.

The statement in italics is true for torque about any axis whatsoever. It is proved in (3) only for an axis A through the left end of the rod, but we can readily show that the statement will hold for any other horizontal axis. Consider, for example, an axis B a distance b further to the left, as in Fig. 4. Add the constant Wb to both sides of (3) to obtain

$$WX_C + Wb = w_1 X_1 + w_2 X_2 + \cdots + Wb.$$

On the right side substitute $w_1 + w_2 + \cdots$ for W:

$$W(X_C + b) = w_1(X_1 + b) + w_2(X_2 + b) + \cdots. \tag{4}$$

The expressions in parentheses are the lever arms of W, concentrated at the center of mass, and of $w_1, w_2, \cdots$, about the axis B; therefore (4) proves the statement given above in italics for any axis such as B. The statement may be similarly proved for any horizontal axis whatsoever.

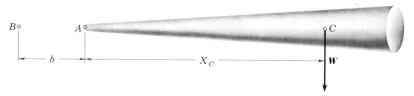

Fig. 4. The whole weight may be concentrated at the center of mass in computing the torque resulting from gravity.

According to (3), the distance X_C of the center of mass from the left end of the rod of Figs. 3 and 4 is given by

$$X_C = \frac{w_1 X_1 + w_2 X_2 + \cdots}{W}. \tag{5}$$

This equation is only an approximation when the rod is divided into a finite number of pieces. The center of mass is actually the limit of this expression as the number of pieces goes to infinity and the size of each piece goes to zero. Thus, the correct expression for X_C is obtained from the integral

$$X_C = \frac{1}{W} \int X \, dw \tag{5'}$$

where dw is the weight of an infinitesimal element at distance X from the axis and the limits of integration include the entire body.

The center of mass plays an important role in the study of dynamical systems, whether or not gravitational forces are involved. We have used weights rather than masses in the previous discussion, but at a given location on the earth, weight is proportional to mass ($w = mg$), so we could substitute masses for weights in the ratio on the right of (5) without altering its value. This leads us to the general definition of the space coordinates of the center of mass of any system of particles, which may or may not constitute a rigid body:

The coordinates X_C, Y_C, Z_C of the center of mass of a system of particles (Fig. 5) of mass m_1 at X_1, Y_1, Z_1; mass m_2 at X_2, Y_2, Z_2; $\cdots$; are given by

$$\left. \begin{array}{l} X_C = (m_1 X_1 + m_2 X_2 + \cdots)/M, \\ Y_C = (m_1 Y_1 + m_2 Y_2 + \cdots)/M, \\ Z_C = (m_1 Z_1 + m_2 Z_2 + \cdots)/M, \end{array} \right\} \tag{6}$$

where $M = m_1 + m_2 + \cdots$ *is the total mass.*

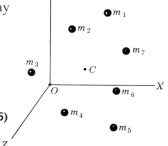

Figure 5

For a continuous body, these expressions can be replaced by integrals similar to the one in (5′).

It is not immediately obvious that expressions (6) determine the *same point C* no matter where the origin of coordinates is placed or what the orientations of the coordinate axes may be. That this property is indeed true is readily demonstrated by writing (6) in vector form:

$$M\mathbf{R}_C = m_1 \mathbf{R}_1 + m_2 \mathbf{R}_2 + \cdots, \tag{6′}$$

where the $\mathbf{R}$'s are radius vectors from the origin O in Fig. 5. From the geometrical definition of vector addition, we immediately see that $\mathbf{R}_C$ is independent of the orientation of the coordinate axes. Also, just as we did in (4) in one dimension, we can add a constant vector $O'O$, leading from a new origin O' to the origin O, to each of the vectors in (6), and thus demonstrate that the point C is independent of the origin chosen for its computation.

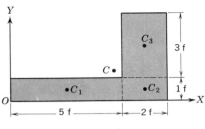

Figure 6

Example. *A piece of plywood is cut in the shape shown in Fig. 6. Find the coordinates of its center of mass. In what orientation will this board hang if suspended by a single string at the corner O?*

As indicated by the broken lines in Fig. 6, we can divide the board into three rectangular pieces each of which has its center of mass at its center. If we denote the mass of one square foot of the plywood by σ, these three pieces have the following masses and centers of mass:

$$m_1 = 5\,\sigma \qquad X_1 = 2.5 \text{ f}, \qquad Y_1 = 0.5 \text{ f};$$
$$m_2 = 2\,\sigma, \qquad X_2 = 6.0 \text{ f}, \qquad Y_2 = 0.5 \text{ f};$$
$$m_3 = 6\,\sigma, \qquad X_3 = 6.0 \text{ f}, \qquad Y_3 = 2.5 \text{ f}.$$

The whole mass $M = 13\,\sigma$. To determine the coordinates X_C, Y_C of the center of mass of the whole board, we can consider the mass of each of these three pieces to be concentrated at its own center of mass and apply the first two relations of (6):

$$X_C = \frac{(5\,\sigma)(2.5 \text{ f}) + (2\,\sigma)(6.0 \text{ f}) + (6\,\sigma)(6.0 \text{ f})}{13\,\sigma} = 4.65 \text{ f};$$

similarly $Y_C = 1.42 \text{ f}.$

The coordinates of this point are marked C in Fig. 6; the point lies outside of the object proper.

To determine X_C and Y_C, it would not have been necessary to divide the board into three pieces as we have done. Division into two pieces, at either of the broken lines, would enable the computation to be carried through. The reader should try this simpler calculation.

When this board is suspended by a string attached at point O, it hangs as indicated in Fig. 7, with the center of mass directly below the point of support. The forces $\mathbf{F}$ and $\mathbf{W}$ are equal and opposite and have the same line of action so their moments about any axis cancel and the condition for rotational equilibrium is fulfilled. We can determine the angle ϕ from the relation

$$\tan\phi = \frac{Y_C}{X_C} = \frac{1.42\ f}{4.65\ f} = 0.305; \qquad \phi = 17^\circ\!.0.$$

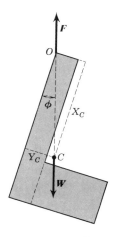

Figure 7

4. Equilibrium of a Rigid Body

We are now prepared to consider both rotational and translational equilibrium of a rigid body acted on by a system of forces. We have seen that a downward force equal to the whole weight of the body, acting at the center of mass, represents correctly the force of gravity in computing resultant torque about any axis. We shall consider the case of coplanar forces, in which all forces acting on the body lie in a common vertical plane. More general cases of noncoplanar forces do not introduce any new principles.

Let us consider first the 'teeter-totter' problem, which illustrates the comparison of masses by means of a *balance*. Let a beam (Fig. 8) be balanced on a knife-edge so that its center of mass is directly over the point of support. Masses m_1 and m_2 are placed on or hung from the beam with their centers of mass at distances S_1 and S_2 on either side of the knife-edge.

The forces acting on the combination of the beam and the two masses, considered as a single rigid body, are the weights w_1, w_2, and the weight of the beam, downward; and the single force F of the knife-edge upward. The condition for rotational equilibrium, if we take an axis at the knife-edge, gives:

$$m_1 g\, S_1 = m_2 g\, S_2 \qquad \text{or} \qquad m_1 S_1 = m_2 S_2. \tag{7}$$

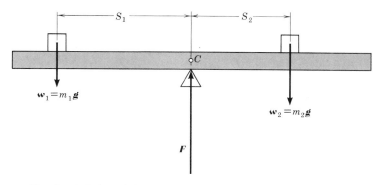

Fig. 8. A balanced beam carrying two masses. For equilibrium, $m_1 S_1 = m_2 S_2$.

This familiar solution, in which the larger mass must be closer to the axis to maintain balance, is known to every child.

Figure 8 illustrates how a balance may be used for accurate comparison of masses. If m_1 is a standard mass, the value of m_2 may be determined by accurate measurement of the distances S_1 and S_2.

In the chemical beam balance, the distances S_1 and S_2 are made equal and standard masses are added to one pan until they balance the unknown mass in the other pan. The physician's scale for determining a person's mass is similar in principle to Fig. 8, but with a more complex system of levers. Essentially, the moment arm of the person's weight is fixed, a standard mass is moved along a beam until balance is obtained, and the position of the standard mass gives a measure of the person's mass.

In spite of loose common terminology, a beam balance does not measure weight. It measures mass and gives, as seen in (7), a reading independent of the value of g. *Honest weight, no springs!* The springless scale is essentially a beam balance and is not measuring *weight* at all—it is giving the *standard weight* (p. 69), which is proportional to the *honest mass*.

Example. *On the earth, at a location where g has its standard value, a man 'weighs'* 150 p *on either a spring scale or a beam balance. What would be his 'weight' on the surface of the moon, as determined by each of these instruments? (Take the acceleration of the moon's gravity at its surface to be* 1/6 *that at the surface of the earth.*)

The beam balance gives standard weight—a measure of mass—and would continue to read 150 p on the moon. On the other hand, the spring scale would show the man's true weight as only 25 p. Although his mass has not changed, his decreased weight would be very apparent to the man; he could, for example, leap about six times as high as on the earth.

Now we shall give an example of a fairly complex problem involving the balance of forces acting in the vertical plane. Careful study of this example should precede the attempt to work the large variety of problems of this character given in the problem set.

Example. *The beam of Fig. 9 is hinged at a frictionless horizontal axle near the wall and is supported by a cable making an angle of* 60° *with the wall. It carries a weight of* 60 p *at its outer end. If the beam weighs* 200 p, *with center of mass* 3.6 f *from the axle, and the cable and weight are attached at* 8 f *from the axle, find the tension in the cable and the force exerted by the axle on the beam.*

In Fig. 9(b), the beam is isolated and all available data concerning the forces acting on the beam are noted. The forces of gravity on the particles of the beam are entirely represented, so far as any quantities entering the *equilibrium conditions*—total forces and torques—are concerned, by a single downward force of 200 p at the center of mass. The

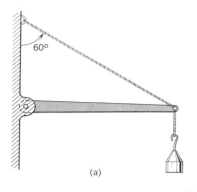
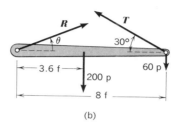

(a) (b)

Figure 9

weight of 60 p is represented by a downward force at the end of the beam, while the unknown tension T in the cable acts upward at an angle of 30° with the beam. The cable is a nonrigid body so it can exert only a force of pure tension in the direction of its length. The pin that attaches the beam to the wall is assumed frictionless, so it cannot exert *torque* that would assist in keeping the beam from swinging down in the absence of the cable. However it can (and must) exert *forces both outward and upward;* the resultant of these forces is represented by the reaction vector **R**, unknown in both magnitude and direction. There are thus three unknowns, T, R, and θ, to be determined from the three equilibrium conditions that we can write. In place of R and θ, it will be simplest first to determine R_H and R_V, the horizontal and vertical components of **R**. We shall proceed to write successively the conditions for equilibrium of the horizontal components of the forces, the vertical components of the forces, and the torques about an axis at the pin:

force components to the right = force components to the left
$$R_H = T \cos 30° = 0.866\ T, \tag{i}$$

force components up = force components down
$$R_V + T \sin 30° = 200\ p + 60\ p, \tag{ii}$$

counterclockwise torques = clockwise torques
$$(T \sin 30°)(8\ f) = (60\ p)(8\ f) + (200\ p)(3.6\ f).$$

The last equation can be solved for T directly to give
$$T = 300\ p.$$

When this value is substituted in (i) and (ii) we obtain $R_H = 260$ p, $R_V = 110$ p. From the values of these components we find
$$R = 282\ p, \qquad \theta = 22°.9.$$

5. The Torque Vector: Vector Product of Two Vectors

In our discussion thus far, we have not referred to torque as a vector, although we have seen that torque is not really a scalar because we must specify whether it is clockwise or counterclockwise. Torque actually has

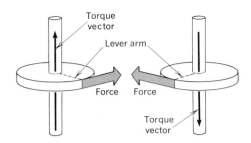

Fig. 10. Sense of the torque vector as given by the right-hand rule: *If one encircles the axis with the fingers of the right hand to indicate the direction of the tendency to rotate, the thumb points along the axis in the direction of the torque vector.*

the properties of a vector. In the case of torque about an axis, this vector is drawn along the axis in the sense shown in Fig. 10. In discussing the rotational motion of bodies that do not have a fixed axis, such as tops, gyroscopes, or airplanes, we need a more general definition of torque, namely, *torque about a point,* which is given in the next chapter.

The torque vector is an important example of the *vector product* of two vectors. Thus, in Fig. 11(a), we draw a 'radius vector' **R** from the axis through O to the point at which force **F** is applied. On the basis of our earlier discussion, we see that the magnitude of the torque is given by

$$L = RF \sin\theta.$$

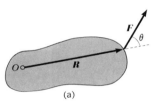

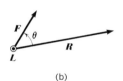

Figure 11

The correct *direction* for the torque vector **L**, determined as in Fig. 10, is given in Fig. 11(b), in which **R** and **F** have been placed tail-to-tail and the direction of **L** is indicated by a point ⊙ representing the point of an arrow pointing toward the reader. This direction of the torque vector corresponds to a counterclockwise torque. The torque vector **L** is perpendicular to both **R** and **F**. The 'product' of **R** and **F** that gives the vector **L** is called *vector product,* symbolized by

$$L = R \times F. \tag{8}$$

The **vector product** $A \times B$ of two vectors is a vector having magnitude $AB \sin\theta$, where θ is the angle less than or equal to 180° between the two vectors when they are placed tail-to-tail. The direction of $A \times B$ is perpendicular to the plane containing A and B in the sense of the advance of a right-hand screw as **A** is rotated through angle θ into **B**, as indicated in Fig. 12.

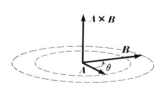

Figure 12

Because the vector product is denoted by a cross, it is sometimes referred to as the *cross product.* It should be noted that the vector product is *not* commutative; rather, we see from the definition that

$$A \times B = -(B \times A).$$

The vector product $A \times B$ is a *vector* quantity of magnitude $AB \sin\theta$ and should not be confused with the scalar or dot product $A \cdot B$, which is the *scalar* quantity $AB \cos\theta$ (*see* p. 104). We shall have occasion to make frequent use of both of these products in later chapters on electromagnetism.

Just as *torque* is a vector directed along the axis of rotation, so are *angular velocity, angular acceleration,* and *angular momentum;* we shall return to these points in the next chapter.

PROBLEMS

1. A horizontal bar hinged at a horizontal axis at the left end can be supported by a vertical force of 6 p acting at a point 18 i to the right of the hinge. What is the lever arm of this force about an axis coinciding with the hinge? What is the torque associated with the force?

Ans: 1.5 f; 9.0 p·f.

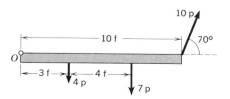

Problem 2

2. In the figure, a uniform horizontal bar 10 f long, weighing 5 p, is hinged about a horizontal axis at O and is acted on by three additional forces. Find the total torque tending to rotate the bar about O.

3. How large a force, applied 18 i to the right of the hinge, is required to support the bar of Prob. 1 in a horizontal position if the line of action of the force is in a vertical plane perpendicular to the axis at the hinge but makes an angle of 30° with the bar? What is the lever arm of this force? Show that your answers are independent of whether the 30° is measured from the left [like α in Fig. 2(a), p. 167] or from the right [like θ in Fig. 2(b)].

Ans: 12 p; 0.75 f.

4. If the uniform stick in the figure has weight w and bears a load F, show that

$$F_1 = \tfrac{1}{2} w + Fl_2/(l_1+l_2);$$
$$F_2 = \tfrac{1}{2} w + Fl_1/(l_1+l_2).$$

using these values of the forces, verify the balance of clockwise and counterclockwise torques about A; about B (at the center of the bar); about C; about D; and about E (a point at an arbitrary distance l_3 beyond the right end of the bar).

5. Demonstrate the statements in italics in the footnote on pp. 88–89 that discusses the tension in a string passing over a pulley.

6. A horizontal 20-f uniform plank weighing 40 p rests across two trestles each 4 f from an end of the plank, as in the figure. How close to the end of the plank can a 130-p painter walk?

Problem 6

7. On a uniform plank 10 f long weighing 20 p is placed a concentrated weight of 10 p, at 1 f from one end, and another concentrated weight of 7.5 p, at 3 f from this same end. Find the distance of the center of mass of plank plus weights from this same end.   Ans: 3.53 f.

8. A piece of plywood 4 f × 8 f weighing 20 p is lying horizontally. A brick weighing 4 p is placed near one corner of the piece with its center 1 f in from each edge. At what point should a vertical cord be attached to the plywood sheet so that it would 'balance' in a horizontal position when lifted by this single cord?

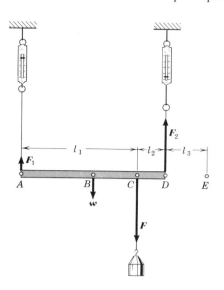

Problem 4

9. In the figure, a rigid stick 32 i long, of negligible weight, carries weights $F_2 = 10$ p, $F_3 = 10$ p, and $F_4 = 6$ p at points 8 i, 21 i, and 24 i from the left end. It is supported by vertical strings 2 i from the left end and 3 i from the right end. Find F_1 and F_5.

Ans: 11.9, 14.1 p.

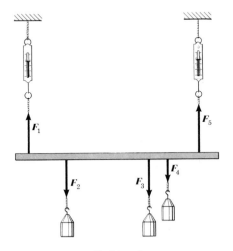

Problem 9

10. In the figure, a uniform stick 80 cm long weighing 20 N carries two weights and is supported in a horizontal position by two strings. If $\phi_1 = 60°$, find F_1, F_2, and ϕ_2 for equilibrium.

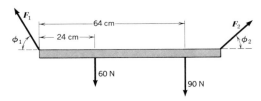

Problem 10

11. Find the center of mass of a circular sheet of metal 3 f in diameter which has a 1-f square cut out of it, two sides of the square lying along diameters of the circle. Ans: 0.116 f from center of the circle—in what direction?

12. A uniform bar 9 f long is bent so that a 3-f arm makes an acute angle of 45° with a 6-f arm. If the bent bar is hooked over a horizontal wire, what angle with the vertical will the long arm assume?

13. When the front wheels of an empty truck are placed on a scale platform, the scale reads 5000 p; with the back wheels on the platform, the scale reads 3000 p. If the distance between the front and rear axles is 10 f, what is the weight of the truck and where is its center of mass?

Ans: 8000 p; 3.75 f behind the front axle.

14. After the truck of Prob. 13 has been loaded, the scale reads 7000 p with the front wheels on the platform and 9000 p with the rear wheels on the platform. What is the weight of the load? Where is the center of mass of the load?

15. A flexible rope having a total weight of 7 p is attached to two rings on the ceiling, as indicated in the figure. At the mid-point of the rope is hung a weight of 3 p. If the angle the rope makes with the ceiling at the points of support is 45°, find the angle ϕ the rope makes with the horizontal at its mid-point. What is the tension in the rope near the points of support and near the mid-point?

Ans: 16°7; 7.07 p, 5.2 p.

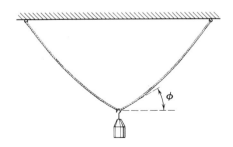

Problems 15–16

16. When the weight is removed from the rope in Prob. 15, the angle between the rope and the ceiling becomes 50°. Determine now the tension in the rope near the points of support and at the mid-point.

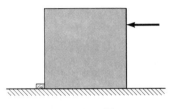

Problem 17

17. A cubical box 4 f on a side rests on the floor with one edge against a small cleat, as in the figure. If the box weighs 150 p and has its center of mass at its center, at what height must a 100-p horizontal force be applied to just tip the box? Ans: 3.00 f.

18. A uniform meter stick with a mass of 200 g has a 400-g mass suspended at the 20-cm mark and a 1200-g mass at the 60-cm mark, and is supported in a horizontal position by laboratory spring scales attached to the ends, as in Prob. 4. What is the reading of the spring scale at the 0-cm mark? the reading of the scale at the 100-cm mark? What would be the scale readings if the scales read in newtons, as they properly should, instead of in grams?

19. Answer all questions in Prob. 18 for a similar experiment performed on the surface of the moon at a place where $g = \frac{1}{6} g_s$. Ans: 150, 150 g; 1.47, 1.47 N.

20. Two identical ladders each 12 f long and each weighing 50 p, with center of mass 5 f from the bottom, are hinged together at the top and stand on a smooth horizontal floor. To prevent slipping, they are joined by links 2 f long fastened at points 3 f from the top ends of the ladders. A 140-p man stands on one ladder, 5 f from the top, measured along the ladder. Find (a) the push of the floor on each ladder, (b) the tension in each of the two links, (c) the force exerted by each ladder on the other at the hinge. *See* figure.

Problem 20

21. In the derrick illustrated, if the weight of the boom is $w = 1250$ p, the load F is 400 p, the center of mass of the boom is 15 f from its axle, the point of attachment of the supporting cable is 20 f from the axle and that of the load is 25 f from the axle, find the tension T in the cable and the magnitude and direction of the force R on the axle.
 Ans: $T = 1050$ p; $R = 1440$ p; $\theta = 51°.0$.

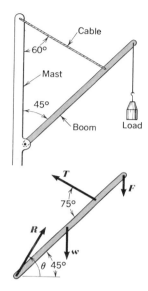

Problems 21-22

22. In the figure, if the boom weighs 800 p, the load F is 2400 p, the center of mass of the boom is 15 f from the axle, the cable is attached 30 f from the axle, and the load is attached 45 f from the axle, find the tension T in the cable and the magnitude and direction of the force R on the axle.

23. In the figure, the ladder is 20 f long, its weight is 150 p, and its center of mass is 7.5 f from its lower end. Its lower end rests on rough ground and its upper end leans at an angle of 32° against a smooth vertical wall. A man whose weight is 225 p stands 12 f up the ladder and exerts force F on the ladder. The smooth vertical wall can exert only a push P perpendicular to its surface against the top end of the ladder. Find the magnitude P of the push of the wall and the magnitude and direction of the force R of the ground.
 Ans: $P = 119$ p; $R = 393$ p; $\phi = 72°.3$.

24. In the ladder problem preceding, find the horizontal component of the ground reaction R as a function of the slant height S (measured along the ladder) to which the 225-p man has

178 STATICS CHAP. 9

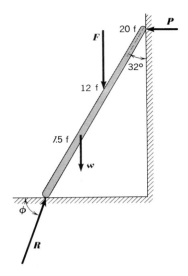

Problems 23–24

climbed. In particular, if 135 p is the greatest horizontal force that the ground can exert to keep the ladder from slipping, what distance S can the man climb before the ladder starts to slip?

25. The figure represents a ladder 20 f long weighing 80 p, with its center of mass 7 f from the bottom end. The base rests on a floor with $\mu_s = 0.6$. The top leans against a smooth wall (no friction). The force F is due to the weight of a 180-p man. (a) What is the greatest angle α the ladder can make with the wall and still permit the man to climb to a rung 19 f from

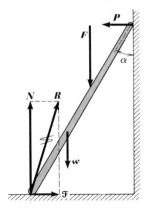

Problems 25–26

the base, measured along the ladder? (b) What is the ratio $\mathcal{F}/N$ when $\alpha = 20°$ and the man climbs to a rung 19 f from the base?

Ans: (a) 38°1; (b) 0.279.

26. Apply the condition of rotational equilibrium to the upper pulley of the differential chain hoist in Prob. 36, p. 126, to determine the ideal force ratio.

27. The box shown in the figure is 4 f high and 2 f wide. The weight of the box is 100 p, and its center of mass coincides with its geometrical center. The coefficient of static friction is 0.4. What horizontal force F is required to set the box in motion? What is the maximum height h at which this force can be applied without causing the box to tip over?

Ans: 40 p; 2.5 f.

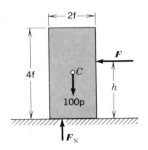

Problems 27–28

28. If no horizontal force is applied to the box in the figure, where does the effective normal force F_N of the floor act on the box? Suppose that a horizontal force $F = 40$ p acts on the box. Locate the position of F_N when the horizontal force is applied at heights $h = 1$ f, 2 f, 3 f, 4 f above the floor. At what height h would the force of 25 p have to be applied to make the box tip over?

29. Two *equal and opposite* forces acting on a body are said to constitute a *couple*. The torque exerted by a couple is given by the particularly simple rule:

> The torque exerted by a couple about any axis perpendicular to the plane containing the forces is the product of the magnitude of either force by the perpendicular distance between the lines of action of the forces.

Derive this rule for the situation shown in the figure.

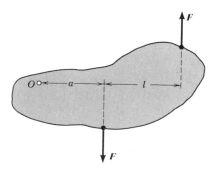

Problem 29

30. A Prony brake is a water-cooled brake used for loading a motor or an engine in tests of its power output. The brake is so arranged that the torque that the motor is exerting on the brake can be measured by means of a lever arm and a scale, as in the figure. If the lever arm $l = 4$ f, the scale reads 5 p with the motor at rest and 20 p when the motor is turning at 2000 rev/min, what horsepower is being delivered by the motor? The electrical input to this motor is measured by electrical instruments as 20 kW. What is the efficiency of the motor?

31. In the figure, a uniform bar rests on two identical blocks which in turn rest on a table. Show, from the laws of friction, that if either block is pushed toward the other, stopping

Problem 31

every centimeter, the bar will not tip over until the blocks are 1 cm or less apart. (Try the experiment with your two index fingers and a ruler.)

32. From the definition of vector product, show that this product is not commutative, but satisfies the relation

$$A \times B = -(B \times A) = (-B) \times A = B \times (-A).$$

Show that the magnitude of $A \times B$ is the area of a parallelogram with A and B as adjacent sides. Show that $(A \times B) \cdot C$ is similarly the volume of a parallelopiped.

33. Prove the following distributive property of the vector product:

$$A \times (B + C) = A \times B + A \times C.$$

This property is proved most simply by showing first that $A \times B = A \times$ (vector projection of B on a plane perpendicular to A). Then taking the paper as a plane perpendicular to A, draw the vector projections of B and C; also draw the vectors $A \times B$ and $A \times C$. The proof is then readily completed.

34. By logical arguments from the results of Probs. 32 and 33, show that the distributive property applies to any number of vectors:

$$(A + B + \cdots) \times (W + U + \cdots)$$
$$= A \times W + A \times U + \cdots + B \times W + B \times U + \cdots.$$

35. By expressing vectors A and B in terms of their vector components as in (6) and (7) on p. 105, expressing the cross products of these vector components in terms of scalar components, and employing the result of Prob. 34,

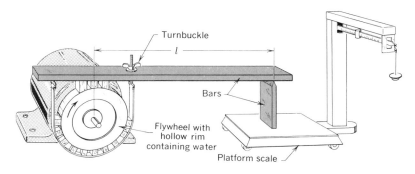

Problem 30

show that the scalar components of the vector $A \times B$ are

$$(A \times B)_X = A_Y B_Z - A_Z B_Y,$$
$$(A \times B)_Y = A_Z B_X - A_X B_Z,$$
$$(A \times B)_Z = A_X B_Y - A_Y B_X.$$

36. In the case of Fig. 11(a), where F is in a plane perpendicular to the axis, show that the torque about the axis is correctly given by expression (8), $L = R \times F$, for *any* vector R in the plane perpendicular to the axis, that terminates at *any point on the line of action of* F. The relation stated on lines 4 to 6 of Prob. 33 should be helpful.

37. In view of the fact (Prob. 35) that the Z-component of $A \times B$ involves only the projections of A and B on the XY-plane, show, using the result of Prob. 36, that if F *is not perpendicular to the axis,* the torque of F about the axis is given by the *component* of $R \times F$ in the direction of the axis, where R is a vector leading from *any point* on the axis to *any point* on the line of action of F.

DYNAMICAL SYSTEMS 10

The general description of the kinematics and dynamics of an extended body—even a rigid body—in three-dimensional motion turns out to be very complex mathematically and is usually treated only in advanced books on analytical dynamics. The complexity arises from the fact that a rigid body has *six degrees of freedom;* its motion involves translation along three perpendicular axes and rotation about each of these axes. No attempt at any general treatment is made in this text. We can, however, give a general treatment of the motion of a rigid body in two dimensions and consider the special case of three-dimensional gyroscopic motion.

We shall first derive certain general relations concerning the momentum of a dynamical system, as defined on p. 133; these relations determine the motion of the center of mass of the system, and hence the motion of the center of mass of a rigid body. Two-dimensional motion of a rigid body is then treated as a combination of translation of the center of mass and rotation about an axis through the center of mass; the special case of rolling motion is considered in some detail.

The principle of conservation of angular momentum is stated in more general terms than it was in Chapter 8, as is the relation between torque and angular momentum, both treated as vectors. Use of this relation enables us to give a brief discussion of a special case of three-dimensional motion: the kinematics and dynamics of *gyroscopes*.

The chapter closes with a discussion of the orbits of *earth satellites* and the trajectories of *ballistic missiles*, based on the joint use of the principle of conservation of energy and the principle of conservation of angular momentum.

1. The Momentum of a Dynamical System

We shall first derive a set of relations that are very important in the study of dynamical systems in general and of rigid bodies in particular. The first such relation is the following:

The total momentum of a dynamical system is the total mass of the system times the velocity of the center of mass of the system.

To prove this statement, let the particles of the system have masses m_1, m_2, $\cdots$, with total mass $M = m_1 + m_2 + \cdots$. Then the position of the center of mass is defined (*see* p. 170) by

$$M \mathbf{R}_C = m_1 \mathbf{R}_1 + m_2 \mathbf{R}_2 + m_3 \mathbf{R}_3 + \cdots = \Sigma\, m_n \mathbf{R}_n, \qquad (1)$$

with respect to the origin O in Fig. 1. By differentiation of (1) with respect to time, we obtain

$$M \frac{d\mathbf{R}_C}{dt} = m_1 \frac{d\mathbf{R}_1}{dt} + m_2 \frac{d\mathbf{R}_2}{dt} + \cdots,$$

or

$$M \mathbf{v}_C = m_1 \mathbf{v}_1 + m_2 \mathbf{v}_2 + \cdots, \qquad (2)$$

where $\mathbf{v}_C$ is the velocity of the center of mass. Since the right side of (2) is the total momentum of the system, this equation is the relation stated in italics above. From this relation, we draw the important conclusion that

When no external forces act on a dynamical system, the center of mass moves at constant velocity.

This conclusion follows from the fact that the total momentum is conserved; the total momentum equals the total mass times the velocity of the center of mass; hence the velocity of the center of mass does not change. Thus an object in free space, away from appreciable gravitational forces, may be tumbling or spinning, or changing

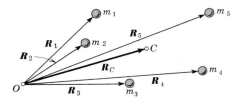

Figure 1

shape or size as a result of internal collisions or explosions or internal gravitational forces, but its center of mass moves in a straight line at constant speed.

2. Effect of External Forces on a Dynamical System

We now turn to the more general case in which *external* forces *do act* on a dynamical system. Here, again, the center of mass plays an important role, since we can prove that:

Irrespective of the internal forces that may act between the particles or bodies constituting a dynamical system, the center of mass of the system moves with acceleration F/M, where F is the vector sum of all external forces acting on all the particles of the system, and M is the total mass of the system.

The proof follows readily by differentiation of (2) with respect to time:

$$M\frac{dv_C}{dt} = m_1\frac{dv_1}{dt} + m_2\frac{dv_2}{dt} + \cdots,$$

or
$$M a_C = m_1 a_1 + m_2 a_2 + \cdots, \tag{3}$$

where a_C is the acceleration of the center of mass. Now $m_1 a_1$ is the resultant force acting on particle 1; let us write this as $(F_1)_{\text{Internal}} + (F_1)_{\text{External}}$, where these terms represent the total internal and the total external forces acting on particle 1; and write similar expressions for the forces acting on the other particles. Then we can write (3) as

$$M a_C = (F_1 + F_2 + \cdots)_{\text{Internal}} + (F_1 + F_2 + \cdots)_{\text{External}}.$$

But the internal forces between any two particles of the system are equal and opposite, by Newton's third principle. Hence, as we pointed out on p. 134, these forces cancel out in pairs and the whole sum of all internal forces is zero. If we let $F = (F_1 + F_2 + \cdots)_{\text{External}}$, we obtain the relations

$$M a_C = F, \qquad a_C = F/M, \tag{4}$$

which express the statement in italics above.

One important case is that in which the only external forces are due to gravity. In this case the external force F in (4) is the weight W of the system, and $a_C = W/M = g$.

Irrespective of the internal forces that may act between the particles or bodies constituting a dynamical system, the center of mass moves, under the force of gravity, with downward acceleration g.

Hence when a rigid body (for example, a chair) is thrown or projected, the *center of mass* moves in a parabolic trajectory like a single particle.

An interesting application of this result concerns the trajectory of

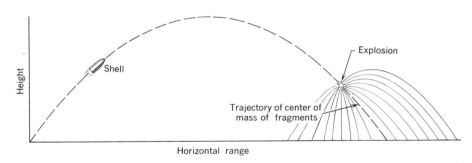

Fig. 2. Trajectory of an explosive shell, neglecting air resistance.

an exploding shell in the absence of air resistance. The first part of the trajectory has the familiar parabolic form shown in Fig. 2. When the shell explodes and shell fragments are blown out in all directions, the forces of the explosion are *internal* forces. Hence we can state that the center of mass of the shell fragments must complete the parabolic trajectory.

Gravitational potential energy. We shall now demonstrate an important theorem, again involving the center of mass, that enables us to compute the change in gravitational potential energy of any dynamical system:

The change in gravitational potential energy of a dynamical system near the earth may be computed by assuming that the whole mass is concentrated in a single particle that is located at the center of mass and moves with it, and computing the change in potential energy of this particle.

The change in potential energy is the negative of the work done by the forces of gravity. Hence the rate of change of potential energy is the negative of the rate at which gravity does work. To demonstrate the above theorem we need merely show that the rate at which gravitational forces do work on all the particles of any dynamical system is equal to $M\mathbf{g}\cdot\mathbf{v}_C$, the rate at which the total weight $W = Mg$ would do work on a single particle of mass M moving with the center of mass. The rate at which the gravitational forces do work on the particles is [*see* equation (28b), p. 121] $m_1 \mathbf{g}\cdot\mathbf{v}_1 + m_2 \mathbf{g}\cdot\mathbf{v}_2 + \cdots = (m_1 \mathbf{v}_1 + m_2 \mathbf{v}_2 + \cdots)\cdot\mathbf{g}$. But, by (2), this expression equals $M\mathbf{v}_C\cdot\mathbf{g} = M\mathbf{g}\cdot\mathbf{v}_C$, and the relation is demonstrated. Thus, if the center of mass of a body of mass M moves from height Y_{C1} to greater height Y_{C2}, the increase in potential energy is

$$\text{increase in } E_P = Mg\,(Y_{C2} - Y_{C1}). \tag{5}$$

3. Equilibrium of a Dynamical System

The dynamical system consisting of the pendulum of Fig. 3 and the earth is in stable equilibrium when the axis of the pendulum makes an angle $\theta=0$ with the vertical. This is its position of *minimum* potential energy because it is the position of *lowest* center of mass of the pendulum. When it is in this position, the pendulum-earth system will not tend by itself to change its configuration, and any stray external forces that might tend to change the configuration would have to do work to increase the potential energy; when such stray forces were removed, the internal forces would tend to return the pendulum to the position of minimum potential energy. The external force might result from a breeze or from jarring of the supports. This is an example of a dynamical system in stable equilibrium.

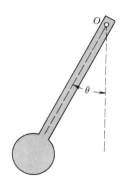

Fig. 3. A pendulum free to rotate about an axis through O.

> A dynamical system at rest is in **stable equilibrium** when any small change in configuration resulting from forces external to the system *increases* the potential energy of the system, and the resulting internal forces tend to return the system to its original configuration.

A second example of stable equilibrium is a cone *resting on its base;* a third is the sphere in Fig. 4 in the position marked 'stable.'

> A dynamical system at rest is in **unstable equilibrium** when any small change in configuration resulting from forces external to the system *decreases* the potential energy of the system, and the resulting internal forces tend to move the system far from its original configuration.

An example is the pendulum of Fig. 3 at $\theta=180°$. At this position, the system is in equilibrium—there is no torque—but any small disturbance from this position will result in torques internal to the pendulum-earth system that tend to send it toppling toward $\theta=0$. Another example of a system in unstable equilibrium is a cone *balanced on its point;* a third is the sphere of Fig. 4 in the position marked 'unstable.'

> A dynamical system at rest is in **neutral equilibrium** when any small change in configuration resulting from forces external to the system *does not change* the potential energy of the system, and no net internal force results.

A wheel mounted on a frictionless horizontal axle exactly through its center of mass is one example of neutral equilibrium; a sphere resting on a perfectly flat horizontal plane is another; so is a circular cone *resting on its side.*

We see that configurations of *stable, neutral,* and *unstable* equilibria

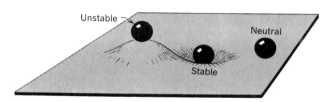

Figure 4

correspond to configurations of *minimum, constant,* and *maximum* potential energy, respectively. These configurations are illustrated in Fig. 4.

4. Two-Dimensional Motion of a Rigid Body

We shall now consider the two-dimensional motion of a rigid body that is *not* constrained to rotate about a fixed axle. The kinematical and dynamical description of this important type of motion is not difficult.

In **two-dimensional motion,** each particle of the body moves in a plane.

For example, each particle moves parallel to the XY-plane of Fig. 5; it has X- and Y-components of velocity but no Z-component. The motion of the body can best be described by the motion of the center of mass C plus a rotation about an axis through the center of mass and perpendicular to the XY-plane, as indicated in Fig. 5. A line that is painted on the body, such as CA in Fig. 5, makes an angle θ with the X-direction. The angle θ will in general vary with time; the angular velocity of the body is defined by $\omega = d\theta/dt$, while the angular acceleration is similarly defined by $\alpha = d\omega/dt$.

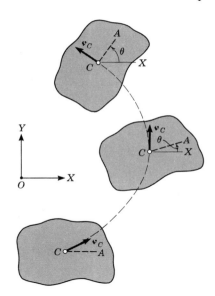

Fig. 5. Two-dimensional motion; C is the center of mass; the line CA is painted on the rigid body.

The *dynamical* behavior of rigid bodies in two-dimensional motion is determined by two relations that are derived directly from Newton's principles. The first, which governs the motion of the center of mass (even in the case of three-dimensional motion) was demonstrated in Sec. 2:

The vector sum of all forces acting on the body equals the mass times the acceleration of the center of mass.

The second relation governs the rotational motion about the center of mass. This relation, a generalization of the equation $L = I\alpha$ to the case where the center of mass is moving, will be demonstrated after we have derived the expression for kinetic energy. The relation is the following:

In two-dimensional motion, the resultant torque about an axis through the center of mass and normal to the plane of motion equals the rotational inertia about this axis times the angular acceleration.

We can apply these relations to the motion of a body of any shape that is cut out of a flat board and thrown so that the plane of the board coincides with the XY-plane of Fig. 5, the Y-axis being vertical, and so that the initial angular velocity is about an axis perpendicular to the XY-plane. (Or we might imagine similarly throwing a wheel, a disc, or a hoop into the air.) Air resistance neglected, the only external force acting is the force of gravity; hence the center of mass moves in a parabola like a projected particle. Since the forces of gravity have no net torque about the center of mass, which is the point at which the body would balance, there is no angular acceleration and the body continues to rotate at constant angular velocity equal to its initial angular velocity.

Kinetic energy. In two-dimensional motion, the kinetic energy is given by a simple relation:

The kinetic energy of a rigid body in two-dimensional motion equals the kinetic energy of a particle containing the whole mass and moving with the center of mass, plus the rotational kinetic energy computed as if the body were in pure rotation about the center of mass.

Thus the kinetic energy can be written as the sum of two terms: The first, called the *translational kinetic energy*, is the same as that of a particle moving with the speed v_C of the center of mass:

$$E_{\text{KT}} = \tfrac{1}{2} M v_C^2. \tag{6}$$

The second, called the *rotational kinetic energy*, is the same as the kinetic energy the rigid body would have if it were mounted on a fixed axle through its center of mass:

$$E_{\text{KR}} = \tfrac{1}{2} I_C \omega^2, \tag{7}$$

Where I_C is the rotational inertia about the axis through the center of mass. The total kinetic energy is thus

$$E_K = E_{\text{KT}} + E_{\text{KR}} = \tfrac{1}{2} M v_C^2 + \tfrac{1}{2} I_C \omega^2. \tag{8}$$

This relation is most readily derived by using vector analysis. In Fig. 6(a), let $\mathbf{R}_n$ be the vector from the origin to a representative particle of the body in Fig. 5. Then, since $v_n = d\mathbf{R}_n/dt$, as in (2), p. 35, and $v_n^2 = \mathbf{v}_n \cdot \mathbf{v}_n$, we can write the kinetic energy of the body as

$$E_K = \sum \tfrac{1}{2} m_n \frac{d\mathbf{R}_n}{dt} \cdot \frac{d\mathbf{R}_n}{dt},$$

where the summation is over all particles of the body. Since we can write $\mathbf{R}_n = \mathbf{R}_C + \mathbf{r}_n$, we find

$$E_K = \sum \tfrac{1}{2} m_n \left(\frac{d\mathbf{R}_C}{dt} + \frac{d\mathbf{r}_n}{dt} \right) \cdot \left(\frac{d\mathbf{R}_C}{dt} + \frac{d\mathbf{r}_n}{dt} \right),$$

$$E_K = \sum \tfrac{1}{2} m_n \frac{d\mathbf{R}_C}{dt} \cdot \frac{d\mathbf{R}_C}{dt} + \sum \tfrac{1}{2} m_n \frac{d\mathbf{r}_n}{dt} \cdot \frac{d\mathbf{r}_n}{dt} + \sum m_n \frac{d\mathbf{R}_C}{dt} \cdot \frac{d\mathbf{r}_n}{dt}. \tag{9}$$

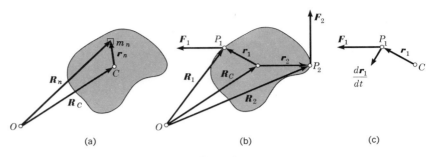

Figure 6

In (9), the last term can be written as

$$\frac{d\mathbf{R}_C}{dt} \cdot \frac{d}{dt}(\Sigma\, m_n\, \mathbf{r}_n),$$

in which the expression in parenthesis *vanishes* since, by definition (1), it is M times the vector leading from the origin of the $\mathbf{r}$'s to the center of mass—but the origin of the $\mathbf{r}$'s is itself the center of mass. The first term in (9) is

$$\tfrac{1}{2}(\Sigma\, m_n)\, \mathbf{v}_C \cdot \mathbf{v}_C = \tfrac{1}{2}\, M\, v_C^2 = E_{\mathrm{KT}}.$$

The second term in (9) is the kinetic energy E_{KR} as it would be computed if there were a fixed axle at the center of mass, since, by definition (2), p. 35, $d\mathbf{r}_n/dt$ is the velocity of the particle of mass m_n *relative to* C [cf. Fig. 6(c)]. Thus (8) is proved.

Derivation of $L_C = I_C\,\alpha$. From (8) we can prove the relation stated on p. 186, that *the resultant torque about the center of mass equals the rotational inertia I_C times the angular acceleration.* Let the external forces be $\mathbf{F}_1, \mathbf{F}_2, \cdots$, acting at points $P_1, P_2, \cdots$, as in Fig. 6(b). We can assume that these forces are in the plane of motion because, if there are forces normal to this plane, as in the two-dimensional case of a body sliding on the ice of a lake, these normal forces will merely balance since no particle of the body has an acceleration normal to this plane. Then the power delivered by the external forces is

$$P = \mathbf{F}_1 \cdot \mathbf{v}_1 + \mathbf{F}_2 \cdot \mathbf{v}_2 + \cdots = \mathbf{F}_1 \cdot \frac{d\mathbf{R}_1}{dt} + \mathbf{F}_2 \cdot \frac{d\mathbf{R}_2}{dt} + \cdots,$$

where $\mathbf{v}_1$ is the velocity of point P_1, etc. Just as above, we can write this expression as

$$P = \mathbf{F}_1 \cdot \frac{d\mathbf{R}_C}{dt} + \mathbf{F}_2 \cdot \frac{d\mathbf{R}_C}{dt} + \cdots + \mathbf{F}_1 \cdot \frac{d\mathbf{r}_1}{dt} + \mathbf{F}_2 \cdot \frac{d\mathbf{r}_2}{dt} + \cdots$$

$$= (\mathbf{F}_1 + \mathbf{F}_2 + \cdots) \cdot \mathbf{v}_C + \mathbf{F}_1 \cdot \frac{d\mathbf{r}_1}{dt} + \mathbf{F}_2 \cdot \frac{d\mathbf{r}_2}{dt} + \cdots.$$

By (4), the first parenthesis is $M\mathbf{a}_C$, so the first term is $M\mathbf{a}_C \cdot \mathbf{v}_C$. Now notice, Fig. 6(c), that since $d\mathbf{r}_1/dt$ is the velocity of P_1 in the rotational motion about C, it is a vector normal to $\mathbf{r}_1$ of magnitude $r_1 \omega$. The dot product of $\mathbf{F}_1$ and this vector is the tangential component F_{1T} times $r_1 \omega$. But $F_{1T} r_1 = L_1$, the torque of $\mathbf{F}_1$ about C. Hence $\mathbf{F}_1 \cdot d\mathbf{r}_1/dt = L_1 \omega$, and we can write the power in the form

$$P = M\mathbf{a}_C \cdot \mathbf{v}_C + (L_1 + L_2 + \cdots)\omega = M\mathbf{a}_C \cdot \mathbf{v}_C + L_C \omega,$$

where L_C is the resultant torque about the center of mass C. The final step is to equate this power to the rate of change of the kinetic energy (8):

$$E_K = \tfrac{1}{2} M\mathbf{v}_C \cdot \mathbf{v}_C + \tfrac{1}{2} I_C \omega^2, \qquad dE_K/dt = M\mathbf{v}_C \cdot \mathbf{a}_C + I_C \omega\alpha.$$

When we equate this expression to the power as given by the preceding equation, we see that the first terms cancel and we obtain the desired expression $L_C = I_C \alpha$. This completes the rigorous discussion of the dynamics of two-dimensional motion.

The parallel-axis relation. From (8), we can also derive an expression for the rotational inertia I_O of a rigid body about any axis O at a distance r from the center of mass, if we know the rotational inertia I_C about a parallel axis through the center of mass. Let the body in Fig. 7, mounted on a fixed axle at O, be rotating at angular velocity ω. Its kinetic energy is then, by (24), p. 153

$$E_K = \tfrac{1}{2} I_O \omega^2. \tag{10}$$

But we can also compute its kinetic energy from (8) by inserting the velocity of the center of mass as $v_C = r\omega$:

$$E_K = \tfrac{1}{2} M r^2 \omega^2 + \tfrac{1}{2} I_C \omega^2 = \tfrac{1}{2} (I_C + M r^2) \omega^2. \tag{11}$$

Comparison of (10) and (11) gives

$$I_O = I_C + M r^2. \tag{12}$$

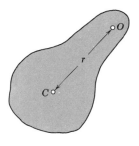

Figure 7

The rotational inertia of a rigid body about any given axis equals the rotational inertia about a parallel axis through the center of mass plus the mass times the square of the distance between the two axes.

This relation, called the *parallel-axis relation,* again indicates the important role played by the center of mass. We also see, from (12), that a rigid body has *minimum* rotational inertia about an axis through its center of mass and that its rotational inertia about any other parallel axis is greater.

Example. *A meter stick is freely pivoted on a horizontal nail at the 90-cm point, as in Fig. 8. If the stick is held horizontally and released, what will be its angular velocity after it has swung down through 90° and is vertical? How far will it swing up on the other side? Neglect friction.*

During the motion indicated in Fig. 8, the center of mass falls 0.4 m; hence, by (5), the decrease in potential energy is $mg\,(0.4\text{ m})$. The initial kinetic energy is zero. The kinetic energy at the bottom must therefore equal the decrease in potential energy. If the angular velocity is ω at the bottom, the kinetic energy is, by (10) and (12),

$$E_K = \tfrac{1}{2} I_O \omega^2 = \tfrac{1}{2} [I_C + m\,(0.4\text{ m})^2]\,\omega^2.$$

From Fig. 8, p. 154, I_C, about the center of a slender 1-m rod, is $\tfrac{1}{12} m \times (1\text{ m})^2 = (0.0833\,m)\text{ m}^2$. Hence

$$E_K = \tfrac{1}{2}\,[(0.0833\,m)\text{ m}^2 + (0.16\,m)\text{ m}^2]\,\omega^2$$
$$= (0.122\,m\,\omega^2)\text{ m}^2.$$

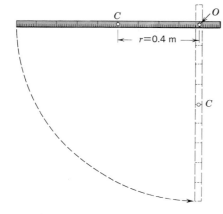

Figure 8

Equating this to the decrease in potential energy as determined by the relation on p. 184, in terms of the 0.4-m change in height of the center of mass, gives

$$0.122\,m\,\omega^2\text{ m}^2 = 0.4\,mg\text{ m}.$$

or
$$\omega^2 = 3.28\,g/\text{m}.$$

For $g = 9.8$ m/s², this becomes

$$\omega^2 = 32.1/\text{s}^2,$$
$$\omega = 5.67 \text{ rad/s}.$$

If there is no friction and hence no loss of mechanical energy, the meter stick will not come to rest until all of this kinetic energy has been reconverted into potential energy, which means that it must swing up into the horizontal position on the other side of the pivot before its angular velocity becomes zero.

Example. *An automobile traveling on a straight and level road brakes with a deceleration of* 16 f/s². *Assume that the car is a rigid body with a weight of* 3000 p, *that its wheelbase is* 10 f, *and that its center of mass is centered between the four wheels and is* 2 f *above the road. Determine the 'weights' N_1 and N_2 on the front and rear wheels, as in Fig.* 9.

Since the braking deceleration is $\tfrac{1}{2} g$, the forces of friction, $\mathcal{F}_1 + \mathcal{F}_2$, must total half the weight of the car, or 1500 p. The sum

$$N_1 + N_2 = 3000 \text{ p}.$$

These conclusions come from the relation stated on p. 186, that the total external force equals the mass times the acceleration of the center of mass.

Since the automobile is in two-dimensional motion, and has no angular acceleration, we conclude, again from p. 186, that the torques of the external forces about the center of mass must be in equilibrium:

$$(\mathcal{F}_1 + \mathcal{F}_2)(2\text{ f}) + N_2\,(5\text{ f}) = N_1\,(5\text{ f})$$
$$3000\text{ p} + 5N_2 = 5N_1$$
$$600\text{ p} = N_1 - N_2.$$

Combining this relation with $N_1+N_2=3000$ p, we find

$$N_1=1800 \text{ p}; \quad N_2=1200 \text{ p}.$$

The weight, which would be 1500 p on each set of wheels when the car is unaccelerated, is shifted to the front when the car is braked. This shift has an important bearing on the design of car brakes to insure lateral stability during braking operations. The decreased weight on the rear results in lower rear-wheel friction, which tends to cause the rear wheels to lock more readily than the front; with locked rear wheels, the car definitely becomes unstable and will swerve.

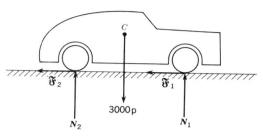

Figure 9

5. Rolling Bodies

One of the interesting cases of two-dimensional motion is that of bodies that roll on a surface without slipping. We shall confine our attention to the rolling of circular bodies such as wheels, cylinders, and spheres, in which the center of mass is at the center of the circle.

The condition that the body roll without slipping imposes a definite kinematical relation between the linear motion and the angular motion of the body. Figure 10 shows a cross section of a wheel rolling along a surface (not necessarily horizontal). Let the initial position of the axis through the center C pass through the origin of the coordinate system as shown. If the wheel rotates in a clockwise direction until one complete clockwise rotation has taken place, the center experiences a displacement $2\pi R$ in the positive X-direction; that is, the wheel must make one complete revolution in the time that its center advances a distance equal to the circumference of the wheel. Thus, when the clockwise angular displacement θ is 2π, the linear displacement of the center in the X-direction is $2\pi R$. Hence, for the rolling wheel, we see that

$$X=R\theta. \tag{13}$$

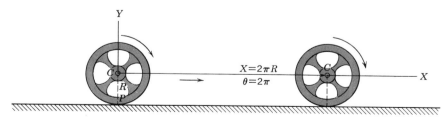

Fig. 10. When the rolling wheel turns through an angle of 2π rad, the center, C, experiences a displacement $2\pi R$ in the X-direction.

From this relation, we find by differentiation that

$$dX/dt = R\, d\theta/dt, \quad \text{or} \quad v_X = R\omega, \tag{14}$$

where v_X is the X-component of the velocity of the center of the wheel and ω is the angular velocity. A second differentiation gives

$$dv_X/dt = R\, d\omega/dt, \quad \text{or} \quad a_X = R\alpha, \tag{15}$$

where a_X is the X-component of the acceleration of the center and α is the angular acceleration. Note that θ, ω, and α are *positive in a clockwise sense* in these equations, rather than in the usual counterclockwise sense. This change of convention is useful for problems of rolling bodies.

Example. *Find the translational acceleration that results when a force $\mathbf{F}$ acts through the center of mass of the solid cylinder of radius R shown in Fig. 11, and find the force of friction $\mathcal{F}$.*

We first note that $w = N$, since the center of mass has no acceleration in the Y-direction. The equation of motion of the center of mass in the X-direction is

$$F - \mathcal{F} = ma_X. \tag{i}$$

The clockwise torque about the center determines the angular acceleration:

$$\mathcal{F}\, R = I_C\, \alpha, \tag{ii}$$

where I_C is the moment of inertia about the center of mass. Remembering that $I_C = \tfrac{1}{2}\, mR^2$ for a solid cylinder, and that $\alpha = a_X/R$, we may rewrite (ii) in the form

$$\mathcal{F} R = (\tfrac{1}{2}\, mR^2)(a_X/R), \quad \text{or} \quad \mathcal{F} = \tfrac{1}{2}\, ma_X.$$

Adding this equation to (i) and solving the resulting equation for a_X, we obtain

$$a_X = \tfrac{2}{3}\, F/m.$$

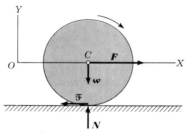

Fig. 11. A solid cylinder that rolls without slipping. The resultant of $\mathbf{F}$ and $\mathcal{F}$ causes a linear acceleration of the center of mass, C. The torque of the frictional force $\mathcal{F}$, about C, causes an angular acceleration.

The rolling cylinder accelerates two-thirds as fast as a particle having the same applied force. The angular acceleration is

$$\alpha = a_X/R = \tfrac{2}{3}\, F/mR.$$

The force of friction is $\mathcal{F} = \tfrac{1}{2}\, ma_X = \tfrac{1}{3}\, F$.

Example. *Using the results of the previous example show that if the cylinder starts from rest and moves a distance S to the right, the work done by $\mathbf{F}$ equals the increase in kinetic energy.*

We note that no work is done against the force of friction $\mathcal{F}$ since there is no sliding—no relative motion of the surfaces; $\mathcal{F}$ is of the nature of a force of static friction. In the ideal case of a rolling body, no energy is lost because of friction. In the real case, some energy is lost—a wheel rolling on a horizontal surface will gradually slow up, but much less rapidly than a sliding body—because there is some *distortion* of the surfaces. We shall consider only the ideal case.

The work done by F is $\quad W=FS$.

The translational velocity acquired is given by $v_X^2 = 2\, a_X S = \tfrac{4}{3}\, FS/m$; hence the translational kinetic energy is

$$E_{KT} = \tfrac{1}{2}\, m v_X^2 = \tfrac{2}{3}\, FS.$$

By (14), $\omega = v_X/R$; hence $\omega^2 = v_X^2/R^2 = \tfrac{4}{3}\, FS/mR^2$. The rotational kinetic energy is thus

$$E_{KR} = \tfrac{1}{2}\, I_C\, \omega^2 = \tfrac{1}{2}\, (\tfrac{1}{2}\, mR^2)(\tfrac{4}{3}\, FS/mR^2) = \tfrac{1}{3}\, FS.$$

The total kinetic energy $E_K = E_{KT} + E_{KR} = FS = W$, as was to be shown.

Example. *Consider a solid cylinder that starts from rest and rolls without slipping down the inclined plane of Fig. 12. Determine the acceleration of the center, the force of friction, the minimum coefficient of static friction, and the velocity at the bottom of the plane.*

The forces acting on the cylinder will be the force $\boldsymbol{w} = m\boldsymbol{g}$ of gravity, the normal reaction $\boldsymbol{N}$ of the plane, and whatever frictional force $\mathfrak{F}$ is required to prevent slipping. The center of mass moves down the plane *in a straight line*, with velocity and acceleration which we denote by v and a. The forces normal to the plane will balance, since there is no acceleration of the center of mass in this direction. Therefore

$$N = mg\,\cos\theta. \tag{i}$$

The net force down the plane will be related by (4) to the acceleration a of the center of mass:

$$mg\,\sin\theta - \mathfrak{F} = ma. \tag{ii}$$

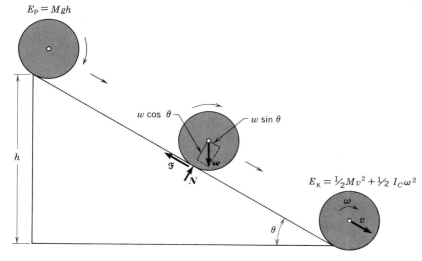

Fig. 12. A solid cylinder of radius R rolls without slipping.

The clockwise torque about the center of mass will equal the moment of inertia times the clockwise angular acceleration:

$$\mathcal{F}R = I_c \alpha. \tag{iii}$$

As before $\alpha = a/R$ and $I_c = \tfrac{1}{2} mR^2$, so (iii) gives $\mathcal{F} = \tfrac{1}{2} ma$. When this value is substituted in (ii), we find that

$$a = \tfrac{2}{3} g \sin\theta,$$

which is a constant acceleration two-thirds as great as the acceleration of a particle *sliding* down a *frictionless* plane of the same inclination. We also find that

$$\mathcal{F} = \tfrac{1}{3} mg \sin\theta,$$

so that we need a coefficient of *static* friction of at least

$$\mathcal{F}/N = \tfrac{1}{3} \tan\theta$$

to prevent slipping.

We can obtain the velocity of the cylinder at the bottom from the kinematical laws of uniform acceleration. The center of mass moves a slant distance $S = h/\sin\theta$ with acceleration $a = \tfrac{2}{3} g \sin\theta$. It therefore acquires velocity given by

$$v^2 = 2\,aS = \tfrac{4}{3} gS \sin\theta = \tfrac{4}{3} gh, \qquad v = 2\sqrt{\tfrac{1}{3} gh}.$$

Example. *Compute the speed v of the solid cylinder in Fig. 12 from considerations of conservation of energy.*

The cylinder at rest at the top of the hill has potential energy mgh. At the bottom of the hill this has been converted into translational kinetic energy $\tfrac{1}{2} mv^2$ plus rotational kinetic energy $\tfrac{1}{2} I_c \omega^2$:

$$mgh = \tfrac{1}{2} mv^2 + \tfrac{1}{2} I_c \omega^2.$$

Remembering that $I_c = \tfrac{1}{2} mR^2$ and $\omega = v/R$, we rewrite this equation as

$$mgh = \tfrac{1}{2} mv^2 + \tfrac{1}{4} mv^2 = \tfrac{3}{4} mv^2,$$

which gives the value

$$v = 2\sqrt{\tfrac{1}{3} gh}$$

for the magnitude of the final translational velocity of the center of mass. It will be noted that the principle of conservation of energy is a powerful tool that has enabled us to solve this problem with much less work than is involved in the direct application of the principles of motion in the preceding example.

Now, let us consider the instantaneous velocity of various points on a wheel rolling in the X-direction as in Fig. 10. We can regard the rolling as a combination of *translation* and *rotation about the axis through the center*. If we consider *translation only*, all points in the wheel have the same velocity v as the center; this is shown in Fig. 13(a). If we consider *rotation only*, the center is at rest, whereas point P at the top of the wheel has X-velocity $+R\omega$ and point O at the bottom of the wheel has X-velocity $-R\omega$; these are indicated in Fig. 13(b). Combining these X-components

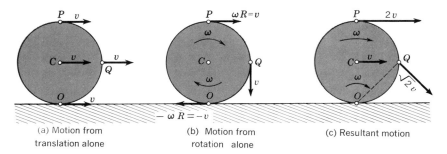

Fig. 13. Velocity components for four points in a rolling wheel. The combined effects of translation and rotation about an axis through C give the same result as a pure rotation with the same angular velocity about O.

of velocity, and noting that $R\omega = v$, we have

for point P: $\quad v_X = v + R\omega = 2v, \quad v_Y = 0;$

for point C: $\quad v_X = v \quad\quad\quad = v, \quad v_Y = 0;$

for point O: $\quad v_X = v - R\omega = 0, \quad v_Y = 0.$

These results are shown schematically in Fig. 13(c).

We note that the point that at any instant is at the bottom of the wheel is at rest. This must be so because this point is in contact with, and not slipping on, a surface at rest. Since the wheel is a rigid body, the velocities given in Fig. 13(c) are sufficient to determine that at any instant the whole velocity pattern is one of pure rotation about the point of contact, with angular velocity $\omega = v/R$, the same as the angular velocity in the rotation about C. A point such as Q therefore has velocity perpendicular to the line OQ, of magnitude ω times the distance OQ.

6. Angular Momentum

On p. 157, we defined the angular momentum J of a rigid body rotating about a fixed axis as

$$J = I\omega, \tag{16}$$

where I is the rotational inertia about the axis. In the more general case of two-dimensional motion, we define the *angular momentum about the center of mass* by the same equation, where now I is the rotational inertia *about the center of mass*. In either case, we may write

$$dJ/dt = I\, d\omega/dt = I\alpha = L. \tag{17}$$

We have now studied two cases for which the equation $L = I\alpha$ is valid—the cases of rotation about a fixed axis and of two-dimensional motion of rigid bodies. In the case of *pure rotation,* the torque and the rotational inertia must be taken *about the fixed axis of rotation;* in the case of *two-dimensional motion* these quantities must be taken *about the center of mass.* We shall assume these conventions as understood in the following statements of a definition, a relation, and a principle.

> The **angular momentum** of a rigid body in pure rotation or in two-dimensional motion is defined as the product of its rotational inertia by its angular velocity.

Equation (17) states that

> *The time rate of change of the angular momentum of a rigid body in pure rotation or in two-dimensional motion is equal to the resultant externally applied torque.*

If there is no applied torque, the angular momentum is constant:

> PRINCIPLE OF CONSERVATION OF ANGULAR MOMENTUM: *If no external torque acts, the angular momentum of a body rotating about a fixed axis, or in two-dimensional motion, is constant.*

In the statement of the principle of conservation of angular momentum, we have deliberately omitted the word 'rigid,' because the principle turns out to be more general than our derivation implies and is found (*see* Prob. 39) to apply also to a nonrigid body that may change its shape and hence its rotational inertia as the rotation is taking place. If, in the absence of external torque, a rotating body changes shape so that the rotational inertia changes from I to I', the angular velocity will change from ω to ω' in such a way that the angular momentum is unchanged:

$$J = J', \quad \text{therefore} \quad I\omega = I'\omega'. \tag{18}$$

As examples of the application of this relation to the case of rotation about a fixed axis, consider the case of a man standing on a small rotating platform, or a toe dancer spinning on one toe, or a spinning skater. If the person extends his arms so that his rotational inertia about the vertical axis of rotation increases, his angular velocity of rotation will decrease, while if he presses his arms close into his body to decrease his rotational inertia, his angular velocity will increase. The observed effects can be striking, especially if the person has weights in his hands, because of the large effect of the r^2-factor in the rotational inertia.

As an application to two-dimensional motion, consider a tumbler or a diver; these athletes essentially execute a two-dimensional motion in a vertical plane. When the diver desires a high angular velocity of tumble, he curls up into a ball to minimize his rotational inertia; just before hitting the water he straightens out and his angular velocity decreases greatly.

There are many powerful and useful theorems concerning the angular momentum of dynamical systems. We shall prove one such theorem, applying to the case of a single particle, which is very useful in discussing the motion of planets, satellites, and ballistic missiles.

If a particle is moving in a plane, as in Fig. 14, we define its angular momentum about an arbitrary origin O when it is at a point P as

$$J = I\omega = mR^2\omega = mRv_T, \qquad (19)$$

where ω is the rate of change, $d\theta/dt$, of the angle θ at the origin O when the particle is at P, and $v_T = R\omega$ is the component of the velocity perpendicular to the radius R.

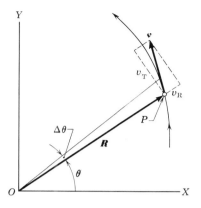

Figure 14

We note, as mentioned on p. 174, that ω and α can be considered as vectors; so can $\boldsymbol{J}$. In fact, from (19), we see that in the case of a single particle we can define

$$\boldsymbol{J} = m(\boldsymbol{R} \times \boldsymbol{v}), \qquad (20)$$

a vector pointing out of the plane in Fig. 14 if ω is positive. If we differentiate this relation with respect to t, we obtain

$$\frac{d\boldsymbol{J}}{dt} = m\boldsymbol{R} \times \frac{d\boldsymbol{v}}{dt} + m\frac{d\boldsymbol{R}}{dt} \times \boldsymbol{v}.$$

In the second term, $d\boldsymbol{R}/dt = \boldsymbol{v}$; and since $\boldsymbol{v} \times \boldsymbol{v} = 0$, the second term vanishes. Hence

$$\frac{d\boldsymbol{J}}{dt} = \boldsymbol{R} \times m\frac{d\boldsymbol{v}}{dt} = \boldsymbol{R} \times (m\boldsymbol{a}) = \boldsymbol{R} \times \boldsymbol{F},$$

where $\boldsymbol{F}$ is the resultant force acting on the particle. But, as we have seen on p. 174, $\boldsymbol{R} \times \boldsymbol{F}$ is the torque $\boldsymbol{L}$ about the origin O. Hence

$$\frac{d\boldsymbol{J}}{dt} = \boldsymbol{L}. \qquad (21)$$

If a particle is moving in a plane and the forces acting on the particle all lie in that plane, the rate of change of the angular momentum of the particle about any point O in the plane equals the resultant torque about O of all the forces acting on the particle.

In particular, if there is no torque, there is no change in angular momentum. This is the case with a planet if we take the point O at the center of the sun. The only force acting on the planet is the *radial* gravitational attraction of the sun, which exerts no torque, so the angular momentum of the planet in its elliptic orbit about the sun is *conserved* (constant). In the case of an *earth satellite* with some air resistance, the air resistance exerts a torque in a sense opposite to the angular momentum, so *the angular momentum decreases*. That this decrease occurs, in spite of the fact that the combination of air resistance and the pull of the earth causes the *speed* and hence the *kinetic energy* to *increase*, will be demonstrated in Probs. 47–49.

We shall now derive a geometric relation that we shall find useful in Sec. 10 in determining the period of revolution of a satellite. Consider the area of the small triangular region in Fig. 14 bounded by the two radii and the path. In the limit $\Delta\theta \to 0$, this region will have the same area as an isosceles triangle with base $R\,\Delta\theta$ and altitude R. This area is $\frac{1}{2}$ (base)$\times$(altitude)$= \frac{1}{2} R^2 \Delta\theta$. If the radius turns through angle $\Delta\theta$ in time Δt, this area is said to be '*swept out*' *by the radius* in time Δt, and the limit of $(\frac{1}{2} R^2 \Delta\theta)/(\Delta t)$ is called the *rate of sweeping out area*. From (19), we see that we can write this rate of sweeping out area as

$$\tfrac{1}{2} R^2 (d\theta/dt) = \tfrac{1}{2} R^2 \omega = \tfrac{1}{2} J/m.$$

If J is constant, the rate of sweeping out area is constant.

Indeed, one of Kepler's laws of planetary motion stated that, for a given planet, the rate at which its radius to the sun swept out area was a constant. This observation is consistent with the fact that the planet experiences no torque about an origin at the sun.

Example. *Verify that the angular momentum, about an arbitrary point, of a particle acted on by no forces, is constant, and that the rate of sweeping out area is constant.*

A particle acted on by no forces moves in a straight line at constant speed. Let us arrange a coordinate system so the arbitrary point is at the origin O and the particle is moving in the XY-plane parallel to the Y-axis at a distance B from that axis, as in Fig. 15. When the radius to the particle makes angle θ with the X-axis, we have $v_T = v\cos\theta$ and $B = R\cos\theta$, or $R = B/\cos\theta$. By (19), the angular momentum about O is

$$J = mRv_T = m\,(B/\cos\theta)(v\cos\theta) = mBv.$$

This value is independent of the angle θ; B and v are constants so J is constant.

In unit time, the height of the triangle in Fig. 15 grows by an amount numerically equal to v, the base remains fixed at the length B, so the area increases by $\frac{1}{2} Bv = \frac{1}{2} J/m$, as in the last equation in the text above. Since B, v, J, and m are all constant, the area swept out by the radius in each unit of time is the same.

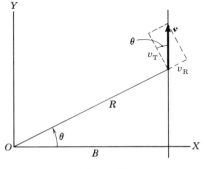
Figure 15

7. Gyroscopic Motion

So far we have discussed rotational motion only in the case where there is a fixed axis or where the motion takes place in two dimensions.

The general problem of rotational motion in three dimensions is extremely complex and far beyond the mathematical level of this text. However, there is one case for which a comparatively simple discussion can be given; this is the case of a top or gyroscope spinning *very rapidly* about its symmetry axis.

Since gyroscopes have been developed for the performance of very important functions in connection with the control and guidance of airplanes, ships, missiles, and satellites, we feel that it will be worthwhile to give a brief introduction to the behavior of tops and gyroscopes, making no attempt to give rigorous derivations of the relations used.

A top or gyroscope can rotate independently about three axes. It is convenient to take these as the nonorthogonal axes shown in Figs. 16 and 17. In the familiar top (Fig. 16), the *spin axis* is the axis of symmetry around which the top spins with *spin angular velocity* ω_S. The *precession axis* is the vertical axis. As the spin axis moves around the vertical axis in a cone of semiangle θ, the top is said to

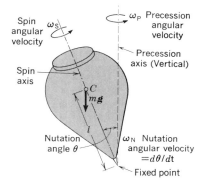

Fig. 16. Top.

precess with a certain *precession angular velocity* ω_P. If the angle θ, called the *nutation angle*, changes so that the spin axis *nods* toward and away from the vertical, the top is said to undergo *nutation*; $d\theta/dt$ is called the *nutation angular velocity* ω_N. The *nutation axis* is a horizontal axis perpendicular to the plane containing the spin axis and the precession axis. The point at the end of the top is supposed to be a fixed point in this discussion; with this restriction, any conceivable motion of the top can be considered as compounded of spin, precession, and nutation.

If we now look at the freely mounted gyroscope of Fig. 17, we can understand the system of three gimbals if we note that rotation in the bearings of each of the three gimbals corresponds precisely to one of the three modes of motion of the top discussed above. Thus the gyro wheel rotates in the bearings of the innermost gimbal at the spin angular velocity ω_S; this gimbal rotates in the bearings of the next at the nutational angular velocity ω_N; and this gimbal rotates in the bearings of the outermost fixed gimbal at the precession angular velocity ω_P.

Although the precession axis is drawn vertically in Fig. 17 to correspond to Fig. 16, the gyro of Fig. 17 is supposed to be a *free* gyro, which means that it is perfectly balanced so that gravity can ex-

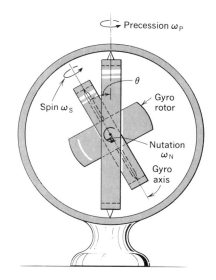

Fig. 17. Gyroscope in three-gimbal mounting.

ert no torques whatsoever; consequently, which way is up is immaterial, and there is no significance to taking the precession axis vertical rather than in some other direction. The center of mass of the gyro wheel in Fig. 17 is supposed to be at its exact center of symmetry, and each of the gimbals is supposed to be perfectly balanced in frictionless bearings.

We now note one important point which a little careful study of Fig. 17 will convince us is true, namely, that for any position of the outer fixed gimbal in Fig. 17, the gyro axis is capable of taking on any direction in space whatsoever by proper rotation of the two inner gimbals in the precessional and nutational bearings.

The top or gyroscope can have, in addition to its spin angular velocity ω_S, angular velocities of precession and nutation ω_P and ω_N. It turns out that these angular velocities can be considered as vectors directed along the respective axes and can be added vectorially. Angular momentum J is of the nature of moment of inertia times angular velocity and is a vector. But since the rotational inertias associated with the different angular velocities are different, the expression for the angular-momentum vector in the general case is complex. However, *in the case where the spin angular velocity is very large compared with the other angular velocities, the angular-momentum vector J will to a good approximation be directed along the spin axis and will have magnitude $I\omega_S$, where I is the rotational inertia of the gyro rotor about the spin axis.* The direction of the vector J is related to the sense of turning by the right-hand rule which we have already discussed on p. 174 in connection with torque. This direction is illustrated in Fig. 18.

If L is the torque acting on the top (about the fixed point), or on the gyro rotor (about the center of mass), the equation of motion is the vector equation

$$d\mathbf{J}/dt = \mathbf{L}, \qquad (22)$$

which says that, in the time dt, the change in the angular momentum vector J has magnitude $L\,dt$ and direction the same as the direction of the torque vector. This equation is outwardly identical with (21), but we are no longer talking about two-dimensional motion, so we must define *torque about a fixed point*, rather than torque about an axis whose direction is fixed in space. Torque about a fixed point is still defined as $\mathbf{L} = \mathbf{R} \times \mathbf{F}$, where R is the radius vector from the fixed point to the point of application of F. Thus, in Fig. 16, if R is the vector from the point of the top to C, the torque of gravity, $\mathbf{L} = \mathbf{R} \times m\mathbf{g}$, is seen to be a horizontal vector perpendicular to the vertical plane containing R and g. For example, if at the particular instant for which Figs. 16 and 19 are drawn,

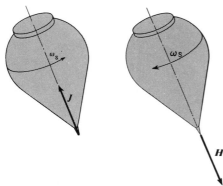

Fig. 18. Relation between direction of angular momentum vector and sense of spin.

the axis of the top lies in the YZ-plane, the torque of gravity tends to cause rotation about the X-axis and hence L is a vector pointing along the X-axis, of magnitude $mgl\sin\theta$.

Now let us derive the equation for the angular velocity of precession of the familiar rapidly spinning top. Let the top be spinning in the sense of Fig. 16, so that the angular momentum is a vector along the axis in the direction shown in Fig. 19. Equation (22) shows that $d\mathbf{J}$ is parallel to the X-axis, which means that it is perpendicular to $\mathbf{J}$ and directed tangent to the circle of radius $J\sin\theta$ shown in Fig. 19. $\mathbf{J}$ changes only in direction, not in magnitude. Furthermore, as the end of $\mathbf{J}$ moves around the circle of Fig. 19, the torque L changes direction continuously so that the direction of $d\mathbf{J}$ remains continuously perpendicular to that of $\mathbf{J}$.

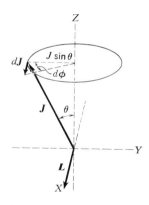

Fig. 19. Rate of precession of the top shown in Fig. 17.

Since $dJ = L\,dt = mgl\sin\theta\,dt$, in time dt the end of $\mathbf{J}$ turns through angle $d\phi = dJ/J\sin\theta = mgl\sin\theta\,dt/J\sin\theta = mgl\,dt/J = mgl\,dt/I\omega_S$. Since $d\phi/dt$ is the precession angular velocity ω_P, this gives

$$\omega_P = mgl/I\omega_S \qquad (23)$$

for the rate of precession. This relation is only valid provided that ω_P comes out small compared with ω_S so that the assumption that the angular momentum vector is directed along the spin axis is justified. The precession angular velocity (23) decreases in proportion to the angular momentum; consequently, the faster a given top spins, the less its precession angular velocity. It will be noted that the precession velocity is independent of the angle θ between the axis of the top and the vertical.

The above discussion shows why the torque of gravity does not make a top tip over. So long as the spin angular velocity does not decrease because of friction, the torque of gravity can add increments $d\mathbf{J}$ of angular momentum *only* in the direction indicated in Fig. 19 and such increments do not tend to change θ. The situation is analogous to the action of centripetal force on a body moving in a circular path. The centripetal force adds increments of velocity that continuously change the direction of the velocity, but not its magnitude.

8. Applications of Gyroscopes

The important property of the free gyroscope in Fig. 17 is that it is impossible by means of gravity, or by means of torques applied to the outer gimbal, to exert any torque whatsoever on the gyro wheel. Consequently, the angular-momentum vector remains fixed in magnitude and direction in space. No matter what motion may be given to the frame of the outer gimbal, the direction in space of the gyro axis will not vary.

Such universally mounted gyros are useful for maintaining a fixed direction of reference in a body which may be undergoing changes in direction. The axis of the steering gyro of a torpedo is initially given the desired direction of motion of the torpedo, and the gyro serves to steer the torpedo in this direction. An essentially free gyro, known as the *directional gyro,* is used in the automatic pilot of an airplane to maintain constant direction of flight. A free gyro with its axis horizontal tells a ballistic missile the correct way to turn after a vertical launch, in spite of the fact that the missile itself may have rotated about its axis since leaving the launching platform.

Foucault first used a free gyroscope to demonstrate the rotation of the earth, making use of the fact that the axis of a free gyroscope maintains constant direction *in space;* i.e. with reference to the distant stars.

There are many applications of gyroscopes that employ a two-gimbal mounting, so that torques can be applied to the gyro wheel about some axes are not about others. Perhaps the most important application of such a gyro is in connection with the shipboard *gyrocompass.* In the gyrocompass, the gyro axis is constrained to remain horizontal, as indicated in Fig. 20. Analysis of the torques on this gyro wheel, whose second gimbal is fixed to the rotating earth, shows that the angular-momentum vector turns and points to the north.

Gyroscopic principles are involved in the explanation of why a rapidly rolling wheel does not tip over; how it is possible to balance a bicycle or motorcycle; how the spin of a rifle bullet prevents it from tumbling under the action of the forces of the air; and why the axis of the earth maintains a practically fixed direction in space, undergoing only a very slow precession because of the torque exerted by the gravitational attraction of the moon and the sun on the equatorial bulge of the earth (the precession of the equinoxes).

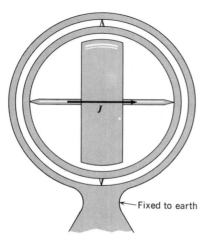

Fig. 20. Gyrocompass.

9. Motion of a Particle Relative to a Spherical Earth

In our treatment of the trajectories of projectiles in Chapter 4, we assumed a flat earth and a gravitational acceleration that did not vary with altitude. These two assumptions are both valid for most projectiles fired from guns, since the range and maximum height are both *very small compared with the* 4000-mi *radius of the earth.* But for rocket-propelled, ballistic missiles, with ranges up to half of the circumference of the earth, and altitudes as high as the planets, these assumptions are certainly not valid.

In considering the motion of ballistic missiles and satellites, it is no longer convenient to use a coordinate system attached to the earth and rotating with it, as we have done previously. Rather, in this section, we use an inertial frame fixed relative to the stars and take into account the rotation of the earth by a separate computation. We denote the radius of the earth by R and the true force of gravity (devoid of the effects of centripetal acceleration) by $F_R = mg_R$, at the surface of the earth. At any arbitrary radius r from the center of the earth, we denote the force of gravity by $F_r = mg_r$. From Newton's principle of gravitation, assuming a spherically symmetrical earth, we have

$$\frac{g_r}{g_R} = \frac{1/r^2}{1/R^2} = \frac{R^2}{r^2}, \quad \text{or} \quad g_r = \frac{R^2}{r^2} g_R. \tag{24}$$

The motion of a body projected in the earth's gravitational field is completely determined by the principles of *conservation of energy* and *conservation of angular momentum*. The angular momentum mrv_T, given by (19), must be expressed relative to the center of the earth so that the force of gravity exerts no torque. We shall first consider conservation of energy, then of momentum.

Conservation of energy. The kinetic energy is given by $E_K = \frac{1}{2} mv^2$. For the potential energy, we cannot use the 'flat-earth' value, but must explicitly consider the inverse-square decrease in gravitational attraction with increasing radius r. The gravitational force of the earth on a body of mass m is, by (24), $F = mg_r = mg_R R^2/r^2$. Hence, the work done in raising the body from R to some greater distance r is given by

$$W = \int_R^r F\,dr = mg_R R^2 \int_R^r \frac{dr}{r^2} = mg_R R^2 \left(\frac{1}{R} - \frac{1}{r}\right).$$

Hence, the potential energy of a body at radial distance r *relative to its potential energy at the earth's surface* is

$$E_P = mg_R R^2 \left(\frac{1}{R} - \frac{1}{r}\right) = mg_R R \left(1 - \frac{R}{r}\right). \tag{25}$$

The principle of conservation of mechanical energy then states that the total energy

$$E = E_K + E_P = \frac{1}{2} mv^2 + mg_R R \left(1 - \frac{R}{r}\right) = \text{constant}. \tag{26}$$

We may use this relation to determine the velocity, as a function of radial distance, of a projectile launched *radially* outward (relative to our nonrotating coordinate system) from the earth's surface. The trajectory is clearly a straight line in our nonrotating coordinate system. The initial energy is $\frac{1}{2} mv_0^2$. Relation (26) then shows that the velocity v when the projectile has reached radius r can be obtained from the equation

$$\tfrac{1}{2} mv^2 + mg_R R \left(1 - \frac{R}{r}\right) = \tfrac{1}{2} mv_0^2. \tag{27}$$

If we set $r = \infty$ in this relation, we find

$$v^2 = v_0^2 - 2 g_R R. \qquad (r = \infty) \tag{28}$$

The velocity v is still real at $r = \infty$ provided that

$$v_0 \geq \sqrt{2 g_R R}.$$

The value $v_0 = \sqrt{2 g_R R}$ is called the *escape velocity* because it is the minimum velocity that will permit the projectile to move completely away from the earth and never return. If $v_0 < \sqrt{2 g_R R}$, equation (28) has no real solution for v; rather the projectile reaches a maximum radius r_{max} at which $v = 0$ and then returns to the earth. Setting $v = 0$ and $r = r_{max}$ in (27) and solving for r_{max} gives

$$r_{max} = \frac{R}{1 - v_0^2/2 g_R R}$$

for the maximum radius attained when the initial velocity is less than $\sqrt{2 g_R R}$. We shall see later that $\sqrt{2 g_R R}$ *is the velocity of escape for any angle of launch*, not necessarily radial. The value of this velocity of escape is 36 800 f/s = 11.2 km/s.

Two points concerning this discussion should be emphasized. The first point is the obvious one that we are ignoring the frictional effect arising from air resistance, which is extremely important close to the earth's surface but becomes negligible for large values of r. The second point is that the velocities given above *are not relative to the surface of the rotating earth. Velocities and trajectories in the preceding and subsequent discussion are given in an inertial frame of reference fixed relative to the stars but having its origin at the center of the earth.* The angle of launch in this coordinate system is not the angle measured with respect to the horizon at the launching site on the rotating earth, but must be corrected for the velocity of the earth's surface so that the initial velocity is radially outward in a fixed coordinate system, not one rotating with the earth.

Equation (26) determines the changes in the magnitude of the velocity of a ballistic missile or satellite as it changes its distance from the center of the earth. But it gives no information, in general, about the direction of the velocity. But the direction is determined if we know the component v_T, whose changes are determined by conservation of angular momentum, which we shall next discuss.

Conservation of angular momentum. We have discussed conservation of angular momentum, relative to an origin at the center of the earth, on p. 197. The angular momentum is given by (19) as $J=mrv_T$, and since angular momentum is conserved,

$$J = mrv_T = \text{constant}. \tag{29}$$

Here v_T is the component of the orbital velocity perpendicular to the radius r. Relation (29) will determine how the tangential velocity component changes as the radius r changes.

Orbit of a satellite or ballistic missile. From (26) and (29), it is shown in more advanced texts that the orbit is a *conic section* in a plane fixed in space, with the focus of the conic section at the center of the earth.

Let us recall the definition of a conic section. As shown in Fig. 21, a conic section is the locus of a point whose distance r from a point F called the *focus* is e times its distance from a straight line called the *directrix*, where e is a constant called the *eccentricity*. From simple trigonometry we see that, in Fig. 21,

$$\frac{r}{e} = \frac{l}{e} + r \cos\theta.$$

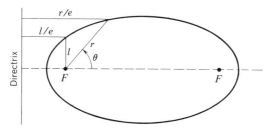

(a) Ellipse ($e = 0.8$ in this example).

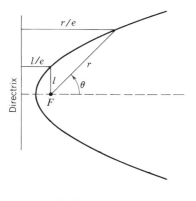

(b) Parabola ($e = 1$).

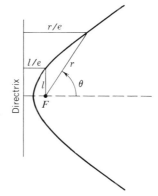

(c) Hyperbola ($e = 1.25$ in this example).

Fig. 21. Conic sections.

Hence the equation of the conic section in polar coordinates r, θ is

$$r = \frac{l}{1 - e\cos\theta}. \tag{30}$$

The shape and size of the conic are completely determined by the value of e and the length l, called the *semi latus-rectum*.

For $e=0$, (30) is the equation of the circle $r=l$.

For $0<e<1$, (30) is the equation of an *ellipse* with $r=l/(1-e)$ at $\theta=0$, $r=l$ at $\theta=90°$, and $r=l/(1+e)$ at $\theta=180°$.

For $e=1$, (30) is the equation of a *parabola* since $r \to \infty$ as $\theta \to 0$.

For $e>1$, (30) is the equation of a *hyperbola* with asymptotes at $\cos\theta = 1/e$.

The orbital parameters l and e are completely determined by the energy E and the angular momentum J and are given by the following relations:

$$l = \frac{J^2}{m^2 g_R R^2}, \qquad e = \sqrt{1 + \frac{2(E - m g_R R) J^2}{m^3 g_R^2 R^4}}. \tag{31}$$

If we know the position and velocity of a missile or satellite at any *one point* on its trajectory, we have enough information to calculate E and J and hence the orbital parameters from (31).

We note from (31) that for $E \geqq m g_R R$, we have $e \geqq 1$ and hence a hyperbolic or parabolic orbit that escapes from the earth. Thus (with air resistance neglected) we see that a projectile launched from the surface of the earth with initial speed v_0 at any angle will escape from the earth if $E = \frac{1}{2} m v_0^2 \geqq m g_R R$. This relation reduces to $v_0 \geqq \sqrt{2 g_R R}$, the velocity of escape. The parabolic trajectories of projectiles launched at exactly the escape velocity and at various angles with the radius are shown in Fig. 22.

10. Earth Satellites

From (31), we can see that any projectile launched from the earth's surface at speed less than the escape velocity travels in an elliptical orbit ($e<1$). Since the orbit is closed, the projectile must return to its starting point. It is prevented from doing so by the interposition of the earth itself. It is in an elliptical orbit *that intersects the earth*. Thus this projectile always returns to earth. A *satellite*, traveling in an elliptical orbit that does not intersect the earth, cannot be launched by a simple one-stage rocket but requires at least a two-stage rocket.

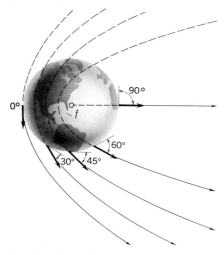

Fig. 22. Parabolic trajectories of missiles launched with the escape velocity $v_0 = \sqrt{2 g_R R}$ at various angles with the horizontal. The equation of these trajectories is $r = l/(1 - \cos\theta)$, with, from left to right, $l = 2R$, $\frac{3}{2}R$, R, $\frac{1}{2}R$, and the limiting case $l \to 0$.

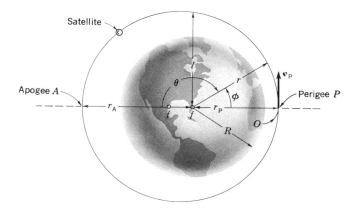

Fig. 23. A satellite orbit.

The orbit of a typical earth satellite is shown in Fig. 23. The usual method of launching a satellite in an elliptical orbit of this type is to start a missile along a ballistic trajectory OP. Then at P, when the velocity is approximately parallel to the earth's surface, the last-stage rocket is fired to give the satellite sufficient energy and angular momentum to keep it in an orbit not intersecting the earth's surface. If the velocity is boosted to exactly that required to give a circular orbit, we see, by equating force to mass times centripetal acceleration that the required velocity is

$$mv^2/r_P = mg_P = m\left(\frac{R^2}{r_P^2}\right)g_R \quad \text{or} \quad v = \sqrt{g_P\, r_P}. \tag{32}$$

If we boost the speed of the satellite to a value somewhat greater than (32), the satellite enters an elliptical orbit with the center of the earth at the *near* focus, as in Fig. 23. If the velocity v_P is exactly perpendicular to the radius r_P, the equation of the orbit is

$$r = \frac{l}{1 - e\cos\theta} = \frac{l}{1 + e\cos\phi}. \tag{33}$$

The constants l and e are determined by substituting the initial values

$$J = m\, v_P\, r_P, \qquad E = \tfrac{1}{2}\, m v_P^2 + m g_R\, R\left(1 - \frac{R}{r_P}\right),$$

in (31) and simplifying by using the relation $g_R R^2 = g_P r_P^2$ given by (24). These constants have the values

$$l = \frac{v_P^2}{g_P}, \qquad e = \frac{v_P^2}{g_P\, r_P} - 1. \tag{34}$$

We see that so long as v_P^2 is greater than the value $g_P r_P$ given by (32) for a circular orbit, l is greater than r_P; hence the apogee radius, r_A, will be greater than the radius at perigee, r_P, and there is no danger that the orbit will intersect the earth. The eccentricity e is zero for $v_P^2 = g_P r_P$, the circular orbit of (32), and is positive for greater v_P. However, $e = 1$ for $v_P^2 = 2 g_P r_P$, the velocity of escape from radius r_P. The radius at apogee, given by setting $\phi = \pi$ in (33), is $r_A = l/(1-e)$, and this radius approaches infinity as e approaches 1.

Example. *A satellite has a mass of* 10 sl, *radius* 4400 mi *at perigee*, 6500 mi *at apogee. Compute its speeds at perigee and apogee and verify the principle of conservation of energy.*

From (33), with $\phi = 0$ and $\phi = \pi$, we find

$$r_P = \frac{l}{1+e} \quad \text{and} \quad r_A = \frac{l}{1-e}.$$

From these equations, we find that

$$e = \frac{r_A - r_P}{r_A + r_P} = \frac{2100}{10\,900} = 0.193.$$

The value of g_P, the gravitational acceleration at perigee, is given by the inverse-square law (24) as

$$g_P = \frac{(4000)^2}{(4400)^2}(32 \text{ f/s}^2) = 26.4 \text{ f/s}^2.$$

We see from (34) that

$$v_P^2 = g_P r_P (e+1) = (26.4 \text{ f/s}^2)(4400 \times 5280 \text{ f})(1.19) = 7.30 \times 10^8 \text{ f}^2/\text{s}^2,$$
$$v_P = 2.70 \times 10^4 \text{ f/s} = 27\,000 \text{ f/s}.$$

The velocity at apogee is obtained directly from the principle of conservation of angular momentum: $v_A r_A = v_P r_P$, or $v_A = v_P r_P/r_A$, which gives

$$v_A = (27\,000 \text{ f/s})(4400/6500) = 18\,300 \text{ f/s}.$$

The kinetic energies at perigee and apogee are

$$E_{KP} = \tfrac{1}{2}(10 \text{ sl})(27\,000 \text{ f/s})^2 = 3.65 \times 10^9 \text{ fp},$$
$$E_{KA} = \tfrac{1}{2}(10 \text{ sl})(18\,300 \text{ f/s})^2 = 1.67 \times 10^9 \text{ fp},$$

so the loss in kinetic energy between perigee and apogee is 1.98×10^9 fp. Let us now verify that this loss in kinetic energy equals the gain in potential energy. From (25), we can write the potential energies at apogee and perigee as

$$E_{PA} = mgR - mg\frac{R^2}{r_A}, \quad E_{PP} = mgR - mg\frac{R^2}{r_P},$$

so
$$E_{PA} - E_{PP} = mgR^2\left(\frac{1}{r_P} - \frac{1}{r_A}\right) = mgR\left(\frac{R}{r_P} - \frac{R}{r_A}\right)$$
$$= (10 \text{ sl})(32 \text{ f/s}^2)(4000 \times 5280 \text{ f})\left(\frac{4000}{4400} - \frac{4000}{6500}\right)$$
$$= (320 \text{ p})(2.11 \times 10^7 \text{ f})(0.909 - 0.615) = 1.98 \times 10^9 \text{ fp},$$

exactly the loss in kinetic energy.

Now we should say a few words about the *period of revolution* of a satellite in an elliptical orbit. The period in a circular orbit, with speed given by (32), is of course $T = 2\pi r_P/v = 2\pi r_P/\sqrt{g_P r_P} = 2\pi \sqrt{r_P/g_P}$. The period in an elliptical orbit with perigee at r_P is larger than this value because the distance to be traveled is greater and the speed is always less than that at perigee. However, from the geometrical expression for the area of an ellipse, and from the expression on p. 198 for the constant rate of sweeping out area, we can derive (*see* Prob. 57) the following simple expression for the period:

$$T = 2\pi \sqrt{a^3/g_R R^2} = 2\pi (a/R) \sqrt{a/g_R}, \qquad (35)$$

where a is the semi-major axis of the ellipse:

$$a = \tfrac{1}{2}(r_P + r_A). \qquad (36)$$

Since $g_R R^2 = g_P r_P^2$, this formula reduces to the previous value for the case of a circular orbit, in which $r_A = r_P$ and $a = r_P$.

We point out again that these are *sidereal* periods, *relative to the fixed stars*—not relative to either the rotating earth or the apparent position of the sun.

Example. *Determine the period of the satellite described in the preceding example.*

From (36), we find $a = \tfrac{1}{2}(4400 + 6500)$ mi $= 5450$ mi. Substitution in (35) then gives

$$T = \frac{2\pi a}{R}\sqrt{\frac{a}{g_R}} = \frac{2\pi \times 5450}{4000}\sqrt{\frac{5450 \times 5280 \text{ f}}{32 \text{ f/s}^2}}$$

$$= 2\pi \times 1.36 \times 948 \text{ s} = 8100 \text{ s} = 135 \text{ min}.$$

As expected, this time is seen to be considerably greater than the period 98.2 min of a satellite in a circular orbit of radius 4400 mi.

11. Ballistic Missiles

We now turn to a consideration of the trajectory of a ballistic missile. A ballistic missile is usually launched vertically, then programmed to turn to a precomputed angle with the vertical in a particular direction, and programmed to cut off its rocket propulsion when it reaches a precomputed 'initial' velocity v_0. From then on, it flies freely in an elliptical 'ballistic' trajectory with the center of the earth at the 'far' focus. Such an ellipse, with foci at *ff*, is illustrated in Fig. 24.

The vertical launch is designed to minimize the slowing-down effect of air resistance by minimizing the length of path in the dense part of the atmosphere. If we neglect the effect of air resistance, we can consider the missile to be launched with velocity v_0 at the surface of

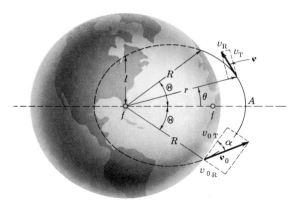

Fig. 24. The trajectory of a ballistic missile.

the earth; that is, at $r=R$. For the sake of simplicity, we assume this to be the case. Then the values of J and E are

$$J = m v_{0T} R, \qquad E = \tfrac{1}{2} m v_0^2. \qquad (37)$$

Substitution of these values in (31) gives, for the constants of the elliptical trajectory,

$$l = \frac{v_{0T}^2}{g_R}, \qquad e = \frac{1}{g_R R} \sqrt{g_R^2 R^2 + v_0^2 v_{0T}^2 - 2 g_R R v_{0T}^2}. \qquad (38)$$

We can readily determine the range for a given initial velocity v_0 by computing l and e directly from (38). Equation (30), with these values of l and e, r set equal to R, and θ set equal to Θ, then determines the value of $\cos\Theta$:

$$R = \frac{l}{1 - e\cos\Theta}, \qquad R - Re\cos\Theta = l, \qquad \cos\Theta = \frac{R-l}{Re}. \qquad (39)$$

Since all quantities on the right of the last equation are known, Θ is readily determined, as well as the range $2 R\Theta$.

We see again that in (38), if $v_0 \geq \sqrt{2 g_R R}$ (the velocity of escape), we have $e \geq 1$, so the trajectory is parabolic or hyperbolic and the missile does not return to earth.

Example. *Find the initial speed at which a ballistic missile must be fired at an angle $\alpha = 45°$ with the horizontal to travel exactly one-quarter the way around the earth ($2\Theta = 90°$ in Fig. 24). Find the maximum height attained, and the speed at this height. Neglect air resistance and assume that the initial velocity is acquired at $R = 4000$ mi, where $g_R = 32$ f/s².*

In the second equation in (39), substitute $\Theta = 45°$, $\cos\theta = 1/\sqrt{2}$, to obtain the relation

$$l = R - \frac{Re}{\sqrt{2}}, \qquad \text{(i)}$$

between l and e. Because $\alpha = 45°$ in Fig. 24, we have $v_{0T} = v_0 \cos\alpha = v_0/\sqrt{2}$. Substitution of this relation in (38) gives

$$l = \frac{v_0^2}{2 g_R}, \qquad e = \frac{1}{g_R R} \sqrt{g_R^2 R^2 + \tfrac{1}{2} v_0^4 - g_R R v_0^2}. \qquad \text{(ii)}$$

Substitution of these expressions in (i) will determine v_0:

$$\frac{v_0^2}{2 g_R} = R - \frac{1}{\sqrt{2} g_R} \sqrt{g_R^2 R^2 + \tfrac{1}{2} v_0^4 - g_R R v_0^2}.$$

Shift R to the left side and multiply by $2 g_R$:

$$v_0^2 - 2 g_R R = -\sqrt{2} \sqrt{g_R^2 R^2 + \tfrac{1}{2} v_0^4 - g_R R v_0^2}.$$

Square both sides to obtain

$$v_0^4 - 4 g_R R v_0^2 + 4 g_R^2 R^2 = 2 g_R^2 R^2 + v_0^4 - 2 g_R R v_0^2,$$

which simplifies to $\quad v_0^2 = g_R R, \quad v_0 = \sqrt{g_R R}.$ \hfill (iii)

For $g_R = 32$ f/s^2 and $R = 4000$ mi $= 21.1 \times 10^6$ f, we obtain as the required initial speed:

$$v_0 = \sqrt{6.75 \times 10^8 \text{ f}^2/\text{s}^2} = 2.60 \times 10^4 \text{ f/s} = 26\,000 \text{ f/s}.$$

Substitution of the value (iii) in (ii) gives, for the parameters of the ellipse,

$$l = g_R R / 2 g_R = \tfrac{1}{2} R, \quad e = \frac{1}{g_R R} \sqrt{g_R^2 R^2 + \tfrac{1}{2} g_R^2 R^2 - g_R^2 R^2} = \frac{1}{\sqrt{2}},$$

so the equation of the ellipse is

$$r = \frac{\tfrac{1}{2} R}{1 - \cos\theta/\sqrt{2}} = \frac{R}{2 - \sqrt{2}\cos\theta}. \tag{iv}$$

To determine the maximum radius r_A, we substitute $\theta = 0$, $\cos\theta = 1$, in (iv):

$$r_A = \frac{R}{2 - \sqrt{2}} = 1.707\, R. \tag{v}$$

The maximum height above the earth's surface is

$$r_A - R = 1.707\, R - R = 0.707\, R = 0.707\,(4000 \text{ mi}) = 2830 \text{ mi}.$$

To determine the speed v_A at this maximum height, we use the conservation-of-angular-momentum relation. Since, at point A, the missile is moving exactly perpendicular to the radius vector, we have $v_{AT} = v_A$; hence from (29),

$$v_A r_A = v_{0T} R = \frac{v_0}{\sqrt{2}} R, \quad v_A \frac{R}{2 - \sqrt{2}} = \frac{v_0}{\sqrt{2}} R,$$

$$v_A = v_0 (\sqrt{2} - 1) = 0.414\,(26\,000 \text{ f/s}) = 10\,800 \text{ f/s}. \tag{vi}$$

Now let us check these answers by computing the total energy (26) at maximum radius and at firing, and show that the values are equal. Initially

$$E = \tfrac{1}{2} m v_0^2 = \tfrac{1}{2} m g_R R,$$

From (iii). At maximum radius we have, from (26), (v), and (vi),

$$E = \tfrac{1}{2} m v_A^2 + m g_R R \left(1 - \frac{R}{r_A}\right)$$

$$= \tfrac{1}{2} m g_R R (\sqrt{2} - 1)^2 + m g_R R [1 - (2 - \sqrt{2})] = \tfrac{1}{2} m g_R R,$$

equal to the initial energy.

The above example represents a computation of the more difficult type—determination of the firing parameters necessary to achieve a specified range. It is considerably easier to determine the range for a given initial velocity $\boldsymbol{v}_0$ directly from equations (38) and (39).

PROBLEMS

1. A 2-kg uniform steel rod 2 m in length lies at rest on the surface of a frozen pond. Find the initial acceleration of the center of the rod, and the initial angular acceleration, if a force of 6 N is applied to one end of the rod in a direction normal to the long dimension of the rod. Assume that friction is negligible.
Ans: 3 m/s^2; 9 rad/s^2.

2. Solve Prob. 1 for cases in which the transverse force is applied at a point ½ m from one end; at a point ⅔ m from one end.

3. A flywheel rotating freely on a horizontal shaft is lopsided, its center of mass being 5 mm from the axis of rotation. The flywheel has a mass of 10 kg and rotational inertia of 0.036 kg·m^2 (relative to the axis of rotation). If friction is neglected, and the speed of the flywheel is 20 rad/s when the center of mass is directly above the axis, what is the speed when the center of mass is directly below the axis?
Ans: 21.3 rad/s.

4. A trapeze artist is swinging in a vertical circle on a rigid trapeze. If his angular velocity is 2 rad/s at the top of the swing, what is his angular velocity at the bottom of the swing? Treat the combination of trapeze and artist as a rigid body with mass 5 sl, rotational inertia 500 sl·f^2, and center of mass 10 f from the axis of rotation.

5. A 2-kg steel rod 2 m in length is initially balanced in a vertical position with one end on the floor of a building. A slight building vibration disturbs the balance and the rod falls so that it lies flat on the floor. There is sufficient friction so that the end of rod that is initially resting on the floor does not move from its initial position. What was the initial gravitational potential energy of the rod relative to the floor? What was the total kinetic energy of the rod just before it struck the floor? How much kinetic energy can be associated with rotation of the rod about its center of mass? with translation of the center of mass?
Ans: 19.6 J; 19.6 J; 4.9 J; 14.7 J.

6. Referring to Prob. 5, find the final translational velocity of the center of mass of the rod and the translational velocity of the upper end of the rod, just before the rod struck the floor.

7. An automobile equipped with rear-wheel drive and weighing 4000 p has a 9-f wheelbase and has its center of mass 4 f ahead of the rear wheels and 2 f off the ground. Assuming no change in location of the center of mass, what is the maximum acceleration the car could attain on a level road without having the front wheels leave the ground?
Ans: 64.0 f/s^2.

8. Show that if a spool of thread lies on a horizontal surface and one pulls the thread in the manner indicated in the figure, the thread will *wind up* if there is no slippage between the rim of the spool and the surface. (Try it.) By equating the work done by the force to the increase in kinetic energy, find an expression for the constant force F that will start the spool

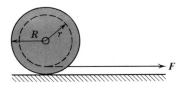

Problem 8

from rest and turn it through 1 rev in 1 s. Express your answer in terms of the inside radius r, the outside radius R, the mass m, and the rotational inertia I of the spool.

9. We can approximate a heavy steel vault door by assuming it to be a uniform sheet of steel 7 f high, 3 f wide, and 4 i thick, hinged at the left edge with one hinge 2 f down from the top, the other 1 f up from the bottom, centered in the 3-i thickness. (*a*) With what horizontal component force does the door act on each hinge? (*b*) If the hinges are frictionless, with what force must one pull on a handle at the right edge of the door in order to swing the door open 45° in 2 s, if the force has constant magnitude but is always perpendicular to the plane of the door?   Ans: 1270 p; 41.5 p.

10. A steel disc 1 i thick and 1 f in diameter is mounted on a solid steel axle that protrudes 6 i from each side of the disc and is 2 i in diameter. Strings are wound around the two ends of the axle as in the figure, and unwind as the disc falls. What are the velocity and the angular velocity of the disc after it has fallen 9 f? What fraction of the energy is rotational? Compare the velocity of fall with that of a particle falling 9 f freely.

Problem 10

11. A straight rod stands on end on a *perfectly smooth* floor, and then falls. What is the path of the center of mass of the rod?

12. The impact-testing machine in the figure has a pendulum weighing 96 p, with rotational inertia 18 sl·f² about the axle, and center of mass 2.8 f from the axle. The pendulum is allowed to swing down from a horizontal position, strike the rod to be tested at the bottom of the swing, and to coast past on the far side. Find the relation between the maximum angle θ made by the pendulum on the upswing and

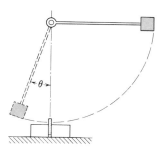

Problem 12

the energy, in fp, used in rupturing the specimen.

13. A lawn roller consists of a 6-sl uniform solid cylinder 3 f in diameter equipped with a light handle. Find the translational acceleration of the center of mass of this roller when a force of 64 p is exerted on the handle as in the figure. What will be the velocity of the center of mass of this roller after the force has been applied for 3 s if the roller was initially at rest? What will be the velocity of the uppermost points of the cylinder? of the points in contact with the ground? of the points Q at the front? Neglect loss of energy because of 'rolling' friction.   Ans: 6.16 f/s²; 18.5 f/s; 37.0 f/s; 0; 26.2 f/s; in what directions?

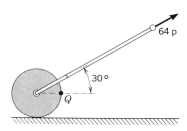

Problems 13–14

14. Referring to Prob. 13, find the angular acceleration of the roller. At the end of 3 s, how much kinetic energy is associated with translation of the center of mass? with rotation of the cylinder about an axis through its center of mass?

15. A small boy has a 6-p hoop 2 f in diameter and wishes to accelerate it by exerting a horizontal force through the center of mass as in the figure. If the coefficient of static friction between the hoop and sidewalk is 0.3, what is the maximum acceleration he can give the rolling hoop without causing it to slide? What horizontal force must he exert to produce maximum acceleration? Ans: 9.6 f/s²; 3.6 p.

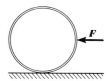

Problems 15-16

16. Answer the questions in Prob. 15 for a 4-p hoop 2 f in diameter; for an 8-p hoop 3 f in diameter. What is the maximum acceleration the boy could give to a rolling solid disc if the coefficient of static friction were 0.3? to a solid wooden sphere such as a croquet ball?

17. A solid wooden ball weighing 12 p rolls from rest down an inclined plane making an angle of 30° with the horizontal. Find the acceleration of the center of mass and the magnitude of the frictional force acting on the sphere. What must be the coefficient of static friction between the sphere and the plane, if no slipping is to occur?
Ans: 11.4 f/s²; 1.72 p; at least 0.165.

18. A solid cylinder, a hoop, and a solid sphere roll from rest down an inclined plane. Which is the first to reach the bottom? the last to reach the bottom? Find the ratios of their final translational velocities and of their times of descent.

19. A 1-kg solid cylinder 0.3 m in diameter rolls from rest down an inclined plane, without slipping. If the initial position of the center of mass is 4.9 m above its final position, what is the decrease in the potential energy of the cylinder as it rolls down the plane? What is its final total kinetic energy as it reaches the bottom of the incline? How much of this energy is associated with translation of the center of mass? with rotation about a horizontal axis through the center of mass?
Ans: 48.0 J; 48.0 J; 32.0 J; 16.0 J.

20. What are the final velocities of translation and rotation of the solid cylinder in Prob. 19?

21. Derive an expression for the final translational speed v of a solid sphere after it has rolled from rest down an inclined plane through a vertical height h.

22. We have asserted that forces of static friction do no work. We have also asserted that the force that accelerates an automobile is exerted by the road and is of the nature of static friction. How then does the kinetic energy of the automobile increase? Analyze the forces and torques acting on the driving wheels and show that, if the wheels have negligible inertia, the net power input to the wheels is zero, but that the wheels absorb power from the rotating axle and transmit this same power to the car body at a rate that accounts for its increase in kinetic energy. [NOTE. There is a general principle (4) that relates the automobile's acceleration to the sum of the *external* forces; there is no corresponding principle relating increase in kinetic energy to work done by *external* forces; *internal* forces are quite permitted to supply the work needed to increase kinetic energy.]

23. An automobile is equipped with wheels 30 i in diameter. Find the angular velocities of these wheels when the car is traveling at 30 mi/h. Which part of the wheel has the greatest instantaneous velocity and what is the value of this velocity? What is the minimum velocity of any part of the wheel? (State velocities relative to the road.)
Ans: 35.2 rad/s; 88 f/s; 0.

24. If the mass of each wheel of the car in Prob. 23 is 1 sl and its rotational inertia is 0.6 sl·f², how much kinetic energy is associated with the rotation of the wheel about the axle? How much kinetic energy is associated with translation of the center of mass of the wheel? If the car has total mass $m = 120$ sl, by what percentage does the expression $\frac{1}{2} mv^2$ have to be corrected to include the rotational kinetic energy of the wheels?

25. A particle slides down a frictionless track forming the loop-the-loop in the figure. The center of mass of the particle initially descends the vertical distance h, then rises the distance D as it traverses the circular loop. Determine the minimum height h required in terms of the diameter D so that the particle will remain in contact at the top of the loop. (To work this problem, assume that the loop exerts such a normal force on the particle that it does remain in contact. If the required normal force turns out to be a force of attraction at the top of the loop, in the absence of this force the particle would have left the loop before reaching the top. The limiting case is when the normal force is exactly zero at the top.) Ans: $h = 5/4\, D$.

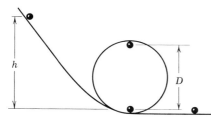

Problems 25–26

26. Solve Prob. 25 for the case where a sphere of diameter small compared with D rolls without slipping around the loop-the-loop. Repeat for a solid cylinder and a thin hoop.

27. A 'yo-yo' has the string wrapped around a reduced section of radius 0.5 cm, mass 200 g, and rotational inertia 900 g·cm². The string is held in the hand and the yo-yo with its axis horizontal is allowed to unwind the string as it falls vertically. What fraction of the kinetic energy that it acquires is rotational?

Ans: $36/37$.

28. A man stands on a small platform that is rotating about a vertical axis at a speed of 1 rev/sec in frictionless bearings; his arms are outstretched and he holds weights in each hand. With his hands in this position, the total moment of inertia of the man and the platform is 6 kg·m². If by drawing in the weights the man decreases the moment of inertia to 3 kg·m², what is the magnitude of the resulting rotational velocity of the platform?

29. A man stands at the edge of a high-diving platform, extends his arms straight over his head, and allows himself to fall forward. His feet lose contact with the end of the board when he makes an angle of 30° with the horizontal. What is his angular velocity at the moment that his feet lose contact? Assume for purposes of computation that the man in this position is dynamically equivalent to a uniform stick 6 ft long and that his initial fall is equivalent to the fall of this stick pivoted at the bottom end. Ans: 2.84 rad/sec.

30. The man in Prob. 29, who is dynamically equivalent to a stick 6 ft long when leaving the diving board at 2.84 rad/sec, curls up and becomes dynamically equivalent to a solid cylinder 2½ ft in diameter. What is then his angular velocity of tumble?

31. Kepler used the astronomical data of Tycho Brahe to formulate the laws of planetary motion. Kepler's first law stated that planets move in elliptical orbits with the sun at one focus; Kepler's second law stated that the line joining the planet and the sun sweeps out equal areas in equal times. Show that Kepler's second law can be restated as follows: *The angular momentum of a planet, about an axis at the sun perpendicular to the plane of the orbit, is constant.* From this restatement, show that the force exerted by the sun acts along the line joining the sun and the planet.

32. When the earth is at aphelion (most distant from the sun), about July 1, its distance is 94.5×10^6 mi and its orbital speed is 18.2 mi/s. From the principle of conservation of angular momentum, compute the orbital speed of the earth at perihelion (the closest point), about January 1, when its distance is 91.4×10^6 mi. Show that these variations in speed result in variations in the length of the solar day.

33. The earth and the moon may be considered together as a dynamical system acted on by an external force, which is the gravitational attraction of the sun. Since the distance of the moon from the earth, 236 000 mi, is small compared with the distance to the sun, the force of the sun can be assumed to be the same whether the moon is new or full. To this approximation, show that it is the center of mass of the system that would move around the sun in the smooth, elliptic, nearly circular orbit, but that the centers of the earth and moon would be expected to 'wobble' with a 29.5-day period (the synodic month). In this wobble, the center of the earth is observed to be a maximum of 2850 mi outside the smooth orbit when the moon is new, and a corresponding distance inside when the moon is full. What is the ratio of the mass of the moon to that of the earth? How far is the center of mass from the center of the earth? Ans: 0.0122; 2850 mi.

34. From the result in the preceding problem, and the fact that the diameter of the moon is

0.270 times the diameter of the earth, determine the ratio of the surface acceleration of gravity on the moon to that on the earth.

35. In the text we derived the 'parallel-axis relation' (12) from energy considerations. This relation can also be derived from the definition of rotational inertia. In the figure, the rotational inertia I'_n of a typical particle of mass m_n, about an axis through O', is $m_n r'^2_n$. By means of the law of cosines, r'_n can be expressed in terms of r_n, b, and $\cos\theta$. By summing I'_n over all particles in the body, derive the parallel-axis relation.

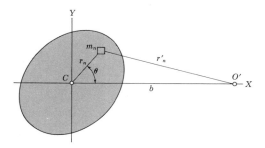

Problem 35

36. Derive equation (5), for the rigid body shown in the figure, by considering the change in the potential energy of a typical particle and then summing over all particles in the body.

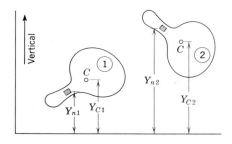

Problem 36

37. A 12-sl wheel is rotating at constant angular velocity of 180 rev/min about a fixed vertical axle. Its rotational inertia about this axis is 48 sl·f². The rim of the wheel is slightly thicker on one side than the other so that the center of mass of the wheel and axle is 1 i from the axis of rotation. Determine the magnitude and direction of the horizontal force that must be exerted by the bearings.

Ans: 355 p, in what direction?

38. In (8), the total kinetic energy of a body in two-dimensional motion is set equal to kinetic energy of translation plus the kinetic energy of rotation about an axis through the center of mass. Demonstrate trigonometrically that this expression is true for a solid disc that rolls without slipping by finding the X- and Y-components of the velocities of a typical particle, computing the kinetic energy of the particle, and then summing over all particles in the disc.

39. Going back to Fig. 15 and equation (21), show that if there are *two* particles moving in a plane and the *only* forces acting on these particles are the forces of attraction or repulsion they exert *on each other,* then the sum of the angular momenta of the two particles (the angular momentum of the *system* of two particles) remains constant. Generalize to the case of a nonrigid body and hence prove equation (18) for rotation about a fixed axis. The further generalization to two-dimensional motion is not difficult.

40. Discuss the manner in which 'reaction wheels' are used to orient satellite vehicles. Specifically, if an electric motor whose rotor has rotational inertia of 0.001 kg·m² is mounted parallel to the axis of a satellite vehicle about which the total rotational inertia is 10 kg·m², through how many revolutions must the rotor be turned to rotate the satellite by ¼ revolution?

41. A top of rotational inertia 0.0175 sl·i² is spinning at 25 rev/s at a nutation angle $\theta = 20°$. The top weighs 14 ounces and has its center of mass 1.5 i from its point. The spin is clockwise as seen from above. What is the angular velocity of precession of the top axis, and is it clockwise or counterclockwise as viewed from above? Ans: 0.913 rev/s; clockwise.

42. A top of rotational inertia 8000 g·cm² is spinning at 18 rev/s at a nutation angle $\theta = 30°$. The top has 900 g mass and has its center of mass 5 cm from its point. The spin is counterclockwise as seen from above. What is the angular velocity of precession of the top axis, and is it clockwise or counterclockwise as viewed from above?

43. A suitcase, with a rapidly turning gyro rotor mounted with its axis fixed horizontally parallel to the long dimension of the suitcase, is carefully handed to a porter.

(a) When the porter, in turning, attempts to turn the suitcase about a vertical axis, how does the suitcase actually move?

(b) When the porter attempts to swing the suitcase about a transverse horizontal axis, how does the suitcase actually move?

In the above show by means of vector diagrams the relation between the directions of the gyro-spin-velocity vector, the torque vector, and the vector that expresses the angular velocity acquired by the suitcase.

44. In Fig. 20, let the plane of the paper be a vertical E–W plane at the earth's equator, and look at the figure from the south. Show that the rotation of the earth occasions a torque on the gyro wheel that tends to make $\boldsymbol{J}$ swing toward the north, into the paper. Show that when $\boldsymbol{J}$ has swung to the north, there is no further tendency for the direction of $\boldsymbol{J}$ to change, so that the gyro is in equilibrium with $\boldsymbol{J}$ pointing north.

45. If an earth satellite of mass m is in a circular orbit of radius r, and the radius of the earth (assumed spherical) is designated by R, and the acceleration of gravity at the surface by g_R, derive the following expressions for the satellite's velocity, kinetic energy, and angular momentum about the center of the earth:

$$v = R\sqrt{g_R/r}; \qquad E_K = \tfrac{1}{2}\, mg_R\, R^2/r;$$
$$J = mR\sqrt{g_R\, r}.$$

46. In the notation of the preceding problem, take $R = 4000$ mi, $g_R = 32$ f/s², and $m = 5$ sl. Find v, E_K, and J for $r = 4500$ mi and for $r = 4400$ mi; that is, for satellites 500 and 400 mi above the surface. Find also the periods of revolution in minutes, for these two cases.

47. In the preceding problem, let the air resistance represent a force of 0.000 15 p in a direction opposite to the velocity. This force causes a gradual shrinking of the orbit from 4500-mi radius to 4400-mi radius, but the change is so gradual that, to a good approximation, the orbit can always be assumed circular for purposes of computation. Compute, to three-figure accuracy, the increase in velocity, the increase in kinetic energy, and the

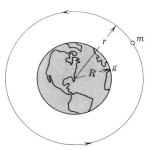

Problems 45–49

decrease in angular momentum as the orbit shrinks by this amount. To make this computation you will need the difference between $\sqrt{4500}$ and $\sqrt{4400}$ to a 3-figure accuracy, and hence these square roots to greater accuracy; 6-figure values are $\sqrt{4500} = 67.0820$, $\sqrt{4400} = 66.3325$. Ans: $\Delta v = 277$ f/s; $\Delta E_K = 3.41 \times 10^7$ fp; $\Delta J = -3.25 \times 10^{10}$ sl·f²/s.

48. In the preceding problems, find the approximate time Δt for the orbital radius to shrink 100 mi under the action of the 0.000 15-p air resistance by using the relation $\overline{L} = \Delta J/\Delta t$ with the average torque computed at the mean radius of 4450 mi. Express your answer in days. Compute also the approximate number of revolutions in this time, using a mean period of revolution.

49. In the preceding problems, compute the work done by the air resistance, using mean values, and the work done by the gravitational pull, using mean values, as the orbit shrinks from 4500 to 4400 mi. Verify that the sum of these 'works' is approximately equal to the increase in kinetic energy computed in Prob. 47. Ans: -3.41×10^7 fp; 6.83×10^7 fp; sum $= 3.42 \times 10^7$ fp.

50. The U. S. earth satellite Explorer III had a very eccentric orbit with perigee at a height of 109 mi. The velocity at perigee was 27 600 f/s (in a direction exactly perpendicular to the radius to the center of the earth). Show that this speed is too great for a circular orbit at the radius of 4109 mi, so the satellite described an elliptic orbit whose apogee was at the height of 1630 mi. Compute the speed at apogee and show that this speed is too *small* for a circular orbit at radius 5630 mi.

51. A ballistic missile whose true range is ¼

the circumference of the earth ($2\Theta = 90°$ in Fig. 24), has a time of flight of 60 min. What is its apparent range on the earth's surface if it is fired toward the east at the equator? toward the west at the equator? at the pole? Consider the earth as a sphere of 4000-mi radius, and be sure to use the sidereal, not the solar period of rotation of the earth.

Ans: 5230, 7330, 6280 mi.

52. On December 7, 1958, the rocket Pioneer III reached the height of 66 000 mi (radius 70 000 mi). At what speed must a rocket be fired radially outward to reach this radius? Would the rocket reenter the atmosphere at the geographical location at which it left?

53. A satellite is said to be in a 'polar orbit' if the plane of its orbit includes the axis of the earth. Show that to an observer on the rotating earth, the plane of the orbit of a satellite in a polar orbit makes one rotation about the earth every sidereal day of 23.9 hours.

54. Referring back to the description of the orbit of Explorer III given in Prob. 50, use the given radius and speed at perigee to *compute* the expected radius at apogee, and compare with the observed value.

55. Compute the period of Explorer III, whose orbit is described in Prob. 50. Ans: 121 min.

56. On p. 209, the period of a satellite in a circular orbit is computed by dividing the circumference by the speed. Compute this same period, as we shall do for the case of an elliptic orbit in Prob. 57, by dividing the area by the rate of sweeping out area as given on p. 198.

57. (*a*) The area of an ellipse is $\pi l^2/(1-e^2)^{3/2}$. From the relations $r_P = l/(1+e)$ and $r_A = l/(1-e)$ in the notation of Fig. 23, show that we can write this area as $\pi a \sqrt{al}$, where a is the semi-major axis (36). (*b*) Show that the rate at which the radius to a satellite sweeps out area, as given on p. 198, can be written as $\frac{1}{2} R \sqrt{gl}$. (*c*) From the results in (*a*) and (*b*), derive equation (35) for the period of a satellite.

58. Discuss 'synchronous' satellites that remain continuously at, or close to, the same longitude relative to the earth's surface. What is the period of a synchronous satellite? What is its altitude if it is a circular equatorial orbit? Synchronous satellites have been more and more widely used for relaying communications since the first SYNCOM was launched on July 26, 1963. Why are synchronous satellites especially suited to this purpose?

59. A ballistic missile is launched with initial radial and tangential speeds, *relative to the fixed stars,* that are each $(26\,000/\sqrt{2})$ f/s = 18 400 f/s. If this missile is fired toward the east at the equator, at what speed and angle should it be fired relative to an observer on the rotating earth?

Ans: 25 000 f/s, 47°.5 above horizontal.

60. We have discussed missiles that move radially, relative to the fixed stars. At what angle and speed relative to the earth's surface should a rocket be fired so that it goes out radially in this sense and reaches a maximum height of 1000 mi from the earth's surface, if it is fired at the equator? at the pole?

61. Show that the expression (25) for potential energy in the earth's gravitational field can be written in the form

$$E_P = mgh\left(1 - \frac{h}{R} - \frac{h^2}{R^2} - \cdots\right),$$

where $h = r - R$, and hence that for $h \ll R$, the flat-earth formula $E_P = mgh$ is valid. At what height h does the true potential energy differ from the flat-earth approximation by one per cent? Ans: 40 mi.

ELASTIC PROPERTIES OF SOLIDS AND LIQUIDS 11

In our treatment of mechanics thus far, solid bodies have been assumed to be rigid. This assumption is not strictly valid in any real physical situation, but can in many cases be justified when deformations are small or when the effects of the deformations are unimportant. In this chapter we shall be concerned with the changes in shape or volume of solids and liquids that occur when external forces are applied. The changes in shape of solids are of great importance in the design of structures of all types, since it is necessary to know how much the structural members will bend, or stretch, or twist, under the loads that the structure will be called on to bear, and whether perhaps the strain will be so great that a structural member might actually break.

The closing sections of the chapter deal with the arrangement of atoms in solids and present a semi-quantitative account of the observed elastic forces in terms of interatomic forces.

1. Hooke's Law, Elastic Potential Energy

Everyone has observed the bending of a piece of wood such as a diving board when a load is added; when the load is removed, the board regains its original shape. Likewise, a helical spring increases in length when a small load is added but regains its original length when the load is removed. These observed effects are examples of the *elasticity* of matter.

Recovery of the original configuration after application of forces is practically perfect for many kinds of materials, provided the distorting forces are not too great. If the distorting forces are too great, the *elastic limit* is exceeded and the recovery of the original configuration is incomplete. In this case, the body is said to have acquired a *permanent set* or permanent deformation. As we shall see later, the term *elastic limit* can be defined in such a way that it has a definite value for every *material*.

Materials for which the elastic limit is extremely small are called *inelastic materials;* for example, dough, putty, and lead solder are inelastic materials, since bodies composed of these materials are permanently deformed when acted upon by relatively small forces. Steel is a highly elastic material, since relatively enormous forces are required to produce permanent distortion of bodies composed of steel.

Nearly three centuries ago, the English experimental physicist ROBERT HOOKE (1635–1703) discovered an important relation:

HOOKE'S LAW: *The deformation of an elastic body is directly proportional to the magnitude of the applied force, provided the elastic limit is not exceeded.*

This law is valid for bodies composed of most metals and other structural materials but *is not valid for plastic materials.*

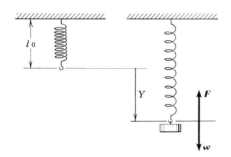

Fig. 1. The force of the weight on the spring is w. The equal and opposite reaction of the spring on the weight is the elastic force F.

The correctness of Hooke's law in a special case can be verified by adding weights to a helical spring in the manner indicated in Fig. 1. Before weight w is added, the length of the spring is l_0. The added weight produces elongation denoted by Y. The equal and opposite force F exerted by the spring is called the *elastic force*. By adding different weights and noting the elongation, we find that the observed elongation Y is proportional to the deforming force w, as in Fig. 2. In accordance with Hooke's law, the relations between the deforming force w, the elastic force F, and the elongation Y can be written

$$w = KY, \qquad F_Y = -KY, \qquad (1)$$

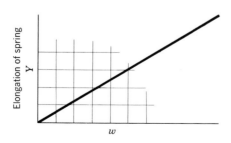

Fig. 2. The deformation Y in Fig. 1 is proportional to the applied force w.

where the constant K is called the *force constant* of the spring.

The **force constant** of a spring is the force per unit elongation.

The force constant is a measure of the 'stiffness' of the spring. The negative sign in (1) is used to indicate that the vertical component of the elastic force has a sign opposite to that of the displacement Y, which in this case is taken as *positive downward*. Thus, if the displacement is downward as indicated in Fig. 1, the elastic force F is directed upward. If an upward external force acted in such a manner as to produce a negative elongation (compression) of the spring in Fig. 1, the elastic force F exerted by the spring would act downward and be regarded as positive.

We shall now consider the potential energy of a spring. We have already defined elastic potential energy on p. 108. In the elastic case, changes in potential energy are associated with changes in configuration of the elastic body. It is only these *changes* in potential energy that have physical significance.

> The change in **elastic potential energy** of a body deformed from configuration A to configuration B is the *negative* of the work done *by the elastic forces* during the change in configuration.

We see that the above definition is consistent with the general definition of energy given on p. 107, because the elastic potential energy as defined above equals the work that could be done *by the elastic forces* as the body returns from configuration B to configuration A.

As the spring of Fig. 1 changes from the condition on the left to the deformed condition on the right, the elastic force F_Y changes linearly from 0 to $-KY$. Therefore it does work $F_Y dY = (-KY) dY$ during the displacement dY. The potential energy relative to the initial condition is by definition

$$E_\mathrm{P} = -\int_0^Y F_Y\, dY = K\int_0^Y Y\, dY.$$

Thus, $\quad \begin{cases} E_\mathrm{P} = \tfrac{1}{2} KY^2. \end{cases}\quad \begin{cases} E_\mathrm{P} \text{ in J} \\ K \text{ in N/m} \\ Y \text{ in m} \end{cases}$ or $\begin{cases} E_\mathrm{P} \text{ in fp} \\ K \text{ in p/f} \\ Y \text{ in f} \end{cases}$ **(2)**

We note that this potential energy is the work that the elastic force F would *do* as the spring returned to the configuration of zero potential energy. We note that it is also the work that an external force would have to do in order to effect the deformation.

Example. Consider again the example of a railroad bumper on pp. 106–107. The work, $\tfrac{1}{2} KX^2$, done by the car on the spring is the negative of the work done by the spring on the car and is, by definition of potential energy, the elastic potential energy of the spring when the car is brought to rest. Demonstrate that this quantity equals the initial kinetic energy of the car.

The spring exerts on the car the force $F_X = -KX$. If the car has mass m, it experiences the acceleration

$$\frac{d^2 X}{dt^2} = a_X = \frac{F_X}{m} = -\frac{K}{m} X. \qquad \text{(i)}$$

The function of time whose second derivative is minus a constant times the function itself is a sine or cosine; in this case we have

$$X = X_1 \sin(\sqrt{K/m}\ t), \qquad \text{(ii)}$$

which may be verified by direct substitution to satisfy (i). In (ii), we have chosen the sine rather than the cosine so that X will be zero when $t=0$. We have chosen the coefficient of the sine as X_1 so that the maximum value of X will be X_1. From (ii), we see that the velocity of the car, as a function of time, is given by

$$v_X = \frac{dX}{dt} = \sqrt{K/m}\ X_1 \cos(\sqrt{K/m}\ t). \qquad \text{(iii)}$$

Hence the initial velocity of the car that is required to give the spring the distortion X_1 is given by

$$v_{X(t=0)} = \sqrt{K/m}\ X_1.$$

Hence the initial kinetic energy of the car must be

$$\tfrac{1}{2}\ mv_{X(t=0)}^2 = \tfrac{1}{2}\ m\ (K/m)\ X_1^2 = \tfrac{1}{2}\ KX_1^2,$$

which is seen to equal the work done by the car on the spring as the car is brought to rest.

This example shows that the car is able to do work equal to its kinetic energy as it is brought to rest by the spring.

Hooke's law can also be easily verified for many other simple cases. Figure 3 shows a wire clamped at the upper end. To the lower end is applied a torque $L_{Ext} = 2\ F_{Ext}\ R$ which tends to twist the wire. The angle of twist can be taken as a measure of the deformation produced in the wire. Experiment shows that the deformation θ is directly proportional to the applied torque. In a notation similar to that used for the helical spring, we may write

$$L_{Ext} = C\theta, \qquad L = -C\theta, \qquad \text{(3)}$$

where L is the elastic reaction of the wire. By an argument similar to that given previously, we find the expression

$$E_P = \tfrac{1}{2}\ C\theta^2 \qquad \text{(4)}$$

for the elastic potential energy stored in the twisted wire. The proportionality constant C in these equations is called the *torsion constant*.

The **torsion constant** of a wire or rod is the torque required per radian of twist.

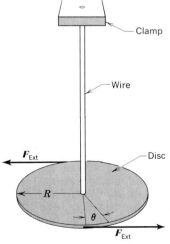

Figure 3

Whereas Hooke's law holds well for most bodies composed of metal, wood, glass, and many other common materials, there are some elastic materials for which it does not hold at all. Rubber is such a material. If we take a rubber band of initial length l_0 and deform the band by suspending weights w from the lower end in the manner indicated in Fig. 4(a), a plot of the resulting elongations Y has the form shown in Fig. 4(b). The curve shown in this figure bears little resemblance to the straight line to be expected on the basis of Hooke's law. *Elastic* elongations of five or six times the original length can easily be obtained with a rubber band, whereas a wire or rod composed of a material like metal, wood, or glass can only be stretched *elastically* a small fraction of its original length before the elastic limit is reached. Hence it is not surprising that these two types of materials have quite different relations between force and elongation. Rubber is a typical example of a *plastic* material; for such materials Hooke's law is not valid.

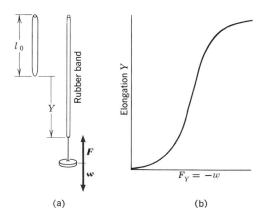

Fig. 4. Elastic properties of rubber.

2. Longitudinal Stress and Strain: Generalization of Hooke's Law

In the above discussion, we have described experiments that verify Hooke's law for particular *bodies*, but we have not considered explicitly the elastic properties of the *materials* of which the bodies are composed. Fortunately, it is possible to generalize Hooke's law in terms of elastic constants for *materials*. The elastic constants of *bodies* such as structural members can then be calculated from the elastic constants of their *materials*. In the following paragraphs, we shall discuss the concepts of *stress* and *strain*, in terms of which the general form of Hooke's law is stated.

Stress is related to the *force* causing deformation; *strain* is related to the *amount of deformation*. In order to understand the ideas involved, consider Fig. 5, which shows a long elastic rod of initial length l_0 and cross-sectional area A. If forces F are applied to the ends of the rod, the rod experiences an elongation Δl and we can say that the rod has undergone a *longitudinal strain*.

> **Longitudinal strain** of a wire or rod under tension or compression is defined as the change in length per unit length.

In Fig. 5 the longitudinal *strain* σ is the ratio of the elongation Δl to the initial length l_0:

$$\sigma = \Delta l / l_0. \qquad \text{(strain)}$$

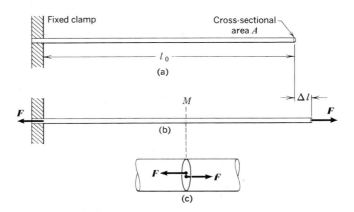

Fig. 5. Longitudinal stress is F/A. Longitudinal strain is $\Delta l/l_0$.

The strain is accompanied by *internal forces* between adjacent parts of the rod. If the cross section of the rod at any point such as M in Fig. 5 be considered as a dividing plane, the material to the left of M exerts a force of magnitude F toward the left on the material to the right of M, whereas the material to the right exerts a force of magnitude F toward the right on the material to the left of M. These forces must have the same magnitude as the external applied forces, since each part of the rod is separately in static equilibrium. When such forces exist in the interior of a body, the body is said to be under *stress*.

Longitudinal stress of a wire or rod under tension or compression is defined as the internal force per unit cross-sectional area.

In the case of the uniform rod shown in Fig. 5, the longitudinal *stress* S is the same at all cross sections:

$$S = F/A. \qquad \text{(stress)}$$

By performing experiments on rods of different sizes but composed of the same material, we find that the longitudinal stress S is proportional to the longitudinal strain σ, the proportionality constant being independent of the size of the rod. Thus we may write

$$S = M_Y\, \sigma,$$

where the proportionality constant M_Y is called *Young's modulus** and is a *constant* for a given *material*.

Thus we see that for a rod or wire of any size,

$$\frac{F}{A} = M_Y \frac{\Delta l}{l_0}, \qquad (5)$$

---

*Named for Thomas Young, English experimental physicist (1773–1829).

in the notation of Fig. 5. Since Δl and l_0 are measured in the same length units, M_Y has dimensions of force/area, such as p/f², p/i², or N/m².

Young's modulus is the ratio of longitudinal stress to longitudinal strain in the case of a rod or wire under tension or compression.

Thus, by introducing the quantities stress and strain we have been able to generalize Hooke's law in the case of stretched rods or wires to the statement:

Stress is proportional to strain, the proportionality constant depending only on the material and not on the particular body.

The values of Young's modulus for various materials are given in Table 11-1. The same values apply to compression and to tension.

Example. *A steel piano wire will withstand a tensile stress of* 100 000 p/i² *and still obey Hooke's law since the elastic limit is about* 120 000 p/i². *How much does a* 100-i *length of wire stretch when this stress is applied? If the wire is* 1/25 *inch in diameter, how much load must be applied to give this stress?*

From Table 11-1, M_Y for steel is 29×10^6 p/i². We substitute this value and $F/A = 10^5$ p/i² in (5) to get the strain

$$\frac{\Delta l}{l_0} = \frac{F/A}{M_Y} = \frac{10^5 \text{ p/i}^2}{29 \times 10^6 \text{ p/i}^2} = 3.4 \times 10^{-3} = 0.0034.$$

Table 11-1 TYPICAL ELASTIC CONSTANTS

Material	Young's modulus M_Y		Shear modulus M_S	
	p/i²	N/m²	p/i²	N/m²
Aluminum...........	10 × 10⁶	6.9 × 10¹⁰	3.8 × 10⁶	2.6 × 10¹⁰
Brass................	13 × 10⁶	9.0 × 10¹⁰	5.1 × 10⁶	3.5 × 10¹⁰
Copper..............	16 × 10⁶	11 × 10¹⁰	6.0 × 10⁶	4.1 × 10¹⁰
Nickel...............	30 × 10⁶	21 × 10¹⁰	11 × 10⁶	7.6 × 10¹⁰
Steel.................	29 × 10⁶	20 × 10¹⁰	11 × 10⁶	7.6 × 10¹⁰
Tungsten............	51 × 10⁶	35 × 10¹⁰	21 × 10⁶	14 × 10¹⁰
Glass................	7.8 × 10⁶	5.4 × 10¹⁰	3.3 × 10⁶	2.3 × 10¹⁰

Material	Bulk modulus M_B		
	p/i²	p/f²	N/m²
Brass................	15 × 10⁶	22 × 10⁸	10 × 10¹⁰
Copper..............	20 × 10⁶	29 × 10⁸	14 × 10¹⁰
Steel.................	25 × 10⁶	36 × 10⁸	17 × 10¹⁰
Glass................	5.2 × 10⁶	7.5 × 10⁸	3.6 × 10¹⁰
Ethyl ether..........	0.9 × 10⁵	1.3 × 10⁷	0.6 × 10⁹
Ethyl alcohol........	1.6 × 10⁵	2.3 × 10⁷	1.1 × 10⁹
Water................	3.1 × 10⁵	4.5 × 10⁷	2.1 × 10⁹
Mercury.............	40 × 10⁵	58 × 10⁷	28 × 10⁹

From this, we find that
$$\Delta l = 0.0034\, l_0 = 0.0034 \times 100 \text{ i} = 0.34 \text{ i}.$$

The 100-inch length of wire stretches by about ⅓ inch.

The area of a circle of diameter $\frac{1}{25}$ i $= 0.04$ i is $A = \pi (0.02)^2 = 12.5 \times 10^{-4}$ i². Hence
$$F = (10^5 \text{ p/i}^2) \times A = (10^5 \text{ p/i}^2) \times 12.5 \times 10^{-4} \text{ i}^2 = 125 \text{ p}.$$

This is the force required to stretch the wire by 0.34 inch.

It is important to distinguish *isotropic* from *anisotropic* solids. An *isotropic* solid material is one whose physical properties are independent of direction; an *anisotropic* solid material is one whose physical properties vary with direction.

Wood is the most familiar example of an anisotropic solid material—it has a grain and a structure that makes its physical properties quite different in the three principal directions. A ball or cube of wood would have different compressive properties in the different directions of applying the compressive stress; these properties could only be expressed by giving three different Young's moduli, not a single value. In our previous discussion we introduced only a single value of Young's modulus for a material—this discussion clearly applies only to isotropic materials. Discussion of the elastic behavior of anisotropic materials is much more complex than for isotropic materials; in this chapter we shall confine our attention to isotropic materials.

Single crystals are (with certain exceptions such as rock salt and diamond) anisotropic. They have different elastic, optical, and electrical properties in different directions. Metals are in general composed of large numbers of individual crystals but, while each individual crystal is anisotropic, so many individual crystals are jammed together with random orientations in a piece of metal that the metal as a whole ordinarily behaves isotropically. The elastic constants given in Table 11–1 are for such ordinary multicrystalline metals. A broken piece of cast iron usually furnishes a surface on which the individual crystals are large enough to be visible without magnification.

3. Volume Elasticity: Bulk Modulus

Let us next consider the problem of bulk or volume elasticity. Figure 6 shows a body that has volume V_0 when subjected to a uniform external pressure P_0, and also shows the same body with smaller volume V when the pressure is increased to $P = P_0 + \Delta P$. By pressure, we mean normal compressive force per unit area of surface.

In the case of bulk compression, **volume stress** is defined as the pressure increase ΔP; **volume strain** is defined as the *decrease* in volume per unit volume, or $(V_0 - V)/V_0 = -\Delta V/V_0$.

Again it is found experimentally that stress is proportional to strain:

$$\Delta P = M_B(-\Delta V/V_0), \tag{6}$$

where the proportionality constant M_B is a characteristic of the material called the *bulk modulus*. The bulk modulus has the dimensions of pressure, that is, force/area.

The **bulk modulus** is the ratio of stress to strain in the case of volume compression.

Values of the bulk modulus for several materials are listed in Table 11-1.

Equation (6) enables us to compute the change ΔV in the volume of a body when the pressure increases by a given amount ΔP, provided we know the initial volume of the body and the bulk modulus. The change in volume will be negative if the change in pressure is positive.

Since no question of rigidity is involved in volume elasticity, equation (6) applies to liquids as well as to solids. The volume elasticity of gases will be considered in Chapter 17.

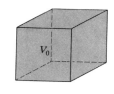
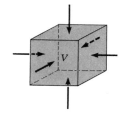

Fig. 6. Decrease in volume associated with increase in pressure from P_0 to $P_0 + \Delta P$.

4. Elasticity of Shape: Shear Modulus

The third type of elasticity we shall consider is one in which the *shape* of a body is changed without change in the *volume* of the body. The type of deformation involved, called a *shear*, is illustrated in Fig. 7. In this figure, it will be noted that the shape of the book has been altered without change in the volume of the book and that a rectangular element of surface area becomes a parallelogram when shear occurs. The angle ϕ in the figure is called the *angle of shear*.

A book is an anisotropic body that is relatively inelastic to deformation of the type indicated in Fig. 7. In order to understand how a shear is produced in an isotropic body, consider the volume element shown in Fig. 8; this element of volume has been taken from the interior of a body experiencing a shear. The forces F_1 are exerted on the element in the directions tangential to the faces of area A_1, and the forces F_2 are exerted in directions tangential to the faces of area A_2; these forces are

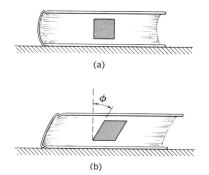

Fig. 7. Shearing deformation.

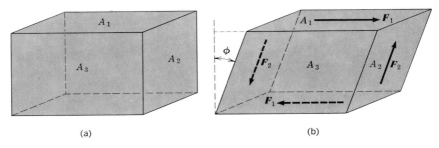

Fig. 8. Shearing stress and strain within a material medium: (a) unstressed parallelepiped, (b) the same parallelepiped under shearing stress.

exerted *by* the material outside the volume element *on* the material inside. The resultant force and the resultant torque on the material in the volume are zero; this must be so, since the material is not experiencing linear or angular acceleration. When the surface forces F_1 and F_2 act in the manner indicated, the face of area A_3, which was a rectangle in the undeformed body, becomes a parallelogram.

Shearing strain is defined as the shear angle ϕ, in radians.

Over the elastic range, the shear angle ϕ is always a very small angle; it is greatly exaggerated in Fig. 8.

Shearing stress is defined as the tangential force per unit area.

In Fig. 8, the shearing stress is F_1/A_1 or F_2/A_2; the condition for rotational equilibrium shows that these two ratios are always equal.

It is found experimentally that shearing stress is proportional to shearing strain:

$$S = F/A = M_S \phi. \tag{7}$$

The proportionality constant M_S, called the *shear modulus* or *modulus of rigidity*, depends only on the type of material of which the body is composed. The shear modulus has dimensions of force/area.

> The **shear modulus** is the ratio of stress to strain when shearing forces act on an isotropic body.

Values of the shear modulus for several materials are given in Table 11-1.

To see how the shear modulus may be determined experimentally, consider a thin-walled hollow cylinder that is twisted about its geometrical axis by clamping one end of the cylinder in a fixed position and applying a torque to the free end; this operation results in a uniform shear. A thin-walled cylinder before deformation is shown in Fig. 9(a); a small 'square' surface element is cross-hatched and a long 'rectangular' surface element is shown by the dashed lines passing along the cylinder surface from end to end. When the cylinder is twisted as shown in part (b) of the figure, the small 'square' becomes a 'rhombus' and the 'rectangle' becomes a 'parallelogram' of the indicated shape; therefore, the material experiences a pure shear. Now let us find expressions for strain and stress from the labeled diagram in Fig. 9(c). The strain can be expressed as the shear angle ϕ. However, the shear angle ϕ is not as easy to measure as θ, the angle of rotation of the twisted end. In the elastic range ϕ is small, and we can write

$$\text{shearing strain} = \phi = a/l = R\theta/l, \qquad (8)$$

where l is the length of the cylinder. The torque L must appear across any section such as AA, since the part of the cylinder above AA is in rotational equilibrium. If we call the shearing stress S at section AA, this stress acts across a total cross-sectional area $2\pi Rt$, so that it exerts torque $L = 2\pi RtS \cdot R$. Hence,

$$\text{shearing stress} = S = L/2\pi R^2 t. \qquad (9)$$

The ratio of shearing stress (9) to shearing strain (8) is by (7) the shear modulus. Hence we have

$$M_S = \frac{L/2\pi R^2 t}{R\theta/l}, \quad \text{or} \quad M_S = \frac{Ll}{2\pi R^3 \theta t}. \qquad (10)$$

Equation (10) enables us to determine the shear modulus M_S for a material from measurements of the rotation θ produced by applying a torque L to a thin-walled cylinder, as in Fig. 10. If the modulus M_S is known, equation (10) can be used to predict the rotation θ produced by the application of a torque L to the end of a thin-walled tube. A similar equation can be derived for a twisted *solid* rod or wire of radius R and length l. This relation is derived by dividing the rod up into cylindrical shells of radius r and thickness dr. According to (10), the torque dL required to twist such an element through angle θ is $dL = (2\pi\theta M_S/l) r^3 dr$. Integration of this expression from $r=0$ to R gives

$$L = \frac{\pi\theta M_S R^4}{2l}, \quad \text{or} \quad M_S = \frac{2lL}{\pi R^4 \theta}. \qquad (11)$$

The ratio L/θ gives the *torsion constant* C of the tube or rod, as defined on p. 222. This constant can be computed from (10) or (11). We note from (8) that even if ϕ is very small, θ can be large if l/R is large so that θ can be an easily measured angle.

The force constant of a *helical spring* is determined by the torsion constant for a unit length of the wire of which it is made. We do not give details here, but merely point out that when a spring elongates, the wire of which it is formed does not stretch, it merely *twists*. Any small

(a)

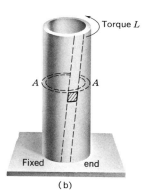

(b)

(c)

Fig. 9. (a) A thin-walled hollow cylinder. (b) The cylinder is twisted by applying a torque. (c) The angle of shear ϕ and the angle of twist θ.

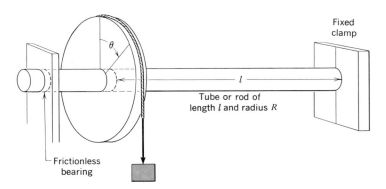

Fig. 10. Measurement of torsion constant of a tube or rod.

section of the wire behaves exactly like a piece of wire in torsion, and hence the torsion constant governs its elastic behavior.

Example. *The elastic limit for shearing stress of a particular steel is 40 000 p/i². An automobile drive shaft made of this steel has 2-i diameter and 0.1-i thickness. How much torque can a 6-f shaft of this tubing transmit without acquiring a permanent set? What will be the angle of twist and the angle of shear for this shaft at the elastic limit?*

From (9), the torque, when the shearing stress is 40 000 p/i², is

$$L = 2\pi R^2 tS = 2\pi (1\text{ i})^2 (0.1\text{ i})(40\,000\text{ p/i}^2) = 25\,000\text{ p}\cdot\text{i}.$$

Using the shear modulus of steel in Table 11–1, we find the shear angle ϕ from (7):

$$\phi = S/M_S = (40\,000\text{ p/i}^2)(11 \times 10^6\text{ p/i}^2) = 3.64 \times 10^{-3}\text{ rad} = 0°208.$$

The angle of twist θ is obtained from (8):

$$\theta = (l/R)\,\phi = (6/\tfrac{1}{12})\,3.64 \times 10^{-3}\text{ rad} = 0.262\text{ rad} = 15°0.$$

5. Relations among Elastic Constants

We have thus far introduced three elastic moduli: Young's modulus, the bulk modulus, and the shear modulus. Of all the elastic moduli that can be defined for an isotropic elastic material, only two are independent. All elastic moduli can be expressed in terms of any two of them. For example, Young's modulus M_Y can be expressed in terms of M_B and M_S by $M_Y = 9M_B M_S/(3M_B + M_S)$. Actually, the stretching of a wire involves a change in both shape and volume of the wire, and consequently both volume elasticity and shape elasticity are involved. As the length increases, the diameter d decreases. *The ratio of the relative lateral contraction $-\Delta d/d$ to the relative longitudinal extension $\Delta l/l$ is called Poisson's ratio ρ.* This ratio is dimensionless. For most metals, Poisson's ratio has a value in the neighborhood of 0.3.

The experimental determination of the elastic moduli is easiest for Young's modulus and the shear modulus. The direct determination of the bulk modulus and of Poisson's ratio are considerably more difficult. The values of the bulk modulus M_B and Poisson's ratio ρ are determined from M_Y and M_S by the relations $M_B = \tfrac{1}{3} M_S M_Y/(3 M_S - M_Y)$ and $\rho = (M_Y/2M_S) - 1$. These relations do not hold for single crystals of material; single crystals are anisotropic and require more than two independent constants to describe their elastic behavior. The relations do, however, hold for polycrystalline materials that are composed of a large number of small crystals oriented at random, as in the usual case for the solid metals of engineering practice. They also hold for amorphous materials such as glass.

The only elastic modulus applicable to a liquid or gas is the bulk modulus, since a fluid in equilibrium will not withstand shearing stress.

6. Elastic Limit and Ultimate Strength of Materials

When an elastic material is subjected to increasingly large tensile stresses, Hooke's law is followed until the elastic limit is attained. Certain *brittle* materials such as glass and phosphor bronze rupture at the elastic limit; for such materials no 'permanent set' can be produced. Other materials such as steel, copper, and brass are said to be *ductile*. Unlike brittle materials, ductile materials can be cold-rolled (as is steel), drawn into wire through dies (as is copper), or beaten into thin sheets (as is gold).

When a wire of a ductile material is subjected to a stress slightly greater than its elastic limit, the *yield point* is reached; at the yield point the wire appears to flow like a viscous liquid and a considerable elongation can be produced without increasing the stretching forces appreciably. When the yield point has been reached, the wire acquires a 'permanent set' and does not regain its original length even after the applied stresses are removed.

Once the yield point has been reached, it is convenient to describe the subsequent behavior of a wire or 'test sample' under increasing distorting forces in terms of an *apparent stress* defined as the ratio of the stretching force to the *original* cross-sectional area of the sample. Since the wire material appears to flow at the yield point, considerable increase in strain is produced in the wire without any increase in apparent stress. However, the apparent stress can be increased beyond the yield point with accompanying still further increase in strain. The maximum apparent stress that can be applied to a material is called the *ultimate strength*. Values of the tensional elastic limits and ultimate strengths of several ductile materials are listed in Table 11–2.

The ultimate strength for shear is usually smaller than the corre-

Table 11-2 Typical Elastic Limits and Ultimate Strengths for Materials in Tension

Material	Elastic limit		Ultimate strength	
	p/i²	N/m²	p/i²	N/m²
Aluminum	19×10^3	13×10^7	21×10^3	14×10^7
Brass...................	55×10^3	38×10^7	67×10^3	46×10^7
Copper.................	22×10^3	15×10^7	49×10^3	34×10^7
Steel, medium..........	36×10^3	25×10^7	72×10^3	50×10^7
Steel, spring............	60×10^3	41×10^7	100×10^3	69×10^7

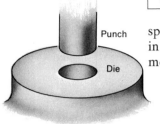

sponding stress for tension. This relative weakness to shear is utilized in the operation of a punch and die or a pair of shears in cutting sheet metal.

Example. *Medium steel requires, typically, a stress of* 50 000 p/i² *for rupture in shear. Determine the force required to punch a* 1-i *diameter hole in a steel plate* ¼-i *thick.*

The circumference of a circle of diameter $d = 1$ i is $\pi d = 3.14$ i. Hence the area of the rim $AAAA$ of the disc (Fig. 11) is $3.14 \times \frac{1}{4}$ i = 0.785 i². If a force of magnitude F is exerted on the punch, the shearing stress (force per unit area) across the rim is $F/(0.785$ i²). The steel sheet ruptures in shear when this stress becomes equal to 50 000 p/i², that is, when

$$F = 0.785 \times 50\,000 \text{ p} = 39\,200 \text{ p}.$$

This is the force (19.6 tons) required to punch the 1-inch hole. Punches and dies are operated by what is known as a *press;* in this case one would have to use a press with a rated capacity of 20 tons or more.

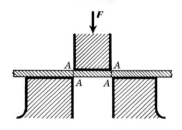

Fig. 11. Punch and die.

7. The Atomic Structure of Solids and Liquids

The elastic properties of a material are directly related to *internal* forces within the material—that is, to the forces between the *atoms* of the material that act to resist changes in the positions of the atoms.

(a) Top view of first layer

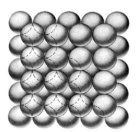

(b) Top view of first and second layers

(c) Front view of three layers

Fig. 12. The face-centered cubic arrangement of atoms.

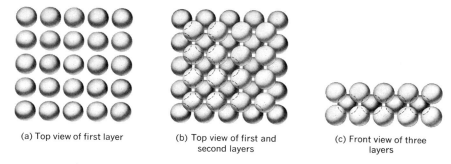

(a) Top view of first layer   (b) Top view of first and second layers   (c) Front view of three layers

Fig. 13. The body-centered cubic arrangement of atoms. The distance between the first and third layers in (c) is the same as the distances between atoms in each layer.

The atoms in a crystal of gold, for example, have a regular arrangement that can be determined by means of X-ray diffraction (Chapter 41). This arrangement, illustrated in Fig. 12, is known as a *face-centered cubic lattice*. We could build up this arrangement out of marbles by laying down a layer in square array as in (a); piling on a second layer as in (b); and so on, as illustrated in the three-layer front view in (c). This same crystalline arrangement occurs for many metallic elements. Other metals such as iron crystallize in the body-centered cubic arrangement of Fig. 13, in which it will be seen that each 'marble' touches only eight others instead of the 12 in Fig. 12. Still other metals such as magnesium crystallize in the hexagonal lattice of Fig. 14, in which each atom also touches 12 others. Most of the nonmetals crystallize in still other lattice arrangements.

But, atoms are not hard spheres like marbles. An atom consists of an electrically positive nucleus of diameter about 1 fm = 10^{-6} nm that contains most of the mass (Chapter 22). Around the nucleus are negative electrons in orbital motion. The outermost electrons are at about 0.1 to 0.2 nm from the nucleus. It is this radius of the orbits of the outermost electrons that determines the apparent size of the atoms in a crystal. (*See* Chapter 42.)

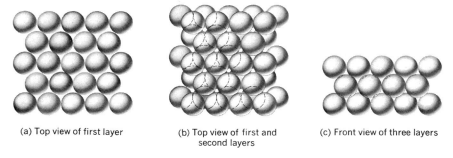

(a) Top view of first layer   (b) Top view of first and second layers   (c) Front view of three layers

Fig. 14. The hexagonal arrangement of atoms.

There is a system of electric forces between any two different atoms. When the nuclei of two atoms are further apart than about one diameter (0.2 to 0.4 nm), the resultant force between the atoms is one of *attraction*. These attractive forces are the binding forces that hold the atoms of a solid together. There is an equilibrium separation, of about 0.2 to 0.4 nm, where the net force is zero. If we try to push two atoms closer than this distance, so that the electronic orbits of the two atoms begin to overlap, repulsive forces come into play. The *centers* of the 'marbles' in Figs. 12–14 should be considered as equilibrium positions of the *nuclei* of the atoms.

The attractive and repulsive forces that we have just discussed are very like those which would result if the nuclei were connected by spiral springs as indicated schematically in Fig. 15. This figure shows nine atoms of the bottom layer, four of the middle layer, and nine of the top layer of Fig. 12 or 13. These springs should be considered as having their equilibrium lengths, so that the forces the springs exert are zero. To compress the solid, one would have to exert a force to compress the springs; to expand the solid, one would have to exert a force to stretch the springs. Because of the girder-like cross bracing, the structure in Fig. 15 will also resist shear, but a detailed analysis of this model shows that resistance to shear is not so strong as resistance to extension or compression; as seen in Table 11–1, most solids require only about one-third the force per unit area to effect the same relative motion of the atoms in shear as is required for extension or compression.

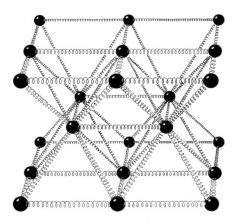

Fig. 15. The lines of action of the spring-like forces between adjoining atoms in the cubic lattice of Fig. 12 or Fig. 13.

The comparative weakness of a crystalline model to shear has immediate importance in the following discussion of the difference between the structures of solids and liquids.

The atoms of a solid are not standing still— they are vibrating, as one would expect from the spring-like nature of the interatomic forces. The higher the temperature, the greater is the average energy associated with this vibration, and the greater is the distance the atoms move. Because the forces resisting shear are less than those resisting extension or compression, the distances in the motions involving shear are greater than the distances in the motions involving extension and compression.

At some definite temperature, the shearing type of motion becomes so violent that the atoms keep right on going and do not return to their previous equilibrium positions. The crystal structure breaks down completely, and the solid melts and changes to a *liquid*. But the forces tending to prevent expansion or contraction are still sufficiently great that the atoms remain at approximately their equilibrium distances apart. *The liquid has a definite volume, but not a definite shape;* the atoms slide around relative to each other, much like the particles in a barrel of flour stirred with the hands. The liquid is said to have an *amorphous* structure; it has a definite density, a definite average separation between neighboring atoms, but not a definite position for each atom. *The liquid resists expansion or contraction, but the liquid at rest has no resistance whatsoever to static shear.*

Before closing this discussion of the atomic structure of solids, we must mention two ways in which real solids usually differ from our models: *First,* real solids are not, in general, pure elementary substances. Impurities are always present and are sometimes deliberately introduced, as in the case of metallic alloys. The presence of impurities merely complicates the pattern of interatomic forces but does not change its general nature. *Second,* a piece of metal is seldom a *single* crystal but, as can be observed under a microscope, is a collection of small crystals oriented in different ways but tightly bound together at the crystalline boundaries. Each such microcrystal contains an enormous number of atoms. This microcrystalline structure affects the magnitude, but not the character, of the internal forces that resist deformation. We can most simply understand the character of these forces by visualizing the forces that would arise in single crystals in the discussion that follows.

8. Internal Forces in Solids

Let us consider (Fig. 16) a wire that is supporting a weight w. Let us draw across this wire an imaginary horizontal plane AA and consider separately the equilibrium of the top part and the bottom part of the wire. In these considerations we shall suppose, for simplicity, that the weight of the wire itself is negligible in comparison with the weight w.

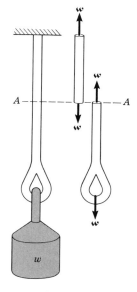

Fig. 16. Internal forces across a plane AA when a wire is in tension.

The bottom part of the wire, shown at the right, is acted on by the downward force w. In equilibrium, it must also be acted on by an upward force w, and this must be the force with which the top part of the wire acts on the bottom part across the imaginary plane. This must be the force exerted *by the atoms* of the lowest layer of the top part *on the atoms* of the uppermost layer of the bottom part. As indicated in the center portion of the figure, an equal and opposite force must be exerted by the uppermost atoms in the lower part on the atoms in the bottom layer of the top part. These are the *internal* forces that might act across the plane AA in Fig. 17, in which the forces between some of the atoms of the two layers are represented schematically by a set of spiral springs.

Let us suppose that in Fig. 17 the atoms are in their equilibrium positions; they exert no forces

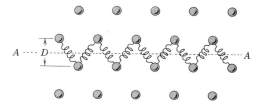

Fig. 17. The internal forces across the plane AA are the forces between the atoms on either side of AA. In this first of a set of four figures, the atoms are supposed to be in their equilibrium positions and these forces are zero.

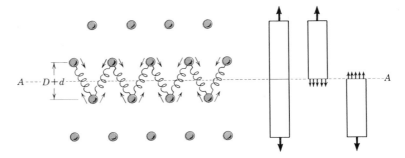

Fig. 18. There is a tensile force across the plane AA when the springs are stretched.

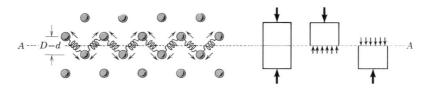

Fig. 19. There is a compressive force across the plane AA when the springs are compressed.

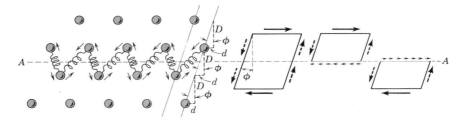

Fig. 20. There are horizontal (shearing) forces across the plane AA when the layers of atoms are displaced horizontally relative to one another.

on each other; the springs are neither stretched nor compressed. This cannot then correspond to the situation in Fig. 16, in which there is a tensile force (a pull) across this plane. In order to have a pull across the plane in Fig. 17, the layers of atoms *must* be further apart and the springs *must be stretched* as in Fig. 18. This means that the *wire* of Fig. 16 *must be stretched* by the tensile force. *A tensile force causes an elongation of a rod or bar.*

If the springs of Fig. 17 are *compressed,* the internal force across the layer is a compressive force, as illustrated in Fig. 19.

The third type of internal force that can be exerted by the atoms of Fig. 17 across the plane AA is a *shearing* force, illustrated in Fig. 20. The springs going down to the left are stretched, while the springs going down to the right are compressed. The net result is that the layer of atoms above AA feels a force to the left while the layer of atoms below AA feels a force to the right.

Figures 17–20 illustrate the fact that *the existence of internal forces implies movement of the atoms of the solid away from their equilibrium positions, and hence distortion of the solid as a whole.*

Moreover, each stretched or compressed 'spring' in Figs. 18–20 has elastic potential energy corresponding to the work required to displace the atoms relative to each other. Thus the elastic potential energy of a strained body is distributed throughout its volume and is of the nature of the potential energy associated with interatomic forces.

Thus, we see that the atomic models presented in Sections 7–8 give a satisfactory qualitative explanation of all the phenomena discussed in Sections 1–6. The models presented even give satisfactory quantitative results when optical data to be discussed in later chapters are used to provide values for the force constants of the interatomic springs.

PROBLEMS

1. When a body weighing 2 p is attached to the end of a spring hanging vertically, the spring is stretched 1.0 i. What is the force constant of this spring?   Ans: 24 p/f.

2. When a 2-kg block is attached to the end of a spring hanging vertically, the spring experiences an elongation of 4 cm. Find the force constant of this spring in N/m.

3. Assuming that the spring in Prob. 1 obeys Hooke's law, find the elongation produced when a body weighing 8 p is suspended from the end of the spring.   Ans: 4 i.

4. Assuming that the spring in Prob. 2 obeys Hooke's law, find the elongation produced (*a*) when a 5-kg block is suspended from the end of the spring and (*b*) when a force of 50 N is applied to the end of the spring.

5. What is the potential energy of the stretched spring mentioned in Prob. 3 when it supports the 8-p weight? when it supports a 16-p weight?   Ans: 1.33 fp; 5.33 fp.

6. What is the potential energy of the stretched spring for each case mentioned in Prob. 4? Express this energy in joules.

7. The end of a long wire hanging vertically and clamped at the upper end is twisted through an angle of 30° when a torque of 1.5 p·f is applied to the free end. What is the torsion constant for this wire? What torque is exerted by the clamp on the upper end of the wire? What is the potential energy of this twisted wire?
    Ans: $9/\pi$ p·f/rad; 1.5 p·f; 0.393 fp.

8. One end of a long brass rod is clamped tightly so as to prevent rotation. When a torque of 9 N·m is applied to the free end, the free end experiences a rotation of 5°.7. What is the torsion constant of this rod? What is the potential energy of the twisted rod?

9. A projectile weighing 5 p is fired horizontally from a gun weighing 2000 p at a muzzle velocity of 1200 f/s. The initial recoil energy of the gun is all transformed into potential energy of a spring. What must be the force constant of the spring if the recoil is to be limited to 2 f?   Ans: 141 p/f.

10. A railroad car weighing 40 tons rolls at 5 f/s into a 'bumper' at the end of a track. The kinetic energy of the car is all transformed into potential energy in a coil spring in the bumper. What must be the force constant of the spring in tons/inch if the motion of the spring is to be restricted to 4 i?

11. A mass of 5 kg is supported by a steel wire 8.0 m in length and 1.0 mm in diameter. What is the resulting elongation?   Ans: 2.50 mm.

12. A mass of 4 kg is supported by a copper wire of length 8 m and diameter 2 mm. What is the resulting elongation?

13. A 2-ton weight is supported by a steel rod 20 f long and 1.0 i in diameter. What is the resulting elongation?   Ans: 0.0426 i.

14. An elevator cage weighs 8500 p and is supported by two steel cables 80 f long. If each of the cables has an effective cross-sectional area of 1 i², what elongation is produced by the elevator cage? What additional elongation does each cable experience when the cage has an upward acceleration of 8 f/s²?

15. A brass rod is 2 f long and has a cross-sectional area of 1.5 i². What compressional force must be applied at the ends of this rod in order to produce a decrease of 0.01 i in the length of the rod?   Ans: 8100 p.

16. What compressional stress is involved when the length of a steel rod 20 f long is decreased by 0.01 i?

17. When an elastic rod is stretched, work is done against the elastic forces and the rod acquires elastic potential energy. Show that the potential energy per unit volume can be expressed as $\frac{1}{2}\sigma S = \frac{1}{2} M_Y \sigma^2 = \frac{1}{2} S^2/M_Y$, where σ is the longitudinal strain, S is the longitudinal stress, and M_Y is Young's modulus for the material of which the rod is composed. What units should be used for S, M_Y, and the potential energy per unit volume in the SI and British systems?

18. If a vertical wire carrying a weight is considered as a spring, express the force constant of the spring in terms of the dimensions of the wire and Young's modulus.

19. What increase in pressure is required to decrease the volume of a cubic meter of water by 0.005 percent?   Ans: 1.05×10^5 N/m².

20. What increase in pressure is required to decrease the volume of a cubic meter of ether by 0.1 per cent?

21. Show that so long as the volume change in (6) is small compared with the original volume, we can write

$$\Delta P = M_B \, \Delta\rho/\rho_0,$$

where ρ_0 is the original density and $\Delta\rho$ the increase in density.

22. What pressure would be required to increase the density of brass by 0.1 per cent? the density of glass by 0.1 per cent? the density of ether by 0.1 per cent? See Prob. 21.

23. What pressure would be required to increase the density of steel by 0.1 per cent? of copper by 0.1 per cent? of water by 0.1 per cent? See Prob. 21.   Ans: 1.7×10^8 N/m²; 1.4×10^8 N/m²; 2.1×10^6 N/m².

24. When equal and opposite forces of magnitude F_1 are applied to the top and bottom faces of a parallelopiped as in Fig. 8, show that to maintain rotational equilibrium forces must be applied to the right and left faces of such magnitude F_2 that the stresses F_2/A_2, and F_1/A_1 are equal. Show that this statement is true for arbitrary magnitude of the angle ϕ, although only small angles are involved in elastic shear.

25. A copper tube 2.0 cm in radius is 8.2 m in length and has a wall thickness of 1 mm. One end of this tube is firmly clamped and a torque tending to twist the tube is applied to the other end. What is the torsion constant for this tube? What angular displacement is produced by a torque of 32 N·m?
Ans: 251 N·m/rad; 0.127 rad.

26. Find the torsion constants of steel and brass tubes having the same dimensions as the copper tube described in Prob. 25.

27. A derrick is equipped with a 'stranded' steel cable consisting of 50 steel wires, each of which is 0.02 i² in area. What is the maximum load that can be lifted without exceeding the elastic limit of the cable material? (Assume the elastic limit for such wire is 95 000 p/i².) What will be the increase in the length of a 20-f length of cable when the maximum load is being supported? Discuss practical reasons for the use of 'stranded' cables.
Ans: 47.5 ton; 0.786 i.

28. A structural-steel 'I-beam' is 12 f long and has a cross-sectional area of 2 i². This beam is to be used as a supporting member in a building. What is the maximum compressive load the beam can support if the maximum permissible change in length is 0.05 i?

29. In Fig. 3, the magnitude of the external forces is 2 p and the radius of the disc is 4 i.

What is the magnitude of the associated torque? If the disc rotates through an angle of 115° when this torque is applied, what is the torsion constant of the supporting wire?
Ans: 1.33 p·f; 0.667 p·f/rad.

30. A solid brass rod is 1 i in diameter and 20 f long. Calculate the torsion constant for this rod. One end of this rod is firmly clamped in a vise. What is the maximum angle through which the free end of the rod can be twisted without producing a permanent deformation of the rod if the elastic limit in shear is 20 000 p/i²?

31. A 20-f length of ½-i diameter steel rod attached to a 1-hp electric motor shaft provides the connection between the shaft and the load. If the motor shaft has a rotational speed of 1800 rev/min, what torque is transmitted by the rod when the motor is running at full load? Through what angle does the rod twist under these conditions? What is the torsion constant of the rod? How much energy is stored in the rod as elastic potential energy?
Ans: 2.91 fp; 0.124 rad; 23.4 p·f/rad; 0.181 fp.

32. What is the minimum wall thickness of a tubular drive shaft 2 i in diameter that will transmit 200 hp at 400 rev/min without exceeding the elastic limit in shear, 40 000 p/i², of the steel of which it is made?

33. The rupture strength of rolled copper for shear is typically 23×10^3 p/i². What forces **F** must be applied to the metal shears shown in the figure in order to cut a strip of sheet copper 2 i wide and 0.05 i thick?   Ans: 2300 p.

34. The figure shows a stick protruding over the edge of a table. Show that in order for

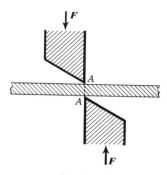

Problem 33

equilibrium to exist there must be both compressive and tensile stresses at section AA. Find the total torque associated with these compressive and tensile forces in terms of the weight and length of the protruding portion. What is the magnitude of the shearing stress at section AA in terms of the weight of the protruding portion and the linear dimensions of the stick?

35. A uniform 5-kg steel rod 2 m long is supported in a horizontal position by vertical wires attached to the ceiling and connected to the ends of the rod. The wire at end A is copper and has a cross-sectional area of 3 mm². The wire at end B is steel and has a cross-sectional area of 2 mm². Find the stresses and strains in each wire.   Ans: A, 8.17×10^6 N/m², 7.43×10^{-5}; B, 12.2×10^6 N/m², 6.10×10^{-5}.

36. At what position on the rod of Prob. 35 would one have to suspend a 5-kg object (*a*) in order to produce equal stresses in the two wires and (*b*) in order to produce equal strains in the two wires?

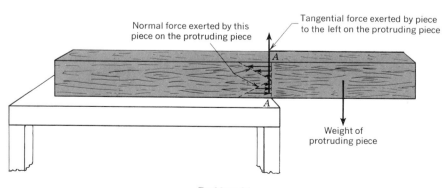

Problem 34

37. A copper wire with a cross-sectional area of 3 mm² is attached to one side of a 24-kg box lying on a horizontal floor. The coefficient of sliding friction between the box and the floor is 0.3. If the rupture strength of copper in tension is 23×10^7 N/m², what is the maximum acceleration that can be given to the box by pulling horizontally on the wire?

Ans: 27.3 m/s².

38. A 4-ton elevator cage is suspended by two stranded steel cables, each of which has an effective cross-sectional area of 0.25 i². If the elastic limit of the steel in the cables is 36×10^3 p/i², what is the maximum upward acceleration that can be given to the elevator without producing a stress of more than one half that corresponding to the elastic limit?

Ans: 4 f/s².

39. From Figs. 17–19, show that we can define the *longitudinal strain* across a plane as the relative change in distance ($\pm d/D$) between atomic layers on the two sides of the plane.

40. From Figs. 17 and 20, show that we can define the *shearing strain* across a plane as the transverse displacement of the layer of atoms on one side of the plane relative to that on the other, divided by the distance between the layers, that is, by d/D in Fig. 20.

41. Show that the distance between the layers in the face-centered cubic lattice of Fig. 12(c) is $0.707\,d$ (exactly $\frac{1}{2}\sqrt{2}\,d$), where d is the diameter of the 'marbles.'

42. Show that, in the face-centered cubic lattice of Fig. 12, the relation between the diameter d of an atom, considered as a ball, and the number N of atoms per unit volume, is $N = \sqrt{2}/d^3$.

43. For the body-centered cubic lattice of Fig. 13, find the number N of atoms, considered as balls, per unit volume, in terms of the diameter d. What is the ratio of this number to that in Prob. 42? Ans: $\sqrt{3}/d^3$; 0.919.

44. For the hexagonal lattice of Fig. 14, find the number N of atoms, considered as balls, per unit volume, in terms of the diameter d. Explain why this lattice and the face-centered cubic are appropriately said to be *close-packed*.

45. Copper has a density of 9.0 g/cm³, and the individual copper atom has a mass of 1.06×10^{-22} g. Copper has the face-centered cubic lattice of Fig. 12. From the result of Prob. 42, compute to two significant figures the 'diameter' of the copper atom. Ans: 0.26 nm.

46. The very light alkali metal, sodium, has a density of 0.97 g/cm³, and the individual sodium atom has a mass of 0.38×10^{-22} g. Sodium has the body-centered cubic structure of Fig. 13. From the result of Prob. 43, compute to two significant figures the 'diameter' of the sodium atom.

47. Explain the terms 'face-centered cubic' and 'body-centered cubic' used in connection with Figs. 12 and 13 by showing that the crystal 'lattice' of Fig. 12 is composed of a continuous succession of contiguous cubes, each with an atom at each of the eight corners and one centered on each of the six faces (these cubes have two faces parallel to the plane of the paper in Fig. 12(a) and (b), but the other faces are at 45° to the axes in the drawing), while Fig. 13 is similarly composed of cubes with atoms at the corners and at the centers of the cubes.

48. The figure again shows the atoms of Fig.

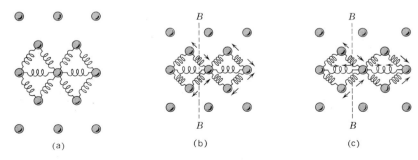

Problem 48

17 in part (a), and those of Fig. 19 in part (b). Show that if the solid is compressed as in Fig. 19 *without* swelling laterally, the forces in (b) across a plane such as *BB* are unbalanced, since there are no forces on the lateral surfaces of the solid. Hence show that the solid must swell sidewise as in (c) and that then these lateral forces are balanced by the forces of the horizontal springs shown in this figure but omitted from previous drawings. Give a similar discussion of the lateral contraction of a rod in tension.

12 PERIODIC MOTION

In this chapter we shall consider some types of motion in which the resultant force or torque acting on a body varies periodically. One extremely important type is the vibratory motion of a particle about an equilibrium position when the restoring force is proportional to the displacement. Vibratory motion results when a body hanging from a spring is pulled downward and then released; other common examples of this type of motion are the vibrations of strings and air columns of musical instruments, the vibrations of bridges and buildings, and the oscillation of the balance wheel of a watch or of the pendulum of a clock.

Vibratory motion, like motion in a circle at constant speed, is a *periodic* motion:

> A **periodic motion** is one in which the motion of a body is identically repeated in each of a succession of equal time intervals.

For example, a particle moving at constant speed in a circle traverses the same path over and over again with the same velocities; a pendulum repeats its to-and-fro motion again and again.

A **cycle** is one complete execution of a periodic motion.

The **period** of a periodic motion is the time T required for the completion of a cycle.

The **frequency** of a periodic motion is the number f of cycles completed per unit time.

From their definitions, it is readily seen that the period T of a periodic motion is the reciprocal of the frequency f, that is, $T=1/f, f=1/T$. For example, a pendulum with a *period* $T = \frac{1}{5}$ s has a *frequency* of five cycles per second, $f = 5 \text{ s}^{-1} = 5$ Hz.

Frequency is measured in a unit* called the **hertz (Hz),** which corresponds to *one cycle per second*, or $1 \text{ Hz} = 1 \text{ s}^{-1}$.

The principles we shall learn in studying periodic motion in this chapter will have applications not only in mechanics but also in the study of sound, light, and electricity. Sound waves, light waves, radio waves, and alternating currents all involve periodic phenomena.

1. Simple Harmonic Motion

Simple harmonic motion is the vibratory motion of a particle about an equilibrium position when the restoring force (the force tending to return the particle to its equilibrium position) is directly proportional to the displacement of the particle from its equilibrium position.

Elastic forces of the type involved in Hooke's law have the character that gives rise to simple harmonic motion. As an example, consider the idealized situation depicted in Fig. 1, where a block of mass m sliding

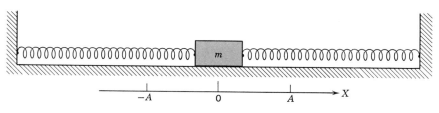

Fig. 1. A body sliding on a frictionless horizontal plane, between two identical stretched springs.

on a frictionless plane is attached to two identical stretched springs. At the central position, $X=0$, the springs exert equal and opposite forces and the resultant force on the block is zero. If the block is pulled a

---

*Named after HEINRICH HERTZ, the German physicist who first observed radio waves in 1886.

distance X to the right, the left-hand spring exerts a greater force to the left, while the right-hand spring exerts less force to the right. The resultant force acting on the block is therefore directed toward the left and, by Hooke's law, is proportional to the displacement X but is opposite in direction. Therefore, the X-component of this resultant force is

$$F_X = -KX, \qquad (1)$$

where K is the *effective force constant*. From Newton's second principle, this force is equal to $ma_X = m\, d^2X/dt^2$, that is,

$$m\, d^2X/dt^2 = -KX, \quad \text{or} \quad d^2X/dt^2 = -(K/m)\, X. \qquad (2)$$

Since K and m are both positive constants, we note from (2) that X must be a function of the time that has a second derivative equal to a negative constant times the function itself.

In our search for such a function, we think immediately of $\sin\omega t$ and $\cos\omega t$, where ω is some constant, since the second derivatives of these trigonometric functions are equal to $-\omega^2$ times the functions themselves; for example, if $X = \cos\omega t$, then $d^2X/dt^2 = -\omega^2 \cos\omega t = -\omega^2 X$, and a corresponding relation holds for $X = \sin\omega t$. We note that throughout this chapter *the arguments of trigonometric functions are expressed in radians*, as they are in the standard differentiation formulas.

In order to make a definite choice of a function, consider a case in which the block in Fig. 1 is moved a distance A to the right, held at rest, and then released at time $t=0$; this description defines the *initial* condition: $X = A$ at $t = 0$. After the block is released, it moves back toward its equilibrium position but, since it has momentum at the time it reaches the equilibrium position, it will 'overshoot' and continue moving to the left until it attains a displacement $-A$. At this position, $X = -A$, the velocity will again be zero and, owing to the restoring force, the block will again move toward the equilibrium position, overshoot, and return to its initial position at $X = A$. This motion will be repeated over and over at some definite frequency f. We shall attempt to express the displacement as $X = A \cos 2\pi ft$, since we know from our discussion of (2) that such a cosine will be a satisfactory function and the one we have selected gives the correct initial conditions $X = A$ at $t = 0$. By selection of the arbitrary constant ω as $2\pi f$, we have introduced the frequency f in a satisfactory manner since $\cos 2\pi ft$ goes through its entire range of values f times per unit time.

Having selected $X = A \cos 2\pi ft$ as a solution, we now attempt to determine the frequency f (as yet unknown) in terms of K and m. We do this by determining d^2X/dt^2 and substituting its value in (2). Thus, $dX/dt = -2\pi f A \sin 2\pi ft$ and $d^2X/dt^2 = -4\pi^2 f^2 A \cos 2\pi ft$. Substitution of this value of d^2X/dt^2 along with our assumed value $X = A \cos 2\pi ft$ in (2) gives

$$-4\pi^2 f^2 A \cos 2\pi ft = -(K/m) A \cos 2\pi ft.$$

Solution for frequency f in terms of K and m gives

$$f = \frac{1}{2\pi}\sqrt{\frac{K}{m}}. \quad \begin{Bmatrix} f \text{ in Hz} \\ K \text{ in N/m} \\ m \text{ in kg} \end{Bmatrix} \text{ or } \begin{Bmatrix} f \text{ in Hz} \\ K \text{ in p/f} \\ m \text{ in sl} \end{Bmatrix} \quad (3a)$$

Because of the relation $T = 1/f$ between the period T and the frequency f, the *period* of the motion is given by

$$T = 2\pi \sqrt{m/K}. \quad \begin{Bmatrix} T \text{ in s} \\ m \text{ in kg} \\ K \text{ in N/m} \end{Bmatrix} \text{ or } \begin{Bmatrix} T \text{ in s} \\ m \text{ in sl} \\ K \text{ in p/f} \end{Bmatrix} \quad (3b)$$

It should be noted that the maximum displacement of the block from its equilibrium position is given by A, which in our particular case also corresponds to the initial displacement; A is called the *amplitude* of the motion:

> The **amplitude** of a simple harmonic motion is the magnitude of the maximum displacement of the vibrating particle from its equilibrium position.

We conclude from (3) that the frequency of the simple harmonic motion executed by a given mass and spring system is independent of the amplitude of the motion. And in general this independence of amplitude is true for other physical systems for which a differential equation of the form $B\, d^2X/dt^2 = -CX$ governs the motion.

A differential equation like (2), which contains no derivatives higher than the second and has no terms involving the function X or its derivatives to any power greater than unity is called a 'linear, second-order differential equation.' Its *general* solution must contain *two* arbitrary constants. One of these constants is the arbitrary amplitude A; the second constant does not appear in our solution $X = A \cos 2\pi ft$ because of our special choice of initial conditions: $X = A$ at $t = 0$. The general solution of (2) can be written as

$$X = A \cos(2\pi ft + \phi) \quad (4)$$

where ϕ is the required second constant and is called the *initial phase angle;* the value of ϕ is determined by the *conditions at* $t = 0$. In our special case, the initial phase angle was zero.

> **Example.** *What initial phase angle ϕ would be needed in the solution (4) if the block in Fig. 1 is released from rest at $X = -A$ at $t = 0$?*
>
> We note that ϕ must be chosen to give the correct displacement $X = -A$ at $t = 0$; hence, $\cos(2\pi ft + \phi)$ must have the value -1 at $t = 0$, that is, $\cos \phi = -1$. Hence, the required value of the phase angle ϕ is π.

Since (4) gives a general solution of (2), we may use it to obtain general expressions for the velocity $v_X = dX/dt$ and the acceleration $a_X = dv_X/dt$:

$$v_X = -2\pi f A \sin(2\pi ft + \phi) = -\sqrt{K/m}\, A \sin(2\pi ft + \phi), \quad (5)$$

$$a_X = -4\pi^2 f^2 A \cos(2\pi ft + \phi) = -(K/m) A \cos(2\pi ft + \phi). \quad (6)$$

Hence the magnitude of the maximum velocity $v_{X\text{Max}}$ is $2\pi f A$ and that of the maximum acceleration $a_{X\text{Max}}$ is $4\pi^2 f^2 A$; both are directly proportional to the amplitude of the motion. Comparison of (4) and (5) shows that v_X attains its maximum positive and negative values when $X = 0$. Comparison of (4) and (6) shows that the magnitude of a_X is directly proportional to the displacement X and that the direction of a_X is always opposite that of the displacement. Figure 2 gives graphical representations of v_X and a_X at various points in the path of a particle executing simple harmonic motion.

Velocity   Acceleration

Fig. 2. The position, velocity, and acceleration of a particle executing simple harmonic motion, shown at intervals of $\frac{1}{16} T$, where T is the period. The long arrows above and below indicate the direction of motion.

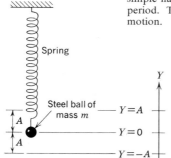

Fig. 3. A mass supported by a spring will execute simple harmonic motion if moved away from its equilibrium position and then released.

The experimental arrangement shown in Fig. 1 is by no means the simplest arrangement we could have chosen to illustrate simple harmonic motion. Perhaps the simplest experimental arrangement is that shown in Fig. 3, in which a steel ball of mass m is suspended from a light spring. We avoided treating this case first, because gravitational forces as well as elastic forces are directly involved. However, as will be shown in Prob. 9, the frequency of the motion in Fig. 3 is still given by (3), where K is the force constant of the spring. Since the motion is in the Y-direction, the displacement is given by $Y = A \cos(2\pi ft + \phi)$, and the velocity v_Y and acceleration a_Y will be given by expressions similar to (5) and (6). The resultant *restoring force* is $F_Y = -KY$.

Example. *A spiral spring* 3 m *long hangs from the ceiling. When a mass of* 1 kg *is suspended from the spring, the spring lengthens by* 40 cm *in the*

equilibrium configuration. The mass is then pulled down an additional 10 cm *and released. Determine the characteristics of the subsequent motion, neglecting the mass of the spring.*

The weight of the 1-kg mass is 9.8 N. Since this weight lengthens the spring 0.4 m, the force constant is

$$K = (9.8 \text{ N})/(0.4 \text{ m}) = 24.5 \text{ N/m}.$$

The period of the simple harmonic motion is given by (3b):

$$T = 2\pi \sqrt{(1 \text{ kg})/(24.5 \text{ N/m})} = (2\pi/4.95) \text{ s} = 1.27 \text{ s},$$

while the frequency is

$$f = 1/T = (1/1.27 \text{ s}) = 0.787 \text{ Hz}.$$

As in the preceding example, we use the phase angle $\phi = \pi$ to satisfy the initial condition, $Y = -10$ cm at $t = 0$. We insert $A = 10$ cm and $2\pi f = 4.95$ s^{-1}. Since

$$\cos(2\pi ft + \pi) = -\cos 2\pi ft,$$

we can write $\qquad Y = -10 \cos[(4.95 \text{ s}^{-1}) t]$ cm.

Since the period is 1.27 s, we compute the fourth column of the accompanying table by inserting the values $t = 0, 0.1$ s, $\cdots$, 1.3 s, to cover the motion during the first period. These values of Y are plotted in Fig. 4.

t (s)	$[4.95 \text{ s}^{-1}] t$ (rad)	Equiv. degrees	Y (cm)	v_Y (cm/s)	a_Y (cm/s^2)
0	0	0	−10	0	+245
0.1	0.495	28°.4	− 8.80	+23.6	+216
0.2	0.990	56°.7	− 5.49	+41.4	+135
0.3	1.485	85°.1	− 0.85	+49.3	+ 21
0.4	1.980	113°.4	+ 3.97	+45.4	− 97
0.5	2.475	141°.8	+ 7.86	+30.6	−193
0.6	2.970	170°.2	+ 9.85	+ 8.4	−241
0.7	3.465	198°.5	+ 9.48	−15.7	−232
0.8	3.960	226°.9	+ 6.83	−36.1	−167
0.9	4.455	255°.2	+ 2.55	−47.9	− 62
1.0	4.950	283°.6	− 2.35	−48.1	+ 58
1.1	5.445	312°.0	− 6.69	−36.8	+164
1.2	6.940	340°.3	− 9.42	−16.7	+231
1.3	7.435	368°.7	− 9.88	+ 7.5	+242

Differentiation of the preceding equation gives

$$v_Y = dY/dt = (49.5 \text{ s}^{-1}) \sin[(4.95 \text{ s}^{-1}) t] \text{ cm}$$
$$= 49.5 \sin[(4.95 \text{ s}^{-1}) t] \text{ cm/s}.$$

A second differentiation gives

$$a_Y = dv_Y/dt = 245 \cos[(4.95 \text{ s}^{-1}) t] \text{ cm/s}^2.$$

Values of v_Y and a_Y are tabulated and plotted along with those of Y.

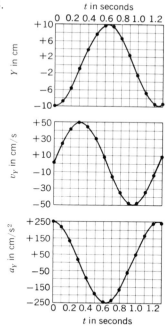

Figure 4

2. The Reference Circle

We have shown that the kinematics of simple harmonic motion can be described by equations like (4), (5), and (6), which give displacement, velocity, and acceleration, respectively. As an aid to understanding these equations and visualizing the motion, we shall compare the motion of a vibrating particle with the already familiar motion of a particle traversing a circular path at constant speed. We shall show that the X-component of the circular motion is the same as that of a particle in simple harmonic motion parallel to the X-axis, as in Fig. 1.

Consider the particle shown in Fig. 5, which moves at constant speed v in a circular path of radius A. If the angular displacement of the particle at $t=0$ is $\theta=\phi$, its angular displacement at any time t is

$$\theta = \phi + \omega t,$$

in radians, where $\omega = v/A$ is the angular velocity in rad/s. We may also write θ in terms of the frequency f; since $\omega = 2\pi f$,

$$\theta = \phi + 2\pi f t.$$

The X-coordinate of the particle as a function of time is

$$X = A \cos\theta, \quad \text{or} \quad X = A \cos(2\pi f t + \phi).$$

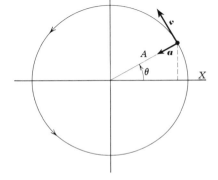

Fig. 5. Circular motion.

Since this expression is the same as (4), the X-components of the velocity and the acceleration are given by (5) and (6).

Hence we see that

The simple harmonic motion of the particle in Fig. 1, with amplitude A and phase angle ϕ, is exactly the projection on the X-axis of the motion of the particle in Fig. 5, moving at constant speed and with the same frequency in a circle of radius A, and having the initial position $\theta = \phi$ at $t=0$.

The validity of this conclusion can be tested experimentally by casting on the same screen the shadow of a small body executing simple harmonic motion and the shadow of a small body in uniform circular motion in a plane parallel to the line of motion of the vibrating particle. The shadows will have identical motions, provided the amplitude of the vibration and the radius of the circle are the same, the frequencies are the same, and the initial phase angles are the same. The circular motion 'matching' a given simple harmonic motion provides a useful means of visualizing the vibratory motion; the 'matching' circle is called 'the reference circle.'

3. Energy Relations in Simple Harmonic Motion

Now let us consider the energy relations involved in the simple harmonic motion of the particle in Fig. 1 or Fig. 3. To give the body its initial displacement, we must do *work* against the restoring force. As a result of the performance of this work, the system acquires potential energy. When the body is released, this initial potential energy is converted into kinetic energy as the body moves back to its equilibrium position; as a result of its momentum at the equilibrium position, the body moves past the equilibrium position and its kinetic energy is converted into potential energy, as work is done against the restoring force. Hence, the energy of the system is alternately all in the form of potential energy when the displacement is a maximum and all in the form of kinetic energy as the body passes through its equilibrium position. However, if there is no friction, *the mechanical energy* (kinetic plus potential) *is at all times constant and is equal to the initial work done on the system*. Let us now see how this fact can be used in determining the velocity of the particle at various points in the path traversed.

The total external work we should have to do against the restoring force to give the body in Fig. 1 a displacement X represents the potential energy. Since the restoring force (1) is exactly of the form considered on p. 220, we can use the result (2), p. 221, and write

$$E_P = \tfrac{1}{2} KX^2. \tag{9}$$

If the body is given an initial displacement A and is released from rest,

$$\text{initial potential energy} = \tfrac{1}{2} KA^2, \quad \text{initial kinetic energy} = 0. \tag{10}$$

This initial potential energy is equal to the total energy E of the system, and at all times after the body has been released the sum of the potential energy and kinetic energy of the system is equal to E:

$$E_P + E_K = E = \tfrac{1}{2} KA^2. \tag{11}$$

The kinetic energy of the system is $\tfrac{1}{2} mv_X^2$, provided the mass of the springs is negligible. Hence, when the body has displacement X, (11) becomes

$$\tfrac{1}{2} KX^2 + \tfrac{1}{2} mv_X^2 = \tfrac{1}{2} KA^2. \tag{12}$$

In order to obtain the magnitude of the velocity, we may solve equation (12) for v_X; thus,

$$v_X^2 = (K/m)(A^2 - X^2), \quad v_X = \pm \sqrt{K/m}\sqrt{A^2 - X^2}. \tag{13}$$

This equation gives the velocity component v_X as a function of the position X. When $X = 0$, the velocity has its extreme values $\pm v_{X\text{Max}} = \sqrt{K/m}\, A$, in agreement with the value obtained in (5). This expression for $v_{X\text{Max}}$ gives $\tfrac{1}{2} mv_{X\text{Max}}^2 = \tfrac{1}{2} KA^2$, which expresses the fact that the value of the energy at the equilibrium position, when it is all kinetic, equals the value at the maximum excursion, when it is all potential.

One other point should be mentioned concerning mechanical energy associated with a particle of mass m executing simple harmonic motion. From (11) and (3a), we can write the total energy in the form

$$E = \tfrac{1}{2} KA^2 = 2\pi^2 f^2 A^2 m. \tag{14}$$

The total energy of a particle of mass m executing simple harmonic motion is proportional to the square of the amplitude and to the square of the frequency of motion.

Example. *For the oscillating mass whose motion is described in the example on pp. 246–247, determine the maximum kinetic and potential energies, the total energy, and the velocity as a function of position.*

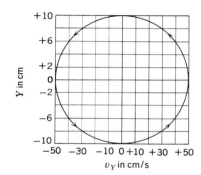

Figure 6

The maximum kinetic energy is

$$\tfrac{1}{2} m v_{Y\text{Max}}^2 = \tfrac{1}{2} (1 \text{ kg})(0.495 \text{ m/s})^2 = 0.122 \text{ J}.$$

The maximum potential energy is

$$\tfrac{1}{2} KA^2 = \tfrac{1}{2} (24.5 \text{ N/m})(0.1 \text{ m})^2 = 0.122 \text{ J}.$$

The total energy has the same value, 0.122 J. From (12), we see that

$$v_Y^2 + (K/m) Y^2 = (K/m) A^2,$$

or $\quad v_Y^2 + (24.5 \text{ s}^{-2}) Y^2 = (24.5 \text{ s}^{-2})(10 \text{ cm})^2.$

This is the equation of the ellipse plotted in Fig. 6. The scales of Y and v_Y in this plot are chosen so as to make the ellipse a circle.

Returning to (12), we note that the potential energy $E_P = \tfrac{1}{2} KX^2$ can be plotted as the parabola in Fig. 7, in which the ordinate is energy and the abscissa is displacement X. The total mechanical energy E, which is constant, is shown by the horizontal line intersecting the parabola at $X = +A$ and $X = -A$. At the points of intersection, $E = E_P$, and the total energy of oscillation is entirely potential; at $X = 0$, $E_P = 0$, and the energy of oscillation is entirely kinetic: $E_K = E$. At all other values of X, the kinetic energy $E_K = E - E_P$ is represented by the length of the vertical line between the horizontal line at ordinate E and parabolic plot of E_P.

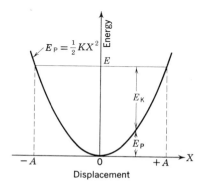

Fig. 7. The parabola $E_P = \tfrac{1}{2} KX^2$ shows potential energy as a function of displacement. The total mechanical energy E is at all points equal to the sum of the potential and kinetic energies: $E = E_P + E_K$.

Since gravitational potential energy is proportional to height above a reference level, the simple harmonic motion can be visualized in terms of Fig. 7 by imagining a particle sliding without friction along a surface shaped like the parabola for E_P. Such a particle would slide rapidly through the bottom of the 'valley' and would spend most of its time high on the 'hillsides.' It would have exactly the same kinetic and potential energies as the body in Fig. 1 at each value of Y, but it would not have the same period, since it has farther 'to go.' The total energy E associated with the motion of the particle would determine the maximum *level* attained on the curve in the figure. Thus, it is convenient at times to speak of the total energy E as corresponding to a certain *energy level*. This concept of energy levels will prove useful in later treatment of problems in atomic structure.

4. Angular Simple Harmonic Motion; Torsional Oscillation

As might be expected, there is a rotational analog to the simple harmonic motion of a particle. Consider a body such as the disc shown in Fig. 8. The body is suspended by a vertical wire that obeys Hooke's law for torsional deformations. If the body is rotated through an angle θ from its equilibrium position, the wire exerts a *restoring* torque proportional to θ which we can write (see p. 222) in the form

$$L = -C\theta. \tag{15}$$

The restoring torque produces an angular acceleration $\alpha = d^2\theta/dt^2$ of the body, given by the relation

$$L = I\alpha = -C\theta, \tag{16}$$

where I is the moment of inertia of the body. Hence,

$$\alpha = d^2\theta/dt^2 = -(C/I)\theta. \tag{17}$$

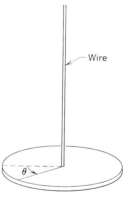

Fig. 8. Torsional oscillation.

The mathematical form of (17) is identical with that of equation (2). Hence the solution to the problem of the motion when the body is released from rest at angle Θ from equilibrium will be analogous to (4), (5), and (6):

$$\left.\begin{array}{l} \theta = \Theta \cos 2\pi ft, \\ \omega = d\theta/dt = -2\pi f \Theta \sin 2\pi ft, \\ \alpha = d\omega/dt = -4\pi^2 f^2 \Theta \cos 2\pi ft = -(4\pi^2 f^2)\theta. \end{array}\right\} \tag{18}$$

Comparison with (17) shows that the frequency f is given by the relation $4\pi^2 f^2 = C/I$; hence

$$f = \frac{1}{2\pi}\sqrt{\frac{C}{I}}. \qquad \begin{Bmatrix} f \text{ in Hz} \\ C \text{ in N·m/rad} \\ I \text{ in kg·m}^2 \end{Bmatrix} \text{ or } \begin{Bmatrix} f \text{ in Hz} \\ C \text{ in p·f/rad} \\ I \text{ in sl·f}^2 \end{Bmatrix} \tag{19}$$

These equations are like those for linear simple harmonic motion, except that all quantities involved in translational motion have been replaced by their rotational analogs; the motion is called *rotational* or *angular simple harmonic motion*. The maximum angle Θ is called the *amplitude* of the angular simple harmonic motion.

We can readily write down the energy relations for a body executing angular simple harmonic motion. The external work done on the system in giving the body its initial angular displacement Θ is equal to $\tfrac{1}{2} C\Theta^2$; this is the value of the initial potential energy of the system and therefore the value of the total mechanical energy E of the system. If the body is released and permitted to oscillate about its equilibrium position, the mechanical energy E of the system remains constant, provided there are no frictional effects. At any angle θ in the subsequent motion the potential energy is $\tfrac{1}{2} C\theta^2$ and the kinetic energy is $\tfrac{1}{2} I\omega^2$; therefore, we may write

$$E_P + E_K = E, \quad \text{or} \quad \tfrac{1}{2} C\theta^2 + \tfrac{1}{2} I\omega^2 = \tfrac{1}{2} C\Theta^2. \tag{20}$$

From this equation, we obtain an expression analogous to (13):

$$\omega = \pm \sqrt{C/I}\sqrt{\Theta^2 - \theta^2}. \tag{21}$$

The system of Fig. 8 is frequently called a *torsion pendulum* and is said to execute *torsional oscillations*.

5. The Motion of a Pendulum

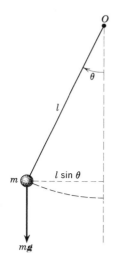

Fig. 9. The simple pendulum.

The rotational motion of a pendulum about a horizontal axis through its point of support approximates simple harmonic motion very closely, provided the amplitude of oscillation about the equilibrium position is small compared with one radian. As the first example of pendulum motion, let us consider the motion of the *simple pendulum* shown in Fig. 9, which consists of a particle of mass m, called the *bob*, supported by a light wire or thread of length l and negligible mass. A simple pendulum is an idealization in which the whole mass is considered as concentrated at a point.

We can consider the motion as the rotational oscillation of a rigid body about a horizontal axis through O, the fixed point of support. The equilibrium position of the pendulum is the position in which the center of mass of the bob is immediately below the point of support O. If the pendulum has an angular displacement θ from the equilibrium position as shown in Fig. 9, the force of gravity exerts a torque about the axis through O that tends to restore the pendulum to the equilibrium position. The magnitude of this torque is $mgl \sin\theta$. For small angular displacements, $\sin\theta$ may be replaced by θ itself measured in radians. Hence, for small values of θ, we may write

$$L = -mgl\theta, \tag{22}$$

where the negative sign is used because the torque is counterclockwise when θ is clockwise, and vice versa. This torque gives the pendulum an

angular acceleration α expressed by the relation $L = I\alpha = -mgl\theta$,
or
$$\alpha = d^2\theta/dt^2 = -(mgl/I)\theta, \qquad (23)$$
where I, the rotational inertia about the axis through O, is $I = ml^2$.

Equation (23) is identical in form with equation (17). Therefore, we conclude that the angular motion of the pendulum is similar to that of the body in Fig. 8. The frequency f of the simple pendulum, for *small* angular oscillations, is given by (19), with C replaced by mgl, as

$$f = \frac{1}{2\pi}\sqrt{\frac{mgl}{ml^2}} = \frac{1}{2\pi}\sqrt{\frac{g}{l}}. \qquad (24)$$

The frequency of a simple pendulum is seen to be independent of its mass.

A rigid pendulum, such as the old-fashioned clock pendulum, is called a *physical pendulum*. A physical pendulum also executes rotational simple harmonic motion when the amplitude of the motion is small. Figure 10 represents a physical pendulum, pivoted about a horizontal axis perpendicular to the drawing and passing through point O located at a distance r from the center of mass of the body. If the body has an angular displacement θ from the equilibrium position, the restoring torque has the value $L = -mgr\sin\theta$, or, for small displacements, $L = -mgr\theta$. The body therefore has angular acceleration $d^2\theta/dt^2$ given by $I_0 d^2\theta/dt^2 = -mgr\theta$, or

$$\alpha = d^2\theta/dt^2 = -(mgr/I_0)\theta,$$

where I_0 is the rotational inertia of the body about the axis through O. This equation is similar to equation (23), and hence the frequency and period of oscillation are given by

$$f = \frac{1}{2\pi}\sqrt{\frac{mgr}{I_0}}, \qquad T = 2\pi\sqrt{\frac{I_0}{mgr}}. \qquad (25)$$

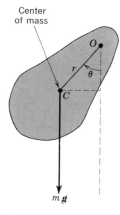

Fig. 10. The physical pendulum.

Example. *Find the frequency of oscillation of a meter stick pivoted at one end as in Fig. 11.*

Let l be the total length of the stick; then $r = \frac{1}{2}l$. From the value $I_C = \frac{1}{12}ml^2$ and the parallel-axis relation (p. 189), we find

$$I_0 = \frac{1}{12}ml^2 + mr^2 = \frac{1}{12}ml^2 + \frac{1}{4}ml^2 = \frac{1}{3}ml^2.$$

Hence, the oscillation frequency is given by

$$f = \frac{1}{2\pi}\sqrt{\frac{\frac{1}{2}mgl}{\frac{1}{3}ml^2}} = \frac{1}{2\pi}\sqrt{\frac{3g}{2l}}.$$

Taking $g = 9.8$ m/s², $l = 1$ m, we obtain

$$f = \frac{1}{2\pi}\sqrt{\frac{3 \times 9.8 \text{ m/s}^2}{2 \times 1 \text{ m}}} = 0.610 \text{ s}^{-1} = 0.610 \text{ Hz}.$$

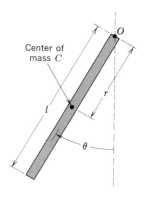

Figure 11

Our discussion of the motion of pendulums has been an approximate one for the case of 'small amplitudes.' In this case the motion is simple harmonic, and the frequency is independent of the amplitude. For large amplitudes the motion is periodic but not simple harmonic. The small-amplitude treatment gives the frequency correctly to within one per cent for an amplitude as large as 20° (on each side of the vertical).

It might be noted that if g is known, the rotational inertia I of a body of mass m and any shape whatever can be determined by suspending the body and allowing the body to swing as a pendulum about an axis through some point O. By noting the frequency of oscillation, the rotational inertia I_0 about the axis can be obtained from (25). The rotational inertia about a parallel axis through the center of mass can then be determined by the parallel-axis relation given on p. 189.

Because of the occurrence of g in (25), this relation can be used to determine gravitational acceleration by measuring the frequency of a pendulum of accurately-known dimensions, for which I_0 can be calculated. This method furnishes one of the most accurate means of measuring g.

6. Forced Oscillations

The periodic motions discussed thus far have been free simple harmonic oscillations. Once oscillations of this idealized type have been initiated, they continue with constant amplitude, since mechanical energy is conserved as expressed in (11). In any real mechanical oscillation, the amplitude gradually decreases as mechanical energy is gradually dissipated by friction. The resulting motion with gradually decreasing amplitude is called *damped oscillatory motion*. One important source of damping involves frictional forces that are proportional to the velocity of the oscillator; as we shall see in Chapter 13, fluids can produce damping of this type as a result of their viscosity. Thus, velocity-dependent frictional forces are exerted on a simple pendulum by the surrounding air, and the amplitude of the pendulum motion decreases logarithmically with time. The amplitude A at time t is given by the expression

$$A = A_0 e^{-\beta t}, \tag{26}$$

where A_0 is the initial amplitude and β is a measure of the size of the damping forces; the constant β is called the logarithmic decrement of the motion.

If a damped harmonic oscillator is subjected to a periodic 'driving force' $F = F_0 \sin 2\pi f_D t$, the frequency of the resulting oscillation is the same as the frequency f_D of the driving force but the amplitude is strongly dependent on the difference between the driving frequency f_D and the natural frequency f_N of the damped oscillator, which is very nearly equal to the frequency given in (3a) for small values of β in (26). The maximum amplitude is attained when $f_D = f_N$; at this 'resonance condition,' oscillations of large amplitude can be produced by driving forces having even small values of F_0. Thus, dangerously large vibrations in structures can be produced by long-continued driving forces with frequencies equal to the natural oscillation frequencies of the structures.

There is a close analogy between mechanical and electrical oscillations; resonance phenomena in the latter case will be treated in some detail in Chapter 32.

PROBLEMS

NOTE: *Consider the masses of springs as negligible, and the amplitudes of oscillation of pendulums as small compared with 1 rad, unless otherwise instructed.*

1. A car moves at a constant speed of 60 mi/h around a circular race track having a diameter of 1 mi. What are the period and frequency of its periodic motion?
Ans: 188 s, 5.31 mHz.

2. The moon moves around the earth in a nearly circular path of radius 240 000 mi, once every 27.3 days. Calculate the frequency of the moon's orbital motion and the magnitudes of the velocity and the centripetal acceleration of the moon.

3. Consider a steel ball dropped from height h and making perfectly elastic collisions with a horizontal plane. (*a*) Show that following its initial release the ball will execute periodic motion with frequency $f = \frac{1}{2}\sqrt{g/2h}$. (*b*) If the collision between the ball and plane is inelastic, derive an expression for the time between collisions in terms of h and the coefficient of restitution e defined in Prob. 19, p. 141. Show that the resulting motion is *aperiodic*.

4. Show by substitution that $X = A \sin(2\pi f t + \phi')$ is a solution to (2), provided f has the value given in (3). What is the relation between ϕ' and the angle ϕ employed in (4)?

5. A 2-kg mass is attached to two identical springs and placed on a frictionless plane as shown in Fig. 1. The mass is displaced 10 cm to the right and then released. An observer notes that the mass executes 15 oscillations in 30 s. What is the amplitude of the motion? the period? the frequency? What is the effective force constant of the spring system? Write an equation describing the motion. Ans: 0.1 m; 2 s; 0.5 Hz; 19.7 N/m; $X = (0.1 \text{ m}) \cos[(\pi \text{ s}^{-1})t]$.

6. A mass of 0.25 sl is mounted in the arrangement indicated in Fig. 1. When the mass is displaced 3 i to the right and then released, it begins an oscillatory motion. If it executes 15 complete oscillations each minute, find the amplitude, period, and frequency of the motion and determine the effective force constant of the spring system; write an equation describing the motion.

7. Referring to Prob. 5, calculate the total energy of the system and the magnitudes of the maximum velocities and accelerations attained by the vibrating mass. Write expressions giving elastic potential energy and kinetic energy as functions of time. Ans: 0.0985 J; 0.313 m/s; 0.981 m/s^2; $E_P = (0.0985 \text{ J}) \cos^2[(\pi \text{ s}^{-1})t]$; $E_K = (0.0985 \text{ J}) \sin^2[(\pi \text{ s}^{-1})t]$.

8. Find the maximum values of kinetic and potential energies of the oscillating system described in Prob. 6. Using the same set of springs, how could you set up a system that would have a period of oscillation of exactly 1 Hz?

9. For a small bob suspended from a light helical spring as in Fig. 3, show that there is a restoring force $F_Y = -KY$, where Y is the vertical displacement from the equilibrium position. Note that when the bob is initially attached to the spring, the spring experiences an initial elongation b, and that at equilibrium the upward force Kb exerted on the bob by the spring is just equal to the weight mg of the bob.

10. Considering both gravitational and elastic potential energy, show that the change in the potential energy of the oscillator in Fig. 3 is given by $E_P = \frac{1}{2} KY^2$, where Y is the vertical displacement of the bob from its equilibrium position.

11. A bob having a mass of 200 g is suspended from a light vertical spring. When displaced from its equilibrium position and then released, the bob executes simple harmonic motion at a frequency of 3 Hz. What would be the frequency of the motion executed by a 1-kg bob if it were suspended from this spring, displaced from its equilibrium position, and then released? Ans: 1.33 Hz.

12. When a 0.2-sl body is suspended from a vertical spring and then set into oscillatory motion along a vertical line, the frequency of the motion is 4 Hz. What is the magnitude of the force one would need to apply to the end of this spring to cause an elongation of 1 i?

13. Most methods of determining masses involve the use of gravitational forces near the earth's surface—either by means of a beam balance or by means of a spring scale. One method not involving gravity is the so-called 'inertia balance,' which consists of a set of springs arranged to permit a supported body to execute simple harmonic motion in a *horizontal* plane, as illustrated in principle in Fig. 1. If a standard 1-kg body is mounted on such an inertia balance, it executes simple harmonic motion at a frequency of 1 Hz. Calculate the frequencies at which bodies of the following masses would oscillate on this balance: 200 g, 400 g, 600 g, 2 kg, 4 kg, 10 kg. Plot a line on log-log paper, showing mass as a function of frequency, for use in determining unknown masses from measured frequencies.
Ans: 2.24, 1.58, 1.29, 0.710, 0.500, 0.316 Hz.

14. Show that the Y-component of the circular motion of the particle in Fig. 5 is correctly obtained from $Y = A \sin(2\pi ft + \phi)$. By differentiation, obtain expressions for v_Y and a_Y. From your expression for v_Y and the expression for v_X given in (5), show that $v_X^2 + v_Y^2 = v^2$ at all times, where v is the speed of the particle.

15. The disc in Fig. 8 has mass 0.125 sl and radius 6 i. When an external torque of $L = 2$ p·f is applied, the disc turns through an angle $\theta = 30°$. What is the torsion constant of the wire? If the external torque is removed, the disc oscillates about its equilibrium position. What is the frequency of the oscillation? What is the total mechanical energy associated with the oscillatory motion?
Ans: 3.82 p·f/rad; 2.50 Hz; 0.524 fp.

16. A slender uniform rod of mass 0.125 sl and length 4 f is attached perpendicular to the same wire as in Prob. 15. The point of attachment is at the midpoint of the rod. What is the frequency of oscillation of the rod? If the amplitude of the oscillation is 30°, what is the maximum angular velocity of the rod?

17. The expression (3a) for the frequency of a simple harmonic oscillator was obtained from (2). An expression for f can also be obtained from the conservation of mechanical energy as expressed in (12) and (13). Since $v_X = dX/dt$, (13) gives

$$\frac{dX}{dt} = \sqrt{\frac{K}{m}} \sqrt{(A^2 - X^2)},$$

or

$$\frac{dX}{\sqrt{A^2 - X^2}} = \sqrt{\frac{K}{m}} \, dt.$$

By integration of the last equation, show that $f = (1/2\pi) \sqrt{K/m}$.

18. The balance wheel of a watch is supposed to have a period of 1 s when it oscillates as a torsion pendulum. If the balance wheel has mass 1.5 g and rotational inertia 13.5 g·mm², what must be the torsion constant of the hairspring?

19. A 'seconds pendulum,' which beats seconds, has a period of 2 s. What is the length of a simple pendulum that beats seconds at a place where the gravitational acceleration is 32 f/s²? Ans: 3.24 f.

20. A simple pendulum consists of a small bob of mass 120 g at the end of a thread 100 cm long. What is the rotational inertia of this pendulum about an axis passing through the point of suspension of the thread? What is the frequency of oscillation of the pendulum?

21. A flat circular disc of 1-f diameter is pivoted for rotation about a horizontal axis perpendicular to the face of the disc and passing through the periphery of the disc. Find the frequency of oscillation of the resulting pendulum. What is the length of the simple pendulum of the same period?

Ans: 1.04 Hz; 0.748 f.

22. What is the period of a pendulum formed by pivoting a meter stick so that it is free to rotate about a horizontal axis passing through the 100-cm mark? through the 75-cm mark? through the 60-cm mark?

23. Show that the three periods in Prob. 22 are in the ratio $\sqrt{40} : \sqrt{35} : \sqrt{56}$ exactly.

24. A bicycle wheel has diameter 26 i and weighs 6 p. When hung by its rim across a horizontal knife-edge 12.5 i from the center, it executes small vibrations with a period of 1.53 s. What is its rotational inertia about an axis through the center of the wheel?

25. To determine the rotational inertia of a flywheel weighing 200 p, it is hung with the inside of its rim over a horizontal knife-edge, and found to oscillate at 0.6 Hz. If the knife-edge is 11 i from the center of the flywheel, determine the rotational inertia about the axis through the center of the wheel.

Ans: 7.70 sl·f².

26. Show that if a thin circular hoop is hung over a knife-edge, it oscillates with the same frequency as a simple pendulum of length equal to the diameter of the hoop. If the lower half of the hoop were removed, what would be the frequency of the resulting pendulum?

27. If the lower end of the oscillating stick of Fig. 11 has speed v when $\theta = 0$, show that the upward lift of the pivot at this point of the swing is $mg + \frac{1}{2} mv^2/l$. State carefully the reasoning involved in all steps of the solution of this problem.

28. A block rests on a horizontal plane which is itself vibrating horizontally with simple harmonic motion at a frequency of 2 Hz. If the coefficient of static friction is 0.6, what is the maximum amplitude that the vibratory motion can have without the block slipping?

29. Prove that the period of a conical pendulum (*see* Fig. 11, p. 95) is the same as the period of a simple pendulum of the same length when the 'cone' angle ϕ is small enough to permit use of the approximation $\sin\phi = \phi$.

30. A solid steel sphere of 6-i diameter is supported by a 20-f length of steel wire, the upper end of which is attached to the ceiling. The diameter of the wire is 0.1 i. (*a*) If the sphere is rotated and then released, it executes simple torsional oscillations. Find the frequency of these oscillations. (*b*) If the sphere is displaced vertically from its equilibrium position, it executes simple harmonic oscillations. Find the frequency of these oscillations.

31. A small block is placed on top of a piston that is executing simple harmonic motion along a vertical line. (*a*) At what position of the piston is the block most likely to leave the piston? (*b*) If the frequency of the piston is 5 Hz, what is the maximum amplitude of its motion for which the block will remain continuously in contact with the piston?

Ans: 1.00 cm.

32. If the amplitude of the piston in Prob. 31 is 10 cm, what is the maximum frequency at which the block would remain continuously in contact with the piston? If the small block has a mass of 20 g, find the magnitudes of the maximum and minimum forces exerted on the block during the oscillation of the piston at this frequency.

33. The 'ballistic' pendulum in the figure is a cubical block of wood suspended at the corners by four parallel strings in such a way that if the block is given a horizontal impulsive force, it will move in pure translation, without rotation. If the strings are 4 f long, the block has mass ⅜ sl, and a bullet of mass 0.006 sl fired into the block from the right causes the pendulum to swing to $\theta = 15°$, what was the speed of the bullet?

Ans: 735 f/s.

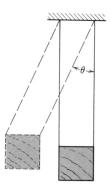

Problem 33

34. Using the crude but phenomenologically realistic model of a solid illustrated in the figure, show that Young's modulus would be given by $M_Y = N^{1/3} K$, where N is the number of atoms per m³ and K is the force constant of each spring.

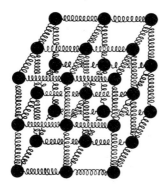

Problems 34–36

35. From optical studies, there is evidence that the vibration frequencies of atoms in solids are of the order of 10^{13} Hz. Assume that this is the frequency of vibration to right and left of a single copper atom in the simple crystal model of the figure, when all other atoms are at rest. Compute the force constant of an individual spring. Use the fact that 64 kg of copper atoms (one kilomole) contain 6×10^{26} atoms (Avogadro's constant). Ans: 210 N/m.

36. From the results of the two preceding problems, compute the order of magnitude of Young's modulus for copper and compare with the value given in Table 11–1, p. 225.

37. What is the maximum tension in the string when a simple pendulum of mass 100 g supported by a string 120 cm long is oscillating with an amplitude of 0.2 rad on each side of the vertical? Ans: 1.02 N.

38. If, as in the figure, the earth were a homogeneous nonrotating sphere, the weight of a body in a cavity inside the earth would be proportional to the distance r of the body from the center of the earth. (This result arises from the fact that the whole gravitational pull results from the sphere of radius r since it can be demonstrated that the gravitational pull of the spherical shell of material at greater radius on a particle *inside* the spherical shell is zero.) Assuming that the earth is a homogeneous sphere of 4000-mi radius and that there is a frictionless shaft passing diametrically through the earth, show that an object dropped into this shaft would execute simple harmonic motion, and find its period.

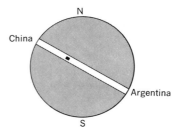

Problem 38

39. It has been suggested that tunnels bored in a straight line through the earth could be used to provide gravitational rapid transit between cities. From the information given in Prob. 38, demonstrate that in the absence of friction a particle in such a tunnel would execute simple harmonic motion with a period that is independent of the length of the tunnel. This can be done by showing that, during its motion along line AOB in the figure, the particle is subject to a restoring force directed toward O that is directly proportional to the particle's displacement X from point O. The transit time from point A to point B is half the period of the harmonic motion and turns out to be approximately 43 minutes, regardless of the length of line AB; thus, the transit times between Boston and New York, between London and Montreal, and between Washington and Moscow would be identical!

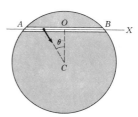

Problem 39

40. Show that the prototype beam balance illustrated in Fig. 8, p. 171, is in *unstable* equilibrium and hence would be a very unsatisfactory laboratory instrument. Show that the common playground 'teeter-totter' is also in unstable equilibrium. Design a 'teeter-totter' whose equilibrium would be stable if properly balanced.

41. Discuss the character of the equilibrium of the typical chemical beam balance. Show how you would compute the period of oscillation of a chemical beam balance, when balanced. The period is made deliberately long; how is this accomplished?

13 MECHANICS OF FLUIDS

The term *fluid* refers to a substance that does not have a fixed *shape* but that is able to *flow* and take the shape of its container; in other words, to a *liquid* or a *gas*. Although there is no sharp line of demarcation between solids and liquids (witness such materials as gelatin, heavy grease, and cold tar), we shall restrict our attention in this chapter to substances that are obviously liquid because they flow with reasonable rapidity, and to gases.

It is perhaps somewhat surprising to find that the same basic laws govern the static and dynamic behavior of both liquids and gases, in spite of their very different appearance. We shall discuss the distinction between liquids and gases in some detail in Chapter 18. In discussing the mechanical behavior of fluids we need make use only of properties that liquids and gases have in common and that are associated with their ability to flow. The first part of this chapter will be concerned with the static behavior of fluids, the latter part with their dynamic behavior.

1. Fluid Pressure

It is a matter of common experience that fluids exert forces on solid bodies with which they are in contact; examples are the forces that tend to burst a high-pressure air tank and those that support a floating boat or balloon. The liquid in the pail in Fig. 1 exerts a force on the bottom which is equal and opposite to the force exerted by the solid bottom on the liquid. That the latter force exists is apparent if we consider what would happen if we took away a piece of the bottom; the liquid would no longer be in static equilibrium but would flow downward.

Similarly, the liquid *below* the imaginary horizontal plane shown by the dashed lines in Fig. 1 must be exerting an upward force on the liquid *above* this plane, because this force is required to hold the latter part of the liquid up; remove this force and the liquid above the broken line would fall to the bottom of the pail. By Newton's third principle, the fluid above this imaginary plane must be exerting an opposite downward force on the fluid below the plane. Thus we conclude that not only does a fluid exert a force on a solid surface but also that if we 'draw' an imaginary plane in the fluid, the fluid on one side of this plane exerts a force across the plane on the fluid on the other side. The plane need not be horizontal. Consider the small imaginary cube sketched in Fig. 1. The liquid in this cube must exert forces on the surrounding liquid across all six of the walls of the cube, since such forces are needed to prevent the surrounding liquid from flowing in and occupying the space occupied by the liquid within the cube. The same arguments apply to imaginary surfaces drawn in a gas such as the air of the atmosphere.

Figure 1

Thus we see the general necessity for the existence of forces across any imaginary surface in the fluid as well as for forces between the fluid and solid surfaces. Experience completely justifies the assumption that *in static equilibrium these forces are normal to the surfaces* in question and are usually of the nature of a *push* rather than a *pull*. This statement is illustrated in Fig. 2. Thus, there are generally *compressive stresses* within a fluid. Such stresses are described by giving the *pressure P*, which is defined as the magnitude of the *normal force per unit area*. The pressure is almost always a positive quantity.*

Our description of the forces across any imaginary surface in a substance in static equilibrium applies to a solid substance as well as to a fluid, *up to the point* where we assert that the force is normal to the surface and almost always compressive. We have seen that in a solid there may be both shearing stresses and tensile stresses. The properties

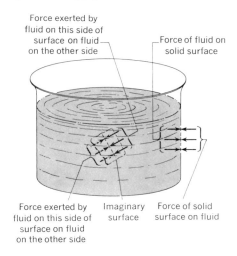

Figure 2

---

*Under certain very exceptional circumstances a liquid (never a gas) can withstand a tensile stress, which would correspond to a *negative* pressure, but the assumption that pressure is positive is valid for nearly all practical purposes.

peculiar to a fluid medium are obtained when we require that the force across any plane be entirely of the nature of a *pressure* normal to the plane. Obviously, a fluid could not have the configuration of a stick protruding across the edge of a table as in Prob. 34, p. 239.

For our purposes we adopt the following definition:

> A **fluid** is a material substance which in static equilibrium cannot exert tangential forces across a surface (either an imaginary surface internal to the fluid or a solid surface bounding the fluid), but can exert only pressure.

From this definition we shall be able to derive all the laws that govern the observed behavior of ordinary liquids and gases in static equilibrium.

We note that the pressure exerted on a real surface or across an imaginary surface in the fluid may in general vary from point to point on the surface. Hence, to define the pressure at a point Q on the surface in Fig. 3, we should take a small area ΔA surrounding Q, on which force of magnitude ΔF acts, and write

$$P = \lim_{\Delta A \to 0} \frac{\Delta F}{\Delta A}. \tag{1}$$

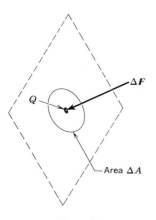

Figure 3

Suitable units for measuring pressure are p/i^2, p/f^2, and N/m^2.

From the discussion so far, one would be tempted to treat pressure as a vector quantity having the direction of the force ΔF in Fig. 3. We might expect the pressure to be different for every different choice of the plane through Q. But it is very convenient that in the case of fluids the pressure at Q is *independent* of the orientation of the plane through Q, so that we can treat P as a scalar function associated with the point and from it derive the vectorial force across any plane whatever through Q. That the pressure is independent of the direction of the plane is proved as follows:

Let the slant face of the prism in Fig. 4 be any plane whatsoever

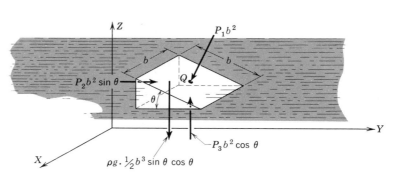

Figure 4

through Q; let this plane make a dihedral angle θ with the horizontal plane. Choose a rectangular coordinate system with the Z-axis vertical and the X-axis parallel to the horizontal line in the plane, and construct the small prism shown in Fig. 4.

Let the average pressure on the slant face be P_1; then the force on the slant face is $P_1 b^2$. Let the average pressure on the left face be P_2; the area of this face is $b^2 \sin\theta$, and hence the force on this face is $P_2 b^2 \sin\theta$. Let the average pressure on the bottom face be P_3; the area of this face is $b^2 \cos\theta$, and hence the force on this face is $P_3 b^2 \cos\theta$. Denote the density of the fluid in the prism by ρ; the volume of the prism is $\tfrac{1}{2} b^3 \sin\theta \cos\theta$; the *weight* per unit volume is ρg; therefore the weight of the prism is $\rho g \cdot \tfrac{1}{2} b^3 \sin\theta \cos\theta$.

Now consider the equilibrium of forces on the prism. In the Y-direction we have

$$P_2 b^2 \sin\theta = P_1 b^2 \sin\theta, \quad \text{or} \quad P_2 = P_1. \tag{2}$$

In the Z-direction we have

$$P_3 b^2 \cos\theta = P_1 b^2 \cos\theta + \rho g \cdot \tfrac{1}{2} b^3 \sin\theta \cos\theta,$$

or
$$P_3 = P_1 + \tfrac{1}{2} \rho g b \sin\theta. \tag{3}$$

Equations (2) and (3) are true for any size prism; in particular they must remain true in the limit as $b \to 0$. In the limit, P_1 approaches the pressure on the slant face at Q; P_3 approaches the pressure on a horizontal surface at Q; and P_2 the pressure on a vertical surface at Q. Also, in (3), the last term approaches zero as $b \to 0$ and can be dropped. Hence in the limit we conclude that $P_1 = P_2 = P_3$, and hence that the pressure on any slant surface or any vertical surface through Q equals the pressure on a horizontal surface through Q. In other words,

At a given point in a fluid, the pressure on any surface through the point is independent of the orientation of the surface.

Although the convenient instruments for measuring pressure employ fluids and hence depend for their operation on laws of fluid statics which we have not yet derived, one could conceive of constructing a little 'gadget' that would measure fluid pressure by compression of a spring, as in Fig. 5. A very flexible rubber diaphragm is backed by a plate attached to a spring. *The inside space is evacuated.* The force of the fluid on the diaphragm is measured by the compression of the spring.

Such a device, in the normal atmosphere, would give a pressure reading of 14.7 p/i², independent of whether the diaphragm were pointed up, down, sidewise, or in any other orientation. Similarly, 35 f down in a lake, the gadget would give a constant reading of about 30 p/i², independent of the way it was pointed.

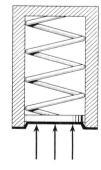

Fig. 5. 'Gadget' for pressure measurement.

2. Fluid Statics

From Newton's first and third principles, we can easily derive two relations that determine the variation of pressure with position in a fluid in static equilibrium. In this analysis, we shall not assume that the fluid is homogeneous; rather, it is important to consider the case where more than one type of fluid is present, as in the case of oil floating on water or the ever-present case of air 'floating' on water.

The first relation compares pressure at two points in the same horizontal layer:

The pressure at every point in a continuous horizontal layer of a fluid at rest is the same.

Imagine a thin square prism to be 'drawn' within the fluid with its axis horizontal as in Fig. 6. Let the Y-axis of a coordinate system be parallel

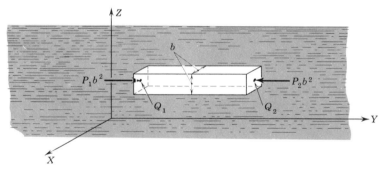

Figure 6

to the axis of the prism. The only forces acting in the Y-direction on the fluid within this prism are the normal forces on the ends. If P_1 is the average pressure on the left end and P_2 the average pressure on the right end, equilibrium requires that

$$P_1 b^2 = P_2 b^2, \quad \text{or} \quad P_1 = P_2.$$

If now b approaches zero, P_1 approaches the pressure at Q_1 and P_2 that at Q_2; therefore the pressures at Q_1 and Q_2, two arbitrary points in the same horizontal plane, are equal. This identity proves the statement above for two points Q_1 and Q_2 that can be joined by such a prism of fluid as in Fig. 6. If, because of a solid obstruction, a straight line cannot be drawn in the fluid from Q_1 to Q_2, but if a succession of straight horizontal lines in the fluid can connect Q_1 and Q_2, the pressures are clearly still equal. Thus, when a solid object is completely *surrounded* by fluid, the pressure of the fluid on the object is the same at all points on the same horizontal level.

The second relation compares pressure at two different heights:

The pressure at a height h_1 in a fluid at rest is greater than the pressure at a greater height h_2 by the weight of a column of fluid of unit cross section and height $h_2 - h_1$ lying between these two levels. For a homogeneous fluid this pressure difference is $\rho g (h_2 - h_1)$, where ρ is the density.

Consider fluid levels a vertical distance $h_2 - h_1$ apart, as in Fig. 7. Connect them by a prism of cross-sectional area A. If the fluid in this prism has weight w, equilibrium of vertical forces requires that $F_1 - F_2 = w$. Dividing this equation through by the area A gives $F_1/A - F_2/A = w/A$, or

$$P_1 - P_2 = \begin{Bmatrix} \text{weight per unit cross-} \\ \text{sectional area of prism} \end{Bmatrix}. \quad (4)$$

If the fluid is homogeneous, in that it has the same density ρ at all heights Z, it is seen that the weight of a column of fluid of unit cross section is ρg times the volume $h_2 - h_1$.

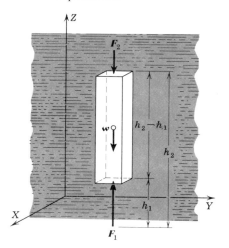

Figure 7

Combination of the above two relations enables us to relate the pressures at *any* two points in a fluid, and to show, for example, that the pressures at two points at the same horizontal level in a *homogeneous* fluid, such as points B and C in Fig. 9, are the same, whether or not these points can be connected by a *horizontal* line in the fluid.

We shall now consider the practically important case of a foreign object such as a ship, a submarine, or a balloon immersed in a fluid. This case was apparently first considered by the Greek philosopher ARCHIMEDES (c. 250 B.C.), who is said to have determined the density, and hence the purity, of a gold crown by comparing its weight in water with its weight in air. In any case, Archimedes demonstrated a fundamental relation, called

ARCHIMEDES' RELATION: *A fluid acts on a foreign body immersed in it with a net force that is vertically upward and equal in magnitude to the weight of the fluid displaced by the body.* (*This upward force is called the buoyant force.*)

For a foreign body of the shape of the prism of Fig. 7, we can immediately demonstrate Archimedes' relation since $F_1 - F_2 = w$, the weight of the fluid that would fill the prism if the foreign body were not present.

For the more general case of an irregular body, such as that of Fig. 8, immersed in a fluid, the proof of this relation can be given in an equally simple fashion. The fluid pressures on the surface of the body would be unchanged if the body were not present and this surface were con-

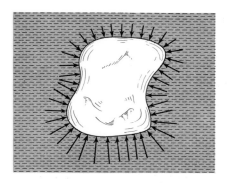

Fig. 8. Cross section of an irregular body immersed in a fluid.

sidered as an imaginary surface drawn in the fluid itself. When the body is replaced by a region of fluid of the same shape, the laws of static equilibrium require that the surrounding fluid exert forces whose resultant is vertically upward and equal in magnitude to the weight of the fluid in the region. Since the surrounding fluid acts with the same system of forces on the foreign body, this body also experiences a net upward buoyant force equal to the weight of the fluid that would occupy the region occupied by the body; that is, equal to the weight of the fluid displaced, as stated in Archimedes' relation.

From this relation, it is seen that a body immersed in a homogeneous fluid will sink if its own average density is greater than the density of the fluid, will rise if its density is less than that of the fluid, and will 'float' at the interface of two fluids if the density of one fluid is greater and that of the other fluid less than the average density of the body, like a log partly in water, partly in air.

The 'body' mentioned in the preceding paragraphs may be a quantity of fluid as well as a quantity of solid. A quantity of denser fluid immersed in a lighter fluid with which it does not mix will sink; similarly, lighter fluid immersed in a denser fluid will rise; until an equilibrium is reached

Table 13-1 TYPICAL STANDARD SPECIFIC WEIGHTS ρg_s AT 68° F

Liquids	N/m³	p/f³	Solid Metals	N/m³	p/f³
Water (32–50° F)	9 807	62.43	Aluminum	26 500	169
Sea water	10 100	64.4	Cast iron	70 600	449
Benzene	8 620	54.9	Copper	87 200	555
Carbon tetrachloride ...	15 630	99.5	Gold	189 300	1205
Ethyl alcohol	7 740	49.3	Lead	111 200	708
Gasoline	6 670	42.5	Magnesium	17 100	109
Kerosene	7 850	49.9	Nickel	86 800	553
Lubricating oil	8 830	56.2	Silver	103 000	656
Methyl alcohol	7 770	49.4	Steel	76 500	487
Sulfuric acid, 100% ...	17 960	114.3	Tungsten	186 000	1190
Turpentine	8 560	54.5	Zinc	70 000	446
Mercury (32° F)	133 300	848.7	Brass or bronze ...	85 300	543
Nonmetallic Solids	N/m³	p/f³	Woods	N/m³	p/f³
Ice	9 040	57.5	Balsa	1 270	8
Concrete	22 600	144	Pine	4 700	30
Earth, packed	14 700	94	Maple	6 300	40
Glass	25 500	160	Oak	7 100	45
Granite	26 500	169	Ebony	11 800	75

with all the fluid of lesser density above the fluid of greater density, the two fluids being separated by a horizontal interface.

The quantity ρg appearing in expressions involving hydrostatic pressure is sometimes called *specific weight,* since it gives the weight of a unit volume of material of density ρ. The standard specific weights ρg_s of various materials are listed in Table 13-1. The standard specific weight, which is proportional to density, gives the specific weight at a location where g has its 'standard' value g_s.

> **Example.** *What is the area of a block of floating ice 2 f thick that will just support an automobile weighing* 3000 p?
>
> The minimum-size block will be just on the point of being sunk by the weight of the automobile; hence it will be just awash and will displace water equal to its whole volume. From Table 13-1, water has specific weight 62.4 p/f^3; ice has specific weight 57.5 p/f^3. Hence a cubic foot of ice entirely immersed in water will experience a buoyant force of 62.4 p. Since its own weight is 57.5 p, it will be able to support an additional weight of only 62.4 p − 57.5 p = 4.9 p. To support 3000 p will require 3000/4.9, or 612 f^3 of ice. Since the ice is 2 f thick, a block of area 306 f^2 will be required, for example, a block of size 20 f × 15.3 f.

3. The Barometer; Pressure Gauges

EVANGELISTA TORRICELLI, in 1643, first devised a method for measuring the pressure of the atmosphere, by means of the mercury barometer. Knowledge of atmospheric pressure is fundamental to most other measurements of pressure because most pressure gauges use the atmosphere as a reference level and measure the difference between an actual pressure and atmospheric pressure.

> **Absolute pressure** is the actual pressure at a point in a fluid.
>
> **Gauge pressure** is the difference between absolute pressure at a point in a fluid and the pressure of the atmosphere.

To make a simple mercury barometer, take a straight glass tube about a meter long, closed at one end. Fill the tube *completely* with mercury, close the open end with the finger, and invert the tube in a dish of mercury as in Fig. 9. The mercury column in the tube will fall to a height h of about 76 cm above the level of the mercury in the reservoir if the experiment is done at sea level.

If the barometer of Fig. 9 is properly constructed, no air will be trapped in the space above the mercury column. Hence this space must be free from substance—a vacuum.* Since a vacuum certainly exerts no

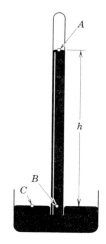

Fig. 9. The mercury barometer.

---

*Actually, there will be a small amount of mercury *vapor* in this space; but since at ordinary temperatures the pressure this vapor exerts will be quite negligible, we can ignore it here. See Chapter 18 for a discussion of vapor pressure.

pressure, the pressure at point *A* is zero. The pressure at point *B* is numerically equal to the weight of a unit column of mercury of height *h*. The pressure at *C* equals that at *B*, since these are two points at the same horizontal level in a homogeneous fluid. But the pressure at *C* is the pressure of the atmosphere. Hence the pressure of the atmosphere is the same as that at the base of a column of mercury of height *h*.

The pressure of the atmosphere is numerically equal to the weight of a unit column of air extending all the way to the top of the atmosphere (99 per cent of the air lies within 30 km of the earth's surface). The total weight of the atmosphere, down to sea level, has exactly the weight of a layer of mercury about 76 cm thick. Atmospheric pressure decreases with altitude because there is less total weight of atmosphere above a point at higher altitude than above a point at sea level. The usual airplane altimeter utilizes this pressure variation to determine altitude. At a given location, there are day-to-day variations in atmospheric pressure which may amount to as much as five per cent. These variations have important meteorological significance.

Pressures are frequently specified in *standard atmospheres* (atm). By definition,

$$1 \text{ atm} = 101\,325 \text{ N/m}^2, \tag{5a}$$

exactly. This number was selected because it represents, within experimental accuracy, the pressure of a column of mercury exactly 76 cm high when the temperature is 32° F and the gravitational acceleration has its standard value g_s. Thus we can write

$$\begin{aligned} 1 \text{ atm} &= 76 \text{ cm of Hg} = 29.92 \text{ i of Hg} = 406.8 \text{ i of water} \\ &= 10.33 \text{ m of water} = 33.90 \text{ f of water} \\ &= 101\,325 \text{ N/m}^2 = 14.70 \text{ p/i}^2 = 2116 \text{ p/f}^2, \end{aligned} \tag{5b}$$

where the water has density 1.000 Mg/m³ and is under standard gravity.

The pressure of the atmosphere may be measured (less accurately) by means of an *aneroid barometer,* which contains a sealed evacuated metal box, flat and circular in shape, with a corrugated top as in Fig. 10. The top is bent in by the pressure of the atmosphere, the amount of bending being a measure of the pressure. Motion of the top is amplified and transmitted to a pointer which reads pressure on a scale that has been calibrated against a mercury barometer. Our 'gadget' of Fig. 5 is essentially an aneroid barometer.

The normal density of air is so small that variations in atmospheric pressure with a change in height of a few feet are ordinarily negligible, so that if one reads barometric pressure in a room, the value obtained will ordinarily apply to apparatus at any height in the room. The decrease in pressure in going up 10 f is only 0.0004 atm.

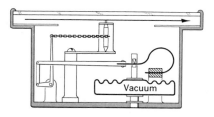

Fig. 10. Schematic cross section of an aneroid barometer.

The barometers illustrated in Figs. 9 and 10 are gauges that measure the *absolute* pressure of the fluid in which they are immersed, since the vacuum exerts no opposing pressure. We shall now describe three common types of pressure gauges that measure *gauge* pressure. One is the *manometer,* illustrated in Fig. 11. The manometer is essentially a U-tube open to the atmosphere on one side and to the fluid whose pressure is being measured on the other, and filled with a suitable liquid. From the difference in heights of column in the two sides, the difference between the fluid pressure and the atmospheric pressure is readily determined.

Another type of pressure gauge, used in many technical applications, is the *Bourdon gauge.* This gauge contains a sealed spiral of flat metal tubing as in Fig. 12. The inside of the tubing is filled with the fluid whose pressure is being measured (or with air in communication with the fluid and hence at the same pressure); the outside of the tubing is exposed to the atmosphere. The elastic properties of a spiral tube are such that when the pressure inside is increased, the tube tends to 'unwind' and straighten out. This unwinding effect is communicated to a pointer. It turns out that the configuration of the tube and hence the position of the pointer depends only on the difference between the pressure inside the tube and the atmospheric pressure outside.

The hand gauge ordinarily used to measure the pressure in automobile tires is of still another type, in which the high-pressure air pushes back a piston against the force of a spring. Since the atmosphere acts on the spring side of the piston, the compression of the spring is again a measure of gauge pressure.

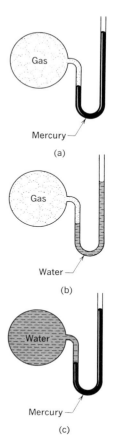

Fig. 11. Manometers: (a) would be used for high gas pressures, (b) for low gas pressures, (c) for water pressure.

Example. *When the mercury manometer shown in Fig. 11(a) is connected to a gas main, the mercury stands* 40 *cm higher in the right-hand tube than in the left-hand. A barometer at the same location reads* 74 *cm. Determine the absolute pressure of the gas in cm of Hg, in atm, and in p/i².*

The pressure at the top of the left-hand mercury column is the gas pressure. This equals the pressure *at the same horizontal level* in the right-hand column. But the pressure at this level in the right-hand column is the atmospheric pressure (74 cm of Hg) plus the pressure exerted by the additional 40 cm of mercury, or a total of 114 cm of Hg. Hence the absolute pressure of the gas is 114 cm of Hg. Using conversion factors obtained from (5b), we find for the same pressure in atm and p/i²:

114 cm of Hg = 114 ($\frac{1}{76}$ atm) = 1.50 atm = 1.50 (14.7 p/i²) = 22.0 p/i².

4. Fluid Dynamics; Bernoulli's Law

So far we have discussed the behavior of fluids at rest—*fluid statics.* We now give a brief introduction to the behavior of fluids in motion—*fluid dynamics.* This subject furnishes the foundations for *hydrodynamics,* which is concerned with the motion of liquids such as water, and for

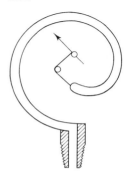

Fig. 12. The Bourdon gauge (schematic).

aerodynamics, which is concerned with the motion of air and with the motion of bodies such as airplanes, rockets, and shells through the air.

The subject of fluid dynamics is mathematically and physically very complex. Bernoulli's law, which we shall derive, is basic to the whole subject. This law applies only to an *ideal, nonviscous* fluid. For an ideal fluid, the only resistance to acceleration is inertial; the absence of viscous (frictional) forces means that there is no resistance to the motion of one layer of fluid past another, nor any resistance to the motion of the fluid along a solid surface. Since the viscosities of air and water are fairly low, Bernoulli's law furnishes a good *first* approximation in the study of their flow. During the past two centuries, empirical corrections to the results given by this law have been determined experimentally. The interest in powered flight in the twentieth century resulted in the first intensive effort to develop *theory* to account for the behavior of *real* fluids.

Our derivation of Bernoulli's law will also assume the fluid to be *incompressible*. This assumption is valid for water, and gives satisfactory results for bodies moving through air at speeds below about 350 mi/h, but as speeds approach that of sound (about 740 mi/h) a more complex form of the law, which takes into account the variation in density with variations in pressure, must be used. One subject under intensive study today is that of *supersonic flow*—motion at speeds greater than that of sound. Supersonic flow is physically very different from subsonic flow; for example, *shock waves* are created in supersonic but not in subsonic flow.

We shall consider only the important special case in which the flow is *steady*. By *steady flow,* we mean that *the velocity and other characteristic properties of the fluid at a given point of space are constant, independent of time*. Thus in steady flow through the pipe of Fig. 14, the velocity at a given point in the pipe is constant in time, so that a definite constant volume of fluid passes a given section in the pipe in each unit of time. In order that the fluid should not 'pile up' anywhere in the pipe, or leak

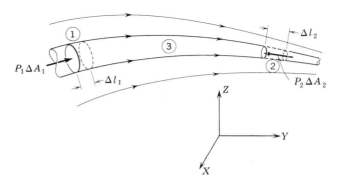

Fig. 13. Streamlines and a stream tube in the flow of an incompressible fluid.

out, the same volume per unit time must pass every section in the pipe. The same volume per unit time enters one end of the pipe as leaves the other.

In more general steady flow, such as the flow of air around a model in a wind tunnel, we can represent the velocity directions by means of *streamlines,* as in Fig. 13. At each point X,Y,Z of space the velocity has a definite magnitude and direction, neither of which vary with time. The streamlines of Fig. 13 are drawn in such a way that at each point the velocity vector is tangent to the streamline. *A tubular region whose generators are streamlines,* such as that indicated in Fig. 13, is called a *stream tube.* Because of the way in which streamlines are defined, the fluid contained within a particular stream tube remains within this tube, never crossing the boundary, because the fluid does not have a velocity component normal to the boundary. Hence the fluid in a particular stream tube behaves in all respects like the fluid flowing in the pipe of Fig. 14, and the same volume of flow crosses each section of the stream tube in each unit of time (if the density of the fluid does not change).

The relations derived in Sec. 2 do not apply to a fluid in motion, but the result derived in Sec. 1 does apply:

At a given point in an ideal fluid in motion, the pressure on any surface through the point is independent of the orientation of the surface.

This relation can be derived from Fig. 4 in the same manner as in the static case, except that terms representing *mass × acceleration,* of the form $\rho \cdot \tfrac{1}{2} b^3 \sin\theta \cos\theta \, a_Y$ and $\rho \cdot \tfrac{1}{2} b^3 \sin\theta \cos\theta \, a_Z$, will appear. However, like the weight term in (3), these terms will disappear in the limit as $b \to 0$. Hence we can still treat P as a scalar function of position.

Bernoulli's law is essentially a formulation of the principle that work equals change in mechanical energy for the case of ideal fluids. Consider the forces that act on a definite quantity of fluid located within a stream tube, such as the quantity marked out by solid lines in Figs. 13 and 14.

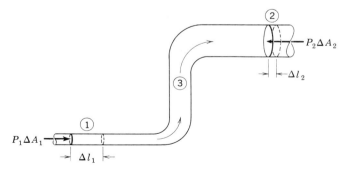

Fig. 14. Flow of an incompressible fluid in a pipe of variable cross section.

On the 'ends' act forces $P_1 \Delta A_1$ and $P_2 \Delta A_2$ arising from the pressure of the adjoining fluid. Similarly, on the 'sides' act normal forces arising from the pressure of the adjoining fluid in Fig. 13, or from the walls of the pipe in Fig. 14. *If the viscosity of the fluid is neglected,* there will be no tangential forces on any of these surfaces. In addition, there will be the force of gravity.

We shall now compute the work done on this quantity of fluid in the time interval Δt and equate this work to the increase in gravitational potential energy plus the increase in kinetic energy. In either Fig. 13 or Fig. 14, in time Δt, one end of our portion of fluid moves distance Δl_1, the other end Δl_2, related in such a way that $\Delta l_1 \Delta A_1 = \Delta l_2 \Delta A_2 = \Delta V$, where ΔV is the volume of fluid passing any point in the stream tube in time Δt. The work done by the pressure P_1 is $P_1 \Delta A_1 \Delta l_1 = P_1 \Delta V$; that done by P_2 is $-P_2 \Delta A_2 \Delta l_2 = -P_2 \Delta V$, negative because the force is opposite to the direction of the motion as the fluid moves from the position shown by the solid outline to that shown by the broken outline. The forces on the sides of the tube do no work because they are perpendicular to the velocity. Hence, in the time Δt,

$$\text{work done} = (P_1 - P_2) \Delta V. \tag{6}$$

Now we compute the change in energy in the same time Δt. Here the assumption of steady flow greatly simplifies the discussion, since in steady flow the energy of the fluid that happens to occupy any given region of space at one time is the same as the energy of the fluid that occupies this same region of space at a later time. Hence, the fluid that occupies a given region of space, such as the region denoted by ①, ②, or ③ in Figs. 13 and 14, has an energy that is characteristic of that particular region. Initially, the fluid bounded by solid lines has energy characteristic of regions ①+③. After time Δt, the same fluid, which has moved into the space outlined by broken lines, has energy characteristic of regions ②+③. The change in energy is the energy characteristic of region ② minus the energy characteristic of region ①. Let us denote the density of the fluid by ρ, the speeds in regions ① and ② by v_1 and v_2, and the vertical heights of regions ① and ② by Z_1 and Z_2. (In the final analysis, the tube of Fig. 13 is supposed to have infinitesimal area, so that the assignment of a single speed and height to these regions is justified. The speeds and heights assigned in Fig. 14 are some kind of average over the cross section.) The fluid in region ② has potential energy $\rho \Delta V g Z_2$ and kinetic energy $\frac{1}{2} \rho \Delta v v_2^2$. The fluid in region ① has potential energy $\rho \Delta V g Z_1$ and kinetic energy $\frac{1}{2} \rho \Delta V v_1^2$. Hence,

$$\text{change in energy} = \rho \Delta V g Z_2 + \tfrac{1}{2} \rho \Delta V v_2^2 - \rho \Delta V g Z_1 - \tfrac{1}{2} \rho \Delta V v_1^2 \tag{7}$$

in time Δt. Equating work done (6) to change in energy (7) and dividing by ΔV gives

$$P_1 - P_2 = \rho g Z_2 + \tfrac{1}{2} \rho v_2^2 - \rho g Z_1 - \tfrac{1}{2} \rho v_1^2,$$

or
$$P_1 + \tfrac{1}{2}\rho v_1^2 + \rho g Z_1 = P_2 + \tfrac{1}{2}\rho v_2^2 + \rho g Z_2. \tag{8}$$

The equality in (8) expresses the law first stated by DANIEL BERNOULLI, a Swiss mathematical physicist, in 1738:

BERNOULLI'S LAW: *At any two points along the same streamline in a nonviscous, incompressible fluid in steady flow, the sum of the pressure, the kinetic energy per unit volume, and the potential energy per unit volume has the same value.*

In applying (8) we must use a consistent system of units, such as

P	ρ	v	g	Z
N/m²	kg/m³	m/s	m/s²	m
p/f²	sl/f³	f/s	f/s²	f

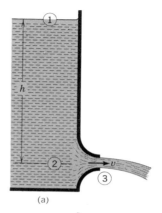

Each term in (8) is seen to have the dimensions of energy/volume.

A fluid at rest is a special case of steady flow, since any two points in the fluid can be connected by a virtual streamline; if we set $v_1 = v_2 = 0$ in (8), we get our previous relation $P_1 - P_2 = \rho g (Z_2 - Z_1)$, which gives the change of pressure with height for a homogeneous fluid.

In the case of a fluid that has frictional (viscous) forces acting, particularly in the case of flow through pipes as in Fig. 14, some of the work done goes into heat rather than into increase in mechanical energy. In this case the change in *mechanical* energy will be *less* than the work done. Expressing this inequality leads to the conclusion that *if there are frictional forces, the value of the left side of* (8) *is greater than the value of the right side, which is evaluated further downstream.* In the case of flow through pipes, Bernoulli's law is ordinarily used with empirical corrections for the conversion of mechanical energy to heat.

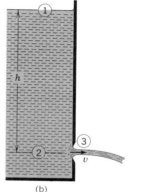

5. Applications of Bernoulli's Law

As a first application of Bernoulli's law, we shall derive *Torricelli's law* for the velocity of efflux of a liquid from an orifice in a tank. Figure 15 shows three types of orifices discharging at height h below the water level in a *large* tank. We can apply Bernoulli's law to points ①, ②, and ③, since some streamline will connect ① and ③, also ② and ③. We shall call $Z = 0$ at ② and ③; $Z = h$ at ①. Since the pressure appears on both sides of Bernoulli's law, an arbitrary reference level for pressure may be used. Normally, it is most convenient to use *gauge pressure*. The gauge pressure at ①, at the free surface at the top of the tank, is zero. The gauge pressure at ③, in the stream just beyond the opening, is also zero because the sides of the stream are open to the atmosphere. We shall write P for the gauge pressure at point ② in the tank at the level of the orifice. The speeds at ① and ② can be taken as zero if the tank

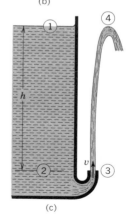

Fig. 15. Efflux from (a) a well-rounded orifice, (b) a sharp-edged orifice, showing the *vena contracta*, (c) a vertical nozzle.

is large. We desire to find the speed v at ③. Writing the terms in (8) for points ①, ②, and ③ successively gives

$$0+0+\rho gh = P+0+0 = 0+\tfrac{1}{2}\rho v^2 + 0.$$

This equation gives $\qquad v=\sqrt{2gh} \qquad \begin{pmatrix}\text{Torricelli's}\\ \text{law}\end{pmatrix}$ (9)

for the velocity of efflux from a tank under 'head' h, or

$$v = \sqrt{2P/\rho} \tag{10}$$

for the velocity of efflux of liquid from an orifice when the tank gauge pressure is P at the same elevation as the orifice. It will be noted that the velocity given by (9) is just equal to the velocity acquired in free fall from rest through a height h. Equation (10) is useful for flow from a closed tank when the liquid is under greater pressure than just the hydrostatic head.

In the case of the vertical jet in Fig. 15(c), if we extend the application of Bernoulli's law to point ④, where the speed is again zero, we see that the vertical jet should reach just the height h. In practice, the jet falls a little short of h because of frictional losses. This result is in accord with our remark that the expression evaluated in Bernoulli's law will in actual cases decrease (but never increase) as one goes *downstream*. Torricelli observed that the height reached in Fig. 15(c) was approximately the level of the liquid in the tank and was the first to recognize that this observation implies a jet velocity given by (9), which is just the velocity of vertical projection required for a body to reach height h.

To find the volume flow out of the tank, one can get a reasonably accurate result by multiplying the velocity v by the area of the opening if the opening is well rounded as in (a) and (c). If the opening is sharp-edged as in (b), the fluid has not completed its acceleration by the time it passes through the opening. It continues to accelerate for a short distance in the jet, and, as the fluid speeds up, the jet contracts because the product of velocity by area must remain constant. The jet takes on a characteristic shape called the *vena contracta*. Torricelli's law gives the velocity after the contraction is complete. For a sharp circular opening, the contracted area is approximately 62 per cent of the area of the opening, and correction for this contraction must be applied in finding the volume flow.

An interesting example of the application of Torricelli's law is furnished by the *siphon* (Fig. 16). We see from arguments similar to the above that the velocity of efflux will be given by the Torricelli formula with h equal to the difference in height between the water surface in the tank and

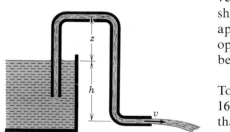

Fig. 16. Siphon.

the discharging orifice. The depth of the entrance to the siphon is irrelevant. Of course the siphon will not work unless it is filled with liquid. The liquid column will break at the top of the siphon if the absolute pressure there falls to zero.* This condition imposes a limitation on the value of the height z in Fig. 16.

As an important application of Bernoulli's law to the flow of gas or liquid through a pipe, we shall discuss the measurement of flow by means of the Venturi flowmeter illustrated in Fig. 17. The flow velocities in Fig. 17 will be related to the cross-sectional areas by

$$v_1 A_1 = v_2 A_2. \tag{11}$$

Bernoulli's law gives $\quad P_1 - P_2 = \tfrac{1}{2} \rho (v_2^2 - v_1^2). \tag{12}$

The pressure difference $P_1 - P_2$ is read directly on the differential manometer in Fig. 17. If we solve (11) for v_1 and substitute in (12), we obtain

$$P_1 - P_2 = \tfrac{1}{2} \rho v_2^2 (A_1^2 - A_2^2)/A_1^2. \tag{13}$$

This equation enables us to obtain the velocity of flow, v_2, and hence the volume rate of flow, $v_2 A_2$, merely from a measurement of the differential pressure. The Venturi flowmeter is widely applied in the measurement of flows of liquids and gases in pipe lines.

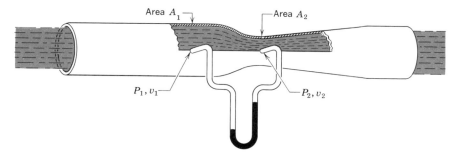

Fig. 17. Venturi flowmeter.

If the flow of liquid through a constricted tube such as that of Fig. 17 is rapid, it is not at all difficult to get conditions in which Bernoulli's law gives a value for absolute pressure in the constricted section that is negative. Since a liquid will not ordinarily sustain a negative pressure, a phenomenon called *cavitation* ensues, in which cavities or holes are formed in the interior of the liquid. Not only does cavitation introduce a large energy loss, but it produces serious pitting on metallic surfaces, apparently owing to sudden collapse of the cavities. Such cavitational

---

*More rigorously, in the case of most liquids, the absolute pressure can fall only to the vapor pressure of the liquid, as discussed in Chapter 18. On the other hand, certain cohesive liquids such as mercury will even flow in a siphon when they are under *tension*—at a slight *negative* absolute pressure.

corrosion can cause destruction of turbine blades and ship propellers.

The low pressure attained in fluid flow through a constricted section is utilized in the design of suction pumps and atomizers.

Example. *In Fig. 17, if $A_1 = 1$ f^2, $A_2 = 0.5$ f^2, the differential mercury height is 15 i, and the fluid is water, determine the flow in* f^3/s.

We have all the data necessary to find v_2 from (13), but we must be careful to use the consistent system of British units given on p. 273.

The differential *pressure* is *not* 15 i of Hg, because the water in the left-hand manometer tube does not have density negligible compared with that of mercury, and hence the pressure resulting from the weight of the 15 i of water in the left-hand tube must be considered. Since mercury is 13.6 times as dense as water, the differential pressure is

$$P_1 - P_2 = (15 \text{ i of Hg}) - (15 \text{ i of water})$$
$$= (15 \times 13.6 - 15) \text{ i of water} = 189 \text{ i of water}.$$

From (5b), or the pressure conversion table in Sec. 5 of the Appendix, we find that 1 i of water = 5.20 p/f^2 and hence that

$$P_1 - P_2 = 189 \ (5.20 \text{ p/f}^2) = 983 \text{ p/f}^2.$$

From the table on p. 70, the density ρ of water is 1.94 sl/f^3.

We are now ready to substitute in (13); from this equation,

$$v_2^2 = \frac{2(P_1 - P_2)}{\rho} \frac{A_1^2}{A_1^2 - A_2^2} = \frac{2 \times 983 \text{ p/f}^2}{1.94 \text{ sl/f}^3} \frac{1}{1 - \frac{1}{4}} = 1350 \frac{\text{f}^2}{\text{s}^2}; \qquad v_2 = 36.8 \text{ f/s}.$$

Hence the volume of flow is

$$A_2 \, v_2 = (0.5 \text{ f}^2)(36.8 \text{ f/s}) = 18.4 \text{ f}^3/\text{s}.$$

6. The Lift of an Airfoil

If we set up an airplane model in a wind tunnel and blow air past it at a constant oncoming speed, we get a *steady* flow that is representable by streamlines as in Fig. 13, and Bernoulli's law is applicable.

But for a plane in flight, the flow is not at all *steady* if we use a coordinate system attached to the ground. However, if the plane is flying at constant velocity and we use *a coordinate system attached to the plane,* the flow is (approximately) steady.* In a coordinate system attached to the plane, the plane *is* standing still and air *is* blowing past it at high speed. *Bernoulli's law is applicable to this steady flow in this coordinate system* and furnishes a very useful first approximation in the computation of forces on the surfaces of airplanes or missiles.

---

*In actual flight, the air passing close to the surface of a plane acquires an eddying motion called *turbulence,* as a result of viscous forces. This turbulence can be seen at the right of the cylinder in Fig. 19. The discussion of lift given in this section is still basically valid in spite of the generation of this turbulence.

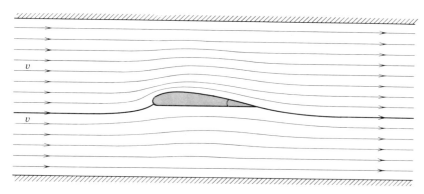

Fig. 18. Flow past an airfoil in a wind tunnel. The heavy streamline divides the flow above the airfoil from the flow below.

Let us now consider the origin of the *lift* of an airplane wing. A wing is called an *airfoil,* and the lift is entirely associated with the characteristics of the flow pattern that is set up around a section having the distinctive characteristic shape of an airfoil, with its sharp trailing edge and greater curvature of the top surface than the bottom surface. This shape results in a greater flow velocity past the top surface of the wing than past the bottom surface, as can be seen from the pattern of streamlines in Fig. 18. As the figure shows, the oncoming air divides in such a way that more than the proper share passes above the wing and less than the proper share below the wing. This inequality results in the flow velocity past the top of the wing being greater than the oncoming velocity v, and hence, by Bernoulli's law, in the pressure on the top of the wing being *less* than the free-stream pressure P. The reverse happens on the bottom of the wing; there the velocity is in general lower than the free-stream velocity, and the pressure is greater than the free-stream pressure. The pressure difference between the top and bottom surfaces results in a lift which is proportional to ρv^2. The lift can be increased by lowering the trailing edge of the wing. Rotation of the trailing section, called the *aileron,* about a transverse axis is used to control the lift. A different amount of rotation of the two ailerons on right and left wings causes the plane to 'roll' about a longitudinal axis, and hence to 'turn.'

A similar phenomenon accounts for the 'curve' of a spinning baseball or tennis ball. Look first at Fig. 19, in which the velocity of flow past the surface of the rotating cylinder is greater above the cylinder than below just as for the airfoil of Fig. 18, and for the same reason a *lifting force* results. A spinning ball rotating clockwise and *moving to the left* would develop a pattern of streamlines similar to those in Fig. 19, from the point of view of an observer moving with the ball; the ball would have a 'back spin' and would experience a lift tending to raise its normal trajectory.

Fig. 19. Streamlines in the flow from left to right past a cylinder rotating clockwise. A net lift develops. (From L. Prandtl and O. Tietjens, *Hydro-* and *Aeromechanik*, vol. 1: Springer, Berlin; reproduced by permission of J. W. Edwards, publisher of the American edition.)

7. Fluid Viscosity

Whenever the velocity of a real fluid changes in magnitude as one moves in a direction perpendicular to the streamlines, there are viscous forces. To understand these forces, consider Fig. 20, in which we have two very large horizontal plates, the lower one at rest and the upper one moving in the X-direction at speed V. The space between the plates is filled with a fluid such as air or water. In such an arrangement, a pattern of horizontal streamlines results, with the fluid velocity entirely in the X-direction, but with magnitude varying linearly from 0 to V as Y increases. There is no slip of the fluid at either solid surface.

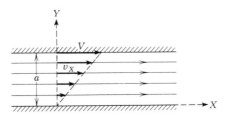

Fig. 20. A layer of fluid between a moving plate and a stationary one.

The last statement is a rigorous observational conclusion. *There is no slip between a fluid and the surface of a solid.* The velocity of the fluid immediately adjacent to the solid is always the same as the velocity of the solid surface; that is, this fluid is at rest *relative to* the solid surface.

In the case of Fig. 20, the fluid in any given layer exerts a tangential force on the faster-moving fluid above it, tending to retard the motion of this faster-moving fluid. Reciprocally, in accordance with Newton's third principle, the faster-moving fluid exerts a force in the forward direction on the slower-moving fluid below it, tending to accelerate this fluid. The fluid in contact with the upper plate exerts a backward force on this plate, tending to retard it; whereas the fluid in contact with the lower plate exerts a forward force on this plate, tending to drag it along with the fluid.

The force per unit area, F/A, which is of the nature of a *shearing stress*, is found to be directly proportional to the velocity gradient in a direction perpendicular to the streamlines, that is, to dv_X/dY in the case of Fig. 20. The constant of proportionality, called the *coefficient of viscosity* μ, is characteristic of the fluid; thus

$$F/A = \mu \, (dv_X/dY) \tag{14}$$

for the case of Fig. 20. This law was first stated by Newton. In principle, we could measure the coefficient of viscosity with the apparatus of Fig. 20 by measuring either the force per unit area required to keep the upper plate moving, or the forward drag force per unit area on the lower plate. In practice, we cannot construct the apparatus of Fig. 20, but we can use two cylinders of large diameter with fluid in the thin annular space between. When one of the cylinders rotates, conditions in the annular space approximate the conditions in Fig. 20.

The dimensions of μ are seen from (14) to be those of (force·length)/(velocity·area), which, since force has dimensions of mass·acceleration, reduce to mass/(length·time). In MKS units, μ is expressed in kg/m·s. The viscosities of water and air at 68° F are

$$\mu_{\text{Water}} = 1005 \times 10^{-6} \text{ kg/m·s},$$

$$\mu_{\text{Air}} = 18.1 \times 10^{-6} \text{ kg/m·s}.$$

For comparison, we might note that olive oil has a viscosity about 100 times, and castor oil a viscosity about 1000 times, that of water.

It turns out that in the flow of air around an object such as an airplane, *the viscous forces have an entirely negligible effect on the air motion except in a thin layer immediately adjoining the solid surface;* this layer is called the *boundary layer* (see Fig. 21). In the boundary layer, which is typically only a fraction of an inch thick, the velocity gradients are sufficiently large that the viscous forces exert an appreciable viscous drag on the solid surface.

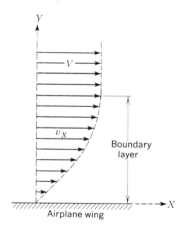

Fig. 21. Velocity distribution in a boundary layer, relative to the material surface.

PROBLEMS

NOTE: *Because of the low density of air in comparison with the densities of solids and liquids, the buoyant force of the air can be neglected in working most of these problems, just as it is neglected in ordinary weighing.*

1. If the U-tube shown in the figure contains mercury, how many centimeters of water must be poured into the right arm to depress the mercury level 0.5 cm in this arm and raise it 0.5 cm in the other arm? Ans: 13.6 cm.

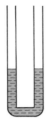

Problems 1–2

2. If the U-tube in the figure contains water and if a 20-cm column of kerosene is poured over the water in the right arm, by how many cm is the water level depressed in this arm and raised in the other arm?

3. In the figure, the right column contains carbon tetrachloride, the center column water, and the left column kerosene. If the water stands at a height of 10 i, at what heights do the other two columns stand?
Ans: 6.25 i; 12.5 i.

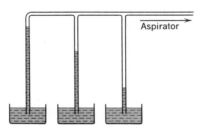

Problems 3–4

4. In the figure, the center column contains water and stands at a height of 15 cm. The other two columns contain liquids of unknown densities and stand at heights of 20.5 cm and 11.5 cm, respectively. Determine the densities and standard specific weights of these liquids.

5. When an iceberg floats on sea water, what fraction of the iceberg is beneath the surface?
Ans: 89.5 per cent.

6. When a block of oak floats in water, what fraction of the block is beneath the surface? What is the magnitude of the vertical force that must be applied to keep a block of balsa wood beneath the surface of a fresh-water lake if the block has a volume of 0.4 f^3?

7. Find the ratio of the density of water at a point 250 f below the surface of a fresh-water lake to the density of water at the surface. Neglect temperature differences.
Ans: 1.000 328.

8. A solid heavier than water is weighed in air and then suspended by a string and weighed in water. The difference between these weights is called the 'loss of weight in water.' This is a convenient method of determining the density of the solid. Show that

density of solid
$$= \frac{\text{weight of solid in air}}{\text{loss of weight in water}} \times \text{density of water}.$$

9. The density of a liquid can be determined by weighing a solid in air, then in water, then in the liquid. Show that, with definitions as in Prob. 8,

density of liquid
$$= \frac{\text{loss of weight of solid in liquid}}{\text{loss of weight of solid in water}}$$
$$\times \text{density of water}.$$

10. Devise a scheme for determining the density of a solid lighter than water by tying it to a solid heavier than water, of known density, and weighing the combination under water.

11. The figure shows a *hydrometer* used to measure the density of a liquid by a determination of the depth at which the hydrometer

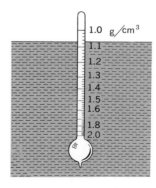

Problems 11–12

floats. If it is desired to construct a hydrometer in which the calibration marks for 1.0 g/cm³ and 2.0 g/cm³ are 15 cm apart on a stem of 1 cm² cross section, what must be the volume of the bulb below the 2.0 mark and what must be the total mass of the hydrometer? How far from the 1.0 mark will the 1.5 mark fall?

Ans: 15.0 cm³; 30.0 g; 10.0 cm.

12. A hydrometer (*see* figure), for fluids lighter than water, has the calibration marks for 0.7 g/cm³ and 1.0 g/cm³ 15 cm apart on a stem of 1 cm² cross section. What must be the volume of the bulb below the 1.0 mark and what must be the total mass of the hydrometer? How far from the 1.0 mark will the 0.9 and 0.8 marks fall?

13. What buoyant force is exerted by water on an iron casting weighing 449 p if the casting lies on the bottom of a lake 42 f deep? What vertical force should be exerted in order to pull the casting toward the surface at constant speed? Ans: 62.4 p; 386 p.

14. What buoyant forces would fresh lake water exert on the following objects dropped into the water following a boat wreck; a bronze bell weighing 60 p, a steel bar weighing 80 p, a maple table weighing 40 p, and a block of ice weighing 100 p?

15. The bag of a balloon is a sphere 25 m in diameter filled with hydrogen of density 0.090 kg/m³. What total mass of fabric, car, and contents can be lifted by this balloon in air of density 1.29 kg/m³? Ans: 9840 kg.

16. Prior to inflation, the total mass of a plastic balloon including 'gondola' is 100 kg. What volume of helium of density 0.178 kg/m³ is required to lift the balloon in air of density 1.29 kg/m³?

17. The operation of the hydraulic press or lift is illustrated in the figure. A liquid is enclosed in a containing vessel equipped with two pistons of areas a_1 and a_2, respectively. If a force F_1 is applied to the small piston of area a_1, what is the increase in pressure at any point in the enclosed liquid? What is the magnitude of the force F_2 exerted by the liquid on the large piston? Ans: F_1/a_1; $F_1 a_2/a_1$.

18. If the area of the small piston in a hydraulic lift is 0.5 i² and that of the large piston is 20 i², what force must be exerted on the small piston in order to raise a total load of 1 ton?

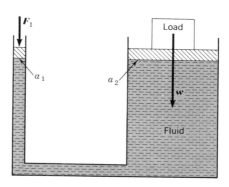

Problems 17–18

How far must the small piston be moved in order to raise the load 4 i? Neglect friction.

19. The hydraulic lift in an automobile service station is operated by compressed air applied directly to the end of a piston 8 i in diameter. If the weight of the lift plus automobile is 2250 p, what gauge pressure of compressed air is required? Ans: 44.7 p/i².

20. Why do bottles containing liquids tend to leak if taken up in an airplane? Do they tend to leak when they are right-side-up or when they are upside-down? What is the gauge pressure in p/i² in a closed bottle in a plane pressurized at 630 mm of mercury (said to be pressurized to an altitude of about 5000 f) if the bottle was closed at sea level?

21. Show that the force F required to pull apart the evacuated hemispheres (Magdeburg hemispheres) in the figure is $\pi R^2 P$, where P is the pressure of the atmosphere and R the radius of the hemispheres to the point at which the vacuum seal occurs.

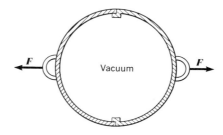

Problems 21–22

22. In 1654, OTTO VON GUERICKE, Burgomaster of Magdeburg and inventor of the air pump, gave a demonstration before the Imperial Diet at Ratisbon in which two teams of eight horses each could not pull apart two evacuated brass hemispheres. If the hemispheres were 24 inches in diameter and imperfectly evacuated to 0.1 atm in the demonstration, what force would the horses have had to exert to pull them apart (*see* figure)?

23. A beaker of water stands on a scale which registers its weight as 2 p. An object weighing 1.2 p with a voluem of 0.0033 f³ is lowered into the water by a string. By how much will the reading of the scale change? Account for this change by computing the rise of the water level and the resultant increase in pressure on the bottom of the beaker. Ans: 0.206 p.

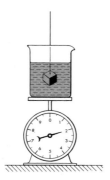

Problem 23

24. In each of the three vessels in the figure, called *Pascal's vases,* the water has the same depth and the bottom the same area. Hence the pressure and the force on the bottom is the same in all cases. If we imagine the bottom of each vessel to be a frictionless watertight piston, we would have to exert the same force F to hold up each piston. This is sometimes considered as a paradox, because while in (a) this force is just equal to the weight of water, in (c) the force is less than the weight of the water and is sufficient to hold up only the vertical cylinder of water indicated by broken lines, while in (b), more remarkably, the force is greater than the weight of water present and is sufficient to hold up all the water which would fill the vertical column indicated by broken lines. Resolve the paradox by analyzing the forces acting on the material of the sides of the vessels. Show that, over and above the weight of the vessel, in (c) a lifting force H is required that is exactly equal to the weight of the excess fluid not supported by F; whereas in (b) a downward force G is required that is equal to the weight of the fluid apparently supported by F but not present.

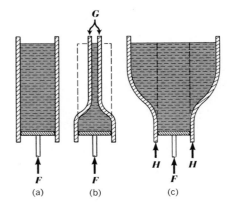

Problem 24

25. If the liquid in a barometer is water, at what height will the water stand on a day when the ordinary mercury barometer reads 30 i? (Neglect the pressure of vapor above the water column.) Ans: 33.9 f.

26. A small building has a flat roof measuring 10 m by 20 m. What total downward force is exerted on the roof by the atmosphere on a day when atmospheric pressure has its standard value? Does this enormous force cause the roof to cave in? Why not?

27. What is the gauge pressure at a point 2 f below the surface in a tank containing lubricating oil if the surface is exposed to the atmosphere? Ans: 112 p/f².

28. What is the gauge pressure at a point 19 cm below the mercury surface inside a barometer if the barometer stands at 76 cm? What is the absolute pressure?

29. Water stands at a height h behind the upstream face of a dam as indicated in the figure. Show that the total horizontal force F exerted on the dam by the water is $\frac{1}{2} \rho g l h^2$, where l is the length of the dam. Show that the total torque exerted by the water, about an axis through O, is $\frac{1}{6} \rho g l h^3$. Show that the effective line of action of the total force F exerted by the water is at a vertical distance $\frac{1}{3} h$ above O.

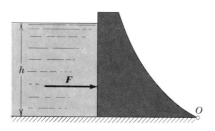

Problem 29

30. From the density of mercury given in Table 4–2 on p. 70, and the value of g_s given on p. 66, compute the pressure exerted by 76 cm of Hg and compare with (5a).

31. If the earth's atmosphere were homogeneous and of normal density 1.292 kg/m³ throughout, what would be the thickness of the atmosphere in km when the barometer reads 1 atm? *Ans: 8.00 km.*

32. An aluminum casting weighing 338 p lies on the bottom of a mill pond, 16 f below the water surface. How much work must be done in raising the casting to the top of the dam over which the water flows? After the casting has been raised to the top of the dam, what is its potential energy with respect to the surface of the creek below the dam if the vertical distance from the top of the dam to the surface of the creek is 16 f? Do the results of your computation violate the conservation-of-energy principle? Discuss.

33. A piece of balsa wood weighing 16 p is pulled downward to a depth of 10 f below the surface of the sea. What is the potential energy of the piece of wood relative to objects located at the surface? If the piece of balsa were released and allowed to move upward, how far would it rise above the surface? (Neglect viscosity and the fact that kinetic energy is actually imparted to some of the water.)
Ans: 1120 fp, 70 f.

34. If the glass tube of the barometer in the figure is supported by a string attached to the upper end, so that the lower end just dips below the level in the reservoir, determine the tension in the string. Analyze the forces carefully and determine whether this tension is (a) just the weight of the glass tube, (b) the weight of the tube plus the weight of the mercury it contains, or (c) some other weight.

Problem 34

35. Explain why a thin-walled metal pipe will withstand more pressure differential between inside and outside if the excess pressure is *inside*. What is the bursting strength in p/i² of a hard-copper tube 1 i in diameter with wall thickness 0.01 i if the tensile stress for rupture is 50 000 p/i². *Ans: 500 p/i².*

36. Water flows through the horizontal pipe in the figure, which has an internal diameter of 20 cm at the right and left and 16 cm in the constricted section. The water level in the right and left tubes is at 2 m above the center line of the pipe. The barometer reads 76 cm of Hg. If the speed of flow in the large section is 2 m/s, at what height does the water stand in the center tube? At what speed of water flow would air be sucked in through the center tube?

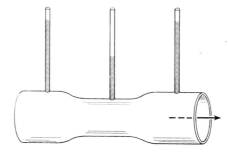

Problem 36

37. What is the speed of flow of water from an open tank through an orifice located 16 f below the water surface? Find the volume rate of flow if the orifice is well rounded and has an area of 2 i². *Ans: 32.0 f/s; 0.444 f³/s.*

38. Answer Prob. 37 for a case in which the water flows from a closed tank in which the region above the water surface contains air at a gauge pressure of 60 p/i².

39. 'Old Faithful' geyser periodically ejects a stream of water that attains a height of 120 f. What excess of gauge pressure is required inside the geyser to cause the stream to reach this height? What is the magnitude of the stream velocity as the water emerges?
Ans: 52.0 p/i²; 87.6 f/s.

40. The pump of a fire engine discharges 18 f³ of water per minute at a gauge pressure of 160 p/i². To what height can this pump send a stream of water? Assuming negligible frictional losses, find the horsepower rating of the engine required to operate the pump.

41. A Venturi flowmeter in a water line has $A_1 = 4$ f² and $A_2 = 3$ f² in the notation of Fig. 17. If the differential height reading is 10.8 i of mercury, what is the volume flow rate of water? Ans: 122 f³/s.

42. In Fig. 17, if absolute pressure $P_1 = 20$ p/i² and $v_1 = 100$ f/s, at what area ratio A_1/A_2 will cavitation ensue in water? Neglect vapor pressure.

43. In a perfume aspirator, air of density 1.25 kg/m³ is blown across the top of a glass tube that dips down into a bottle of perfume of density 0.9 times that of water. What is the minimum air velocity that can lift the perfume 10 cm? Ans: 37.6 m/sec.

44. The hydraulic press considered in Probs. 17–18 can be modified to provide a pump for shooting a stream of liquid into the air. This can be done by replacing the small piston by a small well-rounded nozzle. If the nozzle has an area of 0.1 i² and that of the large piston is 50 i², what will be the velocity of the jet if the large piston moves downward at a speed of 3 i/s? How high will the stream rise if it is ejected vertically (neglect friction)? If the liquid is water, calculate the gauge pressure of the enclosed liquid just below the large piston.

45. The air speed of a plane is measured by a Pitot-static tube whose schematic cross section is shown in the figure. Such tubes can be observed protruding from the wing or the fuselage of an aircraft. If we consider motion of the air relative to the tube, so that Bernoulli's law is applicable, the free-stream pressure and velocity in front at ③ are P and v. In the Pitot tube (which is connected to the right side of the differential manometer), at ① the velocity is reduced to zero and the full 'ram' pressure P_1 is developed. In the static tube, connected to the left side of the manometer, the entry holes are set far enough back so that the velocity and pressure outside the holes have the free-stream values. The pressure inside the static tube will also be the free-stream pressure. Show that if the differential manometer reads $(P_1 - P)$, the air speed is $v = \sqrt{2(P_1 - P)/\rho}$, where ρ is the air density at the altitude of the plane.

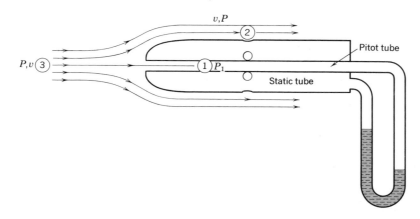

Problems 45–46

46. If a plane is flying at 400 mi/h at an altitude of 10 000 f, where the air density is 0.9 kg/m³, determine the differential pressure indicated by its Pitot-static tube in mm of mercury (*see* Prob. 45).

47. Describe the spin a pitcher should give to a baseball if it is to curve to the right. Justify your answer by a diagram of the streamlines and application of Bernoulli's law.

48. A viscosimeter of the two-cylinder type described on p. 279 has the annular space filled with oil whose viscosity is being measured. The length of the cylinders is 15 cm, their mean radius 6 cm, the thickness of the annular space 2 mm. The inner cylinder is fixed. It is found that a torque of 0.12 m·N is required to rotate the outer cylinder at 30 rev/min. Determine the viscosity of the oil and its ratio to that of water.

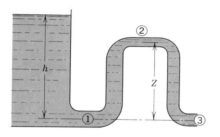

Problem 49

49. Neglecting friction, find the gauge pressures and speeds at ①, ②, and ③ in the figure if $h = 100$ f, $Z = 70$ f, $A_1 = 1$ f², $A_2 = 1$ f², $A_3 = 0.75$ f², and the water discharges into the free atmosphere at ③. Use $g = 32.2$ f/s².
Ans: ①: 19.0 p/i², 60.2 f/s; ②: -11.4 p/i², 60.2 f/s; ③: 0, 80.2 f/s.

50. If water discharges in a stream ½ inch in diameter from a faucet, with a speed of 30 f/s, what is the gauge pressure in the 1-in diameter pipe which the faucet terminates? Neglect losses.

51. Describe the force system acting on the wings of an airplane in a 'rolled' orientation. Show that there is a large transverse force component that causes the plane to travel in a curved path. A plane must roll in order to turn; the rudder is capable of exerting a strong torque to stabilize the plane's orientation, but is ineffective in exerting a transverse force at the center of mass; discuss.

52. Neglecting friction, show that the liquid in an open U-tube of uniform bore will execute simple harmonic motion at a frequency that is $\sqrt{2}$ times the frequency of a simple pendulum of length equal to that of the liquid column.

53. Neglecting friction and the kinetic energy that is actually imparted to the water, show that a cylindrical log, weighted at one end so that it floats upright, would execute vertical simple harmonic motion of frequency $f = (1/2\pi)\sqrt{\rho g A/m}$, where ρg is the specific weight of water, A the cross-sectional area of the log, and m the mass of the log.

54. Consider the stability of a uniform rectangular parallelepiped of wood, such as a plank or a beam, floating in water. Compare the potential energy of the plank, water system in the three orientations in which the faces are horizontal and vertical. Show that the configuration of *least* potential energy is that in which the sides of largest area are horizontal, and hence that a plank will float flat in the water, as observed.

55. If the plank or beam of Prob. 54 is weighted internally near one end, so that it remains a rectangular parallelepiped and still floats, formulate the condition on the location of its center of mass that will cause it to float upended, as in Prob. 53, rather than flat.

56. Show that the speed of rotation at which a submarine's propellor first begins to cavitate increases with the depth of the submarine.

PART II

HEAT AND MOLECULAR PHYSICS

TEMPERATURE; THERMAL EXPANSION 14

With this chapter we begin our study of the branch of physics known as *heat*, which deals with certain phenomena that cannot be described completely in terms of the theories presented earlier under the heading of *mechanics*.

It is the aim of physics to develop a single consistent body of theory that is sufficiently general to permit a complete description of physical phenomena in terms of *fundamental physical quantities*, the number of such fundamental quantities being kept as small as possible. Thus, the whole of mechanics can be described in terms of *length, mass,* and *time*. For these quantities we chose quite arbitrary units—the meter, kilogram, and second. It was appropriate to take length, mass, and time as fundamental quantities because these quantities cannot be defined in terms of a simpler set of physical quantities, or in terms of each other. We then proceeded to define *derived quantities*, such as *force, momentum, energy, acceleration,* and *density*, in terms of the three fundamental quantities that we had selected.

In describing *thermal* phenomena we find that we need one quantity that cannot be defined in terms of the three fundamental quantities of mechanics. This quantity is *temperature*. Until the middle of the last century still another quantity—*the quantity of heat*—was regarded as fundamental. Then the work of Rumford and Joule showed that heat simply represents the kinetic and potential energy associated with the random motions of the atoms and molecules of a substance, so that quantity of heat can be expressed in the energy units defined in mechanics. However, *temperature* is a physical quantity of a type different from *quantity of heat,* and we still have to introduce temperature as a new *fundamental quantity.* Our definition of temperature will consist of a description of the *operations* that must be performed in its measurement.

1. The Common Temperature Scales

Temperature is determined by measurement of some mechanical, electrical, or optical quantity whose value has a one-to-one correlation with temperature. Usually the temperature of a substance is not determined by a measurement made on the substance directly, but by measurement on a *thermometer* that is brought into intimate contact with the substance and is assumed to acquire the same temperature.

Thermometers based on the expansion of a liquid were invented early in the seventeenth century. At first they had completely arbitrary scales, so that each thermometer gave readings peculiar unto itself. The desirability of standardizing the readings was recognized late in the seventeenth century, and our common scales were devised during the first half of the eighteenth century.

In defining the common temperature scales, two conveniently reproducible temperatures called *fixed points* are used:

> The **lower fixed point** (the **ice point**) is the temperature of a mixture of pure ice and water exposed to the air at standard atmospheric pressure.

> The **upper fixed point** (the **steam point**) is the temperature of steam from pure water boiling at standard atmospheric pressure.

The temperature scale used in most scientific work and in common use in most of the countries of the world is the *Celsius scale,* in which the fixed points are taken as 0° C and 100° C. This scale is named for ANDERS CELSIUS (1701–1744), a Swedish astronomer who was one of the first proponents of this scale. It is also, particularly in English-speaking countries, called the *centigrade scale.*

The common Fahrenheit scale was devised in 1714 by GABRIEL DANIEL FAHRENHEIT, who was a scientific-instrument manufacturer in Danzig and Amsterdam. On this scale the fixed points are 32° F and 212° F; Fahrenheit arrived at these curious figures by taking 0° as the temperature of a freezing mixture of ice and salt in his laboratory (actually a temperature as low as $-9°$ F may be obtained with the proper mixture of ice and salt), and choosing 96°, for some unexplained reason, to represent the temperature of the human body (his scale missed the actual average body temperature by about 2.6°, but this discrepancy can be attributed to experimental inaccuracy).

The original thermometers were of the type still in common use; they make use of the volumetric thermal expansion of liquid mercury from a reservoir into an evacuated glass capillary tube of uniform bore. The positions of the end of the liquid column in the capillary tube at the temperatures of the fixed points are determined and marked. The distance between the marks is divided into 100 equal spaces for the Celsius scale, or 180 for the Fahrenheit scale, to give the individual degrees. Above and below the fixed points the scale may be extended by marking off degrees of the same size. This calibration is illustrated in Fig. 1. Because there are 100 Celsius degrees (C deg) and 180 Fahrenheit degrees (F deg) between the ice point and the steam point, we see that 100 C deg = 180 F deg; hence

$$1 \text{ C deg} = \tfrac{9}{5} \text{ F deg}, \qquad 1 \text{ F deg} = \tfrac{5}{9} \text{ C deg}. \tag{1}$$

Note that because of the way the common temperature scales are set up, with origins at different points, we should not write *equations*, such as $0° \text{ C} = 32° \text{ F}$, $5° \text{ C} = 41° \text{ F}$, in comparing a *temperature value* on two different scales, because such equations cannot be manipulated algebraically (try subtracting the first equation from the second!). Rather we should use the symbol $\sim$, standing for 'corresponds to,' and write

$$0° \text{ C} \sim 32° \text{ F}, \qquad 5° \text{ C} \sim 41° \text{ F}.$$

Temperature intervals, on the other hand, are properly expressed by a number and a unit. To avoid confusion with temperatures themselves, we write the unit of temperature difference as C deg or F deg, and we can properly write

$$0 \text{ C deg} = 0 \text{ F deg}, \qquad 5 \text{ C deg} = 9 \text{ F deg}.$$

The units C deg and F deg can be handled in equations like ordinary units by using relations (1).

We now turn to the problem of converting temperatures from one scale to the other. Rather than trying to remember an equation for this, it is best to remember the fixed points of each scale, and hence the relative size of the degrees, and to use the scheme of the following examples:

Example. *What is the Celsius temperature corresponding to* $68°$ *F?*

We start by noting that $68°$ F is 36 F deg above the ice point, which, from (1), is $36 \times \tfrac{5}{9}$ C deg = 20 C deg above the ice point, and hence corresponds to $20°$ C.

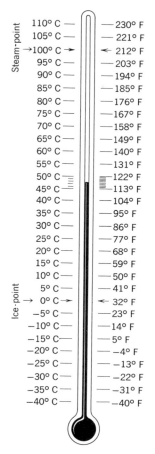

Fig. 1. Celsius and Fahrenheit temperature scales.

Example. *What is the Fahrenheit temperature corresponding to* $-20°$ C?

We note that $-20°$ C is 20 C deg below the ice point, which is $20 \times \frac{9}{5}$ F deg $= 36$ F deg below the ice point. Since the ice point is at $+32°$ F, this temperature corresponds to $-4°$ F.

2. Thermal Equilibrium

No matter what type of thermometer is used to measure temperature, the following principle is found experimentally to be universally true:

PRINCIPLE OF THERMAL EQUILIBRIUM: *If objects having different temperatures are placed in an insulated enclosure, all the objects eventually come to the same temperature.*

In an insulated enclosure, objects that were initially hot will become cooler and initially cold objects will become warmer. For example, if we put a pot of hot coffee and a bucket of ice water into an insulated room, the coffee, the water, the air, the walls, and the room furnishings will eventually reach the same temperature. At this stage *thermal equilibrium* is said to be established. No rearrangement of the objects inside the enclosure will cause any further changes in temperature.

The above principle shows that the assumption we have made that the thermometer comes to the same temperature as that of the object whose temperature is being measured is valid, provided thermal equilibrium is established. A thermometer is a device capable of giving a physical indication of *its own* temperature. When a thermometer is placed in intimate contact with another body—a solid, liquid, or gas—it will come to the same temperature as this body and hence give a reading of the temperature of the body.

3. The Absolute Temperature Scale

Although there is no *upper* limit on temperature, there is a natural *lower* limit called the *absolute zero* of temperature. Temperature changes of a substance are accompanied by corresponding changes in the random energy of atomic motion, called *thermal energy* (Chapter 15). This thermal energy is finite; if all the available thermal energy is extracted, the body cannot be further cooled. Furthermore, no other body can be colder, as we shall see in Chapter 15. The common lower limit of temperature, called absolute zero, can be determined on the basis of Lord Kelvin's work discussed in Chapter 19. It has the value

$$\text{absolute zero} \sim -273.1500° \text{ C} \sim -459.67° \text{ F.} \tag{2}$$

In the laboratory it has been possible to reach temperatures within a few millionths of a degree of absolute zero.

Absolute zero is the temperature of a body that is incapable of giving up any thermal energy.

As would seem reasonable, much of the theory of heat is greatly simplified if temperatures are measured on what is called an *absolute scale,* in which absolute zero is taken as the zero of the scale.

The absolute scale in which temperatures are measured from absolute zero in Celsius-size degrees is known as the *Kelvin scale.* Since absolute zero is $-273°$ C, a temperature may be expressed in °K by adding 273 to the value in °C. Thus,

$$0°\text{ C} \sim 273°\text{ K}, \quad 10°\text{ C} \sim 283°\text{ K}, \quad -10°\text{ C} \sim 263°\text{ K, etc.}$$

Temperature differences on the Kelvin scale are the same as on the Celsius scale and are measured in units that can be called either C deg or Kelvin degrees. The unit of temperature difference on the Kelvin scale is, in the SI system, called the *kelvin* and abbreviated K.

An absolute scale employing Fahrenheit-size degrees, in use in engineering practice, is known as the *Rankine scale.* Since absolute zero is $-460°$ F, a temperature may be expressed in °R by adding 460 to the value in °F.

Any property of matter that varies with temperature in a measurable way can be made the basis of a thermometer. The various properties that are used for thermometric purposes are:

(*a*) Expansion of a liquid.
(*b*) Expansion of a solid (*see* next section).
(*c*) Variation of pressure or volume of a gas (*see* Chapter 17).
(*d*) Variation of electrical resistance (*see* Chapter 25).
(*e*) Thermoelectricity (*see* Chapter 27).
(*f*) Variation of quantity of radiated energy (*see* Chapter 16).
(*g*) Variation of color of radiated light (*see* Chapter 41).
(*h*) Variation of vapor pressure (*see* Chapter 18).
(*i*) Variation of magnetic susceptibility (*see* Chapter 30).
(*j*) Variation of speed of sound in a gas (*see* Chapter 21).

The method of interpolating and extrapolating the common temperature scales between and beyond the fixed points, depending as it does on the properties of a particular substance, mercury, lacks fundamental significance. Also it is impossible to extend the scale in this way below the freezing point ($-39°$ C) or above the boiling point (about $357°$ C) of mercury. Other physical phenomena in the above list can be used to extend the scale indefinitely, but would be equally lacking in fundamental significance.

These difficulties were resolved by LORD KELVIN, who devised a *thermodynamic scale* of temperature that is independent of the physical properties of any particular substance. The thermodynamic scale is the one that would result if temperature measurements could be made with a constant-volume gas thermometer employing an *ideal* gas, with absolute temperature defined as proportional to gas pressure. The properties of real gases can be corrected to correspond to those of an ideal gas by methods described in Chapter 19, and the scale can be extended to indefinitely high temperatures by using the laws of thermal radiation discussed in Chapter 41.

The *absolute thermodynamic scale* requires the definition of only *one* fixed point to determine the size of the degree to be used, since the origin of temperature is now fixed at absolute zero. One could take this one fixed point as the ice point, defined, according to (2), as 273.15° K. Actually, because of its better reproducibility, the fixed point is taken as the *triple point of water* (*see* Chapter 18), which is assigned the temperature 273.16° K exactly; this definition gives the ice point the experimental value 273.1500° K.

By international agreement, *the thermodynamic temperature scale defined as above is the standard scale of temperatures.*

In the region in which it is usable, the mercury thermometer gives readings very close to those of the thermodynamic scale. The standard practical temperature scale, defined by international agreement to give, as nearly as possible, the same values as the absolute thermodynamic scale, makes use of a platinum resistance thermometer up to 903° K, a platinum-rhodium thermocouple up to 1336° K, and an instrument to measure radiated energy at higher temperatures. These instruments are calibrated in terms of a considerable number of secondary fixed points whose thermodynamic temperatures have been accurately determined—for example the value 1336.0° K for the melting point of gold.

4. Thermal Expansion of Solids

When the temperature of a solid body is raised, the body expands. However, the increase in dimension of a solid with increasing temperature is small. The order of magnitude is easy to remember: *a meter length of solid lengthens by about 1 millimeter for a temperature rise of 100 Celsius degrees.*

Thermal expansion is of sufficient magnitude, however, to be an important factor in many practical problems. Thermal expansion makes necessary the provision of expansion joints in buildings, bridges, and pavements; it makes possible the shrink-fitting of collars on shafts; and it results in the breakage of glass when heat is irregularly applied.

The change in any linear dimension of a solid, such as the length, width, height, radius, or distance between two marks, is known as the *linear expansion*. We shall denote the length in question by l and the change in length that arises from a change in temperature of amount ΔT by Δl.

Experiment shows that, over a wide temperature range, the change in length, Δl, is proportional to the change in temperature, ΔT. It is of course also proportional to the length l itself, so we can write

$$\Delta l = \alpha \, l \, \Delta T, \tag{3}$$

where α is called the *coefficient of linear expansion*. This coefficient has different values for different materials. Since we can write

$$\alpha = \frac{\Delta l}{l \, \Delta T}, \tag{4}$$

The **coefficient of linear expansion** of a solid is the change in length per unit length per degree change in temperature.

The dimensions of α are deg^{-1}; its value does not depend on the particular unit of length used, but does depend on the size of unit used to measure ΔT. Thus, if a meter bar lengthens 1 mm for a temperature increase of 100 C deg, we have, from (4),

$$\alpha = \frac{10^{-3} \text{ m}}{1 \text{ m} \times 100 \text{ C deg}} = \frac{10^{-5}}{\text{C deg}}.$$

Since, from (2), 1 C deg = $\frac{9}{5}$ F deg, this same coefficient of linear expansion is

$$\alpha = \frac{10^{-5}}{\text{C deg}} = \frac{10^{-5}}{\frac{9}{5} \text{ F deg}} = \frac{\frac{5}{9} \times 10^{-5}}{\text{F deg}}.$$

Since the F deg is only $\frac{5}{9}$ as large as the C deg, the solid expands only $\frac{5}{9}$ as much per F deg as per C deg.

To the accuracy with which most expansion coefficients are known for commercial materials (two significant figures), it does not matter whether the l in (3) and (4) is taken to be the length at the initial temperature or at the final temperature, since these differ only by the quantity Δl, which is very small compared with l itself.

Values of the coefficient of linear expansion of various commercial materials are given in Table 14-1.

When an isotropic solid expands thermally, the distance between every two points increases in the ratio α per degree temperature rise, just as in the case of a photographic enlargement (Fig. 2) except that the solid is three-dimensional. For example, the diameter of a hole in the solid *enlarges* in the same ratio as an external dimension. Every line drawn on the solid, whether straight or curved, lengthens in the ratio α per degree temperature rise. For example, if C is the circumference of a circle, $\Delta C = \alpha \, C \, \Delta T$.

The difference in expansion coefficients of different metals, usually

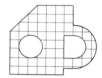
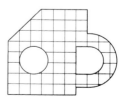

Fig. 2. Thermal expansion of a solid is like a photographic enlargement.

Table 14-1 TYPICAL COEFFICIENTS OF LINEAR EXPANSION OF COMMERCIAL MATERIALS NEAR ROOM TEMPERATURE (PER C DEG)

Aluminum	24 $\times 10^{-6}$		Magnesium	26 $\times 10^{-6}$
Bakelite	28		Nickel	13
Brass or bronze	19		Oak (across fiber)	54
Brick	9		Oak (parallel to fiber)	5
Copper	17		Pine (across fiber)	34
Glass (ordinary)	9		Pine (parallel to fiber)	5
Glass (Pyrex)	3		Platinum	8.9
Gold	14		Quartz (fused)	0.4
Granite	8		Silver	19
Ice	51		Solder (soft)	25
Invar (Ni 36%, Fe 64%)	0.9		Steel	12
Iron (cast)	11		Tin	20
Lead	29		Tungsten	4.3

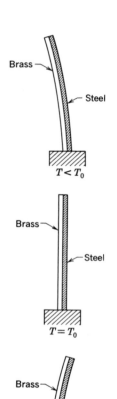

Fig. 3. Principle of the bimetal thermometer or thermostat element. In the thermostat, the motion of the strip opens or closes electrical contacts.

brass ($\alpha = 19 \times 10^{-6}$/C deg) and steel ($\alpha = 12 \times 10^{-6}$/C deg), is utilized in thermometry and more particularly in the common *thermostat* by welding strips of these materials together to form a *bimetal strip*. As illustrated in Fig. 3, if the strip is straight at a certain temperature, it will bend one way at higher temperatures, the other way at lower temperatures, because brass has a greater tendency than steel to lengthen or shorten with temperature change. Bimetal strips are also used in the rim of the balance wheel of a watch to vary the rotational inertia in such a way as to keep the period of the balance wheel constant as the temperature changes.

Let us now consider what happens to the *area* of a figure drawn on the surface of an isotropic solid, or to the area of a sheet of solid. We consider first the change in area of a rectangle (Fig. 4) of sides a and b, area $A = ab$. When the temperature increases by ΔT, a lengthens by $\Delta a = \alpha\, a\, \Delta T$, b lengthens by $\Delta b = \alpha\, b\, \Delta T$, and the area increases by

$$\Delta A = a\, \Delta b + b\, \Delta a + \Delta a\, \Delta b,$$
$$= a\, \alpha b\, \Delta T + b\, \alpha a\, \Delta T + \alpha^2\, ab\, (\Delta T)^2 = \alpha\, ab\, \Delta T\, (2 + \alpha\, \Delta T).$$

Now we see from Table 14-1 that $\alpha \sim 10^{-5}$/C deg; so even if ΔT is as large as 1000 C deg, $\alpha\, \Delta T \sim 10^{-2}$, which is small in comparison with 2. Therefore, we can drop the $\alpha\, \Delta T$ term on the right and write $\Delta A = 2\alpha\, ab\, \Delta T$, or

$$\Delta A = 2\alpha\, A\, \Delta T. \tag{5}$$

What we have neglected is the area $\Delta a\, \Delta b$ of the small rectangle in Fig. 4 in comparison with the area of the two strips $a\, \Delta b$ and $b\, \Delta a$. In equation (5), 2α expresses the change in area per unit area per degree.

The **coefficient of area expansion** of a solid is the change in area per unit area per degree change in temperature.

The coefficient of area expansion of an isotropic solid is twice the coefficient of linear expansion.

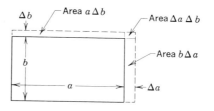

We note that in the limit we can write (3) in the form $dl/dT = \alpha l$. Hence for a square of side l and area $A = l^2$, we have, by differentiation,

$$\frac{dA}{dT} = 2\, l\, \frac{dl}{dT} = 2\, l \times \alpha l = 2\alpha\, A. \tag{6}$$

Fig. 4. Thermal increase in area. The sizes of Δa and Δb are highly exaggerated.

Since an area of any shape can be divided into small squares, each of which expands according to this equation, the whole area so expands. The argument that led to (5) shows that α is sufficiently small that even for fairly large values of ΔT it is permissible to replace dA/dT by $\Delta A/\Delta T$ in (6). We see that (5) applies equally well to plane or curved surfaces; for example it will give the change of surface area of a sphere or of a hole in a solid.

Now consider a cube of side l and volume $V = l^3$. By differentiation,

$$\frac{dV}{dT} = 3\, l^2\, \frac{dl}{dT} = 3\, l^2 \times \alpha l = 3\alpha\, V. \tag{7}$$

Since any solid can be considered as divided into small cubes, (7) applies to the volume of a solid of any shape. It also applies to the volume of a hole in a solid—for example, to the capacity of a container or bottle.

An argument similar to that which led to (5) shows that over a considerable range it is permissible to replace dV/dT by $\Delta V/\Delta T$ in (7) and write

$$\Delta V = 3\alpha\, V\, \Delta T. \tag{8}$$

Here, 3α expresses the change in volume per unit volume per degree.

The **coefficient of volume expansion** of a solid or liquid is the change in volume per unit volume per degree change in temperature.

The coefficient of volume expansion of an isotropic solid is three times the coefficient of linear expansion.

5. Thermal Expansion of Liquids

In the case of a *fluid* (a liquid or a gas) we are not concerned with linear or area expansion because the *shape* of a fluid is not well-defined; only the *volume* of a fluid is of significance. The ways in which liquids and gases respond to changes in temperature or pressure are quite different. Gases respond strongly, whereas the change of volume of liquids with changes in temperature or pressure is very small, only slightly greater than that of solids. Discussion of the thermal expansion of gases will be postponed until Chapter 17.

Experiment shows that for liquids it is possible to define a coefficient of volume expansion of the same type as we defined for a solid. In the case of a solid we denoted the coefficient of volume expansion by 3α, since it was directly related to the coefficient of linear expansion. In the case of a liquid we have no coefficient of linear expansion, so we shall denote the *coefficient of volume expansion* by β, writing

$$\Delta V = \beta \, V \, \Delta T. \tag{9}$$

For most liquids the value of β is relatively independent of the temperature at which the temperature change ΔT takes place. Typical values of β are given in Table 14-2; it is seen that, except for the metal, mercury, these values are of the order of 10 times the coefficient 3α for solids.

Table 14-2 Coefficients of Volume Expansion and Densities of Liquids Near Room Temperature

Liquid	β	Density at 20° C
Alcohol, ethyl	112 $\times$ 10^{-5}/C deg	0.791 Mg/m^3
Alcohol, methyl	120	0.792
Benzene	124	0.877
Carbon tetrachloride	124	1.595
Ether, ethyl	166	0.714
Glycerin	51	1.261
Mercury	18.2	13.546
Turpentine	97	0.873

The use of an equation such as (8) or (9) implies that the change in volume is so small that it does not matter whether the volume inserted on the right is the value at temperature T or that at $T + \Delta T$. This condition is not satisfied over so large a temperature range for volumetric expansion of liquids as it is for solids. When this condition is not satisfied, one must consult detailed data given in handbooks and physical tables.

Example. *Two thermometers made of ordinary glass have exactly the same shape and size and contain, at 0° C, identical volumes of methyl alcohol and mercury respectively. Compare the intervals between the degree marks on the two thermometers.*

When the temperature rises one degree, the internal volume of the glass container increases by an amount proportional to the coefficient of volume expansion of glass; the volume of the liquid increases by an amount proportional to the coefficient of volume expansion of the liquid. Only insofar as the coefficient for the liquid is greater than that for the glass, so that new volume needs to be occupied, will the liquid rise in the stem.

It is easy to see that for the two geometrically identical thermometers in question, the amount of rise per degree will be proportional to the *difference* between the volume expansion of the liquid and that of the glass; that is, proportional to the difference in coefficients of volume expansion. Hence if we denote the intervals between the degree marks by l_{Alcohol} and l_{Mercury}, respectively,

$$\frac{l_{\text{Alcohol}}}{l_{\text{Mercury}}} = \frac{\beta_{\text{Alcohol}} - 3\alpha_{\text{Glass}}}{\beta_{\text{Mercury}} - 3\alpha_{\text{Glass}}} = \frac{120 \times 10^{-5} - 2.7 \times 10^{-5}}{18.2 \times 10^{-5} - 2.7 \times 10^{-5}} = \frac{117}{15.5} = 7.61.$$

The alcohol-in-glass thermometer will have degree intervals 7.61 times as large as those of the mercury-in-glass thermometer.

Anomalous behavior of water. The volume of a given mass of water does not increase at all linearly with temperature according to an equation like (9). Rather, the volume of water decreases slightly from 0° C to 4° C, where the density of water has its maximum value of 1.0000 Mg/m³ or 1.0000 g/cm³. Above 4° C, the volume increases, at first slowly and later more rapidly. The volume of 1 gram of liquid water is plotted against temperature in Fig. 5. *Between 0° C and 4° C, water contracts as the temperature rises.* Such a contraction with increasing temperature is extremely exceptional among known liquids. Salt solutions, such as sea water, behave in the same way for a few degrees above the freezing point. This behavior is critical to the manner in which lakes and the polar ocean freeze from the top down rather than from the bottom up, since water cooler than 4° C rises rather than sinks and the water about to freeze—that at 0° C—is at the top rather than at the bottom.

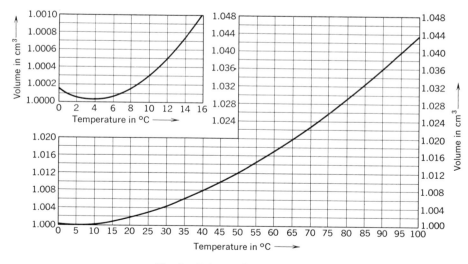

Fig. 5. Volume of one gram of water.

PROBLEMS

1. What is normal human body temperature, 98.6° F, on the Celsius scale? the Kelvin scale? the Rankine scale?
Ans: 37.0° C, 310° K, 558° R.

2. What is the boiling point of ethyl alcohol, 174° F, on the Celsius scale? the Kelvin scale? the Rankine scale?

3. At what temperature will the reading of a Fahrenheit thermometer be the same as that of a Celsius thermometer? Ans: −40.0° C.

4. At what temperature will the reading of a Fahrenheit thermometer be twice that of a Celsius thermometer? three times that of a Celsius thermometer?

5. What is the difference between 40° C and 0° F in C deg? in F deg?
Ans: 57.8 C deg, 104 F deg.

6. What is the difference between 212° F and 40° C in C deg? in F deg?

7. A metal rod 2 m long increases in length by 3.10 mm when heated from 0° C to 92° C. What is the coefficient of linear expansion of this metal in (C deg)$^{-1}$? in (F deg)$^{-1}$?
Ans: 16.8×10^{-6}/C deg; 9.33×10^{-6}/F deg.

8. A glass rod 50 cm long increases its length by 0.34 mm when heated from 0° C to 82° C. What is the coefficient of linear expansion of this glass in (C deg)$^{-1}$? in (F deg)$^{-1}$?

9. If steel railroad rails are placed when the temperature is 10° F, how much gap must be left between each standard 39-f rail section and the next if the rails should just touch when the temperature rises to 110° F? Ans: 0.312 i.

10. A locomotive wheel is 36 inches in diameter. A steel tire, with diameter 0.03 inch undersize at 20° C is to be shrunk on. To what temperature must it be heated to make the diameter 0.03 inch oversize?

11. Shrink fits are frequently made by cooling the male part in solid carbon dioxide, which has a temperature of −78.5° C. How many ten-thousandths of an inch oversize can a steel shaft be made at 20° C and still fit a 1.0-i hole if it is first cooled with Dry Ice? Ans: 11.8.

12. If a steel tape is accurate at 60° F and is used to measure off a mile at 100° F, how many feet and inches should be read on the tape to get a true mile?

13. If a steel tape is accurate at 20° C and is used to measure off a mile at 0° F, how many feet and inches should be read on the tape to get a true mile? Ans: 5282 f, 5 i.

14. A quantity of mercury occupies 250 cm³ at 20° C. What is its change in volume when cooled to 0° C?

15. A hollow spherical aluminum tank has a diameter of exactly 1 m at 0° C. Find the increase in diameter, surface area, and volume when heated by allowing steam at 100° C and 1 atm pressure to enter the sphere. Ans: 24.0 $\times 10^{-4}$ m, 1.51×10^{-2} m², 0.376×10^{-2} m³.

16. A narrow-necked brass bottle is just filled with 500 cm³ of ethyl alcohol at 0° C. How many cm³ spill over when the bottle is brought to a temperature of 30° C?

17. A narrow-necked bottle of ordinary glass is just filled with mercury at 20° C. The volume of mercury is then 150 cm³. How many cm³ spill over when the bottle is placed in steam at 100° C? Ans: 1.86 cm³.

18. Derive a general relation that gives the volume of a liquid that spills over in problems such as 16 and 17 in terms of β for the liquid, α for the container, and the temperature rise.

19. In 1816, Dulong and Petit devised an apparatus (*see* figure) that furnishes the most accurate method of determining coefficients of volume expansion of liquids. If the left column is at temperature T_0 and the right column at T_1, show from the laws of fluid statics that $\beta = \Delta h/(h \, \Delta T)$, where $\Delta T = T_1 - T_0$. Show that

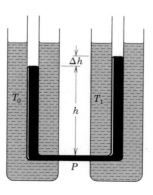

Problems 19–21

this relation does not depend on the material of the containing U-tube, nor on whether the tube is of uniform bore.

20. In the figure, if the water bath on the left is at 5° C, that on the right at 30° C, $h = 153$ cm, and $\Delta h = 2.38$ cm, determine the coefficient of volume expansion of the liquid in the U-tube from the relation in Prob. 19.

21. In the figure, the U-tube is filled with a liquid whose volumetric expansion is to be determined. The water bath on the left is kept at 0° C by melting ice. The bath on the right is at a temperature of 20.0° C, h is 116 cm, and Δh is observed as 2.00 cm. Determine β as in Prob. 19. Ans: $86 \times 10^{-5}/C$ deg.

22. Prove that the change in rotational inertia, I, of a solid object is given by $\Delta I = 2\alpha\, I\, \Delta T$.

23. Prove (using the result of Prob. 22) that the change in period, t, of any physical pendulum is given by $\Delta t = \frac{1}{2}\, \alpha\, t\, \Delta T$.

24. A clock with an uncompensated brass pendulum is adjusted at 68° F. How many seconds will the clock gain or lose per day at 95° F? *See* Prob. 23.

25. A thermometer is made of a capillary tube of ordinary glass of 0.015 mm² cross section at 0° C. At 0° C, there are 1.5 cm³ of mercury in the reservoir and capillary up to the 0° mark. How far does the mercury move up the tube at 10° C? Ans: 15.5 cm.

26. A thermometer is made of a capillary tube of ordinary glass. At 0° F there is 0.400 cm³ of methyl alcohol in the reservoir and capillary up to the 0° mark. The distance between the 0° F and the 100° F marks is 25 cm. What is the cross-sectional area of the capillary?

27. A brass cube 5 cm on an edge at 0° C is weighed in water at 0° C and at 50° C. Find the difference in the weights.
Ans: 1.11×10^{-2} N.

28. What pressure in atmospheres is required to keep water from expanding when it is heated from 20° C to 100° C?

29. A steel bar 1 i² in cross section and 18 i long is inserted with no clearance between two fixed supports when the temperature is 0° C. If the bar is heated to 50° C, what force does the bar exert on each support?
Ans: 17 400 p.

30. A steel surveyor's tape is accurate at 70° F. With what tension in p/i² would it have to be stretched at 30° F in order to read correctly?

31. In modern practice, railroad rails are sometimes butt-welded into a continuous length. If rails are laid and welded in the summer at 90° F, what is the tensile stress in the winter at −10° F? Will the rails acquire a permanent set? Ans: 19 400 p/i².

32. An entrepreneur buys 1500 gallons of carbon tetrachloride when the temperature is −15° F and sells this liquid the following summer when the temperature is 80° F. Assuming no loss by evaporation, find how many gallons he has for sale.

33. A steel tank is filled with turpentine when the temperature is 15° F, at which temperature its volume is 100 gallons. How much liquid will overflow when the temperature is 105° F?
Ans: 4.67 gallons.

34. Show that the coefficient of volume expansion of an isotropic solid is 3α. By similar methods, find the volume expansion coefficients for oak and pine wood from the data given in Table 14–1. If an anisotropic solid has linear coefficients α_1, α_2, and α_3 for three mutually perpendicular directions in the solid, what is the volume expansion coefficient for the solid?

35. It is frequently important in engineering and scientific work to know how the density of a liquid varies with temperature. Recalling that density ρ is defined by the expression $\rho = m/V$, show that we can write

$$\Delta \rho = -\beta \rho\, \Delta T. \qquad (8)$$

36. Show that when the temperature of a liquid in a barometer changes by ΔT, the pressure remaining constant, the height h changes by

$$\Delta h = \beta\, h\, \Delta T, \qquad (9)$$

where β is the coefficient of volume expansion.

37. The volume of the hot-water heating system of a house is equivalent to that of 800 f of 1-i inside diameter steel pipe. How much room must be allowed for expansion in the reservoir if the system is filled at 4° C and is heated to 90° C? Ans: 1.44 f³.

38. A brass wire 12 i long and a steel wire 12 i long of the same diameter are joined end to end and the combination is fastened to rigid supports 24 i apart, as in the figure. If the stress in these wires is zero and the wires are just taut at a temperature of 100° C, determine the stress and strain in each wire when the temperature falls to −50° C, assuming that the distance between supports does not change.

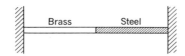

Problem 38

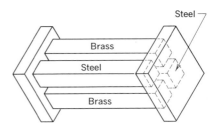

Problem 41

39. If a closed steel 'bomb' is completely filled with water at 0° C, what is the pressure in the bomb at 100° C? Assume that the bomb is so strong that it expands only because of the temperature rise, not because of the increased water pressure. Ans: 12 500 p/i².

40. A bimetal strip to be used in a thermostat is made by fastening a strip of steel to a strip of brass by closely spaced rivets. Each strip is 3 mm thick and 20 cm long at 0° C, and the strip is straight at this temperature. Compute the approximate radius of curvature of the bimetal strip when its temperature is 50° C. If one end of the bimetal strip is fixed, what is the movement of the free end when the temperature is raised from 0° C to 50° C?

41. Two steel bars and two brass bars, each of 1-i² cross section and 12-i length, are firmly attached to two heavy plates, as in the figure. The plates are free to move toward or away from each other. If the stresses in these bars are zero at 0° C, find the stresses at 200° C, assuming that there is no bending of bars or plates. Ans: 12 600 p/i².

42. Show that if a thread of liquid of length l is contained in a capillary tube of uniform bore and the temperature of tube and liquid is raised by ΔT, then the increase in length of the thread will be $\Delta l = (\beta - 2\alpha) l \Delta T$, where α refers to the material of the tube. What will be the *apparent* change in length as measured by graduations on the tube?

43. A shop worker takes a copper rod exactly 1 m long at 20° C and bends it so that it forms a nearly complete circle, with a gap of only 2.4 mm between the ends of the bent rod. By how much should the worker change the temperature of the bent rod in order to cause its length to increase by 2.4 mm? After raising the temperature does the worker obtain a complete circle? Discuss. Ans: 141 C deg; no, why?

15 HEAT AND THERMAL ENERGY

The work of Rumford and Joule demonstrated clearly that *heat* is actually a form of energy—*thermal energy*—in process of transfer from one body to another. The thermal energy of a body is the kinetic and potential energy associated with the random motion of the atoms and molecules making up the body. Ordinary mechanical energy units can therefore be used to measure heat and thermal energy. Since *thermal energy* is associated with motions of the particles internal to a substance, rather than with motion of the substance as a whole, it is frequently called *internal energy*.

For historical reasons, thermal energy is commonly specified in *thermal units* that were introduced before the nature of heat was clearly understood. We define these units and relate them to the ordinary mechanical units.

Calorimetry is the laboratory science of making measurements of 'quantities of heat,' that is, of quantities of thermal energy in the process of transfer from one body to another. We shall describe calorimetric measurements and close the chapter with a discussion of the conversion of *chemical* energy to thermal energy during combustion of fuels.

1. Quantity of Heat

As stated in the preceding chapter, when bodies at different temperatures are placed inside an insulated enclosure, the hotter bodies become cooler and the colder bodies become warmer, as all bodies reach the same temperature. The bodies come into *thermal equilibrium.* One can make quantitative studies of this phenomenon by means of a thermometer and a *calorimeter,* an enclosure designed to provide thermal insulation. The results of such quantitative studies indicate that the final equilibrium temperature of the system depends on the initial temperatures of the bodies, their masses, and the materials of which they are composed. Figure 1 on p. 308 shows a simple calorimeter in which three 'bodies,' a copper block, a quantity of water, and a copper container, are in an insulated enclosure.

Early experiments showed that all observed results could be accounted for by introducing a physical quantity called 'quantity of heat,' usually denoted by Q. The net result of the heat transfers that take place while thermal equilibrium is being established is given by the calorimetric relation:

$$\text{heat lost by hot bodies} = \text{heat gained by cold bodies.} \tag{1}$$

Before discussing the applications of this relation, we shall define the thermal units ordinarily used for specifying quantities of heat.

The metric thermal unit is the *kilocalorie.* Historically, the kilocalorie was defined as *the quantity of heat required to raise the temperature of one kilogram of water from* $14.5°$ C *to* $15.5°$ C. The kilocalorie (kcal) is 1000 calories; hence the *calorie* is the energy required to raise the temperature of one gram of water from $14.5°$ C to $15.5°$ C.* The modern, official definition of the kilocalorie is given later in this section.

In the British system, the thermal unit, called the *British thermal unit* (BTU), is defined with the *standard pound,* rather than the slug, as the mass unit:

One **standard pound** (lb) is the *mass* whose standard weight is 1 p.

We defined standard weight on p. 69 and remarked there and on p. 172 that standard weight actually expressed mass rather than weight. While the standard pound is not a convenient mass unit to use in dynamical problems, the fact that it is universally used in thermal problems when British units are employed makes it necessary now to formalize its defini-

---

*This is the usage among physical scientists. We note, however, that physiologists, in discussing metabolism, use the term *calorie* for what we call the kilocalorie. Popular cookbook tables that say that there are "120 calories in one wedge of angel cake" or "250 calories in two codfish balls" are really referring to the approximate number of *kilo*calories of heat developed when one wedge of angel cake or two codfish balls are burned in a bomb calorimeter (Fig. 2, p. 313), in pure oxygen, after thorough drying.

tion as a mass unit and to give it its customary abbreviation, lb, which readily distinguishes it from the pound as a unit of force, abbreviated p. From the definition of the pound weight on p. 66, we see that 1 lb=0.4536 kg. The BTU is defined as the quantity of heat required to raise the temperature of 1 lb of water by one Fahrenheit degree centered around 15° C (59° F):

> The **British thermal unit** (BTU) is the quantity of heat required to raise the temperature of 1 lb of water from 58.5° F to 59.5° F.

To relate the BTU and the kilocalorie, we use the experimental fact that for small temperature intervals the heat required to raise the temperature of a given quantity of water is proportional to the temperature interval. Since 1 C deg = $\tfrac{9}{5}$ F deg, we see that

$\tfrac{9}{5}$ BTU will raise the temperature of 1 standard pound, or 0.4536 kg, of water from 14.5° C to 15.5° C; and thus that

1 BTU will raise the temperature of $\tfrac{5}{9} \times 0.4536$ kg = 0.2520 kg of water from 14.5° C to 15.5° C.

But it requires 0.2520 kcal to do just this; hence

$$1 \text{ BTU} = 0.2520 \text{ kcal} \tag{2}$$

The thermal units were first defined in the days when it was believed that heat was an invisible, weightless substance known as 'caloric fluid,' which was pictured as flowing from bodies at high temperature to bodies at low temperature somewhat as water flows from high elevations to low elevations. Indeed, it can be seen that equation (1) is entirely consistent with the conservation of a substance. Later, the work of Rumford and Joule, to be described in Sec. 4, demonstrated that heat actually represents transfer of *a form of energy* from hot bodies to cold bodies and that quantity of heat can therefore be measured in conventional energy units such as the J and the fp. The thermal units, kilocalorie and BTU, are still used extensively in specifying the quantities that determine the thermal properties of matter in spite of the fact that modern techniques of electrical calorimetry always determine these quantities initially in joules.

Direct electrical measurement of the energy required to accomplish the same temperature rise as that obtained from one kilocalorie (for example, heating one kilogram of water from 14.5° C to 15.5° C) gives the relation

$$1 \text{ kcal} = 4184 \text{ J.} \tag{3}$$

This value is called the *mechanical equivalent of heat*.

Equation (3), with 4184 treated as an exact number, is now the official *definition* of the kilocalorie. Defined in this way, the unit is sometimes called the *thermochemical* kilocalorie. The current, modern definition is thus:

One **kilocalorie** is exactly 4184 joules.

All accurate calorimetric measurements made since about 1910 have been directly or indirectly based on measurements of electric energy in terms of joules. Nevertheless, the measured data on physical and chemical thermal properties of matter have generally been reported in terms of kilocalories after a conversion that utilized the factor expressing the mechanical equivalent of heat. This was clearly an unsatisfactory situation because the size of the older kilocalorie is tied to the thermal properties of water and the conversion factor was subject to change every time the mechanical equivalent of heat was remeasured to increased accuracy. For this reason, beginning about 1930, the older definition of the kilocalorie given on p. 304 was discarded and use was made of a conventional kilocalorie that agreed with the older kilocalorie to within the then-known accuracy of the mechanical equivalent of heat.

Relation (3) gives the conversion factor by which all thermal properties of matter, determined initially in joules, are now converted to kilocalories. Thus, the definitions of the thermal units of energy have now been divorced from the properties of a particular substance.

2. Heat Capacity; Specific Heat Capacity

We define the *heat capacity* of a body as follows:

The **heat capacity** of a body is the quantity of heat required to raise the temperature of the body by one degree.

Heat capacity is measured in units such as kcal/C deg or BTU/F deg. The *specific heat capacity* is the heat capacity per unit mass:

The **specific heat capacity** of a substance is the quantity of heat required to increase the temperature of unit mass of the substance by one degree.

Specific heat capacity is ordinarily denoted by the symbol c; it is expressed in kcal/kg·C deg, cal/g·C deg, or BTU/lb·F deg. Since

$$\frac{1 \text{ BTU}}{1 \text{ lb} \times 1 \text{ F deg}} = \frac{0.2520 \text{ kcal}}{0.4536 \text{ kg} \times 5/9 \text{ C deg}} = 1\frac{\text{kcal}}{\text{kg} \cdot \text{C deg}} = 1\frac{\text{cal}}{\text{g} \cdot \text{C deg}},$$

the numerical value of the specific heat capacity is the same in any of these three units. We notice that the kcal and the BTU were defined in such a way that the specific heat capacity of water would be exactly one unit in any of the three systems; since the specific heat capacity of any other substance must bear a constant ratio to that of water, the specific heat capacity of any substance must have the same numerical value in any of the systems.

While the specific heat capacity of a substance varies slightly with the temperature at which the temperature change takes place, it will be adequate for our present discussion to assume that the specific heat capacity is a constant independent of temperature. Then we can determine the heat Q necessary to raise the temperature of a mass m of a substance by ΔT degrees by multiplying the specific heat capacity c by m and by ΔT. Hence,

$$Q = c\, m\, \Delta T. \quad \begin{cases} Q \text{ in kcal} \\ c \text{ in kcal/kg·C deg} \\ m \text{ in kg} \\ \Delta T \text{ in C deg} \end{cases} \text{ or } \begin{cases} Q \text{ in BTU} \\ c \text{ in BTU/lb·F deg} \\ m \text{ in lb} \\ \Delta T \text{ in F deg} \end{cases} \quad (4)$$

Values of the specific heat capacities of various solids and liquids are given in Table 15-1; the specific heat capacities of gases will be considered in Chapter 17. It will be noted that water has the highest specific heat capacity of any of the common substances listed in Table 15-1.

Table 15-1 SPECIFIC HEAT CAPACITIES OF SOLIDS AND LIQUIDS
(In kcal/kg·C deg or BTU/lb·F deg)

Metallic solids		Nonmetallic solids		Liquids	
Aluminum	0.212	Ice	0.48	Water	1.00
Brass	0.090	Clay	0.22	Ethyl alcohol	0.58
Copper	0.094	Coal	0.3	Gasoline	0.5
Gold	0.031	Concrete	0.16	Mercury	0.033
Iron and steel	0.11	Glass	0.12–0.20	Mineral oil	0.5
Lead	0.031	Limestone	0.22	Methyl alcohol	0.60
Platinum	0.032	Marble	0.21	Olive oil	0.47
Silver	0.056	Paraffin	0.69	Petroleum	0.51
Tin	0.055	Rubber	0.48	Sea water	0.93
Zinc	0.094	Wood	0.3–0.7	Turpentine	0.41

Specific heat capacities can be determined calorimetrically by the method of mixtures. The following example will illustrate the principle (1) on which all calorimetric computations are made. In this example, it is assumed that the calorimeter is perfectly insulated.

Example. *A copper container (Fig. 1) of 0.25-kg mass contains 0.4 kg of water. Container and water are initially at room temperature of 20° C as measured by the thermometer. A block of copper of 1-kg mass is heated to 100° C by placing it in the steam from water boiling at normal atmospheric pressure. It is then removed from the steam and quickly placed in the water of the calorimeter. The copper block cools, the water and the container become warmer, and the final temperature, as read on the thermometer, is found to be 34.5° C. From these data determine the specific heat capacity c_{Cu} of copper.*

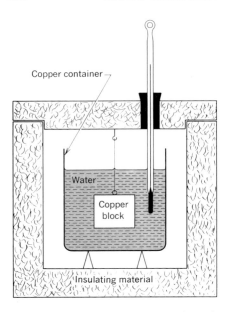

Fig. 1. A simple calorimeter for determination of the specific heat of copper by immersion of a heated block in water.

The fundamental equation (1) becomes

$$\begin{Bmatrix} \text{heat lost by} \\ \text{copper block} \end{Bmatrix} = \begin{Bmatrix} \text{heat gained} \\ \text{by water} \end{Bmatrix} + \begin{Bmatrix} \text{heat gained} \\ \text{by copper} \\ \text{container} \end{Bmatrix}$$

The value of each of these terms is given by (4) as $mc\,\Delta T$. Substituting the given values of m and ΔT in the metric units listed in (4), and setting $c = 1$ kcal/kg·C deg for water, we have

(1 kg) c_{Cu} (65.5 C deg)
$= 0.4 \cdot 1 \cdot 14.5$ kcal $+ (0.25$ kg$) c_{Cu}$ (14.5 C deg)

whence $\quad c_{Cu} = 0.094$ kcal/kg·C deg

In the above substitution, 14.5 C deg is the temperature increase of water and container; 65.5 C deg is the temperature decrease of the copper block.

3. Latent Heats

We now turn to consideration of *latent heats*. If we start with a piece of ice at $-20°$ C, we must add heat to it (0.48 kcal/kg·C deg according to Table 15-1) in order to raise its temperature to $0°$ C. At this point further addition of heat causes some of the ice to melt. *During the melting process no temperature rise takes place.* For each 80 kcal of heat added, one kilogram of ice melts. Only after sufficient heat has been added to melt *all* the ice does any temperature rise of the water take place. The value, 80 kcal/kg, is known as the *latent heat of fusion* of ice.

> The **latent heat of fusion** of a substance is the heat that must be added to unit mass of the solid at its melting point to change it to liquid at the same temperature and pressure.

If we put a pan of water on the stove and turn on the gas, the temperature of the water rises fairly rapidly to its boiling point of $100°$ C (at normal atmospheric pressure). After this temperature has been reached, we must continue to supply heat for a very much longer time if we want to evaporate all of the water. Supplying heat at a constant rate causes the water to evaporate at a constant rate. *During the process of evaporation no temperature rise takes place;* a thermometer in the steam and one in the water will each read $100°$ C continuously. But we have to add 540 kcal to evaporate each kilogram of water. This value, 540 kcal/kg, is known as the *latent heat of vaporization* of water.

The **latent heat of vaporization** of a substance is the heat that must be added to unit mass of the liquid at its boiling point to change it to vapor at the same temperature and pressure.

The boiling point of a liquid varies with pressure, and the latent heat of vaporization varies also. Furthermore, evaporation of a liquid takes place at temperatures below its boiling point and a latent heat is involved in this process also. The melting point of a solid also varies very slightly with pressure. These matters will be treated in Chapter 18. For the present we shall confine our study to processes taking place at normal atmospheric pressure.

Either of the two latent heats can be expressed in kcal/kg or BTU/lb. The relation between these units is given by

$$1 \frac{\text{BTU}}{\text{lb}} = \frac{0.2520 \text{ kcal}}{0.4536 \text{ kg}} = \frac{5}{9} \frac{\text{kcal}}{\text{kg}}, \quad \text{or} \quad 1 \frac{\text{kcal}}{\text{kg}} = \frac{9}{5} \frac{\text{BTU}}{\text{lb}}. \tag{5}$$

Hence, the latent heat of fusion of water is 80 kcal/kg = 144 BTU/lb, and the latent heat of vaporization of water is 540 kcal/kg = 970 BTU/lb.

The factor $9/5$ ($= 180/100$) occurs in converting latent heats from metric to English units because the calorie and the BTU are defined in terms of degrees whose sizes are in the ratio 9:5.

In calorimetric problems involving melting, freezing, vaporization, or condensation, one must include the latent heats among the quantities

Table 15-2 TEMPERATURES AND LATENT HEATS OF FUSION AND VAPORIZATION AT ONE-ATMOSPHERE PRESSURE

Substance	Melting point (°C)	Latent heat of fusion (kcal/kg)	Boiling point (°C)	Latent heat of vaporization (kcal/kg)
Water	0	79.70	100	539.2
Ammonia	−75	108.0	−34	327.1
Helium	...	...	−269	5.97
Hydrogen	−259	15.0	−253	106.7
Methane	−182	14.5	−161	138
Nitrogen	−210	6.2	−196	47.8
Oxygen	−219	3.3	−183	51
Ethyl (grain) alcohol	−115	24.9	79	204.3
Methyl (wood) alcohol	−98	22.0	65	262.8
Aluminum	660	93.0	2056	2000
Copper	1083	50.6	2595	1760
Gold	1063	16.1	2966	446
Iron	1539	65	2740	1620
Lead	327	6.3	1744	222
Mercury	−39	2.7	357	71
Platinum	1774	27.1	4407	640
Silver	960	24.3	2212	552
Tin	232	14.4	2270	650
Tungsten	3400	44	5927	1180
Zinc	419	24.1	907	362

of heat lost or gained. Table 15-2 gives the melting and boiling points and the corresponding latent heats for various substances. The following example will illustrate the application of equation (1) when latent heats are involved:

Example. *How many kilograms of ice at 0° C must be added to 5 kg of water at 100° C in an insulated, 2-kg, aluminum container in order to cool the container and its contents to 25° C?*

We denote the unknown mass of ice by m and use the basic relation (1). The ice gains heat: (80 kcal/kg) $m = 80\ m$ kcal/kg are required to melt the ice and (1 kcal/kg·C deg) (m) (25 C deg) = 25 m kcal/kg are required to raise the temperature of the water resulting from the melted ice to the final temperature of 25° C. The hot water loses (1 kcal/kg·C deg)(5 kg) (75 C deg) = 375 kcal in cooling from 100°C to 25° C, while the aluminum container loses (0.212 kcal/kg·C deg)(2 kg)(75 C deg) = 31.8 kcal. Equating heat gain to heat loss gives

$$80\ m\ \text{kcal/kg} + 25\ m\ \text{kcal/kg} = 375\ \text{kcal} + 31.8\ \text{kcal},$$

$$m = 3.87\ \text{kg},$$

as the required mass of ice.

4. Thermal Energy

When we add heat to a body by placing it in thermal contact with a body of higher temperature, we may raise its temperature or may cause it to melt or vaporize. It is easy to show that we can produce these same changes by performing mechanical work that results in the dissipation of mechanical energy in friction.

In 1798, COUNT RUMFORD observed that, when cannon-boring machinery was operated with a very blunt boring tool, the cooling water surrounding the boring tool continuously boiled away, but little or no drilling was accomplished. The amount of heat that could be generated by doing work against friction appeared to be inexhaustible; the longer the machinery was operated, the more the heat that was added to the water. Hence, Rumford concluded that the results of his experiments could not be explained on the basis of conservation of a caloric fluid that was released, according to the then-current theory, as material was more finely subdivided. Rather, Rumford proposed that heat must somehow be associated with 'motion.'

It was not firmly established until nearly a half century after this observation that there was a definite relation between the amount of work done against friction and the amount of heat produced. In 1843, JAMES PRESCOTT JOULE employed an apparatus in which water was stirred violently by a set of rotating paddles and in which the mechanical energy supplied to rotate the paddles could be accurately measured. The thermal effect of the mechanical work done on the water was a rise in temperature. Joule's experiment showed that the rise in temperature was directly *proportional* to the amount of work done on the water. Thus, the performance of work in stirring the water was found to be equivalent to adding heat to the water. The exact equivalence between work done and heat added is given by the experimental relation:

$$1 \text{ kcal} = 4184 \text{ J}, \quad \text{or} \quad 1 \text{ BTU} = 778 \text{ fp}. \tag{6}$$

This relation gives the so-called *mechanical equivalent of heat,* and is now used as the *definition* of the kilocalorie.

Since adding heat to a body and doing work on the body in such a way as to dissipate mechanical energy are equivalent so far as thermal effects are concerned, we may ask what becomes of the mechanical energy being added. The answer is that the energy does not disappear, but remains inside the body in the form of kinetic and potential energy associated with random motion of the atoms within the body. The word *random* is used to distinguish these thermal motions from the motions of the atoms when the body moves as a whole, in which case there is a resultant momentum. Superposed on the motion of the atoms associated with motion of the body as a whole is the thermal motion, in which the directions of motion of the different atoms is sufficiently random that the vector sum of the momenta of all the atoms adds up to zero, since no momentum of the body as a whole is associated with the thermal motion.

Thermal energy is the potential and kinetic energy associated with the random motion of the atoms of a substance.

Thermal energy is sometimes called *internal energy.*

The term *heat* has the following restricted technical meaning:

Heat is thermal energy in the process of being added to, or removed from, a substance, or in the process of being transferred from one portion of substance to another.

We note that *heat* and *thermal energy* bear the same relation to each other as *work* and *mechanical energy. Heat* and *work* represent energy in transition. A review of mechanical principles shows that we used the word *work* to represent mechanical energy in the process of being added to or removed from a body or substance.

We have seen that the thermal energy of a body can be increased either by adding heat to the body or by doing work on the body. For a given body, we can include both thermal and mechanical effects by writing the principle of conservation of energy in the form:

heat added + work done = increase in (thermal energy

+ mechanical kinetic energy + mechanical potential energy), **(7)** where the last two terms refer to gross mechanical motion or configuration, not the random atomic motions or configurations included in the thermal energy. In using the relation (7), the same units must be used for all terms; all terms must be stated in mechanical energy units such as the J or fp or all must be given in thermal units such as the kcal or BTU.

Although an increase in temperature of a given substance always necessitates an increase in its thermal energy, the converse is not always true, as we see from consideration of latent heats. *Temperature is not a measure of thermal energy, but it is a measure of the ability of a body to transfer thermal energy to another body:*

> Two bodies have the **same temperature** if, when placed in thermal contact, no heat flows from one to the other. Body A is at **higher temperature** than body B if, when they are placed in thermal contact, heat flows from A to B.

This definition shows why, as stated on p. 292, all bodies that have given up all available thermal energy are at one *common* temperature—absolute zero. If a body A is incapable of giving up any thermal energy, no other body B can be colder because, by the definition above, A would have to be able to give up energy to such a colder body B.

5. Solids, Liquids, and Gases

At sufficiently low temperatures, the atoms of all elements except helium, and the molecules of all compounds, are coalesced into crystalline solids. *The atoms of a solid are not ordinarily standing still—they are vibrating.* The higher the temperature, the greater the average amplitude of this vibration and the greater the energy associated with it. At some definite temperature, the shearing type of motion becomes so violent that the molecules slide past each other and do not return to their previous equilibrium positions. The crystal structure breaks down completely, and the solid *melts* and changes to a *liquid*. In the process of melting, no temperature rise takes place. Ice and water can remain indefinitely in thermal equilibrium, both at 0° C. But the water has much greater thermal energy, per kilogram, than the ice. The additional thermal energy that must be added to cause a unit mass of ice to melt, or that must be extracted to cause a unit mass of water to freeze, is the *latent heat of fusion*—'latent' because no temperature change is occasioned by this heat.

The molecules in a liquid are all in violent vibratory motion relative to their neighbors. Some of the molecules at the top of the liquid have sufficient upward velocity so that, in spite of the 'springs' attracting them to the molecules below, they can jump clear of the liquid surface. Thus *vaporization* or *evaporation* takes place, not atom by atom, but gas molecule by gas molecule, since the forces holding the atoms together into molecules are stronger than the forces holding the molecules together into a liquid. As the temperature is raised, the rate of evaporation increases until, at the boiling point, no further increase in temperature of the liquid is possible, and any additional energy added will go into turning the liquid into a gas. In a gas, the molecules fly about independently, colliding with each other, but not sticking together when they collide; the gas fills the whole volume of its container. A given mass of gas has neither a definite shape nor a definite volume. Again, in the process of evaporation, no temperature change takes place although there is a significant energy change.

We shall conclude by carrying this discussion one step further. In a diatomic or polyatomic gas, the molecules are not only rotating freely, but the atoms in the molecule are vibrating relative to each other as the molecule moves through space. As the temperature increases, the violence of this vibration increases, until an energy of vibration is reached at which the atoms fly off independently. At such a temperature, a diatomic gas changes to a monatomic one. This change does not take place at a definite temperature, but over a range of temperatures; oxygen is almost entirely composed of O_2 molecules at 2500° K, it is largely monatomic at 5000° K.

6. Heat of Combustion

When a chemical reaction takes place, it is ordinarily either *exothermic*, meaning that it produces heat, or *endothermic*, meaning that it absorbs heat. A certain amount of chemical binding energy is changed into thermal energy, or vice versa. The amount of this energy, per unit mass of material reacting, is known as the *heat of reaction*.

The principal chemical reaction that is used for heating purposes is the process of combination with oxygen—burning or *combustion*. In this case the heat of reaction is known as *heat of combustion* and is specified in kcal/kg, or BTU/lb. The mass in the denominator refers to the mass of material burned but does not include the mass of the oxygen that enters the reaction. Thus, the statement that the heat of combustion of anthracite is 8000 kcal/kg means that one kilogram of anthracite, when it reacts completely with whatever mass of oxygen is necessary to ensure complete combustion, gives out 8000 kcal of heat.

Heats of combustion are ordinarily meas-

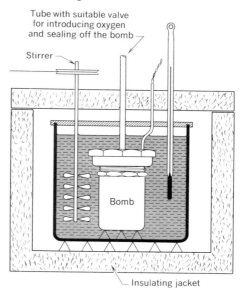

Fig. 2. Bomb calorimeter (schematic).

Table 15-3 TYPICAL VALUES OF HEAT OF COMBUSTION

Substance	kcal/kg	BTU/lb
Solid fuels		
Anthracite......	8 000	14 400
Bituminous coal.	7 500	13 500
Coke...........	6 000	11 000
Pine wood......	4 500	8 000
Liquid fuels		
Gasoline........	11 400	20 500
Kerosene.......	11 200	20 000
Diesel oil.......	10 500	19 000
Alcohol........	6 400	11 500
Foodstuffs		
Proteins........	4 000	7 200
Carbohydrates...	4 000	7 200
Fats............	9 500	17 000

ured in a *bomb calorimeter*. The 'bomb' (Fig. 2) is a massive steel cylinder, fitted with a gas-tight screwed cover, and capable of withstanding the high pressures of the gaseous combustion products In this bomb, a measured mass of material burns in an excess of pure oxygen under pressure so that complete combustion is assured. The ignition is by an electric current through a fine wire, and the combustion takes place with explosive violence. The heat of combustion is computed from the final temperature rise, after thermal equilibrium is reached.

Table 15-3 gives the heats of combustion of various solid and liquid fuels. The relations (5) connect the metric and British units, for the same reason as in the case of latent heats. Heats of combustion of gaseous fuels are ordinarily given in BTU per f^3 of gas, the volume being measured at 1 atm and 0° C. Typical values for gaseous fuels are

Manufactured gas   *Natural gas*
500–550 BTU/f^3 1000–1100 BTU/f^3

Manufactured gas (called *coal gas* or *steam gas*) is made by blowing steam over incandescent coal, resulting in the formation of carbon monoxide and hydrogen. The heat of combustion of the resulting mixture is only about half that of natural gas, which is almost pure methane.

Example. *If all the heat produced were used effectively, how many f^3 of natural gas would be required to heat* 8 lb *(approximately one gallon) of water in a 2-lb aluminum saucepan from* 72° F *to the boiling point?*

From (4), we find that (1 BTU/lb·F deg)(8 lb)[(212−72) F deg]=1120 BTU are required to heat the water; (0.212 BTU/lb·F deg)(2 lb)[(212−72) F deg]=59.4 BTU to heat the saucepan. Thus, the total quantity of heat needed is 1120 BTU+59.5 BTU=1180 BTU. If the heat of combustion of the natural gas is 1000 BTU/f^3, then we see that (1180 BTU)/(1000 BTU/f^3)=1.18 f^3 are required.

PROBLEMS

1. How many kcal are required to raise the temperature of 4 kg of copper from 35° to 53° C? from 35° to 53° F? Ans: 6.75, 3.76 kcal.

2. How many BTU are required to raise the temperature of 30 lb of aluminum from 25° F to 43° F? from 25° C to 43° C?

3. A copper vessel of 6.3-kg mass contains 8.5 kg of a liquid at 20° C. Into this container is placed a 12-kg block of copper at 102° C. The final temperature is 44° C. What is the specific heat capacity of the liquid?
Ans: 0.251 kcal/kg·C deg.

4. A copper vessel of mass 80 g contains 212 g of water at 12° C. A 200-g nugget of gold at a temperature of 96° C is placed in the water. The final temperature is found to be 14.6° C. What value does this experiment give for the specific heat capacity of gold?

5. A 3.0-kg platinum ball is removed from an oven and placed in a copper container of 1.0-kg mass containing 3.2 kg of water. The temperature of the water rises from 20° to 64° C. What was the temperature of the oven?
Ans: 1570° C.

6. A blacksmith throws a red-hot horseshoe of mass 2 lb, at a temperature of 1900° F, into a pail containing 50 lb of water at 50° F. If all the heat goes into warming the water, what is its final temperature?

7. A 1-kg aluminum vessel contains 2.5 kg of water at 20° C. How much steam at 100° C must be added to raise the temperature of the water to 100° C? Ans: 0.400 kg.

8. A 1-kg steel ball is heated in a steam bath at 100° C and is then placed on a large cake of ice at 0° C. How much ice will be melted?

9. If a total of 15 kg of molten lead at 327° C is dropped into 1.5 kg of water initially at 20° C, how much water is evaporated, if no heat loss occurs? Ans: 148 g.

10. Steam initially at 120° C and ice initially at −20° C come into thermal equilibrium as water at 60° C. Assume no heat loss, and take the specific heat capacity of steam as 0.5 kcal/kg·C deg. What is the ratio of the mass of ice to the mass of steam?

11. To determine the latent heat of fusion of water, 1 lb of ice at 32° F is dropped into 5 lb of water in a 1-lb copper container at 100° F. The final temperature is observed as 65° F. What value does this experiment give for the latent heat of fusion of water?
Ans: 142 BTU/lb.

12. How much heat must be added to 3 kg of ice at −30° C in order to convert the ice to steam at 100° C? How much heat must be added to 20 lb of snow at 0° F in order to change it to steam at 212° F?

13. The mechanical output of an electric motor is 1.5 hp. This is 85 per cent of the electric power input, the balance of the input being 'lost' as heat. How many BTU of heat are developed per minute?
Ans: 13.6 BTU/min.

14. The electrical input to a motor is 2 kW, of which 90 per cent is delivered as useful mechanical work, the balance being wasted as heat. At what rate in cal/s is heat being liberated?

15. An electric motor is loaded by means of a Prony brake (see Prob. 30, p. 179) until it is delivering 30 hp. The Prony brake initially contains 20 lb of water at 72° F. Assume that the whole 30 hp delivered by the motor goes into heating the water. How many pounds of water are boiled away during the first 5 min of operation of the motor? Ans: 5.1 lb.

16. A 1000-W electric heater is immersed in 3.5 lb of water at 62° F contained in a 1.2-lb aluminum vessel. Neglecting loss of heat and the heat capacity of the electric heater, find how long it will take to heat the water to boiling and boil off 0.25 lb.

17. A continuous-flow steam generator takes in 500 g of water per minute at 10° C and changes it to steam at 100° C. How many kilowatts of electrical power are necessary to furnish the required heat? Ans: 21.9 kW.

18. Bismuth melts at 271° C. Its specific heat capacity is 0.032 kcal/kg·C deg. To determine its latent heat of fusion, 1 kg of molten bismuth at its melting point is dropped into a 0.5 kg copper vessel containing 1.5 kg of water at 15° C. The temperature rises to 22.1° C. What is the latent heat of fusion of bismuth?

19. At what rate will ice be melted by a 400-W electric heater submerged in a mixture of ice and water? Ans: 72.0 g/min.

20. At what rate will water at 100° C be evaporated by a 500-W electric heater if all the energy supplied goes into latent heat?

21. When delivering 100 hp, a Diesel engine burns 1 lb of oil per minute. Find the over-all efficiency of the engine. Ans: 22.2%.

22. Compare the cost of heating with bituminous coal at $10 per ton, natural gas at 65 cents

per 1000 f³, and electricity at 1.5 cents per kWh.

23. In a determination of the heat of combustion of gasoline, 10 g of gasoline are burned in a bomb calorimeter containing 3.3 kg of water and 7.5 kg of steel. The resulting rise in temperature is observed to be 24 C deg. What value does this give for the heat of combustion?
Ans: 9690 kcal/kg.

24. At what price per kWh would the cost of heating with electricity be the same as the cost of heating with anthracite at $20 per ton if the reasonable assumptions are made that all the electrical energy but only 50 per cent of the heat of combustion of the anthracite are delivered as useful heat?

25. A mixing faucet is supplied with cold water at 60° F and hot water at 140° F. The cold-water tap supplies a flow of 0.5 gal/min out of a total flow of 1.5 gal/min through the faucet. What is the temperature of the resulting warm water? Ans: 113° F.

26. Four pounds of snow at 20° F are placed in a pan on a 1-kW hot plate. What is the minimum time required to produce boiling water?

27. How many gallons of kerosene would be equivalent in heating value to 1 ton of bituminous coal? How long would a 3-kW electric heater have to be operated in order to produce an amount of heat equivalent to that obtained from the combustion of 1 ton of bituminous coal? Ans: 203 gal; 109 d.

28. How much heat must be added to a 250-g block of aluminum initially at 20° C in order to increase the volume of the block by one per cent? If an equal quantity of heat were added to 250 g of mercury initially at 20° C, what would happen?

29. A 'continuous-flow calorimeter' is used to measure the specific heat capacity of a liquid by having a stream of the liquid pass through the calorimeter at a measured rate. While the liquid is inside the calorimeter, heat is added at a known rate, usually by an electric heating element. From the resulting temperature difference between the input and the output points in the stream, the specific heat capacity of the liquid can be obtained. In a continuous-flow calorimeter equipped with a 500-W heating element, it is found that with a flow rate of 1000 cm³/min for a liquid of density 0.9 g/cm³, a steady-state difference of 15 C deg exists between the liquid temperatures at output and input points. Calculate the specific heat capacity of the liquid.
Ans: 0.538 kcal/kg·C deg.

30. If water flows through the continuous-flow calorimeter described in Prob. 29 at a rate of 500 cm³/min, what will be the steady-state temperature difference between input and output?

31. A 2-lb block of ice at 32° F is placed in a bucket capable of holding 8 lb of water. Water is added until the bucket will hold no more, with the block of ice floating freely, and it is found that half the ice melts. Neglecting heat transfer to the bucket, find the initial temperature of the water. Will the bucket overflow, or the water level drop, as the ice melts? Ans: 56.0° F.

32. Compare the quantity of ice fresh from the freezer of a refrigerator at 0° F with the quantity of ice that has been allowed to warm up to 32° F, that is required to cool a given volume of beverage to 32° F.

33. An industrial process requires that 10 000 lb of water per hour be heated from 50° F to 180° F. To do this, steam at 212° F is passed from a boiler into a copper coil immersed in the water. The steam condenses in the coil and is returned to the boiler as water at 190° F. How many pounds of steam are required each hour? Ans: 1310 lb.

34. A lead bullet moving at 800 f/s strikes a log and comes to rest. If half the initial kinetic energy of the bullet goes into heating the bullet, how much will the temperature of the bullet rise?

35. Water moving at a speed of 20 f/s approaches the brink of a waterfall 200 f high. After passing over the fall the water comes to rest in a pool at the base of the waterfall. How much does the temperature of the water rise when the water comes to rest, if all the heat produced goes into heating the water?
Ans: 0.265 F deg.

36. The electrical utilities of the U.S. produce an average of 1 kWh of electric energy per pound of bituminous coal; the best producers obtain 1 kWh for ¾ lb of coal. What are the average and the best over-all efficiencies of this production of electrical energy?

37. An electric pot is advertised to bring to a boil one quart (2 lb) of water in 3 min. Taking the initial temperature as 50° F, find the power required if all energy goes into heating the water. *Ans: 1.89 kW.*

38. An electric power plant develops an average of 4 MW of power, and burns 50 tons of coal per day. The coal has heat of combustion of 13 000 BTU/lb. What is the over-all efficiency of the power plant?

39. How many BTU are developed in the brakes of a 4000-lb automobile brought to rest from 25 mi/h if all the energy goes into heating the brakes? If the four brakes are equivalent to 40 lb of steel, what is the average temperature rise if all heat remains in the brakes? *Ans: 108 BTU; 24.5 F deg.*

40. The energy release in a nuclear reactor is 8.2×10^{13} joules per kilogram of uranium consumed. Express this value in BTU/lb and compare with bituminous coal.

41. Using the data of Prob. 40, determine the number of pounds of uranium consumed daily in a 50-MW nuclear power plant operating at an over-all efficiency of 20 per cent. Compare with the number of tons of bituminous coal consumed in a conventional power plant of the same output that operates at 30 per cent over-all efficiency. *Ans: 0.580 lb; 505 ton.*

42. Crushed ice at 0° C is dropped into a concentrated sodium-chloride solution also at 0° C. As soon as the ice is introduced, some of the ice melts and the temperature of the mixture begins to drop. What is the source of the thermal energy required to melt the ice?

43. How long will it take a 750-W blender to raise the temperature of 1 kg of water by 5 C deg if 50 percent of the electric energy goes into heating the water? *Ans: 55.8 s.*

44. Why are violent storms always accompanied by heavy rains? Where does the energy come from that 'drives' the storm winds?

16 HEAT TRANSFER

In this chapter we consider rather briefly the methods by which thermal energy is transferred from one body to another or from one material medium to another. The computation of heat transfer in practical cases is frequently very complex and mathematically difficult; we confine ourselves here to an introductory discussion of fundamental principles and to simple examples of practical importance. The subject of *thermal radiation* is introduced only briefly; it will be discussed more fully in Chapter 41 after we have described the nature of radiant energy.

1. Methods of Heat Transfer

There are three distinct methods by which thermal energy is transferred from one point to another when temperature differences exist:

Conduction is a process of heat transfer within a material medium, in which thermal energy is passed from molecule to neighboring molecule in the course of the purely thermal motions, no mass motion of the medium being involved.

Convection is the process of transfer of heat from one place to another by the actual mass motion of heated liquid or gas from the one place to the other.

Radiation is the transfer of thermal energy by means of electromagnetic waves, no material medium playing an essential role in the process of transmission.

Whenever two different portions of material media have different temperatures, heat will be transferred from the hot portion to the cool, so that the temperature tends to become equalized. By two portions of material media we can mean various things, such as the hot sun and the cool earth, the hot oven and the cool roast, the hot radiator and the cool furniture, the hot air in a room and the cold air outside in winter, or the hot inside portion of the wall of a house and the cold outside portion.

Radiation involves emission, transmission, and absorption of *radiant energy* in the form of electromagnetic waves similar to the waves that constitute visible light. Because the thermal motions involve electrically charged particles—electrons and atomic nuclei—all bodies at temperatures above absolute zero are continuously radiating energy. Like light, radiant energy is transmitted *best* through a vacuum, but fairly well through a gaseous medium and through *transparent* liquids and solids. The transmission of heat from the sun, through the near vacuum of outer space to the earth, is primarily by radiation. Where there is a material medium available to transmit the heat and where temperatures are low (less than about 200° C), radiation is usually of little importance in comparison with conduction and convection.

Within a *nontransparent solid* medium, heat can be transferred from point to point only by conduction. In liquid and gaseous media, convection by means of circulating currents of fluid is ordinarily the most important process of heat transfer.

In any given problem in heat transfer, two or three of these processes may be active simultaneously. The total heat transferred will be obtained by adding the amounts transferred by the different processes. It is convenient to consider the different processes one at a time.

2. Laws of Heat Conduction

Suppose that we have a slab of solid material (Fig. 1) in which there is a *temperature gradient,* by which we mean a space rate of change of temperature from one point to another in the material. It is found experimentally that inside the slab there is a *uniform* space rate of change of temperature, that is, a constant temperature gradient. This uniformity is indicated schematically by the constant difference ΔT between the equally spaced thermometers in Fig. 1. The constant temperature gradient is given by the ratio $\Delta T/\Delta X$, in C deg/m, for example.

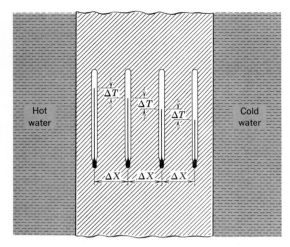

Fig. 1. The rate of heat flow is proportional to the temperature gradient $\Delta T/\Delta X$.

The temperature gradient in Fig. 1 is associated with a heat flow through the solid slab, from left to right (from the region of higher temperature to the region of lower temperature). The heat flow can be measured by calorimetric methods. The heat Q passing through the slab is found experimentally to be proportional to the temperature gradient $\Delta T/\Delta X$, the surface area A (see Fig. 2), and, of course, the time t. Thus we can write

$$Q = k\frac{\Delta T}{\Delta X} A t, \quad \text{or} \quad \frac{Q}{t} = kA\frac{\Delta T}{\Delta X}, \qquad (1)$$

where the proportionality constant k is called the *thermal conductivity* of the material of the slab.

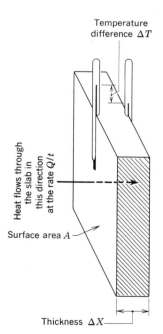

Fig. 2. Definition of the symbols in equation (1).

The **thermal conductivity** of a material is the *rate* of heat flow *by conduction* per unit area per unit temperature gradient.

The thermal conductivity k has the same dimensions as $Q \Delta X/tA \Delta T$. Various systems of units are used in specifying k. If Q/t is in kcal/s, ΔX in m, A in m², and ΔT in C deg, the unit of k is the kcal/s·m·C deg. In British units, the thermal conductivity of structural and insulating materials is usually specified with Q/t in BTU/h, ΔX in i, A in f², and ΔT in F deg. In this case k is in BTU·i/h·f²·F deg. Conversion of k from one system of units to another can be done in the usual way by using the conversion factors for each unit. Thus,

$$1 \frac{\text{BTU} \cdot \text{i}}{\text{h} \cdot \text{f}^2 \cdot \text{F deg}} = \frac{(0.252 \text{ kcal})(0.0254 \text{ m})}{(3600 \text{ s})(0.0929 \text{ m}^2)(0.556 \text{ C deg})} = \frac{1}{29\,000} \frac{\text{kcal}}{\text{s} \cdot \text{m} \cdot \text{C deg}}. \quad (2)$$

Table 16-1 gives typical values of thermal conductivity. It is seen that metals have much larger thermal conductivities than nonmetals, greater by a factor of 100 or more. The thermal conductivity of metals is closely related to the electrical conductivity. The best electrical conductors, and also the best heat conductors, are silver, copper, and gold, in that order.

The thermal conductivities of fluids are so low that, in general, the heat transferred by conduction in liquids and gases is negligible compared with that transferred by convection.

If air is confined in very small spaces so that it cannot circulate, as in the pores of corkboard or wool clothing, convective processes are inhibited; and, since conduction is also low, we have a poor heat transmitter called a *heat insulator*. Such an insulator is said to contain 'dead-air' spaces, meaning that the air cannot circulate readily so that it cannot readily transmit heat by convection.

In determining the amount of heat conducted through a wall or a window, it is not correct to give the outer surface the outdoor temperature and the inner surface the indoor temperature. On a winter day when the room temperature is 60° F, the inner surface of a pane of single window glass may be below 32° F, as witnessed by the familiar frosting of the inside of windows in cold climates. Similarly, the outer surface of the glass may be well above the outdoor temperature. Only part of the temperature drop occurs through the glass, the rest occurs through the layers of air close to the glass inside and outside. In applying (1) to conduction through the glass, one must use the actual temperature difference between the inner and outer surfaces as ΔT; the difficulty of determining this actual temperature difference makes the practical engineering computation of heat losses subject to a good many semi-empirical rules.

The thermal conductivity of a solid is usually determined by measuring the rate at which heat enters or leaves a rod or slab. This is equal to the rate at which heat flows through the rod or slab. If the heat is supplied by circulating hot water, the temperature drop and rate of flow of the water determine the amount of heat entering the solid. If the heat is extracted by means of cold water, the amount of heat leaving can be similarly determined. Or, the heat may be supplied and measured electrically. The temperature gradient is usually computed from the readings of thermocouples (*see* Chapter 27) buried in the material in the manner indicated by the thermometers of Fig. 1. In the measurement of thermal conductivity of a liquid or gas, the sample is in the form of a horizontal sheet which is heated from above and cooled from below so that no heat is transferred by convection.

Table 16-1 TYPICAL VALUES OF THE THERMAL CONDUCTIVITY k NEAR 20° C

Substance	$\dfrac{\text{kcal}}{\text{s·m·C deg}}$	$\dfrac{\text{BTU·i}}{\text{h·f}^2\text{·F deg}}$
Metals		
Silver	0.101	2930
Copper	0.092	2680
Gold	0.070	2030
Brass	0.026	750
Iron and steel	0.011	320
Aluminum	0.048	1390
Mercury, liquid	0.0015	44
Nonmetallic solids		
Brick, common	1.7×10^{-4}	5.0
Concrete	4.1×10^{-4}	12.0
Wood (across grain)	0.3×10^{-4}	0.9
Glass	1.4×10^{-4}	4.0
Ice	5.3×10^{-4}	15.4
Porous materials		
Fiber-blanket insulation	0.09×10^{-4}	0.27
Glass wool or mineral wool	0.09×10^{-4}	0.27
Sawdust	0.14×10^{-4}	0.41
Corkboard	0.10×10^{-4}	0.30
Liquids		
Water	1.43×10^{-4}	4.15
Ethyl alcohol	0.42×10^{-4}	1.23
Gases		
Air	0.056×10^{-4}	0.16
Hydrogen	0.400×10^{-4}	1.16

Example. *A typical household electric refrigerator is equivalent to a box of corkboard* 3 i *thick and of* 50 f^2 *area. During a period when the door is not opened, assume that the interior wall has the average temperature* 40° F *and the exterior wall* 80° F. *If the refrigerator motor is to run only* 20 *per cent of the time during such a period, at what rate in* BTU/h *must heat be extracted from the interior while the motor is running?*

To find the average rate of heat flow from the room into the refrigerator, we use (1) with the mixed British system of units we have described, taking the value of k for corkboard from Table 16-1:

$$\frac{Q}{t} = kA\frac{\Delta T}{\Delta X} = 0.30 \frac{\text{BTU} \cdot \text{i}}{\text{h} \cdot \text{f}^2 \cdot \text{F deg}} \times 50\ \text{f}^2 \times \frac{40\ \text{F deg}}{3\ \text{i}} = 200 \frac{\text{BTU}}{\text{h}}.$$

If the refrigerator motor is to run only one-fifth of the time when the door is unopened and no warm food is introduced, heat must be extracted at five times this rate while the motor is running, or at the rate of 1000 BTU/h.

3. Convection and Radiation; Newton's Law of Cooling

At ordinary temperatures, the principal method of heat transfer in liquids and gases is *convection*. Liquids and gases are comparatively poor heat *conductors*, as will be seen from Table 16-1; and unless temperatures are high, radiation does not play an important role. The mechanism by which heat is transferred from a hot-water or steam 'radiator' to the air, walls, and objects in a room is almost entirely convection. The term 'radiator' is a misnomer. The open fireplace depends almost entirely on radiation for heat transfer to the room, but it is very inefficient because most of the heat is lost by *convection* up the chimney to the out-of-doors. One gets much more effective heating from the same fire burning in a stove, where heat can be transferred to the room by convection.

The process of heating a room by a radiator can be described as follows: The air molecules that collide with the radiator leave with more energy than they had before impact because the radiator is at higher temperature than the air. This process raises the temperature of a thin layer of air in immediate contact with the hot radiator, and the density of this air layer decreases. The hydrostatic balance in the room is thus upset. Archimedean buoyant forces come into play to cause the rarefied hot air to rise to the top of the room. Similarly, when room air comes into contact with the cooler outside walls of the room, the room air transfers heat to these walls; its temperature decreases, and it becomes more dense than the surrounding air. Again Archimedean forces come into play to cause this cooled air to sink to the floor. There is a continuous circulation in which cool floor air is heated by the radiator and rises, while warm ceiling air is cooled by the walls and sinks.

The same process of natural convection is depended on to effect the circulation of hot water from furnace to radiator in older installations. In installations employing a circulating pump, the heat is still transferred by convection—mass motion of material. In this case, or in the case of use of a fan in hot-air circulation, the transfer is said to be by *forced* convection, as distinguished from *natural* or *free* convection.

The laws governing the rate of heat transfer by convection are very complex because complex fluid-dynamic phenomena are involved. One law which was discovered empirically by Newton relates to the rate of cooling (or warming) of a given body when it is at a temperature different from that of the surrounding air:

NEWTON'S LAW OF COOLING: *The rate of heat transfer to the surrounding air is proportional to the difference in temperature between the body and the air.*

This law is found to agree with observed behavior very well when the heat-transfer process is predominantly convective, even though a certain amount of the heat may be transferred by conduction and radiation.

Example. *When a 25-W lamp bulb is placed inside a closed metal can, the temperature of the can rises until it is 60 F deg above that of the surrounding air. To what would the temperature of the can rise if the bulb were changed to one of 50 W?*

The power generated in the lamp bulb can only be dissipated by transfer from the can to the surrounding air. Hence the rate of this heat transfer, expressed in watts, just equals the wattage of the bulb. According to Newton's law of cooling, if the rate of heat transfer is to be doubled, the temperature difference must be doubled. Hence for the 50-W bulb, the can would have to reach a temperature 120 F deg above that of the surrounding air before heat would be transferred as fast as it is generated.

4. Thermal Radiation

As we have noted before, all materials at temperatures above absolute zero are continually emitting radiation. As the temperature of a solid is increased, the energy radiated from the solid increases rapidly and there is also a variation in the character of the emitted radiation, which is at first apparent only as a sense of warmth, then becomes visible to the eye. The nature of this variation in character will be discussed quantitatively in Chapter 41; here we shall discuss the total quantity of radiation and its variation with temperature.

The amount of radiant power emitted by a solid depends very significantly on the character of the surface of the solid. In beginning the discussion of radiation, it is helpful to define a 'perfect radiator,' whose rate of radiation is the maximum possible for its temperature. That such a maximum exists can be shown by considering the inverse process, *absorption*.

Consider Fig. 3, which shows in cross section a solid object maintained at a uniform temperature T throughout. Within this solid there are two identical evacuated spherical cavities containing opaque bodies of the same size but of different ma-

Fig. 3. Two spherical bodies of the same size but of different materials suspended within evacuated spherical cavities.

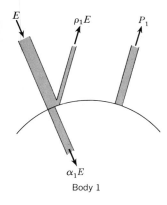

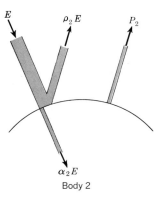

Figure 4

terials; for example, body 1 may be made of wood and 2 may be made of polished metal. It is found by experiment that as a result of radiative interchanges of heat, the temperatures of bodies 1 and 2 eventually become equal to the temperature T of the enclosing walls and remain at that temperature, in accordance with the general principle of thermal equilibrium.

To simplify our discussion we make the convenient (but not necessary) assumption that the inner walls of the cavities are perfectly absorbing. Then the only radiation reaching bodies 1 and 2 is that *radiated* by the inner walls of the cavities; these cavity walls do not reflect any radiation and return it to the bodies. Under these circumstances, the radiant energy incident per second on unit area of each body will be the same; call it E, in W/m². Of the incident radiation E, a certain fraction will be reflected and the remainder will be absorbed. As indicated in Fig. 4, let ρ denote the fraction of the incident radiation that is reflected and α denote the fraction that is absorbed; ρ is called the *reflectance* and α the *absorptance*. These quantities are dimensionless and their sum is unity for the surface of any opaque body: $\alpha + \rho = 1$. The product $\rho_1 E$ gives the radiant power reflected from unit area of body 1, while $\rho_2 E$ gives the radiant power reflected from unit area of body 2, in W/m². Similarly, $\alpha_1 E$ and $\alpha_2 E$ give the radiant power absorbed per unit area of bodies 1 and 2.

Now, let the radiated power per unit area, in W/m², be P_1 for body 1 and P_2 for body 2. If the temperatures of the bodies in Figs. 3 and 4 are to remain constant, as much energy must be lost per second by radiant emission as is gained by absorption and we may write

$$\text{rate of absorption} = \text{rate of emission},$$

or
$$\alpha_1 E = P_1, \quad \alpha_2 E = P_2.$$

Dividing the first equation by the second, we find that

$$\frac{\alpha_1}{\alpha_2} = \frac{P_1}{P_2}, \quad \text{or} \quad \frac{P_1}{\alpha_1} = \frac{P_2}{\alpha_2}. \tag{3}$$

Equation (3), and the observed temperature equality, give

> KIRCHHOFF'S PRINCIPLE OF RADIATION: *The ratio of the rates of radiation of any two surfaces at the same temperature is equal to the ratio of the absorptances of the two surfaces.*

Qualitatively, we can say that '*good radiators are good absorbers.*'

Now we return to the problem of defining a perfect radiator. There *is* a maximum value of the absorptance α; since no surface can absorb more than *all* of the incident radiation, the maximum value α can have is unity. In view of equation (3), we may say that a surface having the maximum rate of radiation is one that has the maximum absorptance and is therefore one that absorbs all radiation incident upon it; such a surface is *black* to all types of radiation. Therefore, we may define a perfect radiator as follows:

> A **perfect radiator** is a body that absorbs all incident radiation and is therefore called a ***black body.*** A perfect radiator is a perfect absorber.

Such a body would appear black unless its temperature were high enough for the body to be self-luminous.

No material surface absorbs *all* of the radiation incident upon it; even lampblack reflects about one per cent of the incident radiation. In practice, a perfectly black surface can be most closely approximated by a very small opening in the wall of a large cavity such as the one shown schematically in Fig. 5; radiation may enter or leave the cavity through the opening. Of the radiation entering through the opening, a part is absorbed by the interior walls of the cavity and a part is reflected. Of the part reflected, only a small fraction escapes through the opening, and the remainder is again partially absorbed and partially reflected by the walls. After repeated reflections, all of the entering radiation is absorbed except for the small portion that escapes through the opening. The opening therefore approximates a *black surface* or *perfect absorber.*

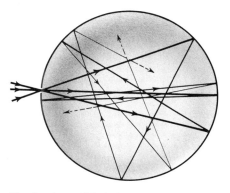

Fig. 5. A small hole in the wall of an enclosure, showing complete absorption of several representative rays. If the cavity walls are rough or dull, the reflection from the cavity walls is largely diffuse, rather than the specular reflection indicated in this drawing.

The inside walls of the cavity are radiating as well as absorbing, and a part of this radiation escapes through the opening. It can be shown that if the walls are at a uniform temperature T, the radiation that escapes is almost identical with that which would be emitted by a perfect radiator at temperature T. The hole closely approximates in all respects the surface of a black body emitting so-called *black-body radiation.* To indicate the accuracy of this approximation, we note that computation shows that even if the interior surface of a sphere has an absorptance of only $1/2$, a 10-inch sphere with a 2-inch hole will absorb 99 per cent of diffuse radiation (coming equally from all directions) incident on the hole, and hence will radiate through the hole 99 per cent of the radiation of a perfect radiator. A smaller hole will do correspondingly better.

The total radiation emitted from the surface of a body increases rapidly as the temperature of the surface is increased. The quantitative relation between rate of radiation and surface temperature of an ideal radiator or black body is called the *Stefan-Boltzmann law* and has the form

$$P_{\text{Black}} = \sigma T^4, \qquad \text{(black body)} \quad (4)$$

where P is the radiated power per unit area. The rate of radiation increases as the *fourth power* of the absolute temperature T. The proportionality constant σ has the value 5.670×10^{-8} W/m² · K⁴.

Example. *A small hole in the wall of an electric furnace used for refining steel behaves to a good approximation as a perfect black body. The radiant power emerging from such a hole can be measured by means of a radiation pyrometer. If the hole has an area of* 1 cm², *and it is desired to maintain the molten steel at* 1650° C, *at what value of radiant power should the electrical controls be set?*

In (4) we substitute $T = (1650 + 273)°$ K $= 1923°$ K. Then

$$P_{\text{Black}} = \sigma T^4 = 5.67 \times 10^{-8} \frac{\text{W}}{\text{m}^2 \cdot \text{K}^4} (1923 \text{ K})^4 = 77.7 \times 10^4 \text{ W/m}^2,$$

which gives 77.7 W as the desired radiant power from the 1-cm² hole.

The total radiation from many surfaces that are definitely not black is also very nearly proportional to the fourth power of the absolute temperature. This is true of surfaces composed of platinum, iron, tungsten, carbon, and many other materials. In every case, however, the proportionality constant is less than that for an ideal-radiator surface. Such a radiator is called a *gray body*. Since the absorptance α of a gray body is independent of its temperature, we see, by comparison with a black body in (3), that its rate of radiation is

$$P = \alpha \, P_{\text{Black}} = \alpha \, \sigma T^4.$$

Because of this relation, α is also called the *emissivity* of the surface.

On the other hand, many surfaces have absorptances that depend significantly on temperature. These are surfaces that have absorptances α_λ that depend on the wavelength λ of the incident radiation. In these cases it can be shown that the rate of radiation at any wavelength λ is α_λ times the rate of black-body radiation *at that wavelength*. The wavelength dependence of black-body radiation will be described in Chapter 41.

PROBLEMS

1. How much heat is conducted in 12 h through a pane of glass 3 f by 4 f in size and ⅛ i thick if the surface temperatures are 60° F and 10° F? Ans: 230 000 BTU.

2. How much heat is conducted in 24 h through a solid concrete building wall 6 i thick and 400 f² in area if the surface temperatures are 60° F and 30° F?

3. A domestic refrigerator has a wall area of 75 f² and is insulated with corkboard 3 i thick. The room temperature is 72° F, the inside temperature is 38° F, and it may be assumed that the temperature drop all occurs through the corkboard. Twenty pounds of food at room temperature, with average specific heat 0.9 BTU/lb·F deg, are placed in the refrigerator and cooled to the inside temperature in one hour. How much heat must be extracted during this hour by the cooling coils in the refrigerator? How many pounds of ice would have to melt (from ice at 32° F to water at 32° F) to extract this same heat?

Ans: 864 BTU; 6.00.

4. A certain freezing-plant building is 60 f long, 25 f wide, and 15 f high. Walls and roof are insulated with corkboard 1 f thick. Let the inside temperature of the corkboard average 0° F and the outside temperature average 60° F over a 24-h period. The heat conduction through the floor is negligible. Let 25 tons of produce of average specific heat and latent heat the same as water and ice be placed in the plant at 70° F to be frozen to 0° F during this period. What must be the 'tonnage' of the refrigerating machine required for this plant, if a 'one-ton' refrigerating machine will extract in one day as much heat as one ton of ice at 32° F melting to water at 32° F? (This is the unit actually used in rating refrigerating machines.)

5. A partition between two parts of a pressure vessel is made of sheet steel 0.50 i thick. What thickness of (a) copper, (b) brass, and (c) aluminum sheet would be equivalent to the steel sheet so far as conductive transfer of heat is concerned. Ans: 4.18, 1.18, 2.16 i.

6. What thicknesses of (a) wood, (b) solid concrete, and (c) cast iron have the same insulating value as 1.5 i of corkboard?

7. If a body cools at the rate of 0.3 F deg/s when it is 30 F deg above its surroundings, how fast does it cool when it is 50 F deg above its surroundings? Use Newton's law of cooling.

Ans: 0.5 F deg/s.

8. If a body cools at the rate of 5 C deg/s when it is at a temperature of 40° C and ambient temperature is 20° C, how fast does it warm up when it is at 5° C and the ambient temperature is 15° C?

9. The wall of a freezing plant is composed of 6 i of corkboard inside 8 i of solid concrete. If the temperature of the inner wall of the corkboard is −10° F and that of the outer wall of the concrete is 75° F, find the temperature of the corkboard-concrete interface and the heat flow in BTU/f^2·h.

Ans: 72.3° F; 4.10 BTU/f^2·h.

10. A large cast-iron steam pipe 0.3 i thick is covered with 1.5 i of asbestos insulation of conductivity 0.27 BTU·i/h·f^2·F deg. If the temperature of the inner wall of the pipe is 230° F and that of the outer wall of the asbestos is 30° F, find the temperature of the iron-asbestos interface and the heat loss per square foot per hour.

11. The walls and roof of a small house have an area of 3410 f^2 and an average heat loss of 0.3 BTU/h·f^2·F deg, equivalent to that through 1 i of corkboard. The house contains a typical small gas furnace rated at 80 000 BTU/h output. What is the maximum temperature difference that this furnace can maintain between the inner and outer surfaces of the walls?

Ans: 77.4 F deg.

12. A one-room building without windows measures 20 f × 30 f × 9 f, and is covered with a 3-i layer of glass-wool insulation. The inside walls of the room are to be maintained at 70° F when the outside wall temperature is 30° F. How much heat must be supplied to the room each hour in order to maintain the desired temperature? If this heat is to be supplied by an electric heater, what should be the power rating of the heater? Neglect heat loss through the floor, and assume that the whole of the temperature drop occurs through the glass-wool insulation.

13. A cubical thin-walled metal box measuring 3 f on an edge contains 75 lb of ice at 32° F. The interior of the box is to be maintained at 32° F in a room whose temperature is 72° F. Approximately what thickness of corkboard attached to the outer walls of the box will make the ice last two days? Ans: 2.88 i.

14. A solid iron sphere is 6 cm in diameter. At a certain temperature, the sphere loses thermal energy by radiation and convective cooling at a rate of 90 W. At what rate is its temperature decreasing?

15. A layer of ice on a pond is 2 cm thick. When the upper surface of the ice is at −10° C and the temperature of the water just below the ice is 0° C, at what rate does the ice become thicker? Ans: 0.0215 cm/min.

NOTE: *Thus far, we have considered only the case of unidirectional conduction through extended slabs in which the temperature gradient is constant. Frequently this is not the case and* it is necessary to express (1) in differential form, $Q/t = KA\, dT/dX$, and determine heat flow and other parameters by integration. The next four problems deal with two simple cases; flow through a sphere and flow through a cylinder.

16. The figure shows the cross section of a hollow sphere whose inside surface, of radius R_1, is maintained at temperature T_1, and whose outside surface, of radius R_2, is maintained at temperature T_2. (a) Show that for the heat flow through a typical spherical shell of radius r and thickness dr, $Q/t = k \cdot 4\pi r^2\, dT/dr$. (b) Since Q/t must be a constant independent of the radius r, show by integration that

$$\frac{Q}{t} = 4\pi k\, (T_1 - T_2) \frac{R_1 R_2}{R_2 - R_1}.$$

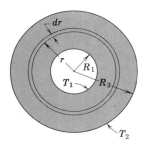

Problems 16–17

17. (a) From the results of Prob. 16, find an expression for the radius r at which the temperature $T = \tfrac{1}{2}(T_1 + T_2)$, the mean of the outside and inside temperatures. (b) A certain hollow sphere has an inner radius of 2 i and an outer radius of 4 i. If the temperature of the inner surface is 100° C and that of the outer surface is 0° C, at what radius is the temperature 50° C?   Ans: (b) 2.67 i.

18. The figure shows a typical section of a hollow circular cylinder of length l, inner surface of radius R_1 maintained at temperature T_1, and outer surface of radius R_2 maintained at temperature T_2. (a) Show that for steady heat flow through a typical cylindrical shell of radius r and thickness dr, $Q/t = k \cdot 2\pi r l\, dT/dr$. (b) By integration show that

$$\frac{Q}{t} = 2\pi k l \frac{T_1 - T_2}{\log_e(R_2/R_1)}.$$

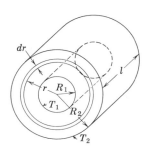

Problems 18–19

19. A stream of superheated steam at 300° C passes through a special tube, consisting of a hollow copper cylinder with inside radius ½ i and outer radius 1 i surrounded by an outer casing of steel with its inner surface in intimate contact with the copper and an outer surface with 2-i radius. If the temperature of the outer surface of the steel casing is 100° C, what is the temperature of the copper, steel interface? Use the results of Prob. 18.   Ans: 278° C.

20. From Newton's law of cooling, show that one may write, for a hot body at temperature T cooling in air of ambient temperature T_{Am}, $dT/dt = -K(T - T_{\text{Am}})$, where K is a constant for the body in question. Show by integration that the temperature of the body at any time t is given by the expression

$$T = T_{\text{Am}} + (T_0 - T_{\text{Am}})\, e^{-Kt},$$

where T_0 is the temperature of the hot body at $t = 0$.

21. A hot block of iron cools from 700° to 600° C in 4 min when the ambient temperature is 20° C. How long is required for this body to cool from 600° C to 500° C? from 500° C to 400° C? from 400° C to 300° C? Assume that Newton's law of cooling applies (see Prob. 20).
   Ans: 4.79, 5.86, 7.65 min.

22. The specific heat of a liquid can be determined by the 'method of cooling.' The scheme involved is the following: A certain mass of water m_W is placed in a closed container of mass m_C composed of a metal of specific heat c_C and heated to temperature T_1. A mass m_L of the liquid of unknown specific heat c_L is placed in an identical container and heated to temperature T_1. The containers are allowed to cool to some lower temperature T_2 and the times t_W and t_L required for the cooling processes for the containers of water and liquid are measured. From Newton's law of cooling derive the relation

$$\frac{m_L c_L + m_C c_C}{m_W c_W + m_C c_C} = \frac{t_L}{t_W},$$

and hence

$$c_L = \frac{m_W c_W t_L + m_C c_C (t_L - t_W)}{m_L t_W}.$$

23. A 200-g sample of water at 80° C is placed in a small covered copper vessel. In 15 min the temperature of the water has dropped to 60° C. The water is then removed from the vessel, and a 300-g sample of olive oil at 80° C is placed in the container. Neglecting the heat capacity of the container, find how long it will take the olive oil to cool to 60° C. (*see* Prob. 22.) Ans: 10.6 min.

24. In the example on p. 323, let T be the temperature of the can *above* the ambient temperature, which is taken as $T=0$. Let C be the heat capacity of the can (see p. 306) in J/deg. Neglect the small heat capacity of the bulb. Let the rate of heat transfer in watts be given by Newton's law of cooling in the form BT, where B is a constant measured in W/deg. If the bulb of power P, in watts, is turned on at time $t=0$, when $T=0$, write the differential equation that governs the rise of temperature of the can. Verify that this differential equation has the solution

$$T = (P/B)(1 - e^{-t/\tau}), \quad \text{where} \quad \tau = C/B.$$

Sketch this curve for a numerical example for which you supply your own constants, and show that τ represents a *time constant* that gives the time required for the fraction $(1 - e^{-1})$ of the total temperature rise.

25. A closed copper can of heat capacity 15 cal/C deg contains 1 kg of water. Can and contents are 20 C deg above ambient and require 20 min to cool to 10 C deg above ambient. How long would it take the empty copper can to cool this same amount? Ans: 17.7 s.

26. What is the rate of radiation of a black body in W/m² at temperatures of 27° C, 327° C, and 727° C?

27. Radiant energy is incident on a certain opaque body at the rate of 80 W/m². The surface absorbs 30 W/m². (*a*) What is the reflectance of the surface? (*b*) What is the absorptance? (*c*) If the body is in thermal equilibrium with its surroundings and can exchange energy with its surroundings only by radiation, what is the rate of radiation of the body? Ans: 0.625; 0.375; 30 W/m².

28. Radiant energy is incident on the surface of a black body at the rate of 80 W/m². (*a*) At what rate is radiant energy reflected from each unit area of the surface? (*b*) At what rate is radiant energy absorbed by each unit area of the surface? (*c*) If the black body is in radiative equilibrium with its surroundings, what is its rate of radiation?

29. What would be the rate of radiation of a black body at the same temperature as the opaque body described in Prob. 27? Ans: 80.0 W/m².

30. The rate of radiation of the surface of a certain opaque body is 50 W/m². The rate of radiation of a black body at the same temperature is 60 W/m². What is the absorptance of the surface of the opaque body?

31. A silver surface has a reflectance of 0.96 at 627° C. What is the rate of radiation of this surface at this temperature? What is the *emissivity* of a silver surface at this temperature? Ans: 1490 W/m²; 0.04.

32. A radiation pyrometer is used to measure the radiant power emerging from a hole of area 1 cm² in the wall of a furnace. If 110 W emerge, what is the temperature of the interior of the furnace?

33. From measurements of solar radiation received on the earth, it is concluded that the sun radiates at the rate of 6250 W/cm² of surface area. Assuming that the sun is a perfect black body (such a gaseous mass has close to zero reflectance), compute its surface temperature. Ans: 5760° K.

34. Show that as the surface of a lake is cooled in the fall and winter, convective processes will tend to mix the water and keep it at constant temperature throughout until the temperature drops to 4° C, and that thereafter the surface cooling will not result in further mixing, but will result in the formation of ice. Water transfers heat readily by convection but poorly by conduction. Hence show that the tendency for the main body of the water to cool further by conduction can readily be overcome by the heat supplied by the warm earth at the bottom, which can be distributed by convection.

35. What diameter should a tungsten lamp filament, with absorptance 0.33, have in order that it will come to a temperature of 3000° K when it is radiating 24 W per cm length?

Ans: 0.252 mm.

36. Why is a Dewar flask or Thermos bottle (*a*) made double-walled? (*b*) evacuated? (*c*) silvered?

37. It is found that the 'surface' of a bundle of highly polished steel needles (*see* figure), viewed end-on, is almost black—that is, the absorptance is almost 100 per cent. Explain this fact. Why must the needles be highly polished and very sharply pointed?

38. In the style of carpet called 'tip-shear,' the carpet loops are sheared in some small areas and left unsheared in others (*see* figure). In spite of a uniform dye color, the sheared areas appear significantly darker or lighter than the unsheared areas. Which are they, darker or lighter, and why?

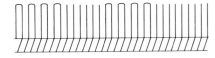

Problem 38

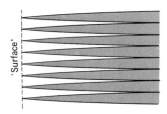

Problem 37

IDEAL GASES 17

The laws that govern the behavior of most of the common gases under ordinary conditions are remarkably simple. They are called the *ideal-gas laws*. These laws are obeyed extremely closely by such gases as air, oxygen, nitrogen, hydrogen, and helium under all conditions except those of extremely low temperature or extremely high pressure. They are very closely obeyed, too, by carbon dioxide under ordinary conditions of temperature and pressure. However *solid* CO_2 (Dry Ice) can exist at *normal* pressure but at the low temperature of $-80°$ C; and *liquid* CO_2 can exist in tanks at *normal* temperature but at the high pressure of 65 atmospheres or more. Clearly, under these conditions the CO_2 is not behaving like an ideal gas. It turns out that *the vapor of any substance behaves like an ideal gas if the pressure is sufficiently low and the temperature sufficiently high*. If the temperature is decreased, or the pressure raised, to a value near to that required for condensation to a liquid, then significant departures from the ideal-gas laws are observed.

The ideal-gas laws furnish such a good approximation to the behavior of most gases under such a very wide range of conditions that it will be profitable to devote this chapter to a study of *the ideal gas*. In the succeeding two chapters we shall discuss cases of departure of gases from ideal behavior. We shall not continue to repeat the word *ideal* in this chapter.

As we shall see, Joule's free-expansion experiment shows that the thermal energy of a gas depends only on the temperature and is independent of the volume. From this fact we conclude that there are no forces between the molecules of a gas except for the impulsive forces that act during actual collisions. Hence the thermal energy of a gas consists solely of energy internal to the molecules (rotation and vibration) and kinetic energy of translation of the molecules—there is no need to introduce a potential energy representing forces between molecules. This simple behavior leads to a very simple *kinetic theory* of gases.

We begin this chapter with a review of certain properties of atoms and molecules that are doubtless familiar to the reader from his study of chemistry. The ideas and terminology involved will prove useful in our treatment of gases.

1. Atoms and Molecules

During the eighteenth century, chemists such as Lavoisier recognized the existence of *elements* and discovered two important laws governing the manner in which elements combine to form *compounds,* namely, the *law of definite proportions* and the *law of multiple proportions*. These discoveries led JOHN DALTON, in 1803, to formulate the hypotheses that each element is composed of atoms, that the atom of each element has a definite fixed mass, and that the molecule* of each type of compound is made up of a definite number of atoms of each of the constituent elements. All later work in chemistry and physics has confirmed Dalton's hypotheses except in certain minor details which we shall discuss presently.

In addition to the 90 elements occurring naturally, 15 others have been made artificially (*see* Chapter 44), for a total of 105. To each element is assigned a *symbol* which, by international agreement, is universal. To each of these 105 elements is also assigned an atomic number Z, ranging from 1 to 105, which, as we shall see later, represents the number of electrons surrounding the positive nucleus of the atom, and which gives the ordering of the elements in the periodic table (*see* Chapter 41). Section 4 of the Appendix contains a complete list of the names, symbols, and atomic numbers of the elements.

> The **atomic number** of an element is an integer that gives the number of electrons in the atom.

---

*The term *molecule* (little mass) was actually introduced by Avogadro at a slightly later date. Avogadro was the first to recognize that Gay-Lussac's volumetric relationships (Sec. 2) made it necessary to assume that even elements such as H_2, O_2, and N_2 were composed of *molecules* containing more than a single atom.

The lightest atom is that of hydrogen. The masses of most of the other atoms were found to be approximately integral multiples of that of hydrogen. It was thought for a time that the masses of the atoms of all elements would turn out to be integral multiples of the mass of the hydrogen atom, but certain well-confirmed exceptions—for example, the mass of chlorine remained obstinately at 35.5 times that of hydrogen—forced the abandonment of this hypothesis. Eventually (*see* Chapter 28) the physicists of the twentieth century learned to measure the masses of individual atoms and found, surprisingly, that the abandoned hypothesis was in fact almost exactly correct. What was incorrect was Dalton's hypothesis that all atoms of an element have the same mass. It was found that, typically, the atoms of a naturally occurring element have a small number of different masses. The atoms of the same chemical element but of different masses are called *isotopes,* or *isotopic nuclides.* Thus the mass 35.5 for Cl relative to H arises from the fact that chlorine has two isotopic nuclides, of relative masses 35 and 37, occurring in natural chlorine in the proportions 75 per cent and 25 per cent, giving an *average* mass of 35.5. Thus the atomic mass of the chemists represented an *average* mass of the isotopes in the proportion in which they occurred naturally. The masses of the nuclides are all *very close* to integral multiples of the mass of the hydrogen atom. That they are not *precisely* integral multiples is responsible for the whole subject of *nuclear energy,* as we shall see in Chapter 45.

A **nuclide** is an atom of a particular mass and of a particular element.

Isotopes, or **isotopic nuclides,** are those nuclides of different masses that belong to the same element.

The different isotopic nuclides of an element are designated by appending to the chemical symbol for the element a superscript giving the (approximate) mass in terms of the mass of hydrogen. This statement is loose, because even hydrogen turns out to have two isotopes. Most hydrogen atoms have the same mass, but one in 10 000 has double the mass of the rest. The two stable isotopes of hydrogen are designated as H^1 and H^2, of which H^1 is greatly predominant. The balance of the ten lightest elements are represented in nature by the following nuclides, of which the one written in bold letters is greatly predominant: He^3, **He^4**; Li^6, **Li^7**; **Be^9**; B^{10}, **B^{11}**; **C^{12}**, C^{13}; **N^{14}**, N^{15}; **O^{16}**, O^{17}, O^{18}; **F^{19}**; **Ne^{20}**, Ne^{21}, Ne^{22}.

The scheme that has been agreed upon for accurate specification of the masses of atoms is one in which the mass of the predominant carbon isotope C^{12} is called 12 *nuclidic mass units* (12 u). In this scheme the mass of H^1 is 1.01 u; that of H^2 is 2.01 u; and the masses of the nuclides of the other ten lightest elements, as listed above, are: 3.02, **4.00;** 6.02,

7.02; 9.01; 10.01, **11.01; 12,** 13.00; **14.00,** 15.00; **15.99,** 17.00, 18.00; **19.00; 19.99,** 20.99, 21.99 u.

The **nuclidic mass unit** (u) is $1/12$ the mass of the lightest naturally occurring isotope of carbon.

Since the masses of all nuclides are very close to integers when expressed in u, it is convenient to identify a nuclide by an integer called the *mass number,* which is the integer nearest its actual mass in u. Thus the superscripts attached to the chemical symbols in the preceding paragraph represent mass numbers.

The **mass number** of a nuclide is the integer closest to its actual mass expressed in nuclidic mass units.

The nuclidic mass unit furnishes a much more convenient unit in terms of which to specify the masses of atoms or molecules than the kilogram, but of course these different units have a definite relation, determined by various methods that will be described later. The relation is

$$1 \text{ u} = (1.6604 \pm 0.0001) \times 10^{-27} \text{ kg.} \tag{1}$$

So far we have been discussing the *masses of atoms.* We now introduce the term *atomic mass,* which is a *dimensionless* number, and should more properly be called the *relative* atomic mass. Until the twentieth century, the *masses* of atoms were unknown, but the *relative masses* were well-known. Chemists called these relative masses simply atomic weights. Until the discovery of isotopes, the (relative) atomic masses were given on a scale in which the *average* mass of the nuclides in naturally occurring oxygen was called 16. With the need for more precision, this has been changed* to a scale in which the *lightest carbon isotope* is assigned the nuclidic mass 12. The term *atomic mass* is still reserved for the average of the relative masses of the isotopes as they occur in nature, and the term *nuclidic mass* is used when an individual nuclide is concerned. Thus we have the following definitions:

The **nuclidic mass** is the ratio of the mass of a nuclide to $1/12$ the mass of the lightest isotope of carbon. It is *dimensionless.*

---

*The $C^{12} \sim 12$ scale was adopted in 1961 by international agreement by both the International Union of Pure and Applied Chemistry and the corresponding International Union of Pure and Applied Physics. It replaces the $O \sim 16$ scale formerly used by chemists, but does not differ from it within the usual accuracy of stoichiometric measurements. It replaces the $O^{16} \sim 16$ scale formerly used by physicists and does differ from it significantly; nuclidic masses are lower on the C^{12} scale by a factor of 1.000 318, and physical constants such as Avogadro's constant, the volume of a kilomole of a gas, and the universal gas constant are lower by the same factor, while the size of the nuclidic mass unit is correspondingly increased.

The **atomic mass** of an element is the *average* of the nuclidic masses of the isotopes of the element in the proportion in which they occur in nature. It is dimensionless.

The **molecular mass** of a compound or of the molecules of a gas is the sum of the atomic masses of the atoms constituting a molecule of the compound or of the gas. It is dimensionless.

The last term is one that we have not introduced previously. Let us consider water, whose molecule is represented by the symbol H_2O. Hydrogen consists of two isotopes of nuclidic masses 1.008 and 2.014. Since there is only one of the latter to each 10 000 of the former, the (average) atomic mass of hydrogen still works out as 1.008. Oxygen contains three isotopes of nuclidic masses 15.995, 16.999, and 17.999. These three isotopes occur in natural oxygen in the relative proportions 2500:1:5. The (average) atomic mass of oxygen works out to be 15.999. The *molecular mass* of the water molecule H_2O is thus $2(1.008) + 15.999 = 18.015$.

The nuclidic, atomic, or molecular mass is ordinarily denoted by the symbol M. It is convenient to consider M as a *dimensionless* quantity, but we note from the definitions that:

The mass of a nuclide of nuclidic mass M is M u; the average mass of the nuclides in a natural element of atomic mass M is M u; the average mass of the molecules in a natural compound of molecular mass M is M u.

2. Avogadro's Law; the Kilomole; the Avogadro Constant

In determining the molecular masses of the molecules of a gas, it was initially very helpful to make use of the discovery, by GAY-LUSSAC in 1808, that when gases react chemically to produce gaseous products, there are simple ratios between the volumes of the reactants and those of the reaction products, provided the volumes are measured under the same conditions of temperature and pressure. To explain this behavior, COUNT AMADEO AVOGADRO, in 1811, advanced the hypothesis, now called

AVOGADRO'S LAW: *Equal volumes of any two gases under the same conditions of temperature and pressure contain equal numbers of molecules.*

Avogadro's law has been demonstrated to be correct to a very good approximation for most gases at ordinary temperatures and pressures (*see* Chapter 18); it is assumed to be rigorously true for our hypothetical *ideal* gas. By its use, one can determine the relative masses of molecules by simply weighing equal volumes of gases under the same conditions of temperature and pressure.

It is convenient, in molecular and atomic physics, to treat *amount of substance* as a basic quantity and to define a unit of amount of substance as a quantity containing a specified number of particles (molecules, or atoms, or ions, or electrons, or other particles, as the case may be). The unit is called the *kilomole* (kmol), and is defined as follows:

> A **kilomole** is an amount of substance containing the same number of particles as there are atoms in 12 kg of the pure carbon nuclide C^{12}.

We note that this amount is 12 kg of C^{12}, where 12 is the nuclidic mass, and 12 u is the mass of an individual nuclide. In view of the fact that the average mass of the atoms or molecules in a naturally occurring element or compound is M u, where M is the atomic or molecular mass, we see that M kg of such an element or compound will contain the same number of particles as 12 kg of C^{12} and hence constitute a kilomole. Thus,

> *One kilomole of an element having atomic mass M is M kilograms of the element; one kilomole of a compound substance or a gas having molecular mass M is M kilograms of the substance.*

Thus, one kmol of oxygen gas, O_2, is 32.0 kg of the gas; one kmol of hydrogen gas, H_2, is 2.02 kg of the gas; one kmol of water, H_2O, whether in the form of ice, liquid, or vapor, contains 18.0 kg; and one kmol of gold is 197 kg.

In chemical usage, the more common unit is the *mole* (mol), which is 10^{-3} kmol and thus is an amount of substance containing the same number of particles as there are in 12 *grams* of the nuclide C^{12}.

Since a kilomole of any substance contains the same number, say N_A, of particles, the number of particles in n kmol of any substance can be written as

$$N = N_A\, n, \tag{2}$$

where N_A is a universal constant called the *Avogadro constant*. The Avogadro constant is, numerically, the number of particles in M kg (one kilomole) of a substance in which the particles each have mass M u. This number is, by (1),

$$\frac{M \text{ kg}}{M \text{ u}} = \frac{1 \text{ kg}}{1.6604 \times 10^{-27} \text{ kg}} = 6.0225 \times 10^{26}.$$

Hence the Avogadro constant has the value, noting the experimental error, of

$$N_A = (6.0225 \pm 0.0003) \times 10^{26}/\text{kmol}. \tag{3}$$

We see now that Avogadro's law can be restated in the form:

> *One kilomole of any gas, under the same pressure and temperature, occupies the same volume.*

The volume per kilomole of an ideal gas at 0° C and 1 atm pressure is observed to be

volume per kilomole $= (22.414 \pm 0.003)$ m³/kmol.   (0° C, 1 atm)   **(4)**

3. The Gas Laws

The three quantities used to specify the *'state'* of a confined gas sample are its *volume, pressure,* and *temperature.* The relations among these quantities were first determined experimentally in the seventeenth and eighteenth centuries.

ROBERT BOYLE demonstrated experimentally in 1660 that the volume of a sample of gas is inversely proportional to the pressure, provided that the temperature of the gas is kept constant:

BOYLE'S LAW: *If the temperature of a gas sample is kept constant, the volume of the gas is inversely proportional to the absolute pressure.*

Thus, if a gas is confined in a cylinder, as in Fig. 1, p. 340, at some initial pressure P_0 and volume V_0, and the pressure is changed to P_1 without change in temperature, the volume will change to V_1 such that

$$P_1 V_1 = P_0 V_0. \qquad (T_1 = T_0) \quad \textbf{(5)}$$

We see that if the pressure is doubled, the volume changes to one-half its initial value; if tripled, the volume becomes one-third; etc.

In 1787, JACQUES CHARLES first determined the relation between the volume and the temperature of a gas when the pressure is kept constant. Although he did not know about absolute temperature, he found a linear relation between volume and temperature that can now be most simply expressed by the statement that at constant pressure the volume of a gas sample is directly proportional to its absolute temperature. For example, if the temperature of a gas sample confined in a cylinder like that in Fig. 1 is raised from 0° C to 273° C (from 273° K to 546° K), with no change in pressure, the volume doubles. In general, if V_0 and T_0 represent some initial values of volume and absolute temperature, the volume V_1 at some other absolute temperature T_1 is given by the relation

$$V_1/T_1 = V_0/T_0. \qquad (P_1 = P_0) \quad \textbf{(6)}$$

CHARLES'S LAW: *If the pressure of a gas sample is kept constant, the volume of the sample is directly proportional to its absolute temperature.*

Equations (5) and (6) provide the experimental basis for a more general relation between volume, pressure, and temperature known as the *general gas law for a given sample*. Relations (5) and (6) are each consistent with the more general relation:

$$\frac{P_1 V_1}{T_1} = \frac{P_0 V_0}{T_0}, \qquad \left\{\begin{array}{l}\text{general gas}\\ \text{law for a}\\ \text{given sample}\end{array}\right\} \quad (7)$$

where P_0, V_0, and T_0 represent some initial set of conditions and P_1, V_1, and T_1 represent a second set of conditions for the same sample of gas.

From (7) it can be seen that the product PV is directly proportional to the absolute temperature T, since we can consider $[P_0 V_0/T_0]$ as a constant for the particular sample, based on the measurement of V_0 under some initial set of conditions P_0, T_0. Thus, we may write (7) in the general form

$$PV = [P_0 V_0/T_0] T \qquad \text{or} \qquad PV = \text{constant} \times T. \qquad (8)$$

The constant in (8) depends, of course, on the *mass* of the gas sample, and on the *kind* of gas. However, from the fact that *one kilomole* of any gas occupies the *same volume* (22.414 m³) at 0° C and 1 atm pressure, we see that *the constant has the same value for one kilomole of any gas*. If we denote the value of the constant for one kilomole of gas by R, and our sample contains n kilomoles, we can write (8) in the form

$$PV = nRT. \qquad \left\{\begin{array}{l}P \text{ in N/m}^2\\ V \text{ in m}^3\\ n \text{ in kmol}\\ R \text{ in J/kmol} \cdot \text{K}\\ T \text{ in K deg}\end{array}\right\} \quad (9)$$

The constant R is called the *universal gas constant*. It can be evaluated by setting, in (9),

$$n = 1 \text{ kmol}, \qquad P = 1.01325 \times 10^5 \text{ N/m}^2, \qquad T = 273.15 \text{ K},$$

corresponding to a pressure of 1 atm and a temperature of 0° C, in which case we know from (4) that the volume will be $V = 22.414$ m³. Substitution of these values in (9) gives

$$R = \frac{(1.01325 \times 10^5 \text{ N/m}^2)(22.414 \text{ m}^3)}{(273.15 \text{ K})(1 \text{ kmol})} = (8314 \pm 1) \frac{\text{J}}{\text{kmol} \cdot \text{K}}. \qquad (10)$$

The relation (9) can be considered as the 'definition' of an *ideal gas*. It is found that all gases are ideal when the number of kmol per unit volume (the 'number density') $n/V = P/RT$ is small, that is, when the pressure is sufficiently low or the temperature sufficiently high.

Example. *What is the mass of helium gas in a cylinder of* 0.15-m³ *volume if the gas is at a pressure of* 120 atm *and a temperature of* 27° C?

We use the fact that one kilomole of He has a mass of 4.00 kg. We shall determine the number of kilomoles in the tank in two ways: first, by substitution in (9) to determine the number n; second, by using (7) to determine the volume the gas would occupy at 0° C and 1 atm and dividing by 22.4 m³ to determine the number of kilomoles.

To determine the number of kilomoles from (9), we substitute $P = 120 \,(1.013 \times 10^5 \,\text{N/m}^2) = 1.216 \times 10^7 \,\text{N/m}^2$, $V = 0.15 \,\text{m}^3$, $T = (273 + 27) \,\text{K} = 300 \,\text{K}$, and R from (10), to obtain

$$n = \frac{PV}{RT} = \frac{(1.216 \times 10^7 \,\text{N/m}^2)(0.15 \,\text{m}^3)}{(8314 \,\text{J/kmol} \cdot \text{K})(300 \,\text{K})} = 0.731 \,\text{kmol}.$$

Alternatively, we use (7) to determine the volume V_0 at 1 atm and 273° K:

$$\frac{120 \times 0.15 \,\text{m}^3}{300} = \frac{1 \times V_0}{273}, \qquad V_0 = 16.38 \,\text{m}^3.$$

Dividing by the volume per kilomole under standard conditions gives

$$n = \frac{16.38 \,\text{m}^3}{22.4 \,\text{m}^3/\text{kmol}} = 0.731 \,\text{kmol}.$$

We now multiply the number of kilomoles by the mass of one kilomole to obtain

$$m = 0.731 \,\text{kmol} \times 4.00 \,\text{kg/kmol} = 2.92 \,\text{kg}$$

as the mass of gas in the cylinder.

We must point out that, in the above discussion, n is a pure number. The unit 'kmol' is added gratuitously to remind us that we are using the kmol as our unit of quantity of gas, but it is physically dimensionless, just as is 'apple' or 'chair' if we were counting apples or chairs.

We now state one other important experimental law that applies when we have a mixture of ideal gases:

DALTON'S LAW OF PARTIAL PRESSURES: *If several types of gas are put into the same container, the total pressure exerted is the sum of the partial pressures that each type of gas would exert if it alone occupied the container.*

For example, if n_A kilomoles of a gas A are alone in a container of volume V at temperature T, the pressure would be

$$P_A = n_A RT/V, \qquad (11)$$

by (9). A sample consisting of n_B moles of a gas B alone in this container would exert pressure

$$P_B = n_B RT/V. \qquad (12)$$

Dalton's law asserts that if the two gases are both placed in this container, the total pressure will be

$$P = P_A + P_B = (n_A + n_B) RT/V, \qquad (13)$$

where P_A and P_B are known as *partial pressures;* hence the general gas laws (7) and (9) are valid for mixtures of gases as well as for single gases.

Comparison of (11) with (12) shows that *the partial pressures of the different types of gas are proportional to the numbers of molecules of each type in the container,* since n_A and n_B are proportional to these numbers.

In *dry air,* 78.09 per cent of the molecules are N_2, 20.95 per cent are O_2, 0.93 per cent are Ar (argon), and the other 0.03 per cent are CO_2, Ne, Kr, Xe, He, and H_2. This composition gives an average molecular mass of 28.97. *For all ordinary purposes, dry air behaves like an ideal gas of molecular mass* 29.0. At 0° C and 1 atm, 1 kmol (29.0 kg) of dry air would occupy 22.4 m³, so dry air would have the density

$$\rho = \frac{m}{V} = \frac{29.0 \text{ kg}}{22.4 \text{ m}^3} = 1.29 \frac{\text{kg}}{\text{m}^3}. \qquad \left(\begin{array}{c}\text{dry}\\\text{air}\end{array}\right) \quad (14)$$

It is noted that the density of air at 0° C and 1 atm is a little over 1/1000 the density of water, which is 1000 kg/m³.

The considerations involved in (14) can be generalized to give the density of any gas of molecular mass M in terms of the temperature and pressure. Consider a 1-kmol sample of gas; the mass of the gas sample is $m = M$ kg, while from (9) the volume of a 1-kmol sample is $V = RT/P$. The density is therefore given by the relation

$$\rho = \frac{m}{V} = \frac{M \text{ kg}}{(RT/P)} = \frac{P}{RT} \times (M \text{ kg}). \qquad (15)$$

This expression gives density in kg/m³ if the metric units given with equation (9) are used. It is noted that the density is directly proportional to the molecular mass, as required by Avogadro's law. Equation (15) is the most convenient form of the general gas law for use with unconfined gases such as those occurring in the earth's atmosphere.

4. External Work; Thermal Energy

If a gas expands by pushing apart the walls of its container, as in the case of the expansion of the gas in a balloon or in the cylinder of Fig. 1 when the piston moves to the right, the gas does mechanical work on the walls of the vessel. This work is called the *external work* done *by* the gas. If the piston moves to the *left* in Fig. 1, external work is done *on* the gas; in this case the external work done *by* the gas is considered as a negative quantity.

We shall now discuss the relation between *heat* added to a gas, *external work* done by the gas, and changes in the *thermal energy* of a gas. If one adds *heat* to a gas, this heat all goes into increasing the thermal energy of the gas, *provided* the gas

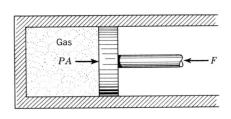

Fig. 1. Gas confined in a cylinder. The force F acting on the right side of the piston is the sum of the force exerted by the pressure of the external atmosphere and any force that may be applied directly to the piston.

does no external work while the heat is being added. But if the gas pushes back the walls of the container and hence does external work while the heat is being added, conservation of energy requires that some of the added heat be used to supply the energy needed for this external work. This is the case, for example, if the pressure of the gas is kept constant while heat is being added. Conservation of energy requires that

(heat added to a gas) = (increase in thermal energy)
+ (external work done by the gas). (16)

It is noted that each of the three terms in equation (16) may be either positive or negative. Heat removed from the gas is considered a negative amount of heat added. A decrease in thermal energy is a negative increase. External work done *on* the gas when the volume contracts is expressed as negative external work done *by* the gas.

We now consider the *thermal energy* of the gas, which we denote by the symbol U. We note that our meaningful term 'thermal energy' is not widely employed in the case of gases; the more usual but less meaningful term is *internal energy*.

Since, as we have seen, the state of (an ideal) gas is completely determined when *temperature, pressure,* and *number of kilomoles* are specified, the thermal energy can depend only on these three quantities, and on the *kind of gas*. We now state an important experimental law that is found to hold *extremely closely* for all real gases for which the general gas law is closely obeyed, and which is assumed to hold *exactly* for our ideal gas. This law states that:

The thermal energy of a gas depends only on the temperature, the number of kilomoles, and the kind of gas; it is independent of pressure or volume.

This experimental law means that there is no variation of thermal energy of a given sample of gas with variations in pressure, and consequent variations in volume, so long as the temperature does not change. It means that the thermal energy of a given quantity of a given kind of gas is known once the temperature is given, no matter what the pressure or volume may be.

The simplest demonstration of this law is by means of Joule's free-expansion experiment, which employs the apparatus shown in Fig. 2. Two vessels are connected by a pipe fitted with a stopcock. Initially one vessel is filled with a gas and the other vessel is evacuated. The two vessels are then immersed in water in a carefully insulated calorimeter

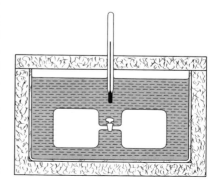

Fig. 2. Joule's free-expansion experiment (schematic).

and allowed to come into thermal equilibrium with the water. After thermal equilibrium has been established, the stopcock is opened and the gas is allowed to rush from the full vessel into the empty one. After thermal equilibrium has been established again throughout the system, it is found experimentally that *the temperature of the water has not changed.* Since the gas is in thermal equilibrium with the water both before and after the expansion, *the temperature of the gas has not changed.* Since the temperature of the water does not change, no heat is gained or lost by the water, so *no heat has been added to or removed from the gas.*

The rest of the argument consists in applying the principle of conservation of energy (16) to the overall effects of this expansion process. As noted above, the experimental result indicates that *no heat is added* to the gas. It is also clear that *no external work is done* by the gas, because the walls of the vessel do not move. Hence, according to (16), *the thermal energy of the gas does not change.* We have noted above that *the temperature of the gas does not change.* So here we have the same mass of gas at two quite different states of pressure, volume, and density but at the same temperature, and have demonstrated experimentally that the thermal energy in these two states is the same. It is by this argument that one derives from Joule's experiment the law stated in italics above.

5. Heat Capacity per Kilomole at Constant Volume

If we place a sample of gas in a container of *fixed* volume and add heat, all the heat added goes into increasing the thermal energy of the gas, since no external work is done. In considering gases, it is convenient to specify thermal energy per kilomole, rather than thermal energy per unit mass; hence we define:

> The **heat capacity per kilomole at constant volume** is the amount of heat that must be added to one kilomole of gas to increase its temperature by one degree when the volume is held constant.

The heat capacity per kilomole at constant volume is denoted by C_V, where we use the capital C to distinguish this quantity from the specific heat capacity (per unit mass) which was denoted by small c in Chapter 15. The amount of heat that must be added to n kilomoles of gas to increase the temperature ΔT degrees at constant volume is thus

$$Q = nC_V \Delta T. \quad \begin{cases} Q \text{ in kcal} \\ n \text{ in kmol} \\ C_V \text{ in kcal/kmol·K} \\ \Delta T \text{ in K} \end{cases} \quad (17)$$

Since this heat all goes into thermal energy, the increase in thermal energy, which we denote by ΔU, is given by

$$\Delta U = nC_V \Delta T. \quad (18)$$

This equation is of particular importance because *it gives the increase in thermal energy that is associated with a temperature rise ΔT whether or not the volume changes while the temperature is rising.* This conclusion follows from the law stated on p. 341, that the thermal energy is dependent only on the temperature. Thus, by a measure of the heat capacity per kilomole at constant volume, where we know that all the heat goes into thermal energy, we can determine the coefficient of ΔT in the expression for the change in thermal energy for any type of process.

6. Work Done by an Expanding Gas; Heat Capacity per Kilomole at Constant Pressure

If the gas in the cylinder of Fig. 3 is at pressure P, it exerts a force F on the piston equal to PA, where A is the area of the piston. We imagine the piston to be frictionless but still gastight. Then, for equilibrium, a force F, equal and opposite to that exerted by the gas, must be exerted externally on the piston. By keeping the external force F constant, and adding or removing heat to raise or lower the temperature, we can cause the gas to expand or contract at constant pressure P.

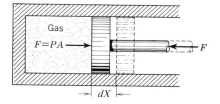

Fig. 3. Work done by an expanding gas.

In Fig. 3, if the piston moves to the right a small distance dX, the work that the gas does on the piston is $dW = F\, dX = PA\, dX$. But $A\, dX$ is the increase of volume of the gas when the piston of area A moves to the right a distance dX. We write $A\, dX = dV$ and obtain $dW = P\, dV$. It can be readily shown that this relation gives the work done whenever a gas exerting a pressure P increases its volume by a small amount dV, whether the volume changes in the geometrical configuration of Fig. 3 or in any other geometrical configuration, such as, for example, the expansion of the gas in a balloon. If, then, a quantity of gas expands from volume V_1 to volume V_2, the external work done by the gas is

$$W = \int_{V_1}^{V_2} P\, dV. \tag{19}$$

We note that (19) gives the work done by the gas in the expansion from volume V_1 to volume V_2 whether or not the pressure P is constant. The evaluation of the integral is particularly simple in the case of an *expansion at constant pressure*, in which case P can be taken outside of the integral sign and we obtain

$$W = P(V_2 - V_1). \qquad \begin{pmatrix}\text{constant}\\ \text{pressure}\end{pmatrix} \tag{20}$$

We are now in a position to compute the amount of heat that must be added to a gas to increase its temperature from T_1 to T_2 *when the pressure is kept constant*. If there are n kilomoles of gas in the cylinder of Fig. 3, then its volume at T_1 will be, by (9), $V_1 = nRT_1/P$, and that at T_2 will be $V_2 = nRT_2/P$. During the process of adding heat, the volume will increase by

$$V_2 - V_1 = (nR/P)(T_2 - T_1).$$

Hence, according to (20), the gas will do external work of amount

$$W = nR(T_2 - T_1).$$

The heat that must be added to the gas is, according to (16), the sum of the increase in thermal energy and the external work done by the gas. According to (18), the increase in thermal energy is $nC_V(T_2 - T_1)$, so the total heat that must be added is

$$Q = nC_V(T_2 - T_1) + nR(T_2 - T_1). \tag{21}$$

The heat capacity per kilomole at constant pressure, denoted by C_P, is defined as follows:

> The **heat capacity per kilomole at constant pressure** is the amount of heat that must be added per kilomole per degree increase in temperature when a gas is heated and the pressure is kept constant.

Thus to increase the temperature of n kilomoles of gas by ΔT degrees at constant pressure, we must add heat

$$Q = nC_P \Delta T. \tag{22}$$

From equation (21), we can compute the value of C_P as

$$C_P = \frac{Q}{n\Delta T} = \frac{Q}{n(T_2 - T_1)} = C_V + R.$$

The heat capacity at constant pressure is greater than the heat capacity at constant volume: in heating a gas at constant pressure we must not only add heat to supply the increase in thermal energy that is associated with the temperature rise but also must add heat to supply the energy for doing external work in the expansion.

In the equation $\qquad C_P = C_V + R, \tag{23}$

we must use the same units in all three terms. If we use R in mechanical energy units, J/kmol·K, we must express C_P and C_V in the same units. It is frequently more convenient to use C_P and C_V in thermal units, kcal/kmol·K, and to change R to these thermal units. If the value of R given by (10) is converted to thermal units by using the relation 1 kcal = 4184 J, it becomes

$$R = 1.987 \text{ kcal/kmol} \cdot \text{K} \tag{24}$$

We shall now discuss the magnitudes of C_V and C_P. First we note that since C_V is a measure of the rate of increase of thermal energy with increase of temperature, and since the thermal energy is independent of pressure or volume and depends only on the temperature of the gas, C_V must be *independent of pressure or volume.* Since, for a given gas, the difference between C_P and C_V is a constant, C_P must also be *independent of pressure or volume.*

Over a very wide range of temperature from values close to the boiling point of the liquid up to at least 500° C, C_V and C_P are very nearly independent of temperature for monatomic and many diatomic gases and have values given by remarkably simple expressions that are explained by the kinetic theory of gases (Sec. 7). *For a monatomic gas, C_V has the value $\tfrac{3}{2} R$; hence, from (23), C_P has the value $\tfrac{5}{2} R$.* This statement holds for gases such as helium (He), neon (Ne), mercury vapor (Hg), etc. *For a light diatomic gas, C_V has the value $\tfrac{5}{2} R$, C_P the value $\tfrac{7}{2} R$.* This statement holds for gases such as hydrogen (H_2), nitrogen (N_2), carbon monoxide (CO), and oxygen (O_2). For heavy diatomic gases such as chlorine (Cl_2) and bromine (Br_2), the heat capacities at room temperature are slightly greater than those given by the above expressions, and for these gases, as well as for polyatomic gases, the heat capacities increase with increasing temperature. (*See* Sec. 7 for further discussion of these points.)

Heat capacities of gases are rather difficult to measure directly, because the mass of gas and hence the quantity of heat involved are necessarily small. Continuous-flow methods are applicable to the determination of the heat capacity at constant pressure. The ratio C_P/C_V may be accurately determined by indirect methods involving measurement of temperature change in an adiabatic expansion or compression (*see* Chapter 19) or the measurement of the speed of sound in the gas (*see* Chapter 21). This ratio alone is sufficient to determine both C_P and C_V if the gas is assumed ideal so that relation (23) is satisfied.

Example. *The air in a room of 150 m³ volume is heated from 0° C to 20° C when the atmospheric pressure out-of-doors is 1 atm. (a) Assuming normal venting to the atmosphere, determine the amount of heat that must be added. (b) Assuming that the air is at 1 atm initially and that the room is hermetically sealed, determine the heat that must be added. In your computations, ignore the fact that there will be heat transfer to and through the walls while the air is being warmed, so that, in fact, additional heat must be added to compensate this loss.*

Since air is almost entirely composed of the diatomic molecules N_2 and O_2, its molar specific heats are given by

$$C_V = \tfrac{5}{2} R = \tfrac{5}{2} (1.99 \text{ kcal/kmol} \cdot \text{K}) = 4.98 \text{ kcal/kmol} \cdot \text{K}$$
$$C_P = \tfrac{7}{2} R = \tfrac{7}{2} (1.99 \text{ kcal} \cdot \text{kmol} \cdot \text{K}) = 6.97 \text{ kcal/kmol} \cdot \text{K}$$

Since 1 kmol occupies 22.4 m³ at 1 atm and 0° C, the number of kilomoles of gas initially in the room is given by

$$n = \frac{150 \text{ m}^3}{22.4 \text{ m}^3} = 6.69 \text{ kmol}.$$

(b) If the room is hermetically sealed, the same amount of air will remain after the temperature is raised to 20° C. Since the volume of the air in the hermetically sealed room does not change, the heat that must be added is given by

$$Q = nC_V \Delta T = (6.69 \text{ kmol})(4.98 \text{ kcal/kmol} \cdot \text{K})(20 \text{ K}) = 666 \text{ kcal}.$$

(a) Normal venting to the atmosphere is actually only slight venting, through crevices and porous materials. A very slight 'overpressure' results as the air in the room is heated, and forces the air through these crevices to the outside, at the same time preventing entry of outside air. Part of the heat added goes into the external work required to push aside the external atmosphere. Since the expansion takes place at constant pressure we can write $dQ = nC_P dT$ as the heat that must be added to effect the temperature rise dT. In integrating this expression we must note that the number of kmol being heated decreases as the temperature rises; in fact, from Charles's law, we can readily see that the number of kmol in the room varies inversely as the absolute temperature, so that we can write $n = 6.69 \ (273°/T)$. With n of this form, we find

$$Q = \int_{273°}^{293°} nC_P \, dT = (6.69 \text{ kmol})(273 \text{ K})\left(6.97 \frac{\text{kcal}}{\text{kmol} \cdot \text{K}}\right) \log_e \frac{293}{273}.$$

We could evaluate the logarithm from the tables; it is somewhat more meaningful to use the rapidly converging series expansion, $\log_e (1+x) = x - \frac{1}{2} x^2 + \frac{1}{3} x^3 - \cdots$, which gives

$$\log_e \frac{293}{273} = \log_e \left(1 + \frac{20}{273}\right) = \frac{20}{273} - \frac{1}{2}\left(\frac{20}{273}\right)^2 + \cdots = \frac{20}{273}(0.963).$$

Substitution in the expression for Q indicates that the average number of moles being heated (averaged over temperature) is 0.963 times the initial number of moles, and gives

$$Q = 897 \text{ kcal}$$

as the heat that must be added when the room is normally vented.

7. The Kinetic Theory of Gases

During the latter half of the nineteenth century, it was realized by JAMES CLERK MAXWELL, LUDWIG BOLTZMANN, and others that the extreme simplicity of the experimental behavior of gases that we have sketched so far in this chapter implied an extreme simplicity in the structure of gases on a molecular scale. Only by making a very simple picture of an ideal gas can one expect to derive such very simple laws. This picture and the derivation of the experimental laws from it constitute the subject matter of the *kinetic theory of gases*.

In order to derive results that are consistent with the fact that the thermal energy of a gas depends on temperature alone, it is necessary to assume that the molecules of a gas are essentially *free particles,* that is, that most of the time they exert no forces on each other. If the molecules exerted forces on each other, the magnitude of the forces and hence the associated thermal potential energy would be expected to be distance- and hence volume-dependent, in contradiction to Joule's free-expansion experiment. The molecules are assumed to move at high speed, and collide with each other and with the walls of the vessel. During these collisions, large impulsive forces are assumed to act to change the direction of the molecular motion.

The following picture is convenient and accurate: The molecules of a monatomic gas behave as if they were hard rigid spheres with diameters of the order of 0.2 nm. At 0° C and 1 atm, there are 2.7×10^{19} of them per cm^3, and they are on the average about 15 diameters apart. They rush about madly in truly random directions at speeds of the order of 1 mi/s, making collisions each time they have gone about 500 diameters along a given path. The molecules of a monatomic gas have no energy of rotation—only energy of translation. The molecules of a diatomic gas behave like rigid dumbbells moving in the same frantic manner except that now rotational energy may be interchanged at each collision. Of course the molecules are not *really* hard spheres or dumbbells, but the intramolecular forces hold the electrons and nuclei so firmly in position that the molecules behave during collisions as if they were rigid bodies.

The pressure of the gas on the walls of the container arises from the collisions of the molecules with the walls. In Fig. 4, P represents the average force that a unit area of the wall exerts on the gas molecules, which is equal and opposite to the average force the gas molecules exert on a unit area of the wall (the gas pressure). In accordance with the discussion of Chapter 7, the pressure P equals the total change of the X-component of momentum of all molecules colliding with unit area of the wall in unit time. Since the number of collisions is enormous ($\sim 10^{14}$ per cm^2 per second), the force P appears as a steady pressure.

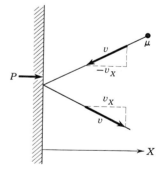

Figure 4

When an individual molecule of mass μ that has X-component of velocity $-v_X$ collides with the wall and rebounds with reversed X-component of velocity, as in Fig. 4, the X-component of momentum before the collision is $-\mu v_X$, after the collision it is $+\mu v_X$, so the change in the X-component of momentum is $2\mu v_X$. We have assumed the collision to be elastic, that is, that the molecule does not lose kinetic energy in the collision. This assumption will hold true on the average if the gas is in thermal equilibrium with the wall and hence is not gaining or losing thermal energy.

Now let us compute the pressure of a gas with N molecules, each of mass μ, per unit volume. The principal purpose of this computation is to show that *the average kinetic energy of the molecules is proportional to the temperature.* We shall also get a measure of the speed with which the molecules must be traveling and an expression for pressure in terms of the absolute temperature and the number of molecules per unit volume.

To take account of the different values of v_X (Fig. 4) that will occur, let us split the N molecules per unit volume up into groups according to their X-component of velocity at a particular time. Let us say that there are dN molecules per unit volume with X-component in a small velocity range dv_X in the vicinity of either $+v_X$ or $-v_X$, that is, $\tfrac{1}{2} dN$ near $+v_X$ and $\tfrac{1}{2} dN$ near $-v_X$. We shall compute the partial pressure of these molecules and then integrate to include molecules of all values of v_X. *This partial pressure will be given by the product of the number of collisions, per unit area of wall, per second, of these molecules by the change $2\mu v_X$ in X-component of momentum per collision.* How many of these molecules collide per second? In the volume $v_X \Delta t$ of Fig. 5 there will be $dN\,v_X\,\Delta t$ of these molecules, half of which will be moving to the left, half to the right. We can take the time Δt sufficiently small so that all the molecules in this volume that are moving to the left may be validly assumed to reach the wall and collide with it without colliding with any other molecule. Since in time Δt a molecule moves a distance $v_X\,\Delta t$ in the X-direction, we see that the half of the molecules contained in the sheet of thickness $v_X\,\Delta t$ at any given instant that are moving toward the wall strike the wall during the succeeding time interval Δt. Hence, the number of molecules striking unit area of the wall in time Δt will be $\tfrac{1}{2}\,dN\,v_X\,\Delta t$, so the number of molecules striking unit area *in unit time* will be $\tfrac{1}{2}\,dN\,v_X$.

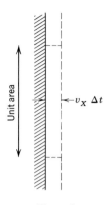

Figure 5

Consequently, if we refer to the sentence in italics above, we see that the partial pressure will be $\tfrac{1}{2}\,dN\,v_X \times 2\,\mu v_X = dN\,\mu v_X^2$. To get the total pressure of the N molecules per unit volume, we must integrate this expression over molecules of all velocities. Since this expression is $\mu \times$(number of molecules of a given X-velocity)$\times$(square of X-velocity), the sum will be $\mu \times$(total number of molecules)$\times$(*average* square of X-velocity). That is,

$$P = N\mu\,\overline{v_X^2}. \tag{25}$$

If v is the magnitude of the velocity of a molecule, $v^2 = v_X^2 + v_Y^2 + v_Z^2$. If we average over all the molecules, we obtain $\overline{v^2} = \overline{v_X^2} + \overline{v_Y^2} + \overline{v_Z^2}$. But since no direction is preferred over any other in a truly random motion, we must have $\overline{v_X^2} = \overline{v_y^2} = \overline{v_Z^2}$. Hence, $\overline{v^2} = 3\,\overline{v_X^2}$, or $\overline{v_X^2} = \tfrac{1}{3}\,\overline{v^2}$, and from (25), $P = \tfrac{1}{3}\,N\mu\overline{v^2}$. Noting that the average kinetic energy of a single molecule is $\tfrac{1}{2}\,\mu\overline{v^2}$, we may rewrite this relation in the form

$$P = \tfrac{2}{3}\,(\tfrac{1}{2}\,\mu\overline{v^2})\,N, \tag{26}$$

a relation that states that the pressure is equal to two-thirds of the total translational kinetic energy of the molecules in a unit volume. If we multiply both sides of (26) by the total volume of the gas, we obtain

$$PV = \tfrac{2}{3} (\tfrac{1}{2} \mu \overline{v^2}) NV,$$

where NV gives the total number of molecules in the sample. If there are n kmol of gas in the sample, we may set $NV = nN_A$, where N_A is Avogadro's constant, to obtain the relation

$$PV = \tfrac{2}{3} (\tfrac{1}{2} \mu \overline{v^2}) nN_A. \qquad (27)$$

So far, our discussion has involved only the principles of mechanics and applies to any gas, whether or not it is ideal. Comparison of (27) with the general gas law as stated in (9) shows that for an *ideal gas*,

$$\tfrac{2}{3} (\tfrac{1}{2} \mu \overline{v^2}) nN_A = nRT$$

or
$$\tfrac{1}{2} \mu \overline{v^2} = \tfrac{3}{2} (R/N_A) T = \tfrac{3}{2} kT, \qquad (28)$$

where the ratio of the gas constant R to Avogadro's constant is set equal to k, a universal constant known as *Boltzmann's constant*. Equation (28) states that *the average translational kinetic energy of a molecule in an ideal gas is directly proportional to the absolute temperature and is independent of the kind of gas*. The value of Boltzmann's constant is

$$k = \frac{R}{N_A} = \frac{8314 \text{ J/kmol} \cdot \text{K}}{6.0225 \times 10^{26}/\text{kmol}} = 1.3805 \times 10^{-23} \text{ J/K}. \qquad (29)$$

Since the average translational kinetic energy is independent of the kind of gas, the average square of the speed must vary inversely as the mass of the molecules, the molecules in a lighter gas moving faster than those in a heavier gas. Equation (28) gives

$$\overline{v^2} = 3kT/\mu \qquad (30)$$

as the average square of the speed. The square-root of $\overline{v^2}$ is called the *root-mean-square* speed.

Example. *Determine the root-mean-square speed of* H_2 *molecules of* $M = 2.017$ *at* $T = 0°$ C.

From (1), we find the mass of a hydrogen molecule as $\mu = 2.017$ u $= 2.017 \times 1.66 \times 10^{-27}$ kg $= 3.35 \times 10^{-27}$ kg. Hence from (30) and (29),

$$\overline{v^2} = \frac{3 \times (1.38 \times 10^{-23} \text{ J/K})(273 \text{ K})}{3.35 \times 10^{-27} \text{ kg}} = 3\,380\,000 \frac{\text{m}^2}{\text{s}^2}.$$

The root-mean-square speed is thus $(\overline{v^2})^{1/2} = 1840$ m/s,

or 1.14 mi/s. For higher temperatures this value would be greater, for heavier molecules it would be less, in accordance with (30).

Substitution of (28) in (26) gives the equations

$$P = NkT, \qquad N = P/kT, \qquad (31)$$

the latter of which gives the number of molecules per unit volume as a function of pressure and temperature; in accordance with Avogadro's law this number does not depend on the kind of gas involved.

From (28) we can obtain an expression for the total translational kinetic energy E_{KT} of a kilomole of gas by multiplying by Avogadro's constant:

$$E_{KT} = \tfrac{3}{2} RT.$$

From this equation we can obtain an expression for the heat capacity per kilomole of a monatomic gas at constant volume by recalling that the addition of heat results merely in an increased thermal energy—in the case of a monatomic gas, translational kinetic energy. Therefore,

$$C_V = dE_{KT}/dT = \tfrac{3}{2} R, \tag{32}$$

in accord with the result stated on p. 345.

Purely classical (nonquantum) mechanics results in a *principle of equipartition of energy* that states that the energy per 'degree of freedom' is $\tfrac{1}{2} RT$. The monatomic molecule has three 'degrees of freedom,' corresponding to the three terms in the expression for its kinetic energy, $\tfrac{1}{2}\mu v_X^2 + \tfrac{1}{2}\mu v_Y^2 + \tfrac{1}{2}\mu v_Z^2$, and hence has energy $\tfrac{3}{2} RT$. A diatomic molecule, on our 'dumbbell' picture, has two more degrees of freedom, corresponding to rotations about the two axes perpendicular to the axis of the dumbbell (the rotational inertia about the dumbbell axis itself is assumed negligible) and hence would have energy of $\tfrac{5}{2} RT$ per kilomole, in agreement with the experimental results stated on p. 345 for light diatomic molecules. The degrees of freedom associated with energy of vibration of the atoms relative to each other results in higher values for heavy diatomic molecules and for polyatomic molecules. The details of the computation of energy and hence specific heat are accurately determined by quantum mechanics, but not by classical mechanics except in the simple situations that we have described.

We should add one refinement to the above discussion. Many gases such as air contain more than one type of molecule; in fact, even simple gases contain molecules of different masses because of differences in nuclidic mass. For such a mixture of molecules, the discussion that led to (26) will give the partial pressure P_a of a given type of molecule, of which there are N_a per unit volume. Since each type acts independently, (28) shows that each type has the same average translational kinetic energy. Hence the relation (31) remains valid, where P is the total pressure and N the total number of molecules per unit volume; and (32) also remains valid.

The above simple picture of a gas, which leads to the general gas law and the correct specific heat, can be extended to explain correctly the observed diffusion, viscosity, heat conduction, and other properties of ideal gases. Modifications of this picture that take into account the small forces that exist between gas molecules when they are close together account in detail for the departures from the general gas law that will be discussed in the following chapter. Hence we feel that we know in complete detail how gas molecules behave. Gases are so much simpler than liquids and solids that it is only in recent years that we have begun to acquire a knowledge of the kinetic behavior of the molecules in liquids or solids that is comparable to the knowledge of the kinetic theory of gases that was acquired almost a century earlier.

We add one more very significant point, namely, that the general gas law, plus knowledge of where departures from it can be expected, extends the range of accurate thermometry far beyond the range between the freezing and boiling points of mercury. By measuring the pressure $P = NkT$ of a quantity of helium held at constant volume, so that N is constant, temperatures can, in principle, be determined accurately from close to the liquefaction point of He ($-269°$ C) to the many-thousand degree temperature at which helium begins to become self-luminous.

8. Pressure in the Interior of a Gas

The kinetic theory gives us a clear picture of the meaning of pressure at a point in the interior of a gas, in basic terms involving momentum transfer. Our earlier definition of pressure as the force that *would be exerted* on a solid surface if one were placed in the gas is unsatisfactory when we want to talk about forces acting on imaginary surfaces in the gas. Our assertion that pressure is the force per unit area exerted by the molecules on one side of a plane on those on the other is inconsistent with the kinetic picture which shows molecules by and large passing *through* the plane.

Consider, in Fig. 6, a right cylinder cut out of a gas and bounded on the left by unit area of a wall, as in Fig. 4, and bounded on the right by an imaginary plane in the gas. As we have seen, the wall is furnishing momentum P, per unit time, to the gas in this cylinder. The momentum added by the wall is to the right. However, even though the particular molecules that inhabit this cylinder change with time, the state of the gas within this cylinder does not change with time, and, in particular, the gas in this cylinder is *not* acquiring an increased momentum to the right. Hence, in addition to the momentum furnished by the wall, this gas must be acquiring an equal and opposite momentum to the left; this momentum to the left is acquired through the right-hand (imaginary) face of the cylinder by actual molecular transfer.

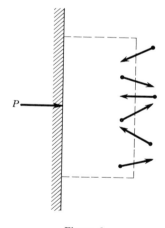

Figure 6

By arguments similar to those used in the preceding section, we see that the molecules passing through this right face *into* the cylinder bring in momentum to the left equal to $\frac{1}{2} P$, per unit time; the molecules passing *out* of the cylinder take out momentum $\frac{1}{2} P$ to the right, which is equivalent to adding $\frac{1}{2} P$ to the left. Thus, through the right-hand surface of the cylinder, momentum equal to *P*, per unit time, to the left is transferred into the cylinder, and this momentum balances the momentum to the right that is furnished by the rigid wall. This *rate of transfer of momentum from right to left through the imaginary plane* is the *physical* quantity that is called the pressure exerted *by the gas to the right* of the plane *on the gas to the left*. Correspondingly, the momentum *leaving* the cylinder is equal and opposite; it is to the right and is called the pressure exerted by the gas to the left on that to the right.

Thus, we see that when we balance 'forces' acting on imaginary planes in a gas, what we are actually balancing are momentum transfer rates. The same remark applies to any fluid, and, in fact, to solids as well.

9. The Maxwellian Distribution; Brownian Motion; Avogadro's Constant

Because of the random way in which thermal energy is imparted to them, and the random way in which they exchange energy with each other and with the walls of the vessel, the molecules of a gas do not all have the same speed. By treating the mechanical collision processes in statistical fashion, Maxwell was able to show that the molecules do, however, have a definite probability distribution of speeds, called the *Maxwellian distribution,* which is given by the curve of Fig. 7. The shape of this curve is given by a fairly complex equation involving exponentials.

There has been ample experimental check that the molecules of a gas do have this Maxwellian distribution of speeds. One type of check

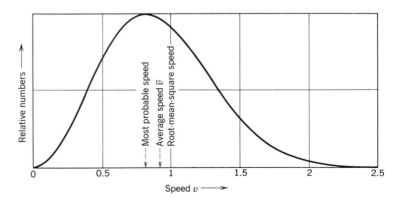

Fig. 7. The Maxwellian distribution. The relative numbers of molecules having speeds in unit speed range at various values of speed are proportional to the ordinates of this curve. The unit of speed is the root-mean-square speed given by (30) as $\sqrt{3\,kT/\mu}$.

uses rotating slits to sort out the molecules of different speeds in a stream of gas that proceeds through a small hole into a vacuum. After such velocity sorting, a gas such as heated mercury vapor can be made to condense on a cold surface so that the relative numbers of molecules in different speed ranges can be measured. Another check involves direct observation of Brownian motion, which we shall describe presently.

In the Maxwellian distribution of Fig. 7, the average speed turns out to be $\sqrt{8/3\pi} = 0.921$ times the root-mean-square speed. So, for H_2 at 0° C, the average speed $\bar{v}$ is 0.921×1840 m/s $= 1690$ m/s $= 1.05$ mi/s. The molecules are distributed over a wide range of speeds, as indicated in Fig. 7. However, only 1 in 10 000 of the molecules has a speed over $3\bar{v}$, but there is a definite but very small probability that a molecule may have a very much greater speed.

An interesting application of the kinetic theory occurs in connection with *Brownian motion*. In 1827 an English botanist, ROBERT BROWN, observed that very fine particles of dust, smoke, or pollen contained in air (or in a liquid) seem to execute very irregular motions when observed with a high-power microscope. The correct explanation of these random motions, which almost make the particles seem to be alive as they perpetually dart hither and thither, was given three quarters of a century later by Einstein. The motion is explained by the kinetic theory of gases. The particle of dust behaves as if it were just a very large molecule. In air it has the same average translational kinetic energy as the gas molecules. It is repeatedly interchanging energy with the air molecules that collide with it, and in the course of time has velocities ranging over the whole of the Maxwellian distribution with relative probabilities given by the curve of Fig. 7. But unlike the air molecules, the particle is large enough to be watched with a microscope, and its average velocity is comparatively slow because of the very large value of its mass μ, which is to be inserted in (30) to obtain the average square of its speed. Hence, details of the kinetic-theory predictions are subject to direct experimental check by observation of the Brownian motion of small particles, and have been accurately confirmed in a number of careful experimental investigations, particularly by JEAN PERRIN. *Such detailed observations of Brownian motion furnish the most direct and most convincing check of the validity of the kinetic theory.*

By prolonged observation of the Brownian motion of particles whose masses can be determined by direct means, the mean kinetic energy of the particles can be determined. Knowing the mean kinetic energy at a given absolute temperature T, we can then use (28) to get a value of Boltzmann's constant k. The value obtained from such experiments is in agreement with that given in (29). Having now determined both k and R, we are in a position to compute Avogadro's constant $N_A = R/k$. Knowledge of N_A enables us to compute the value of the mass of an individual molecule by dividing the mass of one kilomole by Avogadro's constant. We can then proceed to determine the size of the nuclidic mass unit u, as given by (1).

This is only one of a number of ways of determining Avogadro's constant. Optical and electrical methods that give higher precision will be discussed in later chapters.

PROBLEMS

NOTE: *Assume that atmospheric pressure is* 1 atm *unless otherwise specified. Pressures in* cm *refer to* cm *of mercury unless otherwise specified. Use the values* 273°K~0°C; 460°R~0°F, *and* $N_A = 6.02 \times 10^{26}$/kmol. *The tables of conversion factors in the Appendix may prove useful.*

1. Air at a gauge pressure of 29.4 p/i² fills an automobile tire having a total volume of 120 i³. What volume would this air occupy if it were released at standard atmospheric pressure without change in temperature?
Ans: 360 i³.

2. Show that the general gas law (7), when stated in terms of densities, assumes the following form: $\rho_1 T_1/P_1 = \rho_0 T_0/P_0$.

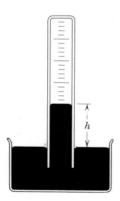

Problem 3

3. The measured volume of a quantity of hydrogen collected over mercury as in the figure is 600 cm³ when the temperature is 18°C and the mercury stands $h = 5$ cm above the reservoir. The barometer reads 72 cm. What is the volume of H₂ at 0°C and 1 atm, and what is its mass?   Ans: 496 cm³; 0.0447 g.

4. A diving bell contains 100 f³ of air at 1-atm pressure when it is at the surface of the water. After it has been lowered to a depth of 200 f, what are the pressure and volume of the air in it? What volume of additional air, measured at 1-atm pressure, must be pumped in to again fill the bell? Neglect temperature change and water-vapor content.

5. A carelessly made barometer has some air trapped above the mercury. The mercury stands at 74.2 cm when the air space is 8 cm long. When the barometer tube is pushed down into the reservoir so that the air space is only 4 cm long, the mercury stands at 72.2 cm. What is the true barometric reading?
Ans: 76.2 cm.

6. Because there is some air trapped in a barometer tube, the mercury stands at 72 cm, with a 10-cm air space above it, when the atmospheric pressure is 76 cm. What is the actual atmospheric pressure when our faulty barometer stands at 68 cm?

7. At what depth below the surface of the ocean would the density of the air in a bubble at 4°C be 50 times its density at 4°C and 1 atm? Neglect the change in density of the water with depth.   Ans: 500 m.

8. In his experimental work, Charles determined the volume expansion coefficient at constant pressure. If V_0 is the volume of a gas sample at 0°C, the volume at T_c°C was found to be $V_1 = V_0(1 + \beta_0 T_c°)$, where the expansion coefficient was measured as $\beta_0 = 0.003\ 660$ per C deg. Show that these results can be stated in the form given in (6).

9. A sample of carbon dioxide at 27° C occupies a volume of 3 m³ at a pressure of exactly 2 atm. How many kmol of gas does the sample contain? What is the total mass of the sample?
Ans: 0.244; 10.7 kg.

10. What is the density of oxygen gas, O_2, at 0° C and 1 atm? What is the density of oxygen at 100° C and 8 atm?

11. Air at 20° C and at a pressure of 76 cm of mercury is admitted to a steel storage cylinder at 20° C. The cylinder is then sealed. What is the pressure of the air in the cylinder when it is placed in a steam bath at 100° C? Does the expansion of the cylinder significantly affect the answer, to three-figure accuracy?
Ans: 96.8 cm.

12. A schematic diagram of a constant-volume gas thermometer is shown in the figure. The volume of gas in the bulb is kept constant by raising or lowering the mercury reservoir so that the mercury in the right-hand tube always stands at the etched mark. In an experiment, the mercury level in the reservoir is 7.8 cm below the etched mark when the bulb is in melting ice; it is 17.2 cm above the etched mark when the bulb is in steam from boiling water. The barometer reads 76.0 cm. From these data and the definitions of the fixed points of the Celsius scale, determine the value of the absolute zero of temperature on the Celsius scale.

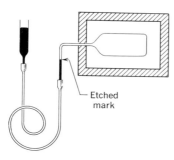

Problems 12–13

13. Hydrogen behaves like a perfect gas down to liquid-air temperatures. If a constant-volume hydrogen thermometer (*see* figure) has mercury level in the reservoir 28.5 cm above the etched mark when the temperature of the bulb is 0° C, 41.4 cm below the etched mark when the bulb is immersed in boiling liquid oxygen, and 46.4 cm below the etched mark when the bulb is immersed in boiling liquid nitrogen, and if the barometer reads 76.0 cm, determine the boiling points of oxygen and nitrogen. Ans: −183° C; −196° C.

14. In order to combine Charles's law and Boyle's law to get (7), consider a gas sample with an initial state P_1, V_1, T_1 and a final state P_2, V_2, T_2, as indicated in the figure. By considering first a change from the initial state to an intermediate state P_2, V', T_1 and then a change from the intermediate state to the final state, show that $P_1 V_1/T_1 = P_2 V_2/T_2$. Note that the curves in the figure, which are called *isothermals*, are rectangular hyperbolas.

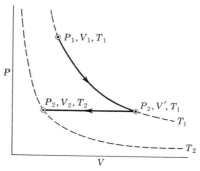

Problem 14

15. What is the density of air on the top of a mountain where the temperature is −13° C and the pressure is 0.9 atm?
Ans: 1.21 kg/m³.

16. The airtight bag of a radiosonde balloon is partially inflated with 50 m³ of He at the ground at 15° C and 1 atm pressure. What is the maximum load (mass of bag plus suspended gear) that can be lifted?

17. As the balloon of Prob. 16 rises, the helium expands to fill the bag more and more completely. If the bag is not yet filled out taut at a height of 10 km where the pressure is 20 cm and the temperature is −55° C, what is the maximum mass that can be supported at this altitude?
Ans: 53 kg.

18. A vessel contains 20.2 g of neon (Ne) and 4.034 g of hydrogen (H_2) under a total pressure of 100 cm. Find the partial pressures of the neon and the hydrogen and the volume of the vessel.

19. A vessel contains 16.03 g of methane (CH_4)

and 66.03 g of carbon dioxide (CO_2). The total pressure is 100 cm of Hg, the temperature 100° C. Find the partial pressures of the methane and of the carbon dioxide and the volume of the vessel.

Ans: 40 cm; 60 cm; 58 200 cm³.

20. A bottle is corked when it is partly filled with water at 0° C, the balance of the bottle containing air at 1 atm pressure. The bottle is then placed in a freezer and the water frozen at 0° C. The cork will blow if the gauge pressure exceeds $\frac{1}{8}$ atm. What is the maximum fraction of the volume that can be filled with water without the cork blowing? (Ignore the negligible change in volume of the bottle or of the ice resulting from increase in pressure.)

21. Two glass bulbs, of volumes 250 cm³ and 100 cm³, are connected by a capillary tube. This container is sealed when it contains dry air at 15° C and 76 cm. The larger bulb is then immersed in steam at 100° C, the smaller in ice water at 0° C. Neglecting thermal expansion of the glass, find the resulting pressure.

Ans: 88.4 cm.

22. An insulated flask containing 250 cm³ of hydrogen gas at 3 atm and 20° C is connected to one containing 500 cm³ of argon gas at 1 atm and 20° C and the two gases are allowed to mix. Show that the temperature remains unchanged. What is the resulting pressure?

23. A compressed air tank has a volume of 2 m³ and is filled with air at a pressure of 2 atm. If the temperature is 27° C, what is the mass of the air enclosed in the tank?

Ans: 4.68 kg.

24. The volume of an oxygen storage tank is 0.3 m³. When the temperature is 50° F, the pressure gauge reads 194 p/i². How many kilomoles of oxygen are in the tank? What is the density of the enclosed oxygen?

25. Two identical compressed-gas tanks are placed on the two platforms of a large beam balance. One tank is 'empty' and open to the atmosphere. The second tank is evacuated and then filled with hydrogen until the two tanks exactly balance. What is the pressure of the hydrogen? Ans: 14.5 atm.

26. A man desires to collect 6 kg of air in a tank having a volume of 0.5 m³. At what gauge pressure should he stop his compressor if the temperature of the compressed air in the tank is 30° C?

27. How much heat must be added to 3 kg of O_2 gas to raise its temperature from $-100°$ C to 25° C at constant pressure? What is the increase in thermal energy? How much external work is done? How much heat would be needed to accomplish the same temperature rise at constant volume?

Ans: 81.6, 58.2, 23.4, 58.2 kcal.

28. How much heat must be added to 3 kg of air to raise its temperature from 0° C to 100° C at constant pressure? What is the increase of thermal energy? How much external work is done by the gas? How much heat would be needed to accomplish the same temperature rise at constant volume?

29. One mole of neon gas at 0° C and 1 atm is placed in a cylinder equipped with a movable piston as in Fig. 1. If the external pressure is exactly 1 atm, to what temperature must the gas be raised in order for it to expand to twice its initial volume? What is the increase in the thermal energy of the gas? How much external work does the gas do in expanding? How much heat must be supplied to the gas?

Ans: 273° C; 3400 J; 2270 J; 1360 cal.

30. As stated in the example on pp. 345–346, show that the number of kmol of air in the ventilated room (fixed volume, fixed pressure) varies inversely as the absolute temperature.

31. Whenever a monatomic gas expands at *constant pressure,* what percentage of the heat supplied goes into increasing the thermal energy of the gas? What percentage of the heat supplied is used in doing the work involved in expansion? Answer the same questions for the case of expansion at *constant temperature.*

Ans: 60%, 40%; 0%, 100%.

32. Answer all questions in Prob. 31 for a light diatomic gas. Generalize your results to apply to *any* gas by writing them in terms of $\gamma = C_P/C_V$.

33. One mole of argon gas (Ar) at 0° C and 1 atm is heated at constant pressure until its volume is doubled, then heated at constant volume until its pressure is doubled. Calculate the heat added to the gas, the work done by the gas, and the increase in thermal energy in each stage of this process, and in the overall process. **Ans: 5670, 2270, 3400 J; 6800, 0, 6800 J; 12 500, 2270, 10 200 J.**

34. Repeat the preceding problem for the case in which the gas is *first* heated at *constant volume*, then at constant pressure. Compare your answers in detail with those in the preceding problem and give a physical explanation of the result of each comparison.

35. An argon (Ar) sample at 20° C and 1 atm is placed in a closed steel container. If the container and gas are heated to 100° C, what is the increase in pressure inside the container? If a similar container were completely filled with mercury at 20° C and sealed, what would be the increase in pressure inside the container if the temperature were raised to 100° C? (Assume that the container expands only because of the temperature increase.)
Ans: 0.27 atm; 3010 atm.

36. What volume will be occupied by 10 g of methane gas (CH_4) at a pressure of 2 atm when the temperature is 27° C? How much work would be done in compressing this gas to one-half its initial volume without change in pressure? How much heat must be removed from the gas during compression? ($\gamma = C_P/C_V = 1.31$.)

37. Prove that the external work done by a gas in expanding from pressure P_0 and volume V_0 to pressure P_1 and volume V_1 *isothermally* (at constant temperature) is

$$W = P_0 V_0 \log_e(V_1/V_0) = nRT \log_e(V_1/V_0).$$

How much heat must be added to the gas during this expansion? **Ans: W.**

38. Return to the problem of the ventilated room on pp. 345–346. Compute the heat that must be added to raise the temperature from −30° C to +30° C. What is the average number of kmol of air in the room, averaged with respect to temperature? What percentage error would be made in computing the heat added by assuming this average number to be the mean of the initial and final numbers? Explain the reason for this error.

39. Show that one would compute the correct pressure of a gas on the walls of a cubical box whose edges are oriented in the X-, Y-, and Z-directions by assuming that one-third of the molecules are moving back and forth parallel to the X-axis, one-third parallel to the Y-axis, and one-third parallel to the Z-axis, all molecules having speed $\sqrt{\overline{v^2}} = \sqrt{3kT/\mu}$ as obtained from (30).

40. Xenon is a monatomic rare gas. Compute its specific heat capacity (per kg) at constant pressure and at constant volume.

41. Compute the specific heat capacity (per kg) of N_2 gas at constant volume; at constant pressure **Ans: 0.177, 0.248 kcal/kg·C deg.**

42. Calculate the average translational kinetic energy of a molecule of benzene vapor (C_6H_6) at a temperature of 27° C; at a temperature of 127° C.

43. Calculate the average translational kinetic energy of an argon atom at 25° C. At what temperature does the average energy have half this value? **Ans: 6.18×10^{-21} J; −124° C.**

44. Calculate the root-mean-square speed and the average speed of benzene-vapor molecules (C_6H_6) at 127° C.

45. Calculate the root-mean-square speed and the average speed of neon atoms of atomic mass 20.2 at 25° C. **Ans: 607, 560 m/s.**

46. Calculate the root-mean-square speed of oil droplets of diameter 2 μm and density 0.9 g/cm³ in their Brownian motion in air at 0° C and 1 atm.

47. Calculate the root-mean-square speed of smoke particles of mass 5×10^{-17} kg in their Brownian motion in air at 0° C and 1 atm.
Ans: 1.50 cm/s.

48. Show that equation (31), $P = NkT$, gives all the information contained in the general gas law (9).

49. Calculate the total kinetic energy of translation of all the molecules in 2000 cm³ of an ideal gas at 25° C and 2 atm pressure.
Ans: 608 J or 145 cal.

50. How many molecules per cm³ does a gas contain at 1 atm and 20° C? The best 'vacuum' yet obtained in the laboratory is 10^{-9} mm of mercury. How many molecules are there in each cm³ of this 'vacuum' at 20° C?

51. Use (31) to find the number of molecules in a cubic meter of any gas at 0° C and 1 atm.
Ans: 2.69×10^{25}.

52. If one makes the crude assumption that

the temperature of the air does not vary as one ascends vertically, the variation of pressure with altitude h is given by the equation

$$P = P_0 \, e^{-g\mu h/kT},$$

where P_0 is the pressure at sea level where $h=0$, μ is the average mass of a molecule of the air, and T and g are assumed constant. Verify this equation by showing that the decrease in pressure $-dP$ when h is increased by dh equals the weight $\rho g \, dh$ of a unit column of air of thickness dh. By using (31) find the relation between N at height h and N_0 at sea level.

53. On p. 204, we computed the *velocity of escape* from the *earth* as 36 800 f/s = 11 200 m/s. Discuss the escape of gas molecules from the top of the atmosphere assumed, as recent evidence indicates, to be at about 0° C. Show that the rate of escape of H_2 will be much faster than for N_2 or O_2. It is believed that the absence of hydrogen in the atmosphere is a result of the ease with which H_2 can escape. The *sun* has a great deal of hydrogen. Compute the velocity of escape from the sun and compare with the speed of H atoms at the solar surface temperature of 6000° K. Why is it impossible for the *moon* to have an atmosphere?

SOLIDS, LIQUIDS, AND GASES

18

In this chapter we shall discuss the three *phases* of matter—solid, liquid, and gaseous—and the conditions that govern transitions from one to the other. We shall confine our attention to the case of pure substances, such as H_2O or CO_2, or of pure substances in the presence of a foreign gas, such as water vapor in air. We shall further restrict our attention to substances that are *crystalline* in the solid phase, in which case there is a definite melting temperature at which a sharp transition from solid to liquid phase occurs. This restriction eliminates from consideration waxlike and glasslike substances in which the solid is not crystalline but *amorphous,* with the molecules arranged in random positions as in a liquid rather than in a regular array as in a crystal. Such amorphous substances do not make a sharp transition from solid to liquid at a single definite temperature. Instead, they make a gradual transition over a range of temperatures, with the solid first softening as does candlewax.

1. Fusion; Freezing

In a crystalline solid, the atoms are arranged in a regular array. For example, in common salt (NaCl), the Na and Cl atoms (actually Na^+

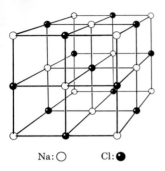

Na: ○   Cl: ●

Fig. 1. Structure of NaCl.

Fig. 2. Photographs of snow crystals.

and Cl⁻ ions) are arranged at the corners of cubes as in Fig. 1. As a result of this cubic atomic arrangement, most of the individual crystals of table salt are perfect cubes, as can be readily observed with a little magnification. In ice, the molecules are arranged in a pattern with hexagonal symmetry, giving rise to the hexagonal patterns of snow crystals in Fig. 2.

As we have already discussed in Chapter 11, the atoms in a crystal are held together by electric forces which act exactly like a collection of little springs (*see* Fig. 15, p. 234). The atoms have definite *equilibrium positions* but are capable of *vibrating* about these positions. The thermal energy of a solid consists principally of the energy of such vibrations. The average energy and amplitude of vibration increase with the increase in temperature as heat is added to a solid until, at a certain definite temperature (0° C for ice), molecules at the surface of the solid that happen to have unusually high energy move completely away from their equilibrium positions and enter the *liquid phase. The molecules of the liquid have a greater energy than the molecules of the solid at the same temperature* (*the melting temperature*).

The temperature of a solid cannot be raised above the temperature of fusion (melting). After the solid has reached this temperature, all added heat goes into melting more solid, the added heat supplying the energy difference between the liquid and the solid. The temperature will not again commence to rise until *all* of the solid has been melted. There is thus only one temperature, the temperature of fusion, at which solid and liquid can coexist in thermal equilibrium. Above this temperature the substance is necessarily all liquid; below this temperature it is necessarily all solid.* When solid and liquid coexist at the fusion temperature and no heat is added, there is equilibrium between the number of particularly energetic molecules that leave their positions on the surface of the solid and pass into the liquid state and the number of particularly slow molecules that recrystallize from the liquid onto the solid. If heat is added, more molecules pass into the liquid; if heat is taken away, more molecules pass from liquid to solid.

The fusion temperature depends slightly on pressure; for example, the fusion temperature of water *decreases* 0.007 50 C deg for each atmosphere increase in pressure. Thus we have not a definite fusion temperature but a *fusion curve* on a *PT*-diagram, such as that shown for water in Fig. 3. It is seen that we can make ice melt not only by increasing the temperature but also by increasing the pressure. Because ice shrinks upon melting, it seems reasonable that increased pressure should tend to make it melt, since this results in a smaller volume. In fact, all substances that

---

*More accurately, below the fusion temperature the substance is *all* solid if there is *any* solid. It is possible to *supercool* a liquid below the fusion temperature.

shrink on melting (the only known pure substances are water, gallium, and bismuth) have fusion curves that slant to the left like that of Fig. 3, so that increased pressure tends to make them melt. On the other hand, most substances expand on melting and have fusion curves that slant to the right, so that increased pressure tends to make them freeze.

The extremely low friction encountered in skating can be attributed to this tendency for ice to melt under pressure. The sharp blades of the skates make contact only over a very small area, where the pressure is truly enormous (*see* Prob. 2).

The latent heat of fusion is the heat that must be added to change unit mass of the substance from solid to liquid *with no change in pressure or temperature*, that is, *at a particular point on the fusion curve*. The value of the latent heat of fusion varies slightly from point to point on the fusion curve.

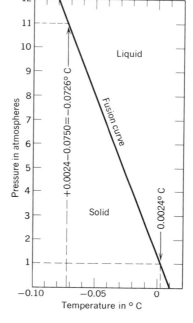

Fig. 3. Fusion curve for pure air-free H_2O. The melting point of pure *air-free* water is not 0° C, but 0.0024° C, at 1 atm. It is the melting point of ice in contact with water saturated with air by being open to the atmosphere that is exactly 0° on the Celsius scale at 1 atm pressure.

2. Vaporization; Condensation

Consider the vessel of Fig. 4, which contains only H_2O, liquid and vapor. The vessel is at a definite temperature T and we are interested in the factors that determine the absolute pressure P.

A molecule in the interior of the liquid experiences, on the average, no resultant force from the other liquid molecules, since it has, on the average, equal numbers of molecules on all sides. It will have an average translational kinetic energy that depends on the temperature and which within wide limits is independent of the pressure in the liquid. But a molecule that approaches the surface of the liquid experiences, as indicated in Fig. 5, a large resultant force directed back into the liquid. The existence of this force is manifested in the phenomenon of *surface tension*, which we shall discuss in Sec. 8. The surface molecules are continually being pulled back into the liquid and new ones are coming out to take their places. One can think of a molecule that approaches the surface from inside as 'hitting' the surface and 'bouncing' back, because as soon as it gets to within a few molecular layers of the surface it begins to experience a force tending to retard its outward motion and to pull it back into the liquid.

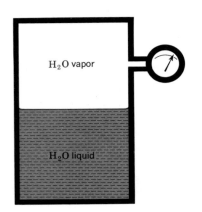
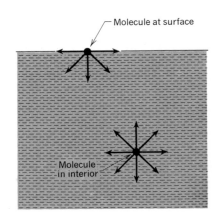

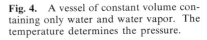

Fig. 4. A vessel of constant volume containing only water and water vapor. The temperature determines the pressure.

Fig. 5. Forces from neighboring molecules that act on a molecule at the surface and on a molecule in the interior of a liquid.

But the velocity distribution among the molecules of the liquid is such that there are some that approach the surface so fast that in spite of the retarding forces they escape through the surface and enter the vapor. These molecules 'evaporate' or 'vaporize.' *The number of molecules that have sufficient kinetic energy to evaporate (per second per unit area of liquid surface) increases rapidly as the temperature of the liquid is raised and depends only on the temperature of the liquid.*

On the other hand, *every vapor molecule that strikes the liquid surface enters the liquid,* since it immediately experiences large forces pulling it into the liquid. *At any given vapor temperature, the number of molecules per second striking unit area of a surface is proportional to the vapor pressure.* This statement is true whether or not the vapor behaves like an ideal gas.

If we start with liquid in the vessel of Fig. 4 and a vacuum in the space above it, the liquid will start evaporating at a rate determined entirely by the liquid temperature. The vapor pressure will increase until there are as many molecules in the vapor striking the liquid surface and condensing as there are molecules evaporating from the surface of the liquid. When this condition is reached, there is no further increase in vapor pressure. The rate of evaporation depends only on the temperature; the rate of condensation is proportional to the vapor pressure; since these two rates must be equal, we see that, *in Fig. 4, the vapor pressure is determined by the temperature.*

The vapor pressure as a function of temperature is readily measured by an apparatus similar to that shown in Fig. 4. In the case of water this pressure is given by the curve of Fig. 6 and the data of Table 18–1.

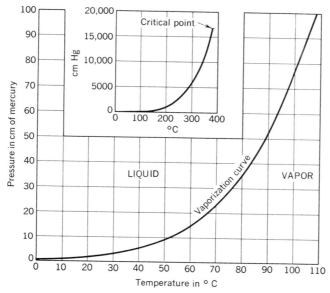

Fig. 6. Vaporization curve for water. This curve gives the conditions under which liquid and vapor can coexist, and hence gives the pressure of vapor saturated as in Fig. 4. No vapor can exist (in stable fashion) to the left of this curve; no liquid to the right. Vapor to the right of this curve is called *superheated steam*. This is also the boiling-point curve.

Now let us consider the cylinder of Fig. 7, which contains only H_2O. Let us assume that we maintain the pressure constant, say at ½ atm (38 cm of Hg) absolute. If we start with ice at a temperature below 0° C and add heat, H_2O will remain solid until we reach $+0.0061°$ C (Fig. 3). At this temperature the ice will change to liquid water as we add sufficient heat to cause fusion. After all the ice has melted, the temperature of the liquid will rise as we continue to add heat. Not until we reach

Table 18-1 Pressure of Saturated Water Vapor

Temp. (° C)	Pressure (mm Hg)	Temp. (° C)	Pressure (mm Hg)	Temp. (° C)	Pressure (atm)
−60	0.008	12	10.52	100	1
−40	0.097	14	11.99	110	1.414
−20	0.776	16	13.63	120	1.959
−10	1.950	18	15.48	140	3.566
0.01	4.579	20	17.53	150	4.697
2	5.29	25	23.76	200	15.35
4	6.10	30	31.82	250	39.26
6	7.01	50	92.52	300	84.79
8	8.04	70	233.7	350	163.2
10	9.21	90	525.9	374.15	218.4

the temperature of 81.6° C given by the curve of Fig. 6 will any vapor be formed. For, suppose that at some lower temperature, say 72° C, an incipient bubble of vapor were formed. The vapor in this bubble would have to be at a pressure of ½ atm if it were to be formed at all. But water at 72° has a surface rate of evaporation which equals the rate of condensation of vapor at a pressure of ⅓ atm and is much less than the rate of condensation of vapor at ½ atm. So the vapor in the bubble would condense much faster than the liquid would evaporate into the vapor bubble, and the incipient bubble would collapse immediately.

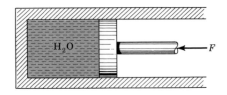

Fig. 7. A cylinder *filled* with H_2O (or other pure substance).

Not until we reach a temperature of 81.6° C will the liquid evaporate from a surface fast enough to maintain a bubble of vapor at ½ atm pressure. At this temperature liquid and vapor can coexist in equilibrium. If we add heat to the liquid so that the liquid exceeds this temperature ever so slightly, the liquid will evaporate faster than the vapor (at ½ atm) condenses, and the vapor bubbles will grow. The liquid will be topped by a layer of vapor. The rate of change of liquid to vapor will be governed purely by the rate at which heat is added since the latent heat of vaporization must be furnished by the added heat. Since it is only the most energetic liquid molecules than can evaporate, every molecule that evaporates in excess of those that condense results in a decrease in the average energy per liquid molecule which must be made up by the addition of heat if the temperature of the liquid is not to fall.

If the *pressure* exerted by the piston in Fig. 7 is kept constant, no temperature rise above 81.6° C will take place until *all* the liquid has evaporated. During this evaporation process, the volume of course increases enormously and the piston of Fig. 7 must be imagined to move about 200 feet to the right on the scale of this figure.

After all the liquid has evaporated, further addition of heat will result in further increase in temperature and volume of the vapor. At 81.6° C, this vapor is said to be *saturated steam;* above 81.6° C it is said to be *superheated steam.*

We now see that we could make the same arguments at any pressure, and that a cylinder containing H_2O under pressure and temperature conditions falling to the left of the curve of Fig. 6 will contain only liquid. To the right of this curve it will contain only vapor. Only for values of P and T falling *on* the vaporization curve can liquid and vapor coexist; at these values of P and T the vapor is said to be *saturated*.

Of the latent heat of vaporization, most goes into the increased thermal energy of the vapor, but an appreciable amount goes into mechanical work in pushing back the piston to increase the volume, as we can show by a sample computation:

Example. *At a pressure of* 1 atm $(1.013 \times 10^5 \text{ N/m}^2)$ *and a temperature of* 100° C, *the latent heat of vaporization of water is* $L = 539.2$ kcal/kg; *the vapor has specific volume of* 1.673 m³/kg; *the liquid has specific volume of* 0.001 m³/kg. *Determine the percentage of the latent heat that goes into performance of external work and the percentage that goes into increase of thermal energy.*

The change in volume per kg is $\Delta V = 1.672$ m³/kg, and the external work done per kg is, by (20), p. 343,

$$P \Delta V = (1.013 \times 10^5 \text{ N/m}^2)(1.672 \text{ m}^3/\text{kg})$$
$$= 1.694 \times 10^5 \text{ J/kg} = 40.5 \text{ kcal/kg}.$$

Since the latent heat L is the total heat that must be added to 1 kg of water to change it to vapor at the same pressure, we conclude from the basic relation (16), p. 341, governing the energy balance, the $L = \Delta U + P \Delta V$, where ΔU is the increase in thermal energy per kilogram. Hence

$$\Delta U = L - P \Delta V = (539.2 - 40.5) \text{ kcal/kg} = 498.7 \text{ kcal/kg}.$$

Of the latent heat, the fraction 40.5/539.2, or 7.5 per cent, goes into doing external mechanical work; the other 92.5 per cent represents the difference between the thermal energy of 1 kg of vapor and 1 kg of liquid.

Exactly similar considerations apply to the vaporization of any pure liquid. Vaporization curves for other familiar liquids are given in Fig. 8.

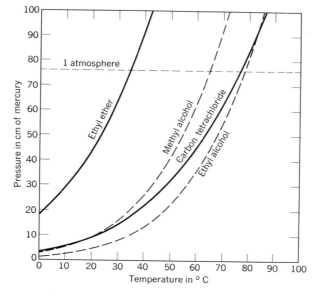

Fig. 8. Vaporization curves for various liquids.

3. Sublimation; The Triple Point

A comparison of the fusion curve of Fig. 3 and the vaporization curve of Fig. 6, both for water, shows that these curves intersect. The point of intersection is at

$$P = 4.579 \text{ mm}, \qquad T = 0.0100° \text{ C.} \qquad (1)$$

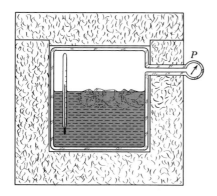

Fig. 9. Triple-point determination (schematic). The vessel contains *only* H_2O, at the pressure and temperature given by (1).

Since solid and liquid can coexist along the fusion curve, and since liquid and vapor can coexist along the vaporization curve, the point of intersection of these curves must give a value of P and T at which solid, liquid, and vapor can all three coexist. This point (1) is therefore called the *triple point* for water.

It is easy to determine the triple point of water, and in practice this point is reproducible to within 0.0001 C deg. Take a thermally insulated vessel (Fig. 9) containing only ice, liquid water, and water vapor. All of the air has been carefully removed by means of a vacuum pump. This vessel will be at the temperature and pressure of the triple point. *There is only one temperature and one pressure at which solid, liquid, and vapor phases can coexist in stable equilibrium.*

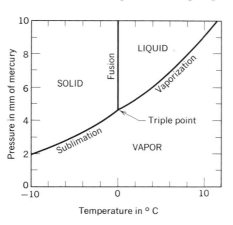

Fig. 10. Triple-point diagram for water.

Because of its accurate reproducibility, the triple point of water is taken as the one fixed point of the absolute temperature scale and assigned the *exact* value $T = 273.16°$ K, as stated on p. 294.

What is the phase of the substance at pressures below that of the triple point? It is seen (Fig. 10) that the fusion and vaporization curves intersect in such a way that no liquid can exist at lower pressures. But both solid and vapor can exist. Below the pressure of the triple point we have a third curve, called the *sublimation curve*, which represents conditions for equilibrium between solid and vapor. This curve is shown for water in Fig. 10; it is not a direct continuation of the vaporization curve, but has a slightly different slope at the triple point.

Sublimation represents a direct removal of molecules from a crystalline solid into the vapor phase. At the low pressures and temperatures occurring on the sublimation curve, solid and vapor can coexist, with equilibrium between the rate of sublimation and the rate of condensation. There is a latent heat of sublimation which is analogous to the other latent heats. The sublimation and condensation that would take place if a piece of ice at −10° C were placed in an insulated evacuated enclosure are similar to the evaporation and condensation discussed in connection with Fig. 4. The first four entries of Table 18-1 represent points on the sublimation curve; the fifth represents the triple point.

While the triple point for H_2O lies well below ordinary pressures, the triple points for some substances, notably CO_2 and iodine, lie well above ordinary pressures, so these substances are not commonly observed in the *liquid* phase.

The triple-point diagram for CO_2 is shown in Fig. 11. The triple point is at −56.6° C and 5.11 atm. We are acquainted with solid CO_2 (Dry Ice) in equilibrium with its vapor at 1 atm pressure and a temperature of −78.5° C, and with its sublimation. The latent heat of sublimation is 137.9 kcal/kg at 1 atm pressure. We are also acquainted with liquid CO_2 in tanks; this liquid is at room temperature but must be at a pressure above 56.5 atm or it could not exist as a liquid at 20° C.

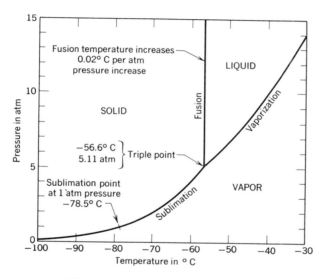

Fig. 11. Triple-point diagram for CO_2.

4. The Critical Point

Figure 6 shows the vaporization curve for water *ending* at a temperature of 374.15° C and a pressure of 218.4 atm. The end of the vaporization curve is called the *critical point*. As we follow up the vaporization curve, the difference in density between liquid and vapor becomes less and less, and the latent heat of vaporization less and less, until both vanish at the critical point. For other substances the vaporization curve ends similarly at a critical point. Values of pressure, temperature, and specific volume at the critical points of various substances are tabulated in Table 18-2.

Table 18-2 CRITICAL CONSTANTS

Substance	Critical temperature (°C)	Critical pressure (atm)	Critical volume (m³/kg)
Water	374.15	218.4	3.1×10^{-3}
Carbon dioxide	31.1	72.9	2.15×10^{-3}
Oxygen	−119	50	2.33×10^{-3}
Argon	−122	48	1.88×10^{-3}
Nitrogen	−147	33.5	3.21×10^{-3}
Neon	−229	27	2.07×10^{-3}
Hydrogen	−240	13	3.23×10^{-3}
Helium	−268	2.3	14.5×10^{-3}

Above the critical pressure, no liquid-vapor phase transition occurs for any temperature. Above the critical temperature, no liquid-vapor phase transition occurs for any pressure. Above the critical pressure only two phases occur, instead of the usual three. These are the crystalline solid phase and the *amorphous* phase. As the temperature is raised with the pressure above critical, the amorphous phase changes *continuously* from a state which behaves much like a liquid to a state which behaves like a perfect gas, as we shall see in the next section.

We now see that we can make a *continuous* change from liquid (say water at 20° C and 1 atm) to vapor (say steam at 110° C and 1 atm) by so altering the pressure and temperature that we go *around the end* of the vaporization curve. This possibility emphasizes the fact that there is no *fundamental* difference between a liquid and a vapor, whereas each of these differs fundamentally from a crystalline solid, in that the structure of each is amorphous rather than crystalline, with a random rather than a regular molecular arrangement.

Gases for which a phase transition cannot be observed when the pressure is increased at room temperature, because their critical temperature lies below room temperature, are sometimes called *permanent* gases. All except the first two substances of Table 18-2 are of this type.

5. Real Gases

Real gases do not obey the general gas law perfectly. This discrepancy is best illustrated by plotting *isothermals* on a $P\rho$-diagram as we do in Fig. 12 for dry air and in Fig. 13 for CO_2. These isothermals are curves in which density is plotted against pressure at a constant temperature; they show the increase of density with increase of pressure as we

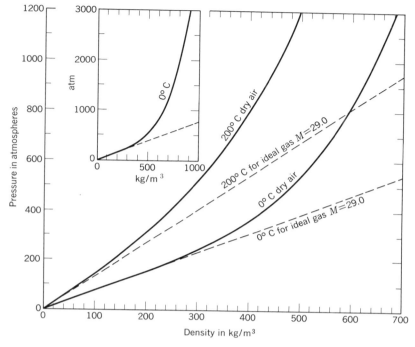

Fig. 12. Isothermals at 0° C and 200° C for dry air.

move up a vertical line in the triple-point diagram.

On such a $P\rho$-diagram, the isothermals for an ideal gas of molecular mass M are straight lines passing through the origin, of slope given by the general gas law (15), p. 340, as

$$\frac{P}{\rho} = \left(0.08206 \frac{\text{atm} \cdot \text{m}^3}{\text{kg} \cdot \text{K}}\right) \frac{T}{M}, \tag{2}$$

where the coefficient has been expressed in units convenient when P is in atmospheres, ρ in kg/m³, and T in ° K.

The isothermals for a real gas always start out from the origin with the slope given by (2); *the vapor of every substance behaves like an ideal gas if the pressure is sufficiently low.*

At a temperature above the critical temperature, the density increases along a smooth curve as the pressure is increased. As shown in Fig. 13, at a temperature below the critical temperature, the density increases smoothly until the vaporization (or sublimation) pressure is reached; then the density increases by a finite amount with no change in pressure as liquefaction (or solidification) takes place. Then the density of the liquid (or solid) again increases smoothly. Liquefaction (or solidification) occurs along a horizontal portion of the isothermal, on which liquid (or solid) and vapor coexist. As we move to the right along this horizontal portion, the density of the vapor and that of the liquid (or solid) remain constant, but the fraction of the substance that is liquid (or solid) increases from 0 to 100 percent.

Departures from the ideal-gas laws arise from forces between the molecules of a gas. These forces are of two types: (*a*) strong repulsive forces when the molecules come very close together, these forces giving the molecules a more or less well-defined size and making them appear to bounce when they collide; and (*b*) weaker attractive forces of somewhat longer range which are the cause of the cohesion of liquids and solids. These forces are usually called *Van der Waals forces,* after the Dutch physicist who first determined the character of the intermolecular forces required to account for observed departures from the ideal gas laws.

The attractive forces cause the density to be greater than the ideal density and result in the curves of Fig. 13 swinging to the right of the ideal curves. Even the dry-air curves of Fig. 12 at first swing slightly to the right of the ideal curves, but only by a small fraction of one percent.

The short-range repulsive force causes the isothermals to swing sharply upward as a density of the order of that of water (1000 kg/m^3) is approached. This is the density at which the molecules are apparently packed tightly together, and the elastic behavior with changes in pressure becomes like that of a liquid.

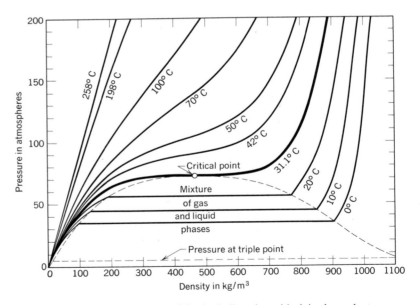

Fig. 13. Isothermals for CO_2, including the critical isothermal at 31.1° C, and showing the region in which vapor and liquid coexist. At the bottom of this region, below the 5-atm pressure of the triple point indicated by the horizontal line, is a region in which vapor and solid coexist.

We can now discuss the properties of the amorphous state, at a pressure above the critical pressure, in more detail. Let us look at Fig. 13 and consider the properties of CO_2 at 100 atm pressure. At 0° C and 100 atm, the substance has all the properties of a liquid. It will start to evaporate and split into two phases if the pressure is lowered to 35 atm isothermally. It is relatively incompressible; doubling the pressure (from 100 to 200 atm) isothermally only results in a 4 per cent density increase or a 4 per cent volume decrease. If now we start with the substance at 0° C and 100 atm and raise the temperature *isobarically* (that is, keeping the pressure constant), we move to the left in Fig. 13 along a horizontal isobar, crossing the isothermals in continuous fashion until we reach a temperature of, say 258° C. At 258° C and 100 atm the substance behaves exactly like an ideal gas, obeying the general gas law very accurately. At this temperature, doubling the pressure isothermally doubles the density or halves the volume. Lowering the pressure isothermally that is, at a constant temperature of 258° C, does not result in any phase change. The transition from a substance that is 'obviously' a liquid to one that is 'obviously' a gas has, however, been continuous as we raised the temperature isobarically, and there is no good criterion for deciding where we should stop calling the substance a liquid and start calling it a gas.

In the ordinary temperature range, water vapor closely approximates an ideal gas at all pressures up to its saturation pressure. *For temperatures below* 60° C, *the density of water vapor never differs from the ideal density by more than 0.5 per cent, even at saturation.* This statement is of importance in connection with the discussion of humidity in Sec. 7.

6. Boiling

So far in this chapter we have confined our attention to a single pure substance in an enclosure. It is important to consider the case in which there is a foreign permanent gas such as air above the surface of a liquid.

Let us start by considering the heating of water in an open vessel. If the barometer reads 76 cm, the pressure on the surface of the water is 76 cm and the pressure within the water is the same except for a slight addition arising from the weight of the water. When the temperature of the water is below 100° C, evaporation takes place only from the surface of the water, and some of the heat added to the water goes into latent heat in this evaporation. The evaporation takes place only at the free surface, since an incipient steam bubble within the water would be unable to support a pressure of 76 cm and would collapse. The amount of evaporation can be restricted by restricting the surface area.

If heat is added at a sufficient rate, the temperature of the water will increase to 100° C. At this temperature, for the first time, steam bubbles can form and grow within the body of the water, because now the vapor pressure within the bubble equals the hydrostatic pressure of the atmosphere. *Ebullition (boiling)* will begin and continue until all the water has boiled away, with no further increase in temperature.

From this type of argument we see that the vaporization curve of Fig. 6 gives the boiling point of water in an open vessel as a function of atmospheric pressure applied to the surface. If the applied pressure is reduced, the boiling temperature goes down. The curves of Fig. 8 similarly give the boiling temperatures of other liquids at various pressures.

Example. *In the vacuum distillation process of drying a pharmaceutical product, it is desirable to have the water boil off at a temperature of not over 50° C in order not to damage the product. What absolute pressure must be maintained in the distillation chamber?*

According to Table 18-1, water will boil at 50° C if the pressure is reduced to 9.25 cm of Hg; hence the vacuum pumps must maintain this pressure or lower.

In our discussion of boiling, we indicated that bubbles are formed in the *interior* of the liquid; the bubbles are usually formed on small, suspended solid particles that serve as nuclei for the bubbles. In the absence of such particles, a pure liquid can be heated to a temperature above its normal boiling point; in such a situation the liquid is unstable and will bubble with extreme violence if particles are suddenly introduced. DONALD GLASER discovered that positive and negative ions can also serve as nuclei for bubbles. Thus, if the pressure is suddenly reduced just prior to the passage of a high-energy ionizing particle through the liquid, bubbles will be formed on the ions along the 'track' traversed by the high-energy particle. This phenomenon is employed in the 'bubble chambers' used in studies of high-energy physics; a photograph of tracks produced in a liquid-hydrogen bubble chamber is shown in Fig. 4 of Chapter 46. A somewhat related bubble track will be formed if a grain of salt is dropped into a carbonated beverage; in this case the bubbles consist of CO_2. Try it!

7. Mixtures of Gases and Vapors; Hygrometry

Let us now return to Fig. 4, p. 362, and suppose that the space above the liquid water contains air or some other permanent gas as well as water vapor. Most permanent gases are only slightly soluble in liquid water, so that only a comparatively few of the gas molecules will enter the water and be dissolved in it. Unlike water-vapor molecules, most of the air or gas molecules that strike the liquid surface will bounce off as if it were a solid surface.

The water seems to ignore the presence of the air, as expected from Dalton's law of partial pressures. The rate of evaporation depends *only* on the temperature of the water and is independent of the pressure of the air. The rate of condensation is the same function of the partial pressure of the water vapor and the temperature as if the air were not present. Hence, *in equilibrium the partial pressure of the water vapor is given by the vaporization curve of Fig. 6.* The total pressure in the space above the liquid water will be the sum of the partial pressure of the air and the partial pressure of the water vapor.

The case we have discussed is that in which the air or other gas is *saturated* with the vapor of the liquid. Another important case is that in which there is no liquid present but only vapor, which may have less partial pressure than the saturation vapor pressure. A vapor cannot have *greater* partial pressure than the saturation pressure corresponding to the temperature of the vapor because then condensation of the vapor to liquid or solid would take place.* Thus, the water vapor in the atmosphere can have any partial pressure up to that of saturation. If the partial pressure exceeds that of saturation, condensation occurs, and formation of clouds, fog, dew, or frost results.

The term *humidity* is used to describe the water-vapor content of the atmosphere.

The **absolute humidity** is the mass of water vapor per unit volume of atmosphere.

The **relative humidity** is the ratio of the partial pressure of water vapor in the atmosphere to the partial pressure that would cause saturation at the temperature of the atmosphere.

The relative humidity is also, to a close approximation, the ratio of the actual density of water vapor to the saturation density at the same temperature, since water vapor below 60° C obeys the ideal gas law very closely, as we have pointed out on p. 371.

The measurement of humidity is called *hygrometry*. Absolute humidity is measured by drawing a known volume of atmosphere through a drying agent and measuring the increase in weight of the agent. Relative humidity is most accurately measured by determining the *dew point*. In the dew-point apparatus, a polished metal surface is slowly cooled, and the temperature is noted at which dew first begins to cloud the surface. In this procedure, the atmosphere is cooled locally in the vicinity of the metal surface. Since the atmosphere in the region near the metal surface is in pressure equilibrium with the balance of the atmosphere, the partial pressures of air and water vapor in this region

---

*Dust particles and electric charges serve as 'condensation nuclei' on which droplets of liquid or small ice crystals form. They will also form on solid surfaces. If few condensation nuclei are present, *supersaturation* may exist for a short period.

will be the same as in the remainder of the atmosphere near the station where the measurement is being made. But when the temperature of the atmosphere near the metal surface drops so that the partial pressure of the water vapor equals the saturation pressure, liquid water will begin to condense on the metal surface.

> The **dew point** of the atmosphere is that temperature at which water vapor would begin to condense, and hence that temperature at which the partial pressure of the water vapor present in the atmosphere equals the saturation pressure.

Since the saturation pressure at the dew point is the actual pressure of water vapor in the atmosphere, we see from the definition of relative humidity that

$$\left\{\begin{array}{c}\text{relative}\\ \text{humidity}\end{array}\right\} = \frac{\text{saturation pressure at dew point}}{\text{saturation pressure at actual atmospheric temperature}}. \quad (3)$$

Another method of measuring humidity is by means of wet- and dry-bulb thermometers. The 'wet' bulb is surrounded by a wick saturated with water, and air is blown at a specified rate past the wick. Evaporation cools the wick and lowers the thermometer reading, but the rate of evaporation and hence the reading is dependent on the humidity of the air, the dryer the air the greater the evaporation. Handbooks contain empirical tables that give the relative humidity as a function of dry- and wet-bulb readings for a standard air velocity. A less accurate form of hygrometer depends on the fact that human hair expands in length in approximate proportion to relative humidity. There are also recording hygrometers that depend on the fact that the electrical conductivities of certain chemicals vary in definite ways with variations in relative humidity.

> **Example.** *In the winter, an air-conditioning system takes in outside air at 4° C and 65 per cent relative humidity and changes it to 25° C and 50 per cent relative humidity. It does so by first heating the air part way, then saturating it with water vapor by means of a spray, then heating the air again to the desired final temperature. What should be the temperature of the air after it leaves the spray saturated with water vapor?*
>
> From Table 18–1 on p. 363, we see that the saturation pressure of water vapor at 25° C is 23.76 mm of Hg. Fifty per cent relative humidity corresponds to 0.50×23.76 mm $= 11.88$ mm water-vapor pressure. Saturated air has this water-vapor pressure at 13.8° C. Hence the air leaving the spray must be at 13.8° C. We note that in the additional heating, the total pressure does not change; hence, since the composition remains constant, the partial pressure of the water vapor will not change, although the whole volume of course increases.

8. Surface Phenomena of Liquids

Let us now return to the subject of *surface tension* mentioned in Sec. 2. As indicated in Fig. 5, a molecule that is located near the surface of a liquid is subjected to a resultant force directed back into the liquid. A molecule at the surface can thus be considered as having potential energy greater than that of a molecule in the interior by the work that has been done against this force in bringing the molecule from the interior of the liquid to the surface. A liquid thus has a *surface potential energy* proportional to its surface area. To increase the surface area requires work equal to the increase in surface energy as more molecules are brought from interior to surface positions. The surface tends, like a stretched membrane, to assume a shape of minimum area, because this shape is the shape having minimum potential energy, the condition for stable equilibrium (*see* p. 185). Thus, a water droplet in the air assumes a spherical form, since a sphere has a minimum surface area for a given volume. In fact, the surface of a liquid behaves in all respects like a membrane stretched with constant tension per unit width.

> The **surface tension** σ of a liquid is defined as the surface potential energy per unit area of surface, which is the same as the work done per unit increase in surface area.

The behavior of a liquid surface can be demonstrated by means of the wire frame equipped with a sliding crosspiece shown in Fig. 14. The frame is dipped into a soap solution and a film is formed as indicated. Since the surface of the film tends to become a minimum, it tends to pull the crosspiece to the left in the figure. In order to keep the crosspiece at rest, it is necessary to exert a force F to the right.

Now let us stretch the film by moving the crosspiece a distance ΔX to the right. This requires work $F\,\Delta X$, while the increase in surface area of the soap film is $2\,l\,\Delta X$, the factor 2 appearing because there are two surfaces. Thus, for the soap film, the *surface tension* σ is given by

$$\sigma = (F\,\Delta X)/(2\,l\,\Delta X) = F/2\,l; \qquad (4)$$

hence

$$F = 2\,l\,\sigma. \qquad (5)$$

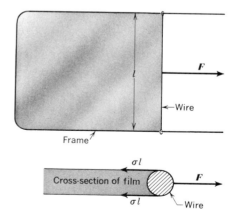

Fig. 14. Surface tension.

The meaning of the last equation is easily visualized from the lower sketch in Fig. 14, which shows the forces exerted on the crosspiece. Each surface of the soap film must exert a force $\tfrac{1}{2}F = \sigma l$ on the wire crosspiece. Careful experiment shows that force F is indeed proportional to l, as indicated

Table 18-3 Surface Tension σ at 20° C (in J/m² or N/m)

Liquid	σ
Acetone	0.0237
Benzene	0.0289
Carbon tetrachloride	0.0268
Ethyl alcohol	0.0223
Mercury	0.465
Water	0.0728

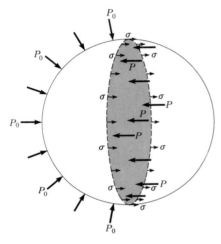

Fig. 15. Forces acting on the left half of a spherical droplet of radius r.

in (5). The surface tension σ thus can be considered as *force per unit width* as well as *work per unit area;* it can be measured in J/m² or in N/m, which are actually identical units. It is like the tension in a stretched membrane such as a drumhead. Table 18-3 gives values of σ for familiar liquids.

Since the surface tension of extremely thin soap films (thickness can be readily determined by *optical* methods involving interference) has the same value as that of thicker films, one is led to the experimental conclusion that *the surface energy is contained in a layer just a few molecular diameters thick.*

Because of surface tension the pressure P inside a liquid droplet is greater than the pressure P_0 outside. The spherical droplet in Fig. 15 is separated into two equal portions by the imaginary vertical plane. Consider the forces acting on the left-hand hemisphere. This hemisphere is subject to two forces directed toward the right: $P_0 \cdot \pi r^2$, the resultant of the forces arising from the external pressure P_0; and $\sigma \cdot 2\pi r$, the resultant force arising from surface tension at the periphery of the shaded plane. (That the resultant of the forces arising from external pressure is $P_0 \cdot \pi r^2$ is readily demonstrated by considering the static equilibrium of a hemispherical volume cut out of a fluid.) In order for the left-hand hemisphere to be in equilibrium, these two forces must be balanced by the force to the left arising from internal pressure at the shaded plane; that is, $P \cdot \pi r^2$, where P is the internal pressure. Thus, the equilibrium condition is

$$P \cdot \pi r^2 = P_0 \cdot \pi r^2 + \sigma \cdot 2\pi r, \quad \text{or} \quad P = P_0 + 2\sigma/r. \qquad (6)$$

The pressure within a spherical droplet is greater than the external pressure by $2\sigma/r$. This pressure difference can become very large for small values of r.

Similar considerations show that the pressure in a gas bubble in a liquid is also $2\sigma/r$ greater than the pressure in the surrounding liquid, while the pressure in a *hollow* spherical bubble such as a soap bubble is $4\sigma/r$ greater than that outside because surface tension acts on both the inside and outside surfaces of the film. This latter relation furnishes one method of measurement of surface tension, as illustrated in the following example.

Example. *A soap bubble* 2 cm *in diameter is formed on a pipe attached to* a *water manometer*, as in Fig. 16. *The differential pressure is observed as* 1.20 mm *of water. What is the surface tension of the soap solution?*

Since the density of water is $\rho = 1000$ kg/m³, the pressure difference $\rho g h$ is

$$(1000 \text{ kg/m}^3)(9.81 \text{ m/s}^2)(1.20 \times 10^{-3} \text{ m}) = 11.8 \text{ N/m}^2$$

We equate this to $4\sigma/r$ to determine σ:

$$4\sigma/r = 4\sigma/(0.01 \text{ m}) = 11.8 \text{ N/m}^2, \qquad \sigma = 0.029 \text{ N/m}.$$

Figure 16

Some of the consequences of surface tension are very familiar. It is because of surface tension that liquid films and bubbles can exist; that liquid droplets assume spherical form; that droplets can 'hang' at the end of a faucet; that a liquid will 'climb' a capillary tube that it wets and be depressed in one that it does not wet, and in general that a liquid meniscus does not meet a solid surface horizontally; that tiny water droplets can remain liquid in atmospheric clouds at temperatures below 0° C because surface tension increases the pressure in the droplet and depresses the freezing point; that small unwetted particles of density greater than water can 'float' on water, the basis of the important flotation process for separation of mineral ores; and that a detergent can displace oil or grease on a surface to be cleaned. Some of these phenomena involve surface energies of a liquid in contact with a solid as well as the surface energy of a liquid in contact with air, which leads to the ordinary surface tension of Table 18-3. It is these different surface energies that determine the behavior of liquids in contact with solids.

PROBLEMS

NOTE: *The saturated vapor pressure of mercury is very low at ordinary temperatures* (0.0001 cm *at* 20° C, 0.028 cm *at* 100° C).

1. What is the *maximum* value that could be obtained for the melting point of pure H_2O? How would it be obtained?   Ans: 0.01° C.

2. If the skating edge of an ice skate has an area of 2 mm², and a skater weighing 200 p rests his entire weight on a single skate, calculate the lowering of the melting point of the ice immediately beneath the cutting edge. Discuss the effect of this variation in melting point on the friction between skate and ice.

3. Compute an accurate height in meters at which a water barometer will stand when the temperature is 20° C, the pressure 1 atm, and the gravitational acceleration is standard.
   Ans: 10.2 m.

4. A mercury barometer reads 760 mm. A drop of water is then introduced at the bottom of the mercury column and rises to the surface of the mercury. After evaporation has taken place, a small amount of liquid water remains. By how many mm is the mercury column depressed if the temperature is 20° C?

5. If it is desired to have water boil at 90° C, what pressure should be maintained in a vacuum cooker?   Ans: 52.6 cm of Hg.

6. If it is desired to have water in a pressure cooker boil at 120° C when the ordinary boiling point is 100° C (the change of 20 C deg about quadruples the cooking speed), to what gauge pressure in p/i² should the valve be set?

7. If it is desired to have water in a pressure cooker boil at 100° C in a mountain location where the atmospheric pressure is only 10 p/i², to what gauge pressure in p/i² should the valve be set? Ans: 4.70 p/i².

8. (*a*) A large vacuum pump continuously exhausts the vapor from a thermally insulated vessel containing water. Why does the water cool down and freeze? (*b*) If you had an ample supply of liquid oxygen at its boiling point under 1 atm pressure, how would you proceed to make solid oxygen? (This procedure has been used to solidify all gases except He.)

9. A vessel of 2000-cm³ volume contains 2 g of H_2O at 100° C. What fraction of the water is liquid? Ans: 0.412.

10. A boiler of 0.3 m³ volume contains 60 kg of H_2O at 200° C. How many kg of steam are in the boiler?

11. At the triple point, the specific volumes of liquid water, ice, and steam are 1, 1.09, and 206 300 cm³/g. The pressure is 4.58 mm. The latent heat of fusion is 79.7 kcal/kg; the latent heat of vaporization is 597.4 kcal/kg. From these data determine the differences in specific thermal energy of water and ice and of steam and water at the triple point, and determine the latent heat of sublimation at the triple point. Show that if heat is added to a vessel of constant volume containing ice, liquid water, and steam, almost all the heat is used to melt ice to form liquid and very little of the heat goes into the formation of more vapor.
Ans: 79.7, 567.2, 677.1 kcal/kg.

12. At 1 atm pressure and the sublimation temperature, solid CO_2 has a density of 1.53 g/cm³. Using the vapor density given by the ideal gas law, find the percentage of the latent heat of sublimation that goes into increase in thermal energy and the percentage that goes into external work.

13. The measured volume of a quantity of hydrogen collected over water is 858 cm³, the temperature being 16° C and the barometer reading 740 mm. The volume is measured with the water level the same inside and outside the hydrogen bottle. Calculate the volume of dry hydrogen at 0° C and 1 atm. Ans: 775 cm³.

14. A sample of oxygen collected over water at 15° C occupies a volume of 500 cm³ when the water level is the same inside and outside the bottle. Find the volume of dry oxygen at 0° C and 1 atm if the barometer reads 750 mm at the time of collection.

15. A flask is partly filled with warm water at 70° C. The air above the water is allowed to become saturated and then the flask is tightly corked. If the barometer reads 760 mm at the time of corking, what will be the pressure in the flask the next day when it is at temperature 25° C and the barometer has dropped to 730 mm? Ans: 481 mm.

16. A vessel containing only air and steam is sealed at 100° C with total pressure of 76.0 cm, partial pressure of water vapor, 21.0 cm. It is then allowed to cool. Compute the total pressure in the vessel at 60° C and at 20° C.

17. What is the dew point when the temperature is 30° C and the relative humidity 75 per cent? Ans: 24.7° C.

18. On a winter day the outdoor temperature is −10° C, and because of snow the relative humidity is 100 per cent. Unless moisture is added to the air, the partial pressure of water vapor will be the same inside a house at 30° C. What will be the relative humidity in the house?

19. If the outdoor temperature is 10° C and the relative humidity 50 per cent, what will be the relative humidity inside a room at 20° C if no moisture is added to the inside air?
Ans: 26.2%.

20. (*a*) If a man enters a room at 25° C from an outdoor temperature of 18° C and his glasses 'steam up,' what is a minimum value of the relative humidity in the room? (*b*) At what relative humidity in a room at 25° C will ice-water glasses collect a coating of moisture?

21. An air-conditioning unit works in the summertime by cooling the air below the desired temperature to condense out excess water vapor, and then reheating the air. If outside air is taken in at 95° F and 70 percent relative humidity, and delivered at 70° F and 50 percent relative humidity, to what temperature must it be cooled to lower the water-vapor content to the desired value?
Ans: 50.4° F.

22. In the air-conditioning installation described in Prob. 21, if air is taken in at 100° F and 50 percent relative humidity, and delivered at 70° F and 60 percent relative humidity, to what temperature must it first be cooled to lower the water-vapor content to the desired value?

23. Find the relative humidity on a day when the temperature is 30° C and the dew point is 16° C. Ans: 42.8%.

24. On a certain day the temperature is 25° C, the relative humidity is 50 percent, and the barometer reads 76 cm of mercury. What portion of the atmospheric pressure is due to dry air?

25. What is the net force exerted on a piston 4 inches in diameter with saturated steam at 150° C on one side and air at 1 atm on the other side of the piston? Ans: 693 p.

26. A water spray is used in an air-conditioning system that changes one cubic meter per second of dry air at 0° C to air at 25° C and 40 percent relative humidity. How many kilograms of water are needed per hour?

27. Show that the ratio of the density of air saturated with water vapor at 30° C to that of dry air at 30° C, the pressure being 760 mm in both cases, is $[31.8 (18.0/29.0) + 728.2]/760$.

28. A barometer at 0° C stands at 76 cm. The vacuum above the merucry has a length of 10 cm and a cross-sectional area of 1 cm². How many grams of ethyl alcohol (C_2H_6O) at 0° C must be allowed to travel up the mercury column into the vacuum to depress the mercury to 66 cm? How many grams of ethyl ether ($C_4H_{10}O$) at 0° C have the same effect? Treat the vapors of these liquids as perfect gases.

29. If a vessel contains only CO_2 at the following temperatures and pressures, what phase or phases (solid, liquid, gas) can be present? (a) −90° C, 1 atm; (b) −78.5° C, 1 atm; (c) 20° C, 1 atm; (d) 20° C, 56.5 atm; (e) 20° C, 75 atm; (f) −56.6° C, 5.11 atm; (g) −56.4° C, 15.11 atm.

30. What changes of phase take place in 200 g of water that is initially at 150° C and is cooled to −150° C under the constant pressure of 1 atm? What changes take place if the cooling is done at 2 mm pressure? Give the approximate temperature at which the changes occur.

31. A sphere 1 f in diameter is made of steel 1 mil (0.001 i) thick that has rupture strength in tension of 100 000 p/i². The sphere is sealed at 0° C and contains air at 1 atm pressure plus a small quantity of water sufficient to keep the air saturated as the sphere is heated. To what temperature can it be heated before the steel ruptures? To solve this problem, consider the forces on half the sphere (including the gas in it) and balance the tension exerted on the steel against the excess gauge pressure of the gas on the diametral area, as in the figure.

Ans: 124° C.

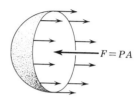

Problem 31

32. In Sec. 8, considerations of hydrostatic equilibrium were used to determine that the external force exerted by the atmosphere on one-half of a small droplet is $P_0 \cdot \pi r^2$, where P_0

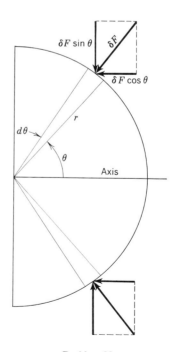

Problem 32

is the atmospheric pressure and r is the radius of the droplet. Verify this statement, in a much more complex fashion, by integration of the forces: (*a*) Note from the figure that for each force δF on a small area on one side of the axis there will be a corresponding force δF acting on the other side of the axis such that the oppositely directed forces $\delta F \sin\theta$ cancel and hence only the components $\delta F \cos\theta$ need be considered. (*b*) Write down $\delta F \cos\theta$ for the force exerted on an element of area consisting of a 'band' of length $2\pi r \sin\theta$ and width $r\, d\theta$. (*c*) Integrate from $\theta = 0$ to $\tfrac{1}{2}\pi$ rad.

33. Determine the diameter of a water droplet that will not freeze until the temperature drops to $-1.5°$ C.   Ans: 7.21×10^{-7} cm.

34. In the last example in the text, what would be the differential pressure if the soap bubble were 4 cm in diameter? 1 cm?

35. Two soap bubbles are connected by means of a glass tube as shown in the figure. At the moment the connection is effected, bubble *A* has a diameter of 1 inch and bubble *B* has a diameter of 3 inches. What happens after the connection is established?

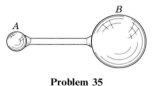

Problem 35

THERMODYNAMICS 19

In this chapter we give a very brief introduction to a powerful discipline called *thermodynamics,* which was developed during the nineteenth century by CARNOT, JOULE, KELVIN, CLAUSIUS, GIBBS, RANKINE, and others.

The principles and theorems of thermodynamics can be stated entirely in terms of directly observable, macroscopic quantities such as pressure, volume, temperature, heat, work, and energy, with no assumptions whatsoever concerning atomic structures or processes. From just two simple postulates, called the first and second principles of thermodynamics, a surprisingly large body of important conclusions and equations can be derived. As the name *thermodynamics* implies, this discipline was initially concerned with the interconversion of thermal and mechanical energy, but the same principles were quickly found to give valuable information on the interconversion of thermal and other forms of energy, such as chemical and electrical.

The first principle of thermodynamics is merely the principle of conservation of energy as it applies to interconversion of thermal and other forms of energy.

The second principle of thermodynamics is a subtle and profound generalization from experience that asserts that many types of conversion of thermal to other forms of energy that would not at all violate the first principle just do not happen and cannot be made to happen. For example, a freely rotating wheel slows down, heating the bearings; it never by itself speeds up, cooling the bearings, gaining mechanical and losing thermal energy. From the first principle, we conclude that someone should invent a machine to extract some of the enormous amount of thermal energy contained in the oceans of the world, turning this energy into mechanical, electrical, or chemical energy, at the expense of merely cooling some of the ocean water. No one has invented such a machine. The second principle asserts that such a machine is impossible. The manifold predictions of the second principle have never been contradicted by observation. In connection with the second principle, we introduce in a very elementary way the concept of *entropy*.

The principles of thermodynamics are fundamental to many branches of physics, chemistry, aerodynamics, and mechanical engineering. In this chapter we confine our attention to certain simple applications in the fields of power and refrigeration, and describe briefly the operation of the steam engine. We also discuss the manner in which the absolute thermodynamic temperature scale is set up. We shall confine attention to the interconversion of thermal and mechanical energy and not attempt to give a more general formulation of thermodynamic theory.

1. The First Principle of Thermodynamics

As we have already noted, the sum of the mechanical energy and the thermal energy of an isolated system is conserved provided only that there are no electrical, magnetic, chemical, or nuclear phenomena that involve forms of energy other than mechanical and thermal. With this restriction taken as understood, the *first principle of thermodynamics* can be stated in the following form:

FIRST PRINCIPLE OF THERMODYNAMICS: *When mechanical energy disappears, an equal quantity of thermal energy appears; and when thermal energy disappears, an equal quantity of mechanical energy appears.*

In the useful processes for converting heat into work or work into heat, there is always a working substance (for example steam or gaseous products of combustion in the case of an engine, ammonia or Freon in the case of a refrigerator). The working substance undergoes certain changes of pressure, volume, and temperature, accompanied by changes in thermal energy. In this case the first principle of thermodynamics gives a relation similar to that discussed on p. 341:

$$\begin{Bmatrix}\text{heat added to}\\ \text{substance}\end{Bmatrix} = \begin{Bmatrix}\text{mechanical work}\\ \text{done by substance}\end{Bmatrix} + \begin{Bmatrix}\text{increase in thermal}\\ \text{energy of substance}\end{Bmatrix} \quad (1)$$

This relation presupposes that changes in *macroscopic* mechanical kinetic and potential energy of the working substance itself are of negligible importance, as is the case in actual engines and refrigerators.

2. The Second Principle of Thermodynamics

Experience tells us that whereas it is very easy to turn mechanical energy completely into thermal energy (as in friction), there are severe restrictions on our ability to effect the reverse transformation. The only way we are able continuously to turn thermal energy into mechanical energy is to have 'heat reservoirs' at two *different* temperatures available, and to introduce *between them* some kind of machine that manages to turn some, *but only some,* of the heat flowing from the hot to the cold reservoir into work.

The *second principle of thermodynamics* is an inference from experience that embodies the above ideas. The second principle can be stated as follows:

SECOND PRINCIPLE OF THERMODYNAMICS: *It is impossible to construct a continuously acting engine that will deliver mechanical work derived purely from heat extracted from a single reservoir, no heat being given out to a reservoir at lower temperature.*

From this principle we can proceed to derive a number of important laws relating to engines and refrigerators. We start with consideration of an *ideal heat engine* in which *isothermal* and *reversible adiabatic* expansions and compressions of an *ideal gas* occur. These considerations will set limits on the efficiencies and performance coefficients of *real* heat engines and refrigerators. Before proceeding with these considerations we must describe isothermal and adiabatic processes.

3. Isothermal Expansion and Compression

Consider a 'working substance,' which might be a gas or might be partly liquid and partly vapor, in the cylinder of Fig. 1. Except for the closed end, the walls of the cylinder are assumed to be insulated perfectly, and the piston of area A is assumed to be nonconducting, gas-tight and frictionless. The closed end of the cylinder is in contact with a large heat reservoir at temperature T. If the force applied to the piston is gradually decreased, the working substance will gradually expand. If the expansion takes place so gradually that all portions of the working substance are

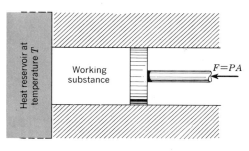

Figure 1

continuously in mechanical and thermal equilibrium at temperature T, the expansion is called an *isothermal expansion*. As shown in (19), p. 343, the work done by the expanding substance is the integral of $P\,dV$. Hence the total work done during the expansion can be represented as the area under the curve giving pressure P as a function of volume V, as in Fig. 2. The work W done during the expansion from A to B is represented by the shaded area $ABEFA$, where point A represents the initial state V_A, P_A, T, and point B the final state V_B, P_B, T. During the isothermal expansion just described, a quantity of heat Q must be supplied by the heat reservoir. This quantity can be obtained from relation (1).

The reverse process in which the applied force F is gradually increased so that the working substance is gradually compressed at constant temperature is called an *isothermal compression*. The work W done *by* the external force F is in this case again equal to the shaded area under the curve on the PV-diagram of Fig. 2. The work done during compression results in a return of heat to the reservoir and may also cause changes in the thermal energy of the working substance if this substance is other than an ideal gas, whose thermal energy depends on temperature only.

Considering the complete cycle of operations just described, we see that we started with a substance in a state characterized by P_A, V_A, and T and caused an isothermal expansion of the substance to a new state P_B, V_B, T, during which process a certain quantity of heat, Q, was supplied to the substance by the heat reservoir, and work W was done on the piston. The isothermal compression caused a return of the working substance to its initial state P_A, V_A, T, with work W done by the piston. Since the initial and final states of the working substance are identical, we conclude that the complete cycle of operations produced no change in the thermal energy of the working substance. Therefore, from the first principle of thermodynamics, we conclude that the heat returned to the reservoir during the isothermal compression is exactly equal to Q, the heat extracted during the isothermal expansion. The isothermal expansion just described is therefore *reversible*, since a corresponding isothermal compression can be carried out with exactly the reverse provess taking place at every stage.

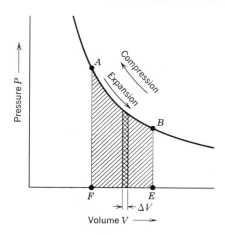

Fig. 2. Work done in an isothermal expansion or compression. The curve shown applies to an ideal gas, with $P \propto 1/V$ at constant T.

An **isothermal process** is a reversible process that takes place without change in temperature and with sufficient slowness that the working substance can be considered to be continuously in mechanical equilibrium and in thermal equilibrium with a heat reservoir.

No real physical process can be strictly isothermal or strictly reversible, since even if we had frictionless devices and perfect thermal insulators, mechanical equilibrium and thermal equilibrium can never be achieved completely at all times.

An isothermal process of special simplicity and importance is one in which the working substance is an *ideal gas.* Let the cylinder of Fig. 1 contain n kilomoles of an ideal gas. Then the general gas law (9), p. 338, is applicable:

$$PV = nRT.$$

Since the thermal energy of an ideal gas depends *only* on its temperature (*see* p. 341), the thermal energy of the gas is unchanged during isothermal expansion. Hence, with an ideal gas as the working substance, only the work done in the expansion must be supplied as heat by the heat reservoir; for isothermal expansion of an ideal gas, (1) becomes

$$Q = W = \text{work done in expansion.} \tag{2}$$

The work done is given by integral

$$Q = W = \int_{V_A}^{V_B} P\, dV = nRT \int_{V_A}^{V_B} \frac{dV}{V} = nRT \log_e(V_B/V_A), \tag{3}$$

where V_B/V_A is called the *expansion ratio*. Thus, the *work done in an isothermal expansion of a given number of kilomoles of an ideal gas depends only on the absolute temperature and on the expansion ratio.* Equation (3) gives an important relation that we shall need to use later, namely: *For the same quantity of gas and the same expansion ratio, the amounts of work done in isothermal expansions at two different absolute temperatures are proportional to the absolute temperatures.*

4. Reversible Adiabatic Expansion and Compression

Consider a substance (which might be a gas or might be partly liquid and partly vapor) in the cylinder of Fig. 3. The cylinder and the face of the frictionless piston are assumed to be perfectly insulated thermally so that *no heat can enter or leave the working substance.* If the force applied to the piston is gradually increased, the working substance will be gradually compressed. At each stage it will be compressed to such a volume that its pressure just balances the force of the piston. Work will be done on the substance by the piston. Since no heat enters or leaves the substance, this

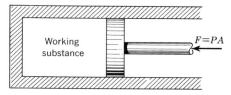

Fig. 3. A working substance in a cylinder with nonconducting walls and a nonconducting piston.

work will go entirely into increasing the thermal energy of the substance, as we see from (1). This increase will result in a gradual increase in the temperature of the substance. Here *gradual* denotes that the work is done sufficiently slowly to give the energy added by the work (in the first instance at the face of the piston) ample time to distribute itself throughout the whole of the substance so that at each instant the working substance can be considered to be in thermal equilibrium, with no more than an infinitesimal temperature gradient needed to carry the heat from the part of the substance near the face of the piston to the balance of the substance. A compression carried out under these ideal conditions is said to be *reversible* and *adiabatic*.

The reverse process, in which the force on the piston is gradually decreased, so that the substance gradually expands, doing work, and hence decreases in thermal energy and temperature is called a *reversible adiabatic expansion*. In a reversible adiabatic expansion the work done by the gas exactly equals the decrease in thermal energy.

The reversible adiabatic expansion is to be contrasted with the free expansion discussed on p. 341, in which the substance, an ideal gas, does no work and suffers no change in thermal energy. One could in principle approximate a free expansion by drawing the piston back so fast that the gas is 'left behind.' The gas then exerts no force and does no work. This process is also called *adiabatic,* since no heat enters or leaves. But it is *not* reversible. There is no way to get the gas in the free-expansion experiment to return spontaneously to the *one* vessel.

Hence the insistence, in defining a *reversible* adiabatic process, that equilibrium conditions exist continuously, so that at each instant the substance exerts a pressure on the piston corresponding to its equilibrium pressure at a given temperature and volume.

> A **reversible adiabatic process** is a process that takes place with no addition or removal of heat, and with sufficient slowness so that the working substance can be considered to be continuously in thermal and mechanical equilibrium.

No entirely reversible adiabatic process occurs in practice, but the expansion of steam in the cylinder of a steam engine, of the hot gases in an internal-combustion engine, and the compression of the air in a Diesel engine or in an air compressor are all approximately adiabatic, as are the pressure variations in sound waves (Chapter 21). These processes take place rapidly enough that only a small amount of heat has time to enter or leave the substance from the cylinder walls, and yet slowly enough so that thermal and mechanical equilibrium is approximately maintained throughout the gas in the cylinder.

Now consider (Fig. 4) *n* kilomoles of an *ideal gas* that are compressed reversibly and adiabatically from a volume V_D to a smaller volume V_A.

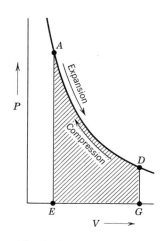

Fig. 4. Work done during a reversible adiabatic expansion or compression of an ideal gas.

Since no heat leaves the gas, all external work done on the gas must be added to the thermal energy of the gas. Thus, at every stage in the compression the work dW is accompanied by a small increase in the thermal energy; by (17), p. 342, the increase in thermal energy is given by $nC_V dT$. Thus, during the compression, every decrease in volume is accompanied by an increase in temperature. The reverse of the above statements applies in a reversible adiabatic expansion.

To obtain the relation between P and V for such a process, we make use of (1) and note that no heat is added. Thus, for a reversible adiabatic expansion of an ideal gas, (1) becomes

$$P\, dV + nC_V\, dT = 0.$$

Substituting, from the general gas law, $P = nRT/V$, we obtain

$$\frac{R}{C_V} \frac{dV}{V} = -\frac{dT}{T}. \tag{4}$$

From (23), p. 344, we have $R = C_P - C_V$; division of this equation by C_V gives $R/C_V = C_P/C_V - 1 = \gamma - 1$, where

$$\gamma = C_P/C_V. \tag{5}$$

Hence (4) becomes

$$(\gamma - 1)\frac{dV}{V} = -\frac{dT}{T}. \tag{6}$$

Integration from the state P_A, V_A, T_A to the state P_D, V_D, T_D (Fig. 4) gives

$$(\gamma - 1)\int_{V_A}^{V_D} \frac{dV}{V} = -\int_{T_A}^{T_D} \frac{dT}{T},$$

$$(\gamma - 1)\log_e V \Big|_{V_A}^{V_D} = -\log_e T \Big|_{T_A}^{T_D},$$

$$\log_e V^{\gamma-1} \Big|_{V_A}^{V_D} = -\log_e T \Big|_{T_A}^{T_D},$$

$$\log_e V_D^{\gamma-1} - \log_e V_A^{\gamma-1} = -\log_e T_D + \log_e T_A,$$

$$\log_e T_D + \log_e V_D^{\gamma-1} = \log_e T_A + \log_e V_A^{\gamma-1},$$

$$\log_e(T_D V_D^{\gamma-1}) = \log_e(T_A V_A^{\gamma-1});$$

finally,
$$T_D V_D^{\gamma-1} = T_A V_A^{\gamma-1}. \tag{7}$$

Equation (7) shows that in a reversible adiabatic process

$$TV^{\gamma-1} = \text{constant}. \tag{8}$$

If we substitute $T_D = P_D V_D/nR$, $T_A = P_A V_A/nR$ in (7) we see that

$$PV^\gamma = \text{constant}; \tag{9}$$

while substitution of $V_D = nRT_D/P_D$, $V_A = nRT_A/P_A$ gives

$$\frac{T^{\gamma/(\gamma-1)}}{P} = \text{constant}. \tag{10}$$

Equations (8), (9), (10) determine the changes in the other two state variables if the change in any one of the three, P, V, or T, is given and the process is known to be reversible and adiabatic. Figure 4 and the broken lines of Fig. 5 are plots of equation (9) for a monatomic gas with $\gamma = 5/3$.

In expanding adiabatically from volume V_A to volume V_D, the gas does work just equal to the decrease in the thermal energy of the gas; thus, in view of (18), p. 342,

$$\text{work done by gas} = nC_V(T_A - T_D). \quad \text{(expansion)}$$

Similarly, in the reverse adiabatic compression all external work results in increasing the internal energy of the gas and hence

$$\text{work done on gas} = nC_V(T_A - T_D). \quad \text{(compression)}$$

In a reversible adiabatic expansion, the pressure drops more rapidly with volume than it does in isothermal expansion, since, as indicated above, in the adiabatic case not only does the volume increase but also the temperature decreases, both of which factors contribute to a pressure drop. This point, which is important for later discussion in this chapter, is illustrated in Fig. 5, which shows a family of PV-curves satisfying the isothermal equation

$$PV = \text{constant}, \quad \text{(isothermal)} \tag{11}$$

and also a family of steeper curves satisfying the reversible-adiabatic equation

$$PV^{1.67} = \text{constant}, \quad \text{(adiabatic)} \tag{12}$$

for the case $\gamma = 5/3 = 1.67$ appropriate to an ideal monatomic gas.

The slope of the isothermal curves (11) is given by

$$P\,dV + V\,dP = 0, \quad \frac{dP}{dV} = -\frac{P}{V}; \quad \text{(isothermal)} \tag{13}$$

whereas the slope of the reversible-adiabatic curves (9) is given by

$$P\gamma V^{\gamma-1}\,dV + V^{\gamma}\,dP = 0, \quad \frac{dP}{dV} = -\gamma\frac{P}{V}. \quad \text{(adiabatic)} \tag{14}$$

At any point of the paper in Fig. 5, the slope of the adiabatic through the point is 1.67 times the slope of the isothermal.

Example. *Air is introduced into the cylinder of a diesel engine at a temperature of 27° C. What is the temperature of the air after it has been compressed to 1/24 of its initial volume?*

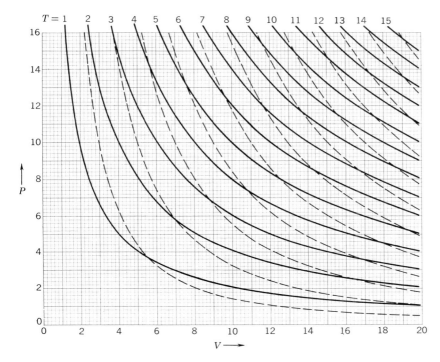

Fig. 5. Isothermals (solid lines) and adiabatics (broken lines) on a PV-diagram for an ideal gas with $\gamma = 1.67$. The units in which P, V, T are measured will vary with the quantity and kind of gas, and must be chosen so that the general gas law is satisfied.

We employ (7) and remember that absolute temperatures must be used. Hence in Fig. 4 we take $T_D = 300°$ K, and write

$$(300°\text{ K}) V_D^{(1.4-1)} = T_A (\tfrac{1}{24} V_D)^{(1.4-1)},$$

$$T_A = (300°\text{ K})(24)^{0.4},$$

since for air $\gamma = 7/5 = 1.4$. To evaluate $(24)^{0.4}$, we take the logarithm of 24, multiply by 0.4, and look up the antilog of the product. The result is 3.57, and hence $T_A = (300°\text{ K})(3.57) = 1071°$ K, or $798°$ C.

Example. *A sample of helium in a cylinder equipped with a movable piston has a volume of* 20 i³ *when the pressure is* 1 atm. *An external force on the piston slowly compresses the gas adiabatically to* ⅕ *its initial volume. What is the pressure of the gas after compression?*

Since the initial pressure and volume are given, it is convenient to use (9) in finding the final pressure. The units to be used are arbitrary provided the same units are used throughout; hence we set $P_D = 1$ atm, $V_D = 20$ i³, $V_A = 4$ i³, and solve for P_A in atm. Thus, since $\gamma = 1.67$,

$$(1 \text{ atm})(20 \text{ i}^3)^{1.67} = P_A (4 \text{ i}^3)^{1.67}, \qquad P_A = 5^{1.67} \text{ atm} = 14.7 \text{ atm}.$$

Note that $V = 20$, $P = 1$ and $V = 4$, $P = 14.7$ are two points on one of the broken curves in Fig. 5.

5. The Carnot Cycle

We shall now discuss a highly idealized heat engine known as the Carnot engine.* This discussion will show how it is possible to obtain mechanical work from heat by using a working substance that is carried through a cyclic process. The Carnot cycle is of fundamental importance because, as we shall demonstrate in the next section, it gives an expression for engine efficiency that furnishes a theoretical limit to the efficiency that a real heat engine can have.

In the Carnot engine shown in Fig. 6, the working substance is contained in a cylinder similar to those in Figs. 1 and 3. We can imagine the insulation at the closed end of the cylinder as removable. With the

*Devised by SADI CARNOT, French engineer, in 1824, in connection with a study of methods for improvement of the efficiency of the steam engine.

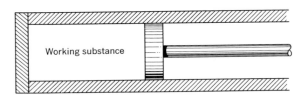

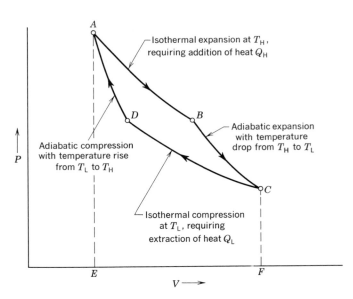

Fig. 6. The Carnot cycle (schematic). All four steps are conducted in reversible fashion.

insulation removed, the cylinder and working substance can be put into thermal communication with heat reservoirs so that isothermal processes may be employed; with the insulation in place, adiabatic processes may be used. Two heat reservoirs are available, one at the higher temperature T_H, the other at the lower temperature T_L. The Carnot cycle involves two isothermal processes, at temperatures T_H and T_L, and two reversible adiabatic processes, which carry the working substance from one of these temperatures to the other.

Consider the initial state of the working substance as characterized by P_A, V_A, T_H, and represented by point A in the PV-diagram of Fig. 6. By removing the insulation at the end of the cylinder and placing the cylinder in contact with a heat reservoir at temperature T_H, an isothermal expansion at T_H can be effected; the state of the working substance at the end of this isothermal expansion is characterized by P_B, V_B, T_H and is represented by point B in the diagram. As indicated in Sec. 3, a quantity of heat Q_H must be supplied from the reservoir in order to accomplish this expansion.

Now the insulation is replaced and the working substance is allowed to expand adiabatically until its temperature drops to T_L, the state of the substance at the end of this expansion being characterized by P_C, V_C, T_L and represented by point C on the diagram. As pointed out in Sec. 4, the work done in this adiabatic expansion is accomplished by reduction of the thermal energy of the working substance, since no thermal energy is supplied from an external source.

The insulation is again removed and the cylinder is placed in contact with a heat reservoir at the lower temperature T_L. The working substance is then compressed isothermally to volume V_D and pressure P_D. This isothermal compression involves delivering heat Q_L to the lower-temperature heat reservoir. Finally, the cylinder is again thermally insulated and the working substance is compressed adiabatically back to its initial state P_A, V_A, T_H, at which point its thermal energy is the same as when the cycle of operations was started.

The work done *by* the working substance during expansion is represented by the area *ABCFEA*. External work done *on* the substance during compression is given by the area *ADCFEA*. The *net work W* performed by the engine during the cycle is therefore presented by the area *ABCDA* on the diagram.

Since in the ideal engine no frictional processes are involved, and since the thermal energy of the working substance has the same value at the end as at the beginning of the cycle, the net work W done by the engine must be just the difference between the heat Q_H supplied from the reservoir at higher temperature T_H and the heat Q_L delivered to the reservoir at lower temperature T_L; i.e., $W = Q_H - Q_L$. This energy balance is represented schematically in Fig. 7. The efficiency of the Carnot engine,

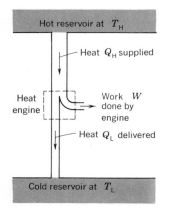

Fig. 7. Energy balance in a heat engine; $W = Q_H - Q_L$.

defined as the ratio W/Q_H of work done to heat supplied by the hot reservoir, is given by

$$\text{efficiency} = \frac{W}{Q_H} = \frac{Q_H - Q_L}{Q_H} = 1 - \frac{Q_L}{Q_H}. \tag{15}$$

In the special case in which the working substance is a perfect gas, we can show that the isothermal expansion ratios V_B/V_A and V_C/V_D are equal. Since points B and C are on the same adiabatic, we have from (7),

$$T_L V_C^{\gamma-1} = T_H V_B^{\gamma-1}. \tag{16}$$

Similarly, since A and D are on the same adiabatic,

$$T_L V_D^{\gamma-1} = T_H V_A^{\gamma-1}. \tag{17}$$

Division of (16) by (17) then demonstrates the equality of the isothermal expansion ratios:

$$V_C/V_D = V_B/V_A. \tag{18}$$

Since the expansion ratios are the same in the two isothermal processes, we see from (3) that these processes involve values of work that are in the ratio of the absolute temperatures and hence that the quantities of heat involved are in this same ratio:

$$\frac{Q_H}{Q_L} = \frac{n R T_H \log_e (V_B/V_A)}{n R T_L \log_e (V_C/V_D)} = \frac{T_H}{T_L}.$$

From (15) we then obtain the following expression for the efficiency of a Carnot engine employing an ideal gas as a working substance:

$$\text{Carnot efficiency} = 1 - \frac{T_L}{T_H} = \frac{T_H - T_L}{T_H}. \tag{19}$$

On the basis of Carnot's theorems given in the next section we can assert that this expression gives the efficiency of *any* Carnot engine working between heat reservoirs at absolute temperatures T_H and T_L regardless of the nature of the working substance employed.

This efficiency depends only on the ratio T_H/T_L of the absolute temperature of the reservoir from which heat is extracted to the absolute temperature of the reservoir to which heat is given. The efficiency is zero at $T_H/T_L = 1$, and increases as T_H/T_L increases. It is 0.5 at $T_H/T_L = 2$; 0.75 at $T_H/T_L = 4$; and so on.

Since all processes in the Carnot cycle are reversible, we can run the heat engine of Fig. 6 'backwards,' meaning that we go around the cycle in the counterclockwise direction rather than in the clockwise, reversing all the arrowheads in Fig. 6. We can start at C with an adiabatic *compression* to B; then an isothermal compression to A, during which heat is *given to* the *higher-temperature* reservoir; then an adiabatic expansion to D; and finally an isothermal expansion back to C, during which heat is *taken from* the *lower-temperature* reservoir. The machine is now acting as a *refrigerator,* taking heat from a cold body and giving it to a hot body. The magnitudes of Q_H, the heat given to the warmer reservoir, and Q_L, the heat taken from the cooler reservoir, are the same as those involved when the engine is used to do work. Furthermore, the net work done in the cycle is again the area *ABCDA,* but now this is net work done *on* the working substance, since more work is done on the working substance during the compression than is done by the working substance during the expansion. This net work is $Q_H - Q_L$, as represented in Fig. 8.

To extract an amount Q_L of heat from the reservoir at T_L and transfer it to the reservoir at T_H, we must do mechanical work of amount $Q_H - Q_L$. The sum, $Q_L + (Q_H - Q_L) = Q_H$, of the heat extracted and the mechanical work done must be added to the higher-temperature reservoir as heat. To accomplish the extraction of the heat Q_L requires mechanical work equal to

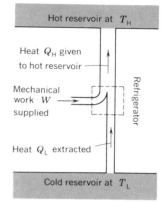

Fig. 8. Energy balance in a refrigerator; $Q_H = W + Q_L$.

$$W = Q_H - Q_L = \left(\frac{Q_H}{Q_L} - 1\right) Q_L = \left(\frac{T_H}{T_L} - 1\right) Q_L = \frac{T_H - T_L}{T_L} Q_L, \quad (20)$$

since $Q_H/Q_L = T_H/T_L$. The greater the temperature ratio T_H/T_L, the more the work required to effect the transfer of a given quantity of heat from a body at T_L to one at T_H.

Example. *On p. 321, we worked out an example in which it was found that a refrigerating system would have to transfer* 1000 BTU/h *from the interior of a refrigerator at* 40° F *to the exterior air at* 80° F. *If this were an ideal Carnot refrigerator, how much work* (in BTU's) *would have to be done by the engine per hour and what would be its required horsepower?*

We get the power from (20) by substituting $T_H = 80°$ F $= 540°$ R, $T_L = 40°$ F $= 500°$ R, $Q_L = 1000$ BTU/h, as

$$P = \left(\frac{40°}{500°}\right)\left(1000 \frac{\text{BTU}}{\text{h}}\right) = 80 \frac{\text{BTU}}{\text{h}}.$$

Since 1 BTU/h $= 3.93 \times 10^{-4}$ hp (*see* the table in Sec. 5 of the Appendix), this represents only

$$80 \times 3.93 \times 10^{-4} \text{ hp} = 0.0314 \text{ hp}.$$

6. Carnot's Theorems

Although the Carnot engine is a highly idealized device, it is an extremely useful concept and was used by Carnot to establish the maximum theoretical efficiencies attainable by real heat engines. His conclusions, which we shall derive from the second principle of thermodynamics, can be stated as follows:

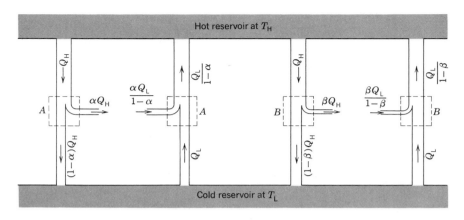

Fig. 9. Energy transformations effected by reversible engines. Left to right: A as engine; reversed A as refrigerator; B as engine; reversed B as refrigerator. The arrows at the center represent mechanical output of the engine or input to the refrigerator.

THEOREM I: *All reversible engines have the same efficiency when they operate between reservoirs at the same two temperatures.*

A reversible engine is one which, like the Carnot engine, will work equally well as a refrigerator, running through a cycle that accomplishes *exactly* the reverse of the engine cycle.

The above theorem is proved as follows: Suppose (Fig. 9) that we have two reversible engines A and B, of efficiencies α and β. As an engine, A takes heat Q_H from the hot reservoir; does work W which, according to the definition (15) of efficiency, equals αQ_H, and gives heat $Q_L = Q_H - W = (1-\alpha) Q_H$ to the cold reservoir. Note that we can write $Q_H = Q_L/(1-\alpha)$. As a refrigerator, A takes $(1-\alpha) Q_H = Q_L$ from the cold body, is supplied with mechanical energy $\alpha Q_H = \alpha Q_L/(1-\alpha)$, and delivers $Q_H = Q_L/(1-\alpha)$ to the hot body. B does similarly, with efficiency β. These relations are illustrated in Fig. 9 in terms of the heat taken from the hot body in the case of the engine and the heat taken from the cold body in the case of the refrigerator.

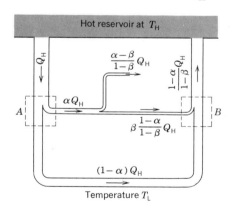

Fig. 10. If $\alpha > \beta$, we could use A to drive B as a refrigerator and get net work $(\alpha - \beta)/(1-\beta)$ for each unit of heat that A took in. The symbolism in the above diagram is the same as in Fig. 9, and the quantities of energy can be verified by comparison with Fig. 9.

Now assume that A is more efficient than B, that is, that $\alpha > \beta$. Then, as indicated in Fig. 10, we can use A as an engine to drive B as a refrigerator. The heat given out by A at the low temperature will be the heat input to B. A will be able to drive B and still furnish *additional* mechanical energy. The combination would be a self-acting engine that takes thermal energy from a single reservoir and changes it into work, in contradiction to the second principle. Therefore A is *not* more efficient than B. Similarly, we prove that B is *not* more efficient than A; therefore, A and B must have the same efficiency. Therefore *every reversible engine has the Carnot efficiency* (19), since it has the same efficiency as a Carnot engine that employs an ideal gas.

THEOREM II: *No actual engine can have a greater efficiency than a reversible engine working between the same two temperatures.*

An actual engine is nonreversible. It always has friction losses that cause nonreversible transformations of mechanical energy into heat, departures from thermal equilibrium that cause nonreversible transfers of heat, loss of thermal energy by conduction to colder bodies, etc.

Let A, Fig. 10, be an actual engine of efficiency α; B a reversible engine of efficiency β. Then if $\alpha > \beta$, we can use part of the mechanical output of A to drive B as a refrigerator to return all the heat given out by A to the hot body and still have some mechanical output from A left over. This result is in violation of the second principle. Therefore, A is *not* more efficient than B. In this case we cannot turn the argument around, since A is not reversible.

THEOREM III: *For any refrigerator to take Q_L units of heat from a cold body at temperature T_L and transfer it to a hot body at temperature T_H, at least $W = Q_L(T_H - T_L)/T_L$ units of mechanical work must be done on the working substance of the refrigerator.*

The mechanical work mentioned in this theorem is the value (20) required by a Carnot cycle, or, from arguments similar to those earlier in this section, by any reversible cycle. The proof of this theorem is similar to those presented above (*see* Probs. 43 and 44).

Carnot's theorems, proved directly from the principles of thermodynamics, can be regarded as exactly correct.

7. The Absolute Thermodynamic Temperature Scale

So far we have not been able to give a satisfactory operational definition of temperature. We are now prepared to define the *absolute thermodynamic temperature scale* already discussed on p. 293. This scale, which does not depend on the properties of any particular substance, was devised by the English physicist WILLIAM THOMSON (LORD KELVIN) about 1850.

An *ideal* gas rigorously obeys the relation $T = PV/nR$. *If we had an ideal gas*, this equation would serve as an excellent definition of absolute temperature. The thermodynamic scale is essentially an ideal-gas scale. But we do not have an ideal gas. However, we shall see that Carnot's theorems will enable us to determine ideal-gas temperatures from the behavior of *actual substances*.

We have seen that for a Carnot cycle employing an ideal gas, the ratio of heat exchanged with the hot body to heat exchanged with the cold body is

$$\frac{Q_H}{Q_L} = \frac{T_H}{T_L}. \tag{21}$$

We have *proved* from the second principle that (21) holds also for *any* reversible cycle employing an *actual* substance. Hence (21), for any reversible cycle working on adiabatics and isothermals, is taken as the *definition* of temperature ratio. Then if, to fix the size of the degree, we arbitrarily assign the temperature of one specified body, (21) in principle determines the temperature of any other body. Because its temperature is more accurately reproducible than that of any other, the body assigned an arbitrary absolute temperature is one at the temperature of the triple point of water, such as the vessel in Fig. 9, p. 366. This vessel is assigned the exact temperature

$$T_{TP} = 273.16° \text{ K}. \tag{22}$$

We shall now show how another temperature, say T_{SP}, the temperature of the steam point, is determined on the absolute thermodynamic scale.

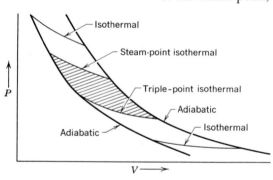

Fig. 11. Isothermals and reversible adiabatics for a real substance (schematic).

Suppose that we have a quantity of actual substance—say a gas—and that we have studied its properties in detail, employing an arbitrary temperature scale, such as the scale depending on the expansion of mercury which we have discussed earlier. In particular, we know the pressure and volume as a function of our arbitrary temperature, and we know C_P and C_V (say in J/kmole per arbitrary degree) for every pressure and temperature. It turns out that this information gives us enough data to *compute* the heat that would have to be added in an isothermal expansion or removed in an isothermal compression at any temperature on our arbitrary scale.

Now consider a cycle (cross-hatched in Fig. 11) operating between the triple point and the steam point and make the computations of Q_{SP} and Q_{TP} mentioned above. From (21) we can write $T_{SP}/T_{TP} = Q_{SP}/Q_{TP}$, from which we can compute the absolute value of T_{SP} on the Kelvin scale.

The second principle of thermodynamics guarantees that we shall get the same value for T_{SP} no matter what substance we use. It is in this way, by careful measurement of the properties of real gases, that the experimental values of the steam-point and ice-point temperatures:

$$T_{SP} = (373.146 \pm 0.004)° \text{ K}, \qquad T_{IP} = (273.1500 \pm 0.0002)° \text{ K}, \qquad (23)$$

have been determined. These two temperatures differ by 100 degrees within the experimental uncertainty of the steam-point value.

The absolute thermodynamic temperature at any other arbitrary fixed temperature point can be similarly determined by computations employing a cycle reaching to an isothermal at that temperature, as indicated in Fig. 11.

Although these ideas are simple in principle, the fixing of the absolute temperature scale has required a formidable amount of experimental work on the accurate determination of the properties of real substances, particularly of hydrogen and helium.

8. Entropy

Let us look again at the isothermal and reversible-adiabatic curves in the diagrams of Figs. 5 and 6. Some of these curves are again represented schematically in Fig. 12. In these so-called *state diagrams,* for a given kind and quantity of ideal gas, each point represents a specific set of values of the state variables P, V, and T. These quantities are called *state variables* because they are constants associated with that particular point in the state diagram, irrespective of how the gas arrived at that point (state).

The state variable P is given by the ordinate of the point, V is given by its abscissa, and T by the value marked on the isothermal curve on which the point lies. We immediately wonder whether it might not be useful to define a fourth state variable associated with values to be marked on the *reversible-adiabatic* curves. It does turn out to be very useful, and this fourth state variable is called the *entropy* and denoted by S. Thus in Fig. 12 we have marked the two adiabatic curves with low and high values S_L and S_H of the entropy.

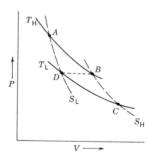

Figure 12

In most cases, the zero value of entropy is taken as arbitrary, just as in the case of potential energy, and we are concerned with entropy *changes*. We therefore ask, how shall we define the entropy difference $\Delta S = S_H - S_L$? We have several clues. If the entropy is to be *constant* along a particular adiabatic curve, we must have $S_B - S_A = S_C - S_D = \Delta S$. Also we must have $S_D - S_A = S_C - S_B = 0$. We also recall, from equation (21), that $Q_H/T_H = Q_L/T_L$. We shall show that these relations are all consistent with the following definition of entropy difference:

If a gas changes from state 1 to state 2 by a reversible process, the **change in entropy** is given by

$$S_2 - S_1 = \int_1^2 \frac{dQ}{T}. \qquad (24)$$

This is a line integral taken along any path in the state diagram that leads from point 1 to point 2; dQ is the heat added to the gas as it moves a distance dl along this line and T is the temperature associated with the point on the line that is contained in dl.

Now let us apply this definition to some of the paths in Fig. 12. Suppose that we change the state from A to D or from B to C along the reversible adiabatic curve. No heat is added so $dQ=0$ in all steps and the entropy does not change; a reversible adiabatic process can therefore be called an *isentropic* process. If we change state from A to B along the isothermal, the temperature is constant, so

$$S_B - S_A = \int_A^B \frac{dQ}{T_H} = \frac{1}{T_H} \int_A^B dQ = \frac{Q_H}{T_H}.$$

Similarly, we could find $S_C - S_D = Q_L/T_L$, which, as we have already noted equals Q_H/T_H.

It can be rigorously proven that the value of the integral (24) is independent of the path taken from 1 to 2. Thus we could compute the entropy change from A to B in Fig. 12 by first following the adiabatic from A to D and then the dotted line, along which volume and temperature are simultaneously increasing, from D to B. Since there is no entropy change along the adiabatic, we conclude that along the path DB, $S_B - S_D = \int dQ/T = Q_H/T_H = Q_L/T_L$.

The definition of entropy is by no means confined to gases. The definition (24) will apply to any material system that moves from a state 1 of thermal equilibrium to a state 2 of thermal equilibrium by a reversible process. It is seen from the above relations that entropy is measured in joules per kelvin, or kilocalories per kelvin.

The real usefulness of the concept of entropy is in connection with the second law of thermodynamics. Without proof, we assert that the second law can be stated as follows: *Any isolated material system that starts in one equilibrium state and ends up in another always suffers an increase in entropy, never a decrease.* The movement of an isolated material system from one equilibrium state to another is always an irreversible process, and entropy always increases. Consider the free-expansion experiment. The gas in the one container is, say, at point A in Fig. 12 at the instant the valve is opened. By itself it changes to point B of double the volume but at the same temperature. During the transition it passes through a series of nonequilibrium states in which pressure and temperature are not constant, so that these states cannot be represented by points on the state diagram. But it eventually settles down at B with, as we have seen, increased entropy.

Again, it may be shown that if two gases at different temperatures, or different pressures, or both, are allowed to mix adiabatically (that is, with no addition or subtraction of heat from external sources), the system composed of the two gases ends up with an entropy greater than its initial entropy—and, of course, the system would never by itself 'unmix' to decrease the entropy back to the starting value.

As has been remarked, *the entropy of the universe is continuously increasing.*

9. The Steam Engine

The Carnot engine is a highly idealized device employing reversible processes. Practical heat engines do not involve reversible processes or reversible cycles; an automobile engine cannot produce gasoline and air from its own exhaust gases, nor will it 'run backwards' to serve as a refrigerator. We shall now consider the steam engine as a typical practical heat engine and show how the Carnot efficiency sets an upper limit on the performance to be expected.

A simple steam power plant is represented schematically in Fig. 13. Water is heated and evaporated in a boiler, usually at a pressure well above atmospheric and hence at a temperature well above 100° C. The pressure is controlled by a feed pump which feeds in water as needed to maintain the pressure. The steam enters the cylinder and is later exhausted to a condenser. The condenser is pumped free from air so that the pressure in the condenser is close to the low vapor pressure at the temperature of the water that circulates through the condenser to condense the steam.

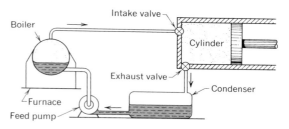

Fig. 13. A simple steam power plant. The circulating-water system that extracts the latent heat in the condenser is not shown.

The conditions *in the cylinder* of the engine are shown by the *indicator diagram* of Fig. 14. On this diagram the actual pressure in the cylinder is plotted as a function of the varying volume of the cylinder. The closed curve on the indicator diagram is traversed clockwise once during each cycle. The net mechanical work done per cycle of the engine, in fp, is seen to be just the area of the closed curve on the indicator diagram in $(p/f^2) \cdot f^3$. This plot is called an indicator diagram because a steam-engine cylinder can be readily equipped with a card that moves back and forth with the piston, and on which a recording pressure gauge connected to the cylinder will write the diagram automatically. The simplest way of determining the actual mechanical power developed by the engine is from the area of an indicator diagram.

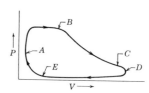

Fig. 14. Typical indicator diagram showing conditions in a steam-engine cylinder.

In Fig. 14, the intake valve is opened at A, and steam from the boiler is admitted during the first part of the stroke until the intake valve is closed at B. During the major portion of this part of the stroke the pressure is essentially constant at boiler pressure. From B to C, both valves are closed and the steam undergoes an essentially adiabatic expansion with resultant drop in pressure and temperature. Since the steam was saturated on admission, some of the steam condenses in the cylinder during this expansion. The exhaust valve is opened at C, a short time before the end of the expansion stroke at D, and the pressure rapidly drops to the condenser pressure, which is maintained during the return stroke until the exhaust valve is closed at E. From E to A the remaining steam is compressed to cushion somewhat the sudden pressure rise occurring when the intake valve is opened at A.

One must not make the mistake of considering the indicator diagram of Fig. 14 as analogous to the Carnot-cycle diagram of Fig. 6. Figure 14 plots conditions in the cylinder for a variable quantity of working substance as steam enters and leaves. The diagram that is analogous to Fig. 6 is the Rankine diagram of Fig. 15, in which one takes a fixed mass of H_2O and follows this same material around the closed circuit of Fig. 13.

At A on the Rankine diagram, the substance is in the form of water which has just been pumped into the boiler but is still at condenser temperature T_L. Between A and B the water has been heated to the boiling point T_H. B to C represents the volume increase (actually enormous in comparison with that shown schematically on Fig. 15) on evaporation. This is an isothermal process in which heat is added at T_H. At the condition C the gas enters the cylinder and accomplishes the first part of the work cycle in the cylinder before the intake valve is closed. CD represents the expansion that accompanies the temperature drop to T_L, partly in the cylinder and partly through the exhaust valve into the condenser; this last part is inherently irreversible. DE represents the volume decrease accompanying complete condensation to water, with latent heat given up at temperature T_L. Finally, EA represents the pressure increase on the liquid when it is pumped back into the boiler.

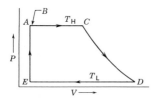

Fig. 15. Rankine diagram for a fixed mass of H_2O passing once around the circuit of the power plant of Fig. 12.

Now let us consider the theoretical limit to the efficiency of the steam engine. Let heat Q_H be added to our mass of H_2O. This heat is added in the boiler, partly at temperatures between T_L and T_H as the water is heated from A to B, mostly at T_H as the water is evaporated. In principle, all this heat could have been derived from a body at temperature T_H. Heat is given up to the cooling water in the condenser at T_L. Our actual engine cannot be more efficient than an ideal reversible engine acting between a reservoir at T_H and one at T_L. So, of the heat added to the water in the boiler, at most only a fraction

$$\frac{T_H - T_L}{T_H} \quad \left\{ \begin{array}{l} T_H = \text{boiler temperature} \\ T_L = \text{condenser temperature} \end{array} \right\} \quad (25)$$

can be converted into work. If we define the *thermal efficiency* of our steam cycle as the ratio: (work done)/(heat added to the water in the boiler), then *expression (25) furnishes a theoretical upper limit to this thermal efficiency*. The value (25) is called the *Carnot efficiency*.

For example, if $T_H = 350°$ F $= 810°$ R and $T_L = 100°$ F $= 560°$ R, the Carnot efficiency is

$$\text{Carnot efficiency} = \frac{T_H - T_L}{T_H} = \frac{250}{810} = 31\%.$$

The thermal efficiency of an actual engine of 250-hp rating operating between these temperatures might be given by:

$$\text{thermal efficiency} = \frac{\text{work done}}{\text{heat added to water}} = 23\%,$$

whereas the overall efficiency, which takes into account waste in heat value of the fuel in heating the water, might be given by

$$\text{overall efficiency} = \frac{\text{work done}}{\text{heat value of fuel burned}} = 17\%.$$

Since the Carnot efficiency furnishes an upper limit, no improvements in this steam plant could possibly bring the thermal efficiency above 31 percent unless T_H and T_L were changed. The advantage of using a high steam temperature and a low condenser temperature is immediately apparent, and accounts for the continuous trend, through the decades, toward higher and higher steam temperatures in the design of new plants.

The more modern *steam turbine* has the same limitation (25) on thermal efficiency. It differs from the reciprocating steam engine of Fig. 13 in that the work accomplished by the steam in expanding and cooling is done on a set of rotating blades.

10. Refrigerators; Heat Pumps

A refrigerator acts as a reversed heat engine in that it extracts heat from a cold body at temperature T_L and gives it to a hot body at temperature T_H, as in Fig. 8. We have seen that to effect this transfer requires mechanical work. For a given pair of temperatures, the more heat the refrigerator extracts from the cold body per unit of work done, the better is the refrigerator, since the cost of operation is principally the cost of supplying the work by means of an electric motor or other type of engine. Hence, we define a performance coefficient as

$$\text{performance coefficient} = \frac{\text{heat extracted}}{\text{work done}} = \frac{Q_L}{W}. \tag{26}$$

The argument of Sec. 6 proves that *the performance coefficient of an actual refrigerator is always less than that of a reversible refrigerator,* which is, from (20),

$$\text{Carnot performance coefficient} = \frac{T_L}{T_H - T_L} = \frac{1}{(T_H/T_L) - 1}. \tag{27}$$

The Carnot performance coefficient depends only on the ratio T_H/T_L;

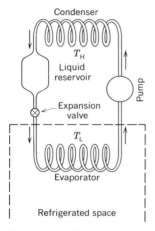

Fig. 16. Refrigeration system (schematic). Heat is extracted at the low temperature T_L and given up at the high temperature T_H.

it is very large when T_H/T_L is close to 1; it is greater than unity when T_H/T_L is less than 2, less than unity when T_H/T_L is greater than 2.

The refrigeration cycle actually used is a vapor cycle, employing a gas that is easily condensed by pressure. Ammonia (NH_3), sulphur dioxide (SO_2), and Freon (CCl_2F_2) are most commonly used. Figure 16 shows a schematic diagram of the equipment. The pump regulates the high pressure in the condenser. The low pressure in the evaporator is regulated by the expansion valve. The working substance evaporates at low temperature, taking in heat, and condenses at high temperature, giving out heat. The work referred to in (26) is that done by the pump. (See Prob. 45 for a PV-diagram.)

The name *refrigerator* implies an apparatus whose purpose is to *cool;* but the *same* apparatus can be used as a *heater*—when so used it is called a *heat pump*. The heat pump extracts heat from a cold body (the earth, underground water, or outside air) and furnishes it to a warmer body (the air in a house or building). To accomplish this heat transfer, mechanical energy must be used, but the mechanical energy can be small compared with the amount of heat transferred, provided the temperature differences involved are low.

Example. *In a heat-pump installation, an ample supply of underground water is available at 50° F to circulate past the evaporator, and it is desired to heat air, circulating past the condenser, to 80° F to use in heating a house. To accomplish the heat transfer, the refrigerant is operated at 40° F in the evaporator and at 110° F in the condenser. The mechanical power is furnished electrically; assume that the installation will give an overall performance coefficient of half the Carnot performance coefficient. Compare the cost of using this heat pump with that of direct electrical heating.*

The Carnot performance coefficient for a refrigerator operating between 40° F and 110° F is

$$\frac{460+40}{110-40} = \frac{500}{70} = 7.1.$$

To account for the fact that a motor is not 100 per cent efficient in converting electric energy to work, we have introduced the term *overall* performance coefficient; by this we mean the ratio of heat extracted to electric energy supplied to the motor. With an over-all performance coefficient of half the Carnot coefficient, or approximately 3.5, the heat extracted from the underground water will be 3.5 times the total electric energy used. The heat given to the house will be 4.5 times the total electric energy used, since, if the motor is in the house, *all* the electric energy will be added to the heat extracted. As compared with house heating by direct expenditure of electric energy in an electric heater, we have gained a factor of 4.5 in heat, 3.5/4.5 = 78 per cent of the heat coming from the cold underground water.

11. Liquefaction of Gases

Gases that can be liquefied by pressure alone at normal temperature, that is, those whose critical temperatures lie well above normal temperature, can easily be obtained in liquid form at their boiling point at normal atmospheric pressure. For example, the critical temperature of ammonia (NH_3) is 132° C. Suppose that we want to obtain a quantity of liquid ammonia in an open flask at $-33.5°$ C, its boiling point at 1 atm pressure. We start with gaseous ammonia and compress it to about 10 atm pressure, at which point it will liquefy at room temperature. We extract the heat of compression and the latent heat by means of circulating water or circulating air until we have liquid ammonia at 10 atm pressure and room temperature. If we then release the pressure, with the liquid in a thermally insulated vessel, the liquid will first evaporate rapidly; but since the latent heat of evaporation must come from the liquid itself, the liquid will also cool rapidly so that soon we shall be left with a large fraction of the liquid at $-33.5°$ C. This liquid can be kept for a considerable time in a double-walled flask. This procedure will not work for a gas whose critical temperature is below room temperature.

A permanent gas (oxygen) was first liquefied in 1877 by a Swiss physicist, RAOUL PICTET, who used a vapor refrigeration cycle similar to the one we have described in the previous section except that the cycle was in two stages. The high-temperature stage employed SO_2, and the evaporator in the SO_2 cycle drew its heat from the condenser in a lower temperature CO_2 cycle. The evaporator in the CO_2 cycle cooled the oxygen gas to its liquefaction temperature. All gases except hydrogen and helium have been liquefied by such multistage vapor refrigeration cycles.

The *Claude process* of liquefaction employs essentially an adiabatic expansion for cooling. In principle, by letting a gas expand and do work, the temperature can be lowered as much as we please.

The *Linde process* of liquefaction depends on the fact that actual gases are not perfect and do suffer some temperature decrease on *free expansion* because of the slight attractive forces that do exist between the molecules. At temperatures not too far above the critical, the temperature suffers a substantial decrease on free expansion through a pressure range of several hundred atmospheres.

By precooling hydrogen with liquid air before expansion, hydrogen is readily liquefied; and by precooling helium with liquid hydrogen before expansion, helium is readily liquefied. The expansion can be either the adiabatic expansion of the Claude process or the free expansion of the Linde process. By using four successive stages, the Claude process can even liquefy helium without precooling. The widely used Collins helium

PROBLEMS

1. During launching, a yacht weighing 400 tons slides down inclined ways through a vertical distance of 10 f before striking the water. If the yacht is then moving at 16 f/s, how many BTU of heat were developed by friction on the ways?   Ans: 6200 BTU.

2. A locomotive exerts a drawbar pull of 30 000 p on a 400-ton train when it is going up an incline of 1° slope at a constant speed of 30 mi/h. How many BTU per second are generated in the wheels and bearings of the cars of the train?

3. The air in the cylinder of a Diesel engine is at 35° C and 1 atm before compression. It is compressed to $\frac{1}{17}$ of its original volume. Find the final temperature and pressure of the air.   Ans: 685° C; 52.8 atm.

4. The air in the cylinder of a Diesel engine is at a temperature of 100° F and a pressure of 14 p/i² at the beginning of compression. It is compressed to $\frac{1}{14}$ of its original volume. Find the final temperature and pressure.

5. Find the work necessary to compress 1000 cm³ of helium at 0° C and 1 atm adiabatically to one-quarter of this volume.
Ans: 232 J.

6. If the isothermals (solid curves) of Fig. 5, p. 389, are plotted for 5 kmol of helium, the unit of volume is 1 m³, and the unit of pressure is 1 atm, what is the unit of temperature?

7. In a *Wilson cloud chamber* (*see* figure), air that is saturated with water vapor is rapidly expanded adiabatically to about 20 per cent greater volume by a sudden compression of the bellows. After the expansion the water vapor is at a partial pressure above that of saturation at the reduced temperature and begins to condense. Electrically charged ions are particularly effective as condensation nuclei. Hence, if a high-speed charged particle passes through the chamber close to the time of the expansion, leaving behind a trail of ionized air molecules, this trail will show up as a track of fog droplets which can be photographed. Such photographs are shown in Figs. 19 and 20 of Chapter 28.

If air saturated with water vapor at 30° C and a total pressure of 1 atm is expanded adiabatically to 20 percent greater volume, find the partial pressure of the water vapor in the supersaturated air just after the expansion, and compare with the pressure of saturated vapor. In computing the temperature after the expansion, neglect the difference between γ for water vapor and that for air and assume that the whole gas has $\gamma = 1.4$. This method will introduce little error because the water vapor, which has a slightly lower γ, constitutes only a small fraction of the whole gas.   Ans: 24.6 mm as compared with 8.4 mm for saturation.

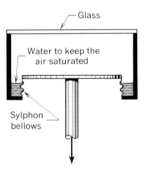

Problems 7–8

8. Repeat Prob. 7 for a 30 percent volume expansion.

9. Compute the efficiency of a Carnot engine operating with reservoir temperatures of 227° C and 27° C.   Ans: 40.0%.

10. Compute the efficiency of a Carnot engine operating with reservoir temperatures of 600° F and 60° F.

11. If the Carnot engine of Fig. 6 contains 0.172 kmol of gas for which $\gamma = 1.4$, and if $P_A = 10$ atm, $P_B = 5$ atm, $T_H = 273°$ C,   and

$T_L = 0°$ C, make a table showing in joules (a) the heat added *to* the gas, (b) the work done *by* the gas, and (c) the *increase* in thermal energy, in each of the four steps of the cycle and for the whole cycle.

Ans:
	A→B	B→C
(a)	542 000	0
(b)	542 000	980 000
(c)	0	−980 000

	C→D	D→A	Whole cycle
(a)	−271 000	0	271 000
(b)	−271 000	−980 000	271 000
(c)	0	980 000	0

12. If the Carnot engine of Fig. 6 contains 0.25 kmol of gas with $\gamma = 1.67$, and if $P_A = 12$ atm, $P_B = 8$ atm, $T_H = 400°$ C, and $T_L = 20°$ C, make a table showing in joules for each step of the cycle and for the whole cycle: (a) the heat added *to* the gas, (b) the work done *by* the gas, (c) the *increase* in thermal energy. How do your answers change if the engine is operated in reverse, as a refrigerator?

13. In Prob. 11, what is the change in entropy in each of the four steps of the cycle and in the whole cycle?
Ans: 1990, 0, −1990, 0, 0 J/K.

14. In Prob. 12, what is the change in entropy in each of the four steps of the cycle and in the whole cycle?

15. What is the increase in entropy of 1 kg of ice at $0°$ C that is melted to water at $0°$ C? Take the latent heat of fusion as 80 kcal/kg. To compute the entropy change we must assume the heat to be added reversibly; discuss in principle how the heat might be added (almost) reversibly.   Ans: 0.293 kcal/K.

16. Show that the increase in entropy of a mass m of water that is heated from absolute temperature T_1 to T_2 is given by

$$\Delta S = cm \log_e(T_2/T_1).$$

To compute the entropy change, we must assume the heat to be added reversibly; discuss in principle how the heat might be added (almost) reversibly.

NOTE: *The next two problems are in British units. Take the specific heat of liquid water as 1 BTU/lb·F deg over the whole range required. The following table will give the required vaporization temperatures and latent heats:*

P (p/i²)......	1.0	2.0	100	165
T (°F)	102	126	328	366
L (BTU/lb)....	1035	1022	888	857

17. A steam engine that exhausts to a condenser at a pressure of 2 p/i² receives steam at 100 p/i² absolute pressure. The steam consumption is 14 lb per indicated horsepower-hour. What is the thermal efficiency of the engine? What is the Carnot efficiency?
Ans: 16.7%; 25.6%.

18. Steam is supplied to an engine at 165 p/i² absolute. The condenser pressure is 1.0 p/i² absolute. The steam consumption is 10 lb per indicated horsepower-hour. What is the thermal efficiency? What is the Carnot efficiency?

19. To extract 2 joules of heat from a body at $0°$ C and give it to a body at $100°$ C, how many joules of mechanical energy are required by a Carnot refrigerator, and how many joules are given to the hot body?   Ans: 0.732, 2.732.

20. To extract 1 kcal of heat from a body at $−73°$ C and give it to a body at $127°$ C, how many kcal of mechanical energy are required by a Carnot refrigerator, and how many kcal are given to the hot body?

21. A 'one-ton' Carnot refrigerator extracts heat from water at $32°$ F to make 1 ton of ice at $32°$ F per day, giving this heat to the air at $90°$ F. What mechanical horsepower is required to drive the refrigerator?
Ans: 0.554 hp.

22. If a Carnot refrigerator which extracts heat from a cold-storage plant at $−20°$ F and gives it to the atmosphere at $100°$ F has 50 'ton' capacity (meaning that the amount of heat extracted in 24 h is the latent heat of 50 tons of ice), what horsepower is needed to drive the refrigerator?

23. If the area of the indicator diagram of Fig. 14 is 25 (p/i²) f³ and the engine operates at 400 cycles per minute, what horsepower does it develop?   Ans: 43.6 hp.

24. If the area of the indicator diagram on each side of a double-acting cylinder is 30 (p/i²) f³, at what speed must the engine run to develop 90 hp? (A *double-acting* cylinder is closed on both ends and steam is admitted alternately on the two sides of the piston.)

25. How many tons of ice can be frozen per day by a refrigerating machine working between 0° F and 100° F if it has 50 percent of the performance coefficient of a Carnot machine working between these same limits and is driven by an 8-hp motor? The ice is frozen to a temperature of 15° F from water initially at 60° F. *Ans:* 3.12.

26. How many horsepower are required to freeze 50 tons per day of ice at 15° F from water initially at 60° F by a refrigerating machine working between 0° F and 90° F which has 70 percent of the performance coefficient of a Carnot machine working between these same limits?

27. Compare the cost of heating a house by coal of 14 000 BTU/lb at $15.00 per ton, with 60 percent overall combustion efficiency, and by electricity at 1.5 cents per kilowatt-hour if a heat pump of overall performance coefficient of 3.5 is used. *Ans:* 11 200 BTU/cent for coal; 10 200 BTU/cent for the heat pump.

28. A sample of CO_2 in a cylinder with a movable piston is heated by a burner. If 60 cal of heat are added to the gas and the gas expands against an external pressure of 1 atm from an initial volume of 1000 cm³ to a final volume of 1600 cm³, what is the change in the thermal energy of the gas?

29. A gas is heated by a burner. If the gas receives 3 BTU and expands from 0.7 f³ to 1.0 f³ against an external pressure of 2 atm, what is the increase in thermal energy of the gas? *Ans:* 1.368 BTU.

30. If we have a reservoir at absolute zero, is it possible to have a 100 percent efficient Carnot engine? By means of a Carnot refrigerator, how much work must be done in order to maintain a reservoir at absolute zero?

31. A certain heat engine uses a diatomic gas as a 'working substance.' The gas has an initial volume of 1 f³ at a pressure of 2 atm. The gas is heated and expands at constant pressure to 2 f³. It is then cooled, the pressure dropping to 1 atm without change in volume. The gas is next cooled further in such a way that its volume decreases to 1 f³, the pressure remaining 1 atm. The gas is finally heated at constant volume until the pressure is again 2 atm. Plot this cycle on a PV-diagram. (*a*) How much work is done by the engine during one cycle? (*b*) At what point of the cycle is the thermal energy of the gas greatest? (*c*) How many complete cycles would this engine make each second if its output is 8 horsepower? *Ans:* (*a*) 2120 fp; (*c*) 2.08.

32. Steam is supplied to a large turbine at a temperature of 800° F and is exhausted at 400° F for use with a lower-pressure heat engine. What is the Carnot efficiency of the turbine? If the lower-pressure heat engine exhausts the steam into the atmosphere at 100° F, what is its Carnot efficiency? Compare the work that would be obtained from these two engines, assumed ideal, with that from a single ideal engine operating between 800° F and 100° F.

33. A certain heat engine has a power output of 2 hp. This heat engine receives saturated steam from a boiler in which the absolute pressure is 39.3 atm, and exhausts into a condenser at 50° C. What is the Carnot efficiency of this heat engine? If the overall efficiency of this engine is 10 %, how much bituminous coal is used each hour in heating the boiler? *Ans:* 38.2 %; 3.77 lb.

34. A gasoline engine consumes 600 lb of gasoline while operating at 124 hp during a test run of 8 h. What mass of fuel does the engine use per horsepower-hour? How many gallons of gasoline are consumed per horsepower-hour? What is the overall efficiency of the engine? (Density of gasoline is 42 lb/f³.)

35. A certain airplane engine develops a torque of 606 p·f when operating a propeller at 2600 rev/min. What is the horsepower developed by the engine? If the overall efficiency of this engine is 20 percent, how many gallons of gasoline of density 42 lb/f³ are burned per hour? *Ans:* 300 hp; 33.1.

36. Energy from the sun reaches the top of the earth's atmosphere at a rate of approximately 1.35 kW/m². Neglecting atmospheric absorption, calculate the rate at which heat is supplied to a solar engine employing a paraboloidal mirror 3 m in diameter. Discuss various ways in which such a solar engine might be constructed. If this solar engine has a Carnot efficiency of 25 percent, what is the maximum power output that might be expected?

37. By using a refrigerator employing a Carnot cycle, we wish to remove heat from various cold reservoirs and give the heat to a reservoir at 27° C. How much work must be done in removing 1 J of thermal energy from a reservoir at 0° C? at −73° C? at −173° C? at −223° C?   Ans: 0.0989, 0.500, 2.00, 5.00 J.

38. The normal temperature decrease of the atmosphere with increasing altitude in the lower levels of the earth's atmosphere is maintained by the fact that in the normal stirring and mixing of the air, air parcels that rise are cooled by adiabatic expansion as the pressure decreases, while those that sink are correspondingly warmed by adiabatic compression. Consider surface air at 70° F in the trade winds that blow across a 5000-f ridge in Hawaii; when the surface air climbs 5000 f, its pressure decreases from 760 mm to 630 mm. What is the corresponding decrease in temperature if the expansion is adiabatic? Express this value in F deg per 1000 f; this value is known as the *normal adiabatic lapse rate*.

39. In the formula $dW = P\,dV$ for the work done by a gas, we can use the kinetic theory value for P only if the gas is continuously in thermal equilibrium, which means, in Fig. 3, that the speed of the piston must be very small compared to the average speed of the gas molecules, so that the energy loss of those particular molecules that hit the piston can be considered as continuously redistributed among all the molecules of the gas. Consider the model of Prob. 39, p. 357, in which one-third of the molecules are moving right or left in Fig. 3, with speed $v = \sqrt{\overline{v^2}}$. Let the piston be moving to the right with speed v_P, where $v_P \ll v$. Using the Newtonian principle of relativity, show that the loss of kinetic energy of each molecule hitting the piston (assuming perfectly elastic collisions) is $2\mu v v_P$, and hence that the total loss of energy per second per unit area of piston is just Pv_P, which is the power developed per unit area.

40. A mixture of water and water vapor is carried through a Carnot cycle between 212° F and 32° F. It is found that 1700 BTU must be added to the mixture at the high temperature and 1240 BTU must be removed at the low temperature. Using these data, determine the value of absolute zero on the Fahrenheit scale.

41. In an attempt to obtain a law that would apply to real gases, Van der Waals suggested the following equation as a replacement for Boyle's law:

$$(P + a/V^2)(V - b) = K,$$

where a, b, and K are constants. The constant a is related to forces between molecules—the so-called *Van der Waals forces*; b is related to the actual volume of the molecules; and K is proportional to the number of moles in the sample and to the absolute temperature. Obtain an expression for the work done by a Van der Waals gas in an isothermal expansion from V_1 to V_2.

42. By consideration of a Carnot engine, attempt to formulate a valid definition of *absolute zero*. Might the Stefan-Boltzmann law be used as the basis for a valid definition of absolute zero? the general gas law?

43. From the figure, show that any reversible engine has the same performance coefficient, when acting as a refrigerator, as the Carnot reversible engine which employs an ideal gas.

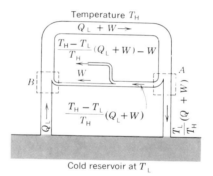

Problems 43–44

44. In the figure, let B be an actual refrigerator and A an ideal reversible engine. Show that the performance coefficient of B cannot be greater than the Carnot performance coefficient.

45. The figure shows isothermals for Freon (CCl_2F_2) on a PV-diagram. The pressure scale is accurate but the volume scale is distorted for the sake of clarity. The diagram is like that for CO_2 on p. 370 except that the abscissa is the inverse, representing volume of a given

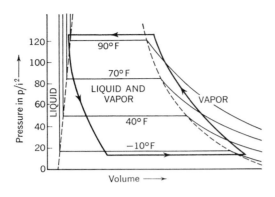

Problem 45

quantity of substance rather than density. Freon is popular for refrigeration because its phase changes, at the temperatures of interest, take place at moderate pressures; its critical temperature lies at 205° F. The heavy line in the figure represents a typical cycle for a refrigerator whose interior is maintained at 0° F and which gives the heat to a room at 80° F. Correlate the points on this cycle with the passage of the working substance around the mechanical system of Fig. 16, p. 402.

PART III

WAVE MOTION AND SOUND

WAVE MOTION 20

The fundamental common characteristic of all the various types of waves that we shall have occasion to describe is that they provide a mechanism for *transfer of energy from one point to another without physical transfer of material between the points.*

The *shock wave* from an explosion is a good example to consider first. The sudden creation of heat in the explosion raises a mass of gas in its immediate vicinity to very high pressure. This pressure is exerted on the surrounding air, which is compressed and increased in pressure. This pressure is in turn exerted on the air farther out. A pressure *wave* travels out from the explosion with a speed of over 1100 f/s; this wave contains the energy required to effect the compression of the air. This energy can be applied, for example, to break windows miles from the explosion. No material travels these miles—the actual motion of any air particle will be comparatively small—it is the *disturbance* that travels rapidly for a great distance and transmits the energy.

An ordinary sound wave involves the transmission of energy in the form of much milder disturbances of the air between a vibrating source and a receiver such as the ear or a microphone. Sound can also be transmitted as motions of the particles of a liquid or a solid. Sound is an example of a *mechanical wave motion*. Water waves are another type of mechanical wave motion; they can transmit energy capable of setting a boat or ship into oscillation. Waves on a string and compressional waves in a spring are other examples of mechanical wave motion.

In later chapters dealing with *light* and *radio* transmission, we shall discuss *electromagnetic waves*. Here the disturbance is not mechanical, but is electric and magnetic. However, the basic ideas that we shall introduce in this chapter are equally applicable to an understanding of these electromagnetic disturbances, which furnish the mechanism of transfer of heat by *radiation,* which we have discussed in Chapter 16.

In the present chapter, we shall introduce the basic ideas of wave motion by considering the simple cases of *transverse* waves in a stretched string and *longitudinal* waves in a spiral spring. Then we shall consider continuous sinusoidal waves and various phenomena that arise when two or more such waves are superposed. In the next chapter, we shall consider the generation and transmission of sound waves.

The speeds of transmission of mechanical waves through a medium can be expressed rather simply in terms of elastic and inertial properties of the medium. The derivation of these expressions involves the solution of equations containing derivatives with respect to time and with respect to spatial coordinates. We present such a derivation for the speed of a transverse wave along a string but content ourselves with a mere statement of the analogous expressions for the speeds of other types of mechanical waves. In Chapter 43, we shall derive, in detail, the speed of electromagnetic waves.

1. Mechanical Waves

In order for mechanical wave motion to occur, it is necessary to have a *source* that produces a displacement or *disturbance* of some kind and an *elastic medium* through which the disturbance can be transmitted. An elastic medium behaves as if it were a succession of adjoining particles with each particle occupying an equilibrium position; if one of the particles is displaced from its equilibrium position, this particle is immediately subjected to a restoring force as a result of attraction or repulsion by neighboring particles, which, in turn, are subjected to reaction forces exerted by the original particle. If one of the particles in a medium is given a sudden displacement by the *source,* this particle exerts forces on its immediate neighbors, which experience displacements; these immediately neighboring particles exert forces on *their* neighbors, which also undergo displacements, and so on. In this way, the initial disturbance at the source causes a displacement 'wave' to travel into the surrounding medium. As a result of the inertia of the particles, the displacements of all the particles do not take place instantaneously; the displacements of the particles far removed from the source occur later than the displacements of particles close to the source.

First consider the motion of a wave along a stretched string. In Fig. 1, part (a) shows a long string with its two ends attached to rigid supports. In part (b), a portion of the string is distorted in the indicated manner by pulling downward at point b while points a and c are held at their initial positions; this distorted portion of the string is called a *trough*. If the string is suddenly released at points b and c, the displaced portion is pulled back to its equilibrium position as a result of the elastic properties of the string; but the parts of the string near c are pulled downward and the trough moves to the right at a definite speed v, as shown in parts (c) and (d) of the figure. A *crest*, or upward displacement, would travel along the string in Fig. 1 at the same speed as that of the trough. The passage of a *crest* or a *trough* along a stretched string is an example of a *transverse* wave motion.

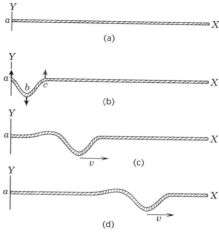

Fig. 1. Production and transmission of a transverse displacement (trough) in a stretched string.

In a **transverse wave** the displacements of the particles of the medium are perpendicular to the direction of propagation of the wave.

The initial work done on the string to effect the distortion in (b) is transmitted with the wave along the string as potential energy of distortion and kinetic energy of the up and down motion of the particles of the string.

Another type of wave, which can occur in a helical spring, is illustrated in Fig. 2. Part (a) of the figure shows a long coil spring with one end supported at point a; the support at the other end is not shown. If the spring is distorted in the manner indicated in Fig. 2(b) so that the coils of the spring are more closely spaced around point b than in other parts of the spring, the region in the vicinity of b is called a *condensation*. If the external forces producing the distortion are removed, the elastic restoring forces accelerate particles near b toward their equilibrium positions. When this acceleration occurs, the coils toward the right of the original condensation position are pushed closer together, and the condensation moves toward the right, at a definite speed v, as shown in (c) and (d).

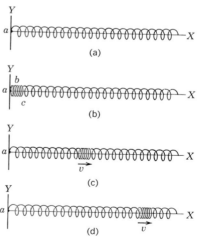

Fig. 2. Production and transmission of a 'condensation' along a helical spring.

WAVE MOTION

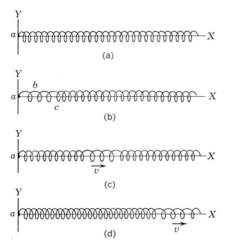

Fig. 3. Production and transmission of a 'rarefaction' along a helical spring.

Similarly, a *rarefaction,* in which the coils are initially pulled farther apart as in Fig. 3, will move to the right along the spring in the manner shown. Figures 2 and 3 illustrate *longitudinal* wave motion. The coils of the spring in Figs. 2 and 3 do not move up and down but only to the right and left, in directions parallel to the X-direction of propagation.

In a **longitudinal wave** the displacements of the particles of the medium are parallel to the direction of propagation of the wave.

Again, in Figs. 2 and 3, the energy of distortion is transmitted to the right along the spring.

The speed with which a wave moves through an elastic medium depends on the rapidity with which a distorted portion of the medium sets adjacent portions in motion. This rapidity depends in direct fashion on the forces that are brought to bear by a distortion, in inverse fashion on the inertia of the material that must be moved. In general, expressions for wave speed are of the form

$$\text{wave speed} = \sqrt{\frac{\text{elastic force factor}}{\text{inertia factor}}}. \quad (1)$$

The elastic force that tends to restore the shape of a stretched string is measured by the *tension* (F) in the string; the inertia is measured by the *mass per unit length* (μ) of the string. The speed of propagation of a transverse wave of small amplitude along a string is given by

$$v = \sqrt{\frac{F}{\mu}}. \quad \left(\begin{array}{c}\text{transverse}\\ \text{wave in string}\end{array}\right) \quad (2)$$

A derivation of (2) is given in Sec. 3 of the present chapter.

The elastic force tending to restore the original configuration of a distorted spring depends on the force constant K and the total length l of the spring; the inertia again depends on μ, the mass per unit length. The wave speed is

$$v = \sqrt{\frac{Kl}{\mu}}. \quad \left(\begin{array}{c}\text{longitudinal wave}\\ \text{in helical spring}\end{array}\right) \quad (3)$$

These equations give v in m/s if F is in N, K in N/m, l in m, and μ in kg/m; they give v in f/s if F is in p, K in p/f, l in f, and μ in sl/f.

Example. *Equation (3) for the speed of a longitudinal wave in a helical spring indicates at first glance a dependence on the length l of the spring. Now it* is physically clear that the speed of a longitudinal wave along a spring can depend only on how the spring is constructed and not on its over-all length since a pulse started as in Fig. 2 must travel with a speed governed by local conditions and not by where the right end of the spring might happen to be. *Compute the speed of travel of a wave on a spring* 16 f *long with* $K=1$ p/f *and* $\mu = \frac{1}{9}$ sl/f. *Then compute the speed of travel on the same spring cut in half so it is only* 8 f *long, and verify that the speeds are the same.*

Direct substitution in (3) gives

$$v = \sqrt{(1 \text{ p/f})(16 \text{ f})/(\tfrac{1}{9} \text{ sl/f})} = 12 \text{ f/s}$$

for the speed in the case of the original spring. The original spring would stretch 1 f under application of a stretching force of 1 p. Each half of this spring would experience the same stretching force of 1 p, but would stretch only ½ f. Hence when the spring is cut to 8-f length, its force constant changes to $K = (1 \text{ p})/(\tfrac{1}{2} \text{ f}) = 2$ p/f. In (3), K doubles if l is halved; the product Kl and hence the wave speed is independent of length.

2. Sinusoidal Wave Motion

In the types of waves we have discussed thus far, a single nonrepeated disturbance, called a *pulse*, is initiated at the source and then travels away from the source through the medium. Another important type of wave motion is the *regular wave train* or *continuous wave*. In this type of wave, a regular succession of pulses is initiated at the source and transmitted through the medium. Thus, if a floating block of wood is pushed up and down regularly on a water surface, a regular train of waves will be propagated outward.

The simplest type of regular wave train is a sinusoidal wave motion, which is illustrated in Fig. 4. Part (a) of this figure shows one end of a long stretched string attached to a weight supported by a spring. The weight is arranged so that it can move freely in the vertical 'ways' of a frame. If the weight is pulled downward a distance A and then released, the weight will move in the vertical direction with simple harmonic motion of a certain period T. Since the end of the string is attached to the weight, the oscillating weight acts as a source of a sinusoidal transverse wave that travels to the right along the string in the manner indicated by the curves of Fig. 4(b). These curves show successive 'snapshots' of the shape of the string during one half-cycle, after the motion has been well established. *The distance between adjacent crests or adjacent troughs in such a wave is called the wavelength;* in the figure the wavelength is denoted by λ. Each time the particle O attached to the weight makes a complete oscillation, the wave moves a distance λ in the X-direction. Hence the wave speed and the wavelength are related by the equation

$$v = \lambda/T, \tag{4a}$$

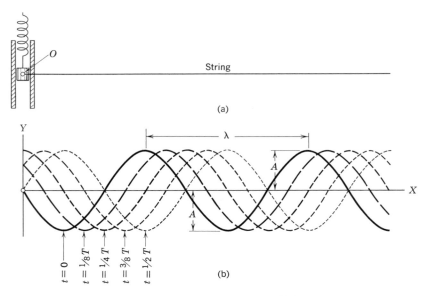

Fig. 4. Production and propagation of a sinusoidal transverse wave in a long string. Part (b) shows the motion of the string during one half-cycle of oscillation of O, from $Y=0$ to $+A$ and back to 0.

where T is the period of oscillation. In terms of the frequency $f=1/T$, this equation can be written as

$$v = f\lambda \tag{4b}$$

The relations (4a) *and* (4b) *are important general relations, between wave speed, wavelength, and frequency or period, that hold for sinusoidal wave motion of any type whatsoever.*

Example. *A fisherman in an anchored boat observes that his float makes* 12 *complete oscillations in* 10 s, *and that it takes* 4 s *for the crest of a wave to move the 36-f length of his boat. What is the wavelength of the waves? How many complete waves are there at any instant along the length of the boat?*

The number of oscillations per second is the frequency $f = 12/(10\text{ s}) = 1.2$ Hz. Since a crest moves 36 f in 4 s, the wave speed is $v = 36\text{ f}/(4\text{ s}) = 9$ f/s. Then, by (4b), the wavelength is

$$\lambda = \frac{v}{f} = \frac{9\text{ f/s}}{1.2/\text{s}} = 7.5\text{ f}.$$

The number of complete waves along the length of the boat would be $(36\text{ f})/\lambda = (36\text{ f})/(7.5\text{ f}) = 4.8$. This number can be obtained in another way: it is the number of waves passing the bow during the 4 s it takes for a crest to move the length of the boat, and since 1.2 waves pass a point per second, this number is $(1.2/\text{s})(4\text{ s}) = 4.8$.

It should be noted that the wave of Fig. 4, as it moves to the right, carries *energy* away from the vibrating body at the left. If the wave is to have *constant amplitude,* as indicated in the figure, the body must be *driven* by a periodic force that supplies this energy loss; in the following discussion we assume that this is done and that the wave has constant amplitude A. Actually, there is some 'internal friction' associated with the 'bending' of the string, so the macroscopic mechanical energy is ultimately converted into thermal energy and the amplitude of the wave gradually decreases as the wave travels along the string.

Now let us give a mathematical description of the transverse wave shown in Fig. 4. Sometime after the wave is established, if we take $t=0$ at the time the weight is moving upward through the origin, the shape of the string at $t=0$ will be given by the equation of the heaviest curve of Fig. 4(b):

$$Y = A \sin(-2\pi X/\lambda). \qquad (t=0)$$

At $t = \frac{1}{8} T$, one-eighth of a cycle later, the equation of the string is

$$Y = A \sin(\tfrac{1}{4}\pi - 2\pi X/\lambda). \qquad (t = \tfrac{1}{8} T)$$

Similarly, at the later times shown in Fig. 4(b), the equations are

$$Y = A \sin(\tfrac{1}{2}\pi - 2\pi X/\lambda), \qquad (t = \tfrac{1}{4} T)$$

$$Y = A \sin(\tfrac{3}{4}\pi - 2\pi X/\lambda), \qquad (t = \tfrac{3}{8} T)$$

$$Y = A \sin(\pi - 2\pi X/\lambda). \qquad (t = \tfrac{1}{2} T)$$

From these, we can write the equation for Y as a function of X and t. Note that for the point O at $X=0$,

$$Y = A \sin 2\pi t/T. \qquad (X=0)$$

This dependence on t at $X=0$ and the above equations for dependence on X at fixed t are consistent if we write

$$Y = A \sin(2\pi t/T - 2\pi X/\lambda). \qquad \textbf{(5a)}$$

If in this expression we substitute $T = 1/f$ and $\lambda = v/f$, we obtain the convenient form

$$Y = A \sin[2\pi f(t - X/v)]. \qquad \textbf{(5b)}$$

This is the equation for a sinusoidal transverse wave traveling in the positive X-direction with amplitude A. From this equation, the Y-coordinate of a particle at any point X can be calculated at any time t, provided we know the amplitude A of the wave, the frequency f of the source, and the speed of propagation v of the wave.

For a wave traveling toward the left, a similar argument gives the equation

$$Y = A \sin[2\pi f(t + X/v)] \qquad \textbf{(6)}$$

for a sinusoidal transverse wave traveling in the negative X-direction.

Sinusoidal *longitudinal* waves in a long coil spring can be produced by the arrangement shown in Fig. 5. One end of the spring is attached to a steel ball supported at the end of a hacksaw blade; the other end of the hacksaw blade is held in a clamp as indicated. If the ball is given a small displacement A to the left and then released, it will executive simple harmonic motion of frequency f and as a result of this motion will set up a sinusoidal longitudinal wave train in the spring. The appearance of the spring at a certain time after the wave train had been started is given in Fig. 5(b); *the wavelength λ is the distance between adjacent condensations or adjacent rarefactions,* as indicated in the figure. Again we must assume that a driving force supplies the energy carried away by the wave, if the wave is to have constant amplitude. As in the case of the string, 'internal friction' will gradually convert the macroscopic mechanical energy to thermal energy; hence the amplitude will gradually decrease as the wave travels to the right along the spring.

It is easy to see that equations (4a) and (4b) apply to a sinusoidal longitudinal wave as well as to a sinusoidal transverse wave. In order to write the equation for the longitudinal wave traveling in the X-direction, we note that all the particles in the spring execute simple harmonic motions about their equilibrium positions and that these motions are parallel to the X-axis. If the amplitude of the oscillation of the steel ball is A, this is also the amplitude of oscillation of point O at the end of the spring and, if there is no friction, of all other points in the spring. The motion of the particle whose equilibrium position is X in Fig. 5 is exactly like the motion of the corresponding particle in Fig. 4, *except that the motions in Fig. 5 are right and left rather than up and down.* If we let x denote the *displacement from the equilibrium position,* then

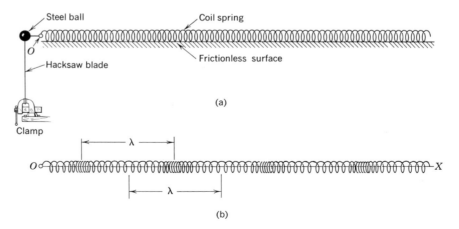

Fig. 5. Production and transmission of a sinusoidal longitudinal wave train in a long coil spring. The wave is moving to the right.

by analogy with (5b), this displacement is given as a function of time by

$$x = A \sin[2\pi f(t - X/v)], \tag{7}$$

where X is the coordinate of the equilibrium position of the particle. This is the equation for *a longitudinal wave traveling in the $+X$-direction*. For *a longitudinal wave traveling in the $-X$-direction*, the equation is

$$x = A \sin[2\pi f(t + X/v)]. \tag{8}$$

Example. *In Fig. 5, if the wave speed is 6 f/s and if the steel ball vibrates through a total distance of 0.1 f at a frequency of 1.5 Hz, what is the wavelength of the longitudinal traveling wave? Write the equation for the wave.*

We get the wavelength from (4b): $\lambda = v/f = (6 \text{ f/s})/(1.5/\text{s}) = 4$ f. Since the wave is traveling to the right, the displacement x at time t of the particle whose equilibrium position is X is given by (7). For the amplitude A we must use *half* the total distance the ball moves, or $A = 0.05$ f. Hence, since $f = 1.5$ Hz and $v = 6$ f/s,

$$x = (0.05 \text{ f}) \sin\left[2\pi (1.5/\text{s})\left(t - \frac{X}{6 \text{ f/s}}\right)\right] = (0.05 \text{ f}) \sin[(3\pi \text{ s}^{-1}) t - (\tfrac{1}{2} \pi \text{ f}^{-1}) X]$$

The argument of the sine is seen to become dimensionless (as all arguments of trigonometric functions must always be) when we insert a *time t* (in seconds) and a *distance X* (in feet), for example $t = 3$ s and $X = 5$ f.

In the above equation, a 'particle' at the origin $X = 0$ has zero displacement at $t = 0$ and is moving to the right. Other initial conditions would be represented by including a phase angle in the argument of the sine as discussed on p. 245.

3. Derivation of the Speed of a Transverse Wave on a String

We shall now write the differential equation satisfied by a simple sinusoidal transverse wave and use this equation in deriving an expression for the speed of a transverse wave on a string.

To obtain the desired differential equation, let us first determine the time derivatives* of Y in (5b), when X is held constant:

$$Y = A \sin 2\pi f(t - X/v)$$

$$\partial Y/\partial t = 2\pi f A \cos 2\pi f(t - X/v)$$

$$\partial^2 Y/\partial t^2 = -4\pi^2 f^2 A \sin 2\pi f(t - X/v)$$

or
$$\partial^2 Y/\partial t^2 = -4\pi^2 f^2 Y. \tag{9}$$

*The symbols $\partial/\partial t$ and $\partial^2/\partial t^2$ denote differentiation with respect to t when X is held constant; similarly $\partial/\partial X$ and $\partial^2/\partial X^2$ denote differentiation with respect to X at a constant value of t. $\partial Y/\partial t$ and $\partial^2 Y/\partial t^2$ are the velocity and acceleration of the particle of the string at X, in its up and down motion. $\partial Y/\partial X$ and $\partial^2 Y/\partial X^2$ are the slope and second derivative of the curve contained in a snapshot of the string at instant t.

Similarly, we may take derivatives of Y with respect to X when t is constant; thus,

$$\partial Y/\partial X = -(2\pi f/v) A \cos 2\pi f(t-X/v)$$

$$\partial^2 Y/\partial X^2 = -(4\pi^2 f^2/v^2) A \sin 2\pi f(t-X/v)$$

or
$$\partial^2 Y/\partial X^2 = -(4\pi^2 f^2/v^2) Y. \tag{10}$$

By combining (9) and (10) we obtain the differential equation

$$\frac{\partial^2 Y}{\partial X^2} = \frac{1}{v^2} \frac{\partial^2 Y}{\partial t^2}, \qquad \left(\begin{array}{c}\text{wave}\\\text{equation}\end{array}\right) \tag{11}$$

which applies to a transverse wave traveling in either the $+X$- or $-X$-direction; proof that (11) applies to waves traveling in the $-X$-direction is left as a problem.

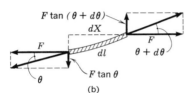

Figure 6

Although we derived equation (11) for a sinusoidal wave, it will be noted that neither frequency f nor wavelength λ appears in the equation. The equation actually applies to *any type* of transverse wave traveling in a direction parallel to the X-axis.

Let us now derive an expression for the speed of a transverse wave in a *perfectly flexible* string *when the amplitude of the motion is small*. Consider Fig. 6, which gives the configuration of the string at a certain instant. An enlarged view of a particular element, of length dl, is shown in part (b) of this figure. If the amplitude is small, we can safely assume that there is no *longitudinal* motion of any part of the string; hence the horizontal components of the tension in the string balance and equal the horizontal pull F furnished by the supports. Since the adjoining parts of a perfectly flexible string can exert forces only along the length of the string, we see that the changing angle in Fig. 6(b) requires a changing vertical component of tension. The net vertical force acting on this element of the string is

$$F \tan(\theta + d\theta) - F \tan\theta. \tag{12}$$

The mass of the element is $\mu\, dX$ (the element is slightly stretched to length dl by the slightly increased tensions on its two ends, but has the same mass as before it was so stretched). The force (12) is to be set equal to the mass times the vertical acceleration, $\partial^2 Y/\partial t^2$:

$$F[\tan(\theta + d\theta) - \tan\theta] = (\mu\, dX)\, \partial^2 Y/\partial t^2. \tag{13}$$

Now $\tan\theta$ is the slope of the curve at X:

$$\tan\theta = \partial Y/\partial X, \tag{14}$$

while $\tan(\theta+d\theta)$ is the slope at $X+dX$, which can be written as

$$\tan(\theta+d\theta)=\frac{\partial Y}{\partial X}+\frac{\partial}{\partial X}\left(\frac{\partial Y}{\partial X}\right)dX=\frac{\partial Y}{\partial X}+\frac{\partial^2 Y}{\partial X^2}dX. \tag{15}$$

Substitution of (15) and (14) in (13) gives directly, after we cancel the dX on both sides:

$$\frac{\partial^2 Y}{\partial X^2}=\frac{\mu}{F}\frac{\partial^2 Y}{\partial t^2}. \tag{16}$$

Equation (16) becomes identical with (11) if we identify F/μ with v^2; that is, to make (11) consistent with Newton's principle we must set

$$v=\sqrt{F/\mu}. \tag{17}$$

Equation (17) for the speed of a transverse wave in a string, derived by applying Newton's principles to the motion of an infinitesimal element of the string, has already been stated in Sec. 1. The above derivation of this speed is rigorous for the case of waves of *small amplitude*.

The above treatment illustrates the type of argument used in deriving the speeds of the various types of mechanical waves considered in this and the following chapter. Since these derivations are slightly above the general mathematical level of this text, we shall omit the proofs for speeds of mechanical waves of other types.

4. Energy of Sinusoidal Wave Motion

Each particle of the string of Fig. 4 is executing simple harmonic motion like the simple harmonic motion of the driven point O. As we have seen in Chapter 12, any particle executing simple harmonic motion must have acting on it a resultant *restoring* force proportional to its displacement, that is, proportional to Y in our case (*see* Prob. 32). The equations for the total energy (kinetic+potential) derived in Chapter 12 are therefore applicable. Equation (14), p. 250, shows that this total energy has the constant value $2\pi^2 f^2 A^2 m$, where m is the mass of the particle. Hence, if a string has mass μ per unit length, it has

$$\text{energy per unit length}=2\pi^2 f^2 A^2 \mu. \tag{18}$$

But this energy is not standing still—it is traveling to the right, actually with the wave speed v. Each element of the string is continuously doing work on the element to its right. To verify this statement, consider Fig. 7, which shows the wave of Fig. 4 traveling to the right in five successive positions 1-1, 2-2, 3-3, 4-4, and 5-5, during one half-cycle of the motion. During this half-cycle, the point at AA is moving down, while the point at BB is moving up. We indicate in Fig. 7 the forces that the

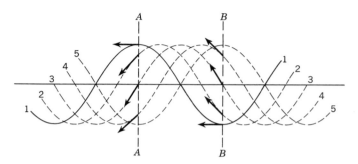

Fig. 7. A wave traveling to the right. The force exerted *by* the string to the left of *AA* or *BB* does work on the portion to the right.

strings to the left of *AA* and *BB* exert on the strings to the right during this half-cycle. Note that during the half-cycle, while the string is changing from configuration 1-1 to configuration 5-5, the force at *AA* has a *downward* component. It acts on a body moving *down*, so the *string to the left does work on the string to the right*. During the succeeding half-cycle, conditions at *AA* would be exactly like those illustrated at *BB* for this half-cycle: the force has an upward component exerted on a body moving up, so again *the string to the left does work on* the string to the right. *Energy is transmitted from left to right* all along the length of a string *when a wave is traveling from left to right*.

We compute the energy transmitted to the right across any section as follows: From Fig. 6(b), we see that the *upward* force exerted *by* the string to the left *on* the string to the right is $-F\tan\theta = -F\partial Y/\partial X$. In time dt, this force acts through a distance $dY = (\partial Y/\partial t)\, dt$, since $\partial Y/\partial t$ is the upward *velocity* of a particle of the string. Hence, the total work done in a finite time interval from t_0 to t_1 is

$$\int_{t_0}^{t_1} -F\tan\theta\, dY = \int_{t_0}^{t_1} -F\frac{\partial Y}{\partial X}\frac{\partial Y}{\partial t}\, dt.$$

Substitution of $\partial Y/\partial X$ and $\partial Y/\partial t$ from pp. 419–420 for the wave traveling to the right gives this work as

$$\frac{F}{v}(2\pi f A)^2 \int_{t_0}^{t_1} \cos^2[2\pi f(t - X/v)]\, dt.$$

The integral is to be evaluated for fixed X. Since the average value of the function $\cos^2$ of any argument over an interval covering many cycles is $\tfrac{1}{2}$ (*see* Prob. 34), the integral above has the value $\tfrac{1}{2}(t_1 - t_0)$, independent of position X, if $t_1 - t_0$ includes many cycles. Since, from (17), $F = v^2\mu$, the work becomes

$$(2\pi f^2 A^2 \mu)\, v\, (t_1 - t_0),$$

and the average *power* (work per unit time) is

$$P = (2\pi^2 f^2 A^2 \mu)\, v, \tag{19}$$

where v is the wave speed. The quantity in parentheses is recognized as the energy per unit length computed in (18). Hence (19) indicates that all of this energy travels along the string with the wave speed v, since the energy crossing any plane per unit time is that contained in a length v of the wave, and this is the length of the wave that passes the plane in unit time. Equation (19) gives the power that must be delivered by the oscillating source in Fig. 4.

We can derive similar relations for the longitudinal wave traveling to the right in the spring of Fig. 5. Each element of this spring executes simple harmonic motion in the longitudinal direction, with a certain amplitude A of displacement to right and left. The energy of this simple harmonic motion is the same as in (18):

$$\text{energy per unit length} = 2\pi^2 f^2 A^2 \mu. \tag{20}$$

Again, a detailed analysis will show that the spring to the left of a point is always exerting a force on the spring to the right *that is in the direction of the instantaneous motion.* Hence, power is transmitted from left to right. The power is that which would be obtained by multiplying (20) by the wave speed v, just as in (19):

$$P = (2\pi^2 f^2 A^2 \mu)\, v. \tag{21}$$

The general ideas contained in the above discussion of energy and power apply to any type of traveling sinusoidal wave in a material medium—in particular to the energy in a sound wave. *The energy represented by the simple harmonic oscillation of the individual particles is transmitted with the velocity of the wave itself.* The velocity of the wave in a string or spring can have only two directions, say *right* or *left,* and power is correspondingly transmitted to the right or left. A sound wave in space can have *any* direction, and power is transmitted in this direction. *Power in a mechanical wave is proportional to the square of the amplitude and the square of the frequency.*

The relations we have given for wave speed, energy, and power are rigorously true only for waves of amplitude small compared with the wavelength ($A \ll \lambda$). Small amplitudes are usually involved in the case of sound, the strings on stringed instruments, and the like. However, for the sake of clarity, Figs. 4 and 7 were not drawn with small amplitude—they would be satisfactory if the vertical measurements were decreased by a factor of about 10.

5. Interference Phenomena; the Superposition Principle

Thus far, we have considered the passage of a *single* wave disturbance—a pulse or a sinusoidal wave train—through a medium. However, it is possible for two or more waves to pass through a medium simultaneously. Let us now consider the effects when *two* transverse wave trains pass simultaneously along a stretched string. In this case, the resultant lateral displacement Y of a point in the string is simply the sum of the displacements the points would have if each wave train traveled along the string by itself. If the first wave train alone would produce displacement Y_1, and if the second wave train alone would produce displacement Y_2, the resultant displacement is given by

$$Y = Y_1 + Y_2. \tag{22}$$

The displacements of the two waves superpose in accordance with the

SUPERPOSITION PRINCIPLE: *When two or more waves move simultaneously through a region of space, each wave proceeds independently, as if the other were not present. The resulting 'displacement' at any point and time is the vector sum of the 'displacements' of the individual waves. This principle holds for mechanical waves on strings, springs, and liquid surfaces, and for sound waves in gases, liquids, and solids, provided the displacements are not too great. It holds rigorously for all light waves and electromagnetic waves in free space.*

To demonstrate this principle, we need merely note that the wave equation (11) is *linear* and *homogeneous* (contains the first power of Y in each term); hence if Y_1 and Y_2 satisfy the equation, so does $Y = Y_1 + Y_2$.

When two waves traveling in the same direction along a string are superposed, the shape of the string at any instant can be determined by the method illustrated in Fig. 8. At a particular instant, the displacements Y_1 arising from the first wave are given by the light solid curve and the displacements Y_2 arising from the second wave are given by the broken curve. The resultant displacements are given by the heavy curve, which is obtained by adding the displacements Y_1 and Y_2 for every point along the string. Since the speeds of the two wave trains in the string are the same, waves with the shape given by the heavy curve will travel along the string with the same speed as that of the component waves.

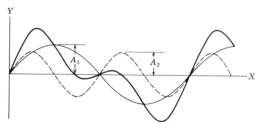

Fig. 8. Superposition of two waves moving in the $+X$-direction. In this example, wave 2 has ¾ the amplitude and ½ the wavelength of wave 1.

One important case involves the superposition of two wave trains of the *same frequency* traveling in the same direction (or in nearly the same direction). This leads to the phenomenon of *interference*. Since interference is of most importance in the case of sound and light, we shall consider the situation in Fig. 9, in which the two identical loudspeakers S and S' are excited by the same electric oscillator E and emit sinusoidal sound waves of the same frequency and amplitude. The solid and broken circular arcs would correspond to the crests and troughs of water waves; however, since sound waves are actually longitudinal, the circular arcs represent the locations of maximum displacement away from the source and toward the source (or, alternatively, of maximum condensation and maximum rarefaction). The wavelength is the distance between successive solid arcs. Surfaces represented by the circular arcs are called *wavefronts;* these surfaces are actually sections of spheres. All points on a wavefront have a common *phase*.

In the region of overlap in Fig. 9, the waves from the two sources are traveling in nearly the same direction, but, at some points in this region the crests arrive together; at certain other points the crest of one wave coincides with the trough of the other.

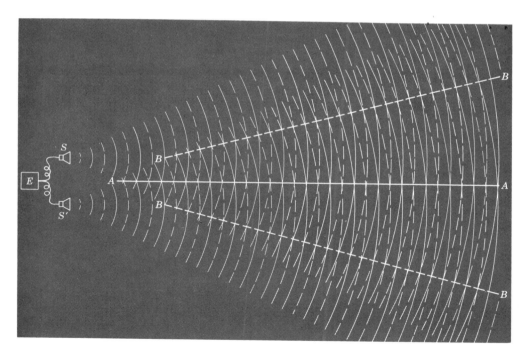

Fig. 9. Interference of sound waves from two speakers. Note that curves *BB* are not straight lines even though no curvature is apparent for the scale employed in the figure. (*See* the example on p. 427.)

Along *AA*, the waves are *in phase;* crests of the two waves are together and troughs are together. The displacements superpose approximately as in Fig. 10(a). If the waves were traveling in exactly the same direction, the *amplitude* of the resulting wave along *AA* would be *twice* that of the wave from either source. As we shall discuss in the next chapter, sound waves are mechanical waves in which power per unit area of wavefront is proportional to the *square* of the amplitude; hence the *power* transmitted along *AA* (per unit cross-sectional area perpendicular to *AA*) would be *four times* the power that would result from either source alone. Along *AA*, the two waves are said to *interfere constructively*.

Along the curves *BB*, the crest of one wave is superposed on the trough of the other, and we have a situation close to that of Fig. 10(b), in which the resultant amplitude is zero. Almost *no sound* would be heard by an ear placed along *BB*, and almost *no power* is transmitted outward along *BB*. Along *BB*, the two waves are said to *interfere destructively*. The reason the two waves along *BB* constitute an intensity *minimum* rather than an intensity *zero,* as in Fig. 10(b), is that the waves are not traveling in exactly the same direction (so that the two displacements are not in exactly opposite directions) and the two amplitudes are not exactly equal (the waves have traveled slightly different distances from the two sources).

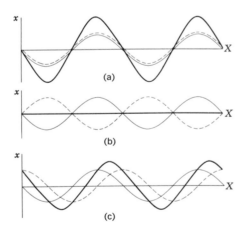

Fig. 10. Interference of two sinusoidal waves of the same amplitude and the same frequency, *traveling in the same direction.* The two waves are represented by the light solid and broken curves, the resultant by the heavy curve. Part (a) represents *constructive interference,* (b) *destructive interference,* while (c) is intermediate.

Between *AA* and *BB*, the situation is intermediate, as indicated in Fig. 10(c). Superposition of the two sinusoidal waves gives a resultant sinusoidal wave of the same frequency. The resultant wave has amplitude that varies from close to zero, along *BB*, to approximately twice that of the individual waves, along *AA*. The power transmitted through the intermediate region varies from close to zero, along *BB*, to approximately four times that of either wave individually, along *AA*.

Note that the superposition principle does not say that the *power* transmitted is additive—rather it says that *displacements* are additive, and since power is proportional to the *square* of the maximum displacement, the powers certainly are not additive. Energy is of course conserved, and in Fig. 9 the total power proceeding outward is the sum of the powers delivered to the air by the two speakers. But interference results in the *concentration* of the power in certain directions. The fact that little power is transmitted in the directions *BB*, when either speaker alone would send significant power in these directions, is compensated for by the fact that the power transmitted along AA is almost *twice* that which one would obtain by merely adding the powers from the two speakers.

Interference is a phenomenon peculiar to wave motion, and it was the existence of interference that finally determined that light was a type of wave.

Example. *If the two speakers in Fig. 9 are 8 f apart and are emitting sound of wavelength 4 f, what is the distance Y between the curve BB and the center line AA on this figure at a horizontal distance $X = 50$ f from the speakers?*

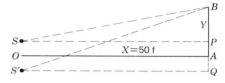

Figure 11

The distances X and Y are indicated on the diagram of Fig. 11, where B is a point on the curve BB of Fig. 9 and A is the point on AA that is 50 f from the origin O halfway between the sources. We are required to find $Y = AB$.

A little study of Fig. 9 will show that the property that characterizes points on the upper curve BB is that they are one-half wavelength farther from source S' than from source S. Hence, in Fig. 11, since one-half wavelength is 2 f,

$$BS' = BS + 2 \text{ f.}$$

We note that $BQ = Y + 4$ f and $BP = Y - 4$ f; hence, applying Pythagoras's theorem to the triangles $BS'Q$ and BSP, we can write the above equation as

$$\sqrt{(Y+4\text{ f})^2 + X^2} = \sqrt{(Y-4\text{ f})^2 + X^2} + 2\text{ f.}$$

Squaring both sides gives

$$(Y+4\text{ f})^2 + X^2 = (Y-4\text{ f})^2 + X^2 + (4\text{ f})\sqrt{(Y-4\text{ f})^2 + X^2} + 4\text{ f}^2.$$

Carrying out the squaring operations and simplifying gives

$$4Y - 1\text{ f} = \sqrt{Y^2 - (8\text{ f})Y + 16\text{ f}^2 + X^2}.$$

Squaring again, we obtain

$$16 Y^2 - (8\text{ f})Y + 1\text{ f}^2 = Y^2 - (8\text{ f})Y + 16\text{ f}^2 + X^2,$$

which simplifies to

$$Y^2 = \tfrac{1}{15}(X^2 + 15\text{ f}^2).$$

This is the equation of the curve BB. Substitution of the value $X = 50$ f gives

$$Y^2 = \tfrac{1}{15}(2500\text{ f}^2 + 15\text{ f}^2) = 168\text{ f}^2, \qquad Y = 13.0\text{ f.}$$

6. Standing Waves

Now let us consider the important case of two sinusoidal transverse wave trains of *equal* amplitude and the same frequency *traveling in opposite directions* along a string. This is a case frequently met in practice, as we shall see in the next section; it results in the production of *standing*

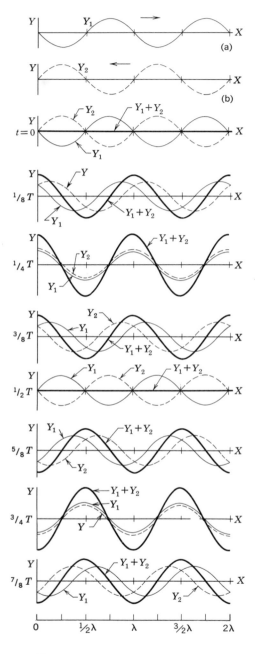

Fig. 12. A standing wave is produced when two waves of the same frequency and amplitude travel *in opposite directions.*

waves, as shown in Fig. 12. At the top of Fig. 12, (a) and (b) show the shapes of two waves of equal amplitude and wavelength at a particular time, say $t=0$. The wave in (a) travels to the right, while that in (b) travels to the left, on the same string. By the superposition principle, we can add the displacements as in the third sketch to find that at $t=0$, there is no resultant displacement. One-eighth of a period later, at $t = \frac{1}{8} T$, the wave (a) has moved $\frac{1}{8}$ wavelength to the right while that in (b) has moved $\frac{1}{8}$ wavelength to the left, and superposition gives the resultant shown in the next sketch. At successive eighth cycles, the resulting patterns are shown in the six following sketches, at times from $t = \frac{1}{4} T$ to $t = \frac{7}{8} T$.

We notice that the resulting motion of the string, indicated by the heavy lines in Fig. 12, is quite different from the motion in the case of a traveling wave. The points of the string at $X = \frac{1}{4}\lambda, \frac{3}{4}\lambda, \frac{5}{4}\lambda$, and $\frac{7}{4}\lambda$ (see scale at bottom of Fig. 12) remain at rest. These points, $\frac{1}{2}\lambda$ apart, are known as *nodes*. Between these points, the particles of the string execute simple harmonic motions, up and down, with frequency f and amplitude varying with position. The maximum amplitude occurs at the points $X = 0, \frac{1}{2}\lambda, \lambda, \frac{3}{2}\lambda$, and 2λ, halfway between the nodes. These points of maximum amplitude are called *loops*, or *antinodes*. The amplitudes of the motion at various points on the string are indicated in Fig. 13, with N denoting the nodes, L the loops. The motion we have described is called a *standing wave*.

It should be noted that the wavelength λ of the component *traveling* waves which produce the standing wave is equal to the distance between *alternate* nodes or between *alternate* loops in the standing wave; that is, *twice* the distance between *adjacent* nodes or *adjacent* loops.

A standing wave may be regarded as a *stationary inference pattern* produced when two traveling wave trains of the *same amplitude and wavelength,* traveling in opposite directions, *interfere.* If two traveling waves of *different* wavelength are traversing a medium, no *stationary* interference pattern can be formed.

In a standing wave, no energy is *propagated* along the string. The energy of each particle of the string is constant, and remains in that particle as it executes its simple harmonic motion. The energy is all kinetic when the string is straight, as at $t=0$ and $t=\frac{1}{2}T$ in Fig. 12, but the particles have their maximum velocities as indicated in Fig. 14; the energy is all potential when the string has its maximum displacement as at $t=\frac{1}{4}T$ and $t=\frac{3}{4}T$, and all particles are momentarily at rest.

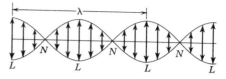

Fig. 13. A standing wave pattern, showing nodes and loops.

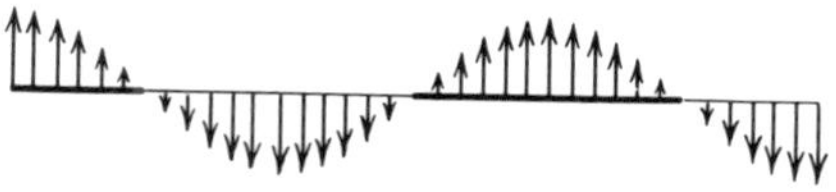

Now that we have given a *physical* description of the origin and behavior of standing waves, let us see how a concise *mathematical* description can be given: Let the first wave, traveling toward the right in Fig. 12(a), be described by the equation

Fig. 14. Velocities of particles in a standing wave.

$$Y_1 = A\sin[2\pi f(t-X/v)],$$

and the second wave, traveling toward the *left* in Fig. 12(b), by

$$Y_2 = A\sin[2\pi f(t+X/v)].$$

Then, from (22), the combination of these two waves is given by

$$Y = A\sin[2\pi f(t-X/v)] + A\sin[2\pi f(t+X/v)]. \quad (23)$$

Recalling the relations for sums and differences of angles:

$$\sin(a+b) = \sin a\cos b + \cos a\sin b, \qquad \sin(a-b) = \sin a\cos b - \cos a\sin b,$$

we may write equation (23) in the form

$$Y = [2A\cos 2\pi fX/v]\sin 2\pi ft,$$

or
$$Y = [2A\cos 2\pi X/\lambda]\sin 2\pi ft. \quad (24)$$

A little reflection will show that (24) gives a complete description of a standing wave. A particle at a given position X executes simple harmonic motion of frequency f; the amplitude of the motion is different for different values of X, becoming a maximum at an antinode, for which $\cos 2\pi X/\lambda = \pm 1$, and zero at a node, for which $\cos 2\pi X/\lambda = 0$. The variation in the sign of the amplitude function $[2A\cos 2\pi X/\lambda]$ as X varies results in the out-of-phase vibration of adjacent regions between antinodes as shown in Fig. 14.

Longitudinal standing waves are similarly produced if two sinusoidal longitudinal wave trains of equal amplitude and the same frequency travel in opposite directions along a helical spring. The most familiar examples of longitudinal standing waves are the standing sound waves in a tube, such as an organ pipe, which we shall consider in the next chapter.

7. Production of Standing Waves

Let us now consider what happens when a transverse pulse advancing along a stretched string arrives at the end of the string attached to a *rigid* support. The end of the string must remain at rest. The arriving pulse exerts a force on the support, and the reaction force exerted on the string by the support sets up a reflected pulse with its displacement in the direction opposite to that of the original pulse. The result is shown schematically in Fig. 15. It will be noted that in reflection the *shape* of the pulse remains unchanged; if, as in Fig. 15, the leading portion of the incident pulse is steep, the leading portion of the reflected pulse will also be steep. However, in the incident pulse the particles of the string are displaced *upward* whereas in the reflected pulse the particles suffer *downward* displacements; in other words, an incident *crest* is reflected as a *trough*. Similarly, an incident *trough* would be reflected as a *crest*.

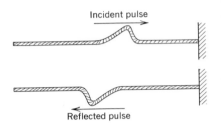

Fig. 15. Reflection of a transverse-wave pulse at a rigid wall.

Just as transverse pulses are reflected at the ends of a string, so also will continuous transverse wave trains be reflected. When a continuous train of sinusoidal waves arrives at a fixed end of a stretched wire or string, a continuous train of reflected sinusoidal waves appears at the end and travels in the opposite direction. Thus, we have *two wave trains of the same wavelength traveling in opposite directions;* as we showed in the preceding section, this is the condition necessary for the production of a standing wave. It should be noted, however, that in the case of a stretched string there are *two* fixed ends; neither end of such a string can move, and therefore *the fixed ends of the string must appear as nodes in any standing-wave pattern that may be formed.* This statement implies that only for certain definite wavelengths will a standing-wave pattern be formed in a given string; some of the possible wave patterns are shown in Fig. 16. The wavelength λ is equal to twice the distance between adjacent nodes; for the standing-wave patterns shown in the diagrams of Fig. 16, the wavelengths are

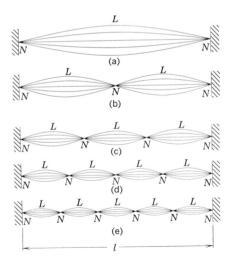

Fig. 16. Standing wave patterns in a stretched string.

(a): $\lambda_1 = 2\,l = 2l/1$,   (c): $\lambda_3 = \tfrac{2}{3}\,l = 2l/3$,   (e): $\lambda_5 = \tfrac{2}{5}\,l = 2l/5$.

(b): $\lambda_2 = l = 2l/2$,   (d): $\lambda_4 = \tfrac{1}{2}\,l = 2l/4$,

Thus, the allowed values for the wavelengths that permit a standing wave to be produced in a string of length l are given by the general equation

$$\lambda_n = 2l/n, \qquad (n = 1, 2, 3, \cdots) \quad (25)$$

where n can be any positive integer. The corresponding frequencies of vibration, given by (4b) as

$$f_n = v/\lambda_n = n(v/2l), \quad (26)$$

will give the frequencies of the sound emitted by the vibrating string, as we shall discuss in the next chapter.

The standing-wave patterns of Fig. 16 are called the *normal modes of oscillation* of the string. The simplest pattern shown in (a) gives the *fundamental mode of oscillation;* the other patterns represent so-called *higher modes of oscillation*. In the case of a string in which the frequency of the fundamental mode is sufficiently low, it is not difficult to excite the fundamental or one of the higher modes by applying alternating forces by hand near the end of the string at one of the frequencies given by (26). If one applies the alternating forces at a frequency different from that of one of the normal modes, one does not obtain a 'regular' motion, but rather a very erratic, irregular motion; that is, no standing wave pattern will be produced.

Example. *A steel piano wire 1 m long has a mass of 20 g and is stretched with a force of 800 N. What are the frequencies of its fundamental mode of oscillation and of the next three higher modes?*

The velocity of a transverse wave on this wire is given by (2) as

$$v = \sqrt{F/\mu} = \sqrt{(800 \text{ N})/(0.02 \text{ kg/m})} = 200 \text{ m/s}.$$

The wavelength of the fundamental mode is $\lambda_1 = 2l = 2$ m; hence the frequency is

$$f_1 = v/\lambda_1 = (200 \text{ m/s})/(2 \text{ m}) = 100 \text{ Hz}.$$

According to (26) the frequencies of the higher modes are 2, 3, 4, $\cdots$ times this, so

$$f_2 = 200 \text{ Hz}, \quad f_3 = 300 \text{ Hz}, \quad f_4 = 400 \text{ Hz}.$$

8. Reflection of Waves

A transverse wave is reflected with reversed phase at the end of a string attached to a rigid support; a crest is reflected as a trough and *vice versa*, as indicated in Fig. 15. A rigid support represents one extreme 'boundary condition.' We now consider reflection of waves for other boundary conditions.

At the opposite extreme from a string with a rigidly fixed end would be a string with a perfectly *free* end. This case has little meaning in the case of a wave along a *stretched* string, since it is difficult to keep a string

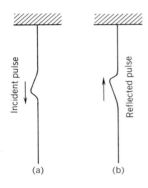

Fig. 17. Reflection of a transverse pulse from the free lower end of a rope hanging vertically. A *crest* is reflected as a *crest* and a *trough* is reflected as a *trough*. (At what part of the rope is the wave speed greatest?)

under finite and uniform tension if one end is *free*. However, reflection of a transverse wave from the free lower end of a rope hanging from a support can be observed, with the results shown in Fig. 17. By giving the rope a sudden sidewise thrust at a point near the support, an observer may produce a transverse pulse which travels down the rope as shown in part (a). After reaching the free end of the rope, the pulse is reflected as shown in part (b) and travels up the rope. It should be noted that the particles in both the incident and the reflected pulse are displaced toward the left.

When the incident pulse is reaching the free end and the reflected pulse is leaving the free end, the combined effects of both pulses result in a displacement of particles near the free end that is twice as large as the displacement arising from either pulse alone. To the observer standing near the upper end of the rope, the particles near the free end appear to 'overshoot' and in so doing set up the reflected wave. In the ideal case, the energy associated with the reflected pulse is just equal to the energy associated with the incident pulse. Whereas the rigidly fixed end of the string experiences *no* displacement when reflection occurs, the free end of a string experiences a *large* displacement. This is the phenomenon that accounts for the 'cracking' of a whip. In reflection of a pulse at the fixed end of a string wave velocity and displacement are both reversed in direction; in reflection at a free end, wave velocity is reversed but displacement is unchanged.

Although reflection of transverse pulses at the free end of a string is not important to our present discussion of waves on *strings,* there are many analogous cases of importance. One of the cases encountered frequently is that of reflection of longitudinal vibrations in a solid. At a 'free' surface, in contact with air, a condensation is reflected in the solid as a condensation; similarly, a rarefaction is reflected as a rarefaction. At a fixed rigid boundary, the reverse is true; a condensation is reflected as a rarefaction, and vice versa. We shall discuss analogous types of phenomena in connection with reflection of sound waves in air columns and with reflection of light waves at boundaries between various media.

In reflection of transverse waves at the fixed end of a string connected to a rigid support, and at the *free* end of a string, reflection is *complete.* These are limiting cases: (a) in the case of a *perfectly rigid* support the elastic force constant involved in the restoring forces *opposing the motions* of the particles of the support is so large (infinite) that the waves traveling down the string can produce no displacement of the support and hence no energy can be imparted to the support; (b) at a *free* end there are *no* elastic restoring forces and hence no wave could be transmitted to the region of space beyond the free end. In these limiting cases, the wave motion is confined to the string and the energy remains confined to the string. However, these cases are never completely realized in practice; usually, some of the energy of the waves is transmitted into the 'medium' to which the end of the string is attached. Thus, some of the energy of the wave reaching the ends of the string may be transmitted by waves produced in the not completely 'rigid' support at the fixed end or by longitudinal waves in the air at the 'free' end; in fact, if the string is in *air*, energy will be transferred from the string to the air along the whole length of the string.

Now consider the composite string shown in Fig. 18, in which two strings of the same material but of different diameters have been joined together at point *P*. Since the *tension is the same in all parts of this composite string*, the speeds of transverse waves in the two parts of the string are different and their ratio is given by the expression

$$v_A/v_B = \sqrt{\mu_B/\mu_A},$$

since the wave speed is inversely proportional to the square root of the mass per unit length. If a wave travels toward the right in string *A* with

Fig. 18. A composite string.

speed v_A, waves are produced in string *B* as a result of the motion of point *P*, and these waves travel along string *B* with speed v_B; thus, a part of the energy associated with the incident wave in string *A* will be transmitted past point *P* into string *B*. However, some of the energy of the incident wave will be retained in string *A* as a result of partial reflection of the incident wave at point *P*. A part of the energy is transmitted and a part is reflected at the boundary at *P* between the two 'media'; this is always the case when any type of wave strikes a boundary between two media in which wave speeds are different. The ratio of reflected energy to incident energy during any time interval is very large when the ratio of speeds in the two media is greatly different from unity; thus, in the case of two strings, a large fraction of the incident energy will be reflected if $v_A/v_B \gg 1$ or $v_A/v_B \ll 1$, and very little reflection will occur if $v_A \approx v_B$. There is also a change of phase to be considered in connection with reflection—ranging from complete reversal at a rigid support to zero at the free end of a string.

PROBLEMS

1. A 40-m flexible string has a mass of 40 g and is stretched between two posts. If the tension in the string is 16 N, what is the speed of propagation of transverse waves in the string?   Ans: 126 m/s.

2. What should be the tension in the string of Prob. 1 in order to make the speed of transverse waves exactly 100 m/s?

3. A 100-f length of flexible cord has a mass of $1/16$ slug. If the tension in the cord is 49 p, what is the speed of transverse waves in the cord?   Ans: 280 f/s.

4. The speed of transverse waves in a 75-f cord is 200 f/s when the tension in the cord is 50 p. What is the mass of the cord?

5. A helical spring of mass $\frac{1}{64}$ slug is suspended in a vertical position. When a 1-p weight is attached to the lower end of the spring, an elongation of 3 i results. If the length of the spring is 10 f, what would be the speed of longitudinal waves in the spring?
Ans: 160 f/s.

6. A 400-g steel helical spring is 5 m in length. When this spring is supported vertically and a 100-g hanger is attached to the lower end, the spring stretches 40 cm. What is the speed of longitudinal waves in this spring?

7. Equation (1) gives a general expression for the speed of waves in elastic media. It can be shown that the speed v of longitudinal waves in a fluid is given by the expression $v = \sqrt{M_B/\rho}$, where M_B is the bulk modulus and ρ is the density. What units should be used for M_B and for ρ if v is to be given in f/s? in m/s?

8. The speed of sound in a gas varies with temperature according to the relation $v = \sqrt{\gamma kT/\mu}$, where the symbols have the meanings assigned in Chapters 17 and 19. Show that this equation can be rewritten as $v = \sqrt{\gamma P/\rho}$ and hence is of the general form given in equation (1). What units must be used for k, T, and μ if v is to be in m/s?

9. A man observes surface waves on a river. If the distance between a crest and the adjacent trough is 7 f and if 12 crests pass in 10 s, what is the speed of the surface waves?
Ans: 16.8 f/s.

10. By rocking a boat, two boys produce surface waves on a pond. One boy notes that the boat makes 8 complete oscillations in 10 s, and that one wave crest moves away from the boat for each oscillation; the other boy notes that it takes 5 s for a given crest to move from the boat to the shore of the pond—a distance of 50 f. What is the average distance between adjacent crests? What is the wavelength of the surface waves? the period?

11. An arrangement like that in Fig. 4 is used to send a train of waves along a long stretched string. The source makes 8 oscillations per second, and the amplitude of the waves is 5 cm. If the string is 50 m long and has a mass of 100 g, how much mechanical energy is associated with each meter of the string through which the waves pass? (Assume negligible internal friction in the string and assume that there is no reflection from the remote end of the string.) If the wavelength of the waves in the string is 1.5 m, how much power must be supplied to the oscillator?
Ans: 6.30×10^{-3} J; 37.8 mW.

12. Answer all questions in Prob. 11 for the situation in which the amplitude is 10 cm and the frequency is 8 oscillations per second.

13. (a) Plot

$$Y = (0.3 \text{ m}) \sin[(120\,\pi\,\text{s}^{-1})\,t - (\pi\,\text{m}^{-1})\,X]$$

against X at $t = 0$ and at $t = \frac{1}{480}$ s. From your plots determine amplitude, wavelength, and speed. (Make the plots neatly and explain your reasoning carefully.)

(b) Plot the same function against t at $X = 0$, and determine the period of vibration. Verify that in this case $\lambda = vT$.

14. Repeat Prob. 13 for the case

$$Y = (0.3 \text{ m}) \sin[(120\,\pi\,\text{s}^{-1})\,t + (\pi\,\text{m}^{-1})\,X].$$

What is the essential difference between this case and that of Prob. 13?

15. A sinusoidal wave train is moving along a string. The equation giving the displacement Y of a point at coordinate X has the following form:

$$Y = (0.09 \text{ m}) \sin[(16\,\pi\,\text{s}^{-1})\,t - 16\,\pi\,X/(40 \text{ m})].$$

Find the following quantities: (a) the amplitude of the wave motion, (b) the frequency of the wave motion, (c) the speed of the wave motion, and (d) the wavelength. (e) In which direction is the wave moving?
Ans: (a) 9 cm; (b) 8 Hz; (c) 40 m/s; (d) 6.25 m; (e) $+X$-direction.

16. Repeat Prob. 15 for a wave motion given by the equation

$$Y = (0.08 \text{ m}) \sin[(20\,\pi\,\text{s}^{-1})\,t + 20\,\pi\,X/(50 \text{ m})].$$

17. Sinusoidal longitudinal waves are sent out in a coil spring, from a vibrating source at one end of the spring as in Fig. 5. The frequency of vibration of the source is 20 Hz. The distance between successive condensations in the spring is 30 cm. Find the speed of a condensation as it moves along the spring. The maximum

longitudinal displacement of a particle in the spring is 5 cm. Write an equation for this wave motion for waves moving in the $+X$-direction if the source is at $X=0$ and the displacement at the source is zero when $t=0$. Ans: 6.00 m/s; $x=(0.05\text{ m})\sin[(40\pi\text{ s}^{-1})t - 40\pi X/(6\text{ m})]$.

18. A source vibrating at a frequency of 16 Hz produces longitudinal waves of amplitude 0.04 m in a coil spring. The speed of propagation of longitudinal waves in the spring is 20 m/s. Write an equation describing this motion.

19. An arrangement like that in Fig. 5 is used to send longitudinal waves along a long helical spring. The spring has a mass of 40 g/m and the wave speed is 6 m/s. If the frequency of the oscillator is 5 Hz and the wave amplitude is 8 cm, how much power must be supplied to the oscillator? Ans: 758 mW.

20. By graphical methods, find the points of constructive and destructive interference on circles of 20-m and 30-m radii drawn about O in the figure if the distance between S and S' is 3 m and sound waves having a wavelength of 2 m are being emitted. Determine the points corresponding to curves AA and BB on Fig. 9, and also the next pair of curves of constructive interference at angles greater than BB.

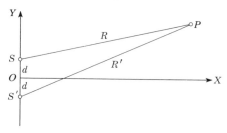

Problems 20–22

21. The speakers in Fig. 9 are 2 m apart and send out sound waves having a wavelength of 1 m. Find the distance between curves AA and BB at a horizontal distance of 25 m from the speakers. Ans: 6.50 m.

22. In order to generalize the results of the example on p. 427 and of Probs. 20 and 21, consider two sources S and S' separated by distance $2d$ as shown in the figure.

(a) Show that the path difference from these sources to point P is given by

$$R' - R = \sqrt{X^2 + (Y+d)^2} - \sqrt{X^2 + (Y-d)^2}.$$

(b) If λ is the wavelength of the disturbances emitted by the sources and the sources are synchronized so as to be exactly 'in phase,' show that the locus of all points for which $R' - R = \tfrac{1}{2}\lambda$, and interference is destructive, is given by the hyperbola

$$Y^2 = \frac{\lambda^2(X^2 + d^2 - \tfrac{1}{16}\lambda^2)}{16\,d^2 - \lambda^2}$$

and that for the case $X \gg d \gg \lambda$, $Y \doteq \lambda X/4d$.

(c) Show that the locus of all points for which $R' - R = n\lambda/2$ is given by

$$Y^2 = \frac{n^2\lambda^2(X^2 + d^2 - \tfrac{1}{16}n^2\lambda^2)}{16\,d^2 - n^2\lambda^2},$$

provided n is sufficiently small that the denominator is positive. Discuss the physical significance of this last restriction.

(d) Give a general method of determining the number of curves, such as AA and BB in Fig. 9, along which constructive or destructive interference occurs.

23. A string 10 f long is attached to two vertical posts. When this string is plucked at the center, a standing wave is produced. If the standing wave has nodes at the points of attachment to the posts and a single loop at the center, what is the wavelength of the component traveling waves? By means of a stroboscope, an observer finds that the string executes 4 complete oscillations per second. What is the speed of transverse waves in the string?
 Ans: 20.0 f; 80 f/s.

24. Find the wavelengths of the next three shorter traveling waves that can set up standing waves in the string of Prob. 23. What would be the frequencies of oscillation of the string when these standing waves are produced?

25. The frequency of the fundamental mode of transverse oscillation of a certain flexible steel wire is 50 Hz. If the length of wire is 4 m and its mass is 16 g, what is the speed of transverse waves along the wire? What is the tension in the wire? Ans: 400 m/s; 640 N.

26. The frequency of the fundamental mode of transverse oscillation of a flexible wire is 60 Hz. The wire is 6 m long and its mass is 40 g. Find the tension in the wire.

27. A helical spring with a mass of 60 g and

a force constant of 1.5 N/m is stretched between two posts 60 cm apart. Describe the fundamental mode of *longitudinal* oscillation of this spring and determine its wavelength λ_1 and its frequency f_1.

Ans: $\lambda_1 = 1.2$ m, $f_1 = 2.5$ Hz.

28. Describe the motion of the spring of Prob. 27 in the next two higher modes of longitudinal oscillation, and determine the wavelengths and frequencies of these modes.

29. The speed of *longitudinal waves in* a slender rod or wire is given by $v = \sqrt{M_Y/\rho}$, where M_Y is Young's modulus and ρ is the density of the material. A copper wire is 1 mm in diameter and 10 m long and is stretched between rigid supports. What is the speed of longitudinal waves in this wire? At what tension would transverse waves have the same speed as longitudinal waves? is this possible?

Ans: 3520 m/s; 102 000 N.

30. Solve Prob. 29 for steel, aluminum, and lead wires of the same size.

31. From the equation for a sinusoidal transverse wave in a string, derive an expression for the *maximum* velocity of a particle in the string. How does the maximum velocity depend upon wave amplitude? upon wave frequency?

32. Verify that the resultant Y-component of force acting on the 'particle' of mass $\mu \, dX$ in Fig. 6 is a restoring force of the form $F_Y = -KY$, where $K = 4\pi^2 f^2 (\mu \, dX)$. This verification can be made by employing equations (13)–(15), (10), and (17). From (3), p. 245, show that the ratio between force constant K and mass is indeed such as to give simple harmonic motion of frequency f.

33. Consideration of Fig. 8 reveals that superposition of two simple sine waves can produce a more complex wave form. This suggests the possibility that *any* periodic function might be represented by proper superposition of simple sinusoidal waves. This was shown to be the case by the French mathematician Fourier in 1807:

> FOURIER'S THEOREM: *Any periodic function can be expressed as the sum of sine and cosine terms.*

Make a plot that shows that a 'saw-tooth' waveform can be closely approximated by even the first three terms of the series

$$Y = A \sin X - \tfrac{1}{9} A \sin 3X + \tfrac{1}{25} A \sin 5X - \tfrac{1}{49} A \sin 7X + \cdots.$$

34. From the trigonometric identities $\sin^2\theta = \tfrac{1}{2} - \tfrac{1}{2} \cos 2\theta$ and $\cos^2\theta = \tfrac{1}{2} + \tfrac{1}{2} \cos 2\theta$, sketch the curves $\sin^2\theta$ and $\cos^2\theta$. Note that these are themselves sinusoidal curves of $\tfrac{1}{2}$ the 'wavelength' of $\sin\theta$ and $\cos\theta$, but with the axis displaced upward by $\tfrac{1}{2}$ unit. From the symmetry of these curves about the displaced axis, show that the average values of $\sin^2\theta$ and $\cos^2\theta$ are $\tfrac{1}{2}$, averaged over a whole period or a large number of periods. Derive the same average values from the identities given in the first sentence by noting that the average value of $\cos 2\theta$ is zero.

35. Demonstrate that the total energy in a loop of a standing wave is the sum of the energies in the corresponding lengths of the two traveling waves, each of amplitude A, of which it may be considered to be composed. The figure shows a portion of any one of the standing waves of Fig. 16. Show that the energy per unit length in the string is a function of X given by $2\pi^2 f^2 \mu (4A^2) [\sin(2\pi X/\lambda)]^2$. Show that the average energy per unit length is $4\pi^2 f^2 A^2 \mu$. Hence demonstrate the proposition posed in the first sentence.

36. Figure 16 shows different standing-wave patterns in the same stretched string. Compare

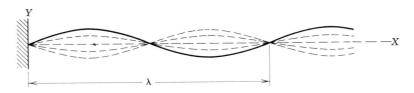

Problems 35–37

the energy of the string in part (a) with that in part (e) if the amplitudes are equal. What should be the ratios of amplitudes if the energies in parts (a) and (e) are to be made equal? (Use the results of Prob. 35.)

37. What is the energy of a piano string of mass 10 g and length 1 m vibrating in its fundamental mode at 440 Hz with an amplitude of 1 cm? (*See* Prob. 35.)   Ans: 7.62 J.

21 SOUND

The term *sound* has two distinct uses. The physiologist or psychologist uses the word *sound* in connection with the sense of hearing and the auditory sensations produced by certain types of disturbances in the air. *In physics the term 'sound' is used to denote the disturbances themselves rather than the sensations produced:*

Sound is a mechanical wave motion in an elastic medium.

The frequencies of *audible* sound waves lie between about 20 Hz and 20 000 Hz, but the term *sound* is used by physicists to include disturbances having frequencies outside the range to which the human ear responds. Waves of frequency below the audible range are termed *infrasonic;* those of frequency above audible frequencies are called *ultrasonic*. Ultrasonic waves are assuming increasing practical importance for such uses as the detection of flaws in the interior of solid bodies (which cause reflection), the production of stable emulsions of normally immiscible liquids, the coagulation of aerosols, the cleaning of surgical instruments, and the remote control of television sets.

The physicist, in his description of audible sound waves, *does* attempt to correlate their psychological effects with the physical properties of the waves; thus, as we shall show, the psychological characteristics of continuous sound—*pitch, loudness,* and *quality*—can be correlated with physical properties—*frequency, intensity,* and *waveform*—of the sound waves.

In order for sound waves to be produced, there must be a *source* that initiates a mechanical disturbance and an *elastic medium* through which the disturbance can be transmitted. A simple experiment will show the necessity of an elastic material medium to transmit the sound. If an ordinary electric doorbell is suspended by fine wires inside a bell jar in such a way that it does not make contact with the walls of the jar, the sound of the ringing bell can be heard when air is inside the bell jar. However, if the air is removed from the bell jar by a vacuum pump, the sound can no longer be heard. That sound waves can be transmitted by solids as well as by air can be shown by tilting the evacuated bell jar so that the bell touches the wall of the jar; as soon as contact is made, the sound can be heard again. That sound waves can also be transmitted by liquids can easily be shown by ringing a bell beneath the surface of oil or water in a large beaker.

1. Production of Sound by Vibrating Solids

We shall first discuss the various ways in which a solid body can vibrate; the ways in which a body can vibrate with all of its parts executing simple harmonic motion of the same frequency are called the 'normal modes of oscillation' of the body. A general motion of vibration is a superposition of normal modes having, in general, different frequencies. The possible normal modes depend upon the shape of the body, the density of the body, the elastic properties of the body, and on the *boundary conditions* imposed at the boundary of the body by restraints imposed by other bodies in contact with the body.

We have already discussed the vibration of a stretched string that has both ends fixed by connection to rigid supports. With these boundary conditions, the normal modes of oscillation correspond to the standing wave patterns shown in Fig. 16, p. 430. If we impart energy to the stretched string by plucking it or by stroking it properly with a violin bow, the string vibrates in one or more of the modes shown in this figure. As the string vibrates, some energy is transferred to the surrounding air in the form of sound waves, and even more energy is transformed into heat as a result of 'internal friction' as the string is deformed. As a result of these processes, the vibration of the string itself gradually dies out.

The frequencies of the sound waves produced in the air are the same as those of the string, since the string is the source of the sound waves.

The frequencies of the normal modes of vibration of the string are given by (26), p. 431, as

$$f_n = \frac{nv}{2l} = \frac{n\sqrt{F/\mu}}{2l}, \qquad (n = 1, 2, 3, \cdots) \quad (1)$$

where F is the tension, μ the mass per unit length, and l the length of the string. The lowest possible frequency corresponds to $n=1$; this is the frequency of the fundamental mode of vibration of the string and is called the *fundamental frequency*. The frequencies of the higher modes of vibration corresponding to $n = 2, 3, \cdots$ are called *overtones*.

Equation (1) shows that for a string we may express the frequencies of all modes of oscillation in terms of the fundamental frequency f_1 by the simple relation

$$f_n = n f_1. \qquad (n = 1, 2, 3 \cdots) \quad (2)$$

The frequencies of the higher modes of oscillation of the string are all integral multiples of the fundamental frequency; a set of frequencies bearing this type of relationship are called *harmonics*. The fundamental frequency, for which $n=1$, is called the *first harmonic;* the frequency for which $n=2$ is called the *second harmonic;* and so on.

For *any* vibrating body, *the lowest frequency is called the fundamental frequency;* the first frequency higher than the fundamental is called the *first overtone;* the second higher frequency is called the *second overtone;* and so on. *The overtones are not always simple integral multiples of the fundamental.* When, as in the case of a drum or a bell, they are not integral multiples of the fundamental, they are called *nonharmonic overtones*.

If a string is plucked at any point selected at random, several modes of oscillation are usually excited simultaneously. However, by employing the proper procedure, we may excite a *particular* mode of oscillation. Thus, if the string is plucked gently at its mid-point, the fundamental mode will be excited. If one plucks the string at a point one-quarter of its length from the end while gently touching the mid-point, the first overtone will be excited. Similarly, by plucking the string at a point one-sixth of its length from the end and gently touching the string at a point one-third of the way from the end, one can excite the second overtone.

We have been considering only the modes of oscillation of a string or flexible wire in which *transverse* motions occur. However, there are also corresponding longitudinal modes of vibration of a wire or rod clamped at the ends. For the longitudinal vibrations also, the ends must be nodes, so the normal modes will again be given by Fig. 16, p. 430, if we imagine these drawings to represent magnitudes of longitudinal, rather than transverse, displacements. The allowed frequencies of longitudinal waves will again be given by (26), p. 431.

The speed of a longitudinal wave in a wire or rod is given by a relation like (1), p. 414, in which the elastic force factor is Young's modulus M_Y and the inertia factor is the density ρ of the material. Thus

$$v_R = \sqrt{M_Y/\rho}. \qquad \binom{\text{longitudinal}}{\text{wave in rod}} \quad (3)$$

The frequencies of the normal modes are thus

$$f_n = \frac{n v_R}{2l} = \frac{n \sqrt{M_Y/\rho}}{2l}. \quad (4)$$

The frequencies of longitudinal modes in a wire will usually be much higher than those of transverse modes because the longitudinal wave speed is usually very high. In musical instruments such as the violin or piano, only the transverse oscillations are desired. In playing the violin, a novice usually manages to excite longitudinal vibrations by moving the bow lengthwise along the string; the resulting sound is an objectionable shrill squeak.

An example of a vibrating body whose overtones are *not* harmonics is the circular membrane used on a drum. The first six normal modes of vibration are shown in Fig. 1, which shows the *nodal lines* as broken lines across the membrane. At these lines and at the periphery there is no motion; at points approximately midway between the nodal lines, displacements of maximum amplitude occur. In the simplest mode of oscillation, the center of the membrane is an antinode and the periphery is a nodal line; this mode of oscillation has the lowest frequency, the fundamental f_1. For the mode giving the first overtone, shown in the second diagram in Fig. 1, the motion of the part of the membrane to the right of the nodal line is 180° out of phase with the motion of the part of the membrane to the left of the nodal line; that is, when the part of the membrane to the right is moving toward the reader, the part to the left is moving away from the reader. It is easy to visualize the motion shown in the other parts of Fig. 1 by remembering that the motions of parts of the membrane on opposite sides of a nodal line are out of phase by 180°.

The frequencies of the various normal modes of oscillation are given in the figure in terms of the fundamental frequency denoted by f_1. The frequencies of the overtones are *not* integral multiples of the fundamental f_1; that is, the overtones are *not* harmonics. The characteristic frequencies may be determined experimentally by subjecting the membrane to vibrations of known frequency by placing a loudspeaker directly above the horizontal membrane and exciting the loudspeaker by means of an audiofrequency oscillator whose frequency can be varied slowly. The drumhead will be set into vibration of large amplitude when the frequency of the sound waves from the loudspeaker coincides with the frequency of some normal mode of oscillation of the membrane; and if the drumhead is covered with a thin layer of fine powder such as chalk dust, the powder will move about over the drumhead and accumulate

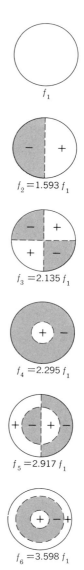

Fig. 1. Normal modes of oscillation of a circular membrane.

along the nodal lines. By noting the pattern of nodal lines and by noting the frequency at which the pattern occurs, one can determine the resonance frequencies for each of the normal modes of oscillation shown in Fig. 1.

The drumhead and the bell are more effective than the vibrating string in the direct production of sound waves of *large amplitude*. If a given amount of energy is imparted to a drumhead, this energy is quickly imparted to the surrounding air by the large, flat surfaces of the vibrating membrane. However, if an equal amount of energy is imparted to a stretched string, this energy is imparted very slowly to the surrounding air because the surface area of the string is small; hence, vibration of the string persists for a long time. In order to make a vibrating string or tuning fork more effective in setting up sound waves in the air, the string can be 'coupled' mechanically to a body of large area or volume which is effective in imparting vibrational energy to the surrounding air. The sounding board of a piano, the case of a violin, and the sound box (Fig. 2) of a tuning fork are familiar examples.

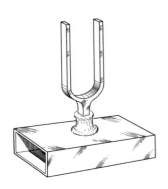

Fig. 2. Tuning fork mounted on a wooden box, which acts as an 'acoustic amplifier.'

2. Speed of Sound in Solids, Liquids, and Gases

We have given in equation (3) the speed of a longitudinal wave in a *rod*. This formula is valid only for rods small in diameter in comparison with the wavelength, because the diametral contractions and expansions that accompany changes in length (Poisson's ratio, p. 230) introduce additional complications if the diameter is not small compared with the wavelength. In the case of an *extended* solid, the speed of longitudinal waves is not given by (3) but by

$$v_\text{L} = \sqrt{\frac{M_\text{B} + 4/3\, M_\text{S}}{\rho}}, \qquad \begin{pmatrix}\text{longitudinal}\\ \text{wave in solid}\end{pmatrix} \quad (5)$$

which involves the bulk modulus, the shear modulus, and the density.

In the case of a *transverse* wave in an extended solid, the shear modulus alone determines the speed, which is given by

$$v_\text{T} = \sqrt{M_\text{S}/\rho}. \qquad \begin{pmatrix}\text{transverse}\\ \text{wave in solid}\end{pmatrix} \quad (6)$$

Comparison of (5) and (6) shows that the speed of the longitudinal wave in an extended solid is always much greater than that of the transverse wave. The difference in arrival time of these two types of waves is used in seismology to estimate the distance of an earthquake from the observing station.

A fluid cannot sustain a shear ($M_\text{S} = 0$), so no transverse wave can be transmitted; the speed of a longitudinal wave is given by

$$v = \sqrt{M_\text{B}/\rho}. \qquad \text{(wave in fluid)} \quad (7)$$

Compressions and rarefactions taking place at audio frequencies must be regarded as *adiabatic* rather than isothermal; hence the M_B to be used in the above formulas is the adiabatic bulk modulus. The distinction is unimportant in the case of solids and liquids, which are relatively incompressible and whose properties vary little with small changes in temperature. Typical experimental values for longitudinal waves in thin solid rods and in liquids are given in Tables 21-1 and 21-2.

Now we turn to the important problem of calculating the speed of sound in gases. Sir Isaac Newton first derived the expression (7) for the speed of sound in a gas, but he made the mistake of using the isothermal bulk modulus; careful measurement revealed that the Newtonian expression gave too low a value. Over a century later, in 1816, Laplace obtained the correct expression when he showed that gas compressions taking place at audio frequencies must be regarded as adiabatic, since the compressions and expansions take place so rapidly that the heat of compression remains in the regions in which it is generated and does not have time to flow to neighboring cooler regions that have undergone an expansion. The correct expression for the speed of sound in a gas is

$$\text{speed} = \sqrt{\frac{\text{adiabatic bulk modulus}}{\text{density}}}. \qquad (8)$$

Table 21-1 SPEEDS OF LONGITUDINAL SOUND WAVES IN THIN SOLID RODS

(From *Smithsonian Physical Tables*)

Material	Speed of sound	
	(m/s)	(f/s)
Aluminum	5104	16 750
Copper	3560	11 680
Iron	5130	16 830
Lead	1322	4 340
Nickel	4973	16 320
Glass (typical)	5550	18 050
Vulcanized rubber	54	177

Table 21-2 MEASURED SPEEDS OF SOUND IN LIQUIDS

(From *Smithsonian Physical Tables*)

Material	Speed of sound	
	(m/s)	(f/s)
Alcohol, methyl	1143	3750
Carbon bisulfide	1060	3477
Ether	1032	3386
Mercury	1407	4614
Turpentine	1326	4351
Water	1461	4794
Sea water	1500	4922

The bulk modulus is defined as in (6), p. 227, as the ratio of stress to strain, where the stress is the pressure increase dP and the accompanying strain is the decrease in volume per unit volume $-dV/V$. Hence

$$M_B = dP/(-dV/V) = -V\, dP/dV. \qquad (9)$$

The value of dP/dV to be substituted in (9) is that obtained in (14), p. 388, for an adiabatic process:

$$dP/dV = -\gamma P/V. \qquad \text{(adiabatic)} \quad (10)$$

Substitution of this value in (9) gives

$$M_B = \gamma P. \qquad (\gamma = C_P/C_V) \quad (11)$$

Hence, the Laplace expression for the speed of sound in a gas is

$$v = \sqrt{\gamma P/\rho}. \qquad \text{(wave in gas)} \quad (12)$$

Experiment shows that the expression given in (12) is correct.

Consideration of (12) reveals the interesting fact that *at a given temperature, the speed of sound in a gas is independent of the pressure.* From Boyle's law, we see that the density ρ is directly proportional to the pressure P, so that P/ρ is a constant if the temperature is constant.

Now consider the effect of temperature variations on the speed of sound. It will be recalled from our discussion of the kinetic theory of gases on p. 349 that $P=NkT$, where N is the number of molecules per unit volume, k is Boltzmann's constant, and T is the Kelvin temperature. Thus, since the density $\rho=N\mu$, where μ is the average mass of a single molecule, $P/\rho=kT/\mu$. Substitution in (12) then gives

$$v=\sqrt{\gamma\, kT/\mu}. \qquad \text{(wave in gas)} \quad (13)$$

Hence, since γ and μ depend only on the particular kind of gas, the speed of sound in a given kind of gas depends only on the temperature. In making use of (13), it is desirable to recall from pp. 334–335 that $\mu = 1.66 \times 10^{-27} M$ kg, where M is the molecular mass of the gas.

The speed is directly proportional to the square root of the absolute temperature and may be calculated for any gas from equation (13), which is in complete agreement with experiment within the accuracy with which gases behave ideally. The measured temperature dependence in air is given in Table 21-3; these values are given by the equation:

$$v = 331\sqrt{T/273°}\ \text{m/s,}$$

with T in °K.

The speed of sound in air is not too fast to be readily determined by direct measurement of the time required for sound to traverse a given distance. Everyone has had the experience of 'seeing a noise made' and then waiting patiently for the sound to arrive—lightning is a common example—and has heard the sound of a jet plane coming from where the plane *was,* and not at all from where it *is.*

Table 21-3 MEASURED SPEED OF SOUND IN AIR

(From *Smithsonian Physical Tables*)

Temp. (C)	Speed (m/s)	Speed (f/s)
0°	331.36	1087.1
20°	344	1129
100°	366	1201
500°	553	1814
1000°	700	2297

Example. *Determine the speed of sound in hydrogen gas at 0°C.*

We shall do this problem in two ways: first by comparing with the speed in air, assumed as known; second, by direct substitution in (13).

The speed in air is 331 m/s at 0° C. Air and hydrogen are both diatomic, so they have the same γ in (13). The speed of sound differs only because of the much lower molecular mass of hydrogen, and is inversely proportional to the square root of the molecular mass. Thus

$$\frac{v_{\text{Hydrogen}}}{v_{\text{Air}}} = \sqrt{\frac{M_{\text{Air}}}{M_{\text{Hydrogen}}}} = \sqrt{\frac{29.0}{2.017}} = \sqrt{14.4} = 3.79,$$

and $\quad v_{\text{Hydrogen}} = 3.79 \times 331\ \text{m/s} = 1250\ \text{m/s}.$

We get the same answer by direct substitution in (13), using $\gamma = 1.40$, $k = 1.38 \times 10^{-23}$ J/K deg, and $\mu = 2.017 \times 1.66 \times 10^{-27}$ kg $= 3.35 \times 10^{-27}$ kg:

$$v_{\text{Hydrogen}} = \sqrt{1.40\ (1.38 \times 10^{-23}\ \text{J/K})(273\ \text{K})/3.35 \times 10^{-27}\ \text{kg}}$$

$$= \sqrt{1.58 \times 10^6\ \text{J/kg}} = \sqrt{1.58 \times 10^6\ \text{m}^2/\text{s}^2} = 1260\ \text{m/s}.$$

We note the remarkable similarity between the expression (13) for the speed of sound in a gas and the expression $\sqrt{3kT/\mu}$ [(30), p. 349] for the root-mean-square speed of the molecules themselves in their random thermal motion. This similarity is not accidental. The molecules of a gas exert forces on each other only when they *collide*. Hence, it is physically impossible for a *disturbance to be propagated* in the gas faster than the molecules themselves move. The molecules have a wide range of velocities, as indicated in Fig. 7, p. 352, but on this same figure we could put a mark at $\sqrt{\gamma}\sqrt{kT/\mu}$ ($=1.2\sqrt{kT/\mu}$ for hydrogen) to denote the speed of sound. *The speed of sound in a gas is of the same order of magnitude as the mean speed of random thermal motion of the molecules.* Thus, on p. 349, we computed 1840 m/s for the root-mean-square speed of hydrogen molecules at 0° C, while the speed of sound is 1260 m/s.

3. Production of Sound by Vibrating Air Columns

We can best approach the subject of the vibrations of an air column by considering a column closed at both ends, as in Fig. 3(a). The air in this column can be set into longitudinal vibration of large amplitude by a small diaphragm vibrating with *very small amplitude* at exactly the resonant frequency.

The fundamental mode of oscillation of this column is one in which the air rushes back and forth from left to right. There is no motion of the air at the closed ends, and a maximum amplitude of oscillation at

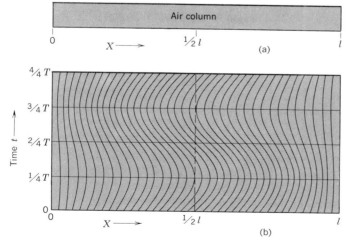

Fig. 3. Fundamental mode of oscillation of an air column closed at both ends. By cutting a narrow horizontal slot in a card, placing the card over part (b) of the figure, and then sliding the card upward along the page, one can 'see' the motion of the air particles.

the center. In Fig. 3(b), the positions of various particles of air are plotted as a function of time. At $t=0$, these particles are in their equilibrium positions and are uniformly spaced along the column. At $\frac{1}{4}T$, they have maximum displacement to the right; at $\frac{2}{4}T$, no displacement; at $\frac{3}{4}T$, maximum displacement to the left; again at $\frac{4}{4}T$, no displacement. The values of this displacement x as a function of X, at these various times, are indicated by the curves 0, 1, 2, 3, 4 of Fig. 4(a), which is the usual way of representing this standing wave.

The velocities of the various air particles are determined by the slopes (from the vertical) of the curves of Fig. 3(b). The particle velocities have their maximum positive values at $t=0$, and go through the cycle indicated in Fig. 4(b).

While the displacement and velocity have maximum values at the center, and nodes at the ends of the column, density and hence pressure changes are largest at the ends and zero at the center; density and pressure changes are represented by the curves of Fig. 4(c). In Fig. 3(b) we see that the set of particles that are uniformly spaced at $t=0$ suffer, at $t=\frac{1}{4}T$, a compression at the right end and a rarefaction at the left end, but that at the center of the column they still have the same spacing as at $t=0$ because neighbors are displaced by equal amounts.

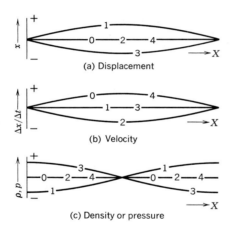

Fig. 4. Variation of displacement, particle velocity, and density or pressure along the column of Fig. 3. Curves 0, 1, 2, 3, 4 refer to successive times $t=0, \frac{1}{4}T, \frac{2}{4}T, \frac{3}{4}T, \frac{4}{4}T$.

The oscillation of Fig. 4 represents a standing wave in which the wavelength is $\lambda=2l$; the frequency would be determined by $f=v/\lambda=v/2l$, where v is the speed of sound in air. Figure 4 shows the fundamental mode of oscillation of the air column of Fig. 3. Higher modes would have additional displacement nodes between the ends of the column, exactly as in the case of a stretched string.

Air columns that are employed in musical instruments are not closed at both ends; they are *open at one or both ends* in order to obtain transmission of the energy of vibration of the column into the atmosphere as sound waves.

If the column of Fig. 3(a) were cut in half at $X=\frac{1}{2}l$ and the right half removed, the left half would continue to have the fundamental mode of oscillation illustrated by the left halves of Figs. 3 and 4. The reason for this is that the center of the column is a *pressure node* at which the pressure remains at atmospheric. Hence, removing half the tube will not change conditions in the other half since the center of the tube does not depart from atmospheric pressure. The only effect will be that air will now rush into and out of the tube from the atmosphere, as is needed for transmission of some of the energy of vibration as a sound wave.

The normal modes of oscillation of air columns or organ pipes are determined by the rules that

A closed end is a pressure antinode or a displacement node.

An open end is a pressure node or a displacement antinode.

Organ pipes are of two types: *closed* (one end closed and one open) and *open* (both ends open). The normal modes of vibration of these two types are illustrated in Fig. 5, in which the curves represent maximum displacements, as in Fig. 4(a). The wavelengths of the sound waves to which the pipes are resonant are given in terms of the length l of the pipe; one can easily verify the correctness of the values given in Fig. 5 by recalling that the wavelength in the standing-wave pattern is equal to the distance between *alternate* nodes or *alternate* antinodes, that is, to four times the distance between a node and the adjacent antinode.

Remembering that the speed of the waves in the pipe is just the speed v of sound in the air, we can write down the frequencies for the various modes of oscillation from the relation $f = v/\lambda$. From Fig. 5 we see that, for an open pipe, $\lambda = 2l/n$, where n is an integer. Thus, the resonance frequencies of an *open* organ pipe can be written as

$$f = nv/2l. \qquad (n = 1, 2, 3, \cdots) \quad (14)$$

If we denote the fundamental frequency of the open pipe by f_0, we may rewrite (14) in the form $f = nf_0$, where $f_0 = v/2l$. We note (a) that the normal frequencies of an open organ pipe are harmonics and (b) that *all harmonics are present*, since n takes on all integral values.

The case of a *closed* organ pipe is somewhat different. From Fig. 5 we see that $\lambda = 4l/n'$, where n' can assume only *odd-integral* values.

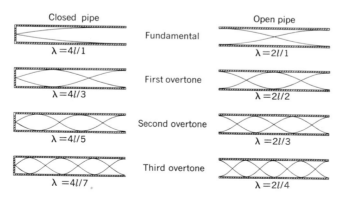

Fig. 5. Normal modes of oscillation of organ pipes. The curves represent maximum displacements.

Hence the resonance frequencies are given by the expression

$$f = n'v/4l. \qquad (n' = 1, 3, 5, \cdots) \quad (15)$$

If we denote the fundamental frequency of the closed pipe by f_C we may write (15) in the form $f = n'f_C$, where $f_C = v/4l$. We note (a) that the normal frequencies of a closed pipe are harmonics but (b) that *only odd harmonics are produced*. Also, we note that the fundamental frequency of a closed pipe is only half that of an open pipe of the same length.

The relations above are not completely accurate since the motion of the air at the open end is not strictly one-dimensional, as we tacitly have assumed. Small 'end corrections' need to be applied; these we shall ignore.

In an organ pipe, vibration of the air column results from oscillations produced in an air jet that is blown upward through a narrow slot against a knife-edge or 'lip' forming a part of the pipe wall as shown in Fig. 6. The jet of air from the slot is deflected alternately into and out of the pipe at the knife-edge. This alternation occurs at the resonance frequency of the pipe, being controlled by air alternately entering and leaving the pipe as the air in the pipe near this end moves alternately up and down at resonance frequency. Usually the fundamental and many of the overtones are excited simultaneously. It should be noted that the lower end of an organ pipe is a region of maximum vibration, and therefore should be regarded as *open*.

Fig. 6. A flue organ pipe.

The knife-edge or lip of a *flue* organ pipe, like that shown in Fig. 6, is a rigid solid which does not vibrate. Other types of organ pipes and certain other wind instruments contain a thin metal or wooden plate called a *reed*. The reed itself is caused to vibrate by the air stream, and, by opening and closing, produces vibrations in the air column. In brass wind instruments the lips of the player act as double reeds. Sounds produced by reeds are modified by the pipe or horn resonator used; in the clarinet and similar instruments the effective length of the pipe is varied by opening and closing small openings in the side.

The human voice organ may be regarded as a double-reed pipe instrument. The double reed is formed by the vocal cords; the oral and nasal cavities act as resonators.

Measurement of the resonance frequency of air columns furnishes a means of determining the speed of sound in a gas directly in terms of the length of the column by equations (14) or (15). If a tuning fork is held near the open end of a tube closed at the other end by a movable rigid disc, it is easy to determine the lengths of tube for which a standing-wave pattern is set up, since the observer hears a much louder sound at these lengths. The tube resonates with the tuning fork and increases the sound transmission to the atmosphere. The frequency of oscillation of the tuning fork can be accurately measured by means of a stroboscope. When a tuning fork is mounted on a sound box (Fig. 2), the 'acoustic amplification' is most effective if the length of the air column in the sound box is chosen so that one of its normal modes of oscillation has the frequency of the tuning fork.

Example. *To determine the speed of sound in hydrogen gas at 20° C, a long glass tube is filled with hydrogen at this temperature. One end of the tube is rigidly closed; the other end is closed by a stiff metal diaphragm driven by a loudspeaker. A small amount of light powder has been dusted into the tube. When a standing wave is set up in the tube, the powder is swept away from the displacement antinodes by the motion of the gas, and collects sharply at the nodes, where there is no motion. Hence the distance between the nodes can be measured. Such a tube is called a Kundt's tube. One of the higher normal modes of oscillation gives nodes 10.5 cm apart when the frequency is 6200 Hz. What value does this give for the sound speed?*

Since the wavelength is *twice* the distance between nodes in a standing wave, the wavelength is $\lambda = 2 \times 10.5$ cm $= 21.0$ cm $= 0.210$ m. Hence

$$v = \lambda f = 0.210 \text{ m} \times 6200/\text{s} = 1300 \text{ m/s}.$$

4. Intensity of Sound Waves

The psychological property of a sound called *loudness* is intimately connected with the *intensity* of the sound wave. Like all traveling waves, a sound wave involves the transmission of energy or power in the direction of propagation. The intensity is defined in terms of this power:

The **intensity** of a sound wave is the power transferred through unit area normal to the direction of propagation. It is measured in W/m^2.

The power transmitted by a sound wave in a gas can be determined by analogy with the power transmitted by a longitudinal wave in a spiral spring, which we discussed on p. 423. A unit cross section of gas along which sound is traveling has mass per unit length equal to ρ, the density of the gas. Hence we get the energy of vibration per unit volume by substituting ρ for μ in equation (20), p. 423:

$$\text{energy per unit volume} = 2\pi^2 f^2 A^2 \rho, \tag{16}$$

where A is the amplitude of the sonic vibration. The power per unit area (*intensity*) is obtained, as before, by multiplying this expression by the speed of sound v:

$$I = \text{intensity} = 2\pi^2 f^2 A^2 \rho v. \tag{17}$$

For a pure tone of given frequency, *loudness* increases with increasing *intensity*, but in general the relation between the loudness of a sound and its intensity is not simple. Loudness cannot be measured in physical terms, since it depends on the ear and judgment of the individual observer. It is relatively easy for two or more observers to agree that two sounds are equally loud, but different observers will not agree that one sound is 'twice as loud' as another. The difficulty of comparing the loudness of two sounds is greatest if the two sounds differ greatly in frequency.

The intensities of sounds that can be heard by the ear vary over an enormous range. A sound that is so loud that it is almost painful may have an intensity as much as 10^{12} times that of a sound that is barely audible. In view of this wide range of intensities, it is convenient to use a logarithmic scale in defining an *intensity level* for use in comparison of sound intensities. One cannot take the logarithm of the intensity I because I has physical dimensions, but one can take the logarithm of the *ratio of two intensities*, I/I_0, since this ratio is dimensionless. The logarithm, to base 10, of this ratio is called the *difference in intensity level* in *bels*.* Ten times this logarithm is called the difference in intensity level in *decibels* (dB), and it is this value that is ordinarily specified in comparing the intensities of two sounds (or of two electromagnetic waves):

The **difference in intensity level** of two sound waves, of intensities I and I_0 is defined as $10 \log_{10}(I/I_0)$ dB.

Since $\log_{10} 10 = 1$, $\log_{10} 100 = 2$, $\log_{10} 1000 = 3$, etc., a sound 10 times as intense as another has an intensity level 10 dB higher; one 100 times as intense has an intensity level 20 dB higher; and one 1000 times as intense has an intensity level 30 dB higher. This logarithmic type of intensity-level scale corresponds roughly with the behavior of the ear. For example, the level of a sound of any intensity must be raised by about 3 dB (a *factor* of 2 in intensity) before the ear perceives a very noticeable change in loudness.

If we take the intensity of the minimum detectable sound as I_0, the intensity I of a sound at the point of painfulness will be approximately $10^{12} I_0$, as mentioned above. Thus, the intensity level of this almost painful sound would be $10 \log_{10}(10^{12} I_0/I_0)$ dB $= 120$ dB above the threshold of audibility.

To facilitate specification of intensity levels, it is desirable to select some standard reference intensity I_0 as the threshold of audibility and to specify the intensity level of any other sound of intensity I in terms of this reference level. By international agreement an intensity of 10^{-12} W/m^2 has been selected as a *standard threshold of audibility* and is used as the value of I_0. If the hearing threshold is taken

Table 21-4 TYPICAL VALUES OF INTENSITY LEVEL ABOVE THRESHOLD

Sound	Intensity (W/m^2)	Intensity level (dB)
Painful	1	120
Riveting	3×10^{-3}	95
Elevated train	1×10^{-4}	90
Busy street traffic	1×10^{-5}	70
Conversation in home	3×10^{-6}	65
'Quiet' radio in home	1×10^{-8}	40
Whisper	1×10^{-10}	20
Rustle of leaves	1×10^{-11}	10
Hearing threshold	1×10^{-12}	0

---

*Named in honor of ALEXANDER GRAHAM BELL (1847–1922), the inventor of the telephone.

as 10^{-12} W/m², it is noted that the pain threshold, 120 dB higher, corresponds to power transmission of 1 W/m². This standard threshold is used in Table 21-4, which is based on a survey made in New York City.

Example. *As we shall see later, the ear is most sensitive to sound of frequency about 3000 Hz. At this frequency, 10 per cent of people can hear a sound of intensity 10^{-12} W/m². What is the amplitude of a sound wave of this frequency and intensity in air at 0° C and 1 atm? What would be the amplitude at the pain threshold of 1 W/m²?*

We substitute, in (17), $I=10^{-12}$ W/m², $f=3000$/s, $\rho=1.29$ kg/m³ [(14), p. 340], and $v=331$ m/s, and solve for A:

$$A^2 = \frac{I}{2\pi^2 f^2 \rho v} = \frac{10^{-12} \text{ W/m}^2}{2\pi^2 (3000/\text{s})^2 (1.29 \text{ kg/m}^3)(331 \text{ m/s})} = 13.2 \times 10^{-24} \text{ m}^2,$$

$$A = 3.63 \times 10^{-12} \text{ m} = 0.003\ 63 \text{ nm}$$

This value is extraordinarily small—about a hundredth of the diameter of an atom. The ear is indeed extraordinarily sensitive!

If we increase the intensity by a factor of 10^{12}, to reach the pain threshold, A^2 is multiplied by 10^{12} and A by 10^6 to give

$$A = 3.63 \times 10^{-6} \text{ m} = 0.003\ 63 \text{ mm},$$

which is less than $1/100$ of a millimeter!

Now consider the velocities of mass motion that are superposed on the random thermal velocities of the molecules of the gas. Consider the sound in the above example at the pain threshold with amplitude $A = 4 \times 10^{-6}$ m. From p. 246 we see that in simple harmonic motion of amplitude A and frequency f, the maximum velocity is $v_{\text{Max}} = 2\pi f A$. Hence, in this case, the maximum velocity of mass motion is

$$v_{\text{Max}} = 2\pi (3000/\text{s})(4 \times 10^{-6} \text{ m}) = 0.075 \text{ m/s},$$

which is extraordinarily small compared with the average speed of the air molecules, about 500 m/s, and with the speed of sound, about 300 m/s. At the threshold of hearing, v_{Max} is lower than the value computed above by a factor of 10^6, and is therefore about 10^{-7} m/s.

5. Pitch and Quality

As noted above, the psychological property of sound known as *loudness* is intimately related to the *intensity* of sound waves. We now discuss briefly two other psychological properties of sound: *pitch* and *timbre*.

Pitch is designated by musicians by letters corresponding to the keys on the piano. Except for extremely loud sounds, there is a one-to-one correspondence between pitch and frequency, the higher the pitch, the higher the frequency. In going *up* the scale one octave, we double the frequency; thus, the A notes on the piano have the following frequencies: 27.5, 55, 110, 220, 440, 880, 1760, and 3520 Hz.

The meaning of the term *quality* or *timbre* can be illustrated by sounding middle C successively on a flute, on a pipe organ, and on a violin, all of which are instruments, unlike a piano, that are capable of giving a reasonably continuous sound of constant volume. If these instruments are properly tuned, the fundamental frequencies of the three notes are the same, but we should have little difficulty in distinguishing between the sounds produced by the three instruments. The property of the three sounds which makes it possible for us to tell them apart is called *quality*. *The quality of a musical sound is determined by the intensities of the different overtones relative to that of the fundamental.* Thus, although the fundamental frequency of the note sounded by the three instruments mentioned above is the same, the note from each instrument has a different and distinctive pattern of overtone intensities which gives the note its distinctive *quality*.

In a good tuning fork only one mode of oscillation is excited, and hence the fork vibrates at only one frequency. The sound waves that are set up are therefore simple harmonic waves, that is, sinusoidal waves of a single frequency. A simple tone like that from a tuning fork is not particularly interesting. An interesting musical tone usually contains many overtones and these overtones are largely harmonics.

Recalling our discussion of the superposition principle in Chapter 20, we see that quality is related to the *waveform* of the sound waves produced by a given source. An auditor has little difficulty distinguishing the different vowel sounds whether these are spoken with the same basic frequency and intensity, or with radically different *pitch* and *loudness*. The difference is one of *quality*, and the radically different characteristic waveforms of the different vowel sounds can be readily demonstrated on an oscilloscope connected to a microphone.

A *noise*, in contrast to a musical sound, has a waveform with no semblance of regularity of any kind; the waveform consists of a series of random displacements.

6. Response of the Ear to Sound Waves

The ear consists of three major portions called the outer ear, the middle ear, and the inner ear (Fig. 7). Sound waves enter the outer ear, travel down the ear canal, and strike a thin membrane called the eardrum, which forms the boundary between the outer ear and an airspace called the middle ear. In the middle ear, a mechanism consisting of three small bones called the hammer, the anvil, and the stirrup transmit vibrations

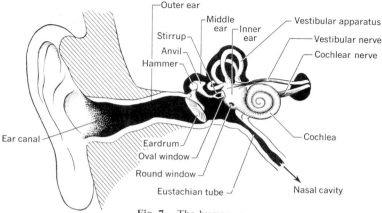

Fig. 7. The human ear.

from the eardrum to the oval window. The oval window is a membrane that transmits vibrations to the inner ear, which is filled with a liquid. In the cochlear spiral, which is a tube about 3.5 cm long and 1 or 2 mm in diameter, are the terminals of the auditory nerve. There are approximately 30 000 nerve terminals in the cochlea, and different frequencies have their greatest effects on nerve terminals located at different positions along the spiral. Sound signals are 'received' by these terminals, and are transmitted electrochemically to the brain. The vestibular apparatus shown in Fig. 7 is not associated with hearing. It consists of three fluid-filled semicircular canals in three orthogonal planes, and controls the sense of balance or equilibrium.

Extensive studies of the sensitivity of the ear to different sound frequencies have been made by the United States Public Health Service and the Bell Telephone Laboratories. Some of the results of these studies are summarized in the graph shown in Fig. 8. Two ordinate scales are shown: the first gives absolute intensities in W/m^2; the second gives the intensity level in dB above the standard reference level of 10^{-12} W/m^2. The uppermost curve shows at various frequencies the intensity of the sound that is loud enough to be almost painful; sounds of this intensity can be felt as cutaneous sensations as well as heard. The height of this curve is almost constant at a level of 120 dB above the standard reference level; this curve does not show extremely wide variations for various individuals. The lower curves represent the thresholds of hearing for various fractions of the group studied. Thus, 99 percent of the individuals studied could hear sounds with intensities shown by the curve labeled 99 percent, 95 percent could hear sounds with intensities given by the next lower curve labeled 95 percent, and so on. The lowest curve shows that 1 per cent of the group could hear sounds of intensity nearly 10 dB

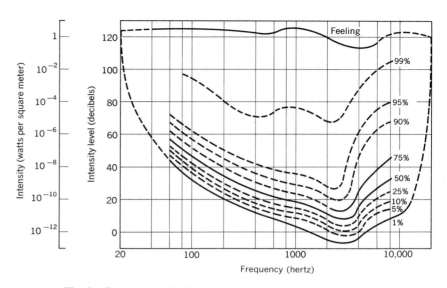

Fig. 8. Percentage of individuals who can hear sounds of various intensities.

below the standard reference level in the region between 2000 and 4000 Hz, the region in which the average ear is most sensitive.

7. Interference of Sound Waves

In the previous chapter, we discussed the interference of sound waves from two different sources. We can also get interference when sound from the same source is propagated along two different paths, as in the arrangement shown in Fig. 9. The diaphragm of a speaker S sends sound waves of a constant frequency into the air inside the tube. These waves travel to the opening O along two paths A and B. Path A has a constant

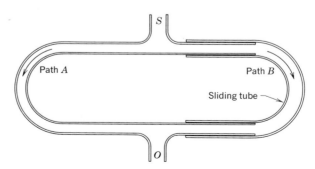

Fig. 9. Schematic diagram of apparatus for demonstrating interference of sound waves.

length, but the length of path B can be varied by means of the 'trombone' arrangement shown in the figure. The observer listens at O.

If the two paths are of *equal* length, the waves arrive exactly in phase at the opening O and sound will be heard by the observer. If the length of path B is now varied slowly, the intensity of the sound wave decreases to a minimum when the difference between paths B and A is one-half wavelength. As the length of path B is increased still further, the intensity of the sound increases to a maximum when the difference between path A and B is equal to a whole wavelength. Further elongation of path B leads to successive minima and maxima of intensity of the sound heard. These observations can be summarized in terms of l_A and l_B, the lengths of paths A and B, respectively, in the following way:

Intensity maxima occur when $\quad l_B - l_A = n\lambda.\quad (n=0, 1, 2, \cdots)$

Intensity minima occur when $\quad l_B - l_A = (n+½)\lambda.\quad (n=0, 1, 2, \cdots)$

The observed effects are readily explained on the basis of the superposition principle. When the path difference is an integral number of wavelengths, the waves arriving at O are effectively *in phase* and reinforcement occurs. When the path lengths differ by an odd number of half wavelengths—that is, by $(n+½)\lambda$—the waves arriving at O are 180° out of phase and destructive interference occurs. This apparatus furnishes a convenient method of measuring the wavelength of sound and hence of determining the speed of sound provided a sound source of known frequency is employed.

If two wave trains of slightly *different frequency* traverse a medium, a stationary interference pattern will not be formed. However, if their frequencies are nearly equal, two sets of waves can interfere in such a way as to produce a sound with pulsating intensity at a given point; these pulsations in intensity are known as *beats*. From the superposition principle it is easy to see the reason for the production of beats. Let the displacement produced at a given point by one wave be $S_1 = A\sin 2\pi f t$, and that produced by the second wave be $S_2 = A'\sin 2\pi f' t$. Then, by the superposition principle, the resultant displacement is

$$S = S_1 + S_2 = A\sin 2\pi f t + A'\sin 2\pi f' t.$$

For the sake of simplicity, consider a case in which $A = A'$; then by using the trigonometric relation for the sum of two sines, we can write

$$S = 2A\cos\left[2\pi\left(\frac{f-f'}{2}\right)t\right]\sin\left[2\pi\left(\frac{f+f'}{2}\right)t\right]. \tag{18}$$

If f is close to f', this can be regarded as a sine wave of frequency $½(f+f')$ modulated by a slowly varying amplitude given by the coefficient of the sine. A little consideration will reveal that the *magnitude* of this amplitude attains a maximum value $f-f'$ times per second. Thus,

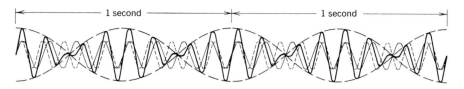

Fig. 10. Beats. The light solid curve has $f = 8$ Hz; the light broken curve has $f' = 10$ Hz. The heavy curve is the sum; the long-dash curve indicates the amplitude variation.

an observer will hear $f - f'$ *beats* per second. Figure 10 illustrates the case of superposition of frequencies 8 and 10 Hz, and the heavy curve representing the sum reaches maximum amplitude 2 times per second. If a tuning fork with a frequency of 256 Hz and one with a frequency of 260 Hz are sounded simultaneously, an observer will hear four beats each second of sound of frequency 258 Hz. By counting beats it is possible to determine accurately the frequency difference between two sounds of nearly equal frequency.

When the difference between the frequencies of two sounds is more than 10 or 15 Hz, the beats become difficult to distinguish. However, when the difference is great enough to correspond to an audible frequency, one may hear a *difference tone*. For example, if two intense sound waves having frequencies of 10 000 Hz and 8500 Hz reach the observer, he may hear a difference tone of 1500 Hz.

> **Example.** *In the natural diatonic musical scale, middle C and G have frequencies* 256 *and* 384 Hz, *in the ratio* 2 *to* 3. *Pianos are tuned to a scale of equal temperament in which C and G are* 256 *and* 386 Hz. *A piano-tuner systematically beats harmonics of notes against each other in order to adjust frequencies exactly. What should he hear when he beats middle G against low C?*
>
> Low C, one octave below middle C, has half the frequency, or 128 Hz. Its third harmonic has three times this frequency, or 384 Hz. If the piano is properly tuned to equal temperament, this should give two beats per second against the middle G of 386 Hz.

There are three other physical phenomena connected with the behavior of sound waves that should be mentioned, namely, *reflection, refraction,* and *diffraction*. Since these phenomena are shared by light waves and are much more important in the case of light, we shall treat them in detail later in Part V. Reflection gives rise to *echoes;* refraction changes the path of sound waves when they pass through a medium in which the wave speed varies with position and gives rise, for example, to anomalous sound transmission over lakes; diffraction gives sound waves the ability to 'bend around corners.'

8. Doppler Effect

When a source of sound is moving with respect to an observer or an observer is moving with respect to the source, the pitch of the sound heard by the observer is different from the pitch heard when the source and observer are both at rest. Observation by a man on a station platform shows that the pitch of a locomotive whistle is *higher* when the locomotive is approaching the station than when the locomotive is standing still; the pitch is *lower* when the locomotive is moving away from the observer than when the locomotive is standing still. The effect is very noticeable when an express train that is blowing its whistle is moving at high speed past the station platform of a local stop. The change in frequency caused by the motion is called the *Doppler effect*.*

We shall consider the simplest case, in which the air is stationary and source and observer are moving directly toward (or away from) each other. If the source is moving toward the right with speed v_S as in Fig. 11, each successive condensation progressing toward the right will be closer to its predecessor than if the source were standing still, and the resulting wavelength will be shorter. In the situation indicated in Fig. 11, the frequency of the source, f_S, is 5 Hz, and five wavelengths are crowded into the distance $v - v_S$, so $\lambda = (v - v_S)/5$. Here v is the speed of sound in air. In general, the sound emitted in the forward direction has wavelength

$$\lambda = (v - v_S)/f_S. \tag{19}$$

Now let the observer at the right of Fig. 11 be moving toward the

---

*Named for the Austrian physicist, CHRISTIAN JOHANN DOPPLER (1803-1853), who first discussed the effect in the case of light waves.

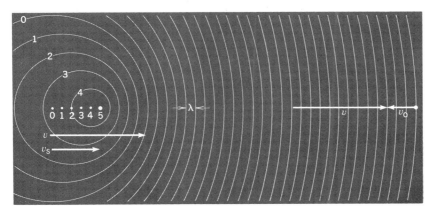

Fig. 11. Doppler effect: positions of successive condensations from a source moving to the right with speed v_S.

source at speed v_O. In unit time the observer gathers up all the waves in the distance $v+v_O$. Hence he hears the frequency

$$f_O = \frac{v+v_O}{\lambda} = f_S \frac{v+v_O}{v-v_S}. \qquad (20)$$

In this equation, v_O and v_S *are positive when they represent speeds of approach.* It is readily seen that the equation also applies when either the source or the observer, or both, are receding, if negative values of the corresponding speeds are used in (20).

Since sound travels at speed v *relative to the air* and not relative to the ground, if there is a wind Fig. 11 should be considered as drawn relative to the air, and v_O and v_S in (20) should be measured relative to the moving air.

Example. Sound of frequency 1000 Hz *from a stationary source is reflected from an object approaching the source at* 31 m/s, *back to a stationary observer located at the source, on a day when the air is at* 0° C. *Determine the frequency heard by the observer.* (*While this situation may be of only academic interest in the case of sound, the corresponding phenomenon for radio waves is the basis of the modern 'radar' method of measuring speeds of moving automobiles and planes, as we shall discuss in the problems in Chapter 34.*)

The moving reflector receives the frequency $f_R = f_S(v+v_R)/v$ given by (20) with the reflector velocity v_R substituted for v_O. This is the frequency reflected by the reflector, which now acts as a moving source, so the observer receives a frequency given by (20) as

$$f_O = f_R\, v/(v-v_R) = f_S(v+v_R)/(v-v_R). \qquad (21)$$

The velocity of the moving reflector enters *twice* into the Doppler-frequency formula. The reflector acts like a moving observer in receiving the sound and then like a moving source in re-emitting it.

Substituting $f_S = 1000/\text{s}$, $v = 331$ m/s, $v_R = 31$ m/s in (21), we find

$$f_O = 1000\ \text{s}^{-1}\,(362/300) = 1210\ \text{Hz}$$

for the frequency heard by the observer.

PROBLEMS

1. A copper wire 0.1 cm in diameter is 4.0 m in length. Find the speeds of transverse and longitudinal waves in this wire when a tensile force of 48 N is applied to each end. Find the frequency of the fundamental transverse and longitudinal modes of oscillation of this wire.
Ans: 82.8 m/s, 3520 m/s; 10.4 Hz, 444 Hz.

2. In a brass wire stressed to its elastic limit, what is the ratio of the speeds of longitudinal and transverse waves?

3. String B has twice the length, twice the diameter, twice the tension, and twice the density of string A. What overtone of B agrees in frequency with the fundamental of A?
Ans: third.

4. A certain string is tuned to give middle C—256 Hz—when under a 40-p tension. How much should the tension be increased in order to produce high C—512 Hz—as the fundamental?

5. An observer sees the flash of a gun and hears the report 3 s later. What is the distance between the observer and the gun if the temperature is 20° C? *Ans: 3390 f.*

6. A man sees the 'steam' from a locomotive whistle 4 s before he hears the sound. What is the distance from the man to the locomotive if the temperature is 20° C?

7. An observer sees the flash of an explosion on the opposite shore of a lake. If the lake is exactly 2 mi wide, how long will it be before he hears the sound reaching him through the air? If the observer uses a hydrophone to receive the sound transmitted through the water, how long after the explosion does he receive the sound? Air and water are at 15° C. *Ans: 9.46 s; 2.20 s.*

8. If a charge of dynamite is set off on a railroad track, how long does it take the sound to travel through the steel rails to an observer exactly 6 mi away? How long after receiving the sound signal transmitted by the rails will the observer receive the sound signal transmitted through the air? Use (3) to compute the speed in the rails. Take air temperature as 20° C.

9. A sound of 306 Hz takes 0.744 s to travel vertically from a point 200 m under water to a point 200 m in the air above the surface of the water. The speed in air is 330 m/s. From these data alone calculate (*a*) the wavelength of this sound in air, (*b*) the wavelength of this sound in water. *Ans: (a) 1.08 m; (b) 5.74 m.*

10. An observer drops a stone down a well, and notes with a stopwatch that 12 s elapse from the time he lets the stone fall until he *hears the sound* of its striking at the bottom. How deep is the well if the temperature of the air in it is 10° C?

11. When the needle on a $33\frac{1}{3}$-rev/min phonograph record is at 4 i from the axis and is picking up a frequency of 10 000 Hz, what is the 'wavelength' of the sinusoidal variations in depth of the record's groove? How many waves are there per inch of groove? *Ans: 11.6×10^{-5} f; 71 700.*

12. A *siren* contains a rotating disc having a circular row of regularly spaced holes through which a jet of compressed air is blown. If there are 90 holes around the circle, at what speed, in rev/min, must the disc rotate to give a 1000-Hz tone? If the tone varies from 900 to 1100 Hz, what is the speed variation of the disc?

13. A dynamite blast is set off on the surface of a lake. A swimmer near the shore receives two shock waves—one through the water and one through the air at 20° C. Four seconds elapse between the times of arrival of the two waves. How far was the swimmer from the explosion? *Ans: 5900 f.*

14. A ship is in a dense fog in a Norwegian fiord, two miles wide. In order to find his location in the channel, the captain fires a gun and hears the first echo after 3 s. (*a*) How far is the ship from the center of the fiord? (*b*) How long after the gun was fired did the second echo arrive? Use 1050 f/s as the speed of the sound.

15. If the Kundt's tube in the example on p. 449 were filled with argon (monatomic, $\gamma = 1.67$) instead of hydrogen (diatomic, $\gamma = 1.4$), what would be the distance between nodes if no other change is made in the experimental arrangement? *Ans: 2.58 cm.*

16. A ship's 'fathometer' measures water depth below the keel by sending out a sharp pulse of high-frequency sound from a source in the keel and measuring time to the first echo. If the chart on which the time is recorded is accurate to 0.005 s, and can read up to 2 s, what is the accuracy of depth measurement in fathoms (1 fathom = 6 f) and what is the maximum depth in fathoms that can be determined?

17. The observed value of γ for nitrogen is 1.40. Using this value, find the speed of sound in nitrogen at 0° C. What is the root-mean-square speed of nitrogen molecules at 0° C? *Ans: 337 m/s; 493 m/s.*

18. The observed value of γ for neon is 1.67. Using this value, find the speed of sound in neon at 0° C. What is the root-mean-square speed of neon molecules at 0° C?

19. On a day when the speed of sound in air is 1080 f/s, what are the frequencies of the fundamental and first three overtones of an organ pipe 8 f long when the tube is operated as a *closed* pipe? *Ans: 33.8, 101, 169, 236 Hz.*

20. Solve Prob. 19 for the case where the pipe is used as an *open* pipe.

21. What should be the length of a closed organ pipe if the fundamental frequency is to be 128 Hz? What should be the length of an open organ pipe to give this fundamental frequency? Assume $v = 1080$ f/s.
Ans: 2.11 f; 4.22 f.

22. Newton's equation for speed of sound was $v = \sqrt{P/\rho}$, derived on the erroneous assumption that the process of transmission of sound was isothermal. Show that this is the equation that would be obtained from (7) on the isothermal assumption.

23. A tuning fork is near the open end of a cylindrical glass tube. The other end of the tube is closed by a movable piston. When the tube is of such length that one of its natural frequencies coincides with that of the tuning fork, a decided increase in volume of sound is heard. If this resonance condition occurs at tube lengths of 20 cm, 60 cm, and 100 cm, what is the frequency of the tuning fork if the temperature is 20° C? Ans: 430 Hz.

24. If a piano is tuned to match an organ on a warm summer afternoon, in which 'direction' will it be out of tune in the cool of the evening?

25. As indicated in Fig. 8, the hearing threshold for sound of frequency 100 Hz is approximately 10^{-9} W/m². What is the amplitude of a sound wave of this frequency and intensity in air at 0° C and 1 atm? in hydrogen?
Ans: 3.43 nm; 6.70 nm.

26. If the amplitude and frequency of a sound wave are both tripled, what is the increase in intensity level in decibels?

27. For sound waves having a frequency of 100 Hz, the intensity at the hearing threshold is approximately 10^{-9} W/m². For sound at this frequency, the pain threshold is reached when the intensity is increased by 90 dB. What is the intensity of sound at the pain threshold? What would be the amplitude at the pain threshold in air at 0° C and 1 atm? in air at 0° C and 4 atm?
Ans: 1 W/m²; 1.08×10^{-4} m; 5.39×10^{-5} m.

28. An open organ pipe at the front of a church is tuned to 256 Hz at 20° C, and an open organ pipe in the echo organ at the rear is tuned to 512 Hz at 20° C. On Sunday the rear of the church is at 25° C, the front at 20° C. How many beats occur between the fundamental of one pipe and the first overtone of the other when they are sounded together?

29. In the arrangement shown in Fig. 9, a vibrating diaphragm at S has a frequency of 1000 Hz. If the length of path A is fixed at 50 cm, for what lengths of path B will intensity maxima be observed? For what lengths of path B will intensity minima be observed? (Take 330 m/s as the speed of sound in air.)
Ans: 50, 83, 116, 149, ⋯ cm; 66.5, 99.5, 132.5, ⋯ cm.

30. If the arrangement in Fig. 9 were set in such a way that the difference in length of paths A and B is 50 cm, what are the five lowest frequencies of S capable of giving intensity maxima?

31. In an arrangement similar to that in Fig. 9, paths A and B are each 2 m in length. If path A is filled with air at 0° C and path B is filled with hydrogen at 0° C, what is the lowest frequency that will produce an intensity maximum? Ans: 224 Hz.

32. A certain note on a piano gives 3 beats per second when sounded with a 256 Hz tuning fork and 7 beats per second when sounded with a 260 Hz tuning fork. What is the frequency of the note?

33. Pure tones of 1000, 1100, 1200, and 1300 Hz are sounded together. To the auditor, a much lower-pitched note seems to predominate. Explain this phenomenon and determine the frequency of the note.

34. A closed organ pipe has a fundamental frequency of 128 Hz under normal conditions. A practical joker fills the pipe with helium. What is the fundamental frequency of the helium-filled pipe?

35. The fundamental frequency of a certain locomotive whistle is 300 Hz and the speed of the locomotive is 100 f/s. On a day when the speed of sound in air is 1100 f/s, what will be the fundamental frequencies of the sounds heard by an observer on a station platform as the locomotive passes the station with its whistle blowing? Ans: approach: 330 Hz; at station: 300 Hz; recession: 275 Hz.

36. If the fundamental frequency of a bell at a railroad crossing is 400 Hz, what frequencies will the engineer of a locomotive hear as the locomotive passes the crossing at 100 f/s if the speed of sound is 1100 f/s?

37. Locomotive A is moving southward at a speed of 100 f/s while locomotive B is moving northward at a speed of 100 f/s. If the whistle on locomotive A has a fundamental frequency of 128 Hz, what will be the frequencies of the sounds heard by the engineer on locomotive B as the two locomotives pass each other? Take the speed of sound as 1100 f/s.

Ans: approach: 153 Hz; adjacent: 128 Hz; recession: 106 Hz.

38. A tuning fork of frequency 1000 Hz is mounted 1 f from the axis of a rotating horizontal table turning at 4 rev/s. What are the highest and lowest frequencies heard by an auditor whose ear is at the same height as the tuning fork?

39. A man walks at 3 m/s on a still day along the line between two fire stations. The fire whistles on the two stations sound simultaneously at a frequency of 300 Hz. Find the apparent frequency of each whistle and the number of beats per second. Take $v = 330$ m/s.

Ans: 297, 303, 6 Hz.

40. A man approaching a stationary wall at speed v_M blows a whistle of frequency f_S. Find the frequency of the echo heard by the man.

41. The sound from a pistol shot at point A is echoed from an object at point B, a distance d from A. Show that the time between the shot and receipt of the echo is

(a) $2\,d/v$ if there is no wind;
(b) $(2\,d/v)/(1 - v_W^2/v^2)$ if there is a wind of speed v_W blowing from A to B;
(c) $(2\,d/v)/\sqrt{1 - v_W^2/v^2}$ if there is a *cross wind* of speed v_W.

Compare with Prob. 16, p. 47, which concerns a related problem in relative velocities and discusses the implications of applying similar reasoning in the case of the velocity of light.

42. In the *sonic anemometer*, invented in 1944, short pulses of sound are emitted simultaneously by sources S and S' and received at R and R' (see figure). The time difference between receipt at R and R' is accurately measured electronically; this measurement permits an accurate determination of the component of wind velocity along the line SR'-$S'R$. Show that a wind perpendicular to this line will not result in any time difference. Show that a wind of speed w parallel to this line will result in a time difference given by the simple formula $2\,Dw/v^2$ if w is small compared with the speed v of sound.

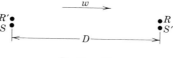

Problem 42

PART IV

ELECTROMAGNETISM

ELECTROSTATICS 22

Electrostatics is the branch of science concerned with electric charges at rest.

It is convenient to begin the study of electricity and magnetism by considering, in this and the following two chapters, the principles of *electrostatics*.

Electricity and magnetism were for thousands of years regarded as separate and unrelated subjects. Certain electrostatic phenomena were familiar to the Greeks, as were the phenomena associated with natural magnets or lodestones. Methods of producing *electric currents,* however, were first discovered less than two centuries ago. Soon thereafter, it was found that electric currents produce magnetic effects; this discovery revealed that electric and magnetic phenomena were related and could be treated in terms of a single theory of *electromagnetism*.

In order to deal with the subject of electromagnetism, it is necessary to introduce one more fundamental physical quantity in addition to *length, mass, time,* and *temperature*. Since we begin with electrostatics, we tentatively select *electric charge* as the additional fundamental quantity and its unit, the *coulomb,* as the additional fundamental unit. Later we shall see that, for operational convenience, the fundamental unit that has been selected by international agreement is the unit of *current,* the *ampere,* which is the coulomb/second. It is interesting to note that *metric* mechanical units are used exclusively in defining the derived electromagnetic units; no system has ever been used that employs English mechanical units.

1. Coulomb's Principle

We can electrify, or charge, a hard-rubber rod by rubbing it with cat's fur, or a glass rod by rubbing it with silk. The electrified rods acquire the ability to attract bits of paper or small balls made of light material such as pith or cork and suspended by silk threads.

By experimenting with very light 'metal' balls such as pith balls coated with metal foil or painted with metal paint, we can easily convince ourselves that there is a difference between the type of electric charge on the hard rubber and that on the glass. The metal ball is attracted by either charged rod until it is permitted to touch the rod; then it bounces away from the rod it has touched and is thereafter strongly repelled by it. A charged metal ball that is repelled by the hard-rubber rod is attracted by the glass rod, and one that is repelled by the glass rod is attracted by the hard-rubber rod. Furthermore, a metal ball that has touched a charged glass rod and another that has touched a charged hard-rubber rod attract each other; but if two metal balls have touched the same charged rod, they repel each other. Evidently the two charged rods must have two different *kinds* of electric charge. Any two dissimilar materials when brought into contact or rubbed together become more or less charged, but always with the same two kinds of electric charge that are produced in considerable quantity on hard rubber and glass when rubbed with cat's fur and silk, respectively.

The two kinds of electric charge are aribtrarily called *positive* and *negative* according to conventions introduced by BENJAMIN FRANKLIN:

Positive electric charge is the kind of charge on glass that has been electrified by rubbing with silk.

Negative electric charge is the kind of charge on hard rubber that has been electrified by rubbing with cat's fur.

We now know that a negatively charged body is one with an *excess of electrons,* while a positively charged body is one with a *deficiency of electrons.*

From the experiments discussed above, we conclude that charged objects exert forces on each other in accordance with the following rule:

Like charges repel each other and unlike charges attract each other.

The electric force between a given pair of charges decreases rapidly as the distance between the charges increases. The law of dependence on distance was first determined experimentally by the French physicist, CHARLES AUGUSTIN COULOMB, in 1785. Using a torsion balance to measure the forces, Coulomb showed that *the mutual force between two charged particles varies inversely as the square of the distance between them*:

$$F \propto 1/d^2. \tag{1}$$

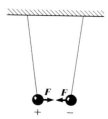

The directions of the force vectors are along the line connecting the charges, as in Fig. 1.

Metals and nonmetals behave quite differently with respect to electric charges. If a piece of metal is brought into contact with a charged rod, the entire surface of the metal becomes charged immediately. Charge can travel easily from one part of a piece of metal to another, and the metal is said to be a *conductor* of electricity. If a piece of nonmetal is brought into contact with a charged rod, charge will be transferred to the nonmetal only at the point of contact. Charge cannot readily travel from one part to another of a nonmetal, which is therefore called an *insulator* or *dielectric*. This difference in behavior is explained by the fact that in metals some of the electrons (the valence electrons) are 'free' to wander from atom to atom; in nonmetals the electrons are not, in general, free to leave the atoms to which they belong.

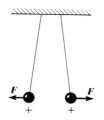

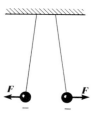

Fig. 1. Unlike charges attract; like charges repel.

The 'free' electrons in a metal move around inside the metal much like the molecules in a gas. In the neutral or uncharged state of a metal, the electrons are uniformly distributed throughout the volume of the metal. If an uncharged metal ball is brought near a positively charged rod as in Fig. 2(a), the electrons in the ball are attracted by the rod and the positive nuclei in the ball are repelled by it. The mobile free electrons concentrate on the side of the ball near the rod. The side of the ball nearer the charged rod acquires a net negative charge; the other side of the ball is left with a deficiency of electrons and hence an *equal* net positive charge. The reverse happens in Fig. 2(b), in which the electrons are repelled by the negatively charged rod. Charges acquired by portions of an uncharged object in this manner are said to be 'induced' by the nearby charged body.

If the metal ball in Fig. 2(a) is made so that the two halves can be separated, one can obtain equal and unlike charges on the two halves by separating them before the charged rod is removed. The sequence of events is shown in Figs. 3(a)–(e), which are intended to be self-explanatory. Instead of a single metal sphere that can be split, two spheres originally in contact can be used as in Figs. 4(a)–(e). The two spheres in Fig. 4(e) have charges that are equal and unlike, since the spheres in Fig. 4(a) were uncharged, and the excess of electrons on the left sphere must exactly equal the deficiency of electrons on the right sphere.

In this way, Coulomb obtained on small spheres charges which he

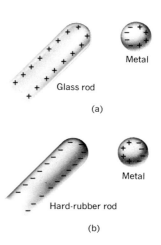

Fig. 2. Induced charges.

468 ELECTROSTATICS CHAP. 22

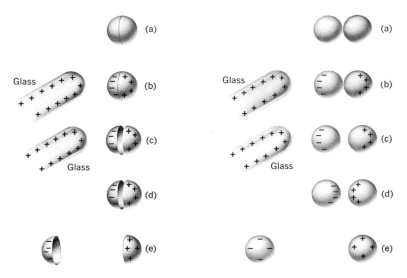

Fig. 3. By induction, equal and opposite charges are obtained on the two halves of a metal sphere.

Fig. 4. By induction, equal and opposite charges are obtained on two metal spheres.

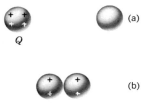

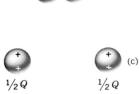

Fig. 5. By touching a charged metal sphere to an uncharged metal sphere of the same size, the charge is divided into two equal parts.

knew were equal and unlike. With his torsion balance he then found that a third charge would exert equal but opposite forces on the equal but unlike charges. In other words, like charges repel each other with the same forces as unlike charges attract, if the charges are numerically equal.

Finally, Coulomb found that the force that either of two charges exerts on the other is proportional to the product of the magnitudes of the charges. He verified this fact by a technique of splitting a given charge into two, four, or more equal parts that is illustrated in Fig. 5. When the two identical spheres in Fig. 5 are separated, they are found to be equally charged (they exert equal forces on any other charge) as would be expected from symmetry. The charge on one of these spheres can again be divided into two, and so on.

Coulomb found that the force between two charges, denoted by Q_1 and Q_2 (Fig. 6), is proportional to Q_1, to Q_2, and to $1/d^2$; that is,

$$F \propto Q_1 Q_2 / d^2.$$

COULOMB'S PRINCIPLE: *The electric force between two point charges is proportional to the magnitude of each of the charges and inversely proportional to the square of the distance between the charges.*

This principle, based on experiment, provides the definition of *magnitude of charge,* once we have selected the proportionality constant. The unit of charge, the *coulomb* (C), rigorously defined later in Chapter 29 in terms of the magnetic forces acting between electric currents, is such that if

F is in newtons and d in meters, the proportionality constant is 8.9875×10^9 N·m²/C². Thus

$$F = (8.9875 \times 10^9 \text{ N·m}^2/\text{C}^2)\frac{Q_1 Q_2}{d^2}. \qquad (2)$$

The reason for the choice of this particular coefficient will be discussed in detail in Chapter 29; suffice it here to note that the numerical value is 10^{-7} times the square of the speed of light in m/s.

It is customary to write the force equation (2) in the form

$$F = \frac{1}{4\pi\varepsilon_0} \frac{Q_1 Q_2}{d^2}, \quad \begin{cases} F \text{ in N} \\ Q \text{ in C} \\ d \text{ in m} \\ \varepsilon_0 \text{ in C}^2/\text{N·m}^2 \end{cases} \quad (3)$$

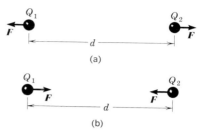

Fig. 6. When Q_1 and Q_2 have like signs, the force F in (2) comes out positive and is to be interpreted as repulsive, as in (a); when Q_1 and Q_2 have unlike signs, F comes out negative in (2), and is to be interpreted as attractive, as in (b).

where ε_0 is a dimensional constant known as the *permittivity of free space*, and, in this so-called 'rationalized' system, the numerical factor 4π is introduced so that many important equations that we shall derive later and employ more frequently than (3) will appear without a factor of 4π. Comparison of (3) with (2) gives the value of ε_0 as

$$\varepsilon_0 = 8.8542 \times 10^{-12} \text{ C}^2/\text{N·m}^2. \qquad (4)$$

For use in problems we give below, to three-figure accuracy, two frequently occurring coefficients involving ε_0:

$$1/\varepsilon_0 = 1.13 \times 10^{11} \text{ N·m}^2/\text{C}^2, \quad 1/4\pi\varepsilon_0 = 8.99 \times 10^9 \text{ N·m}^2/\text{C}^2. \qquad (5)$$

The coulomb is actually chosen to be a unit of convenient size in working with electric *currents*, but, as indicated by the large magnitude of the coefficient in (2), it is an enormous unit from the standpoint of most considerations of electrostatics. Actual static charges are usually conveniently expressed in terms of the submultiples, the *micro-, nano-,* or *picocoulomb* (μC, nC, pC).

Where more than two charges are present, the total force acting on one of the charges is the vector sum of the forces arising from each of the other charges individually. For example, in Fig. 7, we may compute the total force on the -0.2-μC charge by adding vectorially the three forces of repulsion from the other three charges, in the manner indicated.

We are now in a position to see why an *uncharged* metal-coated pith ball is attracted to a charged rod. In Fig. 2, we see that the charges of sign unlike the charge on the rod are closer to the rod than are the charges of like sign. Because of this difference in average distance, the force of attraction experienced by the unlike charges is greater than the force of repulsion experienced by the like charges, and the pith ball experiences a net force of attraction.

The fact that an uncharged *nonmetallic* object, such as an *uncoated*

470 ELECTROSTATICS CHAP. 22

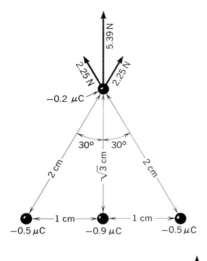

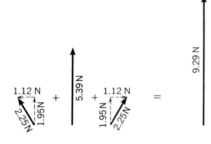

Fig. 7. The resultant electric force on the -0.2-μC charge is 9.29 N upward.

pith ball or a piece of paper, is attracted to a charged rod is associated with a phenomenon called *dielectric polarization*. In spite of the fact that electrons in a nonconductor cannot leave the atom to which they belong and move to the side of the ball, as they do in the conductor of Fig. 2, the electrons in each atom are pulled or pushed so that they lie, on the average, closer to or farther from the charged rod than the positive nucleus. When the charges of the dielectric are thus disturbed, the dielectric is said to become *polarized*. In Fig. 8(a) the electrons in each atom are on the average closer to the positive charges of the rod than is the positive nucleus. Thus, each atom experiences a net force of attraction toward the rod. Similarly, Fig. 8(b) illustrates the polarization resulting from a negative rod, and again we see that there is a net force of attraction.

One of the fundamental principles of physics, for which there is no contrary evidence whatsoever, is the

PRINCIPLE OF CONSERVATION OF ELECTRIC CHARGE: *The total quantity of electric charge in the universe does not change.*

It is reasonable to hypothesize that the total electric charge in the universe is *zero,* but no one has proposed a way to test this hypothesis.

The positive charge of the hydrogen nucleus (the *proton*) and the negative charge of the electron are the smallest (nonvanishing) charges that have ever been observed. These charges have the same absolute magnitudes. Experimental evidence to date indicates that *all* charges are integral multiples of these elementary charges. Hence, in atomic and nuclear physics, it is convenient to employ an 'atomic' unit of charge called the *electronic charge unit* and abbreviated e:

The **electronic charge unit** (e) is an amount of charge equal to the charge of the proton.

The charge of the electron is -1 e; that of the proton is $+1$ e; that of a nucleus of atomic number Z is $+Z$ e. One method of measuring the charge of the electron in coulombs is discussed in Sec. 6. The present best value of the electronic charge unit is

$$1\ e = (1.602\ 10 \pm 0.000\ 07) \times 10^{-19}\ C. \tag{6}$$

2. Electric Fields

An *electric field* is said to exist in any region of space in which an electric charge would experience an electric force, for example in the region around a charged body. The intensity of the electric field at a point is defined in terms of the electric force $\boldsymbol{F}$ that would be exerted on a very small positive test charge $+q$ placed at the point. *The test charge must be imagined to be so small that it does not appreciably disturb the charges on the bodies that set up the field;* that is, it must not cause appreciable redistribution of the charges on conductors or alteration of the polarization of dielectrics in its neighborhood. We should think of the test charge as *infinitesimal*.

The net force $\boldsymbol{F}$ on the test charge $+q$ (*see* Fig. 9, for example) is the resultant of all the forces on q exerted by the various individual charges or elements of charge on the charged bodies in the neighborhood. By Coulomb's law, each of these component forces is proportional to the magnitude of q. Hence, the resultant force $\boldsymbol{F}$ has a magnitude proportional to q, and we may write

$$F = q\,\mathcal{E}. \tag{7}$$

In this equation,
$$\mathcal{E} = F/q, \tag{8}$$

the force per unit of test charge, is known as the *electric intensity*.

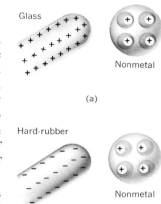

Fig. 8. Dielectric polarization. The darkened circles represent the distributions of electrons around the atomic nuclei.

> The **electric intensity** $\mathcal{E}$ at a point is a vector having the direction of the force that would be exerted on a positive test charge placed at the point and a magnitude equal to the magnitude of this force divided by the magnitude of the test charge, in the limit in which the size of the test charge approaches zero.

Thus, in an electric field there is a vector $\mathcal{E}$ associated with each point of space. In terms of this vector, equation (7) gives the force that would be exerted on a small charge q placed at that point. *If the small charge q is positive, the force $\boldsymbol{F}$ has the same direction as $\mathcal{E}$; if the small charge is negative, the force $\boldsymbol{F}$ has the direction opposite to the vector $\mathcal{E}$.*

In learning to compute electric intensities, let us start with the field in the neighborhood of a single isolated positive point charge of magnitude Q. To measure the intensity at a distance r from the charge Q, we imagine that we place a small test charge $+q$ (Fig. 10) at that distance from the charge Q. At any point on a sphere of radius r drawn around the charge Q, the force on the test charge would be radially outward and of magnitude

$$F = \frac{1}{4\pi\varepsilon_0}\frac{Qq}{r^2}. \tag{9}$$

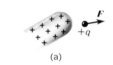
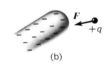

Fig. 9. The electric field has the same direction as the force $\boldsymbol{F}$ acting on a small positive test charge $+q$, and has the magnitude F/q.

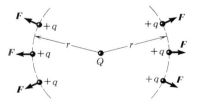

Fig. 10. The field at distance r from a positive charge Q is determined by the force $\boldsymbol{F}$ on a small test charge $+q$.

Fig. 11. Electric intensities $\mathcal{E}$ at points near a positive charge.

The electric intensity is, then, by definition (8), directed radially outward and of magnitude

$$\mathcal{E} = \frac{F}{q} = \frac{1}{4\pi\varepsilon_0}\frac{Q}{r^2}. \qquad \begin{cases}\mathcal{E}\text{ in N/C}\\ Q\text{ in C}\\ r\text{ in m}\\ \varepsilon_0\text{ in C}^2/\text{N}\cdot\text{m}^2\end{cases} \quad (10)$$

The intensity falls off inversely as the square of the distance from the charge, as indicated by the lengths of the vectors in Fig. 11.

If the charge Q were negative in Fig. 10, the force $\boldsymbol{F}$ on a positive test charge would be directed radially inward rather than outward. Since the intensity is in the direction of the force on a positive test charge, the electric intensity would be directed inward as in Fig. 12. Its magnitude would still be given by (10).

It is convenient to visualize a set of lines, called *electric lines:*

> An **electric line** is a directed line whose tangent at each point is in the direction of the electric intensity at that point.

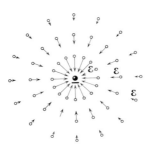

Fig. 12. Electric intensities $\mathcal{E}$ at points near a negative charge.

As so defined, electric lines are not intended to give information about the magnitude of the electric intensity, but only about its direction. Electric lines about the positive and negative point charges of Figs. 11 and 12 are shown in Figs. 13 and 14. Of course these figures can show only sample electric lines in one plane. Electric lines must be imagined to bristle out from the charges in every direction in space like the quills of a curled-up porcupine. Through every point in space there is an electric line starting at the + charge and going out to infinity, or starting at an infinite distance and ending on the − charge. This description assumes that in the whole of space there is only a single positive charge (Fig. 13) or a single negative charge (Fig. 14). This assumption is never strictly accurate, but the pictures of Figs. 11–14 hold in practice if the charge we are considering is well isolated from all other charges so that the forces exerted on our test charge by other charges are negligible.

Fig. 13. Electric lines go radially out from a positive point charge.

Fig. 14. Electric lines go radially into a negative point charge.

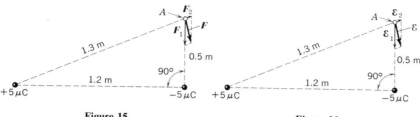

Figure 15 **Figure 16**

Now consider the field of the two charges of opposite sign in Fig. 15. To determine the intensity at point A, we imagine a small test charge $+q$ at this point. The test charge will experience an attractive force F_1 toward the negative charge and a repulsive force F_2 from the positive charge. These two forces must be added vectorially to obtain the total force F:

$$F = F_1 + F_2; \quad \text{hence} \quad (F/q) = (F_1/q) + (F_2/q).$$

But F_1/q is the electric intensity $\mathcal{E}_1$ that the negative charge would set up alone; F_2/q is the intensity $\mathcal{E}_2$ that the positive charge would set up alone; F/q is, by definition, the electric intensity $\mathcal{E}$ at point A. Hence, $\mathcal{E} = \mathcal{E}_1 + \mathcal{E}_2$. This result can be immediately generalized to the field set up by any number of charges, and shows that

The electric intensity arising from any number of individual charges is the vector sum of the intensities that the charges individually contribute.

Example. *Determine the electric intensity at point A for the charges shown in Fig. 15.*

Instead of working with force vectors F, we can work directly with intensity vectors $\mathcal{E}$ as shown in Fig. 16. The intensities $\mathcal{E}_1$ and $\mathcal{E}_2$ have the directions indicated, and magnitudes given by (10) as

$$\mathcal{E}_1 = (8.99 \times 10^9 \text{ N} \cdot \text{m}^2/\text{C}^2)(5 \times 10^{-6} \text{ C})/(0.5 \text{ m})^2 = 180 \times 10^3 \text{ N/C},$$

$$\mathcal{E}_2 = (8.99 \times 10^9 \text{ N} \cdot \text{m}^2/\text{C}^2)(5 \times 10^{-6} \text{ C})/(1.3 \text{ m})^2 = 26.6 \times 10^3 \text{ N/C}.$$

These vectors are added to give $\mathcal{E}$, as follows:

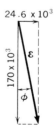

$$\begin{array}{ll} \mathcal{E}_{1X} = 0 \text{ N/C} & \mathcal{E}_{1Y} = -180 \times 10^3 \text{ N/C} \\ \mathcal{E}_{2X} = 24.6 \times 10^3 & \mathcal{E}_{2Y} = 10 \times 10^3 \\ \hline \mathcal{E}_X = 24.6 \times 10^3 \text{ N/C} & \mathcal{E}_Y = -170 \times 10^3 \text{ N/C} \end{array}$$

These are components of a vector $\mathcal{E}$ having, as indicated in the drawing at the right, magnitude and direction given by

$$\mathcal{E} = 172 \times 10^3 \text{ N/C}; \quad \phi = 8°2.$$

It will be noticed from this example that the inverse-square law represents a very rapid decrease of force with distance. The intensity at point A is principally determined by the charge 0.5 m away; the charge of equal magnitude at 1.3 m has a very much smaller effect, so that $\mathcal{E}$ does not differ very much from $\mathcal{E}_1$ in magnitude or direction.

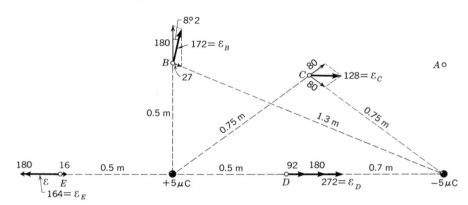

Fig. 17. Computation of the electric intensity at four points near the charges of Fig. 16. Values of the intensity are given in units of 10^3 N/C.

Figure 17 illustrates the computation of the electric intensity at four other points, B, C, D, and E, in the field of the two charges of Fig. 16. Figure 18 shows the system of electric lines for the field of these two charges. Electric lines through the points A, B, C, D, E of Fig. 17 are shown, and a number of other electric lines. All electric lines begin on the positive charge and end on the negative charge. This system of lines shows the direction of the electric intensity at any point.

Figure 19 shows the corresponding picture of electric lines for two equal positive charges. Here all the electric lines start at the positive charges and go to infinity, or in practice probably end on negative charges on the earth or the walls of the room a long distance outside the picture.

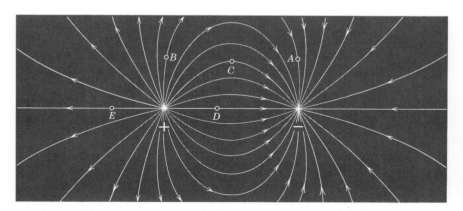

Fig. 18. Electric lines in the vicinity of equal but unlike charges.

SEC. 2   ELECTRIC FIELDS   475

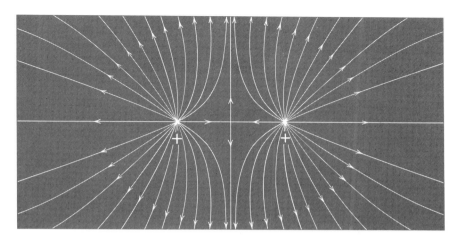

Fig. 19. Electric lines in the vicinity of two equal positive charges.

One important application of the equation $F=q\mathcal{E}$ is the computation of the force on electrons in vacuum tubes so that the motion of the electrons can be determined. The electric field is set up by the charges on the *electrodes* (filament, plate, grids, and so on). Figure 20 shows an *electron-ray tube*, used in 'cathode-ray' oscillographs and in television receivers, in which a beam of electrons is deflected by an amount proportional to the intensity of an electric field between two charged plates. In this electric field, the electron experiences a force

$$F = q\mathcal{E} = (-1\text{ e})\,\mathcal{E} = (-1.60 \times 10^{-19}\text{ C})\,\mathcal{E}.$$

A negative charge is repelled by the lower negative plate, attracted toward the upper positive plate, and experiences an upward electric force, in the direction opposite to the downward electric intensity.

Example. *In the electron-ray tube of Fig. 20, if the speed of the electrons is 10^7 m/s, the length of the plates 2 cm, and the distance from the center*

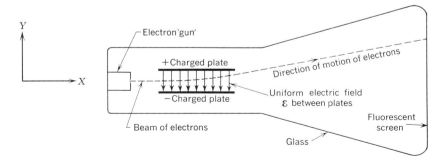

Fig. 20. An electron-ray tube.

of the plates to the screen 30 cm, *determine the deflection of the beam on the screen when a uniform field of intensity* 15 000 N/C *is set up between the plates. Show that the deflection is proportional to the electric intensity if the screen is flat, and that the effect of gravity on the motion of the electrons is negligible.*

Since the electron is initially moving horizontally and neither electric nor gravitational forces act horizontally, the horizontal component of the electron's velocity remains constant: $v_X = 10^7$ m/s. The time t_1 that the electron spends in the uniform field between the deflecting plates is therefore $t_1 = l/v_X$, where l is the length of the plates; the time t_2 required for the electron to traverse the distance d from the center of the deflecting plates to the screen is $t_2 = d/v_X$.

Electric forces. First let us ignore the effect of gravity. We denote the charge of the electron by $-q$ [where $q = 1$ e as given by (6)]. During its passage through the uniform field between the deflecting plates, the electron experiences a constant upward force $F_Y = q\mathcal{E}$ and hence, from Newton's second principle, has a constant vertical acceleration $a_Y = F_Y/m = q\mathcal{E}/m$, where m is the mass of the electron ($m = 9.11 \times 10^{-31}$ kg). The motion is exactly analogous to that of a freely falling particle in a uniform gravitational field. While passing between the deflecting plates, the electron acquires an upward velocity component $v_Y = a_Y t_1 = q\mathcal{E} l/m v_X$. The upward displacement Y_1 of the electron as it leaves the electric field (*see* Fig. 21) is $Y_1 = \tfrac{1}{2} a_Y t_1^2 = \tfrac{1}{2} q\mathcal{E} l^2/m v_X^2$. The tangent of the angle θ in Fig. 21 is $v_Y/v_X = q\mathcal{E} l/m v_X^2 = Y_1/\tfrac{1}{2} l$; therefore, as indicated in Fig. 21, the final trajectory is along a line passing through the midpoint of the plates. The trajectory between the plates is, of course, a parabola. The upward displacement Y at the screen is given by

$$\frac{Y}{d} = \frac{v_Y}{v_X} = \frac{q\mathcal{E} l}{m v_X^2}, \qquad \text{hence} \qquad Y = \frac{q l d}{m v_X^2} \mathcal{E},$$

which shows that the displacement Y is proportional to the electric intensity $\mathcal{E}$. Substitution of the numerical data stated in the example gives

$$Y = \frac{(1.60 \times 10^{-19} \text{ C})(0.02 \text{ m})(0.3 \text{ m})}{(9.11 \times 10^{-31} \text{ kg})(10^7 \text{ m/s})^2} (1.5 \times 10^4 \text{ N/C}) = 0.156 \text{ m} = 15.6 \text{ cm}$$

as the beam deflection on the screen.

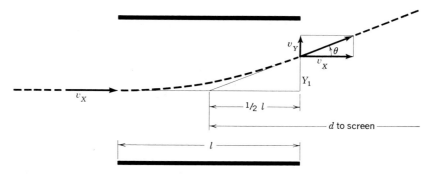

Fig. 21. The region between the plates of Fig. 20. (Vertical distances exaggerated.)

Effect of gravity. The electron has a constant downward acceleration g during the entire flight. The downward displacement resulting from gravity during the time t_2 is

$$\tfrac{1}{2} g t_2^2 = \tfrac{1}{2} g \frac{d^2}{v_X^2} = \tfrac{1}{2} (9.8 \text{ m/s}^2)\frac{(0.3 \text{ m})^2}{(10^7 \text{ m/s})^2} = 4.4 \times 10^{-15} \text{ m} = 4.4 \times 10^{-13} \text{ cm},$$

a displacement negligible compared with that produced by the electric field, and completely undetectable by observation of the screen. Alternatively, we note that the acceleration a_Y arising from electric forces is about 3×10^{14} times the acceleration of gravity; hence, the electric forces have an effect that is *enormous* in comparison with the effect of gravity.

3. Difference of Potential

When a charge moves in an electric field work is, in general, done on the charge by the electric forces, and there is a change in the electric potential energy of the charge; this change is exactly analogous to the change in gravitational potential energy of a mass that moves in a gravitational field (a region in which there are gravitational forces). However, there are several differences between the electrical and the gravitational cases which make it desirable in the electrical case to emphasize *electric potential energy per unit charge in the limit in which the size of the test charge is infinitesimal*. One difference is that the placement of a finite charge in an electric field readily disturbs the distribution of charges that sets up the field and hence alters the field, so we define difference of potential between two points in terms of a *very small* test charge—a charge too small to appreciably disturb the field. Another difference is that charges are of two kinds, so that if the potential energy of a positive charge would increase, that of a negative charge would decrease, in moving between the same two points—the difference of potential is defined in terms of a small *positive* charge. Third, we are vitally interested in *electric currents* (analogous to streams of water moving down- or uphill), and in discussing currents it is most convenient to specify changes in potential energy per unit charge (differences of potential).

In a region in which there is an electric field, an electric charge experiences a force. If the charge moves from one point A to another point B, either the field does work on the charge (like the work that gravity does on a sled when it coasts downhill) or outside energy is required to overcome the electric forces and to produce the motion (like the outside energy required to pull a sled uphill against the force of gravity). We shall presently prove that if the charge is small enough so that it does not appreciably disturb the charges that are setting up the field, *the magnitude of this work is independent of the path taken from A to B*, just as the change in gravitational potential energy is independent of the path.

Since *difference of potential* is defined as *work per unit charge*, the unit of difference of potential is the *joule/coulomb*. This unit is called the *volt** (V); the measuring instrument is called a *voltmeter*, and difference of potential itself is commonly called *voltage*.

> The **difference of potential** between A and B, in **volts**, is a *scalar* quantity defined as the work per unit charge, in joules per coulomb, done by the electric forces when a positive test charge is moved from A to B, in the limit in which the size of the test charge approaches zero.

If this work is positive, A is at a higher potential than B—the forces exerted by the electric field pull the positive charge down the 'potential hill' from A to B. If, on the other hand, this work is negative, indicating that outside energy must be supplied to overcome the forces of the electric field, A is at a lower potential than B. Difference of potential is defined in terms of work done on a *positive* charge; unlike the gravitational case, where all masses tend to move downhill, *negative* electric charges tend to move *up* a potential hill.

> *In an electric field, the electric forces tend to move a positive charge from a region of higher to one of lower potential, and tend to move a negative charge from a region of lower potential to one of higher.*

First consider the difference of potential for two large plates bearing equal and opposite charges, as in Fig. 22. As we shall show later, there will be a uniform electric intensity $\mathcal{E}$ between such plates. At any point between the plates, a small test charge $+q$ experiences a *downward electric force* $F = q\mathcal{E}$, by (7). In the motion from A to B, this force does work $W_{A \to B} = FX = q\mathcal{E}X$. The work done per coulomb of charge is

$$W_{A \to B}/q = \mathcal{E}X. \tag{11}$$

With $\mathcal{E}$ in N/C and X in m, the right side comes out in N·m/C or J/C. By definition, this is the difference of potential between A and B in volts. This difference of potential is usually denoted by $V_A - V_B$ (V for voltage). We have, then,

*Named for ALESSANDRO VOLTA (1745–1827), Italian scientist and professor; inventor of the electroscope and the voltaic cell.

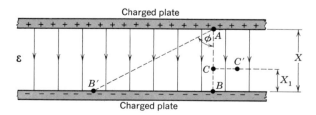

Figure 22

$$V_A - V_B = \mathcal{E} X. \qquad \begin{Bmatrix} V \text{ in V} \\ \mathcal{E} \text{ in N/C} \\ X \text{ in m} \end{Bmatrix} \quad (12)$$

The deflection of the beam in the electron-ray tube was found in Sec. 2 to be proportional to the intensity $\mathcal{E}$; hence, for a given plate spacing X, it is proportional to the difference of potential between the plates. The electron-ray tube acts as a voltmeter to measure the difference of potential between its plates.

Since
$$1 \text{ volt} = 1 \frac{\text{joule}}{\text{coulomb}} = 1 \frac{\text{newton} \cdot \text{meter}}{\text{coulomb}}, \qquad (13)$$

we see that
$$1 \frac{\text{volt}}{\text{meter}} = 1 \frac{\text{newton}}{\text{coulomb}}, \qquad (14)$$

and *volt/meter* is usually used as the name of the unit of electric intensity in place of *newton/coulomb*. Because of relation (14), electric intensity is frequently called *potential gradient;* it is the rate of change of potential with distance. To illustrate, we find from (12) that the electric intensity in Fig. 22 is just the difference of potential divided by the distance:

$$\mathcal{E} = (V_A - V_B)/X. \qquad (15)$$

Example. *If the difference of potential between the plates in Fig. 22 is* 240 V *and the distance between the plates is* 0.8 cm, *determine the electric intensity.*

From (15), $\mathcal{E} = 240 \text{ V}/0.008 \text{ m} = 30\,000 \text{ V/m}.$

Now consider the difference of potential $V_A - V_{B'}$ in Fig. 22. We can imagine a small test charge $+q$ to move from A to B' along the straight line AB'. The component of electric force along this line is $q\mathcal{E} \cos\phi$. The distance AB' is $X/\cos\phi$. The work done is $(q\mathcal{E} \cos\phi)(X/\cos\phi) = q\mathcal{E}X$, and hence

$$V_A - V_{B'} = q\mathcal{E}X/q = \mathcal{E}X.$$

We have obtained the same value for $V_A - V_{B'}$ that we obtained in (12) for $V_A - V_B$. This identity implies that B and B' are at the same potential; that is, that $V_B - V_{B'} = 0$. We can see why this must be so if we imagine moving a charge from B to B' directly, just outside the surface of the plate. Since the motion takes place perpendicularly to the electric intensity, no work is done by the electric forces, so $V_B - V_{B'} = 0$. The surface of the plate is said to be an *equipotential surface.*

Any horizontal plane between the plates is an equipotential surface, since no work is done in moving a charge from one point to another in the same horizontal plane, for example from C to C' in Fig. 22. The potential between the plates may be marked on a whole plane at a time, as illustrated in Fig. 23 for plates 0.8 cm apart at a difference of potential of 240 V. In this case, the electric intensity $\mathcal{E} = 30\,000$ V/m, as in the example above. In Fig. 22, we see that $V_C - V_B$ is proportional to the

480 ELECTROSTATICS CHAP. 22

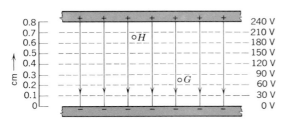

Fig. 23. Equipotential surfaces (broken lines) and lines of force for the plates of Fig. 22, 0.8 cm apart with a difference of potential of 240 V. The potential of the negative plate is arbitrarily called 0 V, but differences between values of potential are the only quantities with physical significance.

distance X_1 since $V_C - V_B = \mathcal{E} X_1$. Alternatively we can argue that since $\mathcal{E}$ is constant, the rate of change of potential with vertical distance must be constant, as indicated in Fig. 23.

Since no work is done in moving a charge along an equipotential surface, we conclude that

Electric lines are normal to equipotential surfaces. The direction of the electric intensity is from higher to lower potential. The magnitude of the electric intensity is the space rate of change of potential along an electric line, in volts per meter.

4. Potential Resulting from a Charge Distribution

In this section we shall first compute the potential resulting from a single point charge; then show how the potential would be computed for any collection of charges. In the course of these computations, we shall demonstrate that potential difference is independent of path.

Consider the difference of potential between two points in space near a single point charge. The equipotential surfaces are clearly concentric spheres surrounding the charge, as indicated in Fig. 24, since the surface of one of these spheres is everywhere perpendicular to the electric lines, and a test charge may be moved about freely on one of these spheres without the electric forces doing any work. The difference of potential, $V_1 - V_2$, between the sphere of radius r_1 and that of radius r_2 in Fig. 24(a) is, by definition, the work per unit charge in joules per coulomb that would be done by the electric forces on a test charge $+q$ moved from radius r_1 to radius r_2. We shall compute this work for the case in which

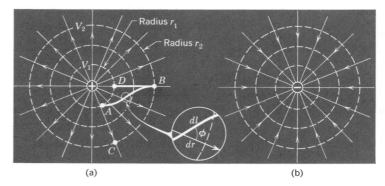

Fig. 24. Equipotential surfaces (broken lines) are spheres surrounding an isolated point charge Q, positive in (a) and negative in (b).

the charge is moved along the line AB, an *arbitrary* path from the sphere of radius r_1 to that of radius r_2.

When the test charge $+q$ is at radius r, it experiences electric force $F = (1/4\pi\varepsilon_0) Qq/r^2$, radially outward. When the test charge $+q$ moves a distance dl along the path AB in Fig. 24(a), the work done by this electric force (*see* inset) is $F\,dl\,\cos\phi = F\,dr = (1/4\pi\varepsilon_0)(Qq/r^2)\,dr$. The total work done when the charge moves from A to B is

$$W = \frac{1}{4\pi\varepsilon_0} \int_{r_1}^{r_2} \frac{Qq}{r^2}\,dr = \frac{1}{4\pi\varepsilon_0} Qq \int_{r_1}^{r_2} \frac{dr}{r^2},$$

and the difference of potential between any point A at radius r_1 and any point B at radius r_2 is

$$V_1 - V_2 = \frac{W}{q} = \frac{1}{4\pi\varepsilon_0} Q \int_{r_1}^{r_2} \frac{dr}{r^2} = -\frac{1}{4\pi\varepsilon_0} \frac{Q}{r}\bigg|_{r_1}^{r_2} = \frac{1}{4\pi\varepsilon_0}\left(\frac{Q}{r_1} - \frac{Q}{r_2}\right). \quad (16)$$

This quantity is seen to be positive if Q is positive and $r_1 < r_2$. The closer we are to the positive charge in Fig. 24(a), the higher the potential. If Q is negative in (16), as in Fig. 24(b), $V_1 - V_2$ comes out negative, so that V_2 is greater than V_1; the closer we get to the negative charge, the lower the potential.

The difference of potential between the sphere at r_1 and a concentric sphere at an infinite distance is obtained by setting $r_2 = \infty$ in (16), to get $V_1 - V_\infty = (1/4\pi\varepsilon_0) Q/r_1$. Frequently it is convenient arbitrarily to call the potential at infinity *zero*, in which case $V_1 = (1/4\pi\varepsilon_0) Q/r_1$, and in general,

$$V = \frac{1}{4\pi\varepsilon_0} \frac{Q}{r}. \qquad (V_\infty = 0) \quad (17)$$

By this convention, $V_\infty = 0$, we can assign a value of *potential* to each point in space near a collection of charges such that the *difference of potential* between any two points is the difference in the values of the potential.

In Fig. 25, the potential values are shown for charges of ± 1 nC, computed from (17). Again we notice that the direction from higher to lower potential is the direction in which the field tends to move a positive test charge; this direction is away from the positive charge, at the top of Fig. 25, but toward the negative charge, at the bottom.

In the above derivation we have rigorously proved that for the field of a single point charge *the work done in the motion of a test charge from A to B is independent of the path taken from A to B*. This proof is still valid as B moves out to infinity.

Now consider, Fig. 26, the electric intensity arising from any number of charges $Q_1, Q_2, \cdots$. At any point P, this intensity can be conveniently written as the vector sum $\mathcal{E} = \mathcal{E}_1 + \mathcal{E}_2 + \cdots$ of the intensities resulting from the individual charges. The force on a small test charge q placed at P

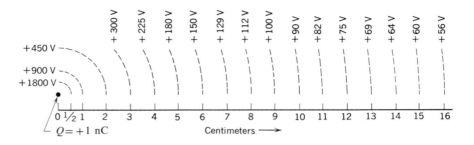

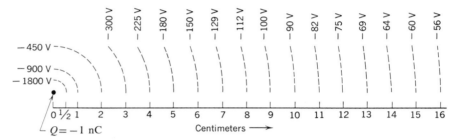

Fig. 25. Potential values near charges of ± 1 nC, with the zero of potential taken at infinity.

is $q\mathcal{E}$, and the work done when this charge suffers the infinitesimal displacement $d\mathbf{l}$ is

$$dW = q\mathcal{E} \cdot d\mathbf{l} = q\mathcal{E}_1 \cdot d\mathbf{l} + q\mathcal{E}_2 \cdot d\mathbf{l} + \cdots = dW_1 + dW_2 + \cdots,$$

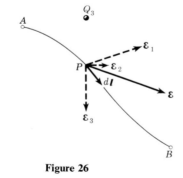

Figure 26

where dW_1 is the work done by the field of charge Q_1 alone, dW_2 that done by the field of charge Q_2 alone, etc.

Hence, the total work done in the motion from A to B is the sum of the values of work computed for the field of each charge separately, and *is independent of the path from A to B* since each of these values is independent of the path. From these arguments we finally conclude that the difference of potential $V_A - V_B$ is the sum of the values $(V_A - V_B)_1$, $(V_A - V_B)_2$, $\cdots$ that would arise from the charges individually. In particular, if B is at ∞, we see from (17) that

$$V_A = V_{A1} + V_{A2} + \cdots = \frac{1}{4\pi\varepsilon_0}\left(\frac{Q_1}{r_{A1}} + \frac{Q_2}{r_{A2}} + \cdots\right), \quad (V_\infty = 0) \quad (18)$$

where r_{A1} is the distance from A to Q_1, etc.

Thus we have shown that *the potentials arising from various charges are additive.* If the charges are discrete, as in Fig. 26, the resultant potential is a sum; if the charges can be regarded as continuous, as on a conductor, the sum is to be replaced by an integral containing the charge density.

5. Line-integral Relation for $\mathcal{E}$

We can now express the difference of potential between A and B in Fig. 26 as a line integral of electric intensity. We note that the work done by the field when a test charge q moves from A to B is given by

$$W_{A \to B} = \int_A^B dW = q \int_A^B \mathcal{E} \cdot d\mathbf{l}.$$

Division by q gives
$$V_A - V_B = \int_A^B \mathcal{E} \cdot d\mathbf{l}. \tag{19}$$

The difference of potential between points A and B is the line integral of the electric intensity along any path from A to B; the integral is independent of the path chosen.

Fields, such as electric and gravitational fields, that have the property that line integrals of field intensity from one point to another are independent of the path chosen between the two points, are called *conservative* fields.

The conservative property of the electric field results in the fact that a potential can be defined as a function that has a definite value at each point in space; an arbitrary assignment of potential to one point, such as $V_\infty = 0$, determines values of potential at all other points such that the difference of potential, which is the integral $\int_A^B \mathcal{E} \cdot d\mathbf{l}$, between points A and B, is just $V_A - V_B$. In particular, if this line integral is computed around a *closed* path starting from point A and returning to the same point, the integral must vanish since it represents a difference of potential of *zero*. Thus we are led to an important property of electric fields (actually of all *conservative* fields) called the

LINE-INTEGRAL RELATION FOR $\mathcal{E}$: *In electrostatics the line integral of $\mathcal{E}$ around any closed curve vanishes:*

$$\oint \mathcal{E} \cdot d\mathbf{l} = 0. \tag{20}$$

Here we use the circled integral sign to indicate that the path of integration is closed. The line-integral relation (20) holds rigorously in all situations encountered in electrostatics, but requires a significant modification in the nonstatic case, as we shall see when we consider electromagnetic induction in Chapter 31. This modified line-integral relation will play an important role in our discussion of electromagnetic radiation in Chapter 43.

6. The Charge of the Electron

Although Thomson, in 1897, initially demonstrated the existence of the electron by determining q/m, the ratio of its charge to its mass (*see* Chapter 28), he did not determine the value of the electronic charge itself. Between 1909 and 1917, R. A. MILLIKAN carefully determined the electronic charge by observations of charged oil droplets in an electric field. A schematic diagram of Millikan's apparatus is shown in Fig. 27. Very tiny oil droplets from an atomizer are introduced into the space between the horizontal plates. X rays passing through the air produce ionization of the air molecules, releasing electrons that become attached to the oil droplets. The vertical motion of one of the oil droplets of mass M is observed by means of a microscope. Let us suppose that the oil droplet has a single electron attached. If the plates are not charged, the oil droplet is subjected to a gravitational force $w = Mg$ and will move downward; owing to the viscosity of the air, the downward motion will be slow. From the rate of fall, the mass of the oil droplet can be determined by known laws of hydrodynamics—a tiny sphere falls at the constant velocity for which the viscous resistance of the air just balances the force of gravity. After the rate of fall of a particular droplet has been determined, the plates are charged so as to exert an upward electric force just sufficient to make the downward motion of the droplet cease. Since q is negative, the magnitude of the *upward* electric force is $-q\mathcal{E} = -qV/d$, where V is the voltage between the plates and d is their separation. When there is no resultant vertical motion, the upward electric force is exactly equal in magnitude to the downward gravitational force. Hence,

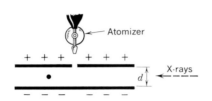

Fig. 27. Schematic diagram of Millikan's oil-droplet apparatus.

$$-qV/d = Mg, \quad \text{or} \quad q = -Mgd/V.$$

The charges observed by Millikan were not the same on all droplets, but the observed value was always a small integer (from 1 to 9) times the least charge observed. This least charge could then be assumed to be the charge of a single electron, with some of the droplets carrying more than one electron. By careful measurements, Millikan was able to obtain a very accurate value of the electronic charge; we have already given the present best value on p. 470.

7. The Electron-volt as a Unit of Energy

The volt was defined in such a way that one joule of work is done when one coulomb moves through a difference of potential of one volt. Thus, the energy unit, joule, is the coulomb-volt, as we see from (13).

When we are discussing the energy of a single particle—electron, atom, molecule, nucleus, or photon—the joule is an inconveniently large and unwieldy unit. It is inconveniently large because the charge is likely to be 1 e $= 1.6 \times 10^{-19}$ C or some small positive or negative integral multiple thereof. It is unwieldly because the charge is usually known *exactly* only in terms of the electronic charge unit.

For the above reasons, the unit of energy most widely used in atomic and nuclear physics is the electron-volt (eV) rather than the coulomb-volt (joule):

> The **electron-volt** is a unit of energy equal to the work done by the electric field when a charge of one electronic charge unit moves through a difference of potential of one volt:
>
> $$1 \text{ eV} = 1 \text{ e} \times 1 \text{ V} = 1.602 \times 10^{-19} \text{ C} \times 1 \text{ V} = 1.602 \times 10^{-19} \text{ J},$$
> $$1 \text{ J} = 6.242 \times 10^{18} \text{ eV}. \quad (21)$$

Example. *If a helium nucleus (alpha particle) is accelerated through a difference of potential of* 600 000 V *in a particle accelerator, what kinetic energy does it acquire?*

The helium nucleus has a charge $Q = +2$ e (atomic number $Z=2$). All of the work done by the electric field in the vacuum chamber of the accelerator goes into increasing the kinetic energy of the nucleus, which becomes

$$W = QV = 2 \text{ e} \times 600\,000 \text{ V} = 1\,200\,000 \text{ eV} = 1.2 \text{ MeV}.$$

Here we have introduced the unit very popular in nuclear physics, 1 MeV $= 10^6$ eV.

8. The Nuclear Model of the Atom

We can now present some of the evidence on which the nuclear model of the atom is based. The approximate sizes of the atoms were known at the beginning of the present century from interpretation of the observed properties of gases on the basis of kinetic theory. After the discovery that electrons formed a part of the structure of atoms, Thomson proposed a model of the atom in which the electrons were embedded in a matrix of positive electric charge "like plums in a pudding." Vibration of the electrons about their equilibrium positions was supposed to result in the emission of light. The Thomson model was not satisfactory, and was discarded in 1911 in favor of a nuclear model proposed by ERNEST RUTHERFORD.

Rutherford proposed the nuclear model to account for results obtained in his experiments on the scattering of high-speed, positively charged helium atoms He^{++} (alpha particles) by matter. A schematic diagram of the experimental arrangement he used in shown in Fig. 28.

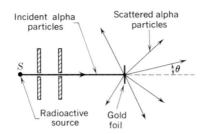

Figure 28

Alpha particles emitted by a naturally radioactive source (Chapter 45) were allowed to pass through a hole in a thick metal shield and strike a thin gold foil which acted as a scatterer. The alpha particles scattered at various angles θ with the incident beam were counted by observing the scintillations produced when the particles impinged upon a fluorescent screen.

When the foil was very thin, most of the alpha particles passed through undeflected, but some were scattered at large angles, even approaching 180°. According to the Thomson model, there should be no electric fields near or within the gold atom intense enough to produce the repulsive forces required to stop the very high-speed alpha particles and send them back. Rutherford was forced to conclude that such an intense field did exist. To provide such an intense field, Rutherford assumed that the entire positive charge of the atom was concentrated in a very small nucleus and that electrons occupied the space outside the nucleus.

For a nucleus of given charge $Q = Ze$ to exhibit a very intense field, the particle on which the field acts must be capable of approaching the nucleus to within a very small distance. Careful scattering experiments performed under Rutherford's direction showed that the angular distribution of scattered particles and its dependence on foil thickness and alpha-particle energy was entirely consistent with that computed from the assumption that the gold nuclei and the alpha particles (helium nuclei) were point charges with masses nearly equal to the entire mass of the atoms. In their motion, these massive points had to be assumed to approach as close as 10^{-15} m in the case of almost backward scattering of very energetic alpha particles. Hence the positive charge on the gold nucleus and that on the helium nucleus (alpha particle) must be assumed concentrated in regions of diameter smaller than this value or the requisite electric intensities would not be generated. This size is only $\frac{1}{100\,000}$ the diameter of the atom as a whole. It is no wonder that most of the alpha particles pass through a thin foil undeviated by the empty space and very light electrons that occupy most of the volume.

In later work with even more energetic particles, where the distance of closest approach would be even less, it was found that the point-charge scattering formulas no longer applied and indeed that a figure of approximately 10^{-15} m represents a true nuclear diameter.

Example. *The original experiments on the scattering of alpha particles from gold foil were done with particles of speed* 1.78×10^7 *m/s from a gaseous radon source. Some of the alpha particles were scattered directly backward. To get an appreciation of the charge concentration required to accomplish this, let us assume that the gold nucleus and the alpha particle each contain spherically* symmetric charge distributions of radius R and calculate the maximum value of R on the assumption that the nuclei do not 'touch,' even in backward scattering. Neglect the motion of the gold nucleus, which is 50 times as heavy as the helium nucleus. Find also the force acting on, and the acceleration of, the alpha particle at the point of closest approach.

As in the gravitational case, the force between two spherically symmetric charge distributions that do not touch each other is the same as the force between two point charges concentrated at the centers of the distributions (*see* Chapter 23). Thus we can find the distance of closest approach by assuming that the nuclei are point charges.

The charge on the gold nucleus is $+79\,e$ (*see* Sec. 4 of the Appendix). Thus the potential in the electric field of the gold nucleus is $V = (1/4\pi\varepsilon_0)(79\,e/r)$. The potential energy of the helium nucleus (charge $+2\,e$) at distance r from the gold nucleus is obtained by multiplying the *potential* (the potential energy per unit charge) by the *charge*:

$$E_P = V(2\,e) = \frac{1}{4\pi\varepsilon_0} \frac{79 \cdot 2\,e^2}{r} = 8.99 \times 10^9 \frac{\text{N} \cdot \text{m}^2}{\text{C}^2} \frac{158}{r}(1.60 \times 10^{-19}\,\text{C})^2$$

or
$$E_P = \frac{3.64 \times 10^{-26}\,\text{m}}{r}\,\text{J}. \quad \text{(i)}$$

The alpha particle has a mass of 6.64×10^{-27} kg (*see* Sec. 2 of the Appendix). It starts its approach to the gold nucleus from a very large distance r with negligible potential energy and kinetic energy equal to

$$E_K = \tfrac{1}{2}\,mv^2 = \tfrac{1}{2}\,(6.64 \times 10^{-27}\,\text{kg})(1.78 \times 10^7\,\text{m/s})^2 = 1.05 \times 10^{-12}\,\text{J}. \quad \text{(ii)}$$

As it approaches the gold nucleus, the repulsive electric force slows it down; it loses kinetic energy and gains potential energy. It comes to rest at such distance r given by (i) that all the kinetic energy (ii) has been converted to potential energy. This distance is found by equating (i) and (ii):

$$\frac{3.64 \times 10^{-26}}{r}\,\text{J} \cdot \text{m} = 1.05 \times 10^{-12}\,\text{J}, \qquad r = 3.48 \times 10^{-14}\,\text{m}.$$

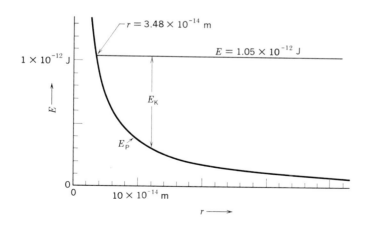

Figure 29

The radius R assumed for the nuclei must be less than half of this distance between centers if the nuclei are not to 'touch':

$$R < 1.74 \times 10^{-14} \text{ m},$$

a radius extremely small compared with the radii of atoms.

Figure 29 shows an energy-level diagram similar to that on p. 250 This diagram shows total energy E, E_P as a function of r, and the distance at which the difference $E_K = E - E_P$ becomes zero.

At the distance of closest approach, the force acting between the charges is

$$F = \frac{1}{4\pi\varepsilon_0} \frac{(79 \text{ e})(2 \text{ e})}{r^2} = \frac{3.64 \times 10^{-26} \text{ N} \cdot \text{m}^2}{(3.48 \times 10^{-14} \text{ m})^2} = 30.1 \text{ N}.$$

The acceleration of the alpha particle when acted on by this relatively enormous force is

$$a = \frac{F}{m} = \frac{30.1 \text{ N}}{6.64 \times 10^{-27} \text{ kg}} = 4.53 \times 10^{27} \text{ m/s}^2.$$

9. Electric Dipoles, Molecular Fields

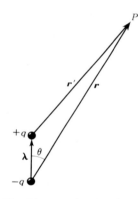

Fig. 30. An electric dipole ($\lambda \ll r$).

A pair of equal unlike charges, such as the pair in Fig. 18, is called an *electric dipole*. The electric field of certain polar molecules such as HCl resembles the electric field of a simple electric dipole.

Consider the electric intensity resulting from a pair of electric charges $\pm q$ separated by a distance λ *small compared with r*, as in Fig. 30. The *electric moment* of the dipole is defined as

$$\mathbf{M}_E = q\boldsymbol{\lambda}, \tag{22}$$

where the vector $\boldsymbol{\lambda}$ is directed from the negative to the positive charge, as indicated in Fig. 30. The electric field for this case is relatively easy to compute. The electric intensity at P is given by (10) as

$$\boldsymbol{\mathcal{E}} = \frac{q}{4\pi\varepsilon_0}\left(\frac{1}{r'^3}\mathbf{r}' - \frac{1}{r^3}\mathbf{r}\right).$$

Since $\mathbf{r}' = \mathbf{r} - \boldsymbol{\lambda}$, this relation can be written as

$$\boldsymbol{\mathcal{E}} = \frac{q}{4\pi\varepsilon_0}\left\{\left(\frac{1}{r'^3} - \frac{1}{r^3}\right)\mathbf{r} - \frac{1}{r'^3}\boldsymbol{\lambda}\right\}. \tag{23}$$

Now, by the cosine law of trigonometry,

$$r'^2 = r^2 - 2r\lambda\cos\theta + \lambda^2 = r^2\left(1 - 2\frac{\lambda}{r}\cos\theta + \frac{\lambda^2}{r^2}\right).$$

Drop the very small term λ^2/r^2, raise to the $-\tfrac{3}{2}$ power, expand by the binomial theorem, and keep terms of only the first order in λ/r:

$$\frac{1}{r'^3} = \frac{1}{r^3}\left(1 - 2\frac{\lambda}{r}\cos\theta\right)^{-3/2} = \frac{1}{r^3}\left(1 + 3\frac{\lambda}{r}\cos\theta\right). \quad (24)$$

In (23), we now substitute (24), and again keep terms of only the first order in λ/r. Writing $q\boldsymbol{\lambda} = \mathbf{M}_E$, we obtain

$$\boldsymbol{\mathcal{E}} = \frac{1}{4\pi\varepsilon_0}\left\{\frac{3\,M_E\cos\theta}{r^4}\mathbf{r} - \frac{1}{r^3}\mathbf{M}_E\right\}. \quad (25)$$

A plot of this field is shown in Fig. 31.

For $\theta = 0$ or π, the intensity is vertically upward and is seen to have magnitude

$$\mathcal{E} = \frac{1}{4\pi\varepsilon_0}\frac{2M_E}{r^3}. \qquad (\theta = 0, \pi)$$

For $\theta = \tfrac{1}{2}\pi$, since $\cos\theta = 0$, the intensity is vertically downward and has the magnitude

$$\mathcal{E} = \frac{1}{4\pi\varepsilon_0}\frac{M_E}{r^3}. \qquad (\theta = \tfrac{1}{2}\pi)$$

In either case, the intensity falls off inversely as the *cube* of the radius.

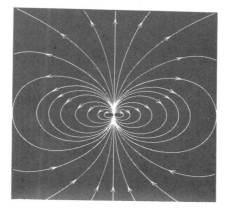

Fig. 31. Field of dipole with electric-moment vector pointing upward.

PROBLEMS

1. An experiment is conducted with three small charged metal balls, A, B, and C. A repels B with a force of 1 N when they are 5 cm apart. B attracts C with a force of $\tfrac{1}{4}$ N when they are 5 cm apart. C attracts A with a force of $\tfrac{1}{2}$ N when they are 5 cm apart. If the charge on A is $+Q$, what are the charges on B and C? Explain each step in your reasoning.
Ans: B, $+\tfrac{1}{2}Q$; C, $-\tfrac{1}{4}Q$.

2. Two small spheres carrying unequal positive charges repel each other with a force of 0.12 N when 3 cm apart. If the charge on each of the spheres is doubled, and the distance between them is tripled, what is the force of repulsion?

3. If two point charges each experience an electric force of attraction of 0.02 N when they are 10 cm apart, what will be the force when they are (*a*) 100 cm apart? (*b*) 50 cm apart? (*c*) 5 cm apart? (*d*) 2.5 cm apart? (*e*) 1 cm apart? Ans: (*a*) 0.0002 N; (*b*) 0.008 N; (*c*) 0.08 N; (*d*) 0.32 N; (*e*) 2 N.

4. Charges of $+5$ and $+10\,\mu C$ and 12 m apart. What is the force between them?

5. Two small metal spheres, initially uncharged, are charged by induction as in Fig. 4 so that, when separated by 3 cm, the force between them is 2×10^{-3} N. How many electrons have moved from the right sphere to the left in the process of charging?
Ans: 12.5×10^{10}.

6. If the two equally charged spheres of Fig. 5(c) would each require the addition of 10^{12} electrons to become neutral, what is the force between these spheres when they are 4 cm apart? (Consider them as point charges.)

7. Two small metal-coated pith balls, each having a mass of 21 g, are held at the ends of silk threads 140 cm long fastened to the ceiling at a common point. Charges are placed on the two balls and they come to equilibrium at a distance apart of 9 cm. If the charge on one ball is $-0.080\,\mu\text{C}$, what is the charge on the other ball?   Ans: $-0.0745\,\mu\text{C}$.

8. In each of the three sketches in Fig. 1, assume that the two strings are of equal length and that the signs of the charges are as indicated. Which of the two strings makes the greater angle with the vertical if (*a*) the charges are equal in magnitude and the balls equal in weight? (*b*) the left-hand charge is greater in magnitude than the right-hand charge and the balls are equal in weight? (*c*) the charges are equal in magnitude but the left-hand ball is heavier than the right-hand one? (*d*) the left-hand charge is greater in magnitude than the right-hand, and the left-hand ball is heavier than the right-hand one? (Give adequate reasons for your answers to the above questions.)

9. How far apart in a vacuum must two electrons be if the force of electrostatic repulsion on each electron is just equal in magnitude to the weight of the electron? (The electron mass is 9.11×10^{-31} kg.)   Ans: 5.09 m.

10. (*a*) Two bodies attract each other electrically. Are they necessarily both charged? (*b*) Two bodies repel each other electrically. Are they necessarily both charged? Explain the reasons for your answers.

11. Two charges, of $+12$ and $-12\,\mu\text{C}$, are 18 cm apart in the air, the $+$ charge being to the left, the $-$ charge to the right. (*a*) What force does each exert on a $+5\text{-}\mu\text{C}$ charge placed halfway between them? (*b*) What is the total force on the $+5\text{-}\mu\text{C}$ charge? (*c*) on the $+12\text{-}\mu\text{C}$ charge? (*d*) on the $-12\text{-}\mu\text{C}$ charge? Show that the total force on the three charges, considered as a dynamical system, is zero.
 Ans: (*a*) 66.7 N to right; (*b*) 133 N to right; (*c*) 26.6 N to left; (*d*) 107 N to left.

12. Two charges of $+15$ and $+5\,\mu\text{C}$ are 6 cm apart. Where must a third charge be placed in order that the resultant force acting on it be zero?

13. Two charges, $+10$ and $-5\,\mu\text{C}$, are 9 cm apart. Where must a third charge be placed in order that the resultant force acting upon it be zero?   Ans: 21.7 cm from the $-5\text{-}\mu\text{C}$ charge (in what direction?).

14. Two charges of $-4\,\mu\text{C}$ each are 6 cm apart. Find the total force exerted by these two charges on a third charge of $+4\,\mu\text{C}$ that is 5 cm distant from each of the two charges.

15. Find the electric intensity at a point 5 cm distant from each of the two $-4\text{-}\mu\text{C}$ charges of Prob. 14.
 Ans: 2.30×10^{7} N/C (in what direction?).

16. What are the magnitude and direction of the electric intensity at a point 18 cm from a point charge of $-50\,\mu\text{C}$?

17. Two charges of $+8$ and $-4\,\mu\text{C}$ are 1.6 m apart. Find the electric intensity at a point that is 2 m from the positive charge and 1.2 m from the negative.   Ans: 20 200 N/C at angle $44°.7$ from line from $+$ to $-$ charge.

18. Two charges of $+2$ and $-1\,\mu\text{C}$ are 1.6 m apart. Find the electric intensity at a point that is 1.2 m from the positive charge and 2 m from the negative.

19. A charged particle of 0.003-g mass is held stationary in space by placing it in a downwardly directed electric field of 24×10^{4} N/C. Find the charge on the particle in μC.
  Ans: $-12.3\times10^{-5}\,\mu\text{C}$.

20. An oil drop has a net negative charge of $-3\,e$, representing an excess of 3 electrons. It remains at rest under the action of the force of gravity and the electric force when it is placed in a downward electric field of 3.0×10^{5} N/C. Find the mass of the droplet.

21. Verify the values of electric intensity given in Fig. 17 at points *B*, *C*, *D*, and *E*.

22. If the charges of Fig. 19 are each $+5\,\mu\text{C}$, separated by 1.2 m, find the electric intensity at points *A*, *B*, *C*, *D*, and *E*, located at the same geometrical positions as in Figs. 16–18. Note that the directions you obtain should agree with the directions of the electric lines shown in Fig. 19.

23. What is the acceleration of an electron in a field of 5×10^{5} N/C? Express this in terms of the acceleration of gravity.
  Ans: 8.80×10^{16} m/s^{2}; $8.95\times10^{15}\,g$.

24. In the field mentioned in Prob. 23, how long would be required for the electron to attain one-tenth the speed of light? What distance would it travel if it starts from rest? Use Newtonian mechanics.

25. What is the magnitude of the force on a charge of $6\,\mu\text{C}$ placed at a point where the potential gradient has a magnitude of 600 000 V/m?   Ans: 3.60 N.

26. If the potential at H, Fig. 23, is 195 V and that at G is 75 V, how much work is required to move a charge of 3 μC from G to H?

27. (a) What is the difference of potential between two points 20 amd 40 cm distant from a charge of -20 μC? (b) Which point is at the higher potential? (c) How much work must be done on a $+2$ μC charge to move it from the point of lower potential to that of higher?
  Ans: (a) 0.45×10^6 V; (c) 0.90 J.

28. What is the potential difference between two points 50 and 20 cm from a $+20$-μC charge? Which point is at the higher potential? How much work does the field do on a 2-μC charge moved from the higher- to the lower-potential point?

29. Two large parallel plates 2 mm apart are held at a potential difference of 300 V. (a) How much work, in joules, is necessary to carry 0.2 μC of charge from the lower- to the higher-potential plate? (b) What are the magnitude and direction of the force acting on this charge when it is between the plates? (c) What is the electric intensity between the plates, in N/C? (d) How much work, in joules, would be necessary to carry a total of 1 C of charge from the lower- to the higher-potential plate, the potential difference being maintained at 600 V? Ans: (a) 6.00×10^{-5} J; (b) 3.00×10^{-2} N; (c) 1.50×10^5 N/C; (d) 300 J.

30. Repeat Prob. 29 for a situation in which the separation of the plates is 3 mm.

31. In an oil-droplet experiment, the horizontal plates are 1.6 cm apart, the radius of the oil droplet is 2.8×10^{-4} cm, and the density of the oil is 0.92 g/cm^3. If a single excess electron is attached to the droplet, what should be the potential difference between the plates in order to counterbalance the weight of the oil droplet?
  Ans: 8.30×10^4 V.

32. With the apparatus described in Prob. 31, what voltage should be applied between the plates in order to support an oil droplet of radius 4.4×10^{-4} cm if two excess electrons are attached to the droplet?

33. When a high-speed alpha particle approaches a calcium nucleus head-on, the calcium nucleus does not remain at rest, but 'recoils,' since its mass (40 u) is only ten times that of the alpha particle (4 u). In an elastic head-on collision between an alpha particle initially moving in the + direction at 1×10^7 m/s and a calcium nucleus initially at rest, what are the velocity of the alpha particle and the recoil velocity of the calcium nucleus after the collision? (All other forces acting on the calcium nucleus may be neglected in comparison with the enormous forces exerted by the alpha particle. Use of nonrelativistic mechanics is justified.)
  Ans: $-\frac{1}{11} \times 10^7$ m/s, $\frac{2}{11} \times 10^7$ m/s.

34. In discussing the detailed dynamics of the collision described in the preceding problem, we cannot use a system of coordinates attached to the calcium nucleus because this nucleus is accelerated and Newton's principles do not apply in an accelerated reference system. But we can use a coordinate system attached to the center of mass of the two particles, because the center of mass moves to the right, relative to the earth at the constant speed of $\frac{1}{11} \times 10^7$ m/s throughout the collision. Demonstrate this fact. What are the initial velocities of the alpha particle and of the calcium nucleus relative to the center of mass? Solve the elastic-collision problem in a coordinate system moving with the center of mass to find the final velocities; change these to velocities relative to the earth and compare with the answers to the preceding problem.

35. Show that in the two preceding problems, the alpha particle and the calcium nucleus *come to rest in the center-of-mass coordinate system at their distance of closest approach*. Hence, employing this coordinate system, set the initial total kinetic energy of our two-particle dynamical system equal to the potential energy when they come to rest, and hence find the distance of closest approach.
  Ans: 3.01×10^{-14} m.

36. Show that the magnitude of the torque exerted on an electric dipole M_E by uniform external electric field $\mathcal{E}$ is given by the expression $L = M_E \mathcal{E} \sin\theta$, where θ is the angle between M_E and $\mathcal{E}$. By writing this result in the form $\mathbf{L} = \mathbf{M_E} \times \mathbf{\mathcal{E}}$ show that the torque always tends to make $\mathbf{M_E}$ parallel to $\mathbf{\mathcal{E}}$.

37. Show by integration that the work done in changing the orientation of dipole M_E from θ_1 to θ_2 with respect to a uniform external electric field $\mathcal{E}$ is given by the expression $W = M_E \mathcal{E} (\cos\theta_1 - \cos\theta_2)$. If we set the potential energy of the dipole $E_P = 0$ when $\mathbf{M_E}$ is

perpendicular to $\mathcal{E}$, show that $E_P = -\boldsymbol{M_E} \cdot \boldsymbol{\mathcal{E}}$. What are the maximum and minimum values of E_P? What are the orientations of the dipole with respect to the field for these limiting values of E_P?   Ans: $M_E \mathcal{E}$, $-M_E \mathcal{E}$.

38. If an electric dipole is parallel to a non-uniform field parallel to the Z-axis and if the electric field has a constant rate of change $d\mathcal{E}/dZ$ with respect to Z, show that the dipole experiences a force $F_Z = M_E (d\mathcal{E}/dZ)$ and that this force tends to move M_E toward 'the stronger electric field.' Generalize the result to show that $F_Z = M_E (d\mathcal{E}/dZ) \cos\theta$, where θ is the angle between M_E and Z; that is, that $F_Z = M_{EZ}(d\mathcal{E}/dZ)$.

39. Molecular electric moments are usually expressed in Debye units, where one Debye unit is the electric moment associated with one electron and one proton 0.1 nm apart; it will be recalled that 0.1 nm is the approximate diameter of a hydrogen atom. Referring to Fig. 30, compute the magnitude of $\mathcal{E}$ at a distance $r = 0.2$ nm from a molecule having $M_E = 1$ Debye unit when $\theta = 0°$ and when $\theta = 90°$. The actual distance λ of charge separations in molecules is small compared with 0.1 nm.
Ans: 3.60×10^{10} V/m; 1.80×10^{10} V/m.

40. The dipole moments of HCl and H$_2$O molecules are 3.46 and 1.84 Debye units, respectively. As in the preceding problem, compute the magnitudes of $\mathcal{E}$ at a distance of 3 nm from these molecules for $\theta = 0°$ and $\theta = 90°$ in Fig. 30.

41. The average translational energy of a molecule of a gas at Kelvin temperature T is $\frac{3}{2} kT$, where k is Boltzmann's constant. What is the average translational energy of a molecule in a room at $T = 300°$ K? Express this energy in electron-volts.   Ans: 0.0388 eV.

42. Chemists commonly express reaction energies in units of cal/mole or kcal/kmol. Show that reaction energies could equally well be expressed in eV/molecule. (a) Express a reaction energy of 1 kcal/kmol in eV/molecule. (b) If the ionization energy of the H atom is 13.6 eV, how many kilocalories of energy would be required to ionize 1 kg of atomic hydrogen? Note that, since hydrogen normally exists in the form of diatomic molecules, still more energy would be required to completely ionize all the atoms in 1 kg of hydrogen gas. Why?

43. The radius of the helium nucleus is approximately 1.5×10^{-15} m. Recalling that the helium nucleus contains 2 protons, calculate (a) the magnitude of the electrostatic force experienced by one of the protons when the other proton is 3×10^{-15} m away and (b) the magnitude of the acceleration a resultant force of this magnitude would give a 1-kg body. Although this electrostatic force is enormous, still larger nonelectric forces of attraction make the nucleus stable!
Ans: (a) 25.6 N; (b) 25.6 m/s^2.

44. The Rutherford scattering experiments indicated that the atom consists of a massive, positively charged nucleus surrounded by the electrons; the hydrogen atom consists of a single proton and a single electron. Consider a model of the hydrogen atom in which an electron of mass m_e moves in a circular orbit around the proton with mass $m_p = 1836\, m_e$. Derive the following expressions for the kinetic energy E_K, the potential energy E_P (taken as zero at $r = \infty$), and the total energy E for the electron:

$$E_K = \tfrac{1}{2}(1/4\pi\varepsilon_0)\, e^2/r,$$
$$E_P = -(1/4\pi\varepsilon_0)\, e^2/r,$$
$$E = -\tfrac{1}{2}(1/4\pi\varepsilon_0)\, e^2/r.$$

45. We shall see later (Chapter 41) that, on the basis of classical physics, the simple atomic model we have proposed is inherently unstable. In attempting to account for the quantum nature of matter and radiation (Chapter 41), Niels Bohr proposed a model of the hydrogen atom involving certain stable circular orbits, the smallest of which has a radius $r_1 = 5.292 \times 10^{-11}$ m. What is the magnitude of the force experienced by the orbital electron when it is in the smallest Bohr orbit? How much work is required to remove the moving electron from this orbit to infinite distance from the proton? Express this work in electron-volts. (This amount of work is called the ionization energy.)   Ans: 13.6 eV.

ELECTROSTATIC FIELDS 23

Before proceeding further with our study of electricity and magnetism, it is desirable to examine the basic ideas involved in the *field concept* and to study some of the general properties of vector fields: gravitational, fluid flow, electric, and magnetic.

In the preceding chapter we introduced the concepts of *electric intensity, electric lines,* and *difference of potential.* The initial observable phenomenon in electrostatics is that charges exert forces on each other in accordance with Coulomb's principle. The electric field is a convenient logical tool for use in visualizing electrical 'conditions' near charged bodies and in computing forces on charges and the work done in their motion.

Later we shall find that the treatment of *magnetic* forces between *electric currents* would be extremely difficult without the introduction of the concept of a *magnetic field;* the magnetic-field concept is an almost indispensable tool in computing forces between currents and in describing the phenomena of electromagnetic induction. Finally, we shall find that oscillating electric and magnetic fields *must both* be invoked in order to account for the transmission of radiation through space. Energy emitted in the form of radiation at the sun's surface resides in the electromagnetic fields in space for eight minutes before reaching the earth's surface!

In the present chapter we first consider certain general properties of vector fields. We then introduce Gauss's relation for the total electric flux through a closed surface,* and use it in describing electrostatic shielding, charging by induction, and other electrostatic phenomena. The following chapter will conclude the formal discussion of electrostatics with the introduction of the concept of capacitance.

1. Vector Lines, Vector Tubes, Vector Flux

In Chapter 13, we defined *streamlines,* which are lines in the direction of the velocity vector in the velocity field arising in fluid flow; in Sec. 2 of the preceding chapter, we defined *electric lines* in a similar manner; later we shall define *magnetic lines*—these are all examples of *vector lines* in a vector field.

In Chapter 13, we also defined a *stream tube.* An *electric tube* is similarly defined:

> An **electric tube** is a tubular region of space bounded by electric lines.

Examples of electric tubes in the fields of a single charged conductor, of a pair of oppositely charged conductors, and of a pair of oppositely charged plates, are illustrated in Fig. 1. Later we shall also define *magnetic tubes* as a third example of a *vector tube.*

The flow or *flux* of a fluid through a stream tube, defined as the volume passing a section of the tube per unit time, is given by the product of the speed by the cross-sectional area of the tube, provided the cross section is so small that the speed can be assumed constant across the section. The flux through a larger tube is obtained by integrating the flux of the infinitesimal tubes of which it is composed.

*In our derivation of Gauss's relation, we make use of solid angles Ω measured in *steradians;* those not familiar with solid angles should refer to Section 7 of Chapter 34 for a formal definition of the steradian.

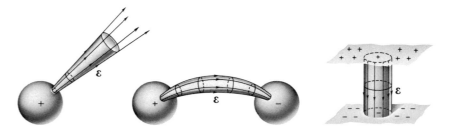

Fig. 1. Examples of electric tubes.

We similarly define *electric flux:*

The **electric flux** through an electric tube is the product of the magnitude of the electric intensity by the cross-sectional area of the tube, provided the cross section is so small that the electric intensity can be assumed constant across the section. The flux through a larger tube is obtained by integrating the flux of the infinitesimal tubes of which it is composed.

Later we shall deal with *magnetic flux* as another example of *vector flux.*

The flux of an incompressible fluid must be constant along a stream tube since no flux enters or leaves the tube and the same volume of fluid must cross all sections. After we have demonstrated Gauss's relation in the next section, we can show that electric flux has this same property of being constant along an electric tube; later we shall find that magnetic flux has this same property.

Next we shall discuss the computation of vector flux through an arbitrary surface, using electric flux as our example. In Fig. 2, the open surface S is bounded by the directed line L. We wish to determine the total flux through the surface, which is the same as the flux 'linking' the curve L. Since a surface has two sides, we need to specify the direction of a line passing through the surface as associated with the assumed 'direction' of the bounding curve L. This direction is always selected in accordance with the *right-hand rule:*

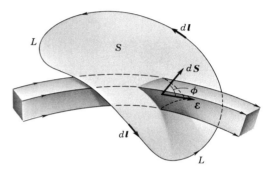

Fig. 2. An electric tube passing through the open surface S bounded by the directed closed curve L.

> RIGHT-HAND RULE: Curl the fingers of the right hand so that they point in the direction of a closed curve; then the thumb will point in the direction to be taken as positive for a line linking that curve.

As illustrated in Fig. 3, this rule may also be expressed in terms of the direction of advance of a right-hand screw thread when a bolt is turned. Thus, in Fig. 2, the positive direction through the surface is from back to front for the direction indicated for L.

Now split the surface in Fig. 2 up into infinitesimal areas dS. Each such area will define an electric tube, as indicated. We cannot determine the flux in this tube by multiplying $\mathcal{E}$ by dS because the cross-sectional area of a tube is measured normal to the tube, whereas the surface S is not, in general, normal to $\mathcal{E}$. However, if we define a vector $d\mathbf{S}$ as an element of area of magnitude dS and direction normal to the surface *in the positive sense,* we see by arguments used previously that the cross section of the tube $dA = \pm dS \cos\phi$, and that the flux $d\Phi$ through the

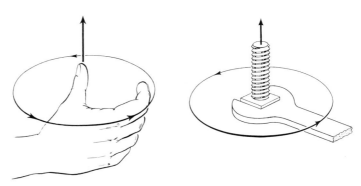

Fig. 3. The *right-hand rule*, relating the direction of a line passing through a closed curve to the assumed direction of the curve.

surface in the positive sense that is contained in this infinitesimal tube is given by

$$d\Phi = \pm \mathcal{E}\, dA = \mathcal{E}\, dS\, \cos\phi = \mathcal{E} \cdot d\mathbf{S}.$$

Note that when ϕ is less than 90°, as in Fig. 2, $\cos\phi$ is positive and the flux through the surface is positive. If $\mathcal{E}$ were reversed in Fig. 2, $\cos\phi$ would be negative and there would be a negative flux through the surface whose positive sense is defined by the given directions of L and of $d\mathbf{S}$.

The total flux through the surface is then the surface integral

$$\Phi = \iint_S \mathcal{E} \cdot d\mathbf{S}, \tag{1}$$

Here we have used the usual notation Φ for *flux*.

We note that, in the next section, we shall also be interested in the flux passing *outward* through a *closed* surface. In this case a similar formula is applicable, provided $d\mathbf{S}$ is in the direction of the *outward* normal. In this case we write

$$\Phi = \oiint_S \mathcal{E} \cdot d\mathbf{S}, \tag{2}$$

using the circled double integral to indicate that the (two-dimensional) surface is closed.

2. Gauss's Relation

We shall now derive a powerful relation due to Gauss* which is extremely useful in the study of electrostatics, and which, in particular,

---

*Karl Friedrich Gauss (1777–1855), a German mathematician who, along with a German physicist, Wilhelm Edward Weber (1804–1891), devised the absolute systems of electric and magnetic units.

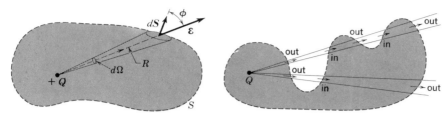

Figure 4 **Figure 5**

will enable us to prove that the flux through an electric tube is constant along its length:

GAUSS'S RELATION: *The total flux passing outward through any closed surface equals* $(1/\varepsilon_0)$ *times the total electric charge inside the closed surface.*

As a first simple illustration of this relation, consider the flux through a spherical surface with a single point charge Q at its center, as in Fig. 24, p. 480. The electric lines are everywhere perpendicular to the spherical surface so the flux Φ given by (2) is equal to the electric intensity times the area of the sphere. If the sphere is of radius R, $\mathcal{E} = (1/4\pi\varepsilon_0)\, Q/R^2$, the area is $4\pi R^2$, and the flux $\Phi = (1/\varepsilon_0)\, Q$, as in Gauss's relation.

Now consider a single point charge Q inside an arbitrary closed surface S, as in Fig. 4. Consider the small conical electric tube of solid angle $d\Omega$. This tube cuts out a surface element of area dS at a radius R. The flux through this surface element is

$$\mathcal{E} \cdot d\mathbf{S} = \mathcal{E}\, dS\, \cos\phi = (1/4\pi\varepsilon_0)\, Q\, dS\, \cos\phi/R^2 = (1/4\pi\varepsilon_0)\, Q\, d\Omega,$$

since $d\Omega = dS\, \cos\phi/R^2 = dA/R^2$, where dA is the cross-sectional area of the tube. Hence, *the flux in a conical vector tube of solid angle $d\Omega$ from a point charge Q is constant along the tube* and equals $(1/4\pi\varepsilon_0)\, Q\, d\Omega$. The flux passing outward through the surface S in Fig. 4 is obtained by integrating $d\Omega$ over the total solid angle to obtain a factor 4π; the flux is therefore $(1/\varepsilon_0)\, Q$, as in Gauss's relation.

Figure 5 shows a surface of a more general type than that in Fig. 4, in which it is seen that the electric flux from a positive charge Q inside the surface that starts out in solid angle $d\Omega$ may pass through the surface more than once. But this flux necessarily passes *out* of the surface *once more* than it passes *in*. Hence, the net outward flux is still given by Gauss's relation.

Now consider a point charge lying *outside* a closed surface, as in Fig. 6. Each flux tube now cuts the surface an even number of times and since the flux is constant along the tube, the *net* flux out of the surface is *zero*, as required by the Gauss relation, since there is no charge within the surface. This observation completes the demonstration of Gauss's relation for a single point charge and an arbitrary closed surface.

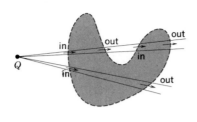

Figure 6

To demonstrate Gauss's relation for a collection of charges, write the electric intensity in the form

$$\mathcal{E} = \mathcal{E}_1^I + \mathcal{E}_2^I + \cdots + \mathcal{E}_1^O + \mathcal{E}_2^O + \cdots,$$

where the intensities $\mathcal{E}_1^I, \mathcal{E}_2^I, \cdots$ arise from charges inside the surface and $\mathcal{E}_1^O, \mathcal{E}_2^O, \cdots$ from charges outside the surface. We can write the flux through the surface as

$$\Phi = \oiint_S \mathcal{E} \cdot dS = \oiint_S \mathcal{E}_1^I \cdot dS + \cdots + \oiint_S \mathcal{E}_1^O \cdot dS + \cdots$$
$$= \Phi_1^I + \Phi_2^I + \cdots + \Phi_1^O + \Phi_2^O + \cdots.$$

Thus the total flux is the sum of the fluxes arising from the individual charges. We can write this relation in the form $\Phi = \Phi^I + \Phi^O$, where the flux Φ^I arises from charges inside the surface and equals $(1/\varepsilon_0)$ times the total charge inside, while Φ^O arises from charges outside, and is zero. Thus Gauss's relation is rigorously demonstrated.

Example. *Determine the electric intensity inside and outside the isolated hollow spherical conductor of Fig. 7, carrying total charge Q.*

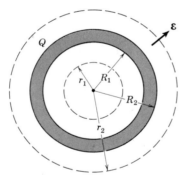

Fig. 7. A hollow charged spherical conductor with internal and external radii R_1 and R_2.

By symmetry, if there are no other charges in the neighborhood, the charge will be uniformly distributed around the sphere and *the electric lines will be radial.*

Consider an imaginary sphere, of radius $r_1 < R_1$, inside the hollow of the conductor. Since there is no charge inside this imaginary sphere, the flux through the sphere, and hence $\mathcal{E}$, must vanish. *There is no field inside the hollow of the charged sphere.*

Consider an imaginary sphere, of radius $r_2 > R_2$, outside the conductor. The flux through this sphere, $\mathcal{E} \times 4\pi r_2^2$, is given by Gauss's relation as

$$4\pi r_2^2 \mathcal{E} = \frac{Q}{\varepsilon_0}; \qquad \mathcal{E} = \frac{1}{4\pi\varepsilon_0} \frac{Q}{r_2^2}.$$

The intensity outside the charged spherical conductor is the same as if the charge Q were concentrated in a point at the center.

We note that in the case of *gravitational* forces, which are also inverse-square forces, there is a relation exactly similar to Gauss's relation. Use of this relation in the same manner as in the example above proves that the gravitational force of a spherical earth on a particle near its surface is exactly the same as if the whole mass of the earth were concentrated at its center, a fact that we made use of on p. 65 (*see* Probs. 19–21 on p. 509).

3. Charges on Conductors; Shielding

In this section we derive a number of important relations applying to the case of conductors. In deriving these relations, we assume charges to be 'smoothed-out,' ignoring the fact that they actually come in discrete units located at discrete points. We must think of the charge densities and electric intensities used here as *averages* over volumes that are very small but nevertheless contain a tremendous number of atoms.

In the static case we are considering, all charges are at rest; if there were a field within the material of a conductor, forces would be exerted on the free electrons and would cause them to move. As we shall see later, in current electricity an electric field is continuously maintained within a conductor and the electrons do continue to move. But in static problems, the electrons in a conductor must have settled down in such a configuration that there is no electric field and therefore no electric lines within the conductor. This settling down takes place very quickly,

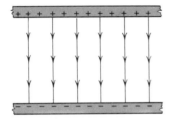

(a) Charged plates.

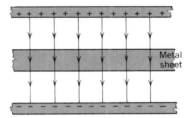

(b) A metal sheet is introduced between the charged plates. Momentarily the field of the charges on the plates acts on the electrons in the sheet.

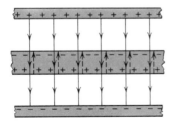

(c) This field causes electrons to move to the top of the sheet. The charges on the sheet then set up a field (broken lines) opposing the original field. When enough electrons have moved so that this field exactly balances the original field . . .

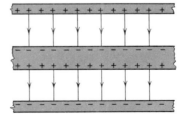

(d) . . . there is no longer any force on the electrons in the sheet and we have electrostatic equilibrium with no electric field within the conducting sheet. The whole process of establishment of equilibrium takes place in about 10^{-17} s.

Fig. 8. In electrostatic equilibrium there is no electric field within a conductor.

as indicated in Fig. 8. We conclude that

There is no static electric field within the material of a conductor.

Since no static electric field can exist within a conductor,

In the static case, there is no difference of potential between two points in the same conductor. The whole of a single conducting body is an equipotential region.

The absence of electric field within a conductor implies that

All the charge on a conductor lies on its surface.

To prove this statement from Gauss's relation, consider an imaginary closed surface that lies entirely within the material of the conductor. Since the electric intensity vanishes everywhere inside a conductor, it vanishes on this imaginary surface, so the net charge inside the surface must vanish. Therefore the net charge must vanish in any volume whatsoever inside the conductor, and so must vanish everywhere. Only the layers of atoms nearest the surface can have an excess or deficiency of electrons; the atoms in the interior must be neutral.

Since the field vanishes inside a conductor, electric lines will begin or end on the surface of a conductor. Because the surface of a conductor is an equipotential surface, and electric lines are perpendicular to equipotential surfaces, we see, as in Fig. 9, that

Electric lines start out perpendicularly from the surface of a conductor.

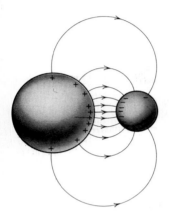

Fig. 9. Electric lines are perpendicular to the surface of a conductor.

Now consider an *electric tube*, starting from an area S_1 of one conductor and ending on an area S_2 of another, as in Fig. 10. We shall show that the magnitude of the positive charge in the area S_1 at the beginning of the electric tube precisely equals the magnitude of the negative charge in the area S_2 at its end:

An electric tube has equal charges of opposite sign at beginning and end.

To prove this statement, apply Gauss's relation to the imaginary surface whose boundaries are the tubular boundaries of the electric tube plus surfaces lying entirely within the materials of the conductors at the ends. By the definition of electric tube, there is *no component* of electric intensity *perpendicular* to the tubular boundaries, and we have seen that there is no field inside the material of the conductors. The surface integral in Gauss's law vanishes, and hence the *net* charge inside this surface vanishes. Therefore there must be *equal and opposite* charges at the ends of the tube. We can now readily demonstrate that

The flux through an electric tube is constant along its length.

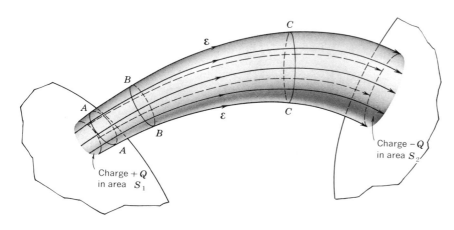

Figure 10

Consider any section of the tube such as that between BB and CC in Fig. 10. Since there is no charge in this section, the net flux leaving the section must be zero. Therefore the *same* flux must enter at BB as leaves at CC.

Finally, we can demonstrate that

The electric intensity immediately outside a point on a conductor is directly proportional to the surface charge density $\sigma = dQ/dS$, according to the relation

$$\mathcal{E} = \pm \sigma/\varepsilon_0, \qquad \sigma = \pm \varepsilon_0 \mathcal{E}. \tag{3}$$

The intensity vector points away from the surface if σ is positive, into the surface if σ is negative.

Imagine the tube in Fig. 10 to be infinitesimal, with positive charge dQ on the left end, in area dS. Imagine the section AA to be just outside and parallel to the conductor, so that it also has area dS and flux $\mathcal{E}\, dS$ if $\mathcal{E}$ is the intensity immediately outside the conductor. Now consider Gauss's relation for a closed surface consisting of the section AA, the walls of the very short tube between AA and the conductor, and a closing surface lying within the substance of the conductor. The flux out of this closed surface is $\mathcal{E}\, dS$; the charge inside the surface is $dQ = \sigma\, dS$. Gauss's relation gives

$$\mathcal{E}\, dS = (1/\varepsilon_0)\, \sigma\, dS, \quad \text{or} \quad \mathcal{E} = \sigma/\varepsilon_0.$$

A similar argument can be applied at the end of a tube where the charge is negative.

The difference of potential between the two conductors of Fig. 10 is given, as in (19), p. 483, by

$$V_1 - V_2 = \int \mathcal{E}\, dl, \tag{4}$$

where the integral is taken *along an electric line* connecting conductors 1 and 2, so that $\mathcal{E} \cdot d\mathbf{l}$ in (19), p. 483, becomes just $\mathcal{E}\, dl$.

The difference of potential between two conductors is the integral, along an electric line, of electric intensity multiplied by the element of length of the electric line.

If we write $\int \mathcal{E}\, dl = \mathcal{E}_{Av} \int dl$, we can state that

The average electric intensity along an electric line connecting two conductors is the difference of potential divided by the length of the electric line.

In general, then, the electric intensity between two conductors is high where the electric lines are short, low where they are long (note Fig. 9). Since the surface charge density is proportional to the electric intensity at the surface, this statement implies that the charges on oppositely charged conductors tend to congregate on the parts of the surface facing each other, where the electric lines are shortest. The extreme example of this congregation is in the case of the parallel-plate capacitor we shall consider in the next chapter, where two large oppositely charged plane-parallel conductors are close to each other. Except near the very edges of the plates, the charge can be considered to be *all* on the *facing* surfaces, uniformly distributed. This distribution is associated (Fig. 11) with the short and constant length of the electric lines between the plates, as compared with the great length of the electric lines connecting charges on the outer surfaces.

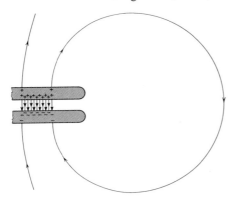

Fig. 11. Schematic diagram of representative electric lines near the edges of oppositely charged parallel plates.

Electric lines either begin at positive charges and end at negative charges, or they may go out to or come in from infinity as in the case of the lines from an isolated point charge or the charged conductor of Fig. 7. Electric lines cannot form closed curves in space because the integral in the line-integral law (20), p. 483, extended once around such a closed curve, would not vanish if there really were such an electric line. To summarize,

Electric lines always begin and end on electric charges, or one end of an electric line may proceed to infinity. Electric lines cannot form closed curves.

The existence of electric lines joining the two conductors in Fig. 9 or Fig. 10 implies that there *must* be a difference of potential between the conductors. Hence, they must be *different* conductors not electrically

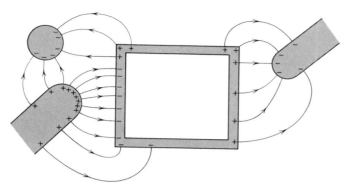

Fig. 12. Charges outside a hollow conductor cannot set up an electric field inside, nor induce charges on the inside walls.

connected, since no difference of potential can exist between two points on the same conductor.

No electric line can begin and end on the same conductor.

This statement furnishes the explanation of the very important phenomenon of *electrostatic shielding*. A closed box made entirely of metal acts as an electrostatic shield. In Fig. 12, the charges outside the hollow conductor can set up no field inside the conductor and can induce no charge on the inner walls of the conductor. Why?—because if there were electric lines in the interior of the conductor, they would begin and end on the same conductor, which is forbidden. Absence of electric field within the conductor implies absence of charge on the inner walls.

If there are charges inside the shield, these will set up their own field, as in Fig. 13, but the field inside the conductor will be unaffected by charges outside, and the charges inside will experience no force resulting from the presence of charges outside. The shield may be as thin as a sheet of foil or a coat of aluminum paint. It makes no difference. Any apparatus inside the shield is completely unaffected by electrified bodies outside. Shielding is familiar to everyone who has looked inside a radio set; it is used extensively to keep electric fields originating in other parts of the set, or in outside electrical circuits, from interfering with the operation of the various vacuum tubes and other field-sensitive components.

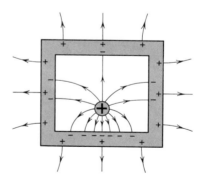

Fig. 13. If a charged conductor is placed inside a shield, an equal and opposite charge appears on the inner wall of the shield.

Example. *Determine the potential at all radii for the charged hollow spherical conductor of Fig. 7, p. 498. Also verify the correctness of equation (3).*

In the example on p. 498, we showed that the electric intensity outside the sphere ($r_2 > R_2$) has exactly the value for a point charge Q. Hence, with $V_\infty = 0$, the computation on p. 481 is applicable, and the potential outside the sphere has the same value as for a point charge:

$$V = (1/4\pi\varepsilon_0)\, Q/r_2 \qquad \text{for } r_2 \geqq R_2.$$

By setting $r_2 = R_2$ in this expression, we find the potential at the outer surface of the conductor. Since the whole of the material of a conductor is an equipotential region, this value also gives the potential throughout the conductor and at its inner surface. Since there is no electric field inside the hollow conductor, the inside of the hollow is also an equipotential region, with potential the same as at the outside of the sphere:

$$V = (1/4\pi\varepsilon_0)\, Q/R_2 \qquad \text{at all radii} \leqq R_2.$$

Once a test charge has been brought in from infinity to the outer surface of the conductor, it encounters no further electric field, and no further work need be done if it is moved to any point within this outer surface.

To verify (3), we note that all the charge Q is uniformly distributed on the *outer* surface of the hollow conductor of area $4\pi R_2^2$—why? Hence the surface charge density is $\sigma = Q/4\pi R_2^2$. $\mathcal{E}$ in (3) is the intensity just outside the surface, which is $(1/4\pi\varepsilon_0)\, Q/R_2^2$. Hence, $\varepsilon_0\, \mathcal{E} = Q/4\pi R_2^2$, which does equal the surface charge density σ.

4. Charging by Induction; the Electroscope

The ideas about electric lines that we have presented in the previous section are useful in explaining the phenomena involved in the process of charging by induction. First, let us consider, as in Fig. 14, what is meant by *grounding* a conductor. At the top of Fig. 14, all the electric lines from the charged conductor are shown ending on the ground; no lines go to infinity. This is the situation when charged objects are near an infinite conducting plane, and the earth is, for practical purposes, such an infinite conducting plane. Equal and opposite charges are induced in the near-by areas of the earth's conducting surface, so that all the electric lines begin or end there. When the charged body in Fig. 14 is 'grounded' (electrically connected to the earth), it and the whole earth become a single conductor. The charge that was on the body is distributed over the whole surface of the earth, which is so enormous that the resulting charge density is negligible, and the body is, for practical purposes, left uncharged (unless there is another ungrounded charge near-by).

The process of *charging by induction* should be clear from Fig. 15, with the possible exception of one point, namely, why *all* the negative charge leaves the ball when it is grounded, in Fig. 15(b). The answer is that if negative charges did remain on the ball, there would have to be positive charges at the beginnings of the electric lines that ended on

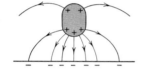

A charged metal body . . .

loses all of its charge when it is grounded . . .

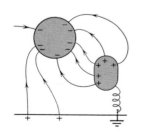

unless there is an ungrounded charge nearby.

Figure 14

these negative charges. But there are no positive charges on the hard-rubber rod. Moreover, the positive charges at the other ends of the electric lines could not be on the ball or the earth because these now form one single conductor, and an electric line cannot begin and end on the same conductor. This absence of necessary positive charges guarantees that all negative charge will leave the ball when it is grounded.

The *electroscope* (Fig. 16) is a semi-quantitative measuring instrument that played a very important role in the development of the laws of electrostatics. It usually consists of two thin gold leaves fastened to a metal rod terminated by a metal ball or plate. The leaves are more or less completely surrounded by a grounded metal case, from which the protruding metal rod is insulated. Grounding the case removes the possibility that the case will have a charge of its own that will influence the ball and hence the leaves. The grounded case will have only such charge as may be *induced* on it. If a positively or negatively charged body is brought near the ball of the electroscope, the leaves diverge because of the repulsion of the charges induced on the two leaves, as in Fig. 16.

If the electroscope is charged, by touching the ball with a charged body, or by induction, the leaves remain permanently deflected, as in Fig. 17(a). If a negatively charged body is brought near the ball of a negatively charged electroscope, as in Fig. 17(b), the leaves diverge further, because more of the negative charge is driven from the ball into the leaves, giving the leaves a greater negative charge and hence greater repulsion. If a positively charged body approaches the negatively charged electroscope, the leaves draw together because negative charge is drawn from the leaves into the ball as in Fig. 17(c). If the body has a sufficiently large positive charge, as it approaches still closer the leaves collapse completely and then diverge again as more and more negative charge moves from the leaves into the ball as in (d) and (e).

Note that the shield prevents any *direct* action of an external charge on the leaves. No electric lines can go directly from an external charge to the

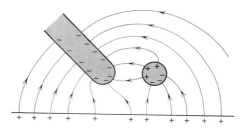

(a) In order to give a metal ball a *positive* charge by induction, a *negatively* charged hard-rubber rod is brought near the uncharged metal ball.

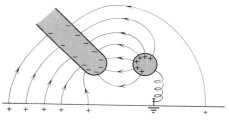

(b) The ball is connected to ground (for example, through the body by touching with the finger). *All the negative charge leaves the ball.*

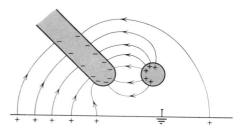

(c) The ground connection is broken.

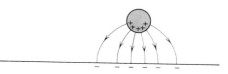

(d) The charged hard-rubber rod is removed, leaving the ball positively charged. The charge acquired by induction is *opposite* in sign to the inducing charge.

Fig. 15. Charging by induction. The electric lines are schematic only.

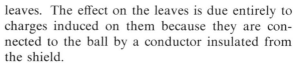

Figure 16

leaves. The effect on the leaves is due entirely to charges induced on them because they are connected to the ball by a conductor insulated from the shield.

The electroscope will serve as the simplest type of electrostatic voltmeter to measure the potential, relative to ground, of a charged conductor. For this purpose the ball is connected by a long, fine, insulated wire to the body whose potential it is desired to measure. Some of the charge flows to the ball and leaves of the electroscope, charging them so that they have the same potential as the conductor. If the amount of charge that flows to the electroscope when it is connected is small compared with the total charge on the conductor, the potential of the conductor is not appreciably altered by connecting the electroscope. If the amount of charge is not small compared with the total charge, at least the electroscope leaves and the conductor come to the same potential, and the electroscope measures the potential that exists after it is connected.

The deflection of the electroscope is a measure of the potential difference between the electroscope leaves and the grounded case.

This statement follows from the fact that the deflection is directly related to the charge on the leaves by Coulomb's law, the charge on the leaves is related to the electric intensity near the leaves by (3), and the electric intensities within the shield are directly related to the difference of potential between the leaves and shield by (4). While the deflection is not exactly proportional to the difference of potential, the deflection increases in regular fashion as the potential difference increases.

5. The Faraday Ice-pail Experiment

All of the discussion in Sec. 3 is based on the accuracy of Gauss's relation. As the reader can readily see, Gauss's relation is valid only if the electric intensity resulting from a point charge is inversely proportional to *exactly the square* of the radius. Crucial experiments based on these considerations can be devised, and have shown that, if we write Coulomb's law in the form $\mathcal{E} \propto 1/r^x$, the exponent x is exactly 2 to within a possible experimental error of only $\pm 10^{-9}$. The prototype of these experiments was performed by Faraday* and is known as the Faraday

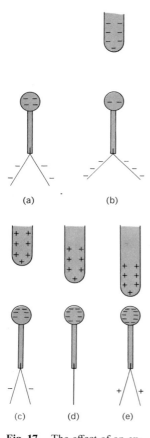

Fig. 17. The effect of an external charge on a charged electroscope. (The electroscope shield is omitted from these schematic drawings.)

*MICHAEL FARADAY (1791–1867), director of the laboratory of the Royal Institution, London, conceived the idea of representing an electrostatic field by electric tubes and formulated its properties in terms of them. He also discovered electromagnetic induction and the laws of electrolysis.

ice-pail experiment. By this experiment, Faraday demonstrated that the charge induced on the inner walls of a closed container is exactly equal and opposite to an inducing charge placed in the container in the manner of Fig. 13. The induced charge is exactly equal and opposite to the inducing charge only if the inverse square law exactly applies, Gauss's relation is exact, and the charges on opposite ends of an electric tube are exactly equal and opposite.

We can describe this experiment in connection with Fig. 18. In (a),

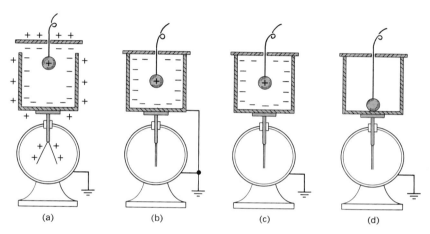

Fig. 18. Faraday's 'ice-pail' experiment.

a charged metal body attached to a nonconducting thread is lowered into a metal 'ice pail.' The container rests on an electroscope. When the ice pail is closed and then grounded as in (b), the electroscope leaves and case are at the same potential, so there is no deflection. Now the following two steps are taken:

1. The ground is removed from the ice pail [Fig. 18(c)].
2. The charge is lowered so as to touch the interior of the ice pail [Fig. 18(d)].

After these operations, the sum of the charges on the body and on the interior of the ice pail is found to be zero because there is still no electroscope deflection. This observation of equality of induced and inducing charge furnishes a very sensitive test of the validity of the inverse-square law of force.

As mentioned earlier, the validity of the inverse-square law has now been confirmed to a few parts in 10^9, by a variety of experiments, over distance ranging from the nuclear radius of 10^{-15} m out to astronomical distances.

PROBLEMS

1. Two large parallel plates 2 mm apart are held at a potential difference of 600 V. (a) How much work, in joules, is necessary to carry 0.2 μC of charge from the lower- to the higher-potential plate? (b) What are the magnitude and direction of the force acting on this charge when it is between the plates? (c) What is the electric intensity between the plates, in N/C? (d) How much work, in joules, would be necessary to carry a total of 1 C of charge from the lower- to the higher-potential plate, the potential difference being maintained at 600 V?   Ans: (a) 1.20×10^{-4} J; (b) 6.00×10^{-2} N; (c) 3.00×10^5 N/C; (d) 600 J.

2. A sphere of radius 6 cm has a total charge of -3 μC uniformly distributed over its surface. What is the electric intensity at the surface of the sphere?

3. What is the electric intensity adjacent to the surface of a sphere uniformly charged with 0.15 μC/cm²?   Ans: $54 \pi \times 10^6$ V/m.

4. Two large parallel plane plates are 0.4 cm apart and have a difference of potential of 800 V. What is the density of charge, in μC/m², on each plate?

5. Two large parallel plane plates are 0.2 cm apart and have a difference of potential of 500 V. What is the density of charge, in μC/m², on each plate?   Ans: 2.21 μC/m².

6. Two large parallel plane plates are 0.4 cm apart and are charged to densities of ± 1.5 μC/m². (a) What is the electric intensity between the plates? (b) What is the difference of potential?

7. Two concentric, thin, metallic spherical shells of radii R_1 and R_2 ($R_1 < R_2$) bear charges Q_1 and Q_2. Using Gauss's law, show that (a) the electric intensity at radius $r < R_1$ is zero; (b) the electric intensity at radius r between R_1 and R_2 is $(1/4\pi\varepsilon_0) Q_1/r^2$; (c) the electric intensity at radius $r > R_2$ is $(1/4\pi\varepsilon_0)(Q_1+Q_2)/r^2$.

8. For the system of charges in Prob. 7, show that the potential, relative to a zero at infinity, is

$(1/4\pi\varepsilon_0)(Q_1+Q_2)/r$ for $r > R_2$;
$(1/4\pi\varepsilon_0)[Q_1/r + Q_2/R_2]$ for $R_2 > r > R_1$;
$(1/4\pi\varepsilon_0)[Q_1/R_1 + Q_2/R_2]$ for $r < R_1$.

9. For the system of charges of Prob. 7, find the charge per unit area on the inner and outer surfaces of each spherical shell. Find the answers in the two different ways given below, and check:
(a) Use the total charge on each shell, the area of the shell, and the theorem that equal and opposite charges reside at the two ends of an electric line.
(b) Use the fact that the electric intensity just outside a conductor is $(1/\varepsilon_0)$ times the charge per unit area. From the electric intensities in Prob. 7, compute the values of charge per unit area.

10. Charge is *uniformly* distributed throughout the interior of a nonconducting sphere of radius R, the *total* charge being Q. Using Gauss's law, show that
(a) the electric intensity at radius $r > R$ is $(1/4\pi\varepsilon_0) Q/r^2$;
(b) the electric intensity at radius $r < R$ is $(1/4\pi\varepsilon_0) Qr/R^3$.

11. Show that the electric intensity at distance r from an infinite straight fine wire with linear charge density q, in coulombs per meter length, is $(1/2\pi\varepsilon_0) q/r$.

12. Show that a charge outside an uncharged hollow conductor experiences a force if a charge is placed inside the conductor, while a charge inside the conductor experiences no force if a charge is placed outside. These statements are in apparent violation of Newton's third principle. Explain the paradox.

13. An isolated metal sphere of 10-cm radius is deprived of 10^{11} electrons. Find the electric intensity at the surface and at 1 m from the center.   Ans: 14 400, 144 V/m.

14. Show that the force between two spherically symmetric charge distributions, not overlapping, is the same as that between two point charges at the centers of the distributions. The proof can be done in three steps by showing (a) that the force of distribution B on distribution A is the same as the force of a point charge on A, and (b) that the force of A on the point charge is the same as the force between two point charges; (c) Newton's third principle can then be used to demonstrate the theorem.

15. Consider the point charge $+Q$ shown in

the figure. Show that $\oint \mathbf{\mathcal{E}} \cdot d\mathbf{l} = 0$ for the two closed paths indicated: (a) a circular path with Q at its center and (b) a path from A along a radius to B, then along an arc of a circle from B to C, then radially inward from C to D, and finally from D to A along the arc of a circle.

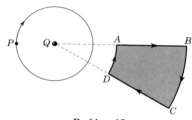

Problem 15

16. Show that $\oint \mathbf{\mathcal{E}} \cdot d\mathbf{l} = 0$ in the field between two large parallel metal plates between which the difference of potential is V. Consider (a) the case of a closed curve in a plane parallel to the plates and (b) the case of a closed rectangle in a plane perpendicular to the plates.

17. Show that if a total charge Q is uniformly distributed along a thin wire forming the perimeter of a circle of radius R lying in the XY-plane and centered at the origin, then the electric intensity at any point on the Z-axis (the axis of the circle) is given by the relation $\mathcal{E} = (1/4\pi\varepsilon_0) Q \cos\phi/(R^2 + Z^2)$. Draw a diagram that properly locates the angle ϕ in this relation, and shows the direction of the electric intensity.

18. Let a charge be uniformly distributed over the whole XY-plane with uniform density σ, in C/m². By splitting this charge into rings and integrating the relation in the preceding problem, compute the electric intensity at any point along the Z-axis and verify that you obtain the same value as is obtained by application of Gauss's relation to this geometrical situation.
Ans: $\sigma/2\varepsilon_0$.

19. The gravitational acceleration g represents force per unit mass and hence is quite analogous to electric intensity, which represents force per unit charge. Show that Gauss's relation for the gravitational case is

$$\oiint_S \mathbf{g} \cdot d\mathbf{S} = -4\pi G \Sigma m,$$

where G is the universal gravitation constant (see p. 61), and the summation is over all masses *interior* to the imaginary surface S.

20. Consider a massive spherical body whose density is not necessarily constant, but is a function only of the radius r from its center (the earth approximates such a body). Using the relation derived in Prob. 19, show that the gravitational attraction of such a body for a particle is the same as if all the mass of the body were concentrated at its center.

21. By simple logical argument, starting with the result of Prob. 20, show that the gravitational force between two massive spherically symmetric bodies of any size is the same as if all the mass of each body were concentrated at its center. Compare with Prob. 14.

22. In the superseded Thomson model of the hydrogen atom, one electron was inside a sphere of uniformly distributed positive charge and of radius R_0. Use Gauss's relation to show that the electron is in equilibrium at the center of the sphere and that the electron is subject to a restoring force $F = -kr$ when it is at a distance r from the center of the sphere, provided $r < R_0$. In view of the restoring force, the electron will oscillate with simple harmonic motion of frequency $f = (1/2\pi) \sqrt{k/m_e}$ if displaced from its equilibrium position and then released; find its frequency of oscillation.

23. A sodium ion Na⁺ is spherical and has a net charge of +e; a chlorine ion Cl⁻ is also spherical and has a charge of −e. In the NaCl crystal (common table salt) adjacent Na⁺ and Cl⁻ ions are 5.63 nm apart. What is the electric intensity midway between the ions? (Neglect the contributions of nonnearest neighbors.)
Ans: 3.65×10^8 V/m.

24. Find the electric intensity $\mathcal{E}$ at a distance of $r_1 = 5.3 \times 10^{-11}$ m from a proton. Remembering that r_1 is the radius of the first Bohr orbit, find an expression for the centripetal force acting on the electron in this orbit. What is the orbital speed of the electron in this orbit? What is the ratio of the orbital speed to the speed of light? What is the orbital frequency of the electron?
Ans: $\mathcal{E} = (1/4\pi\varepsilon_0) e/r_1^2$; $F = (1/4\pi\varepsilon_0)(e^2/r_1^2)$; $v = Fr_1/m_e$; $1/137$; $f = 2\pi r_1/v$.

25. There has been some speculation as to whether the charges of the electron and proton are *exactly* equal. In order to test this point, a large quantity of hydrogen gas at high pres-

sure was placed in a well-insulated tank arranged like the Faraday ice pail in Fig. 18; the tank was first grounded; with ground connection removed, the gas was allowed to escape; then the potential of the "empty" tank was measured. Suppose that 1 kmole of the H_2 gas were initially in the grounded tank; if all the gas were then removed from the insulated tank, what charge would remain on the tank if the proton charge were 1 per cent larger than the charge of the electron? (Note: Careful experiments of the type described have shown that the magnitudes of the charges on proton and electron are equal to within 1 part in 10^{20}!)
Ans: 9.63×10^6 C.

26. What is the electric potential at a distance of 1.5×10^{-15} m from the center of the helium nucleus if $V_\infty = 0$? In making the calculation consider the nucleus as a uniformly charged sphere.

27. In natural radioactive decay the uranium nucleus can emit a helium nucleus and thereby become a thorium nucleus. The original uranium nucleus has a radius of approximately 8.4×10^{-15} m. Recalling that the thorium nucleus contains 90 protons and the helium nucleus contains 2 protons, calculate (a) the magnitude of the repulsive forces between the thorium and helium nuclei and (b) their mutual electrostatic potential energy just as they are separating and the distance between their centers is 9.0×10^{-15} m. (c) Express the potential energy in electron-volts.
Ans: (a) 5.12×10^2 N; (b) 4.61×10^{-12} J; (c) 2.88×10^7 eV or 28.8 MeV.

28. In a nuclear fission process a uranium nucleus divides into a barium nucleus and a krypton nucleus. Recalling that the barium and krypton nuclei contain 56 and 36 protons, respectively, calculate (a) the magnitude of the repulsive forces and (b) the electrostatic potential energy when these nuclei are separating and the distance between their centers is 9×10^{-15} m. (c) If a resultant force having a magnitude equal to that involved in the above fission process acted on a 1-kg body, what would be the acceleration of the body?

NOTE: *Much of our knowledge of the atomic nucleus has been obtained by firing high-energy 'atomic projectiles' into thin 'targets' of various materials and observing the results. As the methods used in accelerating atomic projectiles and the methods of detecting the emerging particles and radiation involve for the most part the elementary principles of mechanics already studied and the elementary principles of electricity and magnetism being introduced in the present chapter and the next few chapters, it will be instructive to examine some of these methods in problems. In working the problems below, the reader should refer to the table in Sec. 2 of the Appendix for data on the masses and charges of the particles.*

29. The most direct method of obtaining high energy projectiles is to produce positive ions in a region of high potential and allow electric forces to accelerate them through an evacuated tube toward a target at ground potential. Such an arrangement is shown schematically in the figure.

Positive ions H$^+$, produced in an electrical discharge in the ion source are directed downward out of a small hole approximately 1 mm in diameter; this direction is accomplished by electrodes inside the source at a potential of +1 or +2 kV relative to the lower plate of the source, which itself has a potential of +6 MV relative to ground. Ions emerging from the hole enter the evacuated accelerator column and are subjected to downward directed electric fields in the spaces between successive metal electrodes which are separated by the glass or ceramic sections forming the walls of the vacuum chamber. By attaching metal rings or 'hoops' outside the evacuated accelerator column it is possible to make each electrode part of an equipotential plane as suggested in the figure. The beam of rapidly moving H$^+$ ions passing through the hole in the final electrode strikes the target at ground potential. In an actual 6 MeV accelerator 100 electrodes are used in a column 3.6 m in length; each electrode is connected to a hoop that surrounds not only the accelerator column but also the rapidly moving charged belt (Prob. 31) used to maintain the ion source at 6 MV.

Suppose that hydrogen atomic ions H$^+$ (protons) are produced in the ion source, which is at a positive potential of 6 million volts with respect to ground. What are the velocity and the kinetic energy of a proton striking the target, which is at ground potential? If the distance from the ion source to the target is 5 m, what average force is exerted on the proton? What is the average acceleration of the proton? Use nonrelativistic formulas.

Ans: 3.38×10^7 m/s, 9.60×10^{-13} J, 2.67×10^{-14} N, 15.9×10^{13} m/s².

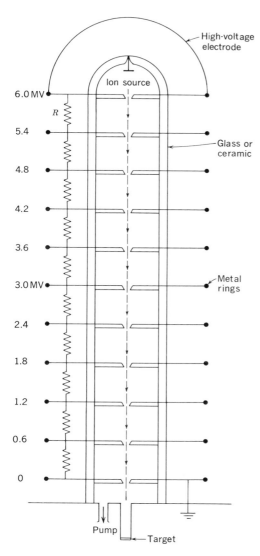

Problems 29–30

30. Answer the questions in Prob. 29 for a case in which doubly charged helium ions He^{++} (alpha particles) are produced in the ion source. Answer the same questions for deuterons (hydrogen atomic ions of atomic mass 2).

31. Although the acceleration of ions by the method in Prob. 29 is simple in principle, it is not a simple matter to obtain and maintain potential differences of several million volts between the ion source and the target. One method of obtaining such a potential difference is that used in the Van de Graaff generator shown schematically in the figure. In this device, static charges are transported on a rapidly moving endless belt from ground potential to a large hollow roughly spherical electrode insulated from the ground. The ion source and focus tube are located inside the sphere, with the accelerator sections between the sphere and ground. The potentials of the accelerator sections are cascaded by a suitable current-leakage arrangement (Prob. 39, Chapter 26). If the spherical electrode has a constant positive potential of 6.0 million volts with respect to ground, how much work is done in placing a charge of 3 μC on the sphere?   Ans: 16.2 J.

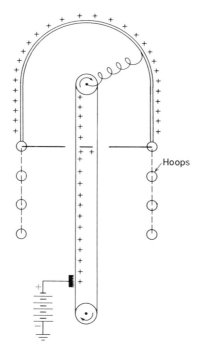

Problem 31

32. What becomes of the kinetic energy of the

ions in an accelerator tube (Prob. 29) when the ions strike a thick target? Why is it necessary to use water cooling on some types of targets?

33. The limiting factor in the design of the conventional Van de Graaff electrostatic device is usually the maintenance of a constant high potential on the large positive electrode; it is possible, however, to make double use of the high-voltage electrode by means of the 'tandem scheme,' initially patented by Willard H. Bennett and now being used in commercially available accelerators. In these devices, negative ions are first produced at ground potential and allowed to move up the potential hill toward the high voltage electrode; while inside the high-voltage electrode, the high-speed ions move through a thin foil and lose several electrons; the resulting positive ions move down the potential hill back to a target at ground potential. In a tandem machine, the high-voltage electrode is usually a large cylinder surrounding the central portion of the accelerator tube.

In a certain tandem accelerator, the high-voltage electrode has a potential of 6 MV. (a) If H^- ions are produced at one grounded end of the accelerator tube, what is their energy when they reach the high-voltage electrode? (b) If an energetic H^- ion loses two electrons in passing through a thin foil, and becomes an H^+ ion, what is the final energy of the H^+ ion after it has moved along the accelerator tube to a target at ground potential.

Ans: (a) 6 MeV; (b) 12 MeV.

34. Repeat the above problem (a) for a system that uses C^- and C^{+++} ions and (b) for a system that uses He^- and He^{++} ions.

35. When a charged particle of high kinetic energy passes through a gas, it collides with the molecules and in many of these collisions tears electrons away from the molecules leaving a trail of 'ion pairs' (electrons and positively charged molecules). In each such collision the high-energy particle loses some of its energy in separating the electron from the molecule and thereby 'fritters' its energy away and eventually comes to rest. By collecting the ions produced in a gas sample, we can detect the passage of high-energy ionizing particles through the gas and if the type of high-energy particles is known, we can get an estimate of the number of such particles passing through the gas sample. The *ionization chamber* shown in the figure is a device for making such measurements. The difference of potential between the central electrode and the container is maintained sufficiently high to collect *all* primary ions formed by the original particle before they recombine but not high enough to give the ions sufficient kinetic energy to produce secondary ions by collision.

It is known that each alpha particle from a certain radioactive element produces 1.4×10^5 ion pairs. If the galvanometer in the figure indicates a current of 1.12×10^{-12} C/s, how many alpha particles of the type mentioned are passing through the gas in the chamber per second? Ans: 50.

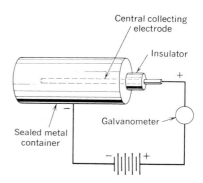

Problems 35–36

36. It will be noted from Prob. 35 that the charge collected in an ionization chamber is extremely small. Larger amounts of charge can be collected if the potential of the inner electrode is raised so that each of the primary ions produced by the original particle is given sufficient kinetic energy to produce secondary ions before being collected; over a limited range of potential the number of secondary ions is directly proportional to the number of primary ions. If the central electrode potential in the chamber mentioned in Prob. 35 is slowly raised while the number of alpha particles passing through the chamber per second remains at 50, the current increases. Find the ratio of the number of ion pairs collected each second to the number of primary ion pairs formed for an electrode potential at which the current is (a) 2.5×10^{-12} C/s, (b) 5.5×10^{-11} C/s, and (c) 0.000 25 C/s. (This ratio is sometimes called the *gas amplification factor*.)

CAPACITANCE 24

The most familiar device whose action is entirely electrostatic is the *capacitor* (or *condenser*), of which considerable numbers usually occur in every electronic device, such as a radio or television set. Electric charge flows through the wires leading to a capacitor but does not flow *through* the capacitor—it merely piles up as a charge on the plates of the capacitor. The relation between the charge on the plates and the difference of potential between the plates is determined by the laws of electrostatics, the charge on the plates being at rest. The ratio of the magnitude of charge on either plate to the difference of potential is a constant, called the *capacitance*. Many circuit elements that are not actually called capacitors behave like capacitors, and a consideration of their capacitance is necessary to an understanding of their performance. Examples of such circuit elements are shielded cables, alternating-current power lines, telephone lines, and the metallic elements of any vacuum tube.

1. Capacitance

A combination of two conductors in proximity (as, for example, in Fig. 1) is called a *capacitor*. The electrical characteristics of a *capacitor* are determined by its *capacitance,* which is a measure of the ability of the conductors to store charge when a potential difference is produced between them, whether by a battery, a radio antenna, or any other source of potential difference. Capacitance, denoted by C, equals Q/V, in coulombs per volt of potential difference. The unit, 1 coulomb/volt, is given the name *farad* (F), after Faraday.

The **capacitance** between two conductors, in **farads,** is the ratio of the absolute magnitude of the charge on either conductor, in coulombs, to the resulting potential difference, in volts, when the conductors have equal and opposite charges.

Thus, capacitance is given by $\quad C = \dfrac{Q}{V}, \quad \begin{Bmatrix} C \text{ in F} \\ Q \text{ in C} \\ V \text{ in V} \end{Bmatrix}$ **(1)**

from which we see that 1 F = 1 C/V. Ordinary capacitances are very small fractions of a farad and are usually specified in *microfarads* (μF), *nanofarads* (nF), or *picofarads* (pF).

The capacitance of a capacitor with air or vacuum between its plates is determined purely by its geometry; as an example we consider the capacitor in Fig. 1, which has two parallel plane plates, each of area A, separated by distance d, and charged with equal and opposite charges $\pm Q$. The separation d is assumed to be small compared with the linear dimensions of the plates. For the reasons that we indicated on p. 502, the charges are almost entirely on the inside surfaces of the plates and, except near the edges, the electric lines go perpendicularly from one plate to the other.

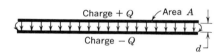

Fig. 1. Parallel-plate capacitor. The area of each plate is A. The plates are in vacuum (or in air).

Then, if the difference of potential between the plates is V, the electric intensity is $\mathcal{E} = V/d$, except near the edges.

We can now employ (3), p. 501, to determine the charge density

$$\sigma = \pm \varepsilon_0 \mathcal{E} = \pm \varepsilon_0 V/d, \qquad (2)$$

positive on one plate, negative on the other. This relation shows that the charge density is constant (except near the edges), so that to a good approximation we can write $\sigma = \pm Q/A$. Substitution in (2) then gives

$$\frac{Q}{A} = \varepsilon_0 \frac{V}{d}, \qquad V = \frac{Qd}{\varepsilon_0 A}, \qquad Q = \frac{\varepsilon_0 A}{d} V. \qquad (3)$$

Comparison of (1) and (3) gives

$$C = \frac{\varepsilon_0 A}{d}. \qquad \begin{pmatrix} \text{parallel-plate} \\ \text{capacitor} \end{pmatrix} \quad (4)$$

On p. 469, we wrote $\varepsilon_0 = 8.85 \times 10^{-12}$ C^2/N·m^2; by rearrangement of (14),

p. 479, we see that $1\ \text{C/m} = 1\ \text{N/V}$, so $1\ \text{C}^2/\text{N}\cdot\text{m}^2 = 1\ \text{C/V}\cdot\text{m} = 1\ \text{F/m}$. Hence, we can conveniently write

$$\varepsilon_0 = 8.85 \times 10^{-12}\ \text{F/m}, \tag{5}$$

and we see that with A in m² and d in m, (4) gives C in farads.

Example. *Find the capacitance of a parallel-plate capacitor formed when two sheets of aluminum, each 1 m² in area, are separated by a 0.1-mm layer of air. What is the charge on either plate of this capacitor when the difference of potential between the plates is 100 V?*

Substitution in (4) gives

$$C = \frac{(8.85 \times 10^{-12}\ \text{F/m})(1\ \text{m}^2)}{10^{-4}\ \text{m}} = 8.85 \times 10^{-8}\ \text{F} = 0.0885\ \mu\text{F}$$

and $\quad Q = CV = (8.85 \times 10^{-8}\ \text{F})(10^2\ \text{V}) = 8.85 \times 10^{-6}\ \text{C} = 8.85\ \mu\text{C}.$

Parallel-plate capacitors are frequently made with a stack of plates connected alternately. The plates are stacked as in Fig. 2. With the exception of the end plates, each plate carries charges on both sides, at the ends of the electric lines going to both adjoining plates. The capacitor of Fig. 2, which has five sets of electric lines, has as much positive and negative charge as five capacitors like that of Fig. 1. Hence, for the same potential difference, it has five times the charge, corresponding to five times the capacitance. In general, if a capacitor has N plates, there are $N-1$ sets of electric lines and the capacitance is $N-1$ times (4), or

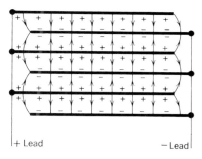

Fig. 2. A six-plate capacitor.

$$C = (N-1)\,\varepsilon_0 A/d.$$

The two conductors forming a capacitor need not, of course, be parallel plates. Any two conductors in proximity have a capacitance determined by the definition above.

2. The Force between Charged Plates

We shall now compute the force per unit area with which the two plates of Fig. 1 attract each other. For this purpose, we shall assume the plates as infinite in area.

We start by using Gauss's relation to compute the electric intensity resulting from an infinite plane that is uniformly charged to area density σ, as in Fig. 3. (Such an infinite charge sheet, with zero thickness is, of course, a mathematical idealization. One could think of a very large plane

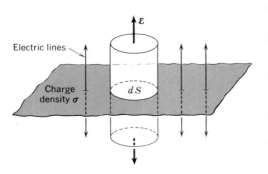

Fig. 3. An infinite plane, uniformly charged.

sheet of thin metal foil, uniformly charged to area density ½ σ on each side, in the limit as the area goes to ∞ and the thickness goes to zero so that the whole charge σ is effectively collected in one sheet.) From symmetry, the electric lines will be straight lines perpendicular to the plane; the electric tubes will be right cylinders. Since the tubes do not change in area as one goes away from the plane, and since the flux through a tube is constant along its length, the intensity $\mathcal{E}$ must be constant everywhere on one side of the plane; on the opposite side, $\mathcal{E}$ will have the same magnitude but the opposite direction. Apply Gauss's relation to the tube of area dS shown in Fig. 3. The flux out of the top is $\mathcal{E}\, dS$; that out of the bottom is $\mathcal{E}\, dS$; the charge within the tube is $\sigma\, dS$. Hence by Gauss's relation, p. 497,

$$2\,\mathcal{E}\,dS = \sigma\,dS/\varepsilon_0, \qquad \mathcal{E} = \tfrac{1}{2}\,\sigma/\varepsilon_0. \tag{6}$$

We contrast this expression with (2), where the field in the capacitor was found to be $\mathcal{E} = \sigma/\varepsilon_0$. The difference of a factor of 2 arises from the fact that the capacitor field is the superposition of the fields of *two* infinite parallel plates, oppositely charged, so that the electric intensities reinforce between the plates and cancel outside, as indicated in Fig. 4.

Now let us denote by f the force per unit area with which the upper plate attracts the lower, and vice versa. An area dS of the lower plate has charge $-\sigma\, dS$ and is in the downward field $\tfrac{1}{2}\sigma/\varepsilon_0$ resulting from the charge on the upper plate. The lower plate feels no force resulting from the field set up by its own charge—this fact can be demonstrated by symmetry; the force would have no way of choosing between up and

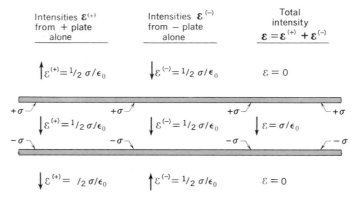

Fig. 4. *Between* two infinite parallel plates, oppositely charged, the intensities superpose to give $\mathcal{E} = \sigma/\varepsilon_0$; *outside the plates*, their vector sum is zero.

down, so it must vanish. The lower plate is therefore acted on by the upward force

$$(\sigma \, dS)(\tfrac{1}{2}\, \sigma/\varepsilon_0) = (\tfrac{1}{2}\, \sigma^2/\varepsilon_0)\, dS.$$

The force of attraction, per unit area, is

$$f = \tfrac{1}{2}\, \sigma^2/\varepsilon_0. \qquad \begin{Bmatrix} f \text{ in N/m}^2 \\ \sigma \text{ in C/m}^2 \\ \varepsilon_0 \text{ in C}^2/\text{N}\cdot\text{m}^2 \end{Bmatrix} \quad (7)$$

Measurement of this force, in an instrument called the *absolute electrometer* that was designed by Kelvin, is a more practical way of determining charge in terms of the definition of the coulomb than is the attempt to measure force between 'point' charges. Problem 36 will illustrate how a section of the capacitor can be isolated for this purpose.

3. Dielectric Constant

The discussion of Sec. 1 assumed that there was air or vacuum between the plates of the capacitor. This condition obtains in the case of variable radio-tuning capacitors, but in most capacitors some solid or liquid insulating material fills the space between the plates. Fixed radio capacitors use mica, a ceramic material, paraffined paper, or a plastic; shielded cables have a plastic compound between the conductor and the sheath; high-voltage capacitors are made with plates of metal foil placed on opposite sides of a sheet of glass.

The reason for using a dielectric material between the plates is not solely *mechanical* convenience. A dielectric improves the capacitor *electrically* in two ways: *it increases the capacitance*, and *it permits the use of higher voltages without danger of breakdown or flashover between the plates*. Breakdown is related to a property of a dielectric called *dielectric strength*, which will be discussed in Sec. 6. We turn here to an explanation of the increase in capacitance.

Consider, Fig. 5(a), a section of a parallel-plate capacitor with charge densities $\pm\sigma$, and vacuum between the plates. The electric intensity between the plates is $\mathcal{E} = \sigma/\varepsilon_0$.

Now insert a sheet of dielectric material between the plates as in Fig. 5(b). Momentarily, the field $\mathcal{E}$ acts on the atoms of the dielectric and polarizes them as in Fig. 8, p. 471. The upward shift of negative electrons relative to the positive nuclei clearly does not produce a *net* charge of either sign in any volume in the interior of the dielectric, but there will be a thin layer at the top of the dielectric where there is a net negative charge and a thin layer at the bottom where there is a net positive charge. Compare with Fig. 6, which illustrates schematically the shift of a smoothed-out distribution of negative charge relative to a distribution of an equal and opposite positive charge.

These layers of so-called 'bound charge' set up their own fields and

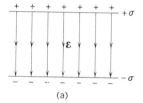
(a)

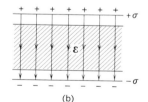
(b)

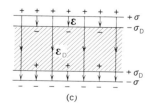
(c)

Figure 5

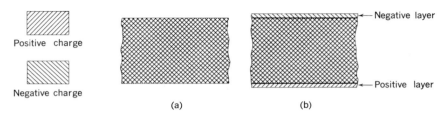

Fig. 6. Parts (a) and (b) illustrate the state of the dielectric in parts (b) and (c) respectively of Fig. 5.

result in an equilibrium field as in Fig. 5(c). Unlike the case of a conducting sheet (Fig. 8, p. 500), the electric intensity within the dielectric does not disappear completely; rather, a residual intensity $\mathcal{E}_D$ must remain in order to maintain the polarized state. As a consequence of experiment, it is found that the surface density σ_D of bound charge is directly porportional to the magnitude $\mathcal{E}_D$ of this electric intensity in the dielectric, say

$$\sigma_D = \alpha \mathcal{E}_D, \tag{8}$$

where α is a positive constant that varies from material to material.

Now, in Fig. 5(c), the electric lines associated with $\mathcal{E}$ end on charge densities $\pm \sigma$, so that $\mathcal{E} = \sigma/\varepsilon_0$. On the contrary, the electric lines associated with $\mathcal{E}_D$ end on the reduced charge densities $\pm(\sigma - \sigma_D)$, so that $\mathcal{E}_D = (\sigma - \sigma_D)/\varepsilon_0$. Substitution of (8) in this last expression for $\mathcal{E}_D$ gives

$$\mathcal{E}_D = \frac{\sigma}{\varepsilon_0} - \frac{\alpha}{\varepsilon_0}\mathcal{E}_D = \mathcal{E} - \frac{\alpha}{\varepsilon_0}\mathcal{E}_D.$$

Solution for $\mathcal{E}_D$ then gives $\quad \mathcal{E}_D = \dfrac{\mathcal{E}}{1+\alpha/\varepsilon_0}.$

The positive constant $1+\alpha/\varepsilon_0$, which is greater than unity, is called the *dielectric constant* and denoted by K:

$$\mathcal{E}_D = \mathcal{E}/K. \tag{9}$$

Since $K = 1 + \alpha/\varepsilon_0$, $\alpha = (K-1)\varepsilon_0$, and substitution in (8) gives the magnitude of the bound charge density in terms of σ as

$$\sigma_D = (K-1)\varepsilon_0 \mathcal{E}_D = (K-1)\varepsilon_0 \frac{\mathcal{E}}{K} = \frac{K-1}{K}\varepsilon_0 \frac{\sigma}{\varepsilon_0} = \frac{K-1}{K}\sigma. \tag{10}$$

In an ordinary capacitor, the dielectric completely fills the space between the plates, as in Fig. 7. The charge that flows through the leads into the metal plates is called the *free charge*. From (10), we see that if free charges $\pm Q$ are on the plates, bound charges $\pm[(K-1)/K]Q$ appear on the dielectric surfaces, as indicated in Fig. 7. The net effective charges setting up the intensity $\mathcal{E}_D$ are

$$Q_E = \pm \left(Q - \frac{K-1}{K}Q\right) = \pm \frac{Q}{K}.$$

For a given charge Q, the presence of the dielectric reduces the electric intensity by a factor K, as in (9). The difference of potential between the plates, which is $\mathcal{E}_D$ times the plate separation, is also reduced by a factor K, so expression (3) must be written as

$$V = \frac{Q_E d}{\varepsilon_0 A} = \frac{1}{K}\frac{Qd}{\varepsilon_0 A}, \qquad Q = K\frac{\varepsilon_0 A}{d}V.$$

The capacitance, which is defined as the ratio of *free* charge (the charge that would flow through the leads) to voltage, is increased by a factor K to

$$C = \frac{Q}{V} = K\frac{\varepsilon_0 A}{d}. \qquad (11)$$

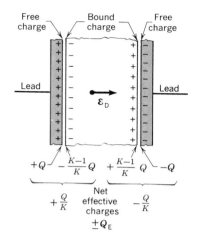

Fig. 7. Bound charges $\pm Q_D$ and effective charges $\pm Q_E$ for free charges $\pm Q$.

The dimensionless constant K, the *dielectric constant* of the material, can then be best defined as follows:

> The **dielectric constant** of a material is the ratio of the capacitance of a capacitor with the material between the plates to the capacitance with vacuum between the plates.

Values of the dielectric constant K for some dielectrics of interest are given in Table 24-1. Values are also given for certain gases. It is not strictly true, as we have assumed so far, that the capacitance of an air capacitor is the same as that of the capacitor evacuated. But the increase of capacitance arising from the polarization of the air is only 6 parts in 10 000, a change that is not of importance for most purposes and that requires apparatus of high precision to detect experimentally. Since the difference between K and unity is due to polarization of mole-

Table 24-1 Typical Values of Dielectric Constant and Dielectric Strength

Material	Dielectric constant	Dielectric strength (kV/cm)
Insulator porcelains	6	100–200
Glass	5–10	200–400
Rubber, vulcanized	3.0	160–500
Transformer oils	2.1	50–150
Mica	4.5–7.5	250–2000
Paraffined paper	2	400–600
Nitrocellulose plastics	6–12	100–400
Dry air at 1 atm	1.0006	30
Carbon dioxide at 1 atm	1.0010	28

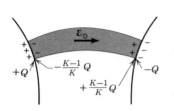

Fig. 8. Bound charges appear at the ends of every electric tube.

cules, the small values of K for gases can be attributed to their low densities; $K-1$ for gases is directly proportional to pressure.

So far, we have discussed the effect of polarization only for the case of parallel plates. However, if the space between plates of any shape be completely filled with a dielectric, the capacitance is increased by a factor K—the same K as given in Table 24-1—no matter what the shape of the plates. This increase arises from the fact that the polarization of the dielectric along any electric tube results in the appearance of bound charges at the ends of the tube, as in Fig. 8. The result is that the net effective charges, the electric intensity at any point, and the difference of potential are all reduced by the factor K.

It can be shown that, *as a result of dielectric polarization, the force between two charged conductors immersed in a fluid dielectric*, such as oil, *is reduced by the factor K*. We do not give the proof, which is rather involved. However, we note that the equation for the force between two point charges Q_1 and Q_2 separated by distance d and immersed in a fluid dielectric becomes

$$F = \frac{1}{4\pi K \varepsilon_0} \frac{Q_1 Q_2}{d^2}. \tag{12}$$

Two ways have been found of substantially increasing capacitance by using dielectric materials different from the older ones listed in Table 24-1. One way, employed in the *electrolytic capacitor,* uses an *extremely thin,* tough, insulating layer of aluminum oxide as the dielectric. The oxide is deposited on aluminum metal, which serves as one plate of the capacitor, and is in intimate contact with a conducting solution of hydrochloric acid, which serves as the other plate. Because d is extremely small in the denominator of (11), an electrolytic capacitor of given size can have a capacitance many times that of a capacitor of the same size employing a conventional dielectric. The other way of greatly increasing capacitance depends on the rather recent development of solid materials designated as *ferroelectric*. The name comes from the fact that such materials have *electrical* properties analogous to the *magnetic* properties of ferromagnetic materials which we shall study in Chapter 30. They can be permanently electrified, exhibit electrical hysteresis, and have extremely high dielectric constant. In particular, barium-titanate ceramic has a dielectric constant of 1400 or more.

The fact that a dielectric increases capacitance is usually, but not always, desirable. In radio capacitors, where the aim is to store charge in as small a space as possible, the increase is desirable. In telephone and power cables, where storage of charge leads to sluggish operation, the increase is undesirable.

4. Energy of a Charged Capacitor

Consider a capacitor that is initially uncharged, as in Fig. 9(a). We start to charge this capacitor by moving electrons, a few at a time, from the right-hand plate to the left. We may imagine that we move the electrons through the space between the plates, although in practice the electrons are usually impelled by a battery or generator through a wire that connects the two plates. This difference is unimportant, since the difference of potential and hence the work is independent of the route followed by the electrons.

At first the work that must be done on the electrons to force them from one plate to the other is small, since the difference of potential is low. But as the charge on the plates builds up, the work that must be done on each electron increases because the difference of potential increases. Finally, the last electron must be moved through practically the full difference of potential V. Since the difference of potential increases in proportion to the charge, that is in proportion to the number of electrons that have been moved, the *average* difference of potential through which the electrons are moved is $½\,V$. If the total charge that is moved is Q, the work required to move this charge, piece by piece, through an average difference of potential $½\,V$ is $W = Q\,(½\,V)$. Alternatively, we can compute this work by an integration. When the charge that has been already moved is q, the difference of potential is q/C, and the work required to move the next element of charge dq is $(q/C)\,dq$. The total work is thus

$$W = \int_0^Q \frac{q}{C}\,dq = \frac{q^2}{2C}\bigg|_0^Q = ½\,\frac{Q^2}{C}.$$

Fig. 9. Charging a capacitor.

From the relation $C = Q/V$, we see that this result is equivalent to the previous value $½\,QV$.

We can write the work required to charge the capacitor in the three equivalent forms:

$$W = ½\,QV = ½\,CV^2 = ½\,(Q^2/C). \qquad \begin{Bmatrix} W \text{ in J} \\ Q \text{ in C} \\ V \text{ in V} \\ C \text{ in F} \end{Bmatrix} \quad (13)$$

This work represents the *electrostatic potential energy* that is stored in the charged capacitor. It is properly considered as stored in the electric field. This energy will play an important role when we consider alternating-current circuits and electromagnetic waves.

5. Capacitors in Parallel and in Series

If a number of capacitors are connected in *parallel* as in Fig. 10 and placed in a (real or imaginary) box, at the terminals on the box they

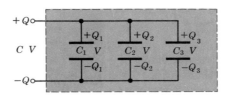

Fig. 10. Capacitors in parallel.

appear indistinguishable from a single capacitor of capacitance equal to the *sum* of the capacitances of the individual capacitors. Each capacitor has the same difference of potential (the positive plates form a single conductor and the negative plates a single conductor), equal to the difference of potential across the terminals. But the charge Q that must enter or leave the terminals on charge or discharge is the sum of the charges on the several capacitors. Hence, in the notation of Fig. 10,

$$Q = Q_1 + Q_2 + Q_3 = (C_1 + C_2 + C_3)V,$$

and
$$C = Q/V = C_1 + C_2 + C_3. \quad \text{(parallel)} \quad (14)$$

If a number of capacitors are connected in *series* as in Fig. 11 and placed in a (real or imaginary) box, the voltages across the individual capacitors add up to $V = V_1 + V_2 + V_3$ between the terminals. The charge on each capacitor has the same value Q because the charges on the two plates of a single capacitor are equal and opposite since they are at the two ends of tubes of force. This same value Q represents the charge entering the $+$ lead and leaving the $-$ lead during charge. We derive the relation for the effective capacitance as follows:

$$V = V_1 + V_2 + V_3 = \frac{Q}{C_1} + \frac{Q}{C_2} + \frac{Q}{C_3};$$

hence, since $V/Q = 1/C$,
$$\frac{1}{C} = \frac{1}{C_1} + \frac{1}{C_2} + \frac{1}{C_3}. \quad \text{(series)} \quad (15)$$

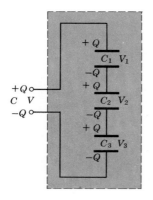

Fig. 11. Capacitors in series.

The reciprocal of the resultant capacitance C equals the sum of the reciprocals of the individual capacitances.

6. Dielectric Strength

We have previously stated that when an electric field exists within an insulating dielectric material, the electrons become slightly displaced relative to the nuclei but still remain bound to the nuclei. This situation obtains so long as the electric intensity is sufficiently low, but there is a certain value of electric intensity that is sufficient actually to pull electrons away from the atoms to which they belong. Exceeding this critical electric intensity usually results in disruptive discharge of a capacitor by sparks that pass through the dielectric material.

The critical electric intensity at which breakdown will take place is called the *dielectric strength:*

> The **dielectric strength** of a material is the maximum electric intensity that the material can withstand without permitting electrical discharge.

SEC. 6   DIELECTRIC STRENGTH   523

Dielectric strength, which depends to some extent on sample history, and on various experimental conditions, must not be confused with dielectric constant. They are essentially unrelated. The dielectric constant determines how much charge a given capacitor will store with a given potential difference; dielectric strength determines how much voltage this capacitor will stand without breaking down.

Dielectric strength is usually expressed in kV/cm. Typical values are given in Table 24–1, p. 519. It will be noticed that the dielectric strengths of most solid and liquid insulators are higher than that of air; hence capacitors with these insulators will withstand a higher voltage, without breakdown, than if air were between the plates. The dielectric strength of a perfect vacuum is infinite; however a vacuum capacitor will break down at the very high value of electric intensity that is sufficient to pull electrons from the surface of the negative plate by a process called *cold emission*. The required intensity is of the order of 2000 kV/cm.

Example. *A high-voltage underground power cable has a wire of 0.5-cm diameter in a lead sheath of inside diameter 1.5 cm. The insulating material is rubber of dielectric strength 400 kV/cm. What is the voltage of the wire, relative to the grounded sheath, at which breakdown occurs?*

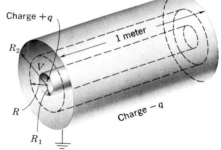

Fig. 12. A 1-m length of a cylindrical capacitor.

The wire and sheath form a cylindrical capacitor. If the wire has charge q per meter length, the net effective charge on the wire and the immediately adjoining dielectric will be $q_E = q/K$. In Fig. 12, the wire has radius R_1, the sheath R_2, and an imaginary cylindrical surface of radius R is also indicated.

By symmetry, the charges must be distributed uniformly around the conductors, and the electric lines must point radially outward. We can find the intensity $\mathcal{E}_D$ in the dielectric, as a function of radius R, by applying Gauss's relation to an imaginary cylinder of 1-m length and radius R, which will enclose total charge $+q_E$. Since the only electric lines penetrating this imaginary cylinder are at its outer surface, of area $2\pi R$, Gauss's relation gives

$$\mathcal{E}_D \cdot 2\pi R = q_E/\varepsilon_0, \quad \text{or} \quad \mathcal{E}_D = (1/2\pi\varepsilon_0)\, q_E/R. \tag{i}$$

The difference of potential between the conductors, which is the integral of $\mathcal{E}_D\, dR$ from R_1 to R_2, is

$$V = \int_{R_1}^{R_2} \mathcal{E}_D\, dR = \int_{R_1}^{R_2} \frac{1}{2\pi\varepsilon_0} \frac{q_E}{R}\, dR.$$

Hence

$$2\pi\varepsilon_0\, V = q_E \int_{R_1}^{R_2} \frac{dR}{R} = q_E \log_e R \Big|_{R_1}^{R_2} = q_E [\log_e R_2 - \log_e R_1] = q_E \log_e(R_2/R_1).$$

Therefore

$$q_E = \frac{2\pi\varepsilon_0\, V}{\log_e(R_2/R_1)}.$$

By substituting this expression for q_E in (i), we find that the relation

$$\mathcal{E}_D = \frac{V}{\log_e(R_2/R_1)} \frac{1}{R} \quad \text{(ii)}$$

gives the electric intensity as a function of radius R in terms of the difference of potential V. The maximum intensity occurs for the smallest R, $R = R_1$, at the surface of the inner conductor. We desire to determine the value of V for which this maximum intensity is 400 kV/cm.

Substitution of $\mathcal{E}_D = 400$ kV/cm, $R_2 = 0.75$ cm, $R = R_1 = 0.25$ cm, in (ii) gives for the breakdown voltage:

$$V = (400 \text{ kV/cm})(0.25 \text{ cm}) \log_e 3 = (400 \times 0.25 \times 1.10) \text{ kV} = 110 \text{ kV}.$$

Note that the breakdown voltage is determined entirely by the dielectric *strength* and the geometry, and is independent of the dielectric *constant*.

The action of a well-grounded and sharply pointed *lightning rod* is primarily to discharge quietly into the air the charge that is induced in the surrounding earth by a charged thundercloud above. The electric intensity in the vicinity of the sharp point is high enough to break down the immediately surrounding air, and the continual discharge prevents building up sufficient voltage over the whole cloud-to-ground path to break down this long path and give rise to a stroke of lightning at this location.

PROBLEMS

1. A capacitor is made of two parallel plates 50 cm square and 0.1 cm apart in air. (*a*) What is its capacitance in farads and in nF? (*b*) What is the charge on each plate when the potential difference is 90 V? Ans: (*a*) 2.20×10^{-9} F, 2.20 nF; (*b*) 1.98×10^{-7} C.

2. What is the maximum capacitance of a radio tuning capacitor consisting of 13 fixed plates and 12 movable plates if the effective area of each plate in the interleaved position is 40 cm² and the air gaps between fixed and movable plates are 1.8 mm? Express your answer in pF?

3. The capacitor of Prob. 2 is charged with a 72-V battery when it has its maximum capacitance. The battery is disconnected, leaving the capacitor charged. The knob is then turned to reduce the effective area to 10 cm². What then is the charge on the capacitor and its voltage? Ans: 34.0 nC; 144 V.

4. A capacitor consists of two large metal plates 0.5 mm apart in air. When it is connected to a 1000-V battery, each plate becomes charged with 42 μC. What is its capacitance in pF?

5. A 100-plate capacitor has plates 50 cm × 40 cm separated by glass plates 1.5 mm thick of dielectric constant 7.0. Find the capacitance in μF. Ans: 0.818 μF.

6. A capacitor is formed from 40 parallel metal plates 30 cm × 30 cm separated by glass sheets ($K = 7.0$) 2.0 mm thick. What is its capacitance in μF?

7. A capacitor with air dielectric ($K = 1.00$) is charged by connecting it across a battery. A meter measures the charge that flows to one of the capacitor plates as 300 μC. Without disconnecting the battery, the capacitor is completely immersed in an insulating oil, and an *additional* charge of 450 μC flows to the

plate. What is the dielectric constant of the oil?
Ans: 2.5.

8. A capacitor with air dielectric ($K=1.00$) is charged by connecting it across a 100-V battery. Without disconnecting the battery, the capacitor is completely immersed in an insulating oil of $K=3.00$. As a result of immersion, the charge on each plate *increases* in magnitude by 300 μC. What is the capacitance with air dielectric?

9. The small capacitors of fixed capacitance used throughout radio sets and other electronic apparatus are frequently constructed as in the figure. An approximate value of their capacitance can be obtained by using the parallel-plate formula, since the dielectric is very thin compared with the radius of curvature throughout most of the capacitor. Consider such a capacitor rolled from two strips of paper and two strips of aluminum foil. Each strip of foil is 1.0 inch wide. If the foil is 0.0005 inch thick and the paper 0.001 inch thick, with $K=2$, compute the approximate length of strips required for a 0.25-μF capacitor, and the approximate outside diameter of the complete roll.
Ans: 278 i; 1.03 i.

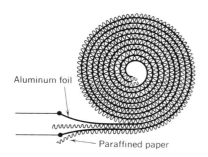

Problems 9-10

10. The type of capacitor described in Prob. 9 has recently been improved by using a strip of 'Mylar' plastic only 0.0005 inch thick with dielectric constant $K=3.8$, in place of paraffined paper. Repeat Prob. 9 using a foil strip 1 inch wide and Mylar dielectric.

11. In making an electrolytic capacitor, the aluminum sheet is *etched* before being oxidized. This doubles its effective area. The oxide has a thickness of 2×10^{-5} i and a dielectric constant of 10, and both sides of the sheet are oxidized and in contact with the acid.

Compare the capacitance of an electrolytic capacitor made from a sheet of aluminum of area A with that of a capacitor with paper dielectric 0.004 i thick each of whose plates has area A.
Ans: 4000 : 1.

12. Find the size of square plates required for a two-plate 1-μF capacitor for the case in which a 1-mm layer of barium titanate with $K=1400$ is the dielectric. Compare with the size for glass 1 mm thick, with $K=5$.

13. What is the energy in a 4-μF capacitor charged to 1200 V?
Ans: 2.88 J.

14. A 50-μF capacitor is charged to 200 V. Calculate the stored energy.

15. A capacitor has a capacitance of 1.5 μF when its plates are separated by a layer of air. It is charged to 1200 V by means of a battery. (*a*) Find the charge on the plates and the energy stored in the capacitor. (*b*) The charged capacitor is first disconnected from the battery and then immersed in an oil having a dielectric constant of 3. What is the difference of potential between the plates? What is the energy of the capacitor? What is the source of the energy change?
Ans: (*a*) 1800 μC, 1.08 J; (*b*) 400 V, 0.36 J.

16. A capacitor consists of two large parallel plates 0.5 mm apart in air. Its capacitance is 80 μF. It is charged to 500 V by means of a battery. (*a*) What is its energy? (*b*) The charged capacitor is disconnected from the battery. The plates are so mounted that they will remain parallel and insulated when a glass rod is used to push them 4 mm apart. What is the energy of the capacitor after this separation is made? What is the source of the extra energy? (*c*) What is the difference of potential between the plates after they have been separated?

17. Three capacitors, of respective capacitances 2, 3, and 6 μF, are connected in *series* across a 120-V battery. Find (*a*) the resultant capacitance; (*b*) the charge that flows from the battery, using the capacitance obtained in (*a*); (*c*) the total energy in joules stored in the capacitor combination, using the results of (*a*) or (*b*); (*d*) the voltage across each capacitor, checking the total; (*e*) the energy of each capacitor, comparing the total with (*c*).
Ans: (*a*) 1 μF; (*b*) 120 μC; (*c*) 7200 μJ; (*d*) 60, 40, 20 V; (*e*) 3600, 2400, 1200 μJ.

18. Three capacitors of respective capacitances

2, 3, and 6 μF are connected in *parallel* across a 120-V battery. Find the same quantities as in Prob. 17.

19. If a 1-μF capacitor charged to 200 V and a 2-μF capacitor charged to 400 V are connected in parallel, + plate to + plate, find the resulting difference of potential, the charge on each capacitor, and the loss of energy.
    Ans: 333 V; 333, 667 μC; 0.0133 J.

20. (*a*) Two capacitors, of 3-μF and 6-μF capacitance, are connected in series and charged from a 300-V battery. What are the charge, voltage, and energy for each capacitor? (*b*) If these two capacitors are now disconnected without discharging and are then connected to each other, + plate to + plate and − to −, what are the new values of charge, voltage, and energy for each capacitor?

21. Capacitors *A*, *B*, and *C* of respective capacitances 4, 3, and 2 μF are connected as shown in the figure. Before the switch is closed, *A* is charged to a potential difference of 300 V, *B* and *C* are uncharged. What will be the charge and voltage of each capacitor after the switch is closed?
    Ans: $V_A = 231$ V, $V_B = 92.4$ V, $V_C = 138$ V; $Q_A = 924$ μC, $Q_B = Q_C = 277$ μC.

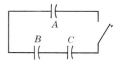

Problems 21-22

22. Capacitors *A*, *B*, and *C* of respective capacitances 4, 3, and 2 μF are connected as in the figure. Before the switch is closed, *B* and *C* are each independently charged to 400 V, the left-hand plate being positive in each case; *A* is uncharged. What will be the charge and voltage of each capacitor after the switch is closed?

23. If a 1-μF capacitor charged to 200 V and a 2-μF capacitor charged to 400 V are connected, + plate of each to − plate of the other, find the resulting difference of potential, charge on each capacitor, and loss of energy.
    Ans: 200 V; 200, 400 μC; 0.12 J.

24. At what voltage will a parallel-plate capacitor with dielectric of dry air 2.0 mm thick break down?

25. How thick must the mica dielectric be in a parallel-plate capacitor built to withstand 30 000 V if the mica has dielectric strength 900 kV/cm?   Ans: 0.334 mm.

26. To what potential can a single isolated sphere 2 cm in diameter in dry air be raised before breakdown occurs? Show that this 'breakdown potential' varies directly as the radius of the sphere and can become very low if the sphere is very small. Why does a spark 'jump' most readily from a sharp point? Why must high-tension transmission lines be made from wire of comparatively large diameter?
    Ans: 30 000 V.

27. A spherical capacitor is formed from two concentric spheres of thin metal, one 0.5 m in radius, the other 1 m, with air dielectric. The outer sphere is grounded, the inner charged with +3 μC. Find the electric intensity and the potential relative to ground at points (*a*) 0.2 m from the center, (*b*) just outside the inner sphere, (*c*) 0.8 m from the center, (*d*) just inside the outer sphere. (*e*) What is the capacitance of the capacitor? (*f*) What is the maximum charge that can be stored on the inner sphere before dielectric breakdown occurs?
    Ans: (*a*) 0 V/m, 27 000 V; (*b*) 108 000 V/m, 27 000 V; (*c*) 42 200 V/m, 6750 V; (*d*) 27 000 V/m, 0 V; (*e*) 111 pF; (*f*) 83.3 μC.

28. Consider the large spherical electrode in the Van de Graaf generator described in Prob. 31, p. 511, used in the open atmosphere. (*a*) What is the maximum potential obtainable with an electrode 2 m in diameter? More frequently, Van de Graaf generators are mounted in tanks and the situation is more nearly approximated by the arrangement described in Prob. 27, with the outer sphere grounded, except for the use of a mixture of gases at high pressure between the electrode and the tank wall. (*b*) Find the maximum difference of potential obtainable between the two spheres of Prob. 27 if the space between the two spheres is filled with a mixture of gases having a dielectric strength 10 times that of air. (*c*) How much charge is on the inner sphere when the sphere has its maximum potential? (*d*) What is the maximum energy stored in the capacitor formed by the high-voltage electrode and the surrounding tank?

29. Show that the capacitance of a spherical capacitor (Prob. 27) formed from two thin metal spheres of radii r_1 and r_2 ($r_2 > r_1$) is given by

$$C = 4\pi\varepsilon_0 r_1 r_2/(r_2 - r_1).$$

30. If the outer sphere of a spherical capacitor (Prob. 29) has a radius of 1 m, what must be the radius of the inner sphere if the capacitance is to be 1 μF? What is the capacitance of a parallel-plate condenser having the same area as the outer sphere and the same distance between plates?

31. Find an expression for the energy stored in a parallel-plate capacitor in terms of the electric intensity $\mathcal{E}$ between the plates. Considering this energy to be stored in the dielectric between the plates, find an expression for the energy per unit volume in the dielectric in terms of $\mathcal{E}$ and the dielectric constant.

Ans: $\frac{1}{2} C\mathcal{E}^2 d^2$; $\frac{1}{2} \varepsilon_0 K\mathcal{E}^2$.

32. A method of obtaining high voltage is the 'Marx circuit,' in which a switching mechanism allows a set of capacitors to be 'charged in parallel' and 'discharged in series.' Design such a circuit. What is the maximum voltage attainable with a 1000-V battery and twelve identical capacitors, each capable of withstanding 1200 V between its plates?

Ans: 12 kV.

33. What are the approximate relative volumes of capacitors made of stacked parallel plates with the following dielectrics? (*a*) Air, $K=1$, dielectric strength 30 kV/cm. (*b*) Paraffined paper, $K=2$, dielectric strength 500 kV/cm. (*c*) Mica, $K=6$, dielectric strength 900 kV/cm. The capacitors are to have the same capacitance and the same breakdown voltage. Neglect the thickness of the metal plates in your computation. Ans: 5400 : 9.7 : 1.

34. Find the capacitance of a parallel-plate capacitor if a dielectric sheet of thickness a is inserted between plates separated by a distance d, where $d \geq a$.

35. With what force per unit area do the plates of a charged parallel-plate capacitor attract each other if they are 1 mm apart in air or vacuum, and their potential difference is 1000 V? Ans: 4.42 N/m².

36. The figure illustrates Kelvin's absolute electrometer in a highly schematic manner. In principle, a circular area A of the upper plate of a parallel-plate capacitor is mechanically but not electrically isolated from the rest of the plate and attached by an insulating rod to one arm of a balance. With the capacitor uncharged, the force F required to hold the circular plate in 'alignment' is measured. With the capacitor charged to a difference of potential V, the force is measured as $F + \Delta F$, where ΔF is the additional force resulting from the attraction between the plates. Show that from the measured value of ΔF, the difference of potential V and the charge Q on the area A can be determined by the relations

$$V = \sqrt{\frac{2\,\Delta F}{\varepsilon_0 A}}\, d, \qquad Q = \sqrt{2\varepsilon_0 A\,\Delta F}.$$

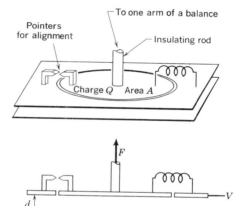

Problems 36–38

37. The absolute electrometer described in Prob. 36 has area $A = 100$ cm² and separation $d = 2$ mm. The balance is adjusted for alignment when the plates are uncharged. If the plates are now charged to a potential difference of 1000 V, what mass in grams must be added to the other pan of the balance to restore alignment? Assume standard gravity.

Ans: 1.13 g.

38. How can one employ the absolute electrometer described in Prob. 36 to isolate an accurately known quantity of charge, which can then be discharged through a convenient charge-measuring instrument (such as a ballistic galvanometer) for calibration purposes?

39. Derive relation (13) for the energy of a

parallel-plate capacitor with vacuum (or air) dielectric by starting with two plates carrying charges $\pm Q$, and almost touching, so that their difference of potential almost vanishes. Then, from the force expression (7), compute the work required to pull the plates apart to a finite separation d and verify that this work agrees with that in (13).

40. From Gauss's relation, find an expression for the capacitance of an isolated metal sphere of radius R. What is the charge on a sphere of 1 m radius when the sphere is at a potential of 1 million volts relative to ∞?

41. Referring to the dielectric constants of gases at 1 atm as listed in Table 24-1 and remembering that the difference between the listed values and unity is due to a shift of charges inside the dielectric, how would you expect the dielectric constant of a nonpolar gas like N_2 or CO_2 to depend on pressure? What value would you predict for the dielectric constants of dry air at 100 atm and carbon dioxide at 40 atm? (Note: Measured values are 1.056 and 1.060 respectively.)   Ans: 1.060, 1.040.

42. The molecules of nonpolar gases like N_2 and CO_2 actually acquire induced dipole moments $M_E = \alpha \mathcal{E}$ when placed in an electric field. The constant α is called the molecular polarizability. On the basis of the values of K listed in Table 24-1, would you say that N_2 or CO_2 has the larger polarizability? Some molecules like water vapor and ammonia have permanent dipole moments; discuss the influence of an electric field on molecules of this type (*see* Probs. 36–40, pp. 491–492).

ELECTRIC CURRENTS 25

In this chapter we make the transition from *static* electricity to *current* electricity—the flow of electric charge in wires. We shall have to introduce the additional concepts of *current* as a rate of flow of charge, *resistance* to the flow of charge, and *electromotive force* as a measure of work done on charges in order to maintain a continuous difference of potential. This chapter lays the groundwork for the detailed discussion of direct-current electric circuits and measuring instruments in the following chapter, and introduces basic concepts useful throughout the study of electricity and magnetism.

We include a brief discussion of the properties of *semiconductors* and their use as rectifiers. We present evidence that the electron is the mobile charge in metallic conductors and show how the electron is removed from metals in thermionic emission processes.

1. Current Arising from a Capacitor Discharge

The electroscope in Fig. 1 gives a measure of the difference of potential between the capacitor plates. If a short, thick wire is connected

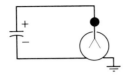

Figure 1

across the capacitor plates, the plates will be discharged almost instantaneously, as indicated by the sudden collapse of the electroscope leaves. But if a very long and very fine wire is connected across the capacitor plates, the electroscope leaves will collapse gradually over a period of seconds or minutes; the longer and finer the wire, the longer the time required for the capacitor to discharge. Although any wire connected across the capacitor plates permits electrons to flow from the negative to the positive plate, a long, fine wire appears to offer more 'resistance' to the flow than a short, thick wire. Furthermore, wires made of certain metals such as nichrome offer much more 'resistance' to the flow than wires of the same size made of other metals such as copper.

In Fig. 2, the conventional symbol for 'resistor' is used to indicate that the wire presents resistance. During the period in which the capacitor is discharging, electrons move from the negative plate through the resistor to the positive plate. *A current* is said to exist. However, by an unfortunate convention originated prior to the discovery of the electron, the *current* is said to have the direction *opposite* to that in which the electrons really move. This convention is so firmly established that no attempt is being made to change it. *We must always specify the direction of current as the direction in which positive charges would move if the charge were transferred by means of positive charges rather than by means of electrons.* Thus, the *current* in Fig. 2 is directed from the positive plate of the capacitor to the negative.

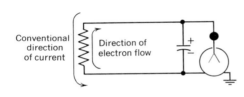

Fig. 2. Rate of decrease of voltage indicates magnitude of current.

If C is the capacitance in Fig. 2 and Q is the charge on either plate, we have, at any instant, $Q = CV$, or, differentiating, $dQ/dt = C\, dV/dt$. The rate of decrease of charge in C/s equals C times the rate of decrease of potential in V/s. But this rate of decrease of charge must represent the rate of flow of charge in C/s through the resistor. The *rate of flow of charge*, in C/s, is called the *current I* in amperes* (A).

A **current** of one **ampere** represents a flow of charge at a rate of 1 C/s past any point. The direction of the current is opposite to that in which the electrons actually move when the current represents a flow of electrons, as in the case of a current in a wire.

Thus, $\qquad\qquad 1\text{ A} = 1 \text{ C/s}.\qquad\qquad$ **(1)**

An instrument called an *ammeter* may be connected into the circuit to measure the current in amperes. We shall discuss the construction of such an instrument in the next chapter.

---

*After ANDRÉ MARIE AMPÈRE (1775–1836), French physicist who formulated the fundamental principles of the magnetic effects of electric currents.

2. Constant Currents

The currents considered in the preceding section were *transient currents* in that charge flowed only during the period required for the capacitor to discharge. If it is desired to maintain a *constant current* for an indefinite period, some means must be found for keeping the capacitor continuously charged, that is, for renewing the supply of electrons on the negative plate as fast as they flow around through the wire. Energy must be supplied to move these electrons from the positive to the negative plate through the difference of potential existing between these plates. This energy could be supplied and the potential difference maintained by an electrostatic generator of some type, but there are more practical and convenient means. In discussing this question, *let us agree to ignore our knowledge that in most cases it is negative electrons, rather than positive charges, that move, and speak as if it were really positive charges that move.* This procedure will lead to no essential error and will avoid much confusion because it is consistent with the conventional current direction.

What we need, then, is a 'charge pump' that will pump charge up the 'potential hill' from the negative plate of the capacitor to the positive plate, as in Fig. 3, as fast as the charge runs back downhill through the resistor. This charge pump must do work equal to V on each unit of charge transferred, if a difference of potential V is to be maintained. If the current is I, charge must be pumped at the rate I, in C/s, and the power that must be supplied is IV, in J/s or watts. In Fig. 3, the pump must supply energy at the rate

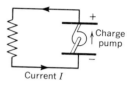

Fig. 3. In order to maintain a continuous current, a 'charge pump' is required.

$$P = IV. \qquad \begin{Bmatrix} P \text{ in W} \\ I \text{ in A} \\ V \text{ in V} \end{Bmatrix} \quad (2)$$

This energy must be supplied to the charges by some energy source, since the charges are being moved *against* the electric forces; the energy reappears as heat in the resistor as the charges flow back down the potential hill through the resistor, *in the direction of* the electric forces.

A 'charge pump,' which converts some other form of energy to electric potential energy by moving charges against an electric field is called a *source of electromotive force*. The first source of *electromotive force* (EMF) suitable for the production of continuous currents was discovered by Volta in 1799 and is called the galvanic pile or voltaic *battery*. This source was the prototype of our modern dry cells and storage batteries, which obtain their power from stored *chemical energy*. A second type of source of EMF, based on the discovery of electromagnetic induction in 1831, converts *mechanical energy* into electric energy. The modern electric generator is of this type. These sources of EMF are illustrated in the circuits of Fig. 4, where of course we can now dispense with the capacitor plates and hook our wires directly to the plates of the battery or to the terminals of the generator.

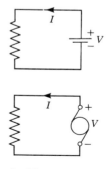

Fig. 4. The most common sources of constant EMF ('charge pumps') are the electric battery and the direct-current electric generator.

The operation of sources of electromotive force will be discussed in detail in later chapters. For the present, we need only understand that EMF is measured in volts and that:

> A **source of EMF** of one **volt** is a source that does one joule of work on each coulomb of charge that passes through it from the low potential side to the high.

To maintain current I, a source of EMF V must furnish power at the rate given by (2).

3. Resistance; Ohm's Law

Whenever there is a difference of potential (voltage) between the ends of a wire, there is a current through it. The current is caused by the electric field (potential gradient) in the wire. If the voltage is continuously maintained by a source of EMF, the current is continuous. The dependence of current on voltage is expressed by *Ohm's law:**

> OHM'S LAW: *For a given conductor at a given temperature, the current is directly proportional to the difference of potential between the ends of the conductor.*

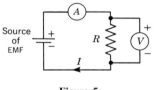

Figure 5

If, in Fig. 5, V is the difference of potential between the ends of a conductor R as measured by a voltmeter V, and I is the current in amperes through the conductor as measured by the ammeter A, and if the temperature is kept constant, Ohm's law states that the current is proportional to the voltage; that is, $I \propto V$. The proportionality constant is called the *conductance G*, and we may write $I = GV$. More frequently used is the reciprocal of the conductance, $R = 1/G$, which is called the *resistance*. In terms of the resistance, we have

$$I = V/R; \qquad V = RI. \tag{3}$$

We see that the greater the resistance of a conductor, the smaller is the current that results from a given voltage. The unit of resistance is the *ohm* (abbreviated by the Greek letter Ω).

> The **resistance** of a conductor is one **ohm** if a difference of potential of one volt is required between its ends to cause a current of one ampere through the conductor.

Thus the ohm is the derived unit,

$$1\ \Omega = 1\ \text{V/A}. \tag{4}$$

---

*Discovered experimentally by GEORG SIMON OHM, a German physicist, in 1826.

The unit of conductance, 1 A/V, which is the reciprocal of the ohm, is called the *mho*.

The value of the resistance R depends on the size, shape, material, and temperature of the conductor, but within wide limits does not depend on the voltage V. Within these wide limits, Ohm's law is obeyed by metallic conductors, but it is not obeyed by nonmetallic conductors, as we shall discuss in Sec. 6.

Since the electric forces do work of amount V on each coulomb of charge passing through the wire, and since all of this work appears as heat in the wire, *the rate of heat generation* is $P = IV$, as in (2). From (3), this expression for power can be written in several convenient forms:

$$P = IV = I^2 R = V^2/R. \qquad \begin{Bmatrix} P \text{ in W} \\ I \text{ in A} \\ V \text{ in V} \\ R \text{ in } \Omega \end{Bmatrix} \qquad (5)$$

The amount of heat (in J) generated in time t can then be written as

$$W = Pt = IVt = I^2 Rt = V^2 t/R. \qquad \begin{Bmatrix} W \text{ in J} \\ P \text{ in W} \\ t \text{ in s} \end{Bmatrix} \qquad (6)$$

Electric energy is usually measured and paid for in kilowatt-hours (kWh) rather than in joules (W·s). We defined the kWh on p. 120.

Example. *A 600-W radiant heater is designed for operation at* 115 V. *What current is drawn by the heater? What is the resistance of the heating coil? How many* BTU*'s are generated in one hour?*

From (5), we find the heater current $I = 600 \text{ W}/115 \text{ V} = 5.22$ A. Then, from (3), the heating-coil resistance $R = 115 \text{ V}/5.22 \text{ A} = 22.0 \, \Omega$. From (6), the heat generated in one hour is $W = 600 \text{ W} \times 3600 \text{ s} = 2\,160\,000$ J $= 2\,160\,000 \; (\frac{1}{1055} \text{ BTU}) = 2050$ BTU.

4. Resistors in Parallel and in Series

A combination of resistances can always be replaced by a single equivalent resistance. Suppose that between two points in a circuit, such as A and B in Fig. 6, three resistors are connected in parallel. The voltage across each resistor is the same, V_{AB}. Hence, $I_1 = V_{AB}/R_1$, $I_2 = V_{AB}/R_2$, $I_3 = V_{AB}/R_3$. The current in the external circuit is the sum of these:

$$I = I_1 + I_2 + I_3 = V_{AB}\left(\frac{1}{R_1} + \frac{1}{R_2} + \frac{1}{R_3}\right).$$

The ratio V_{AB}/I is defined as the effective resistance R of the parallel circuit, so that

$$\frac{I}{V_{AB}} = \frac{1}{R} = \frac{1}{R_1} + \frac{1}{R_2} + \frac{1}{R_3}. \qquad (7)$$

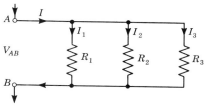

Fig. 6. Resistors in parallel.

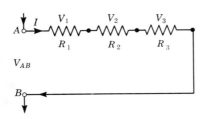

Fig. 7. Resistors in series.

The value of R computed from (7) gives the resistance of the single resistor that will carry the same current with the same voltage as the parallel combination, and will generate the same heat. For any number N of resistors in parallel, the effective resistance is similarly given by

$$\frac{1}{R}=\frac{1}{R_1}+\frac{1}{R_2}+\cdots+\frac{1}{R_N}. \quad \binom{\text{resistors}}{\text{in parallel}} \quad (8)$$

In the same way, the reader can readily demonstrate that one can replace a number of resistors in series as in Fig. 7 by a single equivalent resistance

$$R=R_1+R_2+\cdots+R_N. \quad \binom{\text{resistors}}{\text{in series}} \quad (9)$$

5. Resistivity

Resistivity is the property of the material of which a conductor is composed that enables us to compute the resistance of a conductor of given geometry. Before defining resistivity, let us consider the dependence of resistance on the length and cross-sectional area of a wire of a given material at a given temperature.

If we double the length of a wire, we are effectively putting two equal resistances in series, so that in accordance with (9), we double the resistance. If we triple the length of a wire, we triple the resistance, etc. In general, we see that the resistance varies in direct proportion to the length of a wire:

$$R \propto l \quad (10)$$

If we double the area of a wire, we are effectively putting two equal resistors in parallel so that, in accordance with (8), the resistance is halved. If we triple the area, we are putting three equal resistors in parallel, so that the resistance is cut to one-third, etc. In general we see that the resistance of a wire is inversely proportional to its cross-sectional area:

$$R \propto 1/A. \quad (11)$$

Combining (10) and (11), we have $R \propto l/A$. The proportionality constant, usually written as ρ, is called the *resistivity*:

$$R = \rho \, l/A. \quad (12)$$

The resistivity depends only on the material and the temperature.

In the metric system, l and A are expressed in m and m^2; and since

$$\rho = RA/l, \quad (13)$$

the unit in which ρ is expressed is the $\Omega \cdot$m. Table 25-1 gives the resistivities of various metals and commercial alloys.

In the periodic table (Appendix, p. vi), there is a periodicity of electrical resistivity similar to the periodicity of other properties of the elements. In particular, the three elements of lowest resistivity are Ag, Cu, Au—three metals in the same column of the periodic table—; next comes Al.

The reciprocal of resistivity is called *conductivity*.

We turn now to a consideration of the dependence of resistivity (or resistance) on temperature. The resistivity of a pure metal increases rapidly with temperature, the temperature effect being so pronounced that it must be taken into account in most applications. Figure 8 shows, for example, how the resistivity of copper varies with temperature. The reason for the rapid variation with temperature is that the whole resistance of a metal to electron flow arises from the thermal agitation of the

Table 25-1 RESISTIVITY AND TEMPERATURE COEFFICIENT OF COMMON METALS AND ALLOYS

Alloy	Specification	Resistivity ρ at 20° C		Temperature coefficient α_{20} (per C deg)
		($\Omega \cdot$m)	($\Omega \cdot$circ mil/f)[a]	
Copper	Pure (99.999%)	1.673×10^{-8}	10.06	4.05×10^{-3}
	International standard-annealed ($\sim$99.91%)	1.724×10^{-8}	10.37	3.93×10^{-3}
	Hard-drawn	1.77×10^{-8}	10.6	3.8×10^{-3}
Aluminum	Pure (99.96%)	2.655×10^{-8}	15.97	4.03×10^{-3}
	AIEE standard hard-drawn (99.5%)	2.828×10^{-8}	17.01	4.03×10^{-3}
Iron	Pure (99.99%)	9.71×10^{-8}	58.4	5.76×10^{-3}
	Commercial wire	$11-13 \times 10^{-8}$	66-78	5.5×10^{-3}
	Cast (typical)	60×10^{-8}	360	5×10^{-3}
Steel	Rail	$14-22 \times 10^{-8}$	84-130	4×10^{-3}
Nichrome	60% Ni, 15% Cr, 25% Fe	112×10^{-8}	675	0.16×10^{-3}
Manganin	4% Ni, 12% Mn, 84% Cu	48×10^{-8}	290	$<0.01 \times 10^{-3}$
Mercury	Pure	95.8×10^{-8}	—	0.88×10^{-3}
German silver	18% Ni, 65% Cu, 17% Zn	29×10^{-8}	175	0.27×10^{-3}
Constantan	45% Ni, 55% Cu	49×10^{-8}	294	$<0.01 \times 10^{-3}$
Silver	Pure	1.59×10^{-8}	9.55	3.75×10^{-3}
Gold	Pure	2.44×10^{-8}	14.7	3.4×10^{-3}
Tungsten	Pure	5.50×10^{-8}	33.1	4.7×10^{-3}
Platinum	Pure (99.99%)	9.83×10^{-8}	59	3.64×10^{-3}

[a] In a system convenient for engineering calculations, particularly with round wires, l in (12) is expressed in feet and A in circular mils. A circular mil is defined as the area of a circle one mil (0.001 inch) in diameter. The area of a circular wire of diameter d, in mils, is $A = d^2$, in circular mils. In this system, the unit of ρ is $\Omega \cdot$circ mil/f.

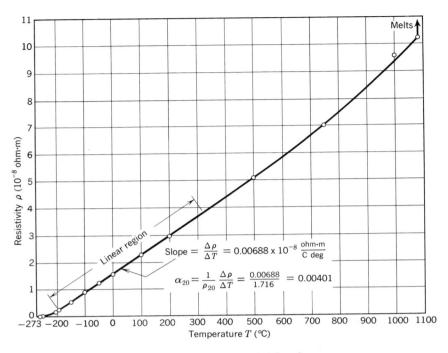

Fig. 8. Typical data on the resistivity of copper.

metallic ions that comprise the lattice structure of the metal. This agitation increases rapidly with increasing temperature. The resistance of all metals approaches zero as the absolute temperature approaches zero. In fact, in the phenomenon known as *superconductivity,* the resistance of a metal drops entirely to zero at a temperature still a few degrees above absolute zero, so that when a current is once started in a metal ring below such temperature it continues indefinitely without necessity of an EMF to maintain it and without causing any heating of the material. The presence of such a persistent current can be detected by means of the magnetic field it sets up.

Over a wide range of the temperatures of practical interest, the resistivity of a metal can be represented by a linear curve. The data of Fig. 8 show this linearity to be very accurate for copper for temperatures from $-200°$ C to $+300°$ C, and to hold approximately for still higher temperatures. It is customary to use the resistivity ρ_{20} at $20°$ C as a reference value, since $20°$ C$\sim 68°$ F is usually taken as the specification of 'room temperature.' The resistivity ρ_T at temperature T (Celsius) is written in the form $\rho_T = \rho_{20} + \text{const} \times (T - 20° \text{ C})$, which is the equation of a straight line for ρ_T as a function of T. The constant is usually written as $\alpha_{20} \rho_{20}$, so that the equation takes the form

$$\rho_T = \rho_{20} + \alpha_{20}\,\rho_{20}\,(T - 20°\,\text{C}),\qquad(14)$$

or

$$\rho_T = \rho_{20}\,[1 + \alpha_{20}\,(T - 20°\,\text{C})].\qquad(15)$$

The constant α_{20} is called the *temperature coefficient of resistance* at 20° C. Since we can write (14) in the form

$$\alpha_{20} = \left(\frac{\rho_T - \rho_{20}}{\rho_{20}}\right) \bigg/ (T - 20°\,\text{C}),$$

we see that α_{20} is the *relative* change in resistivity per C deg change in temperature, and that α_{20} is independent of the units used for resistivity. As seen in Table 25-1, α_{20} has a magnitude of about 0.004 per C deg for most metals. This value corresponds to an increase in resistivity of about 0.4 percent per C deg rise in temperature, or 100 percent for a 250 C deg rise. Such an increase is very large compared, for example, with linear expansion, where the length increases only about 0.001 percent per C deg.

Because changes in physical dimensions are only about $1/400$ the change in resistivity with temperature, we can neglect changes in dimensions in discussing the change in resistance of a conductor of given size and shape. If we multiply (15) on both sides by l/A, where l is the length and A the cross-sectional area of a conductor, we obtain, using (12),

$$R_T = R_{20}\,[1 + \alpha_{20}\,(T - 20°\,\text{C})].\qquad(16)$$

This equation expresses the resistance of any conductor at temperature T in terms of its resistance at temperature 20° C.

Since the change in resistance of a conductor as the temperature changes is large enough to be easily measured, the resistance of a conductor can be used as a thermometer—the so-called *resistance thermometer*. Platinum is used for high-precision resistance thermometers, while tungsten, because of its very high melting point, can be used to measure very high temperatures.

For construction of laboratory apparatus, such as resistance boxes, it is desirable to have a material of very low temperature coefficient so that the resistance will not change when the resistor is heated by current through it. This requirement has been met by certain alloys that have been developed specifically for the purpose—notably manganin and constantan (*see* Table 25-1).

Example. *A standard-annealed copper wire has resistance of 2 ohms per meter length at 20° C. What must be the electric intensity in the wire if the current is 15 A? What electric intensity is required to produce the same current in the wire at 150° C?*

At 20° C, a voltage $IR = (15\,\text{A})(2\,\Omega) = 30$ V is required per meter length. Electric intensity equals voltage gradient so the electric intensity is 30 V/m.

The resistance of the same meter length of wire at 150° C is found by substituting $R_{20}=2\ \Omega$, $\alpha_{20}=0.003\ 93/\text{C deg}$, $T=150°$ C in (16):

$$R_{150}=2\ \Omega\ [1+0.003\ 93\ (130)]=3.02\ \Omega.$$

The voltage gradient for a 15-A current thus increases to $(15)(3.02)$ V/m = 45.3 V/m.

6. Resistivity of Nonmetals

Many nonmetallic materials are called *insulators* because their resistivity is enormous compared with that of a metal; it is greater by a factor of about 10^{20}, and for most practical purposes can be considered infinite. Typical values of resistivity are

Bakelite: $10^{10}\ \Omega\cdot\text{m}$, Mica: $10^{15}\ \Omega\cdot\text{m}$, Shellac: $10^{14}\ \Omega\cdot\text{m}$,
Glass: $10^{12}\ \Omega\cdot\text{m}$, Rubber: $10^{16}\ \Omega\cdot\text{m}$, Sulfur: $10^{15}\ \Omega\cdot\text{m}$.

These numbers are intended only to give the correct order of magnitude, because the resistivity varies from sample to sample, decreases with increasing potential gradient so that Ohm's law is not obeyed, and decreases very sharply with increasing temperature.

The above values apply to current conducted through the body of the material. For exposed insulators, the leakage of current over the surface may be much greater than the current through the body of the material, particularly when the surface is wet or dusty.

Carbon is one nonmetal that is a fairly good electrical conductor. Its resistivity is about $3500\times10^{-8}\ \Omega\cdot\text{m}$ at 20° C, about 2000 times that of copper. Unlike metals, it has a negative temperature coefficient of resistance of about -0.0005 per C deg. Its resistivity drops to $2700\times10^{-8}\ \Omega\cdot\text{m}$ at 500° C, 2100×10^{-8} at 1000° C, and 1100×10^{-8} at 2000° C.

7. Semiconductors, Rectifiers

There is a very important class of crystals called *semiconductors*. This class includes the elements germanium, silicon, and selenium, as well as certain compounds such as CuO, ZnO, PbS, InSb, InP, GaAs, and SiC. Semiconductors have no free electrons at very low temperatures since the atoms are bound together by covalent bonds that involve all the valence electrons. However, the energy required to free an electron is comparatively low (0.7 eV for germanium as compared with 7 eV for diamond, which has exactly the same tetrahedral crystal structure with four valence bonds per atom), so that at room temperature some of the electrons are freed by thermal agitation and the semiconductor has a feeble conductivity. The conductivity increases approximately exponentially with temperature. For example, CuO has a resistivity of about $1 \times 10^3 \, \Omega \cdot m$ at 20° C but only one-tenth this resistivity at 70° C. Since at ordinary temperatures the number of free electrons is very low compared with the number in a metal, the resistivity of a semiconductor is still very high compared with that of a metal. For example, Ge has 2.5×10^{13} free electrons per cm^3 at room temperature, whereas there are 5×10^{22} atoms per cm^3. In the case of a metal there are one or more free electrons per atom, whereas in Ge at room temperature there is only one free electron per 2×10^9 atoms.

It is important to note that, unlike a metal, a semiconductor has both *positive* and *negative* carriers of electricity. A missing binding electron leaves a 'hole'—a region of net positive charge. The hole is mobile and can readily move from atom to atom by being filled by a neighboring electron. The hole acts like a *positive* free electronic charge and contributes to the conductivity by moving in an electric field. In the absence of an electric field, free carriers of both types move randomly at thermal velocities. When an electric field is applied, there is a drift of positive carriers (holes) in the direction of the field, and a drift of negative carriers (electrons) in the opposite direction.

Semiconductors have been used for many decades to form *rectifiers*. Examples are the copper-oxide rectifier and the galena (PbS) crystal and 'cat's whisker' used in the 1920's for radio-frequency rectification. In the copper-oxide rectifier (Fig. 9), a large current of electrons can pass from the copper metal into the CuO semiconductor, greatly supplementing the supply of free electrons in the semiconductor and hence permitting the passage of current through it. In the reverse direction the electron current is necessarily very small, since there are very few free electrons in the semiconductor and a supply cannot enter from the lead sheet because contact conditions at this junction do not happen to permit their entry.

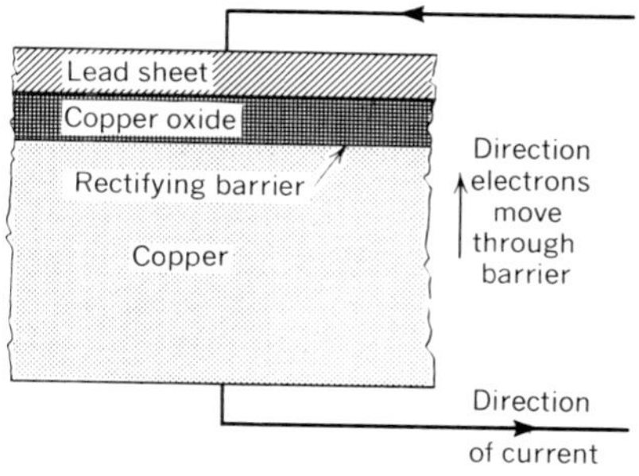

However, the very significant recent advances in the development of semiconductor circuit elements, *diodes* (rectifiers) and *transistors* (see next chapter) have resulted from the controlled 'doping' of semiconductors with impurity elements during crystal growth in order to greatly increase their supply of either negative carriers (free electrons) or positive carriers (holes). We shall consider, as an example, the doping of a Ge crystal.

Fig. 9. The copper-oxide rectifier.

If phosphorus or arsenic, with *five* valence electrons (see Periodic Table), is added during crystal growth as an impurity to germanium, with four valence electrons, each atom of the impurity will substitute for one germanium atom in the tetrahedral lattice but will use only four of its valence electrons to form covalent bonds to neighboring Ge atoms.

The fifth electron will become *free*. Either phosphorus or arsenic is called a *'donor'* for germanium—it donates free electrons; it does not add mobile holes because even though the impurity atom becomes a positive ion there is no vacant valence-bond position into which a neighboring electron can move. The positive charge left at the location of the P^+ or As^+ ion is fixed and not free to move.

Addition of *donor* atoms creates what is called an n-type semiconductor because most of its conductivity is a result of *negative* carriers. By the addition of impurities, this conductivity can be made orders of magnitude greater than the pure-semiconductor conductivity.

If boron or indium, with three valence electrons, is added as an impurity to germanium, with four valence electrons, each atom of the impurity will substitute for one germanium atom but will form valence bonds with only three of the four neighboring germanium atoms. A *hole* will be left; the hole is mobile because it can be filled by an electron from a neighboring atom and can move about as a positive carrier. Either boron or indium is called an *'acceptor'* for germanium—it will accept a neighboring electron and thus cause a movement of a hole.

Addition of acceptor atoms creates what is called a p-type semiconductor because most of its conductivity is a result of *positive* carriers. This conductivity also can be made orders of magnitude greater than the pure-semiconductor conductivity.

Now let us see how rectification takes place at a p-n junction. Figure 10 shows the junction between two semiconductor crystals, one p-type and the other n-type—or, for example, a single crystal of Ge that has been doped with P during the growth of the left end and with B during the growth of the right end. Substantially all of the conduction of the left end is by negative electrons; substantially all of that of the right end is by positive holes. If an electric field is applied from right to left as in Fig. 10(a), the electrons move to the right and the holes to the left, neutralizing each other at the junction (the electrons fill the holes). New electrons move in from the left lead and electrons move out the right lead to form new holes. The diode conducts electricity. With the opposite field direction in (b), both types of carriers are forced away from the junction and no current flows. The diode acts as a *rectifier*. Note that in (b) charge separation will occur in the neighborhood of the junction,

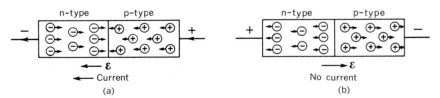

Fig. 10. Rectification at a p-n junction.

leaving the left side of the junction positively charged and the right side negative, until all of the potential difference appears across the junction.

8. The Electron; Thermionic Emission

We have seen that metallic conduction can be satisfactorily explained as the movement of 'free' electrons within the metal. The first indication of the existence of these particles came with the discovery of the phenomenon of *thermionic emission,* in which electrons actually escape from the surface of metals heated in a vacuum.

Thermionic emission was first observed in about 1880 by THOMAS ALVA EDISON in the course of experiments concerned with the development of the incandescent lamp. Edison found that if a plate P were sealed into an evacuated lamp in the manner indicated in Fig. 11, a current could pass from this plate to the heated filament F but could not be made to flow in the reverse direction. In the arrangement shown in this figure, battery A of small EMF (commonly called the *A-battery*) is used to heat the filament to incandescence, and battery B of large EMF (commonly called the *B-battery*) is connected between the plate and the filament. When the plate is positive with respect to the filament as in part (a) of the figure, the meter M indicates a current from the plate to the filament; when the plate is negative with respect to the filament as in part (b), there is no current from filament to plate.

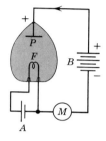

(a) Current from plate to filament

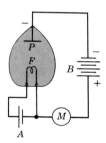

(b) No current from filament to plate

Fig. 11. Edison's discovery.

These observed effects can be explained if we assume that the current consists of electrons that are emitted from the hot filament. When the plate is positive with respect to the filament, the emitted negative electrons are pulled to the plate, and the meter indicates an electric current from the plate to the filament; when the plate is negative with respect to the filament, electrons emitted from the filament are repelled by the plate, and hence there is no current. The cold plate furnishes no electrons for the process of conduction through the tube.

Since conduction in metals takes place by movement of 'free' electrons, it might be thought that a free electron coming toward a metal surface from within the metal could readily pass through the metal surface and escape. The results of the Edison experiment indicate that appreciable numbers of electrons can ordinarily escape from a metal *only when the metal is hot*. Electrons within a metal can be regarded as 'free' so far as electric current *within* the metal is concerned, but they are confronted by a 'barrier' at the surface of the metal. In order to escape from the metal, an electron must have sufficient energy to pass through the surface barrier. The minimum energy an electron must have in order to escape may be called the 'height of the potential barrier' at the metal surface. If an electron has just the minimum energy necessary to pass through the barrier, it will have no kinetic energy when it reaches the outside of the metal; electrons having more than the minimum energy will have appreciable kinetic energies after penetrating the barrier. The escape of electrons from a metal is quite analogous to the escape of the molecules of a liquid in evaporation, where there are also barrier forces tending to prevent the escape (*see* p. 362).

Later experiments (*see* Chapter 28) showed that the negative particles emitted by hot filaments are *all* characterized by a definite ratio of charge to mass. Since the electrons each have a definite charge, -1 e, they each have a definite mass. These unique values of charge and mass are the *distinguishing* characteristics of electrons. Electrons obtained in any manner whatsoever (cold emission, photoelectric effect, Compton effect, ionization, nuclear reactions, etc.) have the same values of charge and mass and are experimentally identical with those obtained in thermionic emission.

PROBLEMS

1. What is the magnitude of the constant current in a wire if it is found that 1200 coulombs of charge flow through the wire in 4 min?
Ans: 5 A.

2. An electroplating tank requires 400 000 coulombs per hour. What current must be fed to the tank?

3. How many coulombs are delivered by a storage battery in 24 h if it is supplying current at the rate of 1.5 A?   Ans: 1.29×10^5 C.

4. There are approximately 10^{29} free electrons per cubic meter of copper (one per atom). For a current of 10 A in a wire of 1 mm^2 cross section, what is the *average* speed of 'drift' of the free electrons along the wire?

5. Within limits, the total amount of charge that a storage battery will deliver without recharging is independent of the rate at which the charge is delivered. A typical automobile battery is guaranteed to deliver 120 ampere-hours, which means that it will deliver 1 A for 120 h, or 2 A for 60 h, and so on. (*a*) How many coulombs is it guaranteed to deliver? (*b*) How many amperes will it deliver for 5 h? (*c*) If the starter draws 500 A, how long would a fresh battery drive the starter?
Ans: (*a*) 4.32×10^5 C; (*b*) 24 A; (*c*) 14.4 min.

6. Why is a coulomb sometimes called an *ampere-second?* How many coulombs are there in an *ampere-hour?* If a storage battery is supplying 15 A and has an EMF of 6 V, what power is it delivering? If a storage battery is rated as a '120-ampere-hour, 6-volt battery,' how much energy will it deliver without recharging?

7. How many electrons per second cross any section of a wire carrying a current of 0.5 ampere?   Ans: 3.12×10^{18}.

8. A 120-V generator delivers 20 kW to an electric furnace. What current is the generator supplying?

9. What voltage should a generator have if it is to supply an 80-kW furnace with a current of 200 A?   Ans: 400 V.

10. What current must a 12-V battery deliver if it is to supply 2.5 hp to an automobile starter?

11. What horsepower must a steam engine have if it is to drive a 2200-V electric generator capable of supplying 75 A, if the generator converts 90 per cent of the mechanical energy supplied to it into electrical energy?
Ans: 245 hp.

12. If a water turbine delivers 1500 hp to an

electric generator of 95 per cent efficiency, what current will the generator deliver at 1400 V?

13. The difference of potential between the ends of a wire is 36 V and the current is 2 A. What is the resistance? Ans: 18 Ω.

14. What power is expended in a 55-Ω resistor connected across a 220-V DC power line?

15. What is the resistance of a 200-W bulb in a 120-V DC power line? Ans: 72 Ω.

16. A current of 5 A flows through a resistance of 200 Ω for 1 h. How much heat is generated in joules? in kcal?

17. What is the resistance of an immersed coil of wire that heats 1000 cm³ of water from 9° C to 89° C in 10 min when 110 V are applied? Ans: 21.8 Ω.

18. It is required to generate 2 kcal of heat per minute in a resistor connected to 120 V. What must be the resistance?

19. At 4 cents per kWh, what is the cost of operating fifteen 40-W lamps for 8 h? Ans: 19.2 cents.

20. At 3 cents per kWh, what is the cost of operating a motor for 24 h if the motor delivers 5 hp and operates at 90 percent efficiency?

21. The coil of a powerful electromagnet is made of copper tubing and is cooled by water flowing through the tubing. The current is 500 A, the resistance 0.24 Ω, and 1.2 f³ of water flow through the coil per minute. Find the temperature rise of the water on the assumption that all the heat generated is carried away by the water stream. Ans: 45.5 F deg.

22. It is desired to design a 220-V water-cooled electromagnet in which the windings carry 400 A. What flow of cooling water, in f³/min, is required to carry off all the heat if the water is to enter at 40° F and leave at 160° F?

23. In Fig. 6, p. 533, let $R_1 = 1$ Ω, $R_2 = 2$ Ω, $R_3 = 6$ Ω, and $V_{AB} = 6$ V. (a) Compute R. (b) Compute I from V_{AB} and R. (c) Compute the total power from V_{AB} and R. (d) Find I_1, I_2, and I_3, and verify that the sum is I. (e) Compute the power developed in R_1, R_2, and R_3 individually, and verify that the sum agrees with (c). Ans: (a) 0.6 Ω; (b) 10 A; (c) 60 W; (d) 6, 3, 1 A; (e) 36, 18, 6 W.

24. Answer the same questions as in Prob. 23 for $R_1 = 100$ Ω, $R_2 = 20$ Ω, $R_3 = 10$ Ω, and $V_{AB} = 60$ V in Fig. 6.

25. In Fig. 7, p. 534, let $R_1 = 1$ Ω, $R_2 = 2$ Ω, $R_3 = 6$ Ω, and $V_{AB} = 24$ V. (a) Compute R. (b) Compute I from V_{AB} and R. (c) Compute the total power from V_{AB} and R. (d) Find V_1, V_2, and V_3, and verify that the sum is V_{AB}. (e) Compute the power developed in R_1, R_2 and R_3 individually, and compare the sum with (c). Ans: (a) 9 Ω; (b) 2.66 A; (c) 64 W; (d) 2.67, 5.34, 16 V; (e) 7.12, 14.2, 42.7 W.

26. Answer the same questions as in Prob. 25 for $R_1 = 100$ Ω, $R_2 = 40$ Ω, $R_3 = 20$ Ω, and $V_{AB} = 120$ V in Fig. 7.

27. Find the single resistance that is equivalent to the four shown in the figure if $R_1 = 10$ Ω, $R_2 = 5$ Ω, $R_3 = 5$ Ω, $R_4 = 10$ Ω. Ans: 4.55 Ω.

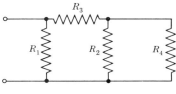

Problems 27-28

28. Find the single resistance that is equivalent to the four shown in the figure if $R_1 = 100$ Ω, $R_2 = 5$ Ω, $R_3 = 100$ Ω, $R_4 = 100$ Ω.

29. At 20° C, what is the resistance of a standard-aluminum bus bar ½ cm × 2 cm in section and 30 m long? Ans: 8.50×10^{-3} Ω.

30. At 20° C what is the resistance of an annealed-copper bus bar ½ cm × 2 cm in section and 10 m long?

31. The density at 20° C of standard-annealed copper is 8.89 g/cm³; that of hard-drawn aluminum is 2.70 g/cm³; and that of silver is 10.5 g/cm³. If 500 f of standard-annealed copper wire has diameter d_{Cu}, weight w_{Cu}, and resistance R, what are the diameters d_{Ag}, d_{Al} and the weights w_{Ag}, w_{Al} of 500 f of silver wire and 500 f of hard-drawn aluminum wire having the same resistance R?
Ans: 0.961 d_{Cu}, 1.28 d_{Cu}; 1.09 w_{Cu}, 0.499 w_{Cu}.

32. (a) What are the ratios $R_{Ag} : R_{Cu} : R_{Al}$ of the resistances of silver, annealed copper, and hard-drawn aluminum wires of the same length and the same *diameter?* Which is the best and which the poorest conductor of the three, volume for volume? (b) What are the ratios $R_{Ag}:R_{Cu}:R_{Al}$ of the resistances of silver, annealed copper, and hard-drawn aluminum wires of the same length and the same *weight?* Which is the best and which the poorest conductor of the three, weight for weight? Use the densities given in Prob. 31.

33. Determine the ratio of thermal conductivity at 20° C to electrical conductivity at 20° C, for pure Ag, Cu, Au, and Hg. The fact that this ratio is approximately constant is known as the *Wiedemann-Franz law*. Heat is conducted in a metal principally by the free electrons; hence thermal conductivity is closely related to electrical conductivity and has the same strong dependence on temperature.

34. The resistance of a conductor is 100.6 ohms at 20° C and 111.8 ohms at 50° C. Calculate the temperature coefficient of resistance of the material.

35. The resistance of the copper field coils of a generator is measured when the room temperature is 25° C and is found to be 500 ohms. What will be the resistance of these coils at the operating temperature of 90° C?
Ans: 624 ohms.

36. What is the resistance at 20° C of the tungsten filament of a lamp bulb which operates at 2000° C and 150 watts on 120 volts DC?

37. To prevent insulation damage, the temperature of the field coils of a certain type of motor should not exceed 105° C when the motor is running continuously under full load. The standard-copper field coils of a particular motor were found to have a resistance of 6.38 ohms at 20° C and 8.56 ohms during a full-load run. Did the motor meet this temperature limitation?

38. A platinum wire has a resistance of 125 ohms at 20° C. When it is placed in a furnace, its resistance is 800 ohms. What is the temperature of the furnace?

39. A capacitor is made by placing sheets of metal foil on the two sides of a glass plate 1 mm thick, of dielectric constant $K=8.0$ and resistivity $\rho=10^{12}$ ohm·m. The foil area is 1.00 m² on each side. The capacitor is charged to 2000 volts and then disconnected from the voltage source. Find the current conducted through the glass immediately after disconnection. Does this current remain constant?
Ans: 2.00 μA.

40. Derive equation (9), p. 534.

41. If the voltage drop across a 1-m length of a straight nichrome wire is 50 V, what is the magnitude of the electric intensity $\mathcal{E}$ in the wire? What is the magnitude of the acceleration of an electron in a field of this magnitude? What would be the magnitudes of the final and average velocities of an electron moving through the wire if the electron were subjected only to the field $\mathcal{E}$?
Ans: 50 V/m; 8.78×10^{12} m/s²; 4.2×10^6 m/s; 2.1×10^6 m/s.

42. We asserted in the text that the energy required to free an electron from a germanium atom is 0.7 eV and the energy required to free an electron from a carbon atom in diamond is 7 eV. Assuming that $\overline{E}_K = \frac{3}{2} kT$, find the temperatures at which the mean thermal energies of electrons would be sufficient to make them 'free' for conduction. Compare these temperatures with the melting points of germanium and diamond.

43. Recalling from Prob. 4 that the actual 'drift' speed of electrons in a conductor is very small, account for the difference between this 'drift' speed and the magnitudes of the velocities as computed in Prob. 41. What becomes of most of the energy given to a conduction electron by the applied field?

It is asserted in the text that the conduction electrons in a metal are free to move about within the metal in much the same manner as the molecules in a gas. Recalling that the mean translational kinetic energy of a molecule is $\overline{E}_K = \frac{3}{2} kT$, compute the root-mean-square velocity of an electron in a metal at 300° K. It should be noted that the 'thermal' speeds of conduction electrons is enormous as compared with the 'drift' speeds even for large currents.
Ans: 1.17×10^5 m/s.

44. Discuss the processes involved in the 'cold emission' of electrons from a metal surface from the standpoint of the forces produced at the surface by a large electric field $\mathcal{E}$ and also from the standpoint of energy considerations. In this connection, is it necessary to make any modifications of the treatment of metallic surfaces given in Chapters 22 and 23?

45. Studies of thermionic emission show that

energy of approximately 3.82 eV must be supplied to an electron in order to move it through the 'potential barrier' at the surface of silver. At what temperature would the mean kinetic energy of an electron be sufficient to allow electrons to escape from the metal?

Ans: 29 400° K.

46. If the beam current in the accelerator tube of a 6-MV Van de Graaf generator is 1 μA when protons are being accelerated, how many protons reach the target each second? If the proton beam strikes a 'thick' target, how much energy is delivered to the target each second? Note: Since the cross section of the beam is ordinarily small, all the energy is delivered to a small portion of the target and can produce large heating effects.

Ans: 6.25×10^{12}; 6 J.

47. Answer the questions in the preceding problem for a case in which the 6-MV Van de Graaf machine produces a beam current of 1 μA consisting of He^{++} ions.

26 DIRECT ELECTRIC CURRENTS

A direct current is a current produced by a source of constant EMF. Except for *transients* that occur when switches are closed or opened, a direct current is a constant current. In this chapter we consider direct-current (DC) circuits, and the instruments used in direct-current measurements. After treating 'simple circuits,' we show how Kirchhoff's rules can be applied to 'multiple-mesh circuits.' We then discuss the transient currents involved in the charge and discharge of a capacitor. This chapter closes with a brief discussion of two 'nonlinear' circuit elements, the *vacuum tube* and the *transistor*.

1. Terminal Voltages

We have defined a source of EMF, V_E, as a device that will do work per unit charge of amount V_E, in joules per coulomb of charge passing through it, deriving the energy from a mechanical, chemical, or other source. This statement regarding the work done is true if the (conventional, positive) charge passes through the source of EMF in a particular direction, from what is called the *negative terminal* to what is called the *positive terminal*. This situation is illustrated in Fig. 1, in which the source of EMF does work on charge passing through it *from the − terminal to the + terminal*. This work raises the potential (potential energy per coulomb) of the + terminal by amount V above that of the − terminal. This difference of potential V across the resistance R causes current in the external circuit from the + terminal to the − terminal, the energy furnished by the source reappearing as heat in the resistor.

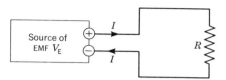

Fig. 1. A source of EMF supplies electric energy that heats a resistor.

When current I passes through a resistance R, *in either direction*, electric energy is converted into heat at the rate I^2R, in watts.

But only when current passes through a source of EMF *from the − terminal to the + terminal* is energy of some other form changed into electric energy. If, by employing a higher EMF in the external circuit, current is forced *backward* through the source of EMF as in Fig. 2, electric energy is changed into energy of the other form, and the charge passing through the source loses energy of amount V, in J/C, because it is going *down* the 'potential hill' just as it does in a resistor. *A source of* EMF *represents a reversible device for changing from electric energy to some other form of energy.* If current is sent backward through a battery, electric energy is converted into chemical energy—in a recoverable form if the battery is a storage battery, in a form mostly useless if the battery is a dry cell. If current is sent backward through a generator, electric energy is converted into mechanical energy—the generator, which is a reversible device, is acting as a *motor*.

Fig. 2. A current is forced backward through a source of EMF.

In contrast, a resistor is a device that converts electric energy into heat no matter what the direction of current. As in (3), p. 532, the potential always *drops* by an amount IR as we traverse a resistor *in the direction* of the current; this change in potential is called the *IR-drop*.

A battery, or a generator turning in a certain direction, has one terminal that can be permanently labeled + and one that can be permanently labeled −. When no charge flows, the + terminal has a potential V_E (E for EMF) above that of the − terminal. When charge flows through the source of EMF from − to +, chemical or mechanical energy is converted to electric energy in the amount V_E, in joules per coulomb of charge. When charge flows in the other direction, the same amount of energy per unit charge, V_E, is converted to mechanical or chemical form. The + terminal would always be at potential V_E above the − terminal except for the fact that all sources of EMF have *internal* resistance in the internal current path *between* their terminals. As a result of internal resistance, there is a conversion of some energy to heat inside the source of EMF, no matter which direction the current has. The copper-wire and carbon-brush path through a generator, or the metal-plate and electrolyte path through a battery, has a certain resistance R_I (I for *internal*), so that, for current I in *either* direction, energy is converted into heat in the amount IR_I, in joules per coulomb of charge passing.

Because of internal resistance, the difference of potential V_T between the *terminals* of a battery, generator, or motor, when charge flows, is not exactly the same as the EMF V_E, which is the difference of potential when no charge flows. We must distinguish two cases:

Generator, or battery on discharge. Internal current passes from − to + as in Fig. 1. The charge gains energy per unit charge in the amount V_E from mechanical or chemical energy but loses energy per unit charge IR_I in heat. Net gain is $V_E - IR_I$, in joules per coulomb. Hence, the difference of potential, V_T, between the + and − terminals is

$$V_T = V_E - IR_I. \quad \begin{pmatrix} \text{terminal voltage of} \\ \text{generator, or of bat-} \\ \text{tery on discharge} \end{pmatrix} \quad (1)$$

Multiplication of this equation by I gives an equation that expresses the power balance:

$$V_T I = V_E I - I^2 R_I.$$

$$\begin{Bmatrix} \text{Electric power} \\ \text{delivered by the} \\ \text{source of EMF} \end{Bmatrix} = \begin{Bmatrix} \text{electric power} \\ \text{generated in the} \\ \text{source of EMF} \end{Bmatrix} - \begin{Bmatrix} \text{electric power con-} \\ \text{verted into heat in} \\ \text{the source of EMF} \end{Bmatrix}$$

Motor, or battery on charge. Internal current passes from + to − as in Fig. 2. The charge loses energy per unit charge in the amount V_E, which is changed to mechanical or chemical energy, *and* loses energy per unit charge in the amount IR_I in heat. Total loss is $V_E + IR_I$, in joules per coulomb. Hence, in this case the + terminal is at higher potential than the − terminal by

$$V_T = V_E + IR_I. \quad \begin{pmatrix} \text{terminal voltage of} \\ \text{motor, or of battery} \\ \text{on charge} \end{pmatrix} \quad (3)$$

Multiplication of this equation by I gives the power balance:

$$V_T I = V_E I + I^2 R_I. \quad (4)$$

$$\begin{Bmatrix} \text{Electric power de-} \\ \text{livered to the motor,} \\ \text{or to the battery on} \\ \text{charge} \end{Bmatrix} = \begin{Bmatrix} \text{electric power} \\ \text{converted to} \\ \text{mechanical or} \\ \text{chemical energy} \end{Bmatrix} + \begin{Bmatrix} \text{electric power con-} \\ \text{verted into heat in} \\ \text{the motor, or in the} \\ \text{battery on charge.} \end{Bmatrix}$$

Example. *A DC motor has an internal resistance of* $2\,\Omega$. *When delivering its rated mechanical power, it draws* 10 A *from* 120-V *power lines. What mechanical power is developed by the motor?*

The situation is that depicted in Fig. 2, with the motor acting as a source of EMF. The motor must be acting as a source of EMF or the current through the motor would be 60 A in place of 10. The motor *generates* an EMF in a manner that we shall discuss later, and it is the loss in energy of the charge passing through this generated EMF from high potential to low that represents the gain in mechanical energy. From (3) we find

$$V_E = V_T - IR_I = (120 - 10 \cdot 2) \text{ V} = 100 \text{ V}$$

for the generated EMF. The mechanical power developed is $V_E I = 1000$ W, while the heat developed is $I^2 R_I = 200$ W, the sum of these representing the total power input $V_T I = 1200$ W.

2. Simple Circuits

The simplest type of electric circuit is illustrated in Fig. 3. Here the terminal voltage of the generator or battery, given by (1), is applied across the resistor and causes current I through it given by $V_T = IR$. Since the terminal voltage itself depends on the current, this relation gives an algebraic equation to be solved for I; substitution of the expression for V_T given by (1) results in the equation

$$V_E - IR_I = IR,$$

or

$$I = V_E/(R + R_I). \tag{5}$$

The current is determined by the EMF and the total resistance in the circuit, both internal and external.

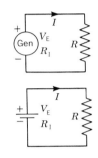

Fig. 3. A resistor connected across a generator or battery.

A more complex circuit, in which a battery is being charged by a generator through wires having different resistances, is shown in Fig. 4. The EMF's and internal resistances of the generator and battery are denoted by V_G, R_G and V_B, R_B. In this case the charge that flows around the circuit gains energy V_G per unit charge in the generator, loses V_B to chemical energy in the battery, and loses $IR_G + IR_I + IR_B + IR_2$ to heat. Since energy gain must equal energy loss,

$$V_G = V_B + I(R_G + R_1 + R_B + R_2);$$

$$I = \frac{V_G - V_B}{R_1 + R_2 + R_G + R_B}. \tag{6}$$

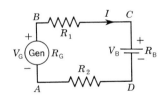

Fig. 4. A generator charging a battery.

It will be readily seen that any simple single-mesh circuit can be solved at once by a generalization of (6). The rule is:

$$\left\{ \begin{array}{l} \text{Current} \\ \text{clockwise} \\ \text{around the} \\ \text{circuit} \end{array} \right\} = \left\{ \begin{array}{l} \textit{algebraic} \text{ sum of EMF's counted posi-} \\ \text{tive if they tend to force current} \\ \text{clockwise, negative if they tend to} \\ \text{force current counterclockwise} \end{array} \right\} \div \left\{ \begin{array}{l} \text{sum of all resist-} \\ \text{ances in the circuit,} \\ \text{both internal and} \\ \text{external.} \end{array} \right\}$$

If the algebraic sum of EMF's in the above expression comes out negative, the clockwise current also comes out negative; the negative sign indicates that the current is counterclockwise instead of clockwise.

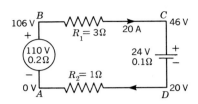

Figure 5

Example. *Determine the current in Fig. 4 and the differences of potential between point A and points B, C, D in Fig. 4 when* $R_1=3\,\Omega$, $R_2=1\,\Omega$, $V_G=110$ V, $R_G=0.2\,\Omega$, $V_B=24$ V, *and* $R_B=0.1\,\Omega$, *as in Fig. 5.*

For these constants, equation (6) gives $I=20$ A. If we arbitrarily call the potential 0 V at point A, the voltage at B is the terminal voltage of the generator given by (1) as 106 V. There is an IR-drop of 60 V in R_1, which makes the potential at C 46 V. The terminal voltage of the battery on charge is given by (3) as 26 V, so the potential drops to 20 V at D. The final drop from D to A is the 20-V IR-drop in R_2. A voltmeter connected between any two of the points A, B, C, D would read the difference between the potential values at these points.

Circuits that are not initially single-mesh can sometimes be changed to single-mesh circuits by replacement of a combination of resistances by a single equivalent resistance, in the manner discussed on pp. 533–534. This procedure is illustrated in the following example.

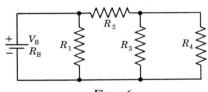

Figure 6

Example. *Find the currents through each resistor in the circuit of Fig. 6, where* $R_1=1\,\Omega$, $R_2=2\,\Omega$, $R_3=3\,\Omega$, $R_4=4\,\Omega$, $R_B=0.2\,\Omega$, *and* $V_B=6$ V.

First note that R_3 and R_4 are in parallel connection between the ends of R_1 and R_2; therefore, from (8), p. 534, R_3 and R_4 can be replaced by a single resistor $R_5=R_3R_4/(R_3+R_4)=(12\,\Omega^2)/(7\,\Omega)=1.71\,\Omega$. Now R_2 and R_5 are in series connection and can be replaced by $R_6=R_2+R_5=3.71\,\Omega$. Resistors R_1 and R_6 are in parallel connection across the terminals of the battery and may be replaced by an $R_7=R_1R_6/(R_1+R_6)=(3.71\,\Omega^2)/(4.71\,\Omega)=0.788\,\Omega$. Since R_7 now represents the entire external resistance, the current I through the battery is given by $I=V_B/(R_B+R_7)=(6\text{ V})/(0.988\,\Omega)=6.07$ A. The terminal voltage of the battery $V_T=6.00$ V $-(0.2\,\Omega)(6.07\text{ A})=4.79$ V. Hence, the current $I_1=(4.79\text{ V})/(1\,\Omega)=4.79$ A. The current $I_6=(4.79\text{ V})/(3.71\,\Omega)=1.29$ A, which is also equal to I_2. The voltage drop across R_2 is $(2\,\Omega)(1.29\text{ A})=2.58$ V; hence the voltage across R_3 and R_4 is 4.79 V -2.58 V $=2.21$ V, and $I_3=(2.21\text{ V})/(3\,\Omega)=0.74$ A, $I_4=(2.21\text{ V})/(4\,\Omega)=0.55$ A. (It is instructive to redraw the circuit diagram after each successive simplification or replacement of resistance combinations by single resistors such as R_5, R_6, and R_7.)

3. Electrical Networks; Kirchhoff's Rules

In this section we introduce briefly a general method for calculating currents in electrical networks of any complexity whatsoever. By a network we mean a circuit like that of Fig. 7 containing several *meshes*. There is a straightforward procedure for the treatment of a network given by two rules formulated by Kirchhoff:

KIRCHHOFF'S FIRST RULE: *At any junction point in a network, the total current arriving at the junction point must equal the total current leaving.*

KIRCHHOFF'S SECOND RULE: *The algebraic sum of the IR-drops around any mesh of a network is equal to the algebraic sum of the EMF's around the mesh.*

The first of these rules is a mere statement that charge does not accumulate at any point, and hence the sum of the currents approaching the point must equal the sum of the currents leaving the point. The second is equivalent to saying that if we start at any point in a mesh and 'walk' around the mesh back to our starting point, we must return to the same electric potential; the same is true, of course, if we start at any point in the network and walk around *any* closed path or loop back to our starting point. The method of applying Kirchhoff's rules is best illustrated by an example:

Example. *Find the current through each of the resistors in Fig. 7.*

In drawing the figure, we have guessed the directions of currents I_1 and I_2 and have already applied Kirchhoff's first rule at point P, since the current leaving the point is labeled I_1+I_2. Now we apply Kirchhoff's second rule to each of the two meshes, proceeding clockwise around mesh 1 and counterclockwise around mesh 2, since by this choice all EMF's and *IR*-drops will appear with a + sign, algebraically:

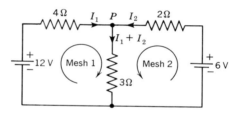

Fig. 7. Two batteries of negligible internal resistance send current through a common resistor.

Mesh 1: $I_1(4\,\Omega)+(I_1+I_2)(3\,\Omega)=12$ V   or $7I_1+3I_2=12$ A,

Mesh 2: $I_2(2\,\Omega)+(I_1+I_2)(3\,\Omega)=6$ V   or $3I_1+5I_2=6$ A,

where, in the simplified equations at the right, we have written 1 A = 1 V/Ω. Solution of this pair of equations gives $I_1=1.62$ A and $I_2=0.231$ A. The current through the 3-Ω resistor is the sum, $I_1+I_2=1.85$ A.

We guessed correctly the directions of I_1 and I_2 in Fig. 7. As an example of a wrong guess, if the 4-Ω resistor in Fig. 7 is replaced by a 1-Ω resistor, and the solution is carried through as above, the results are $I_1=3.82$ A, $I_2=-1.09$ A, $I_1+I_2=2.73$ A. In this case, the 12-V battery supplies 2.73 A through the 3-Ω resistor and, in addition, supplies 1.09 A to *charge* the 6-V battery.

4. Measuring Instruments

We shall now describe three important DC measuring instruments, the *voltmeter*, the *ammeter*, and the *potentiometer*. The indicating element

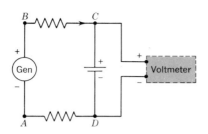

Fig. 8. Connection of a voltmeter to measure the difference of potential between points C and D.

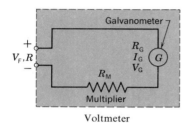

Fig. 9. A DC voltmeter. The symbols V_F, I_G, V_G refer to *full-scale* deflection.

of all three of these instruments is a galvanometer of the d'Arsonval type, whose operation will be described on pp. 593–594. Basically, a galvanometer is an instrument that gives a pointer deflection proportional to the current through itself. It has a certain resistance R_G, and its sensitivity is characterized by the current I_G, or voltage $V_G = R_G I_G$, required for *full-scale* deflection. Good voltmeters and ammeters employ sensitive galvanometers, with I_G and V_G of the order of a thousandth of an ampere or volt, or even less. The galvanometer used in a potentiometer must be even more sensitive.

The DC voltmeter. A voltmeter is intended to measure the difference of potential between two points in a circuit, as in Fig. 8, without appreciably disturbing the currents in the circuit; hence, the voltmeter itself must draw an inappreciable current, and therefore *must have a very high resistance.* A DC voltmeter is constructed by placing a very high resistance, called a *multiplier,* in series with a very sensitive galvanometer, as in Fig. 9. If we want *full-scale* deflection of the voltmeter for voltage V_F, we must add enough resistance R_M so that with voltage V_F across the terminals we get only the small current I_G through the galvanometer. Since

$$I_G (R_G + R_M) = V_F, \quad \text{and} \quad I_G R_G = V_G,$$

we find, by division, that

$$\frac{R_G + R_M}{R_G} = \frac{V_F}{V_G}. \tag{7}$$

Example. *In Fig. 9, if $R_G = 1\,\Omega$, $I_G = 1$ mA, $V_G = 1$ mV, and we want V_F to be 200 V full-scale, find the multiplier resistance.*

Substitution in (7) gives

$$\frac{1\,\Omega + R_M}{1\,\Omega} = \frac{200}{0.001} = 200\,000,$$

so that $\quad R_M = 200\,000\,\Omega - 1\,\Omega = 199\,999\,\Omega$.

The total resistance R of this voltmeter would be $R = R_G + R_M = 200\,000\,\Omega$.

The DC ammeter. An ammeter is intended to measure the current in a circuit without appreciably altering that current. Since the circuit must be broken and the current allowed to pass through the ammeter, as in Fig. 10, the ammeter *must have a very low resistance.* A DC ammeter is constructed by placing a very low resistance, called a *shunt,* in parallel with a very sensitive galvanometer, as in Fig. 11. If we want *full-scale* deflection with current I_F, we must use a shunt resistance low enough that with current I_F at the terminals, there will be only current I_G through the galvanometer. Since $I_F - I_G$ is the current through the shunt, we have, from Kirchhoff's second rule,

$$(I_F - I_G) R_S = I_G R_G,$$

or
$$\frac{R_S}{R_G} = \frac{I_G}{I_F - I_G}. \qquad (8)$$

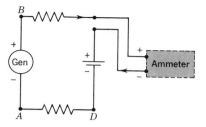

Fig. 10. Connection of an ammeter to measure the current in a circuit.

Example. *In Fig. 11, if $I_G = 1$ mA and $R_G = 1$ Ω, and we want an ammeter with $I_F = 10$ A full-scale, find the proper shunt resistance.*

From (8), $R_S = 1 \, \Omega \times \dfrac{0.001 \text{ A}}{9.999 \text{ A}} = 0.000\ 100\ 01$ Ω.

The resistance R between terminals of this ammeter can be obtained from the equation $R\, I_F = V_G$:

$$R = \frac{V_G}{I_F} = \frac{0.001 \text{ V}}{10 \text{ A}} = 0.0001 \text{ Ω}.$$

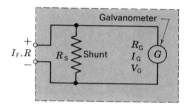

Fig. 11. A DC ammeter. The symbols I_F, I_G, V_G refer to *full-scale* deflection.

The last equation in the example above shows that the ammeter best from the standpoint of low resistance is the one made with the galvanometer of greatest voltage sensitivity (lowest V_G).

The potentiometer. The potentiometer is one of the most important instruments in any electrical laboratory making precision measurements. In principle it is *a device for accurately measuring the ratio of two voltages.* Since one of these is usually the accurately known voltage V_s of a standard cell, the potentiometer serves for precision measurement of the other, unknown, voltage V. It also serves for precision measurement of currents by measuring the voltage across a precision resistor through which the current passes. In this way the potentiometer is used for calibration of both ammeters and voltmeters in terms of the voltage of a standard cell. Standard cells are described in the following chapter.

The basic circuit of the potentiometer is shown in Fig. 12. A constant current, usually furnished by a lead storage battery, passes through the resistor shown in this figure. *The whole resistance R across the storage battery remains constant,* but this resistance is made up of a combination of fixed resistances and a precision slide wire, all in series, in

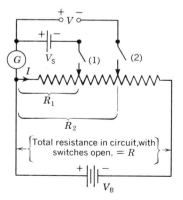

Fig. 12. The basic circuit of the *potentiometer*, arranged to compare the unknown voltage V with the voltage V_s of a standard cell.

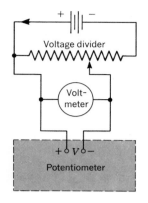

Fig. 13. A potentiometer used to calibrate a DC voltmeter.

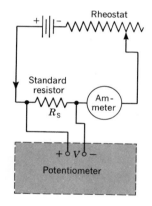

Fig. 14. A potentiometer used to calibrate a DC ammeter.

such a way that *the resistances R_1 and R_2 can be varied and accurately read on knobs and dials.*

In use, first the switch 1 is closed, switch 2 left open, and R_1 adjusted until the very sensitive galvanometer G reads zero. When this adjustment has been made, there is no current through the standard cell so its terminal voltage is just V_S, and we may write, using Kirchhoff's second rule,

$$V_S = IR_1, \quad \text{where} \quad I = V_B/R.$$

Then switch 1 is opened, switch 2 is closed, and R_2 is adjusted until the galvanometer again reads zero. In this case,

$$V = IR_2, \quad \text{where} \quad I = V_B/R.$$

Since V_B and R have not been changed between the two adjustments, I does not change, so by taking ratios we see that

$$V_S/V = R_1/R_2, \tag{9}$$

which determines V in terms of V_S and the known resistances R_1 and R_2.

It is not necessary to know either V_B or I. The fact that they have not changed between settings can be checked by alternately closing switches 1 and 2 and verifying that the galvanometer continues to read zero.

Figures 13 and 14 show how the potentiometer can be used to calibrate a DC voltmeter or a DC ammeter. The potentiometer terminals marked $+ V -$ correspond to the terminals at the top of Fig. 12. The potentiometer is used to read the voltage across these terminals, and gives the correct voltmeter reading in Fig. 13. In connection with Fig. 14, notice that the potentiometer measures the voltage V *without drawing current* (galvanometer reading zero in Fig. 12). Since no current is drawn through the terminals V in Fig. 14, the potentiometer reading is R_S times the correct ammeter reading. We see that in principle *a potentiometer is a perfect voltmeter that draws no current.*

5. Charge and Discharge of a Capacitor

Consider the circuit of Fig. 15 containing a capacitor C, a resistor R, a battery with EMF V_E and negligible internal resistance, and two switches S_1 and S_2. Let the initial charge on the capacitor plates be zero and switches S_1 and S_2 be open. When switch S_1 is closed, the capacitor will be charged by a transient current I through R. The current I will vary with time, but at any instant we can write an equation similar to Kirchhoff's second rule for the loop. If we go around the loop clockwise, starting at point A, the potential first rises by Q/C, where Q is the instantaneous charge on the capacitor. It then rises by an amount

$RI = R\,dQ/dt$ as we traverse the resistor 'uphill.' Finally it drops by the constant V_E in going 'backward' through the battery. When we return to A, we must find the same potential as when we started, so

$$Q/C + R\,dQ/dt - V_E = 0. \tag{10}$$

To solve this differential equation, we multiply by C and rearrange terms to separate the variables, which are Q and t:

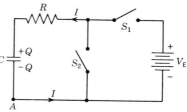

Figure 15

$$\frac{dQ}{CV_E - Q} = \frac{dt}{RC}.$$

Take the indefinite integral:

$$-\log_e(CV_E - Q) = t/RC + \text{const.} \tag{11}$$

At the instant the switch is closed ($t=0$), the charge Q on the capacitor is zero. Setting $t=0$ and $Q=0$ in (11) determines the constant of integration as $\text{const} = -\log_e CV_E$. Hence (11) becomes

$$-\log_e(CV_E - Q) = t/RC - \log_e CV_E$$

or

$$-\frac{t}{RC} = \log_e(CV_E - Q) - \log_e CV_E = \log_e\left(\frac{CV_E - Q}{CV_E}\right).$$

Taking the antilog of both sides gives

$$e^{-t/RC} = (CV_E - Q)/CV_E,$$

which can be solved for Q to give

$$Q = CV_E(1 - e^{-t/RC}). \tag{12}$$

This expression shows how Q varies with time, starting from $Q=0$ at $t=0$ to the final value $Q=CV_E$ at $t=\infty$. Equation (12) is plotted as the solid curve in Fig. 16. It will be noted that the charge rises to $(1 - 1/e)$

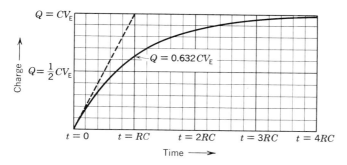

Fig. 16. Growth of charge on a capacitor when connected to a battery of EMF V_E through a resistance R.

times its final value in the time $t=RC$; this time, RC, is called the *time constant* of the circuit.

The charging current I can be obtained as a function of time by differentiation of (12):

$$I=\frac{dQ}{dt}=\frac{V_E}{R}e^{-t/RC}. \qquad (13)$$

As shown in Fig. 17, this current starts with the value V_E/R since at $t=0$

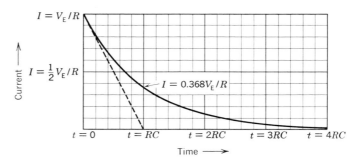

Fig. 17. Capacitor charging current $I=dQ/dt$.

the capacitor is uncharged and the whole battery voltage appears across the resistor. As the capacitor becomes charged, less of the battery voltage appears across R and the current drops—to $1/e$ times its initial value at $t=RC$, and to zero as $t\rightarrow\infty$.

The initial current V_E/R in Fig. 17 is the initial slope of the curve in Fig. 16. As indicated by the broken line in Fig. 16, a straight line of this slope would pass through the point $Q=CV_E$ at $t=RC$, because a line passing through this point has slope $CV_E/RC=V_E/R$, equal to the initial slope of the solid curve. Thus, *if the current continued to enter the capacitor at its initial rate, the capacitor would be fully charged in a time equal to the time constant*. The broken line in Fig. 17 is, similarly, the initial rate of change of current, and has similar properties.

Now consider the question of *discharge*. We first charge the capacitor in Fig. 15 by keeping switch S_1 closed until the full charge CV_E is attained. Then we open switch S_1 and close switch S_2, allowing the capacitor to discharge through the resistance R. In this case, there is no source of EMF in the circuit and (10) assumes the form

$$Q/C+R\,dQ/dt=0. \qquad (14)$$

This differential equation can be solved for Q in a manner similar to that used for (10). The result is

$$Q=CV_E\,e^{-t/RC}. \qquad (15)$$

The charge decays along a curve of the same shape as that in Fig. 17, with the same time constant as in the charging case. The discharge current is

$$I = \frac{dQ}{dt} = -\frac{V_E}{R} e^{-t/RC}, \qquad (16)$$

exactly the same as the charging current given by (13) and Fig. 17, except that it is reversed in direction as indicated by the minus sign.

The time constant RC furnishes a useful order of magnitude estimate of the time required to charge or discharge a capacitor through a resistance.

Example. (a) *What is the time constant for charging a 2-μF capacitor from a dry cell of 1.2-Ω internal resistance through leads of negligible resistance?* (b) *If the capacitor is then disconnected and left charged, what is the time constant for discharge if the leakage path through the dielectric between the plates has a resistance of 5000 MΩ?*

(a) The time constant for charge is

$$RC = (1.2 \ \Omega)(2 \ \mu F) = 2.4 \ \mu s,$$

since $1 \text{ F} = 1 \text{ C/V} = 1 \text{ A} \cdot \text{s/V} = 1 \text{ s/}\Omega$.

(b) The time constant for discharge through the leakage path is

$$RC = (5000 \text{ M}\Omega)(2 \ \mu F) = 10\ 000 \text{ s} = 2.78 \text{ h}.$$

6. Nonlinear Circuit Elements: the Thermionic Vacuum Tube and the Transistor

The circuits we have discussed in this chapter have been made up of so-called 'linear' circuit elements. By a 'linear' circuit element we mean an element having the characteristics of a resistor, in which the current is directly proportional to the applied voltage and Ohm's law applies. There is, however, an important class of 'nonlinear' circuit elements, the most common of which are *thermionic vacuum tubes* and *transistors*.

Thermionic vacuum tubes. The simplest thermionic vacuum tube is of the type shown in Fig. 11, p. 541, used in Edison's experiment. It contains two electrodes, the hot filament F and the plate P. Its nonlinear character is evident from the fact that it carries current when the plate is positive with respect to the filament but that the current is zero when the plate is negative.

The escape of electrons from the surface of a heated metal electrode in a vacuum tube leaves the metal positively charged. The electrons that have escaped are therefore attracted by the metal and form a negative 'cloud,' called a *space charge,* around the electrode. Ordinarily, an equilibrium is quickly established between the number of electrons escaping from a metal surface and the number returning to the surface. If, however, as in Edison's experiment, a second nearby electrode is maintained at a higher potential than the first, electrons are attracted to it; and as long as the potential difference is maintained, there will be a steady movement of electrons from the first electrode to the second. The first electrode is called the *cathode;* the second is called the *anode* or *plate.* A vacuum tube containing two electrodes of this type is called a *diode.* The cathode may be a hot *filament* or a cylindrical metal electrode heated by an internal hot wire called a *heater.*

In order to study the characteristics of a diode, the arrangement shown in Fig. 18 may be used. In this circuit the ammeter gives the plate current I_P for various values of plate voltage V_P, which can be varied by means of the indicated voltage-divider arrangement. If the potential difference V_P between plate and cathode is small, only a few of the emitted electrons reach the plate. Because of repulsion by the negative 'space charge' due to electrons already present, most of the emitted electrons penetrate only a short distance into space and then return to the cathode. As the plate potential is increased more and more of the emitted electrons reach the plate, and with sufficiently high potential difference *all* of the emitted electrons arrive at the plate. Further increase of the plate voltage V_P does not then increase the plate current, which is said to be *saturated.*

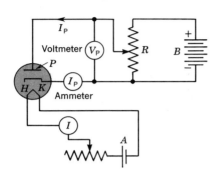

Fig. 18. Circuit for obtaining characteristic curves of plate current vs. plate voltage for a diode in which an indirectly heated cathode K is used.

Typical graphs of plate current I_P as a function of plate voltage V_P are shown in Fig. 19. As indicated in Fig. 19, the magnitude of the *saturation current* I_S increases when the cathode temperature is increased by increasing the heater current. It is noted that there is a small plate current even when the plate voltage is zero and that a small negative voltage must be applied to the plate to stop the current completely; even with zero plate voltage, a few of the fastest electrons emitted by the heated filament are able to penetrate the negative space charge and reach the plate.

Figure 19 shows that the diode can be used as a rectifier, since for positive plate voltages the plate current is appreciable whereas for negative plate voltages the plate current is negligible. If an alternating voltage V_P is applied between plate and cathode, the tube will conduct only when the plate is positive. Thus, the alternating voltage produces a fluctuating current I_P of a constant direction.

Before saturation has been attained, the plate current I_P in a diode can be varied by changing the plate voltage V_P. LEE DE FOREST, in 1907, discovered that if a third electrode called a *control grid* is inserted between the cathode and plate, the potential of this third electrode exerts much greater 'control' over the plate current than does the plate voltage itself. The grid is an open mesh of fine wire, which allows the electrons to pass through the openings. A schematic diagram of a tube containing a control

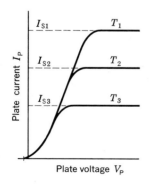

Fig. 19. Characteristic curves of a diode for three cathode temperatures $T_1 > T_2 > T_3$.

grid is shown in Fig. 20. Since small variations in grid voltage produce relatively large variations in plate current, the three-electrode tube or *triode* can be used as an *amplifier*. In use as an amplifier, a small varying 'signal' voltage, obtained from a source such as a microphone or a radio antenna, is impressed between the grid and the cathode. This varying voltage appears in more or less faithfully amplified form across a resistor or other circuit element in the plate circuit. This amplification can be repeated in second and later 'stages.'

Commercial triodes are usually operated at plate voltages much lower than would be required to draw saturation current. The plate current of a triode is a function of cathode temperature, plate voltage, and grid voltage. It cannot be expressed conveniently as an algebraic function of these variables but is customarily expressed by graphs of characteristic curves like those shown in Fig. 21. The filament or heater of a commercial tube is usually operated at a fixed voltage prescribed by the manufacturer, and hence the curves in the graph give the plate current for various grid and plate voltages measured relative to the cathode at a fixed A-battery voltage. The individual curves show plate current as functions of plate voltage when the control-grid voltage is kept constant.

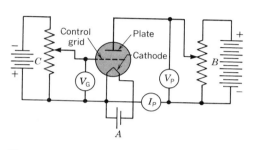

Fig. 20. An arrangement for obtaining the characteristic curves of a triode.

We verify in Fig. 21 that changes in grid voltage have much more

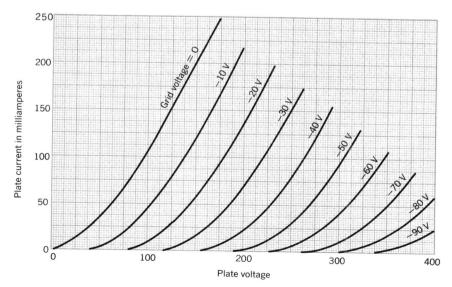

Fig. 21. Average plate characteristics for a particular triode with filament heated by an A-battery of constant voltage.

effect on plate current than do changes in plate voltage. For example, if we start with grid voltage 0 and plate voltage 125 V, the plate current is 155 mA. Dropping the grid voltage by 10 V to −10 V decreases the plate current by 72 mA to 83 mA, whereas dropping the plate voltage by 10 V to 115 V only decreases the plate current by 20 mA to 135 mA. The tube is said to have an *amplification factor* of about $72/20 = 3.6$ when operated in the vicinity of zero grid voltage and 125-V plate voltage.

The transistor. In recent years, it has been shown that the properties of semiconductors can be used to advantage in constructing nonlinear circuit elements. In Sec. 7 of Chapter 25, we discussed the properties of semiconductors and in particular the properties of 'doped' semiconductors of p- and n-types. For example, Fig. 10 on p. 540 showed how a two-element semiconductor containing a p-n junction acts as a *rectifier*.

The desirability of making *three* connections to a semiconductor and having the resulting device behave like a *triode* is immediately apparent. Such a device was developed at the Bell Telephone Laboratories in 1948* and is called a *transistor*.

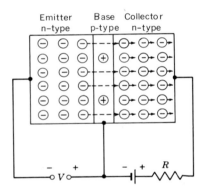

Fig. 22. The n-p-n type of transistor.

The transistor consists of a tiny crystal of doped semiconductor (most commonly germanium or silicon) which contains two rectifying p-n junctions arranged back-to-back, and three leads, as in Fig. 22. This figure illustrates the n-p-n type of transistor; one can equally well make a p-n-p type.

The transistor in Fig. 22 consists of a single crystal in which the two ends are *heavily doped* with a donor impurity and are of n-type, while a *very thin* layer at the center, called the base, is *lightly doped* with an acceptor impurity and is of p-type. In the circuit shown, the left side of the crystal is called the *emitter*, the right side is called the *collector*.

Let us first consider the collector circuit by itself. This circuit is permanently biased by a battery whose voltage is in the indicated direction, which is the direction in which *no current* will occur at the rectifying junction between the base and the collector. Thus any collector current must be the result of effects from the emitter circuit. Now consider application of the voltage V to the emitter circuit in the direction indicated, which is the direction in which the barrier between the emitter and the base will pass current. Electrons from the emitter will enter the base in approximate proportion to the voltage V. Once in the base, these electrons will be strongly drawn to the collector

---

*The Nobel prize in physics for 1956 was awarded to JOHN BARDEEN, WALTER BRATTAIN, and WILLIAM SHOCKLEY for their studies in solid-state physics that culminated in the development of the transistor.

by the collector bias potential. Few of them will fill 'holes' in the base because the material of the base is only weakly doped and the holes are few in number. Hence, with a very thin base layer, practically all the electrons crossing the emitter barrier will also cross the collector barrier and proceed around the collector circuit through the resistance R. With proper choice of circuit constants, the voltage across R will be greater than the applied voltage V, and thus variations of voltage across R can represent an *amplification* of variations of voltage V; the transistor will act as an *amplifier*, just like a thermionic triode.

Transistor circuitry has many advantages over vacuum-tube circuitry for many applications. Transistor circuits can be made small and compact since the individual transistors are very tiny. Transistor circuits do not require the wasted power that is required to heat the thermionic filaments; they remain comparatively cool and respond instantly when the power is turned on. Furthermore, they have proved to be more rugged, reliable, and longer-lived than the corresponding vacuum-tube circuits.

PROBLEMS

1. Consider a 12-V storage battery, with internal resistance of 0.04 Ω, delivering a current of 50 A. (*a*) What is the terminal voltage? (*b*) What is the rate of conversion of chemical to electric energy? (*c*) What is the rate of heat generation in the battery? (*d*) What is the power delivered? Ans: (*a*) 10.0 V; (*b*) 600 W; (*c*) 100 W; (*d*) 500 W.

2. Answer the questions (*a*)–(*d*) of Prob. 1 for a 1.5-V dry cell, with internal resistance of 0.1 Ω, delivering 4.0 A.

3. Consider a 12-V storage battery, with internal resistance of 0.04 Ω, when it is being charged with a current of 50 A. (*a*) What is the terminal voltage? (*b*) At what rate is electric energy being delivered to the battery? (*c*) At what rate is heat being developed in the battery? (*d*) At what rate is electric energy being changed into chemical energy? Ans: (*a*) 14.0 V; (*b*) 700 W; (*c*) 100 W; (*d*) 600 W.

4. Answer the questions (*a*)–(*d*) of Prob. 3 for a 120-V bank of storage batteries, consisting of 60 two-volt cells in series, *each cell* having an internal resistance of 0.008 Ω, when the bank is being charged with a current of 30 A.

5. The terminal voltage of a generator is 120 V at no load (no current output). The internal resistance is 0.1 Ω. What will its terminal voltage be when it is rotating at the same speed and has the same field excitation (in which case it has the same generated EMF) but is delivering 20 amperes? Ans: 118 V.

6. The terminal voltage of a generator of 0.2-Ω internal resistance is 115 V when the generator is delivering full-load current of 30 A. Assuming the same generated voltage, find its terminal voltage (*a*) when it is overloaded and delivering 60 A, (*b*) when lightly loaded and delivering 10 A, (*c*) at no load.

7. A DC motor has an internal resistance of 0.1 Ω. It draws 20 A at full load from 120-V lines.

(*a*) What is the EMF of the motor?
(*b*) What is the power drawn by the motor?
(*c*) What is the heat generated in the internal resistance?
(*d*) What is the mechanical power developed? Ans: (*a*) 118 V; (*b*) 2400 W; (*c*) 40 W; (*d*) 2360 W.

8. Make the same computations as in Prob. 7 for the case of a motor of 0.1-Ω internal resistance that draws 80 A from 240-V lines at full load.

9. In Fig. 5, p. 550, compute (a) the generated electric power; (b) the heat loss in the generator; (c) the heat developed in R_1; (d) the heat developed in the battery; (e) the heat developed in R_2; (f) the rate of creation of chemical energy. Verify that $(a)=(b)+(c)+(d)+(e)+(f)$. Ans: (a) 2200 W; (b) 80 W; (c) 1200 W; (d) 40 W; (e) 400 W; (f) 480 W.

10. In Fig. 4, p. 549, take $V_G=24$ V, $V_B=6$ V, $R_G=0.4\,\Omega$, $R_B=0.2\,\Omega$, $R_1=1\,\Omega$, $R_2=2\,\Omega$. Compute I. Letting the potential at point A equal 0, compute the potentials at B, C, and D. Also compute all the quantities asked for in parts (a)–(f) of Prob. 9.

11. In Fig. 5, p. 550, to what value would R_2 have to be changed to drop the battery-charging current to 10 A? Ans: 5.3 Ω.

12. In Fig. 5, p. 550, to what value would R_1 have to be changed to increase the battery-charging current to 30 A?

13. A man has three 100-Ω resistors. How many different resistances can he obtain by connecting the resistors in any manner he chooses, being free to use any number of the resistors at any one time? Sketch the possible connections, and calculate the resistance of each combination.
Ans: 7; 33.3, 50, 66.7, 100, 150, 200, 300 Ω.

14. In Fig. 6, p. 550, let $V_B=6$ V, $R_B=0.2\,\Omega$, $R_1=4\,\Omega$, $R_2=0.5\,\Omega$, $R_3=2\,\Omega$ and $R_4=3\,\Omega$. (a) Find the currents I_B, I_1, I_2, I_3, and I_4 through the battery and each of the resistors. (b) Find the voltages V_T, V_1, V_2, V_3, and V_4 across the battery and each of the resistors. (c) Find the power P_T delivered by the battery and the amounts of power P_1, P_2, P_3, and P_4 absorbed in each of the resistors. Verify that $P_T=P_1+P_2+P_3+P_4$.

15. Two generators, G_1 and G_2 supply current to two loads I and II in the figure. Calculate the currents in AB, BC, and CD and the voltages V_I and V_{II} across the loads.
Ans: 64⅓ A, A to B; 5⅔ A, C to B; 80⅔ A, D to C; $V_I=114.7$ V; $V_{II}=115.2$ V.

16. In the figure, if the current to load I is 30 A instead of 70 A, with no other change, calculate the currents in AB, BC, and CD and the voltages across I and II.

17. The two generators shown in the figure have nominal ratings of 120 V. Together they supply 100 A to a transmission line. V_{E1}, the EMF of generator 1, is exactly 120 volts. The EMF of generator 2 may be varied by making variations in the exciting field current, in order to make it carry more or less of the load, as indicated by an ammeter measuring I_2. Compute the exact value of V_{E2} if generator 2 is to supply (a) half the line current; (b) one-quarter of the line current; (c) none of the line current (this is the condition to which the power-station attendant would adjust before he pulled a switch to disconnect generator 2 from the line, leaving generator 1 to carry the whole load); (d) all of the line current. (e) What happens if the EMF of generator 2 is raised to 150 V? Ans: (a) 125 V; (b) 117.5 V; (c) 110 V; (d) 140 V; (e) $I_2=133$ A, of which 100 A go into the line and 33 A go backward through generator 1 to drive it as a motor.

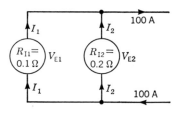

Problems 17–18

18. In the figure, if $V_{E1}=120$ V and $V_{E2}=130$ V, find the currents I_1 and I_2, the voltage across the line, and the effective resistance of the load connected across the line, assuming that this is a pure-resistance load, such as a lighting load.

19. *Wheatstone bridge.* This convenient circuit for the measurement of an unknown resistance is shown in the figure. One of the resistances, say R_1, is unknown. The others are variable

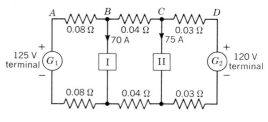

Problems 15–16

but known resistances, which are adjusted until the galvanometer indicates zero current, when the bridge is said to be *balanced*. By applying Kirchhoff's rules show that when the bridge is balanced, the resistances are in the ratio $R_1/R_2 = R_3/R_4$, from which R_1 may be computed if the other three resistances are known.

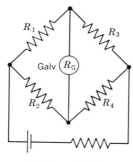

Problems 19–20

20. If the bridge in the figure is balanced with $R_2 = 15.0\ \Omega$, $R_3 = 20.0\ \Omega$, $R_4 = 60.0\ \Omega$, what is the value of the unknown resistance R_1?

21. A two-wire underwater power cable runs across a bay and is 52 000 f long. Each wire has resistance 0.0983 ohm per thousand feet. One wire develops a ground to the lead sheath at some point. A Varley-loop test is applied by connecting the two wires together at the far end and setting up a bridge circuit at the near end as shown in the figure. The bridge is balanced with $R_1 = 15.00\ \Omega$, $R_2 = 7.63\ \Omega$. Which section of the cable should be pulled up to look for the defect? (The resistance of the grounded circuit through the sheath may be assumed negligible.)

Ans: the section 35 000 f from the near end.

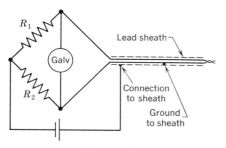

Problems 21–22

22. If the bridge in the figure, used as in Prob. 21, is balanced for $R_1 = 6.00\ \Omega$ and $R_2 = 30.0\ \Omega$, approximately where is the defect?

23. Compute the current through the 50-Ω galvanometer in the Wheatstone bridge in the figure when the bridge is *unbalanced*, with $R = 20\ \Omega$. The battery has an EMF of 6.00 V and negligible resistance.

Ans: 0.005 05 A upward.

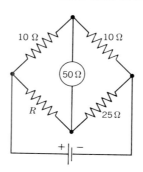

Problems 23–24

24. Compute the current through the 50-Ω galvanometer in the Wheatstone bridge in the figure when the bridge is *unbalanced*, with $R = 30\ \Omega$. The battery has an EMF of 6 V and negligible resistance.

25. Find the single resistance that is equivalent to the combination shown in the figure.

Ans: 5.34 Ω.

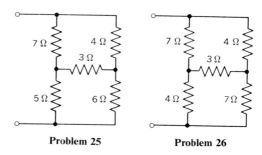

Problem 25   Problem 26

26. Find the single resistance that is equivalent to the combination shown in the figure.

27. A commercial voltmeter reads 150 volts full-scale. The basic element is a d'Arsonval galvanometer of 50 ohms resistance that gives full-scale deflection on 20 mV. Find the resistance of the multiplier. Ans: 374 950 Ω.

28. A laboratory that possesses the instrument of Prob. 27 desires to convert it to one reading 300 volts full-scale. The laboratory writes to the meter-manufacturing company and orders a ×2 external multiplier for this particular instrument. The manufacturer sends a box with two binding posts which is to be connected in series with the voltmeter when in use. What is in the box?

29. What shunt resistance is required in an ammeter whose galvanometer element has a resistance of 5 ohms and gives full-scale deflection on 50 mV, if the ammeter is to read 10 A full-scale? Ans: 0.005 00 Ω.

30. What shunt resistance is required in a milliammeter whose galvanometer element has a resistance of 50 ohms and gives full-scale deflection on 1 mA, if the milliammeter is to read 500 mA full-scale?

NOTE: *In measuring a resistance by the voltmeter-ammeter method, the ammeter is placed in series with the resistance, but the voltmeter can be connected in two ways, as indicated in parts* (a) *and* (b) *of the figure. Connection* (a) *has the disadvantage that the ammeter measures the current through the voltmeter as well as that through the resistor;* (b) *has the disadvantage that the voltmeter measures the voltage drop across the ammeter as well as that across the resistor. We call the ratio* $R_{App} = V/I$ *of voltmeter to ammeter reading the apparent resistance of the resistor being measured. In Probs. 31–34, we contemplate measuring a 20.00-ohm* resistor *by using a dry cell as a current source, together with a 1.5-V full-scale voltmeter and a 50-mA full-scale milliammeter. In each case, the current is adjusted by an external rheostat so that the milliammeter reads full scale.*

31. (*See* note.) If the voltmeter has a resistance of 400 ohms and the milliammeter a resistance of 2 ohms, what is the apparent resistance for connection (a)? for (b)?
Ans: (a) 19.05 Ω; (b) 22.00 Ω.

32. (*See* note.) If the voltmeter has a resistance of 20,000 ohms and the milliammeter a resistance of 2 ohms, what is the apparent resistance for connection (a)? for (b)?

33. (*See* note.) If the voltmeter has a resistance of 400 ohms and the milliammeter a

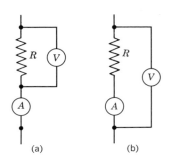

Problems 31–36

resistance of 0.001 ohm, what is the apparent resistance for connection (a)? for (b)?
Ans: (a) 19.05 Ω; (b) 20.00 Ω.

34. (*See* note.) If the voltmeter has a resistance of 20,000 ohms and the milliammeter a resistance of 0.001 ohm, what is the apparent resistance for connection (a)? for (b)?

35. For connection (a) of the figure, show that if I is the ammeter reading, V the voltmeter reading, R_A the ammeter resistance, and R_V the voltmeter resistance, then the true resistance of the resistor is given by $R = V/(I - V/R_V)$, so that $(1/R) = (1/R_{App}) - (1/R_V)$. Hence show that connection (a) is satisfactory only when the resistance of the voltmeter is large compared with the resistance being measured.

36. For connection (b) of the figure, show that if I is the ammeter reading, V the voltmeter reading, R_A the ammeter resistance, and R_V the voltmeter resistance, then the true resistance of the resistor is given by $R = (V/I) - R_A = R_{App} - R_A$, so that connection (b) is satisfactory only when the resistance of the ammeter is small compared with the resistance being measured.

37. Consider the ionization chamber shown in the figure. The central electrode, connected

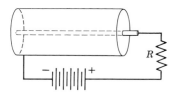

Problems 37–38

through a high resistance R, is initially at a potential just below that at which continuous breakdown occurs. The amplification factor (*see* Prob. 36, p. 512) is about 10^8. Now suppose that a high-energy particle passes through the gas in the tube. The particle produces a very large number of ions. As these ions are collected, a current is produced in the resistor causing an *IR*-drop sufficiently large to reduce the electrode potential enough to stop the incipient continuous discharge. This large sudden voltage change at the electrode can be readily recorded. There will be one such voltage pulse each time a high-energy particle, or a quantum of ionizing radiation, passes through the tube. An ionization chamber operating in this manner is called a *Geiger counter*.

If the battery EMF in the figure is 600 V, and a 1-megohm resistor is used to connect the battery to the central electrode, what is the potential of the electrode when no ions are being collected? If the potential of the electrode must be at least 500 volts in order to produce appreciable ion 'multiplication,' what current in the circuit will cause incipient breakdown to be quenched? Answer the same questions for $R = 1000$ megohms.

Ans: 600 V, 0.1 mA; 600 V, 0.1 μA.

38. If a single high-energy particle produces 10^3 ion pairs in a Geiger-counter tube (Prob. 37) in which the gas multiplication factor is 10^8, how much charge would be collected? If this charge were collected in 1 μs, what would be the average collection current during this period? Would this be possible in the circuit shown in the figure with $R = 1$ megohm?

Ans: 1.6×10^{-8} C; 0.016 A; no, why?

39. As in Prob. 31 on p. 511, the 'sphere' of a Van de Graaff generator is maintained at 6.0 MV relative to ground and an accelerator tube leads from the sphere to ground. Show that if a resistance of 400 MΩ is connected between successive electrodes in the accelerator column and if 100 electrodes are used, the potential of each accelerator section will be 60 kV less than that of the preceding section. What current will flow through these resistors? In an accelerator tube, it is necessary to maintain the difference of potential between sections approximately constant to secure good *focusing* of the ions into a narrow beam. The resistance inserted between sections must be low enough so that the current can make up any losses in charge on the individual sections that arise from leakage along the insulators and from gaseous conduction (the fat 'corona rings' or 'hoops' shown in section on p. 511 help reduce the latter by limiting potential gradients). On the other hand, the resistance must be high enough so that the current drawn is less than the current delivered to the 'sphere' by the charging belt.

Ans: 0.15 mA.

27 ELECTROCHEMISTRY; THERMOELECTRICITY

In the preceding chapter we described sources of EMF as *reversible* devices for changing energy from some other form to electric energy, or from electric energy to the other form. The other form may be *mechanical, chemical, thermal,* or *radiant* energy. In this chapter we discuss briefly the interconversion of chemical and electric energy (*electrochemistry*) and the *reversible* interconversion of thermal and electric energy (*thermoelectricity*).

1. Electrolysis

Pure water is a poor conductor of electricity; aqueous solutions of many organic materials such as sugar, alcohol, and glycerine are also poor conductors. However, aqueous solutions of acids, bases, and salts are comparatively good conductors; these solutions are called *electrolytes*.

Faraday in 1830 performed numerous experiments on conduction of electricity by solutions. In these experiments, he introduced two metallic plates called *electrodes* into the solution being studied, as in Fig.

1. The combination of the two electrodes and the conducting solution is called an *electrolytic cell*. When current is passing through the cell, chemical reactions occur at the electrodes; these chemical reactions are said to occur from *electrolysis*.

When two platinum electrodes are used in a cell containing an aqueous solution of an acid, oxygen gas is liberated at the + electrode and hydrogen gas at the − electrode; these gases are liberated as the result of decomposition of water molecules. When two silver electrodes are placed in a solution of silver nitrate, $AgNO_3$, metallic silver is removed from the + electrode and deposited on the − electrode; the concentration of the solution remains unchanged. Similarly, if two copper electrodes are placed in a solution of copper sulfate, $CuSO_4$, metallic copper is removed from the + electrode and deposited on the − electrode without changing the concentration of the solution. These last two electrolytic processes are examples of *electroplating*.

The principles relating the amount of material deposited, or gas generated, in electrolysis were first determined experimentally by Faraday:

FARADAY'S PRINCIPLES OF ELECTROLYSIS: *The quantity of an element undergoing chemical reaction at an electrode is proportional to the quantity of electric charge passing, proportional to the atomic mass of the element, and inversely proportional to its valence.*

Faraday's principles state that the mass m of an element deposited or dissolved is proportional to the quantity Q of electric charge and the atomic mass M of the element, and inversely proportional to its valence v. Thus we can write

$$m \propto \frac{QM}{v},$$

or, alternatively, the charge required for mass m satisfies the proportionality

$$Q \propto v\frac{m}{M}.$$

The ratio m/M is proportional to the number n of kilomoles undergoing reaction; hence we can write

$$Q = Fvn, \tag{1}$$

where the proportionality constant, F, is called the *Faraday constant*, and has the experimental value

$$F = 96.487 \times 10^6 \text{ C/kmol}. \tag{2}$$

The **Faraday constant** is the charge per kilomole required for electrolysis of a monovalent element.

The explanation of the mechanism of the conduction of electric currents by solutions of electrolytes was given in 1887 by the Swedish chemist, SVANTE AUGUST ARRHENIUS, who suggested that all or part of the electrolyte exists in the solution in the form of separate *ions* carrying positive or negative charges, rather than in the form of neutral molecules. Thus, when electrolytes are dissolved in water, the solute exists largely in the form of ions as indicated in the following reversible reactions:

$$NaCl \rightleftharpoons Na^+ + Cl^-$$

$$MgCl_2 \rightleftharpoons Mg^{++} + 2\ Cl^-$$

$$NaOH \rightleftharpoons Na^+ + OH^-$$

$$HCl \rightleftharpoons H^+ + Cl^-$$

where the positive and negative superscripts indicate charges of $+e$ and $-e$, respectively. It is the availability of ions to carry the current that enables the solution of an electrolyte to conduct electricity; the ions play the role that electrons play in metallic conduction.

2. Charge Transport in Electrolysis

As a first example of electrolysis, consider a copper-plating bath, illustrated in Fig. 1, which accomplishes the transfer of copper from the positive electrode through the solution of copper sulfate ($CuSO_4$) to the negative electrode. Both electrodes are copper, since even if copper is being plated on an electrode of some other metal, the electrode becomes effectively copper as soon as the process is under way and a thin layer of copper has been deposited.

The solution of copper sulfate in water is strongly ionized to form Cu^{++} and SO_4^{--} ions. The source of EMF sets up an electric field that

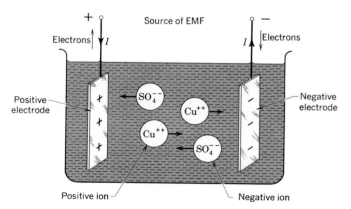

Fig. 1. A copper-plating bath.

causes the Cu^{++} ions to drift (relatively slowly) toward the negative electrode and the SO_4^{--} ions to drift toward the positive electrode.

The Cu^{++} ions that reach the negative electrode combine with two electrons from the metal and are deposited as metallic copper in accordance with the equation (we use e^- as the chemical symbol for an electron):

$$Cu^{++} + 2\ e^- \rightarrow Cu_{Metal}. \qquad (-\ \text{electrode}) \quad (3)$$

The SO_4^{--} ions, on contacting the positive electrode, give up their two electrons to the metal and combine with a Cu atom to form a molecule of $CuSO_4$ which, being readily soluble, goes into solution:

$$SO_4^{--} + Cu_{Metal} \rightarrow 2\ e^- + CuSO_4. \qquad (+\ \text{electrode}) \quad (4)$$

The *net* effect of the pair of reactions (3) and (4) is the removal of a copper atom from the + electrode and its deposition on the − electrode. There is no change in concentration of the $CuSO_4$ in the solution. As ions are removed from the solution, new ions are generated to keep up the ionic concentration.

The combination of the two reactions (3) and (4) involves the gain of two electrons from the solution by the + electrode in (4), and the loss of two electrons from the − electrode in (3). To maintain steady conditions, two electrons must pass through the outside metal circuit from the positive electrode through the source of EMF to the negative electrode. This movement of electrons corresponds to the flow of an equivalent quantity of conventional positive charge in the opposite direction, as shown by the arrows marked *I*. Under steady conditions, the same net amount of negative charge must cross each plane in the electrolyte (from right to left in Fig. 1); otherwise, the net charge of the electrolyte on one side of the plane would increase and that on the other side would decrease. Hence, under steady conditions the same current crosses each plane of a closed circuit that includes the electrolytic cell of Fig. 1. The current is carried by electrons in the metallic parts of the circuit; in the electrolyte it is carried by positive and negative ions.

After the process is started, *exactly the same number of reactions* (3) *and* (4) *will occur*. If this equality does not obtain initially, there will be an initial net transfer of charge between the electrolyte and the wire circuit that will set up an electric field in the electrolyte in such a direction as to speed up the motion of one type of ion, slow down that of the other, and rapidly restore equilibrium between the two rates. In this equilibrium condition, the statement in italics above is true.

We now note that equations (3) and (4) state that in the electrolytic transfer of 2 kmol of electrons, 1 kmol of bivalent copper is transferred. If we consider the similar electrode reactions in a silver-plating bath using silver nitrate ($AgNO_3$) solution, in which the silver is monovalent:

$$Ag^+ + e^- \rightarrow Ag_{Metal}, \qquad (-\ \text{electrode}) \quad (5)$$

$$NO_3^- + Ag_{Metal} \rightarrow e^- + AgNO_3, \qquad (+ \text{ electrode}) \quad (6)$$

we see that only one kmol of electrons is required for the transfer of one kmol of monovalent silver.

In general, *if a metal is plated from a solution in which it has valence v, v kilomoles of electrons are required for the transfer of one kilomole of the metal.* From this result, we can derive *Faraday's principles of electrolysis,* which were discovered empirically before the ionic nature of electrolytic conduction was understood. We see that the value of the faraday constant in (2) must be just the charge per kilomole of electrons. We can compute this charge from the known values of the Avogadro constant and the electronic charge:

$$F = N_A\, e = (6.0225 \times 10^{26}/\text{kmol})(1.602\ 10 \times 10^{-19}\ C) = 96.487 \times 10^6\ C/\text{kmol}.$$

While the value of N_A and the size e of the electronic charge can be determined by independent means, the size of the faraday can also be determined with great accuracy by weighing the deposit in an electrolytic cell and measuring the charge passing through it. Determination of the faraday in this manner has contributed to our knowledge of the value of N_A e and particularly to the value of N_A.

Let us now consider the reactions that take place in the electrolysis of pure water by current that enters and leaves through chemically inert platinum electrodes (Fig. 2). Pure water is only weakly ionized, but it does contain in each mole a definite number of *hydrogen** ions, H^+, and the same number of *hydroxyl* ions, OH^-. The number of ions of each type per kilomole of water molecules is 9×10^{17}, so that only about one in 10^9 of the water molecules is ionized.

In Fig. 2, the hydrogen ions that reach the negative electrode pick up electrons and form hydrogen gas:

$$4\ H^+ + 4\ e^- \rightarrow 2\ (H_2)_{Gas}. \qquad (- \text{ electrode}) \quad (7)$$

The OH^- ions that reach the positive electrode give up their electrons and form oxygen gas and water:

$$4\ OH^- \rightarrow 4\ e^- + 2\ H_2O + (O_2)_{Gas}. \quad (+ \text{ electrode}) \quad (8)$$

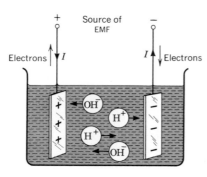

Fig. 2. Electrolysis of water.

The net effect of (7) and (8) is that two molecules of water are converted into two molecules of hydrogen and one of oxygen gas. Transfer of four electrons is required, one for each *atom* of hydrogen that appears as gas and two for each *atom* of oxygen that appears as gas. We see that Faraday's principles again apply to this case, one kilomole

---

*Hydrogen ions are always attached to a neutral water molecule so that the actual ions are $(H_3O)^+ = H^+ + H_2O$, but this detail is unimportant to our discussion.

of electrons being required for each kilomole of monovalent hydrogen *atoms* and two for each kilomole of bivalent oxygen *atoms*.

The dissociation of water into its constituent gases involves a considerable increase in chemical energy; in electrolysis this chemical energy is obtained by conversion of electric energy furnished by the source of EMF. This energy is equal to that converted from chemical energy into heat in the combustion of hydrogen. In the combustion reaction,

$$H_2 + \tfrac{1}{2} O_2 \to (H_2O)_{Liq}, \tag{9}$$

the heat of reaction, determined calorimetrically, is 69 000 *kilocalories per kilomole* of water molecules formed when one kilomole of hydrogen molecules reacts with one-half kilomole of oxygen molecules, both the gases being at 0° C and 1 atm, and the water is cooled to 0° C. Correspondingly, to *dissociate* one kilomole of water molecules, energy of 69 000 kcal would have to be supplied—in the case of electrolytic dissociation, by the source of EMF.

If we divide this energy per kilomole by the Avogadro constant, we get the energy of a single molecular reaction (9) in *kilocalories per molecule*. However, we shall find it much more useful to express the energy in *electron-volts per molecule*. From (21), p. 485, and (6), p. 311, we find that $1 \text{ kcal} = 2.612 \times 10^{22}$ eV; and hence that

$$1 \frac{\text{kcal}}{\text{kmol}} = \frac{2.612 \times 10^{22} \text{ eV}}{6.022 \times 10^{26} \text{ molecule}} = 4.337 \times 10^{-5} \frac{\text{eV}}{\text{molecule}};$$

$$1 \frac{\text{eV}}{\text{molecule}} = 23\,060 \frac{\text{kcal}}{\text{kmol}}. \tag{10}$$

This relation means that if we supply 1 electron-volt of energy to each molecule, we supply 23 060 kcal of energy to each kmol. Hence, if we wish to *reverse* the reaction (9) by electrolysis, we must supply 69 000 kcal/kmol or 3.0 eV/molecule.

3. Voltage Necessary for Electrolysis

When voltage is applied between platinum electrodes in pure water or in a dilute acid solution that electrolyzes to form H_2 and O_2, no charge flows so long as the difference of potential is below 1.7 V. Higher applied voltages result in a current proportional to the difference between the applied voltage and 1.7 V. When charge is flowing, the electrolytic cell acts like a battery on charge, with an EMF of 1.7 V, which, according to what we have learned in the previous chapter, must be related to a change of electric energy into chemical energy. In fact, a chemical reaction is taking place:

$$H_2O \to H_2 + \tfrac{1}{2} O_2,$$

which requires 69 000 kcal of energy per kmol of water decomposed, or

3.0 eV of energy per molecule of water. For each molecule of water decomposed, a charge of 2 electrons passes through the cell; and in order to furnish 3.0 eV of energy to a charge of 2 electrons, the charge must pass through a difference of potential of 1.5 V. This figure is 0.2 V lower than the observed EMF of 1.7 V. The extra electric energy of 0.2 eV/electron goes into thermal energy. We do not expect an *exact* correspondence with the calorimetrically measured heat of reaction because the electrolysis would not be expected to generate the gases at *exactly* the same temperature as the water from which they are formed—the condition under which the heat of reaction is measured.

In the case we have been considering, charge does not start to flow until we apply a potential of 1.7 V to furnish the energy absorbed in the reaction that accompanies the flow. Above this voltage, the amount of current will be governed by the terminal voltage and by the internal resistance just as in the case of a battery on charge. It turns out that thermal energy of 0.2 eV/electron (called *reversible heat*) is produced in addition to the $I^2 R_I$ heat accompanying the internal resistance.

In electrolysis in which copper is being plated from a copper anode to a copper cathode, no chemical energy need be supplied, since for each pair of electrons that traverses the cell, one molecule of solid copper is dissolved, but also one molecule of solid copper is plated, so there is no *net* chemical change. In this case charge will flow with the smallest voltage, and Ohm's law is obeyed, with current proportional to voltage. All the electric energy goes into $I^2 R$ heat.

Now suppose that we arrange a cell so that copper will be dissolved and zinc plated out. This process is accomplished in the old Daniell cell, which was used almost universally in the early days of telegraphy. In this cell, illustrated in Fig. 3, solutions of $CuSO_4$ and $ZnSO_4$ are separated by a porous partition through which diffusion of the solutions is very slow. Although the Daniell cell is no longer used as a commercial source

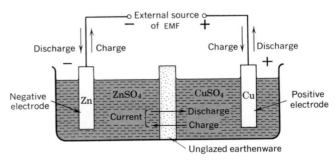

Fig. 3. Schematic diagram of the Daniell cell. Arrows show directions of conventional current in charge and discharge. The cell was usually made with the zinc electrode and $ZnSO_4$ inside a porous cup. The cup in turn was immersed in a glass jar containing $CuSO_4$. A cylindrical copper electrode surrounded the porous cup.

of power, a description of its operation will serve as a good introduction to the use of electrolytic cells as sources of EMF, since the electrode reactions are simpler than those in more modern cells.

If we apply sufficient external EMF to drive current through the cell in the direction marked 'charge,' the electrode reactions put copper in solution and plate out zinc, one atom of each for every two electrons:

$$Cu_{Metal} \rightarrow Cu^{++} + 2e^-, \qquad Zn^{++} + 2\,e^- \rightarrow Zn_{Metal}. \qquad \text{(charge)} \qquad (11)$$

A minimum external voltage of 1.09 V is required to bring about these reactions. Hence, it must be that a minimum energy of 2.18 eV/atom or 50 100 kcal/kmol is required to make Cu displace Zn in solution, since the net result of the above reactions is

$$Cu_{Metal} + Zn^{++} \rightarrow Cu^{++} + Zn_{Metal}. \qquad (12)$$

The reverse reaction, Zn displacing Cu in solution, should then release energy at the rate of 50 100 kcal/kmol. The heat of this reaction,

$$Zn_{Metal} + Cu^{++} \rightarrow Cu_{Metal} + Zn^{++}, \qquad (13)$$

can be measured calorimetrically, because it can be made to take place by merely stirring up flakes of zinc metal in $CuSO_4$ solution. The heat generated in this reaction is found to be 55 200 kcal/kmol. The difference between this figure and the 50 100 kcal/kmol determined from the cell voltage represents thermal energy that is converted into chemical energy while the reactions in the cell are taking place. This thermal energy is called the *reversible heat* because, while it disappears on charge, it will reappear on discharge.

In Fig. 2, when the voltage of the source of EMF drops below 1.7 volts, the current drops to zero and water ceases to be electrolyzed because sufficient energy cannot be supplied. The cell of Fig. 3 behaves differently. In Fig. 3, as the voltage of the external source of EMF is reduced to 1.09 V, the charging current falls to zero right enough, but when the external voltage falls below 1.09 V, the current *reverses* and the cell begins to discharge. Charge is now moving through the cell in the reverse direction to that discussed above, the reactions (11) are proceeding in the reverse direction, copper is being plated, and zinc is being dissolved. The net reaction is now (13) instead of (12). Reaction (13) *releases* chemical energy which in this case does not appear as heat, as in the zinc-flake experiment described above, but principally as electric energy of 2.18 eV per atom, representing an increase of potential energy of + charges moved from the negative to the positive electrode through the cell. *The cell is acting as a source of* EMF. Transformation of 2.18 eV of chemical energy into electric energy for each 2 electronic charges moved represents an EMF of 1.09 V.

To discuss the action of the cell on discharge in another way, we may start by noting that the reaction (13) will go by itself, with release of chemical energy. But in the arrangement of the Daniell cell (Fig. 3), the reaction (13) has no opportunity to take place by itself because there are no copper ions in contact with the zinc plate. This net reaction can take place, however, via two electrode reactions, the reverse of (11):

$$Zn_{Metal} \to Zn^{++} + 2\ e^-, \qquad Cu^{++} + 2\ e^- \to Cu_{Metal}. \qquad \text{(discharge)} \quad (14)$$

But *this pair of reactions can take place only as fast as an external circuit carries the electrons left behind in the zinc plate around to the copper plate to combine with the copper ions.*

The fundamental difference in behavior between this cell, which will act as a *source* of EMF, and the cell of Fig. 2, which will not, is that reactions (11) and (14) are essentially reversible, whereas (7) and (8) are not.

4. Cells in Present Use as Sources of EMF

Many different cells have been devised and used in the past as sources of EMF. These cells fall into three types, which may be designated as *standard cells, primary cells,* and *storage cells.* Of each type, only a few examples, which have proven superior to all others, are presently in use. These examples, which will be described briefly in this section, are

(a) *Standard cells:*   Weston cadmium cell

(b) *Primary cells:*   Leclanché dry cell
                       Mercury dry cell

(c) *Storage cells:*   Lead-acid storage cell
                       Edison alkaline cell
                       Special-purpose storage cells

(a) *Standard cells.* A standard cell is not intended as a source of energy but for use as a secondary standard of voltage. It is always used in a voltage-comparison circuit such as the potentiometer which was described on p. 553. It is designed to maintain a very constant voltage over a long period of time, provided no appreciable current is ever drawn from the cell. The voltage of the Weston cadmium cell, which is about 1.018 V, can be trusted to remain constant to about 1 part in 100 000.

The Weston cell (Fig. 4) has as its positive electrode liquid mercury on which floats a paste of mercurous sulfate, Hg_2SO_4, which is an almost insoluble compound. The negative electrode is an amalgam, 10 to 15 per cent of cadmium dissolved in mercury. The electrolyte is a solution of $CdSO_4$.

The electrode reactions are reversible so long as the current is kept extremely low. On discharge, at the negative electrode cadmium goes into solution as Cd^{++}, in the reaction $Cd_{Metal} \rightarrow 2\,e^- + Cd^{++}$. At the positive electrode, neutral mercury is released from the Hg_2SO_4, the SO_4 going into solution in the reaction $Hg_2SO_4 + 2\,e^- \rightarrow 2\,Hg_{Liq} + SO_4^{--}$.

(b) *Primary cells.* A primary cell is one that cannot be 'charged,' because the cell reactions are not efficiently reversible. All of the energy that the cell will ever deliver is put in as chemical energy when the cell is made; when this energy is exhausted, the cell is dead.

The most widely used primary cell is the common Leclanché dry cell (Fig. 5). This cell is of course not really 'dry'; it is merely *unspillable* because the electrolyte is soaked up in absorbent material. The negative electrode is the zinc can itself. The positive electrode is a stick of carbon, which is inert and does not enter the chemical reaction. The electrolyte is a strong solution of ammonium chloride (NH_4Cl).

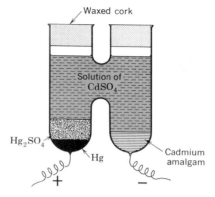

Fig. 4. The Weston standard cell.

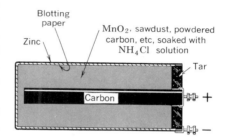

Fig. 5. The Leclanché dry cell.

When this cell delivers current, zinc goes into solution at the negative electrode. At the inert carbon electrode, NH_4^+ ions are discharged, releasing free ammonia and hydrogen: $2\,NH_4^+ + 2\,e^- \rightarrow 2\,NH_3 + H_2$. The ammonia, NH_3, is readily soluble and remains dissolved in the electrolyte. The free hydrogen gas, which would quickly spoil the action of the cell if allowed to collect around the carbon electrode, is taken up and oxidized by manganese dioxide: $H_2 + 2\,MnO_2 \rightarrow H_2O + Mn_2O_3$. The cell, when new, gives an EMF of slightly over 1.5 V.

Another primary cell coming into wide use is the *mercury dry cell*. This cell has a positive electrode consisting initially of solid mercuric oxide, HgO, mixed with graphite. Its negative electrode is zinc. The electrolyte, which is retained in an absorbent pad, is a solution of potassium hydroxide, KOH, saturated with zinc oxide, ZnO. At the positive electrode, the reaction is $HgO_{Solid} + H_2O + 2\,e^- \rightarrow Hg_{Liquid} + 2\,OH^-$, which 'reduces' the HgO to liquid mercury. The reaction at the negative electrode is $Zn_{Metal} + 2\,OH^- \rightarrow ZnO_{Solid} + H_2O + 2\,e^-$, which 'oxidizes' the zinc. The EMF is about 1.35 V. This cell is widely used in hearing aids and automatic cameras because it has a higher ampere-hour output and a lower internal resistance than a Leclanché dry cell of the same size. This cell will also operate at lower temperature than the Leclanché cell, so it is being employed in airborne equipment and in the Arctic.

(c) *Storage cells.* The energy supplied by a primary cell is in general very expensive compared with energy obtained by burning coal or oil. Zinc to 'burn' is extremely expensive compared with fossil fuels to burn. A *storage* cell employs completely reversible chemical reactions so that the chemical energy used up in discharge can be restored by charging, utilizing electric energy obtained from relatively inexpensive sources.

The two types of storage cells in widest use are the *lead-acid cell*, whose electrodes are lead and lead peroxide in a sulfuric-acid solution, and the *Edison alkaline cell*, whose electrodes are nickel and iron in various stages of oxidation, and whose electrolyte is a solution of the alkali, potassium hydroxide (caustic potash, KOH).

The lead-acid cell has the advantage of extremely low internal resistance, which enables it to supply extremely large currents for short periods of time. The Edison alkaline cell is lighter in weight and structurally stronger and more durable than the lead-acid cell. Its high internal resistance (about 10 times that of a comparable lead battery) makes it unsuitable for automobile motor starting.

In the lead-acid cell, the active material of the negative plate is metallic lead. On discharge, the metallic lead changes to solid lead sulfate which remains on the surface and in the pores of the spongy-lead electrode. The active material on the positive plate is lead peroxide, PbO_2, which also changes to solid lead sulfate on discharge. In the overall reaction, for each two electrons that flow, one molecule of lead peroxide is changed to lead sulfate at the positive electrode, one atom of lead is changed to lead sulfate at the negative electrode, and two molecules of H_2SO_4 disappear from solution and are replaced by two molecules of water. This overall reaction is:

$$PbO_2 + Pb + 2\ H_2SO_4 \xrightarrow{\text{discharge}} 2\ PbSO_4 + 2\ H_2O. \tag{15}$$

The density of the sulfuric-acid solution decreases as discharge progresses, since heavy sulfuric-acid molecules are replaced by lighter water molecules, according to (15). Hence, the density of the electrolyte furnishes an indication of the state of charge. The energy released in the chemical reaction (15) varies with the concentration of the sulfuric acid, because of variations in the large heat of solution of H_2SO_4 in water; hence, the EMF of the cell will also vary with charge condition and will furnish an indication of the state of charge. At full charge, the density is 1.285 g/cm³ and the EMF is 2.13 V; whereas at a discharged condition, the density is 1.150 g/cm³ and the EMF has dropped to 2.01 V.

The reactions in the Edison nickel-iron cell are more complex than those in the lead cell. In use, various mixtures of Ni, NiO, Ni_2O_3, Ni_3O_4, $Ni(OH)_2$, and $Ni(OH)_3$ are found on the positive plate, with mixtures of Fe, FeO, and $Fe(OH)_2$ on the negative plate. The chemical energy during discharge is furnished by oxidation of the iron and a corresponding reduction of the nickel.

Other types of storage cells are used for special purposes. For example, a *nickel-cadmium alkaline cell* is employed in the 'solar battery.' In this battery, a barrier-layer, semiconductor type of photocell is used to convert the radiant energy of sunlight directly into electric energy that is used to charge a nickel-cadmium storage cell.

5. Fuel Cells

In recent years there has been a renewal of research and development activity directed toward the perfection of a practical, efficient 'fuel cell.'

A fuel cell differs from a primary cell only in that the substances undergoing reactions are in liquid or gaseous form and can be continuously replenished. The idea is an old one. The first fuel cell was demonstrated by W. R. GROVE in 1839. Grove immersed two strips of platinum in sulfuric acid, supplied one with hydrogen gas, the other with oxygen gas, and generated a current in an external circuit between the platinum electrodes. The electrode reactions, in this type of fuel cell, are

$$2\,H_2 + 2\,SO_4^{--} \rightarrow 4\,e^- + 2\,H_2SO_4, \qquad (-\text{ electrode})$$

$$4\,H^+ + 4\,e^- + O_2 \rightarrow 2\,H_2O. \qquad (+\text{ electrode})$$

In present-day experimental fuel cells, liquid or gaseous fuel and oxidant are allowed to diffuse into the interior of porous electrodes. Simultaneously, a liquid electrolyte diffuses into the electrode and the electrode reactions take place in the pores of the electrode. With this arrangement, the hydrogen-oxygen fuel cell, whose electrode reactions are given above, can be represented schematically as in Fig. 6.

The hope is that, eventually, fuel cells, employing inexpensive conventional fuels, will generate power with higher efficiency than heat engines, since there is no Carnot limit on the efficiency of a fuel cell. This objective has not yet been realized in practice.

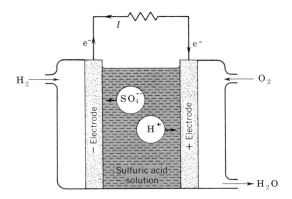

Fig. 6. Schematic diagram of a hydrogen-oxygen fuel cell.

6. Thermoelectric Effects; the Thermocouple

About 1834, JEAN CHARLES ATHANASE PELTIER, in Paris, discovered that when there is a current through a circuit composed of two dissimilar metals, as in Fig. 7, one of the junctions between the metals tends to become warmer and the other junction tends to become cooler. The rates of heat generation and absorption are proportional to the current. When

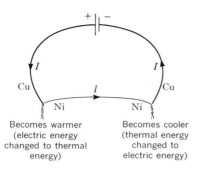

Fig. 7. Peltier effect for copper and nickel.

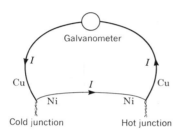

Fig. 8. Seebeck effect. The thermocouple.

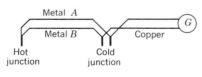

Figure 9

the current is reversed, the roles of the two junctions are reversed. The temperature changes associated with this *Peltier effect* appear in addition to temperature increases resulting from the normal I^2R-heating. By use of low-resistance pieces of metal, it is possible, in spite of the I^2R-heating, to get one of the junctions to cool below room temperature. The dependence of the effect on current direction in the case of the metals Cu and Ni is shown in Fig. 7.

The *Seebeck effect,* discovered in 1821 by THOMAS JOHANN SEEBECK in Berlin, is the inverse of the Peltier effect. Here (Fig. 8), holding the two junctions at different temperatures causes a current when no other source of EMF is present. The current in Fig. 8 is in such a direction as to *tend* to equalize the temperature discrepancy. With the current direction shown, we note from Fig. 7 that the hot junction tends to cool and the cold junction to warm up. The circuit of Fig. 8 is a heat engine which absorbs heat at the hot junction, converts some of the heat into electric energy, and rejects the rest of the absorbed heat at the cold junction.

When used as a thermometer, the circuit of Fig. 8 is called a *thermocouple*. The value of the thermocouple as a thermometer depends on the fact that the net EMF developed is directly related to the temperature difference between the junctions. For small temperature differences, it is approximately proportional to the temperature difference.

In practice, a thermocouple circuit ordinarily consists of at least three metals—the two metals constituting the thermocouple proper, and copper leads to the galvanometer and copper galvanometer windings. The wiring is arranged as in Fig. 9. The two connections to the copper constitute the cold junction; and so long as these two connections are at the same temperature, the EMF developed is exactly the same as if the wires were connected directly together at the cold junction, with no copper circuit.

The thermocouple is one of the most sensitive devices for the measurement of radiant energy. In this usage, the junctions are made small, so as to have low thermal capacity, and blackened. The radiant energy to be measured is focused by a mirror on one of the junctions. Such an instrument is used for the mapping of spectra in the infrared beyond the limits of the photographic plate; for measuring the total amount of radiation of all wavelengths received from a star; and in the *radiation pyrometer,* designed to be pointed at a hole in a furnace and to determine the furnace temperature by measuring the total radiant energy emitted.

The thermoelectric effect is not an unmixed blessing. In any electrical apparatus in which the circuits contain different metals or even different grades of the same metal, temperature differences arising from any cause will set up small 'thermal EMF's' and 'thermal currents,' as they are called. Even when a piece of equipment is constructed of a single grade of metal, small thermal EMF's exist if there are temperature differences between different portions of the equipment. These EMF's appear as a result of a phenomenon known as the *Thomson effect.* If a copper rod is heated at one end and cooled at the other, a difference of potential is observed between the ends. This *Thomson* difference of potential arises from a temperature dependence of the density of free electrons in the metal. These free electrons have properties analogous to those of gas molecules. If a tube containing gas is heated at one end, the density of the gas decreases at that end since the gas pressure remains constant throughout. The 'electron gas' in a metal behaves similarly, and since the electrons are charged, the ends of the rod become charged, negatively at the low-temperature end where the electron density is greatest.

The Thomson effect is particularly striking in semiconductors. A practical *thermoelectric generator* of electricity can be made from the combination of a p-type and an n-type semiconductor arranged as in Fig. 10. (*See* Sec. 7 of Chapter 25 for a discussion of semiconductors.) The *electrons* in the n-type semiconductor concentrate *at the cool end,* charging this end *negative* on open circuit. Correspondingly, the *holes* in the p-type semiconductor concentrate *at the cool end,* charging this end *positive* on open circuit. The difference of potential between the ends of the two conductors can be used to furnish an electric current, as in the figure.

The difference of potential developed by the semiconductor circuit of Fig. 10 is hundreds of times greater than that of any metal circuit employed as in Fig. 8. However, this difference of potential is still only a few tenths of a volt for temperature differences of a few hundred degrees, so for practical energy-generation purposes, a number of circuits must be arranged in series. Practical generators for powering radios, the heat source being a kerosene lamp, are in wide use in remote portions of the U.S.S.R.

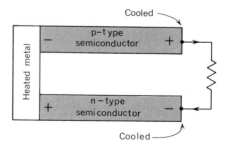

Fig. 10. A thermoelectric generator employing semiconductors.

The generator effect in Fig. 10 is analogous to the Seebeck effect of Fig. 8 in metals. The reverse effect in which a current is forced in the same direction through Fig. 10 can be used to *cool* the metal block. This effect is analogous to the Peltier effect of Fig. 7. Practical refrigerators employing semiconductors are in use.

PROBLEMS

1. The cathode of a silver coulometer increased 0.842 g in mass during an electrolysis. How many coulombs of electricity passed? Ans: 752.

2. How many coulombs are required to purify 1.5 kg of electrolytic copper? If an average current of 125 A is used, how long will it take?

3. Dilute sulfuric acid ionizes to give H^+ and SO_4^{--} ions. On electrolysis with platinum electrodes, hydrogen gas is given off at the cathode and oxygen gas at the anode. Write down the electrode reactions and show that the amount of gas of each type, per faraday, is the same as in the electrolysis of pure water.

4. A current of 0.8 A passes successively through solutions of hydrochloric acid (HCl), sulfuric acid (H_2SO_4), and phosphoric acid (H_3PO_4) via platinum electrodes. Calculate the weight and volume of hydrogen liberated per hour in each solution.

5. What volumes of H_2 and O_2 gas (at 0° C, 1 atm) are liberated if 0.40 A passes through dilute sulfuric acid for 10 min? Ans: 27.8 cm³; 13.9 cm³.

6. Chlorine and caustic soda are prepared commercially by electrolyzing a solution of common salt (NaCl), using inert electrodes. At the cathode, caustic soda (NaOH) is formed in solution, with hydrogen evolution. At the anode, chlorine gas (Cl_2) is evolved. Write down the electrode reactions. What volumes of chlorine and hydrogen gas at 0° C and 1 atm are evolved by passage of a charge of 125 MC?

7. When a ferric-chloride ($FeCl_3$) solution is electrolyzed by using platinum electrodes, chlorine gas is generated at the anode but iron is not plated out at the cathode. Rather, the solution starts changing to one of ferrous chloride ($FeCl_2$) in the neighborhood of the cathode. Write the electrode reactions.

8. Three electrolytic cells with platinum electrodes are connected in series. The first cell contains Ag^+ and NO^- ions; the second, Fe^{++} and Cl^-; the third, K^+ and I^-. In 1 h, 53 g of silver are deposited. In the same time, how many grams of oxygen are released, iron deposited, chlorine released, and iodine deposited? Compare the volumes of O_2 and of Cl_2.

9. A piece of 'costume jewelry' has a surface area of 90 cm² and is to be covered with a layer of silver 40 μm thick. If the plating process is accomplished electrochemically with a silver-nitrate solution, how long will be required for the process if a current of 5 A is used? Ans: 676 s.

10. A smooth, freshly cleaned sheet of aluminum is to be coated with an extremely thin layer of copper. The sheet is 50 cm in length and 20 cm in width and is to be covered on both sides by a layer of copper equal in thickness to 590 nm (the wavelength of yellow light). The plating process is accomplished in a $CuSO_4$ cell. How much electric charge is transported during the plating process? How many Cu atoms are deposited?

11. Outline a method of 'weighing' by means of an electrolytic cell. In particular, show how the loss in mass of the positive electrode in a copper-plating bath can be determined from current and time measurements. What mass of copper is removed from the positive electrode in 1 s by a steady current of 1 A? Ans: 0.329 mg.

12. By means of a stop watch and milliammeter, how accurately can a quantity of copper be 'weighed' if the stop watch can be read to 0.01 s and the meter to 0.1 mA? (*See* Prob. 11.)

13. Why does a lead cell 'gas' when it is 'overcharged'? What gas is given off at the positive electrode? at the negative electrode? Why must rooms in which lead storage batteries are charged be well ventilated to avoid an explosion hazard? A moderate amount of overcharging does not damage the battery; in fact, it is good to overcharge a lead battery occasionally until all cells are gassing freely, to be sure that both plates are fully formed. This procedure is known as 'equalization.' After an overcharge, what needs to be added to bring the electrolyte back to its normal fully charged density? Explain.

14. A lead cell of EMF 2.05 V is connected in parallel to a lead cell of EMF 2.10 V. Assuming an internal resistance of 0.001 Ω per cell, what current will flow initially? Why does the current decrease with time and approach zero?

15. What fractions of the electric energy go

into the chemical energy, reversible heat, and I^2R heat when a cell used to produce hydrogen and oxygen commercially is operated at 2 volts? Use the approximate voltage values given in Sec. 3. Ans: $3/4$, $1/10$, $3/20$.

16. (a) In the lead-acid cell, how many grams of PbO_2 are changed to $PbSO_4$ on the positive plate for each ampere-hour of charge delivered? (b) How many grams of lead are converted to $PbSO_4$ on the negative plate for each ampere-hour? (c) By how many grams does the mass of the electrolyte decrease for each ampere-hour?

17. The capacity of a lead-acid cell in ampere-hours is limited by the amount of PbO_2 on the positive plate. Before *all* the PbO_2 is converted to $PbSO_4$, the voltage of the cell has dropped and the internal resistance has increased to the point where it is desirable to recharge. A cell with a rated useful capacity of 60 A·h has 600 g of PbO_2 on the positive plates. What would be the capacity of this cell if *half* the PbO_2 were converted to $PbSO_4$? Ans: 67 A·h.

18. A storage battery maintains an average terminal voltage of 6.15 V during delivery of 80 A·h. How much energy is delivered to the external circuit? How many coulombs are there in 1 A·h?

19. Assuming that there is adequate $CuSO_4$, calculate how many A·h a Daniell cell will deliver before a 250-g zinc electrode is all dissolved? Ans: 205.

20. The solubility of $CuSO_4$ is 0.14 g per cm³ of solution. If a Daniell cell starts with 800 cm³ of saturated $CuSO_4$ solution and adequate zinc, how many ampere-hours will it deliver before the $CuSO_4$ is used up?

21. In Fig. 9, if metal A is replaced by a copper wire of the same resistance, the galvanometer reads a current I_1; if metal B is replaced by a copper wire of the same resistance, the galvanometer reads a current I_2 in the same direction. Show that for the actual circuit of Fig. 9, the galvanometer will read $I_1 + I_2$.

22. For given hot- and cold-junction temperatures, the *thermoelectric power* of a given metal A is defined as the voltage developed per degree temperature difference in a thermocouple in which the second metal is lead. The thermoelectric power is positive if the current at the hot junction is from the lead to metal A, as in the figure; it is negative if the current is in the reverse direction. If, when the hot junction is at 100° C and the cold junction at 0° C, the thermoelectric power of platinum is -4 μV/C deg and that of nickel is -21 μV/C deg, what will be the voltage developed by a Pt-Ni thermocouple with the junctions at these temperatures? What will be the direction of the current at the hot junction? Explain your reasoning carefully.

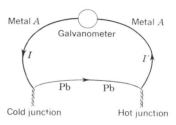

Problems 22-24

23. If the thermoelectric power of iron is $+12$ μV/C deg under the conditions of Prob. 22, what will be the magnitude and direction of the EMF of a Ni-Fe thermocouple with junctions at 0° and 100° C? Ans: 3.30 mV; current from Ni to Fe at hot junction.

24. Near 20° C, the thermoelectric power (*see* Prob. 22) of copper is $+2.7$ μV/C deg, while that of constantan is -38.1 μV/C deg. If a copper-constantan thermocouple is used with a suspended-coil galvanometer that gives a deflection of 0.05 mm/μV when used with a light beam and scale, what is the minimum temperature difference between junctions that can be detected if a 0.1-mm deflection is the minimum observable?

25. A *sonobuoy* is a device, usually dropped from an airplane, that floats in the ocean, listens for the sound of submarines under the water, and broadcasts these sounds by radio to the plane overhead. To power the sonobuoy, a seawater-activated battery is usually employed, with plates of AgCl and Mg. Write the electrode reactions.

28 MAGNETIC FORCES

The word *magnetism* comes from the ancient Greek name for certain naturally occurring iron-oxide stones called *lodestones,* which have the property of exerting forces on each other and on bits of iron or steel. They also have the ability to impart their own distinctive properties to pieces of steel brought near them. A piece of steel (for example, a steel needle) that has thus acquired the properties of the lodestone is said to be *magnetized,* and is called a *magnet.* A lodestone or a steel magnet experiences a torque that tends to orient it in a particular direction on the earth. This property led to the important invention, sometime before the middle of the 12th century A.D., of the mariner's compass.

In 1820, HANS CHRISTIAN OERSTED discovered that forces exist between a magnet and a wire carrying electric current. In the same year, Ampère found that related forces exist between two wires carrying electric currents. Ampère suggested that the forces between magnets arise from the presence of circulating currents within the magnets. This hypothesis has proven to be correct, the circulating currents consisting of electrons in their orbital and 'spin' motions in the atoms. Thus *all magnetic phenomena are explained in terms of forces between electric charges in motion.*

Since the forces between currents are fundamental, we begin the study of magnetism with a description of these forces in this and the following chapter, the study of magnetized materials being left for Chapter 30.

1. Magnetic Forces

As Ampère first discovered, wires carrying currents exert on each other forces proportional to the currents, in addition to any electrostatic forces that may exist between the wires. In addition to the electric forces that exist between charges whether they are at rest or in motion, there are forces associated purely with the motion of the charges; these are called *magnetic forces*. Magnetic forces exist between two charges only if both charges are in motion, the magnitude of the force being proportional to the product of the charges and to the product of the speeds of the charges, and depending in a complex manner on the distance between the charges and on their directions of motion.

Magnetic forces are forces associated with the *motion* of electric charges.

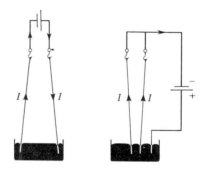

Figure 1

The existence of magnetic forces between parallel wires carrying currents can be easily demonstrated with the apparatus sketched in Fig. 1. Here two stiff wires hang with their lower ends dipping into a pool of mercury so that the lower ends are free to move. If currents are sent through the two wires in opposite directions, the wires are mutually repelled; if currents are sent through the two wires in the same direction, the wires are mutually attracted.

That these forces are fundamentally forces between charges in motion can be easily demonstrated by replacing the left-hand wire of Fig. 1 (whose upward current corresponds to *electrons moving down*) by an electron-ray tube in which a beam of electrons moves down through a vacuum. According to the direction of the current in the fixed parallel wire in Fig. 2, the beam of electrons is transversely deflected in the direction corresponding to the motion of the wire in Fig. 1. The two parts of Fig. 2 are seen to correspond to the two parts of Fig. 1.

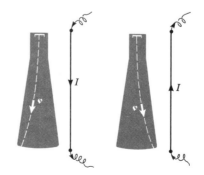

Fig. 2. A stream of electrons in an electron-ray tube is attracted or repelled by a parallel electric current.

The forces between two *parallel* currents constitute a particularly simple case of a fairly complex system of forces which we shall study in detail. This system of forces is best handled by introducing a vector field called the *magnetic field,* the vector being the *magnetic intensity* ℬ. The procedure is similar to the determination of electric forces in terms of the electric intensity ℰ. In the electrostatic case we proceeded in two steps. We first obtained ℰ from the distribution in space of electric charges; similarly, we obtain ℬ from the distribution in space of electric currents. Then as a second step, the force on an electric charge was determined by the magnitude and direction of ℰ at the location of the charge; similarly the magnetic force on a *current element* (a short length of wire carrying current, or a single moving charge) is determined by the magnitude and direction of ℬ at the location of the current element. One should be warned, however, that the laws for computing the magnetic intensity and for computing magnetic forces are much more complex than the corresponding laws in electrostatics. The fundamental reason for this complexity is that a current element is a vector quantity, having magnitude and direction, whereas an electric charge is a scalar quantity, having only magnitude.

In the present chapter, we shall learn to compute the forces exerted on currents in magnetic fields; in the next chapter we shall learn to compute the magnetic fields produced by currents.

2. Magnetic Intensity

We shall define magnetic intensity operationally by specifying a method of measuring the magnetic intensity ℬ at a point of space.

A magnetic field (for example, the earth's magnetic field, the field in the vicinity of magnetized iron, or that in the vicinity of conductors carrying current) can be explored by means of either a small magnet mounted as in Fig. 4 so that it is free to rotate about an axle perpendicular

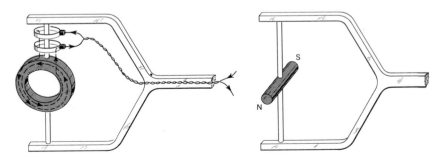

Fig. 3. A small current-carrying coil free to turn about an axle.

Fig. 4. A small magnet behaves like a small coil.

to the magnet, or a small current-carrying coil consisting of many turns of wire mounted as in Fig. 3 so that it is free to rotate about an axle in the plane of the coil. These devices must be carefully balanced so that they will not tend to rotate because of gravitational forces, particularly when the axle is horizontal. The magnet of Fig. 4, when the axle is vertical, is the prototype of the familiar compass needle. As a compass, its ends are *north-seeking* and *south-seeking;* by this criterion its ends are designated as N and S.

At any point in a magnetic field, both the magnet and the coil will tend to orient in a certain direction. We specify the orientation of the magnet by means of a vector **M** (called the *magnetic moment*) along the axis of the magnet pointing in the direction from the south-seeking end toward the north-seeking end, as in Fig. 5. We specify the orientation of a coil by a magnetic-moment vector **M** lying along the *axis* of the coil (not to be confused with the *axle* of Fig. 3) and pointing in the direction related to the current direction by the right-hand rule given on p. 496.

These vectors are called the *magnetic moments* of the magnet and of the small coil. We shall first define the magnetic moment of a coil:

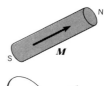

Fig. 5. The magnetic moment **M** of a magnet and of a small coil.

> The **magnetic moment** of a small plane coil is a vector whose magnitude is the product of the number of turns, N, the current in each turn, I, and the area A of the circuit, and whose direction is perpendicular to the plane of the coil in the sense given by the right-hand rule.

Thus
$$M = NIA \qquad (1)$$
and is measured in $\text{A} \cdot \text{m}^2$.

At a given point in a magnetic field, such as that of the earth, it is found experimentally that *a small plane coil and a small magnet behave exactly alike;* either will tend to turn so that its magnetic moment is pointing in the same direction. This direction defines the direction of the magnetic intensity ℬ:

> The **direction** of the **magnetic intensity** ℬ at a point of space is defined as the direction into which the magnetic-moment vector of either a small plane coil or a small magnet tends to turn when the small coil or magnet is placed at that point of space.

Since a small plane coil or a small magnet tends to turn so that its magnetic-moment vector **M** lines up with the magnetic-intensity vector ℬ, there must exist a torque **L** tending to turn **M** into ℬ when these vectors are not aligned. It is found experimentally (*see* Fig. 6) that

1. The magnitude L of the torque is proportional to the sine of the angle between **M** and ℬ.

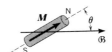

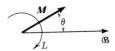

Fig. 6. The torque **L** that tends to turn **M** into ℬ is proportional to $\sin\theta$.

2. In the case of a small coil, the magnitude L of the torque is proportional to the magnitude M of the magnetic moment of the coil.

Finally, as the first step in defining the *magnitude* of $\mathcal{B}$, we postulate that

3. The magnitude L of the torque is proportional to the magnitude $\mathcal{B}$ of the magnetic intensity.

Since $\boldsymbol{L}$ has a direction that tends to turn $\boldsymbol{M}$ into $\boldsymbol{\mathcal{B}}$ and a magnitude proportional to M, to $\mathcal{B}$, and to $\sin\theta$, we see (Fig. 7) that $\boldsymbol{L}$ has magnitude and direction proportional to $\boldsymbol{M}\times\boldsymbol{\mathcal{B}}$. As a final step in the definition of $\mathcal{B}$, we choose the proportionality constant equal to 1, so that we can write

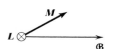

Fig. 7. If M and $\mathcal{B}$ are in the plane of the paper, $\boldsymbol{L}=\boldsymbol{M}\times\boldsymbol{\mathcal{B}}$ points into the paper, as indicated by the tail of an arrow.

$$\boldsymbol{L}=\boldsymbol{M}\times\boldsymbol{\mathcal{B}}, \qquad L=M\mathcal{B}\sin\theta. \qquad \begin{Bmatrix} L \text{ in } \text{N}\cdot\text{m} \\ M \text{ in } \text{A}\cdot\text{m}^2 \\ \mathcal{B} \text{ in } \text{T} \end{Bmatrix} \quad (2)$$

The unit of magnetic intensity defined by this equation is given the name tesla (T).* According to (2),

One **tesla** is the magnetic intensity that will result in a torque of 1 N·m on a coil of magnetic moment 1 A·m² placed with its axis perpendicular to the direction of the magnetic intensity.

No new fundamental unit is required to describe magnetic phenomena. All magnetic units can be expressed in terms of mechanical units and the unit of electric charge or current. Thus, from (2), we note that the unit of $\mathcal{B}$ can be written as

$$1\text{ T} = 1\frac{\text{N}}{\text{A}\cdot\text{m}} = 1\frac{\text{kg}}{\text{C}\cdot\text{s}} = 1\frac{\text{kg}}{\text{A}\cdot\text{s}^2}. \qquad (3)$$

We should also like to have equation (2) applicable to the case of a small magnet by giving a suitable definition of the magnetic moment of the magnet. The following definition accomplishes this purpose:

The **magnetic moment of a small magnet** is equal to the magnetic moment of a small coil that would experience the same torque when placed in the same orientation at the same location in the same magnetic field.

The magnetic moment of a magnet is thus measured in A·m². This unit will seem reasonable when we show in Chapter 30 that the magnetic moment of the magnet is actually the sum of the magnetic moments of the 'electronic' currents circulating in the atoms of the magnet.

---

*Named after Nikola Tesla (1856–1943), the American scientist of Serbian parentage who invented the alternating-current induction motor and the 'Tesla coil' for generation of very high voltages at very high frequencies.

The word 'small' in the preceding discussion requires explanation. In the last analysis, *small* can be taken to mean infinitesimal, or of atomic size, since we shall see later how to compute the magnetic moment of any finite current circuit from that of 'coils' of infinitesimal area, and the magnetic moment of any magnet by summing the magnetic moments of the individual atoms. In practice, the above relations and definitions are valid provided the coil or magnet is sufficiently small that the magnetic intensity $\mathcal{B}$ does not vary significantly over the area of the coil or the length of the magnet.

3. The Earth's Magnetic Field

In the definition of direction of magnetic intensity given above, *one should think of the coil or magnet as somehow cleverly mounted so that the magnetic-moment vector is free to turn in every direction.* In the practical mounting of Figs. 3 and 4, the magnetic-moment vector can rotate only in one plane and define the direction of the *component* of the magnetic intensity in that plane. With such a mounting, successive tests with the axle in various orientations are required to determine the direction of the magnetic intensity in space.

The earth has a magnetic field in which, over most of the inhabited surface, the horizontal component $\mathcal{B}_H$ points generally northward. The earth's magnetic intensity also has a strong downward component throughout most of the northern hemisphere, a strong upward component throughout most of the southern. The magnetic intensity near San Francisco, for example, has a northward component, an eastward component, and a large downward component, so that the actual vector $\mathcal{B}$ points in the direction shown in Fig. 8(a). The direction of this intensity vector is determined in two steps:

First a *compass needle,* mounted to rotate about a vertical axle only, is used to determine the direction of the horizontal component $\mathcal{B}_H$, as in Fig. 8(b). To assure that the needle moves in a horizontal plane, an accurate compass needle is floated, like a ship's card, on the horizontal surface of a liquid. The angle that the horizontal component makes with the geographic north is called the *declination,* and is designated as E or W (east or west of north).

Then a second needle, called a *dipping needle,* which rotates in perfect gravitational balance about a horizontal axle, is used as indicated in Fig. 8(c). The horizontal axle is aligned perpendicular to the horizontal component of the field so that the needle swings accurately in the plane determined by $\mathcal{B}_H$ and $\mathcal{B}_V$, which plane contains the vector $\mathcal{B}$ itself. This needle then points along $\mathcal{B}$ and gives the *dip* (also called *inclination*) of the magnetic field below the horizontal. The dip is designated as N or S according to whether it is the north or south end of the needle that dips below the horizontal.

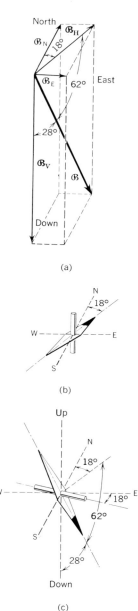

Fig. 8. (a) The direction of the earth's magnetic intensity at San Francisco. (b) A compass needle determines the *declination* as 18° E. (c) A dipping needle determines the *dip* as 62° N.

588 MAGNETIC FORCES CHAP. 28

The earth's magnetic field can be shown to result, principally, from currents inside the earth, but there is as yet no proven theory that explains the origin of these currents. The field is much like that of a current-carrying coil or bar magnet at the center of the earth. The field of such a coil or magnet, mapped by means of iron filings, is shown in Fig. 9.

The earth's field is approximately symmetrical about an axis, called the *geomagnetic axis*, that is tilted 12° with respect to the geographic axis. The great circle in a plane normal to the geomagnetic axis is called the *geomagnetic equator*. Even when plotted in geomagnetic coordinates, the field is not entirely symmetrical but shows localized variations arising from variations in earth structure. Figure 10 is a map, in geographic coordinates, showing the direction of the horizontal component of the earth's magnetic intensity; Fig. 11 shows the angle of dip.

The magnitude of the horizontal component of the earth's magnetic intensity varies from 35 μT at the magnetic equator to zero at the magnetic poles; that of the vertical component varies from zero at the magnetic equator to 70 μT at the magnetic poles. The earth's field is subject to

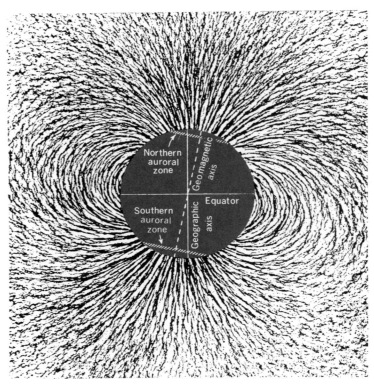

Fig. 9. The magnetic field of the earth. (From J. A. Fleming, *Terrestrial Magnetism and Electricity:* Dover Publications.)

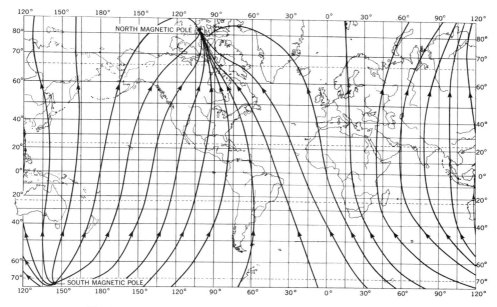

Fig. 10. Lines (called magnetic meridians) whose tangents show the direction of the horizontal component vector $\mathfrak{B}_H$ in the earth's magnetic field; that is, the direction in which a compass needle will point.

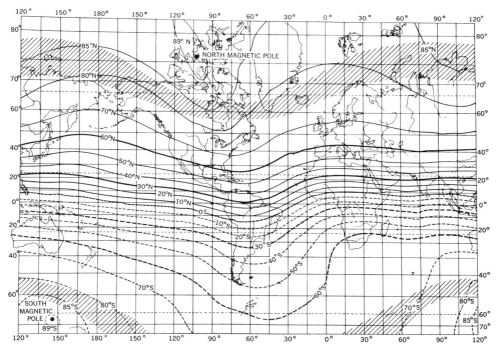

Fig. 11. The dip of the earth's magnetic field. The hatched areas show the zones in which auroras are observed on over 80 per cent of the nights. (*Department of Terrestrial Magnetism, Carnegie Institution of Washington.*)

irregular fluctuations. Large fluctuations of intensity, which may be of the order of 1 μT or more, are called *magnetic storms*. Such storms occur most frequently in the auroral zones shown on Figs. 9 and 11, and most frequently when sunspot activity is high. Magnetic storms and auroras are both believed to be caused by charged particles emitted from sunspots. In traveling toward the earth, these particles are deflected by the earth's magnetic field and concentrated in the auroral zones. When they enter the earth's upper atmosphere, they ionize the air, exciting the auroral radiation and setting up circulating currents of ions whose magnetic fields are the magnetic 'storms.'

Example. *A compass needle experiences a torque of* 5×10^{-4} N·m *when it is pointing geographically north at San Francisco, where the horizontal component of intensity of the earth's field is* 25 μT. *What is the magnetic moment of the compass needle?*

Since the compass needle is mounted on a vertical axle, it is only the horizontal component of the earth's magnetic intensity that tends to rotate it. Hence we can substitute the given values of L and $\mathcal{B}_H$ in (2). The value of θ, the angle between geographical north and $\mathcal{B}_H$, is given in Fig. 8 as 18°. Therefore the magnetic moment is

$$M = \frac{L}{\mathcal{B}_H \sin\theta} = \frac{5 \times 10^{-4} \text{ N·m}}{(25 \times 10^{-6} \text{ T})(0.309)} = 64.7 \text{ A·m}^2.$$

4. Magnetic Force on a Conductor Carrying Current

The torque on a small coil in a magnetic field arises from forces on the different elements of length of the current-carrying conductors in the coil. We now turn to consideration of the forces on such current elements.

By a current element of magnitude $I\,dl$, in A·m, we mean a short piece of conductor of length dl carrying current I. The force on such a current element is determined by the magnetic intensity at the location of the current element. The whole force on a conductor is obtained by properly summing the forces on the current elements. The relation for the force on a current element can be formally derived from (1) and (2), but the derivation is somewhat involved. We shall state the answer and then verify that it gives values of torque in agreement with (2).

A current element has direction as well as magnitude; the force on a current element depends on its direction relative to that of the magnetic intensity. *A current element parallel to the magnetic intensity, as in Fig. 12(a), experiences no force. A current element perpendicular to the magnetic intensity experiences the force* $dF = I\mathcal{B}\,dl$, *in a direction perpendicular to both the magnetic intensity and the current element, in the sense shown by the tail* ⊗ *of the arrow in Fig. 12(b).* In the general case, in Fig. 12(c), when a current element makes an angle ϕ with the magnetic intensity, the magnetic intensity may be resolved into two components: One component is of magnitude $\mathcal{B}\cos\phi$ and parallel to the current; this component occasions no force. The other component, of magnitude $\mathcal{B}\sin\phi$ and perpendicular to the current, gives the force

$$dF = I\mathcal{B}\, dl \sin\phi \qquad \begin{matrix} F \text{ in N} \\ \mathcal{B} \text{ in T} \\ I \text{ in A} \\ l \text{ in m} \end{matrix} \qquad (4)$$

in the direction shown in Fig. 12(c).

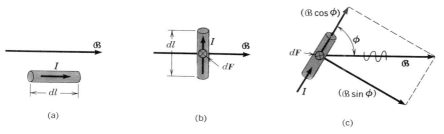

Fig. 12. The force on a current element in a magnetic field. (The ⊗ represents the tail of an arrow.)

Again, we can express these results very conveniently in vector notation. The relations above and in Fig. 12 are summarized by the expression

$$d\mathbf{F} = \mathbf{I} \times \mathcal{B}\, dl, \qquad (5)$$

where $\mathbf{I}$ is the current vector.* Thus, *the force is perpendicular to both $\mathbf{I}$ and $\mathcal{B}$ and has the direction of the vector $\mathbf{I} \times \mathcal{B}$.*

One can determine in which of the two directions perpendicular to the plane containing $\mathbf{I}$ and $\mathcal{B}$ the force acts from the right-hand rule for vector product in which a screw advances in the direction determined by turning $\mathbf{I}$ into $\mathcal{B}$.

Now let us verify that equations (4)–(5) give a value of torque on a small coil that is in agreement with (2). Consider first the N-turn rectangular coil of Fig. 13 with its plane parallel to $\mathcal{B}$. The sides of length l experience forces in the directions indicated by the dot and cross, each of magnitude given by (5) as $I\mathcal{B}l$ per turn or $F = NI\mathcal{B}l$ total. These two forces form a *couple* (see Prob. 29, p. 178) with torque $L = NI\mathcal{B}lw$ about the axis represented by the broken line through the center of the coil. The ends of length w experience no magnetic force. The magnetic

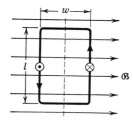

Figure 13

---

*The current $\mathbf{I}$ in a wire is not a proper vector from the standpoint of vector-addition properties, although, as we shall see in Chapter 43, the current density $\mathbf{J}$ (current per unit of area) does constitute a proper vector field. However it is legitimate to treat the current element $\mathbf{I}\, dl$ as a proper vector provided the thickness of the wire is assumed to be small compared with other distances of interest. This assumption is implicitly made in this and the following chapters, and it will be noticed that in expressions such as (5) above and the expressions in the next chapter for the magnetic intensity arising from a current element, the 'vector' $\mathbf{I}$ occurs only in the form of a current element $\mathbf{I}\, dl$.

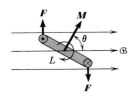

Figure 14

moment of the coil has the magnitude $M = NIlw$ and a direction out of the paper toward the reader. The torque has the magnitude $L = M\mathcal{B}$ and a sense that tends to turn M into $\mathcal{B}$, as we set out to verify.

Now let the magnetic-moment vector of the same coil make angle θ with $\mathcal{B}$, as in the end view, Fig. 14. The forces F have the same magnitude as before, but the distance between them is reduced to $w \sin\theta$. The ends of length w now experience forces (magnitude $NI\mathcal{B}w \cos\theta$ each), but these forces are into and out of the paper in Fig. 14 and do not contribute to the torque. The torque is therefore $L = NI\mathcal{B}lw \sin\theta = M\mathcal{B} \sin\theta$ in a sense that tends to turn M into $\mathcal{B}$, in agreement with (2). It can be shown that similar agreement can be obtained when calculations are made for a coil of arbitrary, nonrectangular shape, with arbitrary orientation in the magnetic field. Such a calculation shows that our original definition (1) of magnetic moment is truly independent of the shape of the coil, which we did not restrict in the definition.

We can now discuss the principle of the DC electric motor in terms of the simple prototype illustrated in Fig. 15, which shows a single rectangular loop arranged so that it can rotate about a horizontal axle in a uniform magnetic field. The axle carries a commutator arranged so that the torque will always have the same sense. As we have already computed in connection with Fig. 14, the clockwise torque is

$$L = I\mathcal{B}lw \sin\theta. \tag{6}$$

The torque (6) vanishes when the plane of the coil is normal to the

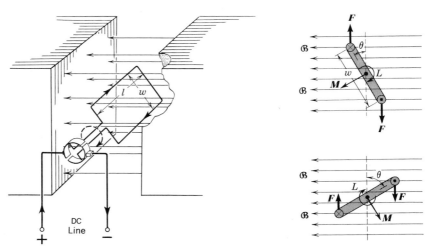

Fig. 15. Prototype DC electric motor. Two positions of the rotating loop are shown at the right. The commutator reverses the current as the plane of the loop passes through the broken center line so that the conductor to the left of this line always has current into the paper and force upward; the conductor to the right always has current out of the paper and force downward; the torque is always clockwise.

magnetic field and is at a maximum when the plane of the coil is parallel to the magnetic field. The torque of such a one-turn motor would come in spurts during each half-revolution, and the motor would depend on rotational momentum to carry it through the no-torque position ($\theta = 0$). An actual motor contains a number of turns arranged at various angles so that only one turn at a time passes through the no-torque position. Such a motor then delivers mechanical energy at a fairly constant torque but requires a complex commutation system to reverse the current in each turn as it passes the no-torque position. Many segments are required in the commutator of a DC motor to do this switching properly.

5. The Moving-Coil Galvanometer

The sensitive element of the DC measuring instruments discussed on pp. 551–554 is a moving-coil galvanometer (frequently called a *d'Arsonval galvanometer*) of the type shown in Fig. 16. The coil of this galvanometer is supported on hardened steel pivots turning in jeweled bearings. Its rotational motion is restrained by a pair of spiral springs (not shown in Fig. 16), one above and one below the coil, which also serve as current leads to the coil. The position of coil and pointer in Fig. 16 is intended to represent the equilibrium position as determined by the springs, with no current in the coil. In use, the current in the coil is into the paper on the right and out of the paper on the left, so that the magnetic forces in the uniform radial magnetic field create a clockwise torque. The coil moves clockwise until the resisting counter-clockwise spring torque equals the clockwise magnetic torque. Since the magnetic torque is proportional to the current in the coil and since the spring torque is porportional to the angle of rotation from the equilibrium position, the angle of rotation is proportional to the current in the coil.

Pivoted galvanometers of the type shown in Fig. 16 that give full-scale deflection on a current as low as 1 microampere can be purchased; greater sensitivity than this cannot be achieved because of friction in the bearings. To achieve sensitivity, the spring torque constant must be made very low, but the spring torque must still be large compared with the frictional torque in the bearings if the instrument is to perform properly. To avoid friction and to achieve a very low torque constant, more sensitive instruments, designed to measure currents in the range 10^{-6} to 10^{-10}

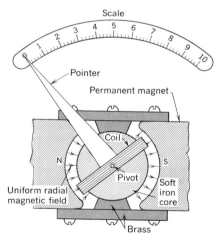

Fig. 16. Moving-coil galvanometer. The drawing is distorted in that the coil and pole pieces would ordinarily be much smaller in comparison with the lengths of the pointer and the scale.

594 MAGNETIC FORCES

amperes, are of the *suspended-coil* type, with the coil suspended from a fiber of phosphor bronze, silver, or quartz.

6. Magnetic Force on a Moving Charged Particle

We shall now derive a fundamental relation which gives the magnetic force acting on a single moving charged particle. We shall derive this relation by considering that a stream of charged particles—for example, the stream of electrons in Fig. 2—represents a current and must be acted on by forces that can be computed from (5). Actually, equation (8), which gives the force on a single moving charged particle, is basically the fundamental relation from which equation (5) for the force on a current element (a stream of charged particles, even in a wire) should be derived. However, since the forces between wires carrying currents can be measured much more conveniently than the forces on individual charges, the approach we have used is the one actually employed *operationally*.

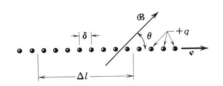

Fig. 17. The particles will experience a force *out of the paper*.

Let us consider Fig. 17, where a stream of particles moves at constant velocity v and with constant spacing δ, each particle carrying charge $+q$. The stream is passing through a constant magnetic field $\mathcal{B}$.

The current is the amount of charge passing any point per second. Since the particles move the distance v in one second, and there are v/δ particles in this distance, the charge passing any point in one second is qv/δ, that is, $I = qv/\delta$. By (5), the magnetic force acting on those particles contained in a length dl of the stream is

$$d\mathbf{F} = \mathbf{I} \times \mathcal{B}\, dl = q\, \mathbf{v} \times \mathcal{B}\, dl/\delta. \tag{7}$$

As we see from Fig. 17, the length dl contains dl/δ particles, so if we divide the right side of (7) by dl/δ, we obtain, for *the force on an individual particle*,

$$\mathbf{f} = q\, \mathbf{v} \times \mathcal{B}. \qquad \begin{Bmatrix} f \text{ in N} \\ q \text{ in C} \\ v \text{ in m/s} \\ \mathcal{B} \text{ in T} \end{Bmatrix} \tag{8}$$

Since the force is perpendicular to the velocity,

A steady magnetic field does no work on a charged particle, neither does it tend to change the speed of the particle.

For accuracy, we have added the word *steady* because, as we shall see in Chapter 31, a magnetic field varying with time *can* do work and increase the speed of a particle.

The path of a charged particle in a uniform magnetic field is actually a helix wrapped around the magnetic lines, as shown in Fig. 18. Figure 19 is a cloud-chamber photograph of electrons moving in a field perpen-

Fig. 18. Path of a *negative* particle in a magnetic field.

dicular to the paper, and corresponds to a top view of the path in Fig. 18. In the cloud-chamber photograph of Fig. 20, the field is from left to right and the sinusoidal path of the electron represents a side view of the helix.

To see the reason for this helical path, consider that at any instant the velocity of the particle can be written as the sum of two components, one parallel and one perpendicular to $\mathcal{B}$, that is, $v = v_\parallel + v_\perp$. The force relation (8) then becomes $f = q\,v_\perp \times \mathcal{B}$, since $v_\parallel \times \mathcal{B} = 0$. *There is no force component parallel to $\mathcal{B}$, hence $v_\parallel$ is constant.* In a plane perpendicular to $\mathcal{B}$, (Fig. 21), the force, normal to $v_\perp$, is of the nature that results in motion in a circle at constant speed. If the particle has mass m, the radius R of the circle is such that

$$f = q v_\perp \mathcal{B} = m v_\perp^2 / R; \qquad R = m v_\perp / q\,\mathcal{B}. \qquad (9)$$

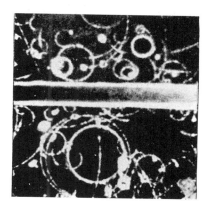

Fig. 19. (From Rasetti, *Elements of Nuclear Physics:* Prentice-Hall, 1937.)

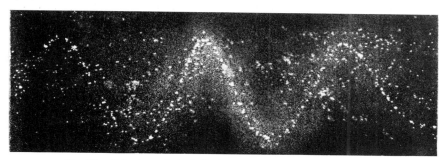

Fig. 20. Side view of the helical path of a stream of electrons in a magnetic field. (Courtesy C. E. Nielsen.)

The radius of the circle increases in proportion to the momentum $mv_\perp$ of the particle and varies inversely as the magnetic intensity and the charge on the particle. A negative particle would have its force and acceleration vectors reversed relative to those of a positive particle. Hence, if q were negative in Fig. 21, the path would curve to the right instead of to the left.

As we shall see in Sec. 9, the operation of the cyclotron depends on an interesting property of an orbit such as that of Fig. 21, namely, that the time it takes a particular type of particle to go once around the circle is independent of the speed $v_\perp$ of the particle. A particle of high speed travels

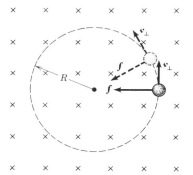

Fig. 21. The force on a positively charged particle in a uniform magnetic field directed into the paper.

rapidly around an orbit of large radius, and one of low speed travels slowly around an orbit of small radius. Since, according to (9), the distance traveled, $2\pi R$, is directly proportional to the speed, the times taken are equal. The time of revolution is $T=2\pi R/v_\perp$, so

$$T=2\pi m/q\mathcal{B}, \tag{10}$$

independent of the speed. Thus the various electrons shown in Fig. 19, traveling at different speeds proportional to the different radii of their circular orbits, all execute these orbits in the same periodic times.

7. The Mass of the Electron

If we know the charge $-q$ on the electron, its mass m can be obtained from an experiment that determines q/m. Electric and magnetic fields are arranged as in Fig. 22 so that the forces they exert on the electrons in the beam are in opposite directions, and the fields are adjusted so as to produce no deflection, as indicated by a maximum reading on the galvanometer G. When there is no deflection, the upward electric force $q\mathcal{E}$ is equal and opposite to the downward magnetic force $\mathcal{B}qv$:

$$q\mathcal{E}=\mathcal{B}qv, \quad \text{or} \quad \mathcal{E}=\mathcal{B}v, \tag{11}$$

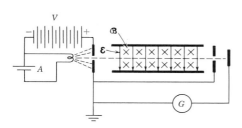

Fig. 22. Schematic diagram of apparatus for measuring q/m. The electron beam, indicated by the broken line, proceeds from left to right in an evacuated enclosure.

where v is the speed of the electrons. The speed is determined by the accelerating voltage V used in the 'electron gun.' Equating kinetic energy to the work done by the accelerating field, we find $\tfrac{1}{2}mv^2=qV$, or $v^2=2qV/m$. Substitution of this value for v^2 in $\mathcal{E}^2=\mathcal{B}^2v^2$ leads to the equation

$$q/m=\mathcal{E}^2/2V\mathcal{B}^2, \tag{12}$$

where the electric field $\mathcal{E}$, the magnetic field $\mathcal{B}$, and the accelerating voltage V are all easily determinable quantities. The first measurement of q/m was made by J. J. THOMSON in 1897.

The experiment we have discussed, Millikan's experiments, and other more recent experiments of various types, give as the present 'best' values of the electron's charge and mass

$$\left.\begin{array}{l} -1\,e=-(1.602\,10\pm 0.0007)\times 10^{-19}\,\text{C}, \\ m=(9.1091\pm 0.0004)\times 10^{-31}\,\text{kg}. \end{array}\right\} \tag{13}$$

8. The Mass Spectrograph; Nuclidic Masses

In the same way that the mass of the electron can be determined by q/m measurements once q has been determined, the masses of atoms can be determined or compared from analogous Q/m measurements on

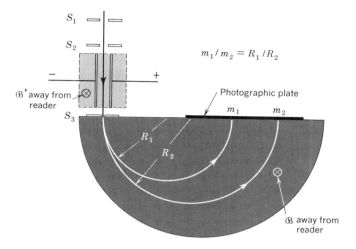

Fig. 23. A mass spectrograph of the Bainbridge type.

ions of known charge Q. A device for making such measurements, called a *mass spectrograph*, is shown in Fig. 23. Positive ions (charge $Q = +1$ e) of the element being studied are formed in a discharge tube; after acceleration, a narrow beam of high-speed ions passes through slits S_1 and S_2 into an evacuated region containing a uniform magnetic field of intensity $\mathcal{B}'$ directed away from the reader, and containing a uniform electric field $\mathcal{E}$ set up by parallel plates. The force $\mathcal{E}Q$ exerted on the ions by the electric field is equal and opposite to the force $\mathcal{B}'Qv$ exerted by the magnetic field only for ions having the particular speed $v = \mathcal{E}/\mathcal{B}'$ given by (11). All ions with speed $v = \mathcal{E}/\mathcal{B}'$ will pass through the third slit S_3 regardless of differences in mass. These ions enter a second uniform magnetic field $\mathcal{B}$ directed away from the reader and traverse a semicircular path of radius $R = mv/Q\mathcal{B}$, as given by (9). Solving this equation for m, we may write

$$m = (\mathcal{B}Q/v) \times R = \text{const} \times R, \qquad (14)$$

which indicates that ions of a given charge Q traverse circular paths of radii proportional to their masses. With the arrangement shown in Fig. 23, the ions strike a photographic plate after traversing one semicircle and produce darkened traces on the photographic plate. From the position of the trace on the plate, R can be accurately determined. If $\mathcal{B}$, Q, and v are also accurately known, the absolute value of the ion mass m can be calculated. With even greater accuracy, a mass spectrograph will determine the *relative* masses of ions whose isotopic or molecular masses are close to each other.

It was with a mass spectrograph that J. J. Thomson discovered in 1913 that not all the atoms of a given element have the same mass, but that most of the elements have two or more isotopes. Isotopes are

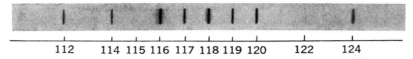

Fig. 24. The mass spectrum of tin, showing the traces of the ten isotopes, and their mass numbers. (Courtesy of K. T. Bainbridge.)

separated in quantity at Oak Ridge by using a device that is in principle a mass spectrograph. Some elements have numerous isotopes; for example, tin has ten isotopes. Figure 24 is a photograph of the mass spectrum of tin. It is impossible to make a very accurate determination of the relative abundances of the different isotopes of an element from a photographic record such as that of Fig. 24, but by adapting the mass spectrograph to electrical recording of the total charge on the ions collected at various radii R, it is possible to make accurate abundance measurements.

Mass-spectroscopic measurements of the nuclidic masses and relative abundances of the naturally occurring isotopes of the elements have become so accurate that all but 20 of the 'best' values of average atomic mass listed in the table in Section 4 of the Appendix are based on such measurements rather than on the older chemical techniques.

9. The Cyclotron

The force exerted on a moving charge by a magnetic field is utilized in many of the particle accelerators employed in studies of nuclear physics. One such accelerator, designed to produce ions of high energy, is the *cyclotron,* invented by E. O. LAWRENCE. In this device, positive ions are made to pass repeatedly through the same accelerating potential and finally to acquire energies corresponding to a fall through a potential many times greater than that actually applied at any one time between the electrodes. The arrangement of the electrodes in the cyclotron is illustrated schematically in Fig. 25. Two semicircular hollow copper boxes, called 'dees' because of their shape, are arranged in the manner indicated. The dees are placed in an evacuated enclosure between the pole faces of a huge magnet so that there is a strong constant magnetic field perpendicular to the flat faces of the dees. An alternating potential of the order of 10 kV is applied between the dees at a frequency of the order of 10 MHz.

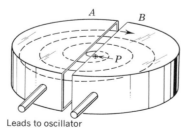

Fig. 25. Schematic diagram of the electrodes (dees) and the ion path in a cyclotron.

A supply of low-energy positive ions is generated at point P midway between the dees by electron bombardment of gas introduced at low pressure; the gas is usually hydrogen, deuterium, or helium. During a half-cycle these ions will be accelerated toward one of the dees, *A*. Once these ions enter the hollow electric-field-free space within the dee, they will no longer be accelerated by the alternating electric field and will move at constant *speed*. However, since they are traveling perpendicular to a constant magnetic field $\mathcal{B}$, they will traverse a semicircular path and will return to the gap between the dees. If the frequency of the alternating potential is such that the time required to describe this semicircle corresponds to one half-cycle, then the ions will arrive at the gap at just the proper time to be further accelerated by the now reversed electric field. The ions are now moving faster after entering the second dee, *B*, and will therefore traverse a circular path of greater radius. However, as we have seen in (10), the *time* required for an ion to traverse a circular path, or half a circular path, in a magnetic field is independent of the speed of the ion. The time for a half-revolution is given by $\pi m/\mathcal{B}Q$, where *m* and *Q* are the mass and charge of the ion.

Thus, for an ion to arrive at the gap between the dees at just the proper instant to experience further acceleration after each half-revolution, the oscillating potential difference must have a period equal to the time of two half-revolutions of the ion, or $2\pi m/\mathcal{B}Q$, corresponding to a frequency of $\mathcal{B}Q/2\pi m$. The ions describe a series of semicircles of increasing radius, gradually spiraling outward and finally emerging at the outer edge of the dees, where they may be concentrated on a target by means of an electrostatic deflecting plate. The ions emerge with an energy equivalent to a fall through a potential many times higher than that used in the accelerating process.

Example. *If protons are accelerated in a cyclotron until the radius of their path is 0.5 m in a field $\mathcal{B}=0.75$ T, and then strike a target, what energy have they acquired? What must be the frequency of the oscillating potential difference between the dees?*

The frequency is given in the text above as $f=\mathcal{B}Q/2\pi m$. The charge Q on the proton is $+1\,e = +1.60 \times 10^{-19}$ C. Its mass m is 1.008 u $= 1.67 \times 10^{-27}$ kg. Substitution and use of relation (3) gives

$$f = \frac{\mathcal{B}Q}{2\pi m} = \frac{(0.75\text{ T})(1.60 \times 10^{-19}\text{ C})}{2\pi\,(1.67 \times 10^{-27}\text{ kg})} = 11.4 \times 10^6 \text{ Hz} = 11.4 \text{ MHz}.$$

The energy acquired by the protons is independent of the magnitude of the oscillating potential. If this magnitude is small the proton will make many trips between the dees, acquiring a little energy on each trip; if the magnitude is greater, it will make fewer trips and increase its path radius faster. But in any case, when its radius has reached 0.5 m, its energy will be that given by the speed v obtained from equation (14), $v=\mathcal{B}QR/m$. Straightforward substitution gives

$$E = \tfrac{1}{2}\,mv^2 = \tfrac{1}{2}\,\mathcal{B}^2 Q^2 R^2/m = 1.08 \times 10^{-12} \text{ J}.$$

Energies of high-speed particles are usually expressed in millions of electron volts (MeV), where, in accordance with (21), p. 485,

$$1 \text{ MeV} = 1.602\ 10 \times 10^{-13} \text{ J}. \tag{15}$$

Thus, in the example above, we find $E=6.75$ MeV as the proton energy. Since the proton charge is 1 e, this is the energy the protons would acquire in a single acceleration through a difference of potential of 6.75 million volts. Because of relativistic phenomena (Chapter 44), the practical energy limit for protons from a standard cyclotron is about 10 MeV.

PROBLEMS

1. If a straight horizontal conductor carrying 20 A from south to north passes through the magnetic field of a large magnet, the magnetic field being 1.5 T vertically upward over a length of 1.2 m of the wire, what is the force on the wire in magnitude and direction?

Ans: 36.0 N toward the east.

2. If 50 cm of a straight conductor is at right angles to a uniform magnetic field of 0.7 T, what current must flow in the conductor in order that the force on this section be 3 N?

3. In a region where the earth's magnetic field has a downward component of 50 μT, a northward component of 10 μT, and no east-west component, what is the force on a 2-meter length of wire carrying 75 A: (a) horizontally, from s to N? (b) horizontally, from w to E? (c) vertically upward?

Ans: (a) 0.007 50 N westward; (b) 0.007 65 N, northward and upward at an angle of 11°.3 above the horizontal; (c) 0.001 50 N westward.

4. In a region where the northward component of the earth's field is 20 μT, what current would an aluminum conductor running E and w have to carry in order that the upward magnetic force should equal the downward gravitational force, if the mass of the conductor is 18 grams per meter length?

5. If the single-turn coil of Figs. 14 and 15 is carrying 20 A and has dimensions $l=20$ cm, $w=12$ cm, what are the torques on the coil at $\theta=0°$, 30°, 60°, and 90°, if the magnetic intensity is 0.6 T?

Ans: 0, 0.144, 0.249, 0.288 N·m.

6. Show that the period of oscillation of a compass needle of rotational inertia I, when it is performing small oscillations about the direction of the horizontal component of the earth's magnetic field, is $T=2\pi\sqrt{I/M\mathcal{B}_H}$. A measurement of T in a known field will determine I/M for a compass needle; this needle of known I/M can then be used in the same way to determine the horizontal component of an unknown field.

7. In the galvanometer shown in Fig. 16, let N = number of turns in coil, $\mathcal{B}$ = intensity of radial magnetic field, l = effective length of wires in the field, r = radius from axis of rotation to coil wires, θ = angle of rotation of coil, measured from no-current equilibrium position, C = torque constant of springs, I = current in each turn of coil. Show that $I = C\theta/2N\mathcal{B}lr$.

8. A galvanometer coil is wound of wire of resistance 0.3 Ω/m. Each turn requires 10 cm of wire. What spring torque constant must be used if, in the notation of Prob. 7, $\mathcal{B}=0.4$ T, $l=2.5$ cm, $r=1$ cm, and the meter is to give full-scale (1 radian) deflection on 1 milliampere and have 1-ohm resistance.

9. A galvanometer is wound of wire of resistance 0.6 Ω/m. Each turn requires 10 cm of wire. What spring torque constant must be used if, in the notation of Prob. 7, $\mathcal{B}=0.5$ T, $l=2.5$ cm, $r=1$ cm, and the meter is to give full-scale (1 radian) deflection on 1 microampere and have 30-Ω resistance?

Ans: 1.25×10^{-7} N·m/rad.

10. (a) Show that a small bar magnet with magnetic moment pointing geographically north, and mounted on a vertical torsion fiber, can be used to measure fluctuations in the eastward component of the earth's magnetic intensity. (b) How can a small bar magnet be mounted on a *horizontal* torsion fiber so as to measure fluctuations in the northward component of the earth's magnetic intensity?

11. An electron moving at a speed of 5×10^7 m/s in a vacuum chamber enters a uniform magnetic field in which $\mathcal{B} = 2.26 \times 10^{-3}$ T. The velocity of the electron is at all times perpendicular to $\mathcal{B}$, and hence the electron describes a circular path like those in Fig. 19. What is the radius of the path?   Ans: 12.6 cm.

12. Determine the time of revolution of the electron in Prob. 11. What is the frequency of the circular motion of the electron?

13. A stream of protons and deuterons in a vacuum chamber enters a uniform magnetic field. Both protons and deuterons have been subjected to the same accelerating potential; hence the kinetic energies of the particles are the same. If the ion stream is perpendicular to $\mathcal{B}$ and the protons move in a circular path of radius 10.0 cm, find the radius of the path traversed by the deuterons.   Ans: 14.1 cm.

14. Singly charged ions of the chlorine isotopes 35 and 37 of equal kinetic energy enter a magnetic field in the manner described in Prob. 13. Find the ratio of the radii of curvature of the paths traversed. Repeat the calculation for singly charged ions of bromine isotopes 79 and 81, and of uranium isotopes 235 and 238. In all these cases the mass in u is numerically equal to the mass number to within three significant figures.

15. Chlorine has two isotopes of masses 34.980 and 36.978 u. If the radius of the path described by the lighter isotope in the mass spectrograph in Fig. 23 is 12 cm, find the separation between the traces produced by the two isotopes on the photographic plate.
       Ans: 1.37 cm.

16. Find the plate separation for the lithium isotopes of mass numbers 6 and 7 if the radius of the path of the lighter isotope in a mass spectrograph (Fig. 23) is 12 cm. The nuclidic masses are 6.017 and 7.018.

17. A proton in a cyclotron (Fig. 25) moves in a direction normal to a magnetic field of intensity $\mathcal{B} = 0.656$ T. (a) Find the time required for the proton to traverse a complete circular path in the magnetic field. (b) Determine the frequency of this motion. (c) If a high frequency alternating potential difference of peak value 20 kilovolts is applied between the dees, how much does the proton's kinetic energy increase each time it makes a complete 'round trip' from one dee to the other dee and back? (d) Does the proton return to its starting point in making such a round trip? (e) If the magnetic field is constant over a circular region of diameter 1.5 m, and if a proton starts near the center of the field, what is the kinetic energy of the proton when it traverses a path of maximum radius 0.75 m? (f) How many times has the proton traversed the gap between the dees by the time its path has maximum radius? (g) How long a time is required for the proton to achieve its maximum kinetic energy?   Ans: (a) 10^{-7} s; (b) 10^7 s^{-1}; (c) 4×10^4 eV $= 6.40 \times 10^{-15}$ J; (e) 18.5×10^{-13} J $= 11.1$ MeV; (f) 555; (g) 2.77×10^{-5} s.

18. If a cyclotron with a magnetic field of the same intensity and extent as that in Prob. 17 were used to accelerate deuterons, what should be the frequency of the alternating high voltage? What would be the maximum deuteron energy attainable? Answer the same questions for alpha particles.

19. In a cyclotron, energy is given to charged particles by a high-frequency oscillator. In the microwave oscillator tube called the *magnetron*, the process is reversed; energy is transferred from a large number of electrons to a set of electrodes. The scheme can be visualized from the figure, which shows a 'bunch' of electrons moving in a circular path normal to a magnetic field $\mathcal{B}$. In traversing the circular path, the electrons pass close to electrodes A and B.

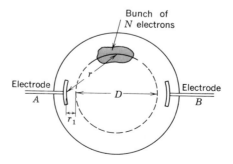

Problems 19–20

Write an expression for the potential V_A of electrode A that arises from a 'bunch' of N electrons at mean distance r from A. Write an expression for the maximum value of the potential difference $V_A - V_B$ in terms of the distance of nearest approach r_1 of the electron bunch to the electrodes and the diameter D of the electron orbit. How many times per second does the potential difference V_{AB} go through a complete cycle? Denote the charge on the electron by $-q$. Ans: $-(1/4\pi\varepsilon_0) Nq/r$; $(1/4\pi\varepsilon_0) NqD/(r_1^2 + r_1 D)$; $\mathcal{B}q/2\pi m$.

20. What should be the value of $\mathcal{B}$ in Prob. 19 if the frequency is to be 1 MHz, 1000 MHz, 10 000 MHz, 25 000 MHz? These latter frequencies are used for radar work, the requisite power being produced by magnetrons operating on the principles just described. Energy is derived from electron 'bunches' in klystron oscillators by a somewhat different method.

21. There are frequent discussions of research on practical power production by means of a *magnetohydrodynamic* (MHD) *generator*. In this concept, a gas is heated by chemical energy to a very high temperature (as it is in a rocket motor). Such a gas becomes conducting by virtue of thermal ionization—the temperature required for appreciable ionization is 2500° K or greater. An ionized conducting gas is called a *plasma*.

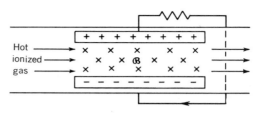

Problem 21

When a plasma moves through a magnetic field between two metal plates, as in the figure, positive ions are forced upward, charging the upper plate positively, while negative ions are forced downward to the lower plate. A difference of potential is established between the two plates that can be used to drive current through an external circuit.

(*a*) If v is the speed of the plasma toward the right, show that the magnetic force acting on charge q is given by $f = q\mathcal{B}v$.

(*b*) Show that the electric intensity in the field is $\mathcal{E} = \mathcal{B}v$.

(*c*) If the separation of the collector plates is l, show that the difference of potential between the plates is $\mathcal{B}vl$.

(*d*) What is the relation between the power output of the MHD generator and the rate of charge collection on the plates?

MAGNETIC FIELDS 29

In the preceding chapter we devoted out attention almost exclusively to a consideration of forces on currents or moving charges in a magnetic field, without discussing the origin of the field. In this chapter we first give *Ampère's principle,* which permits the calculation of the magnetic field produced by any steady electric current. Then, by application of *Ampère's line-integral relation,* we proceed to describe further the general properties of the magnetic fields associated with steady electric currents.

While Ampère's principle is the basic principle involved in the calculation of magnetic fields from current distributions, magnetic-field calculations are facilitated by the introduction of *fictitious magnetic poles* having properties analogous to those of electric charges. By use of the pole concept, it becomes easy to compute the magnetic fields of solenoidal coils, and to determine forces acting on solenoidal coils, by the methods of *magnetostatics.* These methods are formally identical with the already familiar methods of electrostatics. It should be remembered, however, that the pole concept is introduced purely for the purpose of convenience in making calculations and that *the poles have no physical reality.*

We also point out that the *ampere*, rather than the *coulomb*, is taken as the fundamental unit in the international system of weights and measures. The ampere is defined in absolute fashion in terms of the magnetic forces between currents.

1. The Magnetic Field of an Electric Current

We have not yet learned how to compute the magnetic intensity at a point in terms of the currents that set up the magnetic field. Here we must first have recourse to experiment. Since we can measure magnetic intensities, we can determine experimentally how they are related to the current pattern. Let us start with the case of a long straight wire carrying current. At a point in the vicinity of such a wire, a small coil or magnet tends to turn so that its magnetic moment has a direction perpendicular to the wire and perpendicular to the radius from the wire to the point, as in Fig. 1. The magnetic lines are circles surrounding the wire, as in

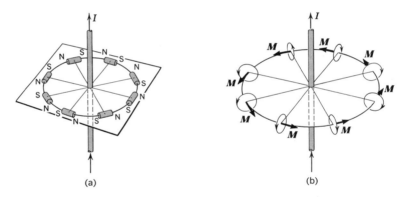

Fig. 1. Exploration of the field near a long straight wire by means of (a) a small magnet, (b) a small coil.

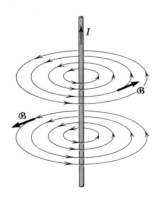

Fig. 2. Representative magnetic lines in the field of a long straight wire.

Fig. 2. The direction of these lines can be obtained from the ubiquitous right-hand rule—*if the thumb of the right hand points in the direction of the current, the fingers will curl in the direction of the circular magnetic lines.*

BIOT AND SAVART, in about 1820, concluded from experiment that the magnetic intensity at distance a from a long straight wire carrying current I is proportional to the current and inversely proportional to the distance. We write this proportionality relation in the form

$$\mathcal{B} = (\mu_0/4\pi)\, 2I/a, \qquad \begin{pmatrix}\text{infinite}\\ \text{straight wire}\end{pmatrix} \quad (1)$$

where μ_0 is a dimensional constant having the value

THE MAGNETIC FIELD OF AN ELECTRIC CURRENT

$$\mu_0 = 4\pi \times 10^{-7} \text{ T} \cdot \text{m/A} = 12.57 \times 10^{-7} \text{ T} \cdot \text{m/A}, \qquad (2)$$

so that
$$\mu_0/4\pi = 10^{-7} \text{ T} \cdot \text{m/A} = 10^{-7} \text{ N/A}^2. \qquad (3)$$

We are now in a position to compute the force between two infinite parallel wires carrying current. If the two wires carry the same current I, as in Fig. 3, we see from (1) and from relation (4), p. 591, that the force that the left wire exerts on a length dl of the right wire is

$$dF = I\mathcal{B}\, dl = I\,(\mu_0/4\pi)(2I/a)\,dl = (10^{-7}\text{ N/A}^2)(2I^2/a)\,dl,$$

where a is the distance of separation of the wires. The force per unit length, which we denote by f, is

$$f = dF/dl = (\mu_0/4\pi)\,2I^2/a = (10^{-7}\text{ N/A}^2)\,2I^2/a. \qquad (4)$$

The force per unit length acting between two infinite parallel wires carrying equal current is proportional to the square of the current and inversely proportional to the distance between the wires.

The preceding statement of proportionality is a direct generalization from experiment—a *generalization* because one cannot in practice use *infinite* wires. The fact that the proportionality constant is a nice round number $(2 \times 10^{-7} \text{ N/A}^2)$ is not a consequence of experiment—this number was chosen in the internationally adopted definition of the ampere:

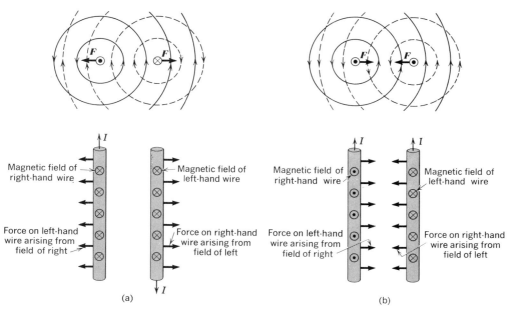

Fig. 3. The forces between parallel wires carrying currents arise from the fact that each wire is in the magnetic field of the other.

The **ampere** is that current which, if flowing in each of two infinite parallel wires one meter apart, would result in a magnetic force of 2×10^{-7} newton per meter length.

We have previously, in order to make progress, defined the ampere in terms of the coulomb. Actually, the above definition of the ampere is taken as fundamental by international agreement, and *the coulomb is defined as the ampere-second*. The constant ε_0 in Coulomb's law, which is not a nice round number, then becomes a matter for experimental determination (*see* Sec. 7).

The proportionality constant in (1) was written in the indicated form purely for later convenience. The system of units in which the proportionality constant takes the form $2\,(\mu_0/4\pi)$ is said to be 'rationalized.'

2. Ampère's Principle

The principle that enables computation of the magnetic intensity resulting from any arbitrary distribution of currents is a profound generalization from experiment made by Ampère:

AMPÈRE'S PRINCIPLE: *The contribution $d\mathcal{B}$ that a current element makes to the magnetic intensity at a point P has a direction perpendicular to both the current element and the line joining the current element to P, in the sense shown in Fig. 4, and has magnitude proportional to the product $I\,dl\,\sin\theta$ and inversely proportional to the square of the distance r, where θ and r are defined as in Fig. 4. Contributions from the various current elements are to be added vectorially.*

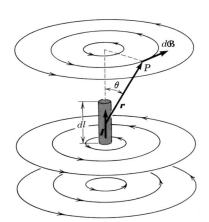

Fig. 4. The field of a current element.

As we shall see in a moment, to reach agreement with the definition of the ampere given above, we must write the proportionality constant as $(\mu_0/4\pi)$. Thus, in the notation of Fig. 4,

$$d\mathcal{B} = (\mu_0/4\pi)\,I\,dl\,\sin\theta/r^2. \qquad (5)$$

We can express both the magnitude and direction of $d\mathcal{B}$ if we note that $d\mathcal{B}$ has the direction of $\boldsymbol{I} \times \boldsymbol{r}$. Since the magnitude of $\boldsymbol{I} \times \boldsymbol{r}$ is $Ir\sin\theta$, we can write $d\mathcal{B}$ in the form

$$d\mathcal{B} = (\mu_0/4\pi)\,\boldsymbol{I} \times \boldsymbol{r}\,dl/r^3. \qquad \begin{Bmatrix}\mathcal{B}\text{ in T}\\ I\text{ in A}\\ r, l\text{ in m}\end{Bmatrix} \quad (6)$$

Let us first verify that this relation gives a value of intensity for an infinite straight wire that agrees with (1). To obtain a somewhat more general result, we shall compute the intensity in Fig. 5 at a point at distance a from a straight wire of finite length at the ends of which the angles between I and r are Θ_1 and Θ_2 respectively. We see that each element of length contributes a component given by (5) and that these components are all in the same direction, so that they can be added like scalars. If we take the origin opposite the point at which we are computing the intensity, we can write

$$\mathcal{B} = \frac{\mu_0}{4\pi} \int_{-L_1}^{L_2} \frac{I}{r^2} \sin\theta \, dl.$$

Now $r = a \csc\theta$, $l = -a \cot\theta$, $dl = a \csc^2\theta \, d\theta$; hence

$$\mathcal{B} = \frac{\mu_0}{4\pi} \int_{\Theta_1}^{\Theta_2} \frac{I}{a^2 \csc^2\theta} \sin\theta \, a \csc^2\theta \, d\theta = \frac{\mu_0}{4\pi} \frac{I}{a} \int_{\Theta_1}^{\Theta_2} \sin\theta \, d\theta,$$

or
$$\mathcal{B} = \frac{\mu_0}{4\pi} \frac{I}{a} (\cos\Theta_1 - \cos\Theta_2). \qquad \left(\begin{array}{c}\text{finite}\\ \text{straight wire}\end{array}\right) \quad (7)$$

In particular, for the infinite wire, $\Theta_1 = 0$, $\Theta_2 = \pi$, and $\mathcal{B} = (\mu_0/4\pi) 2I/a$, in agreement with (1).

By using the relation (7), we can find the field of any circuit composed entirely of straight runs of wire. For example, the intensity at the center of a square coil composed of N turns of wire (Fig. 6) is obtained by using (7) four times over, with $\cos\Theta_1 = \sqrt{2}/2$ and $\cos\Theta_2 = -\sqrt{2}/2$, and has the value

$$\mathcal{B} = \frac{\sqrt{2}}{\pi} \mu_0 \frac{NI}{a}. \qquad \left(\begin{array}{c}\text{center of}\\ \text{square coil}\\ \text{of side } 2a\end{array}\right) \quad (8)$$

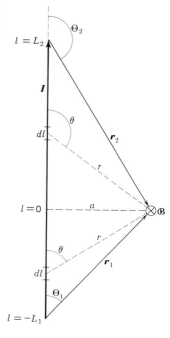

Figure 5

As another application of Ampere's principle, we shall compute the intensity at any point on the axis of a circular turn of radius a (Fig. 7). Here the current element $I \, dl$ makes an intensity contribution $d\mathcal{B}$ that lies in the axial plane and is perpendicular to $\mathbf{r}$. Its magnitude is given by (5) with $\theta = 90°$. This intensity contribution has a component $d\mathcal{B} \sin\phi$ along the axis and a component $d\mathcal{B} \cos\phi$ normal to the axis. We see from Fig. 7 that the elements $d\mathcal{B}_1$ and $d\mathcal{B}_2$ arising from equal lengths dl_1 and dl_2 on opposite sides of the circle have components normal to the axis that cancel in pairs, so that the result of integration around the circle is an intensity in the axial direction. Since $\sin\phi = a/r$, the axial component arising from dl is

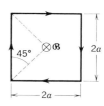

Figure 6

$$d\mathcal{B}_{\text{Axial}} = (\mu_0/4\pi) I \, dl \sin\phi/r^2 = (\mu_0/4\pi) I a \, dl/r^3.$$

Integration over dl merely introduces the circumference $2\pi a$ of the circle in place of dl and gives as the total intensity

$$\mathcal{B} = (\mu_0/4\pi) 2\pi I a^2/r^3 = \tfrac{1}{2} \mu_0 I a^2/r^3 \qquad (9)$$

in the axial direction. The maximum value of this intensity is obtained when r takes on its smallest value, $r = a$, at the center of the coil. The

608   MAGNETIC FIELDS   CHAP. 29

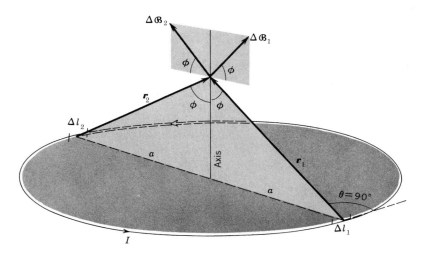

Figure 7

intensity at the center of a circular coil of N turns has the value

$$\mathcal{B} = \tfrac{1}{2}\,\mu_0\,NI/a, \qquad \begin{pmatrix}\text{center of}\\\text{circular coil}\end{pmatrix} \quad (10)$$

slightly greater than the intensity (8) for a square coil of half-side a.

The computation of the field of a square coil at points off the axis involves only the use of relation (7) and a certain amount of geometry and trigonometry. The computation of the field of a circular coil at points off the axis involves complex integrals (elliptic integrals) beyond the scope of a first course in calculus. The magnetic lines in a plane including the axis of a circular coil are plotted in Fig. 8.

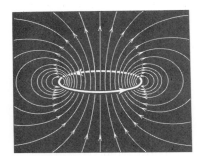

Fig. 8. Magnetic lines linking a circular turn of wire.

3. Ampère's Line-Integral Relation; Magnetic Poles

In this section we shall derive a relation of fundamental importance:

AMPÈRE'S LINE-INTEGRAL RELATION: *In the case of constant currents, the line integral of $\mathcal{B} \cdot d\mathbf{l}$ around an arbitrary, directed, closed path equals μ_0 times the net current linking the path in the positive sense.*

Thus, in Fig. 9, the current linking the path L is that passing through an arbitrary open surface S having L for its boundary. In this case, the relation states that $\oint \mathcal{B} \cdot d\mathbf{l} = \mu_0\,(I_1 + 2I_2 - I_3)$, since I_1 passes through the surface S once, and I_2 twice, in the *positive* sense, while the current I_3

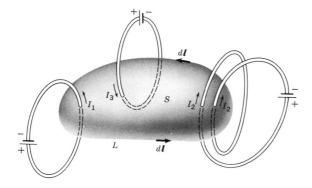

Figure 9

passes through S in the *negative* sense. We can readily convince ourselves that the total current passing through the surface S is independent of the manner in which S may be drawn, provided that its boundary L is fixed.

In general, Ampère's line-integral relation can be written as

$$\oint \mathcal{B} \cdot d\mathbf{l} = \mu_0 (\Sigma I), \tag{11}$$

where the currents I are taken as positive or negative according to the sense of their linkage with the closed path L. The case of the infinite straight wire furnishes a specific verification of the validity of (11). Take as L any one of the circular magnetic lines, of radius a, shown in Fig. 2. Then the line integral is just $\mathcal{B} \cdot 2\pi a$, which, according to (1), has the value $\mu_0 I$, as given by (11).

The line-integral relation (11) for $\mathcal{B}$ is to be compared with the similar line-integral relation for $\mathcal{E}$, p. 483:

$$\oint \mathcal{E} \cdot d\mathbf{l} = 0. \tag{12}$$

We have recalled this relation here because it will be of assistance in the derivation of (11), since this derivation is most easily performed in terms of fictitious entities called *magnetic poles* having properties similar to electric charges. The correspondence is demonstrated by the following argument.

One can show directly from Ampère's principle that the field of a small plane circuit (Fig. 10) of arbitrary shape and magnetic moment $\mathbf{M}$, at a point P at distance r large compared with the circuit dimensions, is given by

$$\mathcal{B} = \frac{\mu_0}{4\pi} \left\{ \frac{3 M \cos\theta}{r^4} \mathbf{r} - \frac{1}{r^3} \mathbf{M} \right\}, \tag{13}$$

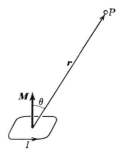

Fig. 10. A small current circuit.

expressed as the sum of a component in the **r** direction and one in the **M** direction. The derivation of (13) is straightforward but extremely tedious so we shall omit it. We note that for $\theta = 0$, $\cos\theta = 1$, **r** has the same direction as **M**, so $\mathcal{B}_{(\theta=0)} = (\mu_0/4\pi)\, 2M/r^3$, in agreement with the circular-coil formula (9) when we substitute $\pi I a^2 = M$ and take $r \gg a$. Also, for $\theta = \pi$, $\cos\theta = -1$, **r** has the direction opposite to **M**, so $\mathcal{B}_{(\theta=\pi)} = (\mu_0/4\pi)\, 2M/r^3$, again as expected. For $\theta = \tfrac{1}{2}\pi$, the first term vanishes and $\mathcal{B}_{(\theta=\pi/2)} = -(\mu_0/4\pi)\, M/r^3$, in the direction opposite to **M**, as is consistent with the magnetic lines in Fig. 8.

There is perfect correspondence between the expression (13) for the magnetic intensity of a small coil of magnetic moment **M** and the expression (25), p. 489:

$$\mathcal{E} = \frac{1}{4\pi\varepsilon_0}\left\{\frac{3\, M_\mathrm{E}\cos\theta}{r^4}\, \mathbf{r} - \frac{1}{r^3}\mathbf{M}_\mathrm{E}\right\}, \qquad (14)$$

that we obtained for the electric intensity of an electric dipole of moment $\mathbf{M}_\mathrm{E} = q\boldsymbol{\lambda}$. The vector lines for either case are shown in Fig. 31, p. 489, repeated here in Fig. 11. We must remember, however, that at the very center of this figure, in a region too small to be seen in the drawing, there is a fundamental difference between the electric and magnetic cases: *electric lines start on the charge $+q$ and end on the charge $-q$* (*see* Fig. 18, p. 474); *magnetic lines are continuous and thread the small coil* (*see* Fig. 8 of the present chapter).

Figure 11

However, so far as the visible parts of Fig. 11 are concerned, the magnetic intensity *could be computed* as the field of the pair of (fictitious) *magnetic poles of strength $\pm m$* shown in Fig. 12 if we imagine a magnetic pole to set up a coulomb-like inverse-square magnetic field, and define the magnetic moment of the *magnetic dipole* in Fig. 12 as

$$\mathbf{M} = m\boldsymbol{\lambda}. \qquad (15)$$

The coefficient in (14) is the coefficient in the Coulomb relation (p. 472), $\mathcal{E} = (1/4\pi\varepsilon_0)q/r^2$, for the electric intensity of a point charge. To get complete correspondence between (13) and (14), we must take the magnetic intensity associated with a point pole as

$$\mathcal{B} = (\mu_0/4\pi)\, m/r^2. \qquad (16)$$

Since M in (15) is measured in $A \cdot m^2$, the pole strength m must be measured in $A \cdot m$.

Fig. 12. A magnetic dipole.

As the next step toward the derivation of Ampère's line-integral law, and also to learn more about the pole concept, which is very useful in the computation of magnetic fields, we shall show how any finite circuit can be replaced by infinitesimal circuits like that of Fig. 10 and, hence, how the field of a finite circuit can be computed from a distribution of magnetic dipole moments. Consider the circuit of Fig. 13, carrying current I. Any surface having this circuit as its boundary can be subdivided into smaller circuits, each carrying current I, such that the field of the original circuit is the sum of the fields of the smaller circuits. This conclusion follows from the fact that the equal and opposite currents flowing along the interior paths in Fig. 13 represent no net current and hence make no contribution to the magnetic field. Hence, we can compute the field for the circuit of Fig. 13 by adding the field of the 15 smaller circuits.

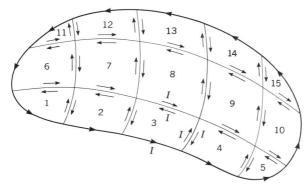

Figure 13

We shall have frequent occasion to make use of both Ampère's line-integral relation, and of the pole concept. The following paragraphs, which complete the rigorous derivation of the line-integral relation from the pole concept, are somewhat difficult and can be omitted on a first reading. A derivation can also be given without employing the pole concept, but this derivation involves very advanced mathematics.

In the last analysis, the subdivision in Fig. 13 can be carried to the point where each area is infinitesimal. A circuit of infinitesimal area dA has magnetic moment $dM = I\, dA$. As indicated in Fig. 14, the field of this infinitesimal circuit is the same as the field of poles $\pm dm$ separated by an infinitesimal distance λ, of the same moment: $\lambda\, dm = I\, dA$, so that the pole density

$$\sigma_p = dm/dA = I/\lambda. \tag{17}$$

Since these poles are uniformly distributed, we conclude that the magnetic field of a current circuit can be computed as the magnetostatic field of two sheets of uniformly distributed poles of density $\pm I/\lambda$, separated by an infinitesimal distance λ, and bounded by the current circuit. This statement is accurate *except at and between the sheets,* where there

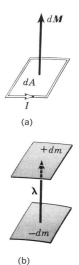

Fig. 14. The magnetic field of the infinitesimal current circuit in (a) is the same as that of the dipole in (b), at distances large compared with λ.

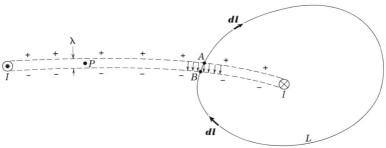

Figure 15

is a discontinuity that does not occur in reality. Thus, in Fig. 15, if one wanted to compute the magnetic intensity at point *P*, one could not use the pole sheets indicated, but would have to use a pair of sheets arranged so that *P* did not lie between them. It is just this discontinuity, however, that will enable us to derive Ampère's line-integral relation.

Consider the cross section of the current circuit shown schematically in Fig. 15, and its representation by pole sheets of density $\pm I/\lambda$. Consider also $\oint \boldsymbol{\mathcal{B}} \cdot d\boldsymbol{l}$ along the path *L*, linking the current in the positive sense. If we let $\boldsymbol{\mathcal{B}}_p$ be the value computed from the pole sheets, we can approximate the line integral, in the limit $\lambda \to 0$, by

$$\oint_L \boldsymbol{\mathcal{B}} \cdot d\boldsymbol{l} \xrightarrow[\lambda \to 0]{} \int_A^B \boldsymbol{\mathcal{B}}_p \cdot d\boldsymbol{l}, \tag{18}$$

where the last integral is over the path *external* to the pole sheets and all that has been omitted is the infinitesimal piece of the integral from *B* to *A* between the pole sheets.

Now, if Fig. 15 represented, instead of fictitious pole sheets, real capacitor plates charged to electric charge densities $\pm \sigma$, the line integral $\int_A^B \boldsymbol{\mathcal{E}} \cdot d\boldsymbol{l}$ would be the difference of potential between the plates, given by (2), p. 514, as $\sigma\lambda/\varepsilon_0$.* Thus,

$$\int_A^B \boldsymbol{\mathcal{E}} \cdot d\boldsymbol{l} = \frac{\sigma\lambda}{\varepsilon_0}. \tag{19}$$

Comparison of the corresponding 'Coulomb' relations,

$$\boldsymbol{\mathcal{E}} = \frac{q}{4\pi\varepsilon_0} \frac{1}{r^2} \quad \text{and} \quad \boldsymbol{\mathcal{B}} = \frac{m\mu_0}{4\pi} \frac{1}{r^2},$$

---

*This statement may be questioned on the grounds that our magnetic 'plates' are, in general, curved, whereas we considered plane plates in the case of a capacitor. A rigorous proof can be given for this statement in the case of curved plates. We omit this proof because it seems intuitively clear that with plates an infinitesimal distance apart (approaching zero as close as we please), plates of 'finite' curvature, and an inverse-square law, the area contributing significantly to the field between the plates can be considered as plane.

and noting that σ corresponds to q and σ_p to m, shows us that to change the integral (19) from a computation of $\boldsymbol{\mathcal{E}}$ to one of $\boldsymbol{\mathcal{B}}$, we must substitute $\sigma_p \mu_0/4\pi$ for $\sigma/4\pi\varepsilon_0$, or $\sigma_p \mu_0$ for σ/ε_0. Making this substitution in (19) gives

$$\int_A^B \boldsymbol{\mathcal{B}}_p \cdot d\boldsymbol{l} = \sigma_p \lambda \mu_0 = \mu_0 I,$$

where the last substitution results from (17). Returning to (18), we see that we have derived Ampère's line-integral relation (11) for the case of a single current linking the path L. For more than one current linking the path, as in Fig. 9, the general relation (11) follows from the fact that the fields of the various currents superpose by vector addition.

Example. *A long, straight, solid-metal wire of radius R carries a total current I uniformly distributed over its circular cross section. Find the magnetic intensity at all radii, internal and external to the wire, treating the wire as of infinite length.*

By symmetry, the magnetic lines must be circles around the center of the wire, as indicated in Fig. 16.

A circular line of radius $r \geqq R$ is linked by the whole current I. Writing Ampère's line-integral relation for such a line gives

$$\boldsymbol{\mathcal{B}} \cdot 2\pi r = \mu_0 I, \quad \text{or} \quad \boldsymbol{\mathcal{B}} = (\mu_0/4\pi)\, 2I/r, \quad (r \geqq R)$$

exactly as if the whole current were concentrated at the center of the wire, as in (1).

A circular line of radius $r \leqq R$ links only the fraction

$$(\pi r^2)/(\pi R^2) = r^2/R^2$$

of the current I, assumed to be uniformly distributed over the cross section. For this line, the line-integral relation gives

$$\boldsymbol{\mathcal{B}} \cdot 2\pi r = \mu_0 \frac{r^2}{R^2} I, \quad \text{or} \quad \boldsymbol{\mathcal{B}} = \frac{\mu_0}{4\pi} \frac{2I}{R} \frac{r}{R}. \quad (r \leqq R)$$

In the interior of the wire, the intensity increases *linearly* with r up to its maximum value $(\mu_0/4\pi)\, 2I/R$ at the surface of the wire.

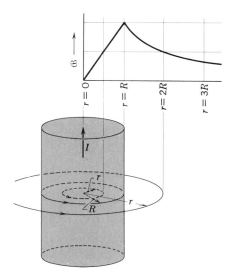

Figure 16

4. Properties of Magnetic Fields

The magnetic flux passing through an open surface S, bounded by a directed closed curve L, is defined, as in the case of electric flux on p. 496, by the surface integral

$$\Phi = \iint_S \boldsymbol{\mathcal{B}} \cdot d\boldsymbol{S}. \qquad \begin{Bmatrix} \Phi \text{ in Wb} \\ \boldsymbol{\mathcal{B}} \text{ in T} \\ S \text{ in m}^2 \end{Bmatrix} \quad (20)$$

The unit of magnetic flux is the *weber* (Wb).*

One **weber** is the magnetic flux through a plane surface of 1 m² area normal to a magnetic field of intensity 1 T.

Thus
$$1 \text{ Wb} = 1 \text{ T} \cdot \text{m}^2 = 1 \text{ N} \cdot \text{m/A}, \tag{21}$$

and we note that we can write

$$\mu_0 = 4\pi \times 10^{-7} \text{ Wb/A} \cdot \text{m}, \quad \mu_0/4\pi = 10^{-7} \text{ W/A} \cdot \text{m}. \tag{22}$$

Since a magnetic field resulting from constant currents can be derived from magnetic-pole distributions, in the manner indicated in Fig. 15, by relations similar to those of electrostatics, magnetic fields will share some, but not all, of the properties of electric fields (*see* Chapter 23). In particular, they share the property of constant vector flux:

The flux in a magnetic tube is constant along the tube.

On the other hand, while electric lines begin and end on electric charges, magnetic lines do not begin or end on poles but proceed continuously; therefore

Magnetic lines do not begin or end but form closed curves in space; these curves link net currents given by Ampère's line-integral relation.

The net magnetic flux passing outward through any arbitrary closed surface is zero.

The last statement is the magnetic equivalent of Gauss's relation, and follows from the closed nature of magnetic tubes (*see* Fig. 17); it expresses the nonexistence of a magnetic 'stuff' analogous to electric charge. Analytically, we can write

$$\oiint \mathcal{B} \cdot d\mathbf{S} = 0 \tag{23}$$

for any closed surface whatsoever.

Magnetic flux through an open surface S is said to link the closed line L bounding the surface (*see* Fig. 2, p. 495). Magnetic flux is denoted by Φ and is expressed in *webers*:

$$\Phi = \iint_S \mathcal{B} \cdot d\mathbf{S}.$$

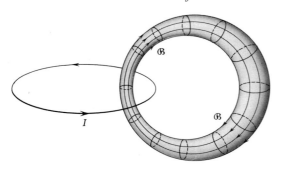

Fig. 17. A magnetic tube in the field of circular current. A magnetic tube is always a region of space that is topologically equivalent to a doughnut.

From the constancy of magnetic flux along a magnetic tube we see that

The flux linking a closed line L is independent of the open surface S, bounded by L, that is used in its computation.

*After WILHELM EDWARD WEBER (1804–1891), the German physicist who, along with KARL FRIEDRICH GAUSS (1777–1855), a German mathematician, devised the absolute system of electric and magnetic units.

5. The Field of a Solenoidal Coil

A solenoidal coil, usually called simply a *solenoid* (Greek *solen*, 'tube') is a single layer of wire wound as in Fig. 18 on the surface of a cylinder, not necessarily circular. A solenoid has approximately the same field as a uniform sheet of current flowing around the cylinder. If the solenoid has n turns per meter length and each turn carries current I, the current around a meter length of the cylinder will be nI, and the current in a length $d\lambda$ will be $nI\,d\lambda$, as indicated in Fig. 19.

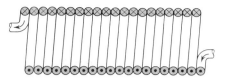

Fig. 18. Cross section of a solenoid with n turns per meter length, each turn carrying current I.

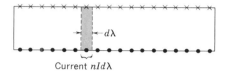

Fig. 19. A current sheet equivalent to the solenoid of Fig. 18.

The magnetic lines of a solenoid whose length is twice its diameter are shown in Fig. 20. A solenoid is said to be *long* if its length is many times (at least 5 times) its diameter. Long solenoids play an important role in magnetic experimentation because a long solenoid provides a very uniform field over its whole central section to within about two 'diameters' from each end. The treatment below will show where the end effects begin to come in along the axis.

We shall start by computing the intensity on the axis at the very end of a circular solenoid; then, by superposition, we can obtain the intensity at any point on the axis. In Fig. 21, the intensity at point P arising from the current $nI\,d\lambda$ in the strip of width $d\lambda$ is given by (9) as

$$d\mathcal{B} = \tfrac{1}{2}\,\mu_0\,nI\,d\lambda\,a^2/r^3.$$

This value is to be integrated over the length of the solenoid. The best integration variable is ϕ, which runs from 0 to α. We write $r = a/\cos\phi$, $\lambda = a\tan\phi$, $d\lambda = a\sec^2\phi\,d\phi$, to get

$$\mathcal{B} = \tfrac{1}{2}\,\mu_0\,nI\int_0^\alpha \cos\phi\,d\phi = \tfrac{1}{2}\,\mu_0\,nI\left[\sin\phi\right]_0^\alpha = \tfrac{1}{2}\,\mu_0\,nI\sin\alpha. \qquad (24)$$

We obtain the intensity on the axis *inside* a solenoid, given in the

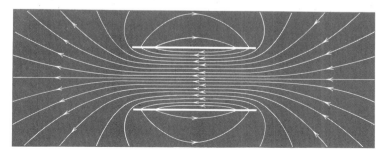

Fig. 20. The field of a short solenoid.

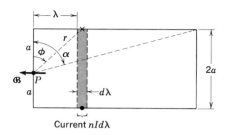

Fig. 21. Intensity on the axis at the end of a circular solenoid.

legend of Fig. 22, by superposition of the intensities from the parts of the solenoid to the right and left of P. Similarly, in Fig. 23, for a point on the axis *outside* of a solenoid, we can *subtract* the intensity of a solenoid subtending angle β from one subtending the whole angle α. We note that on the axis inside and well away from the ends of a long solenoid, where we may take $\beta = \alpha = \frac{1}{2}\pi$ in Fig. 22, we have

$$\mathcal{B} = \mu_0\, nI. \qquad \left(\begin{array}{c}\text{solenoid}\\ \text{relation}\end{array}\right) \quad (25)$$

We shall show that this relation gives the intensity in the interior of a long solenoid of any cross-sectional shape, at any point that is well away from the ends. Away from the ends, the intensity is uniform across the cross section.

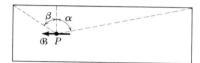

Fig. 22. Intensity on the axis inside a circular solenoid:

$$\mathcal{B} = \tfrac{1}{2}\mu_0\, nI\, (\sin\alpha + \sin\beta).$$

Fig. 23. Intensity on the axis outside of a circular solenoid:

$$\mathcal{B} = \tfrac{1}{2}\mu_0\, nI\, (\sin\alpha - \sin\beta).$$

The intensity on the axis near the end of a long circular solenoid falls off as in Fig. 24. This plot is made by taking $\alpha = 90°$ in Fig. 22 or Fig. 23 and plotting $\mathcal{B}$ against distance from the end of the solenoid expressed in units of the radius a. The magnetic intensity is expressed as a percentage of that inside the solenoid as given by (25). It will be noticed that the intensity at the end of the solenoid is just half the full intensity, whereas the intensity is up to 95 per cent of its full value one diameter ($2a$) inside the solenoid, and down to 5 per cent one diameter outside. The general characteristic of the field of a long solenoid is that the intensity is large and uniform inside, small outside.

A very informative way of studying the field of a solenoid is to consider the volume inside the solenoid to be filled with a distribution of magnetic dipoles. According to the discussion on pp. 609–613, we can consider the magnetic field of the current strip $nI\, d\lambda$ in Fig. 25 to be the field that would arise from $+$ poles on the left and $-$ poles on the right of two planes separated by $d\lambda$, with pole density $\sigma_p = nI$ per unit area of the sheets. The field arising from these poles will be the same as that arising from the current ring *everywhere except between the pole sheets.*

Now consider, Fig. 26, what happens if we slice the whole solenoid into current sheets of the same thickness $d\lambda$ as in Fig. 25. The surface

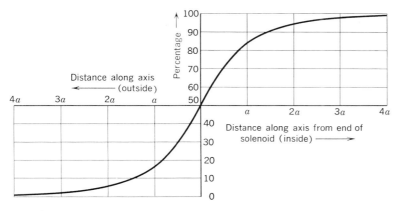

Fig. 24. Magnetic intensity on the axis near the end of a long circular solenoid. Ordinates are percentage of the magnetic intensity given by the solenoid formula. Abscissas are distances from the end of the solenoid in terms of the solenoid radius a. The left end of the solenoid lies at the central vertical line of this chart.

of each slice bears poles of density $+nI$ per unit area, arising from the current strip to its right, and of density $-nI$ per unit area, arising from the current strip to its left. These densities add to zero and we are left in Fig. 26 with merely a distribution of $+$ poles across the left end of the solenoid and one of $-$ poles across the right end. The field of these pole sheets will be identical with the field of the solenoid winding, at points *external* to the solenoid, but *not,* of course, at *internal* points.

The external field of a solenoid of any cross section is the field arising from a sheet of poles of surface density $+nI$ distributed over one end of the solenoid and a sheet of density $-nI$ distributed over the other end.

Since we have obtained this result by dividing the solenoid into *infinitesimal* slices, the result will be good as close to the solenoid as we please, just so we are definitely outside the current sheet that represents the winding and outside the imaginary plane across the end of the solenoid where we put the imaginary pole sheet.

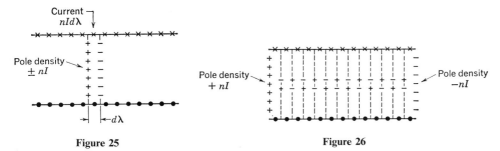

Figure 25   **Figure 26**

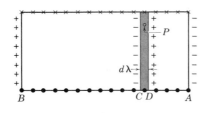

Figure 27

We may now readily extend this treatment to find the internal field of the solenoid. To find the field at point P (Fig. 27), cut a thin strip out of the solenoid by means of planes on either side of point P. The field at P is then the field of two solenoids, to both of which P is external, plus the field of the current $nI\, d\lambda$ in the strip of thickness $d\lambda$. As $d\lambda \to 0$, the current in this strip goes to zero and hence the field it sets up goes to zero. By considering $d\lambda$ as very small, then, we can neglect the field of the strip and consider the field at P as the field of two solenoids BC and DA, and hence as the field of four pole sheets, each of density $\pm nI$, uniformly distributed over the cross section of the solenoid at B, C, D, and A. The magnetic intensity at P arising from the two very close pole sheets at C and D can be computed, as on p. 612, by analogy with the electric field between the plates of the plane-plate capacitor; this computation gives

$$\mathcal{B}_0 = \mu_0\, nI, \qquad (26)$$

from right to left in Fig. 27. Note that this is the field given by the solenoid relation (25). The field at any point P *inside* the solenoid is then obtained by superposing, on the uniform field $\mathcal{B}_0$, the field $\mathcal{B}_p$ of the pole sheets on the ends of the solenoid. In general, this latter field is from left to right inside the solenoid and tends to reduce the intensity below the value $\mathcal{B}_0$ given by the solenoid relation. The field outside the solenoid is just $\mathcal{B}_p$. These two fields are shown in Fig. 28, and lead to the resultant field $\mathcal{B} = \mathcal{B}_0 + \mathcal{B}_p$ already shown in Fig. 20 for a short solenoid of circular section.

Remember that the employment of the pole picture and the splitting of the field $\mathcal{B}$ into these two parts is completely artificial and is just a useful computational trick. The $\mathcal{B}$ obtained in this way is the same as would be obtained by an integration over all current elements of the

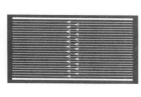

(a) The field $\mathcal{B}_0$.

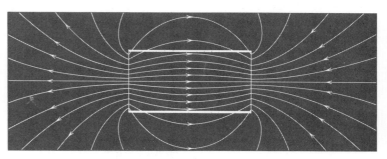

(b) The field $\mathcal{B}_p$.

Fig. 28. The field of the short solenoid whose lines are shown in Fig. 20 may be computed by superposing the fields of (a) and (b).

solenoid, using Ampère's principle. However, such an integral is much more difficult to set up than the computation we have sketched.

We see from the inverse-square law (16), p. 610, that within a long solenoid of cross-sectional area A, at a distance x from one end, $\mathcal{B}_p \sim (\mu_0/4\pi)\,nIA/x^2$, which decreases rapidly with increasing x. For example, when $x^2 \sim A$, $\mathcal{B}_p \sim \mu_0 nI/4\pi \sim \tfrac{1}{12}\mathcal{B}_0$, so that we need not go far from the ends of a solenoid before the pole field $\mathcal{B}_p$ is small compared with the field $\mathcal{B}_0$. As soon as we get much over a 'diameter' away from the ends, *the field inside a solenoid becomes uniform over the cross section and equal to $\mu_0 nI$*. We have now proved this statement rigorously.

The above arguments show that for a very long solenoid, whose ends we may imagine to be at infinity, the intensity inside the solenoid is $\mathcal{B}_0 = \mu_0 nI$, while the intensity outside becomes vanishingly small as the ends go further and further away. It is instructive to check the line-integral relation for this case. Figure 29 shows three rectangular paths, only one of which links current. On path ①, $\mathcal{B} \cdot dl$ has a value only on the bottom leg, and the line-integral relation gives $\mathcal{B}_0 l = \mu_0 nIl$, or $\mathcal{B}_0 = \mu_0 nI$, as in the solenoid relation. Around path ②, there is no magnetic intensity, and no current is linked. Path ③ links no current but has a contribution to the line integral of $\mathcal{B}_0 l$ on the bottom leg and $-\mathcal{B}_0 l$ on the top leg, which cancel.

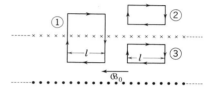

Figure 29

Example. *A solenoid* 3 m *long and* 20 cm *in diameter is wound with* 8 turns/cm *of wire carrying* 90 A. *Determine the flux threading the center of the solenoid.*

The center of the solenoid is 15 radii away from the ends, so according to Fig. 24 the center is in the region where the intensity is 100 per cent of that given by the solenoid relation, with high accuracy. Under such circumstances, the intensity is uniform across the cross section. We substitute $\mu_0 = 4\pi \times 10^{-7}$ T·m/A, $n = 800$ turns/m, and $I = 90$ A in (25) to find

$$\mathcal{B} = (4\pi \times 10^{-7}\ \text{T·m/A})(800/\text{m})\,90\ \text{A} = 0.0905\ \text{T}.$$

The area of a circle 0.2 m in diameter is 0.0314 m²; hence the flux is

$$\Phi = \mathcal{B}A = (0.0905\ \text{T})(0.0314\ \text{m}^2) = 0.002\,84\ \text{Wb}.$$

6. Forces on Solenoids

We have seen that the external magnetic field of a solenoid is correctly given as the field of a certain distribution of magnetic poles, the intensity being computed from the pole strength m by the inverse-square law (16). We can now complete our present discussion of *magnetostatics* by showing that the forces and torques acting on a solenoid placed in a magnetic field arising from some external source are correctly given by assuming that the force on a magnetic pole m in a field of intensity $\mathcal{B}$ is

$$F = m\,\mathcal{B}, \qquad \begin{cases} F \text{ in N} \\ m \text{ in A·m} \\ \mathcal{B} \text{ in T} \end{cases} \quad (27)$$

and hence, from (16), that the forces and torques acting between two (rigid) solenoids are correctly given by assuming that the pole sheets at their ends attract or repel with inverse square forces like those in Coulomb's principle:

$$F = \frac{\mu_0}{4\pi}\,\frac{m_1 m_2}{r^2}. \qquad \begin{cases} F \text{ in N} \\ \mu_0 \text{ in T·m/A} \\ m \text{ in A·m} \\ r \text{ in m} \end{cases} \quad (28)$$

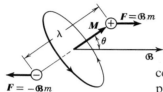

Figure 30

To demonstrate (27), we need only show that this equation gives the correct torque on a small coil, when this coil is replaced by the equivalent pair of poles. If the equation is valid for the small coil, it is valid for the pole sheets that are obtained, as in Fig. 26, by summation over all the small coils in a solenoid. Figure 30 shows a small coil of magnetic moment M in the field $\mathcal{B}$. The torque is, by definition of $\mathcal{B}$, $L = M\mathcal{B}\sin\theta$. The pole strength m is, by (15), p. 610, $m = M/\lambda$. The forces exerted on the poles according to (27) form a couple with an associated torque $L = \mathcal{B} m\lambda \sin\theta = \mathcal{B} M \sin\theta$. This torque has the correct magnitude and is also seen to have the correct direction, along $M \times \mathcal{B}$, to turn M into $\mathcal{B}$. Hence the torque computed from (27) by applying forces on the magnetic poles is accurate.

In a uniform field, such as that of the earth, a thin rigid solenoid behaves as if it were acted on by equal and opposite forces on its two ends, which result in a torque tending to turn it into the direction of the field (Fig. 31). It must be emphasized that this is not at all the system of forces that actually exists. The actual system consists of forces to the right on all current elements going into the paper in Fig. 32, and forces to the left on all current elements coming out of the paper. These forces are to be computed from (4), p. 591. However, the systems of forces in Fig. 31 and Fig. 32 must give the same resultant torque for a solenoid in a *uniform* magnetic field.

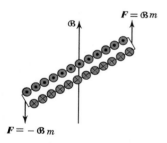

Fig. 31. Forces on a solenoid in a uniform field as given by the pole picture.

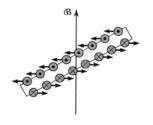

Fig. 32. Actual system of forces on a solenoid in a uniform magnetic field.

Example. *What are the strengths of the 'conceptual' poles at the ends of the solenoid described in the example on p. 619? If this solenoid is in a uniform external magnetic field of intensity 1 T perpendicular to the axis of the solenoid, what is the torque tending to turn the solenoid?*

The pole strength is given by

$$m = nIA = (800/\text{m})(90 \text{ A})(0.0314 \text{ m}^2) = 2260 \text{ A·m}.$$

As in Fig. 31, there will be a 'conceptual' force on the $+$ pole in the direction of the external field, and one on the $-$ pole in the opposite direction. The magnitude of each of these forces is given by (27) as

$$F = m\mathcal{B} = (2260 \text{ A} \cdot \text{m})(1 \text{ T}) = 2260 \text{ N}.$$

Since the solenoid is perpendicular to the field, we get the torque by multiplying the force by the length (3 m) of the solenoid, as

$$L = (2260 \text{ N})(3 \text{ m}) = 6780 \text{ N} \cdot \text{m}.$$

We can verify this result very simply by computing the torque on each turn and multiplying by the number of turns. Each turn has magnetic moment $M = IA = (90 \text{ A})(0.0314 \text{ m}^2) = 2.83 \text{ A} \cdot \text{m}^2$. The torque on this turn is $L = M\mathcal{B} = 2.83 \text{ N} \cdot \text{m}$. There is a total of 2400 turns, so the total torque is $2400 \times 2.83 \text{ N} \cdot \text{m} = 6790 \text{ N} \cdot \text{m}$.

7. The Current Balance; Determination of ε_0; Electrical Standards

It is not convenient to measure the force per unit length between two infinite wires, as in the definition of the ampere on p. 606, but it is convenient to measure the force acting between two coils carrying the same current, as indicated schematically in Fig. 33. If these coils are accurately constructed, it is possible to *compute* the force in terms of the current in amperes by methods indicated earlier in this chapter. Alternatively, it is possible to compute the current in terms of the force and the dimensions of the coils. In this way, the *current balance* of Fig. 33 measures current in terms of the mechanical units, newton and meter.

Very accurately constructed current balances at national standardizing laboratories can determine currents in this way to an accuracy of one part in a million.

Since the coulomb is defined in terms of the ampere, as the charge carried past a point by one ampere in one second, the value of ε_0 in Coulomb's principle, $F = (1/4\pi\varepsilon_0) Q_1 Q_2 / d^2$, becomes a matter for experimental determination. On p. 469, we gave $1/4\pi\varepsilon_0$ the 'best' experimental value, $8.9875 \times 10^9 \text{ N} \cdot \text{m}^2 / \text{C}^2$.

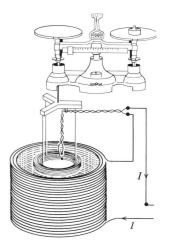

Fig. 33. Schematic diagram of a *current balance*.

There are two ways of determining the value of ε_0. In Chapter 43, we shall show theoretically that the speed of an electromagnetic wave in free space (the speed of light) is given by

$$c = \frac{1}{\sqrt{\varepsilon_0 \mu_0}}; \qquad c^2 = \frac{1}{\varepsilon_0 \mu_0} = \frac{1}{4\pi\varepsilon_0} \times \frac{4\pi}{\mu_0}. \tag{29}$$

The measured value of c^2 is $8.9875 \times 10^{16} \text{ m}^2/\text{s}^2$. Hence, since $\mu_0/4\pi = 10^{-7} \text{ T} \cdot \text{m/A}$ by definition (p. 605),

$$\frac{1}{4\pi\varepsilon_0} = 10^{-7} \frac{\text{T} \cdot \text{m}}{\text{A}} \times 8.9875 \times 10^{16} \frac{\text{m}^2}{\text{s}^2} = 8.9875 \times 10^9 \frac{\text{N} \cdot \text{m}^2}{\text{C}^2}, \tag{30}$$

where the units can best be checked by means of the relations in Sec. 1 of the Appendix.

The above argument gives one way of determining ε_0. However, one would like to see a direct electrostatic measurement of ε_0 as a convincing proof of the validity of the electromagnetic theory of light that leads to relation (29). An electrostatic determination of ε_0 is made most accurately, not by a direct measurement of the force between charges, but rather by measurement of the capacitance of an accurately constructed capacitor, since the constant ε_0 occurs in all expressions for capacitance in Chapter 24. Thus relation (4), p. 514, for a parallel-plate capacitor gives $C = \varepsilon_0 A/d$. It is possible to construct a capacitor whose capacitance can be accurately computed in terms of ε_0 and its physical dimensions. Its capacitance can be measured in terms of the definition $Q = It$ of charge by a method that repeatedly charges the capacitor to a known voltage and permits it to discharge a large but known number of times per second through a galvanometer calibrated in terms of the definition of the ampere. Thus, C is measured in terms of the ampere and the volt. By putting the measured capacitance C in the equation for the capacitance, ε_0 is determined. The most accurate value determined by measurement of capacitance is

$$1/4\pi\varepsilon_0 = 8.9872 \times 10^9 \text{ N} \cdot \text{m}^2/\text{C}^2. \tag{31}$$

The almost perfect agreement between the values (30) and (31) is overwhelming evidence for the validity of the electromagnetic theory of light.

We note that the electrostatic method just described requires an accurate measurement of voltage. We have defined the volt in terms of the ampere and the watt by means of the relation $P = IV$ for the rate in watts at which heat is generated by a current passing through a resistance. *In principle,* the relation $P = IV = I^2R$ can be used, in connection with the current balance, to construct an accurate ohm that will generate heat at the rate of one watt with one ampere of current; the accurate ohm will then have one volt of potential difference per ampere of current. *In practice,* the above procedure does not give a sufficiently accurate ohm for the determination of voltages to the precision required in (31); the lack of precision arises because calorimetric measurements are difficult to make with high accuracy. It turns out that a more accurate standard of voltage can be obtained by methods employing a standard mutual inductance or a standard capacitance (*see* Chapter 31).

8. Magnetic Trapping of Charged Particles

The magnetic force $f = qv \times \mathcal{B}$ exerted on a moving charge in a magnetic field provides a means of confining charged particles to a region of space sometimes called a 'magnetic bottle.' We recall that a positively charged particle moving in a uniform magnetic field moves in a helical path; Fig. 34(a) gives the relation of the helix to the direction of $\mathcal{B}$ for positively charged particles moving to the right and to the left. One type of magnetic bottle shown schematically in Fig. 34(b) and (c) is a magnetic tube in which $\mathcal{B}$ is larger near the ends of the tube than in the central portion; such a tube might be produced by a solenoid in which the windings are more closely spaced near the ends than in the center or by adding a concentrated coil to each end of a solenoid with uniformly spaced windings.

As suggested in Fig. 34(b) and (c), the path of a positively charged particle within such a magnetic tube consists of a helix with large widely spaced turns in the central portion of the tube and smaller closely spaced turns near the ends of the tube. The axial component v_X of the particle's velocity decreases to zero and then reverses; the surfaces at which reversal occurs are called 'magnetic mirrors.' Fig. 34(b) shows the charged particle leaving mirror AA and moving along a helical path to mirror BB; part (c) of the figure shows the particle leaving mirror BB and moving along a helical path back to AA. A trapped negative particle would execute a similar motion except for a reversal in the sense of the helix about the X-axis.

The reason for the reversal of the axial velocity is complicated, but we can readily demonstrate the existence of magnetic force components tending to reverse v_X. Consider Fig. 34(d), which shows representative points at which the particle passes through the XY-plane; the Z-axis is vertically out of the paper in this figure. At each point we can resolve the velocity into components v_X, v_Y, and v_Z. We can resolve the magnetic intensity into components $\mathcal{B}_X$ and $\mathcal{B}_Y$, with $\mathcal{B}_Z = 0$ in the XY-plane. At each point we show the component forces f_X and f_Y, which can be computed from the equations given in Prob. 35 on p. 179. Forces $f_X = -q\, v_Z\, \mathcal{B}_Y$ are responsible for the periodic variations in u_X. Component forces $f_Y = q\, v_Z\, \mathcal{B}_X$ are

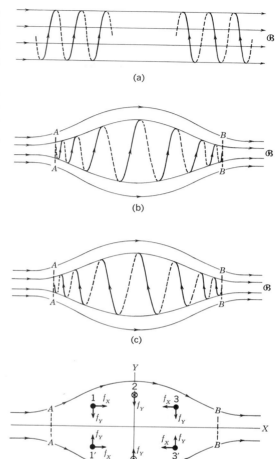

Fig. 34. The magnetic bottle.

directed radially inward toward the X-axis; these can be considered as 'centripetal' since they tend to keep the particle in an approximately 'circular' path around the X-axis. Force components $f_Z = q\, (v_X\, \mathcal{B}_Y - v_Y\, \mathcal{B}_X)$, not shown in the figure, increase the tangential velocity v_Z of the particle as it approaches a magnetic mirror and decrease v_Z as the particle leaves the mirror. Figure 35 is a photograph of an electron beam being reflected at a magnetic mirror like that at BB in Fig. 34. The light used in obtaining the photograph is caused by emission from collisions with the residual gas in the evacuated chamber in which the beam is produced.

Any ion produced inside a magnetic bottle remains permanently

Fig. 35. A stream of electrons in a magnetic bottle (from *International Science and Technology*, September 1965).

trapped inside the bottle provided its energy does not exceed a certain critical value, which depends upon the size and shape of the bottle and on the range of $\mathcal{B}$ values involved. If an ion is produced outside the bottle and has sufficient energy to penetrate the bottle, it can become trapped inside the bottle if it experiences a collision that reduces its energy below the critical value needed for escape.

As we readily see, magnetic tubes in the earth's field are of the form required for magnetic bottles, and do form magnetic bottles responsible for trapping charged particles high above the earth's surface. The presence of these charged particles was discovered in 1958 by JAMES A. VAN ALLEN and his associates. A cross section of the so-called Van Allen radiation belts is shown in Fig. 36; the inner belt consists chiefly of electrons executing helical paths similar to those in Fig. 35, and the outer belt consists chiefly of protons executing similar motions at a greater distance from the earth. The charged particles in the Van Allen belts are, for the most part, initially emitted from the sun and after energy losses become trapped in the earth's magnetic field; some of the particles in the inner belt are of terrestrial origin.

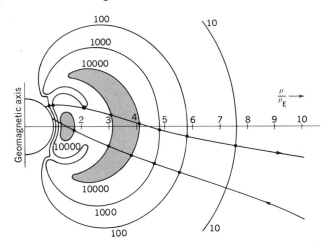

Fig. 36. Van Allen's first map of the earth's "radiation belt" consisting of magnetically trapped charged particles. The radial distance from the earth is given in earth radii; the contours are labeled by the count rate of a Geiger counter with an acceptance area of about 1 cm^2 equipped with a shield of 1 g/cm^2 thickness of lead; the numbers are thus a measure of the number of particles with energies sufficient to penetrate the lead window. The outward and return routes of the rocket probe are indicated by the heavy curves marked with arrows.

When the temperature of a gas is sufficiently high, the molecules of the gas dissociate into electrons and positive ions called a *plasma*. At the temperature of the plasma needed for the production of energy from thermonuclear reactions (Chapter 45), all known solids melt; hence no material containers can be used to enclose the plasma. Much research is currently being done on the design of magnetic bottles for enclosing the plasma required for controlled thermonuclear power production.

PROBLEMS

1. What is the magnetic intensity in T at a distance of 1 cm from an infinitely long straight wire carrying 30 A? at 2 cm? 3 cm? 4 cm?
    Ans: 6, 3, 2, 1.5×10^{-4} T.

2. What current must an infinitely long straight wire carry in order that the magnetic intensity at a distance of 2 cm be 0.005 T?

3. What current must a circular coil of 100 turns and 10 cm radius carry to give a magnetic intensity of 0.004 T at the center?
    Ans: 6.36 A.

4. What current must a square coil of 100 turns and 10 cm half-side carry to give a magnetic intensity of 0.003 T at the center?

5. Show that the magnetic intensity at the center of a rectangular coil of sides a and b is $(\mu_0/4\pi) 8NId/ab$, where d is the length of the diagonal. Find the intensity at the center of a rectangular coil of 100 turns with sides 10 cm and 40 cm long, carrying 10 A.
    Ans: 0.00824 T.

6. In the notation of the figure, show that the magnetic intensity at any point on the axis of a square turn carrying current I is $\mathcal{B} = (\mu_0/4\pi) 8Ia^2/r^2\sqrt{r^2+a^2}$.

7. Show that, at a large distance, the magnetic intensity of Prob. 6 on the axis of a square coil can be written as $\mathcal{B} = (\mu_0/4\pi) 2M/r^3$, where M is the magnetic moment of the coil. Show that (13) gives the same value.

8. The particular arrangement of two coaxial circular coils of radius a with planes separated by distance a shown in the figure is known as a pair of *Helmholtz coils*. This particular coil system is useful because it gives an almost uniform field over a fairly large volume at the center as indicated in the figure. Let X denote the distance from the plane of the lower coil to any point on the axis. If each turn carries current I, compute the intensity (a) at the center ($X=0.5\,a$), (b) at $X=0.4\,a$, $0.6\,a$, (c) at $X=0.3\,a$, $0.7\,a$, (d) at $X=0.2\,a$, $0.8\,a$. The reason for the slow variation of the field in the neighborhood of $X=0.5\,a$ is that with this particular spacing, not only is $d\mathcal{B}/dX=0$ at $X=0.5\,a$, but also $d^2\mathcal{B}/dX^2=0$ at $X=0.5\,a$. Prove this statement.

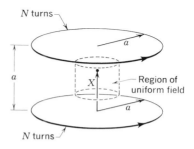

Problem 8

9. From relation (1) for the magnetic intensity close to a long straight wire, verify the line-integral law $\oint \mathcal{B} \cdot d\mathbf{l} = \mu_0 I$, for the two paths

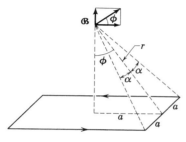

Problems 6-7

shown in the figure, where the current I is directed toward the reader.

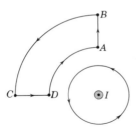

Problems 9–10

10. (a) Show that $\oint \mathcal{B} \cdot dl = 0$ for any closed path *in a plane containing* a long straight wire carrying current. (b) Show from (1) that this line integral is *either* 0 or $\mu_0 I$ for an *arbitrary* closed path in a plane perpendicular to the wire.

11. Show that the magnetic intensity on the axis of the circular coil of Fig. 7 can be written as $\mathcal{B} = (\mu_0/4\pi) \, 2M/r^3$, where M is the magnetic moment of the coil. Show that (13) gives the same value.

12. Show that the pole picture gives magnetic intensity $\mathcal{B} = (\mu_0/4\pi) \, M/r^3$ *in the plane* of a one-turn coil at distances r large compared with the size of the coil. By using (7), verify this relation at at least one point in the plane of a square coil.

13. Let Fig. 21 represent a square solenoid instead of a circular solenoid, with a and r defined as in Prob. 6. Carry out an integration of the intensity given by the equation of Prob. 6 from $\phi = 0$ to $\phi = \frac{1}{2}\pi$, to show that the field on the axis at the end of a very long square solenoid is $\frac{1}{2}\mu_0 nI$, the same as for a circular solenoid. Hence, show that the solenoid relation (25) holds for a square solenoid as well as for a circular solenoid.

14. For testing magnetic mines during World War II, large square solenoids were constructed, 2 or 3 f square and 20 or 30 f long, which could be used to apply a uniform field to the entire mine case. The solenoids were wound with 2 turns per inch length. From the solenoid relation (25), compute the field in T per ampere current. Ships' fields are usually specified in a unit called the milligauss (1 milligauss = $\frac{1}{10}\mu T$). Show that the solenoid constant you have computed is very close to 1 milligauss/milliampere, a very convenient value for test purposes.

15. What is the flux through a circular solenoid 5 cm in radius and 100 cm long wound with 1200 turns of wire carrying 2 A?
Ans: 23.7 μWb.

16. What is the flux through a long solenoid 3 f square wound with 2 turns per inch carrying 4 A?

17. In Fig. 15, p. 592, if $l = 60$ cm, $w = 30$ cm, $\mathcal{B} = 0.6$ T, what is the flux linking the rectangular loop when (a) $\theta = 0°$, (b) $\theta = 30°$, (c) $\theta = 60°$, (d) $\theta = 90°$? Ans: (a) 0.108 Wb; (b) 0.0935 Wb; (c) 0.0540 Wb; (d) 0.

18. Two telephone wires are parallel to a power wire carrying current I. The telephone wires are at distances r_1 and r_2 from the power wire. Show that the total flux from the power wire passing *between* unit length of the telephone wires is $(\mu_0/4\pi) \, 2I \log_e(r_2/r_1)$.

19. Two power wires, each carrying 1000 A but in opposite directions, are 1 m apart on the crossbar of telephone poles. On the same poles, on a crossbar 5 m below, are two telephone wires also 1 m apart, each directly below a power wire. From the relation in prob. 18, compute the magnetic flux from the power circuit that links the telephone circuit, per kilometer of line. NOTE: Telephone wires would never be strung just this way. Such magnetic linkage would be highly undesirable because EMF's would be induced in the telephone circuit whenever the current in the power circuit changed. We shall see in Chapter 31 that the magnitude of the induced current depends directly on the flux linkage you have computed. Ans: 0.007 84 Wb/km.

20. In Fig. 23, let P be a distance X from the left end of the solenoid such that $X \gg a$. Show that the expression for the field at P reduces to
$$\mathcal{B} = (\mu_0/4\pi) \, m \left(\frac{1}{X^2} - \frac{1}{(X+l)^2} \right),$$
as would be given by the pole picture. Here $m = nIA$ and l is the length of the solenoid.

21. Consider the solenoid of Fig. 31 in a uniform magnetic field. Show that the whole torque is correctly given by (2), p. 586, if one uses the integrated magnetic moment NIA, where N is the total number of turns.

22. Show that if the solenoid of Fig. 31 is of length l, and has rectangular cross section of height h in the plane of the paper and width w perpendicular to the plane of the paper, then the force system of Fig. 32 has the same torque as the pole picture of Fig. 31, when the solenoid is in a uniform magnetic field.

23. What is the pole strength of a thin solenoid of area 0.6 cm², length 15 cm, wound with a total of 500 turns carrying 0.2 A?

Ans: 0.0400 A·m.

24. If the thin solenoid of Prob. 23 is placed in the same plane as an infinite straight wire carrying 100 amp, perpendicular to the wire, with one end 5 cm from the wire and the other end 20 cm from the wire, what is the net force acting on the solenoid?

25. Find the total force with which the two thin solenoids shown below repel each other if each solenoid has a cross-sectional area of 0.5 cm².
Ans: 6.25 μN.

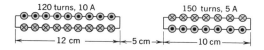

26. Find the total force with which the two thin solenoids shown below attract each other if each solenoid has a cross-sectional area of 0.5 cm².

27. In Fig. 20, p. 475, is shown an electron-ray tube in which the electron beam is deflected by forces exerted by the electric field between two plane plates. Another type of electron-ray tube employs magnetic deflection. In such a tube, two solenoids are employed to set up a substantially uniform field. The two solenoids are end-to-end like those in Prob. 26, except that the gap between the ends would be 1 cm or less; the electron beam then passes through this gap perpendicular to the paper. Show that when such a gap is short compared with the diameters of the two solenoids, the magnetic intensity in the gap is substantially constant. From (9), p. 595, the radius of curvature of an electron path is $R = mv/\mathcal{B}q$. Show that the direction of the beam traversing the field is changed by $\theta = P/R$, where P is the length of the curved path of the electrons in the magnetic field.

28. Using the results of Prob. 27, find the magnetic intensity $\mathcal{B}$ required to give a deflection of 15.6 cm on a screen at 30-cm distance, as in the example on pp. 475–477, if the length P is 1 cm. Sketch this type of electron-ray tube; note that the solenoids are placed outside, not inside, the glass envelope.

29. Show carefully how relation (23) follows from the statements that "the flux in a magnetic tube is constant along the tube" and that "magnetic lines do not begin or end but form closed curves in space."

30. Show how the last statement in italics in Sec. 4 (p. 614) can be rigorously derived directly from equation (23).

31. Show that the magnetic field of a hollow circular cylindrical conductor carrying current vanishes everywhere inside the hollow. Hence, show that the field inside a long circular solenoid wound of sufficiently fine wire is rigorously given by the solenoid formula in spite of the fact that on the current sheet shown in Fig. 19 there is actually superposed a current I from right to left.

32. Show how three perpendicular solenoids capable of carrying controlled currents are used as 'torquers' to control the attitude of the Tiros weather satellite, which flies in an orbit close to the earth.

30 MAGNETIC PROPERTIES OF MATTER

In this chapter, we continue our discussion of the magnetic effects of electric currents, with particular reference to coils wound in toroidal form which, like solenoidal coils, are capable of setting up large uniform magnetic fields inside the coils. When matter is placed in a magnetic field such as that of a toroidal coil, the matter itself sets up a macroscopic magnetic field. This field may be either in the same direction as the applied field, in which case, the matter is said to be *paramagnetic,* or in the opposite direction, in which case the matter is said to be *diamagnetic.*

However, except for those few substances known as *ferromagnetic materials,* the effects of a material medium are always very small. The field set up by most gases, liquids, and solids has an intensity of the order of one-millionth of that of the applied field—always a very small fraction, and usually of little practical importance. Unlike the case in electrostatics, where all materials except gases have a dielectric constant sufficiently large to have a pronounced electrical effect when samples are placed in an electric field, *all except the few ferromagnetic materials can, for most purposes, be treated as magnetically inert and can be considered to behave no differently from a vacuum in the presence of a magnetic field.*

Thus, to a very good approximation, the field of a solenoid may be considered the same whether its core is vacuum, air, brass, or wood. But if the core is iron, or a nickel-iron alloy, the *core* may set up a field that is enormous compared with the applied field of the winding itself.

We begin the chapter with a discussion of permanent magnets and the introduction of a qualitative theory of atomic magnetism. The important macroscopic phenomena involved in ferromagnetism are then discussed quantitatively. A brief treatment of paramagnetism and diamagnetism is also included.

1. Permanent Magnets

When an external magnetic field is applied to a sample of ferromagnetic material such as an iron rod or bar, the sample acquires a magnetic moment. The magnetic moment acquired by the sample, in turn, sets up a magnetic field of its own with magnetic intensities that may be hundreds, thousands, or even hundreds of thousands of times the intensity of the original applied field. The sample is said to be *magnetized*. Once a ferromagnetic material has been magnetized, it tends to some extent to remain more or less magnetized even after the applied external field has been removed.

In the case of pure soft iron or mild steel, the magnetic moment of a test sample decreases to a very small value as soon as the applied field is removed. Such materials are said to be *magnetically soft;* they approximate the ideal magnetic material we shall discuss in Sec. 3. Magnetically soft materials are suitable for use in transformers, dynamos, relays, and similar devices. In contrast, the hard alloy steels, particularly the aluminum-nickel-cobalt (Alnico) steels, are *magnetically hard*. Once a sample of such a steel has been magnetized, it retains most of its magnetic moment indefinitely and constitutes a usable *permanent magnet*. In the present section, we discuss permanent magnets as such without considering the question of how they become magnetized.

Since magnetic fields are associated with electric currents, it seems plausible to attribute the magnetic field of a permanent magnet to some sort of 'internal currents' within the ferromagnetic material itself. Early theories, therefore, attempted to account for the properties of permanent magnets in terms of 'Ampèrian whirls,' or current loops inside the magnetic material. In 1852, Weber proposed that each atom of a ferromagnetic material is itself a microscopic magnet, which can be oriented by magnetic fields. A schematic representation of Weber's model is given in Fig. 1, which shows an array of atoms; each atom has an Ampèrian

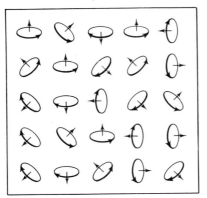

Unmagnetized

Magnetic moment →

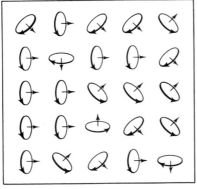

Magnetized, with resultant
magnetic moment to the right

Fig. 1. A schematic representation of the orientation of the microscopic magnetic moments of unmagnetized and magnetized material.

whirl and associated magnetic moment denoted by a small arrow.* When the material is unmagnetized, the atomic magnetic moments are randomly oriented and would give a resultant close to zero if they were summed vectorially over the whole array shown in the upper part of Fig. 1. When the material is magnetized, the atomic magnetic moments are oriented with a preference for a particular direction, as shown in the lower part of Fig. 1; the orientation of the atomic moments in this part of the figure is such that the array has a resultant magnetic moment directed toward the right.

The physical quantity used to describe the 'magnetic state' of a material is the magnetic moment per unit volume. Consider the cube shown in Fig. 2; its volume ΔV is small compared with the volume of a bar magnet, but is large enough to contain many billions of atoms. If the material in the cube is magnetized, the vector sum of the atomic magnetic moments contained in the cube gives a resultant magnetic moment $\Delta \mathbf{M}$ for the volume ΔV. The magnetic moment $\Delta \mathbf{M}$ is proportional to the volume ΔV; by dividing the resultant magnetic moment by the volume, we obtain the *magnetic moment per unit volume,* which we denote by the vector $\mathfrak{M} = \Delta \mathbf{M}/\Delta V$. This vector is called the *magnetization:*

The **magnetization** of a material medium is a vector giving the magnetic moment per unit volume.

Since magnetic moment is measured in A·m² and volume in m³, magnetization is measured in A/m.

Macroscopically (that is, averaged over the violent microscopic irregularities), the field set up by a magnetized cube is the same as the field of a fictitious current, flowing around the edge of the cube in the direction shown in Fig. 2, that has the same magnetic moment. The magnitude ΔM of the magnetic moment of such a current ΔI is $\Delta M = \Delta I \, \Delta A = \Delta I \, (\Delta l)^2$, since the area of the circuit is $(\Delta l)^2$. This quantity is equal to the magnetic moment $\Delta M = \mathfrak{M} \, \Delta V = \mathfrak{M} \, (\Delta l)^3$. Hence

$$\Delta I (\Delta l)^2 = \mathfrak{M} \, (\Delta l)^3, \quad \text{or} \quad \Delta I = \mathfrak{M} \, \Delta l.$$

*Further refinement of the 'atoms' in this representation will be given in Chapter 42; each 'whirl' is actually a 'domain' and contains many atoms.

We need a total current $\mathfrak{M}\,\Delta l$ around the cube, or current $\Delta I/\Delta l = \mathfrak{M}$ per unit width in the current sheet.

We are now prepared to calculate the field of a permanent bar magnet uniformly magnetized in the direction of its length, with magnetization $\mathfrak{M}$. The macroscopic field of each volume element is the field of the current sheet of Fig. 2. In a slice of length Δl across the magnet, these currents will be equivalent to a single current $\mathfrak{M}\,\Delta l$ around the periphery, as indicated in Fig. 3, since the currents in the interior cancel in pairs.

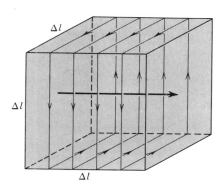

Fig. 2. The magnetic moment $\Delta M = \mathfrak{M}\,\Delta V$ indicated by the heavy arrow sets up the same field as current $\mathfrak{M}\,\Delta l$ in the direction shown by the light arrows.

The whole field of a uniformly magnetized bar magnet, inside and outside, is exactly like the field of a solenoid with current per meter length equal to $\mathfrak{M}$.

The external field is the coulomb field of + poles of total strength $\mathfrak{M} A$ spread across the area A of the N-end of the bar, with equal and opposite negative poles at the S-end of the bar. The internal field has magnetic intensity $\mathfrak{B} = \mu_0 \mathfrak{M}$ at points well away from the ends, as given by the solenoid relation (25), p. 616. The diagram of Fig. 20, p. 615, gives the magnetic lines for a short magnet.

An actual bar magnet is not usually uniformly magnetized. The magnetization usually decreases toward the ends of the bar. Such a magnet has a field like that of a solenoid whose turns are wound more densely near the center than near the ends.

Example. *A long iron bar is uniformly magnetized with magnetization of 10^6 A/m. What is the magnetic intensity $\mathfrak{B}$ just outside the end of the bar?*

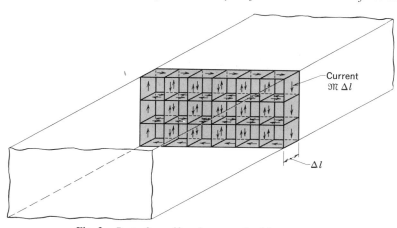

Fig. 3. Part of a uniformly magnetized bar magnet.

The whole field of the bar is the same as the field of a solenoid with current per meter length equal to $nI = \mathfrak{M}$. As we have seen on p. 616, the magnetic intensity just at the end of a long solenoid is $\frac{1}{2} \mu_0 nI$. Hence the magnetic intensity at the end of the long magnetized bar is

$$\mathfrak{B} = \tfrac{1}{2} \mu_0 \mathfrak{M} = (2\pi \times 10^{-7} \text{ T} \cdot \text{m/A}) \cdot 10^6 \text{ A/m} = 0.628 \text{ T}.$$

We now proceed with a quantitative discussion of the processes involved in magnetizing a ferromagnetic material. These processes can be understood by considering an arrangement having a sufficiently simple geometry to permit rigorous calculations.

2. Magnetization of a Ferromagnetic Toroid

There is one type of coil, extensively employed in magnetic measurements, for which the magnetic field can be rigorously computed directly from Ampère's line-integral relation. This is a coil uniformly wound on the surface of a toroid, or ring, of some nonmagnetic material such as wood.

In Fig. 4, the toroid is assumed to be wound with much finer wire than indicated, with the turns so closely spaced that there is essentially a 'current sheet' flowing around the toroidal surface. The total number of turns is N, each carrying current I.

From symmetry, the magnetic lines in Fig. 4 must be circles centered on the axis of the toroid. The only such circles that link current are those lying inside the toroid, that is, inside the body of the ring on which the coil is wound. Hence, *there is no magnetic field except inside the toroid.*

The intensity along a magnetic line of radius r, lying inside the toroid, is readily obtained from the line-integral relation:

$$\oint \mathfrak{B} \cdot d\mathbf{l} = \mathfrak{B} \cdot 2\pi r = \mu_0 NI, \quad \text{or} \quad \mathfrak{B} = \mu_0 NI/2\pi r. \tag{1}$$

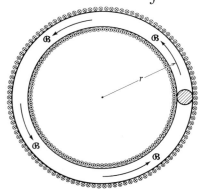

Fig. 4. A toroid wound with N turns, each carrying current I.

Thus the intensity varies (slightly) over the cross section of the toroid, inversely as the radius r. Expression (1) is valid whether the core has a circular cross section, as indicated in Fig. 4, or a cross section of any other shape.

The case of most interest in magnetic measurements is one in which the radius of the cross section is small compared with the radius of the ring. In this case, we can neglect the variation of $\mathfrak{B}$ with r and replace r in (1) by a mean radius R. The mean circumference is then $2\pi R$, and if we define $n = N/2\pi R$ as the number of turns per unit length of the circumference, equation (1) takes the form of the solenoid relation:

$$\mathcal{B} = \mu_0 \, nI. \qquad \begin{pmatrix} \mathcal{B} \text{ in T} \\ \mu_0 \text{ in T·m/A} \\ n \text{ in turns/m} \\ I \text{ in A} \end{pmatrix} \quad (2)$$

Basic measurements of the magnetic properties of ferromagnetic materials are usually made by using a thin toroidal ring of the material with a winding applied completely around the outside, as in Fig. 5. The variation in the flux $\Phi = \mathcal{B}A$ is measured as the current I in the winding is varied. To determine the flux in an iron ring without cutting into the ring seems at first sight impossible; actually, there is an instrument called a *fluxmeter* which, when linked to the ring with a few turns of wire (called a *secondary coil*), will accurately record, in webers, all changes in flux that occur inside the ring. The operation of the fluxmeter depends on the phenomenon of electromagnetic induction, which we shall study in the next chapter.

The flux density $\mathcal{B}$ in the ring is made up of two parts—that arising directly from the 'external' current in the winding, and that arising from 'internal' currents associated with the oriented atoms in the ferromagnetic material. From equation (2), the part of the field arising from 'external' current nI per meter length of winding is $\mu_0 nI$.

As a result of 'internal' currents, the ferromagnetic core has magnetization $\mathfrak{M}$, which, by symmetry, is a vector in the circumferential direction shown in Fig. 5; is the same all the way around the toroid; and for a thin toroid can be assumed constant across the cross section.

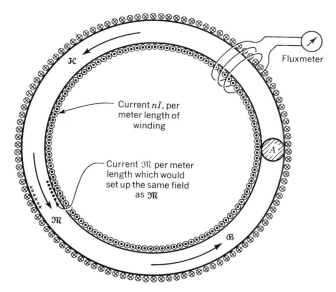

Fig. 5. A toroid with a core of ferromagnetic material, wound with n turns per meter length, each carrying current I.

Such a distribution of magnetization will, by the argument of Sec. 1, set up the same field as a current $\mathfrak{M}$ per meter length around the surface of the core, as indicated in Fig. 5. This fictitious current, from which the field set up by the magnetization of the toroid can be computed, is itself exactly like the current in the winding of a toroid, so it sets up the field $\mu_0 \mathfrak{M}$.

The total magnetic intensity in the core is the sum of these two contributions:

$$\mathfrak{B}_{\text{Core}} = \mu_0\, nI + \mu_0\, \mathfrak{M}, \qquad \left(\begin{array}{c}\text{ferromag-}\\ \text{netic core}\end{array}\right) \quad (3)$$

where $\mu_0\, nI$ is the part resulting from *external* currents and $\mu_0\, \mathfrak{M}$ the part resulting from *internal* currents. The total magnetic intensity in the ferromagnetic core is much larger than the magnetic intensity the same current in the winding would produce in a nonferromagnetic core:

$$\mathfrak{B}_{\text{No core}} = \mu_0\, nI. \qquad \left(\begin{array}{c}\text{no ferromag-}\\ \text{netic core}\end{array}\right) \quad (4)$$

We define the ratio of (3) to (4) as the *permeability:*

$$\mu = \mathfrak{B}_{\text{Core}}/\mathfrak{B}_{\text{No core}}. \qquad (5)$$

> The **permeability** of a material is the ratio of the magnetic intensity in a toroidal core formed of the material to the intensity the same current in the same winding would produce in the absence of the core. Permeability is dimensionless.

The permeability of a ferromagnetic material is not a constant—it depends on the magnetic intensity itself and also somewhat on the previous magnetization history of the specimen. However for 'soft' magnetic materials, it is approximately constant, provided the magnetization is not too great. For such materials it has a value very large compared with unity—of the order of 1000 for the irons and steels and 10 000 for the nickel alloys—so that almost all of the magnetic intensity arises from the last term of (3). The value of μ is very close to 1 in a nonferromagnetic material; it is slightly greater than 1 for paramagnetic materials, slightly less than 1 for diamagnetic materials.

In relation (3), $\mathfrak{B}_{\text{Core}} = \mu_0\,(nI + \mathfrak{M})$, it is the second term that contributes most of the magnetic intensity in a material of high permeability, but it is the first term, representing the current and turns in the winding, that is responsible for the existence of the magnetization represented by the second term. Hence the term nI is called the *magnetizing force* and denoted by the special symbol $\mathcal{H}$:

$$\mathcal{H} = nI, \qquad \text{(toroid)} \quad (6)$$

where $\mathcal{H}$ is measured, as we see from this relation, in *ampere-turns/meter*. We can now rewrite (3) and (4) in the forms

$$\mathcal{B}_{\text{Core}} = \mu_0(\mathcal{H} + \mathcal{M}), \qquad \mathcal{B}_{\text{No core}} = \mu_0 \mathcal{H}; \tag{7}$$

and (5) as
$$\mu = \mathcal{B}_{\text{Core}}/\mathcal{B}_{\text{No core}} = \mathcal{B}_{\text{Core}}/\mu_0 \mathcal{H},$$

or
$$\mathcal{B}_{\text{Core}} = \mu_0 \mu \mathcal{H}. \tag{8}$$

This relation shows that the magnetic intensity in the core is proportional to the magnetizing force $\mathcal{H}$ and to the permeability μ of the material.

We now note that, in the first of relations (7), $\mathcal{B}$ and $\mathcal{M}$ represent the magnitude of *vector* quantities. Hence it is desirable to think of $\mathcal{H}$ as also representing the magnitude of the vector quantity defined by

$$\mathcal{H} = \mathcal{B}/\mu_0 - \mathcal{M}. \tag{9}$$

In the toroidal case, $\mathcal{H}$ has the same circumferential direction as $\mathcal{B}$ and $\mathcal{M}$, as indicated by the vector marked $\mathcal{H}$ in Fig. 5.

We have introduced the concept of a *vector* called *magnetizing force* in the particularly simple case of a uniformly wound toroid, where the magnitude of the vector is given by (6) and the vector can be defined by (9). In the case of more complex arrangements of magnetic materials and windings, no such simple relation as (6) exists for the magnitude of $\mathcal{H}$ but (9) is taken as the definition of the vector $\mathcal{H}$:

> The **magnetizing force** $\mathcal{H}$ is a vector defined by (9). Its magnitude is measured in ampere-turns/meter, and for the case of a uniformly wound toroid has the value (6).

We note from (9) that at any point where $\mathcal{M} = 0$, that is, for practical purposes, at any point in any material except a ferromagnetic material, $\mathcal{H}$ and $\mathcal{B}$ are vectors in the same direction and with magnitudes differing by a constant factor:

$$\mathcal{B} = \mu_0 \mathcal{H} \text{ except in ferromagnetic materials.}$$

Because of this relation, the relations of Chapter 29 for the magnetic effects of electric currents will give $\mathcal{H}$ instead of $\mathcal{B}$ if they are divided by μ_0, provided that no ferromagnetic materials are present.

Example. *An iron toroid of permeability* 1500 *is wound with* 100 *turns/meter each carrying* 8 A. *Determine* $\mathcal{H}$, $\mathcal{B}$, $\mathcal{M}$. *What would be the magnetic intensity* $\mathcal{B}_{\text{No core}}$ *if the core were not ferromagnetic?*

From (6), $\qquad \mathcal{H} = nI = (100 \text{ turn/m})(8 \text{ A}) = 800 \text{ A-turn/m}.$

From (8), $\qquad \mathcal{B}_{\text{Core}} = (4\pi \times 10^{-7} \text{ T·m/A}) \cdot 1500 \cdot 800 \text{ A/m} = 1.51 \text{ T}.$

From (4), $\qquad \mathcal{B}_{\text{No core}} = (4\pi \times 10^{-7} \text{ T·m/A}) \cdot 800 \text{ A/m} = 1.01 \times 10^{-3} \text{ T},$

which is $1/1500$ of $\mathcal{B}_{\text{Core}}$.

To determine the magnetization $\mathcal{M}$, we note that if we substitute (8) in the first equation of (7), we obtain $\mu\mathcal{H} = \mathcal{H} + \mathcal{M}$, from which

$$\mathcal{M} = (\mu - 1)\mathcal{H} = 1499\,(800 \text{ A/m}) = 1.20 \times 10^6 \text{ A/m}.$$

3. The 'Magnetic Circuit'

A closed path of magnetic material (such as the toroid in Fig. 5) is called a *magnetic circuit,* since the magnetic flux runs through the magnetic material like the electric current in an electric circuit. This analogy explains the terminology of some of the following definitions.

If N is the total number of turns in the winding and l is the average length of the core in meters (the length of bar required to make the toroid if an iron bar is bent into a circle and welded), then $n = N/l$. Using (8) and (6), we can write the equation for the total flux as

$$\Phi = \mathcal{B}A = \mu_0 \mu \mathcal{H} A = \mu_0 \mu n I A = \mu_0 \mu N I A/l = \frac{NI}{l/(A\mu_0 \mu)}.$$

The numerator of this expression is called the *magnetomotive force:*

$$\text{MMF} = NI. \tag{10}$$

The denominator is called the *reluctance* of the magnetic circuit:

$$\mathcal{R} = l/A\mu_0 \mu. \tag{11}$$

With these notations the equation for flux through the magnetic circuit can be written as

$$\Phi = \text{MMF}/\mathcal{R}, \tag{12}$$

in exact analogy with the electric-circuit equation $I = \text{EMF}/R$.

If the distributed turns on the highly permeable toroid of Fig. 5 are collected into a concentrated coil, it is an experimental fact that the flux lines continue to follow the toroid around, and the total flux Φ remains unchanged if the total number of ampere-turns remains unchanged. It is possible to develop equations that give the reluctance to be used in (12) for nontoroidal magnetic circuits like the transformer core of Fig. 17, p. 658, and even for 'series' and 'parallel' magnetic circuits. Such equations are used by engineers in magnetic-circuit design.

4. Real Magnetic Materials

If we take a ring specimen of a magnetic material with a toroidal winding as in Fig. 5, we can increase the value of $\mathcal{H} = NI/l$ in small steps by increasing the value of the current I in small steps. For each change in $\mathcal{H}$, we can measure the change in flux, and hence the change in $\mathcal{B}$, by means of the fluxmeter. Hence, we can plot a curve of $\mathcal{B}$ against $\mathcal{H}$. If we start with an unmagnetized specimen, a curve obtained in this way is known as a *magnetization curve*. A magnetization curve for annealed iron is shown in Fig. 6. It will be noticed that this real material corresponds only roughly to the concept of a material of constant permeability (constant ratio of $\mathcal{B}$ to $\mathcal{H}$). In the actual case of annealed iron, μ starts with a value of about 200, determined by the slope of the line marked μ_I in Fig. 6, which is known as the *initial permeability* μ_I. As the magnetizing force increases, the ratio $\mathcal{B}/\mathcal{H}$ at first increases, reaching its maximum at about $\mathcal{B} = 1$ T, as indicated by the slope of the line marked μ_M in Fig. 6. The *maximum permeability* μ_M in this case is 5000. The ratio $\mathcal{B}/\mathcal{H}$ then decreases, and $\mathcal{B}$ slowly approaches the *saturation intensity* $\mathcal{B}_S$, which in this case is 2.15 T. A plot of $\mu = (\mathcal{B}/\mathcal{H})/\mu_0$ is given in Fig. 7.

For many applications of magnetic materials, high flux densities are desired at very low values of magnetizing force. For such applications, which include low-current transformers, low-current relays, inductive 'loading' of telephone cables, and sensitive detectors of small field changes (as in magnetic mines), the best material is the one of highest initial permeability or of highest maximum permeability, depending on the particular application. Tremendous progress has been achieved in the development of magnetic materials for special applications.

If we take an initially unmagnetized sample of iron that has the magnetization curve of Fig. 6, increase the magnetizing force to 160 A-turn/m so that the flux density becomes 1 T, and then start decreasing the magnetizing force, the flux density does not decrease along the same curve but decreases along the curve shown in Fig. 8. When the magnetizing force has dropped to zero (no current in the winding), we still have a *residual flux* of density $\mathcal{B}_R$, which in this case is about 0.7 T. Not until the magnetizing force has assumed the negative value of about -60 A-turn/m (known as the *coercive force* $\mathcal{H}_C$) does the flux density become zero. If the magnetizing force is repeatedly varied between the limits of $+160$ and -160 A-turn/m, the flux density follows around the closed curve of Fig. 8, which is called a *hysteresis loop*. For higher or lower maximum values of cyclically applied magnetizing force, the flux density follows around correspondingly larger or smaller hysteresis loops.

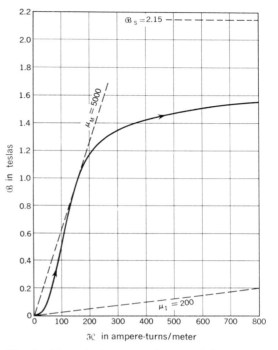

Fig. 6. Magnetization curve for annealed iron of high purity.

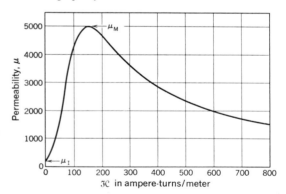

Fig. 7. Permeability vs. magnetizing force for the material of Fig. 6.

A magnetic material with a high residual flux and a high coercive force is a very good permanent magnet material but a very bad material

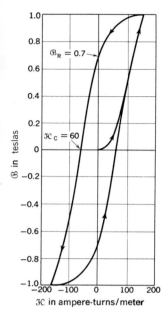

Fig. 8. Typical hysteresis loop for the material of Fig. 6.

for a transformer or a motor where the flux varies continuously. The reason for the latter part of this statement is that the energy of self inductance expended in setting up the magnetic field (see Chapter 31) is not completely recovered when the field decreases to zero if there is hysteresis. Rather, there is an energy loss in each cycle, if an alternating current is applied, that is proportional to the area of the hysteresis loop on a plot such as that of Fig. 8. Hence, for materials subjected to cyclic magnetization, a very narrow hysteresis loop is highly desirable.

For permanent magnets, a high value of residual flux is desired if the magnet is to have a strong field. The residual flux density $\mathcal{B}_R$ when a magnetic material is magnetized to saturation is called the *remanence*. The coercive force required to demagnetize a material magnetized to saturation is called the *coercivity*. It is important for permanent-magnet material to have a high remanence, but it is of even more importance that it have a high coercivity, so that it is hard to 'demagnetize.' Characteristics of permanent-magnet materials have continuously improved from the time of the first introduction of tungsten steel in 1855. This material has a remanence of 1.05 T and a coercivity of 5200 A-turn/m. One of the Alnicos, introduced in 1940, has a remanence of 1.25 T and a coercivity of 44 000 A-turn/m. Practical permanent magnets were first made in 1952 from the pure intermetallic compound manganese bismuthide, which has the enormous coercivity of 290 000 A-turn/m, and remanence of 0.46 T.

5. Magnetic Susceptibility

Returning to our discussion of magnetization of matter in a toroidal field, we recall that the relatively weak magnetizing force $\mathcal{H}$ resulting from the external currents was responsible for the much larger magnetization $\mathcal{M}$ associated with ferromagnetic materials. When nonferromagnetic materials are subjected to a magnetizing force $\mathcal{H}$, they acquire a small magnetization that is directly proportional to $\mathcal{H}$; thus

$$\mathcal{M} = \chi \mathcal{H}, \tag{13}$$

where the proportionality constant χ is called the *magnetic susceptibility*. χ is dimensionless, since $\mathcal{M}$ and $\mathcal{H}$ have the same dimensions. For all nonferromagnetic materials, the value of χ is extremely small compared with unity. Typical values of magnetic susceptibility χ are listed in Table 30-1. Materials having negative susceptibility are called *diamagnetic*; those with positive susceptibility are called *paramagnetic*.

Since, for nonferromagnetic materials, $\mathcal{M} \ll \mathcal{H}$ in equation (9), we can for practical purposes rewrite (13) in terms of the magnetic intensity $\mathcal{B}$ in the form

$$\mathcal{M} = \chi \mathcal{B}/\mu_0. \tag{13'}$$

As we shall see in Section 7, there is a tendency for *all* matter to be diamagnetic, but for certain elements such as the transition elements with atomic numbers $Z=21$ to 30 and 39 to 48, and the rare-earth elements with $Z=57$ to 70 and $Z \geq 89$, paramagnetic effects predominate and give positive values of χ. Compounds of the above elements are also paramagnetic. A few substances such as molecular oxygen, O_2, and nitric oxide, NO, are also paramagnetic. Although we must postpone detailed interpretations of the magnetic properties of matter to Chapter 42, we shall present qualitative interpretations in terms of simple atomic models in Sections 6 and 7.

Table 30-1 MAGNETIC SUSCEPTIBILITIES OF NONFERROMAGNETIC METALS AT 300° K

Metal	χ
Copper	-0.08×10^{-6}
Silver	-0.19
Gold	-0.14
Zinc	-0.15
Antimony[a]	-1.38
Bismuth[a]	-1.5
Vanadium	$+1.5$
Chromium	$+3.0$
Manganese	$+10.0$

[a] The values of χ for antimony and bismuth depend on the orientation of the crystal with respect to the magnetic field.

Magnetic susceptibilities are usually compared by measuring the magnetic forces acting on samples placed in nonuniform external magnetic fields like the one shown schematically in Fig. 9. The magnetization $\mathfrak{M}_z$ is directly proportional to $\mathfrak{B}_z$ but the magnetic force acting on the sample is proportional to $\mathfrak{M}_z$ and to $d\mathfrak{B}_z/dZ$ (*see* Prob. 26, p. 645). The magnetic force on a paramagnetic sample is directed toward regions of larger $\mathfrak{B}$ (stronger magnetic field); for a diamagnetic sample the magnetic force has the opposite direction, toward regions of smaller $\mathfrak{B}$ (weaker magnetic field). These effects can be demonstrated qualitatively by observing that a test tube filled with paramagnetic liquid oxygen is *attracted* by either pole of a strong Alnico magnet while a piece of the diamagnetic metallic bismuth is correspondingly *repelled*.

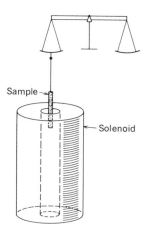

Fig. 9. In an inhomogeneous magnetic field the resultant magnetic force on a paramagnetic sample is directed toward the regions of stronger field; the resultant magnetic force on a diamagnetic sample is directed toward the regions of weaker field.

6. Magnetic Properties of Atoms; Paramagnetism

Returning to Weber's hypothesis that each atom of a ferromagnetic material has a microscopic magnetic moment, we now attempt to interpret this hypothesis in terms of the nuclear model of the atom.

First, let us consider the magnetic moment produced by a positive charge Q associated with a particle of mass m moving at an orbital frequency f in a circular orbit of radius r, as shown in Fig. 10. Since the charge Q passes any point in the orbit f times per second, its orbital motion is equivalent to a current $I=Qf$. By definition of magnetic moment as $M=IA$, the current I produces a magnetic moment $IA = Qf \times \pi r^2$. The angular momentum of the particle has a magnitude given by the product of its rotational inertia mr^2 and its angular velocity $\omega = 2\pi f$, that is, $mr^2 \times 2\pi f$. The magnetic moment and the angular momentum are both vector quantities and in Fig. 10 would be directed toward the reader.

Fig. 10. A particle of charge $+Q$, associated with a mass m, revolves at frequency f in a circular path of radius r.

The ratio of the magnitude of the magnetic moment to the magnitude of the angular momentum is called the *gyromagnetic ratio* and is customarily denoted by γ; thus, for the situation shown in Fig. 10, $\gamma = (Qf \cdot \pi r^2)/(mr^2 \cdot 2\pi f)$, or

$$\gamma = Q/2m. \qquad \begin{Bmatrix} \text{classical gyro-} \\ \text{magnetic ratio} \end{Bmatrix} \quad (14)$$

Although we have derived the classical value of γ for the simple case of a charged particle moving in a circular orbit, our result has much more general application. It is valid for particles moving in elliptical orbits and for any rotating charge distribution in which charge density is proportional to inertial density. Thus, we may write

$$\text{(magnetic moment)} = (\pm Q/2m) \times \text{(angular momentum)}, \quad (15)$$

where $\pm Q$ is the total charge and m the total mass of any system in which charge density is at all points proportional to inertial density. The magnetic moment and angular momentum vectors are parallel for positive charge and antiparallel for negative charge.

We recall that the Rutherford scattering experiments indicated that the atom consists of a massive nucleus of mass M surrounded by electrons each of which has mass m. The total angular momentum of such a system is the vector sum of the resultant angular momenta of the electrons in their motions around the nucleus and the possible angular momentum of the nucleus as a result of spin. In view of the relatively large mass M of the nucleus, the nuclear gyromagnetic ratio $\gamma_n = Q/2M$ of the nucleus is small; we shall neglect the possible effects of nuclear angular momentum in the rest of our present discussion.

Because of the electron's negative charge, the direction of the magnetic moment of a single electron is opposite that of its angular momentum as indicated in Fig. 11(a); its magnitude is simply $(e/2m)$ times its angular momentum. For a complete atom, the resultant angular momentum is given by the expression:

$$\text{(magnetic moment)} = -(e/2m)(\text{resultant electronic angular momentum}) \quad (16)$$

as indicated in Fig. 11(b). In view of the measured values of χ in Table 30-1, we conclude that there is resultant electronic angular momentum only in paramagnetic and ferromagnetic materials. This conclusion is essentially correct and will be considered further in Chapter 42.

When a paramagnetic material is placed in an external magnetic field, the magnetic moments tend to become parallel to $\mathcal{B}$. This partial alignment of atomic magnetic moments in an external field $\mathcal{B}$ results in a resultant magnetization $\mathcal{M}$ parallel to $\mathcal{B}$ in paramagnetic materials. In ferromagnetic materials, all atomic magnetic moments are parallel within small volumes called domains; the interatomic forces responsible for the existence of domains will be described in Chapter 42.

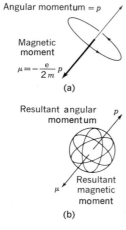

Figure 11

7. Diamagnetism

Paramagnetism and ferromagnetism are properties exhibited by the special class of materials containing atoms whose electrons have a resultant angular momentum. In the case of diamagnetism, the situation is entirely different: *all matter tends to be diamagnetic*. However, this general diamagnetism is always weak and is completely masked by stronger effects in paramagnetic and ferromagnetic materials.

In order to understand the origin of diamagnetism, we recall our earlier discussion of the magnetic force $F = q\,v \times \mathcal{B}$ acting on a charged particle moving in a magnetic field (*see* Sec. 6, Chapter 28). Under the influence of the applied field $\mathcal{B}$ a charged particle, which would otherwise move at constant velocity v, moves instead at constant speed v in a circular or helical path of radius r. The period of revolution is independent of the radius of the path; the frequency f of revolution is $q\mathcal{B}/2\pi m$.

For the sake of simplicity, we restrict our discussion at this point to the magnetic moment of a charged particle moving in a circular path and subject only to the magnetic force $q\,v \times \mathcal{B}$. The magnetic moment of such a system has a magnitude given by the product of a current $I = qf$ and the area πr^2 of the orbit; thus,

$$M = q\,(q\mathcal{B}/2\pi m)\,\pi r^2, \quad \text{or} \quad M = (q^2 \mathcal{B}/2m)\,r^2. \tag{17}$$

Consideration of the direction of the motion of the charged particle shows that *the magnetic moment (17) is antiparallel to $\mathcal{B}$ regardless of the sign of q;* free charged particles such as electrons or ions thus have diamagnetic properties when their motion is restricted to motion in a circular path by the application of a magnetic field. Similar diamagnetic properties are exhibited by charged particles moving in helical paths in a magnetic field.

We note that $q\,v \times \mathcal{B}$ forces act not only on free charges in vacuum but on all moving charges, including conduction electrons in metals and electrons moving in atomic shells. We consider first the case of the relatively 'free' conduction electrons.

Conduction electrons moving at thermal velocities move along circular or helical paths under the influence of a steady magnetic field. However, because of frequent collisions with the lattice, the electrons rarely execute complete circular paths so that (17) cannot be used to compute the magnetic moments involved; nonetheless, the general curvature of the paths is sufficient to make metals feebly diamagnetic.

Much stronger diamagnetism is associated with the phenomenon of electromagnetic induction to be considered in the following chapter. When there is a time-varying external magnetic field in the vicinity of a metal, the induced EMF's, which are proportional to the time rate of change of the magnetic field, produce currents that have associated magnetic fields opposite in direction to the changes in the external field. In ordinary metals with finite resistivity, these currents fall rapidly to zero when the external field becomes constant; thus, ordinary metals exhibit a strong *transient* diamagnetism that disappears when the external field becomes constant. In superconductors the situation is quite different; when an external field is being established, induced EMF's produce currents which persist even after the external field becomes constant. Superconductors are highly diamagnetic; when a superconducting sample is placed in a magnetic field $\mathcal{B}$, currents are produced in the surface layers and the magnetic field in the interior of the sample remains zero.

Now we consider briefly the influence of magnetic fields on orbital electrons within atoms. Consider the simple model in Fig. 12(a), which shows charged particles of mass m moving in opposite senses in circular orbits under the influence of a large centripetal force F_0; the magnetic moments associated with the two motions are of equal magnitude M_0 but are oppositely directed as shown. Now consider the situation in Fig. 12(b), in which a magnetic field $\mathcal{B}$ has been applied in a direction away from the reader. The centripetal force on the particle at the left is increased by the magnetic force $\Delta F = \mathcal{B}qv$, small compared with F_0, and the centripetal force acting on the particle at the right is decreased by the same amount. The changes in the centripetal force are such that the magnetic moment of the orbit at the left is increased by ΔM and the magnetic moment of the orbit at the right is decreased by the same ΔM. The value of ΔM can be calculated if we assume that the radii of

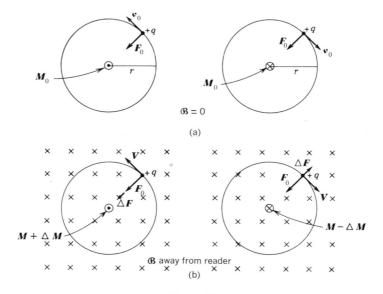

Figure 12

the orbits remain unchanged and that the application of the magnetic field causes changes only in the tangential velocities of the charges by the phenomenon of electromagnetic induction mentioned in the preceding paragraph. The result (*see* Prob. 25) agrees with the results obtained when rigorous methods are applied to more general types of orbital motion.

We now summarize the results shown in Fig. 12. Prior to the application of the field $\mathcal{B}$, the values of the magnetic moments M_0 were equal in magnitude and opposite in direction so that the two orbits gave no resultant magnetic moment. After the magnetic field was applied, the two orbits had a resultant magnetic moment of magnitude $2 \Delta M$ and a direction antiparallel to $\mathcal{B}$; the two orbits were therefore diamagnetic. Analogous diamagnetic effects are associated with each pair of oppositely oriented electron orbits in atoms. Thus, electrons within atoms have diamagnetic properties.

Since 'free' charges such as ions or electrons, conduction electrons, and electrons in atoms all have diamagnetic properties, diamagnetism is a general property of all matter. However, diamagnetic effects are completely masked by the paramagnetism or the ferromagnetism of materials containing atoms with net electronic angular momentum.

PROBLEMS

1. A permanent magnet 15 cm long and 2 cm² in cross-sectional area experiences a torque of 0.030 N·cm when placed perpendicular to a uniform magnetic field of 0.090 T. What is the magnetic moment of the magnet? What is the pole strength? What is the average magnetization of the material? Ans: 3.33×10^{-3} A·m²; 2.22×10^{-2} A·m; 111 A/m.

2. What is the magnetic intensity $\mathcal{B}$ in the central portion of the magnet described in Prob. 1? What is the magnetic flux Φ through the magnet?

3. If we wound 100 turns of wire on a piece of wood 15 cm long and 2 cm² in cross-sectional area, what current in the wire would be required to produce a solenoidal field equivalent to that of the permanent magnet of Prob. 1? Ans: 0.167 A.

4. Compute the magnetic moment of a uniformly magnetized bar magnet by integrating the magnetization throughout the volume. Show that this magnetic moment agrees with that computed by multiplying pole strength by the length of the magnet, and hence that the definition of magnetic moment of a small magnet, given on p. 586, is satisfactory.

5. A long bar magnet has a uniform magnetization of 500 000 A·m. What is the magnetic intensity $\mathcal{B}$ just outside the end of the bar? Ans: 0.314 T.

6. Recalling that the torque exerted on a magnet of moment M by a magnetic field $\mathcal{B}$ is given by the expression $L = M \times \mathcal{B}$, show that the potential energy of a magnet is given by the expression $E_P = -M \cdot \mathcal{B}$ if the zero level of E_P is chosen as the position at which M is perpendicular to $\mathcal{B}$.

7. If the magnet described in Prob. 1 were placed in a uniform external magnetic field of 0.120 T with its magnetic moment parallel to the field, how much work would be required to rotate the magnet from its initial orientation to an orientation in which its magnetic moment is (*a*) perpendicular to the external field and (*b*) antiparallel to the external field?

8. Two permanent magnets like the one described in Prob. 1 are shown in the figure. What are the magnitudes of the resultant forces and torques acting on magnet *B* in the orientations (a), (b), and (c)?

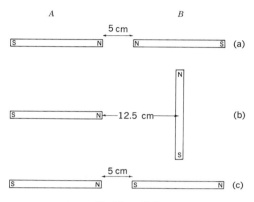

Problems 8-9

9. Using the pole concept and the methods of magnetostatics, compute the magnetic potential energies of the magnets in each of the orientations shown in the figure on the assumption that the magnetic potential at infinite separation is zero. How much work must be done in changing the orientation from (c) to (b)? from (c) to (a)?

10. It is asserted in the text (p. 631) that the external field of a uniformly magnetized bar magnet is the coulomb field of + and − poles of total strengths $\pm \mathfrak{M} A$ "*spread across*" the ends of the bar. This statement does not directly follow from the discussion of the external field of a solenoid in Chapter 29, because in that chapter we considered only thin solenoids, and placed a single point pole at each end to represent the external field. But show how, in Fig. 3, we can slice the magnet longitudinally into 'match sticks,' each of which has an effective current distribution like that of a *thin* solenoid, and that the resulting distribution of poles on the ends of the magnet is equivalent to 'spreading' the pole strength $\mathfrak{M} A$ uniformly over the area *A*.

11. From the results in the preceding problem show that if the N end of one bar magnet is placed opposite the S end of an identical bar magnet with a short air gap between the ends, the field outside the magnets, near the ends, is exactly like the field near a pair of capacitor plates that is illustrated in Fig. 11, p. 502, and Fig. 1, p. 514. By analogy with the electrical case, find the magnetic intensity in the region between the pole faces in terms of the intensity of magnetization $\mathfrak{M}$.   Ans: $\mu_0 \mathfrak{M}$.

12. (*a*) Using the same reference level as that in Prob. 6, show that the potential energy per unit volume of a permanent magnet is given by the expression $-\mathbf{M} \cdot \mathbf{\mathcal{B}}$. (*b*) Recalling our earlier discussion (p. 185) of stable equilibrium, discuss the behavior of a bar magnet in a field where (1) $\mathcal{B}$ varies from point to point and (2) $\mathcal{B}$ is constant in magnitude but rotates.

13. A toroid is made by bending a circular rod of 3-cm diameter and 1-m length into a circle and welding the ends. The material is mild steel of permeability 1100. It is wound with 150 turns uniformly distributed. What current is necessary to set up a flux density of 1.60 T?   Ans: 7.72 A.

14. A cast-iron toroid is made by casting a doughnut shape. The mean radius of the toroid is 12 cm, the radius of the section is 1.5 cm. The permeability is 400. It is wound with 400 turns uniformly distributed. What current is necessary to set up a flux density of 0.2 T?

15. In Prob. 13, what current is necessary to set up a flux of 4.5×10^{-4} Wb?   Ans: 3.07 A.

16. In Prob. 14, what current is necessary to set up a flux of 3.0×10^{-4} Wb?

17. It is asserted in the text that the area of the hysteresis loop shown in Fig. 8 is proportional to an energy loss. By consideration of the physical dimensions of $\mathcal{B}$ and $\mathcal{H}$, show that the area of the loop has the dimensions of energy per unit volume (J/m^3).

18. Compute approximate values of μ_I and μ_M from (8) and the slopes of the dashed lines in Fig. 6, and compare with the values stated along those lines.

19. What is the value of the magnetizing force $\mathcal{H}$ in the air gap between the pole faces of a large electromagnet where $\mathcal{B} = 1.2$ T? Using the values of χ in Table 30-1, compute the magnetization of (*a*) a small sample of silver in this external magnetic field and (*b*) a small sample of chromium.

Ans: (*a*) 2.86 A/m, (*b*) 0.181 A/m.

20. What are the magnetizations of metallic antimony, bismuth, vanadium, and manganese in an external magnetic field where $\mathcal{H} = 1000$ A/m?

21. We can get an approximation of the effective magnetic moment per atom of iron by the following computations: (*a*) If iron has a density of 7870 kg/m³, how many atoms N are there per cubic meter? (*b*) If the saturation intensity of iron (Fig. 6) is observed to be 2.15 T, what is the magnetization $\mathfrak{M}$ of iron when all atoms are aligned? (*c*) By setting $N\mu_{Fe} = \mathfrak{M}$, find the magnitude of the effective magnetic moment μ_{Fe} of the iron atom.

Ans: (*a*) 8.49×10^{28}; (*b*) 16.1×10^5 A/m; (*c*) 1.88×10^{-23} A/m.

22. Using the classical gyromagnetic ratio (14), find the effective angular momentum of each iron atom discussed in Prob. 21. What is the total effective electronic angular momentum in one cubic meter of iron in which all atomic magnetic moments are aligned?

23. By comparison of (17) with (9); p. 595, which gives a value $r = mv_\perp/q\mathcal{B}$ for the radius of the orbit of a charged particle in a magnetic field, show that the orbital magnetic moment M is *inversely* proportional to $\mathcal{B}$. Does this possibly surprising result imply that the *magnetization* is similarly related to $\mathcal{B}$? Discuss.

24. Starting with the model in Fig. 12(a), show that the frequency f_0 of the orbital motion is given by the expression: $f_0 = \sqrt{F_0/mr}/2\pi$. Show also that $M_0 = (q/2)\sqrt{F_0 r^3/m}$.

25. Starting with the model in Fig. 12(b) and assuming that r has the same value as in Fig. 12(a), show that the orbital frequency $f = f_0 \pm \mathcal{B}q/2\pi m$, provided $\mathcal{B}qv \ll F_0$. Show also that $\Delta M = \pm (\mathcal{B}q^2/2m)r^2$.

26. Demonstrate, from the pole picture, the statements made in the second paragraph on p. 639 regarding the magnetic forces on paramagnetic and diamagnetic samples. Assume that $d\mathcal{B}_z/dZ$ is a constant.

31 ELECTROMAGNETIC INDUCTION

We have stated earlier that an electric generator changes mechanical energy into the electric energy associated with an electric current, and that if current is forced through the generator in the opposite direction, the same machine will act as a motor, changing electric energy into mechanical energy. We have described the action of the DC motor in Chapter 28, but we have not yet described the action of the generator.

The action of the motor depends on the fact that a wire carrying current in a magnetic field experiences a force and tends to move. The generator action is the reverse; the motion of a wire in a magnetic field sets up an EMF that tends to cause a current. The fact that an EMF is set up in a wire moving in a magnetic field was discovered in 1831 independently by Faraday in London and by JOSEPH HENRY in America. These men first discovered how to change mechanical energy directly into the energy of current electricity.

The appearance of an EMF in a wire moving in a magnetic field is the simplest example of a much more general phenomenon called *electromagnetic induction*. Whenever the magnetic flux linking a circuit is *changing*, an EMF is induced in the circuit. If the circuit is a *fixed loop of wire* and the flux change arises from the changing magnetic intensity of an electromagnet, one no longer has a case of *motion* in a magnetic field. Yet an EMF appears around the circuit; work is done on the electric charges in the wire; hence these charges experience forces. Such forces must be *electric* forces; the charges are acted on by an *electric field*. If the circuit is an *imaginary closed curve* in space, it is found that the same electric intensities appear; the work that would be done on a unit charge moved once around the closed curve equals the EMF that would appear in the closed loop of wire. No longer is the line integral of $\mathcal{E}$ equal to zero; it turns out to equal the rate of change of magnetic flux linking the closed curve.

The generalizations from experiment described in the preceding paragraph are fundamental to the operation of the *transformer* and of certain types of *particle accelerators,* and also to the theory of *light* and *electromagnetic waves,* as we shall see later. Basically,

A changing magnetic field is always accompanied by an electric field in which the electric lines are closed curves and the line-integral of the electric intensity around a closed curve is no longer zero.

1. Motion of a Wire in a Magnetic Field

Consider a wire, Fig. 1, a length l of which lies in a magnetic field $\mathcal{B}$ and moves with velocity v. The wire, field, and velocity are mutually perpendicular. Although the wire as a whole is macroscopically neutral, each microscopic charge Q in the wire is acted on by a force, in the direction shown, of magnitude given by (8), p. 594, as

$$F = \mathcal{B}Qv.$$

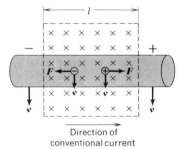

Fig. 1. Forces on the free electrons and positive nuclei in a wire moving downward through a magnetic field directed into the paper.

Unlike the positive nuclei, some of the electrons are free to move and will move to the left. If the ends of the wire are open, the movement of the electrons will cause the left end of the wire to become negatively charged and the right end to become positively charged. Motion of the electrons continues until the ends of the wire reach such different potentials that the force exerted by the resulting electric field on the electrons exactly balances the force F that acts on the electrons because of the motion of the wire. Since the force on a charge Q in an electric field $\mathcal{E}$ is $\mathcal{E}Q$, this requires that an electric field of magnitude $\mathcal{E} = \mathcal{B}v$ be set up in the part of the wire in the magnetic field. The electric field must be from right to left to balance the magnetic forces. An electric field $\mathcal{E}$ acting over the length l corresponds to a difference of potential

$$V = \mathcal{E}l = \mathcal{B}vl. \tag{1}$$

The wire of Fig. 1, if open-circuited, acquires a difference of potential of magnitude $\mathcal{B}vl$ between its ends, the right end being at the higher potential.

On the other hand, suppose the circuit to be closed externally between the ends of the wire. A current will then occur in the direction indicated in Fig. 2. During the passage of the current through the section of wire moving in the magnetic field, work will be done *by* the magnetic forces *on* the charges since the magnetic forces are *in* the direction of motion of the charges. The amount of work will be $Fl = \mathcal{B}Qvl$ on a charge Q. By definition, the EMF in volts is the work per unit charge, in joules per coulomb, or

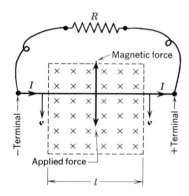

Fig. 2. A prototype electric generator.

$$\text{EMF} = V_E = \mathcal{B}vl. \qquad \begin{Bmatrix} V \text{ in V} \\ \mathcal{B} \text{ in T} \\ v \text{ in m/s} \\ l \text{ in m} \end{Bmatrix} \tag{2}$$

Equation (2) is the fundamental relation for EMF generated in a wire moving in a magnetic field. It is the same as the difference of potential (1) between the ends of the wire in the open-circuit case.

Example. *Verify that in the case where the circuit is closed by an external resistance, as in Fig. 2, the heat generated in the resistance exactly equals the mechanical work that must be done to push the stiff wire through the magnetic field at constant speed v. Take R as the total resistance of the circuit.*

The EMF (2) will cause current $I = \mathcal{B}vl/R$. This current will develop heat in the resistance at the rate $I^2 R = \mathcal{B}^2 v^2 l^2 / R$. This energy must come from mechanical energy expended to push the wire. The need for expenditure of mechanical energy is clear if we note that we now have a wire carrying current in a magnetic field and that therefore there is a magnetic force $\mathbf{I} \times \mathcal{B}l$ acting on the wire. This magnetic force is in the direction opposite to the motion, upward in Fig. 2. The magnitude of this force is $F = \mathcal{B}lI = \mathcal{B}^2 v l^2 / R$. Since the wire has no acceleration, the net force on it must be zero; consequently, our applied force in the direction of motion must have the same magnitude F as this magnetic force. The applied force F then does work at the rate $Fv = \mathcal{B}^2 v^2 l^2 / R$. This power is the same as the rate of development of heat, $I^2 R$, as we were to verify.

From the above example we see that

The induced current is in such a direction as to produce a magnetic force that opposes the force causing the motion.

In Figs. 1 and 2, the magnetic intensity ℬ is perpendicular to the plane in which the wire is moving, that is, to the plane of the paper. Since it is readily seen that neither a magnetic field in the direction of the length of the wire nor one in the direction of motion would result in magnetic forces along the length of the wire, *the value of ℬ to be used in (1) and (2) in the more general case is the component* $\mathcal{B}_\perp$ *of ℬ perpendicular to the plane of motion.* The EMF or the induced current is in the direction of $v \times \mathcal{B}_\perp$. A very convenient alternative rule for determining this direction is illustrated in Fig. 3:

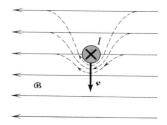

Fig. 3. View of Fig. 2 looking from the left.

Imagine the magnetic lines to be like rubber bands so that they stretch and bend around the wire as it is pushed through them. Then if the fingers of the right hand are curled around the wire, pointing in the same direction as the magnetic lines wrapped around the wire, the thumb will point in the direction of the induced current.

Example. *The field in the air gap between the pole pieces and the armature of an electric generator is* 1 T. *The length of the wires on the armature is* 1 m. *How fast must these wires move in order to generate an* EMF *of* 1 V *in each armature wire?*

From (2) we find that $v = \text{EMF}/\mathcal{B}l = 1 \text{ V}/(1 \text{ T})(1 \text{ m}) = 1 \text{ V}/\text{T} \cdot \text{m}$. This answer does not look at first sight like a *speed,* but we note from Sec. 1 of the Appendix that $1 \text{ T} = 1 \text{ V} \cdot \text{s}/\text{m}^2$, so that we get $v = 1$ m/s. Hence *a wire* 1 m *long moving at a speed of* 1 m/s *in a field of* 1 T *generates an* EMF *of* 1 V.

2. Relation between EMF and Rate of Change of Flux

There is another very useful way of writing the EMF generated in the closed circuit of Fig. 2. Let the symbol Φ stand for the flux in webers that is *linking* the closed circuit; Φ then stands for the amount of flux *above* the straight wire in Fig. 4 and increases in time as the wire moves. In time dt, let the wire move down a distance dX, so that $dX/dt = v$. Then during the time interval dt, Φ increases by the amount of flux in an area $l\,dX$. This amount of flux is $\mathcal{B}l\,dX$, so that $d\Phi = \mathcal{B}l\,dX$. If we divide through by dt, we get $d\Phi/dt = \mathcal{B}l\,dX/dt = \mathcal{B}lv$ for the rate of change of flux. But the expression on the right is just the generated EMF as given by (2). Hence

$$\text{EMF} = V_E = \frac{d\Phi}{dt}. \qquad \begin{Bmatrix} V_E \text{ in V} \\ \Phi \text{ in Wb} \\ t \text{ in s} \end{Bmatrix} \quad (3)$$

The EMF, *in volts, induced in a closed circuit equals the rate of change of flux linking the circuit, in webers per second.*

Since $1\text{ Wb} = 1\text{ V}\cdot\text{s}$, we see that $1\text{ V} = 1\text{ Wb/s}$, and hence that the above relation is dimensionally correct.

In the above derivation of (3) we have been concerned with the magnitude of V_E but not with its direction. It is very desirable to retain the direction convention described on pp. 495–496 that relates, by the right-hand rule, the sign of $\Phi = \iint \boldsymbol{\mathcal{B}} \cdot d\mathbf{S}$ to the sense of traversing the bounding curve. In Fig. 4 we see that the flux traverses the circuit in a *clockwise* sense and is increasing, so $d\Phi/dt$ is also positive in a clockwise sense. But the induced EMF is *counterclockwise*. Hence, to be in agreement with the standard sign convention, we write

$$V_E = -\frac{d\Phi}{dt} = -\frac{d}{dt}\iint \boldsymbol{\mathcal{B}} \cdot d\mathbf{S}. \tag{4}$$

This equation, properly interpreted, will give both the magnitude and the direction of the induced EMF.

A very convenient rule for determining the direction of the induced current and hence of the EMF is known as *Lenz's rule:*

> LENZ'S RULE: *When the flux linking a closed circuit is changing, the flux set up by the induced current is in such a direction as to tend to prevent the change in the flux linkage.*

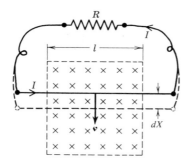

Fig. 4. The circuit of Fig. 2.

Thus, in Fig. 4, the flux Φ linking the circuit is into the paper and increasing in time. Lenz's rule states that the induced current will be in such a direction that the flux it produces is out of the paper through the circuit and therefore *tends to prevent the flux linking the circuit from increasing*. The current in the circuit of Fig. 4 has this property. If v be reversed, Φ will be decreasing. Since the induced current must now set up a flux into the paper to try to prevent Φ from decreasing, it must be clockwise.

The great convenience of equation (4) and Lenz's rule is that they can be shown to apply *whenever* the flux through a closed circuit is changing. For example, in Fig. 5, a single loop of wire is moved upward toward the north pole of a permanent magnet. The flux Φ through the loop is directed downward. The amount of flux is increasing. An EMF is induced whose magnitude is $d\Phi/dt$. The direction of the EMF is determined by Lenz's rule: The flux downward is increasing. The induced current tends to prevent the increase by setting up its own flux *upward*. Hence, the current must be counterclockwise when viewed from above as indicated in Fig. 5.

If the loop of Fig. 5 moves downward, the flux is decreasing; consequently, the induced current is in the opposite direction, tending to set up a downward flux.

If, in Fig. 5, the loop is stationary and the magnet is moved downward it is observed that exactly the same EMF is induced as when the magnet is stationary and the loop is moved upward at the same speed; the rate of change of flux through the loop is the same in the two cases.

If the single loop of Fig. 5 is replaced by a coil wound with N turns and connected into an external circuit as indicated in Fig. 6, the EMF in each of the N turns in series is $d\Phi/dt$, and the whole EMF is

$$\text{EMF} = V_\text{E} = -N\,d\Phi/dt. \tag{5}$$

It is easy to see how to generate an alternating EMF, for example by moving the wire of Fig. 2 up and down or by spinning the coil of Fig. 6 about its diameter in a magnetic field. Such a rotating coil constitutes a simple AC generator, a prototype of which is shown in Fig. 7, in which a rectangular coil of length l and width w is mounted on a shaft between the poles of a magnet in a region of flux density $\mathcal{B}$. The shaft, driven by a source of mechanical power (such as a water wheel or a heat engine), rotates at a frequency f, corresponding to an angular velocity $\omega = 2\pi f$. As the coil rotates in the magnetic field, an EMF is generated and a difference of potential is produced between the terminals A and B which make contact through 'brushes' with the slip rings connected to the coil.

We shall calculate the generated EMF by means of (5). As the coil rotates in the magnetic field, the flux linking the coil varies from zero, when the plane of the coil is parallel to the field ($\theta = 90°$), to $\Phi_\text{Max} = \mathcal{B}wl$

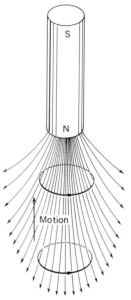

Fig. 5. Motion of a loop of wire in the vicinity of a permanent magnet.

Fig. 6. An N-turn coil.

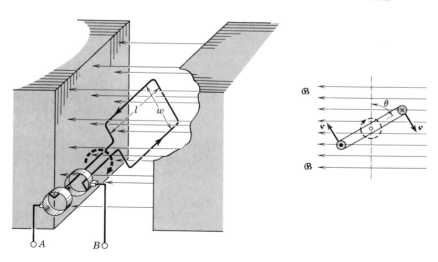

Fig. 7. Prototype AC generator, consisting of a one-turn 'coil.'

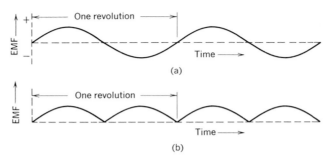

Fig. 8. (a) Alternating EMF obtained from the generator of Fig. 7. (b) Fluctuating DC EMF obtained by adding a commutator.

when the plane of the coil is perpendicular to the field ($\theta = 0°$). The flux linking the coil at angle θ is given by $\Phi = \mathcal{B}wl\cos\theta = \Phi_{\text{Max}}\cos\theta$. The generated EMF is given by (5) as

$$V_E = -d\Phi/dt = \mathcal{B}wl\sin\theta\, d\theta/dt = \Phi_{\text{Max}}\sin\theta\, d\theta/dt,$$

or, since $\theta = \omega t = 2\pi ft$ and $d\theta/dt = 2\pi f$,

$$V_E = 2\pi f\Phi_{\text{Max}}\sin 2\pi ft = V_M \sin 2\pi ft, \tag{6}$$

where $V_M = 2\pi f\Phi_{\text{Max}}$ is the maximum magnitude of V_E reached when the coil is at $\theta = 90°$ and the flux linkage is changing most rapidly. If the coil has N turns, the maximum generated voltage is given by

$$V_M = 2\pi f\, Nlw\mathcal{B}. \tag{7}$$

As shown in Fig. 8(a), V_E is a sinusoidal function of time having the same frequency as the mechanical frequency of rotation of the shaft.

In order to obtain DC from our prototype generator, we replace the slip rings shown in Fig. 7 with a split-ring commutator like the one for the prototype DC motor shown in Fig. 15 on p. 592. Such a commutator can be arranged so that terminal A is always positive with respect to terminal B in Fig. 7; in this case the EMF as measured by the open-circuit difference of potential between A and B would be given by the curve in Fig. 8(b). This EMF has a fluctuating DC value. In the case of a multipolar generator with many sets of armature coils and a many-segmented commutator, the fluctuations are smoothed out and a fairly steady DC EMF is obtained.

3. Eddy Currents

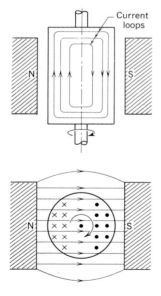

Fig. 9. Side and top views of a solid metal cylinder rotating in a magnetic field. Crosses and dots show the direction of the eddy currents.

Let the solid metal cylinder in Fig. 9 be rotating about its axis in a magnetic field perpendicular to the axis of rotation. Consider the longitudinal elements of this cylinder, parallel to the axis of rotation.

These elements are conductors moving in a magnetic field. The elements in the left portion of the cylinder are moving up in the magnetic field, and hence an EMF is induced in these elements that tends to drive current in the indicated direction. Conversely, the elements on the right side are moving down in the magnetic field, and an EMF is induced in these elements that tends to drive current in the opposite direction. As a result of these EMF's, a system of circulating currents flows in the metal. These currents are called *eddy currents* because they form closed loops within the metal, like eddies within a fluid.

Like the currents in the conductors of a generator, these currents experience magnetic forces in the direction opposing the rotation of the cylinder. *The cylinder experiences a counterclockwise magnetic torque in the end view of Fig. 9.* If the cylinder is spinning freely and the magnetic field is brought around it, this torque will stop the rotation very quickly, the kinetic energy of rotation going into heat created by the eddy currents. If the cylinder is driven at constant speed in the magnetic field, mechanical power equal to the heat produced by the eddy currents must be supplied. For a given magnetic field, this power can be shown to vary as the square of the speed.

Such eddy currents and heat generation would occur in the armature of a motor or generator if it were a solid-iron cylinder. The loss of mechanical energy to heat, called *eddy-current loss,* would be so large as to be intolerable. By constructing the armature from steel sheets or laminations only a few hundredths of an inch thick, as shown in Fig. 10, the eddy-current loss can be kept acceptably small. Lamination of the armature has little effect on the magnetic flux since the flux lines pass through the sheets 'edgewise' and do not need to cross from one lamination to the next. The steel laminations are electrically insulated from each other, either by shellac or by nonconducting iron oxide, so that the eddy currents are confined to circulation within each individual steel sheet. Two factors contribute to the reduction in loss. First, the length of conductor moving perpendicular to the magnetic field is now just the thickness of the lamina, so that the EMF generated is very small. Second, the resistance of the eddy-current path is very large because this path is of small cross-sectional area.

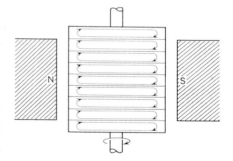

Fig. 10. Eddy-current paths in a laminated armature (schematic).

Figure 9 is just one example of a large class of eddy-current phenomena. In general, *if there is relative motion between a piece of metal and a magnetic field, eddy currents will be set up in the metal in such a direction that the resulting magnetic forces on the eddy-current elements will tend to stop the relative motion.* The only exception occurs in the case of pure translation of a piece of metal all portions of which are in a constant magnetic field, so that there is no change of flux linking any circuit drawn in the metal.

A watt-hour meter, Fig. 11, makes effective use of an *eddy-current brake*. A DC watt-hour meter is essentially a very small DC motor in which the magnetic field is set up by low-resistance coils connected in series with the line, so that the magnetic field is proportional to the line current. The armature, in series with a high resistance, is connected across the line, so that the armature receives a small current proportional to the line voltage. The torque of this motor is thus proportional to the product of voltage and current, that is, to the power taken by the load.

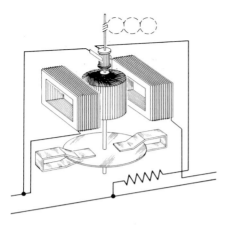

Fig. 11. The DC watt-hour meter. The lines from the power source come in at one side; those to the load go out at the other.

The principal resistance to turning of the motor is furnished by an eddy-current brake, consisting of a metal disc that rotates between the poles of permanent magnets. The eddy currents induced in the disc offer a resisting torque that is proportional to the speed of rotation. Since the motor will run at such speed that the motor torque equals the resisting torque, it is seen that the speed of the motor will be proportional to the motor torque, and hence to the power taken by the load. Dials driven through reduction gears thus turn at a speed proportional to power, so the dial readings are proportional to power × time, that is, to energy. The dials are calibrated in watt-hours or kilowatt-hours.

The AC watt-hour meter is similarly a little AC motor, with the resistance to the torque again furnished by an eddy-current brake.

Another measuring instrument whose satisfactory operation involves eddy currents is the moving-coil galvanometer described on pp. 593–594. The combination of coil and spiral springs or fiber in such a galvanometer forms a torsion pendulum. Such a pendulum would perform harmonic oscillations about its equilibrium position, making the instrument very inconvenient to use, if a special provision were not made to damp out these oscillations. To provide damping, the coil is wound on a light frame, usually of aluminum, which forms a short-circuited turn. When this *damping frame* rotates in the magnetic field, an EMF, and hence a current, is set up in it by the same action as in Fig. 9. The eddy current that circulates around the damping frame is in such a direction as to experience a magnetic force opposing the rotation. The thickness and resistance of the frame are made such as to introduce enough damping to prevent overshooting and harmonic oscillations, but not enough to make the motion of the pointer unduly sluggish. The damping frame has no effect on the position of equilibrium because when the coil has come to rest there is no current in the frame and hence no magnetic force on the frame.

4. Induction by Changing Current

Consider, as in Fig. 12, a coil of wire carrying current I_1 which may vary with time. This coil is called the primary coil and may, for example, be carrying an alternating current. Anywhere within the magnetic field of the primary coil place a one-turn secondary circuit—two examples are shown in Fig. 12. When the primary coil carries current I_1, let the flux *linking* the secondary circuit be Φ_2. By Ampère's principle, Φ_2 will be proportional to I_1, say

$$\Phi_2 = K I_1. \tag{8}$$

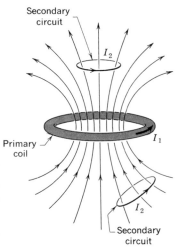

Fig. 12. A changing current in the primary coil induces an EMF in the secondary circuit. The arrows show directions that represent 'the positive sense.'

As indicated in Fig. 12, assign a positive sense to the closed curve representing the secondary circuit such that, by the right-hand rule, if I_1 is positive, Φ_2 will also be positive; then K will be a positive constant.

If now the current I_1 is changing, the flux Φ_2 will be changing and an EMF will be induced in the secondary circuit, of magnitude

$$V_{E2} = -d\Phi_2/dt = -K\, dI_1/dt, \tag{9}$$

in the negative sense in Fig. 12 if I_1 and hence Φ_2 are increasing.

The existence of the EMF V_{E2} around the secondary circuit implies the existence of an electric intensity $\mathcal{E}$ such that

$$\oint \mathcal{E} \cdot dl_2 = V_{E2} = -d\Phi_2/dt, \tag{10}$$

where the line integral is carried once around the secondary circuit.

This EMF is found, by means of experiments with charged particles (*see*, for example, the description of the betatron in the following section), to exist whether the secondary circuit is a wire or merely an imaginary closed curve in space.

A changing magnetic field is accompanied by an electric field such that, when integrated around any closed curve linked by flux Φ,

$$V_E = \oint \mathcal{E} \cdot dl = -d\Phi/dt. \tag{11}$$

It is often a complex matter to derive from (11) the values of the electric intensity $\mathcal{E}$ itself. However, in case there is an axis of symmetry as in Fig. 12, we see that the electric lines must be circles surrounding the axis. The upper secondary circuit in Fig. 12 represents such a circle, and if we write equation (11) for this circle, $\mathcal{E}$ is everywhere tangent to the path of integration and constant in magnitude, so that we have

$$2\pi R\, \mathcal{E} = |d\Phi/dt|, \tag{12}$$

if R is the radius of the circle. In accordance with Lenz's law, the direction of $\mathcal{E}$ is *opposite* to that indicated by the arrow on the upper secondary circuit in Fig. 12 if Φ is increasing, in the direction of the arrow if Φ is decreasing.

5. The Betatron

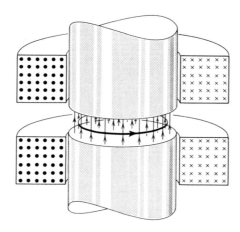

Fig. 13. A secondary circuit, analogous to those in Fig. 12.

Figure 13 indicates schematically an electromagnet with a circle drawn around its axis of symmetry. If this circle is of radius R and a changing flux Φ links the circle, an electric field given by (12) is set up around the circle as indicated in Fig. 14.

The *betatron*, invented by DONALD KERST in 1940, makes use of this induced electric field to give extremely high energies to electrons. In the betatron, Fig. 15, an evacuated, porcelain, doughnut-shaped tube is placed symmetrically between the poles of a magnet. Electrons are injected into the evacuated tube at a speed close to that of light and move in a circular path because of the presence of the magnetic field. A stronger magnetic field increases the flux through the 'hole' of the doughnut and induces an electric field $\mathcal{E}$ to accelerate the electrons. These electrons can acquire many MeV of kinetic energy during the period of increase of the magnetic field. At the end of this period, the electrons are deflected from their circular orbits and strike a target. The details of the operation will be clarified by a study of the specific example that follows.

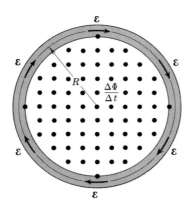

Fig. 14. An *increasing* flux out of the paper sets up a *clockwise* electric intensity.

Example. *In a particular betatron, the electron path in the doughnut-shaped evacuated tube has a mean radius of* 50 cm; *electrons having a kinetic energy of* 1.5 MeV *enter the doughnut; the flux through the 'hole in the doughnut' increases at a constant rate of* 24 Wb/s *for* 1/240 s, *and then the electrons are ejected.* (a) *To what electric intensity is each electron subjected?* (b) *How much work is done on each electron during each revolution around the doughnut?* (c) *Assuming that each electron maintains a speed approximately equal to the speed of light, compute the number of revolutions described before ejection.* (d) *How much does the kinetic energy of each electron increase before ejection and what is the final energy?*

Fig. 15. An early betatron. The evacuated porcelain 'doughnut' described in the text is visible in white at the center of the picture between the poles of the magnet. (Courtesy of Prof. Donald Kerst.)

(a) The electric intensity inside the doughnut is, by (12),

$$\mathcal{E} = (d\Phi/dt)/2\pi R = (24 \text{ Wb/s})/2\pi(0.5 \text{ m})$$
$$= 7.64 \text{ V/m}$$

(b) The work done on each electron during each complete revolution is $V_E e$, which is obtained by multiplying the EMF by the charge $-e$ on the electron, and neglecting signs. From (11), we see that the magnitude of V_E is just the rate of change of flux, so that $V_E = 24$ V. Hence the work per revolution is 24 eV.

(c) The number N of revolutions will be given by dividing the total distance d traversed by an electron in time $t = \frac{1}{240}$ s by the length $2\pi R$ of the orbit. Thus, since an electron of kinetic energy 1.5 MeV already has speed within 3% of the speed of light, we can write $d \approx ct$, where c is the speed of light, and find

$$N \approx ct/2\pi R = (3 \times 10^8 \text{ m/s})(\tfrac{1}{240} \text{ s})/2\pi(0.5 \text{ m}) = 3.98 \times 10^5 = 398\,000.$$

(d) The increase in the kinetic energy of each electron before ejection is given by the product of the work done on each electron during one revolution by the number of revolutions N described before ejection. Thus,

$$\text{increase in } E_K \approx (24 \text{ eV})(3.98 \times 10^5) = 9.55 \text{ MeV},$$

and the final kinetic energy is $(1.5 + 9.55)$ MeV $= 11.0$ MeV.

6. Mutual Inductance

Consider any two coils such that current in one coil will set up a flux linking the other that is proportional to the current in the first. This condition is satisfied by any two circuits where there is no magnetic

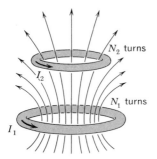

Fig. 16. Arrows indicate 'positive senses.'

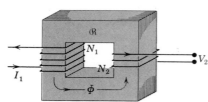

Fig. 17. A transformer. Arrows indicate 'positive senses.'

material present, as in Fig. 16, or when the coils are linked by an ideal magnetic circuit as in the important case of the transformer in Fig. 17.

Current I_1 in coil 1 will set up a flux Φ_2, linking coil 2, that is proportional to I_1; the EMF induced in each turn of coil 2 will be $-d\Phi_2/dt$; hence the total EMF in coil 2 will be

$$V_{E2} = -N_2\, d\Phi_2/dt,$$

which, since Φ_2 is proportional to I_1, may be written in the form

$$V_{E2} = -M_{12}\, dI_1/dt, \tag{13}$$

where the constant M_{12} is called the *mutual inductance* of the two coils.

It can be rigorously demonstrated that the mutual inductance of a pair of coils, where only empty space is involved, as in Fig. 16, or where the coils are linked by an *ideal* magnetic circuit, as in Fig. 17, is independent of which coil is the primary. That is, if current I_2 in coil 2 is changed, an EMF is induced in coil 1 that is given by

$$V_{E1} = -M_{12}\, dI_2/dt, \tag{14}$$

with the same coefficient M_{12} as in (13). For the proof in the case of Fig. 16, reference must be made to more advanced books on electromagnetism; the proof in the case of Fig. 17 is easy (*see* Prob. 18).

> The **mutual inductance** of two coils, in **henrys**, is the EMF, in volts, induced in one coil by a current changing in the other at the rate of one ampere per second.

The unit of mutual inductance is given the name *henry* (H), after Joseph Henry. The mutual inductance is always a positive quantity. As we see from (13),

$$1\ H = 1\ V \cdot s/A = 1\ Wb/A = 1\ \Omega \cdot s = 1\ J/A^2. \tag{15}$$

Example. *Design a 10-millihenry mutual inductance standard by employing a long solenoid as the primary and a winding around its center as the secondary, as indicated in Fig. 18.*

Let us wind the primary long solenoid with No. 19 insulated wire whose diameter is slightly less than 1 mm so that we can wind $n = 1000$ turns/meter. Suppose that from the diameter of the finished primary we compute that the cross-sectional area of the solenoid winding is $A = 106\ \text{cm}^2 = 0.0106\ \text{m}^2$. As we have seen in Chapter 29, such a *long* solenoid sets up a uniform field of intensity $\mu_0\, nI$ in the interior of the solenoid, and a field immediately outside the center of the solenoid that can be made as small as one pleases by lengthening the solenoid. Hence one can wind a multi-layered secondary around the center of the solenoid as indicated in Fig. 18, and compute the flux linking the secondary from the solenoid relation. We must now compute the number of turns, N_2, required for the secondary.

Denote the primary current by I_1. The flux threading the secondary is then given by the solenoid relation as

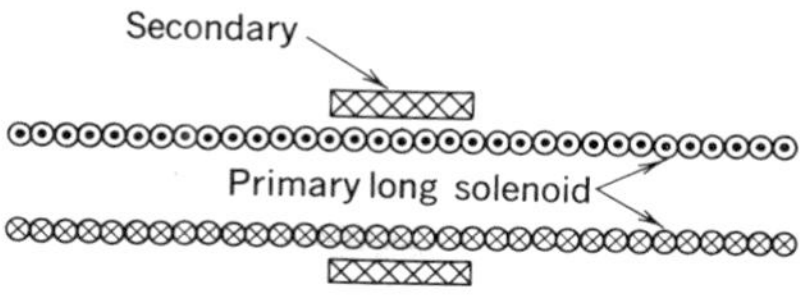

Fig. 18. Schematic diagram of a standard mutual inductance.

$$\Phi = \mu_0 n I_1 A = 10^{-7} \cdot 4\pi \cdot 1000 \times 0.0106\, I_1 \text{ Wb/A} = (13.3 \times 10^{-6} \text{ Wb/A})\, I_1.$$

Hence
$$d\Phi/dt = (13.3 \times 10^{-6} \text{ Wb/A})\, dI_1/dt.$$

The voltage induced in the secondary is given by (13) as

$$V_{E2} = -N_2\, d\Phi/dt = -(13.3 \times 10^{-6} \text{ Wb/A})\, N_2\, dI_1/dt.$$

By the definition of mutual inductance, $M_{12} = -V_{E2}/(dI_1/dt)$; hence

$$M_{12} = 13.3 \times 10^{-6}\, N_2 \text{ Wb/A}.$$

Since Wb/A and H are synonymous, we substitute the desired value $M_{12} = 10 \text{ mH} = 10^{-2} \text{ H}$ to determine

$$N_2 = 10^{-2}/13.3 \times 10^{-6} = 752$$

as the number of turns required on the secondary.

7. Self Inductance

Consider again coil 1 of Fig. 16 or 17. If the current I_1 is changing, an EMF is induced not only in coil 2 but also in coil 1 itself, since the flux through coil 1 is changing and this flux change sets up an EMF around coil 1 itself equal to $-d\Phi_1/dt$ per turn or

$$V_{E1} = -N_1\, d\Phi_1/dt \tag{16}$$

for the N_1 turns. Here Φ_1 is the flux linking coil 1 itself, and is proportional to the current I_1 setting up the flux, so we can write

$$V_{E1} = -L_1\, dI_1/dt, \tag{17}$$

where L_1 is called the *self inductance,* and can, like M_{12}, be measured in henrys. The minus sign indicates that the generated EMF is in a direction to oppose the change of current, in accordance with Lenz's rule.

The **self inductance** L of a coil, in **henrys**, is the EMF, in volts, induced in the coil by the changing magnetic flux it itself sets up when its current is changing at the rate of one ampere per second.

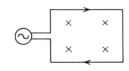

Figure 19

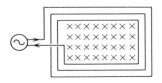

Figure 20

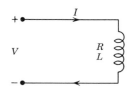

Fig. 21. While the current is increasing, the inductor acts like a battery on charge.

Any circuit, even a simple circuit such as that of Fig. 19, has self inductance. In Fig. 19 a single loop of wire is connected to an AC generator. When the current I has the direction shown, there is a flux linking the circuit in the direction shown. When I increases, Φ increases, and an EMF, $V_E = -d\Phi/dt = -L\,dI/dt$, is generated in the direction opposing the increase of current. The self inductance of a simple circuit like Fig. 19 is very small—often negligibly small. The self inductance is greatly increased if the circuit is an N-turn 'coil' of the same area, as in Fig. 20. For a given current, each turn of this coil sets up the same flux as in Fig. 19, so that the total flux is multiplied by N. Hence, for given rate of change of current, the EMF generated in *each turn*, $d\Phi/dt$, is N times as great as in Fig. 19. But the EMF's in the N turns are additive so, for a given rate of change of current, the generated EMF is N^2 times that in Fig. 19. Consequently, the self inductance is N^2 times as great. This dependence of self inductance on the *square* of the number of turns also applies to a coil wound around an iron or steel core, which will greatly increase the inductance because of the great increase in flux.

The coil in Fig. 21 will have both resistance R and self inductance L. It is called an *inductor*. If we connect a battery of constant voltage V across the terminals of this inductor, we expect the current to reach a constant value, which we denote by $I_F = V/R$. But the current cannot jump *suddenly* from the value 0 to its *final* value I_F, because while the current is increasing there is a *generated voltage* $-L\,dI/dt$ *opposing* the current increase. During the period of current increase, the inductor acts like a battery on charge, with EMF equal to $L\,dI/dt$. Just as in equation (3), p. 548, the terminal voltage must first overcome this *back*-EMF, and only the difference between the terminal voltage and the back-EMF will be available for driving current through the resistance; thus, during the period of current change, we have

$$V = IR + L\,dI/dt. \qquad \text{(inductor)} \quad \textbf{(18)}$$

When we close the switch to connect the battery, the rate dI/dt at which the current can increase is limited by the fact that, in (18), $L\,dI/dt$ cannot exceed V. In fact, just after we close the switch, while I and hence IR are still very small, we can obtain the rate of change of current by equating the first term on the right of (18) to zero, to obtain

$$(dI/dt)_{\text{Initial}} = V/L = (V/R)(R/L) = I_F(R/L). \qquad \textbf{(19)}$$

This is the initial rate at which current starts to build up when a voltage is suddenly applied across an inductor. At later instants, the rate of change is less than that given by (19) because as the IR term in (18) builds up, dI/dt must decrease. The actual manner in which the build-up to I_F occurs is shown in Fig. 22.

To determine the form of the curve of Fig. 22, we solve the differential equation (18) by the standard procedure that we used, in the case of the charging of a capacitor, on p. 555. We divide (18) through by R and rearrange terms to separate the variables, I and t:

$$\frac{dI}{V/R - I} = \frac{dt}{L/R}.$$

Take the indefinite integral of each side to obtain

$$-\log_e(V/R - I) = t/(L/R) + \text{const}.$$

When $t=0$, $I=0$, so the constant of integration is $\text{const} = -\log_e(V/R)$. Hence

$$-\log_e(V/R - I) = t/(L/R) - \log_e(V/R)$$

or

$$-\frac{t}{L/R} = \log_e\left(\frac{V/R - I}{V/R}\right).$$

Taking the antilog of both sides gives

$$e^{-t/(L/R)} = \frac{V/R - I}{V/R},$$

which can be solved for I to give

$$I = (V/R)[1 - e^{-t/(L/R)}], \qquad (20)$$

as the curve of growth plotted in Fig. 22. It is seen that (20) gives the correct values $I=0$ at $t=0$ and $I=V/R$ at $t=\infty$.

As indicated by the broken line in Fig. 22, if the current continued to build up at its initial rate (19), it would reach its final value I_F in a time $t=L/R$. The expression L/R, which has the dimensions of time,

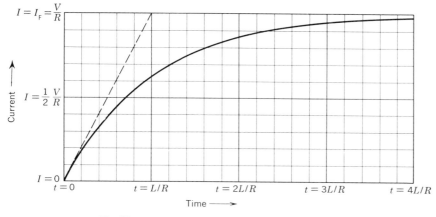

Fig. 22. Growth of current in an inductive circuit.

is called the *time constant* of the inductor. The larger the ratio L/R of inductance to resistance, the slower is the build-up of current.

Example. *The winding of a large electromagnet has an inductance of* 0.8 H *and a resistance of* 0.6 Ω. *What is its time constant? When* 120 V *are applied, what is the initial rate of change of current and the final current? At the time* $t = L/R$, *what is the back-*EMF, *the IR-drop, and the rate of change of current?*

The time constant is $L/R = 0.8\text{ H}/0.6\text{ Ω} = 1.33$ s, where we have made use of (15). The final current is $I_F = V/R = 120\text{ V}/0.6\text{ Ω} = 200$ A. The initial rate of change is given by (19) as $V/L = (120\text{ V})/(0.8\text{ H}) = 150$ A/s. We verify that if the current continued to change at this rate, it would reach its final value in 1.33 s.

At 1.33 s after closing the switch, the current is given by (20) as $(1-e^{-1})I_F = 0.632\, I_F$, or 126 A. The IR-drop is $IR = (126\text{ A})(0.6\text{ Ω}) = 75.6$ V. Hence from (18), the back-EMF is

$$L\, dI/dt = 120\text{ V} - 75.6\text{ V} = 44.4\text{ V},$$

and the rate of change of current is

$$\left(\frac{dI}{dt}\right)_{t=1.33\text{ s}} = \frac{44.4\text{ V}}{L} = \frac{44.4\text{ V}}{0.8\text{ H}} = 55.6\frac{\text{A}}{\text{s}},$$

which is $e^{-1} = 0.368$ times the initial rate of change. The reader can verify that this same rate of change is obtained by differentiation of (20) and substitution of the time $t = L/R$.

While the current in the inductor of Fig. 21 is increasing, work is being done in moving charges through the generated back-EMF; this work represents conversion of electric energy to some other form—in this case to energy associated with the magnetic field that is being set up. If we multiply equation (18) by I, we obtain a power equation analogous to (4), p. 548:

$$IV = I^2R + LI\frac{dI}{dt}. \tag{21}$$

$$\begin{Bmatrix}\text{Power} \\ \text{delivered} \\ \text{to inductor}\end{Bmatrix} = \begin{Bmatrix}\text{power} \\ \text{transformed} \\ \text{to heat}\end{Bmatrix} + \begin{Bmatrix}\text{power being} \\ \text{stored in} \\ \text{inductor.}\end{Bmatrix}$$

When the current is increasing, the last term in this equation must represent a rate at which energy is being stored in the magnetic field of the inductor, since this amount of power is being delivered to the inductor and is not resulting in the production of heat. When the current is decreasing, more power is transformed to heat than is delivered to the circuit; the magnetic field is giving up its previously stored energy.

With the rate of energy storage given by the last term of (21), the energy stored in time dt is $dE_L = LI\, (dI/dt)\, dt = LI\, dI$. While the current

SELF INDUCTANCE

in the inductor increases from 0 to I_F in the manner of Fig. 22 (or in any other manner), the total energy stored in the magnetic field is

$$E_L = \int_{I=0}^{I=I_F} LI\, dI = \tfrac{1}{2} LI^2 \Big|_0^{I_F} = \tfrac{1}{2} LI_F^2. \tag{22}$$

Since I_F is arbitrary, for any current I through an inductance L,

$$E_L = \text{energy stored} = \tfrac{1}{2} LI^2. \qquad \begin{Bmatrix} E_L \text{ in J} \\ L \text{ in H} \\ I \text{ in A} \end{Bmatrix} \tag{23}$$

PROBLEMS

1. A stiff piece of wire 7.2 m long is moved vertically upward in the earth's magnetic field, whose horizontal northward component is 20 μT. The wire is horizontal, points east and west, and moves at 25 m/s. What is the difference of potential between the ends, and which end is positive?   Ans: 0.003 60 V; west end.

2. If the wire of Prob. 1 moves horizontally toward the north at the same speed, and the vertical component of the earth's field is 50 μT downward, what is the difference of potential between the ends, and which end is positive?

3. In Fig. 2, let $l=0.8$ m, $v=3$ m/s, $\mathcal{B}=1.5$ T, and $R=5$ ohms. (a) What is the generated EMF? (b) What is the current? (c) What is the power expended in heating the resistor? (d) What is the magnitude of the magnetic force? of the applied force? (e) What is the rate at which the applied force does work?
Ans: (a) 3.60 V; (b) 0.720 A; (c) 2.59 W; (d) 0.863 N; (e) 2.59 W.

4. In Fig. 2, if $l=0.6$ m, $\mathcal{B}=0.8$ T, $R=8$ ohms, and $F=0.2$ N, find (a) the current, (b) the EMF, (c) the speed, (d) the rate of doing work, and (e) the rate of heat generation.

5. A 50-turn coil of the type shown in Fig. 6 has an area of 1 cm². If this coil is moved toward a magnet as in Fig. 5 in such a way that the mean flux density through the coil changes at the rate of 0.4 T/s, what voltage is induced in the coil?   Ans: 2.00 mV.

6. A 'harbor loop' is a coil of wire of extensive area laid out on the floor of a harbor entrance to detect the entrance of submarines by means of the voltage induced by the submarine's natural magnetism.

(a) If the flux through the loop changes at the maximum rate of 0.006 Wb/s as the submarine passes over the near side of the loop, what is the maximum voltage induced in a 50-turn loop?

(b) If the earth's vertical field may be expected to vary as fast as 0.3×10^{-3} μT/s, what is the maximum area the harbor loop can have if the voltage generated by fluctuations in the earth's field is to be less than 1% of the voltage generated by the submarine in (a)?

7. Show that if the flux linking the coil of Fig. 6 changes at a constant rate $\Delta \Phi / \Delta t$ from Φ_1 to Φ_2, then the *charge* that flows through the circuit of total resistance R is given by $Q = N(\Phi_2 - \Phi_1)/R$. It can be shown that this same amount of charge flows whether or not the change from Φ_1 to Φ_2 is at a constant rate. This relation provides the basis for measurement of flux changes by the ballistic galvanometer—a charge-measuring device.

8. If the coil of Prob. 5 is jerked out of the magnetic field of a magnet and it is found that a charge of 1.5×10^{-4} C passes through the total resistance of 20 ohms, what was the flux density in the field? (Use the relation in Prob. 7.)

9. The 'search coil' of a magnetic mine is a long permalloy rod of 4 cm² cross-sectional area, wound with 25 000 turns of wire. When the external field component parallel to the rod changes by x T, the average flux density in the rod changes by 5000 x T, provided the externally applied field is sufficiently small—a condition adequately satisfied by the earth's field and by ships' fields. When a ship passes over the mine, the field component parallel to the rod changes by 0.5 μT during a period of 20 s. Find the average voltage induced in the search coil during this period.   Ans: 1.25 mV.

10. If the rod of the magnetic mine of Prob. 9 is wound with 100 000 turns of wire and if the field component parallel to the rod changes by 15 μT during a period of 10 s when a high-speed battleship passes over, find the average induced voltage during this period.

11. Let a resistance R be connected between A and B in Fig. 7. Neglect the internal resistance of the generator. For a constant angular velocity ω, determine at each instant (angle) (a) the current I; (b) the torque L required to turn the coil; (c) the power $L\omega$ required. (d) Verify that the power in (c) is equal to IV_E as is required by the definition of EMF.

12. Suppose that, when the field coils of a cyclotron are turned on, the magnetic intensity in the gap is uniform and increases at a constant rate $d\mathcal{B}/dt$. Show that the EMF induced in a circular ring of copper wire lying on one of the pole faces is directly proportional to the square of the radius r of the ring.

13. Consider further the copper ring of radius r in Prob. 12. (a) What is the induced EMF in this ring? (b) Associated with this induced EMF, there is an electric intensity inside the ring; what are the magnitude and direction of the electric-intensity vector? (c) What is the magnitude of the force exerted on a free electron inside the copper ring? (d) How much work would the electric field do in moving an electron entirely around the ring back to its starting point? (e) What becomes of this work?
    Ans: (a) $\pi r^2 \, d\mathcal{B}/dt$; (b) ½ $r\, d\mathcal{B}/dt$; (c) ½ $er\, d\mathcal{B}/dt$; (d) $\pi r^2 e\, d\mathcal{B}/dt$.

14. If the copper ring of Prob. 13 were removed and replaced by a hollow *evacuated* glass or porcelain ring or 'doughnut,' the changing magnetic flux would still produce an electric field inside the doughnut. This electric field would still exert a tangential force on an electron inside the tube. If the electron could be constrained to move in a circular path of radius r, the field would do work as the electron moved completely around the doughnut. (a) Since there is no resistance to the electron's motion inside the doughnut, what becomes of the work done? (b) What is the increase in the kinetic energy of the electron after it makes N traversals of its circular 'orbit' inside the doughnut?

15. If the transformer of Fig. 17 has 400 turns on the primary, 40 turns on the secondary, and an iron path of mean length 110 cm, cross section 10 cm $\times$ 10 cm, and permeability 1300, what is the mutual inductance?
      Ans: 0.238 H.

16. A small permalloy toroid has a cross-sectional area of 1 cm², a mean circumference of 10 cm, and a permeability of 50 000 so long as the flux density is small. It has a primary winding of 25 turns and a secondary winding of 200 turns of very fine wire. What is the mutual inductance? If the primary current changes at the rate of 1 mA/s, what voltage is induced in the secondary?

17. A *long* air-core solenoid has a cross-sectional area of 400 cm² and is wound with 30 turns per cm. Around the center of the solenoid is wound a secondary coil of 200 turns of wire. (a) What is the mutual inductance between the solenoid winding and the secondary coil? (b) If the solenoid current changes at the rate of 4 A/s, what voltage is generated in the secondary winding? (c) If the secondary winding is connected to a source of current and the current in this winding changes at the rate of 4 A/s, what voltage is induced in the solenoid winding?
Ans: (a) 30.1 mH; (b) 120 mV; (c) 120 mV.

18. In Fig. 17, show that $M_{12} = N_1 N_2/\mathcal{R}$, where $\mathcal{R}$ is the reluctance of the ideal magnetic circuit, no matter which of the two coils is taken as primary.

19. In Fig. 17, show that the self inductance of the primary coil is $L_1 = N_1^2/\mathcal{R}$, where $\mathcal{R}$ is the reluctance of the ideal magnetic circuit.

20. Design a toroid, wound on a nonmagnetic core, that will serve as a 1-millihenry self-inductance standard.

21. An induction coil such as is used to set off the spark in an automobile spark plug consists of a primary coil, and a secondary coil of very many more turns of finer wire, both wound around a straight iron core. The primary normally carries a current, which is broken at each instant that a spark is desired. If the primary winding has 200 turns and normally carries a current of 4 A, and if this current falls to zero in 0.001 s when the circuit is broken, what must be the mutual inductance if an average of 20 000 V is to be induced in the secondary? If this current sets up a total flux in the core of 10^{-3} Wb, how many turns are required on the secondary? Ans: 5.0 H; 20 000.

22. How much energy, over and above that expended in heat, is required to set up a current of 15 A through a coil of self inductance 2.0 H?

23. What is the self inductance of the primary winding of the transformer described in Prob. 15? Of the secondary winding?
Ans: 2.38, 0.0238 H.

24. If a current of 10 A in a 120-turn concentrated coil sets up a total flux of 5×10^{-4} Wb linking the coil, find the self inductance of the coil.

25. In the figure, if $R_1 = 2\,\Omega$, $R_2 = 0.5\,\Omega$, and $L = 0.8$ H, what is the time constant for current growth when S_1 is closed? for decay when S_2 is closed? Ans: 0.32 s; 0.40 s.

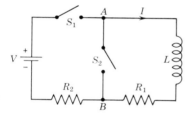

Problems 25–30. The battery, the inductor, and the short-circuiting line from A to B are assumed to have negligible resistance. Let $R = R_1 + R_2$.

26. Repeat Prob. 25 for the case $R_1 = 50\,\Omega$, $R_2 = 1\,\Omega$, and $L = 100$ H.

27. In the circuit described in Prob. 25, and with $V = 6.0$ V, find the current 0.32 s after switch S_1 is closed. At this time, how much energy is stored in the circuit?
Ans: 1.52 A; 0.924 J.

28. In the circuit described in Prob. 25, with $V = 12$ V, find the voltages across R_1, R_2, and L after switch S_1 has been closed for 0.32 s.

29. In the figure, consider the situation when switch S_1 is closed at time $t = 0$, and S_2 is left open. Make a plot of the energy stored in the inductor as a function of time.

30. In the figure, consider the situation in which switch S_1 has been closed for a long time and the current has attained its final value $I_F = V/R$. Then suppose that switch S_2 is closed at a time we now designate as $t = 0$. Show that under these circumstances the current through the inductor satisfies the equation $L\,dI/dt + R_1 I = 0$ and that the solution of this equation that satisfies the initial condition $I = I_F$ at $t = 0$ is $I = I_F\,e^{-R_1 t/L}$. Demonstrate that I drops to $I_F/e \approx 0.37\,I_0$ in time L/R_1 and that, if the current continued its initial rate of decrease without change, the current would become zero at time L/R_1.

31. Consider the circuit shown in the figure; let $R = R_1 + R_2$ and assume all other resistances negligible.

(a) With switch S_2 open, switch S_1 is closed at $t = 0$. To what value does the current I build up, and with what time constant?

(b) After I has built up to its final value, switch S_2 is closed, and then switch S_1 is opened. What is the time-constant involved in the decrease of I to zero?
Ans: (a) V/R_1, L/R_1; (b) L/R.

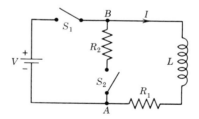

Problems 31–32

32. In the figure, let the inductor represent a large electromagnet; let $L = 120$ H, $R_1 = 6\,\Omega$, and $V = 120$ V. With S_1 closed and S_2 open, the current I has built up to its final value. Then, instead of closing the shunting switch S_2, an attempt is made to open S_1. Why does an arc form between the blades of S_1? Why have persons been electrocuted who have tried this procedure? Why are the windings of the electromagnet sometimes punctured? How much stored energy must be dissipated as heat before the current through L can disappear?

33. Show that the energy per unit volume in a magnetic field is $\frac{1}{2}\,\mathcal{B}^2/\mu\mu_0$, where μ is the permeability of an ideal magnetic material in a circuit such as that of Fig. 5, p. 633. Hence show that in an iron-free toroid,

magnetic energy per unit volume $= \frac{1}{2}\,\mathcal{B}^2/\mu_0$. **(24)**

This is a general relation that we shall use in Chapter 43 in discussing energy transport by electromagnetic radiation. To derive the above relations, use (23), the relation $L = N^2/\mathcal{R}$ of Prob. 19, and the relations between $\mathcal{B}$, Φ, I, and $\mathcal{R}$ given in Chapter 30, p. 636.

34. Starting with equation (2) for the EMF generated in a wire moving in a magnetic field, derive equation (6) for the EMF of the prototype AC generator shown in Fig. 7.

35. Show that the power required to drive the cylinder of Fig. 9 or Fig. 10 at constant rotational speed varies as the square of the speed. Neglect mechanical friction.

ALTERNATING CURRENTS 32

Almost all power circuits are now supplied with alternating voltages, and hence carry alternating currents; they are commonly called AC circuits. The principal reason for the almost universal use of AC is that, by means of a *transformer* with no moving parts, an alternating voltage can be readily and efficiently changed from one amplitude to another; there is no DC counterpart of the transformer.

The usual power frequency is 60 hertz, $f = 60$ Hz; the power in aircraft and guided missiles is usually at 400 Hz. Ordinary telephone circuits operate at the sound frequencies themselves, with emphasis on the range from 100 to 2000 Hz; long-distance telephone circuits frequently superpose the audio signals on a carrier in the radio-frequency range, from 0.1 to 100 MHz. Finally, frequencies in the region 1–10 GHz are used in radar and in 'microwave' communications.

In circuits operating at all of these frequencies, the *inductances* and *capacitances* of the circuit elements play roles just as significant as the *resistances*. In this chapter we shall give an introductory discussion, valid throughout a wide range of frequencies, of the behavior of resistors, capacitors, and inductors when alternating voltages are applied.

1. The Series Circuit

Figure 1 shows a simple circuit containing resistance R, capacitance C, and inductance L, connected in series to a source of alternating EMF indicated by the symbol $\sim$. In general, R, C, and L would be mixed up in individual circuit elements, but to arrive at an understanding of the behavior of the circuit, we assume, in Fig. 1, that all the resistance is lumped in R, all the capacitance in C, and all the inductance in L.

In Fig. 1, the symbols I, V, V_R, V_C, and V_L represent the *instantaneous* values of current, applied voltage, and the voltages across the individual circuit elements. These values all alternate sinusoidally at a frequency f, so they are alternately positive and negative. The arrows in Fig. 1 serve to define the direction of the current *when I is positive;* similarly, the + and − signs in Fig. 1 define the senses of the voltages *when they are positive.*

It will be simplest if we start by specifying the form of the alternating current we desire to have in Fig. 1, and then compute the required voltages. Let us assume that the current has maximum value (amplitude) I_M, and varies with time at frequency f, according to the equation

$$I = I_M \sin(2\pi f t). \tag{1}$$

We shall now proceed to study, one at a time, the voltages V_R, V_C, and V_L across the individual circuit elements when the current is given by (1). Finally, we can combine these voltages to obtain the required applied voltage

$$V = V_R + V_C + V_L, \tag{2}$$

which, *at any instant,* is the sum of the three voltages across the individual circuit elements.

Each of the voltages in (2) will vary sinusoidally at frequency f. We shall denote their amplitudes by V_M, V_{RM}, V_{CM}, and V_{LM}. The complicating factor in AC circuits is that these different voltages *are not in phase*, so that we cannot obtain V_M by adding the amplitudes of the three voltages on the right of (2). In fact, we shall find that V_C and V_L are exactly ½ cycle out of phase with each other, so that we must *subtract* their amplitudes from each other; this is analogous to the case of destructive interference. V_R turns out to be ¼ cycle out of phase with both V_L and V_C; hence the expression for V_M, in terms of V_{RM}, V_{CM}, and V_{LM}, becomes rather complex.

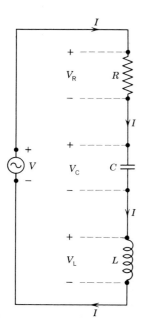

Fig. 1. An AC series circuit. Arrows and +, − signs show the senses *taken as positive.*

2. Resistance

When the current (1) passes through a resistance R, we obtain the instantaneous voltage V_R from Ohm's law

SEC. 2 RESISTANCE 669

$$V_R = RI = RI_M \sin(2\pi ft). \qquad (3)$$

If we write this relation in the form $V_R = V_{RM} \sin(2\pi ft)$, we see that

$$V_{RM} = RI_M. \qquad (4)$$

The current and voltage are *in phase*, as indicated in Fig. 2.

The instantaneous power supplied to the resistance is given by the product of (1) and (3):

$$IV_R = RI_M^2 \sin^2(2\pi ft). \qquad (5)$$

This product is plotted at the bottom of Fig. 2. From the trigonometric identity, $\sin^2\theta = \frac{1}{2} - \frac{1}{2}\cos 2\theta$, we see that the curve $\sin^2\theta$ is itself sinusoidal, of double period, and displaced so that it varies from 0 to 1. By inspection we see that the average value of this $\sin^2$-curve is $\frac{1}{2}$ of its maximum value:

The average ordinate of a sine-squared or cosine-squared curve, during a complete half-cycle, is one-half of the maximum ordinate.

Hence the average power

$$P = \tfrac{1}{2} I_M^2 R = \tfrac{1}{2} V_{RM} I_M = \tfrac{1}{2} V_{RM}^2/R. \qquad (6)$$

These relations are exactly like those for the DC case [(5), p. 533] *except* for the factor $\frac{1}{2}$. We can restore the exact correspondence if we define *effective* values of voltage and current as follows:

The **effective value of any sinusoidal alternating current or voltage** is defined as $1/\sqrt{2}$ times its maximum value.

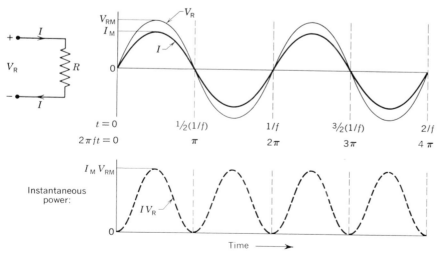

Fig. 2. Instantaneous current, voltage, and power for a pure resistance.

Thus, if we denote effective values by I^{Eff} and V^{Eff}, we have

$$I^{\text{Eff}} = I_M/\sqrt{2}, \qquad V_R^{\text{Eff}} = V_{RM}/\sqrt{2}, \tag{7}$$

and the power relations (6) become

$$P = (I^{\text{Eff}})^2 R = V_R^{\text{Eff}} I^{\text{Eff}} = (V_R^{\text{Eff}})^2/R, \tag{8}$$

exactly the form of the DC relations. Also, the relation (4) between voltage, current, and resistance retains its familiar form: $V_R^{\text{Eff}} = RI^{\text{Eff}}$. For convenience in computing power expended in AC circuits,

When AC voltages and currents are specified without qualification, it is always effective values that are implied.

Thus, when we say that the 'power company' supplies 120 V AC, we always mean 120 V *effective*: $V^{\text{Eff}} = 120$ V; $V_M = 120\sqrt{2}$ V $= 170$ V.

Example. *What current and power are drawn by a 40-W light bulb when connected to* 120 V DC? *Describe the voltage, current, and power when the same bulb is connected to* 120 V AC *at* 60 Hz. (*A light bulb is, to a good approximation, a purely resistive circuit element.*)

For the DC case, $R = V^2/P = (120 \text{ V})^2/(40 \text{ W}) = 360\ \Omega$. Also $I = P/V = (40 \text{ W})/(120 \text{ V}) = \frac{1}{3}$ A. The light bulb has a resistance of 360 Ω and draws $\frac{1}{3}$ A for a power of 40 W.

When this 360-Ω resistor is connected to 120 V AC, $V_R^{\text{Eff}} = 120$ V, $V_{RM} = \sqrt{2}\,(120 \text{ V}) = 170$ V, and the instantaneous voltage is

$$V_R = V_{RM}\sin(2\pi ft) = (170 \text{ V})\sin[2\pi\,(60 \text{ s}^{-1})\,t] = (170 \text{ V})\sin[377 \text{ s}^{-1})\,t].$$

The current has $I_M = V_{RM}/R = (170/360)$ A $= 0.471$ A, so according to (1),

$$I = I_M\sin(2\pi ft) = (0.471 \text{ A})\sin[(377 \text{ s}^{-1})\,t].$$

The effective current is $I^{\text{Eff}} = I_M/\sqrt{2} = (0.471 \text{ A})/1.414 = 0.333$ A. The instantaneous power is

$$IV_R = (0.471 \text{ A})(170 \text{ V})\sin^2(2\pi ft) = (80.0 \text{ W})\sin^2[(377 \text{ s}^{-1})\,t].$$

while the average power is half the maximum, or 40 W, as would be computed also from (6) or (8). The power, *effective* voltage, and *effective* current for AC are the same as the corresponding values for DC.

3. Capacitance

The basic electrostatic relation between the charge Q on a capacitor and the voltage V_C across the plates,

$$Q = CV_C, \qquad V_C = Q/C, \tag{9}$$

must hold at every instant.

Since current cannot pass *through* a capacitor, there is a current into and out of the two terminals only when the charge on the plates is changing, the current equaling the rate of change of charge.

To understand the phase relation between current and voltage across a capacitor, consider the set of little diagrams spaced along the time axis at the top of Fig. 3. The little diagram at the left is at a time when the charge on the capacitor and hence the voltage across it have their maximum values. The current at this instant is zero. Current has been flowing down the wire (I positive) to charge the top plate positively; the charge has reached its maximum value, so the current at this instant is zero and is about to reverse (become negative) to discharge the capacitor. In the second diagram, the capacitor is discharging (I negative), but has not yet lost all the + charge from the top plate, so V_C is still +. Only in the third diagram has all the + charge left the top plate; at this instant the current has its maximum negative value but $V_C=0$. During the succeeding ¼ cycle the negative current continues and charges the *bottom* plate positively (V_C negative).

The reader should continue the above description of the diagrams at the top of Fig. 3, and relate them to the sinusoidal plots of I and V_C, until he has convinced himself that

The voltage across a capacitor lags behind the current by ¼ cycle.

The term *lag* here is used in a time sense. To get a sinusoidal curve of the shape of the voltage curve in Fig. 3, we would have to shift the current curve to the right (which means toward *later* times) by ¼ cycle. The voltage has its positive maximum ¼ cycle *later* than the current, changes

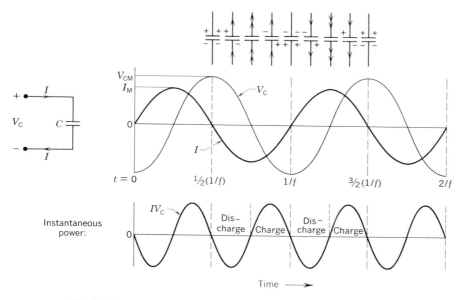

Fig. 3. Instantaneous current, voltage, and power for a pure capacitance. Broken lines are ½ cycle apart.

from positive to negative ¼ cycle *later,* etc; hence the voltage *lags.* Correspondingly, we may say that the current *leads* the voltage.

The V_C curve is seen to be a negative cosine curve; hence if I is represented by the sine curve (1), we must write

$$V_C = -V_{CM}\cos(2\pi ft). \qquad (10)$$

For the alternating current given by (1), we derive the required value of maximum voltage V_{CM} as follows: In the last equation of (9) we substitute $Q = \int I\,dt$ to obtain

$$V_C = \frac{1}{C}\int I\,dt = \frac{I_M}{C}\int \sin(2\pi ft)\,dt = -\frac{I_M}{2\pi fC}\cos(2\pi ft) + \text{const.}$$

We must set the constant equal to zero to obtain a zero *average* value of V_C; hence

$$V_C = -\frac{I_M}{2\pi fC}\cos(2\pi ft). \qquad (11)$$

One can verify the correctness of this equation by differentiation:

$$I = dQ/dt = C\,dV_C/dt = I_M\sin(2\pi ft),$$

which is the current equation we started with in (1).

Comparison of (11) and (10) gives the maximum voltage as

$$V_{CM} = \frac{1}{2\pi fC}I_M. \qquad (12)$$

The coefficient $1/2\pi fC$ in (12) has the same dimensions as resistance and hence is measured in ohms. It is called the *capacitive reactance* and is denoted by X_C:

> The **capacitive reactance,** at frequency f, of a circuit element of capacitance C, is $X_C = 1/2\pi fC$.

Thus we can write (12) in a form analogous to Ohm's law:

$$V_{CM} = X_C I_M, \quad I_M = V_{CM}/X_C, \quad \text{where} \quad X_C = 1/2\pi fC. \qquad (13)$$

The same relations will apply if, instead of maximum values, we use *effective* values of voltage and current.

We note that the higher the capacitance, the lower is the reactance, and the greater the current for a given voltage and frequency. This is reasonable, since a higher capacitance requires a greater charge and hence greater current. We note that the higher the frequency, the lower is the reactance, and the greater the current for given voltage and capacitance. This is reasonable since at higher frequency the capacitor acquires the same charge in a shorter time, so more current is required.

Example. *Compute the capacitive reactance of a 4-μF capacitor, and the current when 120 V is applied, at the following frequencies:* (a) 1 Hz, (b) 60 Hz, (c) 400 Hz, (d) 1 MHz.

We compute the capacitive reactance at $f = 1$ Hz from (13):

$$X_C = 1/2\pi f C = 1/[2\pi (1/\text{s}) 4 \times 10^{-6} \text{ F}] = 39\,800 \text{ }\Omega.$$

The units come out initially in s/F, which is seen to correctly reduce to Ω. When $V^{\text{Eff}} = 120$ V is applied, we find

$$I^{\text{Eff}} = V^{\text{Eff}}/X_C = 120 \text{ V}/3.98 \times 10^4 \text{ }\Omega = 0.003\,02 \text{ A}.$$

The capacitive reactance is large and the current small; we are approaching the DC case ($f=0$), where $X_C = \infty$, $I = 0$.

(b), (c), (d) We can derive the results for these higher frequencies from those at 1 Hz by noting that X_C varies inversely as the frequency, and hence I^{Eff} varies directly as the frequency. Therefore, at 60 Hz,

$$X_C = \tfrac{1}{60}(39\,800 \text{ }\Omega) = 663 \text{ }\Omega, \qquad I^{\text{Eff}} = 60\,(0.003\,02 \text{ A}) = 0.181 \text{ A}.$$

Correspondingly, at 400 Hz, $X_C = 99.5$ Ω, $I^{\text{Eff}} = 1.21$ A;

at 10^6 Hz, $X_C = 0.0398$ Ω, $I^{\text{Eff}} = 3020$ A.

Now consider the power furnished to a capacitor. The instantaneous power delivered by the source of EMF is IV_C; this is plotted at the bottom of Fig. 3. The power is represented by a sinusoidal curve of double frequency and average value *zero*. The capacitor charges for a quarter cycle and energy is delivered to its electric field; it discharges for the next quarter cycle and delivers all its energy back to the line; and so on.

The average power delivered to a capacitor is zero.

This must be the result, because no electric energy is converted to any other form on a permanent basis in a capacitor. No heat is generated, and the electric field has energy only while there is charge on the capacitor, this energy being all delivered back to the line as the charge decreases to zero.

4. Inductance

We have already discussed the voltage across an inductor in some detail on pp. 659–663. In particular, equation (18), p. 660, shows that for a *pure inductance*, the voltage $V_L = L\, dI/dt$ is positive when the current is increasing and negative when the current is decreasing, as shown in Fig. 4, from which we see that

The voltage across a pure inductance leads the current by $\tfrac{1}{4}$ cycle.

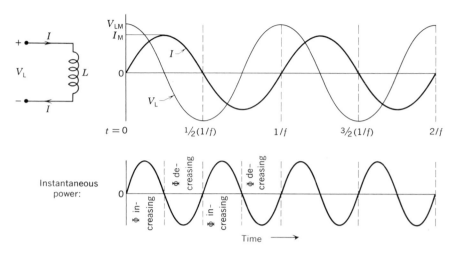

Fig. 4. Instantaneous current, voltage, and power for a pure inductance.

By differentiation of (1), we find

$$V_L = L\, dI/dt = 2\pi f L\, I_M \cos(2\pi f t). \tag{14}$$

Hence, if we write this voltage in the form $V_L = V_{LM}\cos(2\pi f t)$, we find

$$V_{LM} = 2\pi f L\, I_M. \tag{15}$$

The coefficient $2\pi f L$, which is measured in ohms, is called the *inductive reactance* and denoted by X_L:

The **inductive reactance** at frequency f, of a circuit element of self inductance L, is $X_L = 2\pi f L$.

Thus we can write (15) in a form analogous to Ohm's law:

$$V_{LM} = X_L\, I_M, \qquad I_M = V_{LM}/X_L, \qquad \text{where} \qquad X_L = 2\pi f L. \tag{16}$$

The same relations will apply if, instead of maximum values, we use effective values of voltage and current.

We note that the higher the inductance, the higher is the reactance, and the lower the current for a given voltage and frequency. We note that the higher the frequency, the greater is the inductive reactance. In contrast to a capacitor, an inductor passes DC most readily and offers greater and greater reactance as the frequency increases.

Example. *Compute the inductive reactance of a* 4-mH *inductor of negligible resistance, and the current when* 120 V *is applied, at the following frequencies:* (a) 1 Hz, (b) 60 Hz, (c) 400 Hz, (d) 1 MHz.

(a) We compute the inductive reactance at $f = 1$ Hz from (16):

$$X_L = 2\pi f L = 2\pi \cdot (1/\text{s}) \cdot 4 \times 10^{-3}\,\text{H} = 0.0251\,\Omega.$$

The units come out initially as H/s, which is the same as Ω. When $V^{\text{Eff}} = 120$ V is applied,

$$I^{\text{Eff}} = V^{\text{Eff}}/X_L = (120\,\text{V})/(0.0251\,\Omega) = 4780\,\text{A}.$$

The inductive reactance is small and the current large; we are approaching the DC case where $X_L = 0$.

(b), (c), (d) We can obtain X_L at any frequency by multiplying its value at 1 Hz by the frequency, and we can obtain the current by dividing its value at 1 Hz by the frequency. Hence

at 60 Hz, $\quad X_L = 1.51\,\Omega, \quad I^{\text{Eff}} = 79.7$ A;

at 400 Hz, $\quad X_L = 10.0\,\Omega, \quad I^{\text{Eff}} = 12.0$ A;

at 1 MHz, $\quad X_L = 25\,000\,\Omega, \quad I^{\text{Eff}} = 0.004\,78$ A.

We can now look at the lower curve of Fig. 4 and conclude that, just as in the case of capacitance,

The average power delivered to a pure inductance is zero.

The inductance of a solenoid can be accurately computed from its geometrical dimensions; hence its reactance is known at a known frequency. Again, as in the case of a standard capacitor, its reactance can be compared with the resistance of a standard resistor to determine a *standard ohm*. Thus method actually gives the most accurate determination of the *ohm,* and hence, in connection with the current balance, of the *volt*.

5. General Series Circuits

We now return to the circuit of Fig. 1 and to equation (2), which states that, instantaneously, $V = V_R + V_C + V_L$. With the given current (1), we obtain the curve of applied voltage by adding three sinusoidal curves, giving V_R, V_C, and V_L, like those plotted in Figs. 2, 3, and 4.

We first note that V_C and V_L are exactly ½ cycle out of phase, so that when these voltage curves are added the resulting voltage has maximum value $V_{LM} - V_{CM}$ or $V_{CM} - V_{LM}$, depending on whether V_{CM} or V_{LM} is the greater. To this resultant must be added the curve V_R, ¼ cycle out of phase. The result of this addition can be obtained from trigonometry. It turns out that the sum of two sinusoidal curves ¼ cycle out of phase can be obtained by completing a right triangle. The results in our particular case are as follows (*see* Fig. 5):

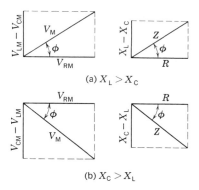

Fig. 5. Amplitude and phase relations in the general series circuit.

When $X_L > X_C$, the amplitude V_M of the applied voltage is the diagonal of the right triangle shown in Fig. 5(a):

$$V_M = \sqrt{V_{RM}^2 + (V_{LM} - V_{CM})^2}. \tag{17}$$

Substitution of $V_{RM} = R\,I_M$, $V_{LM} = X_L\,I_M$, and $V_{CM} = X_C\,I_M$, gives

$$V_M = \sqrt{R^2 + (X_L - X_C)^2}\,I_M. \tag{18}$$

The coefficient of I_M here is called the *impedance* of the circuit; it is measured in ohms and denoted by Z:

$$\left. \begin{array}{c} V_M = Z I_M, \quad I_M = V_M / Z, \\ Z = \sqrt{R^2 + (X_L - X_C)^2}. \end{array} \right\} \tag{19}$$

where

The impedance Z can be obtained from R and $X_L - X_C$ by completing a triangle as shown at the right of Fig. 5. To complete the discussion, we note that the voltage V *leads* the current by the angle ϕ, in radians, given in Fig. 5(a):

$$\tan \phi = (V_{LM} - V_{CM})/V_{RM} = (X_L - X_C)/R \tag{20}$$

and, for current (1), we can write

$$V = V_M \sin(2\pi f t + \phi). \tag{21}$$

When $X_C > X_L$, as indicated in Fig. 5(b), we obtain the same equations except that subscripts L and C must be interchanged; the angle ϕ is now an angle by which the voltage *lags behind* the current, and must appear with a $-$ sign instead of a $+$ sign in (21).

Since no power is expended in either the capacitance or the inductance, the whole power is that used in heating the resistance, and is given by $P = V_R^{Eff} I^{Eff}$. The triangles at the left of Fig. 5 will retain the same shape if effective voltages are plotted instead of maximum voltages, so we see that $V_R^{Eff} = V^{Eff} \cos\phi$. Hence

$$P = V^{Eff} I^{Eff} \cos\phi. \tag{22}$$

The power P read on an AC wattmeter (p. 682) is less than the product of V^{Eff}, read on a voltmeter, and I^{Eff}, read on an ammeter, by the factor $\cos\phi$, which is called the *power factor*.

6. Series Resonance

We now come to an important phenomenon, *resonance,* that is basic to oscillating circuits such as those that generate radio waves, and that is basic to the tuning of receiving circuits. Actually, our principal reason for giving as much detail as we have on AC circuits is to help acquire an understanding of the phenomenon of resonance.

A series circuit is said to be **resonant** if $X_L = X_C$.

In this resonance case, V_L and V_C are exactly equal and opposite at every instant, so, in (2), $V = V_R$. The impedance equals the resistance, current and voltage are in phase ($\phi = 0$), the power factor is 1, and at the terminals the circuit behaves in every respect like a pure resistance.

In a resonant circuit, V_L and V_C may be individually very large, with the capacitance and inductance taking in and giving out large amounts of instantaneous power. But comparison of the instantaneous-power curves of Figs. 3 and 4 shows that, for the case $V_{CM} = V_{LM}$, at each instant the capacitance is absorbing power at the same rate as the inductance is giving it out, and vice versa. Large power can be handed back and forth from capacitance to inductance without influencing the rest of the circuit.

The frequency at which a circuit is resonant is given by setting $X_L = X_C$, $2\pi f L = 1/2\pi f C$, to obtain

$$f = \frac{1}{2\pi \sqrt{LC}}. \qquad \binom{\text{resonance}}{\text{frequency}} \quad (23)$$

We can illustrate the important role that resonance plays in the *tuning* of circuits by means of the following examples.

Example. *Consider the antenna circuit of Fig. 6 when a radio wave of frequency $f = 600$ kHz induces a voltage with $V_M = 0.1$ mV between the antenna and ground. The resistance of the circuit is $R = 100\,\Omega$, the self inductance $L = 10$ mH, and the capacitance C is variable. Compute I_M in μA as a function of the capacitance C in pF, and find the capacitance at resonance.*

The results of the straightforward but somewhat tedious computation of I_M from (19) are plotted in Fig. 6.

The capacitance at resonance is obtained by solving (23) for C:

$$C = \frac{1}{(2\pi f)^2 L} = \frac{1}{(3.77 \times 10^6/\text{s})^2 (10^{-2}\text{ H})} = 7.04 \times 10^{-12}\text{ F} = 7.04\text{ pF}.$$

We notice that in the immediate neighborhood of resonance the current

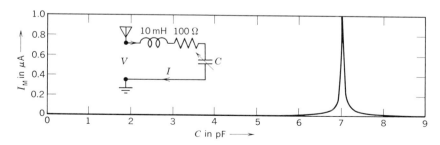

Fig. 6. Current in a series circuit as a function of capacitance, illustrating series resonance at the value of C that makes $X_L = X_C$.

is very large compared with the current for values of C away from resonance, because the impedance Z has its minimum value R at resonance and increases rapidly as the term $X_L - X_C$ in (19) departs from zero either positively or negatively. The sharpness of the peak of a curve such as Fig. 6 depends on the relative values of X_L and R—the larger the ratio X_L/R (called the *quality factor* and denoted by 'Q'), the sharper the peak.

The phenomenon illustrated in Fig. 6 is familiar to everyone in the tuning of the antenna circuit of a radio to an impressed voltage of a given frequency arising from the radio waves broadcast by a given station.

Example. *To see how the antenna circuit of a radio can discriminate between stations of different frequency, plot a curve for the circuit of Fig. 6, except that C is fixed at 7.04 pF and f is varied. Keep V_M fixed at 0.1 mV.*

As we have seen in the previous example, this value of C gives resonance at 600 kHz. Equation (19) gives, at frequency $f = $ f Hz,

$$Z = \sqrt{10^4 + \left(0.0628\, \text{f} - \frac{2.26 \times 10^{10}}{\text{f}}\right)^2}\ \Omega;$$

in this relation f is to be considered as a pure number; the units have all been collected. The impedance Z has its minimum value when the parenthesis vanishes, which is at f = 600 000. At this frequency, $Z = 100\ \Omega$ and $I_M = V_M/Z = 10^{-6}$ A $= 1\ \mu$A. At any other frequency, Z is greater than 100 Ω and I_M is less than 1 μA. Figure 7 shows the curve of I_M as a function of f for a frequency range from 500 to 700 kHz, in the neighborhood of the resonance at 600 kHz. This plot also shows a sharp resonance; if the circuit is turned to a station of one frequency, the response to stations of other frequencies is greatly attenuated.

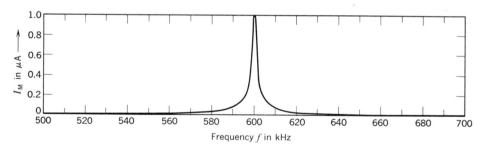

Fig. 7. Current in a series circuit as a function of frequency, illustrating series resonance at the frequency that makes $X_C = X_L$.

Example. *In the antenna circuit of the two preceding examples compute, for the case of resonance, the maximum voltages V_{RM}, V_{CM}, and V_{LM} across the circuit elements, for the applied voltage $V_M = 0.1$ mV.*

At resonance, $R = 100\ \Omega$, $X_C = X_L = 37\,700\ \Omega$, $I_M = 1\ \mu$A $= 0.001$ mA, as computed in the above examples. Hence

$$V_{RM} = R\,I_M = (100\ \Omega)(0.001\ mA) = 0.1\ mV;$$
$$V_{CM} = X_C\,I_M = (37\,700\ \Omega)(0.001\ mA) = 37.7\ mV;$$
$$V_{LM} = X_L\,I_M = (37\,700\ \Omega)(0.001\ mA) = 37.7\ mV.$$

In the last example, we notice that, at resonance, the voltage across the resistance exactly equals the applied voltage, since the voltages across the capacitance and the inductance are equal, ½ cycle out of phase, and exactly cancel each other at any instant. But we also notice that, as in this example, the latter voltages may be much greater than the applied voltage—in our case 377 times as great. Hence the voltage across the capacitor would be very suitable for use on the grid of the triode in the first stage of an electronic amplifier, since it corresponds already to a considerable amplification of the antenna voltage.

7. The Transformer

A transformer consists of two windings around the two legs of a laminated iron 'core,' as illustrated in Fig. 8. In a transformer, AC power is fed into the primary winding, of N_P turns, and taken out at the secondary winding, of N_S turns.

We shall consider the ideal case of a transformer with no losses. In an actual transformer the losses can be kept well under one per cent of the power transmitted. In the ideal case, the windings have zero resistance (no I^2R-losses); the magnetic material is assumed ideal (flux Φ directly proportional to MMF, hence no hysteresis losses); and eddy currents in the core are assumed negligible.

Let us first consider the case in which the switch in the secondary circuit is open, so there is no power transmitted through the transformer. The primary coil then acts like a pure inductance, draws a current ¼ cycle out of phase with the applied voltage, and draws no power. In accordance with (16), p. 659, we can write the instantaneous primary voltage as

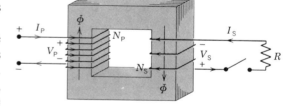

Fig. 8. Diagram of ideal resistanceless transformer with arrows and + and − signs to indicate positive senses of currents, voltages, and flux. The induced EMF's are designated by V_P and V_S.

$$V_P = -N_P\,d\Phi/dt, \qquad (24)$$

where Φ is the alternating flux set up in the core by the alternating primary current. The open secondary will carry no current, but will have the induced voltage

$$V_S = -N_S\,d\Phi/dt, \qquad (25)$$

so that, at any instant, $\qquad V_S/V_P = N_S/N_P. \qquad (26)$

The ratio of instantaneous voltages is proportional to the ratio of turns. Maximum and effective voltages will be in the same ratio. Thus the transformer can be used to step AC voltages up or down in the ratio of the number of turns on the two legs.

Example. *A bell-ringing transformer is connected to a* 120-V, 60-Hz *line and used to furnish* 6 V. *The primary has* 180 *turns, negligible resistance, and self inductance* $L = 0.1$ H. *How many turns are on the secondary? What current and power are furnished to the primary when the bell is not being rung (secondary open)?*

By (26), the number of secondary turns is

$$N_S = N_P (V_S^{Eff}/V_P^{Eff}) = (180 \text{ turns})(6 \text{ V}/120 \text{ V}) = 9 \text{ turns}.$$

The impedance of the primary will equal its inductive reactance

$$Z = X_L = 2\pi f L = (377 \text{ s}^{-1})(0.1 \text{ H}) = 37.7 \text{ }\Omega.$$

Hence $\quad\quad\quad I_P^{Eff} = (120 \text{ V})/(37.7 \text{ }\Omega) = 3.18 \text{ A}.$

Since the current will lag the voltage by ¼ cycle, the power factor will be zero, and the power drawn will be zero. Actual household bell-ringing transformers are sufficiently close to this ideal so that one will not observe the watt-hour meter rotating when the only load on the circuit is the permanently connected bell-ringing transformer, and the bell is not being rung.

Now let us consider briefly the *loaded* transformer, with power being drawn in the secondary, and hence a secondary current I_S. The flux Φ will now result from both the primary and the secondary current. *But* (24) *and* (25), *and hence* (26), *still remain true at every instant.* The primary voltage V_P is fixed at the input voltage; hence $d\Phi/dt$ is fixed and does not change as the secondary is loaded; V_S is also fixed. But since part of the flux is now being set up by the current in the secondary circuit, the primary *current* must change so that it sets up only the balance of the flux. The phase relations are such that the primary current now lags the primary voltage by less than a full quarter cycle, so power is now delivered to the primary. This power will be equal to that drawn from the secondary. The equations that determine the currents and their phases, involving as they do the self-induced voltages in each coil and the voltages mutually induced in each coil by the other, are straightforward but complex and we shall not formulate them here.

When a transformer is *fully loaded,* and furnishing power to an essentially resistive secondary circuit, the secondary current will be in phase with the secondary voltage and the primary current will also be closely in phase with the primary voltage. Under these circumstances (power factor approximately 1), we can write, approximately, $P = E_P^{Eff} I_P^{Eff} = E_S^{Eff} I_S^{Eff}$. Hence, the high-voltage side of the transformer will carry correspondingly low current. It is this property that makes long-distance power transmission feasible. Large currents cannot be transmitted for long distances because I^2R-losses would dissipate much of the power, and prohibitively heavy conductors would be required just to carry the current. But by stepping *up* the transmission voltage by means of a transformer, the current is stepped *down*, and I^2R-losses go down as the square of the current. The situation is illustrated in the following example:

Example. *An* AC *transmission line* 100 mi *long has a resistance of* 50 Ω *and has input power of* 1000 kW *from a* 2000-V *generator. Compare the output power if the input voltage remains at* 2000 V *with the output power if this voltage is transformed to* 200 kV.

To transmit 1000 kW at 2000 V requires a current of 500 A. To safely carry 500 A requires a copper wire about ½ inch in diameter; such a wire has a resistance of about ¼ Ω per mile, or 50 Ω for a 100-mi round trip. But 2000 V applied to 50 Ω gives a current of only 40 A so 1000 kW *cannot be put into this wire at* 2000 V, *even if it is short-circuited at the far end!*
To transmit 1000 kW at 200 kV requires a current of 5 A. The I^2R-loss in a 50-Ω line is $(5\text{ A})^2 (50\text{ Ω}) = 1250\text{ W} = 1.25$ kW, so only slightly over $\frac{1}{1000}$ of the power is lost in the line. Even higher resistance could be readily tolerated.

While discussing long-distance power transmission, we should mention that there is a recent trend toward use of DC lines when the transmission distance is very great—over about 400 miles. Such DC transmission is made feasible by the development, during the last two decades, of mercury-arc rectifiers that will operate at voltages up to 150 kV. The power is generated as AC of comparatively low voltage, transformed to AC of high voltage, rectified, and transmitted as DC; at the receiving end the process is reversed. The advantages of DC transmission are that the line is free of capacitive and inductive reactance and need be insulated only for the actual voltage, not for $\sqrt{2}$ times the effective voltage. The largest DC transmission line that is now in operation carries 2700 MW of power at 800 kV a distance of 840 miles from the Columbia River to Los Angeles.

8. Measuring Instruments

Let us now briefly discuss measuring instruments for use with AC circuits. Those for use with circuits of ordinary power frequencies usually employ a sensitive element called an *electrodynamometer* (Fig. 9).
The electrodynamometer of Fig. 9 differs from the galvanometer of Fig. 16, p. 593, in that the magnetic field in which the moving coil rotates is created by a pair of fixed coils carrying current supplied from outside through a pair of binding posts separate from those that supply current to the moving coil. With this coil arrangement, a given pointer reading

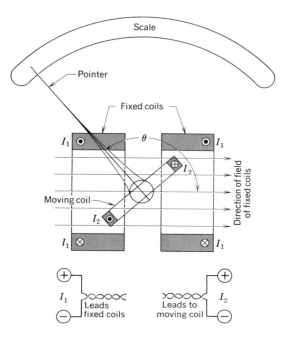

Fig. 9. The electrodynamometer. The pivoted moving coil rotates in the space interior to the fixed coils. A light framework extending through the gap between the fixed coils carries the pointer and the springs supplying the restoring torque. In this diagram the pointer is in the equilibrium position for no current, and the springs are not shown.

corresponds to a definite value of the *product* $I_1 I_2$ of the currents in the fixed and moving coils, and the scale could be calibrated to read $I_1 I_2$.

With a current I_2 in the moving coil in the direction shown, the moving coil will have a magnetic moment represented by a vector perpendicular to its plane and pointing in the same direction as the pointer. This coil, in the magnetic field set up by the fixed coils, experiences a torque tending to turn its magnetic moment into the direction of the field, that is, a clockwise torque tending to swing the pointer to the right. The pointer will move to the right until the countertorque of the springs just balances the magnetic torque.

By (2), p. 586, the torque and hence the pointer deflection will be proportional to the field $\mathcal{B}$ of the fixed coils, to the magnetic moment M of the moving coil, and to the sine of the angle θ between these vectors: $L = M\mathcal{B} \sin\theta$. The intensity $\mathcal{B}$ is proportional to I_1, while M is proportional to I_2. The angle θ varies from about 120° at no deflection, through 90° at half-scale, to about 60° at full-scale deflection, so $\sin\theta$ varies only from about 0.866 to 1. Except for this slight variation in $\sin\theta$, the deflection is proportional to $I_1 I_2$; hence the calibration marks corresponding to $I_1 I_2 = 0$, 1 mA², 2 mA², 3 mA², ⋯ are approximately uniformly spaced on the scale. Note that if we *reverse* the current in *both* coils, we will still get the same torque and hence the *same* pointer deflection. Hence the electrodynamometer is suitable for use with low-frequency (e.g., 60-Hz) AC in both coils.

We have shown (pp. 552–553) how shunts and high-resistance multipliers can be used with d'Arsonval galvanometers to provide DC ammeters and voltmeters; similar uses of shunts and multipliers can be used to convert electrodynamometers into AC ammeters and voltmeters. A wattmeter for use with either DC or low-frequency AC is made by using a shunt with the fixed coils and a multiplier with the moving coil of an electrodynamometer.

By the use of a rectifier (pp. 538–541) in series with a d'Arsonval galvanometer, we can use the d'Arsonval device as the basic element in AC ammeters and voltmeters. This procedure is coming into wide use, particularly for instruments designed for high frequencies, such as radio frequencies, where the high inductive reactance of the coils precludes the use of an electrodynamometer of the type illustrated in Fig. 9.

PROBLEMS

NOTE: *'Current'* and *'voltage'* mean *'effective current'* and *'effective voltage'* unless otherwise qualified.

1. What are the capacitive reactances and the impedances of the circuit in the figure, for DC, and for AC of frequencies 10, 10^3, 10^5, and 10^7 Hz?   Ans: ∞, ∞; 637, 637 Ω; 6.37, 6.68 Ω; 6.37×10^{-2}, 2.00 Ω; 6.37×10^{-4}, 2.00 Ω.

2Ω $25\mu F$

Problem 1

2. What are the inductive reactances and the impedances of an inductor having self inductance of 25 mH and resistance of 3 Ω, for DC, and for AC of frequencies 10, 10^3, 10^5, and 10^7 Hz?

3. (a) What is the current in a circuit containing a 1-μF capacitor in series with a 2000-Ω resistor when an alternating voltage of 500 V is applied, with $f = 6$ Hz? 60 Hz? 600 Hz? (b) Write the equation for the instantaneous current in each case, if $V = (707 \text{ V}) \sin 2\pi ft$.

Ans: (a) 0.0188, 0.150, 0.248 A;
(b) $\begin{cases} (0.0265 \text{ A}) \sin[(37.7 \text{ s}^{-1}) t + 1.50], \\ (0.211 \text{ A}) \sin[(377 \text{ s}^{-1}) t + 0.925], \\ (0.351 \text{ A}) \sin[(3770 \text{ s}^{-1}) t + 0.131]. \end{cases}$

4. A 3-pF capacitor is in series with a 1-MΩ resistor. At 10^4 Hz, at 10^5 Hz, and at 10^6 Hz, (a) what is the reactance? (b) what is the impedance? (c) what is the current when an alternating voltage of 200 V is applied? (d) what is the angle of lead of the current? (e) Write the equation of the instantaneous current in each case if $V = (282 \text{ V}) \sin 2\pi ft$.

5. (a) What is the alternating voltage required to send current $I = (0.4 \text{ A}) \sin 2\pi ft$ through a 1-H coil of 377-Ω resistance, for $f = 6$, 60, and 600 Hz? (b) Write the equation for the instantaneous voltage in each case.

Ans: (a) 107, 151, 1070 V;
(b) $\begin{cases} (151 \text{ V}) \sin[(37.7 \text{ s}^{-1}) t + 0.105], \\ (213 \text{ V}) \sin[(377 \text{ s}^{-1}) t + 0.785], \\ (1510 \text{ V}) \sin[(3770 \text{ s}^{-1}) t + 1.47]. \end{cases}$

6. (a) What is the inductive reactance of a 2-mH inductor of 1-MΩ resistance, at 10^7 Hz? at 10^8 Hz? at 10^9 Hz?

(b) What is the impedance of the inductor at these frequencies? (c) What is the current through the inductor when an alternating voltage of 150 V is applied, at 10^7 Hz? at 10^8 Hz? at 10^9 Hz?
(d) What is the angle of lag of the current in each case in (c)?
(e) Write the equation of the current in each case in (c) if $V = (212 \text{ V}) \sin 2\pi ft$.

7. Consider a series circuit, with $X_L = 3 \Omega$, $X_C = 1.5 \Omega$, and $R = 0.8 \Omega$. Compute I for $V = 48$ volts, and compute the angle of lag or lead.   Ans: 28.2 A; current lags by $61°9$.

8. Consider a series circuit, with $X_L = 3 \Omega$, $X_C = 4.5 \Omega$, and $R = 2.4 \Omega$. Compute I for $V = 60$ volts, and compute the power factor.

9. An inductor, a resistor, and a capacitor are in series. The inductor has self inductance of 0.5 H and resistance of 100 Ω. The resistor has 80 Ω of resistance and the capacitor has 8 μF of capacitance. An applied 60-Hz voltage causes 1 A effective through the series circuit.
(a) Compute the total impedance Z.
(b) From Z and the current, compute the effective total voltage required, and the phase angle between the voltage and the current.
(c) Compute the effective voltage across the terminals of the inductor; the effective voltage across the terminals of the resistor; and the effective voltage across the terminals of the capacitor. In each case obtain the phase angle between these voltages and the current.
Ans: (a) 230 Ω; (b) 230 V, $38°5$ behind current; (c) 213 V, $62°1$ ahead of current; 80 V, in phase with current; 332 V, $90°$ behind current.

10. Repeat Prob. 9 for the case where the inductor has 6 mH of self inductance and 9 Ω of resistance, the resistor has 10 Ω of resistance, the capacitor has 40 μF of capacitance, and an applied 500-Hz voltage causes 1 A effective.

11. What is the resonance frequency for the series circuit of Prob. 9? What voltage is required for a current of 1 A at this frequency?
Ans: 79.6 Hz; 180 V.

12. What is the resonance frequency for the series circuit of Prob. 10? What voltage is required for a current of 1 A at this frequency?

13. In the circuit of Fig. 6, if the capacitance and resistance are fixed at 7.04 pF and 100 Ω, but the inductance is adjustable, what value of inductance will bring the circuit into resonance for a 500-kHz applied voltage?
Ans: 14.4 mH.

14. In the circuit of Fig. 6, if the capacitance and resistance are fixed at 7.04 pF and 100 Ω but the inductance is adjustable, what value of inductance will bring the circuit into resonance for a 700-kHz applied voltage?

15. Show that the curve of Fig. 6 applies to a low-frequency circuit if we let the ordinates represent amperes, the abscissas μF, change L to 1 H and R to 1 Ω, and replace the antenna by a 60-Hz line with $V_M = 1$ V.

16. Show that the curve of Fig. 7 applies to a power circuit if we let the ordinates represent amperes instead of microamperes, the abscissas cover the range from 50 Hz to 70 Hz, and change L to 1 H, R to 1 Ω, C to 7.04 μF, and V_M to 1 V.

17. A current of 5 A, lagging by 30°, exists in an AC circuit when 110 V are applied. What is the power factor? What is the power?
Ans: 0.866; 476 W.

18. A current of 1.0 A, leading by 60°, exists in an AC circuit when 24 V are applied. What is the power factor? What is the power?

19. What is the power in the circuit of Fig. 1 at resonance frequency with $R = 1$ Ω, $L = 1$ H, $C = 7.04$ μF, and $V_M = 1$ V at 60 Hz? What is the maximum energy stored in the self inductance? in the capacitance?
Ans: 1 W, 1 J; 1 J.

20. A series circuit has self inductance, capacitance, and resistance. A fixed AC voltage is applied and the capacitance adjusted until the current is a maximum. The maximum current is found to be 4 A effective. What power is being supplied to the circuit if the total resistance is 15 Ω?

21. A transformer has 50 turns on the primary and 1000 on the secondary. When the primary is connected to 110-V AC mains, what is the voltage of the secondary? Ans: 2200 V.

22. Two circuits are in parallel across a pair of leads in a radio set. One circuit has capacitance 0.1 μF and resistance 50 000 Ω in series; the other has inductance 5 H and resistance 35 Ω. The first circuit is designed to 'pass' both audio and radio frequency, the second circuit to 'choke out' radio frequency.
(a) If an audio-frequency signal of 50 V at 1000 Hz is impressed, what is the current in microamperes in each of the two circuits?
(b) If a radio-frequency signal of 50 V at 1000 kHz is impressed, what is the current in microamperes in each of the two circuits?

23. (a) A feeder line with a total resistance of 2 Ω supplies a group of houses with a nominal 240 V AC for heating and lighting. The transformer supplying the feeder from the transmission line is adjusted so that when the 'nominal' power of 6 kW is used in the houses, the voltage at the houses is 240 V. What must be the output voltage of this transformer? What percentage of the output power of the transformer is wasted in heating the feeder line? Show that the 'regulation' of the voltage at the houses will be very poor in the sense that there will be unacceptably large variations in voltage with the variations in load that are to be expected, say between no load and double the nominal load.
(b) To improve the regulation and reduce the losses, a 10:1 transformer is installed at the house end of the feeder line. What must now be the output voltage of the transformer at the transmission-line end? What percentage of the power is now lost in heating the feeder? Show that the voltage regulation is now satisfactory in that the voltage does not vary more than a few per cent between no load and double the nominal load.
Ans: (a) 290 V, 17.3%; (b) 2405 V, 0.207%.

24. A large step-down transformer has a primary input of 2000 kW at 3300 V. The transformer is 100% efficient. The secondary feeds into a resistive load at 550 V. Find the secondary current.

25. One method of producing 'atomic projectiles' without the use of enormous DC potentials is a *linear accelerator,* in which high frequency AC voltages are applied to the tubular accelerator sections in a tube like that in the figure on p. 511. The focus tube and the second, fourth, sixth, ··· sections are connected to one terminal of the AC 'generator,' while the first, third, fifth, ··· sections are connected to the other terminal. Thus, a proton emerging from the focus tube when the focus tube has its maximum positive potential is accelerated to the first section and moves into the field-free space inside the tubular electrode. If the length of the first section has the correct value, the proton will emerge when the first section has its maximum positive potential and be again accelerated, this time toward the second accelerator section. This process is repeated numerous times before the proton finally reaches the target. If the maximum difference of potential between electrodes is 100 kV, through how many gaps must the proton pass before attaining an energy of 1 MeV? It is desirable to work at high frequencies; in order to see this, make a rough calculation of the length of a 1-MeV linear accelerator with gap voltage 100 kV, when operated at 60 Hz, 1 MHz, and 100 MHz.

26. Why is the core of a transformer laminated in the manner suggested in Fig. 8, p. 679?

33 OSCILLATING CIRCUITS; ELECTROMAGNETIC WAVES

We start this chapter by considering the generation of an oscillating electric current in a resonant circuit. For a hypothetical circuit containing only inductance and capacitance, it is shown that electric oscillations would be sustained once they were initiated. In real circuits containing resistive elements in addition to inductance and capacitance, electric oscillations are damped out as a result of resistive heating; however, by the use of electronic amplifiers and appropriate feedback techniques, it is possible to maintain continuous oscillations by supplying sufficient power to compensate for resistive losses. We describe the conditions under which *additional power* must be supplied to the circuit to compensate for additional power losses resulting from the *radiation of electromagnetic waves*. We postpone our quantitative treatment of Maxwell's theory of electromagnetic radiation to Chap. 43, which follows the part of the book dealing with light. In the present chapter we give a qualitative treatment.

1. Oscillations in a Circuit Containing Inductance and Capacitance

Consider the series circuit of Fig. 1, which is similar to the circuit shown on p. 668 except that a key replaces the source of alternating voltage. If the key is closed in Fig. 1, we must have, at any instant, $V_C + V_L + V_R = 0$. As in the preceding chapter, we can write this equation in the form $V_C + L\,dI/dt + IR = 0$. Now differentiate with respect to t:

$$dV_C/dt + L\,d^2I/dt^2 + R\,dI/dt = 0. \tag{1}$$

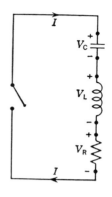

Figure 1

Since $V_C = Q/C$, $dV_C/dt = (1/C)\,dQ/dt = I/C$, where Q is the instantaneous charge on the capacitor; substitution of this expression in (1) gives the *second-order differential equation*

$$L\frac{d^2I}{dt^2} + R\frac{dI}{dt} + \frac{1}{C}I = 0. \tag{2}$$

Before considering the general solution of this equation, let us consider the idealized case where $R=0$ in Fig. 1. In this case, (2) becomes

$$d^2I/dt^2 = -(1/LC)\,I. \tag{3}$$

From our study of simple harmonic motion (*see* pp. 243–247) we can infer that the general solution of the equation $d^2I/dt^2 = -(4\pi^2 f^2)\,I$ is of the form

$$I = A\sin(2\pi ft) + B\cos(2\pi ft), \tag{4}$$

where A and B are arbitrary constants. Hence (4) furnishes the general solution of (3) with

$$4\pi^2 f^2 = 1/LC, \quad \text{or} \quad f = 1/2\pi\sqrt{LC}, \tag{5}$$

exactly the *resonance frequency* discussed in the preceding chapter.

Now return to Fig. 1, with $R=0$, and assume that the capacitor is charged to potential V_0 and that at $t=0$ the switch is closed. Then,

$$I = -V_0\sqrt{C/L}\sin(2\pi ft) \tag{6}$$

is the solution, of form (4), that satisfies the initial conditions $I=0$ at $t=0$ (since the current through the inductance cannot change suddenly) and $dI/dt = -V_0/L$ at $t=0$ (which is obtained from the relation $V_L = -V_C = -L\,dI/dt$, which holds when R and hence V_R are zero).

In the ideal case, the oscillations (6) persist indefinitely, with the energy handed back and forth from capacitance to inductance, just as it is handed back and forth from potential to kinetic form in the case of simple harmonic motion. (This statement is true only if *radiation losses* are negligible, as we shall see in later sections.)

Example. *In Fig. 1, if $L=10$ mH, $C=7.04$ pF, $R=0$, the capacitor is initially charged to 1 V, and the key is closed at $t=0$, determine* (a) *the frequency* of oscillation, (b) *the maximum current*, (c) *the energy of the capacitor as a function of time*, (d) *the energy of the inductor as a function of time*. Show that the sum of (c) and (d) always equals the initial energy.

(a) The frequency of oscillation is the resonance frequency (5) already computed in the examples on pp. 677–679 as $f = 600$ kHz $= 6 \times 10^5$ Hz.

(b) The maximum current is given by the coefficient in (6) as

$$I_M = V_0 \sqrt{C/L} = 2.65 \times 10^{-5} \text{ A} = 26.5 \text{ }\mu\text{A}. \tag{i}$$

Hence,
$$I = -(26.5 \text{ }\mu\text{A}) \sin(2\pi ft). \tag{ii}$$

(c) For the above current, we see from equations (1) and (11) of the preceding chapter (pp. 668 and 672) that the voltage across the capacitor is $V_C = (I_M/2\pi fC) \cos(2\pi ft)$. Since $2\pi f = 1/\sqrt{LC}$, the coefficient of the cosine is $I_M \sqrt{L/C}$, which, from (i), is just 1 V; hence $V_C = (1 \text{ V}) \cos(2\pi ft)$. The energy in the capacitor is

$$\tfrac{1}{2} C V_C^2 = \tfrac{1}{2}(7.04 \times 10^{-12} \text{ F})(1 \text{ V})^2 \cos^2(2\pi ft) = (3.52 \times 10^{-12} \text{ W}) \cos^2(2\pi ft).$$

(d) The energy of the inductor is obtained from (ii):

$$\tfrac{1}{2} LI^2 = \tfrac{1}{2}(10^{-2} \text{ H})(2.65 \times 10^{-5} \text{ A})^2 \sin^2(2\pi ft) = (3.51 \times 10^{-12} \text{ W}) \sin^2(2\pi ft).$$

The sum of (c) and (d) is constant and equal to the initial energy of the capacitor, $\tfrac{1}{2} C (1 \text{ V})^2 = 3.52 \times 10^{-12}$ W.

2. The Damped Oscillating Circuit; Continuous Oscillations

In the resistanceless oscillating circuit described in the preceding section, no energy was lost because no heat was generated. In the case of a low but finite resistance R in Fig. 1, energy will be lost and the oscillating current will die out. The solution of the differential equation (2) in this case is of the form

$$I = A e^{-\alpha t} \sin(2\pi f' t) + B e^{-\alpha t} \cos(2\pi f' t), \tag{7}$$

where α is an exponential damping constant and the frequency f' turns out to be somewhat different from the natural frequency f. The values of these constants are

$$\alpha = \frac{R}{2L}; \quad 2\pi f' = \sqrt{\frac{1}{LC} - \alpha^2} = \sqrt{\frac{1}{LC} - \frac{R^2}{4L^2}}, \tag{8}$$

as may be verified by substitution in the differential equation (2). The first term under the square root gives the *natural* frequency (5). If the resistance is gradually increased the frequency f' will gradually decrease, until, at the value of resistance that makes the radicand vanish, the solution changes from the oscillatory character of (7) and becomes of a purely exponential character. For $R \geq 2 \sqrt{L/C}$, the circuit will not oscillate. Figure 2 shows a plot of the current (7) in the case of the preceding example if $R = 1000 \text{ }\Omega$ rather than zero.

Since a circuit like that of Fig. 1 always contains some resistance, natural oscillations will always die out. However, the invention of the triode made possible the maintenance of *continuous oscillations* in an *LC*-circuit using an electronic amplifier in a 'feed-back' circuit, as indicated schematically in Fig. 3.

Figure 3 shows the circuit of Fig. 1 with the oscillating capacitor voltage fed into an amplifier of high input impedance so that the amplifier draws little current. (The capacitor voltage is usually applied between the grid and cathode of a vacuum tube.) The amplifier output is a current that faithfully reproduces the input voltage, with or without a phase shift. This current is used to 'drive' an inductively coupled tuned antenna and also to induce current in the resonating *LC*-circuit by inductive coupling with L, as indicated. The coupling is in such phase as to aid the passage of the resonating current in L and hence add power to the resonating circuit in order to compensate for the power loss in the resistance.

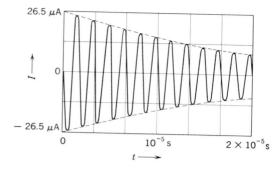

Fig. 2. Oscillatory discharge of a capacitor through inductance and resistance for the case $C = 7.04$ pF, $L = 10$ mH, and $R = 1000$ Ω.

We could imagine 'starting' the oscillator of Fig. 3 by turning the amplifier 'on,' charging the capacitor from a battery, and then closing the switch, in the manner discussed in connection with Fig. 1. In practice, this 'charging' procedure is not necessary; the switch is left closed and the oscillations start by themselves when the amplifier is turned on. The explanation of this behavior involves the fact that there are random *thermal* motions of free electrons in the oscillating circuit (called 'noise' by electronics engineers). These thermal motions will momentarily result in a charge on the capacitor and the oscillations will start. Although they will start with very small amplitude, the amplitude will build up rapidly after the feed-back amplifier is turned on, in just the way that resonant oscillations are built up when air is blown into an organ pipe.

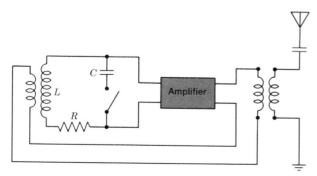

Fig. 3. Continuous oscillations; a simple radio transmitter.

3. Electromagnetic Radiation

We have seen that, in a resonant circuit, relatively large amounts of energy are stored alternately in the electric field of the capacitor and in the magnetic field of the inductor. It turns out that in the course of the transfer back and forth, some of this energy is lost in a form other than I^2R-heat; it leaves the circuit completely and travels out through space as electromagnetic waves of frequency equal to that of the oscillating current. The amount of energy so *radiated* is inappreciable at commercial power frequencies but increases rapidly as the frequency increases. It is also increased if the *physical size* of the circuit approximates the *wavelength* of the radiation in space, as does a broadcast antenna. The radiated waves move with the speed of light, as we shall discuss in detail in Part V of this book. The speed of light is always denoted by c, and has the value

$$c = 3.00 \times 10^8 \text{ m/s}. \qquad \text{(speed of light)}$$

The manner in which this radiation takes place was first described mathematically by Maxwell, in 1864. The radiation was first observed in 1886 by HEINRICH HERTZ, a German physicist, who found that these waves had properties similar to those of light even though his experiments were conducted with waves a meter or more in length; the waves could be reflected, refracted, diffracted, and polarized. By these and later experiments, Maxwell's theory has been confirmed as the correct description of the *transmission* of light waves as well as of radio waves. With respect to *emission* and *absorption* processes, Maxwell's theory gives an accurate description of the long waves produced by oscillating macroscopic charges and currents, but quantum theory must be used in describing the processes of emission and absorption by molecules, atoms, and nuclei involved in waves of visible and shorter wavelength.

The emission of radiation from an oscillating circuit can be visualized by considering Fig. 4, which gives a schematic diagram of an oscillating-dipole antenna. This type of antenna consists of a straight conductor with a source of high-frequency alternating EMF, usually part of a vacuum-tube oscillator, at its center. The two halves of the straight conductor

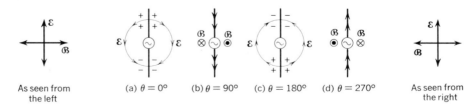

Fig. 4. An oscillating-dipole antenna, showing conditions at successive quarter cycles.

can be considered both as the 'plates' of a capacitor that stores charge and sets up an electric field, and as forming an inductor that carries current and sets up a magnetic field.

The electric and magnetic fields around this antenna are continually changing. When the upper half of the dipole has its *maximum* positive charge, as in Fig. 4(a), electric lines pass from the upper half to the lower half, as indicated schematically by the two $\mathcal{E}$-lines shown; the current is zero, hence there are no $\mathcal{B}$-lines. One-quarter cycle later, there are no $\mathcal{E}$-lines, but since the current is directed downward, magnetic lines form circles about the conductor in the manner indicated schematically by the directions of $\mathcal{B}$ shown in Fig. 4(b). Parts (c) and (d) show the conditions at the two succeeding quarter-cycles. Thus, an observer looking at the oscillating dipole in Fig. 4 from either edge of the paper would observe the electric lines as lines *parallel to the dipole* and directed either upward or downward, and the magnetic lines as lines *perpendicular to the dipole*.

The schematic pictures of the magnetic and electric fields that we have shown in Fig. 4 are only partially correct, since we have shown the electric fields for *static charge distributions* and the magnetic fields for *steady currents*. Actually, *the charge distributions and the currents are both changing*. We have neglected the electric field that results from a changing magnetic field (electromagnetic induction), and the corresponding magnetic field that will be shown in Chapter 43 to accompany a changing electric field. If the dipole oscillated extremely slowly, the simple pictures we have shown would apply fairly well. The electric lines would slowly appear while the ends of the antenna were being charged and would slowly disappear while the charges were being neutralized. Similarly, the magnetic lines would gradually appear while the current was increasing and would slowly disappear while the current was dropping to zero. Actually, the static computations are not accurate for a rapidly oscillating dipole since electric and magnetic fields do not appear instantly at a distant point, but arrive there only after transmission from their source with the speed of light. Similarly, the fields do not disappear instantly when the source disappears, but only after a time governed by the speed of light. Maxwell's theory shows that, particularly at very high frequencies, the electric and magnetic lines do not disappear completely, but some of the lines appear to become 'disengaged' and continue to move away from the dipole, with the speed of light, as *electromagnetic radiation*. For given charge and current distributions, the power so radiated varies as the *fourth power* of the frequency.

The parts of the electric and magnetic fields that do not disappear when the charges or currents decrease to zero, but move outward with the speed of light, still have the same perpendicular relation between the directions of $\mathcal{E}$ and $\mathcal{B}$ as in Fig. 4, but the oscillations of $\mathcal{E}$ and $\mathcal{B}$ are *in phase,* rather than being 90° out of phase as would be concluded from

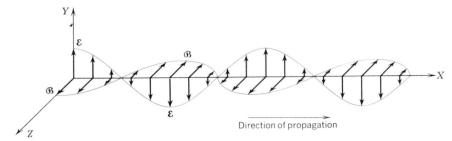

Fig. 5. A plane-polarized electromagnetic wave propagated to the right at a large distance from the antenna of Fig. 4 or Fig. 6. The plane of polarization is, by definition, the XY-plane that contains the $\mathcal{E}$-vector.

the static picture. Thus, an electromagnetic wave consists of a sinusoidal $\mathcal{E}$-wave always accompanied by a similar $\mathcal{B}$-wave at right angles, both $\mathcal{E}$ and $\mathcal{B}$ being perpendicular to the direction of propagation—the wave is *transverse*. Furthermore, as we shall show in Chapter 43, there is a constant ratio between the amplitudes of the $\mathcal{E}$- and $\mathcal{B}$-waves, and for a wave traveling to the right, as in Fig. 5, $\mathcal{B}$ will always be in the $+Z$-direction when $\mathcal{E}$ is in the $+Y$-direction, and in the $-Z$-direction when $\mathcal{E}$ is in the $-Y$-direction. These facts, as well as the equation for the rate of energy transmission, which we shall show to be proportional to the *square* of the $\mathcal{E}$ or $\mathcal{B}$ amplitude, are all given by Maxwell's theory and accurately confirmed by direct observation of radio waves.

Receivers (absorbers) respond variously to the $\mathcal{E}$ or $\mathcal{B}$ part of the wave. Figure 6 indicates the appropriate orientations of the two familiar types of radio-receiver antennas.

In the case of charges or currents oscillating with frequency f and periodic time $T=1/f$, Maxwell's theory shows that *the static relations for*

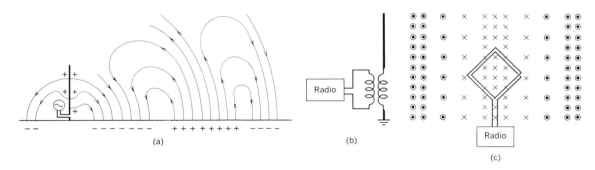

Fig. 6. (a) A 'half-dipole' transmitting antenna, showing the electric lines of the radiation field schematically. (b) A vertical receiving antenna, which responds to the electric component. (c) A 'loop' antenna, which responds to the rate of change of magnetic flux as the magnetic lines indicated by crosses and dots move past it in their progress from the transmitter toward the right.

the intensities $\mathcal{E}$ and $\mathcal{B}$ are valid out to distances from the apparatus such that the time of transmission at the speed of light is small compared with the periodic time T. Since the time of transmission out to distance d is d/c, where c is the speed of light, we see that this condition on distance can be written as $d/c \ll T$, $d \ll cT = c/f$, or $d \ll \lambda$, where λ is the wavelength corresponding to frequency f. For $f = 60$ Hz, the wavelength $\lambda = (3 \times 10^8 \text{ m/s})/(60/\text{s}) = 5 \times 10^6 \text{ m} = 5000$ km. Hence the static relations are applicable at distances small compared with 5000 km from the apparatus, and we could without hesitation use the static relations in our discussion of alternating currents in 60-Hz power circuits.

Again, *the loss of energy through radiation is negligible so long as the dimensions of the apparatus are small compared with the wavelength* λ. Since all the ordinary types of alternating-current apparatus we have considered (including transmission lines) are small in size compared with 5000 km, we can ignore loss of energy by radiation. There is significant radiation only from apparatus comparable in size to the wavelength λ. For radiation of ordinary radio waves of about 300-m wavelength, the transmitting antenna must be of comparable size. At FM and television wavelengths, much shorter antennas suffice; at microwavelengths of a few centimeters, antennas are only a few centimeters long; while at the wavelengths of visible light, single atoms and molecules can radiate (*see* Fig. 12, p. 712, for the wavelengths involved in these examples).

4. Microwaves

It is possible to transmit electromagnetic waves through hollow metal pipes called *waveguides*. A common form of waveguide has a rectangular cross section as shown in Fig. 7. The top view of the waveguide shows the magnetic lines forming closed loops; the side view shows the electric lines, which begin at one wall and end at the opposite wall. $\mathcal{E}$ has no tangential component at a wall of the tube. At places inside the waveguide where they are both perpendicular to the walls, $\mathcal{E}$ and $\mathcal{B}$ are in phase and have the same relation as that shown in Fig. 5 for an electromagnetic field in free space. Both fields are traveling to the right with the speed of light.

The transmission of electromagnetic waves through a waveguide is in some ways analogous to the transmission of sound waves through a tube. Just as standing sound waves can be set up in an

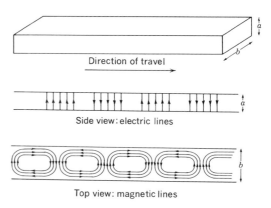

Fig. 7. Rectangular waveguide.

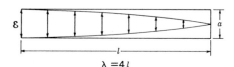

Fig. 8. Standing waves in a waveguide; the length of the arrows is proportional to the amplitude of $\mathcal{E}$ at various points in the waveguide.

organ pipe, so also standing electromagnetic waves can be established in a waveguide.* By making use of the boundary condition that $\mathcal{E}$ has no component tangent to the walls of the guide, we can produce reflection of waves by closing one end of a waveguide by a metal plate as shown in Fig. 8. Just as the closed organ pipe in Fig. 5, p. 447, is resonant for sound waves of length $\lambda = 4l$, our closed or 'shorted' waveguide is resonant to electromagnetic waves of wavelength $\lambda = 4l$. By proper excitation, it is possible to make a waveguide like that in Fig. 8 the *source* of electromagnetic waves just as proper excitation of an organ pipe makes the organ pipe a source of sound waves. One method of excitation is used in the magnetron (*see* Prob. 19, p. 601); by causing bunches of electrons to pass the open end of the waveguide at the proper frequency we can set up standing electromagnetic waves.

The wavelength and hence the frequency of the radiation from a resonant cavity is determined by the dimensions of the cavity; for example, by the *length* of the rectangular cavity in Fig. 8. Practical oscillators have been constructed which produce radiation with wavelengths of a few centimeters; such radiation is called *microwave* radiation. In the klystron, excitation is produced by allowing a beam of electrons to pass through a resonant cavity parallel to $\mathcal{E}$ at a place where there is an antinode for $\mathcal{E}$ in the standing wave pattern of the cavity; the electrons will be 'bunched' by the $\mathcal{E}$ field as the beam passes through the cavity. After passing through the cavity the electron beam approaches a negative 'reflector' electrode, which causes the beam to be reflected back through the cavity; if the bunches of electrons in the reflected beam reach the antinode at the proper phase of $\mathcal{E}$, energy is transferred from the electron beam to the electromagnetic field of the cavity so that electromagnetic oscillations increase in amplitude.

In both magnetron and klystron, the electron beam supplies energy for the production of microwaves; in an analogous fashion a jet of air supplies energy for the production of acoustical waves in an organ pipe (*see* Fig. 6, p. 448). Because of the small wavelengths of microwaves, the techniques used for their transmission and reception more closely resemble optical techniques than commercial radio techniques; thus the parabolic reflectors used as microwave antennas at airport radar systems resemble the mirrors of searchlights more closely than they do the radio antennas of commercial broadcasting stations.

---

* As in the case of our treatment of sound waves in pipes, we shall neglect differences in wave speeds in pipes and in space and shall ignore "end corrections" (*see* p. 448).

PROBLEMS

1. In Fig. 1, if $L=5$ mH, $C=14.08$ pF, $R=0$, the capacitor plates have an initial difference of potential of 1 V, and the switch is closed at $t=0$, determine (a) the frequency of oscillation, (b) the maximum current, and (c) the maximum value of the energy associated with the inductor.

Ans: (a) 600 kHz, (b) 53.2 μA, (c) 7.04 pJ.

2. How can the circuit described in Prob. 1 be modified so as to double the frequency of oscillation (a) by changing the capacitance? (b) by changing the inductance? Assuming an initial voltage of 1 V between the capacitor plates, find the maximum current and the total energy in each of these modified circuits.

3. Show that the total energy of an oscillating circuit is constant provided there is no resistance in the circuit and provided there is no radiation of electromagnetic waves from the circuit. Show that the energy is handed back and forth from capacitor to inductor as oscillations go on.

4. Because of the similarity of the energy relations treated in Prob. 3 and the energy relations encountered in simple harmonic motion, it is possible to set up mechanical analogs of oscillating electric circuits. In such a mechanical analog, *mechanical* quantities are to be associated with *charge, current, inductance,* and *capacitance*. Discuss the analogous quantities carefully.

5. In the example on p. 687, determine the maximum resistance that can be in the circuit if the capacitor discharge is still to be oscillatory. For the case $R=1000$ Ω, determine the frequency of oscillation.

Ans: 107 kΩ; 600 kHz.

6. Determine the maximum value of the resistance R for which oscillations will occur in a circuit with a natural frequency $f=600$ kHz if we employ the following inductors: 100 mH; 10 mH; 1 mH; 100 μH.

7. What is the minimum value of the resistance that can be introduced into the circuit of Prob. 1, that will prevent oscillations from occuring? Ans: 37 800 Ω.

8. With a capacitance of 1.33 pF, what is the inductance in an oscillating circuit that radiates a wavelength of 6.5 m?

9. The magnetic intensity between the pole faces of a cyclotron magnet is 1.5 T. What is the electric intensity between the plates of a vacuum capacitor if the energy per unit volume between the plates is the same as the magnetic energy per unit volume between the pole faces in the cyclotron?

PART V

LIGHT

THE WAVE NATURE OF LIGHT 34

As we have noted earlier in our study of heat, all objects are continually emitting and absorbing radiant energy. The thermal radiation from an object only slightly hotter than the human body is invisible but can be detected by the cutaneous senses. However, if the temperature of an object is above about 525° C, some of the emitted radiation is *visible*. In the restricted sense of the word,

> **Light** is electromagnetic radiation that is capable of affecting the retina of the human eye.

Although visible light is not different in nature from other types of radiation emitted by hot bodies, it is of special significance because it provides one of the most important links involved in our sensory perception of the physical world. The nature of light has therefore long been a subject of great interest.

Newton and others attempted to explain optical phenomena in terms of a theory that pictured light as streams of minute particles emitted by luminous objects. This theory accounted satisfactorily for the straight-line propagation of light in an isotropic medium and for the phenomenon of *reflection*—both regular reflection at smooth mirror surfaces and diffuse reflection at surfaces like the pages of a book. The particle theory was also successful in accounting for *refraction*, the change in direction of light when it passes from one medium to another—for example, from air into water—but led to an incorrect prediction of the ratio of the speeds of light in the two media.

An alternative theory was the wave theory, first elaborated by CHRISTIAAN HUYGENS (1629-1695). The wave theory was able to account satisfactorily not only for rectilinear propagation, reflection, and refraction, but also for other phenomena such as *diffraction, interference,* and *polarization,* which were not readily handled by the particle theory. *Diffraction* is responsible for the departure from strictly rectilinear propagation that is experienced by light that has passed through a very narrow slit. The colors observed in thin layers of oil on water are caused by *interference* between the light reflected from the upper surface and that reflected from the lower surface of the oil film. Light passing through a polarizing film—such as that used in certain types of sunglasses—becomes *polarized;* the effects of *polarization* are such that light transmitted through one polarizing film can be either transmitted or absorbed by a second polarizing film, depending upon the relative orientation of the two films.

Although an anachronism is involved, we can make use of our earlier treatment of radio waves in order to visualize light waves. All electromagnetic waves are transverse waves like the wave in Fig. 5, p. 692; the wave 'displacements' are actually the sinusoidally varying electric intensity $\mathcal{E}$ and magnetic intensity $\mathcal{B}$, which are in phase and mutually perpendicular. Since the electric field $\mathcal{E}$ of the wave usually interacts more strongly with matter than does the magnetic field $\mathcal{B}$, we can, for simplicity, visualize the light waves we shall be discussing as sinusoidally varying $\mathcal{E}$ waves of amplitude $\mathcal{E}_M$. The energy associated with any sinusoidal wave is proportional to the square of the wave amplitude; the energy associated with a light wave is, therefore, proportional to $\mathcal{E}_M^2$. We can give a more meaningful discussion of electromagnetic waves (Chapter 43) after we have become more familiar with the optical phenomena involved.

In the present chapter we shall first consider the phenomenon of interference and show how the *wavelength* of light can be measured. Then we shall describe the methods by which the *speed* of light, *c*, can be determined. As the speed of light in a vacuum represents the maximum speed possible in the universe, we shall discuss its measurement in considerable detail. It should be noted that electromagnetic radiation of all wavelengths is transmitted through free space at the same speed *c*; radio waves have much longer wavelengths than those of visible light, while the wavelengths associated with X rays are much shorter. In view of the common mode of propagation for all wavelengths, the term 'light' is sometimes used in a more general sense than that indicated in the definition given above; it is used to designate all forms of radiation. As in the case of sound, the frequency of light can be obtained from the relation $c=f\lambda$, which holds for any simple sinusoidal traveling wave.

In the chapters immediately following, the phenomena of diffraction, reflection, refraction, and polarization will be treated in terms of light waves. The original Huygens wave theory was refined by THOMAS YOUNG (1773–1829) and by AUGUSTIN JEAN FRESNEL (1788–1827) and formed a background for the classical electromagnetic wave theory of light formulated by JAMES CLERK MAXWELL (1831–1879). Maxwell's theory (Chapter 43) demonstrates conclusively that a light wave is *electromagnetic* in character and describes accurately all aspects of the *transmission* of light. However, it is found that the processes of *emission* and *absorption* cannot be adequately treated in terms of classical electromagnetic waves but must be described in terms of *quanta* or *photons*, entities having certain particle-like characteristics; quantum phenomena will be discussed in Chapter 41. Our treatment of diffraction occurs in two places: Fresnel diffraction, which involves no mirrors or lenses, in Chapter 35; Fraunhofer diffraction in Chapter 39 after we have discussed the necessary properties of mirrors and lenses.

1. The Propagation of Light

That light travels in straight lines in a uniform medium is a fact familiar to everyone. Rectilinear propagation of light has long been utilized in practical ways such as in making land surveys. We rely on rectilinear propagation nearly instinctively in many respects; for example, when, on hearing the characteristic sound of a jet aircraft coming from some direction, we look about the sky to see where the plane 'really is,' we are placing reliance on rapid rectilinear propagation of light to supply the information.

If we observe direct sunlight passing through a window, as in Fig. 1, the bright spot on the floor is fairly 'sharp,' with well-defined boundaries. The sun approximates a point source at an infinite distance and gives a parallel *beam* of light, of constant cross section, between the window and the floor. The slight fuzziness observed at the edges of the beam, on the floor, arises from the fact that the sun is not *accurately* a point source as observed from the earth; a star would be. It is convenient in discussing the propagation of light to speak of light *rays;* a light ray is simply a directed line giving the direction of propagation and thus represents the path along which light travels. Since the rays associated with the beam in Fig. 1 neither converge nor diverge, the beam shown in the figure is called a *parallel* beam; the cross-sectional area of a parallel beam is constant.

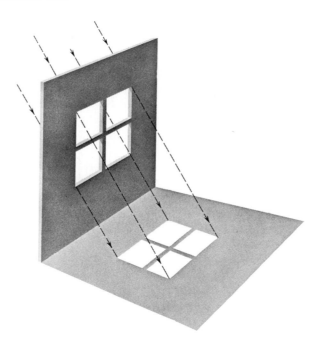

Fig. 1. Rectilinear propagation of sunlight passing through a window, showing a number of light rays.

Light from most familiar light sources spreads as it travels outward and forms *divergent* beams as shown in Fig. 2. If a light source is sufficiently small, it can be considered as a point source S; when light from a point source strikes an opaque object O, a well-defined shadow AA with a sharp boundary is formed. Provided it is sufficiently distant, even an extended source such as a street lamp can be regarded as a point source. In fact, at the earth's surface the sun itself can with fair approximation be treated as a point source, as we have done in Fig. 1.

Most sources such as the frosted lamp bulb in Fig. 3 are too large to be regarded as point sources when the distances to opaque bodies and screens are not large. As indicated in this figure, the part of the screen in the circle AA is in complete shadow; this region is called the *umbra*. The region between circles AA and BB is in partial shadow; this region is called the *penumbra*. It will be noted that the defining rays in Fig. 3 are drawn from two 'point sources' at opposite sides of the lamp. It is quite proper to regard an extended source like the lamp as a collection of point sources.

Example. *The diameters of the sun, earth, and moon are* 864 000 mi, 8000 mi, *and* 2160 mi. *The sun is* 93 000 000 mi *from the earth, while the moon is* 236 000 mi *from the earth. Show that when the earth is on the line between the sun and the moon, the diameter of the umbra of the earth's shadow is*

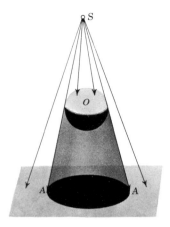 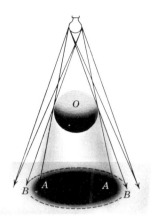

Fig. 2. Shadow of an opaque object illuminated by a point source.

Fig. 3. Shadow of an opaque object illuminated by an extended source.

more than sufficient to completely envelop the moon, so that total eclipses of the moon are possible.

The geometrical situation is shown (out of proportion) in Fig. 4, with the umbra shaded. The length l of the umbra is given by solving the equation $(93\,000 + l)/864 = l/8$, to obtain $l = 868$ (all numbers represent thousands of miles). The moon is at a distance $868 - 236 = 632$ thousand miles from the apex of the cone. The diameter d of the umbra at this distance is computed from

$$d/632 = 8/868; \qquad d = 5.82,$$

which is over twice the diameter, 2.16 thousand miles, of the moon.

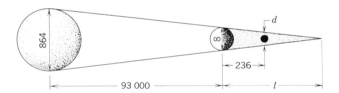

Fig. 4. Total eclipse of the moon (distances in thousands of miles).

Since the light consists of traveling waves, the rays shown in Figs. 1, 2, and 3 represent the normals to the wavefronts. Thus, in Fig. 1 the parallel rays indicate that plane wavefronts are involved, while in Fig. 2 the rays extending radially outward from the point source S indicate that spherical wavefronts are spreading outward from the source. At great distances from a source, the curvature of the spherical wavefronts is so small that the wavefronts are essentially plane, as in the case shown in Fig. 1.

Actually, light is *not* propagated *strictly* rectilinearly—the edge of a shadow from a point source is *not* absolutely sharp. This fact was discovered about 1650 by Francesco Maria Grimaldi, who named the phenomenon *diffraction* (breaking-up). Diffraction was carefully investigated by Newton, who rejected Huygens's wave theory primarily because the sound waves and water waves with which he was familiar bend around obstacles to a much greater degree than light does. The difference in behavior actually arises from the extreme shortness of the wavelength of light as compared with the wavelengths of sound or water waves. The first correct physical explanation of diffraction was not given until 1818, by Fresnel, who went back to the Huygens model. Only after this successful explanation did the wave theory of light gain general acceptance.

2. Interference Resulting from Thin Films

Newton himself was among the first to study a phenomenon that was later satisfactorily explained in terms of *interference* of light waves. He placed a glass lens with a convex spherical surface in contact with a plane piece of glass, illuminated the glass with sunlight, and viewed the reflected light as indicated in Fig. 5(a). In the region in the vicinity of the point of contact between the lens and plate, there appears a series of colored rings, alternately light and dark; these rings are now known as *Newton's rings*. If the arrangement is illuminated with yellow light from a sodium arc light, the rings are alternately yellow and black as shown in the photograph of Fig. 5(b), with a black region near the central point of contact between the convex lens surface and the plane surface of the glass. Newton recognized that these rings indicated the existence of some kind of periodicity in the air film between the lens and plate and attempted to account for this in terms of "fits" of easy transmission and of easy reflection of the light *particles* traversing the air film.

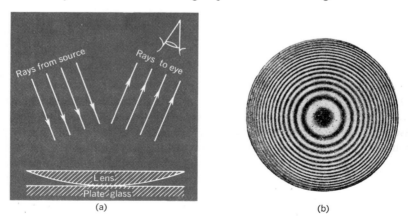

Fig. 5. Newton's rings. (Photograph courtesy of Jemima B. Dutcher.)

Newton's rings can be described elegantly and very simply in terms of interference of light *waves.*

We recall from our treatment of the interference of sound waves (pp. 424-427) that in order to produce a stationary interference pattern it is necessary to employ wave trains having *coherence;* thus, stationary interference patterns like those corresponding to Newton's rings must arise from two wave trains with a constant phase relationship as well as identical wavelengths. In the case of sound waves, it is relatively simple to obtain coherence from two sources such as the loudspeakers in Fig. 9, p. 425. The situation is quite different in the case of light waves; two sodium arcs can be used to produce independent wave trains with identical wavelengths, but there is no constant phase relationship between the wave trains since each is made up of short bits and pieces arising from individual atoms; hence the two light sources are *incoherent.* Although coherent light sources, called *lasers,* have been developed in recent years, it is not really necessary to employ *two* sources in order to produce interference of light waves. We can use a single light source and a bit of trickery; in fact, in this particular bit of trickery, we quite literally 'do it with mirrors'!

Consider, for example, Fig. 6, which shows a thin layer of air between two much thicker glass plates, which have plane surfaces AA and BB parallel to each other and separated by an air film of thickness t. Rays 1 and 2, incident on the upper glass plate, originate from *the same point* of a distant source. Some of the incident light is reflected

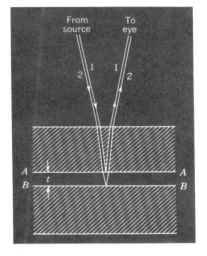

Fig. 6. Interference arising from a thin film of air between plane glass plates.

from glass surface AA; ray 1 represents one of the normals of the set of wavefronts reflected at AA. The portion of the incident light not reflected at AA passes onward through the thin air film to surface BB, at which some additional light is reflected and passes backward through the thin air film and into the upper plate; ray 2 represents one of the normals to the wavefronts reflected at surface BB. Rays 1 and 2 therefore represent normals to wavefronts originating at a single common source traversing different paths; waves of this type are coherent and can therefore give rise to interference effects when rays 1 and 2 are focused at the same point on the retina of the eye or on a photographic plate.

Whether the interference is constructive or destructive depends on the difference in the paths traversed by the two wave trains. In the case of normal, or near-normal, incidence, the only difference in the path lengths traversed by the two sets of waves is the length of the path traversed in the air film by the waves reflected at BB. In the case of normal incidence, it is easy to compute the path difference because there is no change in the direction of light beams passing from air into glass and *vice versa*. For normal incidence, the path difference between the wave train reflected at BB and that reflected at AA is just twice the thickness of the air film; thus,

$$\text{path difference} = 2\,t. \qquad \text{(normal incidence)}$$

From our earlier discussion of interference of sound waves, we might now be prepared to assert that constructive interference between our two wave trains is to be expected whenever the path difference $2\,t$ is equal to an even number of half wavelengths and that destructive interference is to be expected when $2\,t$ is equal to an odd number of half wavelengths. These are expected to be the conditions for the two sets of reflected waves to be exactly 'in phase' and exactly 'out of phase,' respectively.

However, a glance at the photograph in Fig. 5(b) is sufficient to convince us that our hasty conclusion must be wrong. At the very center of the Newton's ring pattern the lens and the plate are actually in contact; this means that the thickness of the air film is actually zero and hence the path difference is also zero. On the basis of our original conclusion, we would expect constructive interference between light reflected at the glass surfaces corresponding to AA and BB at the point of contact; constructive interference means that the point of contact between the lens and the plate should appear bright. Unquestionably, the point of contact in Fig. 5(b) is very black, so if the wave theory of light is valid, we must have neglected something.

Where we have gone wrong is in our tacit assumption that any phase changes occurring during reflection at AA are the same as those occurring during reflection at BB in Fig. 6. Actually, this is not the case; light waves reflected at AA are waves in a 'dense medium' (glass) reflected at the surface of a 'rare medium' (air), while the reverse situation obtains when light waves are reflected at BB. In view of the experimental results shown in Fig. 5(b), we must conclude that there is a phase reversal occurring for reflected waves at one of the surfaces relative to waves reflected at the other surface.

This phase reversal of one of the reflected waves puts the waves out of phase when $2\,t$ is very close to zero, in phase when $2\,t = \frac{1}{2}\lambda$, out of phase again when $2\,t = \lambda$, etc. Thus if we write

$$2\,t = n\,\lambda, \qquad (1)$$

we get *destructive* interference for

$$n = 0, 1, 2, \cdots, \qquad \text{(destructive)}$$

and *constructive* interference for

$$n = \tfrac{1}{2}, \tfrac{3}{2}, \tfrac{5}{2}, \cdots. \qquad \text{(constructive)}$$

A photograph of interference produced in a thin air film between an optically flat glass surface and two gauge blocks is shown in Fig. 7. The parallel equally spaced fringes indicate that the block at the left is plane, while the interference pattern at the right shows that the second gauge block is worn and no longer plane.

It might be remarked that, although we appear to have introduced the phase change on reflection in an *ad hoc* fashion, the introduction can be justified from general principles on which Maxwell's electromagnetic theory is based; meanwhile, the reader can possibly feel somewhat safer regarding the introduction of phase change considerations if he reviews the discussion of reflection of sound waves in closed and open organ pipes, and the reflection of waves at fixed and free ends of a string, where analogous types of phase changes occur.

Returning now to Newton's rings, let us see how the wavelength of light can be determined from the dimensions of the rings and the radius of curvature of the lens surface. The air-film thickness t at a horizontal distance r from the point of lens, plate contact in Fig. 5(a) can be obtained from the so-called *sagitta relation*, which is a relation that approximates a circle by a parabola in the immediate neighborhood of a point O on the circle, as in Fig. 8. Let point P on the circle have rectangular coordinates X_P, Y_P, where the origin of coordinates is the point O, and the X-axis is drawn from O through the center C of the circle. From the Pythagoras theorem, we have

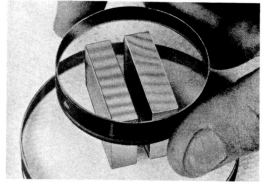

Fig. 7. Interference by thin air films between an optical flat and steel gauge blocks. The optical flat, held by the fingers, is slightly tilted relative to the surface of the blocks, so that there is a thin 'wedge' of air between the flat and the perfect block on the left. (Photograph courtesy of the Do ALL Company.)

$$R^2 = Y_P^2 + (R - X_P)^2 = Y_P^2 + R^2 - 2X_P R + X_P^2,$$

which can be written in the form

$$X_P(2R - X_P) = Y_P^2.$$

If X_P is negligible in comparison with $2R$, we can drop X_P in the parenthesis, to obtain

$$X_P = Y_P^2/2R. \qquad (X_P \ll 2R) \quad (2)$$

This relation, which is the equation of a parabola, is called the *sagitta relation*, or 'arrow' relation, because Fig. 8 is considered to have some resemblance to a bow and arrow.

According to this geometric relation, the film thickness $t = r^2/2R$ at a horizontal distance r from the point of contact in Fig. 5, provided $r \ll R$, where

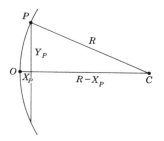

Fig. 8. The sagitta relation approximates the circle by a parabola in the neighborhood of point O.

R is the radius of curvature of the convex surface of the lens. To determine the radius r_n of the nth dark ring *surrounding* the central dark spot we merely substitute $2t = r_n^2/R$ from the sagitta relation in the condition (1) for destructive interference when n is an integer:

$$r_n^2/R = n\lambda; \qquad r_n = \sqrt{n\lambda R}. \tag{3}$$

Since R and r_n can be measured, this relation can be used for the determination of λ.

Although more precise values of λ can be determined by other methods to be described later in this chapter, it seems desirable to discuss the magnitudes of light wavelengths at this point. The wavelength of the yellow light from a sodium arc turns out to be 589 nm, which is about $\frac{1}{2000}$ of a millimeter. When the arrangement in Fig. 5(b) is illuminated with white light, a set of colored rings is observed around the central dark region because white light contains all spectral colors; the wavelengths range from approximately 400 nm for violet light to about 700 nm for red light. The rings observed with white light 'fade out' very rapidly as one proceeds outward from the center of Newton's rings; this rapid fade-out results from overlapping of rings corresponding to various colors or wavelengths. Such a fade-out or blurring of the rings is not observed when a sodium arc is used as a source, because the yellow light emitted from such a source is nearly 'monochromatic' and is thus characterized by a single wavelength* rather than a wide range of wavelengths as in the case of white light.

Thus far we have confined our attention to interference of light resulting from thin films of air. Later we shall consider quantitatively the interference in thin films of other materials such as mica, oil, soap solutions, and metallic oxides. However, one set of experiments involving films other than air deserves qualitative discussion at this point, since certain questions may have been raised in the reader's mind when we spoke of reflection of sound waves at the ends of organ pipes. Acoustical standing waves can be produced in organ pipes; can optical standing waves be produced in thin films?

This question was answered in the affirmative by a clever experiment first performed by WIENER in 1890 and somewhat later by IVES. The experiment utilized photographic techniques, which were of course unknown in the days of Newton and Huygens. A layer of extremely fine-grained photographic emulsion was deposited in the usual manner on a glass plate. The emulsion side of the plate was placed in contact with metallic mercury, and monochromatic light was directed normally upon the glass side of the photographic plate as shown in Fig. 9(a). The light passed through the glass plate into the emulsion; the almost com-

---

*Actually, sodium light consists of two wavelengths, both differing only slightly from 589 nm. Only a single 'average' wavelength would be obtainable by the measurements of Newton's rings.

plete reflection at the surface of the mercury resulted in wave trains of equal amplitude traveling in opposite directions and therefore produced standing waves *within* the emulsion between the mercury and the glass plate as shown in Fig. 9(b). The nodes were formed in planes $\tfrac{1}{2}\lambda_E$ apart, where λ_E is the wavelength of light in the emulsion, with antinodes located between the nodal planes. Since energy, which is proportional to the square of the wave amplitude, is concentrated near the antinodes and is absent from the nodes, the silver salt in the photographic emulsion was reduced in the layers where the antinodes occurred but not in the layers where the nodes were located. Accordingly, when the photographic plate was developed, the emulsion became blackened in the antinodal layers but not in the intervening layers. This alternation was observed by slicing the film at a small angle α to the nodal planes and then examining the film with a microscope. This observation of the optical standing-wave pattern offers very convincing proof of the wave nature of light.

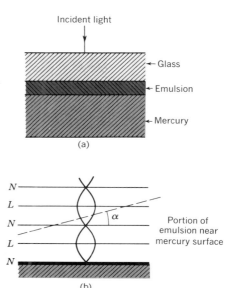

Fig. 9. Optical standing waves in a photographic emulsion (schematic diagram).

3. The Michelson Interferometer

One of the most practically useful methods of producing interference is used in the Michelson interferometer, shown diagrammatically in Fig. 10. The method consists of dividing a beam of nearly parallel rays of light by partial reflection into two beams, sending the two beams over different routes, and finally causing them to recombine and produce interference. Parallel monochromatic light represented by ray S is incident at an angle of 45° on a plane 'half-silvered' mirror M. The silver coating is so thin that half of the light is transmitted as ray 1; the rest of the light is reflected and is shown as ray 2. Ray 1 reaches mirror M_1 and ray 2 reaches mirror M_2. In both cases incidence is normal, and hence the direction of each ray is reversed as indicated in the figure; for the sake of clearness, the reflected rays shown in Fig. 10 have been displaced laterally. On their return journey the two rays reach half-silvered mirror M again and approximately half of the light represented by each ray reaches the observer; the remaining light shown by the broken ray returns to the source. Glass plate N, identical to M but unsilvered, makes the number of extraneous reflections at glass surfaces and the total path in glass identical for rays 1 and 2; hence these rays reach the observer with equal amplitudes.

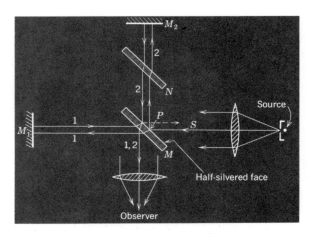

Fig. 10. The Michelson interferometer. Plates M and N must be flat and of uniform thickness to within a small fraction of the wavelength of light, and the front surfaces of mirrors M_1 and M_2 must be flat to this same accuracy.

If the distances of M_1 and M_2 from point P are equal, rays 1 and 2 traverse identical paths; the glass plate N makes the path in glass the same for both rays. Therefore, the light waves associated with rays 1 and 2 emerge in phase and reinforce each other. Reinforcement also occurs when the paths traversed by rays 1 and 2 differ by a whole number of wavelengths. Interference changes from constructive to destructive when one of the mirrors is moved a distance equal to a quarter wavelength, since this movement changes the path by a half wavelength. It follows that the Michelson interferometer can be used for the absolute measurement of wavelength if the mirror M_2 is moved in the direction of ray 2 through a distance that can be measured by a micrometer screw, and if the observed alternations of light and darkness in the field of view are counted while the mirror is being moved.

As the apparatus is actually employed, the mirrors are not aligned with perfect accuracy. Very slight tilts from the perfect alignment discussed above are desirable so that the field is crossed by parallel fringes, which move in a direction normal to their length as mirror M_2 is moved. 'Counting the fringes' that move past the cross hairs of an eyepiece will then give the *number of half wavelengths* that the mirror M_2 is moved. These fringes have an appearance similar to those at the left in Fig. 7, and originate from the slight tilts of the mirrors in essentially the same manner.

It is the Michelson interferometer that is used to calibrate scales in terms of the fundamental definition of the meter, as given on p. 14. The meter is *defined* as 1 650 763.73 $\lambda_{Kr}{}^{86}$, where the last symbol is the wavelength of the orange light emitted by the rare gas nuclide Kr^{86}, measured in vacuum. The light is produced in an electric-discharge tube similar to the familiar tubes that give red light when filled with neon and blue light when filled with argon. A pure nuclide is used because the different isotopes of a naturally occurring element emit light with wavelengths that differ very slightly; the precision attainable in interferometry is enhanced if the light used is accurately monochromatic.

If, then, one of the mirrors of a Michelson interferometer employing krypton light is moved steadily by a screw while 1 650 763.73 fringes are 'counted,' it is known, by definition, that the mirror has moved exactly $\frac{1}{2}$ meter. The varying illumination at a slit in the field of view can be recorded on a moving film as the mirror moves, and it is on this film that the fringes are counted and on which it is feasible to measure to an accuracy of better than 0.01 'fringe spacing.' Such careful measurements, usually conducted at national standardizing laboratories, can be used to calibrate secondary standards such as metal scales and gauge blocks.

The Michelson interferometer provides a highly accurate means of measuring wavelengths. It has been used to measure the wavelengths of light throughout the entire 'visible spectrum'—the range of wavelengths to which the eye is sensitive. As mentioned earlier, these wavelengths range from approximately 400 nm for violet light to approximately 700 nm for red light. The dispersion of these 'spectral colors,' which are all present in sunlight, is familiar to everyone who has seen a rainbow; in the order of increasing wavelength the spectral colors are violet, blue, green, yellow, orange, and red, including, of course, all the gradations in between. Newton was the first to make detailed studies of spectra, as dispersed by a prism; we shall discuss this subject in considerable detail in Chapter 41.

The sensitivity of the eye varies for different wavelengths. Under conditions of moderate or strong illumination the eye is most sensitive to yellow-green light of wavelength 555 nm. The spectral sensitivity curve has a maximum at this wavelength and falls off rapidly for longer and shorter wavelengths, as in Fig. 11, which plots the *reciprocals* of the relative numbers of watts of radiation that must fall on the eye to give equal visual response at different wavelengths. This curve is determined by using the so-called flicker photometer, in which the eye itself is used as the detector to determine equality of visual response to two different colors. In Fig. 11 are also shown typical values of the wavelengths associated with the different colors.

In 1800, the existence of radiation of wavelengths longer than those of red light was discovered by WILLIAM HERSCHEL, who was able to detect this radiation by means of thermometers with blackened bulbs placed at various positions in a dispersed solar spectrum. We now know that the 'thermal radiation' emitted by bodies at temperatures below approximately 525° C consists almost entirely of wavelengths *longer* than those of visible light; this long-wavelength radiation is called *infrared* radiation.

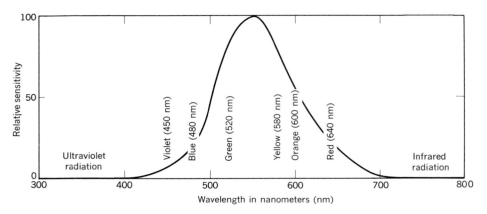

Fig. 11. Relative sensitivity of the human eye to light of various wavelengths.

Radiations of still longer wavelength—*microwaves* and *radio waves*—can be produced and detected by electrical methods.

Shortly after Herschel's discovery of infrared radiation, radiation having wavelengths shorter than those of visible light was discovered by RITTER and WOLLASTON. This radiation can be detected by its high efficiency in promoting certain chemical reactions, notably the reduction of salts of silver in the photographic process, and by the production of *fluorescence* in various materials. Such radiation is called *ultraviolet* radiation. Radiation of still shorter wavelengths—*X rays* and *gamma rays*—was discovered in the closing years of the nineteenth century.

Radiation of all wavelengths—from the shortest gamma rays to the longest radio waves—travels with the same speed in vacuum. The observed wavelengths in this *electromagnetic spectrum* are given in Fig. 12. It is to be noted that visible light—with a spectral range from violet to red—constitutes a very narrow band of wavelengths in the total electromagnetic spectrum. The *frequencies* of radiation in the radio and microwave regions, as listed in Fig. 12, can be measured directly in terms of standard frequencies provided by national standardizing laboratories and broadcast from associated radio stations, such as station WWV associated with the National Bureau of Standards in Washington; these standard frequencies are based on the fundamental time unit. The *frequen-*

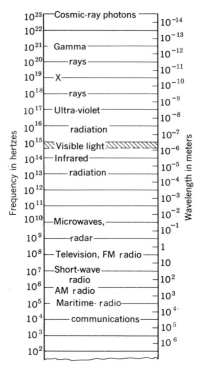

Fig. 12. The electromagnetic spectrum.

cies of radiation having wavelengths shorter than those of the microwaves cannot be measured directly. These higher frequencies listed in Fig. 12, based on careful measurements of *wavelength* λ, are computed from the familiar relation $c=f\lambda$, where c is the speed of light.

4. The Speed of Light

In view of the role in Einstein's relativistic mechanics (*see* Chapter 44) played by the speed of light, c, as a natural upper limit for speed, it is extremely important that this universal physical constant be carefully determined.

The *speed of light* is so great that early attempts at its measurement proved unsuccessful. Galileo conducted an experiment in which two observers stationed in towers some distance apart flashed signals at each other with lanterns. The second observer was supposed to flash a signal as soon as he received a light signal from the first observer. The experiment was inconclusive, and Galileo decided that the speed of light was too great for measurement by this method and that light transmission might indeed be instantaneous. In view of our present knowledge of the speed of light, we recognize that the reaction times of the observers were much greater than the time of transit of the light between the towers, and that Galileo could not correct for these reaction times with sufficient accuracy.

The first successful determination of the speed of light was reported by the Danish astronomer OLAUS ROEMER in 1675, when he announced a calculation of the speed of light from observations of irregularities in the times between successive eclipses of Io, the innermost moon of Jupiter, by that planet. The general argument used by Roemer can be understood by a consideration of Fig. 13, which shows certain relative positions of Jupiter and the earth. For simplicity, we shall consider Jupiter at rest

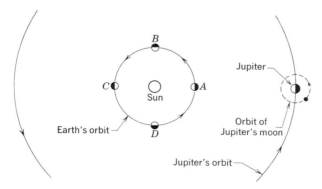

Fig. 13. Roemer's determination of the speed of light (schematic).

in its orbit, since its motion is slow compared with that of the earth and can easily be taken into account in actual calculations. The observed interval between successive eclipses of Jupiter's moon as it passes behind the planet is 42.5 h when the earth is near position A, but the interval between successive eclipses is greater when the earth is near B, and less when the earth is near D. These variations are not due to irregularities in the motion of Jupiter's moon, but are caused by variations in the times required for light to travel from Jupiter to the earth. Because the distance between Jupiter and the earth is increasing when the earth is at B, the time required for light to travel from Jupiter to the earth is increasing, and the measured interval between successive eclipses increases accordingly. A similar line of reasoning accounts for the shorter eclipse intervals noted when the earth is at D. This apparent change in frequency of the eclipses is analogous to the apparent change in frequency of sound in the Doppler effect.

On the basis of his comparatively inaccurate observations, Roemer gave the first value for the velocity of light as 140 000 mi/s; recent measurements of the same eclipse phenomenon give the value 186 000 mi/s. Roemer's conclusions, not credited at the time, were confirmed in 1727 by JAMES BRADLEY, who employed a different astronomical method of determining the speed of light.

The first successful terrestrial measurement was made by A. H. L. FIZEAU in 1849. The principle of his method was the obvious one of sending out a brief flash of light and measuring the time for this light flash to travel to a distant mirror and return to the observer. The equipment used in the measurement is shown in Fig. 14. Light from an arc at S, passing through lens L_1 and reflected by the half-silvered mirror G, is brought to a focus at F. The toothed wheel W rotates at high speed so that it interrupts the light beam passing through the rim at F and produces a series of short flashes. A flash is sent out each time the wheel is in such a position that light can pass through the slot between adjacent teeth. The light beam passing through the slot is rendered parallel by the lens L_2 and, after traversing a large, accurately measured distance, is focused by the lens L_3 on a plane mirror M. The distance D used in Fizeau's experiment was 5.36 mi. After reflection from M, the light retraces its path to L_2 and is again focused on the rim of the wheel by the lens L_2. If during the time required for the light to travel from F to M and back to F the wheel has turned through such an angle that a tooth is interposed at F, this light flash will be intercepted by this tooth.

With the wheel at rest in such a position that light passes through the opening 0, the observer at E will see the image of the source formed by light returning from M. If the wheel is now rotated with increasing speed, a rotational speed will be reached at which the light passing through slot 0 on its outward trip will be stopped by tooth A on its return trip; at the same rotational speed the light transmitted through slot 1

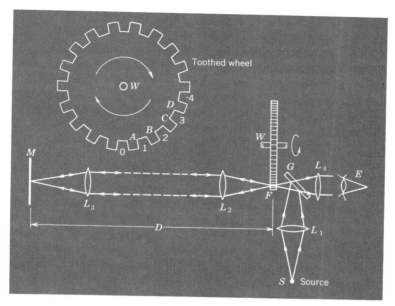

Fig. 14. Schematic diagram of Fizeau's apparatus for the determination of the speed of light.

will be stopped by tooth B on its return trip, and so on. The image will be completely 'eclipsed.' A further increase in speed will cause the image to reappear when these flashes pass through openings 1, 2, $\cdots$; a second eclipse will occur at the speed at which the flashes in question are stopped by teeth B, C, $\cdots$. From the observed rates of rotation of the wheel at which these successive eclipses and reappearances occur, the speed of light can be computed.

In 1850, J. B. L. FOUCAULT devised a second terrestrial method for determining c by observation of the displacement of an image formed by light returning to a rotating mirror after it had traversed a known distance. The distances in Foucault's experiments were relatively short—4 meters in one experiment and 20 meters in another. His work was of importance chiefly in that he measured the speed of light in water as well as in air and found that the speed in water is less than the speed in air, a result which had been predicted by the wave theory of light, but which was in contradiction to Newton's particle theory.

The basic ideas in Fizeau's work were exploited by A. A. MICHELSON (1852–1931) and his associates, who employed an eight-sided rotating mirror in place of the toothed wheel to provide pulses and determine the time for their round-trip passage over a long, accurately measured path of nearly 22 miles; the path was measured by the U.S. Coast and Geodetic Survey in one of the most accurate land surveys ever made. Michelson's results probably cannot be improved by methods involving mechanical devices such as rotating wheels or mirrors. However, recent experiments such as those of Eric Bergstrand (1950) and I. C. C. Mackenzie (1954) have employed electrical methods of creating light flashes and measuring the times of propagation over known paths; these electrical methods give greater precision without the necessity for excessively long path lengths.

Another technique for determination of c is one that involves the determination of frequency f and wavelength λ, separately, and then computing c as the product of these independently measured quantities. It is possible to obtain highly accurate values for both f and λ for microwave radiation. The frequency f is determined in terms of standard radio frequencies, which are linked directly to time standards. The wavelength λ is determined either from the dimensions of standing wave patterns in metallic cavities tuned to the microwaves, as in the experiments of Essen (1950), or from interferometric measurements, like those of K. D. Froome (1954). The value of λ is thus linked directly to length standards.

The microwave interferometer arrangement employed by Froome is shown schematically in Fig. 15. The lengths of the two paths to the receiver are varied by moving the carriage on the lathe bed, just as the lengths of the two paths in Fig. 9, p. 454, are varied by means of the trombone arrangement. Froome used a frequency of about 24 000 MHz, corresponding to a wavelength of about 1 cm. Motion of about $\frac{1}{4}$ cm changes the path difference by about $\frac{1}{2}$ cm and corresponds to a change from constructive interference and maximum receiver signal to destructive interference and zero receiver signal. Froome measured the wavelength accurately by locating the interference maxima and minima as the carriage was moved about one meter along the lathe bed.

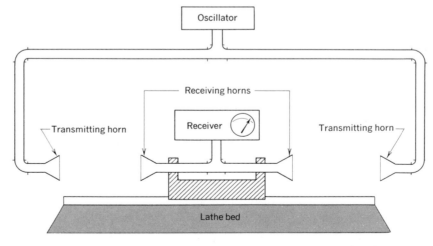

Fig. 15. Froome's interferometer for determination of the wavelength of microwaves.

The presently accepted 'best value' for the speed of light in vacuum is

$$c = 299\ 792\ 500 \pm 300 \text{ m/s}. \tag{4}$$

This value represents a suitably weighted mean of recent values obtained by various methods, largely employing microwaves. We shall discuss later (Chapter 43) the important relation (29), p. 621, between the speed of light and the electric and magnetic constants ε_0 and μ_0—a relation derived from Maxwell's electromagnetic theory of light. This theory provides a means of *computing* the speed of light from laboratory measurements of seemingly quite unrelated electric and magnetic constants. As we have seen on pp. 621–622, the agreement between this computed value of c and the directly measured value (4) is essentially perfect.

5. Index of Refraction

In view of the fact that the speed of light is greater in *vacuum* than in any transparent material medium, it is desirable to refer the 'optical densities' of all media to vacuum by defining an index of refraction:

The **absolute index of refraction** of a medium is the ratio of the speed of light in vacuum to the speed of light in the medium.

If we let c be the speed of light in vacuum, and c_M the speed of light in a medium, the absolute refractive index of the medium is $\mu_M = c/c_M$. The absolute index of refraction is usually called just *the index of refraction* or *the refractive index* of the medium. Note that the indices of refraction of all material media are greater than one because of the way the ratio is defined.

Table 34-1 gives indices of refraction for various gases, liquids, and solids, for yellow sodium light. The speed of light in media other than vacuum varies with wavelength; some of the consequences of this variation will be discussed in Chapter 36. As we see from Table 34-1, the index of refraction of air is very nearly equal to 1; in our discussions of refraction we can usually treat it as unity.

The fact that the speed of light in a material medium is less than that in vacuum implies that the wavelength is also shorter, in the same ratio. Let us consider, in Fig. 16, a light wave that proceeds from vacuum, through a material medium, and again into vacuum. Let the wavelength in vacuum be λ; the frequency is then $f = c/\lambda$. The

Table 34-1 INDICES OF REFRACTION FOR YELLOW LIGHT OF WAVELENGTH 589 nm

Gases at 0° C and 1 atm	
Dry air	1.000 29
Carbon dioxide	1.000 45
Liquids at 20° C	
Benzene	1.501
Carbon disulfide	1.628
Carbon tetrachloride	1.461
Ethyl alcohol	1.362
Water	1.333
Solids at 20° C	
Diamond	2.419
Fluorite	1.434
Glass (typical values)	
Crown	1.517
Commercial plate	1.523
Light flint	1.574
Dense flint	1.656
Quartz (fused)	1.458

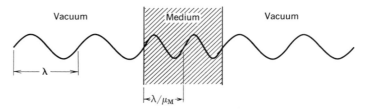

Fig. 16. The wavelength in a material medium is λ/μ_M if λ is the wavelength in vacuum.

frequency (not the wavelength) *is the same in the material medium as in vacuum,* because the frequency is the number of waves passing any point per unit time, and the number of waves passing through the material is the same as the number arriving at its surface—waves cannot pile up and collect anywhere. Hence if f_M is the frequency in the material medium, $f_M = f$. However, the speed in the material medium is less than that in vacuum: $c_M = c/\mu_M$. Hence the wavelength in the material medium is given by

$$\lambda_M = c_M/f_M = c/f\mu_M = \lambda/\mu_M. \tag{5}$$

The wavelength in a material medium is $(1/\mu_M)$ *times the wavelength in vacuum.*

The waves must be shorter in the material because, to get the same number of waves through at reduced speed, the waves must be 'closer together.'

The number of waves in a thickness t of medium is

$$t/\lambda_M = \mu_M\, t/\lambda, \tag{6}$$

which is μ_M times the number that would be contained in thickness t in vacuum because the waves in the material medium are shorter by a factor of μ_M.

We are now in a position to extend the treatment in Sec. 2, of interference arising from thin films, to films other than vacuum (or air) where the refractive index is of importance. The following example discusses the production of 'interference colors' by thin films illuminated by white light:

Example. *A thin layer of oil having refractive index* 1.45 *covers a plate of dense flint glass. If the thickness of the oil layer is* 400 nm, *what is the color of the film if it is illuminated by white light and viewed by reflection at angles near the normal to the surface?*

The problem here is to find which colors in the visible region, having vacuum wavelengths in the range 400 nm to 700 nm, will interfere constructively and which will interfere destructively. However, the situation is somewhat different from that discussed in Sec. 2, since the reflections at the front and back surfaces of the oil film both occur within less dense media; hence, there is *not* a relative phase change introduced by reflection at the surfaces, as was the case with Newton's rings. To determine whether the interference is constructive or destructive we merely need to count the waves contained in the passage through the oil film and back. If this is a whole number of waves, the interference will be constructive; if it is a half-integral number of waves, the interference will be destructive.

If t is the thickness of the oil film, the number of waves contained in the path length $2t$ is, by (6), $n = 2\mu_M t/\lambda$, where λ is the vacuum wavelength. Solving for λ, we obtain $\lambda = 2\mu_M t/n$. *Constructive* interference occurs for those wavelengths for which $n = 1, 2, 3, \cdots$. For $n = 1$, $\lambda = 1160$ nm; for $n = 2$, $\lambda = 580$ nm; for $n = 3$, $\lambda = 283$ nm, etc. Of these wavelengths, only $\lambda = 580$ nm lies in the visible region and, according to Fig. 11, this light is *yellow*. *Destructive* interference occurs when $n = \frac{1}{2}, \frac{3}{2}, \frac{5}{2}, \cdots$. For $n = \frac{1}{2}$, $\lambda = 2320$ nm; for $n = \frac{3}{2}$, $\lambda = 740$ nm; for $n = \frac{5}{2}$, $\lambda = 464$ nm; for $n = \frac{7}{2}$, $\lambda = 331$ nm; etc. Again, of these wavelengths, only $\lambda = 464$ nm, a blue-violet color, lies in the visible region.

Hence, centered at 580 nm, there will be constructive interference in a broad band in the yellow-green, yellow, and orange areas of the spectrum, and, centered at 464 nm, there will be destructive interference in a broad band in the blue and violet regions. If the oil film is illuminated by white light, the yellow part of the spectrum will be enhanced and the blue part of the spectrum will be suppressed. A film of this thickness will have a yellow or golden color.

The reader doubtless wonders how the vacuum value of c in (4) was actually established, since most of the actual measurements were performed in air and since the refractive index of air is not strictly unity. The direct measurements would, of course, give a value for the speed of light in air; this value has been 'reduced to vacuum' in order to obtain (4). This reduction to vacuum involves precise determination of the refractive index of air by means of interferometric techniques. The refractive index of air depends upon density and temperature, as well as wavelength. Though the dependence on these parameters is slight, it is easily measured by interferometric techniques; in fact, refractive index measurements are used in practical measurements of gas densities in such devices as the *Mach interferometer*.

A Mach interferometer is a variant of the Michelson interferometer that is used to measure density differences at various points in a supersonic wind tunnel. A Mach interferogram for the flow past a wedge is shown in Fig. 17. The light beam is split and then reassembled by means of half-silvered mirrors, so that half the light has passed through the glass-sided wind tunnel, half has passed over the tunnel through air of constant density. The fringe shift shown in the photograph arises from the fact that the air flowing past the wedge is compressed and has a higher density, and hence a higher index of refraction, than that in the free stream at the left. The index of refraction of air at 0° C and 1 atm is approximately 1.0003; if the density is doubled, the *difference* between this index and that of vacuum, exactly 1, is doubled, so air of twice normal

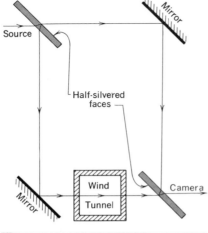

density would have index of 1.0006. The wavelength would correspondingly shorten as indicated in Fig. 16. While these changes are small, the interferometer is so sensitive that a shift of one-half fringe (from constructive to destructive interference) results from a density change of only about 0.1 percent if the path length through the wind tunnel is 1 meter. Note the sharp change in density at the shock wave originating at the point of the wedge in Fig. 17.

6. Transmission of Energy by Light Waves; Radiometry

Fig. 17. Schematic diagram of the Mach interferometer, and photograph of the flow from left to right past a wedge. The mirrors are slightly tilted so that the region of constant density to the left gives uniformly spaced parallel fringes. The fringe shift is to be measured by comparison with this uniform spacing. (*Photograph from U.S. Naval Ordnance Laboratory.*)

We have asserted earlier that energy is transmitted by light waves and have shown how the speed of transmission can be determined. Now we must consider the question of how energy transmission is measured. Actually, two separate sets of phenomena are involved, since the term *light* is used in two ways. In its broader sense, *light* is used to refer to radiation of all frequencies; measurement of transfer of energy by radiation is called *radiometry* and can be accomplished by purely physical instrumentation. In the narrower sense, *light* is restricted to radiation capable of affecting the retina of the human eye and thus involves human physiology and psychology; the quantitative measurement of the visible radiation from light sources is called *photometry* and requires the introduction of a special photometric quantity not related directly to the rest of physics. Radiometry and photometry are both well-established laboratory sciences having specialized techniques and terminologies; here we shall try to give the general ideas involved and attempt to keep specialized terminology to a minimum.

In describing the spreading of light waves from a source, it is necessary to introduce the concept of *solid angle*, which we have not formally defined. A solid angle in *steradians* (sr) is defined in a manner similar to a plane angle in *radians*. Consider a cone whose apex is at the center of a circumscribed sphere of radius R as in Fig. 18. Let the cone cut out an area S of the surface of the sphere.

Since the area S is proportional to R^2, we can use the dimensionless ratio S/R^2 as a measure of the solid angle Ω of the cone, defining $\Omega = S/R^2$, in steradians.

> One **steradian** is the solid angle subtended at the center of a sphere by a portion of the surface of area equal to the square of the radius of the sphere.

The solid angle subtended *by the whole sphere* is the whole area $4\pi R^2$, divided by R^2, or 4π sr.

The transfer of radiant energy from a source involves the *flow* or *flux* F_R of radiation that is properly measured in watts, since it involves the flow of energy per unit time. The *radiant intensity of a source*, I_R, is defined as the radiant flux per unit solid angle,

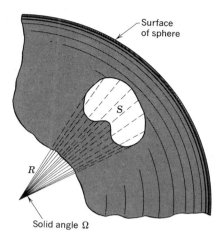

Figure 18.

$$I_R = \frac{\Delta F_R}{\Delta \Omega}, \qquad \left\{\begin{array}{l} I_R \text{ in W/sr} \\ F_R \text{ in W} \\ \Omega \text{ in sr} \end{array}\right\} \quad (7)$$

where ΔF_R is the power passing outward from the source through a small solid angle $\Delta \Omega$. The radiant source intensity may depend upon the direction from which the source is viewed, as indicated in Fig. 19.

The radiant flux reaching unit area of a surface is called the *irradiance*, E_R, and is defined by the relation

$$E_R = \frac{\Delta F_R}{\Delta S}, \qquad \left\{\begin{array}{l} E_R \text{ in W/m}^2 \\ F_R \text{ in W} \\ S \text{ in m}^2 \end{array}\right\} \quad (8)$$

where ΔF_R is the power falling on the small surface area ΔS. If the irradiance is associated with a beam of radiation normal to a surface, it is sometimes called the 'beam intensity.' *The beam intensity is the radiant power per unit area normal to the direction of propagation and is measured in* W/m^2.

In making radiometric measurements, it is necessary to use radiation detectors that are equally responsive to all wavelengths; that is, the detectors must be equipped with *black* receivers. Thermocouples (Chapter 27) equipped with blackened receivers are frequently employed to give an electrical response proportional to radiant power received; calorimetric devices may also be used.

In our earlier treatment of sound waves (Chapter 21), we showed that the sound intensity or acoustic power per unit area is proportional to the square of the wave amplitude. A similar relation exists for electromagnetic waves; the radiant intensity is proportional to the square of the wave amplitude. Thus, for the electromagnetic wave shown in Fig. 5, p. 692, the radiant power per unit area is proportional to the square of the amplitude $\mathcal{E}_M$ of the $\mathcal{E}$ waves. In the case of light waves, the irradiance E_R at a given point is also directly proportional to the square of the amplitude $\mathcal{E}_M$ of the light waves at the point.

This relationship leads to interesting questions regarding interference phenomena: What happens in the case of destructive interference? Does radiant power somehow vanish? The answer is, 'No!' In an interference or diffraction pattern (Chapter 35), the radiant power that disappears in regions of destructive interference reappears in regions of constructive interference in such a way that the total radiant power remains constant. Radiant power 'piles up' at interference maxima so that radiant power is conserved. This relationship is sometimes called the *law of radiometric summation*.

In many optical problems in which we trace rays or optical paths it is convenient to talk about 'point objects' and 'point images.' Although these concepts are very useful, we must realize that our point objects and point images are not really mathematical points; small surface areas must be involved if infinite irradiances (8) and infinite beam intensities are to be avoided. We use the terms point object and point image in somewhat the same way that we used the term particle in our treatment of mechanics; we recall that a point particle would have either zero mass or infinite density!

7. Photometry

In providing a basis for photometry, it is necessary to provide a special unit, the lumen (lm), for the specification of *luminous flux*, F_L, in a measure that corresponds to the effect on the human eye. The corresponding special unit of *luminous intensity*, I_L, the lm/sr, is called the *candela* (cd).* The necessary basis for the presently employed set of photometric units was provided by an international agreement in 1948 that defined a standard light source, in which a black body consisting of a tube of specified size embedded in molten platinum is observed from above as the platinum solidifies. By definition,

> The luminous flux of the standard light source is one **lumen** per steradian in directions very near the vertical, and the luminous intensity of the standard source in this direction is one **candela.**

Thus, on the basis of this definition, we may write

$$I_L = \frac{\Delta F_L}{\Delta \Omega}. \qquad \left\{ \begin{array}{l} I_L \text{ in cd} \\ F_L \text{ in lm} \\ \Omega \text{ in sr} \end{array} \right\} \quad (9)$$

The luminous intensity I_L usually depends upon the direction from which

---

*Candela is the internationally adopted name of the unit that was formerly called the *candle* in English.

a source is viewed, as indicated in Fig. 19. If a source radiates isotropically, its total luminous flux is given by $4\pi I_L$, since there are 4π sr in a sphere.

A third photometric quantity of practical concern is the *illumination*, E_L, which gives a measure of the light flux falling on a surface, per unit area:

> **Illumination** of a surface is defined as light flux per unit area of the surface. It is measured in lm/m² or lm/f².

Thus, the illumination of a surface is given by the relation

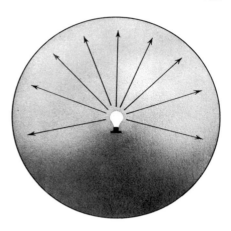

Fig. 19. Luminous intensity may vary with direction.

$$E_L = \frac{\Delta F_L}{\Delta S}. \qquad \begin{cases} E_L \text{ in lm/m}^2 \text{ (or lm/f}^2) \\ F_L \text{ in lm} \\ S \text{ in m}^2 \qquad \text{(or f}^2) \end{cases} \qquad (10)$$

The unit, lm/f², is commonly called the *foot-candle*.

We note that if the surface area ΔS is at distance R from a point source and is perpendicular to the radius, as is the surface of the sphere in Fig. 18, we can write $\Delta S = R^2 \Delta \Omega$. Substitution of this relation in (10) and comparison with (9) gives the *inverse-square law for illumination*:

$$E_L = I_L/R^2. \qquad (11)$$

When we substitute a value of I_L in candelas in (11), E_L comes out initially in candelas/area rather than lumens/area. But since, from (9), 1 lm = (1 cd)×(1 sr), and the sr is dimensionless, we are permitted to substitute lm for cd to obtain lumens/area.

A connection between radiometric and photometric quantities can be established if we employ *monochromatic* light. Thus, if light having a wavelength of 555 nm is employed, it is found that a radiant flux of 1 W is equivalent to a luminous flux of 680 lm. As 555 nm is the wavelength at which the eye is most sensitive, the watt will correspond to fewer lumens at any other wavelengths in proportion to the ordinates of Fig. 11.

> **Example.** *The luminous intensity of the sun is about 2×10^{27} cd. Assuming that none of the light is absorbed by the earth's atmosphere (actually about 10 percent is so absorbed on a clear day), compute the illumination of a horizontal surface on the earth at latitude 40° N on a clear day at noon on June 21–22 (the summer solstice) and at noon on December 21–22 (the winter solstice).*
>
> On a surface perpendicular to the sun's rays (ΔS in Fig. 20) at the distance of the sun from the earth, 92 million miles = 4.86×10^{11} f, the illumination would be, by (11),

$$E_L = \frac{I_L}{R^2} = \frac{2 \times 10^{27} \text{ cd}}{(4.86 \times 10^{11} \text{ f})^2} = 8470 \frac{\text{cd}}{\text{f}^2} = 8470 \frac{\text{lm}}{\text{f}^2} = \frac{\Delta F_L}{\Delta S}.$$

The angle between the plane of the earth's equator and the plane of the ecliptic (the plane of the earth's orbit) is 23°.5 (the latitude of the Tropic of Cancer). At noon at the summer solstice, the sun is at the zenith at the Tropic of Cancer and hence is 40° − 23°.5 = 16°.5 from the zenith at 40° N latitude. Considering ΔA an area on the earth's surface in Fig. 20, we have $\Delta S = \Delta A \cos\theta$ or $\Delta A = \Delta S/\cos\theta$. Thus $\Delta F_L/\Delta A = (\Delta F_L/\Delta S) \cos\theta$, and since $\theta = 16°.5$, the illumination on ΔA is

$$(8470 \text{ lm/f}^2) \cos 16°.5 = (8470 \text{ lm/f}^2)(0.959) = 8120 \text{ lm/f}^2.$$

On the other hand, at the winter solstice, the sun is directly overhead at the Tropic of Capricorn, at 23°.5 S latitude, and the angle θ in Fig. 20 is 40° + 23°.5 = 63°.5 for a point at 40° N. The illumination is then

$$(8470 \text{ lm/f}^2) \cos 63°.5 = (8470 \text{ lm/f}^2)(0.446) = 3780 \text{ lm/f}^2.$$

At these very intense illuminations, the difference in amount of light between the summer and winter noons is not very striking to the eye, but the difference in heat falling on each square foot of earth's surface has a very pronounced effect on climate. The heat varies with angle θ in the same way as the light—actually about one-half of the radiant energy from the sun is in the visible wavelengths.

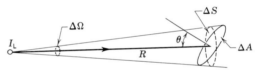

Fig. 20. Light incident at an angle θ to the normal to a surface.

PROBLEMS

1. An incandescent lamp with small filament and clear bulb is suspended 3 f above the center of a table top. If the area of the table top is 6 f² and if the distance from the table top to the floor is 3 f, what is the area of the shadow formed on the floor?   Ans: 24 f².

2. Using the dimensions given in the example on pp. 702–703, compute the approximate width, on the surface of the earth, of the path of a partial solar eclipse. Show that with the moon at 236 000 miles, a total solar eclipse is not possible near January 1, when the sun is 91.4 million miles from the earth, but is possible near July 1 when the sun is at 94.5 million miles. (Note however that the moon's distance varies also, from 222 to 253 thousand miles, with a period of 27.3 d.)

3. A pinhole camera is used to form an image of an object 2 f tall located 10 f in front of the pinhole. If the distance from the pinhole to the film is 4 i, what is the height of the image? (See figure.)   Ans: 0.80 i.

4. The diameter of the sun is 864 000 mi and that of the moon is 2160 mi. A 'pinhole camera,' consisting of a small hole in a window shade and a white card held 10 f from the hole, is used to obtain an image of the sun at a time when the sun is at a distance of 92 900 000 mi from the earth and to obtain an image of the

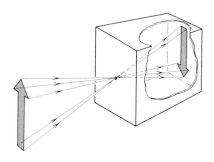

Problems 3-4. The pinhole camera.

full moon at a time when the moon is 236 000 mi from the earth. Find the diameters of the images produced.

5. Explain why a shadow cast by direct sunlight, as in Fig. 1, is conspicuously sharper on a cloudless day than on a day when there are scattered clouds in the sky.

6. The flatness of glass plates such as are used in interferometers can be tested by placing two such plates in contact along one edge, separated by the thickness of a sheet of thin paper along the opposite edge, and observing the reflection of monochromatic light at normal incidence, as in Fig. 7.

(*a*) Show that if the plates are perfect, a system of straight, parallel, uniformly spaced interference fringes will be observed. Determine the spacing of the fringes in terms of the thickness of the paper, the dimensions of the plates, and the wavelength of the light.

(*b*) A plate is perfect except for one low spot. This plate is tested against a perfect plate and the fringe pattern photographed. Sketch the appearance of the fringe pattern. Show how, by measurements on the photograph, the depth of the low spot can be determined to within an accuracy of a fraction of the wavelength of light.

7. A beam of parallel rays of yellow-green light ($\lambda = 500$ nm) falls normally on the plane surface of a plano-convex lens whose convex surface is in contact with a plane glass surface. The radius of curvature of the lens surface is 50 cm. Find the radius of the 50th dark ring observed by reflection, not counting the central dark spot. Ans: 0.323 cm.

8. A plano-convex lens having a radius of curvature of 5 m is placed with its convex surface in contact with a flat piece of plate glass. When this lens is illuminated from above with sodium light of wavelength 589 nm, Newton's rings are observed by reflection. What are the radii of the 5th and 10th bright rings? the 5th and 10th dark rings?

9. Let us suppose that in Galileo's experiment the two observers were stationed in towers 4 mi apart. Using present knowledge of the value of the speed of light, calculate the time of transmission of a light signal from one tower to the other. Ans: 21.4 μs.

10. An observer sees lightning strike a tree on a hill 8 km away. How long after the lightning has struck does the observer see the flash? hear the flash?

11. On a day when the speed of sound is 1100 f/s, a gun is fired 150 f away from an observer. By the time the sound of the shot reaches the observer, how far has the light from the muzzle flash traveled? Ans: 25 400 mi.

12. The explosion of a thermonuclear bomb creates a flash of light that is easily visible from neighboring planets. At a time when Mars is 160 million miles from the earth, how long after detonation of a bomb would the 'news' reach Mars? How long would it take this news item to reach Jupiter at 460 million miles from the earth?

13. Using the value of 92 000 000 mi as the radius of the earth's orbit and 18.5 mi/s as the earth's orbital speed, and considering Jupiter to be very far from the sun in comparison, compare the times between eclipses of Jupiter's moon when the earth is at B and D in Fig. 13 with the time, 42.5 h, observed when the earth is at A.

Ans: 15.2 s longer at B and shorter at D.

14. Fizeau's apparatus for measuring the speed of light employed a toothed wheel, the light flash passing out through one slot in the wheel and returning through a later slot. Fizeau used a total light path length of 17.3 km, and a wheel with 720 teeth. He found that at every increase in speed of the wheel by 1452 rev/min, the light came through to his eye. What value did he obtain for the speed of light?

15. Taking 35 386 m as the distance between stations on Mount Wilson and Mount San Antonio in one of Michelson's experiments, find the lowest rotational speed at which the image will again be observed at the cross hairs after the rotation of the octagonal mirror has begun. (One-eighth of a revolution is required in order to satisfy this condition.)

Ans: 529 rev/s.

16. During the years 1930–1933, Michelson's group made a direct measurement of the speed of light in vacuum by using an evacuated tube approximately 1 mi long. The rotating mirror had 32 faces and the light traveled back and forth between mirrors at the ends of the tube a total distance of 8 mi during the time of replacement of one of the 32 mirror faces by the next one. What is the value of the minimum rotational speed required for the image to be observed in the following mirror face?

NOTE: *One direct application involving the speed of light occurs in radar systems. Microwaves are sent out from radiators backed by parabolic reflectors. If this radiation strikes an airplane, a ship, or a storm cloud, some of the radiation is reflected and detected by a receiver. The radiation is not sent out continuously but in pulses of extremely short duration. The time between transmission of a pulse of radiation and reception of an 'echo' can be accurately measured by electronic methods. This time measurement gives an accurate measurement of the range of the object causing the reflection.*

17. A short pulse of radiation is sent out by a radar set such as that in the figure, and a reflection is received 120 μs later. How far away is the object that produced the reflection?

Ans: 11.2 mi.

18. A reflection of radiation from an airplane is noted by a radar operator 500 μs after a pulse of radiation has been sent outward. What is the distance of the plane from the radar station?

19. Radar pulses have recently been reflected from the moon, which is 236 000 mi from the earth. What is the time interval before receipt of the echo in this case? Ans: 2.54 s.

20. Microwaves can also be used for measurement of the approach speed of an object by means of the *Doppler effect* of electromagnetic waves. In this case a continuous wave is sent out and the frequency shift Δf of the wave reflected by the moving object is measured electronically by determining the beat frequency (exactly analogous to the beat frequency of sound) when the transmitted wave and the reflected wave are superposed. By an argument analogous to that in the example on pp. 457–458 in the case of sound, derive the expression $\Delta f = 2(v/c)f$ for this frequency increase, where v is the approach speed of the object. Note that v is very small compared with c.

21. A microwave radiator emitting radiation of wavelength 0.1 m is used to measure the speed of an oncoming automobile. The observed frequency shift is 456 Hz. What is the speed of the automobile? (*See* Prob. 20.)

Ans: 51.0 mi/h.

22. In the case of the microwave device described in Prob. 21, what frequency shift would be produced by an approaching automobile moving at a speed of 30 mi/h? at a speed of 60 mi/h?

23. A continuous microwave radiator having wavelength 25 cm is used to measure the speed of an approaching airplane at an airport. The Doppler frequency shift is observed as 325 Hz. What is the approach speed of the plane? (*See* Prob. 20.) Ans: 91.0 mi/h.

24. When we observe the sun, we see it at the position it had, relative to the earth, when the light left the sun. What is the angle between this position and its actual position at the moment of observation?

25. What is the frequency of violet light of

Problems 17–19

wavelength 400 nm? red light of wavelength 700 nm? microwaves of wavelength 1 cm? 25 cm? Ans: 7.50×10^{14} Hz; 4.29×10^{14} Hz; 3.00×10^{10} Hz; 1.20×10^9 Hz.

26. A microwave frequency can be measured directly in terms of the unit of time, the second, by electronic techniques that have been developed for counting cycles (just as an electric clock counts the cycles of an alternating current). Show how reflection of microwaves from an object approaching at known speed thus gives another method of measuring the speed of light. (See Prob. 20.)

27. Show how measurements of the interference pattern set up by two microwave sources driven by the same oscillator, whose frequency is known, can be used to determine the speed of light in an arrangement similar to Fig. 9, p. 425. What is the distance AB between the central image and the first-order interference minimum in Fig. 11, p. 427, if the sound sources are replaced by microwave sources of frequency 1.18×10^{10} Hz, the sources are 2 i apart, and the distance $X = 12.5$ i? (Note that you should find the wavelength to be exactly one inch, so the geometry will be similar to that in the example on p. 427.)
Ans: 3.25 i.

28. With the apparatus of Fig. 15, Froome used a frequency of about 24 000 MHz, determined to an accuracy of 1 part in 10^8 by counting cycles. He moved the carriage about 1 m (exactly 81 λ) to count 162 interference minima. He could set the carriage on a minimum with an accuracy better than 1 μm. He could measure the distance of movement of the carriage to an accuracy of ¼ μm. Show that these data correspond to an accuracy in measurement of the speed of light of about 1 part in 1 000 000 as indicated in (3).

29. If we take the speed of light as 3×10^8 m/s and the frequency as 24 000 MHz in Froome's apparatus of Fig. 15, what is the distance the carriage must be moved to start at an interference minimum and proceed to the 160th later interference minimum? Ans: 1.00 m.

30. If the vacuum wavelength of a certain monochromatic light is 500 nm, what is its wavelength in water? in crown glass? in diamond?

31. What is the minimum thickness of a soap film of refractive index 1.33 if the film gives constructive interference of red light of wavelength 648 nm at normal incidence?
Ans: 121 nm.

32. Find the thickness of a soap film that gives constructive second-order reflection of yellow light of wavelength 500 nm. Take $\mu = 1.33$ for the film and assume normal incidence. NOTE: Second-order reflection occurs when the path difference is equal to $\tfrac{3}{2}$ wavelengths.

33. An isotropic point source of light has a luminous intensity of 50 cd. (a) What is the total light flux from this source? (b) Calculate the illumination at a point on a surface 10 f away from the source if the light strikes this surface perpendicularly.
Ans: 628 lm; 0.50 lm/f².

34. An isotropic point source of light emits a total flux of 1256 lumens. (a) What is the source intensity? (b) Calculate the illumination of the inner surface of a sphere of 3-m radius if the point source is located at its center.

35. The full moon, when directly overhead, gives an illumination of 0.03 lm/f². (a) Neglecting absorption in the earth's atmosphere, determine the luminous intensity of the moon. (b) How far away from a 60-W tungsten lamp having luminous intensity of 55 cd in the horizontal direction must a vertical screen be placed to have the same illumination?
Ans: 4.81×10^{16} cd; 42.8 f.

36. The sun, when directly overhead, gives an illumination of 8000 lm/f². (a) Neglecting absorption in the earth's atmosphere, determine the luminous intensity of the sun. (b) How far away from a 75-W tungsten lamp having luminous intensity of 60 cd in the horizontal direction must a vertical screen be placed to have the same illumination?

37. A 600-cd light hangs at the geometrical center of the cubical room 10 f on an edge. Assuming the source to be isotropic and considering only the direct illumination, find: (a) the total number of lumens striking the floor; (b) the average illumination of the floor; (c) the maximum and minimum values of the illumination of the floor.
Ans: 400π lm; 4π lm/f²; 24.0, 4.66 lm/f².

38. If a lamp of luminous intensity 55 cd in the downward direction is 4 f above the surface of a table and is switched on for 7 min, how many lumens fall on a square inch of table top directly below the lamp during the time the light is on? Is the lumen more closely related to the joule or the watt?

39. On March 22 the noon sun at Quito, Ecuador, produces an illumination of 8000 lm/f^2 on level ground when the sun is directly overhead. What is the illumination from direct sunlight at 2 P.M.? At 4 P.M.? (Neglect atmospheric absorption.)

Ans: 6930 lm/f^2; 4000 lm/f^2.

40. A suspended light source has a luminous intensity of 6000 cd when viewed from below. At what distance above a drafting table should this source be suspended in order to give the recommended illumination of 50 lm/f^2? At this distance, what would be the illumination in lm/m^2?

41. A workbench 20 f long is illuminated by three 600-cd lamps suspended at a height of 8 f above the bench. One of the lamps is located above the center of the bench and the other two are located above its ends. What is the direct illumination immediately below the center lamp? (Assume isotropic radiation from the lamps.) Ans: 14.0 lm/f^2.

42. The total energy radiated by the sun is such that 19.4 kcal/min pass through each square meter of area of a sphere of 92 million miles radius surrounding the sun. Neglecting absorption of this energy by the earth's atmosphere, determine the radiant energy in kcal/min falling on a square meter of the earth's horizontal surface at 45° latitude at noon at the summer and winter solstices and at the spring and fall equinoxes.

43. From observation, it is found that the radiant energy (both visible and invisible) from the sun that is received on a square meter of surface perpendicular to the solar rays, at the earth's mean distance from the sun but outside the earth's atmosphere, is 1.35 kW. Compute the total rate of radiation of the sun and the rate of radiation per square meter of solar surface.

Ans: 3.81×10^{23} kW; 6.25×10^4 kW/m^2.

44. A certain light source emits only light of wavelength 555 nm and radiates at the rate of 0.5 W. What is the luminous flux? Estimate the luminous flux from a sodium lamp radiating 0.5 W at 589 nm.

45. A lamp produces direct illumination of 10 lm/m^2 on a screen 75 cm from the lamp. When a sheet of glass is placed between the lamp and the screen, the lamp must be moved 5 cm closer to again give illumination of 10 lm/m^2. What percentage of the light is transmitted by the glass? Ans: 87.1%.

SHADOWS: FRESNEL DIFFRACTION 35

Light from a point source in an isotropic medium travels radially outward; the wavefronts are spherical and the optical paths traversed are accurately given by rays drawn radially outward from the source. There is thus rectilinear propagation along radial lines. However, if we obstruct some portion of the advancing wavefront by an opaque barrier, propagation beyond the barrier is no longer strictly rectilinear; light always penetrates to some extent into regions beyond the barrier that cannot be reached by straight lines drawn from the source. This departure from rectilinear propagation is called *diffraction*. Diffraction always occurs when advancing waves encounter a barrier but is most easily observed when the dimensions of the barrier or the dimensions of openings in the barrier are comparable with the wavelengths involved.

In this chapter, we restrict our discussion of diffraction to the effects involved in *shadow formation*. Since the phenomena involved in shadow formation were first successfully interpreted by Fresnel, they are termed *Fresnel diffraction*. In Chapter 39, we shall consider the diffraction effects involved in image formation by lenses and mirrors. These effects are termed *Fraunhofer diffraction*.

If we place a screen behind a uniformly illuminated opaque barrier in which there is a small circular opening, the light reaching the screen forms a bright circular area when the screen is in contact with the barrier. The size and shape of the uniformly illuminated area of the screen is exactly equal to the size and shape of the opening in the barrier; the pattern of light on the screen is thus a sharp *shadow* of the barrier. If the distance between the screen and the barrier is gradually increased, the shadow initially consisting of the circular bright spot on the screen gradually changes into a *diffraction pattern* consisting of alternately bright and dark concentric circular fringes like those in Fig. 13. Thus, *a Fresnel diffraction pattern is a fringed shadow of a nearby barrier.*

If the reader looks at the bright sky or at a ceiling light through a narrow slit between two fingers, he will see a diffraction pattern similar to those in Fig. 7, consisting of light and dark fringes; the spacing between the fringes depends on the width of the slit and the distance of the slit from his eye.

Diffraction patterns were first carefully studied by Thomas Young early in the nineteenth century and were successfully explained by Fresnel in terms of the wave theory that had been proposed by Huygens in the seventeenth century. We begin this chapter with a discussion of Huygens's hypotheses concerning the behavior of light waves and show how they lead to a correct explanation of the observed phenomena of diffraction of light passing near the edges of material obstructions.

1. Huygens's Hypotheses

The detailed mathematical analysis contained in the electromagnetic theory of light, published by Maxwell in 1873, correctly describes all details of the propagation of light waves. However, this mathematical analysis is extraordinarily complex. Prior to Maxwell, a more intuitive physical model has been developed to explain the observed phenomena; this model gives results consistent with Maxwell's analysis in all except very minor details. It is based on two hypotheses first made by Huygens. In this model, light from a point source is considered to advance as a succession of spherical wavefronts. At a long distance from the source, the spherical wavefronts become essentially plane.

Huygens's most fertile hypothesis, commonly known as the *Huygens principle,* but certainly not a fundamental principle of physics, is the following:

> *Every point on an advancing wavefront can be considered as a source of secondary waves, which in an isotropic medium move forward as*

spherical wavelets. The wave amplitude at any point ahead can be obtained by superposition of these wavelets.

This hypothesis leads to an understanding of the diffraction phenomena that we shall treat in the present chapter and in Chapter 39.

Huygens's second hypothesis relates the successive positions of an advancing wavefront. This hypothesis, which is of particular importance in the consideration of image formation by lenses and mirrors, is the following:

From a given position of an uninterrupted wavefront, a later position of the wavefront can be determined as the envelope of the secondary wavelets.

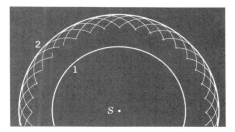

Fig. 1. Huygens construction for the propagation of a spherical wavefront from a source S.

This envelope is illustrated for a spherical wavefront in Fig. 1, where the envelope of the secondary wavelets from wavefront 1 gives the wavefront 2. At large distances from the source S, the wavefront becomes essentially plane and would be associated with a parallel beam of light, as mentioned earlier. Parallel beams and even convergent beams can also be produced by suitable lenses; the Huygens constructions for parallel and convergent beams are shown in Fig. 2. From a *line* source, such as a narrow slit, we also obtain *cylindrical* wavefronts, and we can imagine that, from such a wavefront, cylindrical wavelets proceed outward.

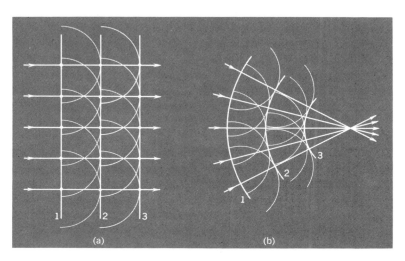

Fig. 2. Huygens constructions for (a) parallel, and (b) convergent beams.

2. Young's Experiment

Although the adherents to the particle theory of light were able to supply a satisfactory—though rather tortured—explanation of interference in thin films, Thomas Young in 1802 performed an experiment that led to results clearly predicted by the wave theory of transmission and clearly at variance with predictions based on the particle theory. Young's experiment involved both diffraction and interference.

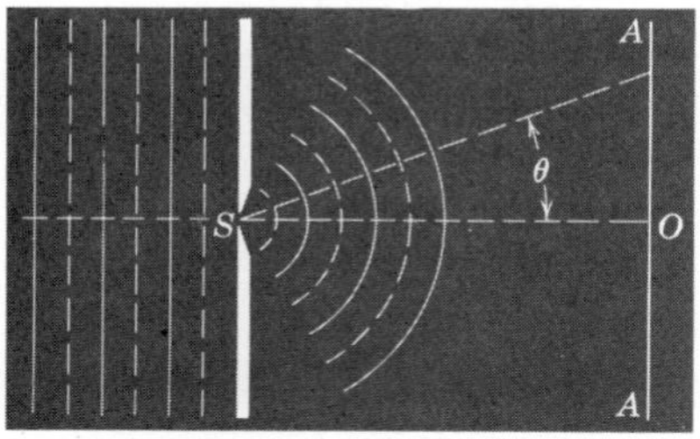

In describing Young's experiment, let us start by considering the situation (Fig. 3) in which a plane light wave meets a slit S having a width small compared with the wavelength. On the basis of the Huygens hypotheses, the light wave would be expected to spread out cylindrically in the manner shown in the figure; on the particle theory, light would pass directly through a slit without change in direction. The effect observed is that on the screen AA there is a broad diffuse region of illumination; the illumination is greatest at O and falls off gradually with increasing angle θ. This spreading of the light beyond the slit is called diffraction.

Fig. 3. Diffraction by a slit of size small compared with the wavelength of the light.

The optical diffraction just described is closely analogous to the more familiar acoustical phenomenon in which a sound wave passes through a hole small in comparison with the wavelength of sound; in the acoustical case, pressure pulses in the air cause the sound wave to spread in all directions after it passes through the hole.

Experimentally, it is found that the central portion of the screen is always brighter than the more remote portions, and hence it must be assumed that the amplitude of the secondary wavelet is greatest in the forward direction; more detailed theory indicates that the amplitude of the wavelet is proportional to $(1+\cos\theta)$, a quantity sometimes called the *obliquity factor*. The obliquity factor was introduced on an *ad hoc* basis by Fresnel and was later justified by Fraunhofer from fundamental considerations of wave propagation.

Young's experimental arrangement is shown schematically in Fig. 4. A monochromatic light source is placed behind the narrow slit S in the first barrier so that the illuminated slit S serves as a source from which a divergent beam spreads beyond the first barrier. The diverging beam strikes a second barrier in which there are two narrow slits, S_1 and S_2, which give rise to diffracted beams that overlap as shown on the screen. Since slits S_1 and S_2 are illuminated by light from a common source S, they can be considered as two coherent sources; interference effects are to be expected in analogy with the acoustical interference phenomena discussed on pp. 424–427. A set of alternately light and dark fringes is

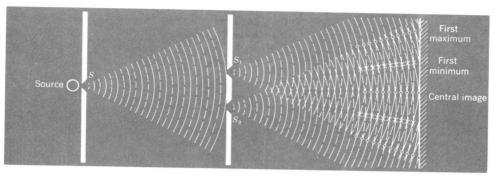

Fig. 4. Young's experiment. Solid and broken arcs correspond to 'crests' and 'troughs,' the distance from solid to broken arc being one-half wavelength. Constructive interference is indicated by circles, destructive interference by crosses.

observed on the screen, but, if either S_1 or S_2 is blocked off, the set of fringes is replaced by a broad, diffuse diffraction pattern like the one discussed in connection with Fig. 3.

For the arrangement shown in Fig. 4, the distances from S to S_1 and S to S_2 are equal. Therefore, we may regard S_1 and S_2 as sources that are equal in phase as well as frequency. Hence, at points on the screen to which the paths from S_1 and S_2 differ by zero or by a whole number of wavelengths, there will be constructive interference and bright fringes will be observed; at points on the screen to which the paths from S_1 and S_2 differ by $\frac{1}{2}\lambda$, $\frac{3}{2}\lambda$, $\frac{5}{2}\lambda$, $\cdots$, there will be destructive interference and dark fringes will be observed. The type of pattern actually observed is shown by the photographic print in Fig. 5; the photograph was obtained by exposing a photographic plate located at the final screen position in Fig. 4, developing the plate, and making a positive print in the usual fashion. Screens for repetition of Young's experiment can be produced conveniently by cutting narrow slits in the opaque emulsion of exposed photographic plates by means of a sharp razor blade.

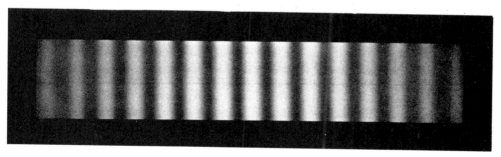

Fig. 5. Photograph of diffraction fringes produced by two slits. (Reprinted from Cagnet *et al.*, *Atlas of Optical Phenomena*, with the permission of Springer-Verlag.)

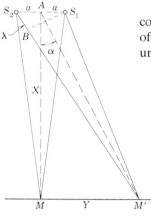

Figure 6

The distances between the fringes in Young's experiment can be computed as in the example below. We note that, so long as the value of α in Fig. 6 is small, as it is in the usual case, the bright fringes are uniformly spaced.

Example. *If the separation of the virtual sources S_1 and S_2 in Fig. 4 is 0.1 mm and the perpendicular distance from the virtual sources to the screen is 50 cm, find the distance between the central bright fringe and the center of the first bright fringe on either side when the slit S is illuminated with yellow sodium light of wavelength $\lambda = 589$ nm.*

The geometry of this problem is illustrated in Fig. 6, except that in this figure horizontal distances $S_2 S_1$ and MM' are exaggerated in comparison with vertical distances by a factor of about 100. The angle α between the central bright fringe, M, and the next, M', turns out to be only about ⅓ of a degree. In Fig. 6, MA is the bisector of $\angle S_2 M S_1$; $S_1 B$ is drawn so that the lengths $M'S_1$ and $M'B$ are equal. The length $S_2 M'$ is greater than $S_1 M'$ by $S_2 B$, and for the first bright fringe this length difference must equal λ, as indicated on the figure.

Let us denote the distances $S_2 A$ and $S_1 A$ by a, the distance AM to the screen by X and the distance MM' between bright fringes by Y. We then have $a = 0.5$ mm $= 5 \times 10^{-5}$ m, $X = 50$ cm $= 5 \times 10^{-1}$ m, and $\lambda = 589$ nm $= 589 \times 10^{-9}$ m $= 5.89 \times 10^{-7}$ m. The distance $M'S_2 = \sqrt{(Y+a)^2 + X^2}$, while $M'S_1 = \sqrt{(Y-a)^2 + X^2}$. Since $M'S_2 = M'S_1 + \lambda$,

$$\sqrt{(Y+a)^2 + X^2} = \sqrt{(Y-a)^2 + X^2} + \lambda.$$

This equation can be simplified, just as in the example on p. 427, to give

$$(4a^2 - \lambda^2) Y^2 = \lambda^2 (X^2 + a^2 - \tfrac{1}{4} \lambda^2),$$

which determines Y in terms of X, a, and λ.

In an example such as that of Fig. 6, we note that, at least to three-figure accuracy, we can neglect a^2 and $\tfrac{1}{4} \lambda^2$ in comparison with X^2 in the parenthesis on the right, and we can neglect λ^2 in comparison with $4a^2$ in the parenthesis on the left. Hence, for cases in which $\lambda^2 \ll a^2$ and $a^2 \ll X^2$, we get the simple relation

$$4 a^2 Y^2 = \lambda^2 X^2, \quad \text{or} \quad Y = \lambda X / 2a.$$

The fringe spacing Y is proportional to the wavelength and the distance to the screen, and inversely proportional to the source separation $2a$.

Substitution of our numerical values in this relation gives

$$Y = \frac{(5.89 \times 10^{-7} \text{ m})(5 \times 10^{-1} \text{ m})}{2 \times 5 \times 10^{-5} \text{ m}} = 2.95 \times 10^{-3} \text{ m} = 2.95 \text{ mm}$$

as the distance between bright fringes. Succeeding bright fringes will be equally spaced at 2.95 mm apart since, to the approximation used above, if the path difference is $n\lambda$ instead of λ, the value of Y will be $n\lambda X/2a$.

We note that the total radiant flux F_R (p. 721) reaching the screen in Fig. 4 must be exactly equal to the sum of the radiant flux F_{R1} from S_1 and the radiant flux F_{R2} from S_2. The irradiance E_R is enhanced in regions of constructive interference and falls to zero in regions of destructive interference in such a way that the integral over the whole screen,

$$\iint E_R \, dS = F_R = F_{R1} + F_{R2},$$

in accord with the law of radiometric summation.

3. Diffraction by a Slit and by a Straightedge

If we start with the experimental arrangement of Fig. 1 and widen the slit from a fraction of a wavelength to several wavelengths, the illuminated region of the screen decreases in size as wavelets from different parts of the slit begin to interfere destructively with each other. The resulting illumination takes on the appearance shown in Fig. 7.

The Huygens picture, carefully applied, gives all details of patterns such as those in Fig. 7. While we shall not attempt a thorough discussion of these patterns, the diagram of Fig. 8 will help us understand the sharpening of the illuminated region and the occurrence of fringes.

The slit in Fig. 8 is five or six wavelengths wide, the wavelength being the distance between the solid 'crests' of the plane wave shown impinging from the left. We have constructed wavelets from two points PP, centered in the upper and lower halves of the slit aperture. These wavelets, and

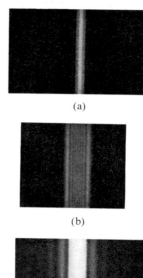

(a)

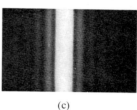

(b)

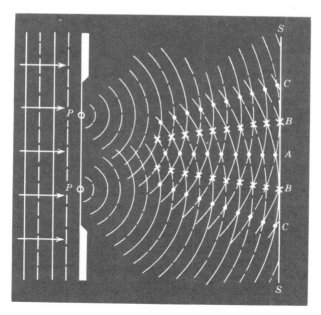

(c)

Fig. 7. Diffraction patterns formed by single slit. (From R. W. Ditchburn, *Light:* Blackie & Son, Glasgow.)

Fig. 8. A wide slit produces a diffraction *pattern* because of the interference of wavelets proceeding from different points of the aperture.

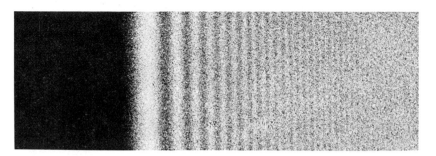

Fig. 9. Photograph of the shadow of a straightedge. (From B. Rossi, *Optics:* Addison-Wesley, Reading, Mass.)

those from other points of the aperture, interfere in a generally constructive fashion along the row of dots leading to point A at the center of the screen, and illuminate this center brightly. Along the rows of crosses leading to points BB, these two wavelets interfere destructively. At points BB, generally speaking, for each point in the bottom half of the aperture, there is a point in the top half that sends out a wavelet that interferes destructively with the wavelet from the lower point. Near BB, there are intensity minima; the above geometrical argument is not perfect so there is not complete darkness at these points. Further out, as at CC, there will be weak intensity maxima. These maxima will be weak because while the wavelets from PP interfere constructively at CC, there will be other wavelets, notably the wavelet proceeding from the point midway between the two points P, that will interfere destructively with both of these wavelets. The strongly illuminated part of the screen will be only that in the region between points BB; outside this region there will be a succession of 'fringes' with comparatively weak intensity maxima.

Similar arguments applied to wavelets arising from a plane wave passing a straightedge show that there will be a series of fringes of intensity maxima and minima in the region that would be uniformly illuminated if propagation were rectilinear, as shown in Fig. 9. The intensity falls off uniformly from the first bright fringe, which lies slightly outside of the geometric shadow, into the shadow proper.

Correspondingly, the geometric shadow of any extended object is surrounded by a series of fringes, as illustrated in Fig. 10. On a screen one meter from the object, the first three bright fringes are at about $\frac{1}{2}$, 1, and $1\frac{1}{2}$ millimeters from the edge of the geometric shadow.

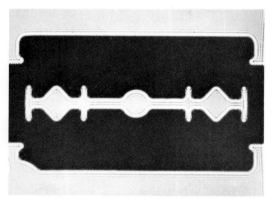

Fig. 10. Diffraction effects in the shadow of a razor blade. (From Francis Watson Sears, *Principles of Physics, III* (*Optics*), 2nd ed.: Addison-Wesley, Reading, Mass.)

Similar fringes are observed when a spherical wavefront, originating from a point source of light, impinges on a slit, a straightedge, or on any other object that interrupts the wavefront.

4. Diffraction by a Circular Aperture

Fresnel developed an interesting approach to an understanding of the effect of a *circular* aperture by considering the effects of wavelets originating in annular *zones* of a plane wavefront. Let us consider a plane wave advancing to the right in Fig. 11 and attempt to find the effect of this wave at a point P by combining the results produced by secondary wavelets from all points of a plane wavefront at AA. Let the wavelength of the incident light be λ and let the distance OP normal to the wavefront be R. Now with P as a common center construct a series of spheres which intersect the plane wavefront and which have radii $R + \frac{1}{2}\lambda$, $R + \lambda$, $R + \frac{3}{2}\lambda$, $R + 2\lambda$, $\cdots$. The intersections of these spheres with the plane wavefront divide the wavefront into a series of zones; the first zone is a circle and the others are rings having O as a center. It is evident from the method of construction that the *average* distance from the point P to successive zones will increase by $\frac{1}{2}\lambda$ in passing from one zone to the next. All the secondary waves from any one zone can be considered to produce a certain resultant amplitude at P, and the amplitudes of the resultants from the successive zones can be represented by M_1, M_2, M_3, $\cdots$. The total amplitude at point P depends on the amplitudes and the phases of the waves reaching P from the different zones. Since the secondary waves start with the same phase from all points on the wavefront, their relative phases at point P will depend simply on the relative dis-

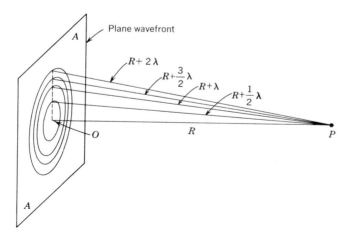

Fig. 11. Construction of Fresnel zones.

tances traveled to P. Since the average distance traveled increases by $\tfrac{1}{2}\lambda$ from zone to zone, the wave from zone 2 will be a half wavelength behind that from zone 1, and so on for successive zone pairs. Consequently, the waves with resultant amplitudes M_1, M_3, M_5, $\cdots$ will be in phase with one another but will be exactly out of phase with waves M_2, M_4, M_6, $\cdots$. Hence, in order to obtain the total amplitude A at point P, we should add the successive M's, using $+$ and $-$ signs alternately:

$$A = M_1 - M_2 + M_3 - M_4 + \cdots . \tag{1}$$

From the sagitta relation (p. 707), it can be shown that *all the zones have the same area* provided their radii are small compared with OP. However, because of the obliquity factor, the amplitude of secondary waves slowly decreases as the angle between the direction of propagation and the normal to the wavefront increases. Therefore the M's slowly decrease:

$$M_1 > M_2 > M_3 > M_4 > \cdots .$$

The expression for A in equation (1) can be rewritten in the form

$$A = \tfrac{1}{2} M_1 + (\tfrac{1}{2} M_1 - \tfrac{1}{2} M_2) - (\tfrac{1}{2} M_2 - \tfrac{1}{2} M_3) + (\tfrac{1}{2} M_3 - \tfrac{1}{2} M_4) - \cdots .$$

Since the M's decrease slowly, the quantities in the parentheses are very small and occur alternately with positive and negative signs; the sum of the series is actually

$$A = \tfrac{1}{2} M_1 . \tag{2}$$

In other words, the resultant amplitude of the plane wave at point P is equal to one-half of that resulting from the first zone alone. This statement can be directly checked experimentally by placing a screen at P in Fig. 11 and a barrier at AA that contains an aperture permitting light to pass through the first zone only; the amplitude at P changes from A to $M_1 = 2A$ when this is done, and the intensity at P quadruples, since the intensity is proportional to the square of the amplitude.

On the basis of the above analysis, it would be expected that blocking out the alternate zones 2, 4, 6, $\cdots$ (or 1, 3, 5, $\cdots$) by opaque rings drawn to the proper size on a transparent plate would considerably increase the amplitude of the disturbance at P, since the effects of all remaining zones would be in phase and the amplitude A could be written

$$A = M_1 + M_3 + M_5 + \cdots .$$

Plates so constructed are called *zone plates*. They produce a strong increase in intensity at point P when placed in a parallel beam of light; in fact, they can be used to produce real images of distant objects in somewhat the same manner as does a converging lens. The fact that a zone plate does increase the intensity at point P is further evidence that the effect of a source at a point is not transmitted simply along the ray connecting the source and the point but does involve *diffraction* effects.

An accurately constructed circular zone plate is shown in Fig. 12; if it is photographed, the photographic negative can be used as an effective zone plate.

Fresnel used arguments based on zones to discuss the diffraction effects observed when a parallel beam of light strikes a small circular opaque object or passes through a screen with a small circular hole.

It turns out that there is always a bright spot at the center of the shadow of a circular disc, and that this spot is almost as bright as would be the direct illumination. Let us assume, for example, that the disc exactly covers the first two zones in Fig. 11. Then, we should have, in place of equation (1), $A = M_3 - M_4 + M_5 - M_6 + \cdots$, which can be transformed, as in equation (2), to $A = \frac{1}{2} M_3$, and since M_3 is very close to M_1, the amplitude is very close to the direct amplitude $\frac{1}{2} M_1$. If the disc does not exactly cover an integral number of zones in Fig. 11, we can start with the distance r from P to the edge of the disc and draw a new set of zones with boundaries at distances $r + \frac{1}{2} \lambda$, $r + \lambda$, $\cdots$ from P; the above argument can then be made, in the same manner, for these zones.

The top photograph in Fig. 13 shows the shadow of a disc, with the bright spot at the center. Around the bright spot, and within the geometrical shadow is a set of very faint circular fringes; outside the geometrical shadow there are more conspicuous fringes. We shall not discuss the

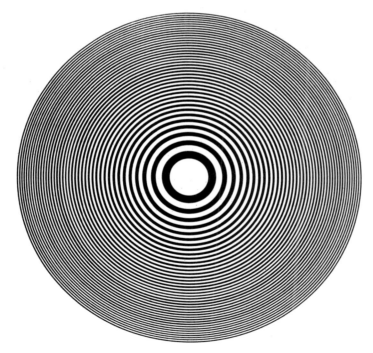

Fig. 12. Zone plate.

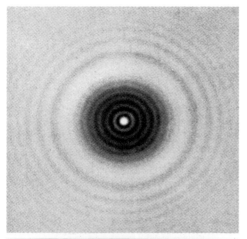

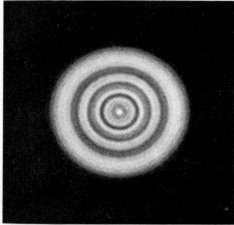

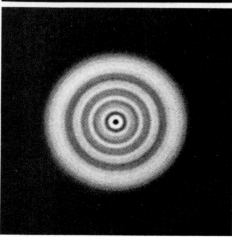

origin of these particular fringes in detail, but will discuss in some detail the similar fringes for the case of a circular opening, as shown in the two lower photographs of Fig. 13.

Consider a circular opening that exposes just the first (central) zone in Fig. 11. As we have already shown, the illumination at P will be four times the direct illumination. For an opening that exposes exactly two zones, the amplitudes for the two zones will be out of phase at P and cancel, leaving P in darkness. For an exposure of three zones, zones 2 and 3 will cancel, leaving just the effect of zone 1, and P will again have four times the direct illumination. The central point P will be alternately bright and dark as successive zones are exposed. Examples of both situations are shown in Fig. 13.

Now let us consider points on the screen that would be in the geometrically illuminated circle but are not at the center of the circle. Figure 14(a) shows the zones as seen from the center of the circle when just three zones are exposed. We can consider that zones 1 and 2 cancel and that bright illumination at the center results from zone 3. In Fig. 14(b), we are looking at the zones exposed by the aperture from the standpoint of a point on the screen slightly to the left of the central point. Zones 1 and 2 are still fully exposed and cancel. Zone 3 is partly covered while zone 4 is partly exposed and substantially cancels the effect of zone 3. We have moved from the center of the bright spot into the dark ring that surrounds it as shown in the photograph of Fig. 14(c). It is in a similar fashion that the dark rings occur in Fig. 13, where the circular apertures expose a greater number of zones.

It is fairly easy to observe the diffraction pattern resulting from a circular aperture by making a small pinhole in a piece of cardboard and viewing the daytime sky through the pinhole.

Fig. 13. Top photograph shows diffraction by an opaque circular disc. Lower two photographs show diffraction by a circular aperture for two positions of the plane of observation. (Photographs reprinted from Cagnet *et al.*, *Atlas of Optical Phenomena*, with the permission of Springer-Verlag.)

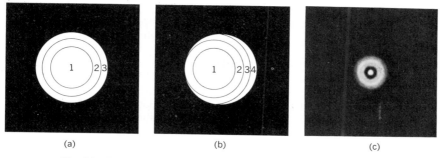

Fig. 14. Demonstrating the occurrence of a dark fringe within the region illuminated by a three-zone circular aperture.

Example. *We are all familiar with diffraction in the case of sound waves, which quite readily 'bend around corners.' Similar easily observed effects exist for electromagnetic waves in the microwave and radiofrequency regions of the electromagnetic spectrum. Referring to Fig. 11, calculate the radius r of the first Fresnel zone if the point of observation P is at a distance $R=1$ m from the plane wave front AA and the microwave wavelength is (a) 1 cm, (b) 3 cm, and (c) 10 cm. Show how a parallel beam of microwaves can be 'focussed' at point P by interposing in the beam either a plane metal sheet with a hole of radius r, or a flat metal disc of radius r.*

From Fig. 11, we note that the radius of the first zone is the short side of a right triangle, the other sides being R and $R + \tfrac{1}{2}\lambda$, respectively. Denoting the radius of the first zone by r and using the Pythagoras theorem, we obtain

$$r^2 = R^2 + \lambda R - \tfrac{1}{4}\lambda^2 - R^2 = \lambda R - \tfrac{1}{4}\lambda^2$$

or $r \approx \sqrt{\lambda R}$. For the wavelengths listed in the problem, we obtain (a) $r=10$ cm, (b) $r=17.3$ cm, and (c) $r=31.6$ cm. Using either the screen or the disc, we obtain at point P a wave amplitude $A = M_1$ in the notation of (2) and (3). Since by (2) the amplitude for the unobstructed wave is $\tfrac{1}{2} M_1$, we have doubled the amplitude at P. Since the irradiance E_R is proportional to the square of the amplitude, we have *quadrupled* the irradiance by inserting the screen or the disc.

Microwave and radio-frequency diffraction patterns analogous to the visible diffractions shown in the photographs of Figs. 5, 7, 9, 10, 13, and 14 can be produced with ease. Such Fresnel-diffraction effects can sometimes lead to difficulties in microwave and radio transmission.

PROBLEMS

1. A narrow slit used in a Young experiment of the type illustrated in Fig. 4 is illuminated by yellow light of wavelength 589 nm. If the separation of the virtual source slits S_1 and S_2 is 0.2 mm and if the perpendicular distance from these virtual sources to the screen is 3.0 m, what is the distance from the central bright fringe to the nearest bright fringes on either side? Ans: 8.84 mm.

2. If light of another color is used in the experiment of Prob. 1, and the distance between the central bright fringe and the next bright fringe is found to be 10.0 mm, what is the wavelength of the light?

3. In Young's experiment, Fig. 4, if the slit separation is 0.2 mm, the distance from the slits to the screen is 70 cm, and the fringe separation (Fig. 5) is 1.75 mm, find the wavelength of the monochromatic light used in the experiment.
Ans: 500 nm.

4. In the example on p. 734, dealing with Young's experiment, a simple expression was obtained for the fringe spacing Y. A more complex expression for Y was used in the example on p. 427, which deals with an analogous acoustical problem. Why does the simple expression not apply in the acoustical problem? Under what circumstances could we apply the simple optical expression to acoustical interference?

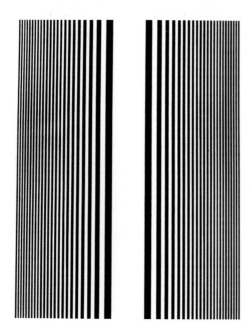

Prob. 5. Cylindrical zone plate.

5. The figure shows a 'cylindrical' zone plate, constructed in a manner analogous to that of the 'spherical' zone plate of Fig. 12. This zone plate will give a bright-line image on a screen placed at an appropriate distance. Show how it is constructed. Show that the successive zones do not have equal area (width) so that the use of these cylindrical zones in the discussion of shadows cast by slits or long narrow obstacles is not as easy as the discussion of shadows cast by circular objects or apertures.

6. Demonstrate, from the sagitta relation (p. 707), that all of the Fresnel zones in Fig. 11, of radii small compared with R, have approximately equal areas.

7. Light of wavelength 500 nm from a distant point source passes through a circular opening, and the resulting Fresnel diffraction pattern is observed on a screen 1 m beyond the opening. Find the radius of the opening if it is just large enough to include the central Fresnel zone.
Ans: 0.646 mm.

8. Compute the radius of a circular zone plate having a 'focal length' $R = 1$ m for light of wavelength 600 nm if the plate has only 10 zones, with zones 1, 3, 5, 7, and 9 open and the even-numbered zones blacked out.

9. If the circular opening described in Prob. 7 is enlarged sufficiently to include the first four Fresnel zones, what will be the diameter of the opening?
Ans: 2.58 mm.

10. Find the radius of the central zone of a Fresnel zone plate that has a 'focal length' of 2.00 m for orange light of wavelength 600 nm.

11. Find the radius of a circular hole in an opaque screen if the hole is just equal in size to the central Fresnel zone for a parallel beam of green light from a mercury lamp (546.1 nm) when the observer's eye is exactly 6 m from the hole.
Ans: 2.56 mm.

12. Show that the hole described in Prob. 11 must have a radius approximately $\sqrt{2}$ times as large to uncover the first two Fresnel zones, $\sqrt{3}$ times as large to uncover the first three Fresnel zones, and so on.

13. From measurements made on Fig. 12, determine the approximate 'focal length' (R in Fig. 11) of this circular zone plate for light of 500-nm wavelength.
Ans: 50 m.

14. Demonstrate that at the geometrical edge of the shadow of a straightedge illuminated by parallel light, as shown in Fig. 9, the intensity is only one-quarter of the intensity of the unobstructed light (see Prob. 5).

REFLECTION AND REFRACTION 36

Light travels in straight lines in an isotropic transparent medium such as air. When a light beam strikes a boundary between two media, such as air and water, some or all of the light may be turned back into the first medium by *reflection;* however, some of the light may pass into the second medium, usually with a change in the direction of propagation—a process known as *refraction*. In the present chapter, we present the experimentally established laws describing these phenomena and show how these laws are interpreted in terms of the wave theory. In treating reflection and refraction, it is usually unnecessary to introduce consideration of the wave nature of light, since these phenomena can be adequately described by tracing *rays*. Treatment of optical problems by ray tracing is called *geometrical optics*. Although we shall make extensive use of the techniques of geometrical optics, it should be borne in mind that rays merely represent directions perpendicular to wavefronts.

1. Reflection at Plane Surfaces

Reflection of light occurs in accordance with two empirical laws known to the early Greek and Egyptian scientists:

FIRST LAW OF REFLECTION: *The reflected ray lies in the plane containing the incident ray and the normal to the reflecting surface at the point of incidence.*

SECOND LAW OF REFLECTION: *The angle of incidence is equal to the angle of reflection.* (The angle of incidence is the angle between the incident ray and the normal; the angle of reflection is the angle between the reflected ray and the normal.)

In Fig. 1, *ABC* represents a plane reflecting surface perpendicular to the plane of the paper. The normal *BN* is in the plane of the paper. If an incident ray *IB* is in this plane, the first law states that the reflected ray *BR* is also in the plane of the paper. The second law states that the angle of incidence $i = \angle IBN$ is equal to the angle of reflection $r = \angle RBN$.

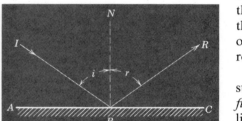

Fig. 1. Specular reflection.

Reflection of light from a smooth polished surface is called *regular reflection* or *specular reflection*. Reflection from a rough or 'mat' surface like that of cement or newsprint occurs in many directions when a parallel beam of light is incident on the surface, as indicated in Fig. 2, although the laws of regular reflection are obeyed by any single ray. The incident beam of light is said to be *diffused* at the rough surface and reflection at a rough surface is called *diffuse reflection*.* It is by diffuse reflection that we are able to see nonluminous bodies when light strikes them. The reflected rays from any small portion of a rough surface travel in so many different directions that the small area is visible to us in essentially the same manner as a similar small area on the surface of a luminous body. On the other hand, we cannot really 'see' a perfect mirror. Diffuse rather than regular reflection is required to make the surface of a body visible.

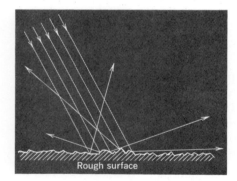

Fig. 2. Diffuse reflection.

The Huygens hypotheses provide a straightforward interpretation of the empirical laws of reflection in terms of light waves. Consider the

---

*The above discussion is not completely satisfactory, because of course all surfaces are rough on an atomic scale. But the scale of atomic size is small in comparison with the wavelength of light. The dividing line between regular and diffuse reflection is not sharp, but if the widths of the hills in Fig. 2 are large compared with the wavelength of the light, the reflection is diffuse as in the figure. On the other hand, if the widths of the hills in Fig. 2 were small compared with the wavelength of the light, the reflection would be regular, as in Fig. 1. Consequently, in order to make a good mirror, a glass or metal surface must be polished so that its irregularities are small in size compared with the wavelength of light.

situation in Fig. 3, which shows the positions of a portion of a plane wavefront before and after striking a reflecting surface; the points A, B, C, D, and E represent points on the incident wavefront. We now wish to find the future position of the wavefront at the end of the time interval t between reflection of the portion of wavefront at point A and the later reflection of the portion of the incident wavefront at point E. During this interval t, the wavefront of the wave traverses the distance EE'', which gives the magnitude of the radius AA'' of the Huygens wavelet spreading out with the speed of light from A. The Huygens wavelet spreading from point B' on the surface has radius $B'B''$ such that $BB' + B'B'' = AA''$; similarly, for the wavelets spreading from C' and D'. By symmetry we see that the reflected wavefront $A''E''$, which by the Huygens hypothesis is the envelope of these wavelets, is a straight line whose normal $C'C''$ makes the same angle with the normal to the mirror, $C'N$, as does the incident 'ray,' CC'. That is, the angle of reflection, r, equals the angle of incidence, i. Thus we have derived the second law of reflection from wave considerations. Realizing that the incident parallel light beam is not all in the plane of the paper in Fig. 3, that the reflected wavefronts are not really circles but are spherical surfaces, and that the envelope is not a straight *line* but rather a *plane,* we can also derive the first law of reflection.

It is interesting to note in connection with Fig. 3 that point A is on the right side of the incident beam, as viewed along the direction of propagation, but point A'' is on the left side of the reflected beam as viewed along the direction of propagation in the figure. Thus, with regard to the wavefronts, the beam is 'folded over' on itself when it is reflected.

Another point to be noted is that when a wave is reflected, the incident wave's velocity component perpendicular to the reflecting surface is reversed in direction. The incident wave's velocity component parallel to the reflecting surface is unchanged when reflection occurs; the tangential velocity components of the incident and reflected waves are equal.

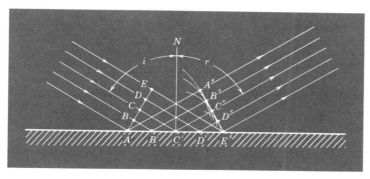

Fig. 3. Huygens construction for reflection at a plane surface.

2. Image Formation by Plane Mirrors

The reversal of the normal component of an advancing wavefront has the interesting result that, when a wavefront strikes a reflecting surface such as a mirror, the position of the reflected wavefront at any time can be obtained by first determining the position the wavefront *would have had* if reflection had not occurred and, second, reversing all the perpendicular distances from the mirror surface. For example, consider Fig. 4, which shows a point source O—called a *point object*—at distance p from a mirror surface. Spherical wavefronts spread out from O in the direction of the mirror surface. Wavefront AA has just reached the mirror. At this time, the center of the wavefront starts traveling to the left, and as time goes on, more and more of the wavefront is reflected and travels to the left. The wavefront AA moves successively to the positions $BB'B$, $CC'C$, and $DD'D$. In the drawing, the broken arcs show the positions the wavefront would have had if the mirror had not been present; the actual positions are just the 'mirror images' of these positions. Because of this symmetry, the wavefronts returning from the mirror and entering the 'eye' are centered at a *point image I* at the distance p behind the mirror. This image *seems* to be the source of the returning wavelets; since the wavelets do not actually originate at I, the image is said to be *virtual*. We have shown that

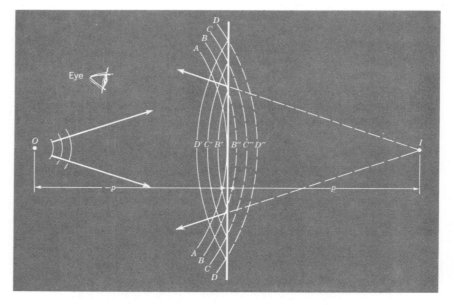

Fig. 4. Formation of a virtual point image by a plane reflecting surface.

A plane mirror forms a virtual image of a point object, the image being behind the mirror on the normal from the object to the mirror, at a distance behind the mirror equal to that of the object in front.

If we call the distance from the object to the mirror the *object distance* and the distance from the image to the mirror the *image distance*, we conclude that for a plane mirror

image distance = object distance.

These conclusions can also be derived quite rigorously from the laws of reflection by means of ray diagrams like the one in Fig. 5. This direct derivation from the empirical laws of reflection is left as a problem.

If the object is not a point object, it is possible to treat it as if it were equivalent to a collection of points as indicated in Fig. 6, in which rays from two points O and O' of the object are traced to the eye. Consideration of Fig. 6 shows that the eye sees a virtual image equal in size to the object, each point of the image being as far behind the mirror as the corresponding point of the object is in front. We also conclude that the image is upright. Thus, we may summarize our discussion of the plane mirror by saying that *it forms a virtual, upright image whose size is equal to that of the object, with image distance equal to object distance.*

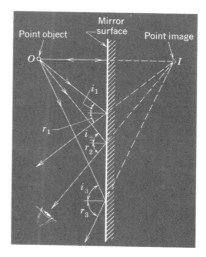

Fig. 5. Formation of a virtual image by a plane mirror.

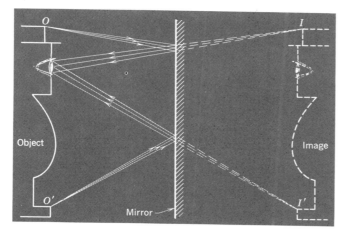

Fig. 6. Formation of the image of an extended object.

3. Refraction of Light

The behavior of light in passing from one medium into another medium is shown in Fig. 7, which uses air and water as examples of the two media. Rays passing from air into water or glass are bent *toward* the normal at the interface; rays passing from water or glass into air are bent *away* from the normal. Refraction takes place in accordance with three empirical laws based on experiment:

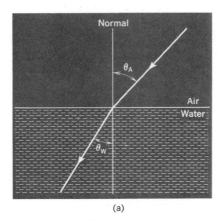

FIRST LAW OF REFRACTION: *The refracted ray lies in the plane containing the incident ray and the normal to the interface at the point of incidence.*

SECOND LAW OF REFRACTION: *The ratio of the sine of the angle of incidence to the sine of the angle of refraction is a constant that is independent of the angle of incidence. (Angles of incidence and of refraction are measured from the normal.)*

THIRD LAW OF REFRACTION: *The path of a ray refracted at the interface between two transparent media is exactly reversible.*

In Fig. 7, the incident ray, the refracted ray, and the normal to the interface are all in the plane of the paper; the plane surface of the water is at right angles to the plane of the paper. The second law of refraction states that

$$\sin\theta_A/\sin\theta_W = \mu_{WA}, \qquad (1)$$

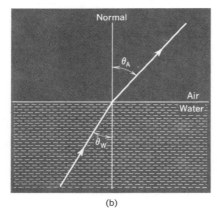

Fig. 7. Refraction at a plane surface. In (a), θ_A is the angle of incidence and θ_W the angle of refraction; whereas in (b), θ_W is the angle of incidence and θ_A that of refraction.

where the constant μ_{WA} is called the *relative index of refraction* of water with respect to air. The third law of refraction enables us to assert that equation (1) holds, with the same index μ_{WA}, whether the incident ray is in the air, as in Fig. 7(a), or in the water, as in Fig. 7(b).

The first and third laws of refraction were known to the early Greeks and Egyptians; the second law of refraction is known as *Snell's law* and was discovered early in the seventeenth century by the Dutch astronomer, WILLEBRORD SNELL (1591–1626).

We shall now 'derive' Snell's law from the wave properties of light. This derivation will show that the refraction results from the fact that the speed c_A of light in air is greater than the speed c_W in water. The relative index of refraction is the ratio of these speeds:

$$\mu_{WA} = c_A/c_W = \mu_W/\mu_A, \qquad (2)$$

where μ_W and μ_A are the *absolute* refractive indices, as defined on p. 717.

In Fig. 8, uninterrupted plane wavefronts strike the interface between two materials, such as air and water, with absolute indices of refraction μ_I and μ_{II}, where $\mu_{II} > \mu_I$. The corresponding incident *rays* make angle θ_I with the normal. From each point on the wavefronts, we draw the positions of wavelets a short time later. The envelope of these positions determines the new wavefront. The wavelets in material II do not travel as far as those in I because the wave speed in II is less than that in I. This fact results in refraction of the wavefronts and hence of the rays, as illustrated. Snell's law is readily obtained from Fig. 8. The time of travel of light from B to D must equal that from A to C, hence

$$BD/c_I = AC/c_{II}, \quad \text{or} \quad BD/AC = c_I/c_{II}, \qquad (3)$$

where c_I and c_{II} are the speeds in I and II. Now

$$BD = AD \sin(\angle BAD) = AD \sin\theta_I,$$
$$AC = AD \sin(\angle CDA) = AD \sin\theta_{II}.$$

Hence
$$BD/AC = \sin\theta_I/\sin\theta_{II}. \qquad (4)$$

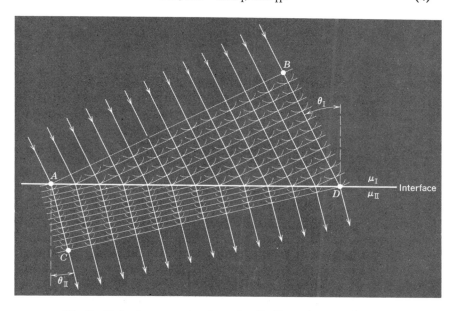

Fig. 8. Refraction at an interface. In this illustration, $\mu_{II} = 2\mu_I$, so the speed of light in II is half that in I. Visualize a column of soldiers arriving at a muddy field, where they take steps only half as long but still keep in time to the band and keep their lines dressed to the best of their ability.

Comparison of (3) and (4) gives

$$c_\mathrm{I}/c_\mathrm{II} = \sin\theta_\mathrm{I}/\sin\theta_\mathrm{II}, \quad \text{or} \quad \sin\theta_\mathrm{I}/c_\mathrm{I} = \sin\theta_\mathrm{II}/c_\mathrm{II}.$$

Since the absolute indices of refraction are defined as $\mu_\mathrm{I} = c/c_\mathrm{I}$, $\mu_\mathrm{II} = c/c_\mathrm{II}$, we can write this relation in the forms

$$\mu_\mathrm{I} \sin\theta_\mathrm{I} = \mu_\mathrm{II} \sin\theta_\mathrm{II}, \quad \text{or} \quad \sin\theta_\mathrm{I}/\sin\theta_\mathrm{II} = \mu_\mathrm{II}/\mu_\mathrm{I}, \tag{5}$$

in agreement with Snell's law.

We now can discuss the phenomenon of *total reflection*. Figure 9 shows the interface between two media I and II, of which medium II is assumed to have the higher index of refraction. According to (5), if $\mu_\mathrm{II} > \mu_\mathrm{I}$, then $\theta_\mathrm{II} < \theta_\mathrm{I}$. A ray passing from I into II is bent toward the normal, and no peculiarities arise. But consider the three rays *within the medium of higher refractive index*, heading toward the interface. Ray 1 has a small angle of incidence and is refracted away from the normal. As the angle of incidence increases, we reach ray 2, whose angle of refraction is 90°. The angle of incidence for which the angle of refraction is 90° is called the *critical angle* θ_C. Substituting $\theta_\mathrm{I} = 90°$ ($\sin\theta_\mathrm{I} = 1$) in (5) gives the relation

$$\sin\theta_\mathrm{C} = \mu_\mathrm{I}/\mu_\mathrm{II}. \quad (\mu_\mathrm{I} < \mu_\mathrm{II}) \tag{6}$$

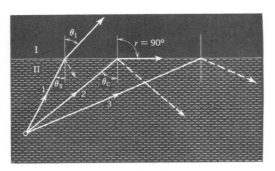

Fig. 9. Total reflection.

Substitution of a value of θ_II larger than the critical angle in (5) gives a value for $\sin\theta_\mathrm{I}$ that is greater than unity, and θ_I does not represent any real angle. No refracted ray is possible. For angles of incidence greater than θ_C, as in the case of ray 3 in Fig. 9, the ray is *totally reflected* at the interface, and no light passes into the other medium. Notice that *total reflection takes place only for light within a medium of higher refractive index at a surface of contact with a medium of lower refractive index*.

Example. *What are the critical angles, with respect to air or vacuum, for the glasses listed in Table 34–1, p. 717?*

In equation (6), we substitute $\mu_\mathrm{I} = 1.000$ for air or vacuum, and for μ_II the refractive index μ_G of the glass, to obtain $\sin\theta_\mathrm{C} = 1.000/\mu_\mathrm{G}$. The results are given in the following tabulation:

Glass	μ_G	$\sin\theta_\mathrm{C}$	θ_C
Crown	1.517	0.659	41°.2
Commercial plate	1.523	0.657	41°.1
Light flint	1.574	0.635	39°.4
Dense flint	1.656	0.604	37°.2

Equation (6) furnishes the basis for a number of types of convenient refractometers designed to determine an unknown index of refraction by

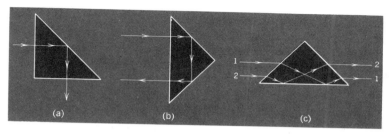

Fig. 10. Total internal reflection in glass prisms.

measurement of a critical angle. For example, the index of refraction of a small drop of liquid can be measured by placing the drop on the surface of glass of known index and measuring the critical angle for total reflection of light traveling within the glass and reflected at the surface of liquid contact, provided that the liquid has lower index than the glass.

We note that in Figs. 7 and 8 there will always be also a reflected wavefront, obeying the laws of regular reflection as in Fig. 3. If we attempt a Huygens construction for a ray like ray 3 in Fig. 9, we find that there are *no* refracted wavefronts, only reflected wavefronts. The reader should try drawing refracted wavelets from the various points of the interface for this case and he will find that *they do not intersect and have no envelope*—hence no wavefront is formed (*see* Prob. 28).

Because total reflection is really *total*, it furnishes the basis for a *perfect* mirror which is utilized in various ways in optical instruments. Because the critical angle for glass is less than 45°, as demonstrated by the above example, a beam of light may be turned through 90° or 180° by a glass prism with 45° and 90° angles, as shown in Fig. 10, (a) and (b). Another utilization of total reflection is illustrated in Fig. 10(c), which shows a glass prism that can be used to invert an image without changing the direction of the light beam. Prisms giving total internal reflection are frequently used in binoculars.

Refraction causes an object immersed in water to appear closer to the surface than it actually is. In order to understand this phenomenon, consider the following example:

Example. *A cylindrical metal cup is 3 i in diameter. A person looks into the cup over the rim at such an angle that he can just see the far edge of the bottom when the cup is empty. The cup is then filled with water and he can just see a spot in the center of the bottom. What is the depth d of the cup?*

The geometry of the situation is shown in Fig. 11. By using Pythagoras' theorem, we see that

$$\sin\alpha = \frac{3\,i}{\sqrt{9\,i^2 + d^2}}; \quad \sin\beta = \frac{1.5\,i}{\sqrt{2.25\,i^2 + d^2}} = \frac{3\,i}{\sqrt{9\,i^2 + 4d^2}}.$$

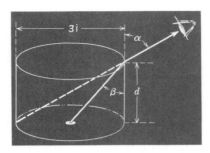

Figure 11

Hence
$$\frac{\sin\alpha}{\sin\beta} = \sqrt{\frac{9\,i^2 + 4d^2}{9\,i^2 + d^2}}.$$

When this ratio equals 1.33, the refractive index of water, the bent line in Fig. 11 will suitably represent a light ray refracted at the water surface. Setting this ratio equal to 1.33, squaring both sides of the equation, and solving for d^2, we find

$$d^2 = 3.11\,i^2; \qquad d = 1.78\,i.$$

This is the required depth of the cup. For any deeper cup, the ray from the spot to the rim is not refracted sufficiently to reach the eye.

4. Fermat's Principle

We introduce at this point a theorem concerning light transmission known as *Fermat's principle of least time,** as illustrative of certain general 'variational principles' that find application in many fields of modern physics:

> FERMAT'S PRINCIPLE: *The path actually traversed by light in going from one point to another is the path for which the time of transmission is a minimum as compared with neighboring paths.*

Fermat's principle can be of practical use in optical work. A trivial case is encountered when the two points are in the same optical medium; in this case Fermat's principle predicts rectilinear propagation along the line connecting the two points. As an illustration of how this quite general principle can be used in two other simple cases, we shall show how the empirically discovered laws of reflection and refraction can be 'derived' from it.

Consider the situation shown in Fig. 12, in which points P and P' are at a height h above the surface of a plane mirror and are separated by a horizontal distance L. Light might conceivably pass from P to the mirror at point O and thence to point P'. The length D of such a path is given by

$$D = \sqrt{h^2 + l^2} + \sqrt{h^2 + (L-l)^2}.$$

According to Fermat's principle, the path actually traversed by the light proceeding from P to P' by reflection is the path for which the time of transmission is a minimum; hence, since the path is entirely in a single medium, the geometrical path

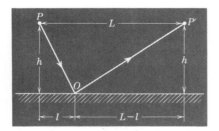

Figure 12

*PIERRE DE FERMAT, French mathematician (1601–1665).

length must be a minimum. Considering the distance l as a variable, we may find the minimum value of D by setting the derivative $dD/dl=0$. Performing this operation leads to a value of $l=\frac{1}{2}L$. Redrawing Fig. 12 with $l=\frac{1}{2}L$ leads immediately to the second law of reflection. The first law of reflection follows from the fact, which is immediately apparent, that movement of point O along the surface out of the plane of the paper increases the path length, so the shortest path lies in the plane of the paper.

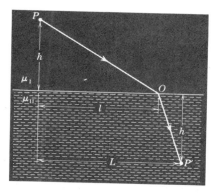

Figure 13

Snell's law can also be derived from Fermat's principle, by means of the diagram shown in Fig. 13, in which point P is located in one optical medium with absolute refractive index μ_I, P' is located in a second optical medium with absolute refractive index μ_{II}, and O represents some point on the interface, in the normal plane containing P and P'. The speeds of light in the two media are c/μ_I and c/μ_{II}, respectively. Hence the time t of transmission from P to P' would be given by the relation

$$ct = \mu_I \sqrt{h^2 + l^2} + \mu_{II} \sqrt{h^2 + (L-l)^2}.$$

Treating the length l as a variable, we set the derivative $dt/dl=0$ to find the actual path traversed. The result obtained is

$$\frac{\mu_I l}{\sqrt{h^2+l^2}} = \frac{\mu_{II}(L-l)}{\sqrt{h^2+(L-l)^2}},$$

which is Snell's law (5), since, in Fig. 13, $\sin\theta_I = l/\sqrt{h^2+l^2}$ and $\sin\theta_{II} = (L-l)/\sqrt{h^2+(L-l)^2}$. The other laws of refraction also follow directly from Fermat's principle.

It is interesting to note that Fermat's principle, a very simple but very general statement, contains within itself the rectilinear propagation of light in an isotropic medium, the laws of reflection, and the laws of refraction, and would thus seem to constitute a 'generalization of higher order' than any of the empirical laws involved. It should be pointed out, however, that generalizations like Fermat's principle are usually firmly supported by empirical laws *prior* to their enunciation. *New* scientific knowledge rarely comes from *ab initio* formulation of such wide generalizations, but, once formulated, they have broad application.

5. Dispersion by Refraction

By *dispersion* of light is meant the separation of light into its component wavelengths or spectral colors. The index of refraction that occurs in Snell's law is not the same for all wavelengths. For most transparent

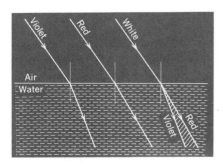

Fig. 14. Dispersion of white light by refraction at a water surface. (The amount of dispersion is exaggerated in this sketch.)

media, the index increases slightly as the wavelength decreases; that is, it is slightly greater for violet light than for red light. As a result of this difference in refractive index, rays of violet light are bent more sharply than rays of red light in passing from air into most transparent media, as illustrated in Fig. 14.

If a ray of white light is incident on the water surface in Fig. 14 at the same angle as the violet and red rays, it splits after passing through the water surface into a group of colored rays with a violet ray at one boundary of the group and a red ray at the other boundary. The explanation of the *dispersion* of the white-light ray into colored rays appears obvious now; we conclude that the original white light is a complex wave motion that can be analyzed into components of different frequencies corresponding to the colored components, which are separated from one another by refraction at the water surface. Before Newton's time it was thought that the colors were somehow *created* in the refracting medium. Using glass prisms to produce dispersion as in Fig. 15, Newton showed that the colored light produced by dispersion could be recombined to produce white light, but that a single color produced by dispersion was not changed into still other colors by a second refraction process. Thus he concluded that "*Light* is not similar or Homogenial, but consists of Difform Rays, some of which are more Refrangible than others."

Table 36–1 gives the refractive indices μ_R, μ_Y, and μ_B of several materials for wavelengths in the red, yellow, and blue. The specific wavelengths are those of three easily reproducible spectral colors, the red and blue wavelengths appearing in the spectrum of hydrogen and the

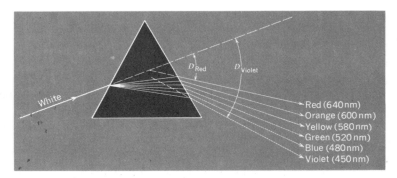

Fig. 15. Dispersion of white light by a prism. The angle D is called the *deviation*.

Table 36-1 REFRACTIVE INDICES AT 20° C FOR VARIOUS WAVELENGTHS

Material	μ_R (656.3 nm)	μ_Y (589.3 nm)	μ_B (486.1 nm)	$\mu_B - \mu_R$
Water	1.3312	1.3330	1.3372	0.0060
Carbon disulfide	1.6182	1.6276	1.6523	0.0341
Crown glass (typical)	1.5145	1.5172	1.5240	0.0095
Flint glass (typical)	1.6221	1.6270	1.6391	0.0170

yellow in the spectrum of sodium. We shall consider the nature of spectra in some detail in Chapter 41.

6. Transmission and Absorption of Light Waves

When a light beam strikes the boundary between two transparent media, there is in general both a *refracted* and a *reflected* beam. The laws of reflection and refraction give a satisfactory account of the *directions* of propagation of these beams but give no information regarding their *intensities*. We shall deal with the *radiant* intensity of a beam of radiation, which represents the energy transferred per unit time per unit cross-sectional area of the beam, measured, for example, in W/m². We note that if the light is monochromatic, the *luminous intensity* is proportional to the radiant intensity, so that our discussion will apply equally well to radiant and luminous intensities. We shall give here only a qualitative discussion of the factors influencing these intensities.

When a light beam strikes the interface between two transparent media, the fraction of the incident light that is reflected, called the *reflectance*, depends upon two factors: (*a*) the relative refractive indices of the two media, and (*b*) the angle of incidence.

For light beams in air ($\mu=1.00$) striking a crown glass ($\mu=1.52$) surface and a water ($\mu=1.33$) surface at normal incidence, the reflectance of the glass surface is greater than that of the water surface. This result is quite general for a given angle of incidence; the more abrupt the change in refractive index at the interface, the greater the reflectance. Stated in another way: for a given angle of incidence, the greater the change in *speed* of propagation at the interface, the greater the reflectance.

For two transparent media, such as air and glass, the reflectance increases with increasing angle of incidence. Since the change in direction of propagation increases with increasing angle of incidence, we may say qualitatively that the greater the change in the *direction* of propagation, the greater the reflectance. When light strikes the surface of a rarer medium, we have seen that the reflection becomes *total* for sufficiently large angles of incidence; i.e. the reflectance is unity. When light strikes the surface of a denser medium, the reflectance increases markedly as the angle of incidence increases toward 90° but reflection never becomes total.

In general, when a light beam is transmitted through a transparent material medium, some of the radiant flux is continuously *absorbed,* and the radiant intensity continuously decreases. The absorption may be slight, as in air, or rapid, as in black glass, but in every case radiant energy is continuously removed from the beam and converted into the thermal energy of atoms and molecules. The decrease in the intensity of a light beam in passing through an absorbing medium is accurately described by a relation known as *Lambert's law.**

Consider a beam of monochromatic radiation of intensity I_0 entering a material medium as shown in Fig. 16; Lambert's law gives the beam intensity I_X after the beam has traversed the distance X in the medium. The law is derived from the reasonable assumptions that the amount of absorption (the change ΔI in intensity) in traversing a small distance ΔX is proportional to ΔX and to the intensity I itself, say $\Delta I = -kI\,\Delta X$, where the proportionality constant, k, which is characteristic of the absorbing medium, is called the *absorption coefficient.* In the limit, we have the differential equation $dI/dX = -kI$, whose solution is

$$I_X = I_0\,e^{-kX}, \quad \text{(Lambert's law)} \quad (7)$$

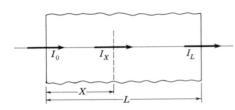

Figure 16

which is Lambert's law. In the whole thickness L of the slab in Fig. 16, the intensity would thus decrease to $I_L = I_0\,e^{-kL}$. This simple exponential expression leads to many useful results; however, it should be remembered that k is usually a function of wavelength, and may be strongly dependent on wavelength as is the case with colored glasses.

Since beam intensity I is proportional to the square of the wave amplitude, it should be noted that the wave amplitude decreases more slowly than the intensity with increasing values of X.

In closing, we might speculate on the situation to be expected if the exponent in (7) were positive instead of negative. This question will be discussed later when we consider the recently developed optical devices known as *lasers* (Chapter 41).

PROBLEMS

NOTE: *In working most of these problems, the refractive index of air can be taken as unity. Take other indices of refraction from Table 34–1, p. 717, or from Table 36–1.*

---

*JOHANN HEINRICH LAMBERT, German-French scientist (1728–1777).

1. A man strikes a match in a dark room and sees the virtual image of the match by light reflected by a plane mirror. If the image *appears* to be 12 f from the lighted match what is the distance from the match to the mirror?
Ans: 6.00 f.

2. A man 6 f tall stands 4 f in front of a large vertical plane mirror. Where will a virtual image of the man be formed? How tall will the image be?

3. A girl's eyes are 4 f from the floor while she is wearing shoes. The top of her hat is 5 f from the floor. At what heights from the floor should the lower and upper edges of a vertical mirror be placed so that she can just see herself, with hat, at full length? Show that the answer is independent of the distance the woman stands from the mirror, and draw a ray diagram to illustrate this point clearly.
Ans: 2.00, 4.50 f.

4. A young woman 5 f, 6 i tall wishes to purchase a mirror just long enough to enable her to see a full-length image of herself. How long should the mirror be?

5. Show that if two adjacent walls of a rectangular room are mirror surfaces, an observer sees exactly three images of himself and of all other objects in the room. Locate the images for an arbitrary position of the object, and trace the rays associated with each image. The rays should be traced from a point of the object to the observer's eye.

6. Show that if the mirror walls of Prob. 5 include an angle of 60°, there are five images. Locate the images and trace rays as in Prob. 5.

7. What does an observer see if two adjacent walls and the ceiling of a rectangular room are mirror surfaces? Explain clearly.
Ans: seven images.

8. Prove geometrically that all rays from O in Fig. 5 that are reflected from the mirror appear to come from a common point I located as described in the text.

9. In Prob. 7 show that if a narrow beam of light strikes near the mirrored corner from any direction, it is reflected back in a direction exactly opposite. How is this property employed in retrodirective reflectors used for highway markers?

10. Sensitive instruments (for example the Cavendish apparatus, p. 61) frequently measure the twist of a fiber by attaching a tiny plane mirror to the fiber and reflecting a narrow beam of light from the mirror to a scale. In this arrangement the light beam acts as an inertialess pointer. Show that if the mirror turns through angle θ, the reflected light beam turns through the angle 2θ. If it is desired that each minute of arc that the fiber twists result in a motion of 2 mm of the reflected beam on the scale, how far must the scale be from the mirror?

11. A ray of light in air strikes a water surface at an angle of incidence of 60°. What is the angle of refraction? Show diagram.
Ans: 40°.5.

12. A ray of light in air strikes the surface of a diamond at an angle of incidence of 45°. What is the angle of refraction? Show diagram.

13. The index of refraction of a certain type of glass is 1.5. Taking $c = 3 \times 10^8$ m/s, calculate the speed of light in this glass.
Ans: 2.0×10^8 m/s.

14. The measured speed of monochromatic light in a liquid is 2.2×10^8 m/s. What is the index of refraction of this liquid?

15. A ray of light in air has an angle of incidence of 30° at the surface of a plate of glass 3.00 cm thick, with $\mu = 1.523$. The ray emerging from the glass is parallel to the incident ray but is displaced laterally with respect to the incident ray. What is the magnitude of the lateral displacement?
Ans: 0.60 cm.

16. A ray of light in water strikes the horizontal surface of the water at angle of incidence of 30°. What is the angle of refraction in air? What is the angle of incidence in water when the angle of refraction in air is 90°? Show diagrams. What is the critical angle for water?

17. What is the angle of deviation D when yellow light is incident at an angle of 50° on the dense flint-glass prism in the figure?
Ans: 52°.7.

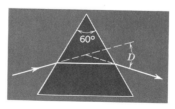

Problems 17–18

18. What is the angle of deviation D when yellow light is incident at an angle of 40° on the dense flint-glass prism in the figure?

19. A piece of plate glass with refractive index 1.50 forms the bottom of an aquarium tank. When the tank is empty, a ray of light coming from above strikes the glass plate at an angle of incidence of 30°. What is the angle of refraction inside the glass? If water is poured into the tank to a depth of 20 cm so that the ray must pass through water before reaching the glass, what will be the angle of refraction of the same ray inside the glass. Draw ray diagrams for the two cases, showing the lateral displacements of the rays. Ans: 19°.5; 19°.5.

20. The index of refraction of air depends only on temperature, pressure, and humidity. An astronomer reads temperature, pressure, and humidity at the earth's surface, and measures the apparent angle of a star from the zenith. He then looks up in a table the correction to be applied to give the true angle. To the approximation in which the curvature of the earth can be neglected and in which the atmosphere can be assumed to have characteristics that vary with altitude but do not vary in any given horizontal plane, show that these tables can be rigorously compiled because the angle of derivation depends only on the index of refraction at the earth's surface and is independent of the manner in which this index varies as one ascends to the top of the atmosphere.

21. Why do stars twinkle? Why does the air over a hot radiator seem to shimmer?

22. What is the minimum index of refraction that glass can have and still give total reflection for the light paths shown through the 90° prism of Fig. 10(c)?

23. What is the minimum index of refraction that glass can have and still give total reflection in the case of the 90° prisms shown in (a) and (b) of Fig. 10? Ans: 1.414.

24. What is the critical angle for diamond? Explain how the high index of refraction of diamond accounts for its extraordinary 'sparkle' when cut as a gem.

25. A large slab of glass of index 1.650 has a small air bubble a short distance below the surface. A dime (diameter 1.75 cm) placed on the surface of the glass is just large enough to completely prevent the bubble from being seen through the surface. How far down is the air bubble? Ans: 1.15 cm.

26. Why is a camera lens with a 180° field of view called a 'fish-eye lens'? What is the angle of the cone in which a fish would see the whole of the world above the level surface of a fresh-water lake?

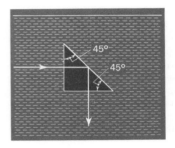

Problem 27

27. What is the minimum index of refraction of the material in a totally reflecting prism immersed in water, as indicated in the figure? Ans: 1.88.

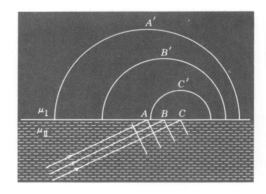

Problem 28

28. As in the figure, a light wave within a medium of high refractive index strikes the interface with a medium of lower refractive index at an angle of incidence greater than the critical angle. Show that, *at a given time,* the wavelets that originated at points A, B, C, of the interface are represented within the medium of low refractive index by *nonintersecting* circles such as A', B', C', which have no envelope.

29. A light ray consisting of wavelengths 656.3 nm and 486.1 nm is incident on the surface of carbon disulfide contained in a beaker. If the angle of incidence is exactly 45°, find the angles of refraction in carbon disulfide.
Ans: 25°.9; 25°.3.

30. Why does a piece of glass of irregular shape become invisible when it is immersed in a liquid of the same index of refraction?

31. A ray of white light is incident at an angle of 60° with the normal to the face of a 60° flint-glass prism as in Fig. 15. Make a large drawing showing the passage of rays of the red, yellow, and blue light of Table 36-1 through this prism, including accurate values of all angles involved.

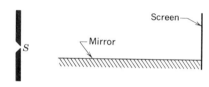

Problem 32

32. Describe the interference pattern formed on the screen in the *Lloyd's mirror* experiment shown in the figure. The mirror is a sheet of black glass—black so the reflection is all from the top surface. S is a narrow slit illuminated from the left by monochromatic light. Show that the interference pattern on the screen might be expected to be that of a pair of slits, S and its image S' an equal distance below the mirror surface, exactly half the pattern of Fig. 5, p. 733, in Young's experiment. Actually the pattern is reversed, with a central *dark* band at the point where the screen meets the mirror surface. Show that this reversal implies *a phase reversal when light rays in air are reflected from glass*. If the light has wavelength 589 nm, the slit S is 1 mm above the glass surface, and the horizontal distance from slit to screen is 50 cm, at what separation are the dark bands on the screen?

33. In the *mica-sheet* interference experiment devised by Pohl, a bright point source of monochromatic light S is placed in front of a thin sheet of mica. The source is imaged in the front surface of the mica at S' and in the back surface at S''. Light from these two images produces a spectacular pattern of cir-

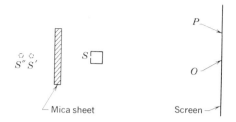

Problem 33

cular interference fringes on a screen across the room. Show accurately on a diagram the actual paths of the two rays of light from S that reach a given point P of the screen, and describe geometrically how you would determine whether they would interfere constructively or destructively. Account qualitatively for the circular interference fringes. If it were not for the fact that it is in the shadow of the light-source housing and hence receives no illumination, give the quantitative condition that would determine whether the center of the pattern at O would be bright or dark.

34. Discuss the luminous intensity of the image of a point source of light, formed by a perfect plane mirror.

35. A room in the shape of a 10-f cube has an isotropic point source of light of 600 cd at the center of the ceiling. (*a*) If the floor, ceiling, and all walls are black, what is the illumination at the center of the floor? (*b*) What is the increase in illumination if one wall is a perfect mirror?
Ans: 6.00 lm/f²; 2.12 lm/f².

36. In Prob. 35, what is the increase in illumination at points near the center of the floor if two adjacent walls are perfect mirrors?

37. After making appropriate corrections for front- and back-surface reflections, an experimenter determines that 2.0 percent of white light is *absorbed* in passing through a 1-cm sheet of a certain glass, that is, that I_L/I_0 in Fig. 16 is 0.980. Determine the absorption coefficient k in Lambert's law (7).
Ans: 0.020 cm⁻¹.

38. It is desired to make a safety shield by using a thickness of 50 cm of the glass described in Prob. 37. Neglecting reflections, determine the percentage of incident white light that will be transmitted through this thickness.

37 MIRRORS AND LENSES

In the preceding chapter we discussed the formation of virtual images by plane mirrors; because the normal component of the velocity of an advancing wave is reversed at the reflecting surface, waves from a point object are reflected in such a way that the reflected waves *appear* to originate at a point behind the mirror called a *virtual image*. By considering an extended object as a collection of points we showed that a perfectly flat reflecting surface produces a perfect image of the object; the image distance and the object distance are exactly equal, as are the sizes of the image and object.

The present chapter considers reflection at curved surfaces. A mirror with a curved surface can produce virtual images that are not necessarily the same size as the original object; in addition, a concave reflecting surface can cause light waves from an object to converge and can thus produce a *real image*, which can be directly observed or can be viewed on a suitably positioned screen. Although curved reflecting surfaces seldom produce perfect images, mirrors with curved surfaces are used to advantage in many optical instruments.

In the preceding chapter we discussed the change in the direction of propagation of light waves that pass from one optical medium into a second optical medium having a different refractive index. In the present chapter we show how this phenomenon can be used in the fabrication of the *lenses* employed in optical instruments to produce real or virtual images. The lens in a motion-picture projector produces a greatly enlarged real image of the film; the lens in a photographic camera normally produces a diminished real image of the object being photographed; the lens used as a 'reading glass' produces an enlarged virtual image of a printed page.

The failures of mirrors and lenses to produce perfect images are called *aberrations*. After we have shown how mirrors and lenses are actually used to produce images of acceptable quality, we shall give a brief discussion of abberations.

1. Ellipsoidal and Paraboloidal Mirrors

In order for a mirror to produce a perfect real image I of a point object O, *all optical paths traversed by light waves traveling from O to the mirror surface and thence to I must be of equal length.* Only in this case will all the waves originating at O interfere constructively when they reach I. Perfect or ideal real images can be produced by curved mirrors in a few special cases, which we shall now consider.

Figure 1(a) shows an ellipse with foci O and I. We recall that the geometrical properties of an ellipse are such that the sum of the distances r_1 and r_2 are the same for all points S on the ellipse. If we rotate the ellipse in Fig. 1(a) about its major axis, we generate an ellipsoid of revolution; if we silver the inner surface of this ellipsoid, all light waves leaving a luminous point object at focus O will, after reflection at the silvered surface, pass through focus I. Since the lengths of *all* optical paths $r_1 + r_2$ are equal, a perfect real image I of the point object O will be produced.

The ellipsoidal mirrors used in optical instruments are not closed surfaces but are similar to the one suggested in Fig. 1(b). Light waves spreading from point object O strike the reflecting surface and are reflected through I along the optical paths shown by the rays in the figure. We note that, if the positions of object O and image I were re-

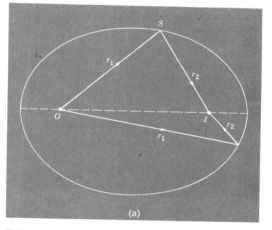

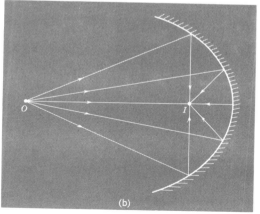

Fig. 1. Ellipsoidal mirrors.

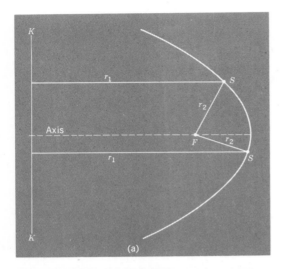

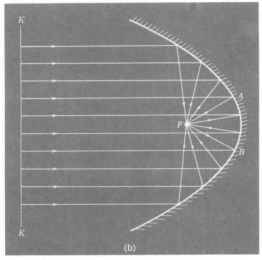

Fig. 2. Paraboloidal mirrors.

versed, the optical path lengths r_1+r_2 would remain the same. We conclude that, when a point object is placed at one focus of an ellipsoidal mirror, a perfect real image is produced at the other focus. However, if a point object is placed at any other position, a perfect image is not produced.

Another situation in which a perfect real image can be produced by a mirror is illustrated in Fig. 2. The geometrical properties of a parabola, such as the one shown in Fig. 2(a), are such that the sum of the lengths r_1+r_2 are the same for all points S on the parabola, where r_1 is parallel to the axis of the parabola and r_2 is the length of the line connecting S and the focus F of the parabola. If we rotate the parabola in Fig. 2(a), we produce a paraboloid of revolution; if then we silver the inner surface of the paraboloid, we produce a paraboloidal mirror. Fig. 2(b) shows parallel light paths from an infinitely distant point object on the axis; since the light paths or rays are parallel, the wavefront KK is plane. Since the optical path lengths r_1+r_2 are the same for all paths between KK and F, a perfect point image will be formed at the focus of the paraboloid. Since light paths are reversible, we may produce a strictly parallel beam of radiation by placing a point object at the focus of a paraboloidal mirror, as is approximated in searchlights and automobile headlights.

Fortunately, we can use mirrors to produce useful, easily recognizable images without meeting the strict requirements we have set up for the formation of perfect images; acceptably good images can be produced even though all optical paths between point object and point image are not *exactly* equal. Thus, a paraboloidal mirror can be used to produce recognizable images of distant point objects even if the point objects are not on the axis of the paraboloid provided the parallel rays from the distant objects make small angles with the axis; these imperfect point images are formed in a plane perpendicular to the axis called the *focal plane*. Thus, the paraboloidal mirror used in an astronomical telescope produces acceptable real images of many stars in the same plane as that in which the perfect image of a star on the mirror axis would be formed. Similarly, we can produce a nearly parallel beam of light by placing a light source of *finite size* at the focal point of a paraboloidal mirror; such an arrangement is employed in searchlights and automobile headlights.

Paraboloidal and ellipsoidal mirrors are difficult to fabricate and are therefore expensive; they are used only for the special purposes we have indicated. In spite of the fact that the easily fabricated spherical mirrors to be described in the next section do not produce perfect images, they are more generally useful and are widely used in optical instruments. Like a paraboloidal mirror, a concave spherical mirror can under certain conditions produce clear images of distant point objects; like an ellipsoidal mirror, a concave spherical mirror can produce an image of a nearby point object. A spherical mirror thus combines desirable properties of both paraboloidal and ellipsoidal mirrors but is, as we shall see, more generally useful than either.

Example. *The solar observatory at Kitt Peak, Arizona, has a paraboloidal mirror with a diameter of* 60 i, *which is used to produce a real image of the sun in its focal plane, which is located* 300 f *from the mirror surface. Using the data given on p. 702, determine the diameter of the image of the sun formed by the Kitt Peak paraboloid.*

The angle subtended by the solar disc is

θ = diameter of sun/distance to sun

= 864 000 mi/92 000 000 mi

= 0.009 29 rad.

If the light from the center of the solar disc is parallel to the axis of the paraboloid, the center of the disc will be imaged *perfectly* at the focus of the paraboloid. In view of the small value of θ, all parts of the disc will be well imaged in the focal plane. The image diameter D will subtend an angle equal to $\theta = D/f$, where f is the distance between the mirror and the focal point; we may verify this by drawing rays from various parts of the sun to the center of the paraboloid. Using this relation, we obtain

$D = f\theta = 300 \text{ f} \times 0.009\ 29 \text{ rad} = 2.78 \text{ f}.$

Fig. 3. Image of the sun obtained at the Kitt Peak National Observatory.

Thus, the diameter of the sun's image is 2.78 f or 33.4 i. As we can see from the photograph in Fig. 3, the image is excellent.

2. Concave Spherical Mirrors

A spherical mirror is a mirror that can be obtained by cutting off a portion of a reflecting sphere, usually by a plane in the manner suggested in Fig. 4. The line AB represents the cutting plane and the arc

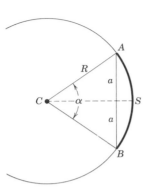

Fig. 4. Construction of a spherical mirror.

ASB represents the portion of the sphere to be used as the mirror. If the *inner* surface of the spherical segment *ASB* is the reflecting surface, the spherical mirror is said to be *concave;* if the *outer* surface is used as the reflecting surface, the mirror is said to be *convex.* The *axis* of a spherical mirror is a line *CS* drawn from the center of the original sphere to the center of the mirror.

The distance $AB = 2a$ is called the *linear aperture* of the mirror and the angle α is called the *angular aperture* of the mirror. If the mirror is to be useful in producing sharp images, the angular aperture must be small compared with one radian. We shall restrict our discussion to mirrors of *small angular aperture,* although our diagrams, for the sake of clarity, will apparently show large apertures.

We first consider the formation of images by a concave spherical mirror. As we have noted in connection with Fig. 2, *all* incident rays parallel to the axis of a paraboloidal mirror converge to a point at the focus of the paraboloid. A small section of a paraboloidal mirror near the axis, such as section *AB* in Fig. 2(b), which has a small angular aperture as measured from the focus, approximates a concave spherical mirror of small aperture. Hence, a concave spherical mirror of small angular aperture also has the property of converging parallel rays to a point *F*, called the *principal focus* or *principal focal point* of the mirror,

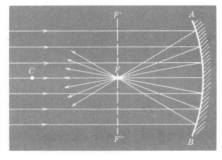

Fig. 5. Parallel light rays converge at the principal focus *F* of a concave spherical mirror of small aperture.

as shown in Fig. 5. Because of the reversibility of light paths, a spherical mirror of small angular aperture produces a parallel beam of light if a luminous point object is placed at its principal focus *F*. For distant point objects slightly off the mirror axis, a spherical mirror, like a paraboloidal mirror, forms point images in the principal focal plane *F'FF''* provided the mirror's angular aperture α is small and provided all rays from the distant objects make small angles with the mirror axis.

A concave spherical mirror also has the property of converging light from a point object *O* on the mirror axis to a point image *I* also on the mirror axis as shown in Fig. 6. Since the angle of incidence of light from *O* at point *P* of the mirror is less than the angle of incidence at the same point in Fig. 5, we note that image point *I*, where the rays cross the mirror axis, is beyond the principal focal point *F* and lies between *F* and the mirror's center of curvature *C*. Because of the reversibility of light paths, we note that if a luminous point object were placed at point *I* in Fig. 6, the mirror would produce a point image at point *O*.

In order to understand the mirror properties just described and also to appreciate the emphasis we have placed on the need for small angular aperture, consider the typical light path shown in the ray diagram in Fig.

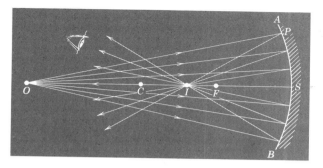

Fig. 6. Formation of a real image of a point object by a concave mirror.

7, in which the object and image positions correspond to those in Fig. 6. In Fig. 7, ray OA represents the path traversed by light waves in passing from the point object O to some typical point A on the mirror surface; ray AI gives the path traversed by light in passing from point A to image point I on the mirror axis. The normal to the mirror surface at point A is given by the line CA drawn from the mirror's center of curvature C to point A; from the laws of reflection, we know that the angle of incidence i and the angle of reflection r at point A are equal. Remembering that the exterior angle of a triangle is equal to the sum of the opposite interior angles, we write for triangle OAC:

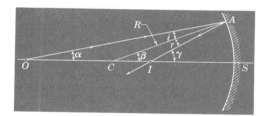

Figure 7

$$\beta = \alpha + i \quad \text{or} \quad i = \beta - \alpha;$$

similarly, for triangle CAI we write:

$$\gamma = \beta + r \quad \text{or} \quad r = \gamma - \beta.$$

Since $r = i$, we obtain the result

$$\gamma - \beta = \beta - \alpha \quad \text{or} \quad \alpha + \gamma = 2\beta.$$

Now angle β in radians equals the arc AS divided by the mirror radius $R = CS$. Angle α equals the arc $A'S$ of a circle centered at O divided by the radius of this circle OS; similarly, angle γ equals the arc $A''S$ of a circle centered at I divided by the radius of this circle IS. We note that if we let angle $\beta \to 0$, the arc lengths $A'S$ and $A''S$ both approach arc length AS; in the limit, we can write $\alpha = AS/OS$ and $\gamma = AS/IS$. In this case, the equation $\alpha + \gamma = 2\beta$ becomes

$$\frac{AS}{OS} + \frac{AS}{IS} = 2\frac{AS}{R} \quad \text{or} \quad \frac{1}{OS} + \frac{1}{IS} = \frac{2}{R}.$$

This is the equation that determines IS in terms of OS and the mirror radius R. It is customary to call OS the *object distance* and to denote it by p; and to call IS the *image distance* and to denote it by q. In terms of these symbols, the above equation becomes

$$\frac{1}{p}+\frac{1}{q}=\frac{2}{R}, \qquad \left(\begin{array}{c}\text{mirror}\\\text{equation}\end{array}\right) \quad (1)$$

which is called the *mirror equation*. We note that the mirror equation holds *exactly* only in the limit when the mirror's angular aperture $2\beta \to 0$. However, the mirror equation holds in close approximation whenever the angular aperture 2β is small compared with one radian; in the remainder of our discussion, we shall assume mirrors of small angular aperture but shall return to a consideration of large angular aperture when we consider aberrations.

We note that equation (1) is symmetric in p and q; that is, values of p and q can be interchanged without altering the equation. This symmetry is associated with the reversibility of optical paths noted in our discussion of Figs. 5 and 6. Thus, if the object were placed at point I in Fig. 7 and a ray were drawn from this point *to* point A on the mirror, the reflected ray would cross the axis at point O.

Equation (1) also states that when $p=R$, q also equals R. This statement is correct because *all* rays from C strike the reflecting surface at normal incidence and are therefore merely reversed at the mirror surface so that they return to point C; in this trivial case alone a spherical mirror of any angular aperture forms a perfect image! Now let us see what happens as the object distance increases in the range $p>R$. As p increases in (1), q decreases. The largest value p can have is ∞; in this case $1/p=0$ and $q=R/2$ as in Fig. 5. 'Object at infinite distance' means physically that rays coming from the object are parallel; the image position in this case is called the *principal focus* of the mirror. The distance of the principal focus from the mirror is called the *focal length* f of the mirror; thus, the focal length of a concave spherical mirror is given by the expression

$$f=\tfrac{1}{2}R. \qquad (2)$$

In terms of the focal length, equation (1) takes the form

$$\frac{1}{p}+\frac{1}{q}=\frac{1}{f}. \qquad (3)$$

Study of this equation shows that as the object distance p decreases from ∞ to f, the image distance q increases from f to ∞. But what happens when the object distance is less than f—that is, when the object is between the principal focus and the mirror surface? Formal solution of (3) gives a *negative* value for q; a negative value for q suggests that the image is *behind* the mirror and is therefore *virtual*. Accurate analysis shows that this supposition is correct as indicated in Fig. 8, which shows selected light paths leading from point object O to the mirror surface; the paths of the reflected light diverge in such a way that they *appear* to come from the virtual point image I behind the mirror.

Thus far, we have considered only the image of a point object located on the mirror axis. We now show that a concave mirror forms an image of an extended object such as the object O lying in a plane normal to the mirror axis in Fig. 9, provided the extended object subtends a small angle at point S. We know that an image of point U is formed at point W in accordance with (1). We know that any image of T must be formed at some point V along a line drawn from T through the center of curvature C to the surface of the sphere, since light traversing this path strikes the reflecting surface at normal incidence and is reversed in direction as indicated by the ray in the figure. By selecting another easily traced light path, we may locate the position of the image at the point at which the two rays intersect; as the second ray we choose the one that strikes the mirror at point S on the axis with angles i and r equal.

We shall now show that p and q, the distances of the off-axis point object T and image V, are related by equation (1), which we derived for object and image on the mirror axis.

For the moment we forget that point W in Fig. 9 is actually the image

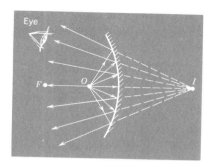

Fig. 8. Formation of a virtual image of a point object by a concave mirror.

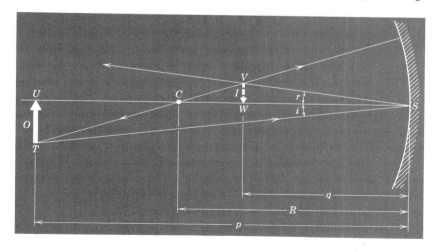

Figure 9

of point U and consider W as merely the foot of the perpendicular dropped from V to the axis. Since the triangles TUC and VWC are similar,
$$TU/VW = CU/CW.$$
Since the triangles TUS and VWS are similar,
$$TU/VW = US/WS.$$
Combining these relations, we obtain
$$CU/CW = US/WS.$$
When these quantities are written in terms of p, q, and R, the last equation becomes
$$\frac{p-R}{R-q} = \frac{p}{q},$$
which by cross multiplication becomes
$$pq - Rq = Rp - pq \quad \text{or} \quad Rq + Rp = 2qp.$$
On dividing each term of this last equation by pqR, we obtain
$$\frac{1}{p} + \frac{1}{q} = \frac{2}{R}.$$
But this is the same equation (1) that applies to point object U and point image W on the mirror axis. Since the same type of derivation can be given for the image of any point object near the axis in a plane perpendicular to the axis, we conclude that *a plane object O at distance p from the mirror has a plane image I at a distance q from the mirror.* This conclusion is correct even for virtual images.

From the equation given above for similar traingles TUS and VWS, we see that
$$\frac{\text{image length}}{\text{object length}} = \frac{\text{image distance}}{\text{object distance}}. \tag{4}$$
This relation also applies to virtual images as well as real images provided we ignore the algebraic sign of q. The ratio of image length to object length is called the *enlargement;* for the situation shown in Fig. 9, the enlargement is less than unity.

The best understanding of the characteristics of image formation by concave mirrors is obtained by drawing 'principal-ray' diagrams. Such diagrams are shown in Fig. 10 for the cases in which the object is (a) beyond the center of curvature, (b) between the principal focus and the center of curvature, and (c) inside the principal focus. The so-called *principal rays* are four rays whose paths, from the tip O of the object in Fig. 10 to the mirror and back, are very easy to trace. The principal rays, numbered as in Fig. 10, are:

1. The ray that leaves the tip of O parallel to the axis and is reflected through the principal focus F (a ray of Fig. 5).
2. The ray that leaves the tip of O along the line through the principal focus and is reflected parallel to the axis (a reversed ray of Fig. 5).
3. The ray that leaves the tip of O along the line through the center of curvature C and is reflected back along itself (as in Fig. 9).
4. The ray that strikes the mirror at the axis and is reflected at an equal angle on the opposite side of the axis (as in Fig. 9).

These four rays intersect at the tip I of the image, and serve to locate the position, size, and orientation of the image graphically. From the three diagrams of Fig. 10 we see at once that:

When the object is beyond the center of curvature, the image is real, inverted, and smaller than the object, and lies between the principal focus and the center of curvature.

When the object lies between the principal focus and the center of curvature, the image is real, inverted, and larger than the object, and lies outside the center of curvature.

When the object is inside the principal focus, the image is virtual, erect, and larger than the object, and lies behind the mirror.

Example. *A concave mirror of 2-f radius of curvature is used to produce an image of a lamp filament on a wall 15 f from the mirror. Where must the filament be placed, and what is the size and character of the image?*

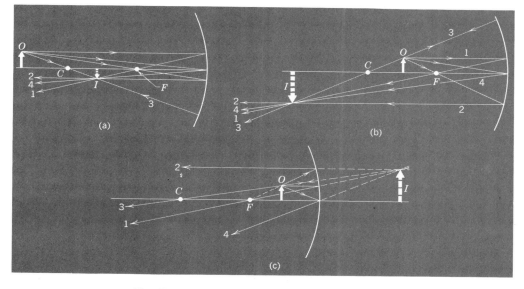

Fig. 10. Principal-ray diagrams for a concave mirror.

A real image beyond the radius of curvature can only occur in the case of Fig. 10(b), in which the object is between the principal focus and the center of curvature. From this figure we see that the image is *real, inverted,* and *enlarged.* We obtain the object distance p by substituting $R = 2\,f$ and $q = 15\,f$ in (1):

$$\frac{1}{p} + \frac{1}{15\,f} = \frac{2}{2\,f}, \quad \text{or} \quad p = {}^{15}\!/_{14}\,f,$$

slightly greater than the focal length. The enlargement is given by (4) as

$$\text{enlargement} = \frac{\text{image length}}{\text{object length}} = \frac{15\,f}{{}^{15}\!/_{14}\,f} = 14.$$

The image is 14 times the size of the filament in height and width.

3. Reflection by Convex Spherical Mirrors

When a point object is on the axis of a *convex* mirror, the rays from the object clearly *diverge* after reflection from the mirror. This statement is true for any position of the object. However, if the angle of incidence of all the rays is small, which implies that the angular aperture of the mirror is small, the rays diverge from a common point as indicated schematically in Fig. 11. This point constitutes a virtual image of the object. A person looking into the convex mirror of Fig. 11 sees a mirrored image of O at I, much as in the case of a plane mirror.

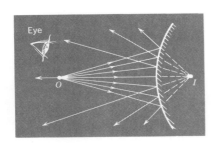

Fig. 11. Formation of a virtual image of a point object by a convex mirror.

A proof that such an image is formed can be given by methods similar to those we used in connection with Figs. 7 and 9. Also, we can show that a plane object such as OA in Fig. 12 has a plane image IB. We shall omit these proofs; but granted that an image of point O in Fig. 12 is formed at some point I, we shall determine the location of I by finding the intersection of the two rays shown in Fig. 12. One of these rays heads from O toward the center of curvature and is reflected back along itself; the other strikes the mirror at the axis and is reflected symmetrically. In the case of convex mirrors, *the radius of curvature R is considered to be a negative number;* hence, the distance from the mirror to the center of curvature is represented by $-R$. As is customary in the case of virtual images, *the image distance q is considered to be a negative number;* hence, the distance from the image to the mirror is represented by $-q$.

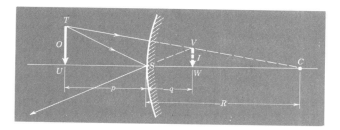

Fig. 12. Image formation by a convex mirror.

From the similar triangles VWS and TUS, $TU/VN = US/WS$. From the similar triangles VWC and TUC, $TU/VW = UC/WC$.

Hence
$$\frac{US}{WS} = \frac{UC}{WC}$$

Inserting the values of these lengths in terms of p, q, and R gives

$$\frac{p}{-q} = \frac{(-R)+p}{(-R)-(-q)} = \frac{p-R}{q-R}.$$

Clearing fractions gives $\quad -pq + qR = pq - pR$

or $\quad qR + pR = 2pq.$

Dividing through by pqR, we again obtain the *mirror equation*

$$\frac{1}{p} + \frac{1}{q} = \frac{2}{R}. \tag{5}$$

For a convex mirror we must remember that R is a negative number; this convention makes q come out negative for any positive value of p, indicating that *the image is always virtual and behind the mirror.*

A study of equation (5) shows that as the object distance p varies from 0 to ∞, the image position varies from $q=0$ to $q = \tfrac{1}{2} R$. Object at infinity corresponds to parallel light incident on the mirror as in Fig. 13. The virtual-image point in this case is called the *principal focus* of the convex mirror. The distance of the principal focus from the mirror, $q = \tfrac{1}{2} R$, is called the *focal length* and is denoted by f. Just as R is taken as a negative number, *the focal length $f = \tfrac{1}{2} R$ is a negative number in the case of a convex mirror.* In terms of focal length the mirror equation takes the form

$$\frac{1}{p} + \frac{1}{q} = \frac{1}{f} = \frac{2}{R}. \tag{6}$$

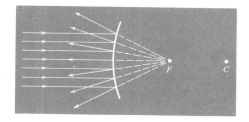

Fig. 13. Parallel light rays diverge from the principal focus F of a convex mirror. The principal focus is behind the mirror, halfway between the mirror and the center of curvature C.

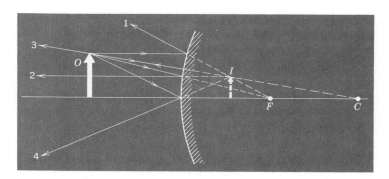

Fig. 14. Principal-ray diagram for a convex mirror.

It is convenient that *equation* (6) *applies to all types of spherical mirrors with the sign conventions we have adopted:*

p is positive for any real object (later we shall define a virtual object for which p is negative).
f and R are positive for a concave (converging) mirror.
f and R are negative for a convex (diverging) mirror.
If q is positive, the image is real (in front of the mirror).
If q is negative, the image is virtual (behind the mirror).

In the case of a convex mirror, we can again make good use of a principal-ray diagram such as that in Fig. 14. The four principal rays shown in Fig. 14 are the same as the four listed on p. 769 in the discussion of the concave mirror.

From Fig. 12, we conclude that the image is always erect and smaller than the object, and we obtain the same analytical expression (4) for image length as in the case of a concave mirror.

4. Treatment of Convex Mirrors by Wave Optics

It is instructive to derive the mirror equation by considering the reflection of wavefronts, much as we did in the case of plane mirrors on pp. 746–747. We shall use a *convex* mirror as our example. (This section of the text may be omitted, if desired, without loss of continuity.)

Figure 15 shows a plane wave advancing toward a convex spherical mirror. At $AA'A$, the wavefront has just reached the mirror, and its center begins to turn back. Successive later positions are shown at $BB'B$, $CC'C$, and $DD'D$. For small mirror aperture, the reflected wavefront is very close to spherical; it is clear from Fig. 15 that it has greater curvature than the mirror itself so it must be centered at some point F between the center C of the mirror and the mirror itself. We assume that the incident plane waves originate from an infinitely distant point source

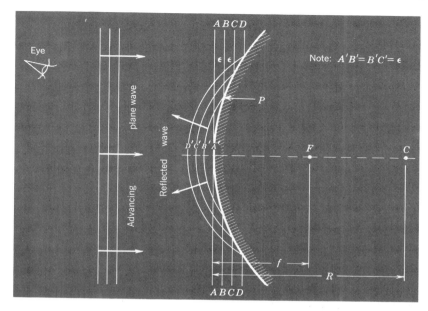

Fig. 15. Reflection of a plane wave from a convex mirror surface.

located on the mirror axis. Point F is therefore the virtual image of the distant point source and is the *principal focus* of the mirror.

From the sagitta approximation, relation (2), p. 707, we can readily show that the distance of the principal focus F from the mirror in Fig. 4 is $f = \frac{1}{2} R$, where R is the radius of curvature of the mirror. In Fig. 16, which matches Fig. 15, we have drawn the mirror surface $A'P$ and the wavefront $B'PB$, and have set up a system of XY-coordinates. The significant point is that P is at the same distance to the right of the Y axis as B' is to the left since the plane oncoming wavefront in Fig. 14 travels from position AA to position BB in the same time as the reflected wavefront on the mirror axis travels from A' to B'. Call this distance ε, since we are going to treat it as small, confining our consideration to the immediate neighborhood of the axis of the mirror. From the sagitta relation, since P lies on the mirror surface $A'P$, we have

$$X_P = Y_P^2 / 2R. \qquad (7)$$

The point P also lies on the wavefront $B'P$ of radius $f + \varepsilon$. Applying the sagitta relation to the circle $B'P$, we find that

$$\varepsilon + X_P = Y_P^2 / 2(f + \varepsilon). \qquad (8)$$

From Fig. 16, we see that $X_P = \varepsilon$; substitution of

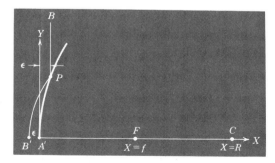

Fig. 16. Derivation of the focal length of a convex mirror.

this value in (7) and (8) gives

$$\varepsilon = Y_P^2/2R; \qquad 2\varepsilon = Y_P^2/2(f+\varepsilon). \tag{9}$$

Division of the first equation in (9) by the second gives

$$\frac{1}{2} = \frac{f+\varepsilon}{R},$$

or, in the limit of small apertures, where $\varepsilon \rightarrow 0$,

$$f = \tfrac{1}{2} R. \tag{10}$$

Thus we have demonstrated that the focal length of a concave mirror is one-half of its radius of curvature.

Now consider a point object O at a *finite* distance from a convex mirror, as in Fig. 17. It is apparent that the reflected wavefront has even greater curvature than in the case of Fig. 15, where the object is at infinite distance. The reflected spherical wavefront is centered at an image point I that is closer to the mirror than is the focus F. We represent the distance of the object from the center of the mirror by p and that of the image by q. We can again find the image distance q by considering the point P in Fig. 17 where the wavefront $BB'B$ contacts the mirror.

We employ the same coordinate system as in Fig. 16, with origin

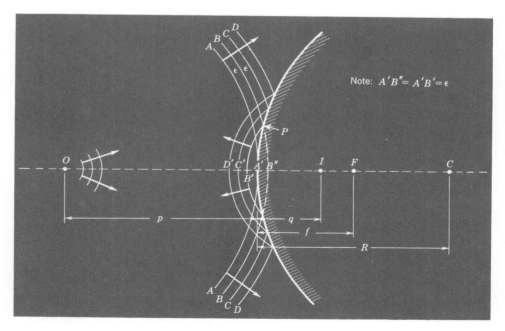

Fig. 17. The image of a point object in a convex mirror.

at the point A', and again denote the distance between the wavefronts AA and BB by ε. The point P lies on three circles. It lies on the mirror surface, where the sagitta relation gives

$$X_P = Y_P^2/2R, \quad \text{or} \quad X_P = Y_P^2/4f. \tag{11}$$

The point P lies on the circle $BB''B$ that represents the oncoming wave in the absence of reflection. The horizontal distance of P, to the left of B'', is $\varepsilon - X_P$. The radius of this circle is $p + \varepsilon$. Hence the sagitta relation applied to this circle gives

$$\varepsilon - X_P = Y_P^2/2(p+\varepsilon). \tag{12}$$

The point P also lies on the outgoing wavefront PB', of radius $q+\varepsilon$. Application of the sagitta relation to this wavefront gives

$$\varepsilon + X_P = Y^2/2(q+\varepsilon). \tag{13}$$

We cannot neglect ε with respect to X_P in these equations because, as we see from Fig. 17, X_P is actually less than ε; but we can, in the limit of small aperture, neglect ε with respect to p and q on the right side of the last two equations. With this neglect we have, by subtraction of equation (12) from equation (13),

$$2X_P = Y_P^2 \left(\frac{1}{2q} - \frac{1}{2p} \right).$$

Then, by substitution in equation (11) and simplification, we obtain

$$\frac{1}{p} - \frac{1}{q} = -\frac{1}{f}. \tag{14}$$

This equation does not exactly agree with our previous mirror equation (6) because we have not introduced the previously adopted convention of taking image distance q and focal length f, for convex mirrors, as negative quantities. Introducing this convention changes the signs of q and f in (14) and we recover equation (6).

A derivation similar to the above will also give the mirror equation for a *concave* mirror.

5. The Ideal Lens

We have seen how the *laws of reflection* can be applied in the design of *mirrors* capable of producing useful images of luminous or illuminated objects. We now turn to a discussion of how the *laws of refraction* can be used to design *lenses* for image formation in optical instruments. We begin again with a discussion of the formation of an ideal or perfect image.

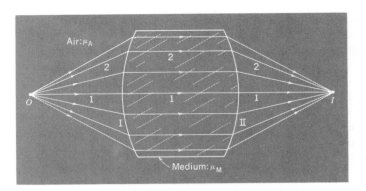

Fig. 18. Ideal lens (schematic).

A lens used in air with refractive index μ_A is usually fabricated from glass or some other transparent material with a refractive index $\mu_M > \mu_A$; at least one of the surfaces of the lens must be curved. If a lens is to form a perfect image I of a point object O, the *number of wavelengths* in *all* paths actually traversed by light waves passing from O to I must be the same. We recall (p. 718) that the *number of wavelengths* in a path length l_M in a medium having refractive index μ_M is equal to $\mu_M l_M / \lambda$, where λ is the vacuum wavelength. Thus, in the situation shown schematically in Fig. 18, *all* paths between O and I must contain the same number of wavelengths as optical path 1. Thus, for path 2,

$$\mu_A l_{2A}/\lambda + \mu_M l_{2M}/\lambda = \mu_A l_{1A}/\lambda + \mu_M l_{1M}/\lambda,$$

where l_{1A} and l_{2A} are the path lengths in air, and l_{1M}, l_{2M} those in the material. Dropping the λ, we see that we must have

$$\mu_A l_{2A} + \mu_M l_{2M} = \mu_A l_{1A} + \mu_M l_{1M} \tag{15}$$

if the lens is to form a perfect image I of point object O. By application of Snell's law at various points on surface I and surface II of the lens, it is possible to determine the proper curvatures of the two surfaces and the thickness of the lens required to give a perfect image for monochromatic light of a given wavelength; this can actually be done! In modern optical-engineering practice, digital-computer techniques can be applied not only to the problem of lens *design* but also to the formidable problem of lens fabrication.

Fortunately, as we have noted earlier, *perfect* images are not required even in optical instruments of high quality. We now turn to a discussion of how lenses can be designed with easily fabricated *spherical* surfaces. Most of our subsequent discussion will be limited to 'thin lenses,' for which $l_A \gg l_M$ for all optical paths.

6. Thin Lenses

The commonest form of simple glass lens designed for use in air has two surfaces that can be considered as parts of spheres as indicated in Fig. 19; the line joining the centers of the spheres is called the *axis* of the lens. A simple lens can be regarded as *thin* if its thickness t is small compared with the radii of curvature R_1 and R_2 of its surfaces. As indicated in Fig. 20, a glass lens thicker in the center tends to *converge* a parallel light beam, while a glass lens thinner in the center tends to *diverge* a parallel light beam. Cross sections of various types of simple lenses are shown in Fig. 21; the first three are converging lenses, the last three are diverging.

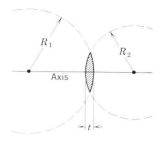

Fig. 19. Geometry of lens surfaces.

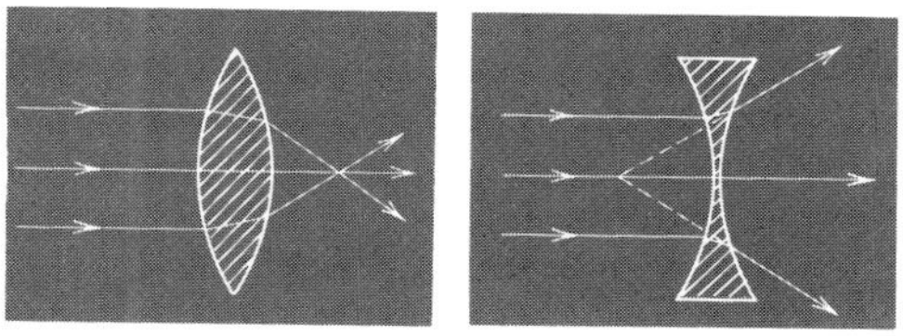

Fig. 20. Deviation of parallel light rays by lenses.

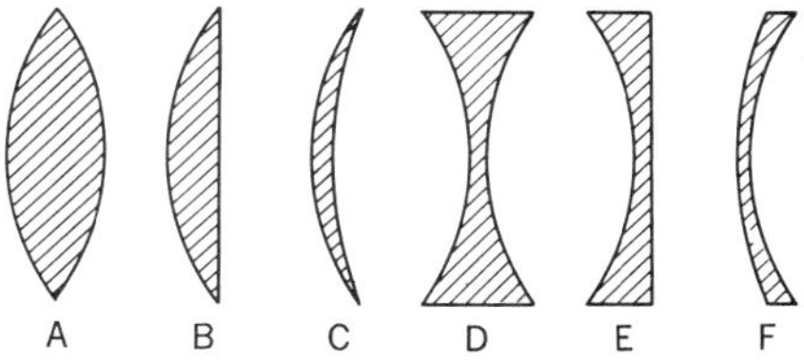

Fig. 21. Common lens types: A, double convex; B, plano-convex; C, concavo-convex; D, double-concave; E, plano-concave; F, convexo-concave.

In order to understand how lenses produce images of luminous or illuminated objects, consider the double convex glass lens in Fig. 22. Part (a) of the figure shows a typical ray leaving a point object O on the lens axis at a distance p from the lens; after passing through the lens, this ray again reaches the axis at point I at a distance q from the lens. Part (b) of the figure shows the refraction of the ray at the first surface. By Snell's law (5), p. 750,

$$\sin i_1 = \mu \sin r_1, \qquad (16)$$

where the refractive index of air is taken as unity and μ is the refractive index of glass. If the angle of incidence i_1 and the angle of refraction r_1 are sufficiently small, the sines can be replaced by the angles themselves, and (16) becomes $i_1 = \mu r_1$. By considering the exterior angles of the triangles in Fig. 22(b), we see that

$$i_1 = \theta + \alpha \qquad \text{and} \qquad r_1 = \alpha - \gamma;$$

hence, for small angles,

$$\theta + \alpha = \mu(\alpha - \gamma). \qquad (17)$$

Fig. 22(c) shows refraction at the second surface of the lens; the ray coming from the left in (c) is the same ray in the glass as that proceeding to the right in (b). Snell's law for small angles i_2 and r_2 at the second surface of the lens is $r_2 = \mu i_2$; consideration of the exterior angles of the triangles in part (c) of the figure leads to the expression

$$\beta + \phi = \mu(\beta + \gamma) \qquad (18)$$

By adding equations (17) and (18), we obtain

$$\theta + \alpha + \beta + \phi = \mu(\alpha + \beta)$$

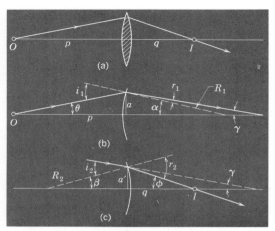

Fig. 22. Formation of a real image of a point object.

or
$$\theta + \phi = (\mu - 1)(\alpha + \beta). \qquad (19)$$

If we now assume that the thickness of the lens is negligible compared with p, q, R_1, and R_2, then it does not matter from what point in the lens we measure p and q; this assumption allows us to set height a in Fig. 22(b) equal to height a in Fig. 22(c). Since the small angles in (19) can be replaced by their sines or tangents, we can now write

$$\theta = a/q, \quad \phi = a/q, \quad \alpha = a/R_1, \quad \text{and} \quad \beta = a/R_2.$$

Substitution of these values in (19) and subsequent division by a gives the *thin-lens equation*:

$$\frac{1}{p} + \frac{1}{q} = (\mu - 1)\left(\frac{1}{R_1} + \frac{1}{R_2}\right), \quad \binom{\text{lens}}{\text{equation}} \quad (20)$$

where p is the object distance, q is the image distance, and R_1 and R_2 are the radii of curvature of the lens surfaces.

Since the ray shown in Fig. 22(a) simply represents a typical optical path between O and I, we conclude that (20) applies to *all* optical paths between O and I, provided all angles are sufficiently small and provided the lens thickness is sufficiently small. Hence, a real image is formed at point I. Although we have derived (20) for a point source on the axis of the lens, the same relation can be used for point objects slightly off the axis. It can therefore also be used for an extended object in a plane perpendicular to the lens axis, provided the light paths from every point in the object make small angles of incidence at the first surface of the lens. For objects of finite size, p in (20) is the distance from the lens to the plane in which the object lies and q is the distance from the lens to the plane in which the image lies.

An incident beam of light parallel to the axis of a converging lens is brought to a point focus F as indicated in the upper diagram in Fig. 23. The point F is called a *principal focus* of the lens, and its distance from the lens is called the *focal length* f. We can determine the focal length by setting $p = \infty$ in (20), in which case $q = f$; thus,

$$\frac{1}{f} = (\mu - 1)\left(\frac{1}{R_1} + \frac{1}{R_2}\right). \quad \binom{\text{lensmaker's}}{\text{equation}} \quad (21)$$

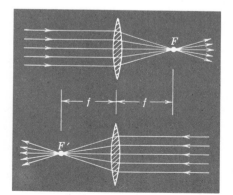

Fig. 23. Action of a converging lens on parallel light rays. Parallel rays intersect at the principal focal points of the lens.

This relation is called the *lensmaker's equation*.

Unlike a mirror, a lens is 'two-sided,' and light can pass through it in either direction. Furthermore, because (20) is symmetric with respect to R_1 and R_2, the image-forming properties of a lens are identical for light traveling in the two directions even though the radii of the two surfaces may be different. Consequently a parallel light beam coming from the right in the lower diagram of Fig. 23 is brought to a focus at point F', which is at the same distance f from the lens as point F. The points F and F' are called the two *principal foci;* they are equidistant from the lens, at the distance given by (21). Because light paths are reversible, we can, by placing a point source of light at either F or F', obtain a parallel beam of light, as shown in Fig. 24.

We can use this knowledge to employ a graphical method of finding the image positions of extended objects, since (*a*) any ray from the left coming to the lens parallel to the axis is refracted in such a manner that it passes through the principal focus F beyond the lens and (*b*) any ray from the left passing through principal focus F' in front of the lens is refracted in such a way as to leave the lens in a direction parallel to the axis. The graphical method of locating images is demon-

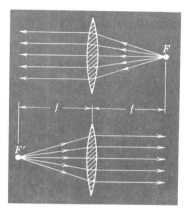

Fig. 24. Action of a converging lens on light rays from point objects at the principal foci.

strated in Fig. 25, in which the end of an extended object is located at point T. We can readily trace three so-called *principal rays* from the tip T of the object to the tip I of the image; these rays are numbered as in Fig. 25:

1. The ray that leaves the tip of O in a direction parallel to the axis and after refraction passes through principal focus F.
2. The ray that passes from the tip of O through principal focus F' and after refraction is parallel to the axis.
3. The ray that passes through the optical center X of the lens and is undeviated.

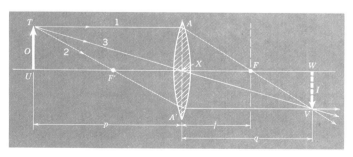

Fig. 25. Principal-ray diagram for a converging lens. The plane through F is called the principal focal plane of the lens.

Principal ray number 3 is defined by a straight line from the tip of O that intersects the axis at point X called the optical center of the thin lens. The undeviated ray through X divides the rays that are 'bent downward' by the lens from those that are 'bent upward.'

The enlargement produced by a lens is defined as the ratio of image length to object length. Consideration of similar triangles TUX and VWX in Fig. 25 shows that

$$\frac{\text{image length}}{\text{object length}} = \frac{\text{image distance}}{\text{object distance}}. \tag{22}$$

Example. *Use the principal-ray diagram of Fig. 25 to demonstrate that the lens equation (20) applies to extended objects.*

In Fig. 25, we let $p = TA$ and $q = A'V$, where V is the image of point T. Then, from geometry, we determine the relation between p, q, and f. In these considerations, U and W are merely the feet of the perpendiculars from T and V. In Fig. 25, the right triangles TUX and VWX are similar; hence,

$$TU/VW = UX/WX.$$

Triangles AXF and VWF are similar; hence,

$$AX/VW = XF/WF.$$

Since $AX = TU$, we can rewrite the second relation as

$$TU/VW = XF/WF.$$

Combining this relation with the first relation gives

$$UX/WX = XF/WF.$$

By rewriting this last relation in terms of p, q, and f, we obtain

$$\frac{p}{q} = \frac{f}{q-f},$$

which can be rewritten in the form

$$\frac{1}{p} + \frac{1}{q} = \frac{1}{f}. \tag{23}$$

Recalling that $1/f$ is given by the lensmaker's equation (21), we have demonstrated that the relation between p and q, as measured from the off-axis points T and V in Fig. 25, is the same as that given by (20) for points on the lens axis, such as U and W. By similar arguments, we can show that (23) applies to any point in the object plane containing O and its image in the plane containing I. Thus, a converging lens produces a plane image of a plane object; in (23), p is the perpendicular distance from the object plane to the lens and q is the perpendicular distance from the image plane to the lens.

Although the lenses in optical instruments are usually thick combinations of lenses, we shall regard the lenses in the following examples as simple thin lenses:

Example. *When a camera is used to photograph a scene, the camera lens must produce a real image of the scene in the plane of the photographic plate or film. If the distance between the lens and the plate in a certain camera is* 10 cm *when a distant skyline is being photographed, what is the focal length of the camera lens? What should be the distance between the lens and the plate if the photographer wishes to photograph an object* 100 cm *away from the camera lens?*

We make use of the lens equation (23). Noting that for a 'distant skyline' $p = \infty$, we know that the real image is formed in the *principal focal plane* and that the center of the image is at the *principal focus* of the lens. Thus, the focal length of the camera lens is 10 cm.

When the photographer photographs an object 100 cm from the lens, $p = 100$ cm. Using $p = 100$ cm and $f = 10$ cm in (23), we write

$$\frac{1}{100 \text{ cm}} + \frac{1}{q} = \frac{1}{10 \text{ cm}} \quad \text{or} \quad \frac{1}{q} = \frac{10}{100 \text{ cm}} - \frac{1}{100 \text{ cm}} = \frac{9}{100 \text{ cm}}.$$

Solution gives $q = (100 \text{ cm})/9 = 11.1$ cm. Since q is the distance between the lens and the plate on which the real image is formed, the photographer must increase the distance between the plate and lens by 1.1 cm in shifting from the distant skyline to the nearby object.

Example. *In a photographic slide projector a converging lens is used to produce a greatly enlarged real image of a well-illuminated photographic transparency. If the focal length of the lens in such a projector is* 10 cm, *what should be the distance between the slide and the lens if the operator wishes to produce an image on a screen* 10 m *away from the lens? If the slide measures* 30 mm *by* 30 mm, *what should be the minimum dimensions of the screen?*

Noting that the image distance $q = 10$ m $= 1000$ cm and $f = 10$ cm, we can immediately use (23) to find $q = (1000 \text{ cm})/99 = 10.1$ cm. Thus, the distance between the slide and the lens should be 10.1 cm; thus, the slide should be 0.1 cm *beyond* the principal focus F of the lens. The enlargement produced by the lens is, by (22), equal to the ratio: (image distance)/(object distance) $= (1000 \text{ cm})/(10.1 \text{ cm}) = 99$. Thus, the minimum screen height should be 3 cm $\times 99 = 297$ cm; the minimum screen width should be the same, since the transparency is square.

We have now treated in considerable detail the formation of real images by converging lenses. We shall treat other types of image formation in less detail, but shall make full use of principal-ray diagrams to provide an understanding of the phenomena and to show how the lens equation (23) can be applied in each situation.

First, we note that the image formed by a converging lens is not necessarily a real image. If we insert a value of p less than f in (23), q comes out negative. As shown in the principal-ray diagram in Fig. 26, this negative value of q implies a virtual image, formed on the same side of the lens as the object; it would be seen at position I by an observer whose eye was at the position indicated in the figure. From the similar triangles in the figure, we see that (22) again gives the ratio of image size to object size.

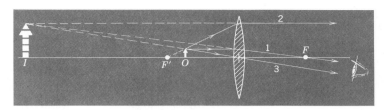

Fig. 26. A converging lens forms a virtual image when the object is inside the focus.

Example. *A converging lens with short focal length used in the manner indicated in Fig. 26 is called a simple microscope; the observer's eye is usually placed just to the right of the lens. If a converging lens with a focal length of* 1 cm *is to be used as a simple microscope, where must a small illuminated object be placed if the observer is to see a virtual image* 25 cm *from the lens? What is the enlargement produced by the lens?*

From the information given we note that $f = 1$ cm and $q = -25$ cm. We find the object distance p by substitution in (23):

$$\frac{1}{p} - \frac{1}{25 \text{ cm}} = \frac{1}{1 \text{ cm}}.$$

Solution gives $p = (25/26)$ cm $= 0.931$ cm. The enlargement, defined as the ratio of image length to object length, is by (22) given by the ratio $q/p = 25/0.931 = 26.9$.

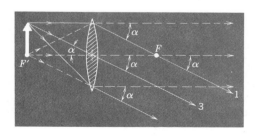

Fig. 27. Object at the principal focus; image at infinity.

We have shown in Fig. 24 that when a luminous point object is placed at the principal focus F of a converging lens, the lens produces a parallel light beam. Figure 27 shows the results obtained when an extended plane object is placed in the principal focal plane; principal rays 1 and 3 drawn from the point of the arrow are parallel by simple geometry. All other rays drawn from the arrow point are parallel to these rays after they pass through the lens and make the same angle α with the lens axis as that the object subtends at the center of the lens. Tracing the rays backward through the lens in Fig. 27 shows that a parallel light beam making an angle α with the lens axis is brought to a point focus in the same plane as the principal focus, with angle α subtended at the optical center.

As mentioned earlier, a glass lens thinner at the center than at the edge is a *diverging* lens. When a parallel light beam like the one in Fig. 28 passes through such a lens, its rays diverge so that to an observer they appear to come from a point focus F or F' on the same side of the lens as the incident light beam. By considering all rays reversed in the lower drawing of Fig. 28, we see that if a converging beam of light is incident on a diverging lens, it is possible for the lens to render the beam parallel.

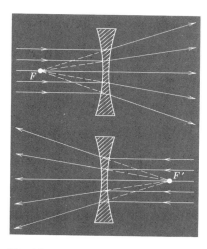

Fig. 28. Action of a diverging lens on parallel light rays.

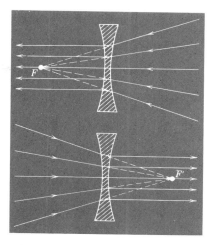

Fig. 29. Rays directed toward the principal focus behind a diverging lens become parallel after passing through the lens.

As indicated in Fig. 29, if a beam of light approaching the lens from the left is converging toward focal point F', a parallel beam emerges from the lens. An incident converging beam like the one in Fig. 29 can be considered as a *virtual object* at F'; the converging beam would have come to a point focus at F' if the beam had not been intercepted by the lens.

We may use principal-ray diagrams to locate the image formed by a diverging lens as indicated in Fig. 30. Principal ray 1 is initially parallel to the axis and is refracted in such a way that it *appears* to come from principal focus F. Principal ray 2 is initially directed toward principal focus F'; after refraction, it leaves the lens in a direction parallel to the axis. Principal ray 3 passes without deviation through the optical center of the lens. After passage through the lens, the three principal rays *appear* to come from a point which defines the end of the virtual image I. From the similar triangles in Fig. 30, we can show that (22) gives the ratio of

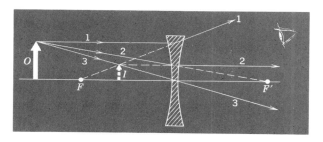

Fig. 30. Principal-ray diagram for a diverging lens.

image length to object length and that (23) gives the relation between p, q, and f provided we make use of certain sign conventions.

The lens equation (20) and (23) and the lensmaker's equation (21) can be applied quite generally to thin lenses, provided we use the following sign conventions:

$$\text{Object distance } p \text{ is } \begin{cases} + \text{ for a real object} \\ - \text{ for a virtual object} \end{cases}$$

$$\text{Image distance } q \text{ is } \begin{cases} + \text{ for a real image} \\ - \text{ for a virtual image} \end{cases}$$

$$\text{Focal length } f \text{ is } \begin{cases} + \text{ for a converging lens} \\ - \text{ for a diverging lens} \end{cases}$$

$$\text{Radii } R_1 \text{ and } R_2 \text{ are } \begin{cases} + \text{ for convex surfaces} \\ - \text{ for concave surfaces} \end{cases}$$

Equations (20) and (21) assume that the lens is in air, for which the refractive index is unity; if this is not the case, these equations can still be used if μ represents the ratio of the refractive index of the lens material to the refractive index of the medium in which the lens is immersed.

7. Treatment of Lenses by Wave Optics

The reader may have noticed a lack of consistency between the arguments of Section 5 and those of Section 6. In Section 5, we demanded that the rays contain the same numbers of wavelengths between object point and image point. In Section 6, we assumed that the image point was where the rays 'crossed,' and we did not mention the numbers of wavelengths. We should extend the argument of Fig. 22 to show that all rays from O to I contain the same numbers of wavelengths and hence indeed are in phase and form an image. This equality can be demonstrated (but the proof is very tedious) in the case of a lens of small angular aperture. Problem 19 calls for such a computation in a particularly simple case.

Rather than follow the above procedure, it will be more illuminating to study the refraction of *wavefronts* at the lens surfaces. This treatment will incidentally solve the problem of *thick lenses* of small aperture, at least in the double-convex case that we shall work through. We consider the lens in Fig. 31(a), but must retain the requirement of small aperture so that we suppose that a diaphragm blocks off all but the very center of the lens. Under these circumstances, all wavefronts can be approximated by sections of spheres and we can use the sagitta approximation (2), p. 707.

When the spherical wavefronts from the object O in Fig. 31(a) reach the first surface of the lens, their radii of curvature are increased because

SEC. 7 TREATMENT OF LENSES BY WAVE OPTICS

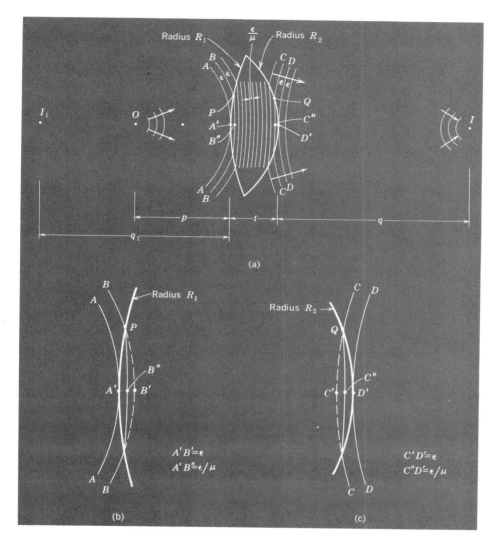

Fig. 31. (a) Image formation by a double-convex lens. (b) First-surface refraction. (c) Second-surface refraction.

the center of the wavefront reaches the lens first, and inside the lens its speed is decreased. Points further out on the wavefront reach the lens and suffer this speed decrease at later times. The wave speed in the lens is c/μ, corresponding to a wavelength of λ/μ, compared with λ in air. Wavefronts a distance ε apart in air will be separated by ε/μ in the glass. The net result, in the case of Fig. 31(a), is that wavefronts centered on the object O, at distance p, before entering the lens, are now centered at point I_1 at distance q_1, greater than p. The rays in the glass are still diverging, but less sharply than before they entered the lens; if the radius of curvature R_1 were sufficiently small, or if the index of refraction μ were sufficiently large, the rays in the glass could actually be converging rather than still diverging.

We can find the distance q_1 of the image formed by the first-surface refraction by the following argument, similar to the argument we used in the case of a spherical mirror. Consider Fig. 31(b), which is an enlargement of Fig. 31(a) in the vicinity of point A'. The circle $BB'B$ shows the location of a wavefront in the glass in the absence of refraction; the curve $BB''B$ shows its actual position. If the distance $A'B'$ is ε, the distance $A'B''$ is ε/μ. Take an origin at point A' and the usual XY-coordinate system. Then the point P, with coordinates X_P, Y_P, lies on three circles. It lies on the lens surface, for which the sagitta relation gives:

$$X_P = Y_P^2/2\,R_1. \tag{24}$$

It lies on the circle $BB'B$ separated from AA by the distance ε, and centered at O. This circle has radius $p+\varepsilon$, and the difference in X-coordinates of P and B' is $\varepsilon - X_P$. Hence the sagitta relation gives

$$\varepsilon - X_P = Y_P^2/2(p+\varepsilon),$$

or
$$X_P = \varepsilon - Y_P^2/2(p+\varepsilon). \tag{25}$$

The point P also lies on the circle centered at I_1, which represents the actual wavefront in the glass; this circle passes through point B'', separated from A' by distance ε/μ. The same argument that led to (25) gives, in this case:

$$X_P = \frac{\varepsilon}{\mu} - \frac{Y_P^2}{2\,(q_1 + \varepsilon/\mu)}. \tag{26}$$

In the limit $\varepsilon \to 0$, we can neglect ε in the denominators in (25) and (26). With this neglect, we can write (26) as

$$\mu X_P = \varepsilon - \mu Y_P^2/2q_1. \tag{27}$$

Subtraction of (25) from (27) gives

$$(\mu-1)\,X_P = Y_P^2\left(\frac{1}{2p} - \frac{\mu}{2q_1}\right). \tag{28}$$

Substitution of the value (24) for X_P in (28) and simplification gives

$$\frac{\mu-1}{R_1} = \frac{1}{p} - \frac{\mu}{q_1} \tag{29}$$

as the equation determining q_1.

Now let us consider the refraction, at the second surface, of the wavefronts traveling in the glass and centered at I_1. Figure 31(c) shows an enlargement of the vicinity of point D' in Fig. 31(a). Take an XY-

coordinate system with origin at point D', distant q_1+t from I_1. Consider point Q, which lies on the lens surface. The sagitta relation gives

$$X_Q = -Y_Q^2/2R_2. \tag{30}$$

From the fact that Q lies on the circle $CC'C$ centered at I, we find

$$X_Q = -\varepsilon + Y_Q^2/2(q+\varepsilon). \tag{31}$$

Since Q also lies on the wavefront in the glass passing through C'' and centered at I_1, we find

$$X_Q = \frac{\varepsilon}{\mu} - \frac{Y_Q^2}{2(q_1+t-\varepsilon/\mu)}. \tag{32}$$

Manipulation of (30), (31), and (32) in a manner similar to our manipulation of (24), (25), and (26) gives the relation

$$\frac{\mu-1}{R_2} = \frac{1}{q} + \frac{\mu}{q_1+t}. \tag{33}$$

Equation (33) determines the final image distance q from the lens geometry and the distance q_1 determined by (29).

It is not easy to eliminate q_1 from (29) and (33) to obtain a direct algebraic relation between p and q; it is best to use (29) and (33) in succession, numerically, as in the following example:

Example. *Consider a thick lens with $\mu = 1.5 = 3/2$, $R_1 = 2$ cm, $R_2 = 4$ cm, $t = 2$ cm. If an object is placed at distance $p = 16/3$ cm in Fig. 31, what is the image distance q in this figure? Compare with the image distance in the 'thin-lens' approximation.*

Substitution in (29) gives

$$\frac{1/2}{4\text{ cm}} = \frac{3}{16\text{ cm}} - \frac{3/2}{q_1},$$

or $q_1 = 24$ cm. Substitution in (33) gives

$$\frac{1/2}{2\text{ cm}} = \frac{1}{q} + \frac{3/2}{26\text{ cm}},$$

or $q = 52/10$ cm $= 5.2$ cm.

For a thin lens of these same radii of curvature, the lensmaker's equation gives $f = 8/3$ cm and $q = 16/3$ cm $= 5.33$ cm. The thin-lens approximation is not far off, even for this lens of extreme thickness.

To recover the lens equation (20) from this wave treatment, we need only assume that t is negligible in comparison with q_1 in (33). Then (29) and (33) become

$$\frac{\mu-1}{R_1} = \frac{1}{p} - \frac{\mu}{q_1} \quad \text{and} \quad \frac{\mu-1}{R_2} = \frac{1}{q} + \frac{\mu}{q_1}.$$

Addition of these two equations gives

$$\frac{1}{p}+\frac{1}{q}=(\mu-1)\left(\frac{1}{R_1}+\frac{1}{R_2}\right),$$

which is just the lens equation (20).

8. Aberrations of Mirrors and Lenses

If we listed all of the approximations made in our derivations of mirror equations (1) and (2) and lens equations (20) and (21), we might have grave doubts that these devices could ever be used to produce images at all! However, surprisingly good images are actually formed by spherical mirrors and by lenses of small aperture with spherical surfaces. We shall now consider briefly lens and mirror *aberrations* or failures to produce perfect images.

First of all, if the requirement of small angular aperture is not fulfilled, parallel incident light paths, parallel to the axis, do not converge to a point focus *F* on the axis; this defect, known as *spherical aberration*, is illustrated in Fig. 32 for a spherical mirror. Spherical aberration also is observed for *lenses* of large angular aperture; the reasons for spherical aberration can be understood by application of Snell's law at various points on the lens surface. The effects of spherical aberration can be minimized by using a 'diaphragm' or 'stop' to limit the angular aperture. Since the use of such a diaphragm or stop reduces the amount of light reaching the image position, the brightness of the image is correspondingly reduced.

Related to spherical aberration is an aberration that applies to light from point sources not on the axis. Light paths striking the mirror or lens away from the center and those at the center do not converge to a common point, and a point object is imaged not as a point but as a *comet-shaped* figure; this aberration is therefore called *coma*. Other aberrations such as *astigmatism* and *curvature of field* are associated with the failure of a lens or mirror to produce a plane image of a plane object. An additional aberration called *distortion* results from a variation of enlargement with distance from the axis; because of this effect a square flat object is not imaged as a square image.

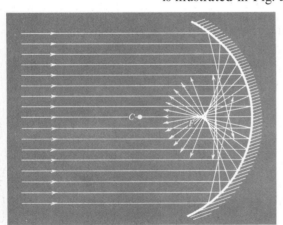

Fig. 32. Reflection of parallel rays by a spherical mirror of large angular aperture.

The aberrations thus far listed are common to lenses and mirrors. Lenses have an additional defect called *chromatic aberration*, which results from the fact that the refractive index μ in (20) and (21) is not the same for light of different wavelengths. Thus, from the refractive indices listed in Table 36-1, we see that the focal length (21) of a glass lens is greater for red light ($\lambda_R = 656.3$ nm) than for blue light ($\lambda_B = 486.1$ nm). The effects of chromatic aberration can be reduced by using compound lenses of the type discussed in the following chapter.

We close this chapter by noting that, even if all the aberrations listed above could be eliminated, there are still certain limitations to perfect image formation that are imposed by the wave nature of light. Thus, even if we use a paraboloidal mirror to produce an on-axis image of a distant star, or use an ideal lens for this purpose, the image of the star is not really a geometrical point but consists of a *diffraction pattern* having a shape determined by the linear aperture of the lens or mirror and the wavelength of the light (*see* Chapter 39).

PROBLEMS

1. An object 4 cm tall is placed 60 cm from a concave mirror with radius of curvature 40 cm. Find the image distance, size, and character (real or virtual, erect or inverted?). Sketch the principal-ray diagram.
Ans: 30.0 cm; 2.00 cm.

2. (*a*) An object 1.5 cm tall is placed 8 cm from a concave mirror of 10 cm radius of curvature. Determine image distance, size, and character. Sketch the principal-ray diagram. (*b*) An object 3 cm tall is placed at a distance of 50 cm from a concave mirror with radius of curvature 20 cm. Find the image distance, size, and character (real or virtual, erect or inverted?). Sketch the principal-ray diagram.

3. An object 2 cm tall is placed 10 cm away from a concave mirror with radius of curvature 50 cm. Find the image distance and the image size. Sketch the principal-ray diagram.
Ans: -16.6 cm; 3.32 cm.

4. An object 2 cm tall is placed 3 cm away from a concave mirror of radius of curvature 10 cm. Determine image distance, size, and character. Sketch the principal-ray diagram.

5. Let R be the radius of curvature and f be the focal length of a concave mirror. Five possible object positions are (1) $p > R$, (2) $p = R$, (3) $f < p < R$, (4) $p = f$, (5) $0 < p < f$. Which ones of these five positions give (*a*) an inverted diminished image? (*b*) an inverted enlarged image? (*c*) an inverted image equal in size to the object? (*d*) an image at infinity? (*e*) an erect image equal in size to the object? (*f*) an erect enlarged image? (*g*) a real image?

6. An object 2 cm tall is placed on the axis at a distance of 10 cm from a convex mirror of radius of curvature -40 cm. Find the image position, size, and character. Sketch the principal-ray diagram.

7. An object 4 cm tall is placed on the axis at a distance of 60 cm from a convex mirror with radius of curvature -30 cm. Find the image distance, size and character. Sketch the principal-ray diagram. Ans: -20.0 cm; 1.33 cm.

8. When an object is placed 60 cm away from a convex mirror, a virtual image one-third as tall as the object is produced. Find the focal length and radius of curvature of the mirror.

9. How far from a convex mirror with radius of curvature -50 cm should an object be placed in order to produce an image one-half as tall as the object? Ans: 25.0 cm.

10. What should be the radius of curvature of the curved surface of a plano-convex crown-glass lens if the lens is to have a focal length of 40 cm?

11. A double-convex lens of dense flint glass has surfaces with radii of curvature of 50 cm and 66.7 cm, respectively. What is its focal length? Ans: 43.5 cm.

12. An object 2 cm tall is placed 30 cm from

a diverging lens of -30-cm focal length. Where will the image be formed? What will be the size of the image? Sketch the principal-ray diagram.

13. An object 2 cm tall is placed near the axis of a thin converging lens at a distance of 66.7 cm from the lens. If the focal length of the lens is 50 cm, where will the image be formed and what will be the size of the image? Sketch the principal-ray diagram.

Ans: 200 cm behind; 6 cm.

14. A diverging lens has a focal length of -15 cm. Locate the images produced when an object is placed at the following distances from the lens: 25 cm, 20 cm, 15 cm, 10 cm, and 5 cm.

15. A converging lens has a focal length of 20 cm. Determine the image distances when an object is placed at the following distances from the lens: 100 cm, 40 cm, 30 cm, 20 cm, and 10 cm.

Ans: $+25.0$, $+40.0$, $+60$, ∞, -20 cm.

16. An image of the sum is formed by a converging lens of 4-m focal length. The sun's angular diameter is 32′ of arc. What is the diameter of the image? Ans: 3.72 cm.

17. A concavo-convex spectacle lens has a focal length of ½ m. If the glass has a refractive index of 1.6 and the concave surface a radius of -1 m, what is the radius of the convex surface? Ans: 23.1 cm.

18. Two distant stars are imaged on a photographic plate by a converging telescope lens of 6-m focal length. The distance between the images is 2 mm. What is the angular separation of the stars in the sky?

19. Verify relation (15) for the two light paths shown in the figure. The lens is assumed 'thin' ($X_1 \ll R_1$ and $X_2 \ll R_2$) and of 'small aperture' ($Y \ll p$, $Y \ll q$). Take $\mu_A = 1$, $\mu_M = \mu$ and compare the number of wavelengths in the path $a+b$, which can be assumed to be essentially all in air, with the number of waves in the straight path through the center of the lens. (a) By using the Pythagoras theorem and the binomial expansion, show that under the above assumptions

$$a+b = p+q+\frac{Y^2}{2}\left(\frac{1}{p}+\frac{1}{q}\right) = p+q+\frac{Y^2}{2f}.$$

(b) By using the sagitta approximation (2), p. 707, show that also

$$(p-X_1)+\mu X_1+\mu X_2+(q-X_2) = p+q+\frac{Y^2}{2f},$$

thus verifying relation (15) for these two rays.

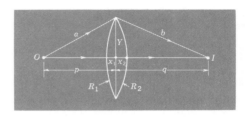

Problem 19

NOTE: *Complex optical instruments can be considered as combinations of the following simple instruments: (1) the camera, which utilizes a lens to form real images of distant objects; (2) the projector, which uses a lens to form an enlarged, real image of a nearby well-illuminated object; and (3) the simple microscope, a lens of short focal length used to produce an enlarged virtual image of a nearby object. These simple instruments are considered in the following problems:*

20. A portrait camera lens has a focal length of 5 i and is to be used in photographing a person 6 f tall. How far from the person should the lens be placed if his photographic image is to be 3 i tall? What should be the distance from the lens to the photographic plate?

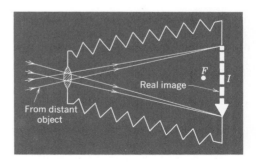

Problems 20–21. Image formation in the camera.

21. A camera has a lens of 4-i focal length and is used to photograph an object 2 f tall. How far from the object should the lens be placed if the photographic image is to be 2 i tall?

What should be the distance from the lens to the film?   Ans: 52.0 i; 4.33 i.

22. The total distance between a slide in a projection lantern and the screen is 40 f. If the projection lens has a focal length of 16 i, what should be the distance between the slide and the lens?

23. A lantern slide has dimensions 3 i×4 i. This slide is to be projected in such a way as to produce an image 4.5×6 f on a screen 20 f from the projection lens. What should be the focal length of the projection lens? What should be the distance from the slide to the lens?   Ans: 12.6 i; 13.3 i.

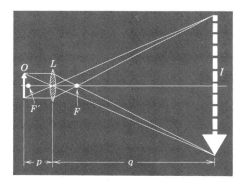

Problems 22-24. Image formation by the lens of a projection lantern.

24. It is desired to produce an image of a 3-i×4-i lantern slide on a screen 40 f from the projection lens. The image on the screen is to be 6 f by 8 f. (*a*) What should be the focal length of the projection lens? (*b*) Where should the slide be placed? (*c*) If the illumination at the slide is 1000 lm/f^2, what is the maximum possible illumination at the screen?

25. In a simple microscope, a single converging lens of focal length less than 25 cm is employed. The object is placed inside the focus and its virtual image is observed by the eye placed close to the lens. It may be assumed that the distance from the eye to the image is the same as the distance from the lens to the image. A converging lens has a focal length of 0.8 cm. If this lens is to be used as a simple microscope, at what distance from the lens should the object be placed in order that an enlarged virtual image may be formed at a distance of 25 cm from the lens?
   Ans: 0.775 cm.

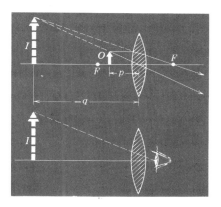

Problems 25-26. Simple microscope.

26. If the lens in Prob. 25 were used in such a way as to produce a virtual image at ∞, where would the object be placed?

27. Consider the same thick lens as in the example on p. 787. What is the object distance p that gives final image at $q = \infty$? Compare p with the focal length, $8/3$ cm, of a thin lens of the same radii of curvature.   Ans: $5/2$ cm.

38 OPTICAL INSTRUMENTS

In the preceding chapter we discussed image formation by lenses and mirrors. The present chapter deals with optical instruments of which lenses and mirrors are the component parts. In fact, the camera lens, the projector lens, and the simple microscope, which we have already discussed, are the basic components of all optical instruments. Although the lenses used in good optical instruments are usually thick compound lenses chosen to minimize aberrations, we shall assume that they are thin simple lenses, since this assumption introduces no essential errors and greatly simplifies the discussion of instrument design. We begin by considering combinations of two lenses.

1. Combinations of Lenses

If light from an object passes through two lenses, one after the other, the combined action of the two can be deduced by considering that the image which would be formed by the first lens is the object for the second lens. In the lens arrangement shown in Fig. 1, lens L_1 forms a real image

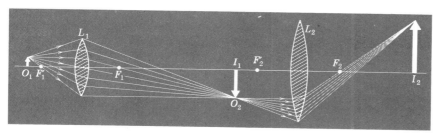

Fig. 1. The image I_1 formed by the first lens acts as a real object O_2 for the second lens.

I_1 of object O_1. Image I_1 can be treated as a real object O_2 for lens L_2. Lens L_2 forms a real image I_2 at the position indicated. Lens L_2 is drawn larger than L_1 so that all rays from O_1 that pass through L_1 will also pass through L_2. However, only one of the rays from the end of O_2 in this drawing is a principal ray. To determine graphically the position of I_2, the other principal rays from O_2 may be sketched, even though no light actually proceeds along them in this particular combination.

In the lens arrangement shown in Fig. 2, lens L_1 would have formed a real image I_1 if the light beam had not been intercepted by the second lens L_2. In finding the position of the final image I_2 formed by the second lens, we must treat I_1 as *virtual object* O_2 for the second lens. As we have noted, the object distance p of a virtual object should be taken as negative when used in the lens equation.

As an example of a compound lens, let us consider the case of two thin lenses in contact as shown in Fig. 3. The focal length of the first lens is f_1, that of the second lens is f_2. Consider parallel rays reaching the first lens from an infinitely distant point object. Then the image distance for the first lens is $q_1 = f_1$. Now let the image I_1 formed by the first lens serve as a *virtual object* O_2 for the second lens. Neglecting the thicknesses of these thin lenses, we can write the following equation for the second lens, with object distance $p_2 = -f_1$:

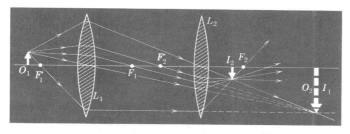

Fig. 2. The image I_1 formed by the first lens acts as a virtual object O_2 for the second lens. In this graphical construction, the three principal rays for lens L_1 are indicated by arrowheads to the left of L_1, those for L_2 by arrowheads to the left of L_2.

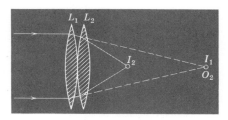

Fig. 3. A compound lens.

$$\frac{1}{-f_1}+\frac{1}{q_2}=\frac{1}{f_2}, \quad \text{or} \quad \frac{1}{q_2}=\frac{1}{f_1}+\frac{1}{f_2}.$$

It is possible to treat the two thin lenses in contact as a compound lens. Remembering that the focal length of a lens is the image distance when the object distance in infinite, we note that q_2 in the above equation is equal to the focal length of the lens combination, and we may write

$$\frac{1}{f_{\text{Comp}}}=\frac{1}{f_1}+\frac{1}{f_2}, \tag{1}$$

where f_{Comp} is the focal length of the compound lens.*

The lens defect called *chromatic aberration* can be partially corrected by the use of a compound lens composed of simple lenses made of different glasses. We note from Table 36–1, p. 755, that the refractive indices for a given wavelength are different for different glasses and also that the values of $\mu_B - \mu_R$ are different for different glasses. These properties make it possible to correct in part the *chromatic aberration* associated with lenses. Consider the case of white light incident on the simple lens in Fig. 4. As a result of the variation of refractive index with wavelength, the focal length of the lens varies with wavelength, so colored images are produced at different points by the lens. Since, from the lensmaker's equation, the focal length varies *inversely* as $\mu - 1$, the focal length is less for blue light than for red, since $\mu_B > \mu_R$.

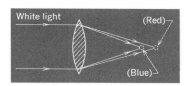

Fig. 4. Chromatic aberration of a simple lens.

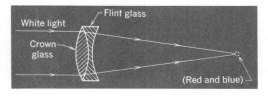

Fig. 5. An achromatic converging lens.

By making a compound lens from two simple lenses of different materials, such as crown and flint glass, it is possible to make the red and blue images coincide, as in Fig. 5. When the lens is so constructed that red and blue images coincide, it is found that the images of colors of intermediate wavelengths also coincide fairly accurately. Such a two-component lens is called an *achromatic doublet*. The diverging lens must have a focal length greater in absolute magnitude than the converging lens if the combination is still to act as a converging lens, and the material of the diverging lens must have greater difference $\mu_B - \mu_R$ than the converging lens if the colors are to be returned to coincidence. From the values given in Table 36–1, p. 755, we see that crown and flint glass are suitable

---

* In specifying lenses, opticians use the term 'power' and measure the 'power' of a lens in a unit called the *diopter;* the 'power' of a lens in diopters is $1/f$, where f is the focal length in *meters*. Converging lenses have positive power; diverging lenses have negative power. The 'power' of a compound lens is, by (1), just the sum of the powers of its components.

for use in making an achromatic combination, since the refractive indices of these materials are not greatly different but for flint glass the value of $\mu_B - \mu_R$ is nearly twice that for crown glass. The details are worked out in Prob. 13.

2. Magnifying Power of an Optical Instrument

We shall give only a brief discussion of optical instruments in this text. We begin with a short discussion of the *human eye* and a definition of *magnifying power*.

Optically, the human eye is similar in many respects to a photographic camera, since it contains a lens system equipped with a shutter (the eyelid), a variable diaphragm (the iris), and a screen (the retina) on which are produced real inverted images of external objects. The eye is very complex physiologically, the complexity arising from the mechanisms required for making various mechanical adjustments and from the incompletely understood processes of detection at the retina and of transmission to the brain.

If the eye muscles are *relaxed*, sharply defined images of *distant objects* are produced on the retina of a normal eye. In order to produce sharp images of nearby objects, the muscles attached to the crystalline lens increase the curvature of the lens surfaces, decreasing the focal length sufficiently to bring the desired images into sharp focus. This adjustment of focus is called *accommodation*. It is possible for the normal eye of the young adult to deform the crystalline lens sufficiently to produce sharp images of objects as close as 25 cm (10 i), which is known as the distance of the *near-point of the normal eye*. Children can see objects clearly at shorter distances, since the crystalline lens is more readily deformable in early life. Normally, the near-point moves beyond 25 cm at about the age of 40.

There are several rather common defects of vision that deserve brief consideration. The normal eye produces sharp images of distant objects when the eye muscles are relaxed, and is capable of sufficient convergence to produce sharp images of objects as close as 25 cm, as in Fig. 6(a). However, the size and shape of the eyeball and the focal length of the lens system are not always properly matched. If the lens is effectively too convergent, the images of distant objects will fall in front of the retina so that on the retina itself the image will be blurred. Only objects close to the eye can be focused sharply on the retina; eyes with this type of defect are said to be *myopic* or *nearsighted*. The *far-point* of a myopic eye is not at infinity, but at some inconveniently close distance. As indicated in Fig. 6(b), this defect can be remedied by using a suitable diverging spectacle lens. A lens incapable of sufficient convergence leads to *far-sightedness*, or *hypermetropia*, which, as indicated in Fig. 6(c), can be remedied by the use of converging spectacle lenses. The *near-point*

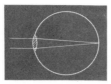

(a) Normal vision; object at ∞, lens fully relaxed; or, object at 25 cm, lens fully converged.

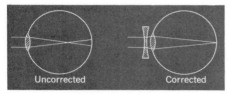

(b) Nearsightedness (myopia); object at ∞, lens fully relaxed.

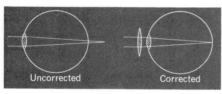

(c) Farsightedness (hypermetropia); object at 25 cm, lens fully converged.

Figure 6

of a hypermetropic eye is beyond 25 cm. Normally, persons over 40 become hypermetropic and reading glasses are required.

A third common type of defect in vision occurs when one or more of the refracting surfaces of the eye, such as the cornea or crystalline lens, are not perfectly spherical. In this case light from a point source will not be focused in a point image. An eye having this characteristic is said to be *astigmatic*. This defect can be corrected by using a cylindrical spectacle lens so placed in front of the eye that its unsymmetric curvature corrects the asymmetric curvature of the crystalline lens or cornea.

Example. *A myopic person has no astigmatism, but has a near-point of 15 cm and a far-point of 50 cm. Can you prescribe a single pair of glasses that will give him normal vision?*

When this person's eyes are relaxed, he sees an object at 50 cm clearly; a normal eye sees an object at infinity. Let us first determine the focal length of the diverging spectacle lens [as in Fig. 6(b)] that will give a virtual image at 50 cm of an object at infinity, so that it can be clearly seen by the myopic person. We substitute $p = \infty$, $q = -50$ cm in the lens formula to obtain $f = -50$ cm. (The optician would call this a -2-diopter lens.)

If vision is to be normal when this person is wearing his spectacles, he should also be able to read a menu held at 25 cm. With his spectacles, an object at $p = 25$ cm gives an image at a distance $-q$ determined by

$$\frac{1}{q} = \frac{1}{f} - \frac{1}{p} = \frac{1}{-50} - \frac{1}{25} = -\frac{3}{50}, \qquad \text{or} \qquad -q = 16\tfrac{2}{3} \text{ cm.}$$

Since the person can clearly see objects as close as 15 cm, he can see this image at $16\tfrac{2}{3}$ cm and will not have to remove his spectacles to read in normal fashion.

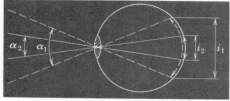

Fig. 7. Magnifying power $= i_1/i_2 = \alpha_1/\alpha_2$, where i_1 and α_1 refer to observation through an optical instrument, i_2 and α_2 refer to direct observation. The angle of the field of view is much exaggerated in this sketch.

Microscopes and telescopes enable us to see objects more clearly than is possible with the unaided eye. The greater clarity is obtained because, with the aid of the optical instrument, a larger image is formed on the retina of the eye than would be formed if the object were viewed directly. In Fig. 7, rays are traced from the ends of the object (or from any two definite points of the object) through the center of the lens of the eye. The distance from the lens of the eye to the retina does not change when the focal length of the eye lens is changed to view objects at different distances. Hence we see in Fig. 7 that the ratio of image lengths on the retina equals the ratio of angles subtended at the eye by rays from the ends of the object, provided these angles are small. The assumption of small angles is always valid, because the eye sees distinctly only the small central portion of the field of view, which subtends a small angle α. These considerations lead to the following useful expression for *magnifying power*:

$$\begin{Bmatrix} \text{magnifying power of} \\ \text{an optical instrument} \end{Bmatrix} = \frac{\begin{Bmatrix} \text{angle } \alpha_1 \text{ subtended at lens of eye} \\ \text{when object is viewed through} \\ \text{the optical instrument} \end{Bmatrix}}{\begin{Bmatrix} \text{angle } \alpha_2 \text{ subtended at lens of eye} \\ \text{when object is viewed directly} \\ \text{in the most favorable manner} \end{Bmatrix}} \quad (2)$$

The denominator of the above expression contains the phrase 'in the most favorable manner,' which we must now discuss. As we have seen, the normal or properly corrected eye has a lens of variable focal length so that it can focus on the retina an image of an object at any distance between 25 cm and ∞. But for a given object size, the size of the image on the retina varies inversely as the object distance. When this page is held at 25 cm, the words on this page form images on the retina that are four times as high as when the page is held at 100 cm. In defining the magnifying power of a magnifying glass or a microscope, we assume that direct observation, in the denominator of (2), is made with the object held at 25 cm, the near-point of the normal or properly corrected eye. On the other hand, a telescope or opera glass is specifically intended for viewing *distant* objects. Hence, in defining their magnifying powers we must assume that, in the denominator of (2), the object is viewed at the distance of its actual location.

Example. *What is the magnifying power of a concave shaving mirror of radius of curvature* 24 i? *The computation should be based on a comparison with a plane mirror, the most favorable manner of shaving in each case being to hold the face* 8 i *from the mirror.*

Consider a line 1 i long drawn on the face. In the case of the plane mirror this line is imaged normal size 8 i behind the mirror or 16 i from the eyes. The angle α_2 in Fig. 5 is thus $\frac{1}{16}$ rad for this 1-i line.

The concave mirror has a focal length of 12 i. Setting $p = 8$ i, $f = 12$ i, in the mirror formula gives $q = -24$ i; the image is 24 i behind the mirror. The enlargement is (image distance)/(object distance) = 3; hence the 1-i line is 3 i long in the image, which is 32 i from the eyes. The angle α_1 in Fig. 5 is $\frac{3}{32}$ rad.

The magnifying power is the ratio $\alpha_1/\alpha_2 = \frac{3}{32}/\frac{1}{16} = 1.5$. When the concave shaving mirror is used, each whisker makes an image on the retina that is $1\frac{1}{2}$ times as long as when a plane mirror is used.

3. The Simple Microscope

A single converging lens can be used as a 'magnifying glass' to form enlarged virtual images of small objects. When the eye is placed close to a converging lens of short focal length and the small object to be viewed is placed at a position such that p is only slightly less than the focal length, the magnifying power is large and the lens act as a *simple microscope*. Examples are the familiar jeweler's lens and the eyepiece of a telescope or microscope.

In order to determine the magnifying power of a simple microscope, consider the arrangement shown in Fig. 8. A small object O is placed between the principal focus and the lens so that a virtual image I is formed at a conveniently large distance D from the eye of the observer. When the observer's eye is close to the lens, the image distance $q \approx -D$.

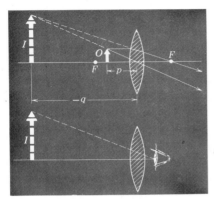

Fig. 8. Simple microscope.

With the arrangement shown in the figure,

$$\left\{\begin{array}{l}\text{angle subtended at lens of eye when} \\ \text{object is viewed through the lens}\end{array}\right\} = \frac{I}{D},$$

$$\left\{\begin{array}{l}\text{angle subtended at lens of eye when} \\ \text{object is viewed directly at 25 cm}\end{array}\right\} = \frac{O}{25 \text{ cm}}.$$

Hence, according to (2), the magnifying power of a simple microscope is given by

$$M = \frac{I/D}{O/25 \text{ cm}} = \frac{I}{O} \times \frac{25 \text{ cm}}{D}, \tag{3}$$

but $I/O = -q/p = D/p$ and hence $M = 25 \text{ cm}/p$. Substitution of $q = -D$ in the lens equation gives $1/p = 1/f + 1/D$. Using this value for $1/p$ in the expression for M gives

$$M = \frac{25 \text{ cm}}{f} + \frac{25 \text{ cm}}{D}. \tag{4}$$

In order to achieve maximum magnifying power, the user should adjust the object position to produce an image such that $D = 25$ cm. Under this condition $M = 25 \text{ cm}/f + 1$. However, for prolonged use it is desirable to choose a larger value of D in order to avoid eyestrain. Since the relaxed eye sees distant objects distinctly, p can be adjusted to make $D = \infty$; in this case $M = 25 \text{ cm}/f$. If f is 5 cm or less, as in the usual case for simple microscopes, M decreases only slightly as D increases from 25 cm to ∞ in (4).

Example. *A nearsighted jeweler has a near-point at 15 cm and a far-point at 50 cm. He employs a jeweler's lens of focal length 5 cm. What is the range of distances from the lens at which he can hold a watch and observe a clear image, and what is the range of magnifying powers he achieves, in comparison with direct observation of the watch at 15 cm, which is his most favorable method of direct viewing?*

To form a virtual image at 15 cm ($q = -15$ cm) from a lens with $f = 5$ cm requires object distance computed from

$$\frac{1}{p} = \frac{1}{f} - \frac{1}{q} = \frac{1}{5 \text{ cm}} + \frac{1}{15 \text{ cm}}, \quad \text{as} \quad p = {}^{15}\!/_{4} \text{ cm} = 3.75 \text{ cm}.$$

Correspondingly, to form an image at 50 cm, the object distance must be $p = 4.55$ cm. The jeweler will observe a clear image at object distances varying from 3.75 cm to 4.55 cm.

Since the jeweler's near-point is 15 cm, rather than the normal 25 cm, we must substitute 15 cm for 25 cm in the numerators of (4) in determining the magnifying powers afforded *this* jeweler. When he places the image at $D = 15$ cm, the magnifying power is

$$M = \frac{15 \text{ cm}}{f} + \frac{15 \text{ cm}}{D} = \frac{15 \text{ cm}}{5 \text{ cm}} + \frac{15 \text{ cm}}{15 \text{ cm}} = 4;$$

while, with the image at $D = 50$ cm,

$$M = \frac{15 \text{ cm}}{5 \text{ cm}} + \frac{15 \text{ cm}}{50 \text{ cm}} = 3.3.$$

The range of magnifying powers is from 3.3 to 4.0.

4. The Compound Microscope

In viewing a small object we can obtain modest magnifying power with a simple microscope consisting of a *single* converging lens of short focal length, but we can obtain much larger magnifying power by using a *compound microscope* consisting of *two* lenses, each of which contributes to the magnifying power. In a compound microscope, shown schematically in Fig. 9, one short-focus converging lens called the *objective*

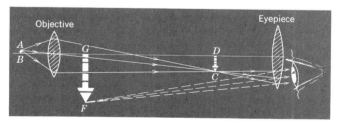

Fig. 9. Image formation in a compound microscope. A compound microscope consists of two essential components: (1) the objective, a short-focus lens that acts as a projector, and (2) the eyepiece, a lens used as a simple microscope.

produces an enlarged real image DC of the small, well-illuminated object AB; the objective thus acts as a projector. The second lens in the compound microscope is also a converging lens with small focal length and is used to produce an enlarged virtual image GF, which is viewed by the observer; the eyepiece thus acts as a simple microscope.

The magnifying power M of a compound microscope is equal to the product of the enlargement E_O produced by the objective and the magnifying power M_E of the eyepiece, because the simple microscope used as the eyepiece forms on the retina an image of CD that is E_O times as long as the image that would be formed if we used the eyepiece to view the object AB directly. The magnifying power can be expressed in terms of the simplified diagram in Fig. 10. In this figure the object AB and the images CD and FG are those shown in Fig. 9; p and q denote object and image distances from the objective; and the final image FG is formed at distance D (between 25 cm and ∞) from the eyepiece. Since the enlargement $E_O = CD/AB = q/p$ as in (22), p. 780, and the magnifying power of the eyepiece is given by (4), the magnifying power of the microscope is

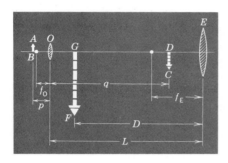

Fig. 10. Positions of object and images in a compound microscope.

$$M = E_O M_E = \frac{q}{p}\left(\frac{25 \text{ cm}}{f_E} + \frac{25 \text{ cm}}{D}\right) \quad (5)$$

where f_E is the focal length of the eyepiece.

We can best appreciate the meaning of (5) by noting that in actual microscopes: (a) f_E is small compared with D so that $(25 \text{ cm})/f_E \gg (25 \text{ cm})/D$; (b) p is negligibly smaller than f_O, the focal length of the objective, so we may in good approximation write f_O for p; (c) f_E is small compared with L, the length of the microscope draw tube in which the objective and eye piece are mounted, so that, to a fair approximation, we can replace q by tube length L. With these approximations, the expression for the magnifying power of the microscope becomes

$$M = L (25 \text{ cm})/f_O f_E.$$

From this expression, we see that, for large magnifying power, the focal lengths of objective and eyepiece should be small and that the microscope tube length should be as large as convenience permits.

There are several limitations to the magnifying power obtainable with a compound microscope. One of these is the illumination that can be used on the object; for biological specimens, the maximum tolerable illumination without damage to the specimen is rather low. A second limitation is involved with lens aberrations, which become increasingly serious as magnifying power is increased. The ultimate limitation is the diffraction limit imposed by the wave nature of light itself; this will be discussed in the next chapter.

5. The Astronomical Telescope

Telescopes are instruments used for the purpose of improving the observer's vision of distant objects. Like the compound microscope, the telescope consists essentially of two components: the objective and the eyepiece. In the case of the telescope, the *objective* forms a real image of a distant object and hence serves the same purpose as a *camera lens;* the *eyepiece* is used to produce an enlarged virtual image of the real image produced by the objective and hence serves as a *simple microscope*.

Figure 11 shows a diagram of a refracting astronomical telescope. The objective lens forms a real inverted image AB of a distant erect object that subtends an angle α at the objective. This real image AB is then viewed through the eyepiece, which produces the enlarged inverted virtual image CD, which subtends the angle β at the eyepiece. Owing to the great distance of the object, the rays from any point of it can be considered parallel on reaching the objective; hence, the real image AB is formed in the principal focal plane of the objective.

In observing a distant object, it is the apparent size that is important, and this is determined by the angle subtended at the eye. Without the telescope, the angle subtended would be α; with the telescope, the angle subtended is β (if the eye is close to the eyepiece). Hence, the magnifying power of the telescope is given by $M=\beta/\alpha$. As in previous discussions, the small angles α and β may be approximated by their tangents, and we may write

$$M = \frac{\beta}{\alpha} = \frac{AB/QB}{AB/PB} = \frac{PB}{QB}.$$

In focusing the telescope, let us assume that the observer arranges the eyepiece position so that the distance QB is the focal length of the

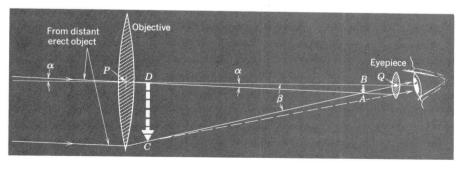

Fig. 11. Image formation in a refracting astronomical telescope.

eyepiece, f_E, and the image CD is at infinity. Since PB is simply f_O, the focal length of the objective, the above expression for the magnifying power can be written as

$$M = f_O/f_E. \tag{6}$$

The magnifying power is equal to the ratio of the focal length of the objective to the focal length of the eyepiece, provided the telescope is focused for final image at infinity. Hence, in order to achieve high magnifying power, the focal length of the objective should be made large and the focal length of the eyepiece should be made small.

In astronomical work, although a certain amount of attention must be paid to magnifying power, the principal emphasis is on seeing fainter and fainter objects by increasing the *light-gathering power* of the telescope. Because of inherent limitations imposed by the wave nature of light, we cannot hope to see any star outside of our own solar system as other than a point; but in order to see fainter and more distant objects, it is desirable to gather as much light as possible from the object, all focused as accurately as possible at a point on a photographic plate. Hence emphasis is placed on increasing the diameter of the telescope objective. There are difficult problems involved in increasing the size of an objective *lens,* since both the production of large pieces of glass of high optical quality and the grinding and polishing of the large compound-lens components needed to correct aberrations involve great difficulties. The largest objective lens that has been made is the one of 40-inch diameter at the Yerkes Observatory of the University of Chicago.

One method of overcoming some of these optical problems is to replace the objective lens by a large concave mirror. With an objective of this type, glass of high optical quality is not needed, since the light is reflected from the front surface and does not traverse the glass; furthermore, there is no chromatic aberration and there is only one optical surface to be ground and polished. The largest objective mirror is the one of 200-inch diameter at Mount Palomar in California.

6. Terrestrial Telescopes

When a telescope is used for astronomical purposes, the fact that the image is inverted is of no inconvenience; but when terrestrial objects are to be viewed, it is necessary to have an erect final image. The erection can be accomplished by introducing a third lens, called the *erecting lens,* between the objective and eyepiece. The arrangement is shown schematically in Fig. 12. In this figure, a bundle of rays from one end of an erect object is shown passing through the objective, the erecting lens, and the eyepiece. If the erecting lens is placed twice its focal length from the real inverted image AB, it forms a reinverted image $A'B'$ of equal

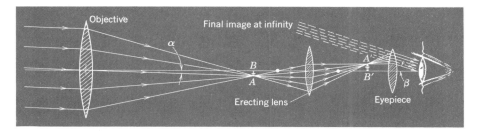

Fig. 12. Terrestrial telescope.

size, which is then viewed through the eyepiece. Since there is no difference in the sizes of AB and $A'B'$, the magnifying power is still given by (6). However, it should be noted that a terrestrial telescope is longer by four times the focal length of the erecting lens than an astronomical telescope of equal magnifying power.

This increased length is sometimes inconvenient and can be avoided by using totally reflecting prisms (*see* Fig. 13) to send the light beam back and forth, as in *prism binoculars;* the prisms can be arranged so that they also reinvert the image, and no erecting lens is needed.

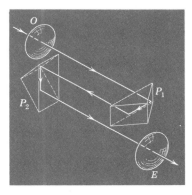

Fig. 13. Arrangement of prisms and lenses in the prism binocular.

Example. *A private detective desires a pair of binoculars that will enable him to read the license plate on a car* 300 f *away when he can normally read such a plate readily at* 20 f. *What magnifying power does he need? If the objective has a focal length of* 12 i, *what should be the focal length of the eyepiece?*

The necessary magnifying power is 300 f/20 f = 15, since the height of the image of a license plate on the retina varies inversely as its distance. From the relation $M = f_O/f_E$, we see that if $M = 15$ and $f_O = 12$ i, f_E must be $12/15$ i = 0.8 i.

PROBLEMS

1. Two converging lenses are placed 20 cm apart. The focal length of the first lens is 30 cm and that of the second is 20 cm. If an object is placed 60 cm in front of the first lens, where will the final image be formed? Sketch the principal-ray diagram.
          Ans: 13.3 cm behind second lens.

2. A converging lens of focal length 10 cm is placed 20 cm in front of a diverging lens of focal length −10 cm. If an object is placed 30 cm in front of a converging lens, where will the final image be formed?

3. Two thin converging lenses of focal lengths 33.3 cm and 50 cm are placed in contact. What is the focal length of this lens combination?
          Ans: 20.0 cm.

4. A thin converging lens with $f = 33.3$ cm and

a thin diverging lens with $f = -66.7$ cm are placed in contact. What is the focal length of the resulting compound lens?

5. A converging lens of focal length 30 cm is placed in front of a converging lens of focal length 3 cm. What is the distance between the lenses if parallel rays entering the first lens leave the second lens as parallel rays?

Ans: 33 cm.

6. A converging lens of focal length 60 cm is placed in front of a diverging lens of focal length -6 cm. What is the distance between the lenses if parallel rays entering the first lens leave the second lens as parallel rays?

7. An object is placed 10 cm in front of a converging lens of focal length 10 cm. A diverging lens of focal length -15 cm is placed 5 cm behind the converging lens. Find the position, size, and character of the final image.

Ans: Erect, virtual image at same location as object; 1.5 times object size.

8. An object is placed 8 cm in front of a converging lens of focal length 8 cm. A converging lens of focal length 16 cm is placed 4 cm behind the first lens. Find the position, size, and character of the final image.

9. A converging lens of focal length 20 cm is placed 20 cm from a diverging mirror of focal length -10 cm. A candle is midway between the lens and the mirror. Where should a screen be placed to catch a real image of the candle flame? What is the image size? Is the image erect or inverted? Ans: 100 cm from the lens; twice the size of the flame; inverted.

10. A converging lens of focal length 10 i is placed 10 i from a converging mirror of focal length 2 i. A candle is placed midway between the lens and the mirror. Describe the two images of the candle flame (position, size, character) seen when one looks through the lens. Sketch the principal-ray diagrams.

11. If a flint-glass lens has focal length 100.0 cm for the yellow light of Table 36–1, what is its focal length for the red and the blue light? Ans: 100.8, 98.10 cm.

12. If a flint-glass lens has a focal length -60.00 cm for the yellow light of Table 36–1, what is its focal length for the red and the blue light?

13. (*a*) Using equation (1), show that the condition for achromatism of the compound lens in–Fig. 5 is

$$\frac{1}{(f_{\text{Comp}})_R} = \frac{1}{(f_{\text{Comp}})_B} \quad \text{(i)}$$

or $\dfrac{1}{f_{\text{CR}}} + \dfrac{1}{f_{\text{FR}}} = \dfrac{1}{f_{\text{CB}}} + \dfrac{1}{f_{\text{FB}}}$

where the subscripts C and F refer to the crown- and flint-glass components, while R and B refer to red and blue light. (*b*) Show by substitution of the refractive indices of Table 36–1 in the lensmaker's equation that for any crown-glass lens the focal lengths for red and blue light are related by

$$f_{\text{CR}} = (0.5240/0.5145)f_{\text{CB}} = 1.0185 f_{\text{CB}}, \quad \text{(ii)}$$

while for any flint glass lens,

$$f_{\text{FR}} = (0.6391/0.6221)f_{\text{FB}} = 1.0273 f_{\text{FB}}. \quad \text{(iii)}$$

(*c*) Show by substitution of (ii) and (iii) in (i) that the condition for achromatism is $f_{\text{FR}} = -1.467 f_{\text{CR}}$ and that the focal length of the compound lens is $f_{\text{Comp}} = 3.102 f_{\text{CR}}$.

14. It is desired to make a diverging achromatic lens of focal length -50 cm from flint- and crown-glass lenses. If the lens surfaces that are in contact are plane, what should be the radii of curvature of the outer curved surfaces? (*See* Prob. 13.)

15. It is desired to make a converging achromatic lens of focal length 80 cm from a flint-glass lens and a crown-glass lens. Each of the component lenses has one plane surface, and it is the plane surfaces that are in contact. What should be the radius of curvature of the curved surface (*a*) of the flint-glass lens and (*b*) of the crown-glass lens? (*See* Prob. 13.)

Ans: (*a*) -23.8 cm; (*b*) $+13.4$ cm.

16. In a compound microscope, the focal length of the objective is 0.6 cm and that of the eyepiece is 4 cm. If the distance between the lenses is 30 cm, what should be the distance from the object to the objective if the observer focuses for image at ∞? What is the magnifying power?

17. In a compound microscope, the focal length of the objective is 0.8 cm, that of the eyepiece is 4 cm, and the distance between the lenses is 20 cm. If the observer places the final image at a distance of 25 cm from his eye, what is the distance from the object to the objective?

What is the magnifying power?

Ans: 0.840 cm; 143.

18. If an astronomical telescope has a magnifying power of 100 when used with an eyepiece of 2-cm focal length, what is the focal length of the objective?

19. The objective of a terrestrial telescope has a focal length of 80 cm and the telescope has a magnifying power of 20 when adjusted in such a way that parallel rays reach the observer. If the erecting lens has a focal length of 20 cm, and image lengths $AB = A'B'$ in Fig. 12, what is the total length of the telescope?

Ans: 164 cm.

20. Show that in the prism binoculars shown in Fig. 13, the prisms not only permit shortening of the telescope tube but reinvert the image, both up and down, and right and left. If the total length of the optical path between objective and eyepiece in a pair of prism binoculars is 18 i, although the actual distance between objective and eyepiece is only 7 i, and the magnifying power of this instrument is 8, what would be the length of a terrestrial telescope using the same objective and eyepiece and having an erecting lens of focal length 4 i?

21. An early form of telescope, invented by Galileo in 1609, survives today as the 'opera glass.' The Galilean telescope employs a diverging lens as eyepiece, as in the figure. The objective would produce a real inverted image at position AB if the eyepiece E were not present. With the eyepiece in position, the observer sees an erect virtual image CD. Show that when the final image CD is at infinity, the magnifying power $M = f_O/(-f_E)$.

22. An opera glass (Prob. 21) measures 3 i between objective and eyepiece. The focal length of the objective is 5 i. What is the magnifying power?

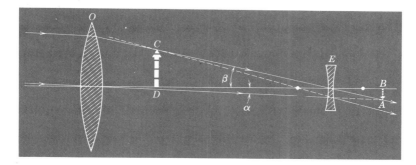

Problems 21–22

39 IMAGE FORMATION: FRAUNHOFER DIFFRACTION

In the preceding chapter we have implied that the real image of a point object, formed by a mirror or lens, is a point image without size or structure. True, we have mentioned spherical aberration. But it is possible to make a perfect mirror from the geometrical standpoint. A paraboloidal mirror focuses parallel rays at a geometrical point. An ellipsoidal mirror accurately images a point object located at one focus as a point image at the other focus. For a particular point object it is also possible to grind a geometrically perfect lens, with slight departures from sphericity, that will form a point image in monochromatic light.

However, quite apart from the question of spherical aberration, the images formed by mirrors and lenses are never geometrical points—the image is spread out into what is called a Fraunhofer diffraction pattern. The fact that the image is 'fuzzy' limits the *resolving power* of a mirror or lens—its ability to focus the images of two nearby points as distinctly separate images. Thus, because of diffraction effects, a microscope cannot separate two points that are too close together nor clearly define the shape of an object that is too small.

In this chapter we shall be concerned with so-called Fraunhofer diffraction effects at the images formed by mirrors or lenses. Actually, we shall discuss only the more important case of images formed by lenses, but our discussion has an obvious translation into the case of images formed by mirrors.

In our treatment of Fresnel diffraction in Chapter 35, we considered plane waves incident on an aperture and were concerned with the fringed *shadow* formed on a screen; no lenses were employed. In this chapter we consider the case in which a lens is placed beyond the aperture and a screen is placed at the focal plane of the lens—we consider the manner in which the *image* is modified by the presence of the aperture. It turns out that the details of these fringed *images* are much easier to compute than those of the fringed *shadows*.

1. Diffraction Resulting from an Aperture

The first arrangement we shall consider is that shown in Fig. 1. Monochromatic light from a source passes from a narrow slit S to a lens L_1 that renders the beam parallel. Before the aperture slit A is introduced, the parallel light beam is focused by lens L_2 in such a manner as to produce a real image of slit S on the screen. When slit A is inserted parallel to slit S, a diffraction pattern is observed on the screen. The bright central portion of the diffraction pattern coincides with the original image of slit S but is considerably broader. If slit A is very narrow, several bright and dark fringes will appear on the screen on each side of the central image. The width of the central image and the spacing of the diffraction fringes are determined by the width of slit A. The width of the central image on the screen is approximately twice that of the fainter bright fringes at the side, as shown in Fig. 2.

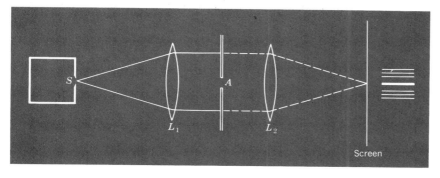

Fig. 1. Arrangement for observing the diffraction pattern resulting from the slit aperture A.

Fig. 2. Fraunhofer diffraction pattern resulting from a single slit.

The explanation of the observed diffraction pattern is relatively simple, since only plane wavefronts reaching the lens L_2 are focused on the screen. The simplified drawing in Fig. 3 will be useful in understanding the observed diffraction patterns; in this figure the magnitude of the wavelength, relative to the slit width, is enlarged for purposes of clarity. In Fig. 3 the lines I represent the path traversed in the forward direction by the secondary Huygens wavelets from the two edges of the slit A of Fig. 1. Since the light approaching the slit consists of a parallel beam, the original wavefront is plane and all sources of Huygens wavelets in the slit opening are in phase. Therefore, the secondary waves traversing the paths I and all similar paths drawn in the forward direction from the plane of the slit are in phase, and the lens produces a bright image in the forward direction; this bright image is the central maximum of the diffraction pattern. The intensity of images produced at other points on the screen depends on the differences in paths traversed by secondary waves from different parts of the slit. If we are interested in the intensity on the screen at a point making angle θ (measured at the slit) with the center of the pattern, we must consider the portions of the Huygens wavelets that leave the slit at angle θ in Fig. 3. The lines II_a and II_b represent paths traversed by such secondary waves. It will be noted that path II_a traversed by waves from the upper limit of the slit is longer than path II_b by an amount Δp. If this path difference Δp is λ, the slit can be divided into two zones as indicated in the figure; the *mean difference* in path length for the secondary waves coming from the two zones is $\frac{1}{2}\lambda$ and the waves from the two zones cancel completely when the lens brings the waves together on the screen. Therefore, at an angle θ for which Δp is λ, complete cancellation occurs, and a dark fringe or region of zero intensity results. Similar reasoning shows that the intensity is zero when Δp is equal to 2λ, 3λ, 4λ, 5λ, $\cdots$.

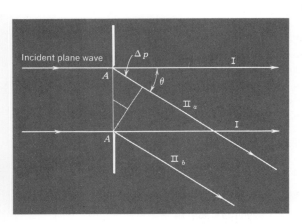

Fig. 3. Enlargement of slit A of Fig. 1.

The situation is different when Δp is equal to an odd number of half wavelengths. For example, consider the case for $\Delta p = \frac{3}{2}\lambda$. The slit can then be divided into three zones, and the mean difference in path traversed by the secondary waves coming from successive zones is $\frac{1}{2}\lambda$. Therefore, the waves from two of the zones will lead to cancellation at the screen, but the secondary waves from the third zone are not canceled and a bright fringe or maximum will appear on the screen. Detailed computation shows that the amplitude of this maximum is $2/3\pi$ times that of the central maximum, so the intensity is $(2/3\pi)^2 = \frac{1}{22}$ that of the central intensity. Similar arguments show that bright fringes of decreasing intensity occur when Δp is equal to $\frac{5}{2}\lambda$, $\frac{7}{2}\lambda$, $\cdots$. Since the first dark fringe occurs at $\Delta p = \lambda$, and $\Delta p = 0$ gives a maximum, the distance between the centermost dark fringes is twice as great as that between succeeding dark fringes; the result is that the observed central maximum is twice as wide as the other maxima, as suggested in Fig. 2.

As for the angles at which the maxima and minima occur, it will be noted that, in Fig. 3, $\sin\theta = \Delta p/w$, where w is the slit width. Fig. 4 shows the way in which the light is focused on the screen. Since parallel light is focused at a point, the distance from the lens to the screen must equal the focal length f of the lens. Parallel light making a small angle θ with the axis of the lens will be focused at the point x for which $x/f = \tan\theta$, as shown by the broken rays drawn through the center of the lens. Since the angle θ is small for those parts of the diffraction pattern that have appreciable intensity, both $\sin\theta$ and $\tan\theta$ are approximately equal to the angle θ in radians. Hence we may replace $\tan\theta$ and $\sin\theta$ by θ and write

$$\theta = \frac{\Delta p}{w} = \frac{x}{f}.$$

Note that the wider the aperture, the narrower is the central image. In principle, if the source slit S in Fig. 1 were small compared with the wavelength of light, so that a true cylindrical wave left the source slit, we could make the central image on the screen as narrow as we pleased

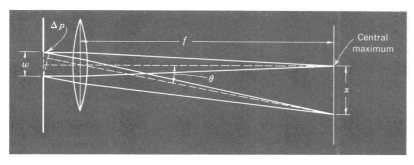

Fig. 4. Formation of a diffraction pattern on a screen.

by widening the aperture A. But we cannot make the aperture wider than the lenses themselves. When the screen containing the slit A is completely removed, the lenses themselves act as circular apertures and there is still a diffraction pattern on the screen.

The calculation of the diffraction pattern of a circular aperture is similar in essentials to the above calculation for a slit aperture. However, since the details of the calculation are considerably more complicated, we shall merely give the expression for the angle at which the first minimum is observed. As we have seen, the first minimum in the diffraction pattern of a slit occurs at an angle $\theta = \lambda/w$. In the case of a circular opening, the first minimum occurs at angle $\theta = 1.22\, \lambda/D$, where D is the diameter of the circular opening (the linear aperture). The diffraction image formed by a circular aperture is illustrated in the photograph in Fig. 8(a) on p. 814.

Example. *If the reflecting mirror at the Kitt Peak Observatory is used to produce the image of a star, the image has the appearance of the diffraction pattern in Fig. 8(a). Recalling that the diameter of the mirror is* 60 i *and its focal length is* 300 f, *calculate the diameter of the first dark ring in the diffraction pattern formed in the focal plane for a wavelength of* $\lambda = 550$ nm.

The diameter of the mirror $D = 60 \times 0.0254$ m $= 1.52$ m. The angular separation between the center of the pattern and the first dark ring is $\theta = 1.22 \lambda/D = 1.22 \times 5.50 \times 10^{-7}$ m$/(1.52$ m$) = 4.42 \times 10^{-7}$ rad. The focal length $f = 300 \times 0.305$ m $= 91.5$ m. In analogy with the last equation above $x = f\theta = 91.5$ m $\times 4.42 \times 10^{-7}$ rad $= 4.94 \times 10^{-5}$ m $= 4.04 \times 10^{-2}$ mm. The pattern could be observed only with a microscope! However, an eyepiece *is* a simple microscope and would reveal a diffraction pattern like the one in Fig. 8(a).

2. Resolving Power of Optical instruments

Let us first consider the diffraction pattern that would result, in the case of Fig. 1, if there were two source slits instead of just one. Such a situation could be realized by placing a second source slit S' just above and parallel to the original source slit S in Fig. 1. Light from slits S and S' would be furnished by different parts of the source and would have no definite phase relationship, so the *intensities* are additive. Two central maxima will be formed on the screen. Each central image will be bordered by a set of bright and dark fringes, which would overlap if the source slits S and S' were sufficiently close together. Indeed, if the two source slits were brought sufficiently close together, the two diffraction patterns would overlap to such an extent that the eye would be unable to distinguish two separate central images. It has been found that the two images can just be seen as separate, or just *resolved,* if the central maximum of one pattern coincides with the first minimum of the other pattern, as illustrated in Fig. 5, which shows a plot of intensity as a function of path difference for the two patterns. According to the discussion on p. 809, the screen separation of the two central maxima in Fig. 5 is $(\lambda/w)f$. This value corresponds to an angular separation of λ/w, where w is the width of slit A in Fig. 1. Hence, we can state that two images can be resolved only if their angular separation θ is at least equal to λ/w, where w is the opening of the slit through which the light passes. The condition for resolution, $\theta \geqq \lambda/w$, is known as the *Rayleigh criterion*.

Note that in the above discussion the 'sources' S and S' are assumed to be very narrow slits, and that the limit of angular resolution of the images of these slits in Fig. 1 is determined by the width w of slit A that limits the lens aperture. *Widening* the slit A increases the resolving power; that is, *widening A permits resolution of the images of more closely spaced* sources S and S'.

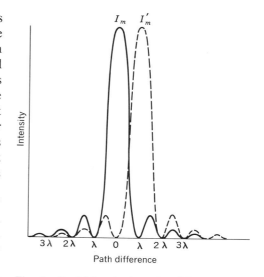

Fig. 5. Rayleigh criterion: the diffraction patterns from two closely-spaced sources at the limit of resolution where the central maximum of one pattern falls at the first minimum of the other.

As we have already noted, even if the diaphragm containing slit A in Fig. 1 were removed, we would still have a *diffraction pattern* of slit S on the screen, because the wave train from S to the screen is still interrupted and the part reaching the screen is limited by the apertures of the lenses themselves. Because of this interruption, even an ideal lens never forms a point image of a point source, but forms an image that is itself a diffraction pattern.

Consider the diffraction pattern in the vicinity of the image I in Fig. 6. It is the aperture AA of the lens itself that divides the spherical wavefront diverging from O into two parts, one of which passes through the lens and is focused at I. Whenever a wavefront is so divided, there is a resulting diffraction pattern that can be computed by considering each point of the partial wavefront (AA in Fig. 7) as a source of Huygens wavelets. In Fig. 7, we draw these wavelets for just two points, P, P, representative of the upper and lower halves of the lens. These wavelets are 'refracted' at the lens surface just like the plane wavefront in Fig. 6, the refraction resulting from the increase in speed and in wavelength as the wavefront passes from glass into air.

The two wavelets from P, P interfere constructively along the row of dots leading to I. Wavelets from *all points* in the plane AA are in phase at I, which is the point of maximum intensity on the screen SS.

The two wavelets are one-half wavelength out of phase along the rows of crosses leading to points B, B and interfere destructively at B, B. Generally speaking, for each point in the lower half of the lens, there is a point in the upper half such that the wavelets are one-half wavelength

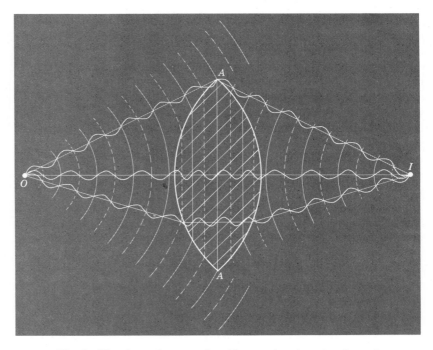

Fig. 6. Wavefronts from a point object passing through a lens of index of refraction 1.5, in the symmetric case where image and object are both at twice the focal length.

out of phase and interfere destructively in the vicinity of B, B. Hence there will be a dark ring around I at about the radius IB. The image intensity falls off *gradually* from a maximum at I to a minimum at B, B. The image is not a *point* but a *region* of radius IB of gradually decreasing illumination.

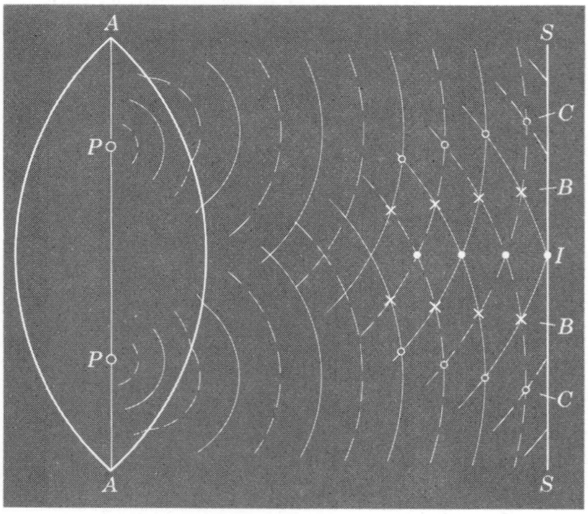

Beyond the circle BB, there will be a weak secondary maximum at about the radius IC where the wavelets from P, P are a full wavelength out of phase and interfere constructively. This secondary bright ring is shown in the photograph in Fig. 8(a). The maximum is weak because there are some wavelets from the plane AA that interfere *destructively* at C with the wavelets from P, P. In particular, since the wavelets from the two points P, P are a *full* wavelength out of phase, the wavelet from the center of the lens, halfway between P and P, will differ from each of these wavelets by one-half wavelength and interfere destructively with both of them. Detailed computation, however,

Fig. 7. Schematic diagram illustrating the origin of the interference pattern forming the image of the point source in Fig. 12.

shows that there will be a weak maximum in intensity on the circle CC, and indeed a succession of weaker maxima farther out, but falling off in intensity so rapidly that they are very difficult to observe.

We now note that the distances IB and IC as drawn on Fig. 7 are large only because we have used, for purposes of illustration, a large wavelength. If Fig. 7 is imagined drawn with the distance between solid and broken arcs equal to half the actual wavelength of light, we see that the radii IB and IC would be very small indeed.

We can get an approximate value of the radius IB to the first minimum by using the relation $Y = \lambda X/2a$ derived in the example on p. 734 concerning the diffraction pattern from two slits separated by a distance $2a$. Y corresponds to IB, X to the image distance q, and $2a$ to $PP = \frac{1}{2}D$, where D is the lens diameter. Hence the approximate arguments above would lead to

$$IB = \lambda q / \tfrac{1}{2} D = 2\lambda q/D,$$

as the approximate radius of the first dark ring.

In the particular case of interest in astronomy, where a star is essentially a point source of light at a very great distance and the image is formed at the focus of the objective lens or mirror, rigorous computation gives the radius of the first dark ring as

$$R = 1.22\, \lambda f/D, \qquad (1)$$

where f is the focal length. The angle subtended by this radius, at the lens, is

$$\theta = R/f = 1.22\, \lambda/D. \qquad (2)$$

Now consider the case in which there are two stars subtending angle α at the objective lens, so that the centers of their images would also subtend

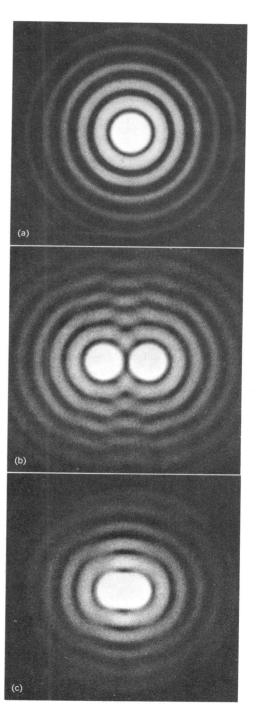

Fig. 8. Fraunhofer diffraction patterns: (a) Diffraction image of a point source in the case of a circular aperture. (b) Separated images of two point sources. (c) Image of a pair of point sources at the limit of resolution. (Photographs reprinted from Cagnet *et al.*, *Atlas of Optical Phenomena* with the permission of Springer-Verlag.)

angle α. Each image will be a diffraction pattern. If the angle α is 1.22 λ/D, the center of one pattern will fall on the edge of the central maximum of the other, as in Fig. 8(c). For a larger angle as in Fig. 8(b), the two images can be clearly distinguished. On the basis of the Rayleigh criterion, the two images can be *resolved* for $\alpha \geqq 1.22\, \lambda/D$. For $\alpha < 1.22\, \lambda/D$, the images cannot be resolved; it is impossible to see that there are actually two images. The wavelength normally used in Rayleigh's criterion when white light is involved is an average value of 500 nm $= 5 \times 10^{-5}$ cm.

The above argument shows how diffraction effects set the ultimate limitation on the usefulness of optical instruments such as telescopes and microscopes. In the problem sets of the preceding chapter, we discussed the magnification obtainable with these instruments; however limits of *useful* magnification are imposed by the wave nature of light itself. It is useless to try to resolve two stars by increasing the magnifying power of a telescope merely by using an eyepiece of shorter focal length if the stars subtend an angle of less than $\alpha = 1.22\, \lambda/D$ at the objective of diameter D. If the angle subtended is less than this, the objective can produce only a blurred image consisting of overlapping diffraction patterns in which separate images of the two stars cannot be distinguished, regardless of the power of the eyepiece.

The ultimate resolution attainable with a microscope is similarly limited. The source is not at a great distance in the case of a microscope, so relation (2) does not give the exact limit of resolution. The coefficient of λ/D depends on the focal length of the objective and the image distance q. But (2) gives the right order of magnitude, and one can say, *approximately,* that a microscope cannot resolve two points that subtend an angle at the objective of less than about λ/D, where D is the diameter of the objective. However, there *is* a way in which the resolving power of a microscope *can* be increased. If ultraviolet light is used to illuminate the object, the limiting angle α is smaller, since ultraviolet light is of shorter wavelength than visible light. It is necessary to use a photographic plate or fluorescent screen to receive the image in an ultraviolet microscope.

We notice that the *larger* the aperture of a telescope or microscope, the *better* is the resolution, that is, the *smaller* is the angle between two points that can just be resolved.

Example. *Using the Rayleigh criterion for angular resolution, compute (a) the distance between two lunar craters that can just be resolved and (b) the distance between two sunspots that can just be resolved with the* 60-i *telescope at Kitt Peak.*

From the example on p. 810, we note that the limit of angular resolution is $\theta = 4.42 \times 10^{-7}$ rad. (*a*) When the moon is 240 000 mi from the earth, the separation x of two points on the lunar surface that would subtend angle θ is $x = 2.4 \times 10^5$ mi $\times 4.42 \times 10^{-7}$ rad $= 0.106$ mi or 560 f. Thus two craters 560 f apart can be resolved. (*b*) Similarly, since the distance to the sun is 93 000 000 mi, $x = 9.6 \times 10^7$ mi $\times 4.42 \times 10^{-7}$ rad $= 41.0$ mi. Two points on the sun's surface 41.0 mi apart can be resolved; sunspots are usually much larger. NOTE: In practice, the resolving power of an astronomical telescope on the earth's surface has its practical limit set, not by diffraction, but by the 'twinkling' effects of variable atmospheric refraction.

3. The Diffraction Grating

The diffraction grating, invented by JOSEPH VON FRAUNHOFER (1787–1826), gives the most startling examples of nonrectilinear propagation of light and the applicability of the wave model. A diffraction grating is a screen containing a very large number of very narrow parallel slits, uniformly spaced at distances only a few times the wavelength of light. This device gives rise to very sharp diffraction maxima at very large diffraction angles—up to almost 90 degrees in some optical arrangements. A transmission grating is formed by ruling grooves on glass with a diamond cutting edge moved by a very accurately constructed 'ruling engine.' It is the clear spaces between the grooves that form the slits.

In Fig. 9(a), a parallel beam of coherent monochromatic light is shown approaching a transmission grating. The source is an illuminated slit parallel to the slits in the grating. Only a few grating slits are shown, but there are actually many thousands of slits per centimeter. The distance between adjacent slits is d. Since the beam is parallel and normal to the grating, all Huygens sources in the plane of the grating are in the same phase. The secondary waves are in phase in the forward direction, since the paths traversed by all waves in this direction are of equal length. If the light waves traveling in the forward direction are focused on a screen, a bright *central image* of the original source slit is formed. In certain other directions, the secondary waves will also be in phase at a plane wavefront. One of these directions is denoted by θ in Fig. 9(b). At this angle the paths traversed by secondary waves from successive slits differ by an amount equal to the wavelength of the light. Hence the condensing lens will produce an image of the original slit on the screen;

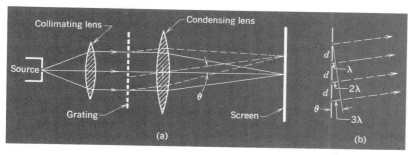

Fig. 9. Transmission grating.

this image is called the *first-order* image. Consideration of Fig. 9 shows that the angle θ for the first order is defined by the relation $\sin\theta = \lambda/d$.

Other directions in which secondary waves from the slits can produce a common plane wavefront are those directions in which the paths traversed by the secondary waves from successive slits differ by 2λ, 3λ, 4λ, $\cdots$. The directions in which diffraction maxima may be produced are accordingly given by the general relation

$$\sin\theta_n = n\lambda/d, \qquad (3)$$

where $n = 1, 2, 3, 4, \cdots$. The diffraction maximum for $n = 1$ is called the *first-order* image, that for $n = 2$ the *second-order* image, etc. Since θ_n can be accurately measured and d is known for a given grating, it is possible to use a grating to measure the wavelength of light. When the angle θ is large it is desirable to turn the condensing lens as in the spectrometer illustrated in Fig. 1 of Chapter 41.

At angles only slightly different from values given by (3), the intensity is sharply reduced since light is received from thousands of Huygens sources covering the complete range of phase differences. Each diffraction image is (for a perfect grating) as sharp as the image of the source that would be formed by the lenses themselves in the absence of the grating. Figures 3-5 of Chapter 41 show examples of the diffraction images obtained with a grating.

> **Example.** *A grating having* 10 000 lines/cm *is illuminated by parallel light of* 590-nm *wavelength, incident normally. At what angles are the diffraction images formed?*
>
> The grating spacing d is 10^{-4} cm $= 10^3$ nm $= 1000$ nm. Hence the first-order image is formed at
>
> $$\sin\theta_1 = \lambda/d = 590 \text{ nm}/1000 \text{ nm} = 0.59, \qquad \theta_1 = 36°\!.2.$$
>
> The second-order image would be formed at $\sin\theta_2 = 2\lambda/d = 1.08$, but this is an impossible angle. The grating spacing is less than 2λ so it is impossible to have a wavelength difference of 2λ from successive slits. There is only one diffraction image on each side of the central image.

The gratings actually used in most research investigations are reflection gratings. Essentially, reflection gratings are metal mirrors on which equally spaced narrow lines have been ruled with a diamond. When the grating is illuminated, the strips between these lines become the sources of Huygens wavelets, just as do the parallel slits in a transmission grating. Since the spacing of grating lines must be constant for all parts of a grating surface and since optical gratings have several thousand lines per centimeter, the fabrication of a grating is a difficult and expensive process. However, it is possible to produce *replicas* of an original grating by coating the grating surface with a plastic film. After the plastic film has dried, it can be removed from the grating surface. The plastic film has ridges corresponding to the ruled lines; hence, the smooth spaces between the rulings can be used in the same manner as slits in a transmission grating. An excellent method for producing replicas that can be used as reflection gratings involves evaporating aluminum on the plastic film. Gratings prepared by this method have proved very satisfactory and are being used in many research laboratories.

PROBLEMS

1. A lens of focal length 60 cm is used to form a diffraction pattern of a slit 0.3 mm wide. Calculate the distance on the screen from the center of the central maximum to the center of the first dark band and to the center of the next bright band when the slit is illuminated by yellow light with $\lambda = 589$ nm.
   Ans: 1.18 mm; 1.77 mm.

2. Parallel light of wavelength 546.1 nm is incident normally on a slit 1 mm wide. If a lens of 200-cm focal length is mounted just behind the slit and the light is focused on a screen, what will be the distance from the center of the diffraction pattern to (a) the first minimum, (b) the first maximum, and (c) the third maximum?

3. In the diffraction pattern of a single slit, the distance between the first minimum on one side of the central maximum to the first minimum on the other side of the central maximum is 6.75 mm. If the wavelength of the light is 546.1 nm and the lens used to form the diffraction pattern has a focal length of 90 cm, find the width of the slit. Ans: 0.146 mm.

4. Light of wavelength 600 nm is incident normally on a slit 0.6 mm wide and is brought to a focus by a lens of focal length 200 cm. Find the distance between the first and second maxima in the resulting diffraction pattern.

5. What is the theoretical angular limit of resolution of the Yerkes telescope, whose objective is 40 inches in diameter? Ans: 0".124.

6. Calculate the limit of angular resolution of the 200-i reflecting telescope at Mount Palomar.

7. The pupil of the eye is approximately 2 mm in diameter under conditions of moderate illumination. Considering the eye as a telescope, compute the eye's limit of angular resolution for $\lambda = 550$ nm. Is the resolving power of the eye greater at night or in bright daylight?
   Ans: 3.4×10^{-4} rad, or approximately 1'.

8. Using the value computed in Prob. 6 for the limit of angular resolution for the Mount Palomar telescope and taking 240 000 mi as the distance from the earth to the moon, find the linear separation of two objects on the moon's surface that can just be resolved by the 200-i telescope.

9. The focal length of the objective of a certain microscope is 3.2 mm and its diameter is 4.0 mm. What is the *approximate* value of the separation of two point objects that can just be resolved when they are illuminated by light of mean wavelength 550 nm?
   Ans: 537 nm. NOTE: Calculation on the basis of detailed theory gives 418 nm.

10. What would be the *approximate* value for the smallest separation of two point objects that could just be resolved if ultraviolet light of wavelength 225 nm were used with the microscope described in Prob. 9 and photographic methods of detection were used?

11. A transmission grating has 5000 lines/cm. Calculate the angular deviation of the second order image for sodium light of wavelength 589.3 nm when a parallel beam of light strikes the grating at normal incidence. Show diagram. Ans: 36°.1.

12. When a parallel beam of light is normally incident, a certain transmission grating produces a diffraction pattern in which the third-order maximum for light of $\lambda = 589$ nm appears at an angle of 60° away from the central image. Find the distance between the lines on the grating.

13. When plane light waves are normally incident, a diffraction grating ruled with 6000 lines/cm forms the first-order diffraction maximum for light of a certain wavelength at an angle of 18°.0. What is the wavelength of the light? Show diagram.   Ans: 515 nm.

14. A diffraction grating is ruled with 8000 lines/cm. If a parallel beam of light consisting of light of wavelengths 500 nm and 550 nm strikes the grating normally, what is the angular separation between the diffraction maxima for 500-nm and 550-nm radiation, in the first order? in the second order?

15. Plane monochromatic waves of wavelength 500 nm are incident normally on a plane transmission grating having 5000 lines/cm. Find the angles at which the first-, second-, and third-order diffraction maxima appear. Show diagram.   Ans: 14°.5, 30°.2, 48°.6.

16. A transmission grating ruled with 6000 lines/cm forms a second-order diffraction maximum at an angle of 54°.0 from the central maximum when a parallel beam of monochromatic light strikes the grating at normal incidence. What is the wavelength of the incident light?

17. What is the grating spacing for a transmission grating giving a first-order image at 45° for microwaves of wavelength 15 cm?
     Ans: 21.2 cm.

18. Compute the angles of diffraction of all orders when microwaves of 6-cm wavelength are incident on the grating in Prob. 17.

POLARIZATION 40

Interference and diffraction effects can be interpreted on the basis of a wave theory of light without the necessity of specifying the type of wave motion involved. Our treatment of these phenomena would hold equally well for transverse or for longitudinal waves. In the present chapter we discuss other phenomena that can be interpreted only on the basis of *transverse waves*.

In longitudinal waves there is always symmetry around an axis in the direction of propagation since the vibrations are parallel to this axis. In transverse waves the vibrations are at right angles to the direction of propagation, and hence symmetry around an axis in this direction may not exist. For example, a simple transverse wave traveling along a horizontal string may involve only up-and-down vertical motions of the particles of the string, or may involve only right-and-left horizontal motions. Such waves are said to have *plane polarization,* since all motions take place in a single plane. If a traveling wave on a string is polarized in a vertical plane, the wave would be undisturbed if the string passed between the pickets of a picket fence, whereas a horizontally polarized wave of sufficient amplitude would be damped by contact of the string with the fence. This difference in behavior of waves of different polarization is exactly analogous to the *selective absorption* of polarized light beams that will be discussed in the first section below.

Light waves can be polarized by various processes such as selective absorption by certain crystals, reflection from nonmetallic mirrors, and scattering by small particles. In this chapter we shall discuss the methods of producing polarized light and various properties of polarized light. The phenomenon of *double refraction* will be treated briefly; double refraction is exhibited by certain crystals in which light waves having different polarizations travel with different speeds. The chapter closes with a discussion of the electro-optical Kerr effect and the magneto-optical Faraday effect; these effects provide evidence that light waves are transverse electromagnetic waves. *The 'light vibrations' we discuss are actually the electric components of the waves.*

Polarization (by double refraction) was discovered by Huygens in 1678, but he found the phenomenon impossible to explain on his wave theory because he assumed longitudinal waves. Newton recognized that polarization implies that "The Rays of Light have different Properties in their different Sides," and used this observation as an argument for the particle theory of light, since particles could have this property but Huygens's *longitudinal* waves could not. It remained for Young and Fresnel, independently in 1816, to suggest that the interference and polarization properties jointly imply that light consists of *transverse waves*.

1. Polarization by Selective Absorption

The production of polarized light by selective absorption (dichroism) is exhibited by crystals of certain minerals and organic compounds. The best known of these minerals is *tourmaline*. When a light beam passes through a single properly cut slab of clear tourmaline, it emerges somewhat diminished in intensity and somewhat colored, but it has no other peculiarity detectable to the eye. However, if *two* slabs of tourmaline are introduced into a light beam, the intensity of the transmitted light depends on the relative orientation of the two slabs, as illustrated in Fig. 1. In (a) the two slabs are oriented with crystalline axes parallel and light is transmitted. In (b) one of the slabs has been rotated through an angle θ with respect to the other slab and less light is transmitted. In (c) one of the slabs is oriented at 90° with respect to the other slab and *no light is transmitted through the pair*. These effects can be interpreted *only* if light waves are assumed to be transverse.

In an ordinary light beam the transverse vibrations have no preferred direction with respect to the direction of propagation; in other words, the different 'pieces' of the light wave that originate in different atoms of the radiating source have their vibrations randomly oriented in all

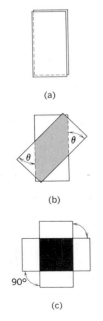

Fig. 1. Transmission of light through tourmaline crystals.

possible planes transverse to the direction of propagation. Then a pencil of light approaching the reader would involve the transverse vibrations shown in Fig. 2(a); in this figure the arrows represent the transverse vibrations associated with the different 'pieces' of the wave and have all possible orientations in a plane perpendicular to the direction of propagation indicated by the arrow point $\odot$. Now let us assume that a properly cut tourmaline slab transmits vibrations having a given orientation with respect to the slab and absorbs all light associated with vibrations at right angles. For example, assume that the tourmaline slab shown in Fig. 2(b) transmits only the light associated with vibration components in the vertical direction and absorbs all the light associated with vibration components in the horizontal direction. If the

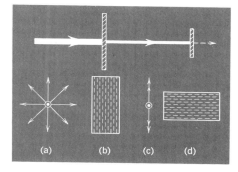

Fig. 2. Transmission of light waves through 'crossed' tourmaline crystals.

tourmaline slab is placed in the light beam approaching the reader, the transmitted light has only vibrations in the vertical direction as indicated in Fig. 2(c). If a second tourmaline slab oriented at right angles to the first is placed in the approaching light beam as shown in Fig. 2(d), it can transmit only light associated with horizontal vibrations. Since light associated with horizontal vibrations has already been removed by the first tourmaline slab, no light reaches the observer. This explanation is necessary to account for the observed phenomena depicted in Fig. 1 and many other experiments, and hence we conclude that light waves are transverse.

Owing to the fact that most tourmaline crystals are colored, they are not very useful in optical instruments. Small needle-shaped crystals of the organic compound quinine iodosulphate having similar optical properties were grown as early as 1852. Because of their small size, they were not useful in optical work until 1935, when EDWIN H. LAND developed a method of orienting large numbers of crystals of this type. The material (called *Polaroid*) that was developed by Land consists of a plastic film in which are suspended large numbers of the minute crystals. These sheets would be useless for polarizing if the crystals were randomly oriented, so the crystals are given the same orientation by giving the film a stretch in one direction during the manufacturing process. For optical use, the films are mounted between glass plates. The development of Polaroid first made possible the construction of *polarizing plates* of any desired size. As indicated in Fig. 3, two polarizing plates can be used to produce effects similar to those produced by tourmaline crystals.

Fig. 3. Photograph of transmission of light through polarizing plates. (Courtesy of the Polaroid Corporation.)

The light transmitted by the first tourmaline crystal or polarizing plate inserted in a light beam is said to be *plane-polarized,* since the vibrations are in a single plane. For example, the vibrations shown in Fig. 2(c) are polarized in a vertical plane. The device, such as the first polarizing plate, that produces the polarization is called the *polarizer*. Since neither the eye nor a photographic film can detect any difference between polarized and unpolarized light, the polarization must be detected by using a second polarizing plate, called the *analyzer*. A polarizing plate is said to be *perfect* if it transmits *all* light polarized along a given transverse axis, called the *axis of the plate,* and absorbs *all* light polarized normal to this axis. Commercial plates only approximate this perfection.

In ordinary light, the vibrations of the different 'pieces' of the light originating in different atoms of the source take place in all directions at right angles to the propagation vector. However, since these vibrations are actually the electric fields of a traveling electromagnetic wave, the amplitude of each vibration can be resolved into components along any two axes normal to the direction of propagation. Since the energy of a light vibration is proportional to the square of the amplitude, this resolution is energetically correct, the sum of the squares of the two rectangular components of a vector being equal to the square of the length of the vector. Thus, if a beam of light is approaching the reader as shown in Fig. 4(a), the vibrations take place in all directions, including those shown. We may resolve each amplitude into components along a pair of transverse X- and Y-axes. The whole unpolarized light wave may be considered as composed of, or equivalent to, two resultant vibrations along the X- and Y-axes as shown in Fig. 4(b).

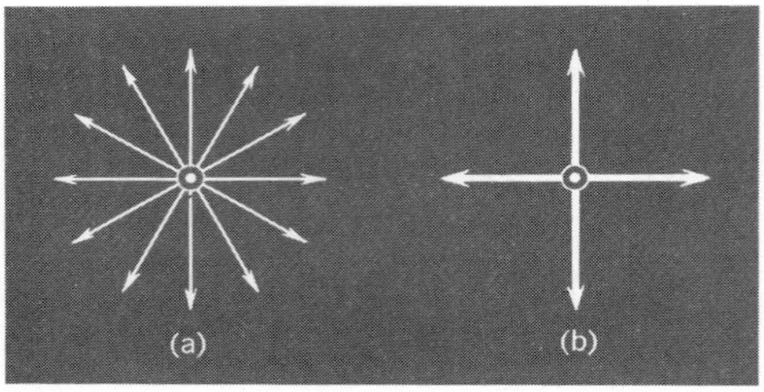

Fig. 4. Resolution of unpolarized light into two plane-polarized components.

One method we shall use for picturing the transverse waves associated with a light beam is shown in Fig. 5. The heavy arrows give the direction of propagation, the vertical arrows represent light vibrations in the plane of the paper, and the circles represent light vibrations perpendicular to the plane of the paper. Thus, the light shown in Fig. 5(a) is polarized in the plane of the paper, that shown in Fig. 5(b) is polarized in a plane at right angles to the paper, and that in Fig. 5(c) is unpolarized. The *plane of polarization* is the plane in which the direction of propagation and the light vibrations lie.

After passing through a perfect polarizer, an unpolarized beam is reduced to half of its original intensity since one component is completely eliminated. If the resulting plane-polarized beam passes through a perfect analyzer making an angle θ with the plane of polarization, the resulting amplitude is $\cos\theta$ and the intensity is $\cos^2\theta$ times that of the plane-polarized beam that leaves the polarizer.

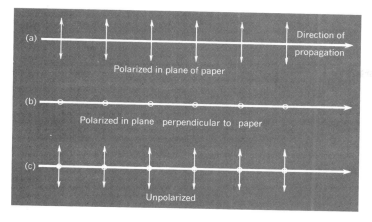

Figure 5

Example. *Four perfect polarizing plates are stacked so that the axis of each is turned 30° clockwise with respect to the preceding plate, the last plate therefore being 'crossed' with the first. How much of the intensity of an incident unpolarized beam of light is transmitted by the stack?*

The first plate transmits $\frac{1}{2}$ of the incident intensity. Each succeeding plate makes a vector resolution at angle 30° and transmits a fraction $\cos 30°$ of the amplitude or $\cos^2(30°)$ of the intensity. Since $\cos 30° = \sqrt{3}/2$ and $\cos^2(30°) = \frac{3}{4}$, the intensity transmitted by the stack is

$$\tfrac{1}{2} \cdot \tfrac{3}{4} \cdot \tfrac{3}{4} \cdot \tfrac{3}{4} = \tfrac{27}{128} = 0.211,$$

or 21.1 percent of the original intensity.

2. Polarization by Reflection

Perhaps the simplest method of polarizing light is the one discovered by ETIENNE LOUIS MALUS in 1808. This method involves reflection of light at the surface of a polished nonmetallic medium such as black glass. Malus discovered that light reflected from a glass surface is partially polarized and that when the light is incident at one particular angle the reflected light is completely plane-polarized.

It is this polarization of light reflected from nonmetallic surfaces that contributes to the effectiveness of polarizing sun glasses, particularly in driving toward the sun over wet road surfaces.

Consider an unpolarized incident light ray SO striking a glass surface at an angle of incidence i, as shown in Fig. 6. There is always a reflected ray OR and a refracted ray OT at such a surface. If the reflected light is examined with an analyzer placed at the indicated position, it is found that the reflected light is partially polarized in such a way that light associated with vibrations parallel to the surface and perpendicular to the plane of incidence is more highly reflected than light associated with

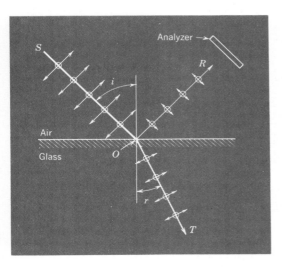

Fig. 6. Partial polarization by reflection.

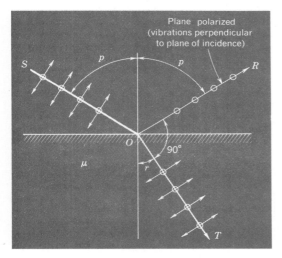

Fig. 7. Complete polarization by reflection: Brewster's law.

vibrations in the plane of incidence. The plane of incidence is the plane of the paper in Fig. 6. When the angle of incidence at a glass surface is about 57°, the reflected light is completely polarized and all vibrations present in the reflected light are parallel to the glass surface and perpendicular to the plane of incidence, as indicated in Fig. 7. The angle p is called the *polarizing angle* (or *Brewster's angle*) for glass. At this angle approximately 8 percent of the incident light is reflected.

DAVID BREWSTER (1781–1868) first noted that when light is incident at the polarizing angle p, the reflected and refracted rays are exactly 90° apart, as shown in Fig. 7. The reflected ray is plane-polarized as indicated. Brewster's discovery that rays OR and OT are 90° apart enables us to write a relation between the polarizing angle and the index of refraction of a medium. Snell's law for the case under consideration is

$$\sin p = \mu \sin r.$$

Since the angle between the reflected and refracted rays is 90°, we may write (Fig. 7)

$$p + r + 90° = 180°, \quad \text{or} \quad p + r = 90°.$$

Hence, $\sin r = \cos p.$

Substituting this value for $\sin r$ in Snell's law, we get

$$\mu = \sin p / \cos p = \tan p. \quad \text{(Brewster's law)} \quad (1)$$

For $\mu = 1.54$, a typical value for glass, $p = 57°$.

Now consider the action of the surface on the refracted light. If the refracted light is examined with an analyzer, it is found to be partially polarized for all angles of incidence, but for no angle of incidence is the refracted light *completely* plane-polarized. Since more light is *reflected* with polarization normal to the plane of incidence than with polarization in the plane of incidence, more light is *refracted* with polarization in the plane of incidence than with polarization normal to this plane. Since the refracted light always includes some light with vibrations in each reference plane, it is said to be *partially plane-polarized*. The ratio of light with vibrations in the plane of incidence to the light with vibrations perpendicular to the plane of incidence is greatest when the initial beam is incident at the polarizing angle.

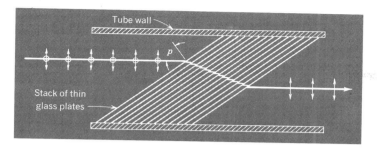

Fig. 8. Polarizer constructed from a stack of glass plates.

Plane polarization of *refracted* light can be approximated by the use of several glass plates as shown in Fig. 8. If a beam of unpolarized light is incident at the polarizing angle on a pile of glass plates, some light with vibrations perpendicular to the plane of incidence is reflected at each surface and all light with vibrations in the plane of incidence is refracted. The net result is that the refracted beam becomes more nearly plane-polarized with vibrations in the plane of incidence as the beam traverses successive glass plates. Although plane polarization is never completely attained by transmission, complete polarization can be approached by using a large number of plates. An effective polarizer can be constructed by mounting a stack of microscope cover glasses in a tube in such a manner that light entering the tube strikes the cover glasses at the polarizing angle as shown in Fig. 8. The reflected light is absorbed by the blackened inner wall of the tube and the refracted light passes through the tube as shown.

3. Polarization by Scattering

When a beam of light passes through a medium in which particles are suspended, its path becomes visible from the side; for example, if a narrow pencil of sunlight enters a darkened room through a small hole in a window shade, its path is plainly visible if there are dust particles in the air. This type of 'scattering' by relatively large particles results from reflection of light by the surfaces of the particles.

However, if the suspended particles are so small that their diameters are comparable to the wavelengths of the incident light, a different type of scattering process takes place. This type of scattering involves not ordinary reflection but a kind of diffraction in which each particle in the light path behaves as if it were a secondary light source. The intensity of the scattered light varies inversely as the fourth power of the wavelength. Hence, if white light enters a medium in which small particles are suspended, more of the light of short wavelength near the violet end of the spectrum is scattered at right angles to the original beam than of the light of long wavelengths near the red end of the spectrum. Of the violet light at 450 nm, a fraction is scattered that is approximately 8 times as great as the fraction of red light at 750 nm that is scattered. The ratio 8 in the previous sentence is $(750/450)^4$. Thus, a pencil of white light traversing a suspension of small particles appears bluish when viewed from the side because a preponderance of the scattered light is of short wavelength; it appears reddish when viewed along the beam in the direction of the source because little of the red and orange light has been scattered out of the beam. This effect can be noted when tobacco smoke is viewed in sunlight; the smoke appears blue-gray but the 'shadow' of the smoke has a red-gray tint.

The red color of the sun at sunrise and sunset is due to selective scattering by small dust particles in the air near the earth's surface and by the molecules of the gases in the atmosphere. When the sun is at the horizon, the sunlight reaching an observer at the earth's surface has an effective atmospheric path that is approximately 50 times greater than the path traversed by light passing vertically through the atmosphere. This longer atmospheric path accentuates the effects of molecular scattering, and, since a larger portion of the actual path is through dust-filled air at low altitudes, the effects of scattering by dust particles are more pronounced than when the sun is high above the horizon. Therefore, since much of the violet, blue, and green in the original sunlight is scattered, the sun appears redder at sunrise and sunset than at midday.

The blue color of the sky can also be explained in terms of scattering. Just as the directly transmitted sunlight appears reddish because the blue end of the spectrum is selectively scattered, when we look away from the sun we see this selectively scattered light as the blue of the sky.

Examination by means of an analyzer reveals that light scattered by small particles is plane-polarized. In order to determine the plane of polarization of the scattered light, let us consider the arrangement shown in Fig. 9. A beam of unpolarized light *NO* enters a liquid at normal incidence. If the liquid holds fine particles in suspension, some of the light will be scattered perpendicularly along a path such as *RS*. By rotating the analyzer *A*, it is found that the scattered light is completely polarized with its plane of vibration perpendicular to the plane of the paper, as indicated in Fig. 9. None of the light of the incident beam with its vibration vector in the plane of the paper is scattered in the direction *RS*. This result is reasonable if one considers that such scattering would necessarily require a 90° change in direction of the vibration vector, since the scat-

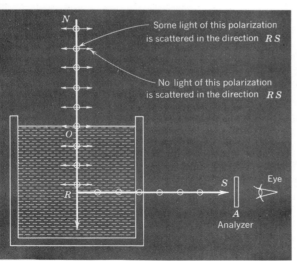

Fig. 9. Polarization by scattering.

tered wave must be transverse. The light is scattered without change in direction of the vibration vector.

Sunlight scattered by the earth's atmosphere (the blue of the sky) is similarly polarized. It is completely polarized when scattered through 90°, partially polarized when scattered through other angles. This effect can also be illustrated by Fig. 9. Let N be the sun more or less directly overhead; let the scattering medium be the earth's atmosphere, S the observer, and R a point in the blue sky near the horizon. Then the blue skylight from R will be polarized in a horizontal plane. Thus, polarizing glasses and polarizing camera filters can be used to reduce the brightness of blue skylight at the same time as they reduce surface glare since the beams RS in Fig. 9 and OR in Fig. 7 both have horizontal polarization.

4. Double Refraction

In our treatment of refraction in Chapter 36, we considered the relatively simple case of the propagation of light through transparent *isotropic* media such as vacuum, air, water, and glass, in which the speed of light is the same in all directions. Many crystals constitute *anisotropic* optical media and possess properties that cannot be interpreted in terms of our earlier discussion. One of these properties is illustrated in Fig. 10, which gives a photograph of a printed word as viewed through a glass plate and as viewed through a calcite crystal; it is evident from the photograph that the word appears double when it is viewed through the calcite crystal. A crystal such as calcite is said to be *doubly refracting* or *birefringent* and the phenomenon exhibited is called *double refraction* or *birefringence*.

A single spot on a piece of white paper viewed through a properly cut calcite crystal appears as *two* spots; if the crystal is rotated about an

Fig. 10. Photograph of a printed word viewed through a calcite crystal, and through ordinary glass. No Polaroid is involved, but the photograph was furnished through the courtesy of the Polaroid Corporation.

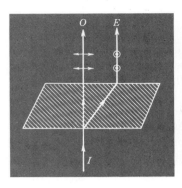
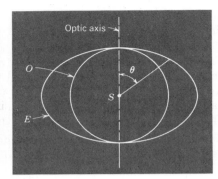

Figure 11   Figure 12

axis normal to the paper, one of the spots appears to move around the other. When a polarizing plate is placed above the crystal, it is found that for a certain orientation only a single spot is observed; when the polarizing plate is rotated through 90°, only the other spot will be observed. This phenomenon is illustrated schematically in Fig. 11, which shows in cross section a narrow pencil or ray of unpolarized light directed vertically upward through the crystal. In passing through the crystal the incident ray is separated into two rays; the ray O passing vertically upward through the crystal is called the *ordinary ray* and the displaced ray E is called the *extraordinary ray*. The light waves associated with the two rays are plane polarized at right angles to each other as indicated in the figure. If the crystal in Fig. 11 is rotated about an axis coinciding with the incident ray I, the emerging extraordinary ray E moves in a circle about the emerging ordinary ray O.

The phenomenon of double refraction can be interpreted on the wave theory if we assume that within the crystal waves associated with the ordinary ray are transmitted with equal speeds in all directions, while those associated with the extraordinary ray are transmitted at different speeds in different directions. The model is shown schematically in Fig. 12, which shows wavefronts emerging from a point source S located inside a doubly refracting crystal. One wavefront labeled O spreads spherically from S in quite *ordinary* fashion. The other wavefront labeled E spreads outward from S in the form of an ellipsoid of rotation in quite *extraordinary* fashion, because its speed of propagation varies with angle θ in the figure. It should be noted that in one direction in the figure the propagation speeds of the O-waves and of the E-waves are equal; this direction, called the *optic axis,* is shown by the broken line in Fig. 12. It should be emphasized that the term 'optic axis' refers to a *direction* within the crystal and not to a single line; thus, any line parallel to the dotted line in Fig. 12 would also represent the optic axis. The speeds of E-waves and O-waves are the same when they travel parallel to the optic axis and differ by the greatest amount when they travel perpendicular to the optic axis.

The propagation of light waves within a doubly refracting crystal can be described in terms of Huygens wavelets provided one uses spherical wavelets for the light of ordinary polarization and ellipsoidal wavelets for the light of extraordinary polarization. Huygens constructions are shown schematically for three situations in Fig. 13. In each part of the figure, plane wavefronts strike a slab of crystal at normal incidence; however, in each part of the figure the slab has been cut in a different manner. The Huygens construction in Fig. 13(a) corresponds to the situation shown in Figs. 10 and 11; in this part of the figure the slab was cut with its face making the indicated acute angle with the optic axis. The Huygens wavelets corresponding to the E-waves spread ellipsoidally but have a plane envelope as shown; the Huygens wavelets associated with the O-waves are spherical and have a plane envelope, which is displaced from that of the E-wavelets. The two sets of emerging wavefronts are parallel but not coincident; the lateral displacements are indicated by the rays. Note that the direction of propagation indicated by the E-rays is *not perpendicular* to the wavefront.

The Huygens construction for a normally incident plane wave traveling in a direction *parallel* to the optic axis of a crystal is shown in Fig. 13(b). In this case the envelopes of the O-wavelets and E-wavelets coincide. Thus, there is no lateral displacement of the E-rays with respect to O-rays; the O- and E-wavefronts coincide at all points. No unusual effects would be encountered in viewing a printed word through a crystal cut in this manner; it would appear as at the right in Fig. 10.

Figure 13(c) gives the Huygens construction for a parallel light beam having normal incidence on a crystalline slab with its optic axis *perpendicular* to the incident beam. In this situation there is no

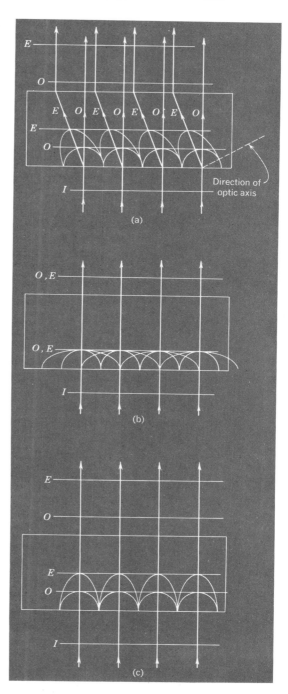

Fig. 13. Passage of a parallel incident beam through doubly refracting crystals. In (a) the incident beam makes an arbitrary angle with the optic axis; in (b) the beam is parallel to the optic axis; in (c) the beam is perpendicular to the optic axis.

lateral displacement of the rays, and no strange effects would be observed visually in reading a printed word through the slab. However, the *wavefronts of the E- and O-waves do not coincide* as they do in the incident wave. The *E*-wavefront arrives at the upper surface of the slab ahead of the *O*-wavefront; this relative displacement of wavefronts that were coincident in the initial incident beam leads to some interesting and important interference phenomena which we shall consider in the following pages.

Some understanding of the reasons for the phenomenon of double refraction can be obtained from a consideration of the atomic models shown in Fig. 14. In Sec. 5 we shall present some direct evidence that the vibrations we have attributed to light waves are actually associated with *electromagnetic* waves. With this in mind, we might consider qualitatively the interaction of light waves with the atoms shown schematically in the isotropic solid in Fig. 14(a). Since the atomic arrangement is similar in the vertical and horizontal directions, the interactions of vertical and horizontal electric vibrations with the atoms should be the same; hence, the speeds of horizontally and vertically polarized light waves would be expected to be equal. The diagram of Fig. 14(b) shows the same solid distorted by the application of the forces FF; as a result of the distortion, the vertical spacing of the atoms is smaller than the horizontal spacing. Hence, horizontal and vertical electric vibrations of a light wave might be expected to interact differently with the atoms in the distorted solid, thus causing a difference in the speeds of propagation of vertically and horizontally polarized light waves. Isotropic media such as lucite or glass become optically anisotropic when they experience elastic strains of the type suggested in Fig. 14(b).

The atomic model of Fig. 14(b) is also appropriate for a doubly refracting crystal with its face parallel to the optic axis. Vertically and horizontally polarized waves would have different speeds if they were traveling toward or away from the reader (perpendicular to the optic axis)

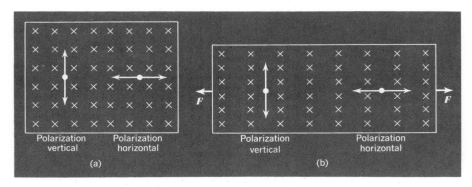

Fig. 14. Transmission of light toward the reader through sheets of (a) isotropic material, (b) anisotropic material.

but would have the same speeds if they were traveling toward the left or the right through the body of the crystal (parallel to the optic axis).

We note that cubic crystals, such as rocksalt, behave isotropically toward light. Other crystals, such as calcite and quartz, have a single optic axis and have *uniaxial* velocity surfaces such as illustrated in Fig. 12. Still other crystals, such as mica, are *biaxial* and have even more complex velocity surfaces for rays of the two polarizations, neither surface being spherical.

Example. *As indicated in Fig. 13(c), when a light beam passes through a doubly refracting crystal in a direction perpendicular to the optic axis, the wavefronts of the E- and O-waves do not coincide when the light emerges; by proper variation of crystal thickness we can control the separation of the emerging wavefronts. If the separation is ¼ λ, the crystal is called a* quarter-waveplate; *if the separation is* ½ λ, *the crystal is called a* half-waveplate, *etc. When plane-polarized light is incident on such a plate, the emerging light has some rather interesting properties. If the light entering a quarter-waveplate is plane-polarized in such a way that the amplitudes of the E- and O-waves are equal, what are the polarization characteristics of the emerging light?*

Let us assume that the incident and emerging beams travel in the Z-direction, that the vibrations of the O-waves are parallel to the X-axis and have an amplitude $\mathcal{E}_{MX}$, and, similarly, that the vibrations of the E-waves are parallel to the Y-axis and have an amplitude $\mathcal{E}_{MY}$. The incident O- and E-wavefronts are *in phase* so that their time variations can be written:

$$\mathcal{E}_X = \mathcal{E}_{MX} \sin(2\pi f t) \quad \text{and} \quad \mathcal{E}_Y = \mathcal{E}_{MY} \sin(2\pi f t). \tag{i}$$

Consideration of (i) indicates that the vibrations of the polarized incident light are in a plane making an angle θ of 45° with the X-axis; also that the amplitude of the incident light

$$\mathcal{E}_M = \sqrt{\mathcal{E}_{MX}^2 + \mathcal{E}_{MY}^2} = \mathcal{E}_{MX}\sqrt{2} = \mathcal{E}_{MY}\sqrt{2},$$

where from the statement of the problem $\mathcal{E}_{MX} = \mathcal{E}_{MY}$.

Since the E- and O-waves emerging from the plate are separated by ¼ λ, their time variations at the plane of emergence can be written as:

$$\mathcal{E}_X = \mathcal{E}_{MX} \sin(2\pi f t) \quad \text{and} \quad \mathcal{E}_Y = \mathcal{E}_{MY} \cos(2\pi f t). \tag{ii}$$

Consideration of (ii) indicates that the resultant intensity $\mathcal{E}$ in the emerging wave moves in a *circle* as time passes; plot the vector (ii) for the range $0 \leq 2\pi f t \leq 2\pi$ and see! The emerging wave is said to be *circularly polarized*. If we identify the amplitudes of our waves as variations in electric intensity, the emerging wave is characterized by an electric intensity $\mathcal{E}$ perpendicular to the Z-axis; in a given plane perpendicular to the Z-axis, the vector $\mathcal{E}$ rotates as the emergent wave passes through the plane.

If the angle θ between the vibrations of the incident polarized light beam and the X-axis is not 45°, the emerging light beam will, in general, be *elliptically polarized*; it will be plane-polarized parallel to the X-axis when $\theta = 0$ and plane-polarized parallel to the Y-axis when $\theta = 90°$. Why?

Let us now consider an arrangement by which polarized light can

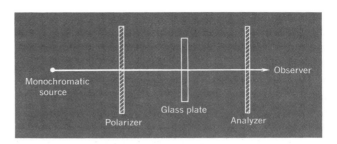

Fig. 15. Photoelastic polariscope (schematic).

be used to detect strains in transparent materials. Figure 15 shows a polarizer-analyzer combination oriented in such a way that the plane-polarized light from the polarizer is not transmitted by the analyzer; the polarizer and the analyzer are 'crossed' and no light is transmitted. Now a plate of glass or some other transparent, normally isotropic material is placed in the light beam between polarizer and analyzer. If there are no strains in the glass plate, the plate will have no effect on the light incident upon it. However, if the glass plate is subjected to a stress as, for example, in Fig. 16, the glass becomes doubly refracting. The incident plane-polarized beam is resolved into two plane-polarized components parallel and perpendicular to the axis of strain as in Fig. 14(b). These two components travel through the glass with different speeds and in general are out of phase on emergence. They are reassembled by the analyzer into a single plane-polarized beam, but in this reassembly the two components will sometimes be in phase, sometimes out of phase, so that fringes will appear, as in Fig. 16. Glass-blowers use such a system for locating the strains appearing in glassware. A photograph of a piece of strained glass observed in the above manner is shown in Fig. 17.

Fig. 16. [Reproduced by permission from *Photoelasticity,* Vol. I, by Frocht (Wiley).]

This method of using polarized light to reveal strains in transparent materials is finding extensive application in the field of mechanical stress analysis. The distribution of internal stresses in structural units or machine parts may be determined by passing polarized light through scale models made of transparent plastic, which are subjected to external forces simulating those in the actual structures or machines. The internal strains can be determined from the patterns observed through an analyzer. The method is illustrated by the photograph in Fig. 16 of the pattern resulting from compression of a disc or cylinder. This important method of solving problems in the theory of elasticity is called *photoelasticity*.

5. Magneto-Optical and Electro-Optical Effects

In earlier chapters we have asserted that light consists of electromagnetic waves; in these waves the electric and accompanying magnetic vectors are at right angles. Hence we should expect the optical properties of materials to be influenced to some extent by externally applied magnetic and electric fields. This is indeed the case.

The first connection between magnetism and optics was discovered by MICHAEL FARADAY in 1845. Faraday found that the plane of polarization of light is rotated when polarized light is allowed to pass through an isotropic medium located in a strong magnetic field if the light travels in a direction parallel to the direction of the magnetic field. An arrangement for demonstrating this *Faraday effect* is shown schematically in Fig. 18. Monochromatic light from a source such as a sodium light passes through a collimating lens L_1 and polarizer P and then through a transparent solid rod or a tube containing a liquid to analyzer A and lens L_2 to the observer's eye. The transparent sample is located inside a large coil of wire like the coils used in electromagnets. Before the direct electric current in the coil is turned on, the analyzer is rotated until no light is transmitted. When the electric current is started, a magnetic field is produced in a direction parallel to the direction in which the light is traveling. Light now comes through the analyzer to the observer, and the analyzer must be rotated through an angle θ before the light is again extinguished. This angle θ represents the angle through which the plane of polarization has been rotated. The magnitude of θ is proportional to the strength of the magnetic field $\mathcal{B}$ and to the length of the light path l through the magnetic field; the relation can be written as

$$\theta = V l \mathcal{B},$$

where the proportionality constant V is called *Verdet's constant*. This constant depends on the material used and on the wavelength of the light. The sense of the rotation is right-handed in some substances, left-handed in others, for the same direction of the field $\mathcal{B}$; it reverses if $\mathcal{B}$ reverses.

A related magneto-optical effect was discovered by A. A. COTTON

Fig. 17. Use of polarized light in detecting strains in glassware.

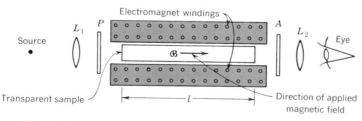

Fig. 18. Rotation of the plane of polarization by a magnetic field: Faraday effect.

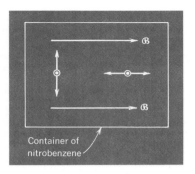

Fig. 19. The Cotton-Mouton effect. Replacement of the magnetic field $\mathcal{B}$ by an electric field $\mathcal{E}$ illustrates the Kerr effect.

and H. MOUTON in 1907. These investigators found that certain isotropic optical media become doubly refracting when placed in a strong magnetic field. For example, if nitrobenzene is placed in a strong magnetic field $\mathcal{B}$ as indicated in Fig. 19, light approaching the reader travels at different speeds when polarized in vertical and horizontal planes.

The electrical analog of this *Cotton-Mouton effect* had been discovered by JOHN KERR in 1875. Kerr found that an *isotropic* optical medium between the plates of an electric capacitor becomes doubly refracting when the capacitor is charged. The *Kerr effect* can be illustrated by the same diagram, Fig. 19, as the Cotton-Mouton effect, by merely replacing the magnetic field $\mathcal{B}$ by an electric field $\mathcal{E}$.

PROBLEMS

1. Show that a polarizing plate absorbs 50 per cent of the intensity of an unpolarized beam.

2. Show that two polarizing plates with axes making an angle θ with each other transmit $\frac{1}{2}\cos^2\theta$ of the intensity of an incident unpolarized beam.

3. Three perfect polarizing plates are stacked. The first and third are crossed; the one between has its axis at 45° to the axes of the other two. Find the fraction of the intensity of an incident unpolarized beam that is transmitted by the stack. Ans: $\frac{1}{8}$.

4. Three perfect polarizing plates are stacked. The first and third have their axes parallel; the one between has its axis at 30° to the axes of the other two. Find the fraction of the intensity of an incident unpolarized beam that is transmitted by the stack.

5. Seven perfect polarizing plates are stacked. Each succeeding plate has its axis inclined at an angle of 15° with that of the preceding plate. What fraction of the light is transmitted by the stack? Comparing the results of this problem with those obtained in Prob. 3 and in the example on p. 823, draw qualitative conclusions concerning the relation between the fraction of the light transmitted and the 'abruptness' of changes in the plane of polarization.

6. Should the axis of Polaroid driving glasses be vertical or horizontal to (*a*) reduce the intensity of sunlight reflected from a road surface? (*b*) reduce the intensity of blue skylight near the horizontal?

7. Calculate the polarizing angles of crown glass and flint glass for sodium *D*-light.
Ans: 56°.6; 58°.4.

8. A beam of sunlight strikes a piece of glass at an angle of incidence of 58° and the reflected beam is completely polarized. What is the angle of refraction of the transmitted light?

9. At what angle with the normal can one view the smooth surface of a lake through Polaroid spectacles and eliminate glare completely?
Ans: 53°.

10. Find the angle of refraction of sodium *D*-light in carbon disulfide when light of this wavelength strikes a carbon-disulfide surface at the polarizing angle.

11. Light of wavelength 550 nm is normally incident on a sheet of quartz crystal cut with its face parallel to the optic axis. If the refractive index for one polarization is 1.553 and for the other is 1.544, where the two polarizations are analogous to those indicated in Fig. 13(c), what are the two wavelengths of the light in the crystal? Ans: 354, 356 nm.

12. A plate of crystal is said to be a 'half-wave' plate if, when light is incident as in Prob. 11, there is one-half wavelength more of one polarization than of the other as the light passes through the plate. Determine the thickness of a quartz half-wave plate for 500-nm light. (*See* Prob. 11.)

13. Show that if a quartz half-wave plate (*see* Prob. 12) is placed between *parallel* polarizing plates oriented at 45° to the orientation of the crystal axis, *no* light is transmitted through the combination. Show that if the polarizing plates are *crossed*, with the same 45° orientation, *all* the light transmitted by the first polarizing plate is transmitted by the combination.

14. By sending an alternating current of frequency f through the coils of Fig. 18, one can produce an alternating magnetic field $\mathcal{B} = \mathcal{B}_0 \sin 2\pi ft$. With the polarizer and analyzer crossed, the alternating magnetic field can be used to 'modulate' the amplitude of the light wave transmitted by the analyzer. Obtain an expression for the time-dependent *amplitude* of the transmitted light. Obtain an expression for the *intensity* of the light transmitted by the analyzer and show that the intensity is modulated at a frequency of $2f$.

15. Draw a diagram showing how a 'Kerr cell' like that shown in Fig. 19 can be placed between a polarizer and analyzer and used as a 'light shutter.' Show that by applying alternating electric fields $\mathcal{E} = \mathcal{E}_0 \sin 2\pi ft$, the intensity of a light beam can be modulated at audio, radio, and even microwave frequencies. Show that by applying an extremely brief electric pulse to the cell, a corresponding brief light pulse can be transmitted through a Kerr-cell shutter; this technique has been used in the measurement of the speed of light.

16. Explain why a wire grid with spacing $d \ll \lambda$ will polarize a beam of microwaves of wavelength λ passing through the grid as in the figure. What will be the plane of polarization?

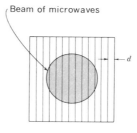

Problem 16

PART VI

QUANTUM AND RELATIVISTIC PROPERTIES OF RADIATION AND MATTER

QUANTUM PROPERTIES OF RADIATION

41

The treatment of optical phenomena in the preceding chapters demonstrates that the *transmission* of light can be satisfactorily described in terms of *waves*. In the present chapter, we consider the processes of *emission* and *absorption* of light, which cannot be described satisfactorily in terms of waves. We shall trace briefly the development, since 1900, of the *quantum theory,* which attributes certain *particle properties* as well as *wave properties* to radiation.

Since the quantum theory was initially developed by MAX PLANCK (1858–1947) to account for the *spectrum* of a perfect radiator—a black body—and was later elaborated by NIELS BOHR (1885–1963) to account for the *spectra* of atoms, we begin the chapter with a brief treatment of *spectroscopy*. Bohr's theory of emission and absorption of light in terms of transitions between 'stationary' energy levels is presented. Bohr's model of the hydrogen atom is included, not only for historical reasons, but also because it provides a useful means of visualizing the electronic structure of atoms, even though it has now been superseded by more satisfactory but less easily visualized models. As further evidence for the existence of 'stationary' energy states in atoms and molecules, we briefly describe the *laser,* which provides a source of intense *coherent radiation*.

According to quantum theory, electromagnetic radiation of frequency f exists only in integral numbers of *quanta* or *photons,* each of which has energy $E=hf$ and momentum $p=hf/c=h/\lambda$, where h is a fundamental physical constant having the dimensions of *energy*$\times$*time,* or, equivalently, of *momentum*$\times$*length,* and known as *Planck's constant.*

1. Emission and Absorption Spectra

The intensity of the radiation emitted by a source is not the same for all wavelengths. A plot of radiant power as a function of wavelength or frequency gives the *spectrum* of the source. Information regarding the spectrum that is characteristic of a given source can be obtained by means of interferometers; however, spectra are usually observed and studied by means of instruments called *spectrometers* or *spectrographs.*

A spectrometer is an instrument that disperses light, and that has an accurate scale for measurement of the dispersed wavelengths or frequencies. In the spectrometers in Figs. 1 and 2, the light of the source illuminates a slit S perpendicular to the plane of the diagram. If the source gives monochromatic light, a single image of the slit is seen in the telescope, at the appropriate angle θ, in light of the monochromatic color. If the source gives several monochromatic wavelengths, several

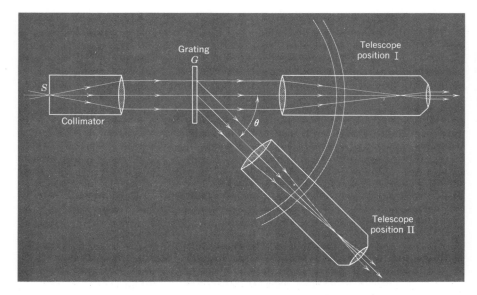

Fig. 1. Schematic diagram of a grating spectrometer. The telescope at position II may be replaced by a camera for photographing the spectrum, in which case the instrument becomes a grating spectrograph.

images of the slit are seen in the light of the different colors as the telescope is moved around the circle. Because the different images of the slit appear to the observer as parallel narrow lines, it is customary to speak of them as *spectral lines*. Figure 3 shows the spectral lines emitted by atomic hydrogen; the photograph was obtained by means of a *spectrograph*, an instrument similar to a spectrometer but employing a camera in place of a telescope.

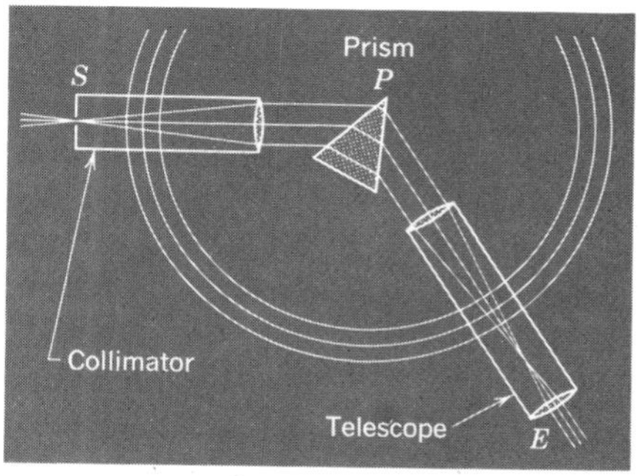

In the grating spectrometer of Fig. 1, the wavelength of a spectral line can be accurately determined by measuring the angle θ that the telescope must be moved from the position I of the central image to the position II of a diffracted image. As the telescope is moved, the first-, second-, third-, $\cdots$ order images will be successively observed. Equation (3), p. 816, gives an absolute measurement of wavelength, for an nth-order image, as $\lambda = \sin\theta_n/Nn$, where N is the number of grating lines per unit width of the grating (the reciprocal of the grating spacing d).

Fig. 2. Schematic diagram of a prism spectrometer.

The prism spectrometer of Fig. 2 does not provide an absolute measurement of wavelength. For any given prism, the angular scale must be calibrated by using a source whose wavelengths are known from grating or interferometer measurements.

The spectrum produced when the light from a luminous source is dispersed is called the *emission spectrum* of the source. Its appearance is determined by the composition and physical state of the source.

Incandescent solids and liquids, and incandescent gases under extremely high pressure produce *continuous spectra*, which include light of all frequencies. Luminous gases and vapors at low pressure have spectra quite different from the spectra of incandescent solids. The emission spectra of such materials consist of distinct lines.

It is found that every chemical element emits a characteristic *line spectrum* when the *atoms* of the element are excited in a flame, a furnace, or an electric discharge. Thus, if different materials containing sodium are introduced into a hot Bunsen flame or an electric arc, two characteristic yellow lines appear in the spectrum near 589 nm, the same position

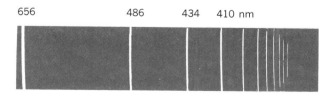

Fig. 3. Spectrum of atomic hydrogen in the visible region. (After Herzberg, *Molecular Spectra and Molecular Structure:* Prentice-Hall.)

as in the spectrum of a sodium-vapor light like those used for highway illumination. Therefore, we might conclude that the mechanism involved in the emission of these yellow lines is to be found in the sodium *atom* itself. This conclusion is correct and can be generalized by the statement that *line spectra originate in the atoms of the chemical elements.* Figure 3 shows the spectrum of the hydrogen atom in the visible and near-ultraviolet regions. With strong excitation, as in an electric spark, line spectra characteristic of positive *atomic ions* are observed.

Other types of emission spectra are sometimes produced by incandescent gases at low pressure. These spectra are called *band spectra* and consist of large numbers of spectral lines closely spaced in groups called *bands.* In the bands, the lines are usually so closely spaced that they cannot be separated by low-dispersion instruments. *Band spectra have their origin in molecules or molecular ions.* Figure 4 shows a portion of the band spectrum of the nitrogen molecule, N_2. Certain nonopaque crystalline solids also emit band spectra when they are heated or otherwise excited.

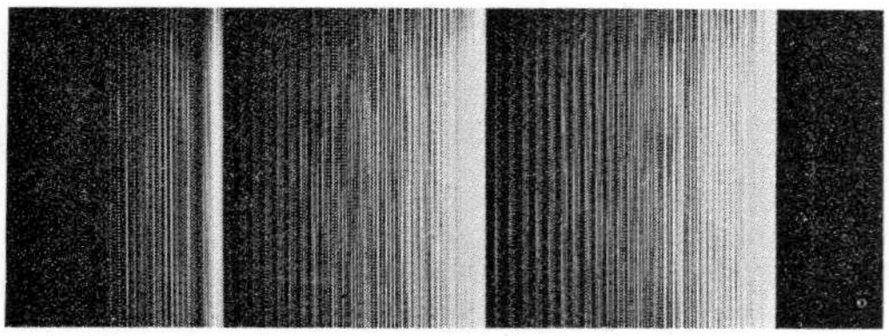

Fig. 4. Bands in the spectrum of nitrogen. (Photograph courtesy of G. H. Dieke.)

When light from a source with a continuous spectrum is allowed to pass through a relatively cool gas or vapor before entering the spectrometer, the observed spectrum consists of the continuous spectrum of the source crossed by dark lines or bands. The dark lines or bands are present as a result of selective absorption by the cool gas or vapor. For example, if light from an incandescent tungsten-filament lamp is allowed to pass through a tube containing sodium vapor, the observed spectrum is continuous except for two dark lines appearing in the yellow region. These lines denote wavelengths *absorbed* by the sodium vapor. The dark lines appear at exactly the same positions as the yellow lines appear in the emission spectrum of sodium. This correspondence between the positions of the lines in emission and absorption spectra is quite general. The wavelengths absorbed by a given type of atom or molecule are identical with wavelengths emitted when the emission spectrum of the atom or molecule is excited. The dark lines or bands in the spectrum, observed when white light is allowed to pass through an absorbing gas, constitute the *absorption spectrum* of the atoms or molecules of the gas.

Dark lines are observed in the otherwise continuous visible spectrum of the sun (*see* Fig. 5). Although these lines were first observed by Wollaston in 1802, they are known as the *Fraunhofer lines,* since Fraunhofer made a careful study of about 600 dark lines in the solar spectrum in the years following 1814.

The light emitted by the sun is generally the continuous spectrum of a gas at a very high pressure, but this light experiences selective atomic absorption in passing through the relatively cooler gases at the surface of the sun. Most of the structure of the otherwise continuous solar

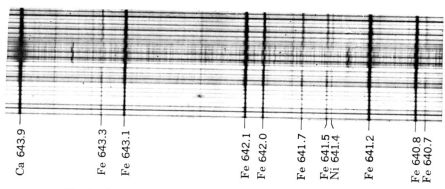

Fig. 5. Part of the spectrum of the sun in the red region, showing Fraunhofer lines arising from absorption by atoms or positive ions of Fe, Ca, and Ni. Wavelengths are given in nm. This is a photographic positive, so the dark lines appear as they would to the eye. (From *Astrophys. Jl.* 75, plate XVIII, 1932.)

spectrum originates in this process, but some dark bands arise from absorption by molecules in the atmosphere of the earth. Comparison of the wavelengths of the Fraunhofer lines with those emitted by atoms in the laboratory enables astrophysicists to determine the composition of the gases in the sun's outer layers. Absorption lines of *helium* were *first* found in the Fraunhofer spectrum of the sun (hence the name helium) and recognized as arising from an element unknown at the time on the earth. This element was later found in the earth's atmosphere.

Nonopaque liquids and solids commonly have broad regions of absorption exhibiting no line structure. For example, the ruby glass used in photographic darkrooms has an absorption region covering the entire visible spectrum except the red and deep orange; the absorption is very intense and shows no line structure.

2. Black-Body Radiation; Planck's Quantum Principle

In our treatment of thermal radiation (Chapter 16), we pointed out that it was helpful to begin the discussion with the consideration of an 'ideal' or 'perfect' radiator, the *black body*. It is interesting to note that the beginning of our understanding of the origin of *spectra* was also based on studies of the *spectrum* of a black body.

The Stefan-Boltzmann law, (4), p. 325, gives the total rate of radiation of a perfect radiator (black body) at absolute temperature T, but gives no information concerning the *spectrum* of a perfect radiator. The spectrum of a perfect radiator is continuous. In order to discuss the relative amounts of energy in radiation of different wavelengths in the spectrum, we introduce the quantity P_λ, which gives the radiated power per unit area in a unit wavelength range at wavelength λ. This quantity,

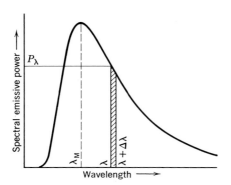

Fig. 6. Power radiated per unit area as a function of wavelength; definition of P_λ.

called the spectral radiance, can be determined by means of a spectrometer. What is actually observed is the amount of radiant power contained in a short wavelength interval between λ and $\lambda + \Delta\lambda$. The radiant power per unit area of source, emitted in this wavelength range, is given by $P_\lambda \Delta\lambda$, represented by the shaded area in Fig. 6. *The unit in which P_λ is measured is* W/m^2 *per unit wavelength interval; for example,* W/m^2 *per nanometer.*

Plots of the distribution of power in the spectrum of a black body at different temperatures are shown by the solid lines in Fig. 7. These curves all have certain basic similarities in form. They do not cross; the curves for higher temperatures are above the curves for lower temperatures at all wavelengths; the maxima of the curves are displaced toward shorter wavelengths as the temperature of the black body is increased. The progressive shift of maximum toward the violet end of the spectrum accounts for the observed change in color of a radiating metal body from red through white to blue as its temperature is increased. Sunlight has the characteristics of black-body radiation corresponding to a temperature of about 6000° K and serves to define

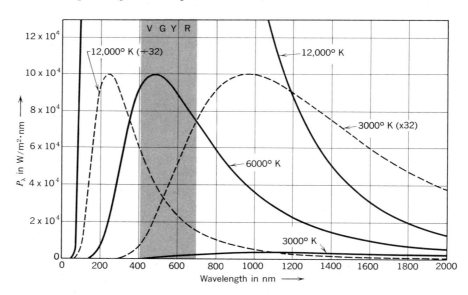

Fig. 7. The solid lines show accurate plots of Planck black-body radiation curves for temperatures of 3000°, 6000°, and 12 000° K. Broken lines show the 3000° curve with ordinates multiplied by 32 and the 12 000° curve with ordinates divided by 32; this adjustment brings the maxima of these curves to the same value as the maximum of the 6000° curve.

'white.' Incandescent-lamp filaments are much cooler (about 3000° K) and give light that is more orange than daylight. Certain stars, such as Vega (12 000° K) are much hotter and appear blue. These points regarding color are best illustrated by the broken curves in Fig. 7, in which the ordinates of the 3000- and 12 000-degree curves have been scaled so that all three curves are plotted with the same maximum.

The area under the radiation curve, $\int_0^\infty P_\lambda \, d\lambda$, is equal to the total rate of black-body radiation and hence varies with the fourth power of the absolute temperature in accordance with the Stefan-Boltzmann law. Hence the total areas under the three curves of Fig. 7 are in the ratios $1:16:16^2$. The wavelength λ_M of the maximum of the curve (see Fig. 6) is found experimentally to vary inversely as the absolute temperature, according to the law $\lambda_M = A/T$, where A is a constant whose value is $A = 2.8978 \times 10^6$ nm·K deg. This relation is called *Wien's displacement law*. Thus with each doubling of temperature in Fig. 7, the value of λ_M is divided by two. Finally, experiment shows that the height of the maximum varies as the fifth power of the absolute temperature, so these heights are in the ratios $1:32:32^2$.

Many attempts were made to derive an analytical expression for the detailed shape of the curves of Fig. 7 by application of *classical* electromagnetic theory to the problem of black-body radiation, but all such attempts ended in failure. The correct form of the equation for P_λ, first given by the German physicist MAX PLANCK in 1900, was based on a hypothesis that represented the very birth of quantum ideas. The correct equation is

$$P_\lambda = \frac{2\pi \times 10^{-9} \, hc^2}{\lambda^5 \, (e^{hc/\lambda kT} - 1)}, \qquad \begin{pmatrix} \text{PLANCK'S} \\ \text{RADIATION} \\ \text{LAW} \end{pmatrix} \quad (1)$$

which gives P_λ in W/m²·nm when

λ is the wavelength in *meters*,
$c = 2.997\,925 \times 10^8$ m/s is the speed of light,
$k = 1.3805 \times 10^{-23}$ J/K is *Boltzmann's constant* (see p. 349),
$h = 6.6256 \times 10^{-34}$ J·s is *Planck's constant*,
$e = 2.718\,28$ is the base of natural logarithms (see tables in Appendix).

The factor 10^{-9} is introduced in (1) merely to convert the wavelength interval in P_λ from the meter to the more appropriate unit, the nanometer.

To derive the expression (1), Planck proposed a radical hypothesis that represented the beginning of *quantum* theory, which has revolutionized our concepts of the nature of the processes taking place in the emission and absorption of radiation. This hypothesis—completely contrary to the ideas of classical electromagnetic theory according to which emission and absorption are essentially continuous processes, but an assumption that has since been well confirmed by many lines of experimental evidence—is the following:

PLANCK'S QUANTUM PRINCIPLE: *Radiation of frequency f cannot be emitted or absorbed in arbitrary amounts but is always emitted or absorbed in a discrete quantity, or quantum, of energy hf, where h is a universal constant. In the radiation field in an enclosure, the radiation of frequency f cannot exist in arbitrary amounts but must consist of an integral number of such energy quanta.*

The complete formulation of the above statement is due partly to Planck and partly to Einstein, who drew his conclusions from an analysis of the photoelectric effect, which we shall discuss in Sec. 3.

Corresponding to the rate of radiation (1), the energy density per unit wavelength, E_λ, in an enclosure whose walls are at temperature T, is given by

$$E_\lambda = 4\, P_\lambda / c. \qquad \left\{ \begin{array}{l} P_\lambda \text{ in W/m}^2\cdot\text{nm} \\ E_\lambda \text{ in W/m}^3\cdot\text{nm} \end{array} \right\} \quad (2)$$

This density of radiant energy exists in any enclosure whose walls are at temperature T, whether or not the walls are black. The radiation is isotropic, with equal amounts traveling in every direction at every point, no matter the shape of the enclosure. The relation (2) between P_λ and E_λ is simply a matter of geometry in the case of an isotropic radiation field; we shall sketch the derivation of this relation in Prob. 10, p. 894.

Example. *Since the electron-volt is a unit of convenient size for use in atomic physics, it is frequently desirable to express Planck's constant in eV·s so that quantum energies $E = hf$ are given directly in convenient units. Express h in eV·s and construct a table expressing quantum energies in eV for radiation in the portion of the electromagnetic spectrum between the far ultraviolet and the far infrared (Fig. 12, p. 712).*

We recall from (21), p. 485, that $1\text{ J} = 6.24 \times 10^{18}$ eV; therefore

$$h = 6.62 \times 10^{-34} \text{ J·s} = (6.62 \times 10^{-34})(6.24 \times 10^{18} \text{ eV})\cdot\text{s} = 4.13 \times 10^{-15} \text{ eV·s}.$$

Using this value for h and typical values of the radiation frequency f in Hz, we obtain the quantum energies $E = hf$ in eV; thus:

Spectral Region	f (Hz)	Quantum energy (eV)
Far-ultraviolet	10^{17}	413
Near-ultraviolet	10^{16}	41.3
Visible	6×10^{14}	2.48
Near-infrared	10^{14}	0.413
Far-infrared	10^{13}	0.0413

We note that quantum energies of visible radiation are in the range 3.1 eV (violet) to 1.6 eV (red), and that 1 eV is the quantum energy of the near-infrared frequency 2.42×10^{14} Hz.

3. The Photoelectric Effect

Now we shall consider *quantum* effects involved in electron emission. Electrons can acquire sufficient energy to escape from *cold* metal if the metal is illuminated with light of sufficiently high quantum energy, that is, of sufficiently high frequency or of sufficiently short wavelength. In this process, the energy needed for escape is furnished by the energy hf of a single light quantum. This phenomenon is known as the *photoelectric effect* and was first discovered in 1887 by HEINRICH HERTZ, who noticed that a spark would jump more readily between two charged spheres when their surfaces were illuminated by light from another spark.

An arrangement that can be used for observing the photoelectric effect is shown schematically in Fig. 8. A beam of light strikes a metal surface S in an evacuated tube. Electrons are emitted by the surface and are drawn to the collector C, normally maintained at a positive voltage with respect to S. The current can be measured by the galvanometer G. It is found that for each particular metal that is used to form surface S, there is a certain frequency, called the *threshold frequency*, that must be exceeded by the light beam before any electrons are emitted at all.

For light frequencies well above the threshold, some of the electrons are emitted from S with considerable speed, as can be demonstrated by removing the battery from the circuit in Fig. 8 and noting that there is still some current. The speed of the fastest electrons can be determined by finding the *negative* voltage that must be applied to the collector C in order to stop the current completely. Experiments show that *the maximum speed of the emitted electrons depends only on the frequency of the incident light*. The number of electrons emitted per second (as determined from current measurements) depends on the intensity of the incident light, but the maximum speed of the electrons is independent of the intensity of the incident light and depends only on its frequency.

The correct explanation of the photoelectric effect was given in 1905 by Einstein, who used the quantum principle that had been proposed earlier by Planck. According to Einstein's explanation, when a photon is absorbed, its total energy hf is imparted to a single electron within the metal. The energy acquired by the electron may enable it to penetrate the potential barrier at the surface of the metal and escape if it is moving toward the barrier with sufficient speed. In penetrating the barrier, the electron loses a certain energy ϕ, called the *work function* of the surface; if the energy hf received from the incident light quantum is greater than ϕ, the electron retains kinetic energy after leaving the surface. Since electrons absorb photons at various depths within the metal and acquire initial velocities in various directions, the electrons emerging from the surface have various energies. But the maximum kinetic energy $\frac{1}{2} mv^2_{\text{Max}}$ of electrons emitted from a metal on which light of frequency f is incident

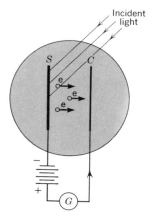

Fig. 8. Photoelectric cell. Light striking the metal surface S ejects electrons that are drawn to the collector C.

is given by the Einstein *photoelectric equation*

$$\tfrac{1}{2} m v_{\text{Max}}^2 = hf - \phi = h(f - \phi/h). \tag{3}$$

The experimental results are described accurately and completely by this expression. For a particular metal, the plot of maximum energy against light frequency is a straight line whose slope determines the quantum constant h, as in Fig. 9. The maximum energy becomes zero at the *threshold frequency* where $hf = \phi$; below this frequency no electrons are emitted because absorption of a photon does not impart sufficient energy to an electron for it to pass the potential barrier at the metal surface. *The photoelectric effect furnished the first direct proof of the quantum principle.*

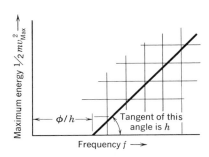

Fig. 9. Photoelectric equation.

The work function of all metals lies in the range from 2 to 5 electron-volts. Those of lower work function, such as potassium ($\phi = 2.2$ eV), have photoelectric thresholds that lie in the visible region, and visible light can cause photoelectric emission. The threshold frequency for most metals lies in the ultraviolet—as in the following typical example—and visible light will not cause photoemission.

Example. *When ultraviolet light of* 300-nm *wavelength falls on a zinc surface, as in Fig. 8, a retarding potential of* 0.5 V *must be applied to keep the most energetic electrons from reaching the collector. Determine the work function, the wavelength of the photoelectric threshold, and the retarding potential required for light of* 200-nm *wavelength.*

In (3), the maximum kinetic energy is 0.5 eV since this is the potential energy increase of an electron passing through 0.5 V of retarding field, and equals its loss of kinetic energy. The frequency f is obtained from

$$f = c/\lambda = (3.00 \times 10^8 \text{ m/s})/(300 \times 10^{-9} \text{ m}) = 1.00 \times 10^{15} \text{ Hz}.$$

Hence (*see* p. 846) $\quad hf = (4.13 \times 10^{-15} \text{ eV} \cdot \text{s})(1.00 \times 10^{15}/\text{s}) = 4.13 \text{ eV},$

and $\quad \phi = hf - \tfrac{1}{2} m v_{\text{Max}}^2 = 4.13 \text{ eV} - 0.5 \text{ eV} = 3.63 \text{ eV}.$

The photoelectric threshold occurs at $hf = \phi = 3.63$ eV, or $f = 0.878 \times 10^{15}$ Hz, corresponding to $\lambda = c/f = 342$ nm. Light of *longer* wavelength than this will give no photoelectric emission since the quantum energy will be less than the work function.

The quantum energy of 200-nm light will be $\tfrac{3}{2}$ that for 300-nm, or $hf = \tfrac{3}{2}(4.13 \text{ eV}) = 6.20$ eV. For this light, $hf - \phi = (6.20 - 3.63)$ eV = 2.57 eV, so a retarding potential of 2.57 V would be required to stop all collection.

4. Atomic Spectra

The quantum theory, initially developed to account for the shape of the black-body radiation curves giving spectral emissive power as a function of wavelength, was next used to explain the photoelectric effect. Now let us see how this theory can be applied to another emission process—the emission of *line spectra* by atoms. In this case, we have to account for the discrete frequencies present in the observed spectra.

The simplest of all atomic spectra is that of hydrogen (Fig. 3); the spectra of other atoms are much less regular (compare the Fe lines in Fig. 5). In atomic spectra, there are no frequency sequences analogous to the fundamental and overtones that might be expected if a radiating atom were comparable in any way to a harmonic oscillator. Much time was spent in looking for sequences of this type until 1885, when JOHANN JAKOB BALMER found a simple equation that gives the frequencies of the hydrogen lines appearing in the visible and near ultraviolet. Balmer's empirical equation may be written in the form $f = cR\,(1/4 - 1/n^2)$, where R is the so-called *Rydberg constant* whose value is 10 967 758 m^{-1}, and n is a number that takes the integral values 3, 4, 5, 6, $\cdots$ for the various lines of the series. The frequencies of the lines in this series increase with increasing n and converge to the limit $1/4\,cR$ as n becomes large. The wavelengths correspondingly become shorter and also converge to a limit, as indicated in Fig. 3, in which the first 10 or 12 Balmer lines are visible, for $n = 3, 4, 5, \cdots$ from left to right.

Further examination of the spectrum of atomic hydrogen in regions outside the visible region revealed the existence of several similar spectral series which have been named for their discoverers:

Lyman series: $\quad f = cR\left(\dfrac{1}{1^2} - \dfrac{1}{n^2}\right)$, where $n = 2, 3, 4, \cdots$ (ultraviolet)

Balmer series: $\quad f = cR\left(\dfrac{1}{2^2} - \dfrac{1}{n^2}\right)$, where $n = 3, 4, 5, \cdots$ (visible)

Paschen series: $\quad f = cR\left(\dfrac{1}{3^2} - \dfrac{1}{n^2}\right)$, where $n = 4, 5, 6, \cdots$ (infrared)

Brackett series: $\quad f = cR\left(\dfrac{1}{4^2} - \dfrac{1}{n^2}\right)$, where $n = 5, 6, 7, \cdots$ (infrared)

Pfund series: $\quad f = cR\left(\dfrac{1}{5^2} - \dfrac{1}{n^2}\right)$, where $n = 6, 7, 8, \cdots$ (infrared)

Humphreys series: $\quad f = cR\left(\dfrac{1}{6^2} - \dfrac{1}{n^2}\right)$, where $n = 7, 8, 9, \cdots$ (far infrared)

The frequencies of *all* lines of atomic hydrogen that are observed in the laboratory are contained in the above series. Certain other lines are observed at radio frequencies by radioastronomers.

It will be noted from these expressions that the frequency of any observed spectral line of hydrogen can be written as the *difference* of two frequencies, called *term frequencies*. The term frequencies are

$$f_n = (1/n^2)\,cR. \qquad (n = 1, 2, 3, \cdots) \qquad (4)$$

The number of term frequencies is smaller than the number of observed lines; every possible difference between two term frequencies gives the frequency of a spectral line. In the case of hydrogen, the term frequencies are represented by the positions of the horizontal lines in Fig. 8, measured down from the line marked 0. The lengths of the various vertical lines in Fig. 10 represent term differences and are proportional to the frequencies of the spectral lines in the various series as indicated.

Similar regularities were found in other spectra more complicated than that of hydrogen, and resulted in the formation by W. RITZ in 1908 of the *combination principle,* according to which each atom may be characterized by a set of numbers called *term frequencies,* such that the actual frequencies of the spectral lines are given by differences between these term frequencies.

It remained for the Danish physicist, NIELS BOHR, in 1913, to suggest that each *term* represents an *energy level* of the atom. This hypothesis represented the very beginning of the *quantum mechanics* of atomic structure, as distinguished from the quantum theory of radiation which we have discussed earlier. According to quantum mechanics, an atom (or a molecule) cannot have an arbitrary amount of internal energy; rather, it must at any time be in one of a discrete set of 'stationary' states, each having a particular value of energy. When it is in a state of higher energy, it will spontaneously make a transition to one of the states of lower energy, the energy released appearing as a single quantum of radiation of frequency f such that the energy released equals hf. This is the energy

$$E = hf \qquad \left(\begin{array}{c}\text{quantum}\\ \text{energy}\end{array}\right) \quad (5)$$

ascribed to a quantum of radiation by Planck's quantum principle. Correspondingly, if irradiated, the atom can absorb a single quantum, but of only such frequencies f that the quantum energy hf is exactly the energy needed to excite the atom to one of its possible states of higher energy.

These ideas can be clarified by returning to the term diagram for hydrogen, Fig. 10. The horizontal lines can now be interpreted as energy levels. At normal temperatures all the hydrogen atoms are in the state of lowest energy at the bottom of the diagram, called the *ground state.* Hydrogen molecules can be dissociated into atoms and the atoms raised to excited states so as to excite the emission spectrum by either (*a*) raising the temperature of the gas sufficiently, (*b*) using an electrical discharge, or (*c*) illuminating the gas with ultraviolet light of sufficiently high frequency.

Emission of radiation corresponds to transitions downward in Fig. 10; absorption of radiation corresponds to transitions upward. In each case the frequency of radiation is exactly proportional to the energy change, the proportionality factor being Planck's constant h. Hence, the frequency of the radiation is proportional to the length of the vertical line in Fig. 10, since this length represents the magnitude of the energy change. If we call the ionization energy zero,* as is most convenient, the term frequency f_n must be associated with the *negative* energy

$$E_n = -hf_n = -(1/n^2)\,hcR. \quad (n=1,2,\cdots) \quad (6)$$

When the atom makes a transition from state n to state n' ($n>n'$ for emission), the energy released is $E_n - E_{n'} = -h(f_n - f_{n'})$. For example, the first Balmer line involves a change from $n=3$ to $n'=2$, releasing energy given by (6) as

$$E_3 - E_2 = -h(f_3 - f_2) = h(f_2 - f_3) = hcR\,(\tfrac{1}{4} - \tfrac{1}{9}),$$

corresponding to the energy of a quantum of the first Balmer frequency.

There is a large body of experimental evidence of various types, all confirming the correctness of the above picture. We shall describe only one type of experiment, first performed by JAMES FRANCK AND GUSTAV HERTZ in 1913, that gives very direct confirmation of the energy-level interpretation of spectral terms. A beam of hydrogen atoms (not molecules) in the ground state is sent down a vacuum tube; the beam is bombarded from the side by electrons of known and controlled energy, and a spectrometer is focused on the beam. A bombarding electron is capable of imparting some or all of its kinetic energy to the electron in a hydrogen atom because of the repulsive force it exerts on that

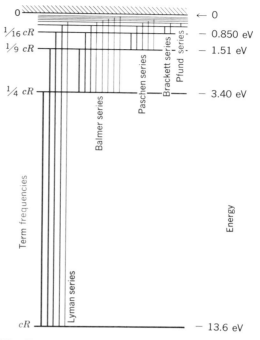

Fig. 10. The spectrum of atomic hydrogen. The horizontal lines represent allowed energy levels of the atom, associated with $n=1, 2, \cdots, \infty$ as we go from the lowest level up to the ionization energy, taken as zero. When the atom makes a transition from one energy level to a lower one in this diagram, it emits radiation of frequency proportional to the energy difference, that is, proportional to the *length* of the vertical line connecting the two energy levels.

electron. It gives the electron an impulse like that required to put an earth satellite in orbit. But when the bombarding electron has energy below the quantum energy hf of the first line of the Lyman series (compare Fig. 10), no spectrum appears, because the hydrogen atom in its ground state is incapable of absorbing *less* than this energy. When the electron energy exceeds the quantum energy of the first Lyman line but is less than that of the second, only the first Lyman line appears in the emission spectrum. When the energy exceeds the quantum energy of the second Lyman line but is less than that of the third, the first two Lyman lines *and* the first Balmer line appear in the emission spectrum. And so on. Similar experiments performed with a large number of different types

---

*The reason for this convention will become clear in Section 5. As in the usual electrostatic situation, the zero of potential energy corresponds to infinite separation of proton and electron; if the electron is at rest the kinetic energy is also zero.

of atoms completely confirm the correspondence of energy levels and term frequencies.

Each element of the periodic table has an energy-level diagram similar to Fig. 10, in which the energy differences correspond to observed spectral lines. In no other case is this diagram so simple as that for the simplest of all atoms, hydrogen. The energy values, and hence the spectral-line frequencies, for any atom are correctly predicted by the *quantum mechanics* formulated by ERWIN SCHRÖDINGER and WERNER HEISENBERG in 1925 (*see* Chapter 42).

Each diatomic or polyatomic molecule also has a characteristic energy-level diagram, in which the energy levels occur in closely spaced groups, so the lines also occur in closely spaced groups called bands—hence the term *band spectrum* (*see* Fig. 4).

5. The Bohr Theory of the Structure of Hydrogen

The discussion of the preceding section represents a radical departure from classical ideas in two respects:

(*a*) *The energy of the atom can have only one of a set of discrete values.*

According to classical mechanics, there is no principle that would *quantize* the energy; the negative electrons in their motion around the positive nucleus would be permitted to have any value of energy whatsoever.

(*b*) *A single quantum is emitted in the transition* (*or quantum jump*) *from one energy level to another.*

According to classical electromagnetic theory, the charged electrons, in their accelerated orbital motion around the nucleus, would be expected to radiate continuously, with loss of mechanical energy as their orbits gradually decreased in size until they fell into the nucleus. In other words, atoms could not exist! The electrons certainly do not fall into the nucleus but have a certain minimum energy in the ground state; this energy they cannot lose.

As an atomic model, Bohr assumed that the hydrogen atom consisted of a single light negatively charged electron attracted to and 'revolving' around a proton. To account for the departures from classical ideas expressed by (*a*) and (*b*) above, Bohr assumed that only certain orbits were 'allowed' and looked for some 'quantum principle' that would determine the permitted orbits and would give the observed values of energy. He noted that Planck's constant h had the dimensions of momentum $\times$ distance. Considering first the circular orbits, he tried the hypothesis that the line integral of momentum, once around the orbit, is restricted to values that are integral multiples of Planck's constant. This line integral is $p \times 2\pi r$, where r is the radius of the orbit. Bohr's quantum assumption was that the values of r are restricted to those values

r_n that satisfy the relation

$$p \times 2\pi r_n = nh, \quad \text{or} \quad pr_n = nh/2\pi, \tag{7}$$

where $n=1, 2, 3, \ldots$. We note that pr_n is the *angular momentum* J_n so in effect we are requiring that *orbital angular momentum* be an *integral multiple* of $h/2\pi$, that is

$$J_n = nh/2\pi. \tag{8}$$

With minor modifications, this principle of quantization of angular momentum applies universally to the electronic structure of all atoms in the periodic table.

Bohr's assumption turned out to give the correct energies for the allowed circular orbits in hydrogen. Let us compute the energies of the circular orbits satisfying the conditions (7) and (8) and show that they agree with the observed energies. The Coulomb force of *attraction* between the electron and the proton (the hydrogen nucleus) is

$$F = \frac{e^2/4\pi\varepsilon_0}{r^2} \tag{9}$$

where e is the electronic charge in C and r is in m. For a circular orbit, we can equate this force (9) to the mass m of the electron times the centripetal acceleration, that is, to $mv^2/r = m^2v^2r^2/mr^3 = p^2r^2/mr^3 = J^2/mr^3$. Writing this equality and putting in the quantum condition (8), $J_n = nh/2\pi$, we have

$$\frac{e^2/4\pi\varepsilon_0}{r_n^2} = \frac{J_n^2}{mr_n^3} = \frac{n^2h^2/4\pi^2}{mr_n^3} = n^2 \frac{h^2}{4\pi^2 mr_n^3}$$

as the equation that can be solved to determine the permitted radii r_n of the allowed orbits:

$$r_n = n^2 \frac{h^2}{4\pi^2 m} \cdot \frac{4\pi\varepsilon_0}{e^2} = n^2 \frac{\varepsilon_0 h^2}{\pi m e^2}. \tag{10}$$

If we enter the known values of ε_0, h, e, and the mass m of the electron, this relation becomes

$$r_n = n^2 (5.292 \times 10^{-11} \text{ m}) = n^2 (0.052\,92 \text{ nm}). \tag{11}$$

Thus, the circular Bohr orbits, shown in Fig. 11, have radii in the ratios $1:4:9:\cdots$, for $n=1, 2, 3, \cdots$, with the radius of the ground-state orbit, $n=1$, approximately 0.05 nm, which represents the 'size' of the normal hydrogen atom.

We now turn to the computation of the energies of these orbits. We take the potential energy as zero when the electron is at infinity, so that the potential energy is negative at all smaller radii r because work must be done against the force F to pull the electron from radius r to ∞. This

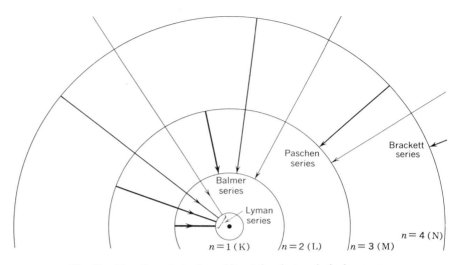

Fig. 11. The allowed circular orbits of the electron in hydrogen on the Bohr model. These orbits have radii in the ratios $1:4:9:16:\cdots$, and energies given by (6). The allowed orbits correspond to the electron's being in the different 'shells.' K, L, M, $\cdots$. The directions of the arrows correspond to *emission*.

work is given by

$$\int_r^\infty F\,dr = \frac{e^2}{4\pi\varepsilon_0}\int_r^\infty \frac{dr}{r^2} = -\frac{e^2}{4\pi\varepsilon_0}\frac{1}{r}\bigg|_r^\infty = \frac{e^2}{4\pi\varepsilon_0 r}.$$

Hence,
$$E_P = -\frac{e^2}{4\pi\varepsilon_0 r}. \tag{12}$$

The kinetic energy is given by $\frac{1}{2}mv^2 = \frac{1}{2}m^2v^2r^2/mr^2 = \frac{1}{2}J^2/mr^2$. Putting in the value of J_n given by (8) and the value of r_n given by (10), we have

$$E_K = \frac{1}{2}\frac{J_n^2}{mr_n^2} = \frac{1}{2}\frac{n^2h^2/4\pi^2}{m(n^4\varepsilon_0^2h^4/n^2m^2e^4)} = \frac{me^4}{8n^2h^2\varepsilon_0^2} \tag{13}$$

Correspondingly, entering the value of r_n in (12) gives

$$E_P = -\frac{me^4}{4n^2h^2\varepsilon_0^2},$$

so the total energy is

$$E_n = E_P + E_K = -\frac{me^4}{8n^2h^2\varepsilon_0^2} = -\frac{2.180\times 10^{-18}}{n^2}\text{J}. \tag{14}$$

This constant is in excellent agreement* with the value hcR in the empirical expression (6) for the observed energy values.

*Actually, as we have performed the computation, the agreement is only exact to the number of significant figures given in (14). To secure exact agreement to a greater

The Bohr theory was later extended to elliptical orbits, again by quantizing the angular momentum, but the details are now unimportant since Bohr's theory has been entirely superseded. Suffice it to say that the allowed elliptical orbits had the same energies (13) as the circular orbits; there were no allowed elliptic orbits for $n=1$, there was one for $n=2$, two for $n=3$, etc.

Bohr's important contribution was the introduction of the idea of *quantization of angular momentum*. His theory was never satisfactorily extended to heavier atoms such as helium, which contains two electrons, or to atoms containing more electrons. But the modern quantum mechanics retains the idea of quantization of angular momentum, and Bohr's picture of an atom with electrons moving in orbits around the nucleus gives a useful qualitative visualization of atomic structure. After discussion of other quantum phenomena in the immediately following sections, we give a brief introduction to modern quantum mechanics in Chapter 42 and give a description of its application to the structure of the atoms of the chemical elements.

6. Emission and Absorption Processes; Lasers

An atom normally in the ground state at room temperature can reach one of its excited states by *radiative* transitions involving absorption of light of the proper frequency or by a *nonradiative* transition involving a collision with a fast-moving electron or an excited atom. An excited atom can lose its excess electronic energy by similar radiative or nonradiative processes. In collisions, an atom in an excited state can transfer all of its excess energy to an atom in the ground state or can have its excess electronic energy transformed into translational kinetic energy shared between the colliding atoms in such a way that momentum is conserved; if an atom is in a highly excited state, it can make a collision that leaves it in a lower excited state so that only a portion of its initial energy is transferred to the other atom or converted into translational kinetic energy.

By radiative transitions an atom can return to the ground state either directly or by several successive steps. For example, if a hydrogen atom is in the state for which $n=4$, it may make a direct transition $n=4 \rightarrow n=1$ by emitting the third line in the Lyman series; or it may return to the ground state by successive transitions $n=4 \rightarrow n=3$, $n=3 \rightarrow n=2$, and $n=2 \rightarrow n=1$, by emitting the first line in the Paschen series, the first line in the Balmer series, and finally the first line in the Lyman series; other

---

number of significant figures, we must include a small correction for the motion of the nucleus, which is assumed stationary in the above computation. By inclusion of this correction, we secure exact agreement within the accuracy of the experimental measurements of wavelength and the accuracies to which the physical constants m, e, and h are known.

possible radiative 'cascades' are $n=4 \to n=3$ followed by $n=3 \to n=1$, and $n=4 \to n=2$ followed by $n=2 \to n=1$. Which radiative transitions are made depends upon the *transition probabilities* (the probability per unit of time that a transition will occur) involved; transition probabilities in agreement with experiment can be determined by quantum-mechanical calculations. If the radiative transition probabilities for transitions from a particular excited state to lower states are all extremely small or even zero, the transitions are said to be 'forbidden' and the excited state in question is said to be *metastable*.

The subject of transition probabilities was studied extensively by Einstein, who showed that, in addition to having *spontaneous* radiative transitions, atoms can undergo transitions to lower states by *induced* or *stimulated* radiative transitions when radiation of the proper frequency is present. The probabilities for induced transitions are proportional to those for spontaneous transitions, but are also proportional to the intensity of the radiation and can become very large when the intensity of the incident radiation is high. It turns out that, so far as induced emission is concerned, the probability of a radiative transition from a given upper state to a given lower state is exactly equal to the probability of absorption of the radiation, leading to a transition from the lower state to the upper state. For example, if radiation of the frequency corresponding to the first line in the Lyman series is incident on a collection of hydrogen atoms, the transition $n=1$ to $n=2$ (absorption) has the same probability as the radiative transition $n=2$ to $n=1$ (emission). Why, then, does the hydrogen sample show a net absorption, resulting in an absorption spectrum?

The answer to this question is obtained by consideration of the *populations* of the various levels. Population differences between the various energy levels of a collection of atoms in thermal equilibrium depends upon the Boltzmann factor introduced earlier (Prob. 52, p. 357) to describe pressure as a function of height in an isothermal atmosphere. For the case of thermal equilibrium at absolute temperature T, the number of atoms, N_2 per unit volume, in an upper level is always smaller than the number, N_1 per unit volume, in a lower level by the Boltzmann factor. Thus,

$$N_2 = N_1 \, e^{-\Delta E/kT} \tag{15}$$

where ΔE is the energy difference between the two levels on the energy-level diagram for the atom in question and k is the Boltzmann constant introduced on p. 349. Since there are more atoms per unit volume in the lower level than in the upper, and since the probabilities of induced absorption and induced emission are equal, there is always a net absorption for a gas in thermal equilibrium. The amount of absorption depends upon the excess $\Delta N = N_1 - N_2$ in the lower state; this excess is given by the relation

$$\Delta N = N_1 - N_2 = N_1 \, (1 - e^{-\Delta E/kT}). \tag{16}$$

In the case of absorption by the atoms of a gas (or of a crystal), we may rewrite our earlier expression for Lambert's law of absorption, given by equation (7), p. 756 in terms of ΔN and an atomic absorption coefficient K' related to the transition probability. The expression is

$$I = I_0\, e^{-Kl} = I_0\, e^{-(K'\Delta N)l}. \tag{17}$$

The negative exponential thus depends upon the probability of induced transitions as measured by K', the excess lower-level population ΔN per unit volume, and the length of absorption path l. It is to be observed that while the beam emerging from the sample of thickness l will be *attenuated*, the emerging radiation remains *coherent* with the incident radiation.

If a light beam can be attenuated by atomic transitions, the question immediately arises as to whether a light beam can be *amplified*—increased in intensity—by similar processes. In view of the Boltzmann factor, $e^{-\Delta E/kT}$, the question can be answered in the negative for samples in thermal equilibrium. However, if by some process the population difference between two levels can be *inverted*, making $N_2 > N_1$ and ΔN in (17) negative, amplification of a light beam becomes possible. The possibility of producing usable population inversions was first demonstrated by CHARLES H. TOWNES in 1952. Townes made use of the phenomenon in a microwave device he called the MASER (an acronym for *M*icrowave *A*mplification by the *S*timulated *E*mission of *R*adiation). Later devices making use of the phenomenon in the visible or near-visible region are called optical masers or LASER's (*L*ight *A*mplification by the *S*timulated *E*mission of *R*adiation).

The general scheme employed in one type of laser is illustrated by the energy-level diagram in Fig. 12. For thermal equilibrium nearly all atoms are in the ground state labeled 1. Now, if the assembly of atoms is irradiated by radiation of frequency f_1, transitions to state 3 will take place. Of the atoms in state 3, some will return to the ground state by spontaneous or induced emission of radiation of frequency f_1; others will go state 2 by spontaneous emission of f_3 or by nonradiative transitions of some sort. If the probability for transitions from state 2 to the ground state is not too great, the population of state 2 will increase. As atoms are 'pumped' out of the ground state and accumulate in state 2, the population of state 2 increases and eventually becomes greater than that of the ground state; i.e., *population inversion* has been produced, N_2 has become greater than N_1, and ΔN in (17) has become negative.

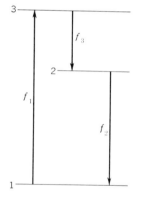

Fig. 12. Atomic energy levels permitting laser action.

Therefore, if a beam of radiation of frequency f_2 and intensity I_0 is incident on the sample in which population inversion exists, the intensity of the beam after transversing path length l in the sample is increased because, with ΔN negative, the exponent in (17) becomes positive. Just as in the case of attenuation, the emerging beam will be *coherent* with the incident radiation. Even though the intensity of the emerging beam has been increased by the addition of single quanta of energy hf_2 from many different atoms, the processes of induced or stimulated emission are such that there are definite phase relationships between the contributions of the various atoms—as contrasted with the random phase relations in spontaneous emission processes of the type involved in most light sources. Another interesting feature of laser action is that the emerging beam has the same *direction* as the incident beam; thus, if a well-collimated narrow pencil of light of frequency f_2 with small beam intensity is incident on a 'pumped' sample, the emerging amplified beam may be extremely intense but will emerge as a narrow pencil with no more divergence than that of the incident pencil.

In addition to the property of amplifying an incident light beam, a laser can be used as a *source* of coherent light provided a suitable arrangement is made for a beam to 'grow.' Such an arrangement is shown schematically in Fig. 13, which shows a sample tube irradiated from the side by radiation 'pumps' having frequency f_1. Beyond the plane ends of the sample tube are two plane mirrors arranged to set up a standing wave of frequency f_2; mirror M_1 is a highly reflecting mirror whereas mirror M_2 is only partially silvered, so that it will allow a beam of f_2-radiation to emerge. If the sample is well-pumped so that there is a sufficiently large population inversion, with ΔN having a large negative value, the spontaneous emission of a single quantum hf_2 is sufficient to initiate a coherent wave provided the emitting atom is located at a non-nodal position in the standing wave pattern and the emitted radiation travels in a direction perpendicular to the plane parallel surfaces of mirrors M_1 and M_2. The standing-wave pattern of light that is set up in the tube is analogous to the pattern of standing sound waves in an organ pipe.

The early lasers employed the chromium *ions* in ruby crystals instead of the atoms of a gas as a light emitting and absorbing source; ruby lasers can provide *pulsed* coherent beams of enormous intensity when large xenon flash lamps are used for radiation pumping. *Continuous* lasers

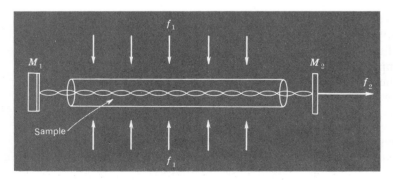

Figure 13

giving coherent beams of considerable intensity can be constructed with gaseous samples. The development of lasers offers additional practical proof of the existence of the quantized energy levels first proposed by Bohr.

7. The Inverse Photoelectric Effect; X Rays

Incident radiation causes the emission of electrons in the photoelectric effect. The inverse process—the emission of radiation by a metal when electrons are incident—results in the production of X rays.

In 1895, WILHELM KONRAD ROENTGEN discovered that penetrating radiation is given off by the walls of tubes carrying an electric discharge. The name *X rays* was applied because the nature of the radiation was not known. Today we know that X rays are electromagnetic radiation like light, but of much shorter wavelengths and higher frequencies than visible light. They are given off when high-speed electrons impinge on the walls of discharge tubes or on metal electrodes within the tubes.

When an electron accelerated through a difference of potential of several thousand volts strikes a metal target, X rays are emitted from the surface. The emitted X rays have various frequencies but the maximum frequency f_{Max}, can be obtained from the photoelectric equation (3) as

$$hf_{\text{Max}} = \tfrac{1}{2}mv^2 + \phi, \tag{18}$$

where v is the speed of the incident electrons. In the case of X rays the magnitude of the work function ϕ is so small compared with the quantum energy hf that it can be neglected in the above equation. The target in an X-ray tube emits X rays of all frequencies below the *maximum* frequency, and the spectrum of the emitted X rays extends indefinitely from the wavelength corresponding to f_{Max} toward longer wavelengths.

X rays are emitted by nonmetals as well as metals, although for convenience the target in a commercial X-ray tube (Fig. 14) is always metallic. The emitted spectrum is continuous, but superimposed on this continuous spectrum are 'bright lines' corresponding to *characteristic* X rays. The frequencies of the characteristic lines are determined by the element of which the target is made. Spectral series of characteristic X rays arise from changes in energy of the *inner* electrons of the atoms, just as spectral series in the visible spectrum arise from changes in energy of the valence electrons (Chapter 42).

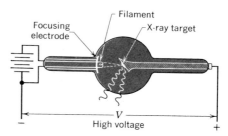

Fig. 14. X-ray tube. Electrons emitted by the filament strike the positively charged metal target. X rays are emitted from the target.

The wavelengths of X rays are much shorter than those of visible light and are small enough to be comparable with the interatomic distances in crystals. Hence, crystal lattices can act as 'gratings' for X rays in much the same manner as ruled gratings are used in the visible region. From the observed X-ray diffraction patterns, interatomic distances in crystals can be accurately measured when the wavelength of the X rays is known, as we shall discuss in Sec. 8.

Example. *If the target in Fig. 14 is at* 3000 V *higher potential than the filament, what are the maximum frequency and the minimum wavelength of the X rays emitted?*

The electrons that are 'boiled off' the heated filament in thermionic emission have negligible kinetic energy compared with that which they acquire in being accelerated through the 3000-V difference of potential. The work done on each electron by the electric field between the filament and the target is

$$W = 3000 \text{ eV}. \tag{i}$$

This work all appears as kinetic energy of the electron, and when the electron strikes the target can, in whole or in part, be transformed into a quantum of X radiation. If the whole energy is so transformed, we have by (18), neglecting the small work function ϕ,

$$f_{\text{Max}} = \frac{W}{h} = \frac{3000 \text{ eV}}{4.13 \times 10^{-15} \text{ eV} \cdot \text{s}} = 7.26 \times 10^{17} \text{ Hz} \tag{ii}$$

as the maximum frequency of the X rays. The corresponding minimum wavelength is

$$\lambda_{\text{Min}} = \frac{c}{f_{\text{Max}}} = \frac{3.00 \times 10^8 \text{ m/s}}{7.26 \times 10^{17}/\text{s}} = 4.13 \times 10^{-10} \text{ m} = 0.413 \text{ nm}. \tag{iii}$$

The above frequency and wavelength appear in Fig. 12, p. 712, near the lower part of the region marked X rays. These would be called 'soft' X rays because the quantum energy is comparatively low and the penetrating power is also low. X rays used for various medical purposes are obtained from electrons falling through differences of potential ranging from about 3 thousand to 3 million volts. For 3 million volts, the maximum quantum energy and frequency would both be 1000 times the values in (i) and (ii), while the minimum wavelength would be $\frac{1}{1000}$ of that in (iii).

The so-called K-lines of the characteristic X-ray spectrum, which represents the most energetic (shortest wavelength) transitions, are produced as follows. The bombarding electron ionizes the atom by removing one of the electrons in the innermost K-shell (*see* pp. 873–876). Subsequently, one of the electrons in an outer L, M, N, O, or P shell jumps into the vacancy in the K-shell, with release of energy that appears as an X-ray quantum. The X-ray lines designated as $K\alpha_1$ and $K\alpha_2$ arise in a jump from the L-shell to the K-shell. HENRY G. J. MOSELY, in 1913, discovered that the frequencies of these lines for the different atoms are, to a close approximation, proportional to Z^2, the square of the atomic number. He was thus, for the first time, able to determine the atomic numbers unambiguously for the various elements in the periodic table.

We can see qualitatively the reason for the variation of frequency with Z^2 by consideration of the Bohr model. From (14), we note that E_n depends on the square of the electronic charge and also the square of the proton's charge in the hydrogen atom; in other atoms the nuclear charge is $+Ze$ and an analogous expression for E_n would thus be proportional to $e^2(Ze^2) = Z^2e^4$. For a one-electron ion, the energy levels would be equal to Z^2 times the values given in (14) for hydrogen. Since most atoms have more than one electron, the dependence on Z^2 becomes an increasingly poor approximation as n increases; i.e., as we progress outward from the K-level.

8. Photons

In our discussion of the quantum theory of radiation, we have pointed out that energy is always emitted or absorbed in discrete units or quanta having energy $E = hf$. When the frequency is low, the energy of a single quantum is so small that it is impossible for any physical instrument to detect a single quantum. Thus, for low radio frequencies a large number of quanta must be received in order for any detecting instrument to give a response. The reception of large numbers of quanta with the energy of each quantum extremely small gives the impression of a *continuous* process of reception, analogous to our feeling that light energy enters our eyes and produces stimuli continuously.

At X-ray frequencies, *a single quantum can easily be detected,* even by a relatively insensitive detector. This fact gives us further evidence that emission and absorption are actually quantum processes. The individual quanta in the X-ray and γ-ray regions of the electromagnetic spectrum have such distinctive properties that they have many of the characteristics of particles, and were given the name *photons*. We now use the term photon more generally:

A **photon** is a single quantum of electromagnetic radiation.

At X-ray frequencies, the interaction of individual photons with matter can easily be studied. ARTHUR H. COMPTON, in 1924, discovered that when a monochromatic beam of X rays of frequency f is scattered by a gas, *two* frequencies are present in the scattered beam observed at a particular angle. One frequency f is equal to the frequency of the original unscattered X rays in the incident beam; this frequency was to be expected on the basis of the classical electromagnetic theory of scattering. The other frequency f', not expected from classical theory, is always less than the frequency of the incident X rays and is dependent on the scattering angle θ.

Compton was able to account for the change in frequency of the scattered radiation by considering the incident photon as a 'particle' with energy hf and with momentum of magnitude hf/c. Scattering is assumed to take place as a result of a collision between the incident photon and an electron in the scattering material. The energy of the incident photon is large compared with the energy binding the electron into the atom, so the electron can be considered as essentially free in this collision. As a result of the collision, a part of the energy and momentum of the incident photon is transferred to the electron, so that the final energy hf' and momentum hf'/c of the scattered photon are less than those of the incident photon. By writing the conservation-of-energy and the conservation-of-momentum relations for a collision of this type, Compton was able to account exactly for the observed frequency changes. We here emphasize that *we can explain the results of Compton's scattering experiment by treating the X-ray photon as a particle of energy hf having momentum of magnitude $hf/c = h/\lambda$*. Observations of the recoil electrons knocked out of the gas have shown that they do indeed have the momentum and energy to be expected from the momentum and energy equations for a collision with such a particle.

On the basis of this *Compton effect,* we might conclude that X rays consist of particles and are not waves at all. However, there are other X-ray phenomena that can be explained only on the basis of wave properties. The wavelength of X rays is of the same order of magnitude as the distance between the atoms or ions in crystals, and crystals can be used as three-dimensional diffraction gratings for X rays. The diffraction pattern produced when a pencil of X rays passes through an aluminum foil is shown in Fig. 15. Such diffraction phenomena give abundant evidence that X rays also have wave characteristics.

Detailed observations of the diffraction patterns resulting when a pencil of monochromatic X rays passes through a single crystal in various orientations determines not only the spatial arrangement of the atoms in the crystal, *but also the distances between the atomic planes.* The phrase in italics is important because this information gives us another means of determining Avogadro's constant, since, if the spacing between atoms is known, the number of atoms per unit volume can be calculated. From density measurements the number of atoms per kmol—Avogadro's constant—can then be readily calculated. Diffraction of a pencil of X rays by a single crystal results in a pattern of spots known as *Laue spots* (Fig. 16), each spot representing the image of a particular order result-

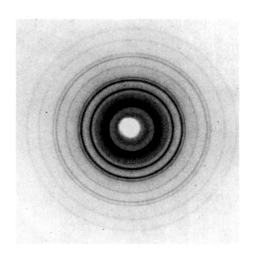

Fig. 15. X-ray diffraction pattern of aluminum foil. (*Courtesy of* Mrs. M. H. Read, Bell Telephone Laboratories.)

ing from atomic planes at a particular orientation acting as a diffraction grating.

The above discussion assumes the wavelength of the monochromatic X rays to be known. The wavelength of X rays of the longer wavelengths can be determined by using a ruled diffraction grating at nearly grazing incidence so that the effective grating spacing is very small. Then the same X rays can be used to determine the 'grating' spacing in a crystal. Finally, the crystal itself can be used like a grating to determine the wavelengths of X rays of still shorter wavelengths.

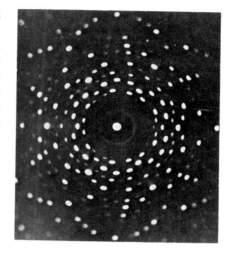

Fig. 16. X-ray diffraction pattern of a single quartz crystal. (From B. Rossi, *Optics:* Addison-Wesley, Reading, Mass., 1957.)

PROBLEMS

1. A prism spectrograph is arranged for photographic observation, with a photographic plate at the focal plane of the objective lens of the telescope, which thus plays the role of a camera lens. What would be the plate separation of the 656.3-nm and the 486.1-nm lines in the hydrogen spectrum if the camera lens has a focal length of 120 cm and the angle between the blue and red rays, as they leave the prism, is $1°55'$? *Ans: 32.5 mm.*

2. What is the angle between the blue and red hydrogen lines of Table 36-1, p. 755, when a 5000-line/cm grating is used at normal incidence and the second order is being employed?

3. If the grating in Prob. 2 were used in a spectrograph in which the camera lens has a focal length of 120 cm, what would be the plate separation of the red and blue hydrogen lines when the second-order spectrum is photographed? *Ans: 252 mm.*

4. Solve Probs. 2 and 3 for a case in which the first-order spectrum is being observed.

5. A *monochromator* is a device used to obtain 'monochromatic' light—actually a narrow band of wavelengths. Such a device can be obtained by placing a slit at the plate position in a spectrograph in such a way as to permit light of the desired wavelength from a source having a continuous spectrum to pass through the slit. Although the dispersion obtainable by grating spectrographs is higher, a prism is usually employed in a monochromator to get adequate light intensity through the slit. Give two reasons why a slit of given width will give a brighter source in a prism monochromator than in a grating monochromator.

6. We have discussed, in Prob. 20, p. 726, the Doppler effect of electromagnetic radiation reflected from a moving object. There is also a Doppler effect whenever source and observer are moving toward or away from each other, which results in a shift of frequency and wavelength *of spectral lines*. Since no medium is involved in the transmission of light through a vacuum, the principle of relativity asserts that it can be only the *relative* velocity of approach v that enters the equations that give the frequency and wavelength shifts:

$$\Delta f = (v/c) f; \qquad \Delta \lambda = -(v/c) \lambda.$$

Show that (20), p. 457, reduces to this expression for the case $v \ll c$, independent of whether it is the source, the observer, or both, that moves.

7. From the Doppler effect of the star's spectrum (Prob. 6), we can obtain the radial velocity of a star relative to the earth. This velocity is typically of the order of 20 km/s, since this is about the speed of the sun relative to the average of the 'fixed' stars; however many stellar radial velocities are more than 100 km/s. Van Maanen's star has a radial velocity of recession of 240 km/s, while μ Cassiopeiæ has radial velocity of approach of 97 km/s. For these two stars, find the wavelength shift of the dark absorption line of hydrogen normally at 656.3 nm. Which shift is 'toward the violet'?
Ans: $+0.525$ nm, -0.212 nm.

8. Doppler effect has demonstrated that a large fraction of the stars are double—two stars rotating around each other with periods as short as a few days. Describe the variation with time of the appearance of a dark absorption line in the spectrum of such a 'binary star,' the variation that enables its recognition as double. (See Prob. 6.)

9. The sun is rotating with a period of 25 days about an axis that is approximately north-south, with the west limb moving away from us and the east limb toward us. If the spectrum of the west limb and that of the east limb are photographed one above the other, what will be the apparent wavelength differences of the Fraunhofer lines in the 642-nm region shown in Fig. 5? Which limb will be shifted toward the violet and which toward the red?
Ans: 0.008 69 nm.

NOTE: *By the angular dispersion of a spectrometer is meant the change in angle on the circles of Fig. 1 or Fig. 2 per unit change in wavelength, that is, $d\theta/d\lambda$. The mean angular dispersion over a wavelength range $\Delta\lambda$ would be $\Delta\theta/\Delta\lambda$. If the spectrometer is arranged for photographic observation, the linear dispersion is defined as $dX/d\lambda$, where dX is the separation on the plate of two spectral lines differing in wavelength by $d\lambda$.*

10. Show that the angular dispersion of a grating spectrometer at angle θ in the nth order is $d\theta/d\lambda = nN/\cos\theta$. Show that the linear dispersion on a photographic plate is $dX/d\lambda = nNf/\cos\theta$, where f is the focal length of the camera lens.

11. What is the angular dispersion of a 5000-line/cm grating in the vicinity of 589 nm in the second order when the grating is used at normal incidence?
Ans: 4.25/nm.

12. A 5000-line/cm transmission grating is mounted as in Fig. 1. Determine its mean angular dispersion in the first-order spectrum in the region between the blue and red wavelengths of Table 36–1, p. 755. Compare with the mean angular dispersion of the prism of Prob. 1 in this same region.

NOTE: *In the following problems, use the rounded values $h = 6.6 \times 10^{-34}$ J·s; $c = 3 \times 10^8$ m/s; and 9×10^{-31} kg for the mass of the electron.*

13. At what wavelength does the black-body emission curve have its maximum at a room temperature of 27° C?   Ans: 9.66 μm.

14. The melting point of tungsten is 3400° C. At what wavelength does the black-body radiation curve for this temperature have its maximum? Explain the relatively 'orange' color of the light from incandescent lamps.

15. The maximum of the radiation curve for the sun occurs at 480 nm. What would be the temperature of a black body for which the radiation curve would have a maximum at this wavelength?   Ans: 6040° K.

16. What would be the temperature of a black body for which the intensity maximum occurred in the infrared at 2000 nm? in the ultraviolet at 200 nm?

17. From Planck's radiation law, compute the relative values of P_λ for black-body radiation at 400 nm (deep violet), 500 nm (green), and 600 nm (orange), setting the value of P_λ at 500 nm equal to 1 in each case, for (a) a tungsten filament at 3000° K, (b) the sun at 6000° K, (c) Vega at 12 000° K.   Ans: (a) 0.28:1:1.99; (b) 0.91:1:0.90; (c) 1.60:1:0.63.

18. Compare Betelgeuse (3000° K) and Vega (12 000° K) with the sun (6000° K) with regard to (a) total rate of radiation, (b) P_λ at 600 nm, (c) P_λ at 100 nm. Assume that all three stars radiate like black bodies.

19. By counting squares under the 6000° curve of Fig. 7, estimate the total rate of radiation and compare with the value 7.4×10^7 W/m² given by the Stefan-Boltzmann relation.

20. If the temperature of a black body is changed so as to triple the wavelength at which the maximum of the radiation curve occurs, by what factor is the total emission multiplied?

21. Find the energy of the quanta associated with radiation of the following frequencies: X rays, $f = 10^{19}$ Hz; visible, $f = 2 \times 10^{15}$ Hz; far infrared, $f = 10^{13}$ Hz; microwave, $f = 10^{10}$ Hz. Recalling that the mean translational kinetic energy of a gas molecule in thermal motion is $\frac{3}{2}kT$, find the frequency of the radiation whose quantum energy is equal to the mean translational kinetic energy of a molecule at 27° C. Ans: 6.62×10^{-15} J; 13.2×10^{-19} J; 6.62×10^{-21} J; 6.62×10^{-24} J; 9.36×10^{12} Hz (far infrared).

22. The maximum of the radiation curve for the sun occurs at 480 nm. What is the frequency of this radiation? What is the energy of a single quantum of this radiation?

23. Find the frequencies of the first three lines in the Lyman series of hydrogen. The lines of this series converge toward a limit; what is the frequency of this limit? Ans: 2.47×10^{15} Hz; 2.92×10^{15} Hz; 3.08×10^{15} Hz; 3.29×10^{15} Hz.

24. Repeat Prob. 23 for the Balmer series in hydrogen. Also determine the wavelengths and compare with Fig. 3.

25. Recalling that hydrogen atoms are normally in their lowest energy level and that the frequency of the first line of the Lyman series is approximately 2.47×10^{15} Hz, calculate the minimum energy of the bombarding electrons required in a Franck-Hertz experiment to excite the first line in the Lyman series. Ans: 1.63×10^{-18} J.

26. What is the minimum energy of the bombarding electrons required to excite the first line of the Balmer series? all the lines in the Lyman series? all the lines in the Balmer series? (*See* Prob. 25.)

27. Planck's constant h is normally expressed as 6.6×10^{-34} J·s. Show that the joule·second can also be considered a unit of angular momentum. Express h in the usual units for angular momentum. Ans: 6.6×10^{-34} kg·m²/s.

28. The orbital angular momentum of the electron in the first orbit of the Bohr model of the H atom is $h/2\pi$. Taking the radius of the orbit as 0.05 nm, calculate the frequency of the orbital motion. Show that the tangential speed of the electron in this orbit is approximately $\frac{1}{137}$ times the speed of light.

29. The diameter of the hydrogen atom is approximately 0.1 nm. What is the frequency of X-radiation having a wavelength of 0.1 nm? What is the energy of a single photon of this radiation? What is the photon's momentum? Ans: 3.00×10^{18} Hz; 1.98×10^{-15} J; 6.62×10^{-25} kg·m/s.

30. The Bohr theory also gives the correct results for the spectra of the ions He⁺, Li⁺⁺, Be⁺⁺⁺, ..., which consist of a single electron in the Coulomb field of a massive nucleus of charge Ze, where $Z=2$ for He⁺, $Z=3$ for Li⁺⁺, $Z=4$ for Be⁺⁺⁺, ... Show that in these cases the energy levels are given by $E_n = -(Z^2/n^2)hcR$. Show that the spectra are exactly analogous to the spectrum of hydrogen shown in Fig. 10, but with all frequencies multiplied by Z^2 or, correspondingly, all wavelengths divided by Z^2. Show that, to our approximation in which motion of the nucleus is neglected, the frequency of the first Balmer line of H is identical with the frequency of the second 'Paschen' line of He⁺.

42 QUANTUM MECHANICS: ATOMIC STRUCTURE

The dual nature of X rays—particle nature in the Compton effect and wave nature in diffraction—is completely contrary to classical ideas, but this dualism is now regarded as characteristic not only of electromagnetic radiation but of matter as well. Just as we were accustomed to think of *light* as consisting of *waves,* so also we were accustomed to think of *matter* as consisting of *particles*. Visible light and radio waves do not have a readily observable particle nature, and particles large enough to be seen do not have a readily observable wave nature. However, tiny 'particles' such as electrons, atoms, and molecules have quantum and wave properties that cannot be completely described in terms of the 'classical' principles of mechanics and electromagnetic theory.

Just as radiation has both particle and wave characteristics, so also does matter. In this chapter, we introduce the de Broglie wavelength $\lambda = h/p$ for a particle of momentum p and show how the Schrödinger equation can be 'derived.' The Schrödinger equation represents one formulation of the modern quantum mechanics, which must be used to describe the behavior of atomic processes.

After a discussion of the Schrödinger equation and the nature of the wave function ψ, we state the Pauli exclusion principle and show how this principle can be used in understanding the existence of electron shells and subshells in atoms. We then interpret the phenomenon of paramagnetism in terms of incomplete electron shells in atoms and discuss briefly the nature of the interatomic forces responsible for the existence of magnetic domains in ferromagnetic materials.

Much of the material in the present chapter will become more meaningful in later physics courses. Meanwhile, if some of the material seems puzzling, the reader can perhaps gain some comfort from the fact that his own plight is not unique. A famous physicist once remarked: "A person never really *understands* quantum mechanics; he merely becomes *used to it!*"

1. Waves Associated with Material Particles

In 1924, Louis Victor de Broglie suggested that, just as a photon of wavelength λ has associated with it a momentum $p = h/\lambda$, *the motion of an electron or of any other material particle is associated with a wave motion of wavelength*

$$\lambda = h/p = h/mv, \qquad (1)$$

where $p = mv$ is the momentum of the particle and h is Planck's constant. An electron with a kinetic energy of 100 eV has, according to de Broglie's relation, a characteristic wavelength of 0.123 nm. This wavelength is comparable with the wavelengths of X rays. Therefore, if de Broglie's hypothesis is correct, an electron beam passing through a thin film of powdered crystal or through a metal film should experience diffraction similar to that experienced by an X-ray beam. Diffraction of an electron beam was first observed in 1927 by C. J. Davisson and L. H. Germer. A photograph of an electron-diffraction pattern is shown in Fig. 1. The essential similarity between this diffraction pattern and the X-ray diffraction pattern shown in Fig. 15, p. 862, is at once apparent. We are led to the conclusion that the de Broglie hypothesis is correct.

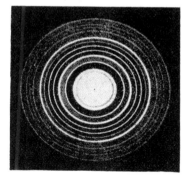

Fig. 1. Electron diffraction pattern of silver. (*Courtesy of* L. H. Germer, Cornell University.)

Consideration of the expression (1) for the de Broglie wavelength shows that particles of macroscopic size would have such large masses that the associated wavelengths would be too small to give rise to observable diffraction phenomena, even with a 'crystal grating.' As a consequence of the short wavelength, the wave nature of particles larger than 'atomic' size has never been detected.

Both matter and photons have properties of the character that we ordinarily ascribe to 'particles,' and properties of the character that we ordinarily ascribe to 'waves.' The 'particle' properties of radio-frequency photons are not apparent because only a very large number of photons can be detected. The 'wave' properties of macroscopic matter are not apparent because the wavelength is undetectably small. In the X-ray region, photons show both 'particle' and 'wave' properties very strikingly; in the 'gamma-ray' region *only* the 'particle' properties of the photons are apparent because again the wavelength is undetectably small. Slow-speed electrons and neutrons have detectably large wavelengths, so their 'wave' properties are apparent.

Both photons and material particles have energy, momentum, and wavelength. The wavelength and momentum are in all cases related by $p = h/\lambda$. 'Wave' properties are apparent only if the wavelength is detectable, which means in practice that it is at least sufficiently large to be comparable to the 'grating' spacing of crystals. Properties ordinarily associated with 'particles' are apparent only if the energy and momentum are sufficiently large so that individual particles can be detected.

The above discussion is not intended to imply that there is not a fundamental difference between photons and material particles. In spite of the fact that they share similar physical properties such as energy, momentum, and wavelength, there is one very fundamental difference: a photon always moves with the speed c of light, whereas a material particle always moves with a speed that is less than c. It is always possible to imagine an observer for whom any material particle would be at rest, but a photon cannot be at rest for any observer, since a photon moves at the same speed c relative to *every* observer (Chapter 44). There are other fundamental differences in the manner of interaction with material charged particles that justify the conclusion that the 'wave' associated with photons is electromagnetic in nature whereas the 'wave' associated with material particles is of quite another character indescribable in terms of ordinary macroscopic physical concepts.

2. Quantum Mechanics

The Bohr theory, which satisfactorily accounted for the spectrum of hydrogen, did not meet with similar success when applied to the spectra of other atoms. Sommerfeld attempted a generalization that included elliptical as well as circular orbits, with the introduction of additional quantum numbers that applied to orbits of different ellipticity and of different spatial orientation. However, the Bohr-Sommerfeld theory was generally unsatisfactory and was never able to give a computation of transition probabilities.

Beginning in 1925, the de Broglie hypothesis was taken over by Schrödinger and Heisenberg, who used it in the development of modern *quantum mechanics,* or as it is sometimes called, *wave mechanics.* The quantum mechanics of Schrödinger and Heisenberg enables us to predict correctly the characteristic energy levels of an atom or molecule from certain perfectly general equations. In the simple case of hydrogen, the wavelength associated with the electron's motion is restricted to a set of discrete values exactly analogous to the set of discrete values of wavelength that gives the normal modes of vibration of a string. To each allowed value of wavelength λ there corresponds a definite momentum computed from de Broglie's relation (1), and hence a definite energy.

When quantum mechanics is applied to a system involving macroscopic bodies, the number of characteristic energy levels is so great and the levels are so close together that the results become equivalent to those obtained by classical Newtonian mechanics.

Although a detailed introduction to quantum mechanics is beyond the scope of the present text, we shall give a brief indication of the general ideas involved in Schrödinger's formulation. Consider a wave of the form

$$\psi = a \sin(2\pi t/T - 2\pi X/\lambda), \qquad (2)$$

where ψ is called the wave function. The nature of ψ will be discussed later, but we can state that (2) 'represents' particles of wavelength λ, hence momentum $p = h/\lambda$, moving in the $+X$-direction. Equation (2) is exactly analogous to the equation giving the electric field in a plane wave of light traveling in the X-direction. From such a plane light wave, we can compute diffraction patterns such as we have discussed. But we remember that a plane light wave represents a 'beam' of photons; in the same manner, equation (2) can be considered to represent a 'beam' of particles moving in the $+X$-direction. We compute the diffraction patterns of the particles, such as the pattern in Fig. 1, by methods exactly analogous to those employed to compute the diffraction patterns produced by light.

Equation (2) is similar in form to the wave equation given in (5a), p. 417. By double differentiation of ψ with respect to X, we obtain the equation

$$\frac{\partial^2 \psi}{\partial X^2} = -\frac{4\pi^2}{\lambda^2}\psi, \qquad (3)$$

in exact analogy with (10), p. 420. Now, substituting de Broglie's relation $\lambda = h/p$ between wavelength and momentum, we write

$$\frac{\partial^2 \psi}{\partial X^2} = -\frac{4\pi^2}{h^2}p^2\psi. \qquad (4)$$

Recalling that kinetic energy $E_K = p^2/2m$ and that the total energy E of a particle is equal to the sum of E_K and the potential energy E_P, we may replace p^2 in (4) by $2m(E-E_P)$ and obtain

$$\frac{\partial^2 \psi}{\partial X^2} = -\frac{8\pi^2 m}{h^2}(E-E_P)\psi.$$

Generalization to three dimensions gives

$$\frac{\partial^2 \psi}{\partial X^2} + \frac{\partial^2 \psi}{\partial Y^2} + \frac{\partial^2 \psi}{\partial Z^2} = -\frac{8\pi^2 m}{h^2}(E-E_P)\psi. \qquad \left(\begin{array}{c}\text{Schrödinger}\\\text{equation}\end{array}\right) \quad (5)$$

This is the equation that determines the allowed energy levels of an electron in an atomic force field—for example, the energy levels of the single electron in hydrogen. Only for certain values of E do single-valued solutions of this equation exist—these are the solutions that are analogous to standing waves on a string—and these values of E represent the energies of the 'stationary' states. In the case of hydrogen, these allowed energies turn out to have the same values (14), p. 854, that we computed on the Bohr theory.

We shall not carry out the detailed treatment of hydrogen, but we can point out, in a general way, the rationale for the Bohr quantization of angular momentum. In the case of the circular orbits, the wavelength of the wave along the orbit must be an integral submultiple of the circumference of the circle, that is,

$$\lambda = \frac{2\pi r_n}{n}. \qquad (n=1, 2, 3, \cdots)$$

Putting in $\lambda = h/p$, we have $h/p = 2\pi r_n/n$, or

$$pr_n = nh/2\pi,$$

which is exactly Bohr's quantum condition (7), p. 853.

Thus far, we have not given any interpretation of the wave function ψ, which was a simple sinusoidal wave in (2). Actually, equation (5) applies much more generally. However, the wave functions involved in applications of (5) to atomic structure must be continuous, single-valued functions with continuous first derivatives and must go to zero at great distances. They are in general complex, rather than real functions. The interpretation of the wave function ψ is that its absolute square, $|\psi|^2$, times a volume element $dX\,dY\,dZ$ represents the probability of finding an electron in the volume element; this *probability* aspect of locating an electron—as contrasted with the certainty implied by Bohr's model—represents an additional radical departure from classical ideas. We note that, just as in the case of light, the *intensity* (number per unit volume) of the beam of particles represented by equation (2) is proportional to the *square* of the wave amplitude.

We now show, in Fig. 2, examples of the electron probability density for hydrogen. The examples are those corresponding to the first four circular orbits in the Bohr theory and represent cross sections in the 'equatorial' plane, which would be the plane of the Bohr orbit. We note that while the probability distributions are 'smeared out' radially, they have their maximum values at the Bohr radii 1, 4, 9, 16, $\cdots \times r_1$. Since the 'equatorial plane' can be any plane through a sphere, it is realistic to think of the electrons as being located in 'shells' near these radii.

We have shown a distribution uniform with angle in Fig. 2. One wonders where the de Broglie wavelength comes in. It turns out that the angular dependence is represented by a *complex* function of the form

$$e^{ik\theta} = \cos k\theta + i \sin k\theta,$$

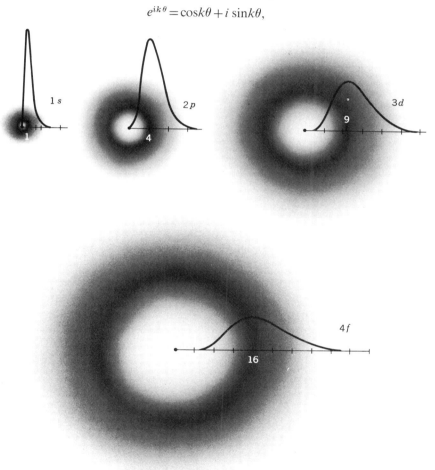

Fig. 2. Radial probability densities corresponding to circular orbits for the first four energy levels of hydrogen, $n = 1, 2, 3, 4$. Radial unit is the first Bohr radius r_1.

with k an integer. This function does exhibit periodic properties, but when we take the absolute square to find the probability distribution:

$$|e^{ik\theta}|^2 = e^{ik\theta}\,e^{-ik\theta} = 1,$$

the periodicity is lost and the electron has equal probability of being found anywhere around the circle, as we would expect by analogy with the Bohr model.

One other quantum-mechanical principle that should be mentioned here is the Pauli exclusion principle, which asserts that *in a multi-electron atom, no two electrons can be described by quantum numbers having identical values*. The wave functions describing the two electrons must differ by at least one quantum number in the complete sets of quantum numbers required for their description. We shall have occasion to employ the Pauli exclusion principle in the next section where we discuss the electronic structure of atoms, and the quantum numbers assigned to the different electrons.

3. Structure of Atoms

The atomic model used as a basis for giving the quantum-mechanical description of atoms is one involving a small positively charged nucleus with a radius of the order of 10^{-15} m and an electric charge of $+Ze$, where Z is the atomic number of the element in the periodic table and e is the magnitude of the electron's charge. Nearly all the mass of an atom is associated with the nucleus. Around the nucleus, Z negative electrons move in various 'orbits,' the mean radius of the outermost electron 'orbits' being of the order of 10^{-10} m. We have written 'orbit' in quotes because, in quantum mechanics, the classical idea of an orbit is replaced by a wave function that gives a probability density of the type depicted in Fig. 2. In using the Schrödinger equation (5) to determine the wave functions ψ and the energy levels E, the value of the potential energy E_P as a function of radial distance from the nucleus can be expressed in terms of the above model and the principles of electromagnetism.

In giving a quantum-mechanical description of an electron in an atom, four quantum numbers are needed. The first of these is the principal quantum number n, which has values equal to the n introduced by Bohr. The second quantum number involved, denoted by l, determines the *orbital angular momentum* of the electron, which has the value $\sqrt{l(l+1)}\,h/2\pi$. If the principal quantum number of an electron is n, the azimuthal quantum number l can take any integral value from 0 up to $n-1$. Recalling that angular momentum is a vector quantity, we might ask whether there are any quantum-mechanical restrictions on the spatial orientations of the vector associated with orbital angular motion. The orientations are indeed restricted and are described by a spatial quantum number m_l, which can take any of the integral values $l, l-1, \cdots, 0, -1, \cdots, -l$, corresponding to a *component of orbital angular momentum* equal to $m_l h/2\pi$ in any arbitrarily selected direction. The selected direction is generally that of an externally applied magnetic or electric field, whose influence on the spectrum we will discuss below.

In addition to angular momentum associated with its orbital motion, every electron has an *intrinsic spin;* with the intrinsic spin is associated an angular momentum that can have only two spatial orientations specified by a *spin quantum number* m_s, which can take only the values $+\frac{1}{2}$ or $-\frac{1}{2}$, corresponding to a component of spin angular momentum $m_s h/2\pi$ in the selected direction.

In a given atom, all electrons with the same principal quantum number n are said to constitute a 'shell'; the shells for which $n=1, 2, 3, 4, 5, 6, 7,$ are called the K-, L-, M-, N-, O-, P-, Q-shells, respectively.

Various 'subshells' are determined by the orbital quantum number l; the number of electrons in any subshell is determined by the restrictions on m_l and m_s imposed by the Pauli exclusion principle: (*a*) Electrons for which $l=0$ are called s-electrons; since m_l must be 0 for $l=0$, an s-subshell can contain only two electrons characterized by $m_s = +\frac{1}{2}$ and $-\frac{1}{2}$. (*b*) Electrons for which $l=1$ are called p-electrons; possible values of m_l for p-electrons are $+1, 0, -1$, and, for each of these values of m_l, m_s can take values of $+\frac{1}{2}$ or $-\frac{1}{2}$; thus, only 6 electrons can be included in a p-subshell. (*c*) Electrons for which $l=2$ are called d-electrons; possible values of m_l are $+2, +1, 0, -1, -2$, with $m_s = +\frac{1}{2}$ or $-\frac{1}{2}$ for each value of m_l; thus, only 10 electrons can be included in a d-subshell. (*d*) Electrons for which $l=3$ are called f-electrons, with possible m_l values of $+3, +2, +1, 0, -1, -2, -3$, and $m_s = +\frac{1}{2}$ or $-\frac{1}{2}$ for each value of m_l; thus, f-subshells can contain at most 14 electrons.*

We are now in a position to understand the periodic table, Table 42-1, in terms of the shells and subshells into which the different electrons are placed as the atoms are built up.

For atoms in the ground energy state, it is found that, in general, electrons occupy levels characterized by the lowest values of n and l that are permitted by the Pauli exclusion principle, since these turn out to be the levels of lowest energy. Thus, for hydrogen with $Z=1$, the single electron is characterized by $n=1, l=0$ (1s); for helium with $Z=2$, the two electrons are both characterized by $n=1$ and $l=0$ and differ only by m_s. The Pauli exclusion principle does not permit additional electrons

---

*The reader will be curious about the historical origin of the series of letters s, p, d, f, g, h, i, $\cdots$. The first four letters of this series derive from adjectives predating Bohr's work relating line frequencies to energy differences. These adjectives refer to four typical series of lines observed in the simple alkali spectra, called the *sharp, principal, diffuse,* and *fundamental* series. The terminal levels of the transitions in these series were later called s, p, d, f. The balance of the letters continue the alphabet.

Table 42-1 The Periodic Table, with the Nonmetallic Elements in Light Type. The Maximum Numbers of Electrons That Can Be in Any Shell or Subshell Are Indicated by the Numbers in Parentheses[a]

Shells		Alkali metals	Alkali-earth metals	Rare earths	Transition elements						Halogens	Rare gases
$n=1$	K (2)	$_1$H										$_2$He
$n=2$	L (8)	$_3$Li	$_4$Be			$_5$B	$_6$C	$_7$N	$_8$O		$_9$F	$_{10}$Ne
$n=3$	M (18)	$_{11}$Na	$_{12}$Mg			$_{13}$Al	$_{14}$Si	$_{15}$P	$_{16}$S		$_{17}$Cl	$_{18}$Ar
$n=4$	N (32)	$_{19}$K	$_{20}$Ca		$_{21}$Sc—$_{30}$Zn	$_{31}$Ga	$_{32}$Ge	$_{33}$As	$_{34}$Se		$_{35}$Br	$_{36}$Kr
$n=5$	O (50)	$_{37}$Rb	$_{38}$Sr		$_{39}$Y—$_{48}$Cd	$_{49}$In	$_{50}$Sn	$_{51}$Sb	$_{52}$Te		$_{53}$I	$_{54}$Xe
$n=6$	P (72)	$_{55}$Cs	$_{56}$Ba	$_{57}$La—$_{70}$Yb	$_{71}$Lu—$_{80}$Hg	$_{81}$Tl	$_{82}$Pb	$_{83}$Bi	$_{84}$Po		$_{85}$At	$_{86}$Rn
$n=7$	Q (98)	$_{87}$Fr	$_{88}$Ra	$_{89}$Ac—$_{92}$U[b]								

Subshells: s(2) $l=0$; p(6) $l=1$; d(10) $l=2$; f(14) $l=3$; g(18) $l=4$; h(22) $l=5$; i(26) $l=6$

[a] A more complete table will be found in the Appendix, Sec. 4.
[b] If the man-made "transuranic" elements are included, this box would read $_{89}$Ac—$_{102}$Nl.

in the K-shell, since the s-subshell is fully occupied and no other subshells exist for $n=1$; with two electrons the K-shell is 'complete.' Thus, for lithium with $Z=3$, only two electrons can enter the K-shell and the third electron is characterized by $n=2$ and $l=0$ (2s). For beryllium with $Z=4$, two electrons are characterized by $n=1$, $l=0$, and two by $n=2$, $l=0$; this completely fills the 2s-subshell of the L-shell. For succeeding elements with Z in the range from 5 to 10, the 2p-subshell characterized by $n=2$ and $l=1$ is gradually filled as additional electrons are added. With neon ($Z=10$), the L-shell becomes complete.

The gradual filling of shells and subshells for elements with Z ranging from 1 to 30 is illustrated by the schematic diagram of Fig. 3, in which s-subshells are represented by dotted circles, p-subshells by solid circles, and d-subshells by dashed circles. The same type of information for the 92 naturally occurring elements is given in the periodic table shown in Table 42-1. In Fig. 3 and Table 42-1, each atom has all of the electrons of the preceding atoms of lower Z *plus* the additional electron that is indicated by its position in the table or figure. The subshells and shells are filled *cumulatively*.

Careful consideration of Fig. 3 and Table 42-1 reveals that the simple building-up procedure, involving the lowest permitted values of n and then l, begins to fail with potassium ($Z=19$). The simple procedure would

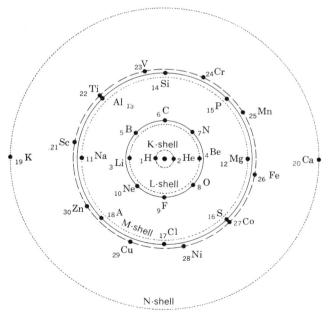

Fig. 3. Cumulative filling of the shells in the first thirty atoms, hydrogen through zinc. In this schematic diagram, s-subshells are represented by dotted circles, p-subshells by solid circles, and d-subshells by dashed circles.

have called for the continued filling of the M-shell by putting the nineteenth electron in the 3d-subshell; actually, the nineteenth electron of potassium is characterized by $n=4$, $l=0$ (4s). For calcium ($Z=20$) the twentieth electron is also characterized by $n=4$, $l=0$; thus, this 4s-subshell becomes completely filled. For elements in the range $Z=21$ to $Z=30$, the 4s-subshell remains complete, and the additional electrons go back into the $n=3$, $l=2$ (3d) subshell of the M-shell; this subshell and the M-shell itself become complete with zinc ($Z=30$).

The elements in the range $Z=21$ to $Z=30$ are called 'transition elements'; they all have two electrons in the outermost M-shell and differ from one another only in the number of electrons in the inner 3d-subshell. As indicated in Table 42–1, there are two other families of 'transition elements' with incomplete inner d-subshells ($l=2$), and two families of 'rare earths' with incomplete inner f-subshells ($l=3$).

The chemical properties of the elements are determined by the *valence* electrons, which are those in the outermost shell (largest n). From his knowledge of chemistry the reader will readily recognize the influence of shell structure on chemical properties. Atoms with completely filled shells—the rare gases—are very stable and are chemically almost inert. Atoms with a single outer s-electron—the alkali metals—can easily lose this electron and become positive ions with completely filled shells. A halogen atom can readily accept an additional electron and as a negative ion achieve completely filled shells. An alkali and a halogen readily combine in the manner illustrated schematically in Fig. 4 to form a molecule that is said to have 'ionic' binding. The chemical properties of the other elements in the periodic table can also be directly related to the electronic arrangements in their atoms. Since the determination of the electron shell structures is based on the quantum-mechanical interpretation of the *spectra* of the atoms, the applicability of these structures to the accurate interpretation of the chemical properties of the elements constitutes confirmation of their physical validity.

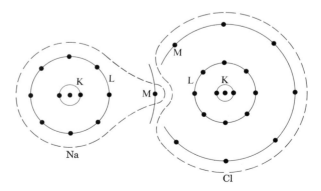

Fig. 4. The structure of the molecule of sodium chloride.

It might be mentioned that the electron arrangements shown in Table 42-1 gave the solution to a mystery that had baffled chemists for many years; this mystery involves the similar chemical properties of the elements in a single 'transition' or 'rare-earth' group. In any one of these groups, the atoms of all elements are similar in that their two valence electrons have the same quantum numbers, and the atoms of the different elements differ only in the number of electrons in partially filled *inner* shells, which do not play a dominant role in chemical behavior. However, as we shall see in the next section, the partially filled inner subshells of the atoms of the transition elements and the rare earths have an important bearing on the magnetic properties of these elements; these are the only strongly magnetic elements.

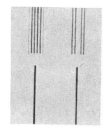

Fig. 5. Zeeman effect. The two yellow lines of sodium are shown in the absence (below) and presence (above) of a magnetic field.

In our earlier introduction of the quantum numbers n, l, m_l, and m_s, we stated that n was related in a general way to the quantum number Bohr introduced to express the characteristic energy levels of hydrogen and we indicated that l was a measure of orbital angular momentum. The spatial quantum numbers m_l and m_s are measures of the components of orbital and spin angular momentum in any reference direction in space; the spatial quantum numbers have no effect on the electron energy levels characterized by n and l unless there is a magnetic or electric field to establish a reference direction.

Spectroscopic evidence for spatial quantization can be obtained by observing the spectrum of an atom in a magnetic or an electric field. Spectral lines that are single lines in the absence of external fields split into several lines when external fields are applied. The 'magnetic splitting' of the yellow lines of sodium is shown in Fig. 5; this type of phenomenon is called the *Zeeman effect*—after its discoverer, PIETER ZEEMAN. The spatial quantum numbers are different in the initial and final states involved in the transitions that result in the different lines. The corresponding 'electric splitting' is shown in Fig. 6 for a line in the helium spectrum; this phenomenon is called the *Stark effect*—after its discoverer, JOHANNES STARK. The electron energies for different values of spatial quantum numbers are the same in the absence of an external magnetic or electric field but become different in the presence of a field.

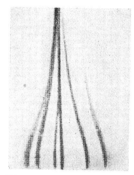

Fig. 6. Stark effect. The strength of the electric field increases from top to bottom of this photograph. [Figures 5 and 6 are reproduced from *Atomic Spectra and Atomic Structure* by G. Herzberg (Prentice-Hall).]

4. Magnetic Properties of Atoms

Now that we have discussed the structures of atoms, we can return to a discussion of paramagnetism and ferromagnetism. Our earlier treatment (Chapter 30) was necessarily incomplete, since these phenomena can be interpreted only on the basis of quantum mechanics.

As an aid in providing a model for use in visualizing atomic magnetic properties, we now use the classical value of the gyromagnetic ratio γ (p. 640) in conjunction with Bohr's model of the hydrogen atom. We recall that the orbital angular momentum of an electron in the first Bohr orbit is $h/2\pi$; the classical γ for an orbital electron is $-e/2m$, where m is the electronic mass. Substitution of these values in (15), p. 640, gives

$$M = -eh/4\pi m = \mu_B, \qquad \text{(Bohr magneton)} \quad (6)$$

where the magnetic moment μ_B is called the Bohr magneton and serves as a convenient unit for atomic magnetic moments. A schematic diagram for an electron in the first Bohr orbit is shown in Fig. 7; the magnetic-moment vector and the angular-momentum vector are oppositely directed because of the negative charge of the electron.

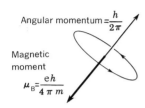

Fig. 7. Electron in first Bohr orbit.

Applying these ideas to the modern quantum mechanical model of the atom, in which the orbital angular momentum of an electron is $\sqrt{l(l+1)}\, h/2\pi$, we get electronic orbital magnetic moments $\sqrt{l(l+1)}\, \mu_B$, where l is the orbital quantum number. However, because of spatial quantization, the components of orbital angular momentum in a given direction such as that established by a magnetic field $\mathcal{B}$ are given by $m_l h/2\pi$ and the corresponding components of magnetic moment are $-m_l \mu_B$, where m_l is the spatial quantum number; as in Fig. 7, the angular-momentum and magnetic-moment components are oppositely directed because of the negative charge of the electron. We recall that, for an electron in a subshell characterized by a given value of l, m_l can take any integral value in the range $l, l-1, \cdots, 0, \cdots, -l+1, -l$ and can thus have any one of the allowed magnetic-moment components $-m_l \mu_B$ in the direction of an applied field $\mathcal{B}$. However, in any completely filled subshell in an atom, there is one electron for each allowed value of m_l and hence, so far as orbital motion is concerned, the resultant magnetic moment for the entire complete subshell is zero, since components characterized by positive values of m_l exactly cancel those characterized by negative values of m_l.

In addition to its orbital angular momentum, each electron also has an intrinsic spin angular momentum $\sqrt{s(s+1)}\, h/2\pi$, where $s = \frac{1}{2}$ for all electrons. The component of spin angular momentum in a given direction such as that of an applied field $\mathcal{B}$ is $m_s h/2\pi$, where $m_s = +\frac{1}{2}$ or $-\frac{1}{2}$. Associated with this angular-momentum component there is a magnetic-moment component of magnitude μ_B directed antiparallel or parallel to the applied field. The magnitude of this magnetic-moment component is actually found to be *twice* the one obtained by use of the classical value of γ. In a completely filled subshell in an atom, the number of electrons with $m_s = +\frac{1}{2}$ is exactly equal to the number of electrons with $m_s = -\frac{1}{2}$; hence, there is no resultant magnetic moment arising from electron spin in a complete subshell.

Since neither orbital motion nor intrinsic spin of electrons can give resultant magnetic moments in complete atomic subshells, we conclude that

Completely filled atomic subshells have no resultant magnetic moments and thus contribute nothing to paramagnetic or ferromagnetic effects.

Hence, for an isolated atom, any resultant magnetic moment must be attributed either to valence electrons or to electrons in incomplete inner shells.

We first consider the magnetic properties of valence electrons. Examination of the periodic table, p. 874, reveals that, with the exception of atoms of the rare gases and of elements like the alkali-earths with valence electrons in complete s-subshells, the atoms of all elements would be expected to have magnetic moments arising from the orbital and spin moments of valence electrons. This is indeed the case for isolated atoms. However, isolated atoms are seldom encountered, since the atoms of most elements either combine to form molecules in gases, or join other atoms in solids or liquids. When atoms combine to form molecules, their valence electrons are either shared between atoms in homopolar bonds or transferred from one atom to another to form ionic bonds. In either case, the valence electrons enter closed molecular or ionic subshells in which their magnetic effects cancel; a few gaseous molecules like O_2 and NO are exceptions and do have magnetic moments as a result of electron motions in molecular orbits, but such molecules are rare. Most non-metallic crystals can be classified as 'molecular' or 'ionic' and have no magnetic moments associated with the valence electrons of their constituent atoms. In the case of metallic solids or liquids, the valence electrons of the constituent atoms become 'conduction electrons,' which are relatively free to move through the metallic lattice but do not contribute to bulk magnetization.

Since complete atomic subshells and valence electrons do not give rise to atomic magnetic moments in solids and liquids, we must regard incomplete inner subshells as the source of the magnetic moments involved in ferromagnetism. If no other effects were involved, we should expect all transition elements and all rare-earth elements to have the high permeabilities characteristic of ferromagnetic materials. Actually, all these elements are *paramagnetic* but only iron, cobalt, nickel, and gadolinium are *ferromagnetic* at ordinary laboratory temperatures. The differences between weakly paramagnetic and strongly ferromagnetic materials are due to *interatomic* effects.

The most important clue to the explanation of ferromagnetism lies in the fact that one can make *permanent* magnets from ferromagnetic materials but not from paramagnetic materials. EWING, in 1890, first pointed out that in order for a permanent magnet to exist, there must be strong interatomic torques which keep the atoms oriented in the permanent magnet, since the thermal motions of the atoms are such as to tend to change any atomic arrangement from one of magnetic *order* to one of magnetic *disorder*. After formulation of the modern quantum mechanics, Heisenberg and others in the years following 1928 were able to compute the torques between the magnetic moments of adjacent atoms; the interatomic torques depend not only on the atoms involved but also on the spacing between the atoms. Calculations for the transition elements show that for Fe, Co, and Ni there are indeed strong forces that keep the magnetic moments of adjacent atoms lined up parallel to one another, whereas for the other transition elements the interatomic forces actually tend to prevent the magnetic moments from lining up. However, in certain alloys it is possible to achieve favorable interatomic distances and to cause manganese and chromium to have ferromagnetic properties; this occurs for manganese in MnBi and the so-called Heusler alloys.

The interatomic torques in ferromagnetic materials are so great that all atomic magnetic moments in whole blocks of atoms have a common orientation. These atomic blocks are of the order of 10^{-6} mm^3 in volume and are called *magnetic domains*. In an unmagnetized sample of ferromagnetic material, the magnetic moments of the various domains are randomly oriented like the 'atoms' shown in the upper part of Fig. 1, p. 630; in fact, each 'atom' in the schematic diagrams shown in this figure can quite properly be considered a domain.

When an external field is applied to an unmagnetized sample, magnetization is accomplished by two kinds of processes: (*a*) wall displacement and (*b*) domain flipping. Domain *wall displacement* is produced by a shift in the wall or boundary between one domain and the next; the domains having a 'favorable' orientation relative to the external field tend to grow while those with 'unfavorable' orientations diminish in size. *Domain flipping* is a *sudden* re-orientation of an entire domain to a new orientation having a larger magnetic-moment component in the direction of the applied field—toward the right in the simple model of Fig. 1, p. 630. If the fluxmeter coil in Fig. 5, p. 633, is attached to an audio amplifier instead of the meter, domain flipping gives rise to audible 'clicks' in a loudspeaker attached to the amplifier output; this phenomenon is called the *Barkhausen effect*.

In closing this discussion, we note that we have omitted discussion of magnetic properties of atomic nuclei. Some nuclei have spin angular momenta and associated magnetic moments (*see* Chapter 45). Nuclear magnetic moments are roughly one-thousandth of those associated with electrons and have little influence on bulk magnetism.

5. Applications of Quantum Mechanics

In this chapter we have attempted to present some of the ideas involved in the formulation of modern quantum mechanics and to show how quantum mechanics gives an understanding of the structure of atoms. We have also described the quantum-mechanical interpretation of the phenomena of paramagnetism and ferromagnetism. Quantum mechanics has also provided a quantitative understanding of chemical bonding, the structure of molecules, and the properties of solids. The electrical conductivity of metals and semiconductors can be interpreted in terms of quantum mechanics, as can the phenomenon of superconductivity. Indeed, if we wish to interpret any of the macroscopic properties of matter in terms of its constituent atoms, we must resort to quantum mechanics. It is interesting to note that classical Newtonian mechanics, which is valid for macroscopic bodies, fails to give valid results when applied to atoms, but that quantum mechanics not only gives valid results for atomic systems but also gives results in agreement with classical Newtonian mechanics when it is applied to macroscopic bodies!

PROBLEMS

1. What is the de Broglie wavelength of an electron having a kinetic energy of 100 eV? of 10 eV? of 1 eV? of 0.1 eV?
Ans: 0.123 nm, 0.317 nm, 1.23 nm, 3.17 nm.

2. What is the kinetic energy of an electron that has a de Broglie wavelength of 0.01 nm? 0.10 nm? 1.0 nm? 10 nm? 550 nm?

3. Show that the de Broglie wavelength of a particle is correctly given by the relation $\lambda = h/\sqrt{2 m E_K}$, where m and E_K are the particle's mass and kinetic energy respectively.

4. According to the example on p. 349, the root-mean-square speed of an H_2 molecule at 0° C is 1840 m/s. What is the magnitude of its linear momentum when it has this speed? What is its de Broglie wavelength?

5. What is the velocity of an electron that has a de Broglie wavelength of 0.1 nm? What is the kinetic energy of this electron?
Ans: 7.25×10^6 m/s; 2.39×10^{-17} J.

6. (a) What is the velocity of a helium atom that has a de Broglie wavelength of 0.1 nm? What is the kinetic energy of this helium atom? (b) What is the de Broglie wavelength of a helium atom moving at a speed equal to its root-mean-square speed at 300° K?

7. Compare the momenta, the energies, and the velocities of a *photon*, an *electron*, and a *proton* if they have the same wavelength λ.

8. Referring to Prob. 28, p. 865, show that the de Broglie wavelength of the electron in the first Bohr orbit is exactly equal to the circumference of the orbit.

9. If an electron in an atom is characterized by an orbital quantum number $l=2$, what is the magnitude of the orbital angular momentum of the electron? What is the magnitude of the maximum component of this electron's angular momentum parallel to an external magnetic field ⓑ?   Ans: $2.46\, h/2\pi$; $2\, h/2\pi$.

10. Solve Prob. 9 for an electron characterized by $l=3$.

11. What is the value of the Bohr magneton in SI units?   Ans: 9.284×10^{-24} A·m².

12. Using the nominal value $r=0.05$ nm as the radius of the first Bohr orbit, calculate the 'current' required in an orbit of area πr^2 to produce a magnetic moment of one Bohr magneton. Calculate the orbital frequency of the electron in the first Bohr orbit.

13. Using the results of Prob. 6, Chapter 30, calculate the potential energy of the magnetic moment M_l associated with the orbital motion of an electron in an atom when the atom is in a uniform magnetic field ⓑ. Express your answer in terms of the spatial quantum number m_l and other pertinent parameters.

14. In the Bohr model, the d-electrons of Fe have angular momentum $3\,(h/2\pi)$ and move in circular orbits at constant speed, but the radius of the orbit is difficult to compute because the electron experiences not only the attraction of the nucleus, but also the repulsion of all the other electrons. Assume that the orbital radius is R and show that the equivalent current I (the charge passing a point per second) is $3 eh/4\pi^2 mR^2$. Hence show that the *magnetic moment* is $3 (eh/4\pi m)$, independent of the value assumed for R.

15. The iron atom has six electrons (out of 10 possible) in the d-subshell of the M-shell that furnishes the ferromagnetic behavior. Each of these electrons has an *orbital* magnetic moment of 3 Bohr magnetons; it also has a *spin* magnetic moment of 1 Bohr magneton. However, the planes of the orbits are necessarily oriented in different directions in space, as are the directions of the electron spins, so the vector sum of these magnetic moments is just a small number of Bohr magnetons. We can get an approximation to the magnetic moment per atom of Fe by the following computations: (a) If iron has a density of 7870 kg/m³, how many atoms are there per cubic meter? (b) If each atom of iron in a bar magnet has a resultant magnetic moment of 1 Bohr magneton, and these magnetic moments are all oriented in the direction of the length of the magnet, what is the value of $\mathfrak{M}$, the magnetic moment per cubic meter? (c) What is the magnetic intensity in the center of a long bar magnetized as in (b)? (d) The saturation intensity (the maximum magnetic intensity that can arise from the oriented atomic moments of the material) of iron is observed to be 2.15 T. To what magnetic moment, in Bohr magnetons per atom, does this value correspond?
Ans: (a) 8.49×10^{28}; (b) 7.87×10^5 A/m; (c) 0.989 T; (d) 2.17 Bohr magnetons per atom.

16. (a) If an atom with a resultant electron spin $s = \frac{1}{2}$ is placed in a strong magnetic field $\mathfrak{B}$, there is one energy level associated with each of the two possible values of the spatial quantum number m_s. Show that the separation of these levels is

$$2 \mu_B \mathfrak{B}.$$

(b) If a sample of a material containing atoms of this type is placed in a strong external field $\mathfrak{B}_0$ and subjected to a small alternating field $\mathfrak{B}_1$ perpendicular to $\mathfrak{B}_0$, show that transitions between the levels occur when $f = 2 \mu_B \mathfrak{B}_0/h$, where f is the frequency of the alternating field. When this 'resonance' relationship between f and $\mathfrak{B}_0$ has been established, the sample absorbs energy from the alternating field. This phenomenon is known as *electron spin resonance* (ESR). Similar resonance conditions are observed for atoms or ions having net orbital as well as spin electronic angular momentum; absorption under these conditions is called *electron paramagnetic resonance* (EPR). Both ESR and EPR have proved useful not only in studies of solid state physics but also in chemical research; for readily attainable fields the frequency is typically in the microwave region of the spectrum.

17. Many nuclides have a characteristic spin angular momentum $\sqrt{I(I+1)} \, h/2\pi$, where I has an integral or half-integral value characteristic of each particular nuclide. Each nuclide also has a magnetic moment μ_N.

(a) Show that in the presence of an external field $\mathfrak{B}_0$ there are $2I+1$ energy levels characterized by

$$E = -\frac{m_I}{\sqrt{I(I+1)}} \mu_N \mathfrak{B}_0.$$

(b) If a selection rule $\Delta m_I = \pm 1$ is assumed, show that nuclear magnetic resonance (NMR) absorption analogous to ESR will occur when a small alternating magnetic field is applied at a frequency

$$f = \frac{\mu_N \mathfrak{B}_0}{\sqrt{I(I+1)} \, h}$$

Nuclear magnetic resonance techniques have proved useful in chemical research as well as in the determination of nuclear magnetic moments. It might be remarked that nuclear magnetic moments are roughly $\frac{1}{1000}$ the size of the Bohr magneton; no valid theoretical values for nuclear gyromagnetic ratios have yet been obtained. In typical laboratory experiments NMR frequencies are in the radio-frequency region.

18. In an electron microscope, electrons diffusely scattered from a surface are focused by electric and magnetic 'lenses' onto a fluorescent screen or photographic film, to form a greatly magnified picture of the surface. The resolving power is limited, just as in the case of a light microscope, by both the 'quality' of the 'lenses,' and by diffraction effects. Show that the limitation set by diffraction, for electrons of the wavelength mentioned on p. 867 is only about $\frac{1}{4000}$ of the limitation for a light microscope of the same equivalent focal length, and hence account for the extreme resolutions that have been obtained with electron microscopes.

ELECTROMAGNETIC RADIATION 43

In this chapter we present a quantitative treatment of the nature of the electromagnetic waves discussed briefly in Chapter 33. We first show how Clerk Maxwell generalized the principles of electromagnetism into four basic relations now known as Maxwell's equations. We then show how these equations can be expressed in differential form for 'free space' and how they form the basis of the classical electromagnetic theory of light; we show that electromagnetic waves have all the properties of light transmitted through free space and that the speed of light can be calculated from the value of μ_0, the permeability of free space, and the value of ε_0, the permittivity of free space.

1. Maxwell's Equations in Integral Form

We have found, in earlier chapters, four integral equations involving $\mathcal{E}$ and $\mathcal{B}$. Maxwell first explored systematically the conclusions that can be drawn from these four equations by logical analysis.

The *first* equation is Gauss's relation (p. 497), which states that the surface integral of electric flux through a closed surface equals Q/ε_0, where Q is the total charge within the surface. This equation can be written in the form

$$\oiint \boldsymbol{\mathcal{E}} \cdot d\mathbf{S} = Q/\varepsilon_0. \tag{M1}$$

The *second* equation [(23), p. 614] says that the corresponding surface integral of magnetic flux is always zero because there is no magnetic stuff analogous to electric charge; that is,

$$\oiint \boldsymbol{\mathcal{B}} \cdot d\mathbf{S} = 0. \tag{M2}$$

The *third* equation is the law of electromagnetic induction stated in (11), p. 655 in the form

$$\oint \boldsymbol{\mathcal{E}} \cdot d\mathbf{l} = -\frac{d\Phi}{dt}, \tag{1}$$

where Φ is the magnetic flux linking the closed curve around which the line integral is taken. By using equation (1), p. 496, which expresses the flux through a closed curve fixed in space as an integral over any open surface S bounded by that curve, we can write

$$\frac{d\Phi}{dt} = \frac{d}{dt} \iint_S \boldsymbol{\mathcal{B}} \cdot d\mathbf{S} = \iint_S \frac{\partial \boldsymbol{\mathcal{B}}}{\partial t} \cdot d\mathbf{S}, \tag{2}$$

where we have used the sign for partial differentiation since $\boldsymbol{\mathcal{B}}$ is a function of X, Y, and Z, as well as of t. Putting (2) in (1) gives Maxwell's third equation:

$$\oint \boldsymbol{\mathcal{E}} \cdot d\mathbf{l} = -\iint_S \frac{\partial \boldsymbol{\mathcal{B}}}{\partial t} \cdot d\mathbf{S}, \tag{M3}$$

where the line integral on the left is taken around any closed curve and the surface integral on the right is taken over any open surface bounded by this same closed curve.

The *fourth* equation is based on Ampère's line-integral relation (p. 608), which states that the line integral of $\boldsymbol{\mathcal{B}}$ around a closed curve equals μ_0 times the current I linking the closed curve:

$$\oint \boldsymbol{\mathcal{B}} \cdot d\mathbf{l} = \mu_0 I. \tag{3}$$

In the general case, where the current may not be in wires, but may be passing through an electrolyte or a gaseous plasma, we can express the right side of (3) in terms of a vector $\boldsymbol{j}$ that represents current density (amperes per square meter), as in Fig. 1. The current through the indicated surface S bounded by the closed curve is now the flux of the current-density vector and can be expressed as a surface integral over the surface S, so that (3) becomes

$$\oint \mathcal{B} \cdot d\mathbf{l} = \mu_0 \iint_S \mathbf{j} \cdot d\mathbf{S}. \qquad (4)$$

It was Maxwell who first realized that *equation (4) is analytically unsatisfactory*. An equation expressing equality of an integral around a closed curve and an integral over a surface bounded by that curve can only be valid provided the value of the surface integral is independent of the particular surface chosen. Thus, as we stated on p. 614, flux through a closed curve can be satisfactorily computed as a surface integral of $\mathcal{B}$ because the flux through a magnetic tube is constant along the tube. Hence, the equation (M3) is analytically satisfactory. But the current through a current tube is not necessarily constant; charge may be accumulating at some location, for example, on the plates of a capacitor.

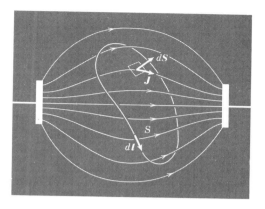

Fig. 1. Current paths in an electrolyte (schematic), and an imaginary closed curve.

Let us consider equation (4) in connection with the alternating-current circuit of Fig. 2. We are interested in the line integral of $\mathcal{B}$ around the closed curve L. If we evaluate the right side of (4) for the surface S_1, we obtain just $\mu_0 I$, since I is the total current through S. On the other hand, we can draw a surface S_2, having the same boundary L, through which there is no current because the surface passes between the plates of the capacitor. Evaluated for this surface, the right side of (4) gives zero. Maxwell noted, however, that while no current crossed S_2, a varying electric field did cross S_2, and suspected that an additional term was needed on the right side of (4), containing a surface integral of $\partial \mathcal{E}/\partial t$, that would give this side of the equation the same value whether S_1 or S_2 is employed. The following argument develops a term suitable for this purpose:

Write Gauss's relation (M1) for the *closed* surface $S_1 + S_2$ (the 'jar' and its 'cover' in Fig. 2, to obtain $\oiint_{S_1+S_2} \mathcal{E} \cdot d\mathbf{S} = Q/\varepsilon_0$, where Q, the charge on the capacitor plate, is the only charge within the closed surface. Next, note that since there is no electric field through S_1, but only through S_2, the integral need only be computed for the open surface S_2; thus $\iint_{S_2} \mathcal{E} \cdot d\mathbf{S} = Q/\varepsilon_0$. Now differentiate with respect to time:

$$\iint_{S_2} \frac{\partial \mathcal{E}}{\partial t} \cdot d\mathbf{S} = \frac{1}{\varepsilon_0} \frac{dQ}{dt} = \frac{I}{\varepsilon_0}. \qquad (5)$$

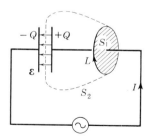

Fig. 2. The surface S_2 is shaped like a narrow-mouthed jar whose rim is L and whose base lies between the plates of the capacitor.

We wanted to add an integral involving $\partial \mathcal{E}/\partial t$ to the right of (4) that would have the same value over S_2 as the integral of $\mu_0 \mathbf{j}$ over S_1, that is, the value $\mu_0 I$. We see from (5) that $\varepsilon_0 \mu_0 \iint_{S_2} (\partial \mathcal{E}/\partial t) \cdot d\mathbf{S} = \mu_0 I$. Therefore, if we rewrite (4) in the form

$$\oint \mathcal{B} \cdot dl = \mu_0 \iint_S \boldsymbol{j} \cdot d\mathbf{S} + \varepsilon_0 \mu_0 \iint_S \frac{\partial \mathcal{E}}{\partial t} \cdot d\mathbf{S}, \qquad \textbf{(M4)}$$

the right side will have the same value whether we use the surface S_1 (where the whole contribution comes from the first integral) or the surface S_2 (where the whole contribution comes from the second). This is the fourth of Maxwell's equations, written in integral form. For our discussion of electromagnetic waves, we shall be interested only in *free space*, where $\boldsymbol{j}=0$ and (M4) takes the form

$$\oint \mathcal{B} \cdot dl = \varepsilon_0 \mu_0 \iint_S \frac{\partial \mathcal{E}}{\partial t} \cdot d\mathbf{S}. \qquad (6)$$

This equation has a considerable resemblance to (M3), with $\mathcal{B}$ and $\mathcal{E}$ interchanged. Equation (M3) shows that a changing magnetic field is accompanied by an electric field; (M4) similarly shows that a changing electric field is accompanied by a magnetic field. Between the poles of a magnet there is an electric field if the magnetic field is varying (*see* pp. 656-659); *there is a similar magnetic field between the plates of a capacitor if the electric field is varying.*

There is no rigorous *derivation* of Maxwell's equations. Like Newton's principles, they represent generalizations from experiment whose validity is established only insofar as logical conclusions drawn from them are consistent with reality. In the case of Maxwell's equations, all such conclusions have proved to be valid *except* in the cases already noted where quantum effects enter the description of emission and absorption of electromagnetic radiation.

For convenience of reference we repeat the four Maxwell equations in the form valid *for free space*, where there are no charges or currents:

$$\oiint \mathcal{E} \cdot d\mathbf{S} = 0, \qquad \textbf{(M1')} \qquad \oint \mathcal{E} \cdot dl = - \iint_S \frac{\partial \mathcal{B}}{\partial t} \cdot d\mathbf{S}, \qquad \textbf{(M3')}$$

$$\oiint \mathcal{B} \cdot d\mathbf{S} = 0, \qquad \textbf{(M2')} \qquad \oint \mathcal{B} \cdot dl = \varepsilon_0 \mu_0 \iint_S \frac{\partial \mathcal{E}}{\partial t} \cdot d\mathbf{S}. \qquad \textbf{(M4')}$$

It is the terms involving rates of change of field intensities in Maxwell's equations that give the departures from the 'static' relations that were mentioned in Sec. 3, Chapter 33, and that result in energy loss by radiation. For example, (M4') shows that the varying electric field between the plates of a capacitor 'induces' a varying magnetic field. But by (M3'), this varying magnetic field in turn induces another varying electric field, so the electric field between the capacitor plates is no longer the same as would be computed from the charges by the static relations. For capacitors with small dimensions, the difference turns out to be negligible, unless the frequency (rate) of variation is extremely high, because $\varepsilon_0 \mu_0$ in (M4') has the extremely small value of about 10^{-17} s^2/m^2.

2. Maxwell's Equations for Free Space in Differential Form

For many purposes, including our discussion of electromagnetic waves, Maxwell's equations are more useful when expressed in terms of the rectangular components $\mathcal{E}_X$, $\mathcal{E}_Y$, $\mathcal{E}_Z$, $\mathcal{B}_X$, $\mathcal{B}_Y$, $\mathcal{B}_Z$ and the derivatives of these components. The method that we shall employ for transformation of the integral relations (M1')–(M4'), valid in free space, into differential equations is typical of the derivation of most of the partial differential equations of mathematical physics.

Let us start with (M1') and compute the surface integral of $\boldsymbol{\mathcal{E}} \cdot d\mathbf{S}$ over the closed surface of a very small cube centered at the point X, Y, Z, and of side 2δ, as in Fig. 3.

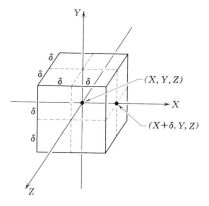

Figure 3

On the right face of this cube, the outward normal points in the positive X-direction, so the contribution of this face to the surface integral of $\boldsymbol{\mathcal{E}} \cdot d\mathbf{S}$ is the area of the face, $4\delta^2$, multiplied by the average value of the X-component, $\mathcal{E}_X$, on this face. For very small δ, it is adequate to take the value of $\mathcal{E}_X$ at the center of the face, that is, $\mathcal{E}_X(X+\delta, Y, Z)$, to represent the average value. From Taylor's theorem we can write, for the case of small δ, $\mathcal{E}_X(X+\delta,Y,Z) = \mathcal{E}(X,Y,Z) + (\partial \mathcal{E}_X/\partial X)\delta$. Hence the contribution of the right face to the integral is

$$4\delta^2 [\mathcal{E}_X(X,Y,Z) + (\partial \mathcal{E}_X/\partial X)\delta]. \tag{7}$$

On the left face of the cube, the outward normal points in the $-X$-direction, so we must multiply the area by the value of $-\mathcal{E}_X$ at the center, that is, by $-\mathcal{E}_X(X-\delta, Y, Z) = -[\mathcal{E}_X(X,Y,Z) - (\partial \mathcal{E}_X/\partial X)\delta]$. This procedure gives, as the contribution of the left face,

$$4\delta^2 [-\mathcal{E}_X(X,Y,Z) + (\partial \mathcal{E}_X/\partial X)\delta]. \tag{8}$$

The sum of (7) and (8) is

$$8\delta^3 \, \partial \mathcal{E}_X / \partial X. \tag{9}$$

In the same way we find that the contributions of the top and bottom faces total $8\delta^3 (\partial \mathcal{E}_Y/\partial Y)$, and the contributions of the front and back faces total $8\delta^3 (\partial \mathcal{E}_Z/\partial Z)$. Therefore, for all six faces constituting the closed surface of the small cube of Fig. 3, we have

$$\oiint \boldsymbol{\mathcal{E}} \cdot d\mathbf{S} = 8\delta^3 \left(\frac{\partial \mathcal{E}_X}{\partial X} + \frac{\partial \mathcal{E}_Y}{\partial Y} + \frac{\partial \mathcal{E}_Z}{\partial Z} \right) = 0, \tag{10}$$

where the integral is set equal to zero in accordance with (M1'). The center expression in (10) is the product of the volume of the cube, $8\delta^3$, times a function of position known as the divergence of $\mathcal{E}$, written div$\mathcal{E}$. Since the volume, although it may be infinitesimal, is nonvanishing, the divergence of $\mathcal{E}$ must vanish at all points in free space:

$$\text{div}\mathcal{E} \equiv \frac{\partial \mathcal{E}_X}{\partial X} + \frac{\partial \mathcal{E}_Y}{\partial Y} + \frac{\partial \mathcal{E}_Z}{\partial Z} = 0. \tag{M1''}$$

The above relation is Maxwell's first equation in differential form.

From the similarity of (M1′) and (M2′) we see immediately that Maxwell's second equation takes the form

$$\text{div}\mathcal{B} \equiv \frac{\partial \mathcal{B}_X}{\partial X} + \frac{\partial \mathcal{B}_Y}{\partial Y} + \frac{\partial \mathcal{B}_Z}{\partial Z} = 0. \tag{M2''}$$

Now consider equation (M3′),

$$\oint \mathcal{E} \cdot d\mathbf{l} = -\iint_S \frac{\partial \mathcal{B}}{\partial t} \cdot d\mathbf{S}, \tag{11}$$

for the case in which the closed circuit is a small square of side 2δ, lying in the XY-plane, and centered at the point (X,Y,Z), while the surface over which the integral on the right of (11) is taken is the plane square lying in the XY-plane. For the counterclockwise direction of traversing the circuit indicated in Fig. 4, the surface element $d\mathbf{S}$ is pointing out of the paper, in the $+Z$-direction for a right-hand coordinate system of the type shown in Fig. 3. (A right-hand coordinate system is one in which a right-hand screw advances along $+Z$ if the head is turned from $+X$ to $+Y$; such a system is universally employed in mathematical physics.) Hence, on the right of (11), we need consider only the Z-component, $\partial \mathcal{B}_Z/\partial t$. The value of the integral on the right is obtained, in the limit of very small δ, by multiplying the value of $\partial \mathcal{B}_Z/\partial t$ at the center of the square by the area, $4\delta^2$:

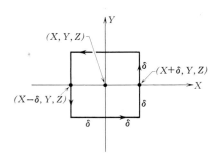

Figure 4

$$-\iint_S \frac{\partial \mathcal{B}}{\partial t} \cdot d\mathbf{S} = -4\delta^2 \frac{\partial \mathcal{B}_Z}{\partial t}. \tag{12}$$

Now consider the line integral on the left of (11) for the circuit of Fig. 4. For the right side of the square, $d\mathbf{l}$ is in the $+Y$-direction and has magnitude 2δ. Along this side, $\mathcal{E} \cdot d\mathbf{l}$ is the average value of $\mathcal{E}_Y$ times 2δ. For small δ, it is adequate to represent the average value of $\mathcal{E}_Y$ by its value, $\mathcal{E}_Y(X+\delta,Y,Z)$ at the center of the side. Therefore, the contribution of the right side to the line integral is

$$(2\,\delta)\,\mathcal{E}_Y(X+\delta,Y,Z) = (2\,\delta)\,[\mathcal{E}_Y(X,Y,Z) + (\partial \mathcal{E}_Y/\partial X)\,\delta]. \quad (13)$$

For the left side of the square, $d\mathbf{l}$ is in the $-Y$ direction, so the dot product is

$$-(2\,\delta)\,\mathcal{E}_Y(X-\delta,Y,Z) = -(2\,\delta)\,[\mathcal{E}_Y(X,Y,Z) - (\partial \mathcal{E}_Y/\partial X)\,\delta]. \quad (14)$$

The sum of (13) and (14) gives the contribution of the right and left sides as

$$4\,\delta^2\,\partial \mathcal{E}_Y/\partial X. \quad (15)$$

By a similar argument, we find that the contribution of the top and bottom sides of the circuit is

$$-4\,\delta^2\,\partial \mathcal{E}_X/\partial Y. \quad (16)$$

Equating the sum of (15) and (16) to (12), we find that (11) becomes, for the circuit of Fig. 4,

$$\frac{\partial \mathcal{E}_Y}{\partial X} - \frac{\partial \mathcal{E}_X}{\partial Y} = -\frac{\partial \mathcal{B}_Z}{\partial t}. \quad (17)$$

Equation (17) was derived for a circuit lying in the XY-plane and relates space derivatives of the X- and Y-components of $\mathcal{E}$ to the time rate of change of $\mathcal{B}_Z$. There are two similar relations obtained by using circuits in the YZ-plane and the ZX-plane. These relations are readily obtained from (17) by cyclical substitution, $X \to Y$, $Y \to Z$, $Z \to X$, a substitution that is always legitimate for a right-hand coordinate system. In this manner we obtain three partial differential equations, which together express Maxwell's third equation (M3'):

$$\left. \begin{array}{l} \dfrac{\partial \mathcal{E}_Z}{\partial Y} - \dfrac{\partial \mathcal{E}_Y}{\partial Z} = -\dfrac{\partial \mathcal{B}_X}{\partial t}, \\[6pt] \dfrac{\partial \mathcal{E}_X}{\partial Z} - \dfrac{\partial \mathcal{E}_Z}{\partial X} = -\dfrac{\partial \mathcal{B}_Y}{\partial t}, \\[6pt] \dfrac{\partial \mathcal{E}_Y}{\partial X} - \dfrac{\partial \mathcal{E}_X}{\partial Y} = -\dfrac{\partial \mathcal{B}_Z}{\partial t}. \end{array} \right\} \quad \text{(M3'')}$$

The right sides of these three equations constitute the three components of the vector $-\partial \mathcal{B}/\partial t$; hence, the left sides must also be the three components of a vector. The vector whose X-, Y-, and Z-components are the three expressions on the left in (M3'') is called the curl of $\mathcal{E}$, written curl$\mathcal{E}$. Hence, with this shorthand notation, the three equations can be written as one vector equation: curl$\mathcal{E} = -\partial \mathcal{B}/\partial t$.

Finally, turning to (M4'), we see by direct analogy with (M3') that the differential form is curl$\mathcal{B} = \varepsilon_0 \mu_0\, \partial \mathcal{E}/\partial t$, or, written out,

$$\frac{\partial \mathcal{B}_Z}{\partial Y} - \frac{\partial \mathcal{B}_Y}{\partial Z} = \varepsilon_0 \mu_0 \frac{\partial \mathcal{E}_X}{\partial t},$$

$$\frac{\partial \mathcal{B}_X}{\partial Z} - \frac{\partial \mathcal{B}_Z}{\partial X} = \varepsilon_0 \mu_0 \frac{\partial \mathcal{E}_Y}{\partial t}, \quad \text{(M4″)}$$

$$\frac{\partial \mathcal{B}_Y}{\partial X} - \frac{\partial \mathcal{B}_X}{\partial Y} = \varepsilon_0 \mu_0 \frac{\partial \mathcal{E}_Z}{\partial t}.$$

3. Character and Speed of a Plane Electromagnetic Wave

By a plane wave, we mean one whose wavefronts are plane. We shall consider a plane wave traveling in the X-direction. Then the physical quantities are the same at a given instant everywhere on an entire YZ-plane, which is perpendicular to the X-axis, and vary only with X and with time (cf. Fig. 5). *Thus, for a plane wave traveling in the X-direction, $\mathcal{E}_X$, $\mathcal{E}_Y$, $\mathcal{E}_Z$, $\mathcal{B}_X$, $\mathcal{B}_Y$, $\mathcal{B}_Z$ are functions of X and t only, and partial derivatives of these quantities with respect to Y and Z vanish.*

We now collect the eight equations (M1″)–(M4″), leaving out all the vanishing terms involving $\partial/\partial Y$ and $\partial/\partial Z$, and rearranging the order for convenience:

$$\partial \mathcal{E}_X/\partial X = 0, \quad \partial \mathcal{B}_X/\partial X = 0; \quad (18)$$

$$\partial \mathcal{E}_X/\partial t = 0, \quad \partial \mathcal{B}_X/\partial t = 0; \quad (19)$$

$$\frac{\partial \mathcal{E}_Z}{\partial X} = \frac{\partial \mathcal{B}_Y}{\partial t}, \quad \frac{\partial \mathcal{E}_Z}{\partial t} = \frac{1}{\varepsilon_0 \mu_0} \frac{\partial \mathcal{B}_Y}{\partial X}; \quad (20)$$

$$\frac{\partial \mathcal{E}_Y}{\partial X} = -\frac{\partial \mathcal{B}_Z}{\partial t}, \quad \frac{\partial \mathcal{E}_Y}{\partial t} = -\frac{1}{\varepsilon_0 \mu_0} \frac{\partial \mathcal{B}_Z}{\partial X}. \quad (21)$$

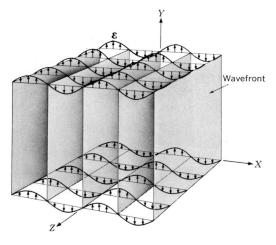

Fig. 5. A plane wave propagated in the X-direction and polarized with the electric vector in the Y-direction. The component $\mathcal{E}_Y$ is independent of Y or Z. The magnetic vector is not indicated on this plot, but is related to $\mathcal{E}$ as in Fig. 5, Chapter 33.

From these equations, we shall proceed to demonstrate that:

1. $\mathcal{E}_X$ and $\mathcal{B}_X$ are constant in space and time; any *wave motion is transverse.*
2. If the wave is plane-polarized with $\mathcal{E}$ in the $\pm Y$-direction, then $\mathcal{B}$ is in the $\pm Z$-direction, as indicated in Fig. 5, Chapter 33.
3. The wave speed is given by $c = 1/\sqrt{\varepsilon_0 \mu_0}$, as already stated in (29), p. 621.
4. The $\mathcal{E}$ and $\mathcal{B}$ waves have the phase relations shown in Fig. 5, Chapter 33, with a constant amplitude ratio.
5. Half the energy is electric, half magnetic, with the total power per unit area proportional to the squares of the electric and magnetic amplitudes.

1. The Wave Is Transverse

From (18) and (19) we see that $\mathcal{E}_X$ and $\mathcal{B}_X$ vary neither with X nor with t; hence, $\mathcal{E}_X$ and $\mathcal{B}_X$ are constant in space and time. There may be any amount of *static* electric or magnetic field in the X-direction in the region in which our wave is traveling, but we are not here concerned with static fields. So far as waves are concerned, the *average* value of the physical quantity that is oscillating must be zero, so we set $\mathcal{E}_X = \mathcal{B}_X = 0$. Only the *transverse* components $\mathcal{E}_Y$, $\mathcal{E}_Z$, $\mathcal{B}_Y$, and $\mathcal{B}_Z$ can contribute to the wave motion.

2. Plane of Polarization

If the wave is plane-polarized with $\mathcal{E}$ in the Y-direction as in Fig. 5, Chapter 33, then $\mathcal{E}_Z = 0$. Setting $\mathcal{E}_Z = 0$ in (20) gives $\partial \mathcal{B}_Y/\partial t = 0$, $\partial \mathcal{B}_Y/\partial X = 0$. Hence, by the same argument as in the preceding paragraph, we must set $\mathcal{B}_Y = 0$. If $\mathcal{E}$ *is in the Y-direction*, $\mathcal{B}$ *is in the Z-direction*.

3. Wave Speed

For the plane-polarized wave described in the preceding paragraph, only two of the eight equations (18)–(21) retain nonvanishing terms—the two equations (21). Differentiation of the first equation of (21) with respect to X and of the second with respect to t gives

$$\frac{\partial^2 \mathcal{E}_Y}{\partial X^2} = -\frac{\partial^2 \mathcal{B}_Z}{\partial X \, \partial t}, \qquad \frac{\partial^2 \mathcal{E}_Y}{\partial t^2} = -\frac{1}{\varepsilon_0 \mu_0} \frac{\partial^2 \mathcal{B}_Z}{\partial X \, \partial t},$$

from which we find that

$$\frac{\partial^2 \mathcal{E}_Y}{\partial X^2} = \varepsilon_0 \mu_0 \frac{\partial^2 \mathcal{E}_Y}{\partial t^2}. \tag{22}$$

This is the familiar *wave equation*. Comparison with (11), p. 420 shows that the wave speed is $c = 1/\sqrt{\varepsilon_0 \mu_0}$. By differentiating the first equation of (21) with respect to t and the second with respect to X, we find that $\mathcal{B}_Z$ satisfies the *same* wave equation (22).

4. Phase and Amplitude Relations

To get further information on phase and amplitude relations between $\mathcal{B}_Z$ and $\mathcal{E}_Y$, we can best specialize to a wave of definite frequency and amplitude, traveling in the $+X$-direction. Let us assume that

$$\mathcal{E}_Y = \mathcal{E}_M \sin[2\pi f(t - X/c)], \tag{23}$$

which, as we know from the discussion of p. 417 is a wave traveling to the right with frequency f, speed c, and amplitude $\mathcal{E}_M$. Then, from the two equations of (21), we can compute

$$\frac{\partial \mathcal{B}_Z}{\partial t} = -\frac{\partial \mathcal{E}_Y}{\partial X} = \frac{2\pi f}{c} \mathcal{E}_M \cos[2\pi f(t - X/c)], \tag{24}$$

and
$$\frac{\partial \mathcal{B}_Z}{\partial X} = -\frac{1}{c^2}\frac{\partial \mathcal{E}_Y}{\partial t} = -\frac{2\pi f}{c^2}\mathcal{E}_M \cos[2\pi f(t - X/c)]. \quad (25)$$

From these two relations we see that $\mathcal{B}_Z$ must also be of the form

$$\mathcal{B}_Z = \mathcal{B}_M \sin[2\pi f(t - X/c)] \quad (26)$$

in order for its partial derivatives with respect to t and X to come out as cosines. To determine the relation between the amplitudes $\mathcal{B}_M$ and $\mathcal{E}_M$, we compute from (26) that $\partial \mathcal{B}_Z/\partial t = 2\pi f \mathcal{B}_M \cos[2\pi f(t - X/c)]$; comparison of this equation with (24) shows that

$$\mathcal{B}_M = \mathcal{E}_M/c, \qquad \mathcal{E}_M = c\,\mathcal{B}_M. \qquad \left\{\begin{array}{l}\mathcal{E}_M \text{ in V/m} \\ \mathcal{B}_M \text{ in T} \\ c \text{ in m/s}\end{array}\right\} \quad (27)$$

From (23), (26), and (27) we conclude that *the $\mathcal{E}_Y$ and the $\mathcal{B}_Z$ waves are in phase* and that *the amplitude of the $\mathcal{E}_Y$ wave is c times the amplitude of the $\mathcal{B}_Z$ wave.*

5. Energy Considerations

In the case of an electromagnetic wave, it can be shown that the power transmitted across a unit area perpendicular to the direction of propagation is the energy per unit volume times the wave speed c, just as in the case of sound waves (*see* p. 449).

The energy per unit volume is given by the same relations that we derived earlier in static situations,

$$\text{electric energy per unit volume} = \tfrac{1}{2}\,\varepsilon_0\,\mathcal{E}^2 \quad (28)$$

(*see* Prob. 31, p. 527), and

$$\text{magnetic energy per unit volume} = \tfrac{1}{2}\,\mathcal{B}^2/\mu_0, \quad (29)$$

as in (24), p. 666. For an electric field varying sinusoidally as in (23), the average value of $\mathcal{E}^2$ is one-half the square of the amplitude, so

$$\text{electric energy per unit volume} = \tfrac{1}{4}\,\varepsilon_0\,\mathcal{E}_M^2, \quad (30)$$

and therefore

$$\text{electric power per unit area} = \tfrac{1}{4}\,c\,\varepsilon_0\,\mathcal{E}_M^2. \quad (31)$$

Similarly,

$$\text{magnetic energy per unit volume} = \tfrac{1}{4}\,\mathcal{B}_M^2/\mu_0 \quad (32)$$

and

$$\text{magnetic power per unit area} = \tfrac{1}{4}\,c\,\mathcal{B}_M^2/\mu_0. \quad (33)$$

We can show that the magnetic power and the electric power are equal. To verify this statement, first introduce into (33) the amplitude ratio (27) and then the relation $c\varepsilon_0 = 1/c\mu_0$, to obtain $\tfrac{1}{4}\,c\,\mathcal{B}_M^2/\mu_0 = \tfrac{1}{4}\,\mathcal{E}_M^2/c\mu_0 = \tfrac{1}{4}\,c\,\varepsilon_0\,\mathcal{E}_M^2$. Hence

$$\text{total power per unit area} = \tfrac{1}{2}\,c\,\varepsilon_0\,\mathcal{E}_M^2, \quad (34)$$

where $\mathcal{E}_M$ is the amplitude of the electric oscillation.

Example. *A typical value of electric-intensity amplitude for good reception in long-distance, short-wave radio transmission is* $\mathcal{E}_M = 100\ \mu V/m$ *at the receiving antenna. At this amplitude, what power in* W/m^2 *is passing the antenna?*

The power is obtained directly by substitution in (34) as

$$\tfrac{1}{2} c \varepsilon_0 \mathcal{E}_M^2 = \tfrac{1}{2} (3.00 \times 10^8\ m/s)(8.85 \times 10^{-12}\ F/m)(10^{-4}\ V/m)^2$$
$$= 1.33 \times 10^{-13}\ F \cdot V^2/s \cdot m^2 = 1.33 \times 10^{-13}\ W/m^2.$$

It is the ability of electronic circuits to amplify signals representing such very low power densities that accounts for the practically world-wide reception obtained under favorable atmospheric conditions even at rather modest total radiated power from the transmitter, for example, 10 to 100 kW.

PROBLEMS

1. Verify the correctness of (5) by using the parallel-plate capacitor equation to compute $\mathcal{E}$.

2. The figure represents a pair of large circular capacitor plates, closely spaced, so that the parallel-plate capacitor equation can be used to a good approximation to compute $\mathcal{E}$ at all radii out to R. From (6), show that, to this approximation, the magnetic field $\mathcal{B}$ at radius r in a plane between the plates has the magnitude

$$\mathcal{B} = \frac{\mu_0}{2\pi} \frac{Ir}{R^2} \quad \text{for} \quad r \leqq R,$$

$$\mathcal{B} = \frac{\mu_0}{2\pi} \frac{I}{r} \quad \text{for} \quad r \geqq R.$$

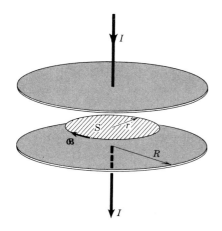

Problems 2-4

Note that the second equation is that for the field of an infinite straight wire.

3. From (6) show that, for the capacitor discussed in Prob. 2, the magnetic intensity can be written as $\mathcal{B} = \varepsilon_0 \mu_0 (r/d)\, \partial V/\partial t$, for $r \leqq R$. Here d is the distance between the plates.

4. Calculate the value of $\mathcal{B}$ at a distance of 10 cm from the center of a large circular capacitor, consisting of parallel plates 1 mm apart in air, when the potential difference between the plates varies sinusoidally at a frequency of (*a*) 10 Hz, (*b*) 10 kHz, (*c*) 1 MHz, and (*d*) 1 GHz.

5. In a region containing charge distributed with volume density ρ, show that Maxwell's first equation (M1) becomes, in differential form, $\text{div}\,\mathcal{E} = \rho/\varepsilon_0$.

6. (*a*) Using the exact value $\mu_0 = 4\pi \times 10^{-7}$ T·m/A given in (2), p. 605, and the experimental value $1/4\pi\varepsilon_0 = 8.9872 \times 10^9$ N·m²/C given in (31), p. 622, compute the velocity of light and show that the relation between c, μ_0, and ε_0 is dimensionally correct. (*b*) Recalling that the speed of light v in a given medium is given by $v = c/\mu$, where μ is the refractive index of the medium, and that the force between charges q_1 and q_2 immersed in a fluid dielectric is given by $F = (1/4\pi\varepsilon_0 K)(q_1 q_2/r^2)$, suggest a possible relation between the refractive index and the dielectric constant of a transparent medium. Check your hypothesis by comparison of tabulated values of μ and K.

7. The magnetic intensity between the pole faces of a cyclotron magnet is 1.5 T. What is the electric intensity between the plates of a vacuum capacitor if the energy per unit volume between the plates is the same as the magnetic energy per unit volume between the pole faces in the cyclotron?

8. What is the total power per square meter of wavefront associated with a plane radio wave that has an electric amplitude of 10 mV/m? Calculate the amplitude of the associated magnetic wave and use the value obtained to check your computation of transmitted power.

9. By means of a device called an '$\mathcal{E}$-meter,' it is found that the amplitude of the electric field associated with a plane radio wave is 5 mV/m. What is the amplitude of the associated magnetic field? Ans: 1.67×10^{-11} T.

10. Show that for an isotropic radiation field having energy density E_λ in a 1-nm wavelength range, the power per unit area impinging on a surface is $P_\lambda = \frac{1}{4} c E_\lambda$. The following paragraph gives a suggested procedure for this proof.

Let N_λ be the number of quanta per unit volume represented by the energy density E_λ. Then the argument proceeds very much like that on p. 348 for impingement of gas molecules. Consider unit area of the surface and those quanta whose direction makes angle θ with the normal to the surface, as in the figure. Of these quanta, those in a layer of thickness

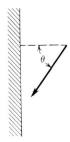

Problem 10

$c \cos\theta \, dt$ will impinge on the surface in time dt. Show that, for isotropic conditions, the number of quanta per unit volume that are heading toward the surface at angles between θ and $\theta + d\theta$ is $\frac{1}{2} N_\lambda \sin\theta \, d\theta$. Determine the number of *these* quanta hitting the surface in time dt and integrate this expression from $\theta = 0$ to $\frac{1}{2}\pi$ to determine the total number of quanta hitting the surface in time dt. From this determine P_λ in terms of E_λ.

11. Show that the total radiation density in any enclosure at thermal equilibrium is given by $E = 4 P_{\text{Black}}/c$, where P_{Black} is the power radiated from unit area of a black body as given by the Stefan-Boltzmann law (p. 325). Show that this density is the same for an enclosure with highly reflecting walls as for one with highly absorbing walls by demonstrating that, if this were not so, the principle of thermal equilibrium would be violated if the interiors of the two enclosures were joined by means of a small hole.

EINSTEIN'S THEORY OF RELATIVITY

44

Maxwell's electromagnetic theory gives a universal value of the velocity c of light in free space without making reference to the frame of the observer. It is thus in apparent contradiction with Newtonian kinematics. However, there is sound experimental evidence that the speed of light *is* the same for all observers. We first discuss some of this evidence.

Einstein formulated the so-called special theory of relativity in 1905. He concluded that the laws of electromagnetic theory were accurate but that Newtonian kinematics and dynamics needed basic revisions. These revisions are observable only for bodies moving at speeds approaching that of light, but nowadays in particle physics we encounter such speeds regularly and have to apply relativistic rather than Newtonian mechanics in their study.

In this chapter we give a very brief description of the profound changes that relativity makes in the kinematics and dynamics of high-speed particles. We state most of the resulting relations without proof, but in sufficient detail so that they may be employed in the discussions of the two following chapters, which are concerned with high-energy physics.

1. The Speed of Light

In the derivation of the speed of 'light' in Chapter 43, we made no mention of the observer relative to whom this speed is measured. Prior to Einstein, there had been much speculation regarding the behavior of light when observers moved relative to light sources, or when there was a moving material medium in the light path.

One hypothesis was that light travels at a speed that is *constant relative to its source,* just as the muzzle velocity of a bullet is fixed *relative* to the gun from which it is fired. This hypothesis is *negated* by the fact that the two components of a double star are observed to move in regular orbits about each other; these orbits are consistent with gravitational laws on the assumption that light travels from each to us *at the same speed.* Double stars are separated by a distance very small compared with their distance from us, and move at comparatively high velocities. If the light from the star that was approaching us at a given moment traveled toward us faster than that from the star that was receding, the light that was emitted by the two stars simultaneously would reach us at *very different times,* since the transit time of light from the nearest star to the earth is over four years. Instead of observing regular orbits of double stars, we should observe a very confused picture indeed, with, in fact, the same star observable simultaneously in more than one place since we would see it both *coming* and *going,* so to speak!

Another hypothesis was that the speed of light as we observe it is *constant relative to the air in the atmosphere,* or even relative to the residual gas in the best vacuum we could produce. This would be the case with sound waves, but Fizeau was able to disprove this hypothesis by direct measurement of the speed of light in moving liquid media.

Another, and the favorite hypothesis for many decades, was that there was an *absolute frame of reference* relative to which light traveled at the speed c. This is the frame of reference that was said to be filled with some hypothesized 'something' called the *luminiferous ether,* the 'something' being conceptually convenient to explain propagation of the electromagnetic wave motion. On this hypothesis, an observer moving relative to the absolute frame would observe different light speeds for different directions of travel of the light, just as a boat on a moving river travels at different speeds, relative to an observer on the bank, according to whether it is headed down, up, or across the stream. This hypothesis was disproved by a brilliant experiment originally performed by MICHELSON AND MORLEY in Cleveland, Ohio, in 1881. They set up a Michelson interferometer (*see* Chapter 34) that could accurately compare the speeds of light in two perpendicular directions, and which was floated on mercury so as to be rotatable. The comparison was exactly like that described in Prob. 16, p. 47. The apparatus was sufficiently sensitive to detect motion relative to the 'ether' of only $1/40$ of the earth's orbital speed around the sun. Even if by chance the earth happened to be at rest relative to the ether at one time of the year, it would certainly be moving at other times of the year, and hence, differences in travel times of light along two equal perpendicular paths should be at least sometimes observed, and one should be able to find the *absolute velocity* of the earth relative to the absolute frame of reference. In fact, no travel-time differences are observed in such experiments, which have since achieved a sensitivity that could detect an ether 'drift' of 30 m/s, or $1/1000$ of the earth's orbital speed; hence one is forced to the assumption that *such an absolute velocity is meaningless.*

From evidence such as the above, Einstein was forced to the assumption that

> *Any observer will always find the same value for the speed of light in vacuum relative to himself, regardless of the direction of travel of the light beam, and regardless of the velocity of the source of light.*

This assumption is consistent with the general

> PRINCIPLE OF RELATIVITY: *The basic principles of physics that are valid in one coordinate system are equally valid in a coordinate system in motion at constant velocity relative to it; hence, there is no way of determining an absolute velocity—only relative velocities can be measured.*

However, the fact that the velocity of light becomes a *preferred* velocity that has the same value relative to all observers is completely inconsistent with Newtonian concepts of space and time and the Newtonian method of combining relative velocities described on pp. 37–39. It was these basic Newtonian concepts and the laws of mechanics that Einstein was forced to modify, rather than electromagnetic theory as formulated by Maxwell.

The concepts of space and time must be modified so that a particle having speed (very close to) c relative to one observer has the *same* speed relative to all observers. Newton's laws must be modified so that the speed c becomes a limiting speed that cannot be exceeded; this limit occurs in relativistic mechanics because mass varies with speed and approaches infinity as the speed approaches c.

2. Relativistic Kinematics

In the following section, we shall discuss the modifications Einstein introduced into *dynamics*. Here we shall first discuss the modifications required in *kinematics* by the existence of the preferred velocity c. We cannot discuss these modifications in detail here; we shall merely state the relativistic equation for the combination of relative velocities and describe some of the interesting consequences of Einstein's theory.

In the case of Fig. 1, which shows an object moving at velocity v'_X relative to an observer on a car moving in the X-direction with the

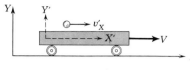

Figure 1

velocity V, Newtonian theory would give

$$v_X = v'_X + V \tag{1}$$

as the velocity relative to an observer on the ground. But if either V or v'_X were close to the speed of light, this equation could give a speed v_X greater than that of light, which is inconsistent with the preceding discussion. The relativistic equation corresponding to (1) is

$$v_X = \frac{v'_X + V}{1 + v'_X V/c^2}, \tag{2}$$

which reduces to (1) for v'_X and V small compared with c, always gives $v_X < c$ provided v'_X and V are both less than c, and gives $v_X = c$ when either $v'_X = c$ or $V = c$.

From the fact that *all laws of physics must remain the same for both the fixed and moving observer*, as required by the *principle of relativity*, and that velocities transform by means of the relation (2), one can logically derive the following interesting conclusions:

1. Consider two observers, one at rest in the 'stationary' X-coordinate system of Fig. 1, the other at rest in the 'moving' X'-coordinate system. *These two observers will not agree as to the time interval between events, nor as to the distance between events. They will not agree as to which events happen simultaneously. They will not even always agree as to which event happened first. They will not agree as to whether a fixed and a moving stick are of equal length.* (The 'events' might be, for example, two strokes of lightning hitting two points on the car or on the ground.)

2. Let the 'moving' observer and the 'stationary' observer be given *identical* meter sticks. *Then to each observer the meter sticks of the other, when placed parallel to the X-axis,* will appear contracted in the ratio $\sqrt{1 - V^2/c^2}$. All distances parallel to the X-axis will be similarly contracted; there will be no disagreement between the observers with respect to distances perpendicular to the X-axis.

3. Let the moving and stationary observers be given *identical* clocks. *Then to each observer the clocks of the other appear to run slow. They appear to mark off too little time in the ratio* $\sqrt{1 - V^2/c^2}$.

4. When a source of light is moving toward (upper sign) or away from (lower sign) an observer, with velocity V relative to the observer, the relation given on p. 726 for the Doppler frequency shift, $\Delta f = (V/c)f$, holds to the first order in V/c, but must be corrected by the slowing-down-of-the-moving-clock effect discussed in **3**. This effect enters twice: If the periodic time in the source frame of reference is $T_s = 1/f_s$, the observed periodic time will be greater, $T_0 = T_s/\sqrt{1 - V^2/c^2}$. In this time, the observer will find that the source has moved a distance $VT_0 = VT_s/\sqrt{1 - V^2/c^2}$. Hence, since light travels at velocity c relative to the observer, he will find that the time between *arrival* of successive cycles will be $T_0 \mp VT_0/c$, the time between emission of successive cycles corrected for the motion of the source in this time interval. This time is the reciprocal of the observed frequency:

$$\frac{1}{f_0} = T_0 \mp \frac{V}{c} T_0 = T_0 \left(1 \mp \frac{V}{c}\right) = \frac{T_s(1 \mp V/c)}{\sqrt{1 - V^2/c^2}}$$

$$= \frac{T_s(1 \mp V/c)\sqrt{1 - V^2/c^2}}{\sqrt{1 - V^2/c^2}\sqrt{1 - V^2/c^2}} = \frac{T_s(1 \mp V/c)\sqrt{1 - V^2/c^2}}{(1 - V/c)(1 + V/c)} = \frac{T_s\sqrt{1 - V^2/c^2}}{(1 \pm V/c)}.$$

Hence, $\quad f_0 = f_s \dfrac{(1 \pm V/c)}{\sqrt{1 - V^2/c^2}} = f_s \left(1 \pm \dfrac{V}{c} + \tfrac{1}{2}\dfrac{V^2}{c^2} \pm \tfrac{1}{2}\dfrac{V^3}{c^3} + \cdots\right).$ **(3)**

The quadratic term in this equation was directly confirmed by H. E. IVES in an experiment in which the frequencies of light emitted in the forward and backward directions by a beam of high-speed hydrogen atoms were recorded on the same plate as the frequency f_s emitted by hydrogen atoms at rest, and the *average* of the two frequencies from the moving atoms was found to differ from the frequency f_s of the atoms at rest by exactly $\tfrac{1}{2}(V^2/c^2)f_s$. *This experiment constitutes a direct confirmation of the slowing-down-of-the-clock conclusion of the theory of relativity.*

5. While the *laws* of electricity and magnetism are the same for all observers, *observers moving relative to each other will measure different electric and magnetic fields.* In particular, $\mathcal{E}$ and $\mathcal{B}$ get 'mixed up,' since a charge at rest relative to one observer appears to have purely an electric field, while, to an observer relative to whom the charge is moving, it will have both an electric and a magnetic field. Again, when an observer sees a charge moving through a magnetic field, he will note that it has an acceleration which he attributes to a magnetic force. However, another observer, relative to whom the charge is momentarily at rest, will also observe the acceleration but must attribute it to an electric force.

3. Relativistic Dynamics

In 1906, KAUFMANN AND BÜCHERER employed a mass spectrograph to measure the masses of the electrons emitted in natural radioactivity. These electrons are emitted with various speeds, and it was found that the mass of the electron increased as its speed approached the speed of light; the experimental data are plotted in Fig. 2. We note that the magnetic-force relation (8), p. 594, is unchanged in Einstein's theory; but that in the relation (9), p. 594, for the radius of curvature of a particle moving perpendicular to a magnetic field, the relativistic mass given by (4), below, must be used. This is the mass given by a mass spectrograph and is the mass that determines the path radius in a particle accelerator, as we shall discuss in Sec. 4.

The departure from the Newtonian idea of an invariant mass, which is illustrated in Fig. 2, had already been predicted by ALBERT EINSTEIN in 1905.

In order to make the principle of relativity applicable simultaneously to mechanical and electromagnetic phenomena, Einstein was forced to modify the principles of mechanics. In particular, Einstein's theory requires a revision of Newton's second principle. The effect of the revision is undetectably small for bodies whose speed is small compared with the speed of light (say less than $\frac{1}{100}c = 3 \times 10^6$ m/s = 6 000 000 mi/h), but the revision is very significant for particles moving at speeds close to that of light. By logical arguments, Einstein not only predicted the observed increase of mass with speed, but made other startling predictions, such as that of the equivalence of mass and energy. This equivalence, which was not verified until much later, is fundamental to an understanding of the transformation of mass to energy in nuclear reactions. We shall first state the expression for the variation of mass with velocity:

The observed mass of a particle is not a constant but increases with increasing speed relative to the observer, in such a way as to approach infinity as the speed of the particle approaches that of light, according to the relation

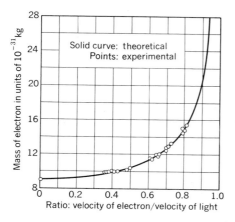

Fig. 2. Electron mass as a function of speed.

$$m = \frac{m_0}{\sqrt{1-(v^2/c^2)}}. \qquad (4)$$

Here m_0, called the *rest mass*, is the Newtonian mass—the mass of the particle when at rest or when moving with a speed small compared with the speed of light. A plot of (4) for the electron, with $m_0 = 9.1 \times 10^{-31}$ kg, is given in the solid curve of Fig. 2.

The first and third of Newton's principles remain unchanged in relativistic mechanics. The *second principle* must be written as the vector equation

$$F = \frac{d(mv)}{dt} \qquad (5)$$

—force equals time rate of change of momentum. This is the same equation as the Newtonian, but now, if the speed is increasing with time, the mass is also increasing with time, and it is the change in the product mv that determines the force.

The force required to give a particle of mass m traveling at velocity v an acceleration a is thus

$$F = \frac{d}{dt}(mv) = m\frac{dv}{dt} + v\frac{dm}{dt} = ma + v\frac{dm}{dt}. \qquad (6)$$

If the force is perpendicular to the velocity, as in the case of motion of a particle in a circle at constant speed, the speed and hence the mass do not change, so the last term of (6) is zero. In this case the force

$$F = ma = m\frac{v^2}{R} = \frac{m_0}{\sqrt{1-(v^2/c^2)}}\frac{v^2}{R}, \qquad (F \perp v) \qquad (7)$$

just as in the Newtonian case, *except that m now depends on the speed.*

On the other hand, *if the force is parallel to the velocity* as in the case of rectilinear acceleration, (6) becomes

$$F = ma + v\frac{dm}{dt} = ma + v\frac{dm}{dv}\frac{dv}{dt} = \left(m + v\frac{dm}{dv}\right)a. \tag{8}$$

From (4), $\quad \dfrac{dm}{dv} = \dfrac{m_0}{[1-(v^2/c^2)]^{3/2}}\dfrac{v}{c^2} = \dfrac{mv/c^2}{1-(v^2/c^2)},$ \hfill (9)

and (8) becomes $\quad F = \dfrac{m}{1-(v^2/c^2)}a = \dfrac{m_0}{[1-(v^2/c^2)]^{3/2}}a.$ \quad $(\mathbf{F} \parallel \mathbf{v})$ \quad (10)

The force is even greater than ma and becomes infinite as $v \to c$, which means that c is a limiting speed that can never be achieved. But, for small values of v/c, (10) reduces to the Newtonian equation. Equation (10) applies whether the force and acceleration are in the *same direction* as the velocity or in the *opposite direction,* since the change in the momentum mv has the same numerical value for a given change in speed whether the speed increases or decreases.

Equation (5) and the third principle show, by arguments given in Chapter 7, that *the principle of conservation of momentum is valid in relativistic mechanics.*

We notice from (6) that the force is no longer parallel to the acceleration, but has a component in the direction of the velocity when the speed is sufficiently great that the value of dm/dt is significant.

Now let us consider the *kinetic energy* of a particle. Just as in Newtonian mechanics, the element of work is defined by $dW = \mathbf{F} \cdot d\mathbf{D}$, and the kinetic energy at speed v_1 is defined as the work that must be done to accelerate the particle from rest to speed v_1. We can then write

$$E_K = \int \mathbf{F} \cdot d\mathbf{D} = \int \mathbf{F} \cdot \frac{d\mathbf{D}}{dt} dt = \int \mathbf{F} \cdot \mathbf{v} \, dt, \tag{11}$$

since $\mathbf{v} = d\mathbf{D}/dt$; this integral is to be taken over the path the particle traverses in starting from rest and reaching speed v_1 under an arbitrarily varying force. Now, using the center expression in (6), we see that

$$\mathbf{F} \cdot \mathbf{v} = m\frac{d\mathbf{v}}{dt} \cdot \mathbf{v} + \mathbf{v} \cdot \mathbf{v}\,\frac{dm}{dt} = m\frac{d\mathbf{v}}{dt} \cdot \mathbf{v} + v^2\frac{dm}{dt}, \tag{12}$$

since $\mathbf{v} \cdot \mathbf{v} = v^2$. We also see, by differentiation of the components in equation (9), p. 106, that we can differentiate the scalar product of two vectors as follows:

$$\frac{d}{dt}(\mathbf{A} \cdot \mathbf{B}) = \frac{d\mathbf{A}}{dt} \cdot \mathbf{B} + \mathbf{A} \cdot \frac{d\mathbf{B}}{dt}. \tag{13}$$

Hence $\quad \dfrac{d}{dt}(v^2) = \dfrac{d}{dt}(\mathbf{v} \cdot \mathbf{v}) = \dfrac{d\mathbf{v}}{dt} \cdot \mathbf{v} + \mathbf{v} \cdot \dfrac{d\mathbf{v}}{dt} = 2\dfrac{d\mathbf{v}}{dt} \cdot \mathbf{v},$ \hfill (14)

since the scalar product is commutative [equation (3), p. 105]. Hence the first expression at the right of (12) reduces to

$$m\frac{d\mathbf{v}}{dt} \cdot \mathbf{v} = \tfrac{1}{2} m\frac{d}{dt}(v^2) = mv\frac{dv}{dt}, \tag{14'}$$

and (12) becomes

$$\mathbf{F} \cdot \mathbf{v} = mv\frac{dv}{dt} + v^2\frac{dm}{dt} = \left(mv\frac{dv}{dm} + v^2\right)\frac{dm}{dt}. \tag{15}$$

The value of dv/dm is the reciprocal of dm/dv computed in (9); insertion of this value in (15) and simplification gives

$$\mathbf{F} \cdot \mathbf{v} = c^2\frac{dm}{dt}.$$

Finally, then, (11) becomes

$$E_K = \int c^2\frac{dm}{dt}\,dt = c^2\int_{m_0}^{m_1} dm = c^2(m_1 - m_0), \tag{16}$$

where the mass is m_0 when the particle is at rest and m_1 when it has speed v_1. This equation expresses the extraordinarily simple relation:

The kinetic energy of a particle equals c^2 times the difference between its mass and its rest mass:

$$E_K = c^2(m - m_0). \tag{17}$$

We can easily verify that at low speeds this expression reduces to the Newtonian value $\tfrac{1}{2} m_0 v^2$ for the kinetic energy. By using the binomial theorem to expand (4) in a series of powers of v^2/c^2, we find

$$m = m_0\left[1 - \frac{v^2}{c^2}\right]^{-1/2} = m_0\left[1 + \tfrac{1}{2}\frac{v^2}{c^2} + \tfrac{3}{8}\frac{v^4}{c^4} + \cdots\right];$$

whence

$$E_K = c^2(m - m_0) = \tfrac{1}{2} m_0 v^2 + \tfrac{3}{8} m_0\frac{v^4}{c^2} + \cdots = \tfrac{1}{2} m_0 v^2\left[1 + \tfrac{3}{4}\frac{v^2}{c^2} + \cdots\right].$$

The Newtonian formula $\tfrac{1}{2} m_0 v^2$ is valid when v/c is sufficiently small that the second term in the last bracket is negligible in comparison with unity.

The fact that the kinetic energy of a particle is directly proportional to the difference between the mass of the particle and its rest mass led Einstein to suspect that mass and energy are just two different measures of the same physical quantity, the conversion factor being c^2. Further consideration, confirmed by direct experimental evidence in nuclear reactions, leads to the conclusion that mass is associated with all forms of energy: kinetic, potential, elastic, thermal, chemical, nuclear, and electromagnetic—and that *whenever a body changes in energy it changes in mass correspondingly.* The conversion factor in all cases is c^2, the change in energy ΔE being related to the change in mass Δm by

$$\Delta E = c^2 \, \Delta m. \qquad \begin{Bmatrix} E \text{ in J} \\ c \text{ in m/s} \\ m \text{ in kg} \end{Bmatrix} \text{ or } \begin{Bmatrix} E \text{ in fp} \\ c \text{ in f/s} \\ m \text{ in sl} \end{Bmatrix} \qquad (18)$$

Thus, *mass and energy are measures of the same physical quantity*, the equivalent changes in mass and energy being

$$1 \text{ kg} \sim 8.987 \times 10^{16} \text{ J}; \qquad 1 \text{ J} \sim 1.113 \times 10^{-17} \text{ kg}. \qquad (19)$$

The extremely small size of the factor 1.1×10^{-17} accounts for the fact that it is impossible to detect the mass changes associated with ordinary energy changes such as occur in thermal heating and chemical reactions. Thus, to heat 1 kg of water from 32° F to 212° F, we must add 418 500 J of energy, which increases the mass of the water by only 5×10^{-12} kg. Again, the energy release in the complete combustion of 3 *tons* of carbon makes the combustion products only 1 *milligram* lighter than the 11 *tons* of carbon and oxygen entering the reaction.

Since 1932, there has been ample direct experimental confirmation of the relation (18), in nuclear reactions in which the release of energy is so large that the accompanying mass changes can be readily measured; thus we know that

There is energy E associated with any mass m, and mass m associated with any energy E, the relation between energy and mass being $E = mc^2$.

By this *identification* of mass and energy, the *law of conservation of energy* and the *law of conservation of mass* become *one and the same physical law*, not two distinct laws. This identification shows that the *total energy* of a particle of mass m is mc^2; from (17) we have

$$mc^2 = m_0 c^2 + E_K. \qquad (20)$$

The term $m_0 c^2$ is called the *rest energy* of the particle (energy associated with the rest mass of the particle). Thus

$$\text{total energy} = \text{rest energy} + \text{kinetic energy}. \qquad (21)$$

Example. *Although the speed of light is enormous, particles such as electrons, protons, and neutrons encountered in atomic and nuclear physics frequently have speeds comparable with that of light. If a particle is moving at a speed of exactly 0.6 c, by what percent do its momentum and kinetic energy differ from the Newtonian values?*

If the Newtonian mass is m_0, the relativistic mass is

$$m = \frac{m_0}{\sqrt{1-(v/c)^2}} = \frac{m_0}{\sqrt{1-(0.6)^2}} = \frac{m_0}{\sqrt{1-0.36}} = \frac{m_0}{0.8} = 1.25 \, m_0.$$

Hence its mass and its momentum mv are 25 per cent higher than the Newtonian values.

The kinetic energy of the particle is $c^2(m-m_0) = c^2(1.25 \, m_0 - m_0) = 0.25 \, m_0 c^2$, while the Newtonian value is $0.5 \, m_0 v^2$. The ratio of true

kinetic energy to the Newtonian is

$$\frac{0.25\, m_0\, c^2}{0.5\, m_0\, v^2} = \frac{0.5}{(v/c)^2} = \frac{0.5}{0.36} = 1.389.$$

Hence the true kinetic energy is 38.9 per cent higher than the Newtonian.

4. Accelerators for High-Energy Particles

The relativistic increase in the mass of a particle with increasing speed places an upper limit on particle energies obtainable with a simple cyclotron (p. 598), since the cyclotron frequency is $f = BQ/2\pi m = BQ\sqrt{1-v^2/c^2}/2\pi m_0$. If the cyclotron operates at resonance frequency $f_0 = BQ/2\pi m_0$ for slow-moving ions in small orbits near the center of the magnetic field, the frequency f_0 becomes too high as the factor decreases for ions as they spiral outward with increasing speeds into orbits of increasing radius. The practical proton energy limit for a simple cyclotron is about 10 MeV.

This relativistic limitation on cyclotron operation can be overcome in two ways: (1) In the *synchrocyclotron*, the frequency f is modulated so as to maintain resonance as a pulse of ions spirals outward to orbits of large radius in a constant magnetic field $\mathcal{B}$ (Fig. 3). (2) In the *spiral-ridge cyclotron* the frequency is kept constant but the magnetic field $\mathcal{B}$ increases with increasing R in such a way that cyclotron resonance is maintained; in such a cyclotron the ion beam is unstable unless a set of spiral ridges is machined into the iron pole faces in such a way as to provide an alternating magnetic field gradient. Although it has been

Fig. 3. The 184-inch synchrocyclotron at the University of California.

used at Berkeley to produce 740 MeV protons, the frequency-modulated synchrocyclotron is wasteful of iron in setting up a strong magnetic field throughout a large volume, only a small part of which is occupied by the pulse of ions at any one time; for this reason synchrocyclotrons are no longer being constructed. However, spiral-ridge cyclotrons are in wide use and provide protons in the 30–150 MeV range; one device of this type for the production of 400 MeV protons is in the planning stage.

More recent machines, designed for still higher energies, keep the *radius of curvature* (9), p. 595, of the ion pulse constant by increasing the magnetic intensity in proportion to the increase of the momentum mv of the particles as they are accelerated. In such machines, called *synchrotrons,* the ion pulse repeatedly follows the same path in a vacuum tube. A pulse of ions is injected into the path at fairly high energy from some other form of accelerator, is then repeatedly accelerated by suitable oscillators at various points in the path, and is finally deflected onto a target. To keep the radius of curvature constant, the magnetic intensity must increase in proportion to the momentum mv of the particles. Since the time of traversing the path is inversely proportional to the speed v of the particles, the frequency of the oscillator must increase in proportion to this speed. However, the speed rapidly approaches the limit c. Figure 4 shows a photograph of a synchrotron.

The most powerful proton synchrotron in operation in the U.S.A. is at the Brookhaven National Laboratory on Long Island and produces 33 GeV protons; comparable machines exist at the international CERN laboratory in Geneva and at Dubna in Russia. The proton synchrotron being constructed at the National Accelerator Laboratory at Batavia, Illinois, is designed to produce 200 GeV protons and is scheduled for completion in 1974. The electron synchrotron at Cornell University

Fig. 4. Synchrotron at the University of California, with 135-f diameter magnet, that produces protons with an energy of 6 GeV.

produces 2.2 GeV electrons, but the largest electron accelerator is the linear accelerator (*see* Prob. 25, p. 684) at Stanford University which produces 20 GeV electrons.

5. Energy and Momentum Transfer; Quantum and Relativistic Considerations

Although classical electromagnetic theory accounts satisfactorily for most of the phenomena involving the transmission of light, we recall that it is necessary to employ quantum theory in order to account for emission and absorption. We now show how quantum theory can be applied to the problem of transmission and how simple relativistic considerations can similarly be applied.

The radiant power per unit area is given by (34), p. 892, which indicates that beam intensity is directly proportional to the square of the wave amplitude. Since the energy of each quantum of radiation of frequency f is given by $E = hf$, we can also express the power per unit area in the form

$$\text{power per unit area} = nhf \tag{22}$$

where n represents the number of photons per unit area per unit time (the photon flux). Thus we note that, for radiation of a given frequency, the number of photons per unit area per unit time is directly proportional to the square of the amplitude of the wave as given by classical theory; for a given power per unit area the required flux of photons is inversely proportional to the frequency of the radiation.

Since energy and mass are equivalent in relativistic theory, the energy of electromagnetic waves, traveling at speed c, has momentum which exerts pressure when it is absorbed or reflected by a material body, just as a stream of water exerts pressure on a water wheel. From the relativistic relation $E = mc^2$, we find that the mass impinging on unit area per unit time is

$$\begin{pmatrix}\text{mass per unit area} \\ \text{per unit time}\end{pmatrix} = \begin{pmatrix}\text{energy per unit area} \\ \text{per unit time}\end{pmatrix} \Big/ c^2 = \begin{pmatrix}\text{power per} \\ \text{unit area}\end{pmatrix} \Big/ c^2.$$

This mass impinges with velocity c, so it exerts a pressure given by

$$P = \begin{pmatrix}\text{mass per unit area} \\ \text{per unit time}\end{pmatrix} \times c = \begin{pmatrix}\text{power per} \\ \text{unit area}\end{pmatrix} \Big/ c. \tag{23}$$

This is the fundamental relation for the pressure exerted by light absorbed by a material body. In the case of reflection of the light, the pressure is doubled, as in the case of the Pelton water wheel.

Example. *Determine the pressure of sunlight absorbed by a material body at the distance of the earth, where sunlight has the intensity* 1350 W/m².

From (23), we find $P = \dfrac{1350 \text{ W/m}^2}{3 \times 10^8 \text{ m/s}} = 4.50 \times 10^{-6} \dfrac{\text{N}}{\text{m}^2}$

The pressure of sunlight, as computed in the above example, is very small. It has, however, been observed to have a measurable effect on the orbits of earth satellites, particularly on that of the large reflecting satellite *Echo 1*, launched in 1960.

The existence of light pressure had been predicted by Maxwell on the basis of classical theory. Light pressure, which is usually masked by much larger thermal effects, was first measured in 1901 by NICHOLS AND HULL prior to the development of relativity and quantum theories. Comparison of (23) with (34), p. 892, shows that the classical expression for light pressure is $\tfrac{1}{2}\varepsilon_0 \mathcal{E}_M^2$ for absorbed radiation and $\varepsilon_0 \mathcal{E}_M^2$ for reflected radiation. Comparison of (23) with (22) gives the corresponding expressions nhf/c and $2nhf/c$; these latter expressions are to be expected in view of the momentum hf/c assigned to a single photon on the basis of the Compton effect (Sec. 8, Chapter 41).

Returning to the relativistic relation $E = mc^2$, we might assign an effective mass $m = E/c^2 = hf/c^2$ to a single photon; we should remember, of course, that photons move only at speed c and have zero rest mass. If we are willing to attribute mass to photons, we might well ask whether photons are influenced by gravitational forces. Gravitational effects on light were indeed predicted by Einstein in 1916 on the basis of his theory of general relativity; two predicted effects have been observed. The first of these is the deflection of light passing close to an extremely massive body; thus, during a total eclipse of the sun there is a slight change in the apparent position of a star when the optical path from the star to the terrestrial observing station passes close to the sun. The second gravitational effect is the so-called gravitational red shift and is observed most readily when the light source is in the strong gravitational field of a star; when a photon of initial energy $E_0 = hf_0$ moves away from the star, a portion of its initial energy is converted into gravitational potential energy and the energy of the photon reaching an observer on earth is $E = hf$, when $f < f_0$. Such a change in frequency is detected by comparing an atomic line in the spectrum of a massive star with the same spectral line produced in a laboratory source.

In experiments involving the Mössbauer effect (Prob. 11), R. V. POUND was able to observe a gravitational red shift in the laboratory. By measuring the frequency of a gamma-ray photon at a height H above its point of emission, Pound found that the observed frequency f was slightly lower than the initial frequency f_0 by an amount suggested by the energy relation

$$\begin{pmatrix} \text{final} \\ \text{energy} \end{pmatrix} = \begin{pmatrix} \text{initial} \\ \text{energy} \end{pmatrix} - \begin{pmatrix} \text{increase in gravita-} \\ \text{tional potential energy} \end{pmatrix}$$

$$hf = hf_0 - mgH = hf_0 - (hf_0/c^2)\,gH.$$

Thus, the value of the frequency f observed at height H is given to a good approximation by the expression

$$f = f_0 (1 - gH/c^2). \tag{24}$$

Consideration of the ratio gH/c^2 indicates that the fractional change in frequency is extremely small, as illustrated by the following example.

Example. *If a Pound experiment were performed with gamma rays of initial frequency $f_0 = 10^{20}$ Hz, what would be the shift $\Delta f = f_0 - f$ if the gamma rays were detected at a height $H = 30$ m above the source? What is the fractional change in frequency, $\Delta f/f_0$?*

From (24) we obtain the relation, $\Delta f = f_0 - f = f_0 \, gH/c^2$. On the basis of this relation, the change in frequency, called the 'red shift,' is $\Delta f = (10^{20} \text{ Hz})(9.8 \text{ m/s}^2)(30 \text{ m})/(9 \times 10^{16} \text{ m}^2/\text{s}^2) = 3.27 \times 10^5$ Hz. The fractional change in frequency $\Delta f/f_0 = (3.27 \times 10^5 \text{ Hz})/(10^{20} \text{ Hz}) = 3.27 \times 10^{-15}$. In view of the extremely small value of $\Delta f/f_0$, it is apparent that gravitational red shifts can be observed in terrestrial laboratories only by extremely careful measurements under very favorable conditions.

PROBLEMS

1. At what speed is the mass of a particle greater than the rest mass by 1 part in 1000? This is the lowest speed at which relativistic effects need to be considered in our usual three-figure accuracy of computation.
Ans: $0.0447 \, c = 1.34 \times 10^7$ m/s.

2. Compare the relativistic and Newtonian values of force required to give the same acceleration a to a particle traveling at a speed of $0.9 \, c$ if the force is applied (a) perpendicular to the velocity, (b) in the direction of the velocity, (c) in the direction opposite to the velocity.

3. Answer the same questions as in Prob. 2 for the case of a particle traveling at a speed of $0.99 \, c$.   Ans: (relativistic force)/(Newtonian force) = (a) 7.09; (b) 357; (c) 357.

4. Show that in relativistic mechanics, the equations replacing (8) and (9) of Chapter 7, for the case of an elastic collision, are

$$m_1' u_1 + m_2' u_2 = m_1'' v_1 + m_2'' v_2,$$
$$m_1' + m_2' = m_1'' + m_2'',$$

where the single and double primes indicate masses before and after the collision, respectively.

5. The equations in Prob. 4 are extremely complex and difficult to solve algebraically after the masses are expressed in terms of the velocities by (4). There is a 'trick' that changes these equations to trigonometric equations readily soluble by 'cut-and-try' methods. Assume that u_1, u_2, and hence m_1' and m_2' are known. Show that if one writes $v_1/c = \sin\theta$, $v_2/c = \sin\phi$, then $m_1'' = (m_0)_1 \sec\theta$ and $m_2'' = (m_0)_2 \sec\phi$, so that the equations in Prob. 4 can be turned into trigonometric equations to be solved for θ and ϕ, from which v_1 and v_2 are readily determined.

6. At what value of v/c is the kinetic energy of a particle greater than that computed from the Newtonian expression by just one part in 1000? Compare your answer with that of Prob. 1.

7. As an example of the computations outlined in Probs. 4 and 5, consider a neutron of rest mass $(m_0)_1 = m_0$ colliding with a deuteron of rest mass $(m_0)_2 = 2m_0$, as in the example on pp. 136–137. Let the neutron have initial speed $u_1 = 0.6 \, c$ and the deuteron be initially at rest. Show that the angles defined in Prob. 5 turn out to be $\theta = 167°3$, $\phi = 26°0$. Show that the neutron's recoil velocity is $v_1 = -0.220 \, c$, in comparison with the value $-0.2\,c$ that would be given by Newtonian theory, and that the deuteron's velocity is $v_2 = 0.438\,c$, in comparison with the Newtonian value $0.4\,c$. Show that the neutron's kinetic energy decreases from $0.250\,m_0 c^2$ to $0.025\,m_0 c^2$, so that it loses approximately 90 per cent of its kinetic energy, about the same as in the Newtonian case. Verify numerically that energy and momentum are both conserved.

8. Repeat the computations of Prob. 7 for the case in which the neutron's initial speed is $0.8\,c$, and make similar comparisons with Newtonian theory.

9. There is much current interest in a subject called 'beam-foil spectroscopy' in which ions are accelerated in a Van de Graaff accelerator and are then allowed to pass through a thin foil. The atoms or ions emerging from the foil emit light, which is studied by means of a spectrograph. On the basis of (3), write expressions for the frequencies f_0 of the observed radiation in terms of the frequency f_s of the radiation the atom would emit if it were at rest, the speed V of the atoms emerging from the foil, and the observation angle θ between the direction of the beam and the light path to the spectrograph. With $\theta = 0$ for beam and light paths parallel and $\theta = 180°$ for beam and light paths anti-parallel, show that $f_0 = f_s\,(1 + V^2/c^2)$ for $\theta = 90°$.

10. In our treatment of emission (p. 851), we set $E_1 - E_2 = hf$, where E_1 and E_2 are energies of 'stationary states' of an atom. Since the emitted photon has momentum hf/c, our treatment (actually, Bohr's treatment) is not strictly correct. Considering an atom at rest prior to emission, use momentum-conservation principles to show that an atom of mass M acquires a recoil velocity of magnitude $V = hf/cM$ when the photon is emitted. From energy-conservation principles show that the proper energy relation for photon emission is

$$E_1 - E_2 = hf(1 + hf/2Mc^2), \quad (25)$$

where the first term on the right represents the photon energy, the second term the energy of recoil.

11. We note that the second term in the parenthesis in (25) represents the ratio of the photon energy to twice the rest energy of the recoil atom; this term is therefore negligibly small for light in the visible range of the electromagnetic spectrum but becomes appreciable in the gamma-ray region. The radioactive nucleus Ir^{191} emits gamma rays with energy $hf = 0.129$ MeV; in Chapter 45, we shall see that the rest energy of $Ir^{191} = 191 \times 931$ MeV. From (25) find the recoil energy $(hf)^2/2Mc^2$. In 1958, R. L. MÖSSBAUER showed that when a radioactive nucleus is part of a crystal lattice at very low temperature the gamma-ray recoil is taken up by the entire crystal, so that the M in (25) is the entire mass of the crystal; in this case $hf/2Mc^2$ is completely negligible in comparison with unity. This *Mössbauer effect* is exhibited strongly by Fe^{57}. Ans: 0.0465 eV.

12. In order to obtain information regarding a structure of dimension d, we must employ waves having wavelength at least this small. ROBERT HOFSTADER at Stanford University has actually used fast electrons to study the structure of the proton and reports that the proton has a fairly definite radius of 0.78×10^{-15} m. What are the momentum and energy of an electron having a de Broglie wavelength of this magnitude?

13. Quasi-stellar objects (QUASARS) are starlike objects in space that are characterized by spectral lines shifted to enormously longer wavelengths. They are classified in terms of a parameter $Z = (\lambda - \lambda_0)/\lambda_0$, where λ is the observed wavelength of a spectral line which has wavelength λ_0 when emitted by a source in the laboratory. One quasar designated as 3C9 is characterized by $Z = 2.01$. Assuming that the quasar's *red shift* is due to the Doppler effect, compute its velocity of recession.

14. Consider the Compton effect for the 'head-on' elastic collision of a photon and a stationary electron, in which the photon is exactly reversed in direction. Let the wavelength of the photon be λ before the collision, λ' afterward. Show that the wavelength increase $\lambda' - \lambda$ is independent of λ and is given by $\lambda' - \lambda = 2h/m_0 c$, where m_0 is the rest mass of the electron. Interestingly, this formula is only an approximation when Newtonian relations are used for the energy and momentum of the electron, but is rigorous in relativistic theory. Hence use relativistic relations. Show that conservation of momentum and conservation of energy lead, respectively, to the two equations

$$\frac{h}{\lambda}+\frac{h}{\lambda'}=\frac{m_0 v}{\sqrt{1-v^2/c^2}},$$

$$\frac{h}{\lambda}-\frac{h}{\lambda'}+m_0 c=\frac{m_0 c}{\sqrt{1-v^2/c^2}},$$

where v is the recoil velocity of the electron. Squaring both equations and subtracting leads directly to the equation for $\lambda'-\lambda$.

15. In Prob. 14, show that the wavelength increase is 0.004 84 nm, as is observed experimentally.

16. In Prob. 14, show that the kinetic energy of the recoil electron is given by

$$hf\,[2\alpha/(1+2\alpha)],$$

where $\alpha = hf/m_0 c^2$, the ratio of the initial energy of the photon to the rest energy of the electron. Discuss the manner in which the fraction of its energy that the photon loses varies with the value of this ratio.

17. In the Compton effect in which the photon is 'scattered' through 90° instead of 180° as in the preceding three problems, show that the wavelength increase is just half as great.

18. A searchlight emits visible radiation at a rate of 2 kW. If this searchlight is adjusted so as to direct the entire radiation on a distant circular mirror 30 cm in diameter, what is the light force exerted on the mirror?

19. A plane radio wave transmits power at a rate of 1 W/m². If this wave is completely absorbed by a plane surface, what pressure is exerted on the surface? What pressure would this wave exert on a surface at which total reflection occurred?

Ans: 3.33×10^{-9} N/m²; 6.66×10^{-9} N/m².

NUCLEAR PHYSICS 45

On the basis of experiments involving the scattering of alpha particles by thin metal foils, Rutherford in 1911 proposed the nuclear model of the atom, which was later used to provide an explanation of atomic properties in terms of the quantized energy levels of electrons in the electrostatic field of the massive, positively charged nucleus. The present chapter deals with *nuclear physics,* which is concerned with the structures and properties of atomic nuclei. Although certain nuclear properties such as charge, mass, spin, magnetic moment, and electric quadrupole moment can be determined by means of the mass spectrograph and by studies of atomic spectra, most of our knowledge of nuclear properties is based on studies of *nuclear reactions*.

The first nuclear reactions studied in detail were those occurring in natural radioactivity, involving the spontaneous disintegration of the nuclei of the heaviest elements. In 1919 Lord Rutherford produced the first 'man-made' nuclear reaction by bombarding nitrogen with high-energy alpha particles produced in natural radioactive decay. In the 1930's, particle accelerators were developed to provide other high-energy charged particles for use in nuclear studies, making possible the determination of nuclear energy levels. With the discovery of the neutron by James Chadwick in 1932, it was recognized that the nucleus was composed of protons and neutrons held together by strong nuclear forces.

Following the discovery of nuclear fission in 1939, large-scale conversion of mass to energy became possible in the nuclear-fission bomb and reactor. Energy from nuclear fusion became available in the thermonuclear bomb in the 1950's, and may someday become available under controlled conditions.

In this chapter we review some of the known facts about atomic nuclei and nuclear reactions and discuss briefly some of the nuclear models that have been proposed. No entirely satisfactory theory of nuclear structure has yet been developed and even the behavior of nuclear force is not yet clearly understood!

1. The 'External' Properties of Nuclei

Certain characteristic nuclear properties can be determined without probing into the interior of nuclei; these properties are sometimes called 'external' properties. In earlier chapters we have discussed two of these properties: nuclear *charge* and nuclear *mass*. The nuclear charge is simply $+Ze$, where Z is the *atomic number*. The nuclidic mass was defined in Chapter 17 (p. 334), and methods of determining nuclidic masses were described in Chapter 29. The integer nearest the value of the nuclidic mass in u is called the *mass number A*. In designating the atomic number and mass number of a nucleus, it is customary to use the notation $_ZX^A$, where X is the chemical symbol for the element. Unless otherwise indicated, *in the present chapter we shall employ such chemical symbols to refer to the nuclei themselves, rather than to the entire atom.*

Another external property of the nucleus is the *nuclear spin*. As indicated previously, the allowed values of the orbital angular momentum of the electrons in atoms are integral multiples of $h/2\pi$; thus, $h/2\pi$ serves as a convenient 'atomic' unit of angular momentum. In discussion of the magnetic properties of matter in Chapter 30, we indicated that, in addition to its orbital angular momentum, the electron has an *intrinsic* angular momentum $\frac{1}{2}(h/2\pi)$ associated with its so-called *spin*. Electron spin was discovered through studies of the *fine structure* of spectral lines; a spectral 'line' actually consists of several closely spaced component lines, associated with different orientations of electron spin relative to the orbital angular momentum of the electron, and is therefore said to have 'fine structure.' Careful studies with spectrographs combined with interferometers show that in the spectra of many atoms each fine-structure component itself consists of several even more closely spaced component lines constituting a *hyperfine structure;* this 'hyperfine structure' results from the effects of *nuclear* spin. Each nucleus has a characteristic intrinsic angular momentum I ($h/2\pi$), called the *nuclear spin*. The observed values of I are always integers when A is even, half an odd integer when A is odd; Table 45-1 lists the spins of selected nuclei.

Since the spinning nucleus is charged, we would expect it also to have a magnetic moment. We have found, in (14), p. 640, that the ratio of magnetic moment to angular momentum for a charge moving in a circular orbit is $q/2m$. For an orbital electron, since the angular momentum is a multiple of $h/2\pi$, the magnetic moment is conveniently expressed in terms of the Bohr magneton $\mu_B = (e/2m_e)(h/2\pi) = eh/4\pi m_e$, where m_e is the mass of the electron. Since the mass of a nucleus is so much larger than the electron mass, nuclear magnetic moments are much smaller than the Bohr magneton. Studies of spectral hyperfine structure and of the behavior of matter in strong magnetic fields show that nuclei with nonzero spin have magnetic moments of the order of $\mu_N = eh/4\pi m_p$, where μ_N is called the *nuclear magneton* and m_p is the mass of the proton. The *observed* nuclear magnetic moments, as typified by those listed in Table 45-1, are *not* simple multiples of μ_N. Positive values are in the *same* direction as the spin, negative values in the *opposite* direction.

It is found that nuclei have no electric dipole moment. However, nuclei with $I > \frac{1}{2}$ have electric fields that are not spherically symmetric;

Table 45-1 'External' Properties of Certain Nuclei

Nucleus	Mass of neutral atom (unit u)	Spin (unit $h/2\pi$)	Magnetic moment (unit μ_N)	Quadrupole moment (unit 10^{-23} m²)
$_0n^1$	1.008 665 4	½	−1.913 15	0
$_1H^1$	1.007 825 2	½	+2.792 76	0
$_1H^2$	2.014 102	1	+0.857 38	+0.002 77
$_1H^3$	3.016 049	½	+2.978 7	0
$_2He^3$	3.016 029 9	½	−2.127 4	0
$_2He^4$	4.002 603 6	0	0	0
$_3Li^6$	6.015 126	1	+0.821 91	+0.000 46
$_3Li^7$	7.016 005	3/2	+3.256 0	−0.042
$_4Be^9$	9.012 186	3/2	−1.177 4	+0.02
$_5B^{10}$	10.012 939	3	+1.800 6	+0.111
$_5B^{11}$	11.009 305	3/2	+2.688 0	+0.035 5
$_6C^{12}$	12	0	0	0
$_7N^{14}$	14.003 074	1	+0.403 57	+0.071
$_8O^{16}$	15.994 915	0	0	0
$_9F^{19}$	18.998 405	½	+2.627 2	0
$_{12}Mg^{24}$	23.985 045	0	0	0
$_{14}Si^{28}$	27.976 927	0	0	0
$_{26}Fe^{56}$	55.934 932	0	0	0
$_{53}I^{127}$	126.904 35	5/2	+2.794 0	−0.69
$_{92}U^{233}$	—	5/2	+0.5	+0.34
$_{92}U^{235}$	235.043 93	7/2	+0.5	+4.0
$_{92}U^{238}$	238.050 76	0	0	0
$_{94}Pu^{239}$	—	½	+0.4	0

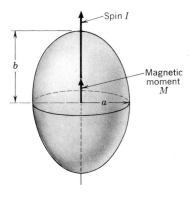

Fig. 1. Nuclear moments. Magnetic moments parallel to spin are called *positive;* magnetic moments anti-parallel to spin are called *negative.* Prolate charge distributions ($b > a$) have positive electric quadrupole moments; oblate charge distributions ($a > b$) have negative quadrupole moments.

they are characterized by an electric quadrupole moment. As indicated in Fig. 1, the electric quadrupole moment measures the departure of a nucleus from spherical symmetry; a positive value corresponds to a prolate charge distribution about the axis of spin I, a negative value corresponds to an oblate distribution. If the nucleus is assumed to be an ellipsoid of revolution whose diameter is $2b$ along the axis of symmetry and $2a$ in the transverse direction, and if charge $+Z\,\mathrm{e}$ is uniformly distributed throughout the volume of the ellipsoid, the electric quadrupole moment is $Z(b^2 - a^2)$. Whether or not the nucleus actually has this form, the observed field asymmetry can be conveniently related to the easily computed asymmetry of this particular ellipsoidal form. We note that *the electric quadrupole moment has the dimensions of area,* since the atomic number Z is dimensionless.

The nuclear quadrupole moment has an effect on the hyperfine structure of atomic spectra, and its value can be determined from a study of such structure. Representative values are listed in Table 45–1.

One additional nuclear property sometimes regarded as 'external' is the so-called *nuclear radius.* The nuclear radius is a measure of the closest approach that can be made by a bombarding particle without experiencing any forces other than those attributable to familiar effects of electric and magnetic fields. It is found experimentally that the nuclear 'volume' is approximately proportional to the mass number A, as would be expected if all nuclei had the same 'density.' Hence, the nuclear 'radius' is proportional to $\sqrt[3]{A}$; it is found to have the approximate value

$$r_0 \approx 1.2 \times 10^{-15}\, \sqrt[3]{A}\ \mathrm{m}. \tag{1}$$

2. Natural Radioactivity

The earliest source of information concerning the internal structure of the nucleus was the phenomenon of nuclear disintegration called *natural radioactivity,* discovered by HENRI BECQUEREL in 1896. The first naturally occurring radioactive isotopes to be discovered were those strongly active isotopes having atomic numbers between $Z = 81$ and $Z = 92$. In more recent years, lighter, naturally radioactive isotopes have been discovered. Two types of natural radioactive-disintegration processes have been observed. One of these types involves the emission of helium nuclei (called *alpha particles*). An example of this type of process is the disintegration of uranium 238; this nucleus spontaneously emits, after an average lifetime of 5×10^9 years, an alpha particle and becomes a thorium nucleus of mass number 234. The reaction involving the disintegration of uranium 238 can be written as

$$_{92}U^{238} \rightarrow {}_{90}Th^{234} + {}_2He^4.$$

The alpha particle is emitted with high kinetic energy, 4.2 MeV in the above reaction. The effect of alpha-particle emission is to decrease the mass number of the parent nucleus by 4 units and the atomic number by 2 units. In reactions of this kind the sum of the mass numbers of the product nuclei on the right is equal to the mass number of the original nucleus on the left; the same is true of the atomic numbers, since the atomic numbers measure charge, and electric charge is conserved. A heavy product nucleus like $_{90}Th^{234}$ may be left in an excited energy state and reach its lowest or ground state only by the emission of one or more high-energy quanta called *gamma rays*.

The second type of naturally occurring disintegration involves the emission of high-energy electrons (called *beta particles*). A typical beta-emission process occurs when the radium nucleus of mass number 228 disintegrates to form an actinium nucleus of mass number 228 and an electron. The reaction may be written as

$$_{88}Ra^{228} \rightarrow {}_{89}Ac^{228} + {}_{-1}e^0,$$

where the electron e is assigned an atomic number of -1 because of its unit negative charge and a mass number 0 because its mass is only 0.0005 u. When a beta particle is emitted, the atomic number of the product nucleus is always 1 greater than that of the parent nucleus, and the mass number is unchanged. The product nucleus may be left in an excited state and reach its ground state by gamma-ray emission.

Observations of the energies of the emitted beta particles show that, if energy and momentum are to be conserved in beta-emission processes, it is necessary to assume that another particle, called a *neutrino*, is emitted along with the beta particle. This particle was first postulated in 1934. The neutrino has no charge and has a mass that is much smaller than the electron mass; according to current theories the rest mass of the neutrino is zero. The above reaction should actually be written as

$$_{88}Ra^{228} \rightarrow {}_{89}Ac^{228} + {}_{-1}e^0 + {}_0\bar{\nu}^0,$$

where the symbol $_0\bar{\nu}^0$ represents the neutrino.

A particle such as the postulated neutrino, extremely light and with no charge, was expected to be extraordinarily difficult to detect. For 23 years, physicists accepted its existence because of several different lines of *indirect* evidence that required either the existence of such a particle or radical revision of some of the most fundamental principles of physics, such as the laws of conservation of energy, momentum, and angular momentum. Finally, in 1957, FREDERICK REINES AND CLYDE COWAN reported unambiguous evidence for the existence of the neutrino, based on experiments on its *direct* interaction with matter. Its existence is now well-confirmed by direct experiment.

The rate at which radioactive processes take place is usually measured by a quantity called the *half-life* of the material.

> The **half-life** of a radioactive material is the time required for the 'activity' of a sample of the material to decrease to one-half of its initial value, or, what amounts to the same thing, the time for half of the nuclei in the sample to disintegrate.

Three series of heavy radioactive nuclides are found in nature, starting with the long-lived nuclides $_{92}U^{238}$, $_{92}U^{235}$, and $_{90}Th^{232}$, which have half-lives of about 10^9 years, and emit alpha particles. After chains of 10 to 14 alpha and beta emissions, all three of these series end in different stable isotopes of lead. The intermediate nuclides in the chains all have half-lives short compared with that of the initial uranium or thorium parent; these half-lives range from 10^{-7} second to 10^4 years.

3. Nuclear Reactions

Further knowledge of the properties of atomic nuclei was next obtained by using the alpha particles from radioactive materials as projectiles for the bombardment of other nuclei. Beginning in 1919, Rutherford used alpha particles to produce *nuclear reactions*.

> A **nuclear reaction** is a reaction involving a change from one nuclear species to another.

Because of their high kinetic energy, naturally occurring alpha particles can approach very close to the atomic nuclei of the lighter elements, in spite of the Coulomb repulsive forces. In certain cases, they can 'penetrate' the nucleus of an atom and cause a change to another type of nucleus. Many nuclear reactions of this type may be thought of as consisting of two steps. The first step involves the entry of the bombarding alpha particle into the target nucleus, with the formation of an unstable compound nucleus. The second step is the immediate or almost immediate breaking up of the compound nucleus into the final products. As an example, the earliest alpha-particle reaction observed by Rutherford involved the nuclei in nitrogen gas:

$$_7N^{14} + {_2He^4} \rightarrow (_9F^{18}) \rightarrow {_8O^{17}} + {_1H^1}.$$

The penetration of the alpha particle into the nitrogen nucleus produces the compound nucleus $_9F^{18}$, an unstable isotope of fluorine, which almost immediately breaks up into a stable isotope of oxygen and a proton. Many reactions of this type, in which an alpha particle is captured and a proton is emitted, have been observed.

It was soon observed that the capture of an alpha particle by a nucleus does not always result in the emission of a proton. In one reaction occurring in the bombardment of beryllium by alpha particles, no charged particle, but a very penetrating type of 'radiation' was emitted. It was at first assumed that gamma rays were being emitted, but it was quickly found that the 'radiation' interacted with matter in quite a different fashion from gamma rays. JAMES CHADWICK showed in 1932 that the emitted 'radiation' consists of *neutral particles, each of mass very nearly equal to that of the proton*. These newly discovered fundamental particles, called *neutrons,* are formed as a result of the reaction

$$_4\text{Be}^9 + {_2\text{He}^4} \rightarrow ({_6\text{C}^{13}}) \rightarrow {_6\text{C}^{12}} + {_0\text{n}^1},$$

where $_0\text{n}^1$ is the symbol for the neutron, which has zero charge and mass number unity (*see* the first line of Table 45–1).

The discovery of the neutron gave further insight into the structure of the nucleus. Prior to its discovery, it had been assumed that the nucleus consisted of protons and electrons, but this assumption was very unsatisfactory for a number of reasons. It is presently accepted that:

A nucleus is composed of protons and neutrons.

The number of protons in a given nucleus is equal to its atomic number Z and the number of neutrons $(A-Z)$ is just sufficient to make up the remainder of the mass number A; for example, the $_2\text{He}^4$ nucleus contains 2 protons and 2 neutrons, while the $_{92}\text{U}^{238}$ nucleus contains 92 protons and 146 neutrons. The neutrons and protons that compose the nucleus are together called *nucleons:*

A **nucleon** is a constituent particle of atomic nuclei, that is, a proton or a neutron.

The mass number is the number of nucleons in a nucleus.

Neutrons behave quite differently from charged particles, and also quite differently from gamma rays. Charged particles interact strongly with the planetary electrons of atoms and usually fritter away their energy by ionizing the atoms or molecules of the materials through which they pass; they are brought to rest after passage through only a few centimeters of air or through thin metal foils. Gamma rays, being electromagnetic in character, can cause excitation and ionization of atoms and can interact with single electrons by Compton processes. Neutrons, on the other hand, have very little interaction with the extra-nuclear portion of atoms and are consequently very penetrating. Neutrons interact strongly *only* with nuclei; a neutron may have elastic collisions with nuclei, but in its passage through matter it is eventually 'captured' by a nucleus.

One type of capture process is that in which a neutron is captured by nitrogen and a proton is emitted, as described in the equation

$$_7N^{14} + {}_0n^1 \rightarrow ({}_7N^{15}) \rightarrow {}_6C^{14} + {}_1H^1. \tag{2}$$

In this process, the product nucleus $_6C^{14}$ is unstable and eventually, with a half-life of 5730 years, achieves stability by the beta-emission process

$$_6C^{14} \rightarrow {}_7N^{14} + {}_{-1}e^0 + {}_0\bar{\nu}^0. \tag{3}$$

These two reactions are illustrated schematically in Fig. 2. The net result of this chain of reactions is that the $_7N^{14}$ is regenerated, but the neutron is transformed into a proton, an electron, and a neutrino. This result is energetically satisfactory, even with an initial neutron kinetic energy close to zero, because, if a neutron does not interact with a nucleus, the spontaneous disintegration,

$$_0n^1 \rightarrow {}_1H^1 + {}_{-1}e^0 + {}_0\bar{\nu}^0,$$

will take place by itself, with a half-life of 13 minutes and an energy release of 0.78 MeV. *The neutron, unless it is bound in a nucleus, is unstable.*

When slow neutrons collide with nuclei, many reactions are observed in which the neutron is captured and the product nucleus achieves stability by emission of a gamma-ray photon. This process is called *radiative capture*. A typical radiative-capture reaction is

$$_{48}Cd^{113} + {}_0n^1 \rightarrow {}_{48}Cd^{114} + hf,$$

where the photon is represented by writing its energy hf. Neutron-capture processes in which an alpha particle is ejected are also observed.

On the presently accepted theory in which nuclei are composed of neutrons and protons, the electrons and neutrinos that are emitted when a nucleus undergoes a beta process must be thought of as *created* when a neutron changes to a proton in the nucleus. When a neutron changes to a proton, an electron must be emitted in order that the total charge may be conserved.

As long as the arsenal of bombarding particles available to the nuclear physicist consisted only of those occurring in natural radioactivity, the study of nuclear reactions was rather limited in scope. However, with the development of vari-

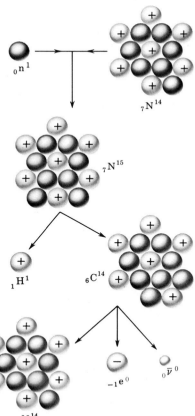

Fig. 2. The neutron-capture reaction of nitrogen.

ous types of accelerators—including the Van de Graff machine, the cyclotron, the betatron, and their modern variants discussed in earlier chapters—it became possible to employ protons, deuterons, electrons, high-energy photons, and various other particles to induce nuclear reactions. More important, it became possible to vary the energies of these particles in a controlled manner. In addition to producing and identifying nuclear reactions, it became possible to determine the minimum energies of bombarding particles required to produce reactions, thereby making it feasible to study 'energy balances' in equations representing the reactions.

By such means, it has been found that nuclei have definite, well-defined energy levels analogous to atomic energy levels (*see* Fig. 10, p. 851). In atoms, the lowest 'excited' levels are of the order of a few electron volts (eV) above the lowest or 'ground' level; the lowest excited energy levels in light nuclei are typically somewhat less than a million electron volts (MeV) above the lowest level. In the case of atoms, transitions between levels are accompanied by the emission or absorption of optical or X-ray photons; in the case of nuclei they are accompanied by the emission or absorption of γ-ray photons. A great deal of effort has now gone into the experimental determination of the energy levels of various nuclei, just as these levels for atoms were assiduously determined and tabulated earlier in the century. Theories are being developed to account in detail for the positions of these levels, as has been done in the case of atomic energy levels.

4. The Positron; Annihilation and Creation of Matter

In our discussion of Einstein's relativistic mechanics in Chapter 44, we pointed out that the relativistic expression for the total energy of a particle is $E = mc^2$ and that its kinetic energy is given by $E_K = m c^2 - m_0 c^2$, where m_0 is the rest mass of the particle. Since energy is relative, the rest energy $E_0 = m_0 c^2$ was at first generally regarded as merely constituting an arbitrary reference level from which observable kinetic energies could be measured, although Einstein himself attributed a 'real' significance to rest energy in the sense that he suspected that it could be transformed to other forms of energy.

However, until 1932, there was no experimental evidence that the *rest* energy of a particle could ever change to another form of energy. With the discovery of the *positron* in 1932, the process of the 'annihilation' of material particles (electrons and positrons) with the complete conversion of their rest energy into the energy of light quanta has become familiar, as has the inverse process, that of the 'materialization' of the energy of photons as the rest energy of electrons and positrons.

The rest masses of the electron and the proton are 0.000 548 6 and

1.007 28 u, respectively. Since the energies of nuclear reactions are ordinarily specified in MeV, we need to know the relativistic equivalence between mass in u and energy in MeV. From (21), p. 485, the definition of u on p. 334, and (19), p. 903, we find the equivalence factor

$$1 \text{ u} \sim 1.492 \times 10^{-10} \text{ J} = 931.5 \text{ MeV}$$

The energy equivalents of the rest masses of the electron and proton are

electron: 0.511 MeV,   proton: 938 MeV.

In 1932, CARL D. ANDERSON discovered a new elementary particle in the cosmic radiation. This particle is called a *positron;* it has a mass equal to that of the electron and a *positive* charge equal in absolute magnitude to the electron's negative charge. The positron has a transitory existence in the presence of terrestrial matter, and therefore had not been previously observed in investigations of atomic structure. Anderson found that positrons are created in a process called *pair production,* in which *a high-energy photon disappears and an electron-positron pair is created.* In other words, *the electromagnetic energy of the photon is converted into matter.* The energy relation for this process is

$$(hf)_{\text{Photon}} = (m_0 c^2 + E_K)_{\text{Electron}} + (m_0 c^2 + E_K)_{\text{Positron}}.$$

Since the rest energies of the positron and the electron are each equal to 0.511 MeV, the energy of the photon must be greater than 1.02 MeV.

When traversing matter, a positron is ultimately annihilated, along with an electron, by a process with the following energy balance:

$$(m_0 c^2)_{\text{Positron}} + (m_0 c^2)_{\text{Electron}} = 2 \, (hf)_{\text{Photon}}.$$

It is observed that *two* photons appear as the result of the annihilation of the positron-electron pair; each photon has an energy of approximately 0.5 MeV, which indicates that the electron and positron have little kinetic energy when they are annihilated. Hence kinetic-energy terms have been omitted on the left of the above equation.

Energy (including rest energy), momentum, and electric charge are all conserved in the processes of pair production and annihilation. *Pair production takes place only in the vicinity of some heavy nucleus* such as that of lead; recoil of the heavy nucleus in connection with the pair-production process is necessary in order to satisfy simultaneously the conditions for conservation of energy and momentum. In annihilation, the two photons travel in opposite directions, as is required for conservation of momentum. (Three-photon annihilation processes also sometimes occur, with the momentum and energy balance involving a recoiling material particle.) Charge conservation requires that electrons and positrons be created or annihilated in pairs, not individually.

Since Anderson's discovery, positrons have been observed to result also from many nuclear reactions. For example, the unstable $_7\text{N}^{13}$ formed in the reaction $_6C^{12} + _1H^1 \rightarrow _7N^{13} + hf$ decays by positron emission:

$$_7N^{13} \rightarrow _6C^{13} + _{+1}e^0 + _0\nu^0,$$

where $_{+1}e^0$ is the symbol for the positron.

It is interesting to note that the existence of the positron, and all of its properties, had been predicted several years before its discovery, by P. A. M. DIRAC, as a necessary consequence of *quantum mechanics*.

5. Transformation of Matter to Energy in Nuclear Reactions

The energies involved in ordinary chemical reactions such as combustion are so small that there is no experimentally detectable difference between the rest mass of the reactants and the rest mass of the reaction products. However, the energies involved in the binding of nucleons into nuclei are so large that mass differences are readily detectable in the case of the nucleon rearrangements that are involved in nuclear reactions. To study these nuclear energies systematically, we start by considering the basic problem of the formation of a stable atomic nucleus from protons and neutrons. Although we cannot, in general, produce stable nuclei from neutrons and protons in the laboratory in a manner comparable with the way in which we can produce carbon dioxide from carbon and oxygen, we can make very accurate comparisons of atomic masses by means of the mass spectrograph; and from these masses we can determine the energy that would be released in the formation of a nucleus from neutrons and protons.

The rest energy $M_{\text{Nucleus}} c^2$ of a nucleus is always *less* than the sum of the rest energies of the constituent nucleons, just as the rest mass of the molecule CO_2 is *less* than the sum of the rest masses of C and O_2, because energy (and hence mass) is released in the formation of CO_2. Correspondingly, energy would be released in the formation of any nucleus from its constituent nucleons, and energy would have to be supplied to break up any nucleus completely into its constituent nucleons. This energy is called the *binding energy*:

> The **binding energy** of a nucleus is the energy that would have to be supplied to break up the nucleus completely into free nucleons.

The term binding energy arises from the fact that this energy would have to be supplied to overcome the forces that *bind* the nucleons together in the nucleus, if the nucleons are to be pulled apart. This energy, which is so large as to be capable of being measured as a mass difference, is analogous to chemical binding energy.

A nucleus of atomic number Z and mass number A is built of Z protons and $A-Z$ neutrons. If we denote the rest mass of this nucleus, in u, by M_{Nucleus}, that of the proton by m_p, and that of the neutron by m_n, the binding energy of the nucleus is given, in u, by

$$E_B = [Z\, m_p + (A-Z)\, m_n] - M_{\text{Nucleus}}.$$

From mass spectroscopy, it is the mass of the neutral atom that is obtained, by adding the mass of the missing electrons to the measured mass of a positive ion that is usually singly or doubly charged. The mass of the nucleus is the mass of the neutral atom, minus the rest mass of the electrons and the mass equivalent of the binding energy of the electrons. The latter binding energy is always small compared with the nuclear binding energy; even in uranium it is only of the order of 1 MeV, while the nuclear binding energy is of the order of 1800 MeV.

If we neglect the binding energy of the extranuclear electrons, we can consider that the neutral atom would have the same mass as Z *neutral* H^1 *atoms* (which contain the same number Z of electrons as the atom) and $A-Z$ neutrons were it not for the nuclear binding energy. If we denote the rest mass of the *neutral atom* by M, that of the H^1 atom by m_H, and that of the neutron by m_n, the binding energy of the nucleus is approximately given by

$$E_B \approx [Z\, m_H + (A-Z)\, m_n] - M.$$

The values of m_H and m_n, and the values of M for selected atoms, were given in Table 45-1.

Example. *Determine the total binding energy and the binding energy per nucleon for the $_6C^{12}$ nucleus.*

From Table 45-1 and the last equation above,

$$E_B = [6\,(1.007\,825) + 6\,(1.008\,665)]\,u - 12\,u = 0.098\,94\,u.$$

The C^{12} nucleus has 0.0989 u or 92.0 MeV *less energy* than its constituent particles—this amount of energy would be released on formation of C^{12} from its constituent particles. Division by 12 gives binding energy of 7.67 MeV per nucleon, as plotted on Fig. 3.

Figure 3 shows the binding energies of various nuclei plotted against mass number. One curve shows the total binding energy in MeV, computed as in the example above. For the other curve, the binding energy is divided by the mass number A to give the binding energy per nucleon. This second curve is of the greatest interest because it shows that those nuclei with mass numbers between 50 and 80 give the highest binding energy per nucleon, and hence represent the most stable arrangements of nucleons. Both lighter and heavier nuclei are relatively less stable.

When two light nuclei combine to form a nucleus with higher mass number, a great deal of energy is released; similarly, by splitting a heavy nucleus into two or more nuclei with intermediate mass numbers, we can release large amounts of energy. Figure 3 is plotted 'upside-down' so

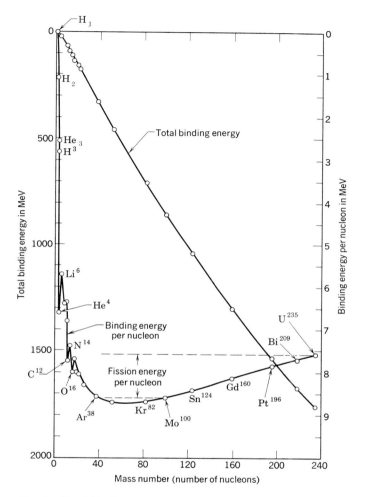

Fig. 3. Total binding energy and binding energy per nucleon as functions of mass number. While the points show only representative nuclei, all nuclei lie close to these curves.

that the most stable elements will be at the bottom of the curve showing binding energy per nucleon, and 'sliding downhill' on the lower curve represents release of energy.

Reactions in which heavy nuclei are split into nuclei of intermediate mass numbers, with the consequent release of energy of about 1 MeV per nucleon indicated in Fig. 3, are called *fission* reactions (*see* Sec. 6). We note that since 1 MeV per nucleon represents 23×10^9 kilocalories per kilomole of nucleons [(10), p. 571], *the fission energy release is of the order of a million times greater than that involved in any chemical reaction.*

Reactions in which lighter nuclei *combine* or *fuse* to form heavier nuclei are called *fusion* reactions. These reactions occur at high temperatures in the interiors of stars, where thermal energies are sufficient to initiate such nuclear reactions. The origin of stellar energy is probably a series of nuclear reactions, such as the following:

$$_1H^1 + {_6}C^{12} \rightarrow {_7}N^{13} + hf, \qquad _1H^1 + {_7}N^{14} \rightarrow {_8}O^{15} + hf,$$
$$_7N^{13} \rightarrow {_6}C^{13} + {_{+1}}e^0 + {_0}\nu^0, \qquad _8O^{15} \rightarrow {_7}N^{15} + {_{+1}}e^0 + {_0}\nu^0,$$
$$_1H^1 + {_6}C^{13} \rightarrow {_7}N^{14} + hf, \qquad _1H^1 + {_7}N^{15} \rightarrow {_6}C^{12} + {_2}He^4.$$

The net result of this set of six reactions is that we start with an ordinary nucleus of C^{12} and four nuclei of ordinary hydrogen and end up again with a C^{12} nucleus and a helium nucleus. The energy release is the difference between the energy of four hydrogen nuclei and one helium nucleus, which Fig. 3 shows to be 6.7 MeV per nucleon or 26.8 MeV total. Astrophysical evidence indicates that the composition of a stellar interior is about 35 per cent hydrogen and one per cent carbon. Using this composition and laboratory measurements of the rates of the above reactions, HANS BETHE showed that under an assumed temperature of 20 000 000° K, the above set of reactions would give a rate of energy release that coincides with that observed from the sun. At the rate at which the above reactions go with a 20 000 000-degree Maxwellian distribution of proton velocities, it is computed that one complete cycle requires an average of 5 million years. Stars are definitely not *exploding*, they are slowly 'burning' hydrogen to form helium.

Recent estimates of the temperature of the interior of the sun give 13 000 000° K. At this lower temperature the most probable set of reactions is

$$_1H^1 + {_1}H^1 \rightarrow {_1}H^2 + {_{+1}}e^0 + {_0}\nu^0$$
$$_1H^2 + {_1}H^1 \rightarrow {_2}He^3 + hf$$
$$_2He^3 + {_2}He^3 \rightarrow {_2}He^4 + {_1}H^1 + {_1}H^1.$$

The net result of this set of reactions—two each of the first two and one of the third—also represents the combination of four nuclei of ordinary hydrogen to form a helium nucleus. In the sun, this set of reactions appears to contribute 90 per cent of the energy. In hotter stars the C–N reactions proposed by Bethe probably predominate.

A high temperature is necessary to initiate such fusion reactions in a thermal source (as distinguished from a particle accelerator) because the positively charged nuclei on the left in the above reactions must travel with sufficient speed to overcome the Coulomb forces of repulsion, and must approach each other to within the short range, $r_0 \sim 10^{-15}$ m, where nuclear forces begin to act. Hence reactions of this type are called *thermonuclear* reactions.

6. Nuclear Energy

Beginning in the year 1934, ENRICO FERMI and his collaborators in Rome began a systematic study of nuclear reactions involving the capture of neutrons by various atomic nuclei. Experiments involving neutron bombardment of all of the elements tested gave reasonable results, *except* that neutron bombardment of uranium led to results that were not immediately interpretable. These results were correctly interpreted as *nuclear fission* in 1939 by OTTO FRISCH AND LISE MEITNER, in Copenhagen, when HAHN AND STRASSMANN, in Berlin, chemically identified *barium* as one of the reaction products appearing when uranium is bombarded by neutrons. The appearance of barium as a reaction product was interpreted as evidence that the uranium nucleus on capturing a neutron split into two nuclei of intermediate mass number. Since the energy per nucleon of nuclei of intermediate mass number is, as indicated in Fig. 3, about 1 MeV less than the energy per nucleon of uranium, energy of more than 200 MeV is released from each U^{235} nucleus that undergoes fission. This enormous quantity of energy is released when a slow neutron with only room-temperature thermal energy (approximately $\frac{1}{30}$ eV) is captured by a uranium nucleus; hence, it was immediately clear that if the fission process also released additional neutrons, this process could be used in power production.

Further studies of uranium fission showed that it was the relatively scarce* isotope $_{92}U^{235}$ that was split by slow neutrons and that *in addition to two elements of intermediate mass number, several neutrons are usually present among the reaction products*. For example, the following fission reaction might occur:

$$_{92}U^{235} + {}_0n^1 \rightarrow {}_{56}Ba^{144} + {}_{36}Kr^{89} + 3\,{}_0n^1.$$

The fission products $_{56}Ba^{144}$ and $_{36}Kr^{89}$ are both unstable since their mass numbers are larger than those associated with atomic numbers 56 and 36 in stable elements; in other words, these two isotopes contain too many neutrons and too few protons. However, by a series of beta-particle emissions, they eventually achieve stability, since the emission of a beta particle results in a reduction of the number of neutrons by 1 and an increase in the number of protons by 1 in a nucleus. It should be noted that the reaction given above is only one of many possible fission reactions. The known fission products are elements near the middle of the periodic table having atomic numbers in the range from $Z=34$ to $Z=58$ and mass numbers in the range from $A=70$ to $A=166$, somewhat to the right of the minimum in Fig. 3. Careful measurements indicate that the

---

*Natural uranium is 99.3 per cent U^{238}, 0.7 per cent U^{235}, with a trace of U^{234}. These isotopes are all naturally radioactive, with half-lives of 4×10^9 y, 7×10^8 y, and 2×10^5 y, respectively.

average initial kinetic energy of fission products is 168 MeV, and that 44 MeV of additional energy appears as kinetic energy of the neutrons and as energy associated with beta and gamma radiation from the fission products, making a total of 212 MeV per fission. Of this energy, 11 MeV is associated with neutrinos and is not available as heat.

The uranium isotope $_{92}U^{235}$ is the most abundant naturally occurring nucleus that undergoes a fission reaction produced by neutrons of very low energy.* However, two other nuclei, which have been produced as a result of radiative capture of neutrons, can also be split by slow neutrons. One of these is the uranium isotope $_{92}U^{233}$ formed when neutrons are captured by $_{90}Th^{232}$ to form $_{90}Th^{233}$, which undergoes two beta-emission process to form U^{233}:

$$_{90}Th^{233} \rightarrow {}_{91}Pa^{233} + {}_{-1}e^0 + {}_0\bar{\nu}^0, \qquad _{91}Pa^{233} \rightarrow {}_{92}U^{233} + {}_{-1}e^0 + {}_0\bar{\nu}^0.$$

The other artificially produced fissionable nucleus is $_{94}Pu^{239}$ (*plutonium*), which can be produced as a result of radiative capture of slow neutrons by the most abundant uranium isotope $_{92}U^{238}$ in the following reactions:

$$_{92}U^{238} + {}_0n^1 \rightarrow {}_{92}U^{239},$$
$$_{92}U^{239} \rightarrow {}_{93}Np^{239} + {}_{-1}e^0 + {}_0\bar{\nu}^0,$$
$$_{93}Np^{239} \rightarrow {}_{94}Pu^{239} + {}_{-1}e^0 + {}_0\bar{\nu}^0.$$

The $_{94}Pu^{239}$ nucleus is an alpha emitter and decays to $_{92}U^{235}$ with a half-life of 24 000 years. Hence, it remains with us for adequate time for use in nuclear reactors or nuclear weapons. The half-lives for the second and third reactions above, which are spontaneous beta emissions, are 23 min and 2.3 d, respectively; hence, once the neutron has been captured in the first reaction, the plutonium is formed fairly rapidly. We note that, even with a half-life of 24 000 years, plutonium is very dangerous to handle compared with U^{235} or U^{238}, both of which have half-lives of about 10^9 years.†

The fission processes occurring when samples containing $_{92}U^{233}$ or $_{94}Pu^{239}$ are irradiated with slow neutrons are similar to the fission processes we have described for $_{92}U^{235}$.

It was realized very quickly that fission could be used for power production, provided an abundant supply of neutrons were available to initiate fission processes. When it was found that two or three neutrons are released in every fission process, it became apparent that nuclear *chain reactions* might be produced by using these released neutrons to produce additional fissions. In order to understand how this process is possible,

---

*Others include protoactinium Pa^{231} and plutonium Pu^{239}, which occurs in trace amounts in uranium ores.

†Neptunium ($Z=93$) and plutonium ($Z=94$) are called transuranic elements. Additional transuranic elements with $Z > 94$ have been produced by nuclear reactions and are listed in the periodic table in Sec. 4 of the Appendix.

consider the schematic diagram in Fig. 4. In part (a) of this figure, a single neutron produces fission of a $_{92}U^{235}$ nucleus in a block of uranium metal, and, in addition to the fission product nuclei, three neutrons are assumed to be released. These neutrons are 'lost' either by escaping from the space occupied by the uranium or in radiative capture by $_{92}U^{238}$ or some other nucleus that may be present. Hence, this reaction is not self-sustaining. In part (b) of this figure is shown a chain reaction that is just self-sustaining. In each fission, on the average, one of the released neutrons produces the fission of another $_{92}U^{235}$ nucleus and two are lost by escape or by radiative capture. In part (c) is shown a chain reaction that involves high 'multiplication', in which every neutron released by a fission process produces additional fission of a $_{92}U^{235}$ nucleus. Hence, assuming 3 neutrons per fission, we would expect 3^N fissions in the Nth 'generation'; remembering that energy of 200 MeV is released during each process, we see that a great deal of energy would be produced in a very short time. The reaction shown schematically in part (b) of Fig. 4 would be the type in a nuclear *reactor* designed for sustained power production or for production of Pu by using the excess neutrons to initiate the last series of reactions listed above. In this case, when the number of fission chains has reached some value corresponding to the desired power level, each chain should be just self-sustaining on the average. The reaction shown schematically in part (c) is the type to be desired in a nuclear explosive device in which enormous amounts of energy are to be released in a very short time.

The neutrons that are created in a fission reaction are initially *fast*, with energies of several MeV. In a nuclear reactor, the neutrons are slowed down before producing additional fissions; if we wish to make a bomb that will detonate with explosive violence, we must utilize a *fast neutron* fission reaction. Not only U^{235} and Pu^{239} but also the common isotope U^{238} undergo fission with *fast* neutrons. If *more than one* of the neutrons generated by fission were, while still fast, to cause a new fission, the rate of energy generation would build up. This build-up will not take place with U^{238} because of competition of the radiative-capture process; either Pu^{239}, separated U^{235}, or U^{233} from thorium, must be used to sustain the

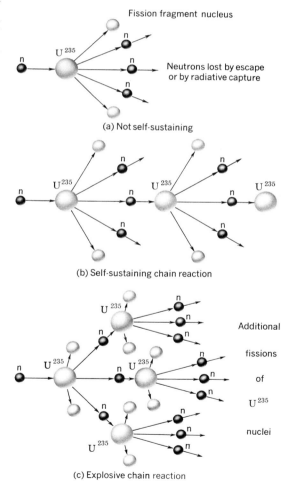

Figure 4

reaction in a bomb. However, U^{238} employed as a casing can add to the total energy released.

When a fission process takes place within the material of a bomb, let us see what may happen to the two or three neutrons that are generated. They may be radiatively captured in the bomb material or in an impurity; they may be elastically scattered and start to slow down; they may escape from the material—none of these processes is useful in generating an explosion. Finally, K of the two or three neutrons may cause another fast-neutron fission. K is known as the reproduction factor; and if K is appreciably greater than 1, the bomb will explode. Let us consider a spherical mass of U^{235} or Pu^{239}, free from impurities, and consider what happens as we gradually increase the radius and the mass of material in this sphere. When the sphere is very small, the neutrons initiated in the fission processes that are continuously being induced by cosmic rays will largely escape. The mean free path of a neutron, before it initiates a fission, will be large compared with the radius of the sphere; under this condition the reproduction factor will be less than unity. As we increase the size of the sphere, the probability of neutron loss by escape decreases; at a certain size, the neutrons generated by each fission will, on the average, produce one additional fission. The mass of fissionable material in such a sphere, which has $K=1$, is called the *critical mass*.

In order to be useful as an explosive device, fissionable material must be assembled into a *supercritical* mass with K substantially greater than 1, so that a great deal of energy can be released by fission processes before the assembly is blown apart. Detonation of the nuclear bomb requires the rapid assembly or compression of subcritical masses into a supercritical mass at the instant detonation is desired.

In a nuclear-fission bomb, sufficient energy is released to convert the uranium or plutonium metal into a highly compressed gas at a temperature of about 10 million degrees Kelvin. In expanding, this compressed gas forms a blast wave of enormous intensity in the surrounding atmosphere. The highly compressed, high-temperature gas expands as a 'ball of fire' which gives off intense ultraviolet, visible, and infrared radiation capable of igniting combustible material and causing severe burns at a considerable distance. Gamma rays and neutrons in lethal quantities are also produced. The energy released in the explosion of a fission bomb is ordinarily expressed in a unit called the *kiloton,* where 1 kiloton $= 3.8 \times 10^9$ J is the energy of 1000 tons of ordinary high explosive (TNT).

In view of the momentary high temperatures produced during the explosion of a uranium or plutonium fission bomb, it is possible to use a *fission* bomb as a detonator for a *nuclear fusion bomb* that utilizes the energy released when the nuclei of very light elements combine or fuse to form a single nucleus of greater mass. Such fusion reactions are called *thermonuclear reactions* because they would depend on a very high temperature to start them off and keep them going. Several such reactions involving hydrogen isotopes were early suggested for a fusion bomb, which therefore is popularly termed a 'hydrogen bomb.' Some of the suggested reactions and the resulting energy releases are:

$$_1H^2 + {}_1H^2 \rightarrow {}_1H^3 + {}_1H^1 + 4 \text{ MeV}, \qquad _1H^3 + {}_1H^3 \rightarrow {}_2He^4 + 2\,{}_0n^1 + 11 \text{ MeV},$$

$$_1H^2 + {}_1H^2 \rightarrow {}_2He^3 + {}_0n^1 + 4 \text{ MeV}, \qquad _1H^3 + {}_1H^2 \rightarrow {}_2He^4 + {}_0n^1 + 18 \text{ MeV}.$$

Other reactions involving heavier nuclei such as those of lithium or boron would also release substantial amounts of energy in a fusion process.

Any of these reactions can, in principle, be used, provided a sufficiently high temperature—about a hundred million degrees—is available to *start* the reaction; once started, the reaction (like a TNT chemical reaction) will be self-sustaining until the bomb blows itself apart. The energy released in a thermonuclear bomb is ordinarily expressed in *megatons*.

Thermonuclear reactions can, in principle, be employed as a source of energy for power production provided they can be initiated and provided a suitable 'container' can be constructed. Much work on these problems is currently in progress but many formidable technical difficulties are being encountered. Early experiments in which a magnetic field is used as a 'container' have given encouraging results (*see* pp. 622–624).

7. Nuclear Models

Thus far, we have described the nucleus only by saying that it is composed of protons and neutrons; we have said very little about the forces that hold these particles together. The nature of these forces is not clearly understood. The known facts, largely established by scattering experiments similar to those of Rutherford, are these: (a) a proton at a relatively large distance $r > r_0 \sim 10^{-15}$ m from the center of a nucleus of charge Ze is acted on only by the force of ordinary electrostatic repulsion, but at a shorter distance the proton is strongly attracted by so-called *strong forces;* (b) a neutron at a large distance $r > r_0$ from the center of a nucleus experiences no appreciable force, but at a shorter distance the neutron is strongly attracted by *strong* forces similar to those experienced by a proton. These short-range *strong* forces, which are the same for protons and neutrons, are different from and much stronger than any forces that could arise from gravitational, electrostatic, or magnetic effects. They must be very strong to hold the protons together in a nucleus in spite of the comparatively large electrostatic forces of repulsion experienced by protons at a very close distance.

We have noted in Sec. 1 that the nucleus is characterized by a definite radius that can be determined from studies of the scattering of fast particles; for example, when fast neutrons are scattered, their observed behavior is very much that to be expected if the scattering nuclei were solid spheres of radius $r_0 = 1.2 \times 10^{-15} \sqrt[3]{A}$ m. As we noted earlier, this value corresponds to a constant value for nuclear density; hence *the volume per nucleon is approximately constant in all nuclei*. Another general property of nuclei is illustrated by the *nearly* linear curve in Fig. 3 for the total binding energy of nuclei; for all but the lightest nuclei, the binding energy per nucleon is *roughly* constant. These facts, along with other evidence, suggest that nuclear forces act only between closely adjacent nucleons. In this case, a nucleon inside the nucleus would experience no resultant force, since forces exerted on it by surrounding nucleons would cancel. When a nucleon moving outward reached the surface, it would experience a strong force pulling it back toward the center of the nucleus. A nucleon approaching the nucleus from outside would experience a strong attractive force as soon as it reached the surface.

The above description of the nucleus bears so much similarity to a description of a liquid droplet that BOHR AND WHEELER proposed a *liquid-droplet model* for the nucleus. The liquid-droplet model accounts very nicely for the formation of a compound nucleus during a nuclear reaction (Sec. 3); when a particle is shot into a nucleus, an excited compound nucleus is formed and persists for about 10^{-14} s, after which time another particle is shot out. The droplet analogy is that of a fast molecule striking a droplet, entering the droplet, raising the energy and temperature of the droplet, and thereby causing the evaporation of another molecule from the surface of the droplet. The various excited energy states of a nucleus are interpreted on the Bohr-Wheeler model as vibrational states of the droplet.

A major success of the liquid-droplet model is in its description of the fission process. Bohr and Wheeler assumed that very heavy nuclei are almost unstable and that, when a slow neutron enters, the nuclear droplet is set into a mode of oscillation that causes a splitting into two smaller droplets. The model indicates that the two large fragment droplets would be of unequal size—as is observed in the case of fission fragments—and furthermore that several very minute droplets would usually be formed—corresponding to the emission of neutrons.

A charged liquid droplet can rotate and vibrate. The quantum-mechanical description of rotation and vibration has been used with considerable success in accounting for the observed excited states of heavy nuclei. The liquid-droplet model is less successful in accounting for the properties of light nuclei containing few nucleons. It also fails to account for 'direct' nuclear reactions in which energy is imparted to a single nucleon and the reaction is completed without the formation of a compound nucleus.

Light nuclei and direct reactions involving single nucleon excitation can be discussed in terms of the *central-field model*, in which each nucleon is considered as moving in a spherically symmetric force field resulting from all the other nucleons. Since a *neutron* experiences *no* force at distances R appreciably larger than r_0, very strong nuclear forces near the nuclear surface $R=r_0$ and again no resultant force at R appreciably smaller than r_0, its potential energy can be approximated by the heavy curve in Fig. 5(a). A *proton* experiences Coulomb repulsion at $R>r_0$, strong nuclear forces at $R=r_0$, and no resultant force at $R<r_0$; the proton's potential energy as a function of R is shown in Fig. 5(b). It is to be noted that protons and neutrons are free to move about inside the nucleus but are restrained from escaping. The allowed energies E_1, E_2, $\cdots$ of the trapped nucleons can be determined in terms of the nuclear radius r_0 and the depth D of the potential 'well,' by using quantum mechancis. Values of r_0 and D are chosen to give the experimentally observed values of one or two energy levels; then attempts are made to calculate the remaining energy levels.

The central-field model leads to the formation of *shells*, analogous to those formed by electrons in atoms. The *completion* of a shell accounts for the remarkable stability of the nuclei formed when the numbers of constituent nucleons are 4, 12, 16, $\cdots$, the so-called 'magic numbers' which had been noted earlier on the basis of purely experimental results, as shown in Fig. 3. This *shell model* accounts fairly well for the values of nuclear spin and, in some cases, for the magnetic moments, but gives no satisfactory account of fission. The complexity of calculations involved in the treatment of 'many-body problems' will probably always limit the accuracy of predictions made on the basis of the central-field model.

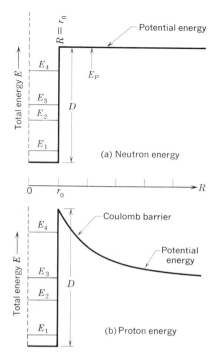

Fig. 5. Energies of neutron and proton in single particle theory. Heavy curves show potential energy as a function of distance; E_1, E_2, $\cdots$ show the allowed energies of a particle inside the nucleus.

PROBLEMS

1. From the fact that both the neutron and the proton have spins of $\frac{1}{2} h/2\pi$ and that the deuteron has a spin of $1\ h/2\pi$, what can be concluded concerning the spins of the nucleons in the deuteron? If there were no orbital motion of the nucleons in the deuteron, what would be the magnetic moment of the deuteron?   Ans: parallel and no orbital motion, or antiparallel and 1 unit of orbital angular momentum; $+0.880\ \mu_N$.

2. From the spins listed in Table 45-1, what can be concluded concerning the spins of the nucleons in $_1H^3$ and $_2He^3$? In view of the great stability of $_2He^4$, what can be concluded as to 'closed shells' of neutrons and protons?

3. Compute the binding energy per nucleon of $_6C^{12}$ and $_7N^{14}$ in MeV. The higher binding energy per nucleon for $_6C^{12}$ is an example of the saturation effects for nuclear forces when the nucleus is composed of a whole number of alpha particles (cf. Prob. 2).
    Ans: $_6C^{12}$, 7.67 MeV; $_7N^{14}$, 7.47 MeV.

4. The isotope $_{83}Bi^{209}$ has mass 208.980 u, $_{36}Kr^{82}$ has 81.913 u. What is the *difference* in MeV of binding energy per nucleon for these two nuclei, one near the right of the curve in Fig. 3 and one near the minimum?

5. Calculate the total energy released in the first series of reactions described on p. 924 if four moles of protons combine to form a single mole of helium. Ans: 2.58×10^{12} J.

6. Calculate the total energy, in J and MeV, released if a proton combines with a $_3Li^7$ nucleus to form two helium nuclei.

7. If $_4Be^9$ is bombarded with 3.00-MeV alpha particles to form $_6C^{12}$ and a neutron in accordance with the reaction on p. 917, find the resulting kinetic energy. This kinetic energy will mostly reside in the lighter neutron.
    Ans: 7.53 MeV.

8. When $_7N^{14}$ is bombarded with 3.00-MeV alpha particles to form $_8O^{17}$ (16.999 13 u) and a proton, as in the reaction on p. 916, find the resulting energy in MeV. This kinetic energy will mostly reside in the lighter proton.

9. Taking 200 MeV as the energy released during the fission of a single U^{235} nucleus, compute the number of fission processes taking place each second in a reactor operating at a power level of 4 MW. Ans: 12.5×10^{16}.

10. How much U^{235} is consumed each day in a nuclear reactor operating at a power level of 12 MW? How much bituminous coal would be required to release this amount of energy?

11. What is the total decrease each day in the total *mass* of the material in a uranium reactor operating at a power level of 6 MW?
    Ans: 5.75×10^{-3} g.

12. If a reactor designed for ship propulsion delivers 60 MW of power continuously, what is the consumption of U^{235} in kilograms per day? What is the equivalent consumption of coal (7500 kcal/kg) in kg/day? Assume 100 percent efficiency in each case.

13. The fuel value of coal is measured by the heat of combustion. Using 200 MeV as the energy release per fission, calculate the 'heat of fission' for pure U^{235} in J/kg and in kcal/kg.
    Ans: 8.21×10^{13} J/kg; 1.96×10^{10} kcal/kg.

14. To how many tons of coal (7500 kcal/kg) is one kilogram of U^{235} equivalent in fuel value?

15. According to official estimates, the nuclear bomb at Hiroshima was equivalent to 20 000 tons of TNT. Assuming 3.8×10^9 J as the energy released by detonation of 1 ton of TNT, find the number of fissions occurring, the total amount of U^{235} consumed, and the total mass decrease involved in the explosion of the bomb.
    Ans: 2.37×10^{24}; 0.925 kg; 8.45×10^{-4} kg.

16. Show that the complete combustion of 3 tons of carbon makes the combustion products only 1 milligram lighter than the 11 tons of carbon and oxygen entering the reaction.

17. A thermonuclear bomb test is rated as a '1-megaton' shot. This term means that the explosion releases as much energy as 1 million tons of TNT. Using 3.8×10^9 J/ton as the heat of detonation of TNT, calculate the total energy release in joules and in kcal.
    Ans: 3.80×10^{15} J; 0.910×10^{12} kcal.

18. How many kg of steam at 100° C would provide an energy release equivalent to that of Prob. 17 if the steam were condensed and the resulting water cooled to 20° C? Find the volume of the water. What area would the water cover if it were deposited in a layer 5 cm thick—corresponding to a '2-inch rain'?

19. From measurements of energy received at the earth, it is estimated that the total energy radiated by the sun is 1.2×10^{34} joules per year. If all of this energy comes from fusion reactions, what is the loss of mass in kg per year? If the sun were to continue to lose mass at this rate, for how many years would its mass last?
    Ans: 1.34×10^{17} kg/y; 1.49×10^{13} y.

20. Attempts at controlled release of thermonuclear energy are being conducted in a 'plasma' of ionized deuterium gas. The deuteron-deuteron reactions

$$_1H^2 + {}_1H^2 \rightarrow {}_1H^3 + {}_1H^1 + 4.0 \text{ MeV},$$
$$_1H^2 + {}_1H^2 \rightarrow {}_2He^3 + {}_0n^1 + 3.2 \text{ MeV},$$

which would occur with equal frequency and be followed by

$$_1H^2 + {}_1H^3 \rightarrow {}_2He^4 + {}_0n^1 + 17.6 \text{ MeV},$$

would release 5 MeV per deuteron. The difficulty is that deuteron energies of at least 10 000 eV are required to overcome the Coulomb repulsion and enable these reactions to take place. Show that to give an average thermal energy of 10 000 eV to a deuteron in a gas requires a temperature of 116 million degrees Kelvin. Once deuterons of this energy or higher are available to 'kindle' the reaction, the high energy of the resulting products would be sufficient to keep the temperature up and continue the reaction under proper plasma conditions. Why is the required temperature much greater than that in the interiors of the stars?

21. In the second reaction listed in Prob. 20, verify that the products of the reaction have 3.256 MeV more kinetic energy than the two deuterium nuclei.

22. For a self-sustaining thermonuclear reaction in a deuteron plasma as described in Prob. 20, it is estimated that the plasma temperature must be 350 million degrees. The problem of a material container for this plasma is insoluble, so experiments are directed toward use of a magnetic field. We have seen that charged particles will move in helices about straight magnetic lines; this motion will serve somewhat to 'contain' charged particles. However, in the case of a dense beam of particles, another effect can be even more important—the 'pinch' effect. The figure shows a plasma current of circular cross section moving into the paper. Near the edge of the circle, there will be circular magnetic lines. Show that the resulting force on the particles near the edge is toward the center of the circle, thus creating the 'pinch.' Show that this force does work in accelerating these particles and thereby assists in raising the temperature. If the diameter of the circle in the figure is 1 cm and the total deuteron current is 1 A, compute the force on a deuteron at the edge of the beam and the resulting acceleration in g's.

23. The measured half-life for alpha emission by radium is 1600 years. The rate of alpha emission, measured by counting scintillations, is 3.7×10^{10} particles per second per gram of radium. From these data, and the known atomic mass of radium, determine Avogadro's constant.

24. Cosmic rays react with the nitrogen in the atmosphere to produce radioactive carbon-14 in accordance with the reaction (2) on p. 918. Since carbon-14 disintegrates with a half-life of 5730 years to recreate stable nitrogen, according to (3), the ratio of C^{14} to C^{12} in the atmosphere has long since stabilized. The isotopes in all plants and animals are stabilized in this same ratio until they die, when no more carbon is assimilated. Thus measurement of the radioactivity of plant or animal specimens (wood or bones) can be used for *dating* their age. If there are 13 disintegrations per minute per gram of living material, how many are there for material 5730 years old? 57 300 years old? The latter age is at about the limit of accuracy of present *radioactive-carbon-dating* techniques.

25. Assuming that a nucleus is a uniform solid sphere of radius r_0 given by (1), p. 914, compute the density of the nuclear material in Mg/m³. Ans: 2.29×10^{14} Mg/m³

26. First observed in 1951 by Martin Deutsch, an 'atom' called *positronium* can be formed from an electron and a positron, in analogy to a hydrogen atom formed from an electron and a proton. However, positronium has a lifetime of only about 10^{-7} s before the electron and the positron undergo mutual annihilation with the emission of gamma-ray quanta.

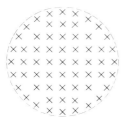

Problem 22

The allowed energy levels of positronium are quantized by the same rule, $J_n = nh/2\pi$ [equation (8), p. 853], as in the Bohr theory of the structure of hydrogen. However, in our treatment of hydrogen we could, to an excellent approximation, assume that the massive proton remained at rest and compute just the angular momentum of the electron in its motion about the proton. In the case of positronium, the two particles have the same mass and rotate about their center of mass, with angular momentum associated with each of the particles. We must take J_n as the *total* angular momentum about the center of mass of the system. If r_n is the distance of separation of the electron and the positron, show that the allowed values of r_n are *twice* the values given by (10) and (11), p. 853, for hydrogen.

46 HIGH-ENERGY PHYSICS: ELEMENTARY PARTICLES

We have described the structure of atoms in terms of electrons bound to positive nuclei by electrostatic forces and the structure of nuclei in terms of protons and neutrons held together by strong short-range forces. We have noted that visible light is emitted when an atom in an excited state a few eV above its ground state makes a transition to a lower state; while X rays are emitted in a transition from an excited state a few keV above the ground state. Similarly, we have noted that nuclei have excited states with energies of a few MeV and may emit particles as well as gamma-ray photons in making transitions to lower states. In the present chapter we discuss phenomena that occur *in the GeV energy range*. Some of these phenomena were originally observed in studies of 'cosmic rays' that come from external sources and continually enter the earth's atmosphere; however, most of the phenomena with which high-energy physics is concerned could be studied systematically only after particle accelerators operating in the GeV range were developed at such centers as the Brookhaven National Laboratory on Long Island, the Lawrence Radiation Laboratory in Berkeley, the European Center for Nuclear Research (CERN) in Geneva, and the Soviet center in Dubna.

Studies of high-energy physics have added to our understanding of the short-range forces acting between nucleons, have revealed the existence of muons and mesons having masses in the intermediate range between the mass of the electron and that of the nucleon, and have shown that nucleons themselves have excited states. Such studies have provided new insights into the nature of the elementary particles of matter.

In this chapter we discuss the nature of the forces that act between particles. As noted earlier, *gravitational forces* between charged particles are negligibly small compared with *electromagnetic interactions,* which, in turn, are small compared with the *strong interactions* between the nucleons inside the atomic nucleus. The *weak interactions* involved in β-decay processes of radioactive decay are usually intermediate between the electromagnetic and the strong interactions. We begin with a reconsideration of the already familiar electromagnetic interactions.

1. Electromagnetic Interactions

When an electron collides with an atom in its ground state, the atom can be raised to one of its characteristic excited states, provided the kinetic energy is sufficiently great. Radiation is emitted when the excited atom returns to its ground state. We can think of the process as occurring in the following way:

$$\text{e} + \text{atom} \rightarrow \text{atom}^\circ + \text{e} \rightarrow \text{e} + \text{atom} + \text{photon.} \tag{1}$$

Momentum, angular momentum, and energy are conserved in the collision but some of the initial kinetic energy of the electron is transformed into the internal energy of the atom, which in its excited state is denoted by atom$^\circ$. When the atom returns to its ground state a photon is emitted. We can thus think of the photon as *created* during the collision; the photon is characterized by energy hf, momentum hf/c, and angular momentum $h/2\pi$.

In our earlier treatment of atomic structure (Chapters 41, 42), we indicated that a moving electron in an atom is subject to electrostatic and magnetic forces that are due to fields produced by the nucleus and the other electrons. Since electromagnetic fields are propagated with the speed of light, the intra-atomic electromagnetic forces can be associated with *virtual photons* within the atom. By adding sufficient energy to the atom, we can create *real photons.* Real photons can be observed by the detection methods discussed in our treatment of electromagnetic radiation; virtual photons cannot be directly observed.

2. Strong Interactions: Pions

In 1935 the Japanese physicist YUKAWA attempted to account for the strong, short-range forces between nucleons in terms of a quantum field theory similar to the one we just described for the electromagnetic interaction. Yukawa's theory involved virtual particles analogous to the virtual photons involved in the electromagnetic interaction; however, the particle required by Yukawa had a *finite* rest mass in contrast to the zero rest mass of the photon. Order-of-magnitude calculations showed that the rest mass of the Yukawa particle was somewhat greater than 200 m_e. Since its mass is greater than the electron mass but less than the proton mass, Yukawa's hypothetical particle was called a *meson*.

The production of an observable meson required a collision between nucleons in which the initial kinetic energy is sufficient for the creation of the meson, that is, we must have $E_K > (200\, m_e)\, c^2$. Thus, in analogy with (1), we write

$$\text{nucleon} + \text{nucleon} \rightarrow (\text{nucleon} + \text{nucleon})^\circ \rightarrow \text{nucleon} + \text{nucleon} + \text{meson}. \quad (2)$$

Since the rest energy of the electron is 0.511 MeV, the energy required in (2) is greater than 200×0.511 MeV $= 102$ MeV. In 1935 there were no particle accelerators capable of providing energies of this magnitude. However in 1947 POWELL and his collaborators in England discovered a cosmic-ray particle—now called the π-*meson* or *pion*—that has a mass of 273 $m_e = 140$ MeV. The pion interacts strongly with the nucleons, as required by Yukawa's theory.

With the development of GeV accelerators in the 1950's, much more has been learned about the pion. When nucleons collide as suggested in (2), pions are indeed produced when the initial kinetic energy exceeds 140 MeV. The pion may have electric charge $+e$, 0, or $-e$; these charge states of the pion are denoted by π^+, π^0, and π^-, respectively. They are produced in the following reactions involving the proton p and neutron n:

$$\left.\begin{array}{lll} p+p \rightarrow p+n+\pi^+ & p+p \rightarrow p+p+\pi^0 & n+p \rightarrow p+p+\pi^- \\ p+n \rightarrow n+n+\pi^+ & n+n \rightarrow n+n+\pi^0 & n+n \rightarrow n+p+\pi^-. \\ & p+n \rightarrow n+p+\pi^0 & \end{array}\right\} \quad (3)$$

Momentum, energy, and electric charge are conserved in all reactions (3); we note also that the total number of nucleons before and after collision remains the same. On the assumption that the pion has zero spin, we note that spin angular momentum is also conserved.

Pions are readily produced in sufficiently energetic collisions; hence we might expect pions to be abundant since they are being produced continuously by cosmic rays. Actually, pions are extremely unstable and decay spontaneously. The neutral pion has a mean life of approximately 10^{-16} s and decays by a process involving the creation of two photons γ:

$$\pi^0 \rightarrow \gamma + \gamma. \quad (4)$$

Since the pion has zero spin, *two* photons with oppositely directed spins

$1\ (h/2\pi)$ are required for spin conservation as well as to provide for momentum conservation. Charged pion decay involves the weak interactions discussed in the next section.

3. Weak Interactions

In addition to strong interactions involving pions, nucleons are subject to the *weak interactions* of the type that produce β-decay of radioactive nuclei. We recall that these β-processes involve no change in atomic mass number and that the β-particle emitted can be either an electron (p. 915) or a positron (p. 921). Such β-processes always involves neutrino emission, and proceed as follows:

$$_Z(\text{nucleus}^*)^A \rightarrow {}_{Z+1}\text{nucleus}^A + {}_{-1}e^0 + {}_0\bar{\nu}^0, \tag{5}$$

$$_Z(\text{nucleus}^*)^A \rightarrow {}_{Z-1}\text{nucleus}^A + {}_{+1}e^0 + {}_0\nu^0. \tag{6}$$

Although the neutrinos are characterized by spin $\frac{1}{2}\ (h/2\pi)$ and by zero rest mass, the neutrino denoted by ${}_0\nu^0$ differs from the one denoted by ${}_0\bar{\nu}^0$ in one extremely important respect to be discussed in Sec. 6.

We note that the end result of (5) is to change a neutron into a proton and to create an electron and neutrino:

$$\text{n} \rightarrow \text{p} + \text{e} + \bar{\nu}. \tag{5'}$$

As we have seen (p. 918), a decay process of this kind is the ultimate fate of a *free* neutron; process (5′) is energetically possible because the mass of the neutron is sufficiently larger than the proton mass to provide for the creation of e and $\bar{\nu}$. The end result of (6) is to change a proton into a neutron

$$\text{p} + \Delta E \rightarrow \text{n} + \bar{\text{e}} + \nu, \tag{6'}$$

where the positron is denoted by $\bar{\text{e}}$. We note that (6′) is energetically possible only if sufficient energy ΔE is added to provide the difference in mass between neutron and proton and to provide the energy required for the creation of $\bar{\text{e}}$ and ν. In the radioactive decay process (6), the required energy ΔE is supplied by the excited parent nucleus. The *free* proton is stable and never spontaneously changes into a neutron.

From the standpoint of high-energy physics, the value of ΔE in (6′) is so small that the neutron and proton can be regarded as merely different charge states of the same particle: *the* nucleon. The similarities of the roles of protons and neutrons in reactions (3) are in accord with this view. Other evidence of the similarity of protons and neutrons subject to strong interactions is supplied by studies of 'mirror' nuclei. Two nuclei constitute a mirror pair if they contain the same total number of nucleons and if the number of protons in one nucleus is equal to the number of neutrons in the other. In the case of the excited energy levels of the mirror nuclei $_3$Li^7 and $_4$Be^7 shown in Fig. 1, the similarity in position and spacing of the levels is striking. Thus, although the neutron and proton are markedly different in weak interactions involving electrons and neutrinos, they are very similar in strong interactions involving pions.

We now return to a consideration of the fate of the charged pions produced in (3). Unless a charged pion reacts with a nucleon or another meson, it experiences spontaneous decay after a mean lifetime of 2.6×10^{-8} s by the weak interactions.

$$\pi^+ \to \mu^+ + \nu_\mu \quad \text{or} \quad \pi^- \to \mu^- + \bar{\nu}_\mu, \quad (7)$$

where μ denotes a particle now called the *muon*. The muon has a rest mass of $207\, m_e = 105.7$ MeV and was discovered independently in 1936 by STREET AND STEVENSON and by ANDERSON AND NEDDERMEYER in studies of cosmic rays. The neutrino created along with the muon *differs* from the neutrino created along with electrons in nuclear β-decay; we note also that the ν_μ associated with μ^+ differs from the $\bar{\nu}_\mu$ associated with μ^- (see Sec. 6).

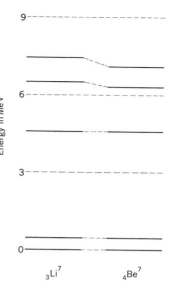

Fig. 1. The energy levels of these mirror nuclei are similar because strong interactions are similar for neutrons and protons.

The muon has very little interaction with nucleons but does interact with electrons and thus loses kinetic energy by ionizing the atoms of matter through which it passes. After a mean lifetime of approximately 2.2×10^{-6} s, muons undergo spontaneous decay involving weak interactions:

$$\mu^- \to e^- + \bar{\nu} + \bar{\nu}_\mu \quad \text{or} \quad \mu^+ \to e^+ + \nu + \nu_\mu, \quad (8)$$

in which two types of neutrinos are created. The muon, like the rest of the particles in (8), has an intrinsic spin angular momentum of $\frac{1}{2}$ $(h/2\pi)$.

The positive muon forms, with the electron, a hydrogen-like 'atom' of transitory existence, whose properties have been observed. This 'atom' is called *muonium*.

4. Mesons and Baryons

Extensive experiments with high-energy particle accelerators show that the pions can be regarded as the lowest energy states of a whole family of mesons. The known mesons are shown in Fig. 2; the energies corresponding to their creation, $E = mc^2$, are plotted vertically. The spin angular momentum of each particle in units of $h/2\pi$ is indicated at the

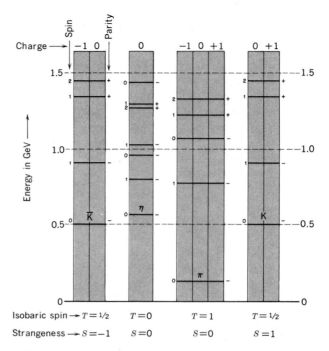

Fig. 2. Energy-level diagram for mesons. Even the lowest level (the pion π) is unstable.

left of the horizontal line representing the particle; we note that the spins of mesons are either zero or integral. The charge of the particles in a given column is given at the top of the column; the charge of a meson is either $+e$, 0, or $-e$. The columns are grouped according to the *strangeness S* and an *isotopic spin T* listed just below each group of columns; strangeness and isotopic spin are quantities employed in the current quantum-mechanical treatment of mesons. The parity of each particle is given by the $+$ and $-$ signs given at the right; parity is connected with the symmetry properties of the wave function used to characterize the particle.

The lowest mesons in each column are subject to decay, either by two-photon emission or by weak interactions. The mesons denoted by K and $\overline{\text{K}}$ are called *kaons* and take part in many reactions. The mesons of higher energy can change to mesons of lower energy by processes involving single-pion emission, two-pion emission, or kaon emission.

The nucleon in its two charge states, the proton and neutron, is the first member of another set of particles called *baryons*. The known baryons are shown in Fig. 3, which gives information similar to that given in Fig. 2. We note that all baryons are characterized by half-integral spin numbers and cover a larger range of strangeness and charge than do the mesons. Baryons of higher energy can reach lower states by pion or kaon emission in all cases except for Λ, Σ, Ξ, and Ω. These latter mesons

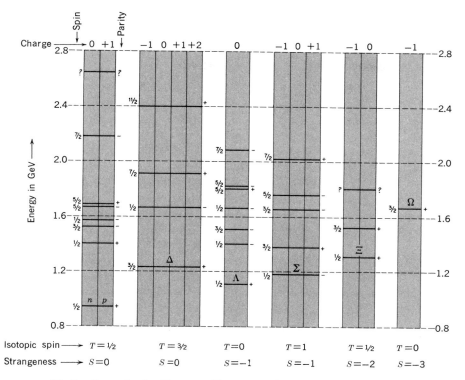

Fig. 3. Energy levels of baryons. The proton p is stable; the neutron n is nearly stable.

achieve stability by one or more weak transitions, eventually attaining stability by becoming nucleons.

We note one important difference between the meson plot in Fig. 2 and the baryon plot of Fig. 3. The *mesons have no stable states;* even the pions eventually disappear by two-photon emission or by weak decay involving muons. The *proton is a stable particle.* It is plotted on Fig. 2 at its rest energy (mass) of 937 MeV. In Table 46-1, p. 942, we give a list of the mesons and baryons that are stable against strong decay, along with their mean lifetimes. We note that longest mean lifetimes listed in the table are of the order of 10^{-8} s; the mean lives of particles subject to strong decay have even shorter mean lives. The decay products of each baryon always include another baryon; this is an example of *baryon conservation.*

5. Detection Devices for High-Energy Particles

A glance at Table 46-1 is sufficient to raise questions about the methods used to detect and identify particles that exist for such brief periods of time, since it is apparent that particles in the GeV energy range cannot be studied in complete detail by many of the conventional instru-

Table 46-1 Mesons and Baryons: Mean Lifetimes[a]

	Particle	Mass (MeV)	Mean life (s)	
Mesons	$\pi^\pm$	139.58	2.60×10^{-8}	
	π^0	134.97	0.89×10^{-16}	
	$K^\pm$	493.8	1.23×10^{-8}	
	K^0	497.7	K_{short}	0.87×10^{-10}
			K_{long}	5.3×10^{-8}
	η	549	—	
Baryons	$p^\pm$	937	stable	
	n	938	1×10^3	
	Λ	1115.4	2.52×10^{-10}	
	Σ^+	1189.5	0.81×10^{-10}	
	Σ^0	1192.5	$< 10^{-14}$	
	Σ^-	1197.4	1.7×10^{-10}	
	Ξ^0	1315	2.9×10^{-10}	
	Ξ^-	1321	1.73×10^{-10}	
	Ω^-	1672	1.1×10^{-10}	

[a] Source: A. H. Rosenfeld et al., University of California Radiation Laboratory Publication 8030.

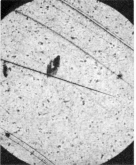

Fig. 4. Cosmic-ray particle ($Z \approx 20$) enters a photographic emulsion and comes to rest; the length of the track is 2100 nm. (Photo courtesy of Dr. Marvin Querry, University of Missouri.

ments of nuclear physics such as ionization chambers, cloud chambers, and Geiger counters.

One of the detection devices used earliest in high-energy physics was the photographic plate. By using stacks of plates with specially prepared thick photographic emulsions, exposing the plates to the particles of interest, and then developing the plates, the investigator can reconstruct the path of an ionizing particle by microscopic examination of the black 'track' as shown by reduced-silver centers in the emulsion, as in Fig. 4. By counting the blackened centers near the track, he can obtain estimates of energy losses by the ionizing particle. A sudden change in the path direction indicates a collision; a sudden change in the ionization per unit path length indicates that a different particle is involved, produced by a decay process or by a collision.

Other detection devices include the *bubble chamber* (see p. 372), in which the path of a charged particle can be traced by the track of bubbles in the liquid (Fig. 5), and the *spark chamber* (Fig. 6), in which the path of an ionizing particle is traced by a series of sparks in the gas between adjacent metal or carbon plates maintained at potentials that provide electric fields in the gas sufficient to produce a spark discharge whenever a high-energy particle produces ions. By placing a detection chamber in an external magnetic field, the observer can determine the momentum of a particle by measurement of the track curvature. Most of the high-energy particles of interest move at speeds nearly equal to the speed of light; hence, the lifetime of a particle can be determined by measuring the total length of its path, provided the entire path can be traced in a bubble chamber or spark chamber. Photographs of the tracks of high-energy charged particles are shown in Figs. 4, 5, and 6.

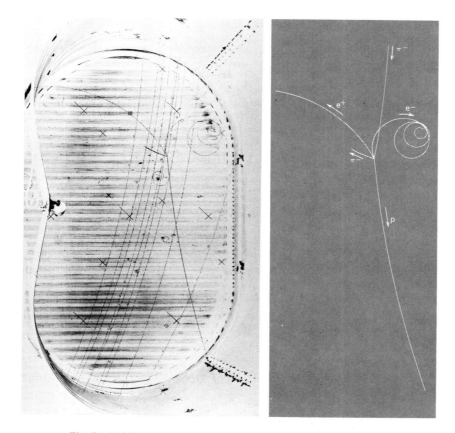

Fig. 5. Bubble-chamber record of a collision between pion π^- and a proton p, which was initially at rest. A neutral pion π^0 was formed during the collision and decayed almost immediately into a positron (e^+) and electron (e^-) pair and a photon γ. The neutral pion π^0 and photon γ left no tracks. (Courtesy of Prof. Owen Chamberlain and Dr. William Chinowsky, University of California.)

We note that the detection of neutral mesons and baryons presents special problems since neutral particles leave no tracks. The properties of neutral particles can be determined from the properties of the particles producing them and from the properties of their decay products. The path of a neutral particle between the point of creation and the point of decay can be determined by use of momentum considerations.

Example. *On the assumption that all particles have nearly the speed of light, compute the total distances d traversed by particles having lifetimes t of* 10^3 s *(neutrons),* 10^{-6} s *(muons),* 10^{-8} s *(charged pions and kaons),* 10^{-10} s *(charged baryons), and* 10^{-16} s *(neutral pions).*

From the relation $d = ct = (3 \times 10^8 \text{ m/s}) t$, we obtain the following results: for $t = 10^3$ s, $d = 3 \times 10^{11}$ m; for $t = 10^{-6}$ s, $d = 3 \times 10^2$ m; for $t = 10^{-8}$ s,

$d = 3$ m; for $t = 10^{-10}$ s, $d = 3 \times 10^{-2}$ m $= 3$ cm; for $t = 10^{-16}$ s, $d = 3 \times 10^{-8}$ m $= 30$ nm. Bubble-chamber measurements could clearly not be used for measuring the half-lives of fast neutrons or muons. Since 30 nm is less than the wavelength of light, the lifetime of neutral pions could not be measured in the way suggested.

NOTE: In this simple calculation we have assumed that the lifetimes apply to the laboratory frame of reference. The mean lifetimes or half-lives listed in tables give these quantities in the rest frames of the particles; time dilatation considerations (p. 898) must be used if we wish to determine path lengths. Thus, the lifetime of the muon is 2.21×10^{-6} s; this is the value determined experimentally for muons that are 'stopped' by ionizing collisions in some material such as iron. However, the muons created by the pions produced by cosmic rays in the atmosphere move at speeds of $0.995\,c$. If we try to measure their half-lives by measuring the numbers present in the upper atmosphere and at sea level, the measured half-life is much greater; the observed half-life is actually increased by a factor of 8.8, in excellent agreement with the value to be expected from relativity considerations: $t_{\text{moving}} = t_{\text{rest}}/\sqrt{1 - v^2/c^2}$.

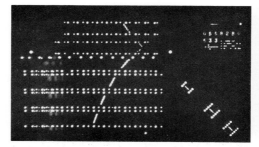

Fig. 6. Spark-chamber photograph. A negative pion π^- collided with a proton in a liquid-hydrogen target just outside the chamber. A neutral kaon K^0 and a Λ-particle were produced in the collision. The neutral Λ-particle decayed to a proton p and a negative pion π^- in the spark chamber; these charged decay products, p and π^-, produced the spark-chamber tracks. Neutral particles produce no tracks. The horizontal arrays of dots represent the location of the charged metal plates in the spark chamber. (Courtesy of Dr. Kurt Reibel, The Ohio State University.)

6. Antimatter

In our earlier discussion of the positron we noted (p. 921) that the existence of the positron and of its properties had been predicted by Dirac on the basis of quantum-mechanical considerations; the positron is the electron's *antiparticle*. Quantum mechanics requires that every particle have an antiparticle. If the particle is charged, the antiparticle has the opposite charge; if the particle is neutral, its antiparticle is also neutral but differs from the particle in some usually distinguishable respect. When a particle collides with its antiparticle, the pair undergoes annihila-

Table 46-2 A Partial List of Particle-Antiparticle Pairs

Group	Particle	Antiparticle
Leptons	electron, e or e^- neutrino, ν negative muon, μ or μ^- neutrino, ν_μ	positron, $\bar{e}$ or e^+ antineutrino, $\bar{\nu}$ positive muon, $\bar{\mu}$ or μ^+ antineutrino, $\bar{\nu}_\mu$
Mesons	positive pion, π^+ neutral pion, π^0 positive kaon, K^+ neutral kaon, K^0	negative pion, π^- neutral pion, π^0 negative kaon, K^- neutral kaon, $\bar{K}^0$
Baryons	proton, p neutron, n	negative proton, $\bar{p}$ antineutron, $\bar{n}$

tion. In the case of the electron-positron pair, we have noted earlier that the annihilation products consist of two photons.

Without mentioning their particle-antiparticle relationship, we have already introduced some of the particle-antiparticle pairs listed in Table 46-2. We note that the electron, together with its neutrino ν, and the muon, together with its neutrino ν_μ, constitute a class of particles called *leptons*, all members of which have a spin of $\frac{1}{2}$ (h/π). The electron-positron pair e, $\bar{e}$ differ in charge, as do the corresponding μ, $\bar{\mu}$ pair. According to current theories the neutrino-antineutrino pairs, ν, $\bar{\nu}$ and ν_μ, $\bar{\nu}_\mu$, consist of particles having zero rest mass and moving with the speed of light; the neutrino and antineutrino can be distinguished by differences in the orientations of their spin-angular-momentum vectors relative (parallel or antiparallel) to their linear-momentum vectors.

In the case of the charged mesons listed in Table 46-2, all particles and antiparticles are subject to spontaneous decay and have very short half-lives but are also subject to annihilation processes if they collide prior to decay. The neutral pion π^0 is its own antiparticle and subject to spontaneous decay with photon emission.

We list two baryons in Table 46-2, the proton and the neutron. On the basis of Dirac's theory there had been many speculations regarding the possible existence of a negative proton. The negative proton was discovered by SEGRÉ, CHAMBERLAIN, WIEGAND, AND YPSILANTIS in 1955 in nucleon-nucleon collisions involving energies somewhat above 5.6 GeV. The proton p and antiproton $\bar{p}$ annihilate in reactions of the type

$$p + \bar{p} \rightarrow 2\pi^+ + 2\pi^- + \pi^0; \tag{9}$$

all pions produced are unstable and subject to further decay processes. The antineutron has also been created in nucleon-nucleon collisions; it can be distinguished from the neutron since its magnetic moment is parallel to its spin whereas the magnetic moment of the antineutron is antiparallel to its spin. Annihilation of a neutron-antineutron pair involves pion creation. According to quantum mechanics, each baryon listed in Table 46-1 must have an antiparticle.

The existence of antiparticles raises some rather interesting questions regarding the nature of the universe. Since particles and antiparticles are equally stable, there is no reason to believe that negative atomic nuclei cannot be formed as a result of strong interactions between negative protons and antineutrons. The antideuteron, negatively charged and consisting of an antiproton and an antideuteron, was first observed in 1965 by the method of mass-spectroscopy. This particle was created in the 33-GeV synchrotron at Brookhaven. Similarly, there is no reason to believe that 'reversed atoms' cannot be formed as a result of electromagnetic interactions between positrons and negative nuclei. If such 'reversed atoms' were excited, they would emit light that would be indistinguishable from the light emitted by what we regard as 'normal atoms.' It is therefore possible that antimatter, consisting of reversed atoms, may exist in remote parts of the universe, which we can observe only by means of radiation reaching the earth.

However, if an appreciable amount of antimatter in the form of a meteorite were to strike the earth, an enormous explosion would occur as a result of annihilation processes. There has been serious speculation by WILLARD LIBBY and others that the meteorite which caused such extensive damage in northern Siberia in 1908 may have consisted of antimatter.

7. Elementary Particles

What are the *elementary* or *fundamental* particles of matter? In answer, we can best paraphrase a statement by ROBERT OPPENHEIMER: *"By calling a particle fundamental we mean only that it has not proven possible, profitable, or useful to regard it as made up of something else."*

On the basis of the then-existing knowledge, the chemists and physicists of the nineteenth century regarded the *atoms* of the chemical elements as the elementary particles. With the discovery of the *electron* and *atomic nuclei* early in the present century, these particles were regarded as fundamental. With the discovery of the *neutron* and with the development of the cyclotron and the Van de Graaff accelerator in the 1930's, it became apparent that the nucleus consisted of neutrons and protons subject to strong and weak interactions; thus, *electrons, protons,* and *neutrons* became the fundamental particles. The discoveries of the *muon* and *pion* and the direct detection of the hitherto hypothetical *neutrino* added these particles to the growing list of elementary particles.

In the 1950's, following the development of particle accelerators operating in the GeV energy range, the at-first-sight baffling arrays of 'particles' shown on Figs. 2 and 3 were discovered. At first, every energy level shown was regarded as an 'elementary particle'! Fortunately, it eventually became 'possible, profitable, and useful' to regard most of the known particles in a different light. As suggested in Fig. 2, we can regard all the more massive mesons as excited states of the pion, remembering that all charge states of the pion itself are subject to spontaneous decay. Similarly, we can regard all more massive baryons in Fig. 3 as excited states of the nucleon, remembering that the neutron is subject to spontaneous β-decay and that only the proton has strict stability.

Although our knowledge of high-energy phenomena is far from complete, much progress is being made in developing theories in terms of such familiar physical quantities as *energy, momentum, angular momentum* (including intrinsic particle spin), along with less familiar quantities such as *parity, isotopic spin, baryon number, lepton number,* and *strangeness*. The existence and physical properties of the Ω particle (Fig. 3) were theoretically predicted by GELL-MANN and by NE'EMAN prior to actual discovery of the particle. An expression correctly giving the charge of every known meson and baryon and its antiparticle in terms of isotopic spin, baryon number, and strangeness has been developed by GELL-MANN AND NISHIJIMA.

Although there is at present no evidence that mesons or baryons have any observable substructure, a theory has been proposed to account for all the known particles in these classes in terms of a few thus-far hypothetical particles known as *quarks,* characterized by spin $\frac{1}{2}$ $(h/2\pi)$ and *fractional* electronic charge. If quarks exist, *they* would be the elementary particles so far as mesons and baryons are concerned. Although no quarks have yet been observed, their hypothesized existence has stimulated further experimental work.

Thus, the great game of observation and experiment, classification, generalization, hypothesis, theoretical prediction, and further experiment goes on in much the same way that it was played several centuries ago by Tycho Brahe and Johann Kepler, by Galileo Galilei, and by Isaac Newton. The game becomes progressively more exciting!

APPENDIX

1. Systems of Mechanical and Electrical Units

The following table contains the derived metric mechanical units and the electromagnetic SI units that have been introduced in this text, expressed in terms of the fundamental units *meter, kilogram, second,* and *ampere*. From these expressions, the *dimensions* (*see* pp. 6–8) of the physical quantities involved can be readily determined.

Length............	1 meter = 1 m	Force....	1 newton = 1 N = 1 kg·m/s^2
Mass.............	1 kilogram = 1 kg	Pressure..	1 N/m^2 = 1 kg/m·s^2
Time.............	1 second = 1 s	Energy...	1 joule = 1 J = 1 N·m = 1 kg·m^2/s^2
Electric current....	1 ampere = 1 A	Power ...	1 watt = 1 W = 1 J/s = 1 kg·m^2/s^3

Rotational inertia.....	1 kilogram-meter^2 = 1 kg·m^2
Torque...............	1 meter-newton = 1 kg·m^2/s^2
Electric charge.......	1 coulomb = 1 C = 1 A·s
Electric intensity......	1 N/C = 1 V/m = 1 kg·m/s^3·A
Electric potential.....	1 volt = 1 V = 1 J/C = 1 kg·m^2/s^3·A
Electric resistance.....	1 ohm = 1 Ω = 1 V/A = 1 kg·m^2/s^3·A^2
Capacitance..........	1 farad = 1 F = 1 C/V = 1 C^2/J = 1 s^4·A^2/kg·m^2
Inductance..........	1 henry = 1 H = 1 J/A^2 = 1 Ω·s = 1 kg·m^2/s^2·A^2
Magnetic flux........	1 weber = 1 Wb = 1 J/A = 1 V·s = 1 kg·m^2/s^2·A
Magnetic intensity....	1 tesla = 1 Wb/m^2 = 1 V·s/m^2 = 1 kg/s^2·A
Reluctance..........	1 ampere-turn/weber = 1 A/Wb = 1 s^2·A^2/kg·m^2
Magnetizing force....	1 ampere-turn/meter = 1 A/m

The preceding units are in the MKSA (meter-kilogram-second-ampere) system, which was initially adopted by international agreement in 1948 and given formal approval in 1960 as the SI system, which also includes the *kelvin* as the fundamental unit of temperature and the *candela* as the fundamental power-like unit in photometry. However, much past and even considerable current work in physics employs a CGS (centimeter-gram-second) system of basic units. For electrical and magnetic units, there are two CGS systems, differing in the definition of the unit of current—one called the *electrostatic system* in which the unit of current, the *statampere,* has the value

$$1 \text{ statampere} = \frac{1}{2.997\,925 \times 10^9} \text{ ampere}$$

(the constant is related in an obvious manner to the speed of light), and one called the *absolute electromagnetic system* in which the unit of current, the *abampere,* has the value

$$1 \text{ abampere} = 10 \text{ ampere.}$$

These two CGS systems employ units derived from the cm, g, s, and unit of current in exactly the manner the MKSA units are derived from the m, kg, s, and A. The unit of force is called the *dyne,* that of energy the *erg;* there is no special name for the unit of power, the *erg/second*. Electrical units in the electrostatic system are given the same names as in the MKS system, preceded by the prefix *stat-,* those in the electromagnetic system are prefixed by *ab-*; the units of magnetic flux and intensity are given the special names *maxwell* and *gauss* in the electromagnetic system but have no special names in the electrostatic system.

All of the relations for derived units in the table on the preceding page are valid in the CGS systems if we make the following substitutions of units and names:

meter→centimeter   newton→dyne
kilogram→gram   joule→erg
second→second   watt→erg/second

Electrostatic   *Electromagnetic*

ampere→statampere   ampere→abampere
coulomb→statcoulomb   coulomb→abcoulomb
volt→statvolt   volt→abvolt
ohm→statohm   ohm→abohm
farad→statfarad   farad→abfarad
henry→stathenry   henry→abhenry
weber→erg/statampere   weber→maxwell
tesla→dyne/statampere·centimeter   tesla→gauss

In this text, the only coherent British system of units that we have discussed is the gravitational foot-slug-second system in which the unit of force is the pound. We have, however, introduced the standard pound (lb) as a unit of mass (*see* p. 304), and employed this unit in problems in heat, but not in mechanics. One sometimes encounters an *absolute* British system of mechanical units in which the standard pound is taken as the unit of mass. In this f-lb-s system, the unit of force, the *poundal,* is defined as the force that will give 1 lb an acceleration of 1 f/s². *No electromagnetic units based on British units of length and mass are ever employed.*

2. Fundamental Physical Constants

The most accurate values of the fundamental physical constants are derived from critical analyses of all precision data that give either the value of a constant directly, or the value of a combination of such constants, such as the ratio of charge to mass of an electron.

The following table presents the most recent, accepted set of consistent values, as recommended by the Committee on Fundamental Constants of the National Academy of Sciences–National Research Council. All data are based on the scale of atomic masses in which the C^{12} isotope is assigned the mass of 12 u exactly. The ($\pm$) values give the uncertainties in the last significant figure, determined in such a way that it is unlikely that the true value of this figure differs from the value given by more than the stated uncertainty.

The first three items in the table are purely *definitions,* included here for reference.

Standard gravitational acceleration (g_s) ...	9.806 65 m/s²
Standard atmosphere (atm)	101 325 N/m²
Thermochemical kilocalorie	4184 J
Speed of light in vacuum (c)	$2.997\ 925(\pm 3) \times 10^8$ m/s
Electronic charge (e)	$1.602\ 10(\pm 7) \times 10^{-19}$ C
Avogadro constant (N_A)	$6.0225(\pm 3) \times 10^{26}$/kmol
Faraday constant (F)	$9.6487(\pm 2) \times 10^7$ C/kmol
Planck constant (h)	$6.6256(\pm 5) \times 10^{-34}$ J·s
	$= 4.1356 \times 10^{-15}$ eV·s
Bohr magneton ($\mu_B = e\,h/4\pi\,m_e$)	$9.2732(\pm 6) \times 10^{-24}$ A·m²
Nuclear magneton ($\mu_N = e\,h/4\pi\,m_p$)	$5.0505(\pm 4) \times 10^{-27}$ A·m²
Universal gas constant (R)	$8314(\pm 1)$ J/kmol·K
	$= 1.987$ kcal/kmol·K
Volume per kilomole of ideal gas at 1 atm and 0° C	$22.414(\pm 3)$ m³/kmol
Boltzmann constant (k)	$1.3805(\pm 2) \times 10^{-23}$ J/K
Wien displacement constant (A)	$2.8978(\pm 4) \times 10^{-3}$ m·K
Stefan-Boltzmann constant (σ)	$5.670(\pm 3) \times 10^{-8}$ W/K⁴·m²
Gravitation constant (G)	$6.67(\pm 2) \times 10^{-11}$ N·m²/kg²
Nuclidic mass unit (u)	$1.6604(\pm 1) \times 10^{-27}$ kg
Rest energy of one atomic mass unit	$931.48(\pm 2)$ MeV
Electron-volt (eV)	$1.602\ 10(\pm 7) \times 10^{-19}$ J

Rest masses of particles

	(u)	(kg)	(MeV)
Electron	$5.485\ 97(\pm 9) \times 10^{-4}$	$9.1091(\pm 4) \times 10^{-31}$	$0.511\ 006(\pm 5)$
Proton	$1.007\ 2766(\pm 2)$	$1.672\ 52(\pm 8) \times 10^{-27}$	$938.26(\pm 2)$
Neutron	$1.008\ 665(\pm 1)$	$1.674\ 82(\pm 8) \times 10^{-27}$	$939.55(\pm 2)$
Deuteron	$2.013\ 553$	3.3433×10^{-27}	1875.58
α-particle	$4.001\ 507$	6.6441×10^{-27}	3727.3

3. Astronomical Data

Earth

Radius: mean	6371 km = 3959 mi
equatorial	6378 km = 3963 mi
polar	6357 km = 3950 mi
Distance from sun: mean	149.5×10^6 km = 92.9×10^6 mi
aphelion	152.1×10^6 km = 94.5×10^6 mi
perihelion	147.1×10^6 km = 91.4×10^6 mi
Period of rotation	86 164 s = 1 sidereal day = 23.94 h
Radiation from sun at earth's mean distance	1.35 kW/m^2

Moon

Radius	1 741 km = 1 082 mi
Distance from earth: mean	384 400 km = 239 000 mi
apogee	407 000 km = 253 000 mi
perigee	357 000 km = 222 000 mi
Period of revolution = period of rotation	27.322 d
Mass	7.343×10^{22} kg
Mean density	3.33 Mg/m^3

Sun

Radius	696 500 km = 432 200 mi
Mass	1.987×10^{30} kg
Mean density	1.41 Mg/m^3

Planets	Distance from sun (10^6 km)			Period of revolution (d)	Mean radius (km)	Mass[a] (10^{24} kg)	Mean density (Mg/m^3)
	Mean	Aphelion	Perihelion				
Mercury	57.9	69.8	46.0	88.0	2 420	3.167	5.46
Venus	108.1	109.0	107.5	224.7	6 261	4.870	4.96
Earth	149.5	152.1	147.1	365.2	6 371	5.975	5.52
Mars	227.8	249.2	206.6	687.0	3 389	0.639	4.12
Jupiter	777.8	815.9	740.7	4 333	69 900	1900	1.33
Saturn	1426	1508	1348	10 760	57 500	568.9	0.71
Uranus	2868	3007	2737	30 690	23 700	86.9	1.56
Neptune	4494	4537	4459	60 190	21 500	102.9	2.47
Pluto	5908	7370	4450	90 740	2 900	5.37	5.50

[a] Excluding satellites.

4. The Elements

In the table on the opposite page, which is arranged alphabetically by the names of the elements, the internationally adopted values of the average nuclidic masses of the elements are given on the C^{12} scale (*see* p. 334). Bracketed values, for elements that have no stable isotopes, give the mass number of the *most stable* isotope. On p. vi of the Appendix appears a periodic table that is more complete than that on p. 874 of the text. The recently produced elements of atomic numbers 104 and 105 are omitted from these tables.

Element	Symbol	Atomic number	Average nuclidic mass	Element	Symbol	Atomic number	Average nuclidic mass
Actinium	Ac	89	[227]	Mercury	Hg	80	200.59
Aluminum	Al	13	26.9815	Molybdenum	Mo	42	95.94
Americium	Am	95	[243]	Neodymium	Nd	60	144.24
Antimony	Sb	51	121.75	Neon	Ne	10	20.183
Argon	Ar	18	39.948	Neptunium	Np	93	[237]
Arsenic	As	33	74.9216	Nickel	Ni	28	58.71
Astatine	At	85	[210]	Niobium	Nb	41	92.906
Barium	Ba	56	137.34	Nitrogen	N	7	14.0067
Berkelium	Bk	97	[249]	Nobelium	No	102	[254]
Beryllium	Be	4	9.0122	Osmium	Os	76	190.2
Bismuth	Bi	83	208.980	Oxygen	O	8	15.9994
Boron	B	5	10.811	Palladium	Pd	46	106.4
Bromine	Br	35	79.909	Phosphorus	P	15	30.9738
Cadmium	Cd	48	112.40	Platinum	Pt	78	195.09
Calcium	Ca	20	40.08	Plutonium	Pu	94	[242]
Californium	Cf	98	[251]	Polonium	Po	84	[210]
Carbon	C	6	12.01115	Potassium	K	19	39.102
Cerium	Ce	58	140.12	Praseodymium	Pr	59	140.907
Cesium	Cs	55	132.905	Promethium	Pm	61	[145]
Chlorine	Cl	17	35.453	Protactinium	Pa	91	[231]
Chromium	Cr	24	51.996	Radium	Ra	88	[226]
Cobalt	Co	27	58.9332	Radon	Rn	86	[222]
Copper	Cu	29	63.54	Rhenium	Re	75	186.2
Curium	Cm	96	[247]	Rhodium	Rh	45	102.905
Dysprosium	Dy	66	162.50	Rubidium	Rb	37	85.47
Einsteinium	Es	99	[254]	Ruthenium	Ru	44	101.07
Erbium	Er	68	167.26	Samarium	Sm	62	150.35
Europium	Eu	63	151.96	Scandium	Sc	21	44.956
Fermium	Fm	100	[253]	Selenium	Se	34	78.96
Fluorine	F	9	18.9984	Silicon	Si	14	28.086
Francium	Fr	87	[223]	Silver	Ag	47	107.870
Gadolinium	Gd	64	157.25	Sodium	Na	11	22.9898
Gallium	Ga	31	69.72	Strontium	Sr	38	87.62
Germanium	Ge	32	72.59	Sulfur	S	16	32.064
Gold	Au	79	196.967	Tantalum	Ta	73	180.948
Hafnium	Hf	72	178.49	Technetium	Tc	43	[99]
Helium	He	2	4.0026	Tellurium	Te	52	127.60
Holmium	Ho	67	164.930	Terbium	Tb	65	158.924
Hydrogen	H	1	1.00797	Thallium	Tl	81	204.37
Indium	In	49	114.82	Thorium	Th	90	232.038
Iodine	I	53	126.9044	Thulium	Tm	69	168.934
Iridium	Ir	77	192.2	Tin	Sn	50	118.69
Iron	Fe	26	55.847	Titanium	Ti	22	47.90
Krypton	Kr	36	83.80	Tungsten	W	74	183.85
Lanthanum	La	57	138.91	Uranium	U	92	238.03
Lawrencium	Lw	103	[257]	Vanadium	V	23	50.942
Lead	Pb	82	207.19	Xenon	Xe	54	131.30
Lithium	Li	3	6.939	Ytterbium	Yb	70	173.04
Lutetium	Lu	71	174.97	Yttrium	Y	39	88.905
Magnesium	Mg	12	24.312	Zinc	Zn	30	65.37
Manganese	Mn	25	54.9380	Zirconium	Zr	40	91.22
Mendeleevium	Md	101	[258]				

Periodic Table[a]

H 1																	He 2
Li 3	Be 4											B 5	C 6	N 7	O 8	F 9	Ne 10
Na 11	Mg 12				Transition elements							Al 13	Si 14	P 15	S 16	Cl 17	Ar 18
K 19	Ca 20	Sc 21	Ti 22	V 23	Cr 24	Mn 25	Fe 26	Co 27	Ni 28	Cu 29	Zn 30	Ga 31	Ge 32	As 33	Se 34	Br 35	Kr 36
Rb 37	Sr 38	Y 39	Zr 40	Nb 41	Mo 42	Tc 43	Ru 44	Rh 45	Pd 46	Ag 47	Cd 48	In 49	Sn 50	Sb 51	Te 52	I 53	Xe 54
Cs 55	Ba 56		Lu 71	Hf 72	Ta 73	W 74	Re 75	Os 76	Ir 77	Pt 78	Au 79	Hg 80	Tl 81	Pb 82	Bi 83	Po 84	At 85	Rn 86
Fr 87	Ra 88																

Rare earths

La 57	Ce 58	Pr 59	Nd 60	Pm 61	Sm 62	Eu 63	Gd 64	Tb 65	Dy 66	Ho 67	Er 68	Tm 69	Yb 70
Ac 89	Th 90	Pa 91	U 92	Np 93	Pu 94	Am 95	Cm 96	Bk 97	Cf 98	E 99	Fm 100	Md 101	No 102

(a) Elements 103 (Lw), 104, and 105 have also been discovered, but their positions in the periodic table have not yet been determined.

5. Tables of Conversion Factors

Angle	°	′	″	rad	rev
1 degree =	1	60	3600	1.745×10^{-2}	2.778×10^{-3}
1 minute =	1.667×10^{-2}	1	60	2.909×10^{-4}	4.630×10^{-5}
1 second =	2.778×10^{-4}	1.667×10^{-2}	1	4.848×10^{-6}	7.716×10^{-7}
1 radian =	57.30	3438	2.063×10^{5}	1	0.1592
1 revolution =	360	2.16×10^{4}	1.296×10^{6}	6.283	1

1 artillery mil = 1/6400 rev = 0.000 981 7 rad = 0°.056 25

Length	m	km	i	f	mi
1 meter =	1	10^{-3}	39.37	3.281	6.214×10^{-4}
1 kilometer =	1000	1	3.937×10^{4}	3281	0.6214
1 inch =	0.0254	2.54×10^{-5}	1	0.0833	1.578×10^{-5}
1 foot =	0.3048	3.048×10^{-4}	12	1	1.894×10^{-4}
1 statute mile =	1609	1.609	6.336×10^{4}	5280	1

1 angstrom = 10^{-10} m
1 X-unit = 10^{-13} m
1 micron (μ) = 1 μm
1 nautical mile = 1852 m = 1.1508 mi = 6076.10 f
1 astronomical unit = 149.5×10^{6} km
1 Bohr radius = $5.291\ 67 \times 10^{-11}$ m

1 millimicron (mμ) = 1 nm
1 light-year = 9.4600×10^{12} km
1 parsec = 3.084×10^{13} km
1 fermi = 10^{-15} m

1 fathom = 6 f
1 yard (yd) = 3 f
1 rod = 16.5 f
1 mil = 10^{-3} i
1 league = 3 naut miles

Area	m²	cm²	f²	i²
1 square meter =	1	10^{4}	10.76	1550
1 square centimeter =	10^{-4}	1	1.076×10^{-3}	0.1550
1 square foot =	9.290×10^{-2}	929.0	1	144
1 square inch =	6.452×10^{-4}	6.452	6.944×10^{-3}	1

1 square mile = 27 878 400 f² = 640 acre
1 circular mil = 7.854×10^{-7} i²
1 acre = 43 560 f²
1 barn = 10^{-28} m²
1 hectare = 10 000 m² = 2.471 acre

Volume	m³	cm³	f³	i³
1 cubic meter =	1	10^{6}	35.31	6.102×10^{4}
1 cubic centimeter =	10^{-6}	1	3.531×10^{-5}	0.06102
1 cubic foot =	2.832×10^{-2}	28,320	1	1728
1 cubic inch =	1.639×10^{-5}	16.39	5.787×10^{-4}	1

1 U.S. fluid gallon = 4 quarts = 8 pints = 128 fluid ounces = 231 i³
1 British Imperial gallon = the volume of 10 lb of water at 62° F = 277.42 i³
1 liter = 1000 cm³

APPENDIX

Mass	g	kg	lb	sl	ton
1 gram =	1	0.001	0.002 205	6.852×10^{-5}	1.102×10^{-6}
1 kilogram =	1000	1	2.205	6.852×10^{-2}	1.102×10^{-3}
1 pound (avoirdupois) =	453.6	0.4536	1	3.108×10^{-2}	0.0005
1 slug =	1.459×10^{4}	14.59	32.17	1	1.609×10^{-2}
1 ton-mass =	9.072×10^{5}	907.2	2000	62.16	1

1 avoirdupois pound = 16 avoirdupois ounces = 7000 grains = 0.453 592 37 kg
1 troy or apothecaries' pound = 12 troy or apothecaries' ounces
= 0.8229 avoirdupois pound = 5760 grains

1 long ton = 2240 lb = 20 cwt   1 stone = 14 lb   1 hundredweight (cwt) = 112 lb
1 metric ton = 1000 kg = 2205 lb   1 carat = 0.2 g   1 pennyweight (dwt) = 24 grains
1 nuclidic mass unit (u) = 1.6604×10^{-27} kg   1 quintal = 100 kg

Time	y	d	h	min	s
1 year =	1	365.2	8.766×10^{3}	5.259×10^{5}	3.156×10^{7}
1 day =	2.738×10^{-3}	1	24	1440	86 400
1 hour =	1.141×10^{-4}	4.167×10^{-2}	1	60	3600
1 minute =	1.901×10^{-6}	6.944×10^{-4}	1.667×10^{-2}	1	60
1 second =	3.169×10^{-8}	1.157×10^{-5}	2.778×10^{-4}	1.667×10^{-2}	1

1 sidereal day = period of rotation of earth = 86 164 s
1 year = period of revolution of earth = 365.242 198 79 d

Density	sl/f³	lb/f³	lb/i³	kg/m³	g/cm³
1 slug per f³ =	1	32.17	1.862×10^{-2}	515.4	0.5154
1 pound per f³ =	3.108×10^{-2}	1	5.787×10^{-4}	16.02	1.602×10^{-2}
1 pound per i³ =	53.71	1728	1	2.768×10^{4}	27.68
1 kg per m³ =	1.940×10^{-3}	6.243×10^{-2}	3.613×10^{-5}	1	0.001
1 gram per cm³ =	1.940	62.43	3.613×10^{-2}	1000	1

Speed	f/s	km/h	m/s	mi/h	knot
1 foot per second =	1	1.097	0.3048	0.6818	0.5925
1 kilometer per hour =	0.9113	1	0.2778	0.6214	0.5400
1 meter per second =	3.281	3.6	1	2.237	1.944
1 mile per hour =	1.467	1.609	0.4470	1	0.8689
1 knot =	1.688	1.852	0.5144	1.151	1

1 knot = 1 nautical mile/hr

SEC. 5   TABLES OF CONVERSION FACTORS

Acceleration: $1 \text{ gal} = 1 \text{ cm/s}^2$

Force	dyne	kgf	N	p	pdl
1 dyne =	1	1.020×10^{-6}	10^{-5}	2.248×10^{-6}	7.233×10^{-5}
1 kilogram-force =	9.807×10^5	1	9.807	2.205	70.93
1 newton =	10^5	0.1020	1	0.2248	7.233
1 pound =	4.448×10^5	0.4536	4.448	1	32.17
1 poundal =	1.383×10^4	1.410×10^{-2}	0.1383	3.108×10^{-2}	1

$1 \text{ kgf} = 9.806\,65 \text{ N}$ $1 \text{ p} = 32.173\,98 \text{ pdl}$

Pressure	atm	inch of water	cm Hg	N/m²	p/i²
1 atmosphere =	1	406.8	76	1.013×10^5	14.70
1 inch of water [a] =	2.458×10^{-3}	1	0.1868	249.1	0.03613
1 cm mercury [a] =	1.316×10^{-2}	5.353	1	1333	0.1934
1 newton per m² =	9.869×10^{-6}	0.004 105	7.501×10^{-4}	1	1.450×10^{-4}
1 pound per i² =	6.805×10^{-2}	27.68	5.171	6.895×10^3	1

[a] Under standard gravitational acceleration, and temperature of 4° C for water, 0° C for mercury.

$1 \text{ bar} = 10^5 \text{ N/m}^2$ $1 \text{ torr} = 1 \text{ mm Hg}$
$1 \text{ cm of water} = 98.07 \text{ N/m}^2$ $1 \text{ f of water} = 62.43 \text{ p/f}^2$

Energy	BTU	fp	J	kcal	kWh
1 British thermal unit =	1	777.9	1055	0.2520	2.930×10^{-4}
1 foot-pound =	1.285×10^{-3}	1	1.356	3.240×10^{-4}	3.766×10^{-7}
1 joule =	9.481×10^{-4}	0.7376	1	2.390×10^{-4}	2.778×10^{-7}
1 kilocalorie =	3.968	3086	4184	1	1.163×10^{-3}
1 kilowatt-hour =	3413	2.655×10^6	3.6×10^6	860.2	1

See also table of relativistic mass-energy equivalents on p. xi.
$1 \text{ kcal} = 2.612 \times 10^{22} \text{ eV}$ $1 \text{ horsepower-hour} = 1.980 \times 10^6 \text{ fp}$
$1 \text{ erg} = 10^{-7} \text{ joule}$ $1 \text{ therm} = 10^5 \text{ BTU}$
$1 \text{ rydberg} = 13.61 \text{ eV} = 2.180 \times 10^{-18} \text{ J}$

Power	BTU/h	fp/s	hp	kcal/s	kW	W
1 BTU/h =	1	0.2161	3.929×10^{-4}	7.000×10^{-5}	2.930×10^{-4}	0.2930
1 fp/s =	4.628	1	1.818×10^{-3}	3.239×10^{-4}	1.356×10^{-3}	1.356
1 horsepower =	2545	550	1	0.1782	0.7457	745.7
1 kcal/s =	1.429×10^4	3087	5.613	1	4.184	4184
1 kilowatt =	3413	737.6	1.341	0.2390	1	1000
1 watt =	3.413	0.7376	1.341×10^{-3}	2.390×10^{-4}	0.001	1

$1 \text{ ton (refrigeration)} = 12\,000 \text{ BTU/h}$

APPENDIX

Electric charge	abC	C	statC
1 abcoulomb (1 EMU) =	1	10	2.998×10^{10}
1 coulomb =	0.1	1	2.998×10^{9}
1 statcoulomb (1 ESU) =	3.336×10^{-11}	3.336×10^{-10}	1

1 franklin = 1 Fr = 1 statC   1 ampere-hour = 3600 C

Electric current	abA	A	statA
1 abampere (1 EMU) =	1	10	2.998×10^{10}
1 ampere =	0.1	1	2.998×10^{9}
1 statampere (1 ESU) =	3.336×10^{-11}	3.336×10^{-10}	1

1 biot = 1 Bi = 1 abA

Electric potential	abV	V	statV
1 abvolt (1 EMU) =	1	10^{-8}	3.336×10^{-11}
1 volt =	10^{8}	1	3.336×10^{-3}
1 statvolt (1 ESU) =	2.998×10^{10}	299.8	1

Electric resistance	abohm	Ω	statohm
1 abohm (1 EMU) =	1	10^{-9}	1.113×10^{-21}
1 ohm =	10^{9}	1	1.113×10^{-12}
1 statohm (1 ESU) =	8.987×10^{20}	8.987×10^{11}	1

Capacitance	abF	F	μF	statF
1 abfarad (1 EMU) =	1	10^{9}	10^{15}	8.987×10^{20}
1 farad =	10^{-9}	1	10^{6}	8.987×10^{11}
1 microfarad =	10^{-15}	10^{-6}	1	8.987×10^{5}
1 statfarad (1 ESU) =	1.113×10^{-21}	1.113×10^{-12}	1.113×10^{-6}	1

Inductance	abH	H	mH	statH
1 abhenry (1 EMU) =	1	10^{-9}	10^{-6}	1.113×10^{-21}
1 henry =	10^{9}	1	1000	1.113×10^{-12}
1 millihenry =	10^{6}	0.001	1	1.113×10^{-15}
1 stathenry (1 ESU) =	8.987×10^{20}	8.987×10^{11}	8.987×10^{14}	1

TABLES OF CONVERSION FACTORS

Magnetic flux	Mx	kiloline	Wb
1 maxwell (1 line or 1 EMU) =	1	0.001	10^{-8}
1 kiloline =	1000	1	10^{-5}
1 weber =	10^8	10^5	1

1 ESU = 299.8 weber

Magnetic intensity $\mathcal{B}$	G	kiloline/i^2	T	mG
1 gauss (line per cm²) =	1	6.452×10^{-3}	10^{-4}	1000
1 kiloline per square inch =	155.0	1	1.550×10^{-2}	1.550×10^5
1 tesla =	10^4	64.52	1	10^7
1 milligauss =	0.001	6.452×10^{-6}	10^{-7}	1

1 T = 1 Wb/m²   1 ESU = 2.998×10^6 T   1 gamma = 10^{-2} mG = 10^{-9} T

Magnetomotive force	abA-turn	A-turn	Gi
1 abampere-turn =	1	10	12.57
1 ampere-turn =	0.1	1	1.257
1 gilbert =	7.958×10^{-2}	0.7958	1

1 ESU = 1 statampere-turn = 3.336×10^{-10} A-turn

Magnetizing force $\mathcal{H}$	abA/cm	A/in	A/m	Oe
1 abampere-turn per centimeter =	1	25.40	1000	12.57
1 ampere-turn per inch =	3.937×10^{-2}	1	39.37	0.4947
1 ampere-turn per meter =	0.001	2.540×10^{-2}	1	1.257×10^{-2}
1 oersted =	7.958×10^{-2}	2.021	79.58	1

1 oersted = 1 gilbert/cm   1 ESU = 3.336×10^{-8} A-turn/m

Mass-energy equivalents	kg	u	J	MeV
1 kilogram ~	1	6.025×10^{26}	8.987×10^{-16}	5.610×10^{29}
1 nuclidic mass unit ~	1.660×10^{-27}	1	1.492×10^{-10}	931.5
1 joule ~	1.113×10^{-17}	6.705×10^9	1	6.242×10^{12}
1 million electron-volts ~	1.783×10^{-30}	1.074×10^{-3}	1.602×10^{-13}	1

6. Natural Trigonometric Functions

sin

	.0	.1	.2	.3	.4	.5	.6	.7	.8	.9	
0°	.0000	.0017	.0035	.0052	.0070	.0087	.0105	.0122	.0140	.0157	89°
1°	.0175	.0192	.0209	.0227	.0244	.0262	.0279	.0297	.0314	.0332	88°
2°	.0349	.0366	.0384	.0401	.0419	.0436	.0454	.0471	.0488	.0506	87°
3°	.0523	.0541	.0558	.0576	.0593	.0610	.0628	.0645	.0663	.0680	86°
4°	.0698	.0715	.0732	.0750	.0767	.0785	.0802	.0819	.0837	.0854	85°
5°	.0872	.0889	.0906	.0924	.0941	.0958	.0976	.0993	.1011	.1028	84°
6°	.1045	.1063	.1080	.1097	.1115	.1132	.1149	.1167	.1184	.1201	83°
7°	.1219	.1236	.1253	.1271	.1288	.1305	.1323	.1340	.1357	.1374	82°
8°	.1392	.1409	.1426	.1444	.1461	.1478	.1495	.1513	.1530	.1547	81°
9°	.1564	.1582	.1599	.1616	.1633	.1650	.1668	.1685	.1702	.1719	80°
10°	.1736	.1754	.1771	.1788	.1805	.1822	.1840	.1857	.1874	.1891	79°
11°	.1908	.1925	.1942	.1959	.1977	.1994	.2011	.2028	.2045	.2062	78°
12°	.2079	.2096	.2113	.2130	.2147	.2164	.2181	.2198	.2215	.2233	77°
13°	.2250	.2267	.2284	.2300	.2317	.2334	.2351	.2368	.2385	.2402	76°
14°	.2419	.2436	.2453	.2470	.2487	.2504	.2521	.2538	.2554	.2571	75°
15°	.2588	.2605	.2622	.2639	.2656	.2672	.2689	.2706	.2723	.2740	74°
16°	.2756	.2773	.2790	.2807	.2823	.2840	.2857	.2874	.2890	.2907	73°
17°	.2924	.2940	.2957	.2974	.2990	.3007	.3024	.3040	.3057	.3074	72°
18°	.3090	.3107	.3123	.3140	.3156	.3173	.3190	.3206	.3223	.3239	71°
19°	.3256	.3272	.3289	.3305	.3322	.3338	.3355	.3371	.3387	.3404	70°
20°	.3420	.3437	.3453	.3469	.3486	.3502	.3518	.3535	.3551	.3567	69°
21°	.3584	.3600	.3616	.3633	.3649	.3665	.3681	.3697	.3714	.3730	68°
22°	.3746	.3762	.3778	.3795	.3811	.3827	.3843	.3859	.3875	.3891	67°
23°	.3907	.3923	.3939	.3955	.3971	.3987	.4003	.4019	.4035	.4051	66°
24°	.4067	.4083	.4099	.4115	.4131	.4147	.4163	.4179	.4195	.4210	65°
25°	.4226	.4242	.4258	.4274	.4289	.4305	.4321	.4337	.4352	.4368	64°
26°	.4384	.4399	.4415	.4431	.4446	.4462	.4478	.4493	.4509	.4524	63°
27°	.4540	.4555	.4571	.4586	.4602	.4617	.4633	.4648	.4664	.4679	62°
28°	.4695	.4710	.4726	.4741	.4756	.4772	.4787	.4802	.4818	.4833	61°
29°	.4848	.4863	.4879	.4894	.4909	.4924	.4939	.4955	.4970	.4985	60°
30°	.5000	.5015	.5030	.5045	.5060	.5075	.5090	.5105	.5120	.5135	59°
31°	.5150	.5165	.5180	.5195	.5210	.5225	.5240	.5255	.5270	.5284	58°
32°	.5299	.5314	.5329	.5344	.5358	.5373	.5388	.5402	.5417	.5432	57°
33°	.5446	.5461	.5476	.5490	.5505	.5519	.5534	.5548	.5563	.5577	56°
34°	.5592	.5606	.5621	.5635	.5650	.5664	.5678	.5693	.5707	.5721	55°
35°	.5736	.5750	.5764	.5779	.5793	.5807	.5821	.5835	.5850	.5864	54°
36°	.5878	.5892	.5906	.5920	.5934	.5948	.5962	.5976	.5990	.6004	53°
37°	.6018	.6032	.6046	.6060	.6074	.6088	.6101	.6115	.6129	.6143	52°
38°	.6157	.6170	.6184	.6198	.6211	.6225	.6239	.6252	.6266	.6280	51°
39°	.6293	.6307	.6320	.6334	.6347	.6361	.6374	.6388	.6401	.6414	50°
40°	.6428	.6441	.6455	.6468	.6481	.6494	.6508	.6521	.6534	.6547	49°
41°	.6561	.6574	.6587	.6600	.6613	.6626	.6639	.6652	.6665	.6678	48°
42°	.6691	.6704	.6717	.6730	.6743	.6756	.6769	.6782	.6794	.6807	47°
43°	.6820	.6833	.6845	.6858	.6871	.6884	.6896	.6909	.6921	.6934	46°
44°	.6947	.6959	.6972	.6984	.6997	.7009	.7022	.7034	.7046	.7059	45°
	.9	.8	.7	.6	.5	.4	.3	.2	.1	.0	

cos

sin

	.0	.1	.2	.3	.4	.5	.6	.7	.8	.9		
45°	.7071	.7083	.7096	.7108	.7120	.7133	.7145	.7157	.7169	.7181	.7193	44°
46°	.7193	.7206	.7218	.7230	.7242	.7254	.7266	.7278	.7290	.7302	.7314	43°
47°	.7314	.7325	.7337	.7349	.7361	.7373	.7385	.7396	.7408	.7420	.7431	42°
48°	.7431	.7443	.7455	.7466	.7478	.7490	.7501	.7513	.7524	.7536	.7547	41°
49°	.7547	.7559	.7570	.7581	.7593	.7604	.7615	.7627	.7638	.7649	.7660	40°
50°	.7660	.7672	.7683	.7694	.7705	.7716	.7727	.7738	.7749	.7760	.7771	39°
51°	.7771	.7782	.7793	.7804	.7815	.7826	.7837	.7848	.7859	.7869	.7880	38°
52°	.7880	.7891	.7902	.7912	.7923	.7934	.7944	.7955	.7965	.7976	.7986	37°
53°	.7986	.7997	.8007	.8018	.8028	.8039	.8049	.8059	.8070	.8080	.8090	36°
54°	.8090	.8100	.8111	.8121	.8131	.8141	.8151	.8161	.8171	.8181	.8192	35°
55°	.8192	.8202	.8211	.8221	.8231	.8241	.8251	.8261	.8271	.8281	.8290	34°
56°	.8290	.8300	.8310	.8320	.8329	.8339	.8348	.8358	.8368	.8377	.8387	33°
57°	.8387	.8396	.8406	.8415	.8425	.8434	.8443	.8453	.8462	.8471	.8480	32°
58°	.8480	.8490	.8499	.8508	.8517	.8526	.8536	.8545	.8554	.8563	.8572	31°
59°	.8572	.8581	.8590	.8599	.8607	.8616	.8625	.8634	.8643	.8652	.8660	30°
60°	.8660	.8669	.8678	.8686	.8695	.8704	.8712	.8721	.8729	.8738	.8746	29°
61°	.8746	.8755	.8763	.8771	.8780	.8788	.8796	.8805	.8813	.8821	.8829	28°
62°	.8829	.8838	.8846	.8854	.8862	.8870	.8878	.8886	.8894	.8902	.8910	27°
63°	.8910	.8918	.8926	.8934	.8942	.8949	.8957	.8965	.8973	.8980	.8988	26°
64°	.8988	.8996	.9003	.9011	.9018	.9026	.9033	.9041	.9048	.9056	.9063	25°
65°	.9063	.9070	.9078	.9085	.9092	.9100	.9107	.9114	.9121	.9128	.9135	24°
66°	.9135	.9143	.9150	.9157	.9164	.9171	.9178	.9184	.9191	.9198	.9205	23°
67°	.9205	.9212	.9219	.9225	.9232	.9239	.9245	.9252	.9259	.9265	.9272	22°
68°	.9272	.9278	.9285	.9291	.9298	.9304	.9311	.9317	.9323	.9330	.9336	21°
69°	.9336	.9342	.9348	.9354	.9361	.9367	.9373	.9379	.9385	.9391	.9397	20°
70°	.9397	.9403	.9409	.9415	.9421	.9426	.9432	.9438	.9444	.9449	.9455	19°
71°	.9455	.9461	.9466	.9472	.9478	.9483	.9489	.9494	.9500	.9505	.9511	18°
72°	.9511	.9516	.9521	.9527	.9532	.9537	.9542	.9548	.9553	.9558	.9563	17°
73°	.9563	.9568	.9573	.9578	.9583	.9588	.9593	.9598	.9603	.9608	.9613	16°
74°	.9613	.9617	.9622	.9627	.9632	.9636	.9641	.9646	.9650	.9655	.9659	15°
75°	.9659	.9664	.9668	.9673	.9677	.9681	.9686	.9690	.9694	.9699	.9703	14°
76°	.9703	.9707	.9711	.9715	.9720	.9724	.9728	.9732	.9736	.9740	.9744	13°
77°	.9744	.9748	.9751	.9755	.9759	.9763	.9767	.9770	.9774	.9778	.9781	12°
78°	.9781	.9785	.9789	.9792	.9796	.9799	.9803	.9806	.9810	.9813	.9816	11°
79°	.9816	.9820	.9823	.9826	.9829	.9833	.9836	.9839	.9842	.9845	.9848	10°
80°	.9848	.9851	.9854	.9857	.9860	.9863	.9866	.9869	.9871	.9874	.9877	9°
81°	.9877	.9880	.9882	.9885	.9888	.9890	.9893	.9895	.9898	.9900	.9903	8°
82°	.9903	.9905	.9907	.9910	.9912	.9914	.9917	.9919	.9921	.9923	.9925	7°
83°	.9925	.9928	.9930	.9932	.9934	.9936	.9938	.9940	.9942	.9943	.9945	6°
84°	.9945	.9947	.9949	.9951	.9952	.9954	.9956	.9957	.9959	.9960	.9962	5°
85°	.9962	.9963	.9965	.9966	.9968	.9969	.9971	.9972	.9973	.9974	.9976	4°
86°	.9976	.9977	.9978	.9979	.9980	.9981	.9982	.9983	.9984	.9985	.9986	3°
87°	.9986	.9987	.9988	.9989	.9990	.9990	.9991	.9992	.9993	.9993	.9994	2°
88°	.9994	.9995	.9995	.9996	.9996	.9997	.9997	.9997	.9998	.9998	.9998	1°
89°	.9998	.9999	.9999	.9999	.9999	1.000	1.000	1.000	1.000	1.000	1.000	0°
		.9	.8	.7	.6	.5	.4	.3	.2	.1	.0	

cos

tan

	.0	.1	.2	.3	.4	.5	.6	.7	.8	.9	
0°	.0000	.0017	.0035	.0052	.0070	.0087	.0105	.0122	.0140	.0157	.0175	89°
1°	.0175	.0192	.0209	.0227	.0244	.0262	.0279	.0297	.0314	.0332	.0349	88°
2°	.0349	.0367	.0384	.0402	.0419	.0437	.0454	.0472	.0489	.0507	.0524	87°
3°	.0524	.0542	.0559	.0577	.0594	.0612	.0629	.0647	.0664	.0682	.0699	86°
4°	.0699	.0717	.0734	.0752	.0769	.0787	.0805	.0822	.0840	.0857	.0875	85°
5°	.0875	.0892	.0910	.0928	.0945	.0963	.0981	.0998	.1016	.1033	.1051	84°
6°	.1051	.1069	.1086	.1104	.1122	.1139	.1157	.1175	.1192	.1210	.1228	83°
7°	.1228	.1246	.1263	.1281	.1299	.1317	.1334	.1352	.1370	.1388	.1405	82°
8°	.1405	.1423	.1441	.1459	.1477	.1495	.1512	.1530	.1548	.1566	.1584	81°
9°	.1584	.1602	.1620	.1638	.1655	.1673	.1691	.1709	.1727	.1745	.1763	80°
10°	.1763	.1781	.1799	.1817	.1835	.1853	.1871	.1890	.1908	.1926	.1944	79°
11°	.1944	.1962	.1980	.1998	.2016	.2035	.2053	.2071	.2089	.2107	.2126	78°
12°	.2126	.2144	.2162	.2180	.2199	.2217	.2235	.2254	.2272	.2290	.2309	77°
13°	.2309	.2327	.2345	.2364	.2382	.2401	.2419	.2438	.2456	.2475	.2493	76°
14°	.2493	.2512	.2530	.2549	.2568	.2586	.2605	.2623	.2642	.2661	.2679	75°
15°	.2679	.2698	.2717	.2736	.2754	.2773	.2792	.2811	.2830	.2849	.2867	74°
16°	.2867	.2886	.2905	.2924	.2943	.2962	.2981	.3000	.3019	.3038	.3057	73°
17°	.3057	.3076	.3096	.3115	.3134	.3153	.3172	.3191	.3211	.3230	.3249	72°
18°	.3249	.3269	.3288	.3307	.3327	.3346	.3365	.3385	.3404	.3424	.3443	71°
19°	.3443	.3463	.3482	.3502	.3522	.3541	.3561	.3581	.3600	.3620	.3640	70°
20°	.3640	.3659	.3679	.3699	.3719	.3739	.3759	.3779	.3799	.3819	.3839	69°
21°	.3839	.3859	.3879	.3899	.3919	.3939	.3959	.3979	.4000	.4020	.4040	68°
22°	.4040	.4061	.4081	.4101	.4122	.4142	.4163	.4183	.4204	.4224	.4245	67°
23°	.4245	.4265	.4286	.4307	.4327	.4348	.4369	.4390	.4411	.4431	.4452	66°
24°	.4452	.4473	.4494	.4515	.4536	.4557	.4578	.4599	.4621	.4642	.4663	65°
25°	.4663	.4684	.4706	.4727	.4748	.4770	.4791	.4813	.4834	.4856	.4877	64°
26°	.4877	.4899	.4921	.4942	.4964	.4986	.5008	.5029	.5051	.5073	.5095	63°
27°	.5095	.5117	.5139	.5161	.5184	.5206	.5228	.5250	.5272	.5295	.5317	62°
28°	.5317	.5340	.5362	.5384	.5407	.5430	.5452	.5475	.5498	.5520	.5543	61°
29°	.5543	.5566	.5589	.5612	.5635	.5658	.5681	.5704	.5727	.5750	.5774	60°
30°	.5774	.5797	.5820	.5844	.5867	.5890	.5914	.5938	.5961	.5985	.6009	59°
31°	.6009	.6032	.6056	.6080	.6104	.6128	.6152	.6176	.6200	.6224	.6249	58°
32°	.6249	.6273	.6297	.6322	.6346	.6371	.6395	.6420	.6445	.6469	.6494	57°
33°	.6494	.6519	.6544	.6569	.6594	.6619	.6644	.6669	.6694	.6720	.6745	56°
34°	.6745	.6771	.6796	.6822	.6847	.6873	.6899	.6924	.6950	.6976	.7002	55°
35°	.7002	.7028	.7054	.7080	.7107	.7133	.7159	.7186	.7212	.7239	.7265	54°
36°	.7265	.7292	.7319	.7346	.7373	.7400	.7427	.7454	.7481	.7508	.7536	53°
37°	.7536	.7563	.7590	.7618	.7646	.7673	.7701	.7729	.7757	.7785	.7813	52°
38°	.7813	.7841	.7869	.7898	.7926	.7954	.7983	.8012	.8040	.8069	.8098	51°
39°	.8098	.8127	.8156	.8185	.8214	.8243	.8273	.8302	.8332	.8361	.8391	50°
40°	.8391	.8421	.8451	.8481	.8511	.8541	.8571	.8601	.8632	.8662	.8693	49°
41°	.8693	.8724	.8754	.8785	.8816	.8847	.8878	.8910	.8941	.8972	.9004	48°
42°	.9004	.9036	.9067	.9099	.9131	.9163	.9195	.9228	.9260	.9293	.9325	47°
43°	.9325	.9358	.9391	.9424	.9457	.9490	.9523	.9556	.9590	.9623	.9657	46°
44°	.9657	.9691	.9725	.9759	.9793	.9827	.9861	.9896	.9930	.9965	1.000	45°
	.9	.8	.7	.6	.5	.4	.3	.2	.1	.0	

cot

SEC. 6   NATURAL TRIGONOMETRIC FUNCTIONS   XV

tan

	.0	.1	.2	.3	.4	.5	.6	.7	.8	.9	
45°	1.000	1.003	1.007	1.011	1.014	1.018	1.021	1.025	1.028	1.032	1.036	44°
46°	1.036	1.039	1.043	1.046	1.050	1.054	1.057	1.061	1.065	1.069	1.072	43°
47°	1.072	1.076	1.080	1.084	1.087	1.091	1.095	1.099	1.103	1.107	1.111	42°
48°	1.111	1.115	1.118	1.122	1.126	1.130	1.134	1.138	1.142	1.146	1.150	41°
49°	1.150	1.154	1.159	1.163	1.167	1.171	1.175	1.179	1.183	1.188	1.192	40°
50°	1.192	1.196	1.200	1.205	1.209	1.213	1.217	1.222	1.226	1.230	1.235	39°
51°	1.235	1.239	1.244	1.248	1.253	1.257	1.262	1.266	1.271	1.275	1.280	38°
52°	1.280	1.285	1.289	1.294	1.299	1.303	1.308	1.313	1.317	1.322	1.327	37°
53°	1.327	1.332	1.337	1.342	1.347	1.351	1.356	1.361	1.366	1.371	1.376	36°
54°	1.376	1.381	1.387	1.392	1.397	1.402	1.407	1.412	1.418	1.423	1.428	35°
55°	1.428	1.433	1.439	1.444	1.450	1.455	1.460	1.466	1.471	1.477	1.483	34°
56°	1.483	1.488	1.494	1.499	1.505	1.511	1.517	1.522	1.528	1.534	1.540	33°
57°	1.540	1.546	1.552	1.558	1.564	1.570	1.576	1.582	1.588	1.594	1.600	32°
58°	1.600	1.607	1.613	1.619	1.625	1.632	1.638	1.645	1.651	1.658	1.664	31°
59°	1.664	1.671	1.678	1.684	1.691	1.698	1.704	1.711	1.718	1.725	1.732	30°
60°	1.732	1.739	1.746	1.753	1.760	1.767	1.775	1.782	1.789	1.797	1.804	29°
61°	1.804	1.811	1.819	1.827	1.834	1.842	1.849	1.857	1.865	1.873	1.881	28°
62°	1.881	1.889	1.897	1.905	1.913	1.921	1.929	1.937	1.946	1.954	1.963	27°
63°	1.963	1.971	1.980	1.988	1.997	2.006	2.014	2.023	2.032	2.041	2.050	26°
64°	2.050	2.059	2.069	2.078	2.087	2.097	2.106	2.116	2.125	2.135	2.145	25°
65°	2.145	2.154	2.164	2.174	2.184	2.194	2.204	2.215	2.225	2.236	2.246	24°
66°	2.246	2.257	2.267	2.278	2.289	2.300	2.311	2.322	2.333	2.344	2.356	23°
67°	2.356	2.367	2.379	2.391	2.402	2.414	2.426	2.438	2.450	2.463	2.475	22°
68°	2.475	2.488	2.500	2.513	2.526	2.539	2.552	2.565	2.578	2.592	2.605	21°
69°	2.605	2.619	2.633	2.646	2.660	2.675	2.689	2.703	2.718	2.733	2.747	20°
70°	2.747	2.762	2.778	2.793	2.808	2.824	2.840	2.856	2.872	2.888	2.904	19°
71°	2.904	2.921	2.937	2.954	2.971	2.989	3.006	3.024	3.042	3.060	3.078	18°
72°	3.078	3.096	3.115	3.133	3.152	3.172	3.191	3.211	3.230	3.251	3.271	17°
73°	3.271	3.291	3.312	3.333	3.354	3.376	3.398	3.420	3.442	3.465	3.487	16°
74°	3.487	3.511	3.534	3.558	3.582	3.606	3.630	3.655	3.681	3.706	3.732	15°
75°	3.732	3.758	3.785	3.812	3.839	3.867	3.895	3.923	3.952	3.981	4.011	14°
76°	4.011	4.041	4.071	4.102	4.134	4.165	4.198	4.230	4.264	4.297	4.331	13°
77°	4.331	4.366	4.402	4.437	4.474	4.511	4.548	4.586	4.625	4.665	4.705	12°
78°	4.705	4.745	4.787	4.829	4.872	4.915	4.959	5.005	5.050	5.097	5.145	11°
79°	5.145	5.193	5.242	5.292	5.343	5.396	5.449	5.503	5.558	5.614	5.671	10°
80°	5.671	5.730	5.789	5.850	5.912	5.976	6.041	6.107	6.174	6.243	6.314	9°
81°	6.314	6.386	6.460	6.535	6.612	6.691	6.772	6.855	6.940	7.026	7.115	8°
82°	7.115	7.207	7.300	7.396	7.495	7.596	7.700	7.806	7.916	8.028	8.144	7°
83°	8.144	8.264	8.386	8.513	8.643	8.777	8.915	9.058	9.205	9.357	9.514	6°
84°	9.514	9.677	9.845	10.02	10.20	10.39	10.58	10.78	10.99	11.20	11.43	5°
85°	11.43	11.66	11.91	12.16	12.43	12.71	13.00	13.30	13.62	13.95	14.30	4°
86°	14.30	14.67	15.06	15.46	15.89	16.35	16.83	17.34	17.89	18.46	19.08	3°
87°	19.08	19.74	20.45	21.20	22.02	22.90	23.86	24.90	26.03	27.27	28.64	2°
88°	28.64	30.14	31.82	33.69	35.80	38.19	40.92	44.07	47.74	52.08	57.29	1°
89°	57.29	63.66	71.62	81.85	95.49	114.6	143.2	191.0	286.5	573.0		0°
		.9	.8	.7	.6	.5	.4	.3	.2	.1	.0	

cot

7. Table of Logarithms to the Base 10

N	0	1	2	3	4	5	6	7	8	9	P.P. 1	2	3	4	5
10	0000	0043	0086	0128	0170	0212	0253	0294	0334	0374	4	8	12	17	21
11	0414	0453	0492	0531	0569	0607	0645	0682	0719	0755	4	8	11	15	19
12	0792	0828	0864	0899	0934	0969	1004	1038	1072	1106	3	7	10	14	17
13	1139	1173	1206	1239	1271	1303	1335	1367	1399	1430	3	6	10	13	16
14	1461	1492	1523	1553	1584	1614	1644	1673	1703	1732	3	6	9	12	15
15	1761	1790	1818	1847	1875	1903	1931	1959	1987	2014	3	6	8	11	14
16	2041	2068	2095	2122	2148	2175	2201	2227	2253	2279	3	5	8	11	13
17	2304	2330	2355	2380	2405	2430	2455	2480	2504	2529	2	5	7	10	12
18	2553	2577	2601	2625	2648	2672	2695	2718	2742	2765	2	5	7	9	12
19	2788	2810	2833	2856	2878	2900	2923	2945	2967	2989	2	4	7	9	11
20	3010	3032	3054	3075	3096	3118	3139	3160	3181	3201	2	4	6	8	11
21	3222	3243	3263	3284	3304	3324	3345	3365	3385	3404	2	4	6	8	10
22	3424	3444	3464	3483	3502	3522	3541	3560	3579	3598	2	4	6	8	10
23	3617	3636	3655	3674	3692	3711	3729	3747	3766	3784	2	4	5	7	9
24	3802	3820	3838	3856	3874	3892	3909	3927	3945	3962	2	4	5	7	9
25	3979	3997	4014	4031	4048	4065	4082	4099	4116	4133	2	3	5	7	9
26	4150	4166	4183	4200	4216	4232	4249	4265	4281	4298	2	3	5	7	8
27	4314	4330	4346	4362	4378	4393	4409	4425	4440	4456	2	3	5	6	8
28	4472	4487	4502	4518	4533	4548	4564	4579	4594	4609	2	3	5	6	8
29	4624	4639	4654	4669	4683	4698	4713	4728	4742	4757	1	3	4	6	7
30	4771	4786	4800	4814	4829	4843	4857	4871	4886	4900	1	3	4	6	7
31	4914	4928	4942	4955	4969	4983	4997	5011	5024	5038	1	3	4	6	7
32	5051	5065	5079	5092	5105	5119	5132	5145	5159	5172	1	3	4	5	7
33	5185	5198	5211	5224	5237	5250	5263	5276	5289	5302	1	3	4	5	6
34	5315	5328	5340	5353	5366	5378	5391	5403	5416	5428	1	3	4	5	6
35	5441	5453	5465	5478	5490	5502	5514	5527	5539	5551	1	2	4	5	6
36	5563	5575	5587	5599	5611	5623	5635	5647	5658	5670	1	2	4	5	6
37	5682	5694	5705	5717	5729	5740	5752	5763	5775	5786	1	2	3	5	6
38	5798	5809	5821	5832	5843	5855	5866	5877	5888	5899	1	2	3	5	6
39	5911	5922	5933	5944	5955	5966	5977	5988	5999	6010	1	2	3	4	6
40	6021	6031	6042	6053	6064	6075	6085	6096	6107	6117	1	2	3	4	5
41	6128	6138	6149	6160	6170	6180	6191	6201	6212	6222	1	2	3	4	5
42	6232	6243	6253	6263	6274	6284	6294	6304	6314	6325	1	2	3	4	5
43	6335	6345	6355	6365	6375	6385	6395	6405	6415	6425	1	2	3	4	5
44	6435	6444	6454	6464	6474	6484	6493	6503	6513	6522	1	2	3	4	5
45	6532	6542	6551	6561	6571	6580	6590	6599	6609	6618	1	2	3	4	5
46	6628	6637	6646	6656	6665	6675	6684	6693	6702	6712	1	2	3	4	5
47	6721	6730	6739	6749	6758	6767	6776	6785	6794	6803	1	2	3	4	5
48	6812	6821	6830	6839	6848	6857	6866	6875	6884	6893	1	2	3	4	4
49	6902	6911	6920	6928	6937	6946	6955	6964	6972	6981	1	2	3	4	4
50	6990	6998	7007	7016	7024	7033	7042	7050	7059	7067	1	2	3	3	4
51	7076	7084	7093	7101	7110	7118	7126	7135	7143	7152	1	2	3	3	4
52	7160	7168	7177	7185	7193	7202	7210	7218	7226	7235	1	2	2	3	4
53	7243	7251	7259	7267	7275	7284	7292	7300	7308	7316	1	2	2	3	4
54	7324	7332	7340	7348	7356	7364	7372	7380	7388	7396	1	2	2	3	4

SEC. 7   TABLE OF LOGARITHMS TO THE BASE 10   xvii

NOTE: $\log_e N = \log_e 10 \log_{10} N = 2.3026 \log_{10} N$
         $\log_{10} e^x = x \log_{10} e = 0.434\ 29\ x$

N	0	1	2	3	4	5	6	7	8	9	P.P. 1	2	3	4	5
55	7404	7412	7419	7427	7435	7443	7451	7459	7466	7474	1	2	2	3	4
56	7482	7490	7497	7505	7513	7520	7528	7536	7543	7551	1	2	2	3	4
57	7559	7566	7574	7582	7589	7597	7604	7612	7619	7627	1	2	2	3	4
58	7634	7642	7649	7657	7664	7672	7679	7686	7694	7701	1	1	2	3	4
59	7709	7716	7723	7731	7738	7745	7752	7760	7767	7774	1	1	2	3	4
60	7782	7789	7796	7803	7810	7818	7825	7832	7839	7846	1	1	2	3	4
61	7853	7860	7868	7875	7882	7889	7896	7903	7910	7917	1	1	2	3	4
62	7924	7931	7938	7945	7952	7959	7966	7973	7980	7987	1	1	2	3	3
63	7993	8000	8007	8014	8021	8028	8035	8041	8048	8055	1	1	2	3	3
64	8062	8069	8075	8082	8089	8096	8102	8109	8116	8122	1	1	2	3	3
65	8129	8136	8142	8149	8156	8162	8169	8176	8182	8189	1	1	2	3	3
66	8195	8202	8209	8215	8222	8228	8235	8241	8248	8254	1	1	2	3	3
67	8261	8267	8274	8280	8287	8293	8299	8306	8312	8319	1	1	2	3	3
68	8325	8331	8338	8344	8351	8357	8363	8370	8376	8382	1	1	2	3	3
69	8388	8395	8401	8407	8414	8420	8426	8432	8439	8445	1	1	2	3	3
70	8451	8457	8463	8470	8476	8482	8488	8494	8500	8506	1	1	2	2	3
71	8513	8519	8525	8531	8537	8543	8549	8555	8561	8567	1	1	2	2	3
72	8573	8579	8585	8591	8597	8603	8609	8615	8621	8627	1	1	2	2	3
73	8633	8639	8645	8651	8657	8663	8669	8675	8681	8686	1	1	2	2	3
74	8692	8698	8704	8710	8716	8722	8727	8733	8739	8745	1	1	2	2	3
75	8751	8756	8762	8768	8774	8779	8785	8791	8797	8802	1	1	2	2	3
76	8808	8814	8820	8825	8831	8837	8842	8848	8854	8859	1	1	2	2	3
77	8865	8871	8876	8882	8887	8893	8899	8904	8910	8915	1	1	2	2	3
78	8921	8927	8932	8938	8943	8949	8954	8960	8965	8971	1	1	2	2	3
79	8976	8982	8987	8993	8998	9004	9009	9015	9020	9025	1	1	2	2	3
80	9031	9036	9042	9047	9053	9058	9063	9069	9074	9079	1	1	2	2	3
81	9085	9090	9096	9101	9106	9112	9117	9122	9128	9133	1	1	2	2	3
82	9138	9143	9149	9154	9159	9165	9170	9175	9180	9186	1	1	2	2	3
83	9191	9196	9201	9206	9212	9217	9222	9227	9232	9238	1	1	2	2	3
84	9243	9248	9253	9258	9263	9269	9274	9279	9284	9289	1	1	2	2	3
85	9294	9299	9304	9309	9315	9320	9325	9330	9335	9340	1	1	2	2	3
86	9345	9350	9355	9360	9365	9370	9375	9380	9385	9390	1	1	2	2	2
87	9395	9400	9405	9410	9415	9420	9425	9430	9435	9440	0	1	1	2	2
88	9445	9450	9455	9460	9465	9469	9474	9479	9484	9489	0	1	1	2	2
89	9494	9499	9504	9509	9513	9518	9523	9528	9533	9538	0	1	1	2	2
90	9542	9547	9552	9557	9562	9566	9571	9576	9581	9586	0	1	1	2	2
91	9590	9595	9600	9605	9609	9614	9619	9624	9628	9633	0	1	1	2	2
92	9638	9643	9647	9652	9657	9661	9666	9671	9675	9680	0	1	1	2	2
93	9685	9689	9694	9699	9703	9708	9713	9717	9722	9727	0	1	1	2	2
94	9731	9736	9741	9745	9750	9754	9759	9763	9768	9773	0	1	1	2	2
95	9777	9782	9786	9791	9795	9800	9805	9809	9814	9818	0	1	1	2	2
96	9823	9827	9832	9836	9841	9845	9850	9854	9859	9863	0	1	1	2	2
97	9868	9872	9877	9881	9886	9890	9894	9899	9903	9908	0	1	1	2	2
98	9912	9917	9921	9926	9930	9934	9939	9943	9948	9952	0	1	1	2	2
99	9956	9961	9965	9969	9974	9978	9983	9987	9991	9996	0	1	1	2	2

8. Tables of Exponentials

$$e^{-x}$$

x		0	1	2	3	4	5	6	7	8	9
0.0		1.000	.9900	.9802	.9704	.9608	.9512	.9418	.9324	.9231	.9139
0.1		.9048	.8958	.8869	.8781	.8694	.8607	.8521	.8437	.8353	.8270
0.2		.8187	.8106	.8025	.7945	.7866	.7788	.7711	.7634	.7558	.7483
0.3		.7408	.7334	.7261	.7189	.7118	.7047	.6977	.6907	.6839	.6771
0.4		.6703	.6637	.6570	.6505	.6440	.6376	.6313	.6250	.6188	.6126
0.5		.6065	.6005	.5945	.5886	.5827	.5769	.5712	.5655	.5599	.5543
0.6		.5488	.5434	.5379	.5326	.5273	.5220	.5169	.5117	.5066	.5016
0.7		.4966	.4916	.4868	.4819	.4771	.4724	.4677	.4630	.4584	.4538
0.8		.4493	.4449	.4404	.4360	.4317	.4274	.4232	.4190	.4148	.4107
0.9		.4066	.4025	.3985	.3946	.3906	.3867	.3829	.3791	.3753	.3716
1.0		.3679	.3642	.3606	.3570	.3535	.3499	.3465	.3430	.3396	.3362
1.1		.3329	.3296	.3263	.3230	.3198	.3166	.3135	.3104	.3073	.3042
1.2		.3012	.2982	.2952	.2923	.2894	.2865	.2837	.2808	.2780	.2753
1.3		.2725	.2698	.2671	.2645	.2618	.2592	.2567	.2541	.2516	.2491
1.4		.2466	.2441	.2417	.2393	.2369	.2346	.2322	.2299	.2276	.2254
1.5		.2231	.2209	.2187	.2165	.2144	.2122	.2101	.2080	.2060	.2039
1.6		.2019	.1999	.1979	.1959	.1940	.1920	.1901	.1882	.1864	.1845
1.7		.1827	.1809	.1791	.1773	.1755	.1738	.1720	.1703	.1686	.1670
1.8		.1653	.1637	.1620	.1604	.1588	.1572	.1557	.1541	.1526	.1511
1.9		.1496	.1481	.1466	.1451	.1437	.1423	.1409	.1395	.1381	.1367
2.0		.1353	.1340	.1327	.1313	.1300	.1287	.1275	.1262	.1249	.1237
2.1		.1225	.1212	.1200	.1188	.1177	.1165	.1153	.1142	.1130	.1119
2.2		.1108	.1097	.1086	.1075	.1065	.1054	.1043	.1033	.1023	.1013
2.3		.1003	*9926	*9827	*9730	*9633	*9537	*9442	*9348	*9255	*9163
2.4	0.0	9072	8982	8892	8804	8716	8629	8544	8458	8374	8291
2.5	0.0	8208	8127	8046	7966	7887	7808	7730	7654	7577	7502
2.6	0.0	7427	7353	7280	7208	7136	7065	6995	6925	6856	6788
2.7	0.0	6721	6654	6587	6522	6457	6393	6329	6266	6204	6142
2.8	0.0	6081	6020	5961	5901	5843	5784	5727	5670	5613	5558
2.9	0.0	5502	5448	5393	5340	5287	5234	5182	5130	5079	5029
3.0	0.0	4979	4929	4880	4832	4783	4736	4689	4642	4596	4550
3.1	0.0	4505	4460	4416	4372	4328	4285	4243	4200	4159	4117
3.2	0.0	4076	4036	3996	3956	3916	3877	3839	3801	3763	3725
3.3	0.0	3688	3652	3615	3579	3544	3508	3474	3439	3405	3371
3.4	0.0	3337	.3304	3271	3239	3206	3175	3143	3112	3081	3050

x		.0	.1	.2	.3	.4	.5	.6	.7	.8	.9
3	0.0	4979	4505	4076	3688	3337	3020	2732	2472	2237	2024
4	0.0	1832	1657	1500	1357	1228	1111	1005	*9095	*8230	*7447
5	0.00	6738	6097	5517	4992	4517	4087	3698	3346	3028	2739
6	0.00	2479	2243	2029	1836	1662	1503	1360	1231	1114	1008
7	0.000	9119	8251	7466	6755	6112	5531	5004	4528	4097	3707
8	0.000	3355	3035	2747	2485	2249	2035	1841	1666	1507	1364
9	0.000	1234	1117	1010	*9142	*8272	*7485	*6773	*6128	*5545	*5017
10	0.0000	4540	4108	3717	3363	3043	2754	2492	2254	2040	1846

$$\log_{10} e^{-x} = -x \log_{10} e = -0.434\ 29\ x$$

TABLES OF EXPONENTIALS

$$e^x$$

x	0	1	2	3	4	5	6	7	8	9
0.0	1.000	1.010	1.020	1.031	1.041	1.051	1.062	1.073	1.083	1.094
0.1	1.105	1.116	1.127	1.139	1.150	1.162	1.174	1.185	1.197	1.209
0.2	1.221	1.234	1.246	1.259	1.271	1.284	1.297	1.310	1.323	1.336
0.3	1.350	1.363	1.377	1.391	1.405	1.419	1.433	1.448	1.462	1.477
0.4	1.492	1.507	1.522	1.537	1.553	1.568	1.584	1.600	1.616	1.632
0.5	1.649	1.665	1.682	1.699	1.716	1.733	1.751	1.768	1.786	1.804
0.6	1.822	1.840	1.859	1.878	1.896	1.916	1.935	1.954	1.974	1.994
0.7	2.014	2.034	2.054	2.075	2.096	2.117	2.138	2.160	2.181	2.203
0.8	2.226	2.248	2.270	2.293	2.316	2.340	2.363	2.387	2.411	2.435
0.9	2.460	2.484	2.509	2.535	2.560	2.586	2.612	2.638	2.664	2.691
1.0	2.718	2.746	2.773	2.801	2.829	2.858	2.886	2.915	2.945	2.974
1.1	3.004	3.034	3.065	3.096	3.127	3.158	3.190	3.222	3.254	3.287
1.2	3.320	3.353	3.387	3.421	3.456	3.490	3.525	3.561	3.597	3.633
1.3	3.669	3.706	3.743	3.781	3.819	3.857	3.896	3.935	3.975	4.015
1.4	4.055	4.096	4.137	4.179	4.221	4.263	4.306	4.349	4.393	4.437
1.5	4.482	4.527	4.572	4.618	4.665	4.712	4.759	4.807	4.855	4.904
1.6	4.953	5.003	5.053	5.104	5.155	5.207	5.259	5.312	5.366	5.419
1.7	5.474	5.529	5.585	5.641	5.697	5.755	5.812	5.871	5.930	5.989
1.8	6.050	6.110	6.172	6.234	6.297	6.360	6.424	6.488	6.554	6.619
1.9	6.686	6.753	6.821	6.890	6.959	7.029	7.099	7.171	7.243	7.316
2.0	7.389	7.463	7.538	7.614	7.691	7.768	7.846	7.925	8.004	8.085
2.1	8.166	8.248	8.331	8.415	8.499	8.585	8.671	8.758	8.846	8.935
2.2	9.025	9.116	9.207	9.300	9.393	9.488	9.583	9.679	9.777	9.875
2.3	9.974	10.07	10.18	10.28	10.38	10.49	10.59	10.70	10.80	10.91
2.4	11.02	11.13	11.25	11.36	11.47	11.59	11.70	11.82	11.94	12.06
2.5	12.18	12.30	12.43	12.55	12.68	12.81	12.94	13.07	13.20	13.33
2.6	13.46	13.60	13.74	13.87	14.01	14.15	14.30	14.44	14.59	14.73
2.7	14.88	15.03	15.18	15.33	15.49	15.64	15.80	15.96	16.12	16.28
2.8	16.44	16.61	16.78	16.95	17.12	17.29	17.46	17.64	17.81	17.99
2.9	18.17	18.36	18.54	18.73	18.92	19.11	19.30	19.49	19.69	19.89
3.0	20.09	20.29	20.49	20.70	20.91	21.12	21.33	21.54	21.76	21.98
3.1	22.20	22.42	22.65	22.87	23.10	23.34	23.57	23.81	24.05	24.29
3.2	24.53	24.78	25.03	25.28	25.53	25.79	26.05	26.31	26.58	26.84
3.3	27.11	27.39	27.66	27.94	28.22	28.50	28.79	29.08	29.37	29.67
3.4	29.96	30.27	30.57	30.88	31.19	31.50	31.82	32.14	32.46	32.79

x	.0	.1	.2	.3	.4	.5	.6	.7	.8	.9
3	20.09	22.20	24.53	27.11	29.96	33.12	36.60	40.45	44.70	49.40
4	54.60	60.34	66.69	73.70	81.45	90.02	99.48	109.9	121.5	134.3
5	148.4	164.0	181.3	200.3	221.4	244.7	270.4	298.9	330.3	365.0
6	403.4	445.9	492.7	544.6	601.8	665.1	735.1	812.4	897.8	992.3
7	1097	1212	1339	1480	1636	1808	1998	2208	2441	2697
8	2981	3295	3641	4024	4447	4915	5432	6003	6634	7332
9	8103	8955	9897	10938	12088	13360	14765	16318	18034	19930

$$\log_{10} e^x = x \log_{10} e = 0.434\ 29\ x$$

9. Table of Square Roots

Numbers *N* from 1.00 to 5.49

N	0	1	2	3	4	5	6	7	8	9
1.0	1.000	1.005	1.010	1.015	1.020	1.025	1.030	1.034	1.039	1.044
1.1	1.049	1.054	1.058	1.063	1.068	1.072	1.077	1.082	1.086	1.091
1.2	1.095	1.100	1.105	1.109	1.114	1.118	1.122	1.127	1.131	1.136
1.3	1.140	1.145	1.149	1.153	1.158	1.162	1.166	1.170	1.175	1.179
1.4	1.183	1.187	1.192	1.196	1.200	1.204	1.208	1.212	1.217	1.221
1.5	1.225	1.229	1.233	1.237	1.241	1.245	1.249	1.253	1.257	1.261
1.6	1.265	1.269	1.273	1.277	1.281	1.285	1.288	1.292	1.296	1.300
1.7	1.304	1.308	1.311	1.315	1.319	1.323	1.327	1.330	1.334	1.338
1.8	1.342	1.345	1.349	1.353	1.356	1.360	1.364	1.367	1.371	1.375
1.9	1.378	1.382	1.386	1.389	1.393	1.396	1.400	1.404	1.407	1.411
2.0	1.414	1.418	1.421	1.425	1.428	1.432	1.435	1.439	1.442	1.446
2.1	1.449	1.453	1.456	1.459	1.463	1.466	1.470	1.473	1.476	1.480
2.2	1.483	1.487	1.490	1.493	1.497	1.500	1.503	1.507	1.510	1.513
2.3	1.517	1.520	1.523	1.526	1.530	1.533	1.536	1.539	1.543	1.546
2.4	1.549	1.552	1.556	1.559	1.562	1.565	1.568	1.572	1.575	1.578
2.5	1.581	1.584	1.587	1.591	1.594	1.597	1.600	1.603	1.606	1.609
2.6	1.612	1.616	1.619	1.622	1.625	1.628	1.631	1.634	1.637	1.640
2.7	1.643	1.646	1.649	1.652	1.655	1.658	1.661	1.664	1.667	1.670
2.8	1.673	1.676	1.679	1.682	1.685	1.688	1.691	1.694	1.697	1.700
2.9	1.703	1.706	1.709	1.712	1.715	1.718	1.720	1.723	1.726	1.729
3.0	1.732	1.735	1.738	1.741	1.744	1.746	1.749	1.752	1.755	1.758
3.1	1.761	1.764	1.766	1.769	1.772	1.775	1.778	1.780	1.783	1.786
3.2	1.789	1.792	1.794	1.797	1.800	1.803	1.806	1.808	1.811	1.814
3.3	1.817	1.819	1.822	1.825	1.828	1.830	1.833	1.836	1.838	1.841
3.4	1.844	1.847	1.849	1.852	1.855	1.857	1.860	1.863	1.865	1.868
3.5	1.871	1.873	1.876	1.879	1.881	1.884	1.887	1.889	1.892	1.895
3.6	1.897	1.900	1.903	1.905	1.908	1.910	1.913	1.916	1.918	1.921
3.7	1.924	1.926	1.929	1.931	1.934	1.936	1.939	1.942	1.944	1.947
3.8	1.949	1.952	1.954	1.957	1.960	1.962	1.965	1.967	1.970	1.972
3.9	1.975	1.977	1.980	1.982	1.985	1.987	1.990	1.992	1.995	1.997
4.0	2.000	2.002	2.005	2.007	2.010	2.012	2.015	2.017	2.020	2.022
4.1	2.025	2.027	2.030	2.032	2.035	2.037	2.040	2.042	2.045	2.047
4.2	2.049	2.052	2.054	2.057	2.059	2.062	2.064	2.066	2.069	2.071
4.3	2.074	2.076	2.078	2.081	2.083	2.086	2.088	2.090	2.093	2.095
4.4	2.098	2.100	2.102	2.105	2.107	2.110	2.112	2.114	2.117	2.119
4.5	2.121	2.124	2.126	2.128	2.131	2.133	2.135	2.138	2.140	2.142
4.6	2.145	2.147	2.149	2.152	2.154	2.156	2.159	2.161	2.163	2.166
4.7	2.168	2.170	2.173	2.175	2.177	2.179	2.182	2.184	2.186	2.189
4.8	2.191	2.193	2.195	2.198	2.200	2.202	2.205	2.207	2.209	2.211
4.9	2.214	2.216	2.218	2.220	2.223	2.225	2.227	2.229	2.232	2.234
5.0	2.236	2.238	2.241	2.243	2.245	2.247	2.249	2.252	2.254	2.256
5.1	2.258	2.261	2.263	2.265	2.267	2.269	2.272	2.274	2.276	2.278
5.2	2.280	2.283	2.285	2.287	2.289	2.291	2.293	2.296	2.298	2.300
5.3	2.302	2.304	2.307	2.309	2.311	2.313	2.315	2.317	2.319	2.322
5.4	2.324	2.326	2.328	2.330	2.332	2.335	2.337	2.339	2.341	2.343

Numbers N from 5.50 to 9.99

N	0	1	2	3	4	5	6	7	8	9
5.5	2.345	2.347	2.349	2.352	2.354	2.356	2.358	2.360	2.362	2.364
5.6	2.366	2.369	2.371	2.373	2.375	2.377	2.379	2.381	2.383	2.385
5.7	2.387	2.390	2.392	2.394	2.396	2.398	2.400	2.402	2.404	2.406
5.8	2.408	2.410	2.412	2.415	2.417	2.419	2.421	2.423	2.425	2.427
5.9	2.429	2.431	2.433	2.435	2.437	2.439	2.441	2.443	2.445	2.447
6.0	2.449	2.452	2.454	2.456	2.458	2.460	2.462	2.464	2.466	2.468
6.1	2.470	2.472	2.474	2.476	2.478	2.480	2.482	2.484	2.486	2.488
6.2	2.490	2.492	2.494	2.496	2.498	2.500	2.502	2.504	2.506	2.508
6.3	2.510	2.512	2.514	2.516	2.518	2.520	2.522	2.524	2.526	2.528
6.4	2.530	2.532	2.534	2.536	2.538	2.540	2.542	2.544	2.546	2.548
6.5	2.550	2.551	2.553	2.555	2.557	2.559	2.561	2.563	2.565	2.567
6.6	2.569	2.571	2.573	2.575	2.577	2.579	2.581	2.583	2.585	2.587
6.7	2.588	2.590	2.592	2.594	2.596	2.598	2.600	2.602	2.604	2.606
6.8	2.608	2.610	2.612	2.613	2.615	2.617	2.619	2.621	2.623	2.625
6.9	2.627	2.629	2.631	2.632	2.634	2.636	2.638	2.640	2.642	2.644
7.0	2.646	2.648	2.650	2.651	2.653	2.655	2.657	2.659	2.661	2.663
7.1	2.665	2.666	2.668	2.670	2.672	2.674	2.676	2.678	2.680	2.681
7.2	2.683	2.685	2.687	2.689	2.691	2.693	2.694	2.696	2.698	2.700
7.3	2.702	2.704	2.706	2.707	2.709	2.711	2.713	2.715	2.717	2.718
7.4	2.720	2.722	2.724	2.726	2.728	2.729	2.731	2.733	2.735	2.737
7.5	2.739	2.740	2.742	2.744	2.746	2.748	2.750	2.751	2.753	2.755
7.6	2.757	2.759	2.760	2.762	2.764	2.766	2.768	2.769	2.771	2.773
7.7	2.775	2.777	2.778	2.780	2.782	2.784	2.786	2.787	2.789	2.791
7.8	2.793	2.795	2.796	2.798	2.800	2.802	2.804	2.805	2.807	2.809
7.9	2.811	2.812	2.814	2.816	2.818	2.820	2.821	2.823	2.825	2.827
8.0	2.828	2.830	2.832	2.834	2.835	2.837	2.839	2.841	2.843	2.844
8.1	2.846	2.848	2.850	2.851	2.853	2.855	2.857	2.858	2.860	2.862
8.2	2.864	2.865	2.867	2.869	2.871	2.872	2.874	2.876	2.877	2.879
8.3	2.881	2.883	2.884	2.886	2.888	2.890	2.891	2.893	2.895	2.897
8.4	2.898	2.900	2.902	2.903	2.905	2.907	2.909	2.910	2.912	2.914
8.5	2.915	2.917	2.919	2.921	2.922	2.924	2.926	2.927	2.929	2.931
8.6	2.933	2.934	2.936	2.938	2.939	2.941	2.943	2.944	2.946	2.948
8.7	2.950	2.951	2.953	2.955	2.956	2.958	2.960	2.961	2.963	2.965
8.8	2.966	2.968	2.970	2.972	2.973	2.975	2.977	2.978	2.980	2.982
8.9	2.983	2.985	2.987	2.988	2.990	2.992	2.993	2.995	2.997	2.998
9.0	3.000	3.002	3.003	3.005	3.007	3.008	3.010	3.012	3.013	3.015
9.1	3.017	3.018	3.020	3.022	3.023	3.025	3.027	3.028	3.030	3.032
9.2	3.033	3.035	3.036	3.038	3.040	3.041	3.043	3.045	3.046	3.048
9.3	3.050	3.051	3.053	3.055	3.056	3.058	3.059	3.061	3.063	3.064
9.4	3.066	3.068	3.069	3.071	3.072	3.074	3.076	3.077	3.079	3.081
9.5	3.082	3.084	3.085	3.087	3.089	3.090	3.092	3.094	3.095	3.097
9.6	3.098	3.100	3.102	3.103	3.105	3.106	3.108	3.110	3.111	3.113
9.7	3.114	3.116	3.118	3.119	3.121	3.122	3.124	3.126	3.127	3.129
9.8	3.130	3.132	3.134	3.135	3.137	3.138	3.140	3.142	3.143	3.145
9.9	3.146	3.148	3.150	3.151	3.153	3.154	3.156	3.158	3.159	3.161

Table of Square Roots (Continued)

Numbers N from 10.0 to 54.9

N	.0	.1	.2	.3	.4	.5	.6	.7	.8	.9
10	3.162	3.178	3.194	3.209	3.225	3.240	3.256	3.271	3.286	3.302
11	3.317	3.332	3.347	3.362	3.376	3.391	3.406	3.421	3.435	3.450
12	3.464	3.479	3.493	3.507	3.521	3.536	3.550	3.564	3.578	3.592
13	3.606	3.619	3.633	3.647	3.661	3.674	3.688	3.701	3.715	3.728
14	3.742	3.755	3.768	3.782	3.795	3.808	3.821	3.834	3.847	3.860
15	3.873	3.886	3.899	3.912	3.924	3.937	3.950	3.962	3.975	3.987
16	4.000	4.012	4.025	4.037	4.050	4.062	4.074	4.087	4.099	4.111
17	4.123	4.135	4.147	4.159	4.171	4.183	4.195	4.207	4.219	4.231
18	4.243	4.254	4.266	4.278	4.290	4.301	4.313	4.324	4.336	4.347
19	4.359	4.370	4.382	4.393	4.405	4.416	4.427	4.438	4.450	4.461
20	4.472	4.483	4.494	4.506	4.517	4.528	4.539	4.550	4.561	4.572
21	4.583	4.593	4.604	4.615	4.626	4.637	4.648	4.658	4.669	4.680
22	4.690	4.701	4.712	4.722	4.733	4.743	4.754	4.764	4.775	4.785
23	4.796	4.806	4.817	4.827	4.837	4.848	4.858	4.868	4.879	4.889
24	4.899	4.909	4.919	4.930	4.940	4.950	4.960	4.970	4.980	4.990
25	5.000	5.010	5.020	5.030	5.040	5.050	5.060	5.070	5.079	5.089
26	5.099	5.109	5.119	5.128	5.138	5.148	5.158	5.167	5.177	5.187
27	5.196	5.206	5.215	5.225	5.235	5.244	5.254	5.263	5.273	5.282
28	5.292	5.301	5.310	5.320	5.329	5.339	5.348	5.357	5.367	5.376
29	5.385	5.394	5.404	5.413	5.422	5.431	5.441	5.450	5.459	5.468
30	5.477	5.486	5.495	5.505	5.514	5.523	5.532	5.541	5.550	5.559
31	5.568	5.577	5.586	5.595	5.604	5.612	5.621	5.630	5.639	5.648
32	5.657	5.666	5.675	5.683	5.692	5.701	5.710	5.718	5.727	5.736
33	5.745	5.753	5.762	5.771	5.779	5.788	5.797	5.805	5.814	5.822
34	5.831	5.840	5.848	5.857	5.865	5.874	5.882	5.891	5.899	5.908
35	5.916	5.925	5.933	5.942	5.950	5.958	5.967	5.975	5.983	5.992
36	6.000	6.008	6.017	6.025	6.033	6.042	6.050	6.058	6.066	6.075
37	6.083	6.091	6.099	6.107	6.116	6.124	6.132	6.140	6.148	6.156
38	6.164	6.173	6.181	6.189	6.197	6.205	6.213	6.221	6.229	6.237
39	6.245	6.253	6.261	6.269	6.277	6.285	6.293	6.301	6.309	6.317
40	6.325	6.332	6.340	6.348	6.356	6.364	6.372	6.380	6.387	6.395
41	6.403	6.411	6.419	6.427	6.434	6.442	6.450	6.458	6.465	6.473
42	6.481	6.488	6.496	6.504	6.512	6.519	6.527	6.535	6.542	6.550
43	6.557	6.565	6.573	6.580	6.588	6.595	6.603	6.611	6.618	6.626
44	6.633	6.641	6.648	6.656	6.663	6.671	6.678	6.686	6.693	6.701
45	6.708	6.716	6.723	6.731	6.738	6.745	6.753	6.760	6.768	6.775
46	6.782	6.790	6.797	6.804	6.812	6.819	6.826	6.834	6.841	6.848
47	6.856	6.863	6.870	6.877	6.885	6.892	6.899	6.906	6.914	6.921
48	6.928	6.935	6.943	6.950	6.957	6.964	6.971	6.979	6.986	6.993
49	7.000	7.007	7.014	7.021	7.029	7.036	7.043	7.050	7.057	7.064
50	7.071	7.078	7.085	7.092	7.099	7.106	7.113	7.120	7.127	7.134
51	7.141	7.148	7.155	7.162	7.169	7.176	7.183	7.190	7.197	7.204
52	7.211	7.218	7.225	7.232	7.239	7.246	7.253	7.259	7.266	7.273
53	7.280	7.287	7.294	7.301	7.308	7.314	7.321	7.328	7.335	7.342
54	7.348	7.355	7.362	7.369	7.376	7.382	7.389	7.396	7.403	7.409

TABLE OF SQUARE ROOTS

Numbers N from 55.0 to 99.9

N	.0	.1	.2	.3	.4	.5	.6	.7	.8	.9
55	7.416	7.423	7.430	7.436	7.443	7.450	7.457	7.463	7.470	7.477
56	7.483	7.490	7.497	7.503	7.510	7.517	7.523	7.530	7.537	7.543
57	7.550	7.556	7.563	7.570	7.576	7.583	7.589	7.596	7.603	7.609
58	7.616	7.622	7.629	7.635	7.642	7.649	7.655	7.662	7.668	7.675
59	7.681	7.688	7.694	7.701	7.707	7.714	7.720	7.727	7.733	7.740
60	7.746	7.752	7.759	7.765	7.772	7.778	7.785	7.791	7.797	7.804
61	7.810	7.817	7.823	7.829	7.836	7.842	7.849	7.855	7.861	7.868
62	7.874	7.880	7.887	7.893	7.899	7.906	7.912	7.918	7.925	7.931
63	7.937	7.944	7.950	7.956	7.962	7.969	7.975	7.981	7.987	7.994
64	8.000	8.006	8.012	8.019	8.025	8.031	8.037	8.044	8.050	8.056
65	8.062	8.068	8.075	8.081	8.087	8.093	8.099	8.106	8.112	8.118
66	8.124	8.130	8.136	8.142	8.149	8.155	8.161	8.167	8.173	8.179
67	8.185	8.191	8.198	8.204	8.210	8.216	8.222	8.228	8.234	8.240
68	8.246	8.252	8.258	8.264	8.270	8.276	8.283	8.289	8.295	8.301
69	8.307	8.313	8.319	8.325	8.331	8.337	8.343	8.349	8.355	8.361
70	8.367	8.373	8.379	8.385	8.390	8.396	8.402	8.408	8.414	8.420
71	8.426	8.432	8.438	8.444	8.450	8.456	8.462	8.468	8.473	8.479
72	8.485	8.491	8.497	8.503	8.509	8.515	8.521	8.526	8.532	8.538
73	8.544	8.550	8.556	8.562	8.567	8.573	8.579	8.585	8.591	8.597
74	8.602	8.608	8.614	8.620	8.626	8.631	8.637	8.643	8.649	8.654
75	8.660	8.666	8.672	8.678	8.683	8.689	8.695	8.701	8.706	8.712
76	8.718	8.724	8.729	8.735	8.741	8.746	8.752	8.758	8.764	8.769
77	8.775	8.781	8.786	8.792	8.798	8.803	8.809	8.815	8.820	8.826
78	8.832	8.837	8.843	8.849	8.854	8.860	8.866	8.871	8.877	8.883
79	8.888	8.894	8.899	8.905	8.911	8.916	8.922	8.927	8.933	8.939
80	8.944	8.950	8.955	8.961	8.967	8.972	8.978	8.983	8.989	8.994
81	9.000	9.006	9.011	9.017	9.022	9.028	9.033	9.039	9.044	9.050
82	9.055	9.061	9.066	9.072	9.077	9.083	9.088	9.094	9.099	9.105
83	9.110	9.116	9.121	9.127	9.132	9.138	9.143	9.149	9.154	9.160
84	9.165	9.171	9.176	9.182	9.187	9.192	9.198	9.203	9.209	9.214
85	9.220	9.225	9.230	9.236	9.241	9.247	9.252	9.257	9.263	9.268
86	9.274	9.279	9.284	9.290	9.295	9.301	9.306	9.311	9.317	9.322
87	9.327	9.333	9.338	9.343	9.349	9.354	9.359	9.365	9.370	9.375
88	9.381	9.386	9.391	9.397	9.402	9.407	9.413	9.418	9.423	9.429
89	9.434	9.439	9.445	9.450	9.455	9.460	9.466	9.471	9.476	9.482
90	9.487	9.492	9.497	9.503	9.508	9.513	9.518	9.524	9.529	9.534
91	9.539	9.545	9.550	9.555	9.560	9.566	9.571	9.576	9.581	9.586
92	9.592	9.597	9.602	9.607	9.612	9.618	9.623	9.628	9.633	9.638
93	9.644	9.649	9.654	9.659	9.664	9.670	9.675	9.680	9.685	9.690
94	9.695	9.701	9.706	9.711	9.716	9.721	9.726	9.731	9.737	9.742
95	9.747	9.752	9.757	9.762	9.767	9.772	9.778	9.783	9.788	9.793
96	9.798	9.803	9.808	9.813	9.818	9.823	9.829	9.834	9.839	9.844
97	9.849	9.854	9.859	9.864	9.869	9.874	9.879	9.884	9.889	9.894
98	9.899	9.905	9.910	9.915	9.920	9.925	9.930	9.935	9.940	9.945
99	9.950	9.955	9.960	9.965	9.970	9.975	9.980	9.985	9.990	9.995

10. A Short Table of Cube Roots

Numbers *N* from 1.0 to 10.9

N	.0	.1	.2	.3	.4	.5	.6	.7	.8	.9
1	1.000	1.032	1.063	1.091	1.119	1.145	1.170	1.193	1.216	1.239
2	1.260	1.281	1.301	1.320	1.339	1.357	1.375	1.392	1.409	1.426
3	1.442	1.458	1.474	1.489	1.504	1.518	1.533	1.547	1.560	1.574
4	1.587	1.601	1.613	1.626	1.639	1.651	1.663	1.675	1.687	1.698
5	1.710	1.721	1.732	1.744	1.754	1.765	1.776	1.786	1.797	1.807
6	1.817	1.827	1.837	1.847	1.857	1.866	1.876	1.885	1.895	1.904
7	1.913	1.922	1.931	1.940	1.949	1.957	1.966	1.975	1.983	1.992
8	2.000	2.008	2.017	2.025	2.033	2.041	2.049	2.057	2.065	2.072
9	2.080	2.088	2.095	2.103	2.110	2.118	2.125	2.133	2.140	2.147
10	2.154	2.162	2.169	2.176	2.183	2.190	2.197	2.204	2.210	2.217

Numbers *N* from 10 to 109

N	0	1	2	3	4	5	6	7	8	9
1	2.154	2.224	2.289	2.351	2.410	2.466	2.520	2.571	2.621	2.668
2	2.714	2.759	2.802	2.844	2.884	2.924	2.962	3.000	3.037	3.072
3	3.107	3.141	3.175	3.208	3.240	3.271	3.302	3.332	3.362	3.391
4	3.420	3.448	3.476	3.503	3.530	3.557	3.583	3.609	3.634	3.659
5	3.684	3.708	3.733	3.756	3.780	3.803	3.826	3.849	3.871	3.893
6	3.915	3.936	3.958	3.979	4.000	4.021	4.041	4.062	4.082	4.102
7	4.121	4.141	4.160	4.179	4.198	4.217	4.236	4.254	4.273	4.291
8	4.309	4.327	4.344	4.362	4.380	4.397	4.414	4.431	4.448	4.465
9	4.481	4.498	4.514	4.531	4.547	4.563	4.579	4.595	4.610	4.626
10	4.642	4.657	4.672	4.688	4.703	4.718	4.733	4.747	4.762	4.777

Numbers *N* from 100 to 1090

N	00	10	20	30	40	50	60	70	80	90
1	4.642	4.791	4.932	5.066	5.192	5.313	5.429	5.540	5.646	5.749
2	5.848	5.944	6.037	6.127	6.214	6.300	6.383	6.463	6.542	6.619
3	6.694	6.768	6.840	6.910	6.980	7.047	7.114	7.179	7.243	7.306
4	7.368	7.429	7.489	7.548	7.606	7.663	7.719	7.775	7.830	7.884
5	7.937	7.990	8.041	8.093	8.143	8.193	8.243	8.291	8.340	8.387
6	8.434	8.481	8.527	8.573	8.618	8.662	8.707	8.750	8.794	8.837
7	8.879	8.921	8.963	9.004	9.045	9.086	9.126	9.166	9.205	9.244
8	9.283	9.322	9.360	9.398	9.435	9.473	9.510	9.546	9.583	9.619
9	9.655	9.691	9.726	9.761	9.796	9.830	9.865	9.899	9.933	9.967
10	10.00	10.03	10.07	10.10	10.13	10.16	10.20	10.23	10.26	10.29

INDEX

Aberration
  chromatic 788
  spherical 764, 788
Absolute
  electrometer 517, 527
  humidity 373
  pressure 267
  temperature 292, 395
  units 57, Ai
Absorber, perfect 325
Absorption 325, 755
  coefficient 756
  spectra 840
Accelerated motion 39–45
Acceleration 39
  angular 147
  centripetal 42
Accelerator (*see also* Cyclotron, Van de Graaff, Betatron)
  linear 684
Achromatic lenses 794, 804
Acoustics (*see* Sound)
Adiabatic process 386
Adiabatic lapse rate 407
Air columns, vibrations of 445
Airfoil 276
Alpha particles 914
  capture of 916
  scattering of 485
Alternating currents (AC) 667 ff
Ammeter 552, DC 681
Amorphous phase of matter 368
AMPÈRE, ANDRÈ MARIE 530
Ampere, definition 530, 606
Ampère's
  line-integral relation 608
  principle 606
Ampèrian 'whirls' 629
Amplification factor 560
Amplitude 245
ANDERSON, CARL D. 839, 920
Anemometer, sonic 461
Aneroid barometer 268
Angle 16
  phase 245
  solid 721
Angular
  acceleration 147
  aperture 764
  displacement 146
  momentum 157, 195
    quantization of 853, 873
  simple harmonic motion 251
  velocity 146
Anisotropic solids 226
Annihilation of matter 919 ff
Anode 558
Antenna 690
Antimatter 944
Antinodes 428, 447

Antiparticles 945
Aperture, optical 764
ARCHIMEDES 265
Archimedes' relation 265
ARRHENIUS, SVANTE AUGUST 568
Astigmatism 788
Astronomical data Aii
Atmospheric pressure 267 ff
Atomic (*see also* Nuclear, Nucleus)
  bomb 925 ff
  energy levels 851, 871
  mass 334, 596
  mass unit (nuclidic) 332
  number 332
  structure of liquids and solids 232 ff
  'weight' 334
Atoms 332 ff
  magnetic properties of 877
  structure of 872
Audibility 452 ff
AVOGADRO, AMADEO 335
Avogadro's
  constant 336, 353, 570, 862
  law 335

Ballistic
  galvanometer 663
  missiles 209 ff
  pendulum 257
Ballistics 75
BALMER, JOHANN JAKOB 849
Band spectra 842
BARDEEN, JOHN 560
Barkhausen effect 881
Barometer 267 ff
Baryons 939
Batteries 574
  charge and discharge 548
Beats 455–6
BECQUEREL, HENRI 914
Bel 450
BENNETT, WILLARD H. 512
BERGSTRAND, ERIC 716
BERNOULLI, DANIEL 273
Bernoulli's law 269 ff
Beta particles 915
Betatron 656
BETHE, HANS 924
Binoculars 803
Biot-Savart law 604
Birefringence 827 ff
Black-body radiation 325, 843
BOHR, NILS 839, 850, 930
Bohr
  magneton 879
  model of atom 852 ff
  model of nucleus 930
  quantum mechanics 850 ff
  radius 853

Boiling 371
BOLTZMANN, LUDWIG 346
Boltzmann's constant 349
Boltzmann factor 358
Bomb calorimeter 313
Boundary layer 279
Bourdon gauge 269
BOYLE, ROBERT 337
Boyle's law 337
BRADLEY, JAMES 714
BRAHE, TYCHO 3
BRATTAIN, WALTER 560
BREWSTER, DAVID 824
Brewster's angle 824
British gravitational units 66 ff
British thermal unit (BTU) 305
BROWN, ROBERT 353
Brownian motion 353
Buoyant forces 265
Bubble chamber 372
Bulk modulus 227

Calorie 304, 306
Calorimeter 304, 307
  bomb 313
  continuous-flow 316
Calorimetry relation 304
Camera
  photographic 790
  pinhole 725
Candela 722
Capacitance 513 ff
  in AC circuits 750
Capacitive reactance 762
Capacitor 515 ff
  charge and discharge 554
  cylindrical 523
  electrolytic 520
  energy of 521
  parallel-plate 514
  spherical 526
Capacitors in parallel and series 521
'Carbon dating' 933
CARNOT, SADI 390
Carnot
  cycle 396
  efficiency 392
  theorems 393
Cathode 558
Cathode-ray tube 475
CAVENDISH, HENRY 61
Cavitation 275
CELSIUS, ANDERS 290
Centripetal
  acceleration 42
  force 60
CHADWICK, JAMES 917
Chain reaction, nuclear 926 ff
CHAMBERLAIN, OWEN 945

Charge (*see* Electric charge)
Charging by induction 504
CHARLES, JACQUES 337
Charles' law 337
Chemical calorie 306
Chromatic aberration 788
CHRZANOWSKI, P. 62
Circle
 motion in 42, 248
Circular polarization 831
Circuits, electric 529 ff
 AC 467, 486 ff
 DC 546 ff
Coercive force Hc 637
Coherence 705, 857
Collisions 134 ff
Combustion, heats of 313 ff
Compression
 adiabatic 386
 isothermal 384
COMPTON, ARTHUR H. 861
Compton effect 862
Condensation 361
 in waves 413
Condenser (*see* Capacitor)
Conduction
 electric 532 ff
 thermal 318 ff
Conductivity
 electric 535
 thermal 320
Conic sections 205
Conservation of
 angular momentum 157, 196
 electric charge 470
 energy 114
 momentum 134, 901
Convection 319, 322
Conversion factors Avii ff
Cooling, Newton's law of 322
Coordinate systems 17, 28
 inertial 76
 moving 75, 117, 898
COTTON, A. A. 833
Cotton-Mouton effect 833
COULOMB, CHARLES AUGUSTIN 467
Coulomb, definition 468, 606
Coulomb's principle 468
Couple 178
COWAN, CLYDE 916
Creation of matter 919 ff
Critical
 angle 750
 constants 368
 mass 928
 point 367
Crystals 226, 827, 862
Current, electric 529 ff
 alternating 667 ff
 balance 621
 direct 529 ff, 546 ff
Cycle 243
Cyclotron 598
 spiral-ridge 904

Cyclotron (*cont.*)
 synchro- 904

DALTON, JOHN 332
Dalton's law of partial pressures 339
Daniell cell 572
DAVISSON, C. J. 867
Day 31
Debye unit 492
DE BROGLIE, LOUIS 867
de Broglie waves 867
Deceleration 40
Decibel 450
Declination angle 587
Definition, operational 4
DE FORREST, LEE 558
Degree
 angle 16
 temperature 290
Degrees of freedom 181
Density 69
 variation with temperature 301
Derived quantities 7, Ai
DEUTSCH, MARTIN 933
Dew point 374
Diamagnetism 628, 641
Dielectric 467
 constant 519
 forces between charges in 520
 polarization of 517 ff
 strength 522
Difference tone 456
Diffraction 729 ff, 807 ff
 electron 867
 Fraunhofer 807
 Fresnel 729
 grating 815
 X-ray 862
Dimensional analysis 6 ff
Diopter 794
Dip angle 587
Dipole
 electric 488
 magnetic 610
DIRAC, P. A. M. 921
Direct electric currents (DC) 723 ff
Direction cosines 28
Dispersion 493 ff, 840
Displacement 13 ff
 angular 146
Dissipation of mechanical energy 117
Dissociation energies 571
Domains, ferromagnetic 881
DOPPLER, CHRISTIAN JOHANN 457
Doppler effect 457, 726, 863
 relativistic 898
Double refraction 827 ff
Dry cells 574
Dulong and Petit 300
Dynamical system 133, 181 ff
Dynamics 51 ff
 fluid 269 ff

Dynamics (*cont.*)
 of particles 51
 of rotation 155 ff
 relativistic 899 ff

Ear 452
Earth
 magnetic field of 587
 mass of 65, Aiv
 satellites 260 ff
EDISON, THOMAS ALVA 541
Edison's discovery 541
Eddy currents 652
Effective values of current and voltage 669
Efficiency 122, 159
 Carnot 392
 of heat engines 400
EINSTEIN, ALBERT 353, 848, 895 ff
Einstein's
 Brownian-motion theory 353
 photo-electric equation 848
 relativity theory 895 ff
Elastic
 collisions 134
 constants 223 ff, 230, 236
 limit 239
 materials 219 ff
 potential energy 221-2
 properties of rubber 223
Electric
 charge 466 ff
  bound 517
  conservation of 470
  in a dielectric 520
  negative 466
  on conductors 499
  positive 466
  test 471
 conductors 467, 534ff
 current 529 ff, 546 ff
  alternating 667 ff
  balance 621
  direct 546 ff
  element 596
 dipoles 488
 fields 493
  energy 892
 flux 495
 insulators 467, 538
 intensity 471
 lines 472
 networks 550
 potential 477, 480
 tube 494
Electrical standards 621, 659
Electricity (*see also* Electric) 565 ff
Electro chemistry 566 ff
Electrodynamometer 682
Electrolysis 566 ff
 Faraday's principle of 567
Electrolytes 566
Electromagnetic
 fields 690, 884 ff

Electromagnetic (*cont.*)
  induction 646 ff
  interactions 936
  radiation 690, 883 ff
  spectrum 712
Electrometer 506, 517, 527
Electromotive force (EMF) 532
Electron 470
  charge of 470, 484
  cold emission 523
  conduction 541
  diffraction 867
  magnetic resonance (EPR) 882
  mass of 596
  microscope 882
  -ray tube 475
  spin 873
  -volt 485
Electronic charge unit 470
Electroscope 506
Electrostatic shielding 503
Electrostatics 456 ff
Elementary particles 946
Elements 332, Aiv
EMF (*see* Electromotive force)
Emission (*see* Light, Radiation, Spectra)
Emissive power 323 ff
Emissivity 326
Endothermic reaction 313
Energy 102, 107 ff
  conservation of 114
  in electric field 521, 527
  in magnetic field 644, 666
  in moving coordinate systems 117
  levels 251, 487, 850
  mass equivalence 903
  mechanical
    dissipation of 115
    kinetic 108, 153, 187
    potential 108, 219
    transformations of 112
  nuclear 920 ff
  radiant 323, 720, 892
  rest 902
  thermal 115, 303 ff
  transmission by waves 421 ff
Entropy 382 ff
Equilibrium
  neutral 185
  of a dynamical system 185
  of a particle 85
  of a rigid body 165-6
  stable 185
  thermal 292
  unstable 185
Equipotential surfaces 480
Escape, velocity of 204
Evaporation 361
Ewing's theory of magnetism 880
Exothermic reaction 313
Expansion, thermal 293 ff
  anomalous (water) 299

Expansion, thermal (*cont.*)
  area 296
  linear 294
  volume 297, 298
Expansion processes
  adiabatic 386
  isothermal 384
Eye 795
Eyepiece (*see* Microscope, Telescope)

FAHRENHEIT, GABRIEL DANIEL 290
Falling bodies 71-75
Farad 514
Faraday, constant 567
  effect 833
FARADAY, MICHAEL 506, 833
Faraday's
  ice-pail experiment 506
  principle of electrolysis 567
Fermat's 'principle' 752
FERMI, ENRICO 925
Ferroelectric 520
Ferromagnetic
  domains 880
  materials 636
Ferromagnetism 629 ff
Fields
  electric 471 ff, 493 ff
  electromagnetic 879, 690, 883
  gravitational 509
  magnetic 584 ff, 603 ff
Fission, nuclear 925
  bomb 927
FIZEAU, A. H. L. 714
Flowmeter, Venturi 275
Fluid 260 ff
  dynamics 269 ff
  pressure 262
  statics 264 ff
Flux
  electric 494
  luminous 722
  magnetic 613
  radiant 721
  vector 494
Focus, principal 764, 778
Foot 15
Foot-pound (fp) 104
Force 56 ff
  -constant 220
  external 133
  impulse 137
  impulsive 131
  internal 133, 235
  normal 84, 88
  reaction 137
  systems 83 ff
  vector properties of 56, 84 ff
FOUCAULT, J. B. L. 715
Fourier's theorem 436
FRANCK, JAMES 851
Franck-Hertz experiment 851
FRANKLIN, BENJAMIN 466

FRAUNHOFER, JOSEPH 815
Fraunhofer
  diffraction 807
  lines 842
Free
  expansion 341
  fall 71-75
Freedom, degrees of 181
Freezing (*see also* Fusion) 359
FRESNEL, AUGUSTIN JEAN 701
Fresnel
  diffraction 729
  zones 737
Frequency 243
Friction 87 ff
  work done by 115
FRISCH, OTTO 925
FROOME, K. D. 716
Fuel cells 577
Fundamental
  frequency 440
  mode 431, 441, 447
  physical constants Aiii
  quantities 6, Ai
Fusion 359
  curve 361
  latent heat 308
  nuclear 924, 928, 929

Galilean telescope 805
GALILEO GALILEI 3, 52, 64, 713
Galvanometer 593
  ballistic 663
Gamma rays, 712, 915
Gas
  constant 338
  external work by 340
  ideal 331 ff
  laws 337
  liquefaction of 403
  molecular nature of 346
  real 368
GAUSS, KARL FRIEDRICH 496
Gauss' relation 497
  for gravitation 509
Geiger counter 564-5
GELL-MANN 947
General gas law 337
Generator 651
  terminal voltage 546
Geomagnetism 587
Geometrical optics 743
GERMER, L. H. 867
GLASER, DONALD 372
Gram 55
Grating, diffraction 815
Gravitation, universal 61
Gravitational
  acceleration 62-66
  deflection of light 908
  energy 108, 203 ff
  red shift 908
  units 66
Gravity

Gravity (*cont.*)
  center of 167
  force of 61 ff
  specific 71
GRIMALDI, FRANCESCO MARIA 704
GUERICKE, OTTO VON 281
Gyrocompass 202
Gyromagnetic ratio 640
Gyroscopic motion 198

HAHN AND STRASSMANN 925
Half-life 916
Harmonic motion (*see* Simple harmonic motion)
Harmonics 440
Heat 289 ff, 303 ff
  capacity 306
  conduction 319
  engines 390 ff, 399
  insulator 320
  latent 308
  of combustion 313
  of fusion 308
  of reaction 313
  of vaporization 308–9
  pump 401
  quantity of 304
  reversible 572
  specific 306, 342, 344
  transfer 318 ff
HEISENBERG, WERNER 852, 869
Heisenberg's theory of ferromagnetism 880
Helmholtz coils 625
Henry, definition 658, 660
HENRY, JOSEPH 646
HERSCHEL, SIR WILLIAM 711
Hertz, definition 243
HERTZ, GUSTAV 851
HERTZ, HEINRICH 690, 847
HEYL, P. R. 62
HOFSTADTER, ROBERT 909
HOOKE, ROBERT 220
Hooke's law 220, 223
Horizontal 63
Horsepower 120
Humidity 373
HUYGENS, CHRISTIAAN 700
Huygens' 'principle' 730 ff
Hydraulic
  press 281
  torque converters 159
Hydrogen
  atom 852, 872
  bomb 929
  spectrum 841, 849
Hydrometer 280
Hygrometry 372
Hypermetropia 795–6
Hysteresis, magnetic 637

Ice point 290, 396
Illumination 723
Image (*see* Lenses and Mirrors)

Impedance 676
Impulse 129 ff
  force 137
  turbine 138
Inclination angle 587
Indicator diagram 399
Induced
  charges 467
  EMF's 646 ff
Inductance 658, 660
  in AC circuits 674
  mutual 658, 659
  self 660
  standard 659
Induction
  charging by 504
  coil 665
  electromagnetic 646 ff
Inductive reactance 674
Inductor 660
Inelastic
  collisions 134
  materials 220
Inertia 54
  moment of 153
  rotational 153
Infrasonics 438
Infrared radiation 711
Insulators
  electric 467, 538
  thermal 320
Intensity
  acoustic 449
  electric 471
  luminous source 722
  magnetic 584 ff
  sound 449
Interference 424
  of light waves 704, 709
  of sound waves 454
Interferometer 709
Internal
  energy 303
  forces 133
  reflection 750
Inverse photoelectric effect 859
Ionization 568, 851
  chamber 512
Irradiance 721
Isothermal atmosphere 358
Isothermal processes 383 ff
Isotopes 333
Isotopic mass 333
Isotropic solids 226
IVES, H. E. 708, 890

Jet propulsion 138, 143
Joule, definition 104
JOULE, JAMES PRESCOTT 104, 311
Joule's
  free expansion experiment 341
  mechanical equivalent of heat 311

Kaufman-Bücherer experiment 899
KELVIN, WILLIAM THOMSON 293
Kelvin
  electrometer 517, 527
  temperature scale 292, 395
KEPLER, JOHANN 3
Kepler's laws 3, 60, 215
KERR, JOHN 834
Kerr effect 834
KERST, DONALD 656
Kilogram 55
Kilomole (*see* Mole)
Kilowatt hour 120
Kinematics 30 ff
  of rotation 146 ff
Kinetic
  energy 108, 153, 187, 902
  friction 90
  theory of gases 346
Kirchhoff's
  law of radiation 324
  rules for electric networks 551
Klystron 694

LAMBERT, JOHANN HEINRICH 756
Lambert's law 756
LAND, EDWIN H. 821
LASER 857
Latent heat
  of fusion 306
  of vaporization 308–9
Lattices, crystal 232, 360, 862
Laue patterns 862-3
LAVOISIER, ANTOINE 332
LAWRENCE, E. O. 598
Length 14 ff
Lenses 775 ff
  achromatic 794, 804
  combinations of 792
  compound 794
  converging 777 ff
  diverging 777, 783
  equation 778
  ideal 775
  optical center of 780
  power 794
Lensmaker's equation 778
Lenz's rule 650
Leptons 945
Lever arm 166
Lift of air foils 276
Light 698 ff
  flux 721–2
  interference 704, 708
  Maxwell's theory of 694, 883 ff
  nature of 699 ff
  Newton's particle theory of 699
  polarization of 819 ff
  pressure 906, 907
  rays 701
  rectilinear propagation of 701
  reflection of 743 ff
  refraction of 748 ff
  scattering of 825

INDEX

Light (*cont.*)
  source, standard 722
  speed of 621, 713 ff, 890 ff
  wavelengths 712
Lines
  electric 472
  magnetic 604, 614
  stream 271
  vector 494
Line-integral 106
  Ampére's relation for $\mathcal{B}$ 608
    Maxwell's modification 885
  relation for $\mathcal{E}$ 483
Liquefaction of gases 403
Liquids (*see also* Fluids)
  atomic nature of 232 ff
  surface tension of 375
  thermal properties of 359 ff
Lloyd's mirror 759
Longitudinal
  strain 223
  stress 224
  wave 414
Loops 428, 447
Loudness 439, 449 ff
Lumen 722
Luminous
  flux 722
  intensity 722

Mach interferometer 719
MACKENZIE, I. C. C. 716
Magdeburg hemispheres 281
'Magic numbers' 931
Magnetic
  'bottle' 623, 933
  circuit 636
  dipole 610
  domains 881
  field 584 ff, 603 ff
    energy 644, 666, 892
    of capacitors 994
    of circular coil 607-8
    of earth 587
    of solenoid 615
    of square coil 607
    of straight wire 607
    of toroid 632
  flux 613
  force 583
    on current 590
    on moving charge 594
    on solenoids 619
  hysteresis 638
  intensity 584
  materials 629, 636
  'mirror' 623
  moments 585
    of atoms 877
  poles 608, 609 ff
  properties of matter 629 ff
  resonance EPR, NMR 883
  storms 590
  susceptibility 638

Magnetic (*cont.*)
  trapping 622, 933
  tubes 614
Magnetism 582 ff (*see also* Magnetic)
Magnetization
  curves 636
  of a material 630
Magnetizing force $\mathcal{H}$ 634
Magnetohydrodynamic generator 602
Magnetomotive force MMF 636
Magneton (Bohr) 878
  nuclear 913
Magnetostatics 603
Magnetron 601-2
Magnets, permanent 629
Magnifying power 795
MALUS, ETIENNE LOUIS 823
Manometer 269
MASER 857
Mass 53
  atomic unit 334
  center of 167
  critical 928
  energy equivalence 903
  isotopic (nuclidic) 333
  number 333
  relativistic 900
  spectrograph 596
Matter, annihilation and creation of 919 ff
MAXWELL, JAMES CLERK 346, 701, 883
Maxwellian distribution 352
Maxwell's
  electromagnetic theory 883
  equations 883 ff
  theory of light 701, 883 ff
Mean solar day 31
Mechanical
  advantage 122
  energy 108 ff
  waves 411
Mechanics 13 ff
  of fluids 260
  relativistic 899
MEITNER, LISE 925
Meridian 31
  magnetic 589
Mesons 937 ff
Meter 14, 710
Metric units 8, Ai
MICHELSON, ALBERT A. 709, 715, 896
Michelson interferometer 709
Michelson-Morley experiment 896
Michelson's measurement of the speed of light 715
Micron (μm) 15
Microscope
  compound 789
  electron 882
  simple 798

Microwaves 693, 712
MILLIKAN, ROBERT A. 484
Millikan's oil-drop experiment 484
Mirror
  concave 763
  convex 770
  ellipsoidal 761
  equation 766
  ideal 761
  Lloyd's 759
  paraboloidal 762
  plane 343 ff
Missiles, ballistic 208 ff
Moderators, reactor 136
Modes of oscillation 431, 441, 447
Molar
  gas constant 338
  specific heat 342, 344
Mole (Kilomole) 336
Molecular
  dipoles 492
  fields 488
  mass 334
  nature of matter 346
  polarizability 528
  weight (mass) 335
Momentum 129 ff
  angular 157
    conservation of 196
    quantization of 853
  conservation of 134 ff, 901
  of a dynamical system 133
MOSELY, HENRY G. J. 860
MÖSSBAUER, R. L. 909
Mössbauer effect 907, 909
Motion (*see also* Dynamics, Kinematics)
  periodic 242 ff
  relative to earth 202 ff
  two-dimensional 186
Motor
  prototype 592
  voltages 548
MOUTON, H. 834
Muon 939
Muonium 939
Mutual inductance 658
Myopia 795-6

NEDDERMYER, SETH 939
Negative proton 945
Networks, electric 551
Neutrino 915
Neutron 917
Newton, definition 57
NEWTON, SIR ISAAC 3, 53, 61, 75, 322, 698, 704, 754
Newton's
  dispersion experiment 754
  speed of sound 460
  law of cooling 322
  particle theory of light 699
  principle of relativity 75

Newton's (*cont.*)
  principle of universal gravitation 61
  principles of mechanics 51 ff, 130, 900
  rings 704, 707
  view of polarization 820
Nodes 428, 447
Noise 452
Non-linear circuit elements 557
Nuclear
  binding energy 921
  chain reactions 926-7
  energy 925
  fission 925
  forces 929
  fusion 924, 929
  magnetic resonance (NMR) 882
  magneton 913
  model of atom 585
  models 929
  physics 911 ff
  properties, 'external' 912
  radius 914
  reactions 916
Nucleon 917
Nuclide 333
Nucleus (*see* Nuclear)
Nutation 199

Objective lens (*see* Microscope, Telescope)
OERSTED, HANS CHRISTIAN 582
Ohm, definition 532
OHM, GEORG SIMON 532
Ohm's law 532
Opera glass 805
Operational definition 4
OPPENHEIMER, ROBERT 946
Optical instruments 792 ff
Optics (*see* Light)
Orbits 3, 60, 202 ff
Organ pipes 447 ff
Oscillation (*see also* Simple harmonic motion)
  forced 254
  modes of 439 ff
  torsional 251
Oscillating circuits 686 ff
Overtones 440

Parallel-axis theorem 189
Paramagnetism 628, 639
Particle
  dynamics 51 ff
  equilibrium 85
  theory of light 699
Particles, elementary 946
Pascal's vases 282
Pauli exclusion principle 872
PELTIER, JEAN-CH. ATHANASE 577
Peltier effect 578
Pendulum 252
  ballistic 257

Pendulum (*cont.*)
  conical 94, 100
  physical 253
  simple 252
  torsion 251
Penumbra 702
Performance coefficient, refrigerators 401
Period 243
Periodic
  motion 242 ff
  table Avi
Permeability μ 634
  of free space μ_0 604
Permittivity of free space ε_0 469
  determination of 621
Phase angle 245
Phases of matter 359 ff
Photoelasticity 832
Photoelectric effect 848
  inverse 859
Photographic camera 790
Photographic-emulsion detectors 942
Photometric quantities 722
Photometry 722
Photons 861
Physical
  constants, fundamental Aiii
  quantities 4-8, Ai
PICTET, RAOUL 403
Pinch effect 933
Pinhole camera 725
Pion 936
Pitch 439, 451
Pitot-static tube 284
PLANCK, MAX 838, 845
Planck's
  constant 845
  quantum principle 845
  radiation law 845
Planetary motion 3, 60 ff, 202 ff
Plasma 602, 933
Pohl's experiment 759
Poisson's ratio 230
Polarization
  dielectric 517 ff
  of light 819
Poles, magnetic 608
Positron 919
Positronium 933
Potential, electric
  difference of 477, 480
Potential energy 108
  elastic 221
  gravitational 180, 203 ff
Potentiometer 553
POUND, R. V. 907
Pound 66
  mass (lb) 304
POWELL, C. F. 936
Power 120
  of a lens 794
  transmission 121, 159

Power, transmission (*cont.*)
  electric 548, 681
Precession 199
Pressure 262
  absolute 267
  atmospheric 267
  fluid 262
  gauge 267
  gauges 267
  of light 906
  partial 339
Principal focus 764, 778
Principal-ray diagram 764, 779
Principles, fundamental
  Ampere's 606
  charge conservation 470
  Coulomb's 468
  Faraday's (electrolysis) 567
  of Newtonian mechanics 51 ff
  of relativity 75, 897
  of thermal equilibrium 292
  of thermodynamics 382-3
  of universal gravitation 61
  Planck's quantum 845
Projectiles (*see also* Missile) 73
Projector 791
Prony brake 179
Proton 470, 917, 941
  negative 945

Quadrupole moment 914
Quality, sound 452
Quantities, physical 4-8, Ai
Quantization 850, 853, 872
Quantum
  condition 853
  energy 846, 850
    in eV 846
  mechanics 866 ff
  principle 846
  theory 846 ff
Quarks 944
Quarter-wave plate 831

Radar 726
Radian 16
Radiant flux 721
Radiation (*see also* Light)
  black-body 325, 843
  electromagnetic 690, 883 ff
  Kirchhoff's law 324
  Planck's law 845
  thermal 319, 323, 843
Radiator, perfect 325, 843
Radioactivity 914 ff
Radiometric summation 722
Radiometry 721
Rankine
  diagram 400
  temperature scale 293
Rarefaction 414
Rayleigh's criterion 811
Rays, light 701
  principal 769, 779

Reactance
  capacitive 672
  inductive 674
Reaction
  endothermic 313
  exothermic 313
  force 137
  heat of 313
  turbine 138
  nuclear 916
Reactor, nuclear 927
Rectifier 539
Rectilinear motion 39 ff
Reflection 743 ff
  laws of 744
  of waves 431 ff
  total internal 750
Refraction 748 ff
  double 827
  laws of 748
Refractive index 717
  for various wavelengths, 753 ff
Refrigerators 401
REINES, FREDERICK 916
Relative velocity 37 ff
Relativistic
  Doppler effect 898
  kinetic energy 902
  mass 900 ff
  phenomena 895 ff
  mechanics 899 ff
Relativity
  Einstein's theory 895 ff
  general 907
  Newtonian- 75
Reluctance 636
Residual flux 637
Resistance 532
  in AC circuits 668 ff
  temperature coefficient of 537
  thermometer 537
Resistivity 534
  of non-metals 538
Resistors 530, 532
  in parallel and series 533
Resolution of a vector 25
Resolving power 811
Resonance in AC circuits 676 ff, 687
Rest energy 903
Restitution, coefficient of 142
Reversible
  engines 390 ff, 393 ff
  processes 384, 385 ff
Right-hand rule 174, 496
Rigid body, equilibrium of 165-6
Rigidity, modulus of 228
RITTER 712
Ritz combination 'principle' 850
Rockets 138, 143
ROEMER, OLAUS 713
ROENTGEN, WILHELM KONRAD 859
Rolling bodies 191 ff
Rotation 145 ff

Rotation (*cont.*)
  dynamics of 155
  kinematics of 146
Rotational
  analogues 158
  equilibrium 166
  inertia 153
    parallel-axis theorem for 189
  kinetic energy 153
  power 152
  work 151
  simple harmonic motion 251
Rubber, elastic properties of 223
COUNT RUMFORD (BENJAMIN THOMPSON) 310
RUTHERFORD, ERNEST 485
Rutherford's
  model of atom 485, 492, 509
  nuclear reactions 916
Rydberg constant 849

Sagitta relation 707
Satellites 79, 206, 217 ff
Scalar 20
  components of a vector 21
  product of vectors 104
Scattering
  of alpha particles 585
  of light 825
SCHRÖDINGER, ERWIN 852, 870
Schrödinger equation 871
Scientific method 2
SEEBECK, THOMAS JOHANN 578
Seebeck effect 578
SEGRÉ, EMILIO 945
Self inductance 660
  energy stored in 663
Semi-conductors 539, 560
Series circuits
  AC 668, 675 ff
  DC 533, 551
Shadows 701, 729 ff
Shear 227
Shielding, electrostatic 503 ff
SHOCKLEY, WILLIAM 560
Sidereal day 31
Simple harmonic motion 243 ff
  angular 251
  energy relations in 249
  reference circle 248
Siphon 274
Slug 67
SNELL, WILLEBRORD 748
'Solar battery' 577
Solar day 31
Solenoid
  magnetic field of 615
  magnetic forces on 619
Solid angle 721
Solids
  atomic structure of 232 ff
  oscillation of 439
Sonic anemometer 461
Sound 438 ff

Sound (*cont.*)
  frequencies 438
  intensity 449
  production 439, 445
  quality 452
  speed 442
  waves, interference of 425, 454
Space charge 557
Spark chamber 942
Specific
  gravity 71
  heat capacity 306
    of gases 342, 344
  weights 266
Spectra 840 ff
Spectrograph 840
  mass 596
Spectrometer 840
Specular reflection 744
Speed 32
  of light 713, 621, 890
  of sound 442 ff
  of wave 414, 419
  terminal 81
Spherical aberration 764, 788
Spin 199
  electron 873
  nuclear 912
Standards (*see* Length, Mass, Time, Light source)
  electrical 621
Standing waves 427 ff, 447, 709
State variables 397
States (Phases) of matter 359 ff
Statics 164 ff
  fluid 264 ff
Steam
  engine 399
  point 290
Stefan-Boltzmann law 325
Stellar energy 924
Steradian 721
STEVENSON, E. C. 939
Storage batteries 574 ff
Strain
  longitudinal 223
  shearing 228
  volume 227
Stream lines and tubes 271
STREET, J. C. 939
Stress
  apparent 231
  longitudinal 224
  shearing 228
  volume 227
Strong interactions 929, 936
Sublimation 366
Superconductivity 536
Superposition principle 424
Supersaturation 373
Surface
  energy 375
  tension 361, 375
Systems

Systems (*cont.*)
  dynamical 131, 181 ff
  of forces 83 ff

Telescope 801 ff
Temperature 289 ff
  absolute (Kelvin and Rankine) 292
  common scales of (Celsius and Fahrenheit) 290
  fixed points of 290, 395
  thermodynamic scale of 395
Tension, surface 375, 361
Term frequencies 849
TESLA, NKOLA 586
Tesla, definition 586
Thermal
  conductivity 320
  energy 31
  equilibrium 292
  expansion 294
  radiation 319, 323, 843
Thermionic
  emission 541
  vacuum tubes 541 ff, 557 ff
Thermocouple 578
Thermodynamic scale of temperature 293-4, 395 ff
Thermodynamics 381 ff
Thermoelectric effects 577 ff
Thermometers 290, 293
  gas 355
  resistance 537
Thermometric properties 293
Thermonuclear reactions 924, 928
Thermostat 296
THOMSON, J. J. 656, 484, 596-7
Thomson model of atom 485
THOMSON, WILLIAM (LORD KELVIN) 293, 395
Thomson effect 579
Threshold frequency 848
Timbre 452
Time intervals 31 ff
Time dilatation 898, 944
Toroid, magnetic properties of 632
Torque 152, 166, 173
  about a point 200
  ratio 159
  vector 173
TORRICELLI, EVANGELISTA 267
Torricelli's law 273
Torsion
  -constant 222
  pendulum 251
Total internal reflection 750
TOWNES, CHARLES H. 857
Trajectories 73 ff, 202 ff
Transformations of energy 112 ff
Transformer 658, 679
Transient currents 554, 660
Transistors 560
Translational
  equilibrium 85, 165

Translational (*cont.*)
  motion 83
Transverse wave 413
Triple point 366
Tube (*see also* Vacuum tubes)
  electric 494
  magnetic 614
  stream 271
  vector 494 ff

Ultimate strengths 231
Ultrasonics 438
Ultraviolet radiation 712
Umbra 702
Units Ai
  absolute 57
  fundamental 8-9, Ai
  gravitational 66
  metric 8
  systems of Ai

Vacuum tubes 541 ff, 557 ff
Valence electrons 875
VAN ALLEN, JAMES A. 624
Van Allen 'radiation belts' 624
Van der Waals'
  equation 407
  forces 369
Van de Graaff generator 510 ff, 512
Vapor pressure 362
Vaporization 361 ff
  curves 363, 365
  latent heat 308
Varley loop test 563
Vector 19 ff
  components 25
  flux 494
  lines 494
  resolution of 25
  scalar components of 21
  torque 173
  tubes 494 ff
Vectors 19 ff
  addition of 22
  cross (or vector) product of 173
  dot (or scalar) product of 104
  subtraction of 24
  sum (or resultant) of 22
Velocity 33 ff
  angular 146
  of escape 204
  relative 37 ff
  terminal 81
Vena contracta 274
Venturi flow meter 275
Vertical 63
Vibrations (*see also* Oscillations, Simple harmonic motion)
  of air columns 445 ff
  of solids 439 ff
Virtual object 783, 793
Viscosity 278
Vision (*see also* Eye) 795
Volt 478
VOLTA, ALESSANDRO 478

Voltage (*see also* Potential)
  effective values 669
  terminal 546
Voltaic cells 574
Voltmeter
  AC 681
  DC 552

Water
  anomalous expansion of 299
  triple point of 366, 395 ff
  vapor pressure 363
Watt 121
Watt-hour meter 654
Wave 411 ff
  equation 417, 420
  -guide 693
  length 415
  longitudinal 414
  mechanics 867 ff
  nature of matter 867 ff
  speed 414, 421
  standing 427
  transmission of energy by 421 ff
  transverse 413
Waves
  de Broglie 867
  electromagnetic 690, 883 ff
  interference of 424 ff
  mechanical 411 ff
Weak interactions 938
Weber, defined 613
WEBER, WILHELM EDWARD 613
Weight 64
  specific 266
  standard 68
Wheatstone bridge 562
WHEELER, JOHN A. 930
Wiedemann-Franz law 544
Wien's displacement law 845
Wiener's experiment 708
Wilson cloud chamber 404
WOLLASTON 712, 842
Work 102 ff
  done by friction 114 ff
  done by gases 340
  function 848
  in moving coördinate systems 117
  in rotational motion 151

X-rays 712, 859 ff

Year 31
Yield point 231
YOUNG, THOMAS 224, 701
Young's
  interference experiment 733
  modulus 224
YUKAWA, HEDEKEI 936

Zero
  absolute 292, 395
  gravity experiment 81
Zone plates 739, 742